REMEDIATION OF PETROLEUM CONTAMINATED SOILS

Biological, Physical, and Chemical Processes

REMEDIATION OF PETROLEUM CONTAMINATED SOILS

Biological, Physical, and Chemical Processes

Eve Riser-Roberts, Ph.D.

LEWIS PUBLISHERS
Boca Raton   London   New York   Washington, D.C.

Library of Congress Cataloging-in-Publication Data

Riser-Roberts, Eve.
    Remediatin of petroleum contaminated soils : biological, physical, and chemical processes / Eve Riser-Roberts.
        p. cm.
    Includes bibliographical references and index.
    ISBN 0-87371-858-5 (alk. paper)
    1. Oil pollution of soils. 2. Soil remediation. I. Title.
TD879.P4R575 1998
628.5′5--dc20   97-46784
                                                                                                                                                              CIP

This book contains information obtained from authentic and highly regarded sources. Reprinted material is quoted with permission, and sources are indicated. A wide variety of references are listed. Reasonable efforts have been made to publish reliable data and information, but the author and the publisher cannot assume responsibility for the validity of all materials or for the consequences of their use.

Neither this book nor any part may be reproduced or transmitted in any form or by any means, electronic or mechanical, including photocopying, microfilming, and recording, or by any information storage or retrieval system, without prior permission in writing from the publisher.

The consent of CRC Press LLC does not extend to copying for general distribution, for promotion, for creating new works, or for resale. Specific permission must be obtained in writing from CRC Press LLC for such copying.

Direct all inquiries to CRC Press LLC, 2000 N.W. Corporate Blvd., Boca Raton, Florida 33431.

Trademark Notice: Product or corporate names may be trademarks or registered trademarks, and are used only for identification and explanation, without intent to infringe.

© 1998 by CRC Press LLC
Lewis Publishers is an imprint of CRC Press LLC

No claim to original U.S. Government works
International Standard Book Number 0-87371-858-5
Library of Congress Card Number 97-46784
Printed in the United States of America   2 3 4 5 6 7 8 9 0
Printed on acid-free paper

PREFACE

This comprehensive technology survey describes and compares the many biological, chemical, and physical processes available for remediating soils contaminated by jet fuels, gasoline, bunker oil, hydraulic and lubricating oils, and related petroleum products. Many details have been collected from the literature and assembled under one cover to provide a convenient and informative reference source for those who must contend with the critical worldwide problem of environmental contamination by these compounds.

The survey was initially conducted for the Naval Civil Engineering Laboratory (NCEL), Port Hueneme, CA, which is now merged with other commands in the new Naval Facilities Engineering Service Center (NFESC). The survey was performed in connection with the installation restoration effort at Twenty-nine Palms, a marine corps base training and staging facility at Twenty-nine Palms, CA, on Purchase Order Number N62583/88 P 2085. It was later expanded and updated for this publication.

Bioremediation is emerging as an important tool for treating petroleum-contaminated soils, whether used as a stand-alone technology or in combination with other physical or chemical methods. Bioremediation was considered to be the desired primary approach for remediating the contaminated soil at Twenty-nine Palms, supplemented, as necessary, by other processes. Also, because of heightened world interest in the phenomenon of bioremediation and its appropriation as a viable treatment option, this book presents an in-depth coverage of its application for contaminated soils.

The results of current research are combined with essential background information to cover all aspects of *in situ* and *ex situ* bioremediation of petroleum-contaminated soils. This information elaborates on the numerous factors affecting biodegradation of petroleum hydrocarbons and describes how they can be enhanced to optimize bioremediation. The susceptibility of individual petroleum components to biodegradation by specific microorganisms is reported, as are the chemical reactions and metabolic pathways involved. All groups of microorganisms are considered for their potential contribution, and the effects of both aerobic and anaerobic conditions are discussed.

This survey also contains an extensive overview of current *in situ* and *ex situ* physical and chemical soil remediation processes for dealing with petroleum contamination, including many innovative approaches. It investigates means of controlling release of volatile organic compounds (VOCs) to the atmosphere and leachate that could migrate to the groundwater during remediation. Methods for collecting and treating VOCs and leachate are included to address these secondary waste streams generated during soil treatment, whether *in situ* or *ex situ*. The importance of selecting appropriate technologies for each contamination incident and the potential value of combining processes for maximum efficiency are discussed. The expansive coverage of these subjects will furnish the reader with a wide range of options for developing treatment strategies and for customizing remediation procedures to the specific site requirements.

Information for this report was obtained through API (American Petroleum Institute), NTIS (National Technical Information Service), DTIC (Defense Technical Information Center), and Dialogue searches, and by extensive use of the library facilities of the University of California at Santa Barbara, in Goleta, CA.

ABOUT THE AUTHOR

Eve Riser-Roberts, Ph.D., received her doctoral degree in microbiology from the University of London, England. She has over 30 years's experience in the life and physical sciences as a consultant, researcher, technical writer, copywriter, and editor. She has conducted and directed research and written for scientists, engineers, and the general public while working in England, Germany, and the United States.

Dr. Riser-Roberts compiled two major reports for the U.S. Navy on remediation of the environment contaminated by petroleum products. Her previous book, *Bioremediation of Petroleum-Contaminated Sites,* was published in 1992 by Lewis Publishers.

At the University of Arizona, Tucson, she wrote for the Lunar and Planetary Laboratory (LPL), Department of Planetary Sciences, Department of Physics, and the Department of Agriculture. She also conducted research in the university's Department of Geosciences, and coordinated the first microbiological research ever performed on hydroponic systems at the Environmental Research Laboratory. Prior to that she conducted medical research at the University Health Sciences Center at the University of Arizona; at the Royal Free and Middlesex Hospitals in London, England; at the Technical University in Munich, Germany; and at the University of Tübingen and Max Planck Institute in Tübingen, Germany.

ACKNOWLEDGMENT

Thanks must be given to all the practitioners and researchers in the many diverse areas related to remediation of soils contaminated by petroleum products. The information they contribute from their work and studies help facilitate restoration of our contaminated world.

DEDICATION

This book is dedicated to Richard M. (Mike) Roberts, my personal, in-house chemical and environmental consultant, whose help in so many ways made this book possible.

CONTENTS

Section 1
Introduction 1
1.1 Background 1
1.2 Biodegradation as a Treatment Alternative 1
1.3 Combined Technologies 3

Section 2
Current Treatment Technologies 5
2.1 On-Site or *Ex Situ* Processes 5
    2.1.1 Physical/Chemical Processes 6
        2.1.1.1 Soil Treatment Systems 6
            2.1.1.1.1 Thermal Treatment 6
            2.1.1.1.2 Incineration 9
            2.1.1.1.3 Soil Washing 10
            2.1.1.1.4 Chemical Treatment 11
            2.1.1.1.5 Chemical Extraction 11
            2.1.1.1.6 Supercritical Fluid (SCF) Oxidation 11
            2.1.1.1.7 Volatilization 12
            2.1.1.1.8 Steam Extraction 12
            2.1.1.1.9 Solidification/Stabilization 12
            2.1.1.1.10 Encapsulation 12
            2.1.1.1.11 Supercritical Fluid Extraction 13
            2.1.1.1.12 Beneficial Reuse 13
        2.1.1.2 Leachate/Wastewater Treatment Systems 13
            2.1.1.2.1 Carbon Adsorption 19
            2.1.1.2.2 Resin Adsorption 20
            2.1.1.2.3 Adsorption with Brown Coal 21
            2.1.1.2.4 Wet Air Oxidation (WAO) 22
            2.1.1.2.5 Supercritical Fluid (SCF) Oxidation 22
            2.1.1.2.6 Chemical/Photochemical Oxidation 23
            2.1.1.2.7 Chemical Catalysis 27
            2.1.1.2.8 Chemical Precipitation 27
            2.1.1.2.9 Crystallization 28
            2.1.1.2.10 Density Separation 28
                2.1.1.2.10.1 Sedimentation 28
                2.1.1.2.10.2 Flotation 28
            2.1.1.2.11 Flocculation 28
            2.1.1.2.12 Evaporation 28
            2.1.1.2.13 Stripping 28
            2.1.1.2.14 Distillation 29
            2.1.1.2.15 Filtration 29
            2.1.1.2.16 Ultrafiltration 29
            2.1.1.2.17 Dialysis/Electrodialysis 29
            2.1.1.2.18 Ion Exchange 29
            2.1.1.2.19 Reverse Osmosis 29
            2.1.1.2.20 Solvent Extraction 29
    2.1.2 Biological Processes 30
        2.1.2.1 Soil Treatment Systems 30
            2.1.2.1.1 Landtreatment/Landfarming 30
            2.1.2.1.2 Composting 41
            2.1.2.1.3 Bioreactors 46

			2.1.2.1.3.1	Bioslurry Reactors 47
			2.1.2.1.3.2	Dual Injected Turbulent Suspension (DITS) Reactor 49
			2.1.2.1.3.3	Gravel Slurry Reactors 49
			2.1.2.1.3.4	Tubular Reactors 49
			2.1.2.1.3.5	Blade-Mixing Reactors 50
			2.1.2.1.3.6	Prepared-Bed Reactors 50
			2.1.2.1.3.7	Enclosed Reactors 50
			2.1.2.1.3.8	Fermenters 51
			2.1.2.1.3.9	Fungal Compost Bioreactors 51
			2.1.2.1.3.10	Combination Reactors 51
			2.1.2.1.3.11	Pressure Reactors 51
			2.1.2.1.3.12	Wafer Reactors 51
		2.1.2.1.4	Biopiles 51
		2.1.2.1.5	Vacuum Heap Biostimulation System 52
		2.1.2.1.6	Vegetation 53
		2.1.2.1.7	Photolysis 53
	2.1.2.2	Leachate/Wastewater Treatment Systems 54
		2.1.2.2.1	Aerobic Systems 54
			2.1.2.2.1.1	Suspended Growth Systems 56
			2.1.2.2.1.2	Fixed-Film Systems 57
			2.1.2.2.1.3	Microbial Accumulation of Metals 66
			2.1.2.2.1.4	Combination Aerobic Reactors/Microbial Adsorption 67
			2.1.2.2.1.5	Bioreactor for Aromatic Solvents 68
			2.1.2.2.1.6	Sequencing Batch Reactor (SBR) 69
			2.1.2.2.1.7	Self-Cycling Fermenter (SCF) 69
			2.1.2.2.1.8	Autothermal Aerobic Membrane Bioreactor (ATA MBR) 70
			2.1.2.2.1.9	Evaporation and Biofilm Filtration 70
			2.1.2.2.1.10	Biocatalyst Beads 70
		2.1.2.2.2	Anaerobic Systems 70
			2.1.2.2.2.1	Anaerobic Bioconversion Process 70
			2.1.2.2.2.2	Suspended Growth Systems 71
			2.1.2.2.2.3	Fixed-Film Systems 74
		2.1.2.2.3	Combined Aerobic/Anaerobic Treatment 76
			2.1.2.2.3.1	Aerobic/Anaerobic Biofilm Reactor 76
			2.1.2.2.3.2	Sequential Anaerobic/Aerobic Treatment 76
2.2	*In Situ* Processes 78
	2.2.1	Physical/Chemical Soil Treatment Processes 79
		2.2.1.1	Shallow Soil Mixing (SSM) 79
		2.2.1.2	Oxidation/Reduction 79
		2.2.1.3	Hydrolysis 80
		2.2.1.4	Neutralization 80
		2.2.1.5	Stabilization/Solidification 80
		2.2.1.6	Mobilization/Immobilization 80
		2.2.1.7	Soil Flushing/Washing/Extraction/Pump and Treat 81
		2.2.1.8	CROW Process 87
		2.2.1.9	Injection/Extraction Process 87
		2.2.1.10	Air Stripping 87
		2.2.1.11	Soil Vapor Extraction (SVE) 89
		2.2.1.12	Air Sparging 93
		2.2.1.13	Detoxifier™ 93
		2.2.1.14	Soil Heating 95

			2.2.1.14.1	Hot Air Injection/Flushing	96
			2.2.1.14.2	Steam Injection/Steam Flushing/Steam Stripping	96
			2.2.1.14.3	Radio Frequency (RF) Heating	98
		2.2.1.15	Vitrification		99
		2.2.1.16	Tensiometric Barriers		99
		2.2.1.17	Electric Fields		99
	2.2.2	Biological Soil Treatment Processes			100
		2.2.2.1	Bioremediation/Bioreclamation		100
		2.2.2.2	Bioventing		103
		2.2.2.3	Bioslurping		105
		2.2.2.4	BioPurge^{SM}/BioSparge^{SM}		106
		2.2.2.5	Hydraulic/Pneumatic Fracturing		107
		2.2.2.6	Deep Soil Fracture Bioinjection™		109
		2.2.2.7	Combined Air–Water Flushing		109
		2.2.2.8	*In Situ* Electrobioreclamation/Electro-Osmosis/Electrokinetics/ Electrochemical Remediation		110
		2.2.2.9	Biopolymer Shields		112
		2.2.2.10	Bioscreens		112
		2.2.2.11	Phytoremediation		113

Section 3
Biodegradation/Mineralization/Biotransformation/Bioaccumulation of Petroleum Constituents and Associated Heavy Metals ... 115

3.1	Chemical Composition of Fuel Oils				115
	3.1.1	Naphtha			115
	3.1.2	Kerosene			115
	3.1.3	Fuel Oil and Diesel #2			115
	3.1.4	Gasoline			115
	3.1.5	JP-5			115
	3.1.6	JP-4			115
3.2	Organic Compounds				115
	3.2.1	Aerobic Degradation			123
		3.2.1.1	Degradation of Alkanes		128
		3.2.1.2	Degradation of Branched and Cyclic Alkanes		130
		3.2.1.3	Degradation of Alkenes		131
		3.2.1.4	Degradation of Aromatic Compounds		131
		3.2.1.5	Degradation of Specific Compounds		131
			3.2.1.5.1	Mononuclear Aromatic Hydrocarbons and Derivatives	134
			3.2.1.5.2	Polycyclic Aromatic Hydrocarbons	141
			3.2.1.5.3	Branched-Chain Aliphatics	155
			3.2.1.5.4	Straight-Chain Aliphatics	156
			3.2.1.5.5	Fatty Acids and Carboxylic Acids	159
			3.2.1.5.6	Alcohols	159
			3.2.1.5.7	Alicyclic Hydrocarbons	159
			3.2.1.5.8	Asphaltenes	160
		3.2.1.6	Nonaqueous-Phase Liquids (NAPLs)		160
	3.2.2	Anaerobic Degradation			165
		3.2.2.1	Anaerobic Respiration		168
			3.2.2.1.1	Denitrification	169
			3.2.2.1.2	Sulfate Reduction	172
			3.2.2.1.3	Methanogenesis	172
		3.2.2.2	Fermentation		173
		3.2.2.3	Anaerobic Photometabolism		173
		3.2.2.4	Specific Compounds		173

			3.2.2.4.1	Mononuclear Aromatic Hydrocarbons	173
			3.2.2.4.2	Polycyclic Aromatic Hydrocarbons	177
			3.2.2.4.3	Straight-Chain Aliphatics	177
			3.2.2.4.4	Branched-Chain Aliphatics	178
			3.2.2.4.5	Alcohols	178
			3.2.2.4.6	Alicyclic Hydrocarbons	178
			3.2.2.4.7	Fatty Acids	178

3.3 Heavy Metals 178
    3.3.1 Specific Elements 184
        3.3.1.1 Arsenic (As) 185
        3.3.1.2 Cadmium (Cd) 185
        3.3.1.3 Chromium (Cr) 187
        3.3.1.4 Iron (Fe) 188
        3.3.1.5 Lead (Pb) 188
        3.3.1.6 Mercury (Hg) 189
        3.3.1.7 Nickel (Ni) 190
        3.3.1.8 Selenium (Se) 191
        3.3.1.9 Silver (Ag) 191
        3.3.1.10 Other Metals 192
3.4 Intermediate Metabolites and End Products of Biodegradation 192

Section 4
Factors Affecting Biodegradation in Soil–Water Systems 199
4.1 Chemical and Physical Factors 200
    4.1.1 Chemical Solubility 200
    4.1.2 Advection 203
    4.1.3 Dispersion and Diffusion 203
    4.1.4 Sorption 203
    4.1.5 Volatility 207
    4.1.6 Viscosity 207
    4.1.7 Density 208
    4.1.8 Chemical Structure 208
    4.1.9 Toxicity 210
    4.1.10 Hydrolysis and Oxidation 212
    4.1.11 Concentration of Contaminants 213
        4.1.11.1 Low Concentrations 213
4.2 Biological Factors 217
4.3 Soil/Environmental Factors 217

Section 5
Optimization of Bioremediation 219
5.1 Variation of Soil Factors 219
    5.1.1 Soil Moisture 221
        5.1.1.1 Irrigation 225
        5.1.1.2 Drainage 226
        5.1.1.3 Additives 226
    5.1.2 Temperature 227
    5.1.3 Soil pH 229
        5.1.3.1 Increasing Soil pH 231
        5.1.3.2 Decreasing Soil pH 231
    5.1.4 Oxygen Supply 232
        5.1.4.1 Ozone 236
        5.1.4.2 Hydrogen Peroxide (H_2O_2) 237
        5.1.4.3 Hypochlorite 241
        5.1.4.4 Other Electron Acceptors 241

		5.1.4.5	Soil Oxygen Delivery Approaches 241
		5.1.4.6	Commercial Soil Oxygen Delivery Approaches 246
		5.1.4.7	Creating Anaerobic Conditions 247
		5.1.4.8	Combination Aerobic/Anaerobic Treatment 248
	5.1.5	Nutrients 249
	5.1.6	Organic Matter 257
		5.1.6.1	Addition of Products to Immobilize Heavy Metals 259
	5.1.7	Oxidation-Reduction Potential 260
	5.1.8	Attenuation 261
	5.1.9	Texture and Structure 261
		5.1.9.1	Texture 262
		5.1.9.2	Bulk Density 262
		5.1.9.3	Water-Holding Capacity 262
5.2	Biological Enhancement 263
	5.2.1	Microorganisms in Bioremediation 263
		5.2.1.1	Aerobic Bacteria 264
		5.2.1.2	Anaerobic Bacteria 265
		5.2.1.3	Oligotrophs 266
		5.2.1.4	Fungi 267
			5.2.1.4.1 Extracellular Enzymes 272
			5.2.1.4.2 Soil Inoculation 273
			5.2.1.4.3 Screening Strategies 273
		5.2.1.5	Phototrophs 274
		5.2.1.6	Higher Life Forms and Predation 276
	5.2.2	Bioaugmentation 277
		5.2.2.1	Acclimated/Adapted Bacteria 281
		5.2.2.2	Mutant Microorganisms 283
		5.2.2.3	Microbial Consortia 286
		5.2.2.4	Emulsifier Producers 287
		5.2.2.5	Microbial Transport 287
		5.2.2.6	Microbial Preservation 289
		5.2.2.7	Encapsulation/Immobilization 290
	5.2.3	Cometabolism and Analog Enrichment 292
		5.2.3.1	Diauxie Effect 297
	5.2.4	Application of Cell-Free Enzymes 297
	5.2.5	Addition of Antibiotics 299
	5.2.6	Use of Aerobic/Anaerobic Conditions 300
	5.2.7	Use of Biosorption/Bioaccumulation/Bioconcentration 300
	5.2.8	Use of Vegetation 300
	5.2.9	Other Microbial Applications 300
5.3	Contaminant Alteration 300
	5.3.1	Use of Surfactants 300
		5.3.1.1	Chemical Surfactants 302
		5.3.1.2	Microbial Surfactants 307
		5.3.1.3	Biodegradation of Surfactants 312
	5.3.2	Photolysis 313

Section 6
Volatile Organic Compounds in Petroleum Products 317
6.1 Emissions Produced from Soil Contamination 317
    6.1.1 Gasoline Vapor Composition 322
    6.1.2 Human Health Criteria 322
6.2 Parameters Affecting Volatilization 323
    6.2.1 Temperature 324
    6.2.2 Operating Surface Area 325

	6.2.3	Wind/Barometric Pressure	325
	6.2.4	Soil Moisture/Volumetric Water Content	325
	6.2.5	Mass Transfer Coefficient/Partition Coefficient	326
	6.2.6	Effective Depth	326
	6.2.7	Mole Fraction of Diffusing Component	327
	6.2.8	Humidity	327
	6.2.9	Solar Radiation	327
	6.2.10	Vapor Pressure	327
	6.2.11	Soil Properties	328
	6.2.12	Adsorption onto Soil	329
	6.2.13	Evaporation	329
	6.2.14	Water Solubility	329
	6.2.15	Henry's Law Constant	330
	6.2.16	Density	330
	6.2.17	Viscosity	330
	6.2.18	Dielectric Constant	330
	6.2.19	Boiling Point	330
	6.2.20	Molecular Weight	331
	6.2.21	Air-Filled Porosity	331
	6.2.22	Retention	332
	6.2.23	Diffusion Travel Times	332
6.3	Control of VOC Emissions	332
	6.3.1	Design and Operating Practices	333
		6.3.1.1 Surface Area Minimization	333
		6.3.1.2 Freeboard Depth	333
		6.3.1.3 Inflow/Outflow Drainage Pipe Locations	334
		6.3.1.4 Operating Practices	334
		6.3.1.4.1 Temperature of Influent	334
		6.3.1.4.2 Dredging, Draining, and Cleaning Frequency	335
		6.3.1.4.3 Handling of Sediments and Sludge	335
		6.3.1.4.4 Collecting Samples for Monitoring	335
	6.3.2	*In Situ* Controls	335
		6.3.2.1 Air-Supported Structures and Synthetic Membranes	335
		6.3.2.2 Vapor Extraction Systems (VES)	340
	6.3.3	VOC Pretreatment Techniques	342
		6.3.3.1 Pretreatment Processes for Organic Liquids	342
		6.3.3.1.1 Distillation	342
		6.3.3.1.2 Steam Stripping	343
		6.3.3.1.3 Solvent Extraction	344
		6.3.3.1.4 Air Stripping	344
		6.3.3.1.4.1 Aeration Devices	346
		6.3.3.1.4.2 Secondary Effects of Aeration	349
		6.3.3.1.5 Carbon Adsorption	350
		6.3.3.1.5.1 Gaseous Carbon Adsorption	350
		6.3.3.1.6 Biological Treatment	357
		6.3.3.1.7 Refrigeration/Condensation	357
		6.3.3.1.8 Evaporation	357
		6.3.3.2 Pretreatment Processes for Sludge with Organics	358
		6.3.3.2.1 Air Stripping with Carbon Adsorption	358
		6.3.3.2.2 Evaporation with Carbon Adsorption	358
		6.3.3.2.3 Steam Stripping	358
		6.3.3.3 Pretreatment Processes for Soils	358
		6.3.3.3.1 Soil Washing/Extraction	358
		6.3.3.3.2 Thermal Desorption	359
		6.3.3.3.3 Soil Venting/*In Situ* Air Stripping	359

		6.3.3.3.4	Soil Vapor Extraction (SVE)	362
		6.3.3.3.5	Soil Vapor Extraction/Shallow Soil Mixing (SSM)	362
		6.3.3.3.6	Detoxifier™	363
		6.3.3.3.7	Photodegradation	363
6.3.4	VOC Posttreatment Techniques			363
	6.3.4.1	Combustion/Incineration		363
		6.3.4.1.1	Thermal Incinerators	364
		6.3.4.1.2	Afterburners	365
		6.3.4.1.3	Catalytic Incinerators	365
		6.3.4.1.4	Flares	369
		6.3.4.1.5	Boiler/Process Heater	369
	6.3.4.2	Condensation		369
	6.3.4.3	Distillation		372
	6.3.4.4	Absorption		372
		6.3.4.4.1	Packed Columns	372
		6.3.4.4.2	Plate Columns	372
		6.3.4.4.3	Polymer-Based Adsorbent	376
	6.3.4.5	Biofiltration		376
		6.3.4.5.1	BIOPUR®	380
	6.3.4.6	Photo-oxidation		380
6.3.5	Recycle			380
6.3.6	Treatment Residuals			381

Section 7
Monitoring Bioremediation 383
7.1 Microbial Counts 383

7.1.1	Methods for Enumerating Subsurface Microorganisms		385
	7.1.1.1	Direct Microscopic Counts	385
	7.1.1.2	Direct Counts with Acridine Orange	385
	7.1.1.3	Direct Viable Counts by Cell Enlargement	385
	7.1.1.4	Direct Viable Counts from Cell Division	385
	7.1.1.5	Dip Slides	385
	7.1.1.6	INT Activity Test	385
	7.1.1.7	ATP Content	386
	7.1.1.8	Direct Epifluorescence Filtration Technique (DEFT)	386
	7.1.1.9	Microcolony Epifluorescence Technique	386
	7.1.1.10	Immunofluorescence Microscopy	386
	7.1.1.11	Plate Counts	386
	7.1.1.12	Enrichment Techniques	387
	7.1.1.13	Fume Plate Method	389
	7.1.1.14	Drop Count Method	389
	7.1.1.15	Droplette Method	389
	7.1.1.16	Broth Cultures	389
	7.1.1.17	Most-Probable-Number (MPN) Method	389
	7.1.1.18	Membrane Filter Counts	390
	7.1.1.19	Rapid Automated Methods	390
	7.1.1.20	Fatty Acid Analysis/Lipid Biomarkers	391
	7.1.1.21	Dehydrogenase-Coupled Respiratory Activity	391
	7.1.1.22	Microautoradiography	391
	7.1.1.23	Protozoan Counts	391
	7.1.1.24	Fungal Counts	391
	7.1.1.25	Opacity Tube Method	391
	7.1.1.26	Turbidimetric Measurement	391
7.1.2	Counts in Uncontaminated Soil		392
7.1.3	Counts in Contaminated Soil		393
7.1.4	Effect of Biostimulation on Counts		394

- 7.2 Other Monitoring Methods ... 396
  - 7.2.1 Biomolecular/Nucleic Acid–Based Methods ... 396
    - 7.2.1.1 Reporter Genes ... 398
    - 7.2.1.2 mRNA Extraction ... 399
    - 7.2.1.3 Chromosomal Painting ... 399
    - 7.2.1.4 rRNA Methods ... 399
    - 7.2.1.5 Polymerase Chain Reaction ... 400
  - 7.2.2 Biomarkers ... 400
    - 7.2.2.1 Carboxylic/Hopanoic Acids ... 400
    - 7.2.2.2 Bicyclic Alkanes, Pentacyclic Terpanes, and Steranes ... 400
    - 7.2.2.3 Phenanthrenes/Anthracenes ... 400
    - 7.2.2.4 Pristane and Phytane ... 401
  - 7.2.3 Hydrocarbon Concentration ... 401
  - 7.2.4 Other Organic Indicators ... 401
  - 7.2.5 Electron Acceptor Concentration ... 402
  - 7.2.6 Soil Gas Monitoring ... 402
    - 7.2.6.1 Carbon Dioxide and Oxygen ... 402
    - 7.2.6.2 Nitrous Oxide ... 403
    - 7.2.6.3 Methane ... 403
  - 7.2.7 Anaerobic By-Products ... 403
  - 7.2.8 Inorganic Indicators ... 403
  - 7.2.9 Stable Isotope Analysis ... 404
  - 7.2.10 Labeled Contaminants ... 404
  - 7.2.11 Enzyme Assays ... 405
  - 7.2.12 Intermediary Metabolite Formation ... 405
  - 7.2.13 Monitoring Conservative Tracers ... 405
  - 7.2.14 Gas Chromatography and Mass Spectrometry (GC/MS) ... 405
  - 7.2.15 Thin-Layer Chromatography–Flame Ionization Detection ... 406
  - 7.2.16 Antibiotic-Resistant Microorganisms ... 406
  - 7.2.17 ELISA ... 406
  - 7.2.18 Biolog® System ... 406
  - 7.2.19 Respirometry/Radiorespirometry ... 407
  - 7.2.20 Microcalorimetry ... 407
  - 7.2.21 Flow Cytometry ... 407
  - 7.2.22 Biochemical Testing ... 408
  - 7.2.23 Modeling ... 408
- 7.3 Rate of Biodegradation ... 408
- 7.4 Differentiating Biotic and Abiotic Processes ... 413

Section 8
Treatment Trains ... 415

- 8.1 Limitations of Soil Treatment Systems ... 415
  - 8.1.1 Physical/Chemical Treatment Systems ... 415
  - 8.1.2 Landtreatment ... 415
  - 8.1.3 *In Situ* Biodegradation ... 415
  - 8.1.4 On-Site/*Ex Situ* Biological Systems ... 417
- 8.2 Remediation Guidelines ... 417
- 8.3 Combined Technologies ... 417
  - 8.3.1 On Site/*Ex Situ* ... 417
  - 8.3.2 *In Situ* ... 419
  - 8.3.3 Processes for Treatment Trains ... 420
- 8.4 Examples of the Use of Treatment Trains ... 420
- References ... 429
- Index ... 503

GLOSSARY

TERM/ACRONYM DESCRIPTION

ABF: Activated biofilters
Abiotic reactions: All reactions not biological in origin, including inorganic, photolytic, surface-catalyzed, sorptive, and transport processes
Absorption: Retention of the solute within the mass of the solid rather than on its surface
Acclimation: The lag time during which organisms acquire the ability to degrade novel compounds
Acetogens: Microorganisms that convert higher volatile acids to acetate and hydrogen
ACGIH: American Conference of Governmental Industrial Hygienists
Acidophilic: Favors acidic conditions
Adaptation: The modification of characteristics of organisms to improve ability to survive and reproduce in a particular environment
Adsorption: Retention of solutes in solution by the surfaces of the solid material
Aerobic: In the presence of oxygen
AFCEE: U.S. Air Force Center for Environmental Excellence
AGP: Attached growth ponds
Alfonic 810-60: A nonionic alcohol ethoxylate surfactant
Alkalophilic: Favors basic conditions
Allochthonous: Nonindigenous microorganisms
Ambersorb 563: Activated carbon
Anaerobes: Microorganisms that require anoxic conditions and oxidation-reduction potentials of less the -0.2 V
Anaerobic: The absence of oxygen
Anisotropic: Exhibiting properties with different values when measured along axes in different directions
Anoxic: Oxygen free
Anthropogenic: Of man-made origin
AODC: Acridine orange direct counting method
API: American Petroleum Institute
Assimilatory: Results in the reduction of nitrate to ammonia for denitrification cellular synthesis
ATAB: Autothermal aerobic bioreactor
ATA MBR: Autothermal aerobic membrane bioreactor
ATF: Automatic transmission fluid
ATP: Adenosine-5′-triphosphate
Attenuation: Mixing of contaminated soil with clean soil to reduce concentration of hazardous compounds
ATTIC: Alternative Treatment Technology Information Center (EPA database for technical information on innovative treatment technologies for hazardous waste and other contaminants)
Autochthonous: Indigenous or native bacteria found in soil in relatively constant numbers that do not change rapidly in response to the addition of specific nutrients
Autotrophic: The ability to use reduction of carbon dioxide as major source of organic compounds needed for growth
Autotrophs: Organisms that can survive autotrophically
Axenic: Free from other living organisms
BAC: Biological activated carbon
BARR: Bioanaerobic reduction and reoxidation; a remedial technique for *in situ* biodegradation in soil and groundwater
BCP: Bacterial chromosomal painting
BDAT: Best demonstrated available technology
Bioaccumulation: Accumulation of organic contaminants or metals by some microorganisms
Bioaugmentation: Supplementation of microorganisms to a contaminated site to enhance bioremediation; see Enhanced biodegradation
BIOCELS: Bioreclamation with Innovative On-Site Controlled Environment Landtreatment Systems

Biodegradation: Breakdown of organic substances by microorganisms by breaking intramolecular bonds; e.g., involving substituent functional group or mineralization. As a result, the microorganisms derive energy and may increase in biomass.
Bioemulsifier: An emulsifier produced by a microorganism
BIOFAST: Biological forced-air soil treatment for biopiles
Biofiltration: Treatment of off-gases using biological filters to remove VOCs
Biolog® system: Measures metabolic potentials to describe bacterial communities (Biolog, Inc.)
Biopiles: Mounds of excavated contaminated soil for controlled *ex situ* treatment
BIOPUR®: A patented, aerated, packed-bed, fixed-film reactor using PUR as a carrier material for microorganisms
BioPurge^SM: Technology using bioventing with a closed-loop concept to regulate soil moisture and release of nutrients, oxygen, and microorganisms into the vadose zone
Biorecalcitrance: Resistance of a compound to biological attack
Bioreclamation: A natural or managed process involving biodegradation of environmental contaminants
Bioremediation: A natural or managed process involving biodegradation of environmental contaminants
Biorestoration: A natural or managed process involving biodegradation of environmental contaminants
BioSparge™: Technology using bioventing with a closed-loop concept to regulate soil moisture and release of nutrients, oxygen, and microorganisms below groundwater level
Biostim: Uses "Tysul" WW H_2O_2 to circulate oxygen in the soil (Biosystems, Inc.)
Biotic reactions: Reactions that are biological in origin
Biotransformation: Microbial or enzymatic alteration of the molecular structure of a chemical; i.e., microbial metabolism
Bioventing: Process of aerating subsurface soils to stimulate *in situ* bioremediation using SVE systems
Bio XL: Process employing stabilized solutions of H_2O_2 to increase level of oxygen in soil (Aquifer Remediation System)
BOD: Biochemical oxygen demand
Brij 30: A surfactant
BR: Butyl rubber, can be used as a liner
Brij 30: A surfactant
Brij 35: A surfactant
BSRR: Rotary reactor
BTEX: Benzene, toluene, ethylbenzene, and xylenes
BTX: Benzene, toluene, and xylenes
C8PE9.5, C9PE10.5: Nonionic alkylphenol ethoxylate surfactants
C12-E4: Nonionic alkylethoxylate surfactant
Catox: Catalytic/thermal oxidation units for controlling VOC emissions
Cedephos FA-600: Anionic surfactant mixture of mono- and diorganophosphate esters
CEQ: Council on Environmental Quality
CERCLA: Comprehensive Environmental Response, Compensation, and Liability Act (Superfund)
CFU: Microbial colony-forming unit
CGAs: Colloidal gas aphrons (foams); e.g. NaDBS
Chemoautotrophic: Derives energy from the respiration of inorganic electron donors
CO-601 carbons: Coal-based or coconut shell carbons for removal of hydrocarbons from gas streams with the Detoxifier™
CoA: Coenzyme A
COD: Chemical oxygen demand
Commensalism: Sequential degradation of a compound by two or more microorganisms in a relationship that may benefit only one partner
Cometabolism or Cooxidation: The indirect metabolism of a recalcitrant substance; the process by which microorganisms, in the obligate presence of a growth substrate, transform a nongrowth substrate
Composting: A form of biodegradation involving mesophilic and thermophilic microorganisms
Conjugation: Reaction between a normal metabolite and a toxicant
Convective Transport: Passive transport of microorganisms through soil by transport addition of water or aqueous nutrient feed solution
Corexit 0600: A surfactant

CPE: Chlorinated polyethylene, can be used as a liner
CR: Neoprene, can be used as a liner
CREAM®: Video image analyzing system for reading ELISA plates (Kem-En-Tec A/S)
Critical micelle concentration: Lowest concentration at which micelles begin to form
CROW: Contained recovery of oily wastes process to recover DNAPLs
Cryo-SEM: Cryoscanning electron microscopy
CSPE: Chlorosulfonated polyethylene, can be used as a liner
Customblen: A slow-release fertilizer containing calcium phosphate, ammonium phosphate, and ammonium nitrate in a vegetable oil coating
Cy3, Cy5: Fluorochrome (fluorescing) DNA label
Cyanobacteria: Blue-green algae, which are actually bacteria; can be present in surface soil
DCE: 1,2-Dichloroethane
DEFT: Direct epifluorescence filtration technique for counting microorganisms
Dehalogenation: Enzymic removal of a halogen
DEHP: Di-2-ethylhexylphthalate
Denitrification: Also called dissimilatory nitrate reduction, where nitrate serves as the terminal electron acceptor in the oxidation of an organic substance, with production of N_2 and energy for the cell; begins when oxygen concentration goes below 10 µmol/L
Denitrifying bacteria: Facultative bacteria that reduce nitrate using the oxygen of nitrate as a hydrogen acceptor; they (denitrifiers) have the ability to oxidize inorganic energy sources (e.g., hydrogen or sodium sulfide)
DETOXIFIER™: Technology/equipment potentially capable of implementing a range of *in situ* treatment methods; e.g., air/steam stripping, neutralization, solidification/stabilization, oxidation (Toxic Treatments U.S.A.)
Diauxie: The opposite of cometabolism; a sparing effect, when a compound is not degraded in the presence of another compound
Dissimilatory nitrate reduction: Nitrate serves as the terminal electron acceptor in the oxidation of an organic substance, producing N_2 and energy for the cells
Dissimilatory sulfate reduction: Strict anaerobes that utilize organic carbon as a source of carbon and energy and use reducible sulfur compounds (e.g., sulfate, thiosulfate) as terminal electron acceptors
DITS: Dual injected turbulent suspension
DNA: Deoxyribonucleic acid
DNAPLs: Dense nonaqueous-phase liquids
DNOC: 4,6-Dinitro-*o*-cresol
DO: Dissolved oxygen
Dobanols 91-5, 91-6, 91-8: Surfactants
DOC: Dissolved organic carbon
DOD: Department of Defense
Dowfax C10L: A sulfonated anionic surfactant
Dowfax 8390: A sulfonated anionic surfactant
DPA: Diphenylamine
EC: Elimination capacity
EDTA: Ethylenediaminetetraacetic acid
Eh: Redox potential
Electro-osmosis: Soil water is induced to flow toward a cathode during *in situ* or *ex situ* electrobioreclamation of low permeability, unsaturated soils
ELISA: Enzyme-linked immunosorbent assay; a monoclonal antibody immunoassay
ELPO: Elasticized polyolefin, can be used as a liner
END: Enhanced natural degradation; *in situ* process to increase amount of H_2O_2 in contaminated soil (Groundwater Technology)
Engineered bioremediation: Any modification or intervention in the bioremediation process
Enhanced biodegradation: Stimulation of microbial degradation of organic contaminants by addition of microorganisms, nutrients, or optimization of environmental factors on-site or *in situ*
Enrichment culturing: Addition of a specific hydrocarbon to a minimal medium to select for degraders of that compound

EO: Ethoxylate (surfactant)
E°: Standard reduction potential
EPA: Environmental Protection Agency
EPDM: Ethylene propylene rubber, can be used as a liner
EPS: Extracellular polymeric substances; exopolysaccharide; polysaccharide produced by microorganisms external to the cell
Eukaryotic: Nucleus is surrounded by a membrane, as in fungi and higher organisms
Ex Situ: Latin for not in its original place. *Ex situ* treatments could be on site or off site
F-1: Controlled-release, hydrophobic fertilizer; modified urea-formaldehyde polymer with N and P
F-400 GAC: Granular activated carbon
Facultative: The ability to adapt to the conditions specified with this term
Facultative anaerobes: Microorganisms that are metabolically active under aerobic or anaerobic conditions
FBR: Fluidized-bed reactor
FDA: Fluorescein diacetate-hydrolyzing activity assay for determining biological potential of pelleted fungi for bioaugmentation
FID: Flame ionization detector
Field capacity: Water-holding capacity of soil
FITC: Fluorescein isothiocyanate
Fluor-X: Fluorochrome (fluorescing) DNA label
F/M: Food-to-microorganism ratio
FyreZyme™: Bioremediation enhancing agent containing extracellular enzymes, microbial nutrients, and bioemulsifiers
GAC: Granular activated carbon
GAC FBR: Integrated biological granular activated carbon fluidized-bed reactor
GAS 3D: Three-dimensional gas flow model to aid in design of soil-venting systems
GC: Gas chromatography
GC/MS: Gas chromatography/mass spectroscopy
GLC: Gas-liquid chromatography
GPMS: Gas-permeable-membrane supported (reactor)
Gram-negative bacteria: Bacteria that do not possess a cell wall and thus do not retain the blue dye that stains cell walls; these cells are enclosed by a cell membrane that will absorb the red counter stain in the Gram stain technique
Gram-positive bacteria: Bacteria that possess a cell wall, which retains the blue dye that stains cell walls in the Gram stain technique
H: Henry's law constant or Henry's coefficient
H_2O_2: Hydrogen peroxide, used to supply oxygen to the subsurface
Half-life: The time required to decrease original concentration by one half
Heterotrophic: The ability to derive energy and carbon for survival and growth from decomposition of organic materials
Hoechst 33342: Allows assessment of macromolecular composition for DNA
Homologous: Identical compounds except for number of repeating units
HPAH: High-molecular-weight PAH
HPCD: Hydroxypropyl-β-cyclodextrin
HRUBOUT®: Hot air injection process for soil flushing (Hrubetz Environmental Services)
HSWA: Hazardous and Solid Waste Amendments
Hydrocarbonoclastic: Ability to degrade and utilize hydrocarbons
Hydrolysis: Chemical reaction involving cleavage of a molecular bond by reaction with water
Hydrophilic: Water attracting
Hydrophobic: Water repelling
Hydrophobization: Conversion to a hydrophobic (water-repellent) state
Hydroxylation: Addition of OH to an aromatic or aliphatic molecule
ICB: Immobilized cell bioreactor (Allied Signal)
ICP/MS: Inductively coupled plasma/mass spectrometry
Igepal CA-720: A surfactant

Igepal CO-603: Nonionic ethoxylated alkylphenol surfactant
Indigenous microorganisms: Microorganisms occurring naturally in a particular region or environment; stable members of a community that have a selective, competitive advantage in that environment
Inipol EAP-22: An oleophilic fertilizer (Elf Aquitaine in France)
***In situ* bioreclamation:** Biodegradation operations taking place in the contaminated soil or groundwater without excavation
INT Activity Test: A dye, 2-(p-iodophenyl)-3-(p-nitrophenyl)-5-phenyl-tetrazolium chloride; identifies bacteria active in electron transport
Intrinsic bioremediation: Lack of intervention in bioremediation process, or natural attenuation; the result of several natural processes (e.g., biodegradation, abiotic transformation, mechanical disperion, sorption, and dilution) that reduce contaminant concentrations in the environment
Intrinsic remediation: Results from natural processes; e.g., biodegradation, abiotic transformation, mechanical dispersion, sorption, or dilution
***In vitro*:** In a test tube
***In vivo*:** In life
***K*:** Soil adsorption constant; the measure of the tendency of a pollutant to be adsorbed and stay on soil; the greater the K value, the stronger the binding
KAX-50, KAX-100: Proprietary rubber particulates as stabilization additives
K_{oc}: This value reflects the impact of organic material to adsorb organic compounds out of solution
K_{ow}: Octanol/water partition coefficient; also P
K_p: The linear partition coefficient
lacYZ: Gene for lactose utilization
Landfarming, Landtreatment: Controlled application of waste materials to soil for immobilization or for degradation or transformation by resident microorganisms
Leachate: Liquids generated by movement of liquids by gravity through a disposal site
LiP: Lignin peroxidase; extracellular fungal enzyme
Lipotin: Glycerophospholipids
Lithotrophic: The ability to obtain energy from oxidation of inorganic compounds
LNAPLs: Light nonaqueous-phase liquids
log K_{OH}^0: The atmospheric reaction rate of a specific compound
log K_{ow}, log P: A measure of the tendency of a compound to dissolve in hydrocarbons, fats, or organic component of soil rather than in water
LPAH: Low-molecular-weight PAH
LPH: Liquid-phase hydrocarbon
luxAB: Gene for bioluminescence
Macrofauna: Soil organisms, such as insects, protozoa, earthworms, and slugs that aid in decomposition of organic material
MARS: Membrane aerobic or anaerobic reactor system
MBR: Membrane biological reactor
MeOH: Methanol
Mesophilic: The ability to grow at temperatures from 10 to 45°C, with optimum growth around 20 to 40°C; most human pathogens grow best at 37°C
Methanogenic consortia: Groups of microorganisms that function under highly reducing conditions and produce methane from degradation of small or low-molecular-weight organic compounds
Methanogenesis: Conversion of short-chain organic compounds by anaerobic microorganisms to methane, carbon dioxide, and inorganic substances
Methanotroph: Microorganisms that break down methane
Methylation: Addition of a methyl group
MF: Microfiltration
Micelles: Surfactant molecules emulsify oily material into fine droplets that form aggregates 10 to 100 Å in diameter, called micelles
Microaerophilic: The ability to survive on very low levels of oxygen
Microbial consortia: Mixed population of interacting microorganisms
Microbial diffusion: Transport of microorganisms in soil as a result of (microbiological) life/death cycle and natural microbial movement; requires expenditure of energy

Mineralization: Complete biodegradation of organic molecules to mineral products; e.g., CO_2, NO_3^-, SO_4^{2-}, PO_4^{3-}. A portion of the carbon from the organic molecule is usually incorporated into biomass
MLSS: Mixed liquor suspended solids
MLVSS: Mixed liquor-volatile suspended solids
MnP: Manganese-dependent peroxidase; extracellular fungal enzyme
MPN: Most-Probable-Number; method for estimating counts of viable microorganisms
MPP: Multiplasmid *Pseudomonas putida*
MRI: Magnetic resonance imaging
mRNA: Messenger RNA
MS: Mass spectroscopy
MTBE: Methyl-*tert*-butyl ether, a gasoline additive
NaDBS: Sodium dodecyl benzosulfonate; colloidal gas aphron foam
Na₅DTPA: Pentasodium salt of diethylenetriaminepentaacetic acid; most effective of the commercially available chelating agents for preventing contact between metals and H_2O_2
NAH plasmids: Naphthalene-degrading plasmids
NALS: Narrow angle light scatter at the cell surface reflects cell size
NAPLs: Nonaqueous-phase liquids (e.g., solvents and fuels); pollutants present in liquids that are immiscible with water
Natural attenuation: Unassisted biochemical degradation, evaporation, adsorption, metabolism, or transformation by microorganisms of subsurface contaminants
ndoB: Naphthalene degradation gene probes
Nitrogen demand: Amount of nitrogen required for degradation of a given amount of contaminant
Normal flora: Mixed population of microorganisms occurring in nature
Novel II 1412-56: A nonionic alcohol ethoxylate surfactant
NRV: Nitrogen requirement value; the amount of nitrogen required by organisms to decompose or degrade a particular organic chemical
NTA: Nitrilotriacetic acid
NVOCs: Nonvolatile organic compounds
Obligate: Strict dependence upon the conditions specified with this term
Obligate anaerobes: Microorganisms that require the absence of oxygen
OCAs: Oil-core aphrons
ODR: Oxygen diffusion rate
Oligotrophic: The ability to survive on very low concentrations of nutrients
Oligotrophs: Organisms that can survive on low organic concentrations (<15 mg carbon/L)
OLR: Organic load rate
On-site bioreclamation: Biodegradation operations that occur above ground at the site of contamination
Operon: A DNA region that codes for several enzymes in a reaction pathway; it enables or prevents repression of structural gene function by controlling synthesis of mRNA by RNA polymerase enzyme
OR: Oxidation-reduction potential; Eh
Orange I: Azo dye; substrate for assaying for fungal manganese peroxidases
Orange II: Azo dye; substrate for assaying for fungal LiP
OTA: Congressional Office of Technology Assessment
P: Octanol/water partition coefficient; also K_{ow}
PAC: Powdered activated carbon
PAHs: Polycyclic aromatic hydrocarbons, also called polyaromatic hydrocarbons and PNAs
PB: Polybutylene, can be used as a liner
PCBs: Polychlorinated biphenyls
PCE: Tetrachloroethylene
PCPs: Polychlorinated phenols
PE: Polyethylene, can be used as a liner
PEL: Polyester elastomer, can be used as a liner
PHB: Poly-3-hydroxybutyrate, an intracellular storage polymer; improves eroding soil, enhances soil strength, reduces soil permeability
PHENOBAC® Mutant Bacterial Hydrocarbon Degrader: Mixture of mutant microorganisms (Polybac)

Photodegradation: Use of light for direct photodegradation or sensitized photo-oxidation to degrade organic compounds
Photolysis: Light-sensitized oxidation of resistant complex compounds
Photo-oxidation: UV light-induced oxidation for destruction of organic contaminants
Phototrophs: Organisms that derive energy from sunlight
PID: Photoionization detector
PISB: Passive *in situ* biotreatment
pKa: Dissociation constant indicates degree of acidity or basicity of a compound and thus the extent of adsorption and ease of desorption
Plasmids: Extra-chromosomal genetic material
Pleomorphs: Bacteria having multiple shapes
pMOL28 (163 kb): Microbial plasmids specifying nickel, mercury, chromate, cobalt, and thallium resistance
pMOL30 (240 kb): Microbial plasmids specifying zinc, cadmium, cobalt, mercury, copper, lead, and thallium resistance
pMOL85 (240 kb): Microbial plasmids specifying zinc, cadmium, cobalt, and copper resistance
PNAs: Polynuclear aromatic hydrocarbons, also called PAHs
Poly B-411: Polymeric dyes that serve as substrates for lignin degrading enzymes
Poly R-478: Polymeric dyes that serve as substrates for lignin degrading enzymes
Poly R-481: Polymeric dyes that serve as substrates for lignin degrading enzymes
Poly Y-606: Polymeric dyes that serve as substrates for lignin degrading enzymes
POLYBAC® E biodegradable emulsifier: Synthetic biodegradable emulsifier (Polybac)
POLYBAC® N biodegradable nutrients: Commercial fertilizer containing balanced nitrogen and phosphorus to enhance bioremediation by soil microorganisms (Polybac)
Pozzolanic: Materials such as Portland cement, fly ash, kiln dust
PP: Polypropylene, can be used as a liner
PRISM: Plasma remediation of *in situ* materials
Procaryotic "Nucleus": Single chromosome without a membrane, as in bacteria
Proppant: Material that props a fracture, as created in the subsurface by hydraulic/pneumatic fracturing
Protocooperation: Sequential metabolism of compounds by two or more microorganisms where both benefit
Pseudomonads: Bacteria belonging to the genus *Pseudomonas*
PSO: Petroleum sulfonate-oil surfactant; commercial Petronate
Psychrophilic: The ability to grow best at temperatures from –5 to 30°C, with optimum growth between 10 and 20°C
Psychrotrophs: Organisms growing optimally at lower temperatures (e.g., <20°C)
PUF: Porous polyurethane foam for immobilizing enzymes and living microorganisms
PUR: Reticulated polyurethane, a carrier material for microorganisms
PVC: Polyvinyl chloride, can be used as a liner
PWEs: Platinum wire electrodes
Q_{10} effect: Decrease in microbial enzymatic activity as a result of low temperature
qO_2: Oxygen consumption rate
RAAS: Remedial action assessment system
RAPD: Randomly amplified polymorphic DNA to characterize bacteria in biodegradation
RAS: Return activated sludge
RBC: Rotating biological contactor
RCRA: Resource Conservation and Recovery Act of 1976
RE: Removal efficiency
Recalcitrant: Resistant to microbial degradation
RF: Radio frequency heating to desorb organic contaminants from soil; improves soil venting by increasing vapor pressure of contaminants
RESOL 30: Solution containing nonionic and anionic biodegradable surfactants for soil washing
Restore 375: Fertilizer with sodium triphosphate combined with orthophosphates
Rexophos 25/97: A phosphate ester blend weak-acid anionic surfactant

RF: Radiofrequency heating
Rhamnolipid R1: A glycosylated, anionic, amphipathic surfactant secreted by *Pseudomonas aeruginosa*
Rhodamine 123: A bacterial stain to demonstrate viability
RNA: Ribonucleic acid
RREL: Risk Reduction Engineering Laboratory (EPA)
rRNA: Ribosomal RNA
RT-PCR: Reverse transcrition-coupled-PCR
SARA: Superfund Amendments and Reauthorization Act
SBR: Sequencing batch reactor
SCF: Self-cycling fermenter
SCF: Supercritical fluid oxidation
SDS: Sodium dodecyl sulfate, an anionic surfactant used for soil flushing
Serqua 710: A surfactant
SITE: Superfund Innovative Technology Evaluation Program
SITE ETP: SITE Emerging Technologies Program
SLB: Signature microbial lipid biomarker indicates viable biomasss
MMO: Soluble form of methane monooxygenase from the methanotroph, *Methylosinus trichosporium* OB3b
Sorption: Refers to both "adsorption", the retention of solutes in solution by the surfaces of the solid material and "absorption", retention of the solute within the mass of the solid
SPR: Single particle reactor
S^{SM}: Shallow soil mixing; can be combined with SVE to extract VOCs from soil
Sulfate-reducing bacteria: Strict anaerobic bacteria that use sulfate as a terminal electron acceptor, converting sulfate to sulfide
Sulfidogens: Organic acids are used as electron donors
Superbugs: Strains of microorganisms developed in the laboratory with the potential of biodegrading a range of contaminants
Superfund: See CERCLA
Surfactants: Surface active agents that promote the wetting, solubilization, and emulsification of organic chemicals
SVE: Soil vapor extraction; also called soil vacuum extraction, soil venting, and soil vapor stripping; reduces vapor pressure in soil and increases volatilization of contaminants, which are then withdrawn by the vacuum
SVOCs: Semivolatile organic compounds
Synergism: Sequential metabolism of compounds by two or more microorganisms where both benefit
Syntrophism: One organism supplying a missing nutritional requirement of another
TAD: Thermophilic aerobic digestion
TAH: Total aliphatic hydrocarbons
TBA: Tertiary butyl alcohol
TCLP: Toxicity characteristic leaching procedure
TEA: See terminal electron acceptor
TEL: Tetra-ethyl lead
Tergitol 15-S-9: An ethoxylated nonionic surfactant
Tergitol NP-10: A surfactant
Tergitol NPX: A surfactant
Terminal electron acceptor (TEA): Chemicals necessary for transfer during metabolic processes while microorganisms biodegrade contaminants; aerobic biodegradation of petroleum hydrocarbons requires the TEA, oxygen; anaerobic biodegradation requires the TEAs iron, sulfate, or nitrate
TESVE: Thermally enhanced soil vapor extraction
Tetren: Tetraethylenepentamine, a chelator
Thermophilic: The ability to grow at temperatures from 25 to 80°C, with optimum growth at 50 to 60°C
TLD-FID: Thin-layer chromatography-flame ionization detection
TLV: Threshold limit value
TOL plasmid: Toluene-degrading plasmid in bacteria

TPH: Total petroleum hydrocarbons
Treatment trains: Use of more than one technology or process, in series or in parallel, to remediate a contaminated site; these will be site and incident specific
Triton N101: A surfactant
Triton X-100: An ethoxylated nonionic surfactant
Triton X-114: A surfactant
TSDFs: Treatment storage and disposal facilities
Turnover time: The amount of time required to remove the concentration of substrate present
Tween-20-80: A surfactant
Tween-80: A surfactant
Tysul WW: H_2O_2 from du Pont, used in Biostim to provide oxygen to the soil
UASB: Upflow anaerobic sludge blanket
UST: Underground storage tank
UTCHEM: Multiphase, multicompositional simulator to model migration and surfactant-enhanced remediation of an NAPL
UV: Ultraviolet
Vadose zone: Unsaturated soil above water table
VES: Vapor extraction system; also called soil vacuum extraction, soil venting, and soil vapor stripping; *in situ* technique for removing VOCs from soil
VOCs: Volatile organic compounds
WALS: Wide angle light scatter reflects internal cell structure as refractility
WIGEs: Wax-impregnated graphite electrodes
Xenobiotic: Compounds that are man-made or are unique in nature; also refers to compounds released in the environment by the action of humans and, thereby, occur in a concentration that is higher than natural
***xylE*:** Toluene, xylene degradation gene probes
Zymogenous: Soil bacteria that increase rapidly when furnished with certain nutrients and then diminish in numbers when the material is exhausted

ILLUSTRATIONS

Figure

2.1	Flameless thermal oxidizer	9
2.2	Wet air oxidation (WAO) process	23
2.3	Biological activated carbon/wet air oxidation combination process schematic	24
2.4	SCF reactor for retrofit application	25
2.5	SCF reactor for stand-alone application	26
2.6	Process schematic of a typical UV/ozone system	26
2.7	Treatment zone definition	32
2.8	Fate of refinery waste at a landfarm site	33
2.9	Evaluation of the economics of options meeting environmentally acceptable performance standards for hazardous wastes	37
2.10	System flowchart of the Ebiox Vacuum Heap™ systems: air circulation and water circulation	53
2.11	Biological water treatment processes	55
2.12	Schematic of rotating biological contactor	60
2.13	Oxitron system fluidized-bed process schematic	62
2.14	Membrane aerobic or anaerobic reactor system (MARS)	62
2.15	Biochemical removal of anthropogenic organic compounds from water	67
2.16	Removal of organic contaminants with photosynthetic microorganisms	68
2.17	Mechanisms for photosynthetic organism growth	69
2.18	Reactor configurations for anaerobic biotechnology	71
2.19	Conventional anaerobic digester	72
2.20	Anaerobic activated sludge process	72
2.21	Upflow anaerobic sludge blanket (UASB)	73
2.22	Diagram of typical anaerobic filter system	75
2.23	Anaerobic expanded/fluidized bed	75
2.24	Schematic diagram of experimental anaerobic activated carbon filter	77
2.25	Reactor schematic for the growth of methylotrophs	78
2.26	Soil flushing	81
2.27	Schematic of an elutriate recycle system	82
2.28	Schematic of air stripping process equipment	88
2.29	A typical process layout for vapor extraction	90
2.30	The *in situ* Detoxifier™	94
2.31	HRUBOUT® process	96
2.32	Steam-enhanced recovery process	97
2.33	Bioreclamation technology for treatment of contaminated soil and groundwater	101
2.34	Schematic diagram of bioventing installation: section view	103
2.35	Well construction detail and slurper tube placement for the skimmer test configuration	106
2.36	BioSparge^{SM}/BioPurge^{SM} schematic	107
2.37	Hydraulic fracturing	108
2.38	Pneumatic fracturing schematic	109
2.39	Electrokinetic remediation process	110
3.1	Differences between the reactions used by eukaryotic and prokaryotic organisms to initiate the oxidation of aromatic hydrocarbons	125
3.2	Hydrocarbon substrate recalcitrance: relative proportions of hydrocarbon fractions	128
3.3	Terminal or diterminal oxidation of alkanes or aliphatics	129
3.4	Pathways utilized by prokaryotic and eukaryotic microorganisms for the oxidation of PAHs	142
3.5	Homocyclic aromatic "benzenoid" nucleus (enclosed) of benzoate (a) and heterocyclic aromatic "pyridine" nucleus (enclosed) of nicotinate (b)	168
3.6	Possible sequencing of soil manipulation	181
6.1	Typical chart trace from a total hydrocarbon monitor	320
6.2	Overall system for treatment, storage, and disposal of VOCs	321

6.3	Approximate ranges of applicability of VOC removal techniques as a function of organic concentration in liquid waste stream	334
6.4	Schematic of soil gas VES	341
6.5	Aeration devices that put water through air	348
6.6	Fixed-bed carbon adsorption system	352
6.7	Schematic of solvent recovery by condensation and distillation	354
6.8	Schematic of VOC recovery by decantation	355
6.9	Schematic diagram of fluidized-bed carbon adsorption system	356
6.10	Schematic illustration of the low-temperature thermal stripping pilot system	360
6.11	Schematic process diagram of the full-scale low-temperature thermal stripping system	361
6.12	Thermal incinerator with primary and secondary heat recovery	366
6.13	Regenerative thermal incinerator	368
6.14	Catalytic incinerator with primary heat recovery	370
6.15	Diagram of an inert gas condensation solvent recovery system	373
6.16	Typical packed absorption column	374
6.17	Bubble cap absorption column	375
8.1	Decision framework for remediation technologies	418
8.2	System integration	419
8.3	Schematic of biological/carbon sorption process train	424
8.4	Process train for leachate containing metals	425
8.5	Air stripping, carbon adsorption, and ion exchange process flow diagram (typical)	426
8.6	Process diagram for the Detoxifier™ II treatment train	427
8.7	Treatment train used at a southern California site	428

TABLES

Table

2.1	Summary of Suitability of Treatment Processes	6
2.2	Technology Applicability	7
2.3	Critical Factors in a Leaching Procedure	15
2.4	Treatment Process Applicability Matrix	16
2.5	Leachate Treatment Process By-Product Streams	17
2.6	Activated Carbon Adsorption of Organics	21
2.7	Average Microbial Degradation of Oily Waste after 1 Year and after 6 Months	40
2.8	Biodegradation Rates Reported from Full-Scale Landtreatment Operations	40
2.9	Rates of Degradation of Phenol and Substituted Phenols	41
2.10A	Operational Characteristics for Landtreatment Facilities 01 to 06	42
2.10B	Operational Characteristics for Landtreatment Facilities 07 to 13	43
2.11	Factors Governing Selection of FBR vs. MARS in the Treatment of Hazardous Water and Wastewater	63
3.1	Composition of Diesel Fuel #2	116
3.2	Composition of Various Gasolines	116
3.3	Components of Gasoline	117
3.4	Composition of Gasoline	118
3.5	Selected Compound Types Occurring in JP-5	121
3.6	Trace Elements in Shale-Derived JP-5	121
3.7	Major Components of JP-5	122
3.8	Composition of JP-4	122
3.9	Major Components of JP-4	123
3.10	Trace Elements in Petroleum-Based JP-4	124
3.11	BOD_5/COD Ratios for Various Organic Compounds	124
3.12	Relationship between Representative Microbial Processes and Redox Potential	125
3.13	Hydrocarbons and Their Relative Levels of Recalcitrance	127
3.14	Aromatic Hydrocarbons Known to Be Oxidized by Microorganisms	132
3.15	Degradation of Anthropogenic Compounds by Different Groups of Microorganisms	132
3.16	Growth of Microorganisms on Components of Gasoline	133
3.17	Biodegradation of Gasoline Components by Mixed Normal Microflora	134
3.18	Fuel Components/Hydrocarbons and Microorganisms Capable of Biodegrading/Biotransforming Them	135
3.19	Microbial Growth on Selected Branched Alkanes	157
3.20	Microbial Mechanisms for Metal Extracting/Concentrating/Recovery	184
3.21	Products/Metabolites Formed from the Oxidation of Petroleum Hydrocarbons by Various Microorganisms	196
4.1	Solubility and Biodegradability of Some Common Organic Contaminants	201
4.2	Log K_{ow} Values for Several Compounds	203
4.3	Soil Adsorption Constants and Water Quality Criteria for Hydrophobic Organics	205
4.4	Soil Adsorption Constants and Water Quality Criteria for Hydrophilic Organics	206
4.5	Summary of Organic Groups Subject to Biodegradation	209
4.6	Relative Persistence and Initial Degradative Reactions of Nine Major Organic Chemical Classes	210
4.7	Problem Concentrations of Selected Chemicals	214
4.8	Biodegradable Concentrations of Compounds in the Field	215
4.9	Biodegradable Concentrations of Compounds in Bioreactors or Small-Scale Studies	216
5.1	Important Site and Soil Characteristics for *In Situ* Treatment	220
5.2	Soil Modification Requirements for Treatment Technologies	222
5.3	Selective Use of Microorganisms for Removal of Different Anthropogenic Compounds	223
5.4	Mulch Materials	229
5.5	Liming Materials	232
5.6	Oxygen Supply Alternatives	234

5.7	Estimated Volumes of Water or Air Required to Completely Renovate Subsurface Material That Originally Contained Hydrocarbons at Residual Saturation	234
5.8	Nutrients Required by Microorganisms	250
5.9	Succession of Events Related to the Redox Potential, Which Can Occur in Waterlogged Soils or Poorly Drained Soils Receiving Excessive Loadings of Organic Chemical Wastes or Crop Residues	261
5.10	Microbial Genera Degrading Hydrocarbons in Soil	263
5.11	Autotrophic Modes of Metabolism	265
5.12	Fungi Capable of Growth on a Variety of Crude Oils	269
5.13	Microorganisms That Can Bioaccumulate Anthropogenic Compounds	275
5.14	Microbial Species Exhibiting the Phenomenon of Cometabolism	295
5.15	Organic Compounds, Analogs/Growth Substrates, Microorganisms, and/or Products of Co-oxidation	295
5.16	Organic Substances Subject to Cometabolism and Accumulated Products	296
5.17	Biosurfactant-Producing Microorganisms	312
5.18	Rate Constants for the Hydroxide Radical Reaction in Air with Various Organic Substances	314
5.19	Atmospheric Reaction Rates and Residence Times of Selected Organic Chemicals	315
6.1	Emissions from Hazardous Waste Treatment, Storage, and Disposal Facilities	318
6.2	Most Frequently Reported Substances at 546 MPL Sites	318
6.3	VOCs Included on the Hazardous Substance List (HSL)	319
6.4	Cumulative Measured Emissions of Selected Individual Compounds	321
6.5	Approximate Gasoline Vapor Components	323
6.6	EPA Water Quality Criteria for Protection of Human Health (10^{-5} Risk Level)	324
6.7	Vapor Pressure for Several Compounds	328
6.8	Henry's Law Constants for Several Compounds	330
6.9	Time to Diffuse L = 1 m through a Soil	332
6.10	Estimated Air-to-Water Ratios Necessary to Achieve Desired Water Quality	346
6.11	Summary of LTTS System Data for Thermal Stripping Systems with and without Flue Gas Scrubbing	362
6.12	The Destructibility of Different Compound Classes	371
6.13	Temperatures for 90% Conversion in a Catalytic Incinerator	371
6.14	Comparison of Packed and Plate Columns	375
6.15	Treatment Process Residuals	381
7.1	Distribution of Microorganisms in Various Horizons of a Soil Profile	392
7.2	Distribution of Aerobic and Anaerobic Heterotrophic Bacteria and Fungi with Depth in a Retorted Shale Lysimeter	392
7.3	Bacterial Populations in Subsurface Soils	393
7.4	Summary of Viable and Direct Counts in Uncontaminated Soils from Several Studies	394
7.5	Summary of Viable and Direct Counts in Contaminated Soils from Several Studies	395
7.6	Summary of Effect of Biostimulation on Counts in Contaminated Soils from Several Studies	397
7.7	Kinetic Parameters Describing Rates of Degradation of Aromatic Compounds in Soil Systems	410
7.8	Turnover Times for Microbial Hydrocarbon Degradation in Coastal Waters	413
7.9	Reported Oil and Grease Biodegradation Rates in Soil	414

Section 1

Introduction

1.1 BACKGROUND

Many of the standard treatment processes used to decontaminate soil and groundwater have been limited in their application, are prohibitively expensive, or may be only partially effective (Nicholas, 1987). Problems associated with the cleanup of leaking disposal sites and spills of toxic substances have demonstrated the need to develop remediation and waste reduction technologies that are efficient, economical, and rapidly deployable in a wide range of physical settings (Catallo and Portier, 1992).

Traditional methods of treating soil and groundwater contamination have relied upon removal or containment (Brown, Loper, and McGarvey, 1986). These were found to be the most common techniques in a survey of 169 remedial actions (Neely, Walsh, Gillespie, and Schauf, 1981). Traditional remediation efforts at hazardous waste sites have been partially effective 54% of the time and completely successful only 16% of the time (Neely, Walsh, Gillespie, and Schauf, 1981; Lee and Ward, 1985). Most of these treatment schemes are not completely effective and do not offer permanent solutions for containment or remediation. Some methods may even create additional uncontrolled hazardous waste.

There are circumstances where excavation of contaminated soils is feasible (FMC Aquifer Remediation Systems, 1986). For instance, in repairing or replacing underground tanks or pipelines, contaminated soil must be removed to gain access to the faulty equipment. Excavation can also be practical in dealing immediately with high concentrations that pose a health or environmental hazard. Excavation can also be cost-effective by preventing groundwater contamination from substances limited to the surface. However, excavation raises the question of what to do with the contaminated soil that essentially transfers the contamination from one site to another.

1.2 BIODEGRADATION AS A TREATMENT ALTERNATIVE

Government, industry, and the public have come to recognize the need for greatly reducing the volume and toxicity of waste and developing safe, effective, and economic alternatives for its disposal (Nicholas, 1987).

In situ and on-site treatment processes avoid the economic and technical disadvantages, as well as environmental risks, incurred by transport of hazardous wastes to alternative treatment facilities (Ahlert and Kosson, 1983). Both *in situ* and on-site biological processes involve the use of microorganisms to break down hazardous organic environmental contaminants (Lee and Ward, 1985). Both employ many of the same procedures and are influenced by the same environmental factors.

Natural attenuation in the subsurface environment is accomplished by biochemical degradation, evaporation, adsorption, metabolism, and transformation by microorganisms (Brown, Loper, and McGarvey, 1986). It is well known that microorganisms are capable of degrading a wide range of organic compounds (Pierce, 1982b; Jhaveri and Mazzacca, 1983; Flathman and Caplan, 1985).

An organic chemical may be subjected to nonenzymatic or enzymatic reactions brought about by microorganisms in the soil (Alexander, 1980a); however, it is the enzymatic reactions that bring about the major changes in the chemical structure of these compounds. Extensive removal of organic materials is accomplished primarily through enzymatically mediated biological reactions, i.e., biodegradation (Thornton-Manning, Jones, and Federle, 1987). Few abiotic processes completely mineralize complex organic compounds in soil, and complete degradation depends upon microbial activity (Alexander, 1981). However, physical/chemical transformation processes may act synergistically with biochemical decomposition in this process.

The term *biodegradation* is often used to describe a variety of quite different microbial processes that occur in natural ecosystems, such as mineralization, detoxication, cometabolism, or activation (Alexander, 1980b). It can be defined as the breakdown of organic compounds in nature by the action of microorganisms, such as bacteria, actinomycetes, and fungi (Sims and Bass, 1984). The microorganisms derive energy and may increase in biomass from most of the processes (Lee and Ward, 1985).

The organisms that occur naturally in almost every soil system (the indigenous microbial populations) appear to be the chief agents involved in the metabolism of the chemicals in waters and soils. Heterotrophic bacteria and fungi are responsible for most of the chemical transformations. Aerobic bacteria, actinomycetes, cyanobacteria, anaerobic bacteria, fungi, and some true algae have all been shown to be capable of degrading a wide variety of organic chemicals (Amdurer, Fellman, and Abdelhamid, 1985).

Many aerobic bacteria found in soil and water utilize biologically produced substances, such as the sugars, proteins, fats, and hydrocarbons of plant and animal wastes, and can also metabolize petroleum hydrocarbons (Brown, Norris, and Brubaker, 1985). During biodegradation, certain anaerobic bacteria commonly produce short-chain organic acids, while other microorganisms further break down these by-products to methane, carbon dioxide, and inorganic substances (Pettyjohn and Hounslow, 1983).

Natural soil bacteria may be present in a dormant or slow-growing state, but when stimulated by optimum environmental conditions, they multiply rapidly and subsequently adapt to the new environment. Some of the more common genera involved in biodegradation of oil products include *Nocardia*, *Pseudomonas*, *Acinetobacter*, *Flavobacterium*, *Micrococcus*, *Arthrobacter*, and *Corynebacterium*. Studies have revealed that cultures containing more than one genus have greater hydrocarbon-utilizing capabilities than the individual culture isolates.

Mineralization occurs when there is complete biodegradation of an organic molecule to inorganic compounds — i.e., carbon dioxide, water, and mineral ions — based on nitrogen, phosphorus, sulfur (Sims and Bass, 1984; JRB Associates, Inc., 1984), cell components, and products typical of the usual catabolic pathways (Alexander, 1994). Under anaerobic conditions, methane may be produced, and nitrate nitrogen may be lost as N_2 or N_2O gas through denitrification. Microorganisms can also transform hazardous organic compounds into innocuous or less-toxic organic metabolic products. Chemical alteration can also be the result of cometabolism (co-oxidation); i.e., growth on another substrate while the organic molecule is degraded coincidentally (Alexander, 1980a). This latter process may be promoted by enzymes that catalyze reactions of chemically related substrates. Contaminants in solution in groundwater, as well as vapors in the unsaturated zone, can potentially be completely degraded or transformed to new compounds (Wilson, Leach, Henson, and Jones, 1986).

Biorestoration is useful for hydrocarbons, especially water-soluble compounds and low levels of other compounds that would be difficult to remove by other means (Lee and Ward, 1985). It is environmentally sound, since it destroys organic contaminants and, in most cases, does not generate problem waste products.

Degradation in the contaminated soil and aquifers may be affected by environmental constraints, such as dissolved oxygen, pH, temperature, toxicants, oxidation-reduction potential, availability of inorganic nutrients (e.g., nitrogen and phosphorus), salinity, and the concentration and nature of the organics. The number and type of organisms present in the environment will also play an important role in this degradation process.

Treatment, therefore, generally consists of optimizing conditions of pH, temperature, soil moisture content, soil oxygen content, and nutrient concentration to stimulate the growth of the organisms that will metabolize the particular contaminants present (Sims and Bass, 1984). Optimum environmental conditions and nutrient application rates generally have to be established in laboratory bench-scale studies and small field pilot tests. Some hazardous compounds may be degraded more readily under aerobic conditions, and some under anaerobic conditions. Anaerobiosis is always present at microsites in soil. Remediation might, therefore, consist of a combination of both aerobic and anaerobic treatment methods.

Biodegradation techniques are versatile and can be used at different stages of treatment or with different approaches (Nicholas, 1987). For instance, microorganisms or their active products (e.g., enzymes) can be released directly into the contaminated environment; microorganisms already present in the environment can be enhanced by the addition of oxygen and suitable nutrients; or microorganisms can be used in contained or in semicontained reactors.

If the locally occurring organisms are not effective for the given set of contaminants, the soil can be inoculated with microbial isolates that are specific for those compounds (Buckingham, 1981). Microorganisms can be selectively adapted by growth on media containing the target chemicals. They can even

INTRODUCTION

be genetically engineered to enhance their ability to degrade these compounds. The use of genetically altered microorganisms may, in the future, expand the range of compounds that can be degraded and/or accelerate the rates at which degradation occurs (Walton and Dobbs, 1980).

Biodegradation is a feasible option for treatment of hazardous chemical spills (Buckingham, 1981). Its use depends upon the availability of necessary equipment and manpower, as well as upon decontamination time restrictions. Biodegradation of massive spills may prove ineffective in terms of percent recovery per unit time, whereas it may be the only treatment alternative for low level quantities or concentrations of contaminants.

According to the Office of Technology Assessment (OTA), once developed and proved, biodegradation is potentially less expensive than any other approach to neutralizing toxic wastes (Nicholas, 1987). Such systems involve a low capital investment, have a low energy consumption, and are often self-sustaining operations. Biological means of decomposition require less energy than physicochemical processes and can be a competitive option under certain circumstances (Ghisalba, 1983).

For commercial applications of biological technologies to achieve more widespread use, they must meet specific criteria (Scholze, Wu, Smith, Bandy, and Basilico, 1986):

1. *Degradation* — The end result of biological treatment should be to destroy the hazardous constituents of concern completely. This will allow the residues to be disposed of in nonhazardous landfills at a much lower cost.
2. *Concentration* — Dilute hazardous wastes can be difficult to manage cost-effectively, since many chemical and thermal treatments are only cost-effective on concentrated waste constituents. Dilute hazardous wastes might be most cost-effectively treated by using biological treatment to concentrate the organic constituents, followed by thermal treatment of the residue.
3. *Diverse Target Constituents* — Treatment often has to include management of recalcitrant compounds or combinations of many hazardous constituents. Organisms in commercial applications should be able to degrade mixtures of organics. It would also be advantageous for the organisms to accomplish multiple tasks, such as degrading organics while they concentrate inorganics for further treatment or recovery.
4. *Consistency* — Variability in degradation efficiency among batches would be costly to a waste management firm; consistency in the composition of the residue is essential. The end product of the biological process should be predictable to keep monitoring costs down. Consistency in the degradation products also allows the biological process to be used in sequence with many other treatment processes, which may be sensitive to changes in constituent concentrations.
5. *Relatively Low Cost* — It has been difficult to reduce the use of landfills because of the high costs associated with alternative options. Biological treatment must be able to compete with other processes (e.g., chemical and thermal applications) in effectiveness and cost.

1.3 COMBINED TECHNOLOGIES

The remediation approach selected for a given contamination incident should be site and incident specific. The many variables associated with the nature of the pollutants, the environmental conditions of the site, and the microorganisms present or required for augmentation must be evaluated. In response to the growing need to respond to environmental contamination, many new technologies have been developed to treat soil, leachate, wastewater, and groundwater contaminated by petroleum products, including both *in situ* and *ex situ* methods. New and innovative approaches are being continuously explored for this application.

The solution to the needs of a particular site may ultimately require a combination of procedures selected to allow optimum remediation for the prevailing conditions. Biological, physical, and chemical techniques may be used in series or in parallel to reduce the contamination to an acceptable level. Use of such treatment trains is further discussed in Section 8, and specific suggestions are given throughout the text.

The following literature and technology survey provides a description of a wide variety of biological, physical, and chemical techniques currently available for dealing with soil contaminated by petroleum and related products. Since some of these processes may generate a contaminated liquid waste stream, methods employed for treating leachate, wastewater, or groundwater are also described and evaluated for potential application to this phase of the problem. Since volatile organic compounds may also be released during remediation, methods for controlling and treating contaminated vapors are also addressed. Remediation of polluted soil is, thus, a multidimensional problem, not only in the complexity of the factors that can affect the restoration efforts, but also in the need to select appropriate technologies for the different phases of treatment required for a specific pollution incident. The appropriate procedures will facilitate the treatment process while keeping costs down. The information in this book should help those involved with remediation of petroleum-contaminated soils in their determination of the best treatment options.

Section 2

Current Treatment Technologies

Soil treatment technologies are often developed and evaluated in order to conform with regulatory demands, which may require or suggest that residual total petroleum hydrocarbon (TPH) concentrations in soil be reduced below 1000 mg/kg or, in some areas, below 100 mg/kg TPH.

There are many technologies available for treating sites contaminated with petroleum hydrocarbons; however, the treatment selected depends upon contaminant and site characteristics, regulatory requirements, costs, and time constraints (Ram, Bass, Falotico, and Leahy, 1993). These authors propose a decision framework that is structured and tiered for selecting remediation technologies appropriate for a given contamination incident. Commonly used technologies can be integrated to enhance performance. Variation in design and implementation of the technologies, with concurrent or sequential configurations, can help to optimize the effectiveness of the treatment.

The American Petroleum Institute (API) developed a petroleum decision framework to facilitate decision making for investigation and cleanup of petroleum contamination of soils and groundwater (API, 1990). Kelly, Pennock, Bohn, and White (1992) of the U.S. Department of Energy Pacific Northwest Laboratories also produced a Remedial Action Assessment System (RAAS) for information on remedial action technologies. The EPA Risk Reduction Engineering Laboratory (RREL) provides a treatability database, which is accessible through the Office of Research and Development network retrieval system, the Alternative Treatment Technology Information Center (ATTIC), the EPA database for technical information on innovative treatment technologies for hazardous waste and other contaminants (*Haztech News,* 1992; Devine, 1994). An expert system for remediation cost information, Cost of Remediation Model (CORA), has been designed by EPA. EPA has also compiled descriptions of technologies for processes that treat contaminated soils and sludges (U.S. EPA, 1988). Emerging and developing technologies being studied in the EPA Superfund Innovative Treatment Evaluation (SITE) Program are also described (U.S. EPA, 1991). The EPA Soil Treatability Database organizes and analyzes treatment data from a variety of technologies, including innovative technologies (e.g., biotreatment, chemical extraction, and thermal desorption), for the applicability and performance in treating hazardous soil (Weisman, Falatko, Kuo, and Eby, 1994).

The successful treatment of a contaminated site depends on designing and adjusting the system operations based on the properties of the contaminants and soils and the performance of the systems, and by making use of site conditions rather than force-fitting a solution (Norris, Dowd, and Maudlin, 1994). Integration of bioremediation with other technologies either simultaneously or sequentially can result in a synergistic effect among the techniques employed (National Research Council, 1993).

Information regarding remediation systems is furnished by Katin (1995) to explain to the practicing plant engineer or small business person how to recognize a good design and the aspects of a good design that will allow ease of operation and maintenance. Remediation systems discussed include air strippers, oil/water separators, vacuum extraction systems, thermal and catalytic incinerators, carbon beds, sparging systems, and biological treatment systems.

Table 2.1 lists a number of unit operations and the waste types for which they are effective (Canter and Knox, 1985). Table 2.2 compares various features and the applicability of a variety of remediation technologies (Ram, Bass, Falotico, and Leahy, 1993).

2.1 ON-SITE OR *EX SITU* PROCESSES

Excavation is a common approach to dealing with contaminated soil (Lyman, Noonan, and Reidy, 1990). The excavated soil may be treated on site, treated off site, or disposed of in landfills without treatment. If treated, it may then be returned to the excavation site. Excavation is easy to perform, and it rapidly removes the contamination from the site in a matter of hours, as opposed to other remediation methods, which may require several months. It is often used when urgent and immediate action is needed.

There are problems associated with excavation (U.S. EPA, 1989). It allows uncontrolled release of contaminant vapors to the atmosphere. Nearby buildings, buried utility lines, sewers, and water mains

Table 2.1 Summary of Suitability of Treatment Processes

Process	Volatile Organics	Nonvolatile Organics	Inorganics
Air stripping	Suitable for most cases	Not suitable	Not suitable
Steam stripping	Effective concentrated technique	Not suitable	Not suitable
Carbon adsorption	Inadequate removal	Effective removal technique	Not suitable
Biological	Effective removal technique	Effective removal technique	Not suitable — metals toxic
pH adjustment precipitation	Not applicable	Not applicable	Effective removal technology
Electrodialysis	Not applicable	Not applicable	Inefficient operation/inadequate removal
Ion exchange	Not applicable	Not applicable	Inappropriate technology — difficult operation

Source: From Canter, L.W. and Knox, R.C., *Ground Water Pollution Control,* Lewis Publishers, Boca Raton, FL, 1985.

could be in the way, and aboveground treatment approaches tend to be more expensive than *in situ* methods. Contaminated soil may be considered a hazardous waste, and disposal is becoming increasingly restricted by regulation. In addition, the excavation site must be filled.

The following physical, chemical, and biological processes are some of the techniques that might be employed to treat the contaminated soil, once it has been excavated and transported to an on-site or off-site location.

2.1.1 PHYSICAL/CHEMICAL PROCESSES
2.1.1.1 Soil Treatment Systems
2.1.1.1.1 Thermal Treatment

Thermal desorption is an innovative, nonincineration technology for treating soil contaminated with organic compounds (Fox et al., 1991). It is a proven method in the field of nonhazardous waste treatment and can be used for treating petroleum-contaminated soils (Molleron, 1994). Contaminated soil is heated under an inert atmosphere to increase the vapor pressure of the organic contaminants, transferring them from the solid to the gaseous phase (Wilbourn, Newburn, and Schofield, 1994). This separates the organics from the soil matrix.

Boehm (1992) describes an on-site/off-site method to treat polluted soil, which is based on a thermal process to remove oxidizable, organic pollutants with low boiling points. The thermal treatment plant consists of a mechanical pretreatment of soil material, a thermal treatment in a rotary kiln, and an outlet-gas treatment. Since 1987, a mobile pilot plant has been in operation and has demonstrated remarkable success by cleaning up more than 70 different kinds of soil.

Low-temperature thermal treatment (low-temperature thermal stripping or soil roasting) can be used on excavated, contaminated soil (Ram, Bass, Falotico, and Leahy, 1993). A mobile thermal processor, which uses low-temperature thermal treatment of soils contaminated by volatile organic compounds (VOCs) is described by Velazquez and Noland (1993). With this method, the soil is heated to 450°C in an indirect heat exchanger. Jensen and Miller (1994) cite the requirement of heating the soil to >600°C for successful thermal treatment of petroleum-contaminated soil.

The effect of thermal treatment by means of a natural gas-fired, batch, rotary kiln; by a single particle reactor (SPR); and by a rotary reactor (BSRR) on toluene, naphthalene, and hexadecane was studied at 300 to 650°C (Larsen, Silcox, and Keyes, 1994). The ease at which the hydrocarbons were removed were toluene > naphthalene > *n*-hexadecane, and increasing the temperature increased their desorption rates. Moisture had a large effect on the desorption rate, which was first order with respect to individual and total hydrocarbon concentrations.

Chern and Bozzelli (1994) showed that a continuous-feed rotary kiln is highly effective in removing volatile and semivolatile organic contaminants from sand and soils. Temperature, residence time volatility, and purge gas velocity are the main parameters affecting the desorption, with higher temperatures and longer residence times resulting in higher removal efficiency. For complete removal (98%) of the organics at 20 min residence time, the temperature should be 100°C for 1-dodecene, 200°C for 1-hexadecene, 150°C for naphthalene, and 250°C for anthracene.

CURRENT TREATMENT TECHNOLOGIES

Table 2.2 Technology Applicability

Technology	Applicability	Soil Type and Saturated Zone Characteristics	Variations	Cost	Permits
LPH recovery LPH withdrawal	All lighter-than-water petrochemicals except for the most viscous fuel and lube oils	Works better with more-permeable soils	Total fluid extraction, passive bailers, dual pump recovery, recovery wells, thermally assisted LPH recovery, mop and disk skimmers	Variable	Groundwater discharge, product storage, and possibly, groundwater withdrawal
Vadose zone Soil vapor extraction	LPH less than about 0.5 ft, contaminants with Vp > 1 mmHg (BTEX, gasoline, MTBE, PCE, TCE, TCA, mineral spirits, MeOH, acetone, MEK, etc.)	Permeable soils, ROI > 10 ft, depth-to-water greater than 3 ft	Thermally assisted venting, horizontal venting, surface sealing, passive vent points, closed loop venting, concurrent groundwater pumping for VOCs in capillary fringe	Low	Air discharge permit may be required
In situ percolation (bioremediation)	Any aerobically biodegradable chemical in the vadose zone	Works better in permeable soils; depth-to-water greater than 3 ft	Oxygen and nutrients need to be supplied to the subsurface	Low to moderate	Air discharge permit may be required when soil venting used to provide oxygen
Excavation	All soils and contaminants	All soil types	Dewatering may be used to expose soils in capillary fringe	High	On-site treatment of excavated soil may require permitting
Saturated zone Sparging	Contaminants in saturated zone with K$_H$ > 0.1 and Vp > 1 mmHg; contaminants: BTEX, gasoline, PCE, TCE, TCA, mineral spirits	Hydraulic conductivity > 10^{-5} cm/s (silty sand or better); at least 5 ft of saturated thickness	Hot air, steam, and cyclic sparging, concurrent groundwater pumping	Low	Air discharge permit; water discharge if concurrent groundwater pumping
In situ bioremediation	Any biodegradable chemical in the saturated zone; inhibited by pH extremes, heavy metals, and toxic chemicals	Nutrients are transported better in more-permeable soil	Oxygen supplied by sparging or peroxide addition; nutrient addition with groundwater recovery and reinjection	Moderate to high	Water discharge for nutrient injection, air discharge if performed with sparging/venting
Excavation	All soils and contaminants	All soil types	Dewatering needed, groundwater containment may be used (slurry walls, sheet piles)	Very high	Permits for dewatering operations

Table 2.2 (continued)   Technology Applicability

Technology	Applicability	Soil Type and Saturated Zone Characteristics	Variations	Cost	Permits
Groundwater recovery and treatment Groundwater recovery	Uses: (1) LPH recovery, (2) provides hydraulic control of contaminant plume, (3) pump and treatment technologies	Transmissivity, depth-to-water and saturated-zone thickness determine optimal strategy	Recovery wells, well points, interceptor trenches	Variable	Well installation, groundwater withdrawal and groundwater discharge
Liquid-phase carbon	Removal of compounds with low solubility/high adsorptivity	See groundwater recovery	High pressure (75 to 150 psi) and low pressure (12 to 15 psi)	Low to high depending on contaminant loading	Water discharge permit
Air stripping	Compounds with $K_H > 0.1$; contaminants with K_H between 0.01 and 0.1 may require an air-water ratio > 100	See groundwater recovery	Packed towers, low profile, heated and closed-loop air stripping; off-gas treatment may be required	Low, if no off-gas treatment required	Air and water discharge permits
Advanced oxidation	Most effective on sulfide cyanide, double-bonded organics (PCE, TCE), BTEX, phenols chlorophenols, PCBs, PAHs, some pesticides	See groundwater recovery	Hydroxy/radicals produced by combinations of UV, ozone, and peroxide	Moderate to high	Water discharge permit
Bioreactors	Any biodegradable compound	See groundwater recovery	Fixed-film and suspended growth reactors	Moderate to high	Water discharge permit
Off-gas treatment Vapor-phase carbon	Adsorptive capacity generally increases with increasing molecular weight	NA	Pretreatment dehumidification; on-site regeneration	Moderate	Air discharge permit
Catalytic oxidation	Conventional units can treat all compounds containing carbon, hydrogen, and oxygen; concentrations should not exceed about 20% of the LEL	NA	Some units can treat chlorinated compounds, exhaust gas scrubbing may be required	Moderate to high	Air discharge permit
Thermal oxidation	Compounds containing carbon, hydrogen, and oxygen; usually not amenable to halogen-containing compounds	NA	Exhaust gas scrubbing may be required	Moderate to high	Air discharge permit

Abbreviations: NA, not applicable; LEL, lower explosion limit; ROI, radius-of-influence; LPH, liquid-phase hydrocarbon; MTBE, methyl *tert*-butyl ether; PCE, perchloroethylene; TCE, trichloroethylene; TCA, trichloroethane; MEOH, methanol; MEK, methyl ethyl ketone; BTEX, benzene, toluene, ethylbenzene, and xylenes; PCBs, polychlorinated biphenyls; PAHs, polyaromatic hydrocarbons.

Source: Ram, N.M., Bass, D.H., Falotico, R., and Leahy, M. *J. Soil Contam.* 2(2):167–189. Lewis Publishers, Boca Raton, FL, 1993.

CURRENT TREATMENT TECHNOLOGIES

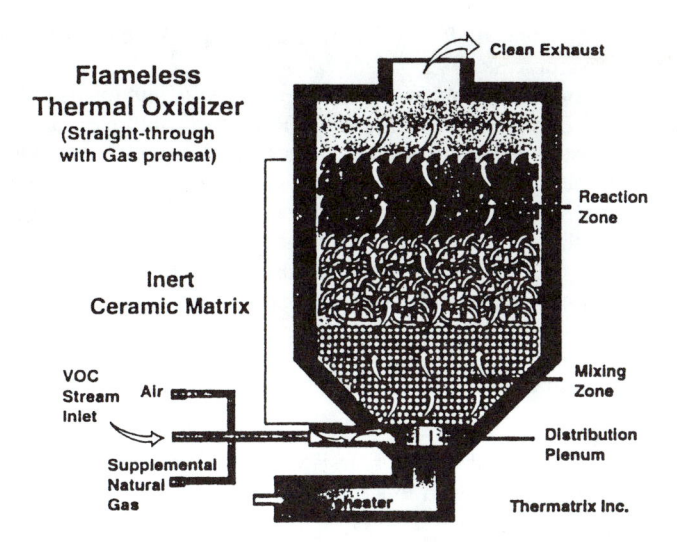

Figure 2.1 Flameless thermal oxidizer (straightthrough with gas preheat). (From Wilbourn, R.G. et al. in *Proc. 13th Int. Incineration Conf.*, University of California, Irvine, 1994. With permission.)

Thermal desorption can be combined with the Thermatrix flameless oxidation process for an integrated waste-processing system offering operational simplicity, near zero emissions, heat recovery and reuse, and reduced costs (Wilbourn, Newburn, and Schofield, 1994). After the organic contaminants are separated from the soil, the Thermatrix unit (Figure 2.1) treats the vapors. The heat produced during operation of the unit can be used to facilitate desorption of organic contaminants from soil matrices. An integrated Thermatrix/thermal desorption system can treat soils contaminated with VOCs at a feed rate of 5 ton/h.

Use of a laboratory-scale quartz furnace enabled researchers to remove BTEX (benzene, toluene, ethylene, and xylene) and BTEX with heavy metals from contaminated soil (Yang and Ku, 1994). The removal efficiency increased with increasing reaction temperature and reaction time. Thermal treatment of heavy metal-contaminated soil would stabilize the heavy metals within, resulting in a lower leaching toxicity.

A bench-scale treatment of soil contaminated with polycyclic aromatic hydrocarbons (PAHs) employed the ReTeC screw auger process for thermal desorption (Weisman, Falatko, Kuo, and Eby, 1994). A pilot-scale treatment of soil contaminated with PAHs, heterocyclic compounds, and phenols utilized the IT Corporation process for thermal desorption. Another thermal desorption treatment for removal of PAHs on a pilot scale employed the WES screw auger-based process. The Chemical Waste Management, Inc., X TRAX process has also been used on a pilot scale for treatment of soil contaminated with solvents, chlorinated pesticides, and cyanide.

A thermal desorption unit has been developed and patented for removing chemical contaminants from soil (Crosby, 1996). Contaminated soil is loaded and hydraulically sealed in a modified, sealable drum of a cement truck. A vacuum is drawn and the soil heated indirectly through a heat transfer plate from the natural gas of a propane-fired burner under the plate. The contaminants are vaporized and flow through the vacuum discharge pipe toward the condenser unit, through a series of refrigerated condensing coils. The vapors are liquidized, collected, recycled, or sent to an appropriate facility. The treated material is then downloaded into a roll-off-type container for posttreatment analysis and cooldown prior to recycling or backfilling. Process time is about 45 min to 1 h for a 6 yd^3 batch. The system is self-contained, mobile, and operable by a two-person crew.

2.1.1.1.2 Incineration

For complete destruction of the contaminants, incineration is one of the most effective treatments available. Greater than 99.99% destruction of carbon tetrachloride, chlorinated benzenes, and polychlorinated biphenyls (PCBs) was achieved by a trial burn with an EPA mobile incinerator (Yezzi, Brugger, Wilder, Freestone, Miller, Pfrommer, and Lovell, 1984). Aqueous waste streams are difficult to incinerate, but contaminated soils can be handled effectively (Absalon and Hockenbury, 1983). However, incineration is a relatively expensive process.

The most common types of incinerators in use are the rotary kiln, multiple hearth, fluidized bed, and liquid injection incinerators (Ehrenfeld and Bass, 1984). Rotary and multiple hearth incinerators can be used with most organic wastes, including solids, sludges, liquids, and gases, while liquid injection incinerators are limited to pumpable liquids and slurries. Fluidized-bed incinerators work well with liquids and can also be used with solids and gases. Incineration may generate incomplete combustion products and a residual ash that may need to be disposed of as a hazardous waste, but it offers one of the best methods for the destruction of organic compounds. Section 6.3.4.1 describes this technology in depth, although mainly in connection with treatment of gaseous emissions.

High-temperature thermal treatment, such as incineration, pyrolysis, and vitrification technologies are generally not considered for treating petroleum hydrocarbon-contaminated soil because of their high costs (Ram, Bass, Falotico, and Leahy, 1993).

2.1.1.1.3 Soil Washing

Soil washing is a variation of the soil flushing process, with similar requirements (Lyman, Noonan, and Reidy, 1990). It is performed above ground in a reactor and has been shown to be more effective than the *in situ* flushing system. This approach overcomes some of the problems that may be encountered with the *in situ* method — low hydraulic conductivity, channeling, and contamination of underlying aquifers. However, tightly bound contaminants are difficult to remove by flushing or washing. See Section 2.2.1.7 for a discussion of *in situ* soil flushing techniques.

A Mobile Soils Washer was built for the U.S. EPA to remove hazardous and toxic materials from soils (Elias and Pfrommer, 1983). The unit includes

- A drum washer operating at rates up to 18 yd^3/h, while separating and washing the stones and other large materials from the drier soils;
- A four-stage countercurrent extraction operation processing up to 4 yd^3/h;
- A mobile flocculation/sedimentation trailer to remove soil fines and inorganic contaminants from water prior to recycle or discharge to additional water treatment equipment.

There are several state-of-the-art soil-washing systems, including the EPA mobile system, two hot water systems for removing oil from sandy soils, and a flotation process (Assink and Rulkens, 1984). The quantity of residual sludge formed in the extraction process can be a problem and, generally, requires additional handling as a hazardous waste.

A multiple-stage, continuous-flow, countercurrent washing system, each stage consisting of a complete mixing tank and clarifier, for soil remediation has been simulated to produce a mathematical model, which can be used to manage a treatability study and assist the operator in determination of the steady state in the system (Chao, Chang, Bricka, and Neale, 1995).

A proprietary soil-washing process has been developed in Germany (Castaldi, 1994). It is a two-step mechanical separation using water, with no detergents, solvents, acids, or bases as an extracting agent. The process concentrates contaminants in a froth, which is discharged during flotation separation, thickened, and dewatered with gravity thickeners and plate-filter presses.

There is another two-stage process for soils containing semivolatile and nonvolatile organic compounds, such as substituted phenols, PAHs, fuel oils, creosote, lubricating oils, and diesel fuel (McBean and Anderson, 1996). The contaminated soil is excavated, piled onto polymer linings, washed to extract the hydrocarbons into an aqueous phase (by slowly flooding and draining from the bottom), and returned to its original site. The next stage involves biological treatment of the leachate with conventional wastewater technologies. The advantage of separating these stages is that conditions for each can then be optimized, without negatively impacting the other. For example, surfactants may be necessary in the initial extraction stage, and they can be added at a concentration that would be inhibitory to microorganisms, if the two steps would not separate. A concentration of at least 1% surfactant is typically necessary, while concentrations greater than 2% reduce the hydraulic conductivity. The wash solution can then be treated on- or off-site by an acclimated mixed microbial culture. This process is especially useful for areas with a cold climate. Hydrocarbons are rapidly removed, and the leachate is treated under optimized conditions. Removal efficiencies of over 90% are possible with sandy soils.

BioGenesis Enterprises, Inc. developed a soil- and sediment-washing process (BioGenesis^{SM}) for cleaning heavy hydrocarbon pollutants, such as crude oil, fuel oils, diesel fuel, and PAHs, from most matrices (Amiran and Wilde, 1994). Controlled temperature, pressure, friction, and duration are combined with proprietary chemical blends tailored to specific site requirements. Synthetic biosurfactants continue remediation after washing is completed.

Washing of tar-contaminated soils (attrition of soil, separation of light particles and soil fines) can be significantly enhanced by using additives (Sobisch, Kuehnemund, Huebner, Reinisch, and Olesch, 1995). To reduce the amount of contaminated soil fractions for disposal, the fraction of soil fines can be cleaned by a subsequent extraction step using surfactant solutions.

Ultrasound-enhanced soil washing with a surfactant (octyl-phenyl-ethoxylate) is being investigated as a means of improving the performance and economics of this method (Meegoda, Ho, Bhattacharjee, Wei, Cohen, Magee, and Frederick, 1995). Results of the preliminary studies indicate that ultrasound energy supplied by a 1500-W probe operating at 50% power rating, applied for 30 min to 20 g of coal tar–contaminated soil with 1% surfactant in 500 mL can enhance the soil-washing process by over 100%. For soil heavily contaminated with coal tar, the surfactant to contaminant ratio of >0.625 and a solvent ratio >10 is needed for near total removal efficiency. The solution pH does not contribute to removal efficiency, and the ultrasound energy increases soil temperatures.

Soil washing can be enhanced by use of solid sorbents and additives (El-Shoubary and Woodmansee, 1996). Hydrocyclone, attrition scrubber, and froth flotation equipment can be used to remove motor oil from sea sand. Sorbents (e.g., granular activated carbon, powder activated carbon, or rubber tires) and additives (e.g., calcium hydroxide, sodium carbonate, Alconox, Triton X-100, or Triton X-114) are mixed with soils in the attrition scrubber prior to flotation. Addition of these nonhazardous additives or sorbents can enhance the soil-washing process, thereby saving on residence time and number of stages needed to reach the target cleanup levels.

Soil washing has been used on a pilot scale to treat soil contaminated with cadmium, chromium, cyanide, and zinc, by use of the Chapman soil-washing process (Weisman, Falatko, Kuo, and Eby, 1994).

2.1.1.1.4 Chemical Treatment

Peroxide spraying can be used to treat excavated, contaminated soil (Ram, Bass, Falotico, and Leahy, 1993).

A new laboratory method for stagnant digestion studied oil release from oil–sand aggregates (Hupka and Wawrzacz, 1996). Oil is released when submerged in an alkaline solution of pH 10.5. The rate of oil release can be two to seven times greater at 50 than at 20°C, depending upon the kind of oil, surfactant concentration, and size of sand grains. The efficiency of oil liberation from sand is inversely proportional to oil–sand-conditioning time and is controlled by surfactant concentration (at least 1 wt%).

Organic substances can be destroyed by indirect electro-oxidation (Leffrang, Ebert, Flory, Galla, and Schnieder, 1995). The oxidation agent, Co(III) is used because of the high redox potential of the Co(III)/Co(II) redox couple (EPV0PV = 1.808 V). Organic carbon is ultimately transformed to CO_2 and to small amounts of CO.

2.1.1.1.5 Chemical Extraction

Chemical extraction, such as heap leaching and liquid/solid contactors, can also be used in the treatment of excavated, contaminated soil (Ram, Bass, Falotico, and Leahy, 1993). Chemical extraction has been employed on a pilot scale for remediating soil contaminated with PAHs, by applying the Resource Conservation Company solvent extraction process (Weisman, Falatko, Kuo, and Eby, 1994).

Multiple regression analysis of solvent extractions of pyrene and benz(a)pyrene from sand, silt, and clay gave an equation for the optimal extraction efficiency and process parameters (Noordkamp, Grotenhuis, and Rulkens, 1995). Soil type and extraction time did not affect extraction efficiency. Acetone, methanol, and ethanol were similar in efficiency, although the optimal extraction efficiency was with 19% water and 81% (vol/vol) acetone, which was surprising because the compounds are more soluble in pure acetone.

2.1.1.1.6 Supercritical Fluid (SCF) Oxidation

Oxidation in supercritical water is fast and can lead to total oxidation of the organic compounds (Brunner, 1994). Supercritical water is an excellent solvent for extraction of mineral oil fractions from soil, even without oxygen, and the effluents are biologically degradable.

A supercritical water oxidation system can clean PAH-contaminated soil by extracting hazardous material from the soil and completely destroying it by an oxidation reaction (Kocher, Azzam, and Lee, 1995). Since most organics dissolve readily in supercritical water, the oxidation reaction proceeds very rapidly, producing a clean soil with residual hydrocarbon contamination of <200 ppm and a top gas stream rich in CO_2 and water. The process can be an effective, *ex situ* remediation technology that can readily be implemented on a mobile unit. See Section 2.1.1.2.5 for a full description of this process.

2.1.1.1.7 Volatilization

Enhanced volatilization refers to any process that removes contaminants from soil by increasing their rate of volatilization (Lyman, Noonan, and Reidy, 1990). This includes the processes of mechanical volatilization, enclosed mechanical aeration, pneumatic conveyer systems, and low-temperature thermal stripping, which is considered to be the most effective. Repeated rototilling with successively deeper levels of excavation results in volatilization of contaminants from greater depths. Enclosed mechanical aeration systems use pug mills or rotary drums to increase turbulence in the reactor, with greater aeration and volatilization. Low-temperature thermal stripping systems are similar but include heat to increase the volatilization rate. Pneumatic conveyers use both increased temperature and high velocity airflow to remove contaminants. Excavated contaminated soil can be treated by surface spreading, soil pile aeration, or soil shredding (Ram, Bass, Falotico, and Leahy, 1993).

2.1.1.1.8 Steam Extraction

Laboratory-scale tests and a semi-industrial-scale plant equipped with vapor condensation and subsequent wastewater treatment capability demonstrated that steam extraction can be easily used to remove soil contamination caused by diesel fuel, solvents, and PAHs (Hudel, Forge, Klein, Schroeder, and Dohmann, 1995). The process is not limited by soil structure (grain size distribution). Treatment costs of about 300 Deutsch marks/Mg soil are expected for an industrial-scale plant with a 5 Mg/h capacity. There is interest in the U.S. and Germany for industrial-scale plants.

2.1.1.1.9 Solidification/Stabilization

This approach incorporates chemical or biological stabilization processes to treat excavated, contaminated soil (Ram, Bass, Falotico, and Leahy, 1993).

Use of carbon-grade fly ash as the only binding agent is a simple, inexpensive method acceptable to the Toxicity Characteristic Leaching Procedure (TCLP) of stabilization/solidification of hazardous wastes (Parsa, Munson-McGee, and Steiner, 1996). Waste and fly ash are mixed and compacted for <3 s at 1.4 to 6.9 MPa to form a monolith. The optimum operating conditions are a waste pH of 9.2 and an applied pressure of 4.65 MPa.

If the effectiveness of stabilization is to be mainly determined by the total constituent analysis rather than the previous TCLP, it will be more difficult to meet the standards by stabilization treatment (Conner, 1995). Thus, new stabilization additives and formulations are being developed. These include cement-based formulations with additives, such as activated carbon, organoclay, and proprietary rubber particulates (KAX-50 and KAX-100). The rubber particulates were superior to the other additives. Stabilization of low-level organic constituents in soils is feasible, even for volatile organics.

Bench-scale studies of soil contaminated with lead, cadmium, zinc, barium, chromium, and nickel have employed either the Risk Reduction Engineering Laboratory process, the TIDE process, or the WES process for stabilization (Weisman, Falatko, Kuo, and Eby, 1994).

2.1.1.1.10 Encapsulation

Other than asphalt blending and other thermoplastic encapsulation methods, most stabilization techniques for fixing organic contaminants in a soil matrix use pozzolanic materials (portland cement, fly ash, kiln dust) as the main ingredient (McDowell, 1992). This process does not work with moderate to high levels of hydrocarbons. The increase in volume and need for pozzolanic materials can be avoided with the Siallon process for microencapsulation of hydrocarbons, which uses two water-based products, an emulsifier, which is specifically selected for different hydrocarbons and soil types, and a reactive silicate. The first stage desorbs and emulsifies the hydrocarbon; the second applies the reactive silicate, which reacts with the emulsifier to form a nonsoluble silica cell measuring <10 µm. The silica cell is essentially pure silica, is nonporous and relatively solid, has a honeycomb or mazelike interior, reduces the mobility and toxicity of hydrocarbons, and does not change the physical characteristics of the soil. It has been successfully applied by *in situ* or *ex situ* remediation of sites contaminated with gasoline, diesel, waste motor oil, crude oil, coal tars, and PCBs.

2.1.1.1.11 Supercritical Fluid Extraction

Use of supercritical CO_2 is a novel technique to remediate contaminated soil, but there is limited information for costs and timing estimates (Zytner, Bhat, Rahme, Secker, and Stiver, 1995). Partition results suggest a weak dependence on the vapor pressure of the contaminant and on soil type. The film mass transfer coefficient appears not to be a rate-limiting kinetic step. Key parameters are axial dispersion and internal aggregate diffusion.

A pilot-plant experiment indicated that SCF extraction was effective for cleanup of hydrocarbon-contaminated soils (Schulz, Reiss, and Schleussinger, 1995). The residual concentration of benzo(a)pyrene after the extraction was <1 mg/kg in the soil at 140°F.

Supercritical CO_2 can be used to extract anthracene and pyrene from soil at conditions ranging from 35 to 55°C and 7.79 to 24.13 MPa (Champagne and Bienkowski, 1995).

Cleanup of soils contaminated with organics by extraction with supercritical carbon dioxide is influenced by additional substances (Schleussinger, Ohlmeier, Reiss, and Schulz, 1996). Both continuous and discontinuous addition of water elevates the extraction yield by altering the adsorption phenomena, which indicates the extraction is limited by adsorption and not by diffusion effects. The contaminant is more accessible and transported faster out of the soil with water.

2.1.1.1.12 Beneficial Reuse

Soil that has been contaminated by petroleum products can be excavated and incorporated into asphalt or other construction applications (Ram, Bass, Falotico, and Leahy, 1993).

Sometimes, the waste can be converted into a useful product, such as a compost for landscaping (Savage, Diaz, and Golueke, 1985). However, the toxic contaminant and toxic breakdown products must first be completely destroyed or reduced to an acceptable level. Also, the residue can be made quite small by using the compost product as a bulking agent and recycling it in the compost system.

2.1.1.2 Leachate/Wastewater Treatment Systems

Contaminated leachate may be released during the process of remediating contaminated soil. It may be necessary to treat any leachate collected, or it may be desirable to prevent a leachate from occurring. Therefore, background information on leachate formation and a variety of leachate, wastewater, and groundwater treatment systems are discussed as possible options for dealing with this phase of the remediation program.

Large concentrations of many organic compounds, both volatile and nonvolatile, can leach through landfill sites into the groundwater (Sawhney and Kozloski, 1984). Leachate is generated as a result of the movement of liquids by gravity through a disposal site (Shuckrow, Pajak, and Touhill, 1982b). The leachate percolating through a particular waste reflects the composition of all the materials through which that leachate has passed and depends upon site characteristics, such as annual rainfall volume and composition, evapotranspiration, biological activity, and the nature of the surrounding soil and wastes (Ham, Anderson, Stegmann, and Stanforth, 1979). It is possible that the liquid could be multiphase, e.g., water, oil, and solvents, with the various phases moving through the solid medium at different rates (Shuckrow, Pajak, and Touhill, 1982b).

Soil batch leaching protocols based on the EPA TCLP for petroleum hydrocarbons were evaluated and refined by Daymani, Forster, Ahlfeld, Hoa, and Carley (1992) for the ability to predict the leaching potential of volatile organic compounds in gasoline-contaminated soils. They substituted deionized water as an extraction fluid, reduced the test time to 2 h, and found that the TCLP was most effective in assessing the leaching characteristics of gasoline constituents with relatively high solubilities and low vapor pressures. They also determined that the relationship calculated from the TCLP ratio study results, between the mass of soil and mass of contaminant leached from the soil, may be used to obtain an indication of the amount of contamination that leaches from an area of homogeneously contaminated soil. Under the new regulatory test methods and treatment standards used by the EPA in the Land Disposal Restrictions, the effectiveness of stabilization is judged primarily by the total constituent analysis rather than, as previously, by the TCLP (Conner, 1995). This approach will likely be extended to remedial actions in the future.

A unique analytical method was developed by GTEL Environmental Laboratories in cooperation with the Shell Development Company Westhollow Research Center (Felten, Leahy, Bealer, and Kline, 1992). The analysis segregates hydrocarbons by their respective elution times, which correspond to molecular weights. Hydrocarbons are segregated into five fractions:

Fraction 1 containing pentane and compounds eluting prior to pentane;
Fraction 2 containing benzene and compounds eluting between benzene and pentane;
Fraction 3 containing toluene and compounds eluting between toluene and benzene;
Fraction 4 containing ethylbenzene and compounds eluting between ethylbenzene and toluene; and
Fraction 5 containing compounds that elute after ethylbenzene.

Fraction 1 contains the most-volatile compounds and Fraction 5, the least volatile.

Leaching ability is related to the proton and electron environments (Lowenbach, 1978; Rai, Serne, and Swanson, 1980) and the presence of solubilizing agents (Means, Kucak, and Crerar, 1980). The proton and electron environments are determined for natural environments and landfill leachates by measuring the pH, redox potential, ionic strength, and buffering capacity (Baas Becking, Kaplan, and Moore, 1960; Chian and deWalle, 1977). Movement of organic pollutants through soil may be increased in the presence of organic solvents (Green, Lee, and Jones, 1981). Solubilizing agents include constituents, such as complexing and chelating agents (hydroxyl ion, ammonia, ethylene diamine tetracetic acid [EDTA]), colloidal constituents (unicelles or surfactants), and organic constituents (melanic materials, humic acids) (Baas Becking, Kaplan, and Moore, 1960; Chian and deWalle, 1977). Some of these agents can affect the mobility of inorganic and organic constituents of the waste, even at low concentrations of the agents.

A number of factors affect the quality of a leachate (Shuckrow, Pajak, and Touhill, 1982b). Solubility is one of the most important factors. Chemical composition of the leachate determines dissolution and reaction rates. Dissolution is directly proportional to the surface contact area. Porosity influences the flow rate of liquid and, thus, the contact time between liquid and solids. Longer contact times permit more-complete chemical reactions until an equilibrium concentration is reached. The pH also has a significant effect on the leachate composition. Soil admixtures also influence solubility. For example, acid soils tend to promote solubilization of waste constituents, whereas the higher pH in alkaline soils likely will retard solubilization. Warmer temperatures increase reaction rates between liquid and solid and improve microbial catalysis. The main physical transformation expected in the leaching process is plugging of pore spaces and the resultant influence on chemical processes and leachate flow rates.

On-site hazardous leachate treatment can be used to accomplish either pretreatment of the leachate with discharge to another facility for additional treatment before disposal or treatment complete enough to meet direct discharge limitations (Shuckrow, Pajak, and Touhill, 1982b). The major difference between complete on-site treatment and pretreatment is likely to be the extent of the treatment. Most leachate treatment processes result in production of by-products, such as sludges, air pollution control residues, spent adsorption or ion exchange materials, or fouled membranes, which also require disposal. Residue disposal considerations may determine selection of a leachate management technique.

One possible approach to on-site leachate management is leachate recycling (Shuckrow, Pajak, and Touhill, 1982b). This technique involves the controlled collection and recirculation of leachate through a landfill to promote rapid landfill stabilization.

Information on leachate composition is used in judging the adequacy of a leachate treatment system (Garrett, McKown, Miller, Riggin, and Warner, 1981). A leachate procedure provides a realistic leachate profile, showing the change in constituent concentration with amount of leaching. It can be site specific and applicable to a variety of solid wastes.

A leaching procedure has been developed to estimate the total amount of leachable species to be released from a unit mass of solid waste (Garrett, McKown, Miller, Riggin, and Warner, 1981). In addition, the profile of the leachate will indicate the concentration or mass of that constituent likely to be present in the leachate and the time period, in terms of total volume of leachate produced, when that constituent will be present at any particular concentration or mass. This information will also indicate the composition of leachate that can be expected in the field under the duplicated conditions.

Ideally, the leaching medium and test conditions used in a leaching test should reproduce the actual leachate and conditions to be encountered at the field disposal site (Garrett, McKown, Miller, Riggin, and Warner, 1981). While no single medium can duplicate field conditions, certain factors have been identified that influence leaching and, thus, determine the leaching medium composition (Table 2.3).

Test Conditions
Distilled, deionized water is used as the leaching medium with a monofilled solid waste (Garrett, McKown, Miller, Riggin, and Warner, 1981). Where environmental conditions warrant, alternate media,

CURRENT TREATMENT TECHNOLOGIES

Table 2.3 Critical Factors in a Leaching Procedure

I. Leaching Medium Composition
    A. Proton and electron environment
        1. pH
        2. Redox potential
        3. Ionic strength
        4. Buffering capacity
    B. Presence of solubilizing agents
        1. Complexing and chelating agents
        2. Colloidal constituents
        3. Organic constituents
II. Leaching Test Conditions
    A. Contact area/particle size
    B. Method of mixing
    C. Mixing time
    D. Temperature control
    E. Number of leachings on the same solid
    F. Number of leachings on the same liquid
    G. Solid-to-liquid ratio

Source: From Garrett, B.C. et al., in *Proc. of 7th Annual Res. Symp.*, Philadelphia, March 16–18, 1981. PB81-173882. With permission.

such as one that duplicates acid rain, might be more appropriate. A solid-to-liquid ratio of one-to-ten (weight/volume on a wet weight basis) may not always reflect field conditions, but is a workable amount for the analysis. The approximate time per leaching is 24 hr. Ideally, the time should allow equilibrium to be reached. The temperature should be close to that expected for the site leachate. Room temperature may be used unless there is a substantial difference between the two. The leaching medium–sample mixture is then mixed with a rotary mixer (Ham, Anderson, Stegmann, and Stanforth, 1979), being careful to prevent stratification and ensuring continuous liquid-solid contact.

Treatability of Leachate Constituents
Once the compounds have been identified in a leachate, treatability tables can be consulted to see which treatment techniques can be applied to each of the hazardous constituents (Garrett, McKown, Miller, Riggin, and Warner, 1981). These techniques can be evaluated for treatment feasibility, and a treatment train can be proposed, based upon a combination of the treatment options for the various constituents.

A number of technologies that have potential application to hazardous waste leachate treatment are described below (Shuckrow, Pajak, and Touhill, 1982b). The applicability of these treatment processes for different classes of chemicals is summarized in Table 2.4 (Shuckrow, Pajak, and Osheka, 1981).

Treatment By-Products
Most leachate treatment processes generate sludges, brines, gaseous emissions, or other by-product streams, which often contain hazardous constituents that must be managed as hazardous waste (Shuckrow, Pajak, and Touhill, 1982b). These streams will probably be of mixed composition and can be divided into two categories, residues and gaseous emissions, which require different methods of treatment. By-products that can be expected from the various treatment processes are given in Table 2.5.

Residues may be managed using most of the techniques available for hazardous wastes on- or off-site (Shuckrow, Pajak, and Touhill, 1982b). There are three basic control measures for gaseous emissions. One is to treat the emission using air pollution control technologies, e.g., scrubbers, precipitators, chemical or thermal oxidation, or gas phase adsorbents. In many cases, these also generate by-product waste streams.

Another approach is a process that produces an emission of less magnitude or severity (Shuckrow, Pajak, and Touhill, 1982b). For example, gravity sedimentation is less likely to strip volatile compounds than dissolved air flotation; the same applies for trickling filtration vs. diffused aeration–activated sludge. The third alternative is a "do nothing" approach, which allows emissions that are within acceptable limits. Dilution of the emission may be a factor in this approach.

Table 2.4 Treatment Process Applicability Matrix

Chemical Classification	Biological Treatment	Carbon Adsorption	Chemical Precipitation	Chemical Oxidation – Alkaline Chlorination	Chemical Oxidation – Ozonation	Chemical Reduction	Ion Exchange	Reverse Osmosis	Stripping	Wet Oxidation
1. Alcohols	E	V		N	G,E	N		V		
2. Aliphatics	V	V		N	P	N		V		
3. Amines	V	V		N	N	N				
4. Aromatics	V	G,E	F	N	F,G	N		V		
5. Ethers	G	V		N		N				
6. Halocarbons	P	G,E		N	F,G	N				
7. Metals	P,F	N,P	E	N		G	E	E	N	
8. Miscellaneous:										
Ammonia	G,E	N	N	N		N	G		G	
Cyanide	F,G	N	N	E	E	N			N	
TDS	N	N	N	N	N	N	E	E	N	N
9. PCB	N	E		N		N				
10. Pesticides	N,P	E		N	E	N		E		
11. Phenols	G	E		N	E	N		V		
12. Phthalates	G	E	G	N		N				
13. Polynuclear aromatics	N,P	G,E	R	N	G	N				

Key for Symbols: E = Excellent performance likely; G = Good performance likely; F = Fair performance likely; P = Poor performance likely; R = Reported to be removed; N = Not applicable; V = Variable performance reported for different compounds in the class. A blank indicates that no data are available to judge performance; it does not necessarily indicate that the process is not applicable.

Note: Use of two symbols indicates differing reports of performance for different compounds in the class.

Source: Shuckrow, A. J. et al. Concentration Technologies for Hazardous Aqueous Waste Treatment. EPA-600/S2-81-019. U.S. EPA. Cincinnati, OH, February, 1981.

CURRENT TREATMENT TECHNOLOGIES

Table 2.5   Leachate Treatment Process By-Product Streams

Treatment Process	Residuals Generated	Gaseous Emissions
I. Biological treatment		
A. Aerobic		
1. activated sludge	Excess biological sludge must be removed — amount of sludge varies with the process configuration	Stripping of volatile compounds during aeration process — use of pure oxygen process may reduce air emissions
2. lagoons	Settled solids will accumulate on lagoon bottom, clean-out frequency depends on performance requirements and lagoon capacity	Stripping of volatile compounds if mechanical or diffused aeration is used
3. trickling filter	Excess biological sludge must be removed — plastic and high-rate filters generate more sludge than low-rate filters	The most volatile compounds may be stripped at the point of waste application; if improperly operated, odor problems may occur
B. Anaerobic		
1. filters	Some anoxic residue may be generated; less sludge than aerobic process	Properly operating system will generate gas composed of methane, carbon dioxide, and water vapor; highly volatile compounds also may be present
2. lagoons	Settled sludge will accumulate in lagoon; need for clean-out depends on lagoon performance and capacity	May create odor problem — some opportunity for stripping of volatile compounds
II. Carbon adsorption		
A. Granular carbon	Spent carbon — may be regenerated and reused; performance may decline with continued reuse and blowdown of some portion of the spent carbon may be required	Emission problems generally associated with spent carbon handling and regeneration operations
B. Powdered carbon (PAC)	When used with activated sludge process a residue containing excess biological sludge and PAC results — may be regenerated thermally or by wet oxidation with some wasting to prevent buildup of inerts; if not regenerated, sludge disposal is necessary	Same as for the activated sludge process
III. Catalysis	Depends on the process in which the catalyst is used	
IV. Chemical oxidation	Small amount of residue may be formed during the oxidation process; residue likely to be less hazardous than raw waste	During the rapid mix phase stripping may occur or gaseous reaction products could be released
	Use of chlorine may result in formation of chlorinated organics in liquid product stream; ozone and hydrogen peroxide add no harmful species to the effluent	Gaseous chlorine and ozone are toxic; however, these should not escape from the system in appreciable quantity
V. Chemical precipitation	Relatively large amounts of inorganic sludge will be generated by lime, ferric chloride, and alum coagulants; polymer addition would increase sludge amounts	Stripping may occur during the rapid mix or flocculation phases
VI. Chemical reduction	As with chemical oxidation, small amounts of residue may be formed; some metal ions or sulfate from the reducing agents may carry over in the liquid effluent	Emissions may occur during rapid mixing
VII. Crystallization	Brines high in organics or inorganics will be formed	Emissions could include lost refrigerant, noncondensable compounds, and water vapor
VIII. Density separation	Either a sludge or a floating scum is produced by these processes; the quantity produced depends on the suspended solids content of the raw wastewater and the use of coagulant chemicals	Gravity separation is not likely to generate emissions; dissolved air flotation may cause stripping of volatile compounds

Table 2.5 (continued) Leachate Treatment Process By-Product Streams

Treatment Process		Residuals Generated	Gaseous Emissions
IX.	Distillation	Still bottoms consisting of tars and sludges will be laden with nonvolatile organics; condensed overhead stream also could contain volatile organics	No emissions if the overhead stream is condensed trapping volatiles in a liquid phase
X.	Dialysis/Electrodialysis	No solid residue is formed; however, the original pollutants will be present in different concentrations in the two product streams	Venting of gases produced at electrodialysis electrodes causes emissions
XI.	Evaporation	Similar to distillation with evaporator liquor laden with less-volatile organics and condensed vapor rich in volatile compounds	Evaporation vapors could contain volatile compounds; these can be condensed and trapped in liquid phase
XII.	Filtration (granular media for aqueous waste)	In the case of granular media filters, the major residue is suspended solids trapped by the filter and removed by backwashing	Emissions generally should not be a problem; if anaerobic conditions are allowed to occur in granular media filter, anoxic odors could occur; during backwashing, turbulence may induce some stripping of volatiles
XIII.	Flocculation	See discussion of chemical precipitation	
XIV.	Ion exchange	Residuals include the (1) concentrated regenerant stream and (2) spent ion exchange materials; unless spent exchange materials are regenerated both types of residues could contain the original hazardous pollutants	Emissions should not occur
XV.	Resin adsorption	One residue will be spent resin which can no longer be used effectively Another will be solutes extracted from the sorbent; These solutes may be separated from the regenerant solvent or discarded with the used regenerant solution Waters used to rinse regenerant solution from resin also require attention	Emission problems generally associated with spent resin handling or regeneration operations; steam regeneration and distillation of solvents used for solvent regeneration are principal emissions sources
XVI.	Reverse osmosis	The primary residual will be a brine stream containing the concentrated pollutants Other residues include solutions which may be used to wash or maintain the membranes and degraded or fouled membranes; these all could contain the original pollutants	Emissions should not occur
XVII.	Solvent extraction	No solid residuals are generated by the process; spent solvent, solvent containing the solutes, or solutes alone will have to be disposed of at some time during process operation	Gaseous emissions from the extraction process should be minimal; however, processes to remove solute from solvent or recover solvent from the treated water could produce emissions of either volatile solutes or volatile solvent since these procedures usually employ stripping or distillation
XVIII.	Stripping		
	A. Air	No solid residue is generated unless chemicals are added to adjust operating conditions; use of lime can result in substantial quantities of sludge	Volatile compounds will be contained in stripper emission by design
	B. Steam	No solid residues are formed; however, stripper bottoms will contain concentrated nonvolatile organics and cannot be discharged directly	No emissions occur if stripped volatile compounds are trapped in the condensed overhead stream
XIX.	Ultrafiltration	Same as reported for reverse osmosis	
XX.	Wet oxidation	Residues are not generated by the process, but solids present in the raw wastewater could remain after treatment; these solids are likely to be more inert than those originally present	Vapors may be released when the high pressure and temperature operating conditions are removed and the waste is exposed to atmospheric conditions

Source: From Shuckrow, A.J. et al. *Hazardous Waste Leachage Management Manual.* Noyes Data Corp., Park Ridge, N.J., 1982. With permission.

CURRENT TREATMENT TECHNOLOGIES

It has been suggested by Shuckrow, Pajak, and Touhill (1982a) that the most practicable leachate treatment operations are chemical coagulation, carbon adsorption, membrane processes, resin adsorption, stripping, and biological treatment. Carbon adsorption is the most frequently employed.

2.1.1.2.1 Carbon Adsorption

When a toxic organic is to be removed from a water stream, which is otherwise relatively clean and free of suspended matter, and the toxic material is present in concentrations of less than about 10%, activated carbon adsorption can be considered (Hackman, 1978). At higher concentrations of the toxic organic, the preferred separation methods would be distillation, extraction, or another method not using relatively large quantities of solids like carbon.

Activated carbon adsorption is well suited for removal of mixed organic contaminants from aqueous wastes (Shuckrow, Pajak, and Touhill, 1982b). Granular activated carbon is the most well developed approach and may be used to provide complete treatment, pretreatment, or effluent polishing. Combined biological–carbon systems also appear promising for leachate treatment. Energy requirements for systems employing thermal reactivation are significant — approximately 14,000 to 18,600 kJ/kg of carbon (6000 to 8000 Btu/lb). Unit costs depend upon the waste, the adsorption system, and the regeneration technique, but have been shown to be economical.

Organic contaminants come into contact with and adhere to an activated carbon surface by physical and chemical forces (Nielsen, 1983; IT Corporation, 1987). The hydrophobic nature of the contaminants and the affinity of the contaminants for the activated carbon are the primary factors and driving forces affecting the quantity of contaminants that can be adsorbed from the groundwater. The physical and chemical characteristics of the contaminants in the water (e.g., solubility, pH, molecular weight, temperature), concentration, carbon properties, and contact time between the carbon and the groundwater all affect the balance between the attraction of the contaminants to the carbon and the forces to keep them in solution. The degree of sorption onto the carbon depends upon (Knox, Canter, Kincannon, Stover, and Ward, 1984):

1. Solubility of the compound, insoluble compounds being more likely to be adsorbed;
2. The pH of the water, which controls the degree of ionization of the compounds — acids are adsorbed better under acidic conditions and adsorption of amine-containing compounds is favored under alkaline conditions;
3. Characteristics of the adsorbent, which are a result of the process used to generate and activate the carbon;
4. Properties of the compound, for example, aliphatic compounds are less well adsorbed than aromatics and halogenated compounds.

Activated carbon sorbs every one of the representative hazardous chemicals, but different activated carbons are selective for different hazardous compounds (Robinson, 1979). A carbon surface can be acidic or basic, hydrophilic or hydrophobic, or oleophilic or liphobic. It can vary in porosity. The surface area per unit weight is a function of the size of the carbon particles and of the area generated by the process of activation. Further, activated carbon is sold with its particles in various states of agglomeration and aggregation. It can come in bead, pellet, rod, sheet, and other forms and shapes.

Activated carbon can be in granular or powdered form (Hackman, 1978). The powdered carbons are much finer in particle size, passing through 325-mesh sieves, as opposed to the retention of granular carbon on 10- to 40-mesh sieves. Granular carbons were specifically designed for use in beds; however, with the need for low-pressure-drop fluid flow through the bed, they also need the ability to be fluidized for transport, and then to be thermally regenerated. Until the present, powdered carbons were not considered good candidates for regeneration, being disposed of when their activity was lost. Granular activated carbon is typically employed in reactors, and the powdered carbon is added to the wastewater, then either settled or filtered for removal with the sludge (Ehrenfeld and Bass, 1984).

F-400 GAC and Ambersorb 563 can be regenerated by impregnating the absorbents with photocatalysts (e.g., Pt-TiO$_2$), which allow them to act as both an adsorbent for capturing the organics and as a photocatalyst for destroying the organics using artificial light during regeneration (Liu, Crittenden, Hand, and Perram, 1996). Increasing the temperature improves the regeneration rate; however, there is a low photo-efficiency compared with photocatalysis alone, because the desorption of the organics may be slow, even at elevated temperatures.

Effluent levels of between 1 and 10 µg/L can be achieved for many organics (Ehrenfeld and Bass, 1984). Partial adsorption of several heavy metals also occurs. Over a wide variety of systems, activated carbon can be expected to adsorb in the range of 1 to 30% of its weight (Hackman, 1978). Simultaneous removal of organics and heavy metals is feasible provided that the organic contaminants do not desorb at the extreme pHs experienced during regeneration for heavy metals (Reed and Thomas, 1995). If desorption does occur, that portion of the column effluent with an acceptable concentration of organics can be recycled through the column. Of the whole spectrum of toxic organics, the larger, more-complex molecules, which are not very soluble in water, and molecules that tend to concentrate at interfaces are all logical candidates for carbon adsorption (Hackman, 1978).

The effectiveness of carbon adsorption is controlled by the tendency of the contaminating species to fit into the micropores on the surface of the carbon (Brubaker and O'Neill, 1982). It is most often used with aromatics (including chlorinated aromatics, phenols, and PAHs), fuels, chlorinated solvents, and high-molecular-weight amines, ketones, and surfactants. Because compounds much larger or smaller (on a molecular level) than these materials do not fit into the pores, they are not generally good candidates for carbon treatment. A mixture of materials might not respond like the sum of its individual parts. There are many compounds that inhibit the adsorption of other contaminants to a carbon surface. In addition, those materials that adsorb most effectively to carbon also adsorb effectively to the soil and are thus difficult to transport into the water in the first place.

There are many factors to consider in selecting a carbon, beyond the prime concern for large, and very active, surface area per pound (Hackman, 1978). A carbon of high bulk density, while maintaining a high specific surface, will tend to minimize the size and cost of filter hardware.

Carbon adsorption systems are sensitive to the composition of the influent, to flow variations, to fine precipitates, to oil and grease, and to suspended solids in the influent water (Lee and Ward, 1985, 1984; Lee, Wilson and Ward, 1987). Activated carbon systems have a finite loading capacity. They may be clogged by biological growth, although this growth may provide additional treatment by destroying organics. They may be regenerated at a high temperature, which is expensive, or by treatment with steam or a solvent. The spent carbon could be placed in secure landfills or other sites that do not allow any desorbed organics to contaminate other environments. Carbon adsorption is the best system for emergency response. Activated carbon systems can be batch, column, or fluidized-bed reactors.

Carbon adsorption systems work at about 95% efficiency. They are effective in removing aromatics but relatively ineffective in removing t-butyl alcohol or methyl t-butyl ether (American Petroleum Institute, 1983). Isotherms for the adsorption of priority pollutants, VOCs, and other hazardous organic compounds in aqueous solutions have been developed and can be used to estimate adsorption capacities for an activated carbon treatment system (Dobbs and Cohen, 1980; Love, Miltner, Eilers, and Fronk-Leist, 1983).

Table 2.6 presents influent and corresponding effluent concentrations of several organics that can be achieved by use of carbon adsorption (Canter and Knox, 1985).

Polluted soil from a gasworks site was converted into a carbonaceous adsorbent using $ZnCl_2$ (Fowler, Sollars, Ouki, and Perry, 1994). Organic pollution, consisting of coal tars, phenols, etc., was converted into a carbonaceous matrix with development of microporosity within the carbons that could entrap metallic pollutants. The complex soil pollutants influenced the adsorption characteristics, and sulfur appeared to play a major part in this development.

Treatment of leachate from a landfill in the U.K. with conventional activated carbon technology proved unacceptable for economic and technical reasons (Sojka, 1984). However, biological pretreatment of the effluent in a sequencing-batch reactor prior to the carbon proved to be cost-effective.

See Section 6.3.3.1.5 for further discussion of carbon adsorption.

2.1.1.2.2 Resin Adsorption

Phthalate esters, aldehydes and ketones, alcohols, chlorinated aromatics, aromatics, esters, amines, chlorinated alkanes and alkenes, and pesticides are adsorbable with resins (Shuckrow, Pajak, and Touhill, 1982b). Resins adsorb certain aromatics better than activated carbon. Resin adsorption has greatest applicability when

> Color due to organic molecule must be removed;
> Solute recovery is practical or thermal regeneration is not practical;
> Selective adsorption is desired;
> Low leakages are required;
> Wastewaters contain high levels of dissolved inorganics.

Polymeric adsorbents are nonpolar with an affinity for nonpolar solutes in polar solvents or of intermediate polarity capable of sorbing nonpolar solutes from polar solvents and polar solutes from

Table 2.6 Activated Carbon Adsorption of Organics

Contaminants	Influent Concentration, µg/L	Effluent Concentration, µg/L
Phenol	63,000	<100
	2,400	<10
	40,000	<10
Carbon tetrachloride	61,000	<10
	130,000	<1
	73,000	<1
1,1,2-Tetrachloroethane	80,000	<10
Tetrachloroethylene	44,000	12
	70,000	<1
1,1,1-Trichloroethane	23,000	ND
	1,000	<1
	3,300	<1
	12,000	<5
	143,000	<1
	115	1
Benzene	2,800	<10
	400	<1
	11,000	<100
2,4-Dichlorophenol	5,100	ND

ND = Nondetectable.

Source: From Canter, L.W. and Knox, R.C. *Ground Water Pollution Control.* Lewis Publishers, Boca Raton, FL, 1985.

nonpolar solvents (Shuckrow, Pajak, and Touhill, 1982b). Carbonaceous resins have a chemical composition intermediate between polymeric adsorbents and activated carbon in a range of surface polarities.

Resin adsorption has a wide range of potential applications for organic waste streams (Shuckrow, Pajak, and Touhill, 1982b). There is a high initial cost. Costs for resins are $11 to 33/kg ($5 to 15/lb, 1980 dollars). If not reused, spent regenerant requires disposal, frequently by incineration or land disposal. Resin sorption is a potentially viable candidate for treatment of hazardous waste leachates; however, the technique is not as well defined or economic as carbon adsorption.

Many polymeric adsorbents will adsorb toxic organics (Hackman, 1978). Ion exchange resins adsorb ionic organics, and the macroreticular resins have an even greater adsorptive capability. Nonpolar adsorbents are particularly effective for adsorbing nonpolar toxic organics from water. Conversely, the highly polar adsorbents are most effective for adsorbing polar solutes from nonpolar solvents. It is desirable to use the adsorbent with the highest surface area available having a suitable polarity. A limitation is the size of the molecule to be adsorbed, since the average pore diameter in the adsorbents decreases as the surface area increases. Thus, for large molecules, it is necessary to use the lower-surface-area adsorbent. The solvents to use for removal of the adsorbate from the adsorbent are

- Methanol or other organic solvents — often most effective
- Base — for weak acids
- Acid — for weak bases
- Water — where adsorption is from an ionic solution
- Hot water or steam — for volatile materials

2.1.1.2.3 Adsorption with Brown Coal

Metal-bearing aqueous streams can be treated by adsorption on lignite and its maceral fractions (Gaydardjiev, Hadjihristova, and Tichy, 1996). The denser coal-refined fraction shows superior performance and resembles to a certain extent the activated carbons.

Felgener, Janitza, and Koscielski (1993) performed studies on five municipal landfill leachates using a two-stage adsorption in a fluidized bed of brown coal coke. The COD (chemical oxygen demand) and BOD (biological oxygen demand) values, the content of organic carbon, and adsorbable chloro-organic compounds in the leachates were decreased below acceptable limit values.

2.1.1.2.4 Wet Air Oxidation (WAO)

This process is similar to the previously discussed SCF oxidation process (Section 2.1.1.1.6), but is it subcritical. It may have potential for treatment of high-strength leachates or those containing toxic organics, especially those waste streams too dilute for incineration but too concentrated or refractory for chemical or biological oxidation, for example, COD in the range of 10,000 mg/L up to 20% by weight (Bove, Lambert, Lin, Sullivan, and Marks, 1984). The process has limited applicability to treatment of groundwater containing low concentrations of organics, due to high energy requirements and high capital and operating costs. Generally, the process involves high capital and operating costs and requires skilled operating labor. It is potentially suitable for hazardous waste leachate treatment, with the area of greatest potential being for treating concentrated organic streams generated by other processes, such as steam stripping, ultrafiltration, reverse osmosis, still bottoms, biological treatment process waste sludges, and regeneration of powdered activated carbon used in biophysical processes. Extensive site-specific treatability studies are required.

Wet air oxidation is used for organic concentrations of less than 1%, but there are, generally, more cost-effective techniques available for the higher concentrations (Allen and Blaney, 1985). The process has had limited application in hazardous waste treatment (Spivey, Allen, Green, Wood, and Stallings, 1986). It is not specific for removal of volatiles, and other nonvolatile or slightly volatile hazardous waste stream constituents may compete with the dynamics of the process. The method is limited in the species of volatiles that it can destroy; for example, it will not readily decompose highly chlorinated organics. As with ozonation, in practice, this technique does not completely oxidize the treated compounds to water and carbon dioxide and may remove limited amounts of some volatiles and produce new volatile species in the process.

This process is kept under pressure between 1500 and 2500 psig (103 to 172 bar) and temperatures of 450 to 600°F (232 to 315°C) (see Figure 2.2; Bove, Lambert, Lin, Sullivan, and Marks, 1984). It typically reduces complex organic compounds to short-chain organics, such as formic and acetic acids, aldehydes, ketones, and alcohols. Therefore, additional polishing treatment, such as biological treatment and carbon adsorption, may be necessary to remove the remaining biodegradable, as well as biorefractory, organic material (see Figure 2.3).

2.1.1.2.5 Supercritical Fluid (SCF) Oxidation

Destruction of hazardous organic wastes in an SCF reactor may be the most attractive of all the SCF technologies (Welch, Bateman, Perkins, and Roberts, 1987). The hydrocarbon compounds and their derivatives may be converted to carbon dioxide and water, and the salts of the inorganic oxides may be precipitated. No new developments in process equipment are required, since there already exists considerable expertise concerning supercritical-steam plants, steam chemistry, ammonia-synthesis reactors, and steam reformers. Advantages of using SCFs are

> The solute may be separated readily from the SCF solvent by decreasing the density of the fluid.
> The contact and separation processes may be conducted at relatively low temperatures, which results in increased safety in the handling of heat-sensitive materials, such as propellants and explosives.
> The solvent may serve as an inert gas cover, thereby reducing the hazard of explosion or fire.
> The solvent does not become part of the waste disposal problem.
> The proper scheduling of solvent density changes permits fractionation, if multiple solutes are present.
> The solvent power of the SCF solvent may be altered in certain cases by the addition of "entrainers," which reduce the pressure change required in the separation step.

A major advantage of using SCF for hazardous waste management is the relative ease of separation of the solute from the solvent (Welch, Bateman, Perkins, and Roberts, 1987). The density of the fluid may be altered by changing temperature, pressure, or both to alter the selectivity and to separate the extract solvent from the solute.

The wide variation in the solvent power of fluids in the supercritical state is an important feature of this technology and allows the SCF to be used as (Welch, Bateman, Perkins, and Roberts, 1987)

> A replacement for an ordinary solvent;
> A solvent for materials that are not usually soluble;
> A medium in which chemical reactions may be conducted.

CURRENT TREATMENT TECHNOLOGIES

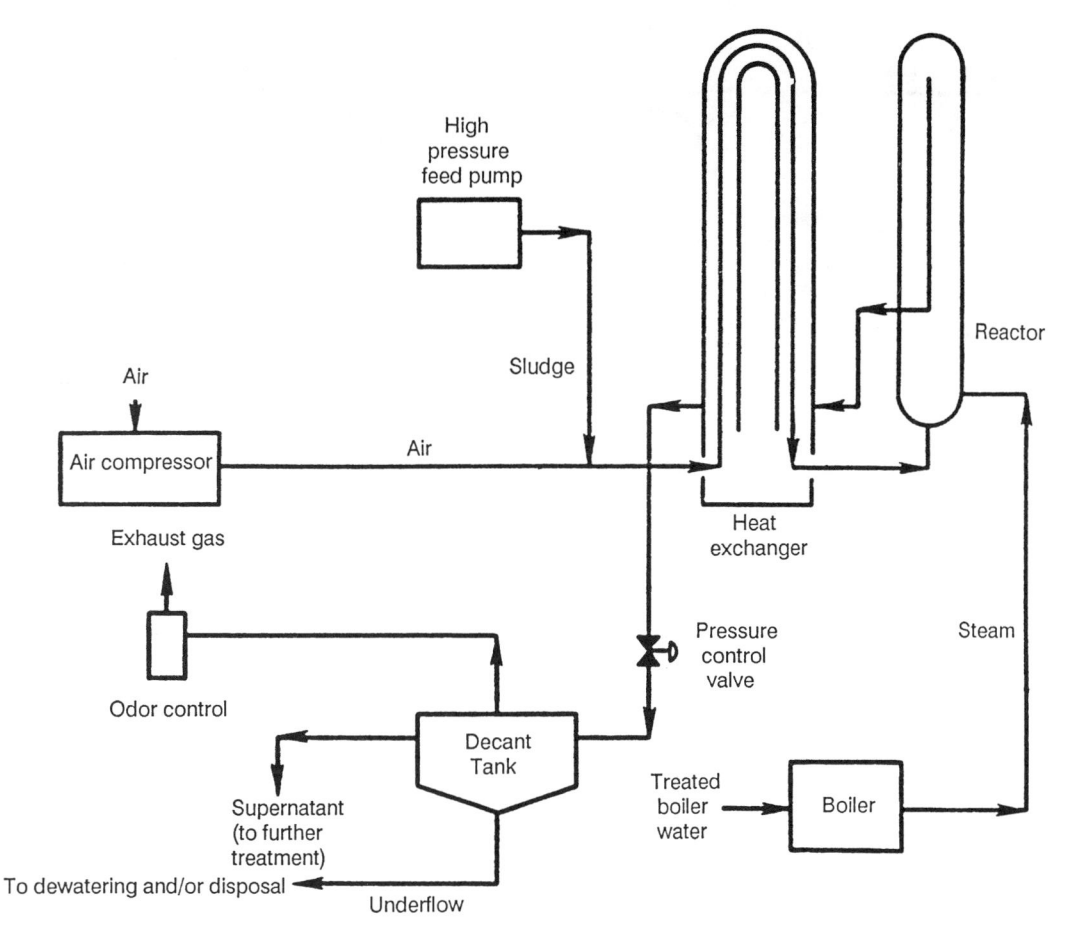

Figure 2.2 Wet air oxidation (WAO) process. (From Bove, L.J. et al. Report to U.S. Army Toxic and Hazardous Materials Agency on Contract No. DAAK11-82-C-0017, 1984. AD-A162 528/4.)

There is an increase in capital cost associated with pressure vessels and an increase in operating expense due to compression work with this technology (Welch, Bateman, Perkins, and Roberts, 1987). However, these costs are irrelevant when the unit operation cannot be accomplished by the use of an ordinary fluid, or when the solute is thermally labile.

Two SCF reactor systems are illustrated in Figures 2.4 and 2.5 (Welch, Bateman, Perkins, and Roberts, 1987). The first is a retrofit to an existing Navy boiler, furnace, or incinerator. The system in Figure 2.5 is a conventional Rankine cycle with supercritical water as the working fluid. This system is able to generate power, as well as destroy the wastes.

A recent patent presents an improved method for initiating and sustaining an oxidation reaction (Mcguinness, 1996). Hazardous waste serves as a fuel and is introduced into a reaction zone in a pressurized container with a permeable liner. An oxidizer, such as oxygen, is mixed with a carrier fluid, such as water, heated and pressurized to supercritical conditions of temperature and pressure. The mixture is added gradually and uniformly to the reaction zone by forcing it radially inward through the permeable liner. The exhausted by-products are then cooled.

2.1.1.2.6 Chemical/Photochemical Oxidation

When organic contaminants are mineralized, i.e., chemically oxidized to completion, carbon dioxide and water will be produced, and halogens will be converted to inorganic salts (IT Corporation, 1987).

Relatively poor removals of most organics are effected by chemical oxidation, although chemical transformations may occur, which could facilitate treatment by other processes (Shuckrow, Pajak, and

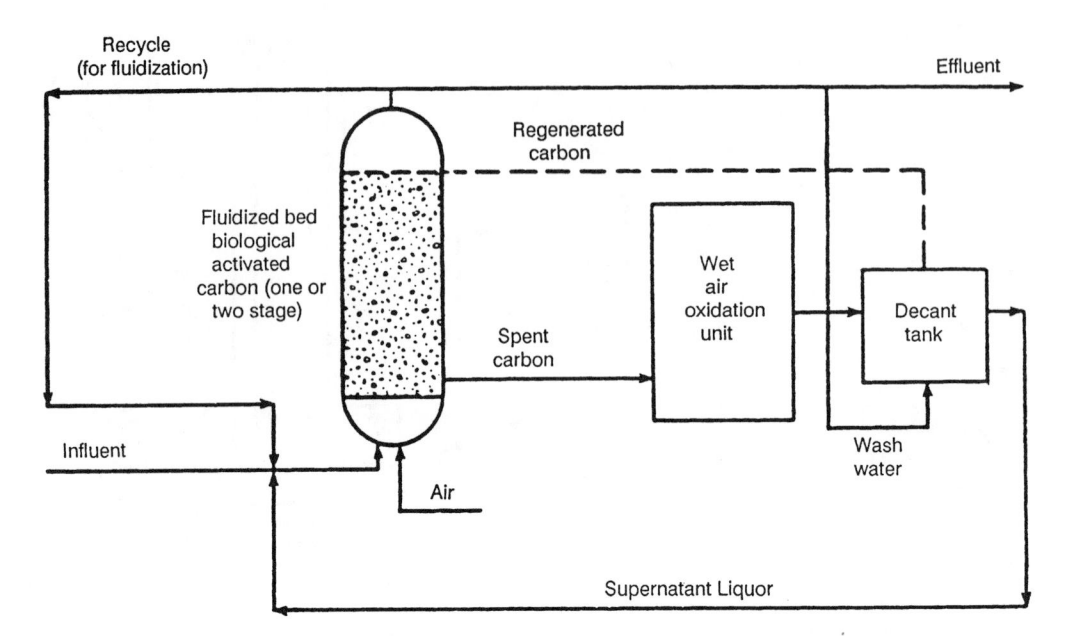

Figure 2.3 Biological activated carbon/wet air oxidation combination process schematic. (From Bove, L.J. et al. Report to U.S. Army Toxic and Hazardous Materials Agency on Contract No. DAAK11-82-C-0017, 1984. AD-A162 528/4.)

Touhill, 1982b). Inorganics can often be transferred to a less toxic or more easily precipitable valence state. Most chemical oxidation technologies (including ozone) are fairly well developed but have, generally, been applied to dilute waste streams.

Wastewaters containing refractory, toxic, or inhibitory organic compounds should be pretreated before being introduced to conventional biological treatment systems (Cho and Bowers, 1991). Pretreatment can remove or destroy these compounds or convert them to less-toxic and more readily biodegradable intermediates. Chemical oxidants can be used as a pretreatment to oxidize these contaminants partially, which reduces their toxicity and improves overall reduction of COD and total organic carbon (TOC).

Ozonation has potential for aqueous hazardous waste treatment (Shuckrow, Pajak, and Touhill, 1982b). It can serve as a pretreatment process prior to biological treatment. It can also be used alone or with ultraviolet (UV) irradiations as the primary treatment. Combination of ozonation and granular activated carbon has had mixed results, with performance depending upon the wastewater composition.

Hydrogen peroxide or ozone as an oxidizing agent with UV light as a catalyst provides a means to degrade or destroy VOCs in groundwater (IT Corporation, 1987). The hydrogen peroxide or ozone is converted into hydroxyl radicals, which are strong oxidants and react with the organic contaminants. The organics also absorb UV light to undergo chemical structural changes, such as dechlorination.

A basic flow system of a UV/hydrogen peroxide treatment process consists of a feed reservoir with heating/cooling for temperature control, a peroxide metering system for mixing peroxide with the contaminated water, and an oxidation chamber (or reactor) equipped with UV lamps to catalyze the reaction (Figure 2.6; Bove, Lambert, Lin, Sullivan, and Marks, 1984). Chemical catalysts may also be added. The reaction rate is controlled by the UV and peroxide doses, pH and temperature, chemical catalyst, mixing efficiency, light transmittance of the water, and concentration of the contaminants. UV peroxide performance will be affected by the hardness of the water. Pilot studies will determine the optimum conditions for the specific situation.

The use of UV/ozone treatment is similar to the UV/hydrogen peroxide process (IT Corporation, 1987). Ozone also forms hydroxyl radicals by UV light catalysis. Ozone is a stronger oxidizing agent, but it must be generated on-site and is more difficult to handle than the peroxide. In addition, each hydrogen peroxide molecule will form two hydroxyl radicals. For many applications, the hydrogen peroxide will be the most cost-effective.

CURRENT TREATMENT TECHNOLOGIES

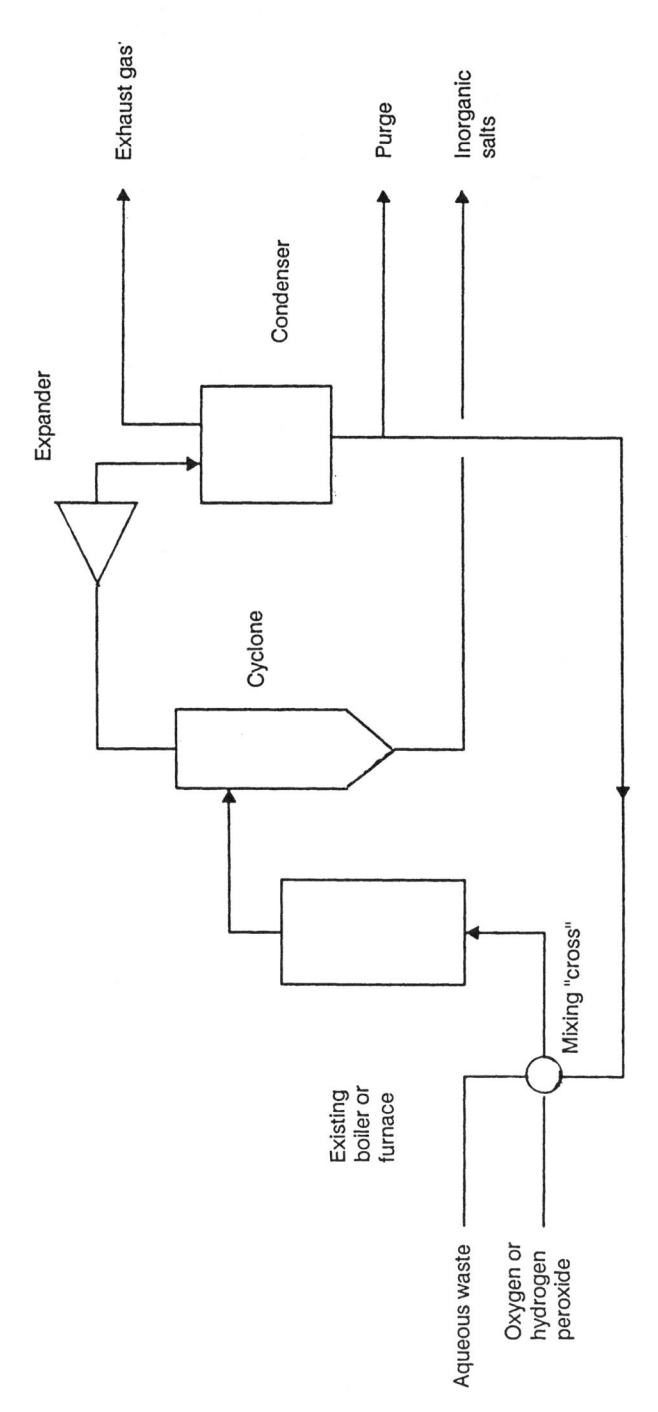

Figure 2.4 SCF reactor for retrofit application. (From Welch, J.F. et al. Report No. TM 71-87-20. Naval Civil Engineering Laboratory, Port Hueneme, CA, 1987.)

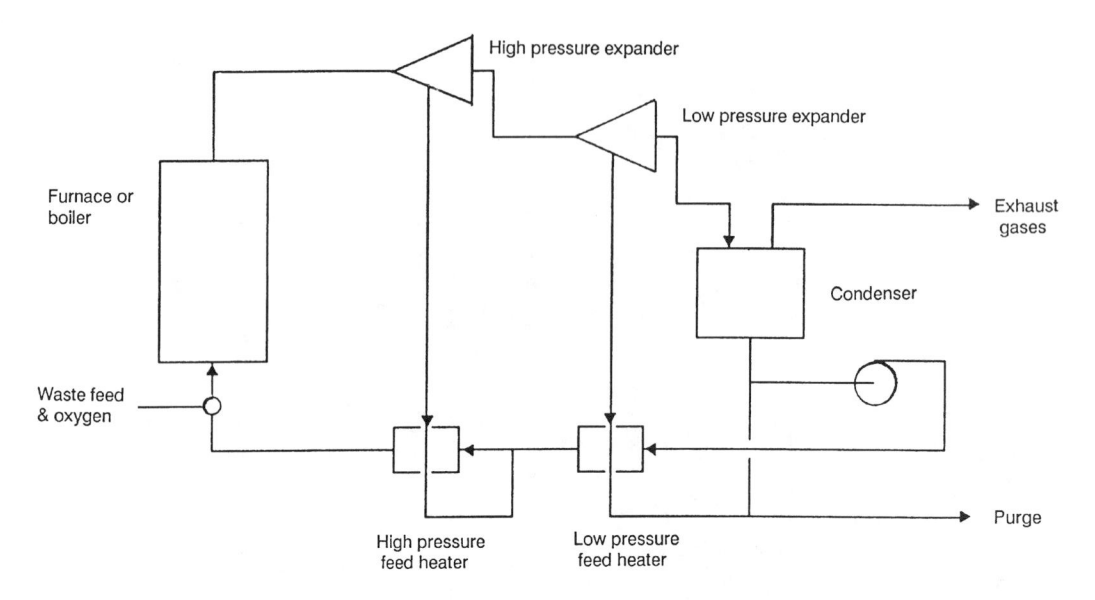

Figure 2.5 SCF Reactor for stand-alone application. (From Welch, J.F. et al. Report No. TM 71-87-20. Naval Civil Engineering Laboratory, Port Hueneme, CA, 1987.)

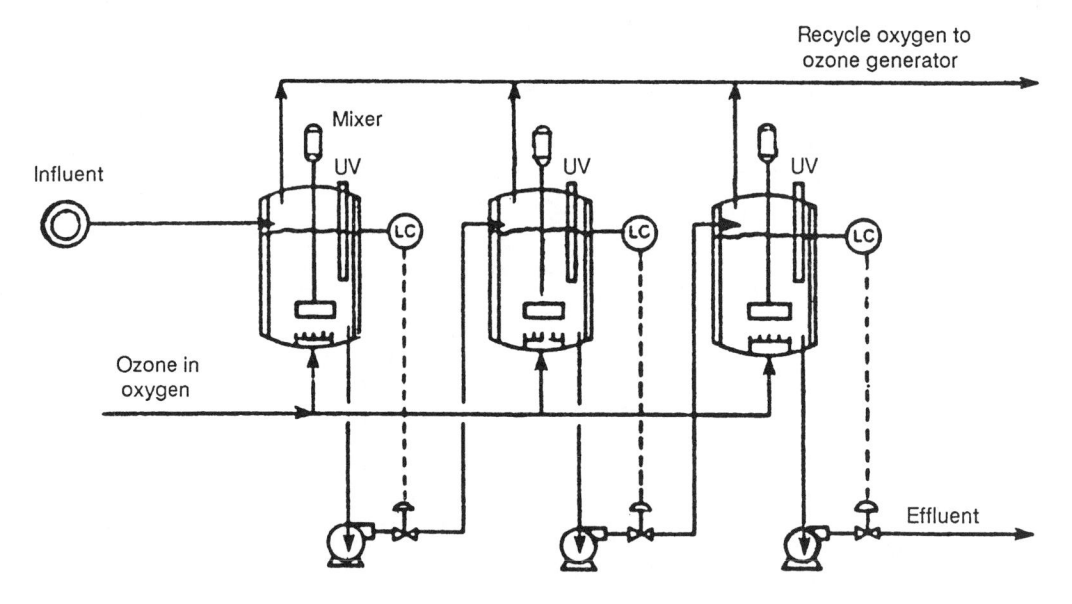

Figure 2.6 Process schematic of a typical UV/ozone system. (From Bove, L.J. et al. Report to U.S. Army Toxic and Hazardous Materials Agency on Contract No. DAAK11-82-C-0017, 1984. AD-A162 528/4.)

A UV/hydrogen peroxide or UV/ozone oxidation treatment system is reported to achieve low effluent concentrations with no air emissions and may be cost-competitive with air stripping and carbon treatment systems that must meet stringent air pollution control requirements for the treatment of VOCs in some situations (IT Corporation, 1987). Cost of treatment depends upon the objectives, concentration, and types of contaminants to be destroyed or removed.

There has been little application of ozonation/UV radiation, except in cleanup of disposal site leachates (Allen and Blaney, 1985). The technique will not specifically oxidize volatiles in hazardous waste streams, since other nonvolatile or slightly volatile stream constituents will compete in the process dynamics. It is limited in terms of the volatile species it can destroy. While the process should potentially result in the complete mineralization of treated compounds to water and carbon dioxide, in practice this does not always occur (Spivey, Allen, Green, Wood, and Stallings, 1986). Some volatiles may be removed to only a very limited degree, and in the process new volatile species may be produced.

Photocatalyzed hydrogen peroxide and ozone are effective oxidants at pH 3.5 (Cho and Bowers, 1991). Optimum oxidation by permanganate may require a different pH. Ozone oxidation reduces TOC toxicity better than H_2O_2 and permanganate, while the percentage reduction with catalyzed hydrogen peroxide gives the highest value in most of the compounds tested. Most of the oxidation products are biodegraded rapidly. While there are no harmful residues generated with ozone or hydrogen peroxide, the intermediate products must be assessed. Off-gases containing residual ozone should be passed through activated carbon to decompose the ozone.

Low concentrations of benzene can be removed from water using UV light–catalyzed hydrogen peroxide oxidation (Weir, Sundstrom, and Klei, 1987). H_2O_2 alone does not reduce the level of contaminant by 50% after 90 min; however, UV light alone does. The combination of UV/H_2O_2 reduces the concentration by 98% in 90 min. Increasing either H_2O_2 concentrations or UV light intensity improves the benzene oxidation rate.

PAHs absorb UV light energy and are subject to photolytic breakdown (Wilson and Jones, 1993). Natural sunlight or UV light (300 nm) in the presence of a dilute oxidant, H_2O_2, can degrade dilute solutions of benzo(a)pyrene (Miller, Singer, Rosen, and Bartha, 1988). Costs for complete breakdown, however, are prohibitively expensive (Wilson and Jones, 1993). Photolysis and photo-oxidation are further discussed in Sections 2.1.2.1.7, 5.3.2, and 6.3.4.6.

2.1.1.2.7 Chemical Catalysis
Catalysts, generally, are very selective and, while potentially applicable to destruction or detoxification of a given component of a complex waste stream, do not have broad spectrum applicability (Shuckrow, Pajak, and Touhill, 1982b).

2.1.1.2.8 Chemical Precipitation
Precipitation of certain waste components can be accomplished by adding a chemical that reacts with the hazardous constituent to form a sparingly soluble product or by adding a chemical or changing the temperature to reduce the solubility of the hazardous constituent (Ehrenfeld and Bass, 1984).

Chemical precipitation with carbonate, sulfides, or hydroxides is used routinely to chemically treat wastewaters containing heavy metals and other inorganics (Knox, Canter, Kincannon, Stover, and Ward, 1984). Sulfides are probably the most effective for precipitating heavy metals; however, sulfide sludges are susceptible to oxidation to sulfate, which may release the metals.

The hydroxide system with lime or sodium hydroxide is widely used but may produce a gelatinous sludge, which is difficult to dewater (Knox, Canter, Kincannon, Stover, and Ward, 1984). Removal of metals by chemical precipitation with lime requires a pH at which a soluble form of the metal is converted to an insoluble form (Stover and Kincannon, 1983). After metals are removed, the characteristics of the water can change significantly.

Soda ash is employed with the carbonate system and may be difficult to control (Knox, Canter, Kincannon, Stover, and Ward, 1984). Alum is another common agent used in chemical precipitation. The effectiveness of these chemical treatments will vary with the nature and concentration of the constituents of the waste stream (Lee and Ward, 1985, 1984; Lee, Wilson, and Ward, 1987). A process design for chemical precipitation must consider the systems for chemical addition and mixing, the optimal chemical dose, the time required for flocculation, and the removal and disposal of the sludge.

Precipitation results in production of a wet sludge, which may be hazardous and require further processing (Shuckrow, Pajak, and Touhill, 1982b). It is the technique of choice for removal of metals (arsenic, cadmium, chromium, copper, lead, manganese, mercury, nickel) and certain anionic species (phosphates, sulfates, fluorides) from aqueous hazardous wastes. This technique can be applied to large volumes of almost any liquid waste stream containing a precipitable hazardous constituent. It is inexpensive, and equipment is commercially available.

In the case of chromium in the hexavalent state, reduction to the trivalent form is necessary in order to promote precipitation. This can be accomplished using sulfur dioxide, sulfite salts, or ferrous sulfate. Precipitation of trivalent chromium as $Cr(OH)_3$ with lime or sodium carbonate usually follows reduction.

2.1.1.2.9 Crystallization
The crystallization process cannot respond to changing wastewater characteristics and is so operationally complex it is not practiced. It has little potential for this application (Shuckrow, Pajak, and Touhill, 1982b).

2.1.1.2.10 Density Separation
2.1.1.2.10.1 Sedimentation
These processes are easy to operate, are low cost, consume little energy, and require simple and commercially available equipment (Shuckrow, Pajak, and Touhill, 1982b). They can be applied to almost any liquid waste stream containing settleable material and have a high potential for leachate treatment. However, sedimentation must be utilized in conjunction with another technique, such as chemical precipitation. Alternatively, it may be used as a pretreatment technique prior to another process, such as carbon or resin adsorption.

2.1.1.2.10.2 Flotation
This is a solids/liquids separation technique for certain industrial applications (Shuckrow, Pajak, and Touhill, 1982b). It has higher operating costs, as well as more skilled maintenance and higher power requirements. It is potentially applicable but probably only in situations where the leachate contains high concentrations of oil and grease.

2.1.1.2.11 Flocculation
This must be carried out in conjunction with a solid/liquid separation process, usually sedimentation (Shuckrow, Pajak, and Touhill, 1982b). Often, it is preceded by precipitation. It is a simple process with low costs and energy consumption, requiring commercially available equipment. The process can be applied to almost any aqueous waste stream containing precipitable or suspended material. Flocculation followed by sedimentation is a viable candidate process for hazardous waste leachate treatment, particularly where suspended solids or heavy metal removal is an objective. It can be used in conjunction with sedimentation as a pretreatment step prior to a subsequent process, such as activated carbon adsorption.

A patented method and equipment for removing oil from oil-contaminated water consists of a flocculation device and a flotation device (Henriksen, 1996). One or more chemicals are added to the liquid in the flocculation device, which is composed of one or more pipe loops with built-in agitators to provide turbulence and plug-type flow through the loop. Purified liquid and pollutants are separated in the flotation fitting or in a sedimentation apparatus.

2.1.1.2.12 Evaporation
Evaporation would not have broad application to treatment of hazardous waste leachate containing moderately volatile organic constituents (BP 100 to 300°C) (Shuckrow, Pajak, and Touhill, 1982b). These organics cannot be easily separated in a pretreatment stripper and will appear in the condensate from the evaporator to some extent, depending upon their volatility. Good clean separation of these organics is not possible without posttreatment of the condensate. Capital and operating costs are high, with high energy requirements. This process is more adaptable to wastewaters with high concentrations of organic pollutants than to dilute wastewaters.

2.1.1.2.13 Stripping
Air stripping has potential for leachate treatment, primarily when ammonia removal is desired and, then, only when the concentrations of other VOCs are low enough not to produce unacceptable air emissions (Shuckrow, Pajak, and Touhill, 1982b). The process would be difficult to optimize for leachate containing a spectrum of volatile and nonvolatile compounds. It is a useful pretreatment prior to another process, such as adsorption, to extend the life of the sorbent by removing sorbable organic constituents. Air emission problems would be most severe from biological treatment processes using aeration devices.

Steam stripping has merit for wastes containing high concentrations of highly volatile compounds (Shuckrow, Pajak, and Touhill, 1982b). It requires laboratory and bench-scale investigations prior to application to leachates containing multiple organic compounds. Energy requirement and costs are relatively high. It has greatest potential as a pretreatment step to reduce the load of volatile compounds to a subsequent treatment process. Organics concentrated in the overhead condensate stream would also require further treatment, possibly by wet oxidation.

2.1.1.2.14 Distillation
This has limited applicability to treatment of complex hazardous waste leachate because of its high cost and energy requirements, unless recovery of useful products can be practiced (Shuckrow, Pajak, and Touhill, 1982b).

2.1.1.2.15 Filtration
Both granular and flexible media filtration are well-developed processes and are commercially available (Shuckrow, Pajak, and Touhill, 1982b). They are economical. Filtration is a good candidate for leachate treatment; however, it is not a primary treatment, but rather used as a polishing step (granular media) subsequent to precipitation and sedimentation or as a dewatering process (flexible media) for sludges generated in other processes.

2.1.1.2.16 Ultrafiltration
This has limited potential for treating a complex leachate (Shuckrow, Pajak, and Touhill, 1982b). Its use would probably be limited to relatively low-volume leachate streams containing substantial quantities of high-molecular-weight (7500 to 500,000) solutes, such as oils. Concentrated organics would require further treatment, possibly by wet oxidation or off-site incineration. Pilot testing is necessary.

Ultrafiltration will remove colloids, and when operated in the cross-flow mode, will stay on-line longer without blinding (needing backwash to reduce the pressure buildup).

2.1.1.2.17 Dialysis/Electrodialysis
Dialysis and electrodialysis are not well suited to mixed constituent waste streams, being most applicable for removal of inorganic salts, and are, therefore, not appropriate for hazardous waste leachate treatment (Shuckrow, Pajak, and Touhill, 1982b).

2.1.1.2.18 Ion Exchange
This process removes dissolved salts, primarily inorganics, from aqueous solutions (Shuckrow, Pajak, and Touhill, 1982b). It is economical, with low energy requirements. It has some potential for leachate treatment where it is necessary to remove dissolved inorganic species. However, other processes, such as precipitation, flocculation, and sedimentation, are preferred. There is an upper concentration limit (around 10,000 to 20,000 mg/L). Ion exchange would be limited to supplying a polishing step for removing ionic constituents that could not be reduced to satisfactory levels by other methods.

2.1.1.2.19 Reverse Osmosis
This process can concentrate inorganics and some high-molecular-weight organics from waste streams (Lee and Ward, 1985, 1984; Lee, Wilson, and Ward, 1987). The contaminated water passes through a semipermeable membrane at high pressure. The resulting clean water leaves behind the concentrated wastes and any particulates. Pretreatment of the waste stream is likely to be required to achieve a constant influent composition (pH is particularly important) to kill any organisms that might form a biological film that would reduce permeability, to remove suspended solids, and to remove chlorine, which might affect the membrane. Microfiltration (MF) is being studied as a pretreatment prior to reverse osmosis to reduce microbes in secondary effluent from municipal wastewater (Ghayeni, Madaeni, Fane, and Schneider, 1996). Bacterial bioadhesion studies of various reverse osmosis membranes show differences between membrane-based reclamation of secondary effluent.

This is a relatively new process for removing inorganic salt from rinse waters (Shuckrow, Pajak, and Touhill, 1982b). It is a relatively costly process, requires pretreatment to remove solids, and may experience membrane fouling due to precipitation of insoluble salts. It also requires extensive bench- and pilot-scale testing, prior to any application. Thus, it has limited potential for leachate treatment.

2.1.1.2.20 Solvent Extraction
This has minimal potential for leachate treatment. Carbon adsorption is more effective and economical (Shuckrow, Pajak, and Touhill, 1982b).

2.1.2 BIOLOGICAL PROCESSES
2.1.2.1 Soil Treatment Systems
2.1.2.1.1 Landtreatment/Landfarming
The limitations, side effects, and high expense of traditional cleanup technology has stimulated interest in unconventional alternatives, such as the use of hydrocarbon-degrading microorganisms for cleanup

of contaminated soils and groundwaters (Shailubhai, 1986). The controlled application of waste materials to soil for immobilization or for degradation or transformation by the resident microflora is called landfarming, or landtreatment, and has become a recognized process technology. Landtreatment is categorized in the Resource Conservation and Recovery Act of 1976 (RCRA) as one of the land disposal options for managing hazardous wastes (Martin, Sims, and Mathews, 1986). Biodegradation allows landtreatment to function both as a treatment mechanism and a disposal process (Huddleston, Bleckmann, and Wolfe, 1986). Such disposal can be effective, provided that application rates and scheduling do not result in conditions that allow undesirable components or degradation products to run off or leach through the soil, and provided that no materials accumulate to toxic levels in the soil (Arora, Cantor, and Nemeth, 1982).

Landfarming is practiced in the U.S., Canada (Loehr, Neuhauser, and Martin, 1984), U.K., The Netherlands, Sweden, Denmark, France, and New Zealand (CONCAWE, 1980). There are some 197 registered hazardous waste landtreatment facilities in the U.S., extending from Alaska to Florida (Brown, 1981), where more than 2.45×10^6 tons of hazardous waste are treated annually (Overcash, Brown, and Evans, 1987). Over half of this amount is associated with petroleum refining and production (Brown and Associates, Inc., 1981). About one half of the disposable volume of oily sludges is landtreated at more than 100 sites across the country, under a variety of soil and climatic conditions (Arora, Cantor, and Nemeth, 1982).

Land application of various wastes has been practiced worldwide for over 100 years (Sprehe, Streebin, Robertson, and Bowen, 1985) and by the petroleum industry for more than 25 years (Martin, Sims, and Mathews, 1986). The objective of hazardous waste landtreatment technology is to dispose of the waste in an environmentally safe manner by designing and operating the system to utilize the natural biological, chemical, and physical processes in the soil for the purpose of assimilating those wastes receiving such treatment (Kincannon and Lin, 1985). These processes include leaching, adsorption, desorption, photodecomposition, oxidation, hydrolysis, and biological metabolism by plants and soil microorganisms; however, microbial processes are usually the dominant soil decomposition mechanisms for organic waste constituents (Loehr, 1986). The relative importance of the different processes will depend upon the specific wastes involved and environmental or site-specific factors acting on the system (Overcash, Brown, and Evans, 1987).

The technology of landtreatment relies on detoxification, degradation, and immobilization of hazardous waste constituents to ensure protection of surface water, groundwater, and air (Martin and Sims, 1986). Landtreatment of petroleum industry wastes involves the immobilization of metal constituents and the immobilization and biodegradation of organic constituents (Martin, Sims, and Mathews, 1986). Diverse populations of soil microorganisms degrade waste oil and other organic compounds through a series of complex reactions to yield carbon dioxide, water, and innocuous by-products (Arora, Cantor, and Nemeth, 1982). An important advantage that landtreatment offers for oil biodegradation is that the soil tends to hold the oil in place and provide large surface areas for its metabolism (Huddleston, Bleckmann, and Wolfe, 1986).

A landtreatment site is a biological–physical–chemical reactor that contains soil particles that filter applied wastewater and transform (adsorb, exchange, precipitate) many of the applied chemicals and bacteria and macroorganisms (e.g., earthworms) that stabilize the applied organics (Loehr, 1986). Landtreatment sites may also include vegetation that can utilize applied nutrients and inorganics during growth. With landtreatment, there is no sludge that requires subsequent treatment and disposal. An increase in biomass undergoes natural degradation in the soil until it is stabilized and becomes part of the soil humus. Movement of the applied constituents and of the net precipitation and any applied water is slow, and detention times in the soil are long. Thus, slow, as well as rapid, degradation and immobilization reactions contribute to controlling the applied organics and inorganics. It is common to have greater mass and percentage removals of waste constituents at a landtreatment site than in conventional waste treatment systems.

Soil disposal of many wastes is effective because the soil has large surface areas in which to absorb and inactivate waste components (Brown, Deuel, and Thomas, 1983). And, if the soil is properly managed, it also presents an ideal medium for microbial decomposition because of the presence of oxygen, water, and the nutrients needed for degradation of organic constituents. The microbes digest the organic matter and recycle the nutrients into the environment (Brown, 1981). Landtreatment can be thought of as a slow oxidizer and is an effective alternative for wastes that have large concentrations of degradable organic constituents.

There is a general trend toward increasing use of landtreatment (Rosenberg et al., 1976). Many untreated wastes currently being landfilled could be treated and rendered less hazardous by landtreatment, often at lower cost (Overcash, Brown, and Evans, 1987). From 1981 estimates, about 1.9 million tons (wet weight) of hazardous wastes were incinerated, while 3.8 million tons were treated by land application (Booz, Allen, and Hamilton, Inc., 1983).

Landfarming is an environmentally attractive alternative for the disposal of petroleum wastes (Arora, Cantor, and Nemeth, 1982) and has proved to be a successful alternative to incineration when energy conservation is considered. This alternative to *in situ* biotreatment may be employed in cases where soil permeability is too low for effective groundwater recirculation. Landfarming has been found to be a successful treatment technology for removal of petroleum hydrocarbons from weathered crude oil–contaminated soils, resulting in biodegradation of 96% of compounds with carbon numbers from 10 to 20 and 85% of compounds with carbon numbers above 44 (Huesemann and Moore, 1993). In landtreatment, a period of 1 to 2 years might be needed to decompose PAHs (Overcash and Pal, 1979a).

For each waste that will be applied to the treatment zone, the owner or operator must demonstrate, prior to the application, that hazardous constituents in the waste can be completely degraded, transformed, or immobilized in the treatment zone and that the process will be protective of human health and the environment (Loehr, 1986). The treatment demonstration is then used to determine permit requirements for the wastes to be applied and the operating principles to be used. A draft manual providing guidance for such treatment demonstrations has been developed (Ward, Tomson, Bedient, and Lee, 1986).

The objectives of landtreatment are (Overcash, Brown, and Evans, 1987)

1. *Treatment* to convert substantial quantities of hazardous wastes containing organics, heavy metals, and other inorganic constituents into materials that are, at a practical level, nonhazardous.
2. *Environmental protection* to maintain a minimal, acceptable effect on the environment and to avoid creating unusable areas of land. Thus, a functional, long-term protection of the environment can be achieved, recognizing that all hazardous waste management alternatives for treatment also have an effect on the environment.

In the process of design and permitting of landtreatment sites, detailed investigations and evaluations are required of the following: the waste to be treated, the site to be used, and the waste–soil interactions (Overcash, Brown, and Evans, 1987). The impact of design specifications, management plans, monitoring activities, and ultimate closure criteria is an integral part of the design.

Procedure

The contaminated soil is spread over the surface of the landfarm and incorporated into the top 6 to 12 in. (10 to 15 cm; Shailubhai, 1986) of clean soil (Loehr, Neuhauser, and Martin, 1984). This incorporation zone, in conjunction with the underlying soils where additional treatment and immobilization of the applied waste constituents occur, is the treatment zone. The treatment zone in the soil may be as great as 5 ft deep. The maximum depth of the treatment zone must be no more than 1.5 m (5 ft) from the initial soil surface and more than 1 m (3 ft) above the seasonal high-water table (Loehr, 1986). Excavated soils are spread over about 0.5 acres/1000 yd^3 soil (Eckenfelder and Norris, 1993). Figure 2.7 describes the different elements of the treatment zone (Overcash, Brown, and Evans, 1987). The dimensions of layers and zones can vary, depending upon natural conditions.

The applied material is allowed to dry for about 1 week (Shailubhai, 1986). The zone of incorporation may be considered a bioreactor operating in a quasi completely mixed mode, in which conversion of substrate (oil) to various end products occurs (Sprehe, Streebin, Robertson, and Bowen, 1985). By optimization of management practices, the biodegradation rates can be maximized (Arora, Cantor, and Nemeth, 1982). Nutrients can be added at this time, and the soil can be tilled to increase the oxygen level for enhanced biodegradation. Because of the considerable amount of carbon from the wastes, the levels of nitrogen and phosphorus would probably be too low to support the growth necessary for bioremediation and should be supplemented (Alexander, 1994). Rototilling equipment vigorously mixes the soil, promoting the aeration and mixing process more effectively than disks or bulldozers (Raymond, Hudson, and Jamison, 1976). The value of cultivating the soil is the redistribution of the oil, nutrients, oxygen, and microorganisms to create new points of attack for the microorganisms (Harmsen, 1991). Adding microorganisms was found to act as a booster at the beginning of the biodegradation, but after 2 months the amount of degraded oil was the same whether organisms had been supplemented or not

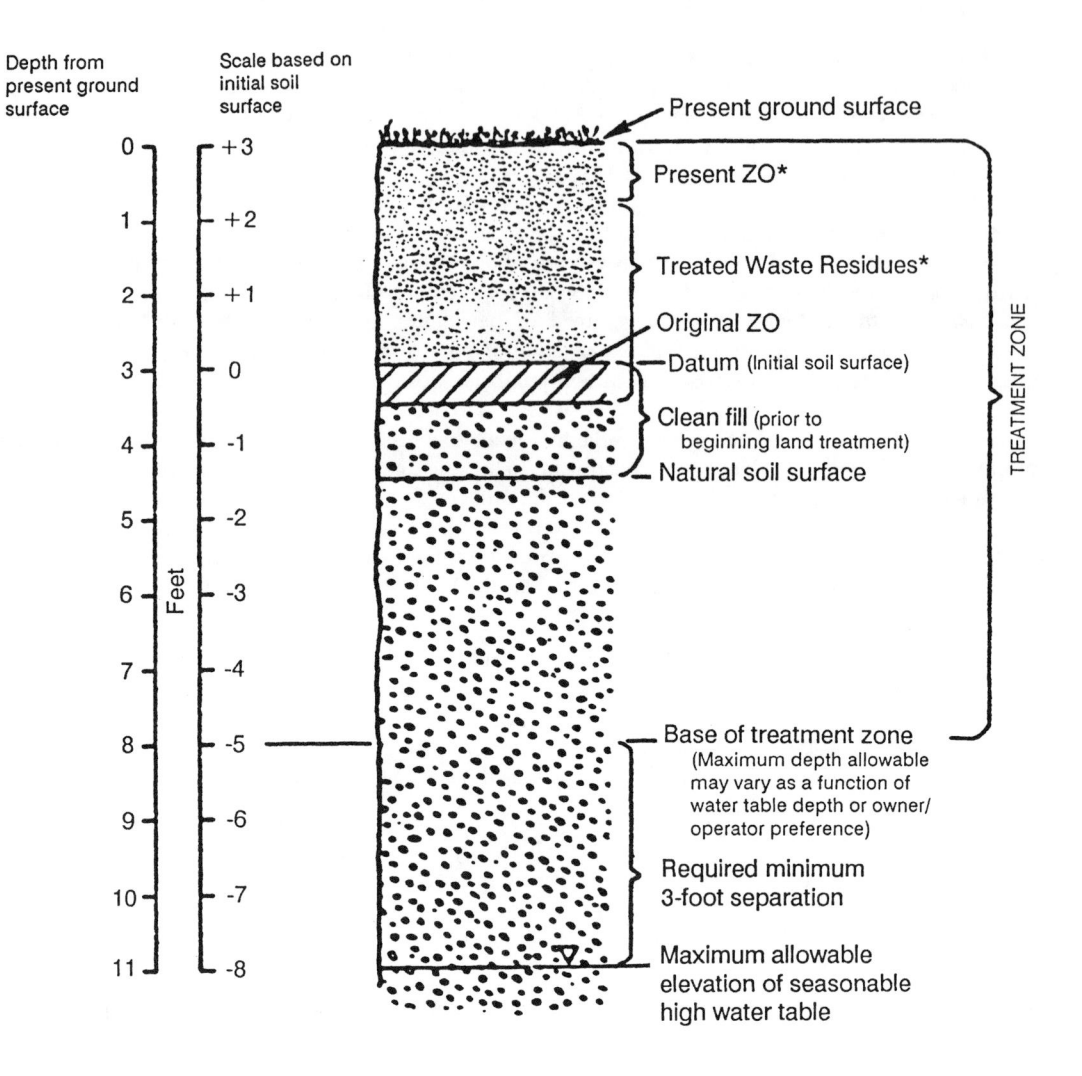

Figure 2.7 Treatment zone definition. (From Overcash, M. et al. Report No. ANL/EES-TM-340. DE88005571. Argonne National Laboratory, Argonne, IL, 1987.)

(IWACO Consultancy, 1989). The area is managed by fertilization, irrigation, and lime addition to maintain optimum conditions of nutrient content, moisture, and pH (Wilson and Jones, 1993).

Tilling the waste material into the soil immediately after application will decrease its chance of migration out of the area (Raymond, Hudson, and Jamison, 1976). Off-site migration of oily waste constituents has not been observed in several field or laboratory studies (Arora, Cantor, and Nemeth, 1982). Metals are immobilized in the top 15 cm of the soil (Loehr, Martin, Neuhauser, Norton, and Malecki, 1985).

The leached residual may be adsorbed, assimilated, or inactivated in the upper soil horizons; however, an individually tailored monitoring system should permit detection of waste transport, as well as allow evaluation of the performance of the biodegradation process (Arora, Cantor, and Nemeth, 1982). Because of the slower movement and longer detention times in the soil, changes in the characteristics of the soil, pore water, and groundwater are not rapid. Monitoring can detect trends in any changes, and management adjustments can be made to minimize developing problems (Loehr, 1986). In a judiciously located and operated landfarm facility, groundwater quality is not likely to be endangered (Dibble and Bartha, 1979b),

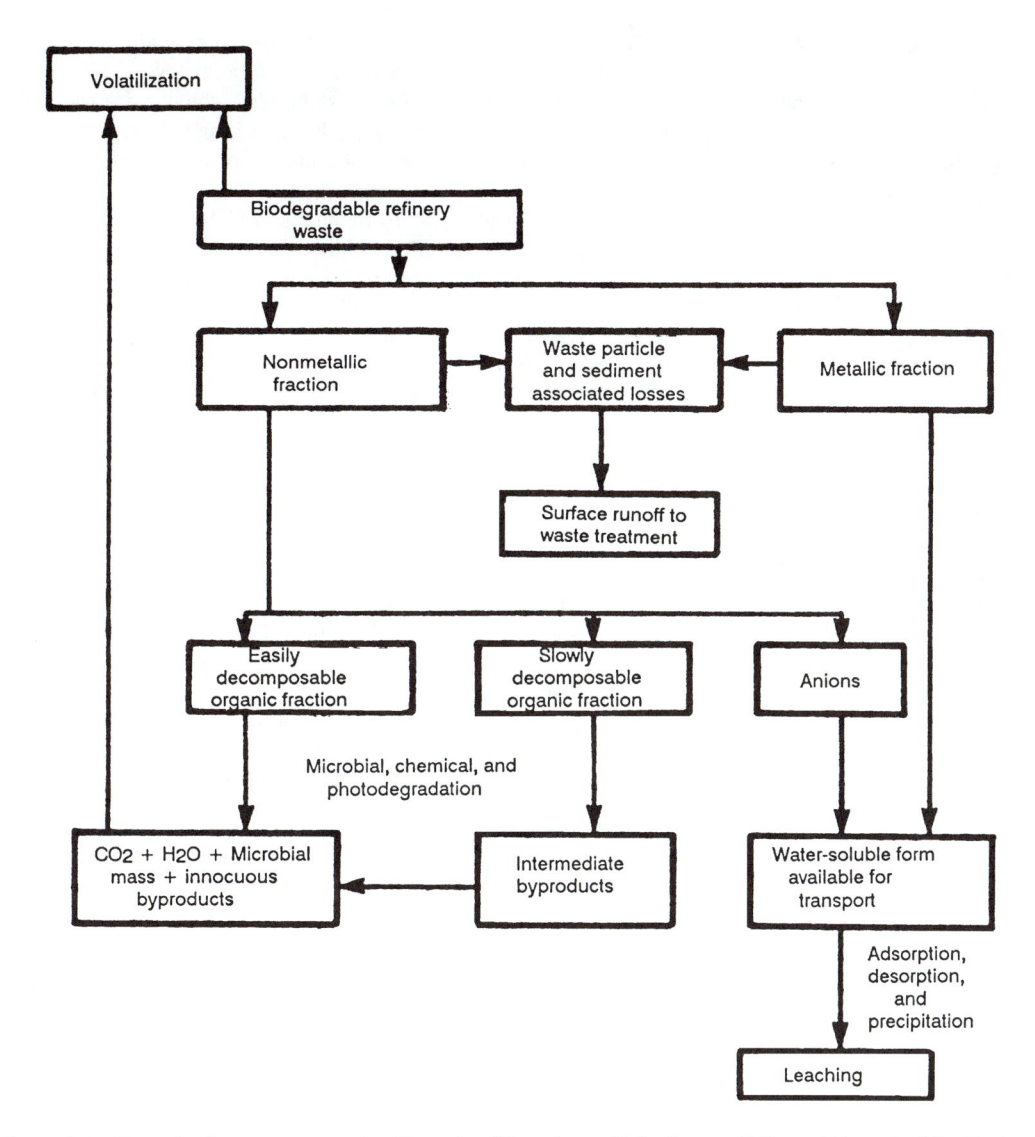

Figure 2.8 Fate of refinery waste at a landfarm site. (From Arora, H.S., Cantor, R.R., and Nemeth, J.C., *Environ. Int.* 7: 285–291, 1982. Elsevier Sci. Ltd. With permission.)

and maintenance and cleanup responsibilities are considerably less than with other waste disposal options (Loehr, 1986). Studying landfarming of weathered oil-contaminated soil by means of a mesocosm, Huesemann and Moore (1993) concluded that leaching was insignificant with this contaminant and that landfarming of weathered soils would not adversely impact groundwater or surface water quality. Figure 2.8 shows the processes that take place concurrently after the incorporation of waste in the soil at a landfarm site (Arora, Cantor, and Nemeth, 1982).

After the easily degraded compounds have been removed by biodegradation, a residual concentration is left in the soil (Harmsen, 1991). Efforts should be directed at increasing the bioavailability of these residuals. Detergents may be useful at this point to create more bioavailable material. High concentrations of detergents may be necessary, however, and there will still be residual concentrations. Also, the detergents producing the highest solubility of oil (the Dobanols) are the most easily degradable. Biodegradation of the detergents consumes oxygen, leaving less for degradation of the oil. It is questionable whether or not these detergents work long enough in landfarming. Time and sufficient water for continuously transporting contaminants into the soluble phase may be the solution for biodegradation of the more recalcitrant compounds.

During the first year of application, the highest total losses occur for the saturates fraction, followed by aromatics, polar compounds, and asphaltenes (Sprehe, Streebin, Robertson, and Bowen, 1985). The pattern is different in the second year. During the winter months, saturates, asphaltenes, and polar compounds show net increases, probably due to anaerobic decomposition. Aromatics are believed to degrade into other fractions, even during winter months. Phenolics (not found in any of the added sludges) are apparently formed at low concentration in the soil matrix during the first year.

Application of refinery effluent sludge to the soil was followed by bursts of CO_2 evolution, and after 25 months 85% of the total PAHs was found to have disappeared (Balkwill and Ghiorse, 1982). Three-ringed PAHs are readily degraded. There is a pattern of increased persistence with increasing molecular weight and condensation.

When landfarming was used to treat weathered Michigan crude oil–contaminated soil, the process was successful in removing up to 90% of the total petroleum hydrocarbons (TPH) in the soil within 22 weeks (Huesemann and Moore, 1993). Up to 85% of heavy petroleum hydrocarbons with carbon numbers above 44 were biodegraded. Approximately 93% of saturated and 79% of aromatic compounds of the TPH were biodegraded during that period. Leachate concentrations of BTEX and PAHs were below detection limits. It was concluded that farming of such weathered soils would be highly successful for removing petroleum hydrocarbons, while not adversely impacting either groundwater or surface water quality.

Landtreatment is a controlled and managed treatment and disposal technology (Loehr, 1986). It is an active and dynamic technology for degrading and immobilizing applied constituents and should not be confused with passive storage technologies, such as waste piles, surface impoundments, landfills, or deep well injection. There have been no demonstrated hazards to workers or local residents during or after treatment. If petroleum wastes could be somewhat degraded or stabilized by a pretreatment process, such as composting, application rates could be increased without environmental problems resulting (Hornick, Fisher, and Paolini, 1983).

After initial biodegradation of organic compounds in the soil, there is a residual concentration left that is adsorbed by organic matter and entrapped in micropores (Harmsen, Velthorst, and Bennehey, 1994). This fraction is only slowly biodegraded, as the contaminant gradually becomes available to the microorganisms. Rather than try to increase the bioavailability of the pollutants, it may be necessary to extend the duration of the treatment to allow time for desorption to occur. If the pollutant had been in contact with the soil over a long period of time, a larger fraction would be sorbed in the micropores and would take more time to diffuse out of the soil pores. Contaminants with larger partition coefficients between soil and water will also need a longer treatment period.

A multiphase biological remediation approach was employed by Turney, Aten, and Zikopoulos (1992) to treat a fuel tank farm impacted by JP-4 jet fuel and aviation gasoline. The site was remediated in 2 years by combining landfarming, *in situ*, and biological reactor remedial methods.

Landtreatment was employed at a site in southern California to remediate 50,000 yd^3 of petroleum-contaminated soil (Graves, Chase, and Ray, 1995). Landtreatment was combined with prescreening of the soil to remove large globules of tar and pieces of asphalt. In less than 8 months, TPH levels went from a range of 4000 to 5000 mg/kg to less than 1000 mg/kg.

Although the material applied to land may initially be phytotoxic and reduce the yield of the vegetation that manages to emerge, the toxicity diminishes with time (Brown, Deuel, and Thomas, 1983). Thus, soils used for landtreatment can eventually be revegetated.

An extensive literature search has been performed on the subject of biological degradation of hazardous waste via landtreatment by Wetzel and Reible (1982).

Environmental Factors
Several factors have been identified as being important to the landtreatment process: the relative composition of the organic fraction of the material to be treated, temperature, soil moisture, availability of nutrients, soil pH, and oxygen availability (Sprehe, Streebin, Robertson, and Bowen, 1985).

Temperature
Temperature has an important influence on degradation rate. Biodegradation declines with temperature, due to reduced microbial growth and metabolic rates (Huddleston, Bleckmann, and Wolfe, 1986). It is essentially zero at the freezing point. Degradation at 10°C is about one third that at 40° (Brown, 1981). This factor has been demonstrated in the laboratory and is, for all practical purposes, a function of

geographic location and climate (Sprehe, Streebin, Robertson, and Bowen, 1985). It is normally refinery/site specific, and not a process control parameter.

In field plots that received higher applications of oily wastes, the temperature of the soil increased by 1 to 10°C, due to increased metabolic activity or the presence of the waste constituents (Huddleston, Bleckmann, and Wolfe, 1986).

Several correlations have been made between soil temperature and moisture:

1. Sensitivity of biodegradation to moisture increases with temperature.
2. Sensitivity of biodegradation to temperature increases with moisture.
3. Sensitivity of biodegradation to temperature increases with decreasing temperature.

Section 5.1.2 discusses in depth the effect of temperature on biodegradation and provides suggestions for controlling soil temperature in the field.

Moisture
Soil moisture is a major control parameter in the landtreatment process (Sprehe, Streebin, Robertson, and Bowen, 1985). It is influenced by local climatic conditions, but is at least somewhat controllable through addition of water, via sludge application or irrigation, and by proper site design to allow for controlled surface drainage. The importance of soil moisture in the landtreatment process is manifested primarily in its impact on the ability to maintain adequate soil aeration. The optimum moisture content for tilling of oiled soil depends upon the oil concentration and soil characteristics. The soil should be tilled only when it is friable, i.e., at moisture contents below the plastic limit.

The optimum moisture content for the highest degradation rate for landtreatment of a refinery waste was found to be 18% (Brown, 1981). At 33% moisture (too wet) or 12% moisture (too dry), the degradation rate was lower. See Section 5.1.1 for extensive background information on soil moisture requirements for biodegradation and methods for providing the optimum moisture levels.

If there is no cover over the soil, runoff water may have to be collected and treated, sprayed back on the soil, or released to a sewer (Eckenfelder and Norris, 1993).

Nutrients
The availability of nutrients, especially nitrogen, within the soil of the zone of incorporation is important to allow the biological processes to proceed efficiently (Kincannon, 1972; Cresswell, 1977; Huddleston and Cresswell, 1976; Dibble and Bartha, 1979b). Availability of nitrogen (N) and phosphorus (P) for soil microbial growth is controlled by three factors: (1) amount of N and P in the soil and rate at which they are mineralized (become available for use), (2) amount of biodegradable carbon and available N and P in the added waste, and (3) rate at which the waste organic carbon is assimilated in the soil environment (Huddleston, Bleckmann, and Wolfe, 1986). By estimating or measuring these factors, fertilization needs can be approximated.

Addition of inorganic fertilizers ensures nutrient availability, and a soil carbon-to-nitrogen (C:N) ratio of 600:1 is recommended as a guideline for nitrogen addition (Kincannon, 1972; Huddleston and Cresswell, 1976; Cresswell, 1977; Dibble and Bartha, 1979b). The most rapid biodegradation of refinery sludge occurs when nitrogen is added to reduce the C:N ratio to 9:1 (Brown, Donnelly, and Deuel, 1983b). Petrochemical sludge is degraded most rapidly when nitrogen, phosphorus, and potassium (K) are added at a rate of 124:1 (C:NPK). Ammonia nitrogen should be used as a parameter for evaluation of nitrogen utilization (Kincannon, 1972; Huddleston and Cresswell, 1976; Cresswell, 1977; Dibble and Bartha, 1979b). The phosphorus source is typically a salt of phosphoric acid, and a nitrogen source may be an ammonium salt, a nitrate salt, urea, or combination (Eckenfelder and Norris, 1993). From 10 to 50% of the total amount of nutrients anticipated is added initially. Subsequent additions are based on consumption.

As a landtreatment site matures, progressively less fertilization is required, because most of the nitrogen and phosphorus added to the site remains for reuse (Huddleston, Bleckmann, and Wolfe, 1986). Overfertilization can lead to groundwater or surface water problems, while underfertilization leads only to slower than optimal biodegradation rates. See Section 5.1.5 for a detailed discussion of the nutrient requirements for optimum biodegradation.

Soil pH
This is an important process control parameter (Sprehe, Streebin, Robertson, and Bowen, 1985). Maintaining the soil pH at or above 6.5 minimizes the solubilization and migration of heavy metals and

provides optimum conditions for biodegradation. Soil treated with sludges containing heavy metals should be medium to fine textured, have a pH above 6.5, and contain 3 to 7% organic matter with a cation exchange capacity of at least 14, in order to be considered acceptable for retention of metals (Leeper, 1978; Huddleston, 1979; Loehr, Tewell, Novak, Clarkson, and Freidman, 1979).

The optimum pH for soil biodegradation lies between 6 and 8; however, effective biodegradation can be found outside this range (Huddleston, Bleckmann, and Wolfe, 1986). The soil pH may increase somewhat during the operation of the landtreatment system, because of the addition of both fertilizer and sludge (Sprehe, Streebin, Robertson, and Bowen, 1985). Application of oily wastes to field plots increased the pH of the acid soils by as much as 1 pH unit for the higher application (Loehr, Martin, Neuhauser, Norton, and Malecki, 1985). Section 5.1.3 describes the effects of pH on biodegradation and how to achieve optimum pH levels in the contaminated soil.

Oxygen
Availability of oxygen throughout the zone of incorporation is important for bio-oxidation of the organic materials (Sprehe, Streebin, Robertson, and Bowen, 1985). Maintenance of aerobic conditions also aids in preventing the migration of heavy metals (Raymond, Hudson, and Jamison, 1976; Fuller, 1977; Dibble and Bartha, 1979b; Frandsen, 1980). This can be accomplished by tilling or cultivating the zone of incorporation and by avoiding saturated soil conditions. Adequate surface and lateral subsurface drainage of the zone of incorporation is essential following rainfall, or tilling operations can be inhibited for long periods of time. Prevention of saturation of the soil with water benefits the soil oxygenation (Huddleston, Bleckmann, and Wolfe, 1986). Oxygen content is also improved by the presence of sand or loam (heavy clay is undesirable), avoidance of unnecessary compaction (heavy trucks, etc.), and limited loading of rapidly biodegradable matter. Lower application rates result in greater bacterial populations, possibly due to decreased aeration from excessive hydraulic loading (Arora, Cantor, and Nemeth, 1982).

Soil microorganisms use oxygen that has been transferred to soil water from the atmosphere (Huddleston, Bleckmann, and Wolfe, 1986). Thus, oxygen available for biodegradation is a function of (1) amount of void space in the soil, (2) partial pressure of oxygen in the soil atmosphere, (3) oxygen transfer rate from soil atmosphere to soil water, (4) rate at which soil microorganisms are using available oxygen, and (5) geometric distribution of oxygen-consuming soil area. Oxygen is not rapidly transferred to water, and it can be quickly depleted by active microbial metabolism (Huddleston, Bleckmann, and Wolfe, 1986). Generally, aerobic bacteria function well at ≥0.2 mg/L dissolved oxygen. See Section 5.1.4 for an in-depth coverage of soil oxygen, the oxygen requirements of microorganisms for biodegradation, and a variety of methods for controlling oxygen levels in soil for optimum degradation. Section 3.2.1 describes the process of aerobic degradation and discusses the microorganisms that can aerobically degrade the different petroleum constituents.

Although anaerobic degradation occurs in soil, it must be limited for effective landtreatment applications since (1) anaerobic biodegradation results in noxious products, such as hydrogen sulfide, ammonia, amines, and mercaptans; (2) anaerobic biodegradation is slower and less complete; and (3) in a reduced state, most hazardous metals are more water soluble (Huddleston, Bleckmann, and Wolfe, 1986). Anaerobic degradation of petroleum constituents is described in Section 3.2.2, along with the different microorganisms that can degrade specific compounds. Utilization of anaerobic conditions for degradation of petroleum compounds in soil is described in Section 5.1.4.7.

Measurements of redox potential (Eh) and oxygen diffusion rate (ODR) can be used to monitor biodegradation of hazardous wastes in soil and determine the effectiveness of landtreatment operations (Shaikh, Hawk, Sims, and Scott, 1985). Wax-impregnated graphite electrodes (WIGEs) perform better than platinum wire electrodes (PWEs) for the measurement of ODR and Eh in soil environments. This method appears to provide an indication for complete biodegradation, and the course of the entire process can be monitored conveniently. The method is quite suitable for a landfarm situation where a large number of probes can be distributed over the entire area. Eh measurements may indicate only the complete biodegradation of organic compounds into end products, such as carbon dioxide and water. This method is not recommended for nonaerobic conditions created by flooding, since "channeling" of the solution may cause different redox environments around the electrodes.

Cost-Effectiveness
Landtreatment of oily wastes has been found to be not only an environmentally sound waste disposal practice but also a very cost-effective treatment option for disposal of this material (Arora, Cantor, and

CURRENT TREATMENT TECHNOLOGIES

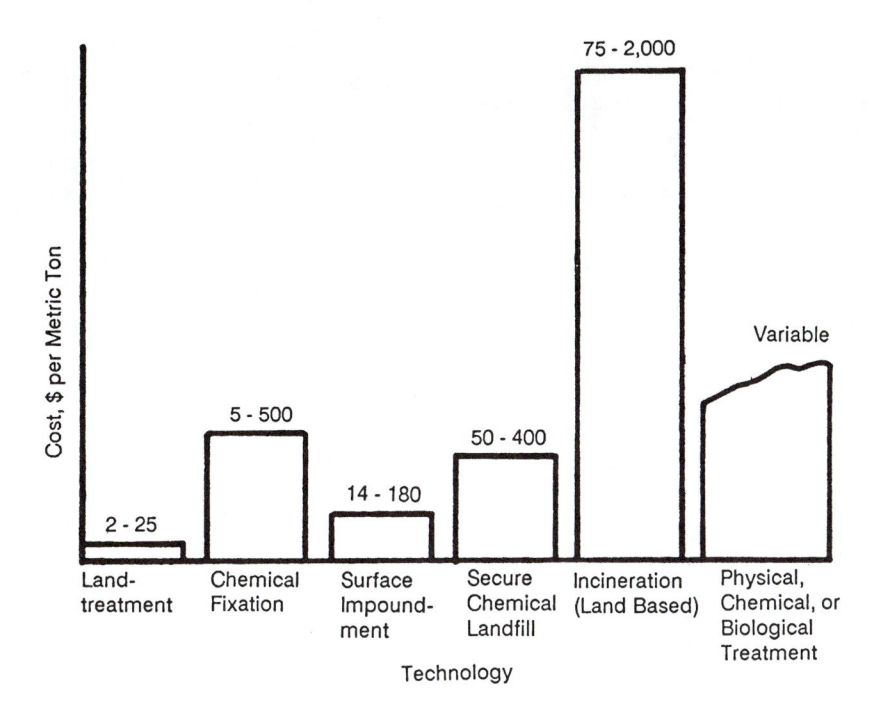

Figure 2.9 Evaluation of the economics of options meeting environmentally acceptable performance standards for hazardous wastes. (From Plehn, S.W., Draft economic analysis (regulatory analysis supplement) for Subtitle C, Resource Conservation and Recovery Act of 1976. Office of Solid Waste, U.S. EPA, Washington, D.C. Reprinted in Monroe, *Am. Biotechnol. Lab.* 3:10–19. 1985. With permission.)

Nemeth, 1982). When land area is readily available, landtreatment is usually more cost-effective than other disposal methods, including landfilling (Sprehe, Streebin, Robertson, and Bowen, 1985). This soil treatment may prove to be the most economical and environmentally sound means of disposing of many of the complex industrial wastes (Brown, Deuel, and Thomas, 1983). In landfilling, oil remains in the landfills with little biodegradation and represents a continuous potential for groundwater pollution. In addition, future land use for landfilling is restricted. Incineration requires energy, and smoke stack residues can spread harmful constituents. Figure 2.9 compares the relative costs per metric ton for the disposal of hazardous wastes by the available technologies (Plehn, 1979).

Land application for wet municipal sludge entails an aggregate cost of about $6 to 8/ton (Overcash, Brown, and Evans, 1987). For hazardous waste, this range is $10 to 50/ton. Incineration has a cost range of $100 to 500/ton, with an average of $150 to 200/ton. The capital costs for incineration are substantial ($1 million to more than $5 million).

Limitations

Landfarming has the virtue of simplicity and, consequently, modest capital and operational costs (Savage, Diaz, and Golueke, 1985). Unfortunately, as is the case with landfilling, it is encumbered by problems of land availability and a serious potential for contaminating water, air, and soil (Grabowski and Raymond, 1984). This process is also sensitive to weather conditions, and it may not be possible to meet optimum sludge application schedules.

Landfarms cannot degrade the heavy components of petroleum oils or chlorinated solvents. In an EPA study, 20 to 50% of applied oily waste was not biodegradable (U.S. EPA, 1985b). Naphthalenes, alkanes, and unsubstituted aromatics were rapidly degraded, with a half-life of less than 30 days, while refractory compounds, e.g., creosote (Mueller, Lantz, Blattmann, and Chapman, 1991) and bunker C oil (Song, Wang, and Bartha, 1990), accumulated in the soil. The regulatory acceptance of long-term disposal of residual oil and grease in landfarms has not been resolved.

The major disadvantage with landfarming is the possibility of contaminant movement from the treatment area (Wilson and Jones, 1993). The criteria for determining the treatability of a waste in a landtreatment facility will involve an evaluation of the degradability, mobility, and potential bioaccumulation of the waste constituents (Loehr, 1986). If waste constituents are sufficiently volatile or mobile or have a propensity to bioaccumulate at a landtreatment site such that there is an adverse impact to human health and the environment, then the wastes should not be landtreated.

Assessment of this technology rated it very unfavorable for contaminant interferences and versatility. It is expected to be adversely affected by the presence of a wide variety of identified contaminants. It is expected to demonstrate, or has demonstrated, inability to remove a wide range of both organic and inorganic compounds of interest from groundwater.

VOC Emissions

Some organics in applied wastes are volatile, and these materials may be emitted by vapor transport from the soil during landtreatment (Loehr, 1986). Volatilization is, of course, not considered an acceptable treatment method in and of itself. The driving force for volatilization results from the vaporization of a chemical within the soil pore space. The volatilization rate of a chemical is strongly affected by its adsorption to soil organic matter and its solubility within the soil water. Volatilization losses are affected by waste-, soil-, and site-specific parameters.

In some areas, there is concern over the air pollution from VOCs released from landfarms (U.S. EPA, 1985b). The maximum percent of applied oil that is lost due to volatilization has been reported to be 23%, within 20 days of application (Sprehe, Streebin, Robertson, and Bowen, 1985). Volatile emissions may account for up to 65% of the total oil losses from a plot during the short term. With less-volatile contaminants, such as diesel fuel and the heavier heating oils, volatile losses may be acceptable (Eckenfelder and Norris, 1993). A cover over the contaminated soil would contain the VOCs. Soils with high levels of VOCs should probably be treated by other means.

Rates of Application

Waste application rate, or loading rate, is a function of oil concentration in the waste and the land area for waste treatment, assuming a conventional 15 cm depth of incorporation (Martin, Sims, and Mathews, 1986). Waste application rates can be determined from the quantity of waste produced, waste oil concentration, and the land area required for actual treatment. Waste application frequency is a function of waste quantity and waste generation frequency. Waste oil degradation can be expressed in terms of stabilized weight percentage of oil in the treatment soil during the active life of the units (oil/soil concentration). It appears that waste degradation half-life and waste application frequency have a greater influence in determining stabilized oil/soil concentrations than waste application rate.

Smaller and more-frequent applications yield higher overall biodegradation rates than does infrequent application of large batches (Dibble and Bartha, 1979a). They also minimize the adverse effects of toxic oil sludge components and keep the hydrocarbon-degrading microbial population in a continuous state of high activity. At most temperate zone landfarming sites, two 100,000-L/ha (255 barrels/acre) or four 50,000-L/ha oil sludge hydrocarbon applications per growing season seem appropriate.

Biodegradation rates for oily sludges from a petroleum refinery and a petrochemical plant were greatest when small applications were made at frequent intervals (Brown, 1981). Comparison of degradation rate with the microbial population indicated that the optimum application rates for both wastes were from 5 to 10% (wt/wt). The highest oxygen uptake rate and the greatest total microbial counts occur at an oily waste concentration of 5% (Jensen, 1974). Other authors, however, report that the population of total soil bacteria is greatest when 1% of these sludges is added to the soil; whereas, 5 and 10% sludge additions result in slightly lower microbial populations (Brown, Donnelly, and Deuel, 1983). In another study, four to six applications of sludge per year were found to be required in order not to exceed the field capacity of the soil (Sprehe, Streebin, Robertson, and Bowen, 1985). The resulting equilibrium oil concentration in the soil was not to exceed 10 to 12% (dry soil weight basis).

An application rate of 1.0 g of hydrocarbon per 20 g of soil–sand mixture was found to be optimal (Dibble and Bartha, 1979a). Application rates that are overoptimal for a rapid removal of saturated hydrocarbons favor removal of the aromatic and asphaltic classes. Biodegradation of the latter compounds may be dependent upon a continued presence of saturated hydrocarbons to support the cometabolic biodegradation of the former classes.

Some of the potential problems caused by overloading a landtreatment system are (Huddleston, Bleckmann, and Wolfe, 1986)

    Toxicity to degrading microorganisms;
    Exceeding the soil-binding capacity and resultant waste migration;

Oxygen starvation caused by demand exceeding transfer rate;
Water exclusion by hydrophobic components;
Creation of cultivation (mechanical) problems;
Waterlogging (oxygen starvation) if the waste has a high water content;
Crust or film formation on the soil surface.

In a 2-year study of oil loadings (2.1 to 26.50% dry soil weight basis), there was no apparent maximum oil loading above which inhibition of biodegradation occurred (Sprehe, Streebin, Robertson, and Bowen, 1985). A maximum practical hydraulic loading for this site was found to be approximately 40 L/m^2 (1 gal/ft^2). At the relatively high oil concentrations of the sludges used in the study (60 to 90 wt%), the maximum hydraulic loading corresponds to an approximately 7% (dry soil weight basis) increase in oil concentration per application.

Because of metal accumulation in a plot with a high loading rate (27%, dry soil weight basis) of oil in 22 months, the useful life of the plot was found to be limited by zinc and cadmium concentrations. Zinc would have reached a critical level in about 17 years, and cadmium, in about 24 years, at that loading rate (Sprehe, Streebin, Robertson, and Bowen, 1985). However, it may be possible to use higher application rates, if the waste to be applied receives preapplication treatment to reduce a specific constituent that may be limiting the application rate (Loehr, 1986).

Oily waste landtreatment systems should be designed for equilibrium conditions (Sprehe, Streebin, Robertson, and Bowen, 1985). Ideally, equilibrium is reached when the amount of degradable material applied is removed (via degradation and volatilization) in the period prior to the next application. Actually, "equilibrium" applies to time intervals over which sufficient loading/loss cycles have occurred, such that process fluctuations are insignificant.

The water-soluble compounds in sludges can be low in degradability, potentially toxic, and extremely mobile in high concentrations (Brown, 1981). This indicates a need for careful management of land treatment sites to avoid groundwater contamination. Gas-liquid chromatography (GLC) combined with column chromatography is recommended for effective monitoring of oily wastes applied to soils.

Degradation Rates

High treatment efficiencies (in terms of TOC and specific chemicals) can be achieved by biodegradation of industrial sludges and waste sludges containing toxic or hazardous chemicals in landtreatment systems (Kincannon and Lin, 1985). Priority pollutants can be removed to very low levels. Removal of the organic pollutants cannot, of course, be described by simple kinetic equations. The removal can be described by one, two, or, in some cases, three first-order rates. This makes use of the first-order reaction, $dc/dt = -KC$ (Loehr, 1986). This indicates that at any one time, t, the rate of degradation is proportional to the concentration, C, of the chemical in the soil. First-order kinetics, generally, apply where the concentration of the chemical being degraded is low relative to the biological activity in the soil. The first-order rates are a function of the type of waste; the site and soil characteristics, including climatic conditions, soil temperature and moisture, soil texture and chemistry; and operation/management practices, such as loading and cultivation (Kincannon and Lin, 1985; Sims, Sorensen, Sims, McLean, Mahmood, and Dupont, 1985).

The overall rate of biodegradation is influenced by the type of oil sludge, by the microorganisms present in the soil, and by the climate (Shailubhai, 1986). The degradation rate is especially affected by the loading rate when the waste exerts a toxic effect on the microorganisms (Sims, 1986). Degradation rates are, generally, expressed as half-lives, or the time required to decrease the original concentration by one half, and may incorporate all of the site/soil factors identified above. Half-lives can be estimated from first-order kinetics, if first-order rate constants are known for waste constituents (Loehr, 1986). The time required for the compound to decrease to any fraction of its initial level also can be estimated, if the appropriate rate constants are known. There is no given optimum loading pattern that fits all or even most wastes for landtreatment (Huddleston, Bleckmann, and Wolfe, 1986). The best policy is to load at the maximum rate that does not result in any problems.

The half-life of organics applied to soil varies and can range from very rapid (half-life of minutes) for readily degradable compounds, such as amino acids and sugars, to very slow (half-life of months or years) for some polynuclear aromatics (Loehr, 1986). The loss of some organics (naphthalenes, alkanes, and certain aromatics) is rapid, especially in the warmer months (Loehr, Martin, Neuhauser, Norton, and Malecki, 1985). The half-life of these compounds is, generally, less than 30 days, while the half-life of the total oil and grease ranges from about 260 to 400 days. Tables 2.7 (Raymond, Hudson, and

Table 2.7 Average Microbial Degradation of Oily Waste after 1 Year[a] and after 6 Months[b]

Oily Waste	Percent Degraded (1 yr)	Percent Degraded (6 months)
Used crankcase oils	67.9	
Crude oils	61.9	
Home heating oil (#2)	87.3	
Residual oil (#6)	57.8	
Refinery sludge		63
Petrochemical sludge		34

Source: [a] Raymond, Hudson, and Jamison, *AIChE Symp.* 75:340–356. American Institute of Chemical Engineers 1980 AIChE. With permission.
[b] Brown, Donnelly, and Deuel, *Microbiol. Ecol.* 9:363–373, 1983.

Table 2.8 Biodegradation Rates Reported from Full-Scale Landtreatment Operations

Waste	Location	Loading Rate g/kg soil/year	Loading Rate lb/ft³ soil/year	Degradation Rate g/kg soil/year	Degradation Rate lb/ft³ soil/year
Refinery oily waste	Montana	11	0.98	6	0.57
	California	148	12.25	114	10.28
	New Jersey	87	7.82	61	5.47
	Illinois	16	1.40	11	0.98
	Louisiana	44	4.00	39	3.52
	Washington	22	1.97	14	1.26
	Texas	29	1.97	22	1.96
	Texas	79	7.16	75	6.73
	Texas	62	5.62	55	4.94
	Oklahoma	67	6.00	53	4.80
	Oklahoma	17	1.54	11	0.98
	Texas	—	—	165	15.00
Heavy oil	Oklahoma	1.2	0.11	0.5	0.05
Sulfite liquor wastes		<150 lb BOD₅/acre/day		100%	
Vegetable-canning wastes		1300 lb COD/acre/day		99%	
Potato-processing wastewater		10–85 T COD/ha/yr		>80%	

Source: Huddleston, R.L. et al. in *Land Treatment: A Hazardous Waste Management Alternative,* Loehr, R.C. and Malina, J.F., Jr., Eds. Center for Research in Water Resources, University of Texas, Austin, 1986. With permission.

Jamison, 1980) and 2.8 (Huddleston, Bleckmann, and Wolfe, 1986) present some biodegradation rates observed in landtreatment operations. Many of the rates for the latter were examined with respect to "degradation months," which are defined as months having an average air temperature above 50°F. This narrows reported oil degradation values from 0.57 to 10.28 lb/ft³/month to 0.09 to 0.86 lb/ft³/month. Table 2.9 summarizes rates of degradation obtained for phenol and substituted phenols, for a given application rate (Overcash and Pal, 1979a).

In a laboratory simulation of landtreatment, a PAH decrease of 3.3 µg/g of soil per day occurred over a 2-year "active" period of waste addition (Bossert, Kachel, and Bartha, 1984). During a subsequent 1-year "closure" period, the PAH decrease was 0.1 µg/g soil/day. However, nearly half the loss appeared to be due to abiotic mechanisms. There was extensive degradation (to 99%) of three-ring and some four-ring PAHs. The condensed four-ring compounds were relatively recalcitrant, and all the five- and six-ring compounds resisted degradation (from 1 to 70% loss). The half-lives of 11 PAHs were found to be 18 to 190 days, when they were added to soil at 1 to 147 ppm (Sims, 1982). The initial concentration had no effect on the degradation rate over the addition range studied.

Table 2.9 Rates of Degradation of Phenol and Substituted Phenols

Compound	Application Rate (mg/kg)	Degradation Rate
Pyrocatechin	500	100% in 1 day
Hydroquinone	500	100% in 1 day
Resorcinol	500	100% in 2 days
Phenol	500	100% in 2 days
Phenol	500	100% in 6 days
Thymol	500	100% in 3 days
2,4-Xylenol	500	100% in 4 days
α-Naphthol	500	100% in 5 days
p-Cresol	500	100% in 7 days
p-Hydroxybenzoic acid	500	100% in 7 days
o-Cresol	500	100% in 8 days
3,4-Xylenol	500	100% in 9 days
1,4-Naphthoquinone	500	100% in 9 days
m-Cresol	500	100% in 11 days
3,5-Xylinol	500	100% in 11 days
2,5-Xylinol	500	100% in 14 days
β-Naphthol	500	Very little in 90 days
Pentachlorophenol	100	Half-life of 40 days
Various mono-, di-, and trihalogenated phenols	—	Ring cleavage in 2 to 4 weeks

Source: Overcash, M.R. and Pal, D. *Design of Sand Treatment Systems for Industrial Wastes — Theory and Practices.* Ann Arbor Science Publishers, Ann Arbor, MI, 1979. With permission.

Design of Landtreatment Systems

It is important to have proper design, management, monitoring, contingency, and closure plans for all landtreatment facilities (Brown, 1981). An understanding of the mechanism of site-specific biodegradation processes is necessary in the development of a design for the landtreatment facility and in establishing operational practices for realizing accelerated biodegradation rates (Arora, Cantor, and Nemeth, 1982). Periodic measurements of oil concentration within a bioreactor facilitate the evaluation of the rate of oil disappearance, and in combination with measurements of percolation, runoff, and volatilization allow calculation of the biokinetic rates needed for rational design (Sprehe, Streebin, Robertson, and Bowen, 1985). The relative simplicity of operation is a major advantage; however, operational simplicity can lead to quick abuse, especially in the absence of rational guidelines for process design and operation.

To design or evaluate a landtreatment system, information is needed on factors, such as waste and soil characteristics, site characteristics (including slope, depth to bedrock, and depth to the groundwater), vegetation to be grown (if any), climatic conditions (precipitation, evaporation, and temperature), and environmental criteria and standards (Loehr, 1986). These factors help establish the proper application rates and scheduling to avoid adverse effects to human health and the environment.

Major emphasis in the design of the treatment facility and the monitoring program must be on avoidance of potential detrimental impacts on ground and surface water, air quality, soil resources, vegetation, biotic life, and human health (Arora, Cantor, and Nemeth, 1982). Areas of potential concern include accumulation of metals, accumulation of resistant organic compounds, and excess salinity. Soil processes, such as ion exchange and adsorption, help mitigate the deleterious effects of these ingredients and protect soil microorganisms from the potentially harmful effects of excessive levels of these compounds in the soil solution.

Section 7 describes a wide variety of methods that can be used for monitoring biodegradation in the field. Table 2.10A and B provides the operational characteristics for several landtreatment facilities (Martin, Sims, and Mathews, 1986).

2.1.2.1.2 Composting

An alternative to landfilling and landfarming is composting of biodegradable wastes (Rose and Mercer, 1968; Willson, Sikora, and Parr, 1983). This method combines many of the good points of incineration

Table 2.10A Operational Characteristics for Landtreatment Facilities 01 to 06

Parameter	01	02	03	04	05	06
			Facility Code			
EPA region	V	VI	VI	V	IX	X
Treatment area (ha)	8	85	2.8	6	4	8
Zone of waste incorporation (depth, m)	0.2–0.3	0.46	0.15	0.66	0.25–0.30	0.30
Reported waste application rates (maximum)	472 m^3/ha/appl.	265 metric t waste/ha/appl.	786 m^3/ha/appl.	5 kg/m^2/appl.	112 t/ha/appl.[a]	70 metric t/ha/appl.
Number of months site actively used per year	6–7	12	7	7–8	9–12	12
Frequency of waste application	Monthly	Yearly per plot	Yearly	Every 2 weeks	Weekly[a]	Monthly
Oil application per[b] degradation month (kg/m^3/month) (max)[c]	167[d]	57	102[d]	14.7	36.8	3.4
Calculated wt% of oil in soil (single application)[e]	13	4.4	8	0.58	10.0[f]	5.0[f]
Reported wt% of oil in soil (percent)	2	4.6	—[g]	—	—[h]	—
Method of waste application	Spraying	Surface spreading by gravity	Surface spreading	Surface spreading and spraying	Subsurface injection	Surface spreading
Amendments added to soil	NPK	None	None	NPK + lime	K, lime	Lime +

[a] Usually approx. 2 acres are injected in any 1 day. Frequency of waste application is reported frequency for maximum application rate.
[b] Per depth of incorporation.
[c] Based on maximum waste application rates reported.
[d] kg/m^3/year.
[e] Based on reported waste application rates.
[f] Reported maximum target percentage of oil in the treatment soil.
[g] Information not available.
[h] Section will not be injected until the oil content is less than 10%.

Source: From Martin, J.P. et al. EPA-600/J-86/264, 1986. PB 87166339.

and landfarming and minimizes their disadvantages (Savage, Diaz, and Golueke, 1985). Composting allows biological decomposition of organic material under controlled conditions.

Composting is a form of prepared-bed type of treatment that can be used to treat highly contaminated material (Wilson and Jones, 1993). The process involves the activity of a succession of mesophilic and thermophilic microorganisms. The soil is piled and mixed with an organic bulking agent, such as straw or wood chips. Aeration can be achieved by forced air or pile turning. Moisture, nutrient, and pH levels are controlled.

The success of landtreatment of hazardous wastes can be validly applied to composting, since the microbiology and biological processes involved in both systems are comparable and, to some extent,

Table 2.10B Operational Characteristics for Landtreatment Facilities 07 to 13

Parameter	07	08	09	10	11	12	13
EPA region	VIII	V	VI	VI	X	VIII	IX
Treatment area (ha)	5.7	4.6	6.1	20	2.8	4.8	2.0
Zone of waste incorporation (depth, m)	0.20	0.20	0.15	0.20	0.30	0.15	0.25
Reported waste application rates (maximum)	20,815 kg waste/ha/appl.	76118 kg waste/ha/appl.	59 m³ waste/ha/appl.	3.5 kg oil/m³/appl.	772 kg/ waste/ha/appl.	1050 kg oil/ha/appl.	2.4 m³ oil/ha/appl.
Number of months site actively used per year	8	6	10	12	12	10	12
Frequency of waste application	Every 3 months	2 to 3 times/week	Monthly	Monthly	Monthly	Monthly	5 times/week
Oil application per[a] degradation month (kg/m³/month) max[b]	0.64	0.18	0.11	4.16	2.31	4.21	1.83
Calculated wt% of oil in soil (single application)[c]	0.05	0.0036	0.8	0.28	0.18	0.34	0.35
Reported wt% of oil in soil (%)	—[d]	—	—	—	—	—	—
Method of waste application	Surface spraying/ spreading	Surface spreading	Surface spreading	Subsurface injection	Vacuum truck hose	Surface spreading	Surface injection
Amendments added to soil	Lime and fertilizer	Lime and fertilizer	Lime and (NPK) fertilizer	None	Lime and fertilizer (once in 3 years)	Lime and (NPK) fertilizer	None

[a] Per depth of waste incorporation.
[b] Based on maximum waste application rates reported.
[c] Based on reported waste application rates.
[d] Information not available.

Source: From Martin, J.P. et al. EPA-600/J-86/264, 1986. PB 87166339.

identical (Savage, Diaz, and Golueke, 1985). For example, pseudomonads most active in landtreatment are among the predominant microorganisms in composting. The main difference between the activities of the organisms in the two media is that environmental conditions of most importance to the microorganisms are subject to greater control in composting than in landtreatment. With composting, the rate and extent of biological degradation can be made to surpass significantly those in landfarming. Most of the PAHs can apparently be decomposed by composting (Epstein and Alpert, 1980).

Another difference between composting and landfarming is that the seeding of the raw wastes is an important feature (Savage, Diaz, and Golueke, 1985). This is done by recycling a portion or even most of the compost product into the new raw waste to be composted. A suitable ratio would be one part seed to nine parts raw waste. This procedure ensures the presence of a population of organisms capable of attacking the hazardous contaminant without incurring the need for a lag period during which the necessary population could develop.

Composting facilitated degradation of 500 mg naphthalene/kg soil and 100 mg/kg of phenanthrene, anthracene, fluoranthene, and pyrene/kg soil within 25 days in soil systems below the water-holding capacity (Kaestner and Mahro, 1996). The degradation seemed to be enhanced by the solid organic matrix of the compost. It was not enhanced by the presence of specific microorganisms, the fertilizers, or the shift of pH of the compost.

Composting of service station soil contaminated with lubricating oil for 5 months in Finland decreased the mineral oil concentration from about 2400 to 700 mg/kg dry wt (Puustinen, Joergensen, Strandberg, and Suortti, 1995). A commercial bacterial inoculant and nutrient addition had no significant effects.

Technology
The technology should fall primarily into three classes or types: turned windrow systems, static windrow systems, and in-vessel (reactor) systems (Savage, Diaz, and Golueke, 1985). Contaminated soil is mixed in a pile with a solid organic material, such as fresh straw, wood chips, wood bark, or livestock bedding straw (Alexander, 1994). Nitrogen, phosphorus, and other inorganic nutrients may be added. This mixture might be shaped into long rows, or windrows, or placed in a vessel. Moisture and aeration are provided. In selecting one of these systems, the advantages and disadvantages of each must be weighed in terms of their effect on the control of emissions from the composting operation (Savage, Diaz, and Golueke, 1985). All three are equally amenable to the effective control of solid and liquid emissions and discharges, but possibly not with gaseous emissions. The likely technology for controlling gaseous emissions would be an in-vessel (reactor) system (Savage, Diaz, and Golueke, 1985). For hazardous gaseous emissions, the suction phase would be the only one to use in forced air systems (e.g., Beltsville system) because all air exits through a single duct and, hence, can be filtered or subjected to some process to remove the hazard. Bracker (1993) employed a sealed system with controlled aeration for the manufacture of compost from municipal biowaste by allowing intensive rotting at about 70°C to produce a hygienic, biologically stabilized compost of high quality.

Parameters to be Controlled
The engineering parameters to be optimized for composting of hazardous wastes are the same as those for composting of nonhazardous wastes (Savage, Diaz, and Golueke, 1985).

Aeration — The amount and thoroughness of aeration determine the rate and extent of the destruction of the waste, since this is, essentially, an aerobic process (Savage, Diaz, and Golueke, 1985). Aeration also determines the level to which the temperature will rise in the composting mass, as the temperature rise is a result of bacterial activity. The chemical and physical makeup of the waste determines the amounts and rates of aeration required. Insufficient aeration leads to anaerobiosis and generation of objectionable odors.

An exception is the use of anaerobic conditions to break down certain pesticides and many haloaromatic compounds (Suflita, Horowitz, Shelton, and Tiedje, 1982). In fact, an anaerobic phase must precede aerobiosis in the breakdown of certain compounds, such as DDT.

See Section 3.2.1 for a discussion of aerobic degradation and Section 3.2.2 for a description of anaerobic degradation, with examples of specific organisms for degradation of particular petroleum constituents. Section 5.1.4 describes oxygen requirements for optimum biodegradation and methods for controlling aerobic and anaerobic conditions.

Moisture Content — Bacterial activity becomes severely inhibited when the moisture content drops below about 40% (Savage, Diaz, and Golueke, 1985). Fungi and actinomycetes are more tolerant of lower moisture contents than are bacteria. Section 5.1.1 discusses moisture requirements of microorganisms for optimal biodegradation. The maximum level of moisture content is a function of the physical structure of the wastes and the ratio of air to water in the soil.

A problem in composting hazardous wastes is the high moisture content and amorphous structure of the wastes (Savage, Diaz, and Golueke, 1985). A bulking agent can be added to provide ample porosity under all moisture conditions. It is absorbent, resists compaction, degrades very slowly, and can be easily recovered from the composted wastes and, subsequently, recycled. This compost product makes the best bulking agent (Savage, Diaz, and Golueke, 1985). An external bulking agent, such as wood chips, is used for the first composting pass, and the product from that can serve as the bulking agent in the following composting. Recycling the compost product not only has the advantage of eliminating the need to import a bulking agent, it also reduces the amount of residue for disposal. A highly matured compost (>6 months old) at a 2:1 ratio of soil:compost enhances biodegradation (Stegmann, Lotter, and Heerenklage, 1991).

It can be important to maintain sufficient moisture content during composting of wastes under severe hot and dry weather conditions (El-Haggar, Hamoda, and Elbieh, 1996). A mobile composting unit has been developed to provide optimum conditions compatible with site considerations and cost-effectiveness. Aquastore hydrogel (Cyanamid, Rotterdam, Netherlands), an insoluble absorbent polymer, is used to maintain compost moisture. It also requires an aeration system and turning the mixture for 10 min every 2 days for 2 weeks. The end product is an environmentally sound soil conditioner that can eliminate some chemical fertilizer costs.

Temperature — Temperatures rise to 50 to 60°C in composting and are often favorable to biodegradation (Alexander, 1994). Optimum temperatures for biodegradation and associated organisms in the different temperature ranges are discussed in Section 5.1.2.

Nutrients — Nutrients must be available in the amounts and proportions needed to meet the nutritional demands of the various microorganisms (Savage, Diaz, and Golueke, 1985). If the waste has the nutrient (C,N,P,K) concentrations and physical characteristics required for the growth and activities of the microbes, no additional materials are required. Section 5.1.5 describes the nutritional requirements of microorganisms for optimum biodegradation.

Compostable Organic Wastes

Essentially, all but the most refractory of organic wastes can be biologically degraded with this process, given adequate time (Savage, Diaz, and Golueke, 1985).

There is a fairly widespread incidence of hazardous waste decomposers in nature (Savage, Diaz, and Golueke, 1985). These can be isolated and developed to workable concentrations through conventional enrichment techniques. However, the biodegradability of a waste should be assessed before a sizable operation to compost it is set up. The possible contributions of physical and chemical forces (e.g., solar radiation and pH) to the waste destruction should also be recognized and not falsely attributed to biodegradation.

Advantages and Disadvantages

Advantages and disadvantages of composting must be compared with those of physical, chemical, and thermal methods (Savage, Diaz, and Golueke, 1985). These are, in part, a function of the waste.

Because the volumes of toxic organics being treated are usually large, composting contaminated residues is less expensive than incineration, in terms of operation and equipment (Savage, Diaz, and Golueke, 1985). The threat to the environment is less severe. Composting can be carried on at generation sites with less effort than is required with portable incineration units. Compared with simple storage and landfarming, composting is safer and is competitively inexpensive, because the rate of biological destruction is more rapid.

The anaerobic phase needed to initiate breakdown of halogenated organic compounds is readily attained by not aerating the composting mass during the first stage (Savage, Diaz, and Golueke, 1985).

Composting is safer than landfarming because the material is contained during the compost process, rather than being spread or worked into the land (Savage, Diaz, and Golueke, 1985). The containment makes it possible to control and treat all emissions before releasing them to the environment. Composting also requires less land than landtreatment. In landfarming, expensive measures must be taken to minimize the generation of leachate and to collect what is generated to avoid contamination of groundwater.

Sometimes, the waste can be converted into a useful product, i.e., the compost product for landscaping (Savage, Diaz, and Golueke, 1985). However, the toxic contaminant and toxic breakdown products must have been completely destroyed or reduced to an acceptable level before they can be safely used for this purpose. Also, the residue can be made quite small by using the compost product as a bulking agent and recycling it in the system.

Experiments on the degradation of anthracene and hexadecane in soil amended with mature compost (20% dry wt/dry wt) resulted in both mineralization of the compounds and the formation of unextractable bound residues (humification) (Kaestner, Lotter, Heerenklage, Breuer-Jammali, Stegmann, and Mahro, 1995). The annual turnover of humic bound residues may range from 2 to 8% (Hsu and Bartha, 1976; Haider and Martin, 1988). These compounds are used more efficiently for biomass formation when supplied to microorganisms on culture plates than in a compost-manured soil (Kaestner, Breuer-Jammail, and Mahro, 1994).

Retention Time — Retention time is much shorter than with landfarming, i.e., weeks rather than months. Conversely, it is much longer than that with incineration, i.e., weeks instead of minutes in an incinerator (Savage, Diaz, and Golueke, 1985).

Emissions — Gaseous emissions are far less in terms of quantity, severity, and toxicity than are those from incineration, and probably on a par with those from landfarming. However, they can be much more easily controlled in composting than in landfarming. The mechanism for this depends upon the nature of the gaseous substance and can be based upon absorption or adsorption, on combustion, or on a chemical reaction (Savage, Diaz, and Golueke, 1985). The conventional forced aeration (static pile) system used for sewage sludge treatment is not acceptable. Most of the current compost reactors (drum, silo, tank) could probably be suitably modified to fit these requirements.

Residues — If the composted product is considered a residue, then incineration would create less to dispose of than composting. The incineration residue might be more toxic than that from composting, depending upon the retention time used for the latter. Recycling the compost product submits the remaining compounds to a longer exposure to destructive agents. The bulk of the residue for disposal is also reduced. Landfarming produces no residues to be removed from the disposal site. The soil is the disposal receptacle.

Land Requirement — Incineration requires much less land than composting and landfarming, and composting less than landfarming (Savage, Diaz, and Golueke, 1985).

Siting Difficulty — This depends upon permission to use a particular site (Savage, Diaz, and Golueke, 1985). There may be greater resistance to siting an incineration facility than a composting facility. Public resistance to composting and landfarming would probably be about the same.

Groundwater Contamination — If proper disposal methods are used, neither composting nor incineration should pose a problem (Savage, Diaz, and Golueke, 1985). The chances are much greater with landfarming.

Control of Process — Incineration allows more control of the process than composting, and composting more than landfarming (Savage, Diaz, and Golueke, 1985). For example, aeration is more complete in composting than in landfarming. Moisture content is readily adjusted in composting but with difficulty in landfarming (at least with respect to lowering the moisture content).

2.1.2.1.3 Bioreactors

If the potential hazards from discharges and emissions are very serious, the likely technology to use would be one involving an in-vessel system (Savage, Diaz, and Golueke, 1985).

Advantages of a Contained Soil Treatment System — *In situ* vadose zone technologies offer the advantages of being relatively inexpensive and versatile, and they may be able to achieve site closure within a desired time period (Ram, Bass, Falotico, and Leahy, 1993). However, these technologies might require long-term monitoring, and they are not implementable at all sites. Very dense soils or those containing nonvolatile contaminants may have to be excavated and treated on- or off-site. Excavation is a traditional treatment solution, which rapidly removes the contaminant source and allows it to be subjected to a more-focused remediation program. In fact, on-site treatment of PAH-contaminated soil may be preferable to *in situ* treatment, especially if the soil is relatively easily excavated (Wilson and Jones, 1993). Bioreactors have advantages in terms of time and effectiveness when compared with on-site and *in situ* bioremediation. Biological, physical, and chemical processes can be combined in a cost-effective, controlled manner in a bioreactor (Annokkee, 1990).

One of the problems facing the use of biodegradation for *in situ* treatment of contaminated soils is the complexity of the process. There are many reactions and interactions occurring in the subsurface, most of which are not fully understood and are difficult to control. Since it is advisable that optimum conditions be maintained during biodegradation, an enclosed reactor would provide an opportunity to maximize the process. An economical on-site approach that employs the principles of biological degradation, without the current limitations of *in situ* systems, could have wide-ranging applications for site remediation.

An enclosed system would permit control of temperature, moisture, pH, oxygen, nutrients, addition of surfactants, supplementation of highly efficient contaminant-degrading microorganisms, and monitoring of reactions and conditions, resulting in a more rapid and greater extent of biological degradation. This method is less dependent on favorable weather conditions, and seasonal fluctuations would not be a limiting factor.

Many remediation methods generate VOCs. Treatment of excavated soil in an enclosed unit would permit all emissions to be contained and treated before being safely released to the atmosphere. Other advantages of the system are computerized process control with biological deodorization of waste gases, negligible odor emission, and controlled wastewater collection. Leachates that might normally be formed under field conditions and potentially carry pollutants to the groundwater could be collected, treated, or recirculated through the soil in the reactor. Runoff water containing high amounts of oil and fertilizer would not occur with a contained system.

It may also be desirable to employ anaerobic biodegradation, alone or in combination with aerobic degradation. Anaerobic biodegradation can take place in native soils; however, an airtight reactor is needed, if the treatment is to be used as a controlled process.

Contaminated soil should remain permeable to water and air during solid-phase bioremediation (NcNabb, Johnson, and Guo, 1994). Tillage is effective to only a depth of about 15 cm and destroys soil structure of wet soil, forming large clods and lowering the rate of bioremediation. A combination of amendments has been determined that can be used to create small aggregates of oil-contaminated soil to increase aggregate stability, decrease bulk density, and reduce the compressibility of the soil (NcNabb, Johnson, and Guo, 1994). The compressibility and water retention characteristics of aggregated soil will determine the maximum depth that material can be placed in a solid-phase bioreactor. Inorganic amendments increase aggregate stability, while organic amendments decrease aggregate stability, with peat moss decreasing stability the most. A combination of straw, hydrated lime, and starch is most effective. The range of 50 to 80% of the maximum water capacity provides high oxygen consumption in bioreactors, but may vary with the particular soil type (Stegmann, Lotter, and Heerenklage, 1991). A hydrocyclone can separate soils and sediments into fine and coarse fractions. Coarse fractions are then treated in a bioreactor or by landfarming, while fine fractions are treated in a bioreactor or aerated lagoon.

In a field situation, the uncertainty of subsurface geochemical and contamination conditions makes it difficult to provide information typically required for a permit. The complexity of an *in situ* treatment technology can cause delays in obtaining a permit, hindering the cost-effective feasibility demonstrations. With excavated soil contained and treated in a controlled environment, these uncertainties could be reduced or eliminated.

2.1.2.1.3.1 Bioslurry Reactors

Slurry bioremediation is a process that treats contaminated soils and sludges in a bioreactor by extraction and biodegradation (Castaldi, 1994). Slurry-phase treatment can be implemented by constant mixing of contaminated soil with a liquid (Alexander, 1994). It can be set up in a contained reactor (e.g., stainless steel EIMCO Biolift slurry reactor; Leavitt and Brown, 1994) or in a lagoon with a liner. Most commercially available soil slurry reactors consist of one to three tanks in series (King, Long, and Sheldon, 1992). A full-scale unit can range from 10 to 50 ft in diameter and from 15 to 25 ft in height and hold from 15,000 to 300,000 gal. The process is similar to that of activated sludge treatment (Alexander, 1994). It provides for aeration and mixing. It allows addition of nutrients, oxygen, surfactants, and microorganisms, and can incorporate a means of capturing VOCs generated. It allows many of the factors that affect biodegradation to be monitored and controlled for rapid destruction of many PAHs, heterocycles, and phenols. Anaerobic conditions can also be established for more-recalcitrant compounds (Kaake, Roberts, Stevens, Crawford, and Crawford, 1992; King, Long, and Sheldon, 1992).

An advantage of this approach is that the bioreactors are technically simple, and pellet formation during dry treatment can be avoided (Parthen, 1992). Application of slurry reactors is considered especially useful when treating problematic substances, such as concentrated residues from soil scrubbing (Bhandari, Dove, and Novak, 1994).

Degradation occurs in the aqueous phase by suspended microorganisms and by immobilized microbiota in the soil phase (Tabak, Govind, Fu, Yan, Gao, and Pfanstiel, 1997). The presence of the soil places a physical limitation on the degradation rate of oil (Geerdink, Kleijintjensm, van Loosdrecht, and Luyben, 1996). The whole soil should be pretreated to separate out the most-contaminated fraction of the soil and reduce the amount to a suspension of soil fines for biotreatment (Ahlert, Black, Bruger, Kosson, and Suenter, 1990; King, Long, and Sheldon, 1992). It is essential that the soil be sized and graded to smaller than 60 mesh to be able to maintain a slurry against gravity. Silts and clays would be more appropriate than sands and gravels. Pretreatment also prepares the soil by adjusting the pH and increasing the surface area for mass transfer (Black, Ahlert, Kosson, and Brugger, 1991).

One pretreatment technique involves use of a commercially available soil-washing process (King, Long, and Sheldon, 1992). For example, a washing process used by BioTrol, Inc., of Chaska, MN, employs froth flotation, attrition scrubbing, vibrating screens, mixing trommels, hydrocyclones, and pug mills, coupled with a countercurrent flow of washing fluid (e.g., water). This process removes surface contamination and helps release strongly adsorbed compounds. Commercially available soil slurry reactors combine aeration and mixing. Extensive, high-power mixing is generally required to suspend solids in a slurry and optimize the mass transfer of organic compounds to the aqueous phase, where biodegradation normally occurs (Castaldi, 1994). Mixing can be mechanical, such as with an innovative process by EIMCO of Salt Lake City, UT, or simply by the action of aeration (King, Long, and Sheldon, 1992). The treated soils are then dewatered by centrifugation or a belt filter press.

Counts of indigenous, nutrient-supplemented microorganisms have been observed to go from 10^5 to 10^7 CFU/g (colony-forming units/gram) solids in 8 weeks in a slurry reactor with very high removal rates (Leavitt and Brown, 1994). Counts of bioaugmented organisms, on the other hand, dropped to 10^5 in that time, with poor removal performance.

Semivolatile organics and total organic carbons are biologically degraded more rapidly in a slurry-phase system than in a solid-phase system (Yare, 1991). As long as the organic carbon is available to microorganisms, it can be degraded rapidly. Sludge can be biotreated in a closed system, such as a slurry reactor, during cold winter months prior to land application (Persson and Welander, 1994).

Bioslurry treatment of PAH-contaminated soil has been demonstrated under the SITE (Superfund Innovative Technologies Evaluation) Emerging Technologies Program (SITE ETP) (Brown, Davila, and Sanseverino, 1995; Brown, Davila, Sanseverino, Thomas, Lang, Hague, and Smith, 1995). The process involved two 60-L bioslurry reactors and a 10-L fermentation unit in semicontinuous, plug-flow mode for 6 months. The first 60-L reactor received fresh feed daily, supplemented with salicylate and succinate to enhance PAH biodegradation. Salicylate was added to induce the naphthalene degradation operon on NAH (naphthalene) plasmids. Effluent from the first reactor was fed to the second, where Fenton's reagent was added to accelerate oxidation of four- to six-ring PAHs. Fenton's reagent is responsible for hydroxylation of multiring aromatic hydrocarbons, since this is generally the rate-limiting step in biological oxidation of PAHs. The third reactor was aerated, nutrient amended, and pH adjusted only to serve as a polishing reactor to oxidize remaining contaminants biologically. This system demonstrated an average removal of 85% of the total PAHs and 66% of the carcinogenic PAHs.

When compared with a static system, with and without nutrient addition, the slurry system promoted the most thorough biodegradation of petroleum, especially for the heavier compounds (Novak, Schuman, and Burgos, 1995). In the soil-slurry reactor, significant degradation of contaminant occurs in the aqueous phase by the suspended soil microorganisms, rather than by the soil immobilized biofilms (Tabak, Govind, Pfanstiel, Fu, Yan, and Gao, 1995). Data from this reactor can be used to derive the biokinetic parameters for the suspended and immobilized microorganisms.

An open-system slurry reactor is a system in which the liquid medium is replaced daily and the solids retained in the reactor for 2 weeks (Blackburn, Lee, and Horn, 1995). It allows removal of small black particulate solids, which are chemically or biologically produced and easily elutriated from the bioreactor as the liquid medium is changed. This approach can achieve 60 to 80% total petroleum hydrocarbon removal and has the potential to be twice as effective as other bioremediation schemes.

Removal of tetradecene in soil slurries in a rotating drum bioreactor was enhanced by addition of oil-degrading bacteria (Krueger, Harrison, and Betts, 1995). Preinoculated polyurethane foam cubes introduced into the system did not increase the rate of tetradecene removal but should not be discounted as a possible means of improving biomass retention and treatment rate in a continuous culture system.

Another modification of the soil-slurry reactor is one with intermittent mixing capabilities (Stormo and Deobald, 1995). It can function with sand or soil with four basic modes of operation. The water and air pumps are controlled by a repeat-cycle timer, which allows either or both pumps to be operated continuously or intermittently, and to run the blade. The stirrer is propelled by the fluid as it exits the nozzle. This keeps solids in suspension. Addition of air can be intermittent or by bubbleless oxygenation with a membrane gas-transfer system to prevent foaming, if necessary. Vertical mixing in anaerobic operations is increased by pumping headspace gases, and volatiles are remixed into the slurry with further degradation.

An acclimated microbial population can use still bottoms (i.e., aromatic and polyaryl species) as a sole source of energy (Black, Ahlert, Kosson, and Brugger, 1991). Lowering the rate of airflow reduces volatility losses by at least 50%, without interfering with biodegradation.

A pilot-scale study with a 30% slurry, an inoculum of indigenous PAH-degraders, and inorganic nutrients provided 97.4% degradation of two- and three-ring PAHs and 90% degradation of four- to six-ring PAHs (Lewis, 1993). Air emissions from semivolatile compounds, such as naphthalene, anthracene, and phenanthrene, and volatile compounds, such as toluene, xylene, and benzene, occurred primarily during loading of the reactors in the first few days of operation. Thus, it may not be cost-effective to have elaborate emission controls in a bioslurry reactor system, unless dealing with large quantities of VOCs.

Another 4.2 m^3 pilot slurry plant demonstrated degradation of around 65% of oil and 92% of PAHs in contaminated clay soil, and the process will be commercialized (Oostenbrink, Kleijntjens, Mijnbeek, Kerkhof, Vetter, and Luyben, 1995). Other pilot-scale studies have been conducted using Eimco™ reactors for treating PAHs, heterocyclic compounds, phenols, and solvents (Weisman, Falatko, Kuo, and Eby, 1994).

As of 1992, the size of available full-scale slurry bioreactor units ranged from 10 to 50 ft in diameter and from 15 to 25 ft in height (King, Long, and Sheldon, 1992). They could hold slurry volumes from 15,000 to 300,000 gal. The general price range for capital equipment, design, and construction at that time was on the order of $125,000 to 2,000,000.

Biodegradation occurs at a rapid rate in these reactors, with typical treatment times ranging from less than 1 month to more than 6 months (King, Long, and Sheldon, 1992). The kinetics will depend upon the nature and concentration of the contaminants and the level of treatment that must be attained. Slurry bioreactors offer considerable potential for reaching currently targeted treatment goals (Glaser, Tzeng, and McCauley, 1995). These authors provide an overview of current slurry bioreactor practice to aid in developing objectives and criteria for use of the technology. Their process evaluation has identified pitfalls and problems associated with this new technique.

2.1.2.1.3.2 Dual Injected Turbulent Suspension (DITS) Reactor

DITS is a modification of the slurry reactor, which combines remediation of part of the contaminated (polydisperse) soil with separation of the soil into heavily and lightly polluted fractions (Geerdink, Kleijintjensm, van Loosdrecht, and Luyben, 1996). In continuous operation, a soil residence time of 100 h, oil degradation rates at the steady state are over 70 times faster than in a comparable landfarm. After DITS treatment, the remaining oil in the contaminated soil fraction is slowly released from the soil (e.g., by landfarming), which takes another 10 weeks to reach the Dutch reference level of 50 mg/kg.

2.1.2.1.3.3 Gravel Slurry Reactors

Slurry reactor technology was adapted to on-site, field-scale bioreactors to handle treatment of waste oil–contaminated gravel from several sources (Wilson, Saberiyan, Andrilenas, Miller, Esler, Kise, and DeSantis, 1994).

Two portable, 20-yd^3 (15.20-m^3), watertight, steel drop boxes were constructed for the study (Wilson, Saberiyan, Andrilenas, Miller, Esler, Kise, and DeSantis, 1994). The rate of airflow was about 200 ft^3/min (5.66 m^3/min). Galvanized well screen along the bottom of the reactors promoted uniform air distribution. A water recirculation system allowed the uniform introduction of nutrients and monitoring of pH and temperature in a flowthrough system. The covered reactors were operated under negative pressure and off-gases filtered through activated carbon units. The temperature averaged 22°C and the pH from 6.5 to 7.0. Microbial counts ranged from 4.0×10^6 to 9.0×10^7 CFU/mL. Gravel contaminant TPH concentrations were reduced from about 34,000 to 1,400 mg/kg in 14 weeks, while the process was still ongoing. Use of a laboratory-scale model, duplicating field conditions, is a useful step for designing field units and interpreting operational parameters during field trials.

2.1.2.1.3.4 Tubular Reactors

These packed-bed reactors packed with biomass are effective for treating gas currents either by changing the oxidation level of the inorganic compound or by transforming toxic VOCs into innocuous biological material (Auria, Christen, Favela, Gutierrez, Guyot, Monroy, Revah, Roussos, Saucedo-Castaneda, and Viniegra-Gonzalez, 1995). Spherical Amberlite beads can be used as a support and aerated columns as tubular reactors.

A mobile revolving tubular reactor, operating on the helical-screw feed principle, can provide optimal conditions for microorganisms to degrade organic contaminants in soil (Kiehne, Berghof, Mueller-Kuhrt, and Buchholz, 1995). Through its continuous operational mode and rapid decomposition, it provides high throughput rates while using small equipment. It is economical, especially for decontaminating 500 to 3000 t. The sealed process prevents uncontrolled release of volatiles and can, thus, be used for physical soil decontamination.

A porous tube reactor can provide data for a quantitative estimation of oxygen diffusivity in the soil matrix (Tabak, Govind, Pfanstiel, Fu, Yan, and Gao, 1995). The rate of oxygen diffusion through the compacted soil affects the overall rate of biodegradation (Tabak, Govind, Fu, Yan, Gao, and Pfanstiel, 1997).

A bench-scale biofilter using peat as a packing material and activated sludge as inoculum demonstrated continuous removal of toluene from an airstream by biofiltration (Morales, Perez, Auria, and Revah, 1994). The bioreactor is now computerized and fully instrumented.

Tubular bioreactors have been used successfully in Mexico by a company that developed a proprietary system of bioscrubbers and microorganisms to remove contaminants by a liquid–gas separation system (Auria, Christen, Favela, Gutierrez, Guyot, Monroy, Revah, Roussos, Saucedo-Castaneda, and Viniegra-Gonzalez, 1995).

2.1.2.1.3.5 Blade-Mixing Reactors

Use of mixing reactors should have a positive effect on the biological turnover of contaminants, especially for cohesive soils (Hupe, Lueth, Heerenklage, and Stegmann, 1995). The effect of the dynamic treatment of contaminated soils with this method depends on the soil type and the contamination. For instance, the turnover of diesel fuel in a soil compost mixture was enhanced, but the opposite occurred with lubricating oil.

Water content must be held at about 50% of the maximum water capacity. When blade-mixing reactors are employed, it is essential to inhibit or reduce pellet formation during the dynamic treatment of soils. It is difficult to avoid pellet formation, which reduces microbial activity and contaminant turnover. This method is expensive, because of high investment and operating costs. Also, the reduction of the treatment period is minimal.

The above authors do not recommend this process for biological treatment of dry soils. Nevertheless, the mixers could be applied to pretreat soils mechanically prior to biological treatment. Plowshare mixers with installed cutter heads can pretreat soils for disagglomeration, homogenization, disintegration (aeration), and mixing of additives. The mixers should probably not be used when mycel-forming organisms are involved.

2.1.2.1.3.6 Prepared-Bed Reactors

At many Superfund sites contaminated with PAHs or BTEX, soil is excavated and placed in a container with a clay or synthetic liner at the bottom of the soil to collect leachate to prevent contamination of the groundwater (Alexander, 1994). An overhead spray-irrigation system can supply water and nutrients to promote biodegradation. Perforated pipes embedded in sand can be used to collect the leachate for treatment in an adjacent bioreactor.

This method was used to treat 115,000 m^3 of soil contaminated with bunker C fuel oil (Compeau, Mahaffey, and Patras, 1991) and 23,000 m^3 of soil contaminated with gasoline and fuel oil (Block, Clark, and Bishop, 1990).

Another demonstration project involved placing soil contaminated with petroleum hydrocarbons (#2 diesel fuel) onto a bermed polyethylene liner to construct an *ex situ* biovault (Huismann, Peterson, and Jardine, 1995). Nutrients were added as the soil was loaded. An automated soil vapor recirculation system circulated gas through the treatment vault, while a closed-loop irrigation system controlled moisture levels. A methane barrier cover was placed over the vault to prevent short circuit of the induced airflow in and around the pile and release of fugitive emissions. In 4 months, the #2 diesel concentration had decreased from 683 to 81 ppm.

2.1.2.1.3.7 Enclosed Reactors

The prepared-bed reactor can be enclosed in a plastic greenhouse to contain hazardous volatile emissions (Alexander, 1994).

The European company, Umweltschutz Nord GmbH, has developed the Terraferm Biosystem Soil, a biological soil regeneration method in which microorganisms degrade contaminants in a closed reaction room under controlled conditions (Porta, 1991).

A new type of closed soil bioreactor was constructed for bioremediation, where the soil is not treated as a slurry, but retains its original water content (Saner, Bollier, Schneider, and Bachofen, 1996). This simulates remediation techniques and allows assessment of the degradation potential of microbial populations and test of the efficiency of commercial bioremediation techniques. Optimization of mass transfer in the soil can be achieved by introduction of a planetoid mixing geometry, which ensures intensive mixing without clogging, regardless of the water content of the soil.

2.1.2.1.3.8 Fermenters

These bioreactors consist of horizontal drums that rotate about their axes like cement mixers to keep the soil loosely packed (Fouhy and Shanley, 1992). Fermenters allow closer control of temperature, oxygen, and nutrient supply and can reduce soil regeneration time by more than 50%, over *in situ* treatment. These reactors can treat either dry soil (dry process) or a slurry formed with aqueous solution (wet process).

2.1.2.1.3.9 Fungal Compost Bioreactors

Bound residue formation is a potential soil detoxification and contaminant containment process for certain hazardous organic compounds (Berry and Boyd, 1985). Indigenous peroxidase enzymes may be stimulated to enhance the rate of bound residue formation.

An uninoculated compost system and one inoculated with *Phanerochaete chrysosporium* were compared with respect to their ability to promote covalent binding of benzo(a)pyrene to soil humus matter (McFarland, Qiu, Sims, Randolph, and Sims, 1992). Mineralization of PAHs is not a major mechanism of contaminant removal in soil compost systems. Although the inoculated system initially had a higher rate of bound residue formation, after 100 days the uninoculated system demonstrated a greater extent of removal of the compound. The fungal system may have become carbon limited. If there is insufficient carbon source, *P. chrysosporium* secretes proteases that degrade the ligninase enzymes responsible for bound residue formation (Dosoretz, Chen, and Grethlein, 1990). Periodic addition of nutrients may prevent this from happening. Berry and Boyd (1985) also report that additions of growth substrates are required to enhance bound residue formation.

Binding of benzo(a)pyrene by fungal composting provides an option for low-cost soil treatment that may be used in the preparation of a no-migration variance to land treatment of a petroleum-impacted soil (McFarland, Qui, Sims, Randolph, and Sims, 1992). It could be used as a pretreatment that meets land disposal criteria. In 1992 dollars, a cubic yard of PAH-contaminated soil could be so pretreated prior to land disposal for about $97, not including costs of capital equipment, permitting, or insurance.

2.1.2.1.3.10 Combination Reactors

Slurry reactor treatment followed by solid-phase bioremediation combines the advantages and minimizes the weaknesses of each treatment method when used alone (Irvine and Cassidy, 1995). Periodic aeration in solid-phase sequencing batch reactors can be less costly than continuous aeration and provides a greater ratio of biological to abiotic contaminant removal than obtained with continuous aeration. Pretreatment by slurrying for 1.5 h greatly increases the rate and extent of contaminant biodegradation achievable in solid-phase bioreactors. Slurrying the soil at or above its saturation moisture content, though, results in lengthy dewatering times that retard biological treatment in the solid-phase stage.

2.1.2.1.3.11 Pressure Reactors

A European company, Rethmann Stadtereinigung GmbH, is developing an alternative microbial approach in which contaminated soil is treated with microorganisms and nutrients in a pressure reactor at 15 to 20 bar (Porta, 1991). Then growth conditions are optimized, and the material is pressed into bricks that are stored for 1 to 2 months to allow complete degradation of the pollutants. The bricks are then crushed and used as earth filler.

2.1.2.1.3.12 Wafer Reactors

In a soil wafer reactor, which consists of a thin layer of soil, biodegradation occurs primarily in the soil phase, as compared with the aqueous phase, which consists of free and bound water in and around the soil particles (Tabak, Govind, Fu, Yan, Gao, and Pfanstiel, 1997). A wafer reactor can provide data on biodegradation rates with no oxygen limitations (Tabak, Govind, Pfanstiel, Fu, Yan, and Gao, 1995).

2.1.2.1.4 Biopiles

Soil can be excavated and placed in a treatment area, in mounds resembling large compost heaps (Fouhy and Shanley, 1992). A covering over the piles prevents release of vapors to the atmosphere during the process. Here, the soil can be periodically turned over for aeration (or ventilated; Hayes, Meyers, and Huddleston, 1995), or a sprinkling system can add water, nutrients, and a source of oxygen. Sometimes the soil is placed in piles that extend laterally for some distance (Hildebrandt and Wilson, 1991). An optional vacuum aeration system would allow volatile compounds to be collected and treated, if necessary (Hayes, Meyers, and Huddleston, 1995). The ability to control moisture can prevent soluble constituents from leaching.

A physical model of the technology involving *ex situ* forced aeration of soil piles, or soil pile aeration, can serve two purposes (Battaglia and Morgan, 1994). It can be used to screen the technology, i.e., to determine if, based on soil characteristics and contamination levels, the technology is applicable. Also, the model can be used as a design tool to optimize the basic design parameters — pile radius and length — and the vacuum blower characteristics. This can help in the preliminary design of aeration piles.

Biodegradation of organic contaminants by indigenous microorganisms can be stimulated by *ex situ* forced aeration of soil piles (Battaglia and Morgan, 1994). The treatment unit requires less area than landfarming, and optimal treatment conditions can be maintained to reduce the treatment time needed. A vacuum blower is attached to slotted pipes extending through the soil pile. When a vacuum is exerted, air is drawn through the soil to provide the microorganisms with oxygen for biodegradation. A bermed liner can be used to contain possible leaching, and a cover to protect against precipitation and volatilization. Soil conditioners, such as straw, sawdust, or manure, can improve soil texture and increase soil permeability. Soils with permeabilities lower than 10^{-8} cm^2 would require too great a vacuum for adequate aeration of the pile. Nutrients and moisture can be added to the soil. The vacuum required at the blower end of the slotted pipe and airflow are an increasing function of the pile radius. Variations on this design could include having more than one slotted pipe under vacuum or installing the blower in the middle of the pipe instead of at the end. Equations relating the vacuum at the blower to the airflow necessary for proper aeration of a soil pile have been derived and can be useful for design purposes.

If a passive pile is used, a longer treatment time and more space would be required, because the height would have to be limited to 0.8 m (Benazon, Belanger, Scheurlen, and Lesky, 1995). However, remediation with a passive system would cost less, since off-gas treatment, as well as monitoring and maintenance, would not be necessary.

A biological forced-air soil treatment (BIOFAST) system treated around 2000 yd^3 of soil polluted with diesel fuel and heavy bunker oil (Anenson, 1995). A soil pile was constructed in layers interspersed with layers of slotted pipes for air circulation and nutrients added. Monitoring the bioremediation parameters provided information for designing a full-scale operation. After a few months, there was >72% reduction to the target level of 1000 ppm.

2.1.2.1.5 Vacuum Heap Biostimulation System

The Ebiox bioremediation system employs natural decontamination processes utilizing biostimulation of natural, contaminant-adapted microorganisms (Eiermann and Bolliger, 1995a). The vacuum heap biostimulation system (Eiermann and Menke, 1993) is appropriate for large-scale treatment of excavated soil (thousands of cubic meters at a time) and best applied for bioremediation projects with short completion deadlines, confined working space, and tough air emission restrictions. Soil structure remains unchanged and allows for high-value reuse.

The contaminated soil is excavated, conditioned by mechanical homogenization (no bulking agents added), and piled up to 4.5 m in several layers on a sealed biobed (Eiermann and Bolliger, 1995a). Figure 2.10 is a flowchart of the vacuum heap air and water circulation systems. Perforated plastic piping is installed between the layers and connected to a vacuum blower system. A vacuum is pulled through the perforated pipes, drawing outside air uniformly through the soil by bioventing to provide oxygen to the microorganisms. Permanent low pressure prevents atmospheric emissions. Volatile compounds are removed by the vacuum system and purified by a biofilter. Leachate is pumped into the bioplant system, where the dissolved contaminants are degraded to water and carbon dioxide. The purified, oxygen- and nutrient-rich solution is resprayed by a sector-controlled irrigation system on top and on the sides of the vacuum heap. Heavier petroleum fractions are subjected to a two-phase biodegradation: *in situ* biodegradation in the vacuum heap and dissolved biodegradation in the bioplant system. A black plastic liner covers the entire heap to retain moisture and passive solar heat.

Application of the process to decontaminate 10,500 m^3 with three vacuum heaps successfully resulted in 75 to 83% degradation of PAHs in the soil and 87 to 98% in the eluates, within 6 to 12 months (Eiermann and Bolliger, 1995b).

CURRENT TREATMENT TECHNOLOGIES

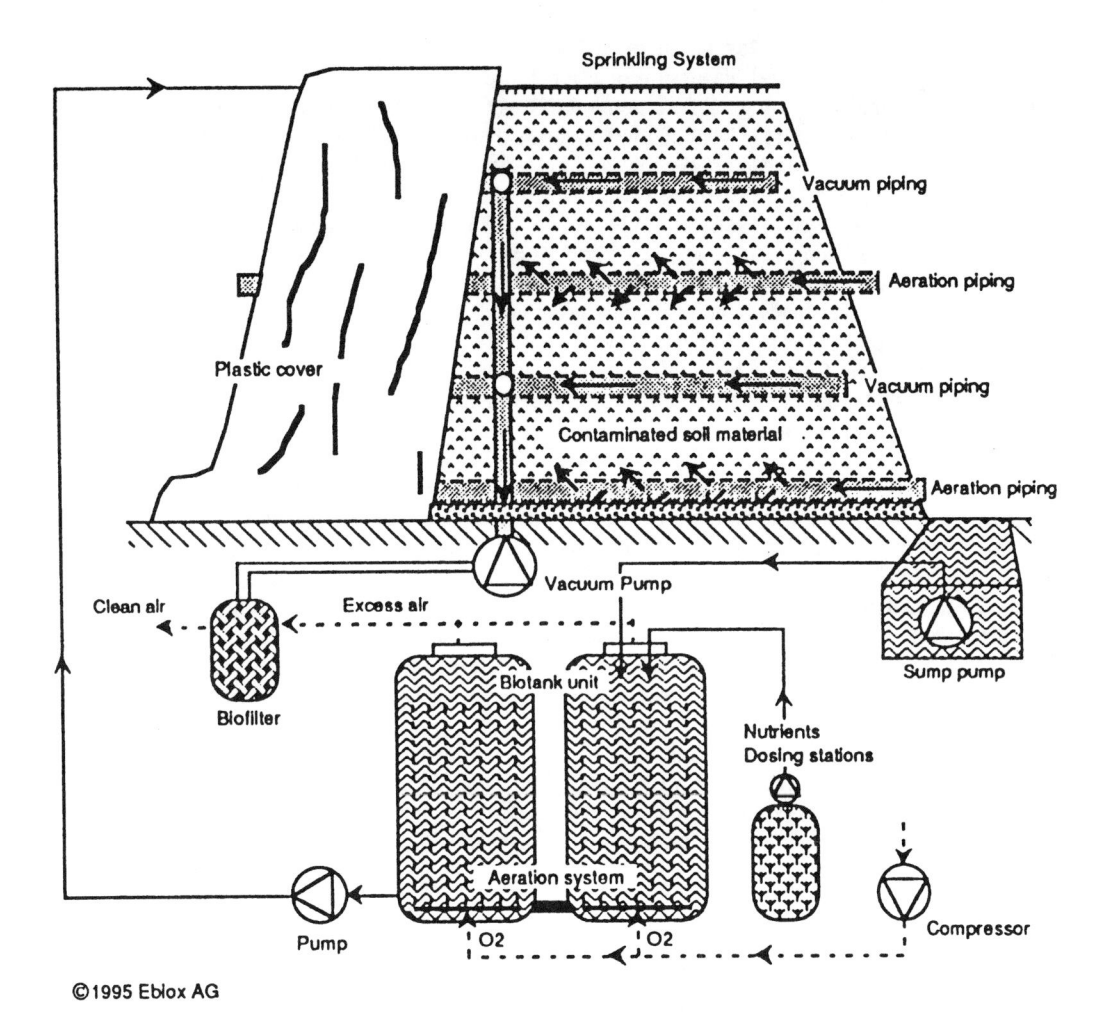

Figure 2.10 System flow chart of the Ebiox Vacuum Heap™ systems: air circulation and water circulation. (From Eiermann, D.R. and Bolliger, R. In *Applied Bioremediation of Petroleum Hydrocarbons*. Hinchee, R. E. et al., Eds. Batelle Press, Columbus, OH, 1995. With permission.)

2.1.2.1.6 Vegetation

Vegetation can enhance the rate and extent of degradation of PAHs in contaminated soil (Santharam, Erickson, and Fan, 1994). Plant roots release exudates capable of supplying carbon and energy to microflora for degrading PAHs. It has been established that the population of microorganisms in the rhizosphere is significantly greater than that in the nonvegetated soil and that these microorganisms are apparently responsible for enhanced biodegradation of PAHs. The plants themselves can also accumulate heavy metals in their roots and shoots (Kumar, Dushenkov, Motto, and Raskin, 1995); however, no uptake of anthracene or benzo(a)pyrene was observed with ryegrass (*Lolium multiflorum*), soybean (*Glycine Max* (L.) Merr.), or cabbage (*Brassica oleracea* var. capitata L.) (Goodin and Webber, 1995).

Prairie grasses have been tested for enhancing degradation of high-molecular-weight PAHs in soil (Aprill and Sims, 1990). Removal mechanisms could involve direct incorporation into humic material, increased abiotic incorporation of biologically generated intermediate metabolites, or increased microbial interaction resulting from the effects of the rhizosphere. This approach might be applicable for cleanup of low PAH levels in soils. See Sections 2.1.1.2.6, 5.3.2, and 6.3.4.6.

2.1.2.1.7 Photolysis

It might be possible to utilize photolysis to render PAHs more susceptible to microbial attack during biodegradation in some contaminated soils (Wilson and Jones, 1993).

2.1.2.2 Leachate/Wastewater Treatment Systems

See Section 2.1.1.2 for a general discussion of leachate treatment.

Figure 2.11 summarizes biological treatment processes typically used on waste streams contaminated with organic hydrocarbons and the level of contamination they can handle (Eckenfelder and Norris, 1993).

Biological processes are the most-flexible and cost-effective techniques for treating aqueous waste streams containing organic constituents (Shuckrow, Pajak, and Touhill, 1982b). The greatest concern with this process is probably the potential release of VOCs to the atmosphere as a result of aeration.

Biodegradation has been found to lower pH conditions, resulting in high heavy metal concentrations in the leachate (du Plessis, Phaal, and Senior, 1995).

If hazardous waste leachates contain organic compounds that are not readily biodegradable, it may be necessary to acclimate a biological system to the waste. The presence of refractory or toxic compounds may preclude biological treatment or may require an additional treatment process.

The activated sludge process, or one of its modifications, appears to have the greatest potential for leachate treatment, because it can be controlled to the greatest extent and allows development of an acclimated culture. However, because of their ease of operation, minimal sludge production, and energy efficiencies, anaerobic filtration or anaerobic lagoons are worth considering in some situations.

Conventional biological wastewater treatment techniques that employ microorganisms can be used to treat polluted water (Lee and Ward, 1985, 1984; Lee, Wilson, and Ward, 1987). Both aerobic and anaerobic biological treatment systems are available (Roberts, Koff, and Karr, 1988). The systems that have used acclimated bacteria to restore contaminated aquifers typically have relied upon biological wastewater treatment techniques, such as activated sludge, rotating biological disks (membrane bioreactors), aerated and anaerobic lagoons, composting, waste stabilization ponds, trickling filters, and fluidized-bed reactors for the treatment of relatively large flow volumes with consistent loading characteristics (Lee and Ward, 1985).

The biological process reactors available for water and wastewater treatment can be classified according to the nature of their biological growth (Sutton, P.M., 1987). Those in which the active biomass is suspended as free organisms or microbial aggregates can be regarded as suspended growth reactors, whereas those in which growth occurs on or within a solid medium can be termed supported growth or fixed-film reactors.

2.1.2.2.1 Aerobic Systems

Biological treatment systems normally are either activated sludge (suspended growth) or fixed-film systems (IT Corporation, 1987). Such a treatment constitutes its own ecosystem (Atlas, 1977). Aerobic treatment systems include conventional activated sludge processes, as well as modifications (such as sequencing batch reactors) and aerobic attached growth biological processes (such as rotating biological contactors; RBCs) and trickling filters (Roberts, Koff, and Karr, 1988). Aerobic processes are capable of significantly reducing a wide range of organic toxic and hazardous compounds; however, treatment is limited to dilute aqueous wastes (usually not exceeding 1%). Genetically engineered bacteria have been recently developed for effective biological treatment of specific hazardous wastes that are relatively uniform in composition. Such systems are typically used to treat aqueous wastes contaminated with low levels (BOD < 10,000 mg/L) of nonhalogenated organic or certain halogenated organics. This treatment requires consistent, stable operating conditions.

Oil wastes are often dumped into municipal sewage systems (Itoh, Ohguchi, and Doi, 1968). Many of the organisms in activated sludge are capable of metabolizing hydrocarbons; however, they do not, generally, extensively degrade petroleum hydrocarbons. Hydrocarbonoclastic microorganisms may be added to activated sludge tanks to degrade the oily wastes. There is an even antagonism between such organisms in activated sludge. Within organic-rich sewage, substrates other than hydrocarbons (e.g., proteins and carbohydrates) may be preferentially attacked, rather than the hydrocarbons (Atlas, 1977). Also, many of the oils found in sewage systems have high concentrations of heavy metals, which may be toxic to microorganisms. On the other hand, activated sewage is a balanced microbial ecosystem that is continuously rich in organic compounds. It is likely that a hydrocarbonoclastic microorganism would be able to compete and survive in this complex ecosystem. Rather than seed microorganisms into an activated sludge treatment facility, it would be better to remove oily wastes separately (Ludzack and Kinkead, 1956; U.S. EPA, 1971) by setting up a specialized secondary tank for oil degradation in a sewage treatment facility that is receiving oily wastes.

CURRENT TREATMENT TECHNOLOGIES

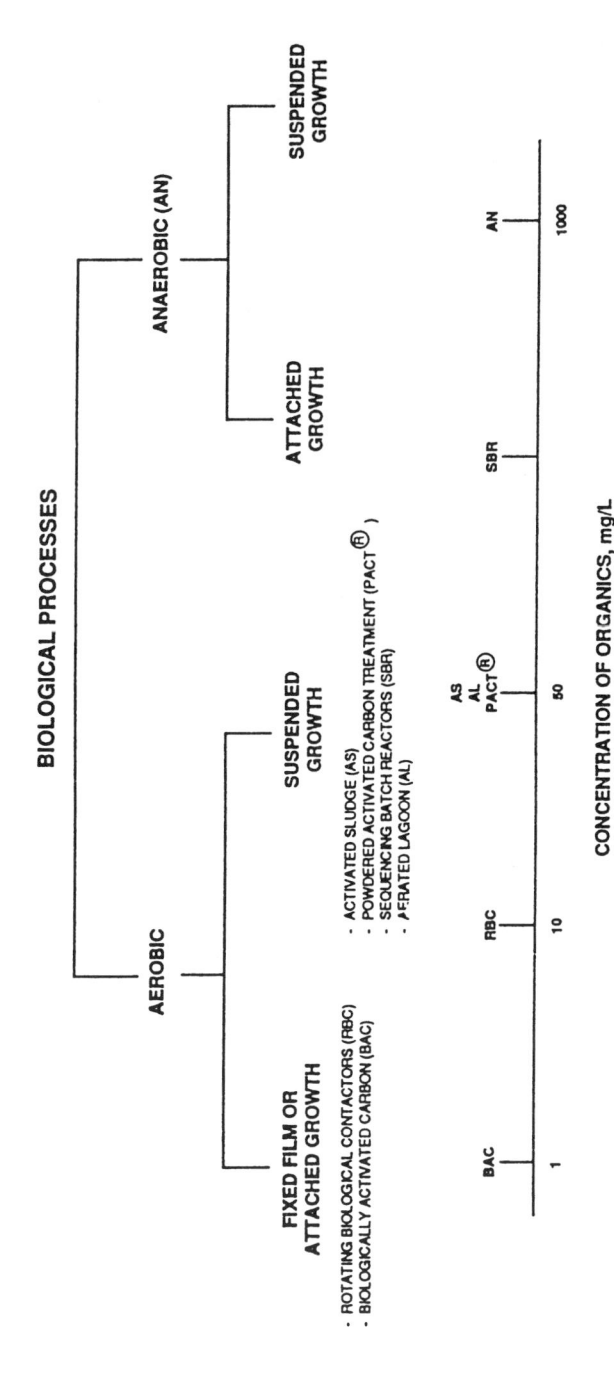

Figure 2.11 Biological water treatment processes. (From Eckenfelder, W. W., Jr. and Norris, R. D. in *Emerging TEchnologies in Hazardous Waste Management III*, Tedder, D. W. and Pokland, F. G., Eds., American Chemical Society, Washington, D.C., 1993. With permission.)

With biological treatment, continuous processes are desirable, since the process would tend to be more stable as the different wastes are blended before introduction to the process and step changes in concentration are avoided (Allen and Blaney, 1985). Equalization will moderate the concentration surges.

2.1.2.2.1.1 Suspended Growth Systems
1. Activated Sludge

There are many variations of the conventional activated sludge process, all of which, basically, use the same principles of unit operation (Roberts, Koff, and Karr, 1988). The first step in the process involves aeration in open tanks, in which the organic biodegradable matter in the waste is degraded by microorganisms in the presence of oxygen. The hydraulic detention time of this unit operation is usually from 6 to 24 h, depending upon the process mode. This is followed by a sludge–liquid separation step in a clarifier. The organisms multiply during the process. A zoogleal sludge is settled out and a portion of the organisms (return activated sludge, or RAS) is recycled to the aeration basin, which allows growth of an acclimated population. The remaining sludge is wasted, while the clarified water is discharged in a manner appropriate to its quality. Organic loading rates can vary from 10 to 180 lb of BOD applied per 1000 cf, depending upon the mixed liquor suspended solids (MLSS) concentration, the food-to-microorganism (F/M) ratio, and the oxygen supply.

Variations of the conventional activated sludge system that utilize pure oxygen, or oxygen-enriched air instead of air, produce more rapid breakdown of chemical solutes (Roberts, Koff, and Karr, 1988). Extended aeration involves longer detention times than conventional activated sludge and relies on a higher population of microorganisms. Contact stabilization involves only short contact of the aqueous wastes and suspended microbial solids, with subsequent settling of sludge and treatment of the sludge to eliminate the sorbed organics. Use of powdered activated carbon is also reported to have excellent pollutant removal capabilities for wastes that are difficult to treat.

Activated sludge has a great potential for treatment of hazardous waste, since it can be easily controlled (Shuckrow, Pajak, and Touhill, 1982a). Treatment of contaminated groundwater in activated sludge systems is sometimes used to remove biodegradable compounds before activated carbon treatment (IT Corporation, 1987). However, nonbiodegradable organics and metals can be adsorbed by the biomass, interfere with metabolism, and cause the generated sludge to be hazardous (Shuckrow, Pajak, and Touhill, 1982a). Activated sludge processes can be applied only when the organic concentration exceeds 50 mg/L (Eckenfelder and Norris, 1993). They can handle organic loadings as high as 10,000 ppm BOD, but are sensitive to shock loads (JRB Associates, 1982). Recalcitrant organics can be removed by adsorption onto powdered activated carbon (PAC).

Variation of initial total solids levels from 1000 to 80,000 mg/L did not significantly impact the reduction of total and volatile solids (i.e., 35 to 40% and 30 to 35%, respectively), but the required detention time varied from 6.5 to 22 days (Datar and Bhargava, 1985; Bryant, 1986). Batch studies at a 9-day detention indicated maximum reduction of BOD and COD at 30 to 35°C (Datar and Bhargava, 1984). Aerobic digestion of activated sludge produced refractory compounds of 15 to 25 mg COD/g input (Chudoba, 1985). The thickening of waste sludge from 0.5 to 5% solids and subsequent aerobic digestion produced a sludge for application to agricultural land (Pizarro, 1985).

During 9-day digestion of activated sludge at 27°C, nitrification proceeded in two steps (Bhargava and Datar, 1984). Nitrite formation occurred for 2.5 days, followed by nitrate formation. A 76% reduction in organic nitrogen was measured. Digestion of activated sludge from the treatment of refinery wastewaters removed 58% and 64.85% solids in batch and semicontinuous reactors (Lopatowska, 1984). Addition of pure oxygen raised performance by 10%, and the product sludge was still high in N and P. A 2-year pilot study resulted in 98% reduction of COD by aerobic digestion of waste coolant from metal-cutting operations (Young, 1985).

Pressure filtration has been found to be the most-effective treatment for dewatering of the sludge (Catalytic, Inc., 1984).

If there are high levels of VOCs in the waste stream, they must be removed by upstream air stripping or by capturing and treating the off-gas from the aeration basin (Eckenfelder and Norris, 1993).

2. Thermophilic Aerobic Digestion (TAD)

A polymer-degrading *Bacillus stearothermophilus* was found to constitute at least 95% of the thermophilic bacteria in a TAD reactor, but acidophilic and alkalophilic thermophiles were always present at lower concentrations (Sonnleitner and Fiechter, 1985; Bryant, 1986). Continuous culture systems were essential for efficient metabolism. Full-scale operation of TAD retreatment of municipal sludge has shown high bacteria removal, exothermic heat production, and intensified efficiency of subsequent mesophilic anaerobic digestion (Zwiefelhofer, 1985).

Increasing oxygen levels in TAD increases the reactor temperature, but not the quality of the product sludge (Booth and Tramontini, 1984). Volatile solids reduction was, normally, 17 to 25%. Pure oxygen TAD has been successfully pilot tested on primary wastewater sludge containing alum sludge (Paulsrud et al., 1984). TAD of wastewater sludge with air reduced COD, BOD, and TOC by 32, 48, and 27%, respectively (Wolinski and Bruce, 1984a). The foam layer above the sludge supported oxidation and enhanced the oxygen utilization. TAD studies of wastewater sludge at retention times of 8 to 10 days produced solids removal of about 30% (Wolinski and Bruce, 1984b). Use of pure oxygen in autothermal digesters produced a stable temperature and effective disinfection in 2 to 4 days, but the final sludge was very difficult to settle and dewater (Trim and McGlashan, 1985). Generally, an 8-day retention time and a diffused aeration system are recommended (Vismara, 1985).

3. Lagoons and Waste Stabilization Ponds

These are similar to activated sludge processes, except for the biomass recycle (Wilkinson, Kelso, and Hopkins, 1978). Lagoons provide dilution and buffering of load fluctuations; however, they require more land, and the operational controls are less flexible (Shuckrow, Pajak, and Touhill, 1982a). Oxygen can be supplied by aerating the lagoons to speed up the degradation (Johnson, 1978).

Some industrial plants with large amounts of oily wastes have set up sewage lagoons for their treatment (McLean, 1971). Proper aeration and additional nutrients will be required for effective petroleum biodegradation in these lagoons. This may require buffering and inclusion of algae in the seed mixture to provide continuous oxygen. Biodegradation of oil in lagoons may even be effective for oil with a large proportion of asphaltenes.

Some commercial mixtures of microorganisms have been marketed for use in degrading oil in such lagoons and other situations. The applicability of seeding selected bacteria and fungi to oil spills has been patented by Azarowicz and Bioteknika International, Inc. (Azarowicz, 1973).

Waste stabilization ponds are principally a polishing technique useful for low organic wastewaters (Johnson, 1978). Since natural biodegradation processes are employed, requirements for energy and chemical additions are low; however, large land areas are needed. The performance of waste stabilization ponds can be improved by incorporating attached growth media, or so-called artificial fibrous carriers, in the pond water (Zhao and Wang, 1996). Better COD, BOD_5, and NH_4–N removal is obtained in the attached growth ponds (AGP). A pilot-scale AGP system has demonstrated superior stability and other advantages.

2.1.2.2.1.2 Fixed-Film Systems

The subsurface environment is generally characterized by low substrate and nutrient concentrations and high specific surface area, which favor predominance of bacteria attached to solid surfaces in the form of biofilms (Kamp and Chakrabarty, 1979). Attached bacteria have an advantage over suspended bacteria, as they can remain near the source of fresh substrate and nutrients contained in water flowing by them.

Fixed-film systems operate by providing a medium onto which the microorganisms attach themselves and grow in a film (IT Corporation, 1987). A solution containing contaminants is passed over this biofilm, which causes a rapid biodegradation of the chemicals. The most common fixed-film systems are trickling filters, biological towers, and RBCs, with removal efficiencies similar to that of the activated sludge process. Immobilized films are also used in fixed-bed reactors, fluidized beds, and airlift bioreactors (Gadd, 1988; Huang and Morehart, 1990). Fixed-film reactors can treat waste streams with an organic content of 5 to 50 mg/L (Eckenfelder and Norris, 1993).

There is evidence that in fixed-film processes, the cell retention time is long compared with suspended growth processes (Switzenbaum and Jewell, 1980; Stratton, 1981). Fixed-film processes foster long cell retention and enhance growth of slow-growing microorganisms. Such populations are particularly advantageous when sorption is the main mechanism for the removal of a compound. One advantage of these processes is that they can provide cell concentrations of an order of magnitude higher than those found in suspended growth systems (Matter-Muller, Gujer, Giger, and Strumm, 1980; Stratton, 1981). A study of the partitioning of organic compounds into biomass indicates that efficient removal is possible only when the biomass concentration is large.

The organic removal efficiency of nine biofilm reactors, which were either packed beds or open channels, showed that a column reactor packed with thin vinyl strings was the best system (Toda and Ohtake, 1985; Kinner and Eighmy, 1986). It removed 75% of the influent TOC at an organic loading rate of 4.0 kg/m^3/day. Modified activated sludge, activated biofilters (ABFs), RBCs, and a deep shaft process were evaluated as alternatives for design of a municipal waste treatment system (Viraraghavan et al., 1985a). RBC and ABF processes were evaluated favorably, but the ABF process was preferred because of its stability, reliability, cold weather operation, longer-lasting media, and uncovered design. In a comparison of activated sludge, trickling filter, biological activated carbon (BAC), and fluidized-bed treatment of industrial wastewater with techniques using immobilized bacterial cultures, immobilization appeared to be more efficient than the more traditional biological processes (Gvozdyak et al., 1985).

Biofilms on porous membranes of polyetherimides for polycyclic aromatic hydrocarbons removal from coking wastewater have shown advantages over suspension reactors or solid-bed reactors with immobilized microorganisms (Kniebusch, Hildebrandt-Moeller, and Wilderer, 1994). Ideal supports have a large surface area but are porous enough to allow high flow rates and minimal clogging (Gadd, 1992). This includes planar surfaces, such as glass, metal sheets and plates, and plastics; uneven surfaces, such as wood shavings, clays, sand, crushed rock, and coke; and porous material, such as foams and sponges (Macaskie and Dean, 1989). Sintered clay was found to be superior to pine wood shavings and peat as the support for bacteria (*Micrococcus varians*, *Staphylococcus saprophyticus*, and *Acinetobacter lwoffi*) in treatment of diesel fuel–contaminated soil (Muszynski, Karwowska, and Kaliszewski, 1996).

A steady-state biofilm has a constant amount of active biomass per unit surface area (Ascon-Cabrera, Thomas, and Lebeault, 1995). The biofilm can be removed by sloughing, abrasion, predator grazing, or erosion by shear forces caused by moving fluid. Cells can also become detached from the biofilm during reproduction. Variation in the physiological activity of the cells causes changes in the cell surface hydrophobicity, which influences initial cell adhesion. During steady state, the detachment rate is equal to the growth rate of the biofilm.

Living cells can be immobilized by entrapment within gels or other matrices (Gadd, 1992). Polyacrylamide gel has been used successfully for this purpose (Nakajima and Sakaguchi, 1986; Macaskie and Dean, 1987).

Dissolved oxygen concentration has little effect on cell yield, but higher concentrations do result in thicker films (Huang, Chang, Liu, and Jiang, 1985). Biofilm density and composition are related to thickness. Substrate removal rate is linearly proportional to biofilm thickness, to a certain limit.

In a continuous-flow, fixed-bed reactor, during xenobiotic biodegradation, it was found that the immobilized, sessile (parent) cells and suspended detached cells grew synchronically at the end and at the beginning of the cell cycle, respectively (Ascon-Cabrera, Thomas, and Lebeault, 1995). The difference in hydrophobicity of the two cell types permitted the cell detachment. Therefore, the synchronized growth and hydrophobicity of cells are probably the main factors permitting the maintenance of a steady-state xenobiotic-degrading biofilm reactor, in which the overall accumulation of biofilm is determined by the growth rate of the biofilm cells minus the rate of detachment of cells from the biofilm.

A bubble interface may be produced in an aerobic, fixed-film, upflow reactor (Shah and Stevens, 1995). A model has been developed to understand the removal of contaminants due to their attachment and partitioning to the air bubbles. Significant portions of PAHs, especially naphthalene, with high Henry's law constant values should be removed from the reactor via volatilization. The residence time of contaminants may be independent of water flow rate but dependent upon the airflow rate and bubble diameter.

Nonliving biomass can also be immobilized, and almost all microbial groups have been used successfully in this manner (Gadd, 1988; Macaskie and Dean, 1989; Harris and Ramelow, 1990; Watson, Scott, and Faison, 1990; Kuhn and Pfister, 1990; De Rome and Gadd, 1991). Cells can be entrapped within alginates, polyacrylamide, and silica gels for small-scale systems (Macaskie and Dean, 1989; Macaskie, 1990). The immobilized biomass should have a particle size similar to that of other commercial absorbents (0.5 to 1.5 mm), with good particle strength, high porosity, hydrophilicity, and chemical resistance (Tsezos, 1990; Brierley, 1990). Particles should contain a maximum amount of biomass and minimal amount of binding agent. The cells can be immobilized with agar, cellulose, alginates, crosslinked ethyl acrylate-ethylene glycol dimethylacrylate, polyacrylamide, toluene diisocyanate, glutaraldehyde (cross-linking reagent), or silica gel (Brierley, 1990). Cell immobilization is further discussed in Section 5.2.2.7.

Biofilms with living and nonliving cells can simultaneously remove heavy metals, while degrading hydrocarbons (see Section 3.3) (Gadd, 1992).

1. Trickling Filters

Trickling filter systems are similar to fixed-film reactors, except they operate in continuous-flow mode (King, Long, and Sheldon, 1992).

These systems involve contact of the aqueous waste stream with microorganisms attached to some inert medium, such as rock, specially designed plastic material (Roberts, Koff, and Karr, 1988), cobbles, or silicate sand (King, Long, and Sheldon, 1992). The original trickling filter consisted of a bed of rocks over which the contaminated water was sprayed. Microbial deposits formed slime layers on the rocks where metabolism of the solute organics occurred. Oxygen was provided with the air being injected countercurrently to the wastewater flow. Present technology suggests that gas-suspended biomass systems are more applicable to treating oily sludges than are fixed-film systems.

Trickling filters require very short retention times and must run in a recirculation mode to produce significant reductions in contaminant concentrations (King, Long, and Sheldon, 1992). They are quite adaptable to fluctuating waste streams and can be especially effective when used in the pretreatment mode. Release of volatiles is not a significant problem. Trickling filters are not as effective as activated sludge, but are less sensitive to shock loads and have lower energy costs (Lee and Ward, 1985, 1984; Lee, Wilson, and Ward, 1987). Their principal use is for secondary treatment or as a roughing filter to even out loading (JRB Associates, 1982).

There are some disadvantages associated with this technique (King, Long, and Sheldon, 1992). If the operation of the system is disrupted for any reason, there can be difficulty in rebounding. Also, when there is not a proper contact of biomass, contaminants, oxygen, and nutrients, dead spots can occur within the unit.

A hospital waste was treated using a trickling filter followed by a chemical coagulation/flocculation process (Ademoroti, 1985; Kinner and Eighmy, 1986). The combined system removed up to 100% and 98.0% of the suspended solids and BOD_5, respectively. Pilot-scale, high-rate trickling filters using stone and plastic media have been designed to achieve partial BOD removal (Yau, 1985). Plastic media filters remove up to 74.9% of the BOD with Filterpak media and 62.5% with Flocor media, although there can be some operational problems with the former media. Modular cross-flow plastic media used in shallow trickling filters (depth less than 3 m) does not deteriorate when wetted normally, but discolors and becomes brittle when improperly dosed (Deis et al., 1985). Use of UV light stabilizers and carbon black has been considered to eliminate this problem. Pulsed-bed sand filters added before and after a trickling filter can increase the capacity of a municipal wastewater treatment plant (Becker and Garzonetti, 1985).

When a trickling biofilter was used to remove toluene from waste gas, the overall liquid mass transfer coefficient, $K_L a$, was a factor of six higher in a system with biofilm than in one without (Pedersen and Arvin, 1995). This may have been due to a high level of water recirculation to keep the liquid-phase concentration constant and to achieve a high degree of wetting, to mass transfer occurring directly from the gas phase to the biofilm, or to enlarged contact area between the gas phase and the biofilm due to a rough biofilm surface. The toluene removal was nearly independent of the gas/liquid mass transfer and was governed by the biological degradation. With loads of toluene above 60 to 70 $g/m^3/h$, the elimination capacity was about 35 $g/m^3/h$.

2. Biological Towers

These are a modification of the trickling filter (Roberts, Koff, and Karr, 1988). The medium (e.g., polyvinyl chloride, polyethylene, polystyrene, or redwood) is stacked into towers, which typically reach 16 to 20 ft. The wastewater is sprayed across the top, and, as it moves downward, air is pulled upward through the tower. A slime layer of microorganisms forms on the medium and removes the organic contaminants as the water flows over the slime layer. Towers are also discussed in Section 6.3.3.1.4.1.

3. Rotating Biological Contactor

RBC consists of a series of rotating disks, connected by a shaft set in a basin or trough (Figure 2.12) (Roberts, Koff, and Karr, 1988). The contaminated water passes through the basin where the microorganisms, attached to the disks, metabolize the organics present in the water. Approximately 40% of the surface area of the disk is submerged. This allows the slime layer to come alternately into contact with

CH₄ and CO₂

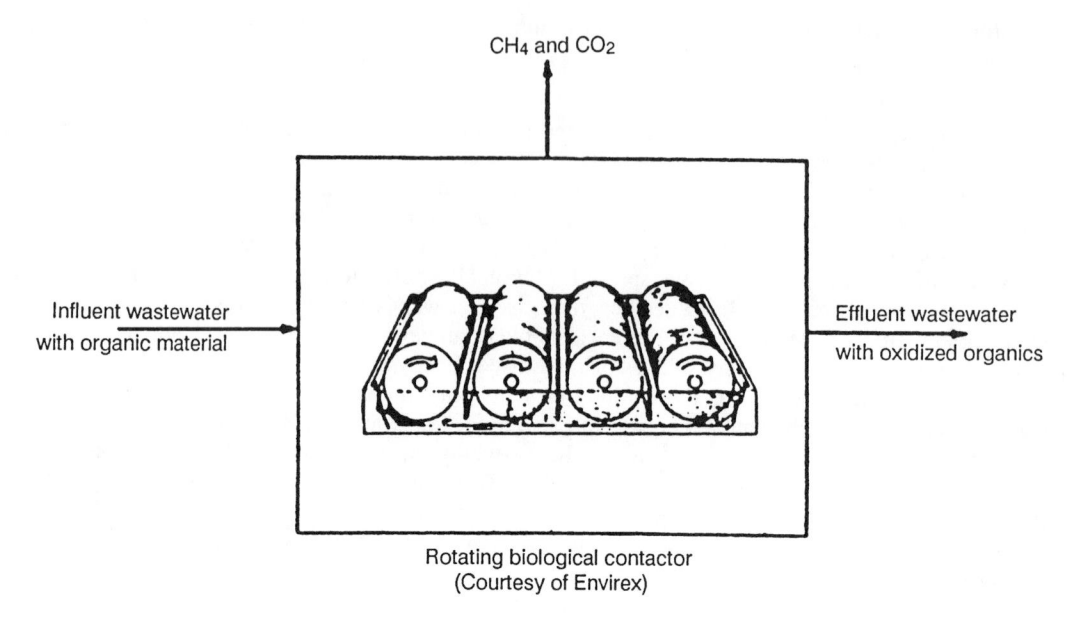

Influent wastewater with organic material → [Rotating biological contactor (Courtesy of Envirex)] → Effluent wastewater with oxidized organics

Figure 2.12 Schematic of rotating biological contactor. (From Roberts, R.M. et al., Hazardous Waste Minimization Initiation Decision Report. TN-1787, 1988.)

the contaminated water and the air, where the oxygen is provided to the microorganisms. These units are compact, and they can handle large flow variations and high organic shock loads. They do not require the use of aeration equipment. Due to the varied composition of oily sludges and high concentrations of solids, oils, and heavy metals, the applicability of the RBC to this problem material is questionable. Reliability of this method is moderate in the absence of shock loads and at temperatures less than 55°F (JRB Associates, 1982). Suspended or colloidal organics are effectively treated.

A full-scale RBC treating municipal wastewater achieved 88 and 70% removal for BOD_5 and ammonia nitrogen, respectively (Norouzian and Gonzales-Martinez, 1984; Kinner and Eighmy, 1986). RBCs are a good choice for treatment in remote locations, but lower than expected ammonia nitrogen removals occur when predators feed on nitrifying bacteria. RBC performance is improved by placing Pall rings between RBC disks (Norouzian, Gonzalez-Martinez, Pedroza-de-Brenes, and Duran-de-Bazua, 1985). This modification yields an average effluent BOD_5 of 5 mg/L and complete nitrification at an organic loading rate of 12 kg TOC/1000 m²/day. Supplemental aeration with high organic loading rates will achieve good effluent quality (Surampalli and Baumann, 1985). With added aeration, dissolved oxygen remains above 1.5 mg/L in the bulk liquid, eliminating growth of *Beggiatoa*.

The most important process variables in predicting RBC performance are influent BOD, substrate type, hydraulic loading, stage number, and wastewater temperature (Del Borghi, Palazzi, Parisi, and Ferrajolo, 1985). Biofilms growing in RBC compartments receiving organic loading rates from 2.0 to 6.0 g TOC/m²/day show different morphological and ultrastructural characteristics (Kinner, Maratea, and Bishop, 1985). At low loading rates, nitrifying bacteria become predominant. Nitrification is oxygen rate limited, which should be considered during RBC design (Gonenc and Harremoes, 1985).

A modified rotating biological contactor with polyurethane foam attached to the disks as porous support media was tested for biodegrading a petroleum refinery wastewater (Tyagi, Tran, and Chowdhury, 1993). Foam of 1 cm thickness was attached to both sides of each disk. The efficiency of the modified system for removal of organics, ammonia nitrogen, phenol, hydrocarbons, and suspended solids was generally better than that of a conventional RBC.

4. Fluidized-Bed and Membrane Biological Reactors (FBR and MBR)

These treatment systems have been developed to remove organics and solvents from contaminated aqueous streams (U.S. EPA, 1983; Bove, Lambert, Lin, Sullivan, and Marks, 1984). They combine the features of activated sludge and fixed-film biological processes. Particles, such as sand or coal, are fluidized by the action of the aeration gas stream and the wastewater stream and are colonized by a dense growth of microorganisms, which provides rapid treatment (McCarty, Rittmann, and Bouwer, 1984). Organics that are resistant to biological treatment can be removed by adsorption on the carbon (Eckenfelder and Norris, 1993).

A fluidized-bed reactor represents a highly efficient fixed-film reactor in which biomass buildup occurs on an inert (sand) or active (activated carbon, resin material) fluidized support medium high in external surface area. A typical system consists of a fluidized-bed reactor and oxygenator; a sand–biomass separator; and feed, recycle, and chemical-addition tanks. Oxygen, required for carbonaceous BOD removal or nitrification, is dissolved in the influent stream prior to entry into the fluidized reactor. The oxygenation system and influent distribution system, both proprietary devices, allow the transfer of at least 50 mg oxygen/L to the fluid phase. If the oxygen demand of the influent exceeds the level of available oxygen, then effluent recycle is required. Pure oxygen eliminates off-gas from the treatment (Eckenfelder and Norris, 1993).

In the fluidized-bed system, the contaminated water or wastewater and recycled effluent pass upward through the bed of medium at a velocity that expands the bed beyond the point at which the frictional drag is equal to the net downward force exerted by gravity. From this point of minimum fluidization, the medium particles are individually and hydraulically supported (Sutton, P.M., 1987). They provide a vast surface area for biological growth, in part leading to the development of a biomass concentration approximately five to ten times greater than that normally maintained in a suspended growth system. A high concentration of active biological organisms can be achieved in the fluidized reactor, i.e., 12,000 to 40,000 mg/L of mixed liquor–volatile suspended solids (MLVSS). This compares with MLVSS values of 1500 to 3000 for conventional activated sludge, and 3000 to 6000 mg/L for pure oxygen–activated sludge treatment.

It may be necessary to control the biofilm thickness to prevent the density of the biofilm-covered medium (bioparticle) from decreasing to the point where bed carryover occurs (Sutton, P.M., 1987). This is accomplished by monitoring the bed expansion optically, carrying out separation of the medium from the biomass (if the maximum specified bed height is reached), and return of the medium to the reactor. To control the fluidized bed at a specific height, a fraction of the biomass coating is separated from the sand particles by means of a vibrating screen. The sand particles are returned to the bed for reseeding, and the biomass is removed as excess sludge.

The system requires less land area and contact time than other biological treatments, because higher volumetric loading rates are possible. Organisms are fixed in the system, as in a trickling filter, and are claimed to be capable of providing greater process stability in handling shock and toxic loads. However, unlike trickling filtration processes, there is minimal sloughing of biological growth. Therefore, no clarifier or recycle stream of sludge is required.

Fluidized beds can be used in secondary treatment for BOD removal or in tertiary treatment processes, such as nitrification of ammonia or denitrification of nitrates. They may not work for dilute contaminated groundwater without the addition of supplemental carbon, since the concentration of organic material in the groundwater would be too low to sustain proper biological growth. However, they are probably comparable with most other biological treatment processes in that removal of most of the nonbiodegradable organic material (particularly highly chlorinated compounds) would be quite limited. This process is particularly suited to treatment of relatively high-strength, biodegradable wastewaters. The presence of metals, however, may hinder this process by destroying the active organisms.

Assessment of this technology demonstrated very favorable removal of an identified contaminant from an aqueous stream. The process was rated as very unfavorable for low concentrations and versatility. The expected difficulties in removal of identified contaminants at low concentrations were found in a typical contaminated groundwater stream. The process is expected to demonstrate or has demonstrated inability to remove a wide range of both organic and inorganic compounds of interest from groundwater. There were no demonstrated hazards to workers or local residents during or after treatment.

Fluidized-bed biological systems have been developed by Ecolotrol, Inc., and full-scale aerobic and anaerobic configurations incorporated, respectively, into the Oxitron (Figure 2.13) and Anitron systems by Dorr-Oliver.

Complete nitrification was obtained within 30 min in a fluidized bed reactor operating with an upflow velocity of 25 m/h (Green and Hardy, 1985; Kinner and Eighmy, 1986). Optimum sand particle diameters were 350 to 600 μm. The nitrifying films were fairly resistant to shearing and dissolved-oxygen starvation. Dissolved oxygen concentrations governed the amount of nitrification.

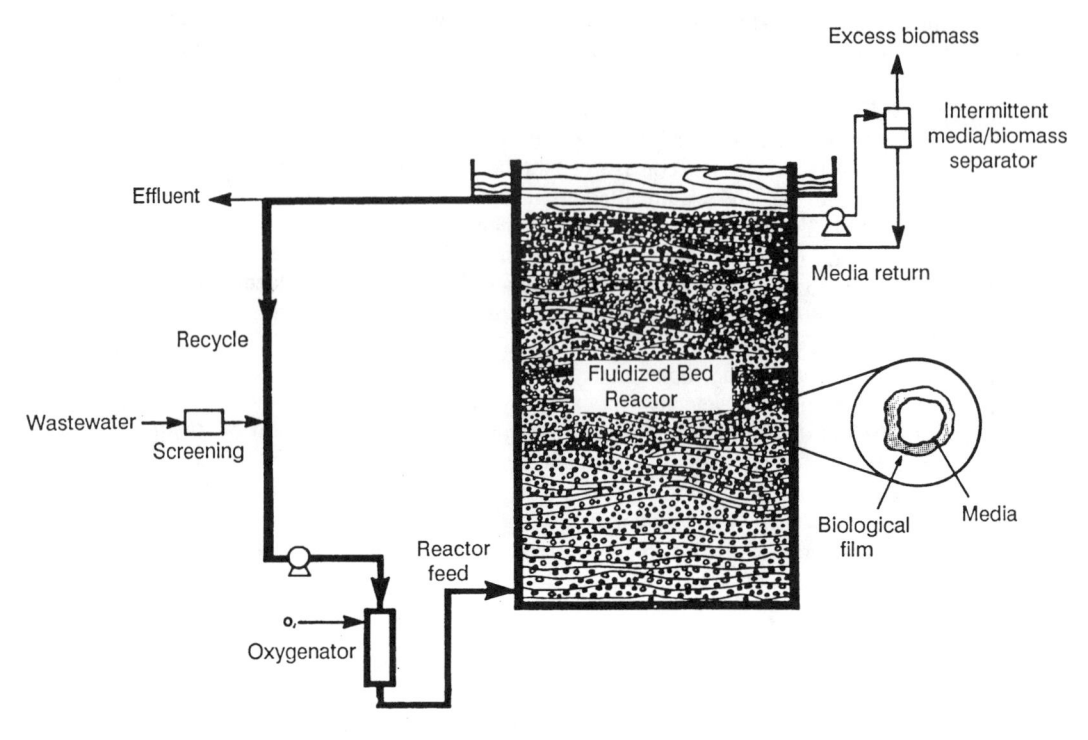

Figure 2.13 Oxitron system fluidized-bed process schematic. (From Sutton, P.M. *Pollut. Eng.* 19:86–89, 1987. With permission.)

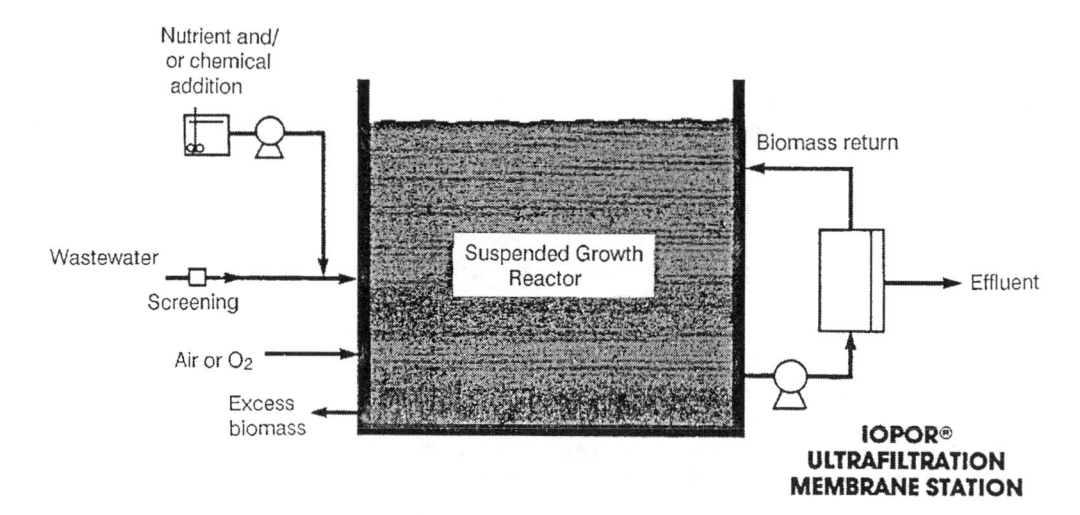

Figure 2.14 Membrane aerobic or anaerobic reactor system (MARS). (From Sutton, P.M. *Pollut. Eng.* 19:86–89, 1987. With permission.)

The membrane bioreactor is an advanced suspended-growth reactor in which a high, active microbial concentration (12,000 to 30,000 mg/L volatile suspended solids) is achieved through the use of ultrafiltration for biomass–effluent separation and subsequent recycle to the biological reactors. An example of a commercial setup of the aerobic or anaerobic configurations is represented by the Membrane Aerobic or Anaerobic Reactor System (MARS) (Figure 2.14) (Sutton, 1987).

Table 2.11 Factors Governing Selection of FBR *vs.* MARS in the Treatment of Hazardous Water and Wastewater

Factor	Fluid Bed Bioreactor (FBR) (Oxitron/Anitron) *vs.* Membrane Bioreactor (MARS)
Effluent quality	MARS normally will provide better effluent quality as effluent will contain no suspended solids
Treatment of highly volatile organics	Little or no volatilization/stripping of organics will occur in Oxitron; more will occur in MARS
Treatment of particulate and soluble organics	Particulate organics are handled better in MARS due to retention by ultrafiltration component and subsequent biotreatment; both reactors handle soluble organics efficiently
Economics	Fluid bed is often more cost-effective than MARS in treatment of high-volume wastewater
Aerobically biodegradable and volatile or semivolatile compounds, such as naphthalene, ethylbenzene, toluene, benzene, methyl chloride	Oxitron

Source: From Sutton, P.M. *Pollut. Eng.* 19:86–89, 1987. With permission.

These technologies are more favorable than alternative biological systems in situations where the contamination must be treated as rapidly as possible. Accumulation of a large biomass concentration in the fluidized-bed or membrane bioreactor will allow removal of complex organic compounds efficiently with a short liquid contact time or hydraulic retention time relative to more conventional suspended growth (activated sludge, sequencing-batch reactors, aerobic/anaerobic lagoons) and fixed-film (downflow or upflow packed-bed reactor) systems. Hydraulic retention times are about 5 min with organic loadings (COD) of up to 3 kg COD/m^3/day (Eckenfelder and Norris, 1993).

Both technologies allow selective development and retention of microbial populations effective against specific complex compounds. The fluidized bed does it with a biofilm. Biomass loss must be less than the rate of growth of new biofilm, which declines as the concentration of the contaminants is reduced in the reactor, with a loss of efficiency. The MARS bioreactor achieves a large biomass density by absolute and controlled retention of all developed microorganisms.

Tolerance to toxic or inhibitory feed inputs can be further achieved by the use of granular activated carbon (GAC) as the fluidizing medium in the fluidized-bed reactor (FBR) and the addition of PAC to the membrane bioreactor. The activated carbon also provides more rapid initial removal upon startup and greater removal of recalcitrant compounds, and it reduces the volatilization of adsorbable compounds.

Bioactivity develops faster in an integrated biological granular activated carbon fluidized-bed reactor (GAC-FBR) than in an FBR without activated carbon (Voice, Zhao, Shi, and Hickey, 1995). The GAC-FBR produces superior effluent quality during step organic load rate (OLR) increases. Partial oxidation products are formed from the degradation of benzene, toluene, and xylene (BTX) with an extremely high step OLR increase. Adsorption capacity gradually decreases over a 6-month period to about 50% of its original value.

Hazardous water and wastewaters containing degradable organic suspended material and emulsified oil and grease are more readily handled in MARS. MARS is often more cost-effective for treatment of lower-volume water and wastewaters. The biodegradability, adsorbability, and volatility of the waste stream will determine which of the two technologies is more appropriate. Table 2.11 compares the factors that govern the selection (Sutton, P.M., 1987).

In situations where the treated water is to be disposed of in a municipal treatment plant, the anaerobic versions of the fluidized-bed and the membrane bioreactor systems may be attractive. Anaerobic pretreatment of a leachate source from a landfill site may be more cost-effective than aerobic treatment. Alternatively, series operation of anaerobic and aerobic fluidized-bed or membrane bioreactors may be the most attractive flow scheme.

Oak Ridge National Laboratory has developed a fluidized-bed digester using immobilized aerobic organisms (Roberts, Koff, and Karr, 1988). In laboratory-scale testing, less than 4 min was needed to reduce phenol levels from 30 mg/L to less than 1 mg/L. This experimental system has handled waste concentrations up to 50%.

A novel membrane bioreactor allowed only 1.5% loss of a contaminant to the exit gas stream (Freitas dos Santos and Livingston, 1995). In this case, the contaminant was 1,2-dichloroethane (DCE). A silicone rubber membrane coiled around a perspex draft tube separates the contaminated wastewater from the aerated biomedium. The contaminant diffuses through the membrane and into a biofilm growing on the surface of the membrane, while oxygen diffuses into the biofilm from the biomedium side. Air stripping is avoided, since the biofilm prevents direct contact between the contaminant and the aerating gas. The biofilm limits the flux of contaminant across the membrane and its accumulation at the membrane–biofilm interface, which would reduce the mass transfer driving force for contaminant extraction from the wastewater.

5. Biological Activated Carbon (BAC) Systems

Carbon adsorption is effective in removing low levels of some organic contamination (Brubaker and O'Neill, 1982). It can handle less than 5 mg/L of organic constituents (Eckenfelder and Norris, 1993). Activated carbon systems can be batch, column, or fluidized-bed reactors (Lee and Ward, 1985, 1984; Lee, Wilson, and Ward, 1987). Batch systems typically use powdered carbon for applications, such as protecting biological treatment systems or when low capital costs and ease of operation are desirable. Column systems can be single or parallel adsorbers, which are useful in high-volume flows or where a pressure drop is expected, and have moderate adsorbent costs. In contrast, adsorbers in series produce a gradual breakthrough curve, can be used continuously, can give lower effluent concentrations, but have higher adsorbent expenses. Expanded upflow systems have been employed for high flows containing considerable quantities of suspended solids, whereas moving-bed reactors provide efficient carbon use for wastewaters with low amounts of suspended solids and minimal biological activity is expected.

If organisms are immobilized on a GAC filter, enhanced biodegradation of industrial aromatic effluents can be promoted (Roberts, Koff, and Karr, 1988). Refractory aromatic compounds, such as indole, quinone, and methylquinone, have been successfully degraded in concentrations over 300 mg/L. With biologically activated carbon beds, oxygen is added to the water by aeration or as hydrogen peroxide in solution with nutrients to provide 1 mg oxygen for each milligram of carbon (Eckenfelder and Norris, 1993).

This system removes biodegradable and biorefractory materials within the same treatment unit (Tsezos and Benedek, 1980; Bove, Lambert, Lin, Sullivan, and Marks, 1984). It is a variation of the carbon adsorption process and can be in either of two forms:

1. Addition of PAC to aeration basins of biological systems;
2. Biological GAC in an FBR.

Form number 1 is relatively simple, since only the carbon feed system is required. Part of the carbon is recycled from clarifiers back to the aeration basins, while other spent carbon is removed from the system together with waste-activated sludge, which can be either disposed of or regenerated in a wet air oxidation unit.

Benefits of this method are a high BOD and COD reduction, despite hydraulic and organic overloading; an aid to solids settling in the clarifiers; a high degree of nitrification due to extended sludge age; substantial reduction in phosphorus; adsorption of toxic compounds and coloring materials, such as dyes; adsorption of detergents; and reduction of foams.

In the case of a fluidized-bed biological reactor (form number 2), the GAC bed is expanded (i.e., fluidized) by the combined action of upward hydraulic flow, together with an air supply. The carbon adsorbs nonbiodegradable organic materials and provides support for growth of the biological film.

When biological activated carbon is used to treat dilute solutions of various organic compounds, most of the initial removal is the result of adsorption (De Laat, Bouanga, Dore, and Mallevialle, 1985; Kinner and Eighmy, 1986). After an acclimation period, the length of which is a function of the compounds present, bacterial activity eliminates biodegradable material and leads to an increase in the adsorption capacity of the carbon caused by regeneration. A gram of activated carbon adsorbs 4×10^9 *Pseudomonas* and 3×10^8 *Candida* in 10 h (Ehrhardt and Rehm, 1985). As the adsorbed contaminant diffuses out of the carbon, it is metabolized by the attached microorganisms.

Laboratory-scale tests found that biologically activated carbon was effective in removing biologically degradable organics, organics that are not normally degraded in a biological treatment process under given treatment conditions (the system provides a large surface area for supporting a very long sludge age), and nondegradable organic substances either existing in the influent or generated during the treatment process (Tsezos and Benedek, 1980).

The powdered carbon/activated sludge system is probably not applicable for treatment of groundwater, since the biodegradable organic content in groundwater would be too low to sustain biological growth of activated sludge (Tsezos and Benedek, 1980). A fluidized bed of biologically activated carbon may be suited for treatment of contaminated groundwater since, unlike the activated sludge system, biological growth will be attached to the carbon and no sludge recycle will be necessary. The process is also adaptable for use with other processes, such as wet air oxidation, which may break complex nonbiodegradable organic compounds into smaller molecules of biodegradable material. Although neither biological activated carbon nor wet air oxidation may be a feasible process by itself, the combination process scheme shown in Figure 2.3 may be technically and economically feasible for treatment of dilute contaminated groundwater.

Aerobic biological activated carbon will have performance results similar to those of a typical granulated carbon treatment system. The treatment is enhanced by the coresident biota. Whether or not the groundwater contaminants will be able to maintain a high enough organic content to sustain the biological growth must be answered on a site-specific basis. Typical retention times range from 5 to 15 min (Eckenfelder and Norris, 1993).

Volatile organics/solvents, explosive-related organics, heavy metals, and miscellaneous organics can be removed effectively by this technology. Biological activated carbon systems have been assessed as being very favorable for removal of an identified contaminant from an aqueous stream. The process unit operation is not rate limiting for removal of contaminants from groundwater. There were no demonstrated hazards to workers or local residents during or after treatment. This technology rated high in technology effectiveness, time for decontamination, and safety. This process was considered conceptually feasible for treating contaminated groundwater. It is cost-competitive with known technologies and was recommended for further consideration and testing at the pilot-study level. See also Section 2.1.1.2.1.

6. Activated Biofilters (ABFs)

ABFs operate as attached and suspended growth systems (Viraraghavan, Landine, Winchester, and Wasson, 1985b; Kinner and Eighmy, 1986). Little is known about ABF kinetics. Potential advantages of ABFs include lack of heat loss, good operational stability, and high BOD removal. A hydroautomatic biofilter with floating polystyrene foam media achieved tertiary treatment of a mixed sanitary and industrial waste (Yakimchuck, Zhurba, Prikhod'ko, and Shevchuk, 1985). Removal rates of 80 to 95%, 50 to 60%, 60 to 70%, and 10 to 30% were obtained for suspended solids, COD, BOD_5, and ammonia nitrogen, respectively.

7. Immobilized Cells

A modification of fixed-film treatment immobilizes microorganisms by firmly attaching them or physically embedding them in a solid matrix (Alexander, 1994). The cells can be sorbed in or on various media, such as alginate beads, diatomaceous earth, hollow glass fibers, polyurethane foam, activated carbon, and polyacrylamide beads. The attached cells can be employed in a variety of different types of reactors to facilitate rapid biodegradation of organic compounds, which often produces microorganisms with greater resistance to the chemicals than is achieved with suspended cells. For instance, *Trichoderma harzianum* can be attached to alginate beads to degrade anthracene (Ermisch and Rehm, 1989), and *Pseudomonas putida* can be sorbed to activated carbon to degrade phenol (Ehrhardt and Rehm, 1989).

The Allied Signal Immobilized Cell Bioreactor (ICB) is a fixed-film, fixed-bed bioreactor that employs a dual support system for microorganisms (Gromicko, Smock, Wong, and Sheridan, 1995). It consists of a patented, carbon-coated, polyurethane foam packing and conventional, random, plastic packing. It showed 86.9% removal for COD, 96.3% for total phenols, 98.9% for total PAHs, and 97.2% for BTEX. A full-scale ICB as a pretreatment step would decrease GAC usage by a factor of 3.5.

8. Innovative Fixed-Film Processes

Microalgal *Scenedesmus* sp. can be immobilized in carrageenan and used to treat secondary effluent (Chevalier and de la Noue, 1985; Kinner and Eighmy, 1986). Results are promising for removal of nitrogen and phosphorus. Immobilized and free algae exhibit similar growth and uptake characteristics.

A new moving-bed biofilm reactor has been developed in Norway (Oedegaard, Rusten, and Westrum, 1994). Biomass is attached to carrier elements that move freely along with water in the reactor, resulting in a very compact reactor and very efficient biomass.

9. BIOPUR®

This is a patented, aerated, packed-bed, fixed-film reactor, which uses reticulated polyurethane (PUR) as a carrier material for microorganisms (Oosting, Urlings, van Riel, and van Driel, 1992). PUR is a foam with a very open structure and large (500 m^2/m^3) specific surface.

Biomass grows as a thin biofilm on PUR, with a long retention time (Marsman, Appelman, Urlings, and Bult, 1994). A variety of microorganisms can colonize the biofilm for mineralization of different contaminants. The reactor has several compartments in series to model a plug flow pattern. Completely different organisms can be established in each compartment at a temperature of 9 to 12°C. Water and air flow concurrently upward through the compartments. Volatile components stripped in the first section are forced into the next. Volatile compounds in the vapor phase will dissolve in the water phase. Soil vapor can also be treated in the biofilm reactor.

A pilot-scale fixed-film system (RBC and the biofilm reactor) removed >99% of BTEX, naphthalene, and chlorobenzene and >70% of phenolic compounds. Stripping of volatile compounds was minimal (<1 to 8%). A full-scale biofilm reactor is used mainly at gasoline stations for treating groundwater contaminated with BTEX and volatile and nonvolatile hydrocarbons, gasoline, kerosene, and diesel fuel, with high removal efficiencies at short hydraulic retention times. Iron oxide formation is not a problem. The PUR is removed eventually to clean off excess sludge with a high-pressure spraying pistol. The reactor can be inoculated with microorganisms present in the groundwater or with selected microbes.

10. Immobilized Enzymes

Entrapment of biocatalysts within a porous polymeric matrix can be employed for immobilization of single or multiple enzymes, cellular organelles, microbial cells, plant cells, and animal cells (Hu, Korus, and Stormo, 1993). The durability of gel immobilization systems is limited. An alternative to gels is porous polyurethane foams (PUF) that can be preformed or formed *in situ*. A matrix of PUF from prepolymer HYPOL FHP 2000 or 3000 can also be used to immobilize living microorganisms for biodegradation of toxic chemicals.

The highest activity was obtained with a β-D-galactosidase from *Aspergillus oryzae* immobilized with PUF prepared by *in situ* copolymerization between enzyme and prepolymer HYPOL 3000 (Hu, Korus, and Stormo, 1993). Adsorption on the surface of macropores was not needed for immobilization, but the enzymes could be entrapped in the PUF micropores.

2.1.2.2.1.3 Microbial Accumulation of Metals

Much recent research has been conducted into the acclimation of microbes to metallic contamination (Johnson, Kauffman, and Krupka, 1982; Pierce, 1982a; Bove, Lambert, Lin, Sullivan, and Marks, 1984). Natural or mutant microorganisms are mixed in with a metal-containing aqueous waste, where they selectively accumulate the metals in their cells. These microbes are subsequently separated from the waste solution as biomass, and the concentrated elemental metals are recovered by burning the microbes.

A variation of this technology is microbial leaching, the prime application being removal of copper from sulfide ore. Microorganisms convert copper into a water-soluble form, which is then extracted from the solution.

Microorganisms are known to bioaccumulate cadmium, copper, iron, lead, molybdenum, radium, and uranium. Polybac and the O'Kelley Company are studying use of microbes for metals removal from wet-scrubber blowdown streams. B.C. Research has been investigating microbial copper leaching. It has been able to remove 95% of the copper from 600-g batches, with recovery of elemental sulfur. This company plans to develop this into a 2- to 10-tpd pilot system (Short and Parkinson, 1983). McGill University has patents pending on a number of microbial formulations that recover metals from dilute aqueous streams.

With this technology, concentrated metals in the biomass must have proper disposal. Metals will have a toxic effect on microbes, if they are allowed to reach a certain concentration in the cell.

Assessment of this technology demonstrated very favorable removal of an identified contaminant from an aqueous stream. It is expected to demonstrate or has demonstrated inability to remove a wide range of both organic and inorganic compounds of interest from groundwater. Expected toxic or hazardous residuals are produced that will require additional treatment. There were no demonstrated hazards to workers or local residents during or after treatment. The technology has been rated high in technology effectiveness and safety and unfavorable in residuals generated and versatility, because it is primarily

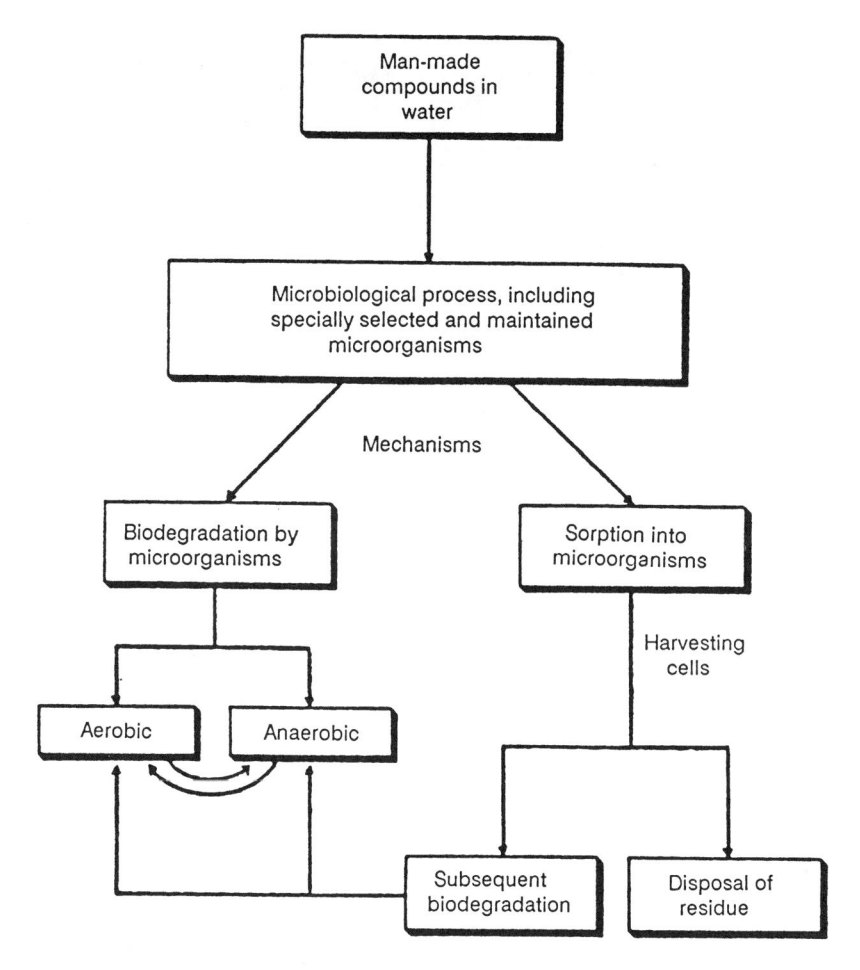

Figure 2.15 Biochemical removal of anthropogenic organic compounds from water. (From Kobayashi, H. and Rittmann, B.E. *Environ. Sci. Technol.* 16:170A–183A. American Chemical Society, Washington, D.C., 1982. With permission.)

effective for treating specific contaminants. This process is considered conceptually feasible for treating contaminated groundwater. It is cost-competitive with known technologies and has been recommended for further consideration.

2.1.2.2.1.4 Combination Aerobic Reactors/Microbial Adsorption

A scheme for separating and treating readily degradable and refractory compounds has been proposed (see Figure 2.15) (Kobayashi and Rittmann, 1982). The degradable compounds follow the left path and are treated directly in reactors, while the relatively refractory compounds are removed by sorption. Depending upon the character of the waste, the reactors will be either aerobic or anaerobic, chemotrophic or phototrophic, or a series of several types.

Removal by sorption to microorganisms of compounds not readily biodegraded can provide an extension of the detention time in a treatment system for relatively refractory compounds, can concentrate the substance, and can be a means to transfer compounds from one environmental condition to another without making it necessary to deal with the entire volume of wastewater. Once concentrated and removed from the water, refractory compounds can be disposed of by other means, such as incineration or burial. The "relatively" recalcitrant compounds can be effectively degraded, if they are first sorbed.

An example of such use is the sorptive removal from aerobic wastewater of compounds that undergo the necessary reductive dechlorination during subsequent anaerobic digestion of the cells.

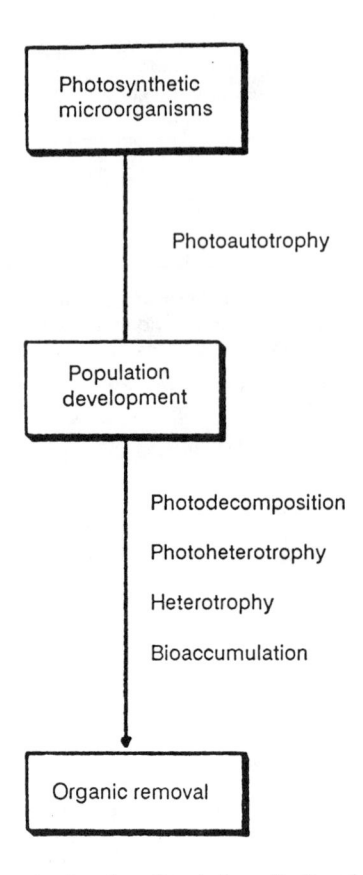

```
                    ┌─────────────────┐
                    │  Photosynthetic │
                    │  microorganisms │
                    └────────┬────────┘
                             │
                             │  Photoautotrophy
                             │
                    ┌────────┴────────┐
                    │    Population   │
                    │   development   │
                    └────────┬────────┘
                             │
                             │  Photodecomposition
                             │  Photoheterotrophy
                             │  Heterotrophy
                             │  Bioaccumulation
                             ▼
                    ┌─────────────────┐
                    │  Organic removal│
                    └─────────────────┘
```

Figure 2.16 Removal of organic contaminants with photosynthetic microorganisms. (From Kobayashi, H. and Rittmann, B.E. *Environ. Sci. Technol.* 16:170A–183A. American Chemical Society, Washington, D.C., 1982. With permission.)

Photosynthetic organisms appear to be potentially valuable when initial sorption is required, because large populations can be readily developed, even under low organic nutrient conditions. A generalized scheme for use of photosynthetic organisms is presented in Figures 2.16 and 2.17. Cell populations would be initially developed by photosynthetic activity; sorptive processes would then operate to bioaccumulate substance. In addition, heterotrophic activity (with or without the mediation of light), or reductive dehalogenation brought about by the phototrophs, could biotransform the compounds. Once the initial, limiting reactions are performed by the phototrophs, products could be further biodegraded by other organisms.

2.1.2.2.1.5 Bioreactor for Aromatic Solvents

A novel bioreactor for biodegradation of toxic aromatic solvents, such as BTX in liquid effluent stream, has been developed (Choi, Lee, and Kim, 1992).

The bioreactor is a baffled, impeller-agitated type (KLF 2000, Bioengineering, Switzerland). Silicone tubing is wound around the baffles at the bottom of the bioreactor and liquid solvent circulated at a fixed flow rate within the tube from a reservoir, using a diaphragm pump. The solvent diffuses out of the tube wall and is transferred at a high rate into the surrounding culture broth, where biodegradation takes place. A dissolved oxygen probe (Ingold, Switzerland) monitors the dissolved oxygen concentration in the bioreactor.

Degradation of toluene with *Pseudomonas putida* ATCC 23973 at 32°C in a mixed and aerated bioreactor was tested by innoculating broth. During continuous operation, the biodegradation rate was considerably higher than rates obtained using conventional methods. The operating parameters, such as dilution rate and toluene transfer rate, should be determined by taking into account both the removal efficiency and the maximum allowable concentration of solvent in the effluent stream.

CURRENT TREATMENT TECHNOLOGIES

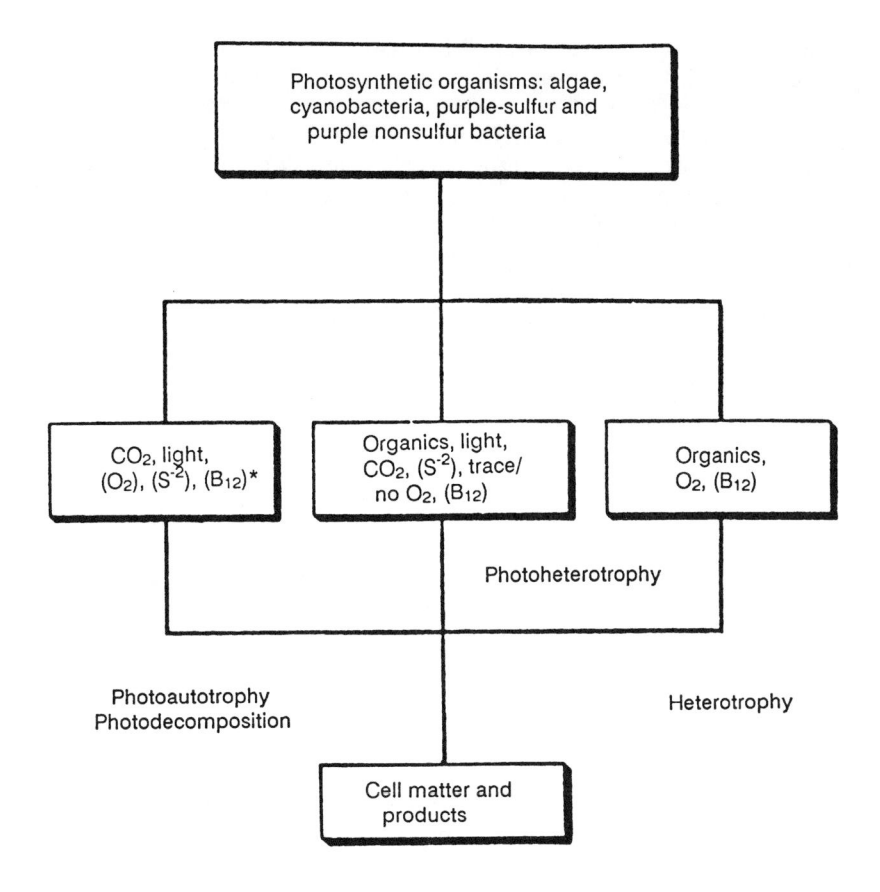

[Flow chart: Photosynthetic organisms: algae, cyanobacteria, purple-sulfur and purple nonsulfur bacteria branches into three conditions — CO₂, light, (O₂), (S⁻²), (B₁₂)*; Organics, light, CO₂, (S⁻²), trace/no O₂, (B₁₂); Organics, O₂, (B₁₂) — labeled Photoautotrophy/Photodecomposition, Photoheterotrophy, and Heterotrophy, all leading to Cell matter and products.]

*( ) Indicates that the substance may or may not be required

Figure 2.17 Mechanisms for photosynthetic organism growth. (From Kobayashi, H. and Rittmann, B.E. *Environ. Sci. Technol.* 16:170A–183A. American Chemical Society, Washington, D.C., 1982. With permission.)

2.1.2.2.1.6 Sequencing Batch Reactor (SBR)

SBRs can be utilized only when the organic concentration is greater than 50 mg/L (Eckenfelder and Norris, 1993). A recent modification of the SBR has several advantages, including a high degradation capacity, over conventional batch and continuous operations for the degradation of aromatic hydrocarbons (Lee, Choi, and Kim, 1993). The process, however, is difficult to operate because it depends upon the operator estimating when one of the nutrients has been depleted and adding fresh medium. An SBR used in this mode does not result in improvement in the rate of degradation.

2.1.2.2.1.7 Self-Cycling Fermenter (SCF)

The SCF is similar to the sequencing batch reactor, but it is a very stable system that can be operated for many cycles (Sarkis and Cooper, 1994). It has the potential to provide as high a degradation capacity as a sequencing batch reactor, but at a faster rate.

The SCF process is a computer-controlled fermentation, which uses the level of dissolved oxygen (DO) in the fermenter as a control parameter (Sarkis and Cooper, 1994). The DO level drops as the organism grows and the respiration rate increases. When the limiting nutrient is depleted, the DO value begins to increase, and the computer allows half the fermentation broth to be harvested and replaced, which takes about 2 min. The carbon source can be used as the limiting nutrient.

The cycle time is equal to the doubling time characteristic of the microorganism under the conditions used. The cells are always growing at their maximum rate, and after a few cycles the growth is synchronous. The cycle time is independent of contaminant when different initial levels are used as the carbon substrate.

2.1.2.2.1.8 Autothermal Aerobic Membrane Bioreactor (ATA MBR)

This is a patented, aerated, hot membrane bioreactor process for treating recalcitrant compounds in wastewater (Tonelli and Behmann, 1996). It utilizes acclimated, thermophilic, or caldoactive living microorganisms (hot cells) that thrive in an ATA reaction zone in an ATA bioreactor (ATAB) operating at ambient atmospheric pressure in combination with a microfiltration (MF) or ultrafiltration (UF) membrane filtration device from which a solids-free permeate can be withdrawn. The ATA MBR operates autothermally with a feed containing biodegradable organic material having a BOD of $\geqq 5000$ mg/L and preferably $\geqq 10,000$ mg/L (10 g/L). If a stable population of live hot cells can be maintained, the ATAB can be operated in the thermophilic range from 45 to 75°C, with HRT from 1 to 12 days.

2.1.2.2.1.9 Evaporation and Biofilm Filtration

This is a patented process carried out by concentration of the organic wastewater in a thin-film downflow evaporator, with double-walled tubes that allow wastewater to flow down along the inner wall of the tube while being heated by steam on the outside. The vapor is separated in a gas–liquid separator, condensed in a condenser, and filtered in a biofilm filtration apparatus.

2.1.2.2.1.10 Biocatalyst Beads

Highly porous, absorbent biocatalyst beads of synthetic organic polymer have PAC dispersed throughout the polymer and a biocatalyst, such as bacteria, in the macropores of the beads (Bair and Camp, 1995). The beads of this patented process are used for remediating contaminated aqueous streams. The biocatalyst consumes adsorbed organic contaminants and converts them into harmless gases, while continuously renewing the adsorptive capacity of the activated carbon.

2.1.2.2.2 Anaerobic Systems

In spite of the present significance and future potential, anaerobic waste treatment processes have not been favorably received (Venkataramani, Ahlert, and Corbo, 1988). These systems promote the reduction of organic matter to methane and carbon dioxide in an oxygen-free environment (Lee and Ward, 1985, 1984; Lee, Wilson, and Ward, 1987). Since the anaerobic process results in a lower cell yield, less sludge is generated and sludge handling costs are lower (Switzenbaum, 1983). This is a major consideration since a large proportion of both the capital and the operation and maintenance costs at a municipal wastewater treatment plant is associated with sludge processing and disposal. Certain energy requirements would also be lower, since aeration is not necessary and methane is produced as a by-product.

To date, no anaerobic process has been developed for treatment of domestic wastewater, since the temperature requirement of 35°C would require more energy than could be recovered from a low-strength waste (Switzenbaum, 1983). However, the expanded- and fluidized-bed, anaerobic filter and upflow anaerobic sludge blanket (UASB) systems have high flow capacity and operate at lower temperatures, and are, thus, being investigated for this application. Figure 2.18 provides an overview of the various reactor configurations available for anaerobic biotechnology (Speece, 1983).

2.1.2.2.2.1 Anaerobic Bioconversion Process

The bioconversion of the organic feedstocks in industrial wastewaters to methane is accomplished by a consortium of bacteria composed of chemoheterotrophic, nonmethanogenic, and methanogenic bacteria (Mah, 1981). Complex organics are first hydrolyzed by the chemoheterotrophic nonmethanogens to free sugars, alcohols, volatile acids, hydrogen, and carbon dioxide. The alcohols and volatile acids longer than two carbons are oxidized to acetate and hydrogen by obligate, proton-reducing organisms (acetogens), which must exist in symbiotic relation with hydrogen-utilizing methanogens (McInerney, Bryant, and Pfenning, 1979). The acetate and hydrogen are then converted to methane by the methanogenic bacteria (Mah, 1982). This latter reaction is the rate-limiting step, as these organisms are very slow growing. Therefore, long retention times are necessary for solids in anaerobic reactors.

An obligate, syntrophic (nutrient exchange between two organisms) relationship exists between the hydrogen-utilizing methanogens and the acetogens, which convert the higher volatile acids to acetate and hydrogen (Speece, 1983). The hydrogen-utilizing methanogens maintain extremely low hydrogen partial pressures in the system to prevent higher volatile acids from accumulating.

Compounds that were previously thought to be nondegradable under anaerobic conditions, such as aromatic compounds, have proved to be degradable anaerobically (Healy and Young, 1979). However, it has been discovered that methanogens have unusual nutritional requirements, such as a dependence

CURRENT TREATMENT TECHNOLOGIES

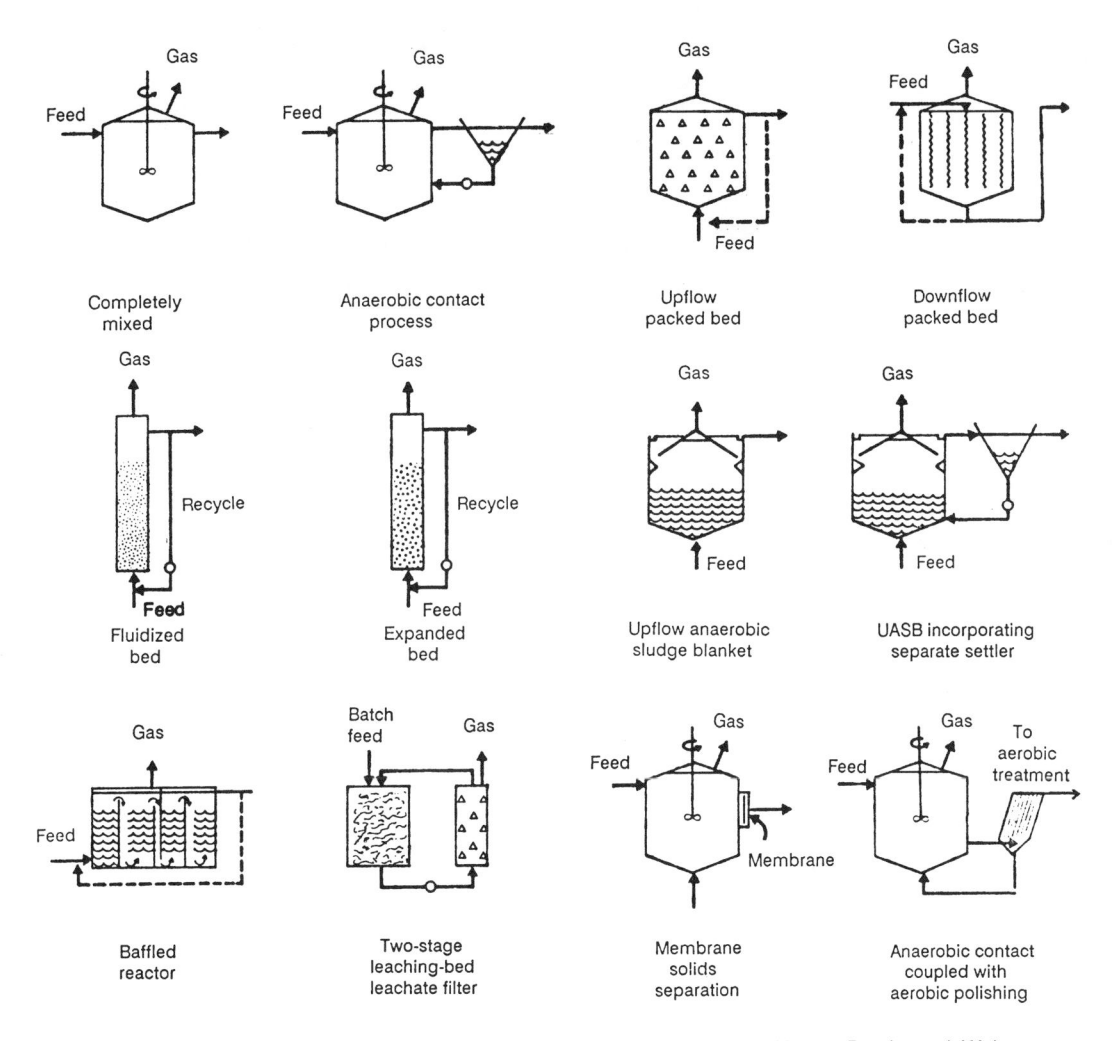

Figure 2.18 Reactor configurations for anaerobic biotechnology. (Zehnder, Huser, Brock, and Wuhrmann, 1980). (Reprinted with permission from Speece, R.E. *Environ. Sci. Technol.* 17:416A–427A. American Chemical Society, Washington, D.C., 1983. With permission.)

upon nickel (Switzenbaum, 1983). The lack of success of anaerobic degradation in the past has, at least partially, been attributed to the absence of nickel in the medium.

2.1.2.2.2.2 Suspended Growth Systems

1. Conventional Anaerobic Digester

Anaerobic reactors have been developed primarily for sludge digestion (Switzenbaum, 1983). Typically, sludges are digested in large holding tanks, which are usually maintained for about 15 days. This conventional digester (Figure 2.19) is usually heated and its contents mixed. Because of the temperature requirement, high-strength wastes are more suitable, since the methane produced is used to heat the reactors. The high solids retention times, which dictate large reactor volumes, generally make the conventional digester more suitable for solids processing than for liquid waste streams. On the other hand, a low hydraulic retention time is desirable for system economy.

An increase of about 10% in the gas production rate occurs when the hydrogen concentration is decreased by 25% (Poels, Van Assche, and Verstraete, 1985). A two-phase digestion process allows higher loading rates with reduced hydraulic retention times and a superior performance than the conventional high-rate anaerobic digestion process (Ghosh, Ombregt, and Pipyn, 1985). Fast anaerobic

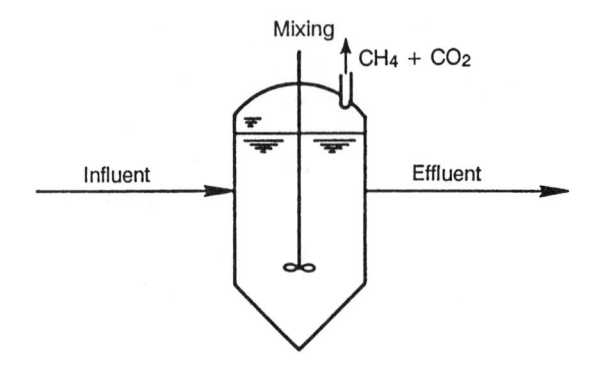

Figure 2.19 Conventional anaerobic digester. (From Switzenbaum, M.S. *ASM News.* 49:532–536, 1983. From the American Society of Microbiology, Journals Division. With permission.)

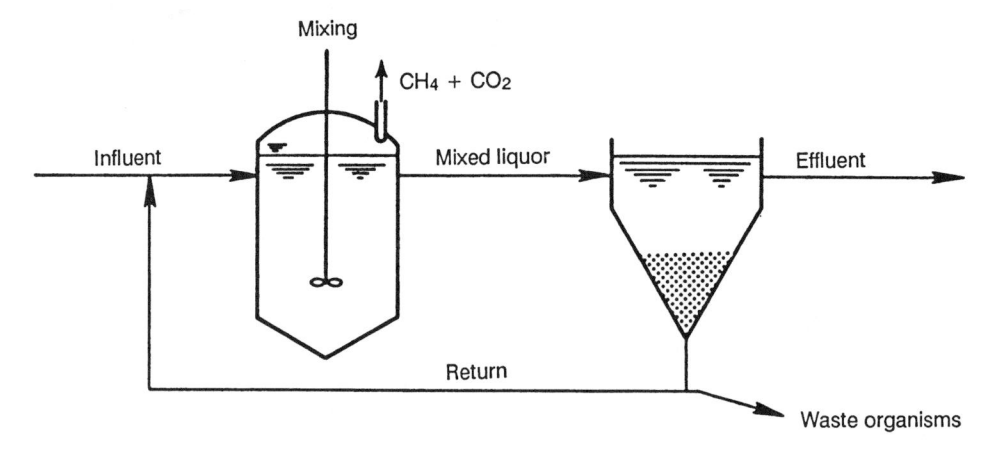

Figure 2.20 Anaerobic activated sludge process. (From Switzenbaum, M.S. *ASM News.* 49:532–536, 1983. From the American Society of Microbiology, Journals Division. With permission.)

digestion of liquid organic wastes is being developed (Auria, Christen, Favela, Gutierrez, Guyot, Monroy, Revah, Roussos, Saucedo-Castaneda, and Viniegra-Gonzalez, 1995). Retention and reuse of biomass in fast anaerobic digesters can be achieved by using the natural tendency of anaerobic bacteria to form aggregates in the form of biofilms or bioactive granules.

Research suggests that bioaugmentation of anaerobic digestion with a biocatalyst does not improve process performance (Koe and Ang, 1992). Comparison of the bacteria in a commercial biocatalyst preparation (obligate aerobes or facultative anaerobes, and no methanogens) with those recovered from a laboratory anaerobic digester showed no significant difference.

2. Anaerobic Activated Sludge Process

The anaerobic contact process was initially developed in the 1950s (Schroepfer, Fullen, Johnson, Ziemke, and Anderson, 1955) from the concept of recycling biological solids to obtain a larger biomass for a longer retention time. Several modifications are currently available (Frostel, 1982; Rippon, 1983). The process is basically an anaerobic activated sludge process (Figure 2.20). The effluent from the bioreactor is pumped to a settling unit, where a portion of the settled sludge is returned to the reactor, enabling the contact unit to maintain a high concentration of active mass. Solids concentrations can be maintained independently of waste flow by the biomass recycling.

3. Upflow Anaerobic Sludge Blanket (UASB)

A sludge blanket is, basically, a dense layer of granular or flocculated sludge placed in a reactor designed to allow the upward movement of liquid waste through the blanket (Switzenbaum, 1983). There are

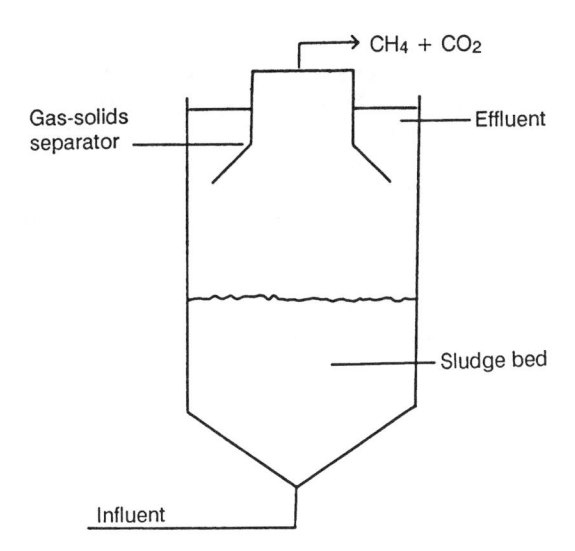

Figure 2.21 Upflow anaerobic sludge blanket (UASB). (Abeliovich, 1985; Bryant, 1986). (From Switzenbaum, M. S. *ASM News.* 49:532–536, 1983. From the American Society of Microbiology, Journals Division. With permission.)

various types of sludge blankets. The UASB is similar to the basic sludge blanket process except that the reactor is equipped with a gas–solids separator in the upper part of the reactor (Figure 2.21). The separator acts to separate the gas provided by the methane reaction and to separate dispersed sludge particles from the liquid flow. This is very important for the retention of sludge in the reactor, and mixing and recirculation are kept at a minimum.

Rod-shaped Methanothrix-like organisms were prevalent in UASB digesters receiving a simulated waste containing acetate and propionate (Ten Brummeler, Hulshoff Pol, Dolfing, Lettinga, and Zehnder, 1985).

Chemical flocculants enhance biomass retention and accumulation and promote granular sludge formation in an upflow floc digester (Cail and Barford, 1985; Fannin, Conrad, Srivastava, Chynoweth, and Jerger, 1986). Substrate concentration has no effect on the values of maintenance coefficient and true biomass yield with an UASB and filter, but has a significant effect on the saturation constant (Guiot and van den Berg, 1985). A reactor combining a UASB with a fixed film demonstrated better performance than the sludge blanket alone, with regard to biofilm development, HRT, and gas productivity (Fiebig and Dellweg, 1985).

4. Anaerobic Lagoons

Anaerobic or facultative lagoons are also available, with their ease of operation and low costs, but these will probably produce a lower-quality effluent than aerobic lagoons (Johnson, 1978; JRB Associates, 1982). They utilize anaerobic degradation pathways, which may be more efficient for the removal of some compounds, and sludge production is minimized.

A significant number of nitrifying bacteria were present in effluent samples from a heavily loaded anaerobic pond, which supports the theory of heterotrophic nutrition in such conditions. In a Canadian storage pond for wastewater produced by bitumen extraction from tar sands, the counts of aerobic and anaerobic heterotrophs and sulfate-reducing bacteria were 10^6/mL, 10^3/mL, and 10^4/mL, respectively (Foght and Westlake, 1985). Only in the upper layers could *n*-hexadecane be degraded.

Hydraulic loading does not affect unaerated lagoons with a specific surface area greater than 40 $m^2/m^3/d$ (Tariq and Ahmad, 1985). Series operation improves bacterial removal, and the operational variables do not affect nutrient removal or the dominant type of phytoplankton. With addition of $Al_2(SO_4)_3$, a pilot-scale, high-rate pond removed 95% of BOD and total P (Kawai, Grieco, and Jureidini, 1984). A 50:50% mixture of raw wastewater and mesophilically digested sludge enhanced treatment in an anaerobic lagoon, providing 50% reduction of organic solids in 14 weeks (Lowe and Williamson, 1984).

2.1.2.2.2.3 Fixed-Film Systems

Fixed-film reactors provide an anaerobic process with a high solids retention time for the methane-producing bacteria, with a short hydraulic retention time for system economy (Switzenbaum, 1983). In these systems, microorganisms grow on a solid support and organic matter is removed from the liquid flowing past them. A large reduction of required reactor volume is possible through application of a fixed-film concept combined with a liquid–solid separation pretreatment (Liao and Lo, 1985; Fannin, Conrad, Srivastava, Chynoweth, and Jerger, 1986). Several types of fixed-film systems have been developed for anaerobic treatment (Switzenbaum, 1983). These include the anaerobic filter, expanded-bed, fluidized-bed, and anaerobic baffled reactors.

Methanogenic isolates that cannot consume hydrogen below partial pressures of 6.5 Pa have been reported, suggesting a threshold for methane production from hydrogen in sediment (Lovley, 1985; Fannin, Conrad, Srivastava, Chynoweth, and Jerger, 1986). The sulfur-reducing bacterium, *Desulfovibrio vulgaris*, grows via interspecies H_2 transfer by utilizing reducing equivalent generated in coculture by *Methanosarcina barkeri*, while growing on acetate or methanol (Phelps, Conrad, and Zeikus, 1985). Hydrogenase may function to vent excess reducing equivalents as H_2, which are consumed to maintain a constant partial pressure in the system.

Overloading of downflow stationary fixed-film digesters is characterized by elevated volatile acids, decreased pH, and biofilm sloughing (Kennedy, Muzar, and Copp, 1985). Recovery from instability is directly related to buffering capacity. Performance of a downflow fixed-film reactor is dependent upon design and operational factors (Van den Berg, Kennedy, and Samson, 1985). COD removals of 70 to 95% have been achieved at loading rates of 5 to 15 kg COD/m^3/day. Reactor performance is stable even with severe hydraulic and organic loading rates. Performance at thermophilic temperatures is similar to that at mesophilic temperatures. COD conversion rates of carbohydrate-containing wastewaters are significantly increased when preacidified with volatile acids (Cohen, Breure, Schmedding, Zoetemeyer, and van Andel, 1985). Loading rates can be increased by two to four times when 3 to 13% volatile acids are added.

A start-up period of 90 days is reported for a downflow stationary fixed-film reactor (Samson, van den Berg, and Kennedy, 1985a). Performance is related primarily to surface-to-volume ratios. Reactors receiving lower-strength feeds exhibit faster start-up rates and greater stability (Kennedy, Muzar, and Copp, 1985). In an anaerobic downflow stationary fixed-film reactor, mixing is a function of horizontal channel separation, feed distribution system, and recirculation (Samson, van den Berg, and Kennedy, 1985b). Digester performance is related to mixing and recirculation.

In fixed-film systems, denitrification, resulting in a loss of oxidized nitrogen, is as high as that in completely anoxic conditions (Strand, McDonnell, and Unz, 1985). Since oxygen does not completely penetrate biofilms of certain thickness, bulk liquid oxygen has little effect on denitrification. Denitrification rates in thick biofilms are at least 50% of those in anoxic films (Strand and McDonnell, 1985).

1. Anaerobic Filter System

The most common anaerobic attached growth treatment process is the anaerobic filter (Lee and Ward, 1985, 1984; Lee, Wilson, and Ward, 1987). This is an upflow fixed-bed (or static-bed) configuration (Figure 2.22) (Switzenbaum, 1983). The filter is composed of one or more vertical beds containing some inert material, such as rocks or plastic media, which acts as a stationary support surface for microbial film attachment. Wastewaters are pumped upward through the support medium, allowing contact between the attached microorganisms and the wastewater. Microbial growth also takes place in the spaces in the support medium.

A number of proprietary anaerobic biotechnology processes are on the market, each with distinct features, but all utilizing the fundamental anaerobic bacterial conversion to methane (Lee and Ward, 1986). The digester gas can be flared or fired in boilers, gas turbines, or reciprocating engines with or without the prior removal of sulfurous gases.

These systems are used to treat aqueous wastes with low to moderate levels of organics. Anaerobic digestion can handle certain halogenated organics better than aerobic treatment. Stable, consistent operating conditions must be maintained. Anaerobic degradation can take place in native soils; however, when used as a controlled treatment process, an airtight reactor is required. Hazardous organic substances that have been found to be amenable to anaerobic treatment include acetaldehyde, acetic anhydride, acetone, acrylic acid, aniline, benzoic acid, butanol, creosol, ethyl acrylate, methyl ethyl ketone (MEK), phenol, and vinyl acetate. The upflow fixed-bed reactor performance is more stable than that of other

CURRENT TREATMENT TECHNOLOGIES

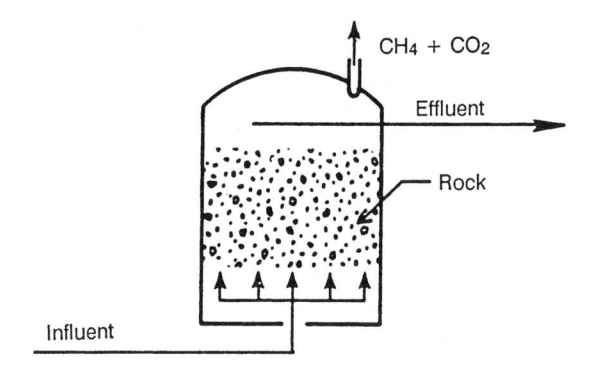

Figure 2.22 Diagram of typical anaerobic filter system. (From Switzenbaum, M. S. *ASM News.* 49:532–536, 1983. From the American Society of Microbiology, Journals Division. With permission.)

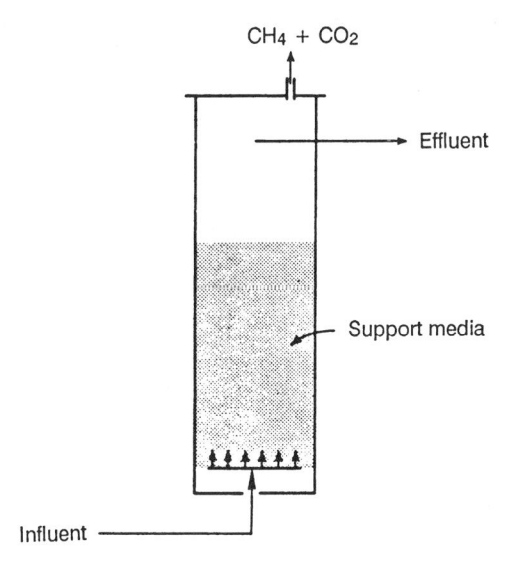

Figure 2.23 Anaerobic expanded/fluidized bed. (Abeliovich, 1985; Bryant, 1986). (From Switzenbaum, M. S. *ASM News.* 49:532–536, 1983. From the American Society of Microbiology, Journals Division. With permission.)

fixed-film reactors in the production and utilization of volatile acids (Dohanyos, Kosova, Zabranska, and Grau, 1985).

There are some filter designs with downflow direction (Switzenbaum, 1983). An example is the Bacardi plant in San Juan, Puerto Rico. Methane gas produced from the rum distillery wastewater provides nearly all of the energy required by the distillery. Approximately 2×10^6 ft^3 of gas is produced per day, which supplies 40% of the energy needed for distillation. This is equivalent to approximately 150 barrels of fuel oil per day, having an approximate value of $1 million/year.

2. Anaerobic Expanded/Fluidized Bed

Anaerobic expanded or fluidized beds consist of inert sand-sized particles, contained in a column, which expand with the upward flow of wastewater through the column (Switzenbaum, 1983). The inert particles act as a support surface for the growth of attached microorganisms. A schematic of the process is shown in Figure 2.23. This system is known for its large surface area for film attachment and excellent mixing characteristics for substrate–biomass contact (Meunier and Williamson, 1981). The Oxitron fluidized-bed system (Figure 2.13) can also be used anaerobically.

The difference between the expanded- and fluidized-bed processes is somewhat ambiguous (Switzenbaum, 1983). In many cases, *fluidized bed* has been used to refer to the more-than-doubled reactor volume achieved by the high flow rate of fluid through the active bed, which is composed of small particles. *Expanded bed* has been used to designate reactors with a smaller degree of expansion of the static volume. In other aspects the systems are quite similar.

A completely mixed, expanded-bed anaerobic GAC reactor (Figure 2.24; Bove, Lambert, Lin, Sullivan, and Marks, 1984) shows steady-state removal efficiencies of acetate and COD exceeding 98 and 97%, respectively (Wang, Suidan, and Rittmann, 1985). The reactor responds well to sudden organic loading rate changes.

In addition to anaerobic fermentation, these systems have been used for denitrification, aerobic oxidation of carbonaceous matter, and nitrification (Switzenbaum, 1983).

2.1.2.2.3 Combined Aerobic/Anaerobic Treatment
2.1.2.2.3.1. Aerobic/Anaerobic Biofilm Reactor

A novel support aerated biofilm reactor for the biodegradation of toxic organic compounds has been developed (Woods, Williamson, Strand, Ryan, Polonsky, Ely, Gardner, and Defarges, 1987). The gas-permeable-membrane-supported (GPMS) reactor differs from traditional biofilm processes in that electron donors or acceptors are provided to the microorganisms by diffusion of gases through a gas-permeable membrane. In addition to supplying substrates directly to the biofilm, the membrane also physically supports the biofilm and the overlying bulk liquid. By this means, an enrichment culture of methane-oxidizing bacteria and an enrichment consortium of aerobic and anaerobic bacteria are produced.

A methylotrophic biofilm is developed by providing methane and oxygen through the membrane (Woods, Williamson, Strand, Ryan, Polonsky, Ely, Gardner, and Defarges, 1987). Because methane and oxygen are sparingly soluble gases, the growth of methylotrophs is facilitated in the GPMS reactor by providing methane and oxygen directly to the microorganisms through the membrane. Figure 2.25 shows a reactor for the growth of methylotrophs. The original bacterial inocula for the selection of methylotrophs consisted of 10 mL of thickened trickling filter effluent from a municipal wastewater treatment plant and 10 mL of thickened sludge from a bench-scale anaerobic digester.

The anaerobic/aerobic consortium is developed by operating the reactor at high organic loading levels in the liquid compartment to generate an anaerobic bulk liquid and a thick biofilm, and by supplying pure oxygen through the membrane at the base of the Goretex support to allow development of aerobes at the surface of the membrane (Woods, Williamson, Strand, Ryan, Polonsky, Ely, Gardner, and Defarges, 1987). The different cultures will degrade different compounds. For example, the anaerobic portion of the biofilm will facilitate reductive dehalogenation of haloaromatic compounds, and the aerobic portion of the biofilm will allow degradation of the dehalogenated metabolic products. By controlling the type of gas transferred through the gas-permeable membrane, the electron donors and electron acceptors and bacterial consortia can be selected for the compounds to be removed.

2.1.2.2.3.2. Sequential Anaerobic/Aerobic Treatment

Anaerobic pretreatment of a leachate source from a landfill site by use of a fluidized bed may be more cost-effective than aerobic treatment. Alternatively, series operation of anaerobic and aerobic fluidized-bed or membrane bioreactors may be the most attractive flow scheme. A number of field-scale studies have shown that sequential anaerobic/aerobic treatment processes are a viable treatment approach for leachates (Venkataramani, Ahlert, and Corbo, 1988).

Since about 60 to 70% of the leachate-derived dissolved organic carbon (DOC) can be removed by anaerobic treatment, a process scheme involving anaerobic pretreatment of the waste residue, followed by final aerobic polishing, would provide the optimal approach (Venkataramani, Ahlert, and Corbo, 1988). Leachate-derived carbon can serve as the sole source of carbon for growth and energy. Aerobic biological studies have shown that a mixed microbial population, acclimated to landfill leachate, can degrade 80 to 90% of the organic species present in the hazardous industrial waste liquor, with or without addition of glucose and other nutrients (Venkataramani, Ahlert, and Corbo, 1988).

Control of pH is important, while added nutrients and the presence of a preferred carbon source seem to be less critical (Venkataramani, Ahlert, and Corbo, 1988). The pH increases to a certain point after which it remains steady. The decrease in TOC and DOC with time indicates that removal of organic carbon from the system is due to biological oxidation and not to sorption effects. Cometabolism is not the sole mode of oxidation of the organic matter present in leachate. It appears that a mixed microbial culture utilizes the fatty acids first, before utilizing other compounds present in the leachate, suggesting a diauxic type of growth, with two distinct growth phases. The point at which the growth shifts coincides

CURRENT TREATMENT TECHNOLOGIES

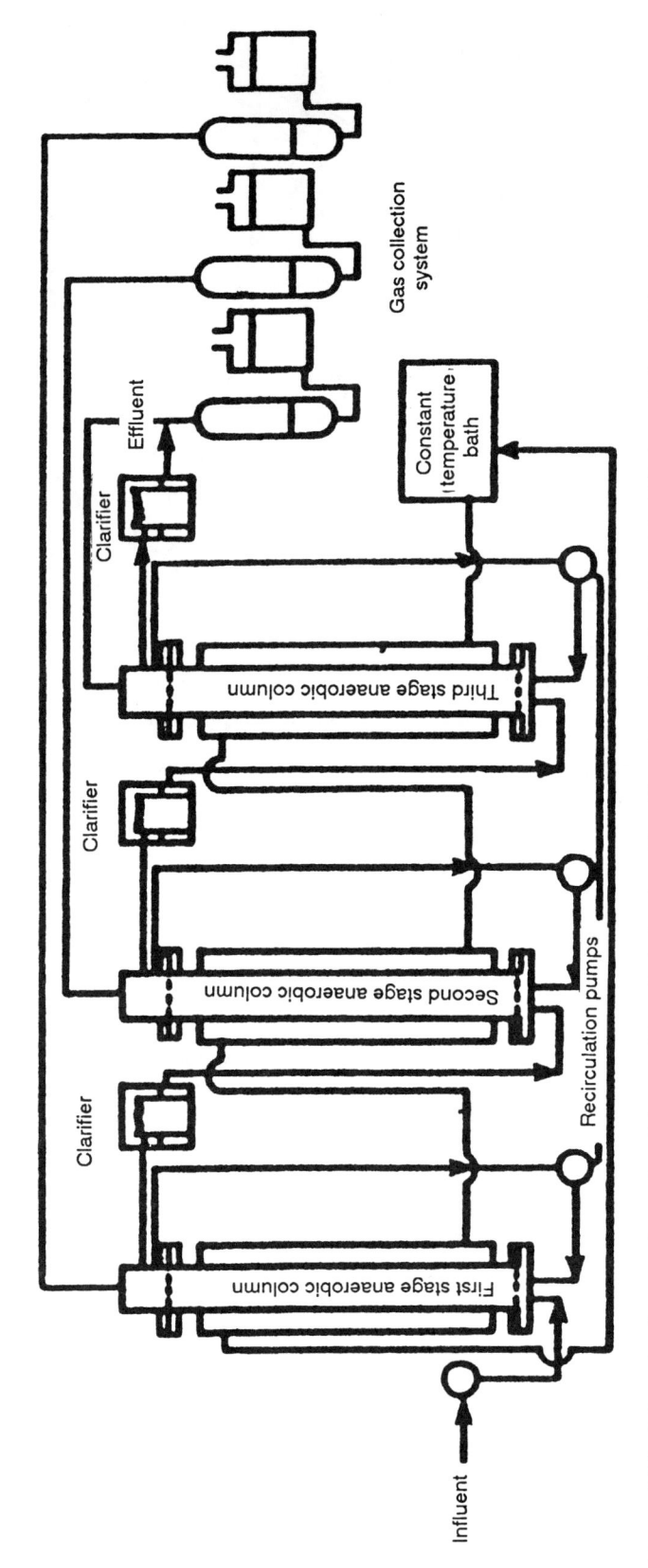

Figure 2.24 Schematic diagram of experimental anaerobic activated carbon filter. (From Bove, L.J. et al. Report to U.S. Army Toxic and Hazardous Materials Agency on Contract No. DAAK11-82-C-0017, 1984. AD-A162 528/4.)

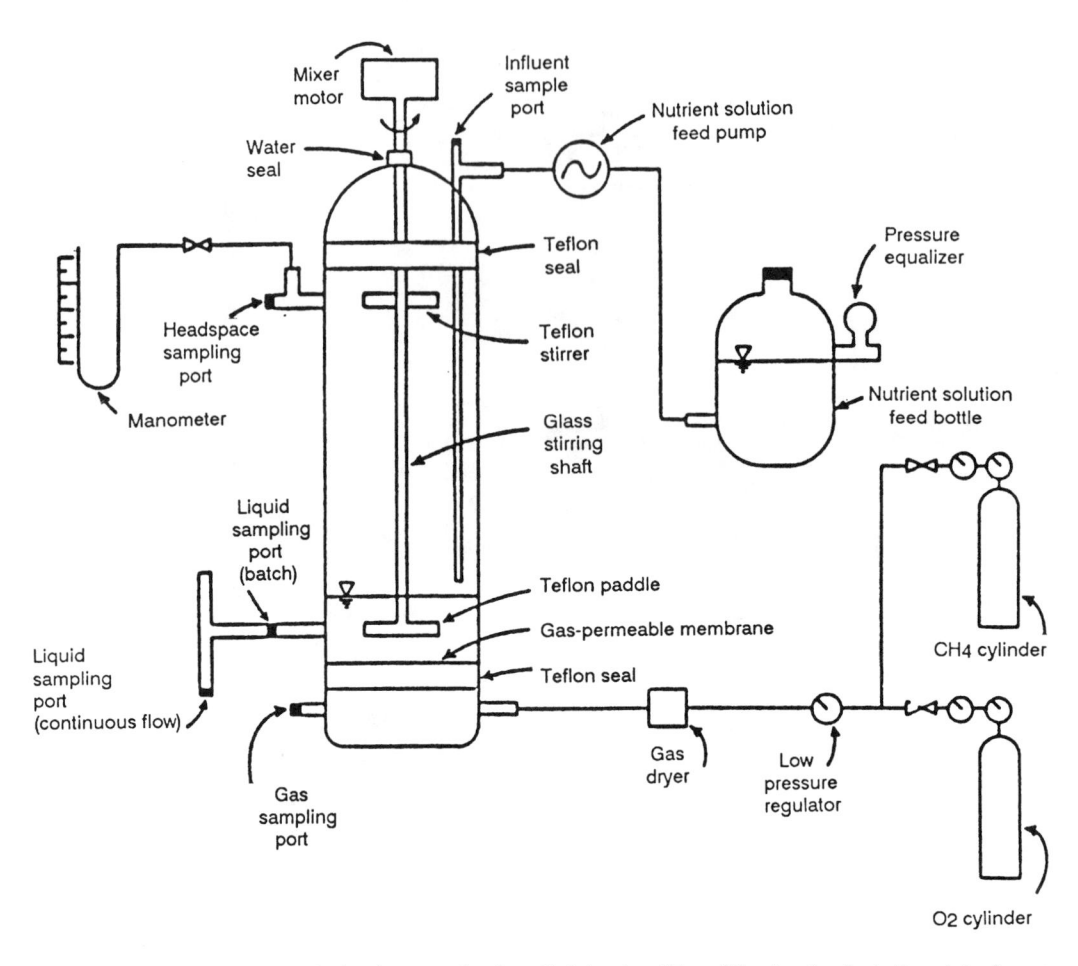

Figure 2.25 Reactor schematic for the growth of methylotrophs. (From Woods, S. et al. Preprint of paper presented at 194th Am. Chem. Soc. Natl. Mtg. Dept. of Civil Eng., Oregon State University, Corvallis, OR, 1987. With permission.)

with the time at which the pH value becomes steady. Enhanced specific substrate uptake rate is observed with a controlled pH of 7.5.

On the other hand, yield and specific growth rate decrease with increasing leachate concentrations, indicating substrate inhibition (Venkataramani, Ahlert, and Corbo, 1988). This may be due to inhibitors of methanogenesis being present at appreciable concentrations or too high a total volatile fatty acid concentration in the reactors, leading to toxicity from the un-ionized portion. A lag with low concentrations of leachate may be due to removal of acetate that acts as the driving force for the removal of other compounds. It is possible to treat highly concentrated waste liquor, i.e., up to 10,000 mg/L of organic carbon, with this method, with low sludge production. Methane is produced at levels of 0.95 to 0.99 L/g (m^3/kg) DOC removed. This option would greatly reduce costs associated with energy requirements and sludge disposal.

2.2 IN SITU PROCESSES

In situ treatment of waste and soil at contaminated sites offers an alternative to the traditional approach to site remediation involving excavation and redisposal or on-site isolation or treatment (Ghassemi, 1988). Based upon the technologies that have been developed and used in conventional water and wastewater treatment and in mining, oil and gas, and chemical process industries, a number of processes and systems have been proposed for *in situ* treatment. These methods use biological, chemical, physical, or thermal methods to degrade, detoxify, extract, or immobilize contaminants. They include biodegradation, air/steam stripping, neutralization, solidification/stabilization, and oxidation and can be used in combination as a treatment train.

2.2.1 PHYSICAL/CHEMICAL SOIL TREATMENT PROCESSES
2.2.1.1 Shallow Soil Mixing (SSM)
This technique is being increasingly relied on in the *in situ* remediation of contaminated soils (Carey, Day, Pinewski, and Schroder, 1995). The primary applications of SSM have been to mix cement, bentonite, or other reagents to modify properties and thereby remediate contaminated soils or sludges. SSM is being combined with soil vapor extraction (SVE) for extracting VOCs from soils. It also enhances *in situ* bioremediation (Johns and Nyer, 1996).

2.2.1.2 Oxidation/Reduction
Hydrogen peroxide, ozone, and hypochlorites are the most useful oxidizing agents available. Ozone oxidizes many organic compounds that cannot be easily broken down biologically (Lee and Ward, 1985, 1984; Lee, Wilson, and Ward, 1987). The addition of ozone, hydrogen peroxide, and hypochlorites to the subsurface for oxidation of organic compounds is discussed in depth in Section 5.1.4.

A patent has been developed for a method for *in situ* oxidation of soils by introduction of an aqueous solution containing hydrogen peroxide and a compound to control the mobility of the aqueous solution by increasing the viscosity and density or modifying the interfacial properties of the aqueous solution within the formation (Brown and Norris, 1986). The mobility control agent is selected from a group consisting of hydratable polymeric materials, interface modifiers, and densifiers.

Takahashi (1994) compared use of ozone for oxidation and destruction of biorefractory organic compounds with a partial oxidation (ozonation as a pretreatment for biological treatment) and combined oxidation (combined use of ozone and other chemical or physical treatment). Combined oxidation uses active radicals (e.g., OH) instead of O_3. Destruction of biodegradation-resistant organic compounds can be improved with this treatment, but the optimum conditions for the combined oxidation must be determined. Recent developments have made ozone usage less expensive and more efficient.

Wet air oxidation involves addition of air at high pressures and temperatures in a form of combustion (Lee and Ward, 1985, 1984; Lee, Wilson, and Ward, 1987). A catalyst promotes the oxidation process. Dilute wastes that cannot be treated with incineration can be handled by this process with greater than 99% destruction. Catalytic oxidation is less expensive than carbon adsorption (DePaoli, Wilson, and Thomas, 1996).

The U.S. Department of Energy's Office of Technology Development *In Situ* Remediation Integrated Program is developing a new technology that uses electric fields to oxidize organic compounds *in situ* by a two-step electrical corona process (Heath, Caley, Peurrung, Lerner, and Moss, 1994). In the first stage, relatively low voltage heats the soil in the vadose zone. Steam formed from intrinsic soil moisture and any volatile and semivolatile soil contaminants are removed through a central soil–vapor extraction vent. In the second stage, higher voltages delivered through the same electrodes create corona directly on soil particles in narrow bands along spreading interfaces between moist and dry soil. Ions, electrons, and secondary oxidants formed in the corona oxidize any residual organic contaminants. *In situ* corona may be applicable to sites with complex stratigraphies, including low-permeability soils contaminated with nonvolatile or bound contaminants; difficult organic contaminants, such as PAHs, that are not removable by other means; or situations where the need for added chemicals or long treatment times precludes use of other *in situ* techniques.

The Thermatrix flameless oxidation process provides >99.99% destruction of a wide variety of organics (Wilbourn, Newburn, and Schofield, 1994). Combining this process with other successful waste contaminant separation and removal technologies can result in effective integrated systems. For example, Thermatrix oxidation coupled with thermal desorption provides an integrated waste-processing system that is technically and economically attractive, due to the need for a very low gas flow rate. The system is simple to operate, allows for heat recovery and reuse, and results in near zero emissions.

Inorganics can also be oxidized in the soil. Chromium can be reduced from the hexavalent state to the trivalent and then precipitated with hydroxide (Lee and Ward, 1985, 1984; Lee, Wilson, and Ward, 1987). Reducing agents for chromium include gaseous sulfur dioxide, iron sulfate, waste pickling liquor from metal-plating industries, and sodium bisulfite with sulfuric acid commonly used to reduce the pH

(Ehrenfeld and Bass, 1984). Levels of less than 1 ppm chromium can be achieved. Arsenic and possibly some lead compounds can be oxidized by use of potassium permanganate.

2.2.1.3 Hydrolysis
Hydrolysis is a chemical reaction involving cleavage of a molecular bond by reaction with water (Amdurer, Fellman, and Abdelhamid, 1985). The rates of hydrolysis for some compounds can be accelerated by altering the solution pH, temperature, or solvent composition, or by introducing catalysts. For *in situ* treatment, alteration of pH, particularly raising the pH (base-catalyzed hydrolysis), is the most-promising approach. Chemicals potentially treated by base-catalyzed hydrolysis include amides, esters, carbamates, organophosphorus compounds, pesticides, and herbicides. The process has been used successfully on surface spills of acrylonitrile.

The primary design concern is the production and maintenance of high-pH (9 to 11) conditions with saturation or high moisture content in the waste deposit (Amdurer, Fellman, and Abdelhamid, 1985). For shallow contamination, surface application of lime, sodium carbonate, or sodium hydroxide followed by surface application of water may be appropriate. For deeper deposits, subsurface delivery or injection of alkaline solutions may be required.

2.2.1.4 Neutralization
Neutralization involves injecting dilute acids or bases into the groundwater to adjust the pH (U.S. EPA, 1985a). This can serve as a pretreatment prior to *in situ* biodegradation to optimize the pH range for the microorganisms. Adjustment of the pH may be required to make the water less corrosive and suitable for other unit processes (Stover and Kincannon, 1983).

2.2.1.5 Stabilization/Solidification
The same technology is applied to *in situ* stabilization/solidification as to the *ex situ* process (see Section 2.1.1.1.9) for immobilizing heavy metals or other inorganic compounds (Johns and Nyer, 1996). It involves mixing the soil with additives to produce a cementlike consistency, to depths of 100 ft. The soil is mixed in place with additives, such as cement, cement or lime kiln dust, fly ash, or lime. A specially designed hollow-stem auger mixes and delivers the reagents.

The process is for inorganics and some organically contaminated soil (Johns and Nyer, 1996). A simple treatability study will determine the best stabilizing agent for the latter. Addition of oxygen and nutrients through the augers can stimulate biodegradation. A bulking factor between 10 and 30% should be used for the stabilizing agent.

2.2.1.6 Mobilization/Immobilization
Soil sorption is perhaps the most important soil waste process affecting immobilization of toxic and recalcitrant fractions of hazardous wastes (Sims and Bass, 1984). Leaching potential and the residence time in soil for constituents that undergo degradation are directly affected by the extent of immobilization. Treatment techniques to enhance immobilization of constituents by controlling or augmenting the sorption process in soils have been developed based upon fundamental principles and applied landtreatment techniques. These techniques are also useful in regulating biological degradation. See Sections 4.1.4 for an in depth discussion of sorption and its effect on biodegradation and 5.1.6 for a wider discussion of the role of organic matter in sorption and how it can be used to aid biodegradation.

Surfactants may be added during soil washing to mobilize the contaminants (Ellis, Payne, Tafuri, and Freestone, 1984). A 4% solution of two nonionic surfactants removed greater than 90% of PCBs and a high boiling distillation fraction of crude oil from test soil columns with 10 pore volume washes.

Contaminants can be immobilized by precipitation or encapsulation in an insoluble matrix (Pye, Patrick, and Quarles, 1983). A spill of acrylate monomer was treated with catalyst and activator in order to produce a solidified polymer, thereby immobilizing an estimated 85 to 90% of the liquid monomer (Knox, Canter, Kincannon, Stover, and Ward, 1984).

Other *in situ* treatments include radio frequency heating (Dev, Bridges, and Sresty, 1984) and vitrification using an electric current to melt the soils and wastes in place (Ehrenfeld and Bass, 1984).

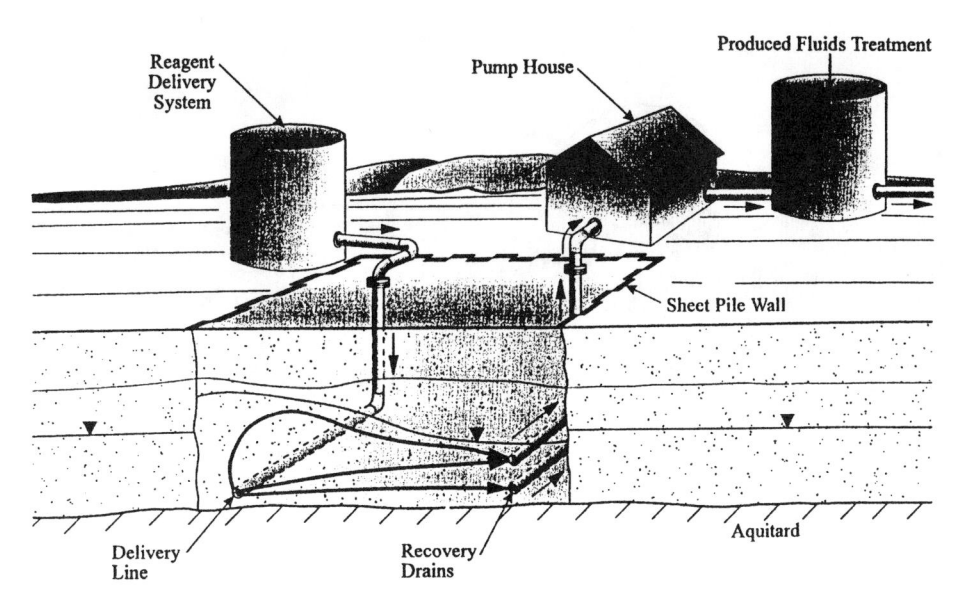

Figure 2.26 Soil flushing. (From Johns, F.J., II and Nyer, E.K. In *In Situ Treatment Technology.* Nyer, E. K. et al., Eds. Lewis Publishers, Boca Raton, FL, 1996.)

2.2.1.7 Soil Flushing/Washing/Extraction/Pump and Treat

Principle/Description — Chemical adsorption of a contaminant to soil particles and the portion of contaminant that is retained or entrained within the pore space surrounding the soil particles must be addressed for soil treatment.

Soil flushing is an extraction process to remove organic and inorganic compounds from contaminated soils (U.S. EPA, 1985a). Essentially, water or an aqueous solution is injected into the area of contamination, and the contaminated elutriate is pumped to the surface for removal, recirculation, or on-site treatment and reinjection (U.S. EPA, 1985a). Groundwater can be continuously recycled by this "pump-and-treat" method (Gruiz and Kriston, 1995). During elutriation, sorbed contaminants are mobilized into solution by reason of solubility, formation of an emulsion, or by chemical reaction with the flushing solution. See Figure 2.26 (Johns and Nyer, 1996).

Soil-flushing methods remove contaminants by dissolving the liquid, sorbed, or vapor phase or by mobilizing contaminants existing as free product in the soil pores and adsorbed to the soil (Lyman, Noonan, and Reidy, 1990). The former processes are controlled by the solubilities of the contaminants and Henry's law constants, while the latter are removed by the pressure gradient of the flushing water and are controlled by the viscosity and density of the contaminants.

The treatment solution for flushing or rinsing a contaminant from soil can be delivered via gravity (e.g., flooding, ponding, surface seepage) or forced systems (e.g., injection pipes) (U.S. EPA, 1985a). Unused treatment agents and by-products are collected via gravity (e.g., open ditches and trenches, porous drains) or forced systems (e.g., well points).

Flushing or mobilizing wastes can serve two purposes: to promote the recovery of wastes from the subsurface for treatment on the surface or to solubilize adsorbed compounds in order to enhance the rate of other *in situ* treatment techniques, such as biodegradation or hydrolysis (Amdurer, Fellman, and Abdelhamid, 1985). The elutriate can also be recycled back into the site (Sims and Bass, 1984). An example of a soil-flushing system with elutriate recycle is given in Figure 2.27.

With extraction, the extraction agent is mixed with the contaminated soil to dissolve or disperse the contaminants from the soil into the extracting agent and separate them from the soil particles (IT Corporation, 1987). The soil may be postwashed with clean extraction agent to remove residual contaminated extraction agent. The extraction agent is cleaned for reuse, and the chemical waste residue further treated or landfilled. Heat may also be used to enhance the extraction process.

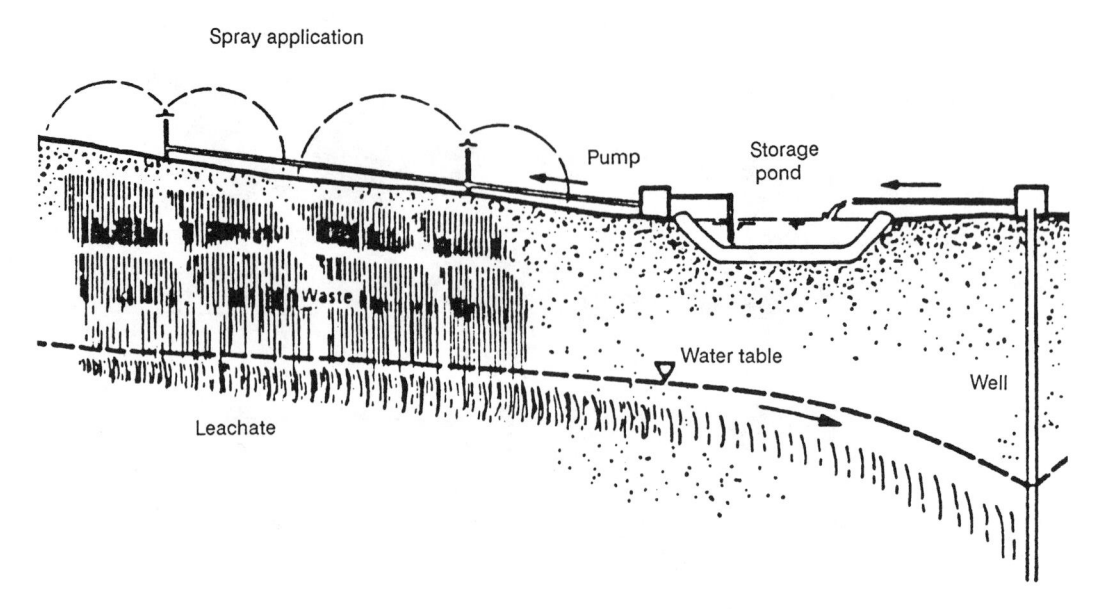

Figure 2.27 Schematic of an elutriate recycle system. (From Sims, R. and Bass, J. EPA Report No. EPA-540/2-84-003a. 1984.)

Solutions with the greatest potential for use in soil flushing would be water, acidic aqueous solutions (e.g., sulfuric, hydrochloric, nitric, phosphoric, and carbonic acid; Sims and Bass, 1984), basic solutions (e.g., sodium hydroxide; Sims and Bass, 1984), complexing and chelating agents, surfactants (e.g., alkylbenzene sulfonate; Sims and Bass, 1984), and certain reducing agents (U.S. EPA, 1985a).

Selection of an aqueous extraction agent is based upon safety to humans and environment, natural presence of water in soil, purification possibilities of the extracting agent, ease of use, and costs (IT Corporation, 1987). Water is appropriate for flushing water-soluble compounds (e.g., phenols) or water-mobile organics and inorganics (Amdurer, Fellman, and Abdelhamid, 1985). Organics amenable to water flushing can be identified according to their soil/water partition coefficient, or estimated using the octanol/water partition coefficient, K_{ow} (Nash, 1987). Inorganics that can be flushed from soil with water are soluble salts, such as the carbonates of nickel, zinc, and copper.

Adjusting the pH with dilute solutions of acids or bases will enhance inorganic solubilization and removal (Nash, 1987). Acids can be added to dissolve impurities, bases to dissolve impurities or disperse insoluble impurities in the extraction phase, surfactants to aid dispersion, or complex-forming agents to increase solubility of the impurities in the extracting agent (IT Corporation, 1987). Weak acids (e.g., sodium dihydrogen phosphate and acetic acid) or dilute solutions of strong acids (e.g., sulfuric) could be used if the soil contains sufficient alkalinity to neutralize it (U.S. EPA, 1985a). Complexing and chelating agents can mobilize metals strongly adsorbed to manganese and iron oxides in soils.

Surfactants (surface active agents) are a class of natural and synthetic chemicals that promote the wetting, solubilization, and emulsification of various types of organic chemicals (Amdurer, Fellman, and Abdelhamid, 1985). Surfactants can improve the solvent property of the recharge water, emulsify nonsoluble organics, and enhance removal of hydrophobic organics sorbed onto soil particles. This is a promising *in situ* chemical treatment method. Some 93% of petroleum hydrocarbons can be removed by adding surfactants, which is orders of magnitude greater than obtained with just water washing (Ellis, Payne, and McNabb, 1985).

Soil-flushing methods involving the use of water surfactants appear to be the most feasible and cost-effective chemical treatment for organics (U.S. EPA, 1985a). Relatively cheap, innocuous treatment reagents can be used to treat a broad range of waste constituents, without producing toxic degradation products. Accelerating the natural leaching process by flushing the contaminated soil *in situ* with an aqueous surfactant solution and recovering the wash effluent from the aquifer is a remedial method being investigated for removing these pollutants from soil (Ellis, Payne, and McNabb, 1985; Nash, 1987). A laboratory experiment can be conducted in order to estimate the number of times groundwater would have to be turned over or filtered through the contaminated soil to achieve the required level of water quality (Stover and Kincannon, 1983).

The removal of PAHs from soil matrices using pure water, by *ex situ* soil washing or *in situ* soil flushing, is quite ineffective because of their low solubility and hydrophobicity (Joshi and Lee, 1996). However, use of a suitable nonionic surfactant for *in situ* flushing can increase the removal efficiency severalfold. The surfactant concentration, the ratio of washing solution volume to soil weight, and the temperature of the washing solution have significant effects on PAH removal. There is an optimal range for each parameter for given washing conditions.

Surfactants will be required for significant solubilization of insoluble (hydrophobic) compounds (Amdurer, Fellman, and Abdelhamid, 1985), such as the gasoline additive, tetraethyl lead, and many of the major constituents in fuel oil (Lyman, Noonan, and Reidy, 1990). A saline solution will displace only limited amounts of mobile gasoline and associated tetra-ethyl lead (TEL) components from soil (Ouyang, Mansell, and Rhue, 1996). However, 95% of the immobile or residual gasoline and 90% of the associated TEL entrapped in the soil pores can be removed by a surfactant/cosurfactant/water solution. Essentially, 1 g of surfactant (sodium lauryl sulfate) can remove 0.6 g immobile gasoline and 2 mg immobile lead from the soil.

Sodium dodecyl sulfate (SDS), an anionic surfactant, is frequently used for soil flushing (Liu and Roy, 1995). The maximum SDS adsorption and precipitation occur when SDS concentration is in the range of the critical micelle concentration. Surfactant injection in a soil matrix decreases hydraulic conditions, due to clay expansion, sodium dispersion, fine particle mobilization, and especially to precipitation of calcium and magnesium dodecyl sulfate.

The efficiency and effectiveness of using surfactants for flushing depends on their mobility in the soil (Allred and Brown, 1996). These authors determined that the concentration fronts of two commercial anionic surfactants, an alkyl ether sulfate and a linear alkylbenzene sulfonate, were one half and one fifth of the advance of the wetting front. This indicated a high degree of sorption of the surfactants, which could be either electrostatic (coadsorption) and/or hydrophobic in nature. At substantial concentrations, both surfactants significantly reduced soil moisture diffusivity values.

A field test of the surfactant-washing method was applied to a site contaminated with PCBs and oils (Abdul, Gibson, Ang, Smith, and Sobczynske, 1992). Over 70 days, 5375 gal of an 0.75% aqueous surfactant solution was applied on the test plot at an average rate of 77 gal/day, and 10,981 gal of leachate were recovered at an average of 157 gal/day. About 10% of the initial mass of the contaminants was washed from the plot during the test.

The question then arises of how to deal with the remaining surfactant after the treatment is complete. Several alternatives for removing surfactant and contaminants from leachate have been considered (Ellis, Payne, and McNabb, 1985): flocculation/coagulation/sedimentation, foam fractionation, sorbent adsorption, ultrafiltration, surfactant hydrolysis/phase separation, centrifugation, and solvent extraction. The surfactant and contaminant together can be separated from the leachate, but it has been difficult to separate the surfactant from the contaminant. Reuse of the surfactant is essential for cost-effective application of this technology in the field.

Experiments with several of these leachate treatment alternatives showed that dilute alkaline hydrolysis of a surfactant in the leachate, followed by neutralization, separated the surfactant components from the leachate solution (Ellis, Payne, and McNabb, 1985). Either activated carbon or foam fractionation could be used to purify the leachate further following hydrolysis, producing a very clean effluent. Hydrolysis was the only treatment method that effectively removed contaminants from the raw leachate; however, in the process, one of the surfactants was destroyed and it was impractical to recycle the other.

The cost of using synthetic surfactants for multiple washing of large volumes of contaminated soil is very high (Nash, Traver, and Downey, 1987). The estimated cost for removing 25,000 gal of oil and fuel would be about $540,000. The high cost is due to the present inability to separate organic contaminants from the surfactant solution and reuse the surfactant economically. A treatment system must be devised to separate organic contaminants from surfactant rinse waters before *in situ* soils washing is advanced to full-scale development. In addition, a biodegradable surfactant that will not reduce soil permeability is needed (Neely, Walsh, Gillespie, and Schauf, 1981). It must not persist in the environment and continue to mobilize contaminants after treatment is stopped.

Ang and Abdul (1994) showed, in an *in situ* washing of contaminated soil, that a Romicon Model HF-Lab-5 ultrafiltration unit with an XM50 membrane can recover 46% of the surfactant, while the membrane retains 89% of contaminant oils. A PM500 membrane recovers 67% of the surfactant and retains >83% of the oils. The significant recovery of the nonionic surfactant makes its application for soil washing more economical.

It would be economically feasible to recycle the elutriate from soil flushing back through the contaminated soil for treatment by biodegradation (Sims and Bass, 1984; U.S. EPA, 1985a). This approach may eliminate the need for separate processes for treatment and disposal of the collected waste solution and be less expensive than using unit operations alone for treatment of elutriate. For soils contaminated with inorganic and organic constituents, a combination of pretreatment land applications, where the metal constituents are reduced or eliminated in the elutriate by precipitation, followed by land application of the elutriate, may be a feasible, cost-effective approach.

BioGenesis Enterprises, Inc., has developed a soil-washing technology using a proprietary surfactant solution to first transfer contaminants from the soil matrix to a liquid phase and then to enhance biodegradation (Gatchett and Banerjee, 1995). Total recoverable petroleum hydrocarbon concentrations decreased by 65 to 73% in washed soils and by 85 to 88% with both soil washing and biodegradation together, after 120 days.

An advanced treatment system has been developed for fine-particle, hydrocarbon-contaminated sludges from the soil-washing process using ethoxylate (EO) surfactants and terminal-blocked EO surfactants (Schmid and Hahn, 1995).

Hydroxypropyl-β-cyclodextrin (HPCD) is a microbially produced compound that can be used to reduce the sorption and to enhance the transport of several low-polarity organic compounds (Brusseau, Wang, and Hu, 1994). Cyclodextrin did not interact with the porous media in the study; thus, there would be no retardation of HPCD during transport. Retardation of anthracene, pyrene, and trichlorobiphenyl was significantly reduced with its use.

Natural surfactant solutions from fruit pericarps of *Sapindus mukorossi* are very effective for solubilizing and flushing hydrophobic organic compounds from soil columns and would probably perform in a similar manner in the subsurface (Kommalapati and Roy, 1996). They are comparable to commercial surfactants. Natural surfactants can serve as both carbon and energy sources and degrade considerably. Growth of microbes in natural surfactants is not inhibited by surfactant–bacterial cell interactions, and growth increases with an increase in surfactant concentration, especially with the inclusion of nutrients.

The soil adsorption constant (K) is a measure of the tendency of a pollutant to be adsorbed and stay on soil (Nash, 1987). Contaminants can be grouped according to a K value and their removal efficiencies (RE) evaluated. Tables 4.3 and 4.4 list the soil adsorption constants and water quality criteria for a number of hydrophobic and hydrophilic compounds, respectively. The greater the value for K, the stronger the binding to the soil. Organic material, generally, has high surface areas and exchange properties ideal for adsorption of organic compounds (Devitt, Evans, Jury, Starks, Eklund, and Gholson, 1987). The K_{oc} value reflects the impact of this organic material to adsorb organic compounds out of solution. Compounds with low water solubility often have higher K_{oc} values (Wilson, Enfield, Dunlap, Cosby, Foster, and Baskin, 1981). The logarithm of the octanol/water partition coefficients (log P) is a measure of the tendency of a compound to dissolve in hydrocarbons, fats, or the organic component of soil rather than in water (Nash, 1987). Log P can be used to estimate the tendency of an organic compound to become (or remain) adsorbed in soil (Nash, 1987).

A simple approach to evaluating the potential use of surfactants in inorganic waste recovery involves consideration of the aqueous solubility or octanol/water partition coefficient K_{ow} (Amdurer, Fellman, and Abdelhamid, 1985). Surfactants would be most effective in promoting the mobilization of organic compounds of relatively low water solubility and high K_{ow} values.

Reasonable estimation of the sorption behavior of organic pollutants can be made from a knowledge of organic carbon content of the biomass and octanol/water partition coefficient of the organic pollutants (Selvakumar and Hsieh, 1988). The pKa, or dissociation constant, of a compound indicates the degree of acidity or basicity that a compound will exhibit and, therefore, should be very important in determining both the extent of adsorption and the ease of desorption (Kaufman and Plimmer, 1972).

Augustijn, Jessup, Rao, and Wood (1994) conducted solvent flushing experiments, where the soil was equilibrated with a solution of naphthalene and anthracene, and found that compounds with different retardation factors are separated at low cosolvent contents, while coelution of the compounds occurs at higher contents. In general, the smaller the retardation factor in water and the higher the cosolvent fraction, the faster the contaminant is recovered. The presence of nonequilibrium conditions, soil heterogeneity, and type of cosolvent will influence the time required to recover the contaminant.

Many of the biological wastewater treatment systems recharge the effluent from biological treatment to an aquifer to create a closed loop of recovery, treatment, and recharge, which flushes the contaminants out of the soil rapidly and establishes hydrodynamic control to separate the contaminated zone from the rest of the aquifer (Lee and Ward, 1985, 1984). Acclimated bacteria can also be added to the aquifer and can act *in situ* to degrade the contaminant. The recharge water can be adjusted to provide optimal conditions for the growth of the acclimated bacteria and of the indigenous populations, which may also act on the contaminants.

The achievable level of treatment with soil washing is variable, depending upon the contact of flushing solution with waste constituents, the appropriateness of solutions for the wastes, and the hydraulic conductivity of the soil (Sims and Bass, 1984). The technology is more applicable to highly permeable soils, such as gravel or sand. Local soil conditions affect the process (Lyman, Noonan, and Reidy, 1990). High silt and clay content hinders movement of the flushing solution. Soils with greater than 1% organic matter and high clay content would have stronger sorption characteristics and, therefore, be less appropriate for flushing. The leaching fluid may not reach all contaminants in a well-compacted soil, and it may follow flow paths in the subsurface. Edwards, Liu, and Luthy (1994b) determined that treating soil with successive surfactant washings results in greater removal of hydrocarbons than a single washing with the same amount of surfactant.

Numerous laboratory studies have shown 80% removal of crude oils and PCBs from soil columns washed with surfactant solutions (Ellis and Payne, 1985). Three synthetic surfactants were tested in the laboratory and in pilot-scale studies (Nash, Traver, and Downey, 1987):

1. A mixture of ethoxylated fatty acids sold by Witco Chemical Corporation (used in agriculture as a soil penetrant)
2. An ethoxylated alkyl phenol (Diamond Shamrock)
3. An anionic sulfonated alkyl ester (Diamond Shamrock)

In spite of the ability of a 50/50 mixture (at 1.5%) of number 1 and 2 surfactants to clean contaminated soils in laboratory tests with 12 pore volumes, there was little evidence that they could clean the contaminated soil (fine sand with a high infiltration rate) *in situ* by using 14 pore volumes. The laboratory results also showed a decreasing ability to remove hydrocarbons from a column with increasing depth. The initial field test of *in situ* soil washing has raised serious doubts over its full-scale feasibility. There was a noticeable reduction in field soil permeability not predicted in the laboratory, which underscores the importance of small pilot-scale testing on all contaminated sites before designing full-scale decontamination technologies. This clogging effect may have been caused by surfactant micelle formation in the pore spaces or a surfactant-enhanced movement of fine particles down the soil structure until they filled in the pore spaces deeper in the soil. It should be noted, however, that heavy rainfall and uncontrolled runoff onto the test site occurred during the treatment.

Another possible explanation for the failure of this test is that the contamination may not have been evenly dispersed throughout the soil (Nash, Traver, and Downey, 1987). There is a difference between contaminated vadose zone soil and uncontaminated vadose zone soil. Because of the greater surface area available in fine soil, an equal mass of fine soil particles will hold more contaminants than coarse particles. The surfactant solution may have followed the paths of least resistance when applied to the top of the undisturbed soil. These paths have less fines and less contamination. If this is the reason for the failure of the field test, this technique would not be effective as a treatment measure for soils of variable composition.

A method of incorporating surfactant in soil washing followed by surfactant-aided bioremediation is being developed in France (Ducreux, Baviere, Seabra, Razakarisoa, Shaefer, and Arnaud, 1995). Over 80% of diesel oil was recovered by flushing the soil with a RESOL 30 solution containing nonionic and anionic biodegradable surfactants at a concentration of 10 g/L and 1.5 g/L sodium chloride. This was followed by pumping air through the porous soil and adding an aqueous solution containing nutrients (nitrates and phosphates) and an 0.05 wt% solution of RESOL 30.

Addition of any flushing solution to the system requires careful management and knowledge of reactions that may adversely affect the soil system (Sims and Bass, 1984); e.g., addition of sodium hydroxide to soil systems may adversely affect soil permeability by affecting the soil sodium absorption ratio. Surfactant/polymer flooding typically leaves behind residual contaminants and flushing agent in the soil, which could then be treated by *in situ* bioremediation (biopolishing) (Brown, Mahaffey, and Norris, 1993).

Surfactants also have other applications in remediation. They may be applied to removing heavy metals from soil. A surfactant system containing Dowfax 8390 and diphenyl carbazide was found to be quite effective in remediation of chromium-contaminated soil, by enhancing the elution of chromate (Nivas, Sabatini, Shiau, and Harwell, 1996).

Soil washing with surfactant is useful for simultaneous *in situ* stabilization of inorganic lead and remediation of organic lead, as the surfactant induces inorganic lead in soil to change into a less-soluble form (Shin and Huang, 1994). The efficiency increases with increasing temperature. Sequential chemical extraction indicates the order of abundance for the chemical forms of lead to be organic bound > carbonate > sulfide > exchangeable > residual > water-soluble > adsorbed. Electrokinetic soil flushing can be used to remove lead from saturated soil (Thompson, Hatfield, and Reed, 1994). Soil washing and flushing have been increasingly selected for U.S. EPA Superfund sites but appear to be primarily directed toward treatment of inorganics (Rubin and Mon, 1994).

Section 4.1.4 describes the process of sorption and relates it to availability of contaminants for biodegradation. Section 5.3.1 covers application of surfactants to enhance bioremediation.

Desirable Features — This technique can provide for recovery of chemicals in cases involving spills of individual chemicals (U.S. EPA, 1985a). Both inorganics and organics are suitable for soil-flushing treatment, if they are sufficiently soluble in an inexpensive solvent that is obtainable in a large enough volume (Sims and Bass, 1984). An advantage of this method is that it removes contaminants more quickly without disturbing the soil (Johns and Nyer, 1996).

The pump-and-treat method can be appropriate when the contamination has reached a high water table and the pollutants are at least somewhat water soluble (Gruiz and Kriston, 1995). If the groundwater is contaminated, this method can be combined with bioventing.

Limitations — Soil flushing is not feasible when complex wastes containing a range of contaminants with different solubility characteristics are involved (U.S. EPA, 1985a). Oil is poorly extracted by water, and intensive washing may deteriorate the soil (Gruiz and Kriston, 1995). Aeration of the soil is limited by water diffusion, as is the water permeability of most types of soils. It is difficult to limit the reaction to target contaminants (i.e., prevent loss of treatment agents through side reactions or sorption/retention by soil) (U.S. EPA, 1985a).

Proper location and design of wells require good knowledge of subsurface characteristics (U.S. EPA, 1985a). Channeling and uneven treatment may result, due to nonhomogeneity of the subsurface. The soil system after treatment is altered from its original state (Sims and Bass, 1984). Its physical, chemical, and biological properties may be altered adversely; for example, the pH may be lowered by the use of an acidic solvent, or the soil may be compacted from being flooded. These soil properties may have to be restored to assure that other treatment processes can occur (e.g., biodegradation). Also, the treatment rate can be very slow (U.S. EPA, 1985a). It is difficult to monitor or control treatment progress and completeness. Limited field experience and operating data are available for this method.

One problem of soil flushing is the large volumes of contaminated elutriate that require treatment (Lyman, Noonan, and Reidy, 1990). The recovered contaminant solution can be very dilute in a large volume and, hence, difficult to treat and dispose of economically (U.S. EPA, 1985a). In addition, the solutions used for the flushing may themselves be potential pollutants (Sims and Bass, 1984). The groundwater flow pattern should be well defined to ensure complete recovery of the elutriate, otherwise physical barriers, such as slurry walls, may be needed (Lyman, Noonan, and Reidy, 1990). There must also be a source of water for the flushing. If this is groundwater, which is often extracted, treated, and recycled as the flushing agent, its geochemistry must first be assessed for problem constituents. If a surfactant is used, the technique will be too expensive until an economical method is developed to separate the surfactant from the water.

Soil flushing is less effective with contaminants that are relatively insoluble or tightly bound to the soil (Lyman, Noonan, and Reidy, 1990). High clay content can cause adsorption of a surfactant. Phenanthrene in contaminated soil is resistant to continuous desorption (Chen and Maier, 1993). A significant fraction of soil phenanthrene requires extended flushing to increase desorption. The slow desorption process will effectively restrict its transport. These authors found that biodegradation of phenanthrene in the soil was rapid and extensive and enhanced the release of the compound from the soil, with the rate and extent of mineralization being similar to the amount of compound desorbed. Although flushing pollutants from a contaminated site is a common remediation practice, bioremediation may be able to give the same extent of treatment without toxic end products.

2.2.1.8 CROW Process

The Contained Recovery of Oily Wastes (CROW) process was developed to recover dense, non-aqueous-phase liquids (DNAPLs) from contaminated industrial sites (Johnson and Leuschner, 1992). It uses hot water displacement to reduce concentration of oily wastes in subsurface soils and underlying bedrock. After hydraulically isolating the contamination, the downward flow of the DNAPLs is reversed by controlled heating of the subsurface. The oily material floating in water is displaced to production wells by sweeping the subsurface with hot water. Temperature and concentration gradients are maintained to control waste flotation and vapor emissions. The oily wastes are immobilized by reducing waste concentrations to residual saturation.

The CROW process is only effective for removing free product from soils (Johnson and Leuschner, 1992). It can then be followed by enhanced *in situ* bioremediation, which is most effective for soils not heavily contaminated with free product, for complete site restoration. The two processes share their requirements for equipment installation and site characteristics, such as soil type, permeability, etc., and would thus be compatible.

In laboratory and small-scale pilot testing of soils contaminated with coal tars or creosote, polychlorinated phenols (PCP), and petroleum products, CROW achieved over 60 wt% removal of coal tars at a temperature of 156°F (69°C) and over 80 wt% removal of creosote-wood treatment waste at a temperature of 120°F (49°C) from contaminated soils (Johnson and Leuschner, 1992). PCP concentrations were reduced by over 99%. Removal efficiency of coal tars was increased to over 80 wt% with surfactant addition at about 1% by volume at 156°F (69°C). The CROW process alone reduced PAHs to a level similar to the end points achieved by slurry reactor treatment of the site soil. CROW process waters were treatable by both aerobes and anaerobes. Both aerobic and anaerobic (denitrification) treatment were successful in degrading CROW process residuals to levels below 5 mg/kg total PAHs.

2.2.1.9 Injection-Extraction Process

This technology was developed by Ecosite, Inc., and applies the basic principles of soil washing with improved distribution of the washing solution and improved hydraulic control using air sparging and vacuum capability to remediate soil to a depth of 10 m (Ross, Tremblay, and Boulanger, 1995). Free-phase hydrocarbons are recovered; then different treatment solutions were injected and recovered through cyclic manipulation of the water table level.

The contaminated zone is isolated, and washing solutions composed of nontoxic, biodegradable surfactants are injected from the main injection network and pumped out through specially designed wells (Ross, Tremblay, and Boulanger, 1995). The recovered washing solutions form emulsions, which are processed in an on-site wastewater treatment system. The volatile compounds pass through a biofilter.

The process achieves high performance under infrastructure without the need for excavation. It can also be extended to biodegradation and can include bioventing, as well as sparging, in a sequential treatment program.

2.2.1.10 Air Stripping

Principle/Description — Under some circumstances, soil containing volatile hydrocarbons or solvents can be decontaminated by air stripping (Nash, 1987). In this process, air is injected into the soil. It is forced into injection wells and pulled out of extraction wells. As it flows through the soil, volatile materials are stripped off into the airstream. The contaminated air is vented to an emission control system or the atmosphere, depending upon contaminant levels. The organics can be removed from the air in a vapor-phase carbon adsorption system or by fume incineration. The success of *in situ* air stripping depends upon relatively unrestricted and uniform flow of air through the soil. Clay soils, packed soils, or soils with a high water table are not good candidates for *in situ* air stripping. The interphase contaminant transport from the sorbed to the vapor phase plays a dominant role in influencing the effectiveness of this process (Mehrotra, Karan, and Chakma, 1996).

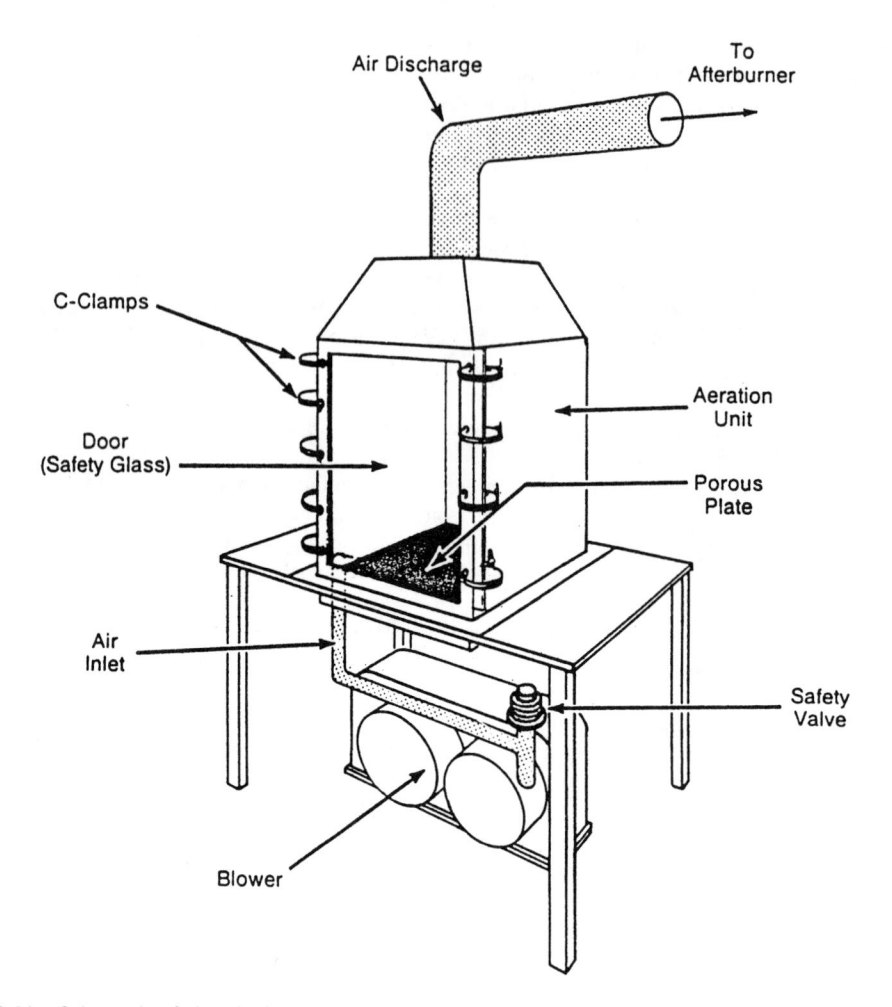

Figure 2.28 Schematic of air stripping process equipment. (From McDevitt, N.P., Noland, J.W., and Marks, P.J. Report No. AMXTH-TE-CR-86092. ADA 178261. U.S. Army Toxic and Hazardous Materials Agency, Aberdeen, MD, 1987.)

An illustration of an aeration unit that is used industrially to supply a constant, low-pressure flow of diffused air is presented in Figure 2.28 (McDevitt, Noland, and Marks, 1987). This unit allows intimate contact between the airstream and contaminated soil, thereby aerating the soil and stripping it of VOCs.

A larger soil grain size results in greater recovery of benzene from the soil (Shah, Hadim, and Korfiatis, 1995). An increase in airflow rate from 5 to 10 L/min improves recovery efficiency from 56 to 70%. Preheating the air to 45°C at the inlet increases the recovery efficiency from 70 to 90%.

Desirable Features — It is a simple, possibly low-cost, and safe method (with proper controls). It has been tested on a pilot scale and can treat significant soil depths in the unsaturated zone.

Limitations — Its applicability is limited to cases involving volatile compounds, low groundwater table, and loose, sandy formation. Proper location and design of wells require good knowledge of subsurface characteristics. Channeling and uneven treatment may result, due to nonhomogeneity of the subsurface. It is not recommended for low hydraulic conductivity soils (requires hydraulic conductivities perhaps exceeding 10^{-3} to 10^{-2} cm/s). It is difficult to monitor or control treatment progress and completeness. Limited field experience and operating data are available for this technique. Off-gas treatment may be required from air-stripping units to limit hydrocarbon discharge to the atmosphere (Ram, Bass, Falotico, and Leahy, 1993). Air stripping is further discussed in Section 6.3.3.1.4.1 as a VOC pretreatment process for organic liquids.

2.2.1.11 Soil Vapor Extraction (SVE)

SVE is also referred to as soil vacuum extraction, soil venting, and soil vapor stripping. SVE employs a blower or vacuum pump connected to small-diameter vertical wells or lateral trenches to reduce the vapor pressure in the soil and increase volatilization of contaminants, which are then withdrawn by the same applied vacuum (Dupont, Doucette, and Hinchee, 1991; Burke and Rhodes, 1995). The performance of SVE systems is based on the rate of mass removal, the time required to achieve cleanup goals, and the cost of cleanup (Ghuman, 1995). These performance parameters depend on physical and chemical factors, such as the rate and pattern of airflow through the affected soils, contaminant type, and the degree of partitioning among the vapor, liquid, dissolved, and adsorbed phases. SVE systems are inexpensive and easy to implement (Dupont, Doucette, and Hinchee, 1991). They have been used successfully to remediate contamination, such as gasoline, in relatively permeable soils.

The U.S. EPA has prepared a guide to support decision making by regional and state corrective action permit writers, remedial project managers, on-scene coordinators, contractors, and others responsible for the evaluation of technologies (U.S. EPA, 1994a). This guide directs managers of sites being cleaned up under Resource Conservation and Recovery Act (RCRA), Underground Storage Tanks (UST), and Comprehensive Environmental Response Compensation and Liability Act (CERCLA) waste programs to SVE resource documents, databases, hotlines, and dockets and identifies regulatory mechanisms that have the potential to ease the implementation of SVE at hazardous waste sites. The guide provides abstracts of representative examples of over 70 SVE guidance/policy and reference documents, overview/program documents, and studies and demonstrations. The "Soil Vapor Extraction Treatment Technology Resource Matrix," which accompanies this guide, identifies the contaminants, soil type, and activities used to support the application of SVE covered in each abstracted document.

There are an optimum flow rate and system size for each design strategy at a particular site (DePaoli, Wilson, and Thomas, 1996). A conceptual design of soil-venting systems provides a means of estimating the cost and schedule of site cleanup for the purposes of technology selection and system design. Estimates of transient off-gas concentration and the vacuum required at the extraction vents for a given set of site and system design conditions are obtained from idealized treatments of contaminant volatilization and flow of gas in the soil. Capital and operating costs of blowers and emission control devices can be estimated, allowing comparison of costs for various design strategies.

Bench-scale studies can be an important stage in the remediation process. Such studies can ascertain whether SVE can successfully remediate soils polluted with VOCs. They provide qualitative results and can be used for model development and characterization.

Johnson, Stanley, Byers, Benson, and Acton (1991) and Hutzler, Murphy, and Gierke (1988) provide additional information on this process. A general approach to the design, operation, and monitoring of *in situ* SVE systems has been discussed by Johnson, Stanley, Kemblowski, Byers, and Colthart (1990). Application of SVE for remediation of vadose zone contamination is described by Brown and Bass (1991). Soil venting is relatively inexpensive and can be implemented in 2 to 4 weeks (Lyman, Noonan, and Reidy, 1990). It typically requires 6 to 12 months for a venting program. Vapor extraction systems are further discussed and their application for bioventing given in Sections 2.2.2.2 and 6.3.2.2.

Principle/Description — Soil venting may be passive (with no energy input) or active (Lyman, Noonan, and Reidy, 1990). Passive venting consists of perforated pipes sunk into the contaminated area, providing an outlet for gases in the subsurface. A slight draft might be provided by a wind-driven turbine at the outlet. Active venting is more effective than passive and uses an induced pressure gradient to move vapors through the soil. With active venting, vacuum is applied to the subsurface to volatilize and remove contaminants. Vacuum may be applied through vertical extraction wells (low water tables) or horizontal extraction systems (high water tables) using a vacuum blower. The resulting pressure gradient forces the soil gas to migrate through soil pores toward the vapor extraction wells. Positive-pressure injection wells increase the removal rate. Plastic sheeting on the ground may avoid short circuiting the airflow. VOCs are volatilized and transported out of the subsurface by the migrating soil gas. The removed vapors may require treatment before discharge to the atmosphere. Figure 2.29 shows a typical process layout for a soil venting operation (Long, 1992). GAC or catalytic combustion is commonly used. As an alternative, the contaminant vapors may be passed through an acclimated biofilter bed, where they are biodegraded.

Contaminants exist in the vapor, liquid, and/or dissolved phase in the unsaturated zone (Lyman, Noonan, and Reidy, 1990). Contaminants in the vapor phase or easily volatilized may be removed by

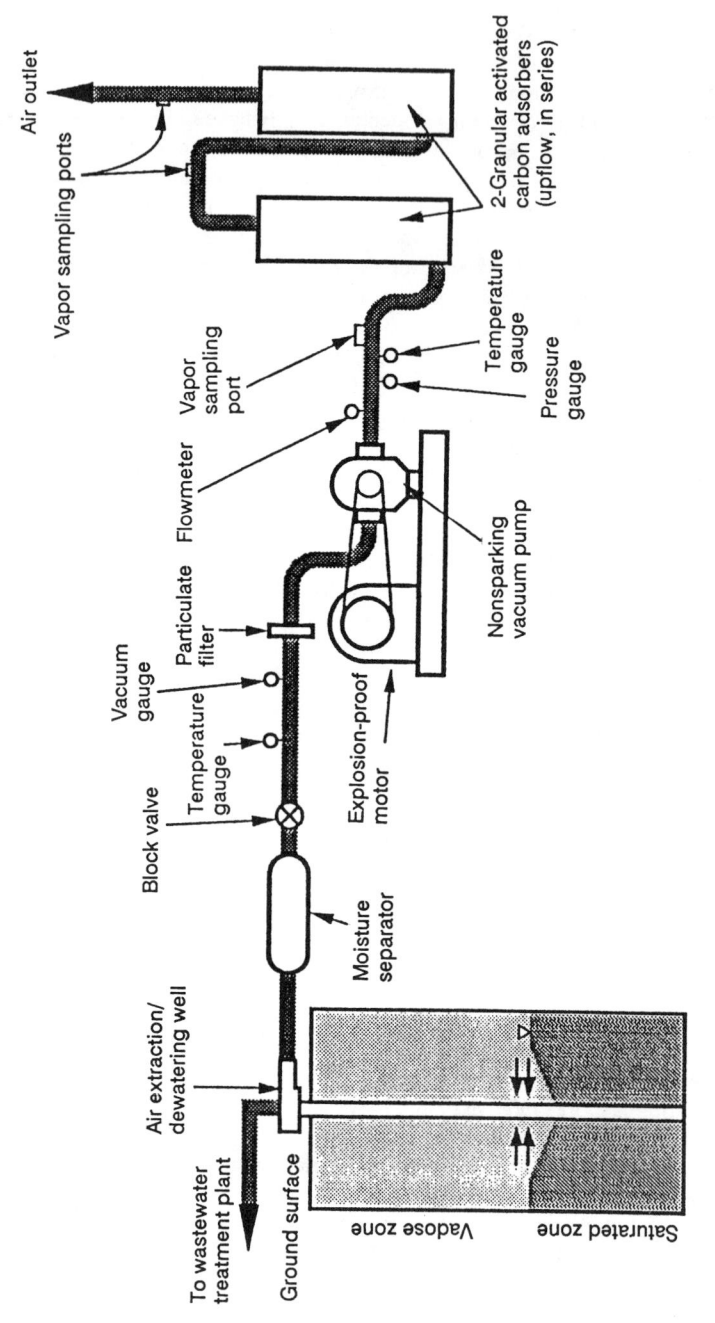

Figure 2.29 A typical process layout for vapor extraction. (From Long, G. J. *Air Waste Manage. Assoc.* 42(3):345–348, 1992. With permission.)

this method. Removing contaminant vapors and furnishing clean air causes the equilibrium to shift and draw out more contaminants from the liquid phase or from the portion dissolved in water. The rate and extent of removal of VOCs from the soil by active or passive vapor extraction systems are affected by factors, such as soil permeability, soil moisture content, applied suction, airflow rate, temperature, vapor pressure, and external boundary conditions (Sepehr and Samani, 1993).

VOCs with high vapor pressures are more likely to be removed by vacuum extraction (Lyman, Noonan, and Reidy, 1990). Course soils, such as sand and gravel, have low soil sorption coefficients and are more likely to be amenable to vacuum extraction than fine-grained materials. Venting is affected by the water solubility of the contaminants and the soil properties, such as air conductivity, temperature, and moisture. Higher temperatures help promote volatilization. In fact, heat may be injected into the vadose zone around the vapor extraction wells to improve soil venting by increasing the vapor pressure of the contaminants (Ram, Bass, Falotico, and Leahy, 1993). The soil temperature can be raised by electrical heating or hot air volatilization (Ghuman, 1995).

VOCs such as BTEX can be removed from unsaturated (vadose zone) soils by use of SVE (Ram, Bass, Falotico, and Leahy, 1993). High concentrations of gasoline vapors become mobile when air is injected into the soil (Downey, Frishmuth, Archabal, Pluhar, Blystone, and Miller, 1995). Contaminants in the vadose zone that exhibit a vapor pressure greater than 1 mm Hg are generally amenable to SVE (Ram, Bass, Falotico, and Leahy, 1993). Bolick and Wilson (1994) advocate the sooner that pumping can begin, even before negotiations are completed, the less chance there is of contaminating groundwater and the more rapid the remediation.

When soil moisture is increased during bioventing of fuel-contaminated arid soils, microbial activity can be significantly enhanced (Zwick, Leeson, Hinchee, Hoeppel, and Bowling, 1995). This can be accomplished with a subsurface drip irrigation system. However, Stenseng and Nixon (1996) have shown that it will take longer to clean soils with high moisture content than similar soils with lower moisture content. Dry air might be injected into the soil to reduce soil moisture content and produce a tensiometric, or dry, barrier to contain liquid-phase transport (see Section 2.2.1.16 for a description of tensiometric barriers) (Thomson, Morris, Stormont, and Ankeny, 1996). If large amounts of water are withdrawn from the soil with the gaseous steam and contaminants during SVE, the pneumatic soil permeability could change, and the temperature could drop up to 10°C (Garcia-Herruzo, Gomez-Lahoz, Rodriguez-Jimenez, Wilson, Garcia-Delgado, and Rodriguez-Maroto, 1994).

In a remediation of gasoline-contaminated soils, total effluent hydrocarbon concentration for an SVE system ranged up to 1121 ppmv, corresponding to a hydrocarbon removal rate of about 1.68 lb/h (Felten, Leahy, Bealer, and Kline, 1992). An estimated total pounds of VOCs removed by SVE over 4 months was 1400 lb. In subsequent combination with air sparging, an additional 600 lb of VOCs were removed over 140 days. This study showed that SVE alone can provide excellent removal of VOCs adsorbed to unsaturated soils, but that fluctuations in groundwater elevations influence the rate of hydrocarbon removal. Air sparging contributed to removal of VOCs from saturated zone soils, as well as from groundwater and removed hydrocarbons from regions of the subsurface that had not previously been affected by the SVE system. The increase in dissolved oxygen (DO) with air sparging increased the population of indigenous bacteria without the addition of supplemental nutrients.

The design of vapor-extraction remedial systems and the analysis of their performance can be improved by using models that can simulate the chemical and physical processes affecting the occurrence and movement of multiple-compound vapor-phase chemical mixtures (Benson, Huntley, and Johnson, 1993). Some models are multiple-compound phase distribution models, which are either nondimensional (no transport) or one dimensional (column experiments). Other models are multidimensional, single-compound transport models. These authors have constructed a model that couples the steady-state vapor flow equation, the advection-diffusion transport equation, and a multiple-compound, multiphase chemical partitioning model.

A three-dimensional gas flow model, GAS 3D, has been developed to aid in the design of soil-venting systems. It takes into account the effect of partial penetration and partial screening of vapor extraction wells, as well as the nonhomogeneity and anisotropy of the soil. In a sandy soil with high clay and organic matter, a model was used to optimize biodegradation (De Wit, Urlings, and Alphenaar, 1995). It calculated that SVE with a low extraction rate (e.g., by intermittent extraction) would be more favorable than extraction at a high flow rate. SVE of VOCs from soils rich in natural humic organic carbons is described by a model (Gomez-Lahoz, Rodriguez-Maroto, and Wilson, 1995). Partitioning of VOCs between the aqueous phase and the natural organic carbon can be handled by two different isotherms, both of which suggest substantially different amounts of time required for the remediation. A method for establishing the cost and schedule of site cleanup using soil-venting systems has been described by Kang and Oulman (1996), using a mathematical model to predict the evaporation rate of VOCs from petroleum-contaminated sand.

A multicomponent, nonisothermal, three-dimensional software model called Thermally Enhanced Soil Vapor Extraction (TESVE) is a powerful tool for evaluating the feasibility of SVE, optimizing system design, predicting system performance, and reducing cleanup costs (Ghuman, 1995). The TESVE model was run for a site at McClellan Air Force Base, CA, and would probably save about $500,000 and 4 years of system operation over the conventional SVE design.

Since all the decision variables considered for maximizing the net marginal mass removal rate per incremental remedial cost are generally related to the air extraction rate, this variable can be used as a surrogate for remedial cost (Farr, McMillan, and Shibberu, 1995). Knowledge of the subsurface response in terms of achievable mass removal rates as a function of operational air extraction rates is critical to optimizing the process. Field testing and transport modeling can be used to characterize the subsurface response and optimize SVE system design and operation.

Parker (1993) discusses application of internal combustion engines in SVE and provides performance data, flow diagrams, vapor extraction curves, air/fuel ratios, soil quality, carburetor vacuum, engine rpm, preventive maintenance, and cost per operating hour. These engines do not require supplemental fuel until the inlet TPH concentration drops to about 8900 ppm.

A successful pilot study of the use of a SVE system was conducted at a gasoline-contaminated site composed of a fine-grained soil with zones of low permeability (Widdowson, Aelion, Ray, and Reeves, 1995). It demonstrated that SVE could rapidly decrease hydrocarbon concentrations to asymptotic values, with elevated levels of CO_2, which indicated increased microbial activity. It appears that intermittent pumping, with relatively short SVE and long standstill phases, are sufficient to contain pollution and minimize risks with this procedure (Spuij, Lubbers, Okx, Schoen, and de Wit, 1995).

A dual vacuum extraction system was successful in remediating hydrocarbon-impacted clays and contaminated groundwater at a former service station in northern California (Dockstader, 1994). Air conductivities ranged from 1.67×10^{-7} to 9.82×10^{-8} cm/s horizontally and 1.02×10^{-7} to 2.38×10^{-8} cm/s vertically. After 76 days of operation, the soil TPH concentration had gone from 2500 ppm to one sample above 100 ppm and 50% of the other samples with no detectable levels. No free product remained on the water table, TPH levels had declined by an order of magnitude, and benzene concentrations had fallen from 5000 to 50 ppb.

The vapor extraction system for *in situ* remediation of soils contaminated with VOCs, weathered organic compounds, and residual organics can be enhanced by a change in the design of the system (Nazarian, 1996). This patented approach consists of air influx wells around the periphery of the contaminated zone, an air extraction well in the contaminated zone, and vacuum pump or suction blower pulling on the extraction well to create an air circulation that flows from the influx wells through the pore spaces of the contaminated soil to the extraction well. The improvement involves increasing the air velocity through the system to increase the shear action of the air as it strips contaminants from the soil surfaces and pore spaces. This is accomplished by tapering the air influx wells from a maximum flow cross-sectional area at ground level to a minimum flow area at the depth of the contaminated soil, where the maximum air velocity is obtained. The taper may be gradual (conical with tubular wells) or stepwise (series of straight tubes of diminishing cross sections). This increases the stripping rate of the contaminants and reduces operating time and costs, with only minor structural changes.

Desirable Features — Same as for air stripping. SVE has been used to remove chemical spills before they reach the groundwater.

Limitations — Same as those for air stripping. Permeability of the contaminated soil will greatly affect cleanup time with this process (Bolick and Wilson, 1994). Marquis (1993) used an *in situ* vapor extraction system with a combined thermal-catalytic oxidizer vapor treatment system to remediate petroleum-contaminated soil with low permeability.

It has been shown that a NAPL can be trapped at the base of a fine-grained lens in a heterogeneous vadose zone (Smith, Reible, Koo, and Cheah, 1996). This should reduce the effectiveness of *in situ* SVE

CURRENT TREATMENT TECHNOLOGIES

for removal of the NAPL. Mass transfer resistances and liquid-phase resistances will then affect the removal rate of the compounds in the NAPL from the low permeability lens.

Gomez-Lahoz, Rodriguez-Maroto, and Wilson (1994) report that it is difficult, if not impossible, to develop models that permit accurate prediction of SVE cleanup times from data taken in short-term, pilot-scale experiments. They also note that the lack of decrease of the effluent soil gas VOC concentration during the course of cleanup may be due to diffusion of VOC from NAPLs through aqueous boundary layers. Many full-scale, *in situ* bioremediation efforts tend to stagnate because of the slow, diffusion-controlled release of the contaminants, which may be caused by stagnant areas without a convective flow (De Wit, Urlings, and Alphenaar, 1995). To overcome this, nonequilibrium phenomena should be considered by using a model that is geared toward time, budget, and data constraints.

2.2.1.12 Air Sparging

Air sparging can be used in conjunction with a vacuum recovery system in a soil remediation strategy. Air sparging is a technology for injecting air below the water table (Brown, Mahaffey, and Norris, 1993). In the process of oxygenating the groundwater, volatile components can be transferred to the unsaturated zone, where they can then be removed by a vapor recovery system.

2.2.1.13 Detoxifier™

The *in situ* Detoxifier™ (by Toxic Treatments U.S.A., Inc., San Mateo, CA) is an innovative technology/equipment potentially capable of implementing a range of *in situ* treatment methods (e.g., air/steam stripping, neutralization, solidification/stabilization, oxidation) (Ghassemi, 1988). It is an adaptation of the drilling technology providing capabilities for *in situ* delivery of treatment agents in dry, liquid, slurry, or gaseous form to the soil, with thorough mixing and homogenization of a vertical column of soil. The treatment agents can be ambient or heated air or aqueous solutions containing surfactants and oxidizing agents. Delivery can be via gravity (e.g., flooding, ponding, or surface seepage) or via forced systems, such as injection pipes. The unused treatment agents and by-products can be recovered via gravity (e.g., open ditches and trenches and porous drains) or via forced or vacuum systems (e.g., well points, induced draft fans, and leachate collection underdrains).

The mobile system was applied in a field demonstration of a full-scale remediation of a site contaminated with hydrocarbons from leaking underground fuel storage tanks (Ghassemi, 1988). Steam and hot air were used to strip hydrocarbons from the soil; the off-gas was processed in a treatment train and recycled to the soil. By adjusting treatment conditions, the TPH concentration in the soil could be reduced from an initial level of 5000 ppm or higher to less than 100 ppm. Based upon field demonstration results, the equipment vendor is developing designs for a more powerful and compact Detoxifier with enhanced off-gas treatment capabilities. The air can also be returned to the ground via a compressor (Mori and Mori, 1992).

The heart of the Detoxifier™ technology is the "process tower" (Figure 2.30), which is essentially a drilling and treatment agent–dispensing system, capable of penetrating the soil/waste medium to depths of 25 ft (7.6 m) or more (Ghassemi, 1988). The process tower consists of an assembly of two cutter/mixer bits connected to separate, hollow Kelly bars. The bits overlap and rotate in opposite directions. The rotating action provides for simultaneous cutting and mixing of the soil/waste material. Treatment agents (in dry, liquid, vapor, or slurry form) can be conveyed through the hollow Kelly bars and ejected through feed jets and orifices to the mixing area. A rectangular shroud covers the mixing area to minimize dust generation and to capture gas and vapor released during the subsurface treatment. The captured off-gas is treated in a process train and recycled through the process tower to the treatment zone.

The treatment train is selected and designed based upon the type and level of pollutants to be removed from the off-gases (Ghassemi, 1988). The off-gas from the shroud is monitored continuously using an on-line TPH analyzer with a flame ionization detector and a strip chart recorder. The readout is used to determine the required dosages of treatment agents and the treatment time. The output is used to adjust the treatment conditions, including the length of treatment, to achieve desired treatment objectives. Off-gas monitoring provides a basis for assessing completeness of *in situ* treatment. An example of a treatment train employing a Detoxifier™ on a site contaminated with hydrocarbons is given in Section 8.

At the site where this was used, the off-gas from the shroud, which contained the exit air, steam, and volatilized hydrocarbons, was cooled and passed through three demisters of differing designs (Ghassemi, 1988). It then passed through a refrigeration coil to condense and remove excess moisture and, subsequently,

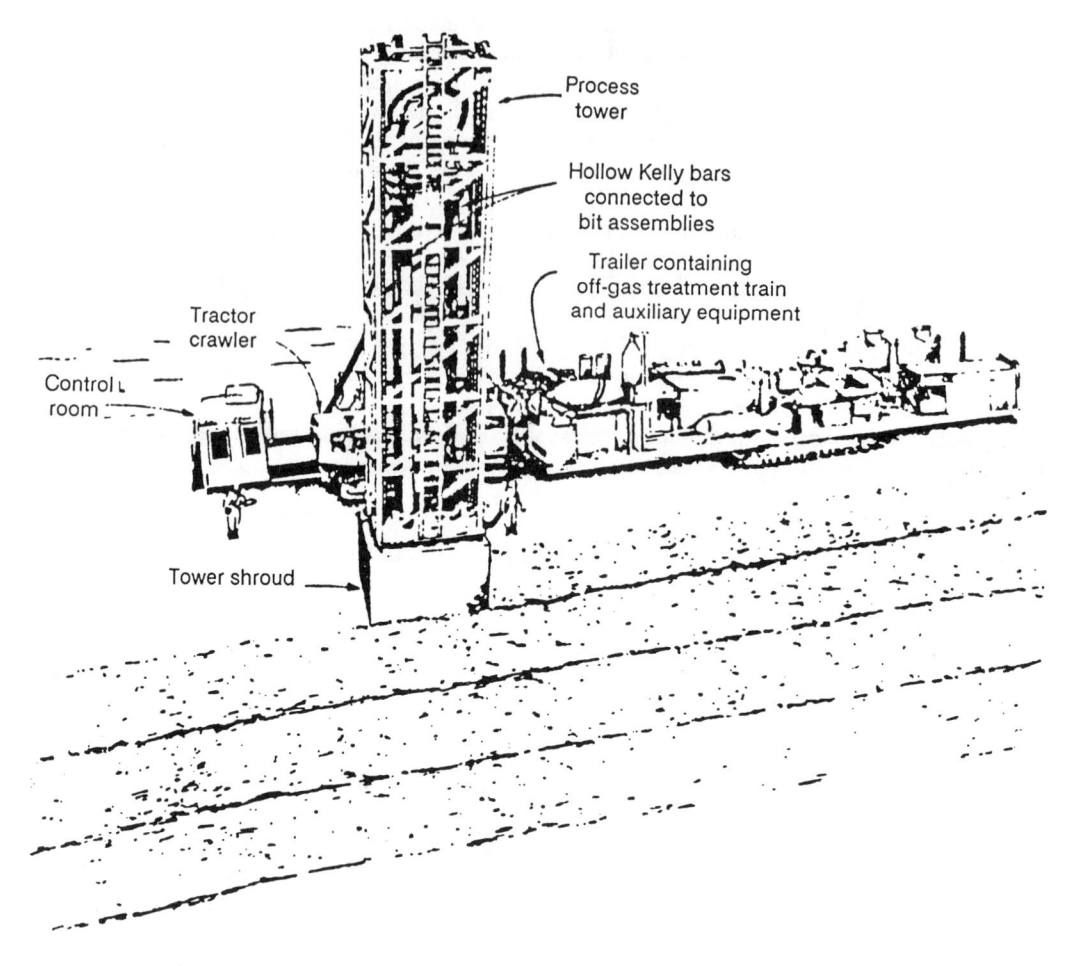

Figure 2.30 The *in situ* Detoxifier™. (From Ghassemi, M. *J. Haz. Mater.* 17:189–206, 1988. Elsevier Scientific Publishers, Academic Publishing Division. With permission.)

was heated, when necessary, before entering the activated carbon adsorption unit. With CO-601 (coal-based) and CC-601 (coconut shell) carbons, the preferred temperature range for the removal of hydrocarbons from gas streams is in the ambient (about 75 to 100°F) range (about 24 to 38°C). Since the temperature of the gas exiting the refrigeration coil is always above 75°F, the gas does not have to be heated before entering the adsorption unit.

After carbon adsorption, the gas is split into two streams (one for each drill bit), compressed, reheated, and recycled to the treatment zone through the Kelly bars in the process tower (Ghassemi, 1988). If powder (solidification agent) addition is used, the heated gas stream is diverted to the powder feed system before entering the Kelly bars. In the center of each Kelly is a separate line, which receives steam and treatment solutions at the top and delivers them to the soil through screw-on-type nozzles along the drill bit mixer assembly.

In actual site cleanup, the area to be treated is divided into rows of blocks, and the process tower treats one block at a time (Ghassemi, 1988). The process train, the control room, the components of the off-gas treatment train, and auxiliary support equipment are all tractor mounted and are, thus, mobile. Each bit assembly is capable of drilling a hole of 4.5 ft (1.4 m) in diameter. The drill is positioned with about 10% overlap of the grid cells, leaving a treatment block of about 3.25 × 7.3 ft (0.99 × 2.23 m).

The Detoxifier™ system (Ghassemi, 1988):

1. Delivers treatment agents directly to the treatment zone;
2. Causes thorough mixing and homogenization for effective contact between the treatment agents and the contaminant;

3. Contains a metal shroud that thoroughly encloses the area above the blades and captures the vaporized hydrocarbons (Mori and Mori, 1992);
4. Creates a closed-loop process train that condenses and removes the liquids and returns the air to the ground via a compressor (Mori and Mori, 1992);
5. Uses a range of treatment agents in liquid, gas, solid, and slurry forms, employing stripping of volatile organics (with hot air and/or steam), oxidation, reduction, precipitation, stabilization/solidification, and neutralization;
6. Is mobile.

A subsequent application of the Detoxifier™ for field remediation resulted in a verifiable and cost-effective posttreatment sampling plan where cleanup standards were met or exceeded in most cases (Mori and Mori, 1992). The process has advantages over passive vapor extraction systems in that it actively mixes the soil with the vapor and steam. The process can be operated at large sites with no above- or below-ground obstructions and requires a fairly level site. It can be applied *in situ* or *ex situ*.

A new design of the Detoxifier™ will be more compact (i.e., the components will be housed in one large trailer), more powerful, and will cover a greater treatment area per block.

Desirable Features — A variety of *in situ* treatment methods (e.g., air/steam stripping, neutralization, solidification/stabilization, oxidation) are potentially possible with this approach (Ghassemi, 1988). It allows thorough mixing and homogenization of soil and treatment agents and, thus, a uniform treatment. On-line monitoring permits good process control and adjustment in treatment conditions to achieve the desired level of treatment. It establishes a closed-loop operation, and the treatment system is completely mobile. The system has been field demonstrated at a site involving decontamination of soil contaminated with hydrocarbons.

Limitations — The treatment depth is possibly limited to 60 ft (current experience is limited to depths of less than 25 ft) (Ghassemi, 1988). A problem with this process is that the carbon adsorption system is quickly overloaded when soils containing very high levels of hydrocarbons are treated. However, the carbon adsorption system can be preceded by an additional treatment step, such as a cryogenic unit, to remove most of the hydrocarbons from the scrubbed off-gas before it enters the adsorbers.

2.2.1.14 Soil Heating

Heat can be applied to soil at a depth to desorb organic contaminants thermally or can be injected into the vadose zone around the vapor extraction wells to improve soil venting by increasing the vapor pressure of the contaminants (Ram, Bass, Falotico, and Leahy, 1993). Heat greatly accelerates the release and transport of contaminants through the soil (Edelstein, Iben, Mueller, Uzgiris, Philipp, and Roemer, 1994). Heat may be applied by the following processes (Ram, Bass, Falotico, and Leahy, 1993):

1. **Hot air injection.** Air is heated and injected under pressure. Hot exhaust gas from catalytic or thermal oxidation of off-gas can be used for this purpose.
2. **Steam injection.** Steam has a much greater heat capacity than air at the same temperature. Steam injected into the subsurface will, however, create some groundwater mounding as the steam condenses, possibly requiring groundwater pumping.
3. **Radio frequency (RF) heating.** Radio waves are directed into the contaminated area from an RF transmitting antenna placed in a well. The RF energy is converted into thermal energy. Costs are significantly higher with this method, but the heat can be directed more evenly and precisely.

As soil is heated, *in situ* water and NAPLs vaporize and are generally transported to the vapor extraction borehole (Phelan and Webb, 1994). Contaminant removal rates are proportional to the vapor pressure of the compound. Proper design of the vapor extraction system will contain the contaminant within the heated zone. Optimized treatment temperatures are based on steam distillation principles, pure component vapor pressures, or adjusted component vapor pressures due to mixtures of contaminants. Off-gas treatment and vapor sampling and analysis must be able to handle the large volume of water removed from the soil.

Resistance Heating — VOCs and semivolatile organic compounds (SVOCs) become more volatile and more easily removed by SVE after the soil is heated (Johns and Nyer, 1996). Resistance heating can also be used in vitrification (see Section 2.2.1.15).

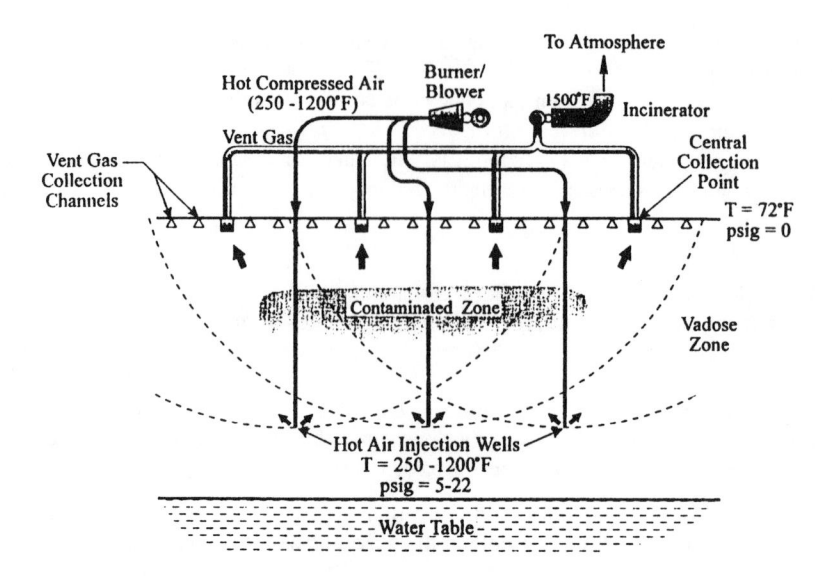

Figure 2.31 HRUBOUT® Process (U.S. EPA, 1994d; From Johns, F. J., II and Nyer, E. K., in *In Situ Treatment Technology,* Nyer, E. K. et al., Eds., Lewis Publishers, Boca Raton, FL. 1996.)

An electrical current is passed through electrodes in the ground, with the soil matrix as a conductor (Johns and Nyer, 1996). Soil water is the major conducting pathway. Heat is generated in response to resistance to the flow of current in the soil. Vapor extraction will remove the organic contaminants in the water vapor formed. If the voltage is increased, nonvolatile organic compounds (NVOCs) can be pyrolyzed or oxidized, which creates more volatile breakdown products. With low soil moisture, conductivity of the soil is decreased, which causes uneven heating. This process is applicable for soil types from sands to clays, with removal rates up to 99% for SVOCs and nonvolatile compounds. It has not yet been demonstrated at full scale.

2.2.1.14.1 Hot Air Injection/Flushing

Like steam flushing, hot air flushing increases the soil temperature to enhance SVE, but it does not require a water supply to operate and water does not condense (Johns and Nyer, 1996). There is enhanced promotion of biodegradation at the elevated soil temperatures. However, very high air temperatures are required.

The HRUBOUT® (Hrubetz Environmental Services, Inc., Dallas, TX) is a patented hot air injection process (Figure 2.31) (U.S. EPA, 1994d). Soil gas collection channels at the soil surface collect the hot air and organic contaminants. The system includes an air blower, an adiabatic burner for heating the injection air, and an incinerator for treating off-gases. It can remove gasoline, diesel, jet fuel, crude oil, lubricants, solvents, and creosotes from contaminated soil.

2.2.1.14.2 Steam Injection/Steam Flushing/Steam Stripping

Steam injection is a successful technology for remediating sites contaminated by heavy fuel oils (Dablow, Hicks, and Cacciatore, 1995). SVE is effective for removing VOCs from the vadose zone, but it is limited to contaminants with vapor pressures of >1.0 mmHg at ambient conditions. Heavy fuels oils (e.g., #2 diesel fuel to #6 fuel oil) have vapor pressures considerably lower. They are also highly viscous, with high residual saturation in the vadose zone, which makes them difficult to remove. These materials can be more easily removed by the use of steam injection to heat the subsurface and manipulate their volatility, viscosity, and residual saturation.

A "steam cleaning" process can effectively remove high concentrations of toxic petrochemicals from soil and groundwater at significantly lower cost than traditional methods of cleaning up sites contaminated with these materials (Baum, 1988). Injection of steam into the ground heats the soil, causing the pollutants to vaporize and flow toward a recovery well that captures and condenses the vapors. Since the steam becomes water as it cools, a steam front must flow uniformly from the point of injection to the point of extraction (Johns and Nyer, 1996).

CURRENT TREATMENT TECHNOLOGIES

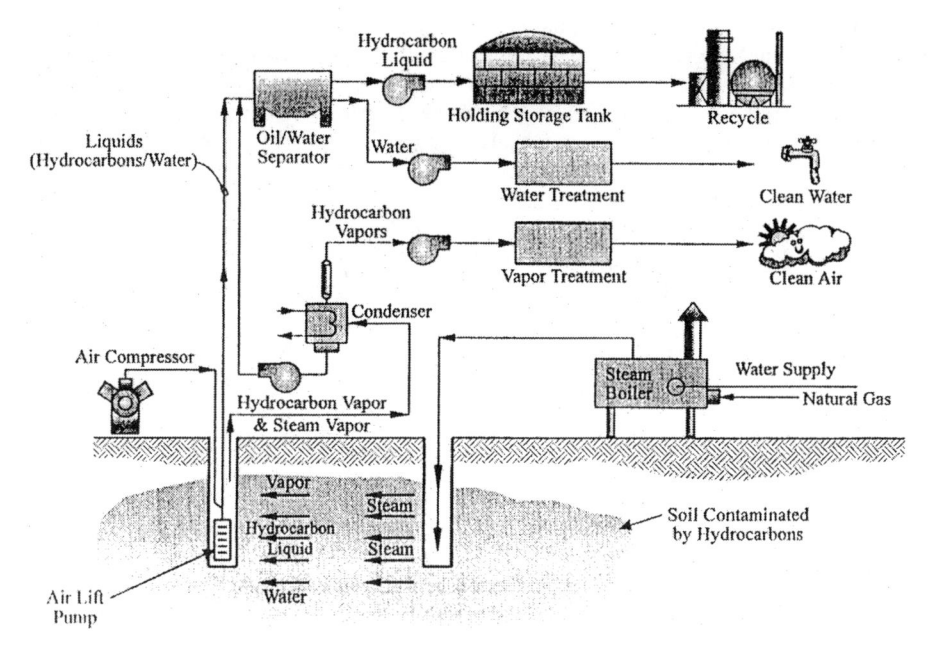

Figure 2.32 Steam-enhanced recovery process. (Courtesy of Hughes Environmental Systems, Inc.) (From Johns, F.J., II and Nyer, E.K. In *In Situ Treatment Technology.* Nyer, E. K. et al., Eds. Lewis Publishers, Boca Raton, FL, 1996.)

There are different methodologies for applying steam stripping (Johns and Nyer, 1996). Steam can be used to enhance the standard SVE system but requires a condenser ahead of the moisture separator. Vacuum is extracted from a well or series of wells to draw the steam through the contaminated soil (U.S. EPA, 1994b). Another technology, developed and licensed by NOVATERRA, Inc., employs a rotating, hollow shaft with drill bits to inject steam into the contaminated soil; vacuum is then applied at the surface to withdraw the steam and VOCs (de Percin, 1991; U.S. EPA, 1994c). A Detoxifier™ unit can be used to implement the process (see Section 2.2.1.13). Both approaches remove not only VOCs but also SVOCs from the soil, which is an advantage over conventional SVE (Johns and Nyer, 1996). Steam can also be used to simultaneously enhance gas and liquid movement (Figure 2.32).

Tests have shown that bacteria present in the poststeam and postthermal treatment can be used in the remediation of heavy fuel oil–contaminated sites (Dablow, Hicks, Cacciatore, and van de Meene, 1995). While the results do not confirm that the surviving microorganisms can degrade the pollutants, they do suggest that steam injection does not itself prevent the subsequent use of bioremediation.

In a pilot test of the innovative process, 99.5% of the petrochemicals polluting soil and groundwater were removed (Baum, 1988). Six 20-ft-deep steam injection wells were installed around a 10-ft-diameter area. The walls of the wells and the surface were cemented to prevent steam from escaping. A recovery well with a vacuum pump and condenser was installed at the center of the area. Steam was injected continuously for 5 days. The level of toxic organic pollutants was reduced from 30,000 ppm to about 10 ppm. Increasing the temperature of the soil increases the equilibrium concentration and, hence, the mass transfer of the compounds in the steam phase by a factor of about 40. Thus, xylenes are completely removed, even though their boiling points are higher than that of water. About 1000 lb of waste chemicals were recovered from the site, making recycling feasible. The steam-cleaning process is about one tenth the cost of conventional methods of hauling contaminated soil to a hazardous waste disposal facility, which can cost up to $300/yd^3 of soil.

Steam injection may be used to remove the residual saturation of NAPLs sorbed to soil particles or occupying soil pore spaces, by reducing the viscosity and increasing vaporization (Dablow, 1992). During the process, low-boiling-point contaminants are vaporized, contaminant evaporation rates are enhanced, NAPLs are displaced at the steam condensation front, initial pore water concentrations are diluted, contaminants are desorbed, and interstitial water and contaminants boil. Contrary to the common assumption of local equilibrium, the interphase contaminant transport from the sorbed to the vapor phase or from the liquid to the vapor phase is mass transfer controlled (Karan, Chakma, and Mehrotra, 1995).

An *in situ*, hot air/steam stripping technology to remove VOCs and SVOCs from soil in the vadose and saturated zone has been demonstrated (La Mori, 1994). A drill tower injects and mixes steam and hot air continuously into the soil and immediately captures all vapors that come to the surface. These are removed with condensation and carbon beds. The air can be recompressed and recycled. The condensed liquid is distilled off and can be destroyed or recycled.

Cyclic steam injection combined with vacuum extraction was effective in remediating a JP-5 jet fuel–contaminated site (Chan, Yeh, and Bialecki, 1994). The second cycle of the process was especially effective in lowering the residual hydrocarbon content of the soil, and further reductions are expected with additional cycles.

A similar steam stripping process was developed in The Netherlands (Dablow, 1992). It utilizes wet or dry steam to strip and vaporize contaminants from the soil, while a vacuum extracts the vapors and condenses them for disposal. It has been used successfully at gasoline-contaminated sites to take the level of hydrocarbons from 10,000 to 15,000 ppm down to 5 ppm in 4 to 9 months. Diesel fuel concentrations have been reduced from 20,000 ppm to <1000 in 2 months. Portable steam systems can be used in place to typically treat from 1000 to 5000 yd^3 of soil. Vapors can be treated in a packed-bed thermal oxidizer. Extracted liquids can be pumped into an oil/water separator, with diesel fuel being collected for recycling. Contaminated groundwater and condensate can be treated by a series of filters and carbon adsorption before discharge to a storm drain.

Some of the factors that affect rates of biodegradation can be optimized immediately after cessation of steam injection, which allows enhanced bioremediation to serve as an effective polishing technique for heavy fuel oil–contaminated soils (Dablow, Hicks, and Cacciatore, 1995). Microorganisms are not killed off during the steam treatment, and nutrient augmentation reactivates the dormant cells.

This process is limited by inhomogeneity in the soil and impermeable layers that could block the flow of steam (Johns and Nyer, 1996). Also, analysis of air-dried or oven-dried soil samples may be subject to error, and thermal desorption processes for treating soil may cause polymerization on the soil surface (Karimi-Lotfabad, Pickard, and Gray, 1996).

2.2.1.14.3 Radio Frequency (RF) Heating

With RF heating, soil is heated to high temperatures, in the region of 140°C, thereby desorbing most organic contaminants (Edelstein, Iben, Mueller, Uzgiris, Philipp, and Roemer, 1994). When the temperature is 150°C, 95 to 99% of the VOCs and 90 to 95% of the SVOCs can be removed (Johns and Nyer, 1996). Pollutants are primarily vaporized, but they may also be decomposed or pyrolyzed to form more-volatile compounds, which are removed by SVE. This method is appropriate for removal of higher-boiling-point organic contaminants in the soil. The selected frequency depends upon the soil properties and the depth of the contamination. It must be altered to allow for moisture loss and change in soil dielectric properties. The frequencies used range from 45 Hz to 10 GHz, primarily from 6.8 MHz to 2.5 GHz.

An RF transmitter is connected to exciter electrodes, which are placed either vertically or horizontally along the ground (Johns and Nyer, 1996). The soil is heated either by resistance heating or dielectric heating and covered by a synthetic membrane to trap vapors for treatment.

A problem with RF heating is that the vacuum tube RF amplifiers are touchy and unreliable (Edelstein, Iben, Mueller, Uzgiris, Philipp, and Roemer, 1994). A simple, inexpensive solid-state amplifier, which can be combined to give multikilowatt and, ultimately, megawatt power levels at lower cost per watt than vacuum tube amplifiers, is needed for large-scale remediation and is being developed by these authors.

A demonstration of the process involved treating 1000 m^3 of soil contaminated with organic wastes ranging from acetone to vacuum pump oils for about 90 days to a final treatment temperature of 200°C (Phelan and Webb, 1994). This technique was also applied in a pilot study in combination with soil vapor extraction to enhance the recovery rate of #2 fuel oil in silty soil at a depth of 20 ft (Price, Kasevich, and Marley, 1994).

There is the possibility that the high temperatures could sterilize the soil or affect the distribution of microorganisms. However, Brombach, Schwabe, and Theissen (1996) report no impairment of bacterial growth or change of morphology of oil-degrading bacteria on exposure to heating by high-frequency fields. Therefore, this method can be applied to heat oil-polluted soil, independent of seasonal influences, when an optimum temperature for bacterial growth and oil degradation is provided.

2.2.1.15 Vitrification

Principle/Description — This is a soil-melting technology whereby electric current is passed between electrodes placed in the ground (Lendvay, 1992). Inorganic contaminants are encapsulated and immobilized by heating the soil to high temperatures (1500 to 2000°C) (Johns and Nyer, 1996). The soil and contained materials are converted to a stable glass. Organic contaminants and naturally occurring organic compounds are volatilized and can be removed by vapor extraction, or they are pyrolyzed and the evolved gases removed. A cover over the treatment area captures the emissions, which are sent to a treatment unit or vented to the atmosphere. The most-expedient vitrification method is direct electric heating (Lendvay, 1992).

Battelle Pacific Northwest Laboratory has developed an *in situ* vitrification process, which is marketed by Geosafe Corp., Richland, WA (U.S. EPA, 1994g). Soil is heated by resistance heating, as described above, but with much higher temperatures (Johns and Nyer, 1996). Flaked graphite and glass frit on the soil surface initiate the melt. Electrodes are lowered 1 to 2 in/h, until the entire contaminated area is molten. The area then cools into a monolith. Off-gases are contained by a cover and treated before release.

Most soils can be treated with this method, with a volume reduction of 20 to 40% (Amend and Lederman, 1991). Johns and Nyer (1996) discuss limitations of this method. The durability of the solidified mass and the removal of organic contaminants are advantages of this technology over conventional *in situ* stabilization. A disadvantage is the treatment cost for the high amount of energy required to melt the soil.

Major advances in plasma arc technology allow *in situ* transformation of all soil, rock, and waste types into a vitrified, rocklike material, similar to obsidian, that is durable, strong, and highly resistant to leaching (Circeo, Camacho, Jacobs, and Tixier, 1994). A plasma arc torch can be lowered into a borehole to any depth, thermally converting a mass of soil into a vertical column of vitrified and remediated material. Plasma arc torches operating at power levels exceeding 1 MW could produce vitrified columns >10 ft in diameter. The plasma remediation of *in situ* materials (PRISM) is rapid, efficient, cost-effective, and simple. This treatment process could be applied to soil containing hazardous materials and heavy metals.

Desirable Features — Experience is available from applications to radioactive waste (Lendvay, 1992). The product glass is inert, and the hazardous compounds are incorporated into a chemically durable, nonleaching solid (Kirts, 1995).

Limitations — There are limited data for cases involving hazardous wastes. Available technology and cost data are from small-scale tests (Lendvay, 1992). Emission control and design requirements are not fully defined for applications involving large sites and are a factor of site and contaminant characteristics. Recent cost analyses on the use of vitrification found previously published cost data to be highly inaccurate and noninclusive (Kirts, 1995).

2.2.1.16 Tensiometric Barriers

A tensiometric, or dry, barrier is produced by injecting dry air into an unsaturated formation to reduce the soil moisture content (Thomson, Morris, Stormont, and Ankeny, 1996). Containment can thus be achieved by reducing the hydraulic condition of the unsaturated media to the point where liquid-phase transport becomes negligible. The air injection could be coupled with a vacuum extraction system to recover soil vapors. The process may have to include a cover design to prevent infiltration from atmospheric precipitation. A properly designed tensiometric barrier is competitive with conventional containment methods. See Section 2.2.1.11 for application with SVE.

2.2.1.17 Electric Fields

The dominant transport process for removing charged species from soils by electric fields is electromigration (Jacobs, Sengun, Hicks, and Probstein, 1994). For heavy metals, the polarity and magnitude of the charge depend on the pH. Positive ions are generally stable at low pH, and negatively charged complexes dominate at high pH. Strong pH gradients can develop in the soil trapping the metals by a process of isoelectric focusing, under some conditions. The focusing effect can be eliminated and high metal removal efficiencies achieved by washing the cathode. A model demonstrates the effect of background ions and electroneutrality in governing distribution of species and shows how concomitant variations in the electric field result in the virtual cessation of the transport process.

2.2.2 BIOLOGICAL SOIL TREATMENT PROCESSES
2.2.2.1 Bioremediation/Bioreclamation

Environmental contamination by petroleum products is widespread, and development of remediation efforts must accommodate the fact that there are vast numbers of sites that need to be treated and that there will be individual requirements for each site (McGugan, Lees, and Senior, 1995). Appropriate measures must also be available for those areas of the world that cannot afford expensive remediation and may not even have the driving legislation to enforce them to clean up the pollution. Site remediation must be simple and cost-effective for both industrialized nations and underdeveloped countries to be able to confront and resolve this enormous problem.

More than $1 million a day was spent in a partially successful attempt to clean up the oil spill at Prince William Sound, Alaska (Atlas, 1991). Neither government nor private industry can afford the cost of physically cleaning up the known U.S. toxic waste sites. Because of the large volumes of unsaturated soils contaminated with metals and organic solvents, *in situ* remediation is often the most economically attractive remediation technique (Lindgren and Brady, 1995). Therefore, bioremediation is now being considered as a viable alternative for this purpose.

Principle/Description — Bioremediation has been defined by Madsen (1991) as "a managed or spontaneous process in which biological, especially microbial, catalysis acts on pollutant compounds, thereby remedying or eliminating environmental contamination."

Harmful hydrocarbon contaminants may be assimilated by microorganisms and converted into biomass or transformed by cells or cell-free enzymes (Babel, 1994). These are normal processes in the environment. Under certain conditions, petroleum hydrocarbons, gasoline, diesel fuel, jet fuel, and motor oil can be bioremediated rapidly and at a low cost (Amdur and Clark-Clough, 1994). Bacteria capable of biodegrading petroleum hydrocarbons may commonly be found in subsurface soils; however, natural breakdown of the compounds will occur too slowly without intervention to prevent accumulation of the pollutants to unacceptable levels (Lyman, Noonan, and Reidy, 1990).

Bioreclamation or bioremediation refers to the enhancement of this native capability of the microorganisms. The indigenous (naturally occurring) microbes can be stimulated, or specially developed microorganisms can be added to the site to degrade, transform, or attenuate organic and organometallic compounds to low levels and nontoxic products (Catallo and Portier, 1992). Petroleum contaminants can be converted by this method to inert or less harmful materials (Ram, Bass, Falotico, and Leahy, 1993). Oxygen and nutrients are added to the system to support biological growth and improve the degradation (Figure 2.33). This is a multidisciplinary approach to site cleanup that involves the input of soil chemists, microbiologists, hydrologists, and engineers.

Unlike other techniques that temporarily displace the problem or transfer the contaminants to another medium, bioremediation attempts to render the contaminants into harmless substances (Fouhy and Shanley, 1992). Bioremediation has been found to restore fuel spill–contaminated soil to the point where it could support plant growth in 4 to 6 weeks, with complete recovery of the soil in 20 weeks (Wang and Bartha, 1990). PAH components of diesel oil were completely eliminated in 12 weeks (Wang, Yu, and Bartha, 1990).

The subsurface environment is characterized by a complex natural physiological and biological heterogeneity (Levine, MacDonald, Rothaus, Ruderman, and Treiman, 1995). In 1985, the U.S. Department of Energy established the Subsurface Science Program to clean up underground contamination by improving the understanding of the biology of the subsurface, in order to make bioremediation a successful approach for environmental remediation.

Bioremediation is currently receiving considerable attention as a remediation option for sites contaminated with hazardous organic compounds (Piotrowski, 1991). There is an enormous amount of interest in this process, with both field applications and laboratory tests being conducted to elucidate and refine the technology. Use of bioremediation is growing around the world (Fouhy and Shanley, 1992). From 1982 through 1986, bioremediation was used in only 3 of 48 site cleanups. From 1987 through 1989, 18 out of 165 sites used the technology. In 1992, the process was being considered or used in over 135 Superfund, RCRA, and underground storage tank sites across the U.S.

In situ bioremediation of unsaturated soils generally involves three very simple, well-documented processes: (1) the ability of microorganisms to degrade petroleum hydrocarbons, (2) the use of above-ground pumps and blowers to move vapors through the unsaturated soil, and (3) the addition of nutrients

CURRENT TREATMENT TECHNOLOGIES

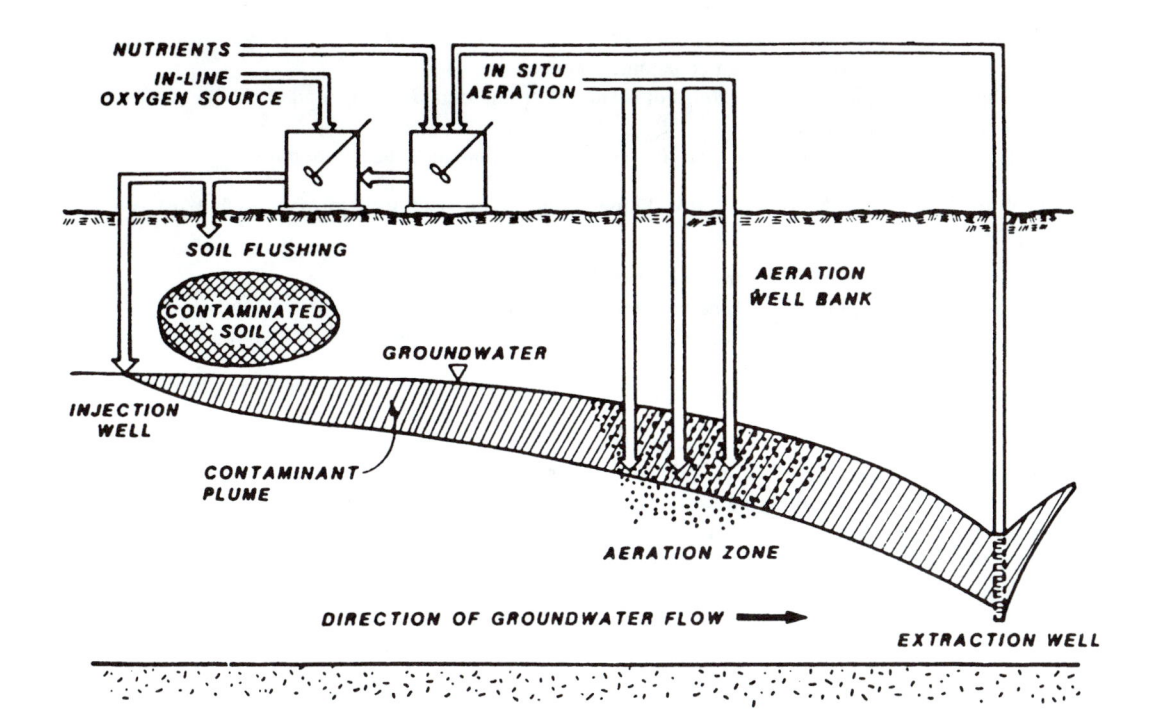

Figure 2.33 Bioreclamation technology for treatment of contaminated soil and groundwater. (From Ghassemi, M. *J. Haz. Mater.* 17:189–206, 1988. From Elsevier Scientific Publishers, Academic Publishing Division. With permission.)

to the subsurface to facilitate growth of the organisms (Dineen, Slater, Hicks, Holland, and Clendening, 1993).

There are two forms of bioremediation: the microbiological approach and the microbial ecology approach (Piotrowski, 1991). The microbiological approach involves supplying microorganisms that have been conditioned to degrade target compounds, along with appropriate nutrients, to the subsurface. These organisms could be prepackaged "superbugs," which are strains developed in the laboratory and shipped to a contaminated area, or they could be site-specific superbugs, which have been isolated from the affected area itself and reintroduced at higher concentrations. The microbial ecology approach, on the other hand, involves altering the environment of the indigenous organisms to optimize their biodegradation of the contaminants.

Bioremediation can also be expressed as being engineered or intrinsic. Any modification of the bioremediation process is considered engineered bioremediation, and the lack of intervention is intrinsic bioremediation, or natural attenuation (Hart, 1996). Intrinsic remediation results from several natural processes, such as biodegradation, abiotic transformation, mechanical dispersion, sorption, and dilution that reduce contaminant concentrations in the environment (Morin, 1997).

For natural attenuation to be a viable approach, the site must have a high natural supply of nutrients and oxygen, and the source of contamination must be small (Hart, 1996). Significant evaluation up front and follow-up monitoring are necessary to ensure removal of contaminants of concern at reasonable rates.

There are various approaches for determining whether or not a site would be appropriate for remediation via biological means. Ogunseitan (1996) advises environmental testing and analysis to characterize the site first as to determination of the chemical nature, concentration, and hydrogeological context of the pollutants; the existence, diversity, and population densities of relevant biodegradative organisms; engineering constraints to implement *in situ* or contained-system bioremediation; as well as microcosm bioassays (remediation cost estimations and establishment of conditions supportive to bioremediation).

Piotrowski (1991) has developed a conceptual framework for bioremediation decision making. The following site aspects would tend to favor the microbiological (engineered) approach for biological treatment of the site:

- A site recently contaminated by a spill event
- A site containing extremely high contaminant concentrations
- A site containing high concentrations of toxic metals and organic compounds
- A site requiring rapid cleanup
- A site containing extremely low, but unacceptable contaminant concentrations

The following site aspects would tend to indicate that the microbial ecology approach would be more appropriate:

- A site that has been contaminated for some time
- A site contaminated for some time with multiple organic contaminants
- A site contaminated with "moderate" concentrations of contamination
- A site that does not have to be cleaned up quickly
- A site containing a contaminated aquifer

A three-tiered approach has been suggested as a course of action for any potential bioremediation project (Forsyth, Bleam, and Wrubel, 1995). Tier I assesses the feasibility of bioremediation and develops an estimation of remediation time and costs. Tier II conducts pilot-scale studies. Tier III includes a complete work plan, budget, and project management plan for full-scale bioremediation conducted in the field. At each tier, the project is evaluated to determine the proper courses of action and options for experimental and remedial alternatives.

In situ bioremediation offers the possibility of a rapid and cost-effective solution to contamination of the environment by hazardous organic compounds (Catallo and Portier, 1992). Additional background on this process can be found in Section 1.2. Further information about the use of *in situ* bioremediation of petroleum-contaminated soil can be found in review articles by Hicks and Brown (1990), Arvin, Godsy, Grbic-Galic, and Jensen (1988), and Litchfield and Clark (1973).

Desirable Features — *In situ* bioremediation is a simple, low-cost, safe method. It has been demonstrated to be effective at many organic spill sites. The objective of biodegradation is to break down organic contaminants to harmless products, instead of temporarily displacing them or transferring them to another medium. Control of environmental parameters can increase the efficiency of the process. Bioremediation is even being used successfully for treating sites contaminated with PAHs and other complex organic compounds. As more data are accumulated and the complex subsurface interactions become better understood, the more refined the technology will become, allowing wider and more successful applications.

Limitations — One of the drawbacks of bioremediation, or any *in situ* remediation effort, is the numerous uncertainties that occur as a result of dealing with a complex and inaccessible subsurface (National Research Council, 1993). Evaluation of bioremediation requires integrating tools and concepts from various diverse disciplines, which may not be standardized for cross-referencing. Uncertainties can be minimized, however, by increasing the number of samples, using models to weight the importance of variables, and building safety factors into the engineering design.

Bioreclamation is, generally, limited to applications involving localized contamination (organic spills). It is not applicable where the contaminants are refractory or present at toxic levels. There are many variables to consider with this process, and the soil conditions and contaminant characteristics are unique for every incident. Oxygen, pH-adjusting chemicals, and nutrients may have to be provided. Effective, proven methods for uniform distribution of oxygen and nutrients in the subsurface (especially to significant depths) do not exist. Channeling and uneven treatment may result, due to nonhomogeneity of soil or to formation of precipitates and biological deposits. It is difficult to ensure contact between the microbes and the contaminant. For contaminated soils, applications seem to be limited to very shallow depths. It is not recommended for low-hydraulic-conductivity soils (may require hydraulic conductivities exceeding 10^{-3} to 10^{-2} cm/s).

There are advantages and disadvantages to both the microbiological and microbial ecology approaches (Piotrowski, 1991). The former may be more effective when immediate results are needed or if there are high concentrations of contaminants, which would be toxic to the indigenous organisms. However, superbugs from other sources may not be able to survive or compete in the new environment. The microbial ecology approach is less costly and less complicated to apply and assess. In addition, it has

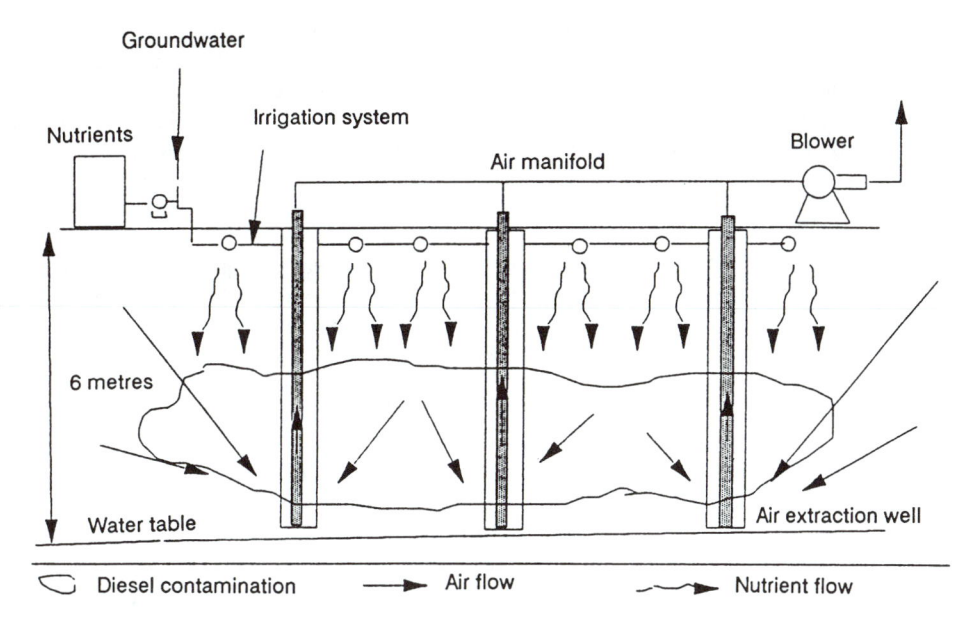

Figure 2.34 Schematic diagram of bioventing installation: section view. (From Bulman, T.L. Newland, M., and Wester, A. *Hydrol. Sci. J.* 38(4):297–308, 1993. With permission.)

been observed that biodegradation rates of prepackaged superbugs have been essentially the same as those for indigenous or nonselected microbes.

While bioremediation is useful for reducing the volume and concentration of organic contaminants in soils and waters, it is uncertain what final contaminant concentrations will be considered acceptable in the future and whether this process would be able to achieve those levels (Piotrowski, 1991). The process can be difficult to monitor and control. Reliable cost and performance data have been difficult to obtain, and remediation can take anywhere from 2 months to 10 years (Fouhy and Shanley, 1992).

While bioremediation is effective for treating soil contaminated with medium-distillate fuels, it is not effective in removal of heavier fuels and has been shown to play a relatively minor role in removing lower-distillate fuels, compared with vaporization (Song, Wang, and Bartha, 1990).

2.2.2.2 Bioventing

SVE technology was originally developed to remove volatiles from the subsurface (see Section 2.2.1.11). SVE alone is not effective for removing heavier material, such as diesel fuel, JP-4 jet fuel, or fuel oils, because of the nonvolatile, high-molecular-weight fractions they contain (Dupont, Doucette, and Hinchee, 1991). These compounds, however, are susceptible to biodegradation. It was noticed that when SVE was applied, biodegradation rates were also being stimulated (Brown, Mahaffey, and Norris, 1993). Active warming of fuel-contaminated soil in areas in cold climates can be used in combination with bioventing to improve biodegradation rates (Leeson, Hinchee, Kittel, Sayles, Vogel, and Miller, 1993). Figure 2.34 illustrates a bioventing installation (Bulman, Newland, and Wester, 1993).

Bioventing now utilizes SVE hardware, vertical extraction wells (placed on the edge of the contaminated area) and/or lateral trenches, piping networks, and a blower or vacuum pump for gas extraction, as well as for delivering oxygen and a gaseous nutrient source to the subsurface (Dupont, 1993). These systems are generally operated at rates that provide far more than stoichiometric amounts of oxygen during the remediation (Dupont, 1992).

Bioventing is appropriate when the water table is deep and the contaminant has not reached the groundwater, leaving the oil absorbed primarily in the solid phase (Gruiz and Kriston, 1995). There are advantages to employing bioventing. The gas diffusion rate is about 10,000 times higher than that of the solutes. In addition to fuel vapor removal, bioventing may increase the rate of air diffusion, providing oxygen to microorganisms in soil pores, in the deeper subsurface, and in the water film of the aggregates. The ratio of the volatilized oil increases, which improves and homogenizes the distribution and sorption of the oil on the surface of soil aggregates.

A disadvantage of the continuous airflow is desiccation of the soil (Gruiz and Kriston, 1995). Soil moisture (30 to 50% field capacity) and nutrient (if required) levels should be maintained to optimize biodegradation (Dupont, 1993). This allows stimulation of *in situ* biodegradation of fuel components by the indigenous microorganisms. Introduction of a gaseous nutrient, such as ammonia, can be a practical option for sites where surface application of liquid nutrients is not possible (Marshall, 1995). Ammonia dissolves readily in soil moisture and sorbs strongly to soil particles, and the ammonium ion is the preferred nutrient form of many microorganisms. Conservative addition of ammonia can promote appreciable increases in evolved CO_2 and rate of oxygen utilization.

If bioventing is conducted in municipal areas, off-gas treatment may be required (Burke and Rhodes, 1995). Also, bioventing/sparging allows little control of the flow direction, while bioventing/extraction requires treatment and monitoring of the off-gas. Treatment of soil venting off-gas can contribute to more than 50% of the remediation costs (Miller, 1990).

Modified SVE or innovative bioventing techniques could be employed to control vapors, while increasing oxygen levels in the soil for the hydrocarbon degraders (Downey, Frishmuth, Archabal, Pluhar, Blystone, and Miller, 1995). Some alternative approaches include low rates of pulsed air injection, a period of high-rate SVE and off-gas treatment followed by long-term air injection, or an innovative approach combining regenerative resin for *ex situ* vapor treatment with *in situ* bioventing to reduce bioremediation costs. Venting operated with cyclic, or surge, pumping could reduce operating costs and the need for off-gas treatment (Dupont, 1993). With this method, air venting is continued on an as-needed basis by the microbes. Vapor retention in the soil encourages microbial degradation of the contaminant vapors. Some researchers believe that SVE can be more efficient than H_2O_2 in supplying oxygen to unsaturated soils (Brown, Mahaffey, and Norris, 1993).

By managing airflow rate, it should be possible to increase degradation to 85% (Miller, 1990; Miller, Vogel, and Hinchee, 1991). Since rate constants for oxygen consumption and carbon dioxide production follow zero-order kinetics for oxygen concentrations above 1%, lower flow rates with longer soil retention times should increase removal of hydrocarbons by biodegradation (Miller, Vogel, and Hinchee, 1991). Airflow rates can be minimized to maintain oxygen levels between 2 and 4% for this purpose. An airflow rate of 0.5 air void volumes per day was optimal at one test site. When vapor extraction rates are reduced, flow path distances are maximized, and vapor retention times in the soil are increased, volatilization is minimized, which promotes greater biodegradation and allows direct discharge of vent gas that does not need off-gas treatment (Dupont, Doucette, and Hinchee, 1991).

Bioventing increases biodegradation of VOCs and SVOCs (Heuckeroth, Eberle, and Rykaczewski, 1995), as well as PAHs (Alleman, Hinchee, Brenner, and McCauley, 1995). Bioventing commonly increases the numbers of TOL plasmid-containing bacteria by two to three orders of magnitude (Brockman, 1995a).

Soil venting was used to stimulate aerobic bioremediation of JP-4 in contaminated soil in an unsaturated vadose zone at Hill Air Force Base, UT (Hinchee, Downey, Dupont, Aggarwal, and Miller, 1991). Although most of the contaminant was removed through volatilization, from 15 to 25% of the jet fuel was biodegraded *in situ*. When moisture (35 to 50% field capacity) and nutrients were supplied to contaminated soil along with vacuum extraction, more than 80% of JP-4 was biodegraded (Dupont, Doucette, and Hinchee, 1991).

At a hazardous waste landfill, bioventing resulted in an increase of microbial populations and an increase in temperature from 19 to 27°C (Heuckeroth, Eberle, and Rykaczewski, 1995). About 99% reductions were observed in specific VOC and SVOC concentrations after 3 months, about 90% of which was attributable to biodegradation.

Bioventing was shown to be suitable for full-scale bioremediation of petroleum in volcanic and marine soils 1.4 to 50 m below ground surface (Ratz, Pierson, Caskey, and Barry, 1995). Another study demonstrated the effectiveness of bioventing with nutrient additions in sandy soil (Breedveld, Olstad, Briseid, and Hauge, 1995). After 1 year, the TPH content was reduced 66%. Without nutrient addition, there was only minor removal of the light-end compounds.

The U.S. Air Force Bioventing Initiative has conducted field treatability studies using bioventing at more than 120 sites to evaluate what parameters correlate with biodegradation rates (Leeson, Kumar, Hinchee, Downey, Vogel, Sayles, and Miller, 1995). *In situ* respiration test data, soil gas permeability test data, and soil chemistry and nutrient data were analyzed. Results showed that some biodegradation was occurring at all sites, and, therefore, all should respond well to bioventing.

CURRENT TREATMENT TECHNOLOGIES

A depletion of O_2 and an increase of CO_2 in contaminated soil, compared with uncontaminated soil gas levels, will indicate increased microbial activity and can be used to determine whether or not the contaminated site is a good candidate for bioventing (Dupont, 1993). *In situ* respiration measurement techniques can then be used to quantify the maximum respiration rates under field conditions (Hinchee et al., 1991). These data will allow estimation of biodegradation rates, required oxygen transfer rates, the feasibility of *in situ* bioventing, and required time for remediation at that site. *In situ* soil air permeability measurements will then help determine the rate of transfer of the electron acceptor to the contaminated soil (Johnson, Kemblowski, and Colthart, 1990).

Site accessibility and power are usually required for bioventing (Graves, Dillon, Hague, Klein, McLaughlin, Wilson, and Olson, 1995). Wind-powered bioventing systems for stimulating air movement through vadose zone soil were designed to operate at remote locations without electric power. The low cost, low maintenance, and simplicity of the biovents make them an attractive option for remediating windy, remote sites.

The technique is becoming increasingly popular in soil remediation for removing VOCs and SVOCs from the vadose zone (Sepehr and Samani, 1993), while supplying oxygen to the soil to enhance the bioremediation process. Like SVE, it is ineffective when the porosity and transmissivity are low, such as in silt and clay soils (Burke and Rhodes, 1995). There may be little capital investment initially for this procedure, but it can be expensive monitoring the site for regulatory compliance. Since microbial populations may be limited at greater depths due to a lack of naturally occurring substrates and nutrients, the effectiveness of *in situ* bioventing is questionable in contaminated soil zones that extend far below the ground surface (Frishmuth, Ratz, Blicker, Hall, and Downey, 1995). Also, since the microorganisms require soil moisture for hydrocarbon degradation, the viability of bioventing in arid climates is questionable.

2.2.2.3 Bioslurping

Bioslurping is an innovative technology for remediation of sites contaminated with the light NAPLs of petroleum hydrocarbons (Baker and Bierschenk, 1996). LNAPLs may be present in the free phase, floating on the water table. They may occur in solution, being dissolved in the groundwater or pore water. LNAPLs may be nondrainable and be retained in soil within the vadose zone, saturated zone, and capillary fringe. They can also exist in the vapor phase as soil gas in the vadose zone. The bioslurping systems apply subatmospheric pressure to remove free-floating product from the water table, and to release non-free-draining LNAPL, additional water, and soil gas from the soil pores.

Bioslurping employs the technology of vacuum-enhanced recovery of free product in combination with the technology of *in situ* bioventing (Kittel, Hinchee, Hoeppel, and Miller, 1994; Keet, 1995). Bioventing is typically employed after NAPL recovery and aerates the vadose zone to enhance biodegradation of low-volatility hydrocarbons, while promoting vapor extraction of the more volatile fractions (see Section 2.2.2.2).

A simple bioslurping installation consists of wells, a negative-pressure vacuum system, and a "slurper spear" near the hydrostatic groundwater level (Keet, 1995). A single, aboveground vacuum pump can service several extraction wells to remove LNAPLs from the surface of the water table and soil gas vapor in the same process stream (Kittel, Hinchee, Hoeppel, and Miller, 1994). The system pulls a vacuum of up to 20 in. Hg on the recovery well, which creates a pressure gradient forcing fuel into the well. The depth of the "slurp" tube can be adjusted to minimize water uptake (Baker and Bierschenk, 1996). Groundwater and oil are then separated so free product can be recovered and recycled. The soil vapor and water can be treated, if necessary. Figure 2.35 illustrates a well construction and slurper tube (Leeson, Kittel, Hinchee, Miller, Haas, and Hoeppel, 1995).

When the free-product recovery stage is complete, the system can be converted to a conventional bioventing system for bioremediation of the vadose zone (Leeson, Kittel, Hinchee, Miller, Haas, and Hoeppel, 1995). Oxygen is drawn into the subsurface as a result of a vacuum created by the extraction of air from the soil (Kittel, Hinchee, Hoeppel, and Miller, 1994). A slow extraction rate allows the soils to be aerated with minimal volatilization.

The technology is proving to be faster, more effective, and much less expensive than conventional LNAPL remediation techniques (Baker and Bierschenk, 1996). Bioslurping has been applied at a number of sites in the U.S., Europe, and Australia for recovery of LNAPL (Kittel, Hinchee, Hoeppel, and Miller, 1994; Leeson, Kittel, Hinchee, Miller, Haas, and Hoeppel, 1995). While it is generally employed for

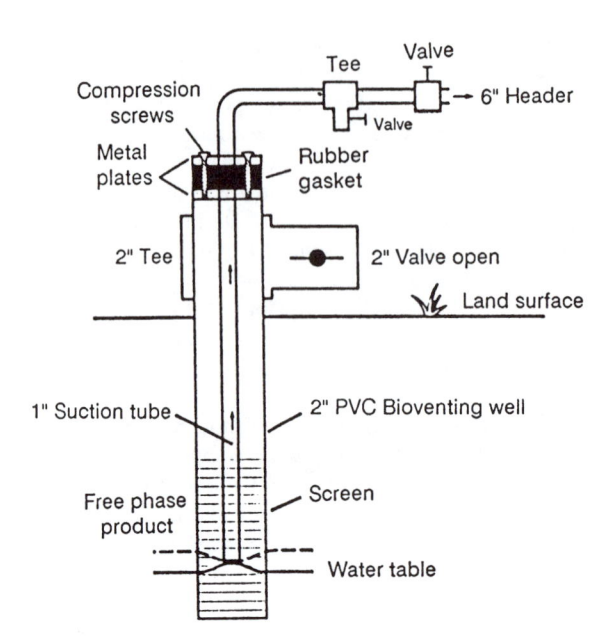

Figure 2.35 Well construction detail and slurper tube placement for the skimmer test configuration. (From Leeson, A., Kittel, J.A., Hinchee, R.E., Miller, R.N., Haas, P.E., and Hoeppel, R.E. In *Applied Bioremediation of Petroleum Hydrocarbons.* Hinchee, R.E. et al., Eds. Battelle Press, Columbus, OH, 1995. With permission.)

recovery of free product from fine to medium-fine sediments, it has also been successful with medium-coarse sands and fractured rock. Bioslurping is especially useful for activating *in situ* biodegradation of the medium- to low-boiling-point hydrocarbon distillates in the unsaturated and capillary zones above the free-product layer (Kittel, Hinchee, Hoeppel, and Miller, 1994). The process is useful during the early stages of remediation and can then later be augmented by other methods (Keet, 1995).

A bioslurper system at Fallon Naval Air Station, NV, employed a 10-hp liquid ring pump to extract 15 to 60 gpd LNAPL (JP-5), 50 cfm soil gas, and minimal groundwater (1 gpm) (Kittel, Hinchee, Hoeppel, and Miller, 1994). Recovery rates of LNAPL appeared to correlate directly with the vacuum placed on the system (Hoeppel, Kittel, Goetz, Hinchee, and Abbott, 1995). Although the low-permeability soil was being aerated by the process, biodegradation of the JP-5 was minimal. This may have been due to the indigenous organisms having a diminished ability to bioemulsify the LNAPL.

The U.S. Air Force Center for Environmental Excellence (AFCEE) Technology Transfer Division Bioslurper Initiative is a multisite program to evaluate the bioslurping technology (Leeson, Kittel, Hinchee, Miller, Haas, and Hoeppel, 1995). The objective of the initiative is to develop procedures for evaluating the potential for recovering free-phase LNAPLs from petroleum-contaminated sites. The *Test Plan and Technical Protocol for Bioslurping* (Battelle, 1995) contains details for materials and methods for testing of the bioslurper process at more than 35 test sites.

There are factors that need to be considered when using this process (Keet, 1995). For instance, emulsions can be formed by the high-velocity pump systems; however, prepump separation or a de-emulsification unit would solve this problem. Also, treatment might be required for emissions that would be released to the air and water.

Pilot studies and field applications have confirmed the potential benefits of this technology, including substantial recovery of LNAPLs (Baker and Bierschenk, 1996). However, this approach should be selected only after a thorough assessment of the site characteristics, including geology and hydrogeology. See Section 3.2.1.6 for a discussion on biodegradation of NAPLs.

2.2.2.4 BioPurge[SM]/BioSparge[SM]

This technology was developed in the late 1980s to provide enhanced subsurface control and maintain consistent moisture levels (Burke and Rhodes, 1995). It uses the low-volume airflow of bioventing with a closed-loop concept to regulate soil moisture and controlled release of nutrients, oxygen, and microorganisms

CURRENT TREATMENT TECHNOLOGIES

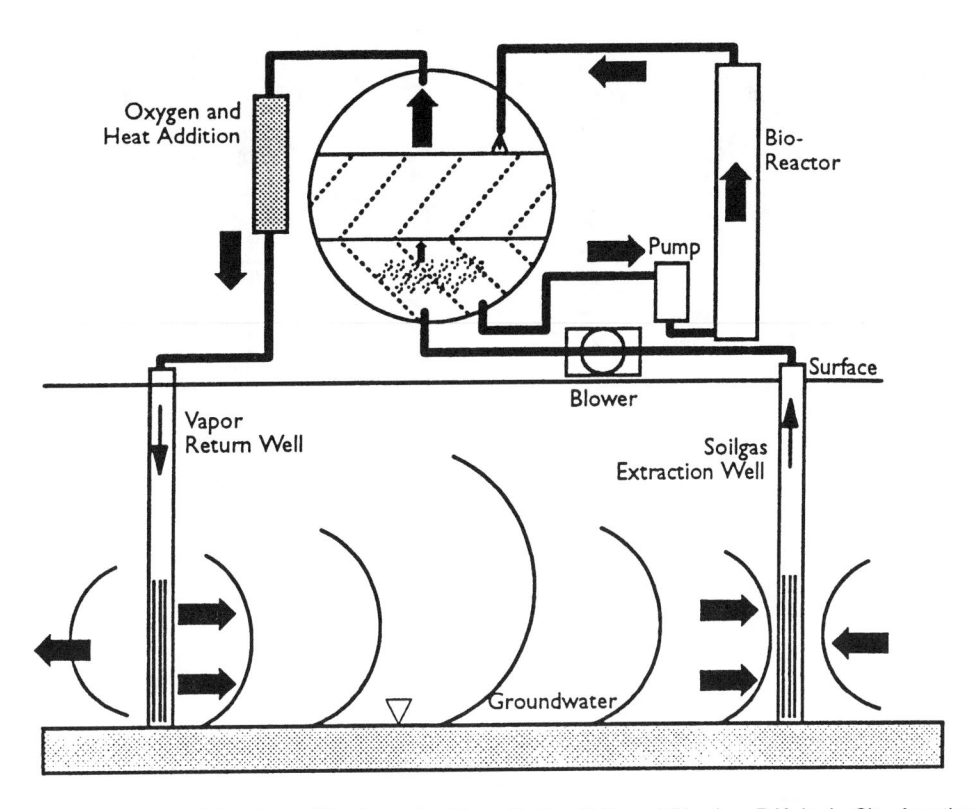

Figure 2.36 BioSparge^SM/BioPurge^SM schematic. (From Burke, G.K. and Rhodes, D.K. In *In Situ Aeration: Air Sparging, Bioventing, and Related Remediation Processes.* Hinchee, R.E. et al., Eds. Battelle Press, Columbus, OH, 1995. With permission.)

(see Figure 2.36). A grid pattern of pressure injection allows good distribution of the materials into the soil mass, accessing soil macropores and micropores. Wells may be placed around the perimeter of the contaminated soil area and soil vapor extracted. This vapor is treated in an on-site unit, which absorbs and biodegrades the volatile compounds. Oxygen, nutrients, and heat are added to the vapor stream and the vapor humidity adjusted. This stream is reinjected into the contaminated soil, with no off-gassing. The system is called BioPurge^SM when the vapor is injected above groundwater and BioSparge^SM when it is injected below groundwater level.

Excellent transmittance of materials is achieved with the bioinjection, even through silts and clays (Burke and Rhodes, 1995). It is more expensive initially, but costs considerably less in reduced sampling and monitoring, decreased treatment time, and ease of regulatory acceptance (Hobby, 1993).

2.2.2.5 Hydraulic/Pneumatic Fracturing

A major obstacle to *in situ* bioremediation is the difficulty of distributing nutrients effectively and uniformly to the subsurface (Anderson, Peyton, Liskowitz, Fitzgerald, and Schuring, 1995). By creating fractures in the subsurface, it is possible to enhance permeability to improve the flow of carrier fluids for contaminant removal or delivery of nutrients or reactive agents (Kidd, 1996). This process might be adapted by combination with other technologies to enhance the bioremediation process (Davis-Hoover, Murdoch, Vesper, Pahren, Sprockel, Chang, Hussain, and Ritschel, 1991).

The following materials are generally treatable by fracturing (Schuring and Chan, 1993): silty clay/clayey silt, sandy silt/silty sand, clayey sand, sandstone, siltstone, limestone, and shale.

There are two types of fracturing: hydraulic (water based) and pneumatic (air based) (Kidd, 1996). The main difference is in the penetrating fluids. The highly viscous hydraulic fracturing fluids (e.g., cross-linked guar) create a wide fracture and transport particles (proppants or propping agents, e.g., silica sand) with minimal fluid loss to the formation. They are also easily broken down in posttreatment, which prevents clogging. The proppants then support the open fractures.

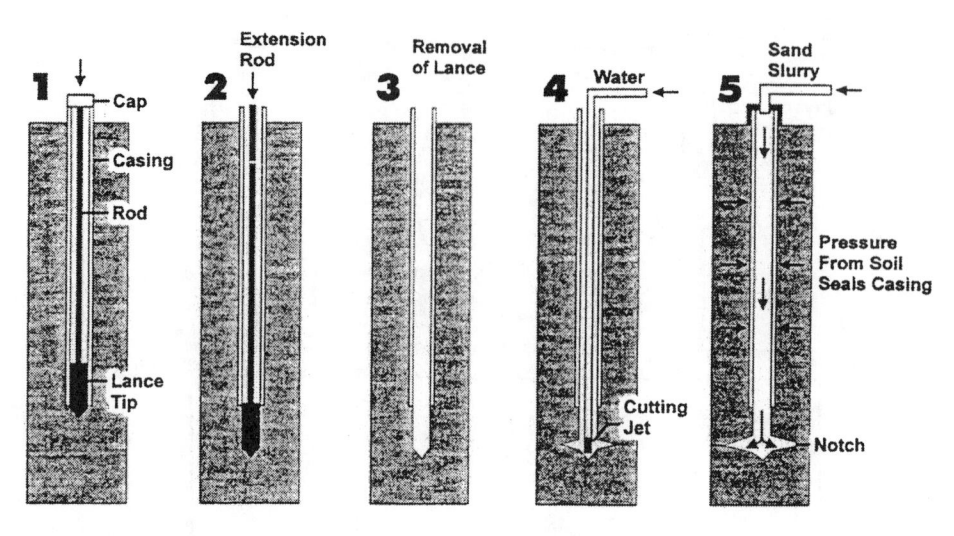

Figure 2.37 Hydraulic fracturing. (Murdoch, 1991). (From Kidd, D. F., *In Situ Treatment Technology.* Nyer, E. K. et al., Eds., Lewis Publishers, Boca Raton, FL, 1996.)

Hydraulic fracturing is accomplished by injecting fluid (high-pressure water) into a well until the pressure exceeds a critical value and a fracture is nucleated in about 10 to 60 min (Davis-Hoover, Murdoch, Vesper, Pahren, Sprockel, Chang, Hussain, and Ritschel, 1991). A fracturing fluid and proppants are injected into the fracture, at the time of formation (Kidd, 1996). This multistage process is illustrated in Figure 2.37 (Murdoch, 1991). The high-viscosity fracture fluid must break down to leave the fractures permeable (Kidd, 1996). Guar gum is often used as a fracture fluid and mixed with an enzyme for rapid decomposition. Vapors and liquids can be withdrawn through a centrally located well.

It may then be possible to pump granules of slow-release nutrients or oxygen-generating compounds in a viscous gel through the permeable sand conduits (Davis-Hoover, Murdoch, Vesper, Pahren, Sprockel, Chang, Hussain, and Ritschel, 1991). Multiple layers of fractures can be created. Whether they are horizontal or vertical depends upon the *in situ* state of stress. The volume of an injection is restricted to minimize migration of the fractures toward the ground surface (Kidd, 1996). Injection pressures are generally less than 100 psig, and fracture intervals are typically 2 to 5 ft apart.

The slow release of oxygen from an encapsulated solid peroxide was found still to attain toxic levels (Davis-Hoover, Murdoch, Vesper, Pahren, Sprockel, Chang, Hussain, and Ritschel, 1991). Nevertheless, this method shows promise for being able to dispense nutrients and oxygen more evenly to contaminated subsurface soils, when a longer-lasting, less-toxic solution is found. It may also be possible to inject microorganisms via this route. The selection criteria for such organisms may then include smaller microbes that can travel larger distances (Fontes, Mills, Hornberger, and Herman, 1991) and those less likely to adhere easily to soil particles near the injection site (Deflaun, Tanzer, McAteer, Marshall, and Levy, 1990; Fletcher, 1994).

Pneumatic fracturing is accomplished with compressed air or other gas source (Kidd, 1996). The compressed air is supplied at a pressure (around 150 psig) and flow (>500 scfm) that exceed the *in situ* stresses and the permeability of the material. The material is fractured in about 20 s, producing conductive channels radiating from the point of injection (Schuring, Jurka, and Chan, 1991). Figure 2.38 shows a pneumatic fracturing schematic (Kidd, 1996). Since there is no fluid involved, these fractures may be more permeable than those of hydraulic fracturing, and there is no fluid to remove.

Sodium percarbonate can be employed as a proppant and time-release (4 month) oxygen source (Vesper, Murdoch, Hayes, and Davis-Hooper, 1993). The "Lasagna Process" used a graphite-based proppant to enhance electroosmotic dewatering and *in situ* resistive heating due to its electroconductivity (Kidd, 1996).

A pilot-scale study of the integrated pneumatic fracturing and bioremediation system demonstrated enhanced removal of BTX from a gasoline-contaminated, low-permeability soil formation (Venkatraman, Schuring, Boland, and Kosson, 1995). Fracturing improved subsurface permeability by over 36 times and established an extended bioremediation zone supporting aerobic, denitrifying, and methanogenic

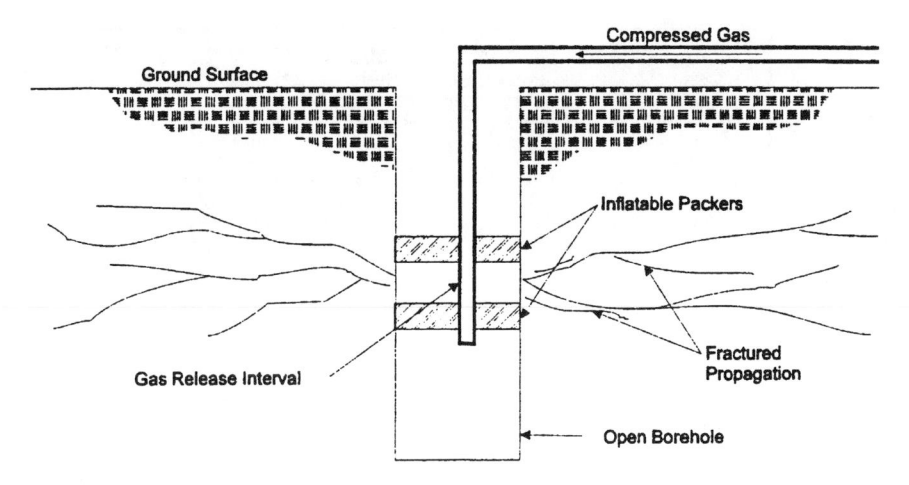

Figure 2.38 Pneumatic fracturing schematic. (From Kidd, D. F., *In Situ Treatment Technology.* Nyer, E. K. et al., Eds., Lewis Publishers, Boca Raton, FL, 1996.)

microorganisms. Phosphate/nitrogen amendments and injections were periodically introduced to the subsurface over 50 weeks, resulting in 79% of soil-phase BTX removal, with more than 85% of the mass removed attributable to bioremediation.

Pneumatic fracturing tests performed at Tinker Air Force Base, Oklahoma City, resulted in significantly improved formation permeability by enhancing secondary permeability and promoting removal of excess soil moisture from the unsaturated zone (Anderson, Peyton, Liskowitz, Fitzgerald, and Schuring, 1995). Postfracture airflows were 500 to 1700% higher after the treatment. Free-product recovery rates were also improved.

Pneumatic fracturing was applied at a site in New Jersey and hydraulic fracturing at a site in Illinois (Kidd, 1996). Surface heave in the siltstone and shale substrate in the former was observed up to 35 ft from the fracturing well. Vapor extraction and contaminant recovery rates were increased manyfold in both cases.

Results were not as successful with pneumatic fracturing of hydrocarbon-impacted clays at a former service station in northern California (Dockstader, 1994). Air conductivities ranged from 1.67×10^{-7} to 9.82×10^{-8} cm/s horizontally and 1.02×10^{-7} to 2.38×10^{-8} cm/s vertically. Fracturing applied to extraction wells increased flow rates but decreased the mass extraction. The vertical fractures appeared to "short-circuit" the extraction wells.

2.2.2.6 Deep Soil Fracture Bioinjection™

This process involves pressure injecting a slurry with a mixture of controlled-release oxygen, nutrients, and (if necessary) cultures into the subsurface, in an overlapping grid pattern that covers the entire contaminated area (Burke and Rhodes, 1995). Bioinjection can permeate the area to access the macropores and micropores and requires much less time for remediation than traditional *in situ* systems. This is an economical alternative *in situ* treatment option.

Hydraulic injection rigs, consisting of rubber-tired, crawler-mounted or truck-mounted vehicles, with a hydraulically powered, vertical injection mast with one to four injection rods to each mast, can treat soils to a depth of 40 ft (12 m) (Burke and Rhodes, 1995). The injection flows through the paths of least resistance in all soil types, e.g., fissures, sand lenses, desiccation cracks, etc., following the same pathways as the contaminants (Burke and Rhodes, 1995). Even silt and clay soils can be permeated. Water-soluble materials will also diffuse even further.

2.2.2.7 Combined Air–Water Flushing

A geochemical treatment has been developed to supply oxygen and moisture to the subsurface soil (Lund, Swinianski, Gudehus, and Maier, 1991). A network of positive (injection) and negative (suction) lances in the soil creates a horizontal airflow to deliver oxygen needed for biodegradation. Water is delivered by a surface sprinkling system to dissolve the contaminants and make them available for biodegradation. Nutrients can also be dispersed with the water, if necessary.

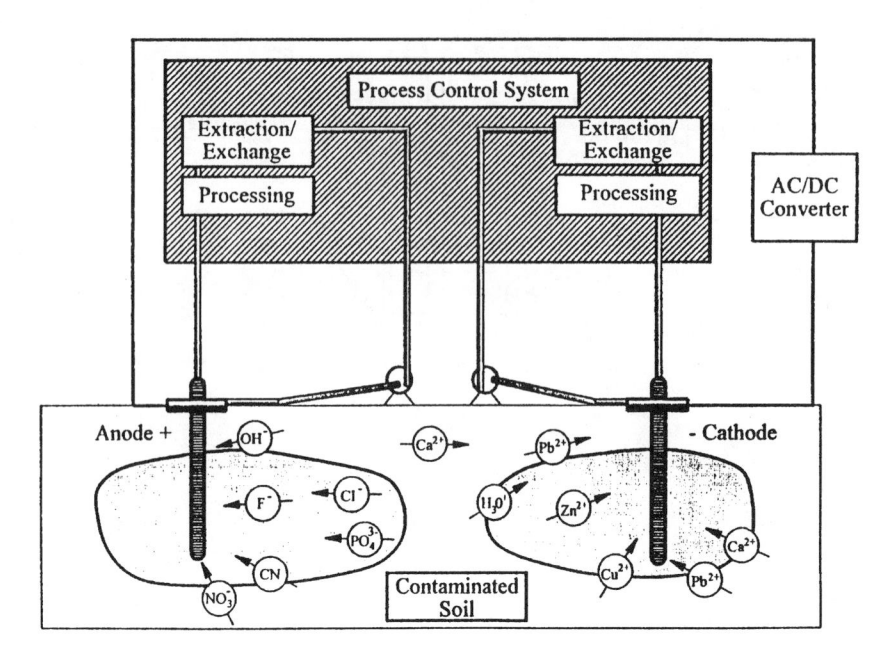

Figure 2.39 Electrokinetic remediation process. (Courtesy of Electrokinetics, Inc., Baton Rouge, LA.) (From Johns, F.J., II and Nyer, E.K. In *In Situ Treatment Technology.* Nyer, E. K. et al., Eds. Lewis Publishers, Boca Raton, FL, 1996.)

Airflow occurs in the larger pores and water flow in the smaller pores. As dissolved hydrocarbons are biodegraded, the depleted oxygen in the water is replenished by diffusion from the air, maintaining a constant concentration of oxygen in the water. Contamination is also reduced by the outflow of hydrocarbons with the airstream and water stream.

2.2.2.8 *In Situ* Electrobioreclamation/Electro-Osmosis/Electrokinetics/ Electrochemical Remediation

The process of electrobioreclamation has been developed for low-permeability, unsaturated soils (Lindgren and Brady, 1995; Lageman, Pool, van Vulpen, and Norris, 1995). Electrokinetic reclamation is based on electrokinetic phenomena that occur when the soil matrix is charged with low-voltage, direct current (DC) electrical power induced into the soil material by means of alternating anode and cathode arrays. Soil/water is also induced to flow toward the cathode (electro-osmosis). The ionic pollutants dissolved in the soil/water solution migrate and collect around the oppositely charged electrode, where they are captured and removed by excavation or by a fluid around the electrode (U.S. EPA, 1994e). This process is illustrated in Figure 2.39. Electro-osmosis is most effective in moist, fine-grained soils, and it is also useful in low-permeability, silt- and clay-type soils (Johns and Nyer, 1996). It can remove both inorganic and organic contaminants.

For *in situ* or *ex situ* electroreclamation, the electrodes can be installed at any depth (Lageman, Pool, van Vulpen, and Norris, 1995). The electrodes are placed either vertically or horizontally in the soil (Johns and Nyer, 1996). The electrode system consists of an upper plastic well casing with a lower porous ceramic section containing the active portion of the electrode system, an electrolyte solution, and a drive electrode (Lindgren and Brady, 1995). The anodes and cathodes are integrated in circulation systems with chemical additives, such as HCl, NaOH, or complexing agents, which capture the contaminants and control the pH and the redox potential around the electrodes (Lageman, Pool, van Vulpen, and Norris, 1995). Without this control, the soil around the anodes will acidify and cause metal hydroxides to precipitate around the cathodes.

Vacuum is applied to the casing headspace to limit the degree of soil saturation (Lindgren and Brady, 1995). Moisture can be added to the soil so the field capacity is never exceeded, which would prevent undesired mobilization of dissolved contaminants by saturated wetting fronts (Lindgren and Brady, 1995). Because unsaturated conditions (typically <50% saturation) are maintained, there is residual void space, which could be utilized to disperse the gaseous phase of nutrients (Hazen, Lombard, Looney, Enzien, Doughtery, Fliermans, Wear, and Eddy-Dilek, 1994).

A current of about 12 A/m of electrode is practical; higher currents may cause electrode and soil heating problems (Lindgren and Brady, 1995). At this current, about 0.45 g NO_3/min is injected to the soil pore water. Around 2.3 g NO_3/min/m is required for *in situ* bioremediation, and 26.5 g NO_3/min is necessary to prevent biofouling. Because of low flux rates, electrokinetic delivery of nutrients alone would probably not be adequate to prevent biofouling near the electrodes. However, electrokinetic injection methods used in conjunction with hydraulic injection methods may increase the uniformity of treatment in heterogeneous soils.

This technique can be used to remove metal ions, which move by means of soil moisture toward the cathode (Lageman, Pool, van Vulpen, and Norris, 1995). In other words, positively charged particles move in the direction of the negatively charged cathode, resulting in a decrease in the concentration of metal ions in the liquid phase (soil moisture). The ion displacement and subsequent ion replacement from the solid phase continues as long as the electrical field is maintained. The captured contaminants in the electrode solutions are periodically run through an electrolytic device to remove heavy metals.

Since contaminated soils may have a low permeability, e.g., clay soils generally have a hydraulic permeability of less than 10^{-8} m/s (Freeze and Cherry, 1979), it is difficult to deliver nutrients (e.g., nitrate, ammonium, phosphate), molecular oxygen, other electron acceptors (e.g., nitrate or sulfate), or electron donors through the vadose zone (Lageman, Pool, van Vulpen, and Norris, 1995). Movement of these materials can be accelerated predictably through the use of electrokinetic transport. With this technique, ion transport does not depend on the hydraulic nature of the soil. Production of electricity heats the soil, which increases hydraulic permeability and the solubility of organic constituents (Lageman, Pool, van Vulpen, and Norris, 1995). An electro-osmotic water flow is also generated.

Bioremediation is also enhanced by the process (Lageman, Pool, van Vulpen, and Norris, 1995). Microbial growth improves with higher voltages; a 50% increase resulted after an application of 22.5 V/m. There may be a change in the dominant species present under these conditions. For instance, *Aeromonas hydrophila* was replaced by a *Pseudomonas* sp. after application of an electrical potential.

Ionic forms of nutrients, electron acceptors, electron donors, and chelating agents can be introduced at the electrodes (Lageman, Pool, van Vulpen, and Norris, 1995). The maximum concentration at which a bionutrient can be distributed through the pore water is roughly equivalent to the ionic strength of the pore water (Lindgren and Brady, 1995). The electrode reactions should be buffered to maintain a neutral pH to prevent an excess of H^+ or OH^-, which would transport most of the current, instead of the contaminants or nutrients (Lindgren and Brady, 1995). For instance, nitrate is best introduced electrokinetically into the soil by using nitric acid to neutralize the hydroxyl ions formed.

The technique has been employed to move both bacteria and chemical nutrients through a low-permeability clay soil to enhance bioremediation (Nowatzki, Lang, Medellin, and Sellers, 1994). A solution of anaerobic, hydrocarbon-degrading bacteria and commercial nutrients (20:20:20 soluble plant food in water; 400 mg/L N) was introduced at the anode, with a 20-V DC power source. Fluid collecting in the cathode well was recirculated back to the anode to maintain electrical currents of 370 to 400 mA at an applied electrical potential of 35 V (DC), and to keep the soil from drying. The organisms were unharmed by the process, but it was inconclusive whether phosphate-based nutrients could be transported under these conditions or not. A phosphate source as a nutrient with this system must be able to form soluble complexes with calcium and other metals; e.g., sodium nitrate and sodium metaphosphate would be acceptable, while orthophosphate salts would not (Lageman, Pool, van Vulpen, and Norris, 1995).

An electrokinetic-enhanced passive *in situ* biotreatment (PISB) system was implemented to treat soil and groundwater contaminated by gasoline, diesel, and kerosene leakage under a desert lake bed in Nevada (Loo, 1994). The PISB consisted of heating the ground to 100°F to promote rapid *in situ* biodegradation. Around 50,000 gal of a nontoxic, nonhazardous solution of nutrients was inoculated through a percolation gallery and distributed into the clayey fraction of the soil by electrokinetic enhancement. In less than 6 months, 3000 yd^3 of soil was remediated to less than 100 ppm TPH.

While the nutrient flux rates achievable with electrokinetics may not be adequate to support *in situ* bioremediation totally, the process could be useful for stimulating bioremediation in unsaturated soils where degradation is limited by soil moisture content (Lindgren and Brady, 1995). During electrokinetic treatment of the soil to remove anionic heavy metal contamination, the soil moisture would be raised to about 15 wt%, kept below field capacity, and liquid pore water nutrients added. After adjustment of the moisture content, air or gaseous nutrients could be injected to stimulate biodegradation further. Biodegradable chelating agents could also be added to help remove strongly adsorbed cationic heavy metal contamination. Electrobioreclamation is also compatible with vapor-phase *in situ* techniques, such as SVE or bioventing.

Several modifications of this process are being developed (Johns and Nyer, 1996). One of these, by Battelle Memorial Institute, uses a patented electroacoustic process between the electrodes to enhance dewatering of sludgelike wastes (U.S. EPA, 1994f).

2.2.2.9 Biopolymer Shields

The soil matrix can be damaged by ecological changes in the soil organic matter (Yang, Li, Park, and Yen, 1994). The strength of soil can be improved by applying slime-forming bacteria to the soil matrix to produce a biopolymer inside it, or by directly applying biopolymer from slime-forming bacteria or commercially available products, such as poly-3-hydroxybutyrate (PHB), xanthan gum, and sodium alginates to the soil matrix. This technique could be applied to improve eroding soil or to enhance soil strength in the foundations of hydraulic systems. The strength of sand, silica, and clay is enhanced by this process. Land subsidence could be reduced.

Biopolymer can be used to seal the pores of sand and soil to reduce the permeability in porous media, serving as a contaminant barrier (Yang, Li, Park, and Yen, 1994). This technology could be used to control leachate migration from polluted areas into an aquifer. It could generate a capsule, such as a slurry wall, to contain a spill, allowing bioremediation in the less-contaminated areas (Li, Yang, Lee, and Yen, 1994).

Alcaligenes faecalis (ATCC 49677) and *A. viscolactis* (ATCC 21698) are bacteria that produce very hydrophilic and viscous slimes consisting of polysaccharides and small amounts of lipids and proteins (Magee and Colmer, 1960b; Punch, 1966). *A. eutrophus* (ATCC 17699) produces a huge amount of the intracellular polyester, PHB, as high as 70% of the cell weight (Li, Yang, Lee, and Yen, 1994).

There are three types of plugging brought about by *A. eutrophus*: living cell suspension, dead cell suspension, and intracellular product (PHB) (Li, Yang, Lee, and Yen, 1994). A living cell suspension of *A. eutrophus* can reduce effluent rates 280-fold and the relative permeability of sand by a factor of a million. Particulate plugging and filter cake formation can be due to accumulation of living and dead bacterial cells. Only living cells can produce plugging by biofilm formation, and their plugging will last for long periods of time. The living organism and its product, PHB, are much more efficient plugging agents than dead cells.

PHB is not an extracellular slime or polysaccharide (Li, Yang, Lee, and Yen, 1994). PHB is an intracellular storage polymer for providing reserve carbon and energy for the cells in which it is produced. It is a highly crystalline thermoplastic with a melting temperature around 180°C and a glass-transition temperature around 4°C (Doi, 1990). Only insoluble polymers with a low glass-transition temperature can be used (Li, Yang, Lee, and Yen, 1994). PHB is produced by many organisms, and it may be possible to find indigenous microorganisms that can produce the material at the intended site. It is also commercially available.

An ideal medium can yield slime with a viscosity up to 50,000 cps (Magee and Colmer, 1960a; Stamer, 1963). Many microbially derived biopolymers behave like humus, which acts as a hydrophilic colloid with many functional groups as donor atoms (Yang, Li, Park, and Yen, 1994). Interfacial chemical interactions in the soil matrix are either adhesion or binding, which combine with the biopolymer to strengthen the soil. Heating this mixture causes cross-linking in the matrix.

2.2.2.10 Bioscreens

In situ bioscreens (patent pending) are an emerging technology for *in situ* isolation and extensive remediation of contaminated sites (Rijnaarts, Hesselink, and Doddema, 1995). The bioscreen is a local zone in a natural porous medium that exhibits a high pollutant retention capacity (isolation) and an increased biodegradation of hazardous organics and immobilization of dissolved heavy metals. There are different types of bioscreens. For example, they can be formed from activated carbon grains coated with microorganisms or by *in situ* biofouling.

2.2.2.11 Phytoremediation

Natural plants can help remediate a contaminated site by accumulating contaminants or enhancing biodegradation (Johns and Nyer, 1996). Heavy metals and organic compounds can be removed from contaminated soil. The root system of plants is an important area for these reactions to take place. The process is especially useful in tight soils.

Section 3

Biodegradation/Mineralization/Biotransformation/ Bioaccumulation of Petroleum Constituents and Associated Heavy Metals

3.1 CHEMICAL COMPOSITION OF FUEL OILS

3.1.1 NAPHTHA
3.1.2 KEROSENE
3.1.3 FUEL OIL AND DIESEL #2

Table 3.1 lists organic compounds found in diesel fuel #2, as reported by Clewell, 1981.

3.1.4 GASOLINE

Tables 3.2 through 3.4 list the organic compounds and trace elements found in gasoline, as reported by different references.

3.1.5 JP-5

Tables 3.5 through 3.7 list the organic compounds and trace elements found in JP-5, as reported by different references.

3.1.6 JP-4

Tables 3.8 through 3.10 list the organic compounds and trace elements found in JP-4, as reported by different references.

3.2 ORGANIC COMPOUNDS

Crude oil is a highly complex mixture, containing hundreds of thousands of hydrocarbons (Cooney, 1980). Compounds in crude oil can be divided into three general classes consisting of saturated hydrocarbons, aromatic hydrocarbons, and polar organic compounds (Huesemann and Moore, 1993). Saturated hydrocarbons can be separated further into straight-chain and branched alkanes, as well as cyclic alkanes with varying numbers of saturated rings and side chains. Aromatic hydrocarbons contain one or more aromatic rings ranging from simple monoaromatic compounds, such as benzene and toluene to polyaromatic compounds, such as pyrene. The polar fraction is made of compounds containing "polar" heteroatoms, such as nitrogen, sulfur, and oxygen.

Refined oils are also complex; for example, kerosene may contain as many as 10,000 different hydrocarbons (Sharpley, 1964). This complexity also extends to other petroleum products (Atlas, 1977). For microorganisms to biodegrade petroleum completely or attack even simpler refined oils, thousands of different compounds must be metabolized. The chemical nature of these petroleum components varies from the simple n-paraffin, monoalicyclic, and monoaromatic compounds, to the much more complex branched chains and condensed ring structures (Horowitz, Sexstone, and Atlas, 1978). Many different enzymes are presumably necessary to biodegrade these types of compounds.

The principal biochemical reactions associated with the microbial metabolism of xenobiotics include acylation, alkylation, dealkylation, dehalogenation, amide or ester hydrolysis, oxidation, reduction, aromatic ring hydroxylation, ring cleavage, and condensate or conjugate formation (Kaufman and Plimmer, 1972). Microbial action is initiated at metal, sulfur, and other functional group sites, and proceeds to breakage of carbon–carbon bonds, even to biocracking of heavy petroleums (Premuzic, Lin, Racaniello, and Manowitz, 1993).

Subsurface screening of petroleum hydrocarbons in contaminated soils can be conducted with laser-induced fluorometry over optical fibers with a cone penetrometer system to identify the contaminants as an initial step in the remediation process (Lieberman, Apitz, Borbridge, and Theriault, 1993). The

Table 3.1 Composition of Diesel Fuel #2

Component	Concentration (% Volume)	Component	Concentration (% Volume)
C_{10} paraffins	0.9	C_{15} paraffins	7.4
C_{10} cycloparaffins	0.6	C_{15} cycloparaffins	5.5
C_{10} aromatics	0.4	C_{15} aromatics	3.2
C_{11} paraffins	2.3	C_{16} paraffins	5.8
C_{11} cycloparaffins	1.7	C_{16} cycloparaffins	4.4
C_{11} aromatics	1.0	C_{16} aromatics	2.5
C_{12} paraffins	3.8	C_{17} paraffins	5.5
C_{12} cycloparaffins	2.8	C_{17} cycloparaffins	4.1
C_{12} aromatics	1.6	C_{17} aromatics	2.4
C_{13} paraffins	6.4	C_{18} paraffins	4.3
C_{13} cycloparaffins	4.8	C_{18} cycloparaffins	3.2
C_{13} aromatics	2.8	C_{18} aromatics	1.8
C_{14} paraffins	8.8	C_{19} paraffins	0.7
C_{14} cycloparaffins	6.6	C_{19} cycloparaffins	0.6
C_{14} aromatics	3.8	C_{19} aromatics	0.3

Source: From Clewell, H.J., III. The Effect of Fuel Composition on Groundfall from Aircraft Fuel Jettisoning. Report. Air Force Engineering and Services Center, Tyndall Air Force Base, FL, 1981.

Table 3.2 Composition of Various Gasolines

Paraffins

Propane	Dimethyl pentanes
Isobutane ($1C_4$)	Methyl hexanes
n-Butane (nC_4)	Trimethyl pentanes
Isopentane ($1C_5$)	Normal heptane
n-Pentane ($1C_5$)	Dimethyl hexanes
Dimethyl butanes (C_6)	Methylethyl pentanes
Methyl pentanes (C_6)	Dimethyl hexanes
n-Hexane	Trimethyl hexanes
n-Octane	

Naphthenes

Methylcyclopentane	Methylcyclohexane
Cyclohexane	Other cyclic saturates

Aromatics

Benzene	Propylbenzene
Toluene	Methylethylbenzenes
Ethylbenzene	Trimethylbenzene
Xylenes	Other aromatics

Olefins

Methylbutene	Methylpentene
Pentene	Other olefins

Source: From Ghassemi, M. et al. *Energ. Sourc.* 7:377–401, 1984. With permission.

system provides the capability for real-time, *in situ* measurement of petroleum hydrocarbon contamination and soil type to depths of 50 m.

A common technique for measuring the biodegradability of an organic compound is calculation of the BOD/COD ratio. This ratio is an indication of the amount of degradation that occurs, or biochemical oxygen demand (BOD), relative to the amount of material available to be degraded, chemical oxygen

Table 3.3 Components of Gasoline

Component	Component
n-Propane	3,3-Dimethylpentane
n-Butane	2,3-Dimethylpentane
n-Pentane	2,5-Dimethylhexane
n-Hexane	2,4-Dimethylhexane
n-Heptane	2,3-Dimethylhexane
n-Octane	3,4-Dimethylhexane
n-cis-Butene-2	2,2-Dimethylhexane
n-Pentane-2	2,2-Dimethylheptane
2,3-Dimethylbutene-1	1,1-Dimethylcyclopentane
Olefins C_4	1,2- and 1,3-Dimethylcyclopentane
Olefins C_5	1,3- and 1,4-Dimethylcyclohexane
Olefins C_6	1,2-Dimethylcyclohexane
Isobutane	2,2,3-Trimethylbutane
Cyclopentane	2,2,4-Trimethylpentane
Cyclohexane	2,2,3-Trimethylpentane
Methylcyclopentane	2,3,4-Trimethylpentane
Methylcyclohexane	2,3,3-Trimethylpentane
2-Methylbutane	2,2,5-Trimethylpentane
2-Methylpentane	1,2,4-Trimethylcyclopentane
3-Methylpentane	Ethylpentane
2-Methylhexane	Ethylcyclopentane
3-Methylhexane	Ethylcyclohexane
2-Methylheptane	Benzene
3-Methylheptane	Ethylbenzene
4-Methylheptane	Toluene
2,2-Dimethylbutane	o-Xylene
2,3-Dimethylbutane	m-Xylene
2,2-Dimethylpentane	p-Xylene
2,4-Dimethylpentane	

Source: From Jamison, V.W. et al. in *Proc. 3rd Int. Biodegradation Symp.* Sharpley, J.M. and Kaplan, A.M., Eds. Elsevier, New York. 1976. With permission.

demand (COD). Table 3.11 presents relative biodegradabilities by adapted sludge cultures of various substances in terms of a BOD/COD ratio, after 5 days of incubation (U.S. EPA, 1985a). A higher ratio represents a higher relative biodegradability. As the ratio approaches zero, the compound becomes less degradable.

Recalcitrant compounds are more difficult to treat with microorganisms. Nevertheless, there are microbes and techniques available for dealing with such contaminants. For example, a novel *Pseudomonas anaerooleophila*, which is resistant to aliphatic, alicyclic, and aromatic carbohydrates, or a mixture thereof, has been patented by Imanaka and Morikawa (1993). This strain, HD-1, can grow in a medium containing n-tetradecane, toluene, cyclohexane, and petroleum and can be useful for treating environmental pollutants. Novotny (1992) has even characterized some thermophilic microorganisms that can utilize recalcitrant substrates.

Many different types of microorganisms are involved in biodegradation. These range from microbes that require oxygen to perform the catabolic reactions to those that require an anaerobic environment. Table 3.12 summarizes several microbial processes and shows an approximate relationship between the process and the environmental redox potential (Berry, Francis, and Bollag, 1987). The more negative the number, the stronger the reducing environment, as can be seen by the strict requirement for anaerobiosis.

A database has been developed to provide rapid, reliable information on the biodegradability of soil pollutants and on possible bioremedial action (Gleim, Milch, and Kracht, 1995). The database was created to facilitate better transfer of scientific results on biodegradation of soil pollutants to those involved in bioremedial action. Information in the database is derived from international scientific publications on biodegradation. It includes information on polycyclic aromatic hydrocarbons (PAHs),

Table 3.4 Composition of Gasoline

Compound	Number of Carbons	Concentration (w%)
Straight Chain Alkanes		
Propane	3	0.01 to 0.14
n-Butane	4	3.93 to 4.70
n-Pentane	5	5.75 to 10.92
n-Hexane	6	0.24 to 3.50
n-Heptane	7	0.31 to 1.96
n-Octane	8	0.36 to 1.43
n-Nonane	9	0.07 to 0.83
n-Decane	10	0.04 to 0.50
n-Undecane	11	0.05 to 0.22
n-Dodecane	12	0.04 to 0.09
Branched Alkanes		
Isobutane	4	0.12 to 0.37
2,2-Dimethylbutane	6	0.17 to 0.84
2,3-Dimethylbutane	6	0.59 to 1.55
2,2,3-Trimethylbutane	7	0.01 to 0.04
Neopentane	5	0.02 to 0.05
Isopentane	5	6.07 to 10.17
2-Methylpentane	6	2.91 to 3.85
3-Methylpentane	6	2.4 (vol)
2,4-Dimethylpentane	7	0.23 to 1.71
2,3-Dimethylpentane	7	0.32 to 4.17
3,3-Dimethylpentane	7	0.02 to 0.03
2,2,3-Trimethylpentane	8	0.09 to 0.23
2,2,4-Trimethylpentane	8	0.32 to 4.58
2,3,3-Trimethylpentane	8	0.05 to 2.28
2,3,4-Trimethylpentane	8	0.11 to 2.80
2,4-Dimethyl-3-ethylpentane	9	0.03 to 0.07
2-Methylhexane	7	0.36 to 1.48
3-Methylhexane	7	0.30 to 1.77
2,4-Dimethylhexane	8	0.34 to 0.82
2,5-Dimethylhexane	8	0.24 to 0.52
3,4-Dimethylhexane	8	0.16 to 0.37
3-Ethylhexane	8	0.01
2-Methyl-3-ethylhexane	9	0.04 to 0.13
2,2,4-Trimethylhexane	9	0.11 to 0.18
2,2,5-Trimethylhexane	9	0.17 to 5.89
2,3,3-Trimethylhexane	9	0.05 to 0.12
2,3,5-Trimethylhexane	9	0.05 to 1.09
2,4,4-Trimethylhexane	9	0.02 to 0.16
2-Methylheptane	8	0.48 to 1.05
3-Methylheptane	8	0.63 to 1.54
4-Methylheptane	8	0.22 to 0.52
2,2-Dimethylheptane	9	0.01 to 0.08
2,3-Dimethylheptane	9	0.13 to 0.51
2,6-Dimethylheptane	9	0.07 to 0.23
3,3-Dimethylheptane	9	0.01 to 0.08
3,4-Dimethylheptane	9	0.07 to 0.33
2,2,4-Trimethylheptane	10	0.12 to 1.70
3,3,5-Trimethylheptane	10	0.02 to 0.06
3-Ethylheptane	10	0.02 to 0.16
2-Methyloctane	9	0.14 to 0.62
3-Methyloctane	9	0.34 to 0.85

Table 3.4 (continued) Composition of Gasoline

Compound	Number of Carbons	Concentration (w%)
4-Methyloctane	9	0.11 to 0.55
2,6-Dimethyloctane	10	0.06 to 0.12
2-Methylnonane	10	0.06 to 0.41
3-Methylnonane	10	0.06 to 0.32
4-Methylnonane	10	0.04 to 0.26
Cycloalkanes		
Cyclopentane	5	0.19 to 0.58
Methylcyclopentane	6	Not quantified
1-Methyl-*cis*-2-ethylcyclopentane	8	0.06 to 0.11
1-Methyl-*trans*-3-ethylcyclopentane	8	0.06 to 0.12
1-*cis*-2-Dimethylcyclopentane	7	0.07 to 0.13
1-*trans*-2-Dimethylcyclopentane	7	0.06 to 0.20
1,1,2-Trimethylcyclopentane	8	0.06 to 0.11
1-*trans*-2-*cis*-3-Trimethylcyclopentane	8	0.01 to 0.25
1-*trans*-2-*cis*-4-Trimethylcyclopentane	8	0.03 to 0.16
Ethylcyclopentane	7	0.14 to 0.21
n-Propylcyclopentane	8	0.01 to 0.06
Isopropylcyclopentane	8	0.01 to 0.02
1-*trans*-3-dimethylcyclohexane	8	0.05 to 0.12
Ethylcyclohexane	8	0.17 to 0.42
Straight Chain Alkenes		
cis-2-Butene	4	0.13 to 0.17
trans-2-Butene	4	0.16 to 0.20
Pentene-1	5	0.33 to 0.45
cis-2-Pentene	5	0.43 to 0.67
trans-2-Pentene	5	0.52 to 0.90
cis-2-Hexene	6	0.15 to 0.24
trans-2-Hexene	6	0.18 to 0.36
cis-3-Hexene	6	0.11 to 0.13
trans-3-Hexene	6	0.12 to 0.15
cis-3-Heptene	7	0.14 to 0.17
trans-2-Heptene	7	0.06 to 0.10
Branched Alkenes		
2-Methyl-1-butene	5	0.22 to 0.66
3-Methyl-1-butene	5	0.08 to 0.12
2-Methyl-2-butene	5	0.96 to 1.28
2,3-Dimethyl-1-butene	6	0.08 to 0.10
2-Methyl-1-pentene	6	0.20 to 0.22
2,3-Dimethyl-1-pentene	7	0.01 to 0.02
2,4-Dimethyl-1-pentene	7	0.02 to 0.03
4,4-Dimethyl-1-pentene	7	0.6 (vol)
2-Methyl-2-pentene	6	0.27 to 0.32
3-Methyl-*cis*-2-pentene	6	0.35 to 0.45
3-Methyl-*trans*-2-pentene	6	0.32 to 0.44
4-Methyl-*cis*-2-pentene	6	0.04 to 0.05
4-Methyl-*trans*-2-pentene	6	0.08 to 0.30
4,4-Dimethyl-*cis*-2-pentene	7	0.02
4,4-Dimethyl-*trans*-2-pentene	7	Not quantified
3-Ethyl-2-pentene	7	0.03 to 0.04

Table 3.4 (continued) Composition of Gasoline

Compound	Number of Carbons	Concentration (w%)
Cycloalkenes		
Cyclopentene	5	0.12 to 0.18
3-Methylcyclopentene	6	0.03 to 0.08
Cyclohexene	6	0.03
Alkyl Benzenes		
Benzene	6	0.12 to 3.50
Toluene	7	2.73 to 21.80
o-Xylene	8	0.68 to 2.86
m-Xylene	8	1.77 to 3.87
p-Xylene	8	0.77 to 1.58
1-Methyl-4-ethylbenzene	9	0.18 to 1.00
1-Methyl-2-ethylbenzene	9	0.19 to 0.56
1-Methyl-3-ethylbenzene	9	0.31 to 2.86
1-Methyl-2-*n*-propylbenzene	10	0.01 to 0.17
1-Methyl-3-*n*-propylbenzene	10	0.08 to 0.56
1-Methyl-3-isopropylbenzene	10	0.01 to 0.12
1-Methyl-3-*t*-butylbenzene	11	0.03 to 0.11
1-Methyl-4-*t*-butylbenzene	11	0.04 to 0.13
1,2-Dimethyl-3-ethylbenzene	10	0.02 to 0.19
1,2-Dimethyl-4-ethylbenzene	10	0.50 to 0.73
1,3-Dimethyl-2-ethylbenzene	10	0.21 to 0.59
1,3-Dimethyl-4-ethylbenzene	10	0.03 to 0.44
1,3-Dimethyl-5-ethylbenzene	10	0.11 to 0.42
1,3-Dimethyl-5-*t*-butylbenzene	12	0.02 to 0.16
1,4-Dimethyl-2-ethylbenzene	10	0.05 to 0.36
1,2,3-Trimethylbenzene	9	0.21 to 0.48
1,2,4-Trimethylbenzene	9	0.66 to 3.30
1,3,5-Trimethylbenzene	9	0.13 to 1.15
1,2,3,4-Tetramethylbenzene	10	0.02 to 0.19
1,2,3,5-Tetramethylbenzene	10	0.14 to 1.06
1,2,4,5-Tetramethylbenzene	10	0.05 to 0.67
Ethylbenzene	8	0.36 to 2.86
1,2-Diethylbenzene	10	0.57
1,3-Diethylbenzene	10	0.05 to 0.38
n-Propylbenzene	9	0.08 to 0.72
Isopropylbenzene	9	<0.01 to 0.23
n-Butylbenzene	10	0.04 to 0.44
Isobutylbenzene	10	0.01 to 0.08
sec-Butylbenzene	10	0.01 to 0.13
t-Butylbenzene	10	0.12
n-Pentylbenzene	11	0.01 to 0.14
Isopentylbenzene	11	0.07 to 0.17
Indan	9	0.25 to 0.34
1-Methylindan	10	0.04 to 0.17
2-Methylindan	10	0.02 to 0.10
4-Methylindan	10	0.01 to 0.16
5-Methylindan	10	0.09 to 0.30
Tetralin	10	0.01 to 0.14
Polynuclear Aromatic Hydrocarbons		
Naphthalene	10	0.09 to 0.49
Pyrene	16	Not quantified

Table 3.4 (continued) Composition of Gasoline

Compound	Number of Carbons	Concentration (w%)
Benz(a)anthracene	18	Not quantified
Benzo(a)pyrene	20	0.19 to 2.8 mg/kg
Benzo(e)pyrene	20	Not quantified
Benzo(g,h,i)perylene	21	Not quantified
Elements		
Bromine		80 to 345 µg/g
Cadmium		0.01 to 0.07 µg/g
Chlorine		80 to 300 µg/g
Lead		530 to 1120 µg/g
Sodium		<0.6 to 1.4 µg/g
Sulfur		0.10 to 0.15(ASTM)
Vanadium		<0.02 to 0.001 µg/g
Additives		
Ethylene dibromide		0.7 to 177.2 ppm
Ethylene dichloride		150 to 300 ppm
Tetramethyl lead		
Tetraethyl lead		

Source: From State of California. *Leaking Underground Fuel Tank Field Manual.* Academic Press, Orlando, FL, 1987. With permission.

Table 3.5 Selected Compound Types Occurring in JP-5

Aromatic	Partial Saturation	Saturated
Benzene	—	Cyclohexane
Indene	Indane (Indan)	Hydrindane (Hydroindane)
Naphthalene	Tetralin (Tetrahydronaphthalene)	Decalin (Decahydronaphthalene)
Acenaphthalene	Acenaphthene	Perhydroacenaphthalene
Phenanthrene	Tetrahydrophenanthrene	Perhydrophenanthrene

Source: From Varga, G.M., Jr. et al. Report to Naval Air Propulsion Center, Contract N00140-81-C-9601, 1985. NAPC-PE-121C.

Table 3.6 Trace Elements in Shale-Derived JP-5

Element	ppm	Element	ppm	Element	ppm
Al	0.048	Cu	<0.02	Si	≦10
Sb	<3	Fe	<0.01	Ag	<0.02
As	<0.5	Pb	<0.06	Na	0.14
Be	<0.01	Mg	<5.3	Sr	≦0.94
Cd	<0.02	Mn	<0.02	Tl	<6
Ca	≦0.6	Hg	<2	Sn	≦0.93
Cl	<2	Mo	≦0.03	Ti	<0.4
Cr	≦0.094	Ni	<3.9	V	≦0.0008
Co	≦0.04	Se	<0.5	Zn	≦0.02

Source: Ghassemi, M., Panahloo, A., and Quinlivan, S. *Environ. Toxicol. Chem.* 3:511–535. Society of Environmental Toxicology and Chemistry (SETAC). 1984. With permission.

Table 3.7 Major Components of JP-5

Fuel Component	Concentration (w%)	Fuel Component	Concentration (w%)
n-Octane	0.12	n-Dodecane	3.94
1,3,5-Trimethylcyclohexane	0.09	2,6-Dimethylundecane	2.00
1,1,3-Trimethylcyclohexane	0.05	1,2,4-Triethylbenzene	0.72
m-Xylene	0.13	2-Methylnaphthalene	0.90
3-Methyloctane	0.07	1-Methylnaphthalene	1.44
2,4,6-Trimethylheptane	0.09	1-Tridecene	0.45
o-Xylene	0.09	Phenylcyclohexane	0.82
n-Nonane	0.38	n-Tridecane	3.45
1,2,4-Trimethylbenzene	0.37	1-t-Butyl-3,4,5-trimethylbenzene	0.24
n-Decane	1.79	n-Heptylcyclohexane	0.99
n-Butylcyclohexane	0.90	n-Heptylbenzene	0.27
1,3-Diethylbenzene	0.61	Biphenyl	0.70
1,4-Diethylbenzene	0.77	1-Ethylnaphthalene	0.32
4-Methyldecane	0.78	2,6-Dimethylnaphthalene	1.12
2-Methyldecane	0.61	n-Tetradecane	2.72
1-Ethylpropylbenzene	1.16	2,3-Dimethylnaphthalene	0.46
n-Undecane	3.95	n-Octylbenzene	0.78
2,6-Dimethyldecane	0.72	n-Pentadecane	1.67
1,2,3,4-Tetramethylbenzene	1.48	n-Hexadecane	1.07
Naphthalene	0.57	n-Heptadecane	0.12
2-Methylundecane	1.39		

Source: From Smith, J.H. et al. SRI Int., Menlo Park, CA. Report No. ESL-TR-81-54. Engineering and Services Laboratory, Tyndall Air Force Base, FL, 1981.

Table 3.8 Composition of JP-4

Component	Concentration (% Volume)	Component	Concentration (% Volume)
C_5 hydrocarbons	3.9	C_{11} paraffins	4.8
C_6 paraffins	8.1	C_{11} cycloparaffins	2.5
C_6 cycloparaffins	2.1	Dicycloparaffins	3.4
Benzene	0.3	C_{11} aromatics	1.1
C_7 paraffins	9.4	C_{11} naphthalenes	0.2
C_7 cycloparaffins	7.1	C_{12} paraffins	2.8
Toluene	0.7	C_{12} cycloparaffins	1.2
C_8 paraffins	10.1	C_{12} aromatics	0.5
C_8 cycloparaffins	7.4	C_{12} naphthalenes	0.2
C_8 aromatics	1.6	C_{13} paraffins	1.1
C_9 paraffins	9.1	C_{13} cycloparaffins	0.4
C_9 cycloparaffins	4.3	C_{13} aromatics	0.1
C_9 aromatics	2.4	C_{14} hydrocarbons	0.2
C_{10} paraffins	7.3	C_{15} hydrocarbons	0.1
C_{10} cycloparaffins	3.7	Tricycloparaffins	1.8
C_{10} aromatics	1.8	Residual hydrocarbons	0.1
Napthalene	0.2		

Source: From Clewell, H.J., III. The Effect of Fuel Composition on Groundfall from Aircraft Fuel Jettisoning, Air Force Engineering and Services Center, Tyndall Air Force Base, FL, 1981.

halogenated compounds, degradation in soils, and use of pure or mixed cultures. The files cover compounds, mixtures and classes of compounds, microorganisms, culture conditions, metabolism/metabolites, soil properties, experimental scale and experimental conditions, bioremediation treatment, degradation data, and bibliography. The language of the database was German until April 1994, after which

Table 3.9 Major Components of JP-4

Fuel Component	Concentration (w%)	Fuel Component	Concentration (w%)
n-Butane	0.12	m-Xylene	0.96
Isobutane	0.66	p-Xylene	0.35
n-Pentane	1.06		
2,2-Dimethylbutane	0.10	3,4-Dimethylheptane	0.43
2-Methylpentane	1.28	4-Ethylheptane	0.18
3-Methylpentane	0.89	4-Methyloctane	0.86
n-Hexane	2.21	2-Methyloctane	0.88
Methylcyclopentane	1.16	3-Methyloctane	0.79
2,2-Dimethylpentane	0.25	o-Xylene	1.01
Benzene	0.50		
Cyclohexane	1.24	1-Methyl-4-ethylcyclohexane	0.48
2-Methylhexane	2.35	n-Nonane	2.25
3-Methylhexane	1.97	Isopropylbenzene	0.30
trans-1,3-Dimethylcyclopentane	0.36	n-Propylbenzene	0.71
cis-1,3-Dimethylcyclopentane	0.34	1-Methyl-3-ethylbenzene	0.49
cis-1,2-Dimethylcyclopentane	0.54	1-Methyl-4-ethylbenzene	0.43
n-Heptane	3.67	1,3,5-Trimethylbenzene	0.42
Methylcyclohexane	2.27	1-Methyl-2-ethylbenzene	0.23
2,2,3,3-Tetramethylbutane	0.24	1,2,4-Trimethylbenzene	1.01
Ethylcyclopentane	0.26	n-Decane	2.16
2,5-Dimethylhexane	0.37	n-Butylcyclohexane	0.70
2,4-Dimethylhexane	0.58	1,3-Diethylbenzene	0.46
1,2,4-Trimethylcyclopentane	0.25	1-Methyl-4-propylbenzene	0.40
3,3-Dimethylhexane	0.26	1,3-Dimethyl-5-ethylbenzene	0.61
1,2,3-Trimethylcyclopentane	0.25	1-Methyl-2-i-propylbenzene	0.29
Toluene	1.33	1,4-Dimethyl-2-ethylbenzene	0.70
		1,2-Dimethyl-4-ethylbenzene	0.77
2,2-Dimethylhexane	0.71	n-Undecane	2.32
2-Methylheptane	2.70	1,2,3,4-Tetramethylbenzene	0.75
4-Methylheptane	0.92	Naphthalene	0.50
cis-1,3-Dimethylcyclohexane	0.42	2-Methylundecane	0.64
3-Methylheptane	3.04	n-Dodecane	2.00
1-Methyl-3-ethylcyclohexane	0.17	2,6-Dimethylundecane	0.71
1-Methyl-2-ethylcyclohexane	0.39	2-Methylnaphthalene	0.56
Dimethylcyclohexane	0.43	1-Methylnaphthalene	0.78
n-Octane	3.80	n-Tridecane	1.52
1,3,5-Trimethylcyclohexane	0.99	2,6-Dimethylnaphthalene	0.25
1,1,3-Trimethylcyclohexane	0.48	n-Tetradecane	0.73
2,5-Dimethylheptane	0.52		
Ethylbenzene	0.37		

Source: From Smith, J.H. et al. SRI Int., Menlo Park, CA. Report No. ESL-TR-81-54. Engineering and Services Laboratory, Tyndale Air Force Base, 1981.

it was continued in English. The database was projected to be available in 1996 on diskettes (MS-DOS) from the DSM-Deutsche Sammlung von Mikroorganismen und Zellkulturen GmbH, Mascheroder Weg 1b, D-38124 Braunschweig, Germany.

3.2.1 AEROBIC DEGRADATION

Some compounds appear to be degraded only under aerobic conditions, others only under anaerobic conditions, and some under either condition, while others are not transformed at all. It has been concluded that hydrocarbons are subject to both aerobic and anaerobic oxidation (Dietz, 1980). The first stage of biodegradation of insoluble hydrocarbons is predominantly aerobic, while the organic carbon content is then reduced by anaerobic action.

Table 3.10 Trace Elements in Petroleum-Based JP-4

Element	ppm	Element	ppm	Element	ppm
Al	NA	Cu	<0.05	Si	NA
Sb	<0.5	Fe	<0.05	Ag	NA
As	0.5	Pb	0.09	Na	NA
Be	NA	Mg	NA	Sr	NA
Cd	<0.03	Mn	NA	Th	NA
Ca	NA	Hg	<1	Sn	NA
Cl	NA	Mo	NA	Ti	NA
Cr	<0.05	Ni	<0.05	V	<0.05
Co	NA	Se	<0.03	Zn	<0.05

NA = not applicable.

Source: From Ghassemi, M. et al. *Environ. Toxicol. Chem.* 3:511–535. Society of Environmental Toxicologicy and Chemistry (SETAC), 1984. With permission.

Table 3.11 BOD5/COD Ratios for Various Organic Compounds

Compound	Ratio	Compound	Ratio
Relatively Undegradable			
Heptane	0	*m*-Xylene	<0.008
Hexane	0.	Ethylbenzene	<0.009
o-Xylene	<0.008		
Moderately Degradable			
Gasolines (various)	0.02	*p*-Xylene	<0.11
Nonanol	>0.033	Toluene	<0.12
Undecanol	<0.04	Jet fuels (various)	0.15
Dodecanol	0.097	Kerosene	0.15
Relatively Degradable			
Naphthalene (molten)	<0.20	Jet fuels (various)	0.15
Hexanol	0.20	Kerosene	0.15
Benzene	<0.39	Benzaldehyde	0.62
p-Xylene	<0.11	Phenol	0.81
Toluene	<0.12	Benzoic acid	0.84

Source: Lyman, W.J. et al. in *Handbook of Chemical Properties Estimation Methods: Environmental Behavior of Organic Chemicals,* McGraw-Hill, New York, 1982. Chap. 16. Reprinted in U.S. EPA Handbook No. EPA/625/6-85/006, 1985.

For most compounds, the most rapid and complete degradation occurs aerobically (U.S. EPA, 1985a). It can be generalized that for the degradation of petroleum hydrocarbons, aromatics, halogenated aromatics, polyaromatic hydrocarbons, phenols, halophenols, biphenyls, organophosphates, and most pesticides and herbicides, aerobic bioreclamation techniques are most suitable. Aerobic degradation with methane gas as the primary substrate appears promising for some low-molecular-weight halogenated hydrocarbons.

Microorganisms have evolved catabolic enzyme systems for metabolism of naturally occurring aromatic compounds (Gibson, 1978). In the oxidation of aromatic hydrocarbons, oxygen is the key to the hydroxylation and fission of the aromatic ring. Bacteria incorporate two atoms of oxygen into the hydrocarbons to form dihydrodiol intermediates. The hydroxyl groups are *cis*-dihydrodiols. Oxidation of the dihydrodiols leads to the formation of catechols, which are substrates for enzymatic cleavage of the aromatic ring. In contrast, certain strains of fungi and higher organisms (eukaryotes) incorporate one atom of molecular oxygen into aromatic hydrocarbons to form arene oxides, which can undergo the

BIODEGRADATION/MINERALIZATION/BIOTRANSFORMATION/BIOACCUMULATION

Table 3.12 Relationship between Representative Microbial Processes and Redox Potential

Process	Reaction (electron donor + electron acceptor)	Physiological Type	Redox Potential (mV)
Respiration	OM + O_2 → CO_2	Aerobes	700–500
Denitrification	OM + NO_3^- → N_2 + CO_2	Facultative anaerobes	300
Fermentation	OM → organic acids (mostly acetate, propionate, and butyrate)	Facultative or obligate anaerobes	
Dissimilatory sulfate reduction	OM (or H_2) + SO_4^{-2} → H_2S + CO_2	Obligate anaerobes	–200
Proton reduction	OM (C_4 to C_8 FA) + H_2 + acetate (propionate) + CO_2	Obligate anaerobes	
Methanogenesis	CO_2 + H_2 → CH_4	Obligate anaerobes	<–200
	Acetate → CO_2 + CH_4		<–200

OM = organic matter; FA = fatty acid.

Source: From Berry, D.F. et al. *Microbiol. Rev.* 51:43–59, American Society of Microbiology, 1987. Journals Division. With permission.

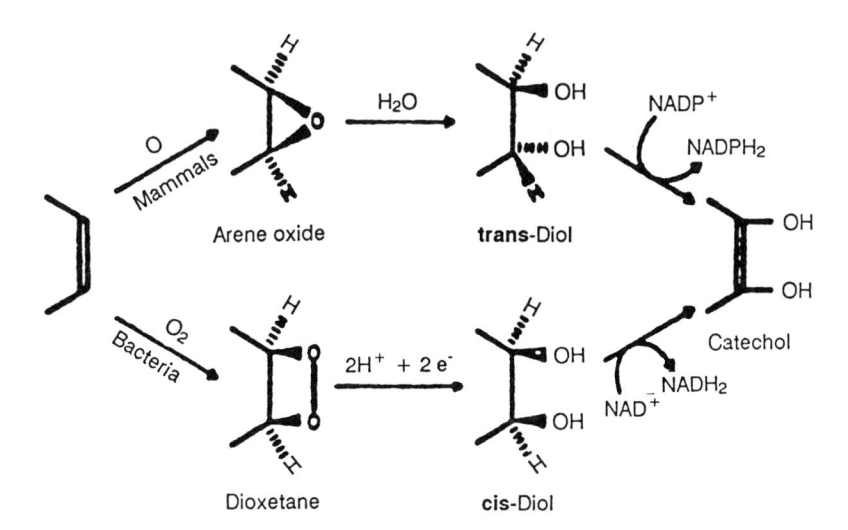

Figure 3.1 Differences between the reactions used by eukaryotic and prokaryotic organisms to initiate the oxidation of aromatic hydrocarbons. (From Gibson, D.T. in *Fate and Effects of Petroleum Hydrocabons in Marine Organisms and Ecosystems.* Wolfe, D.A., Ed. Pergamon Press, Oxford, 1977. With permission.)

enzymatic addition of water to yield *trans*-dihydrodiols. Figure 3.1 shows the reactions used by eukaryotes and prokaryotes for the initial oxidation of aromatic hydrocarbons (Gibson, 1977).

Incorporation of oxygen from the environment requires the action of oxygenase enzymes (Bausum and Taylor, 1986). The oxygenases required for the initial attack on hydrocarbons are typically inducible enzymes, although induction is sometimes accomplished by molecules other than the substrates being oxidized (LePetit and Tagger, 1976). These oxygenases include the following categories (Bausum and Taylor, 1986):

1. *α-Oxygenase.* This enzyme is present only in *Arthrobacter simplex* (Ratledge, 1978), where pentadecane is readily oxidized to pentadecanoic acid. This organism is also able to degrade some fatty acids by oxidative decarboxylation with the evolution of CO_2.
2. *Dioxygenases* (Wood, 1982). Enzymic activation of oxygen from the triplet state to the singlet state is a crucial prerequisite for biological oxidation to occur (Dagley, 1977; Ratledge, 1978). This is accomplished by dioxygenases, where oxygen is fixed directly into organic compounds, e.g., in the oxidation of benzene to catechol, a process commonly found in bacteria, yeast, and fungi (Giger and Roberts, 1978).

3. *Hydroxylases* (Giger and Roberts, 1978; Ratledge, 1978; Wood, 1982). These enzymes accomplish the insertion of an oxygen atom and the transfer of electrons, e.g., in the oxidation of acyclic hydrocarbons with alcohols as the intermediate product. These enzymes are also active in the degradation of PAHs as mixed-function oxygenases of microsomal origin.
4. *Cytochrome oxidases.* These effect the transfer of electrons in the respiratory chain, usually with the formation of water and an organic acid (Bausum and Taylor, 1986). Among the cytochrome oxidases is a group termed cytochrome P-450. This is found in all types of cells, including mammalian, when the cells are stressed by a soluble hydrocarbon molecule or other xenobiotic (Alvares, 1981).

Two hydroxyl groups must be present on the aromatic nucleus for enzymatic fission of the ring to occur, and these may be *ortho* (adjacent) or *para* (opposite on the ring) to each other (Davies and Westlake, 1979). Subsequent metabolic sequences will then vary, depending upon the organism and the site of ring cleavage. Anthropogenic molecules are degraded by these enzymes, if they are structurally similar to the naturally occurring compounds.

Important chemical transformations yielding smaller or more degradable oxygen-containing compounds include the following (Ratledge, 1978):

1. *Terminal oxidation.* The insertion of activated oxygen usually occurs at the free end of an alkyl hydrocarbon chain, with formation of an alcohol, then an aldehyde, and, with further oxidation, a fatty acid. The fatty acid is then degraded by β-oxidation.
2. *Diterminal oxidation.* The insertion of oxygen at both ends of an alkane (α and ω) to give the dioic acid of the alkane.
3. *β oxidation.* This is the degradative pathway for fatty acids. The fatty acids are linked to coenzyme A and degraded to the acid with two fewer carbon atoms and acetyl coenzyme A. ATP (adenosine triphosphate) is required, and the coenzyme A is recycled.
4. *Subterminal oxidation.* Oxidation of the aliphatic hydrocarbons has been shown to occur also at any carbon atom in the hydrocarbon chain. This has been demonstrated in some varieties of bacteria but more often with fungi. Secondary alcohols and subsequently ketones are formed.
5. *Oxidation of double bonds.* Terminal alkenes are readily degraded. These are often attacked at both ends, with the double bond region converted to an epoxide or diol. The utilization of internal alkenes has been reported and can lead to both unsaturated and saturated fatty acids.
6. *Oxidative cleavage of aromatic rings.* The dioxygenases insert active oxygen, forming *cis*-diols or dihydrobenzenoid compounds (*trans*-diols with eukaryotic organisms). The next step is, usually, the formation of an unsaturated acid or dioic acid, by further insertion of oxygen. This breaks the ring structure, and the fatty acids are further oxidized by β oxidation.
7. *ω oxidation.* This has been reported, especially for branched alkanes, which tend to be more resistant (Schaeffer, Cantwell, Brown, Watt, and Fall, 1979). The degradative pathway for a highly branched compound, such as pristane or phytane, may proceed by ω oxidation, forming dicarboxylic acids instead of only monocarboxylic acids, as in normal β oxidation (Markovetz, 1971).
8. *Citronellol pathway.* This pathway occurs in certain *Pseudomonas* sp. and some *Acinetobacter* (Fall, Brown, and Schaeffer, 1979) for the degradation of otherwise recalcitrant, branched hydrocarbons (Bausum and Taylor, 1986). Geranyl-coenzyme A carboxylase is a key enzyme in this pathway, permitting subsequent oxidation to continue by the β oxidation pathway.

Metabolic pathways have been established for the degradation of a number of simple aliphatic and aromatic structures (Atlas, 1978a). The general degradation pathway for an alkane involves sequential formation of an alcohol, an aldehyde, and a fatty acid. The fatty acid is then cleaved (decarboxylated), releasing carbon dioxide and forming a new fatty acid two carbon units shorter than the parent molecule. This process is known as β oxidation. The initial enzymatic attack involves the class of enzymes called oxygenases. The general pathway for degradation of an aromatic hydrocarbon involves *cis*-hydroxylation of the ring structure forming a diol, e.g., catechol. The ring is then cleaved by oxygenases, forming a dicarboxylic acid, e.g., muconic acid.

Degradation of substituted aromatic compounds generally proceeds by initial β oxidation of the side chain, followed by cleavage of the ring structure (Atlas, 1978). Simple alkyl substitution of benzene generally increases the rate of degradation, but extensive alkylation inhibits degradation. This also occurs with polyaromatic hydrocarbons (Cripps and Watkinson, 1978). If side chains are present, oxidation

Table 3.13 Hydrocarbons and Their Relative Levels of Recalcitrance

Easily Degraded		
Volatile aliphatics, *n*-paraffins, aromatics	→	Alkenes, alkadienes, alkynes
Heavy aliphatics, aromatics	→	Saturated alkanes, cyclic hydrocarbons
Phenolic compounds	→	Phenol, cresols, naphthols, xylenols
Intermediate		
Polyaromatic hydrocarbons	→	Mono-, di-, and trinuclear aromatics
Recalcitrant		
Residuum	→	Asphalts, asphaltenes, resinous compounds
Tars, waxes	→	Asphaltic compounds, paraffinic waxes

Data from Lapinskas, J., *Chem. Ind.*, 4 Dec.: 784–789, 1989.

usually occurs at a point next to the ring, but may occur in more than one molecular region (Bausum and Taylor, 1986). A mixed-function oxygenase is the active enzyme, with formation of an arene oxide. This oxide can form a phenolic compound, or, with the addition of water, a diol, which can be further oxidized to a diol epoxide. This is a very potent carcinogen, since it binds with DNA (Harvey, 1982). Species of *Pseudomonas, Beijerinckia,* and *Nocardia* can further degrade these compounds, usually one phenol ring at a time, although the pathways are not clearly defined (Bausum and Taylor, 1986).

Branching generally retards the rate of alkane degradation (Atlas, 1978a). It may also change the metabolic pathway for utilization of a hydrocarbon (Pirnik, 1977). Some long-chain alkanes may be degraded by different metabolic pathways, such as subterminal oxidation (Markovetz, 1971).

The sequence of hydrocarbon degradation in an oil spill is also likely to be determined by the ecological succession of the degrading microorganisms (Bartha and Atlas, 1977). *n*-Alkane degraders with rapid growth rates would out-compete the slow-growing decomposers of the more recalcitrant hydrocarbons for the nutritional resources until the *n*-alkanes are depleted. These organisms would then be replaced by microbes with slower growth rates but greater metabolic flexibility to degrade the more recalcitrant hydrocarbons (Fredericks, 1966).

The relative biodegradability of hydrocarbons has been reported to be (in decreasing order of degradability) (Perry and Cerniglia, 1973):

Linear alkanes (C_{10} to C_{19})
Linear alkanes (C_{12} to C_{18})
Gases (C_2 to C_4)
Alkanes (C_5 to C_9)
Branched alkanes to 12 carbons
Alkenes (C_3 to C_{11})
Branched alkenes
Aromatics
Cycloalkanes

The number of microorganisms isolated that grow on these substrates decreases from top to bottom. Any organism isolated on the compounds further down the list will, generally, grow on those above.

Table 3.13 indicates the relative biodegradability of the various hydrocarbon types, and Figure 3.2 shows the recalcitrance of the relative proportions of hydrocarbon fractions from lubricating oil (Lapinskas, 1989).

A buildup of hydrocarbon substances to larger molecules has also been reported (Ratledge, 1978; Rosenberg, Zuckerberg, Rubinovitz, and Gutnick, 1979b; Atlas, 1981). Very long chain alkanes (waxes) and esters arise during microbial action upon at least some oils. These are highly resistant to further enzymic attack and play a role in the formation of recalcitrant tarry materials.

The diversity of catabolic pathways for the degradation of hydrocarbons between different species and even strains of microorganisms makes it difficult to summarize all of the varied mechanisms and reactions that can occur (Hornick, Fisher, and Paolini, 1983). However, some of the major degradative pathways will be discussed here.

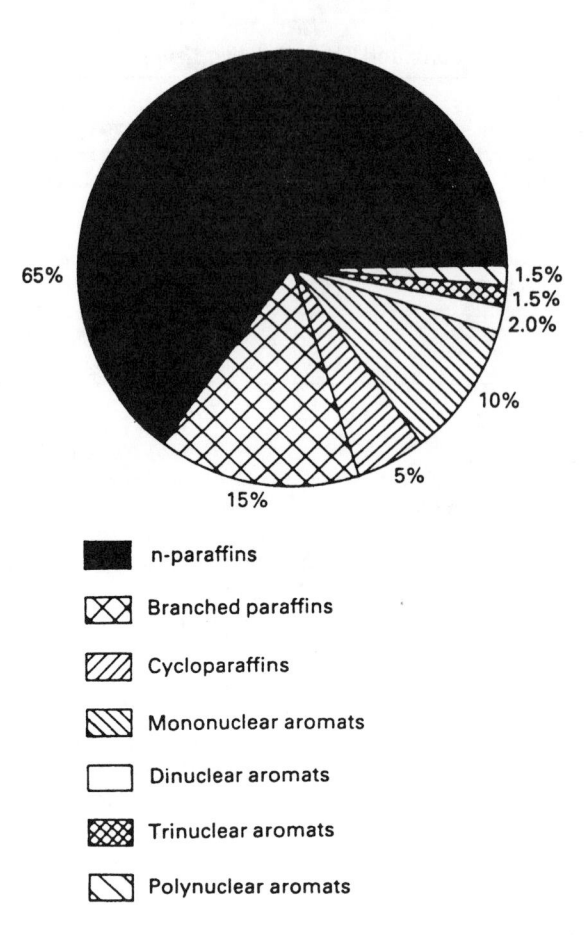

Figure 3.2 Hydrocarbon substrate recalcitrance: relative proportions of hydrocarbon fractions. (From Lapinskas, J. *Chem. Ind.* 4 Dec.:784–789, 1989. With permission.)

In a product, such as gasoline, the most important aliphatic hydrocarbons are the alkanes and the alicyclics, which exhibit important differences in their degradative pathways and their susceptibility to biodegradation.

3.2.1.1 Degradation of Alkanes

Alkanes are completely saturated hydrocarbons; i.e., they contain only carbon–hydrogen and carbon–carbon single bonds. A large amount of structural diversity is represented by the many isomers of alkanes, but only a limited number of these isomers occur in large amounts, and only the *n*-alkanes and the branched alkanes are important environmental contaminants.

In a study of the fate of petroleum aliphatic hydrocarbons in sewage sludge–amended soils, it was found the *n*-alkanes were lost by both abiotic and biotic processes, the former being more effective for short-chain alkanes (Stronguilo, Vaquero, Comellas, and Broto-Puig, 1994).

The *n*-alkanes are the most widely and readily utilized hydrocarbons, with those between C_{10} and C_{25} being most suitable as substrates for microorganisms (Bartha and Atlas, 1977). The *n*-alkanes are the most susceptible components in oil to microbial attack (Westlake, Jobson, and Cook, 1978). The process is similar to the degradation of fatty acids. Biodegradation of *n*-alkanes with molecular weights up to *n*-C_{44} has been demonstrated (Haines and Alexander, 1974), even with a solubility as low as 1 ng/L (Alexander, 1994).

Oxidation of alkanes is classified as being terminal or diterminal (Zajic, 1964). These terms indicate that the initial breach occurs at one of the terminal carbon atoms (Figure 3.3).

Monoterminal oxidation proceeds by the formation of a free radical and then the corresponding alcohol, which is readily oxidized to its respective aldehyde or aliphatic acid (Zajic, 1964). The terminal

A. Terminal Oxidation of Alkanes:

$$RCH_2CH_3 \longrightarrow [RCH_2 \cdot CH_2]$$
$$\downarrow O_2$$
$$RCH_2CH_2OH$$
$$\downarrow$$
$$RCH_2COOH$$

B. Alpha Oxidation (variation of monoterminal oxidation):

$$RCH_2CH_3 \longrightarrow [RCH_2 \cdot CH_2] \longrightarrow [R \cdot CHCH_3]$$
$$\downarrow O_2$$
$$RCHOHCH_3$$
$$\downarrow$$
$$\overset{O}{\underset{\|}{R}}CCH_3$$

methyl ketones

C. Diterminal Oxidation

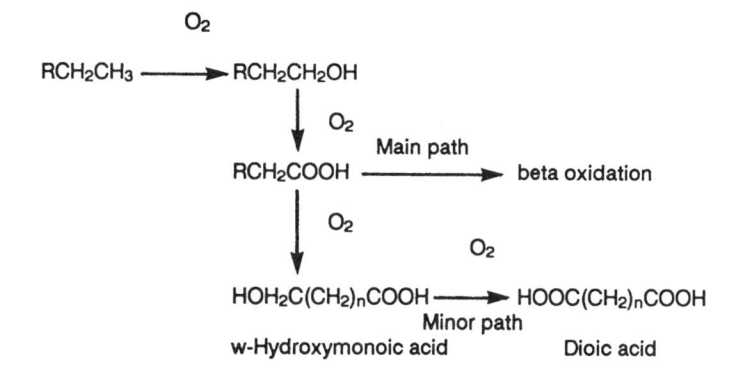

$$\overset{O_2}{\downarrow}$$
$$RCH_2CH_3 \longrightarrow RCH_2CH_2OH$$
$$\downarrow O_2 \quad \text{Main path}$$
$$RCH_2COOH \longrightarrow \text{beta oxidation}$$
$$\downarrow O_2$$
$$HOH_2C(CH_2)_nCOOH \overset{O_2}{\longrightarrow} HOOC(CH_2)_nCOOH$$
$$\text{Minor path}$$
w-Hydroxymonoic acid   Dioic acid

Figure 3.3 Terminal or diterminal oxidation of alkanes or aliphatics. (From Zajic, J.E. *Devel. Ind. Microbiol.* 6:16–27, 1964. With permission.)

methyl group is enzymatically oxidized by incorporation of molecular oxygen by a monooxygenase, producing a primary alcohol, with further oxidation to a monocarboxylic acid (Atlas, 1981; Hornick, Fisher, and Paolini, 1983). β oxidation of the carboxylic acid results in formation of fatty acids and acetyl coenzyme A, with eventual liberation of carbon dioxide. Fatty acids can be toxic and may accumulate during hydrocarbon biodegradation. Fatty acids produced from certain chain length alkanes may be directly incorporated into membrane lipids, instead of going through the β oxidation pathway (Dunlap and Perry, 1967).

α oxidation of alkanes to methyl ketones may proceed with the same initial breech, but at the α carbon (Zajic, 1964). α alcohols are formed, which are further oxidized to methyl ketones. α oxidation has been obtained from propane, butane, pentane, and hexane.

Diterminal (ω) oxidation (Kester and Foster, 1963) of C_{10} to C_{14} alkanes has been obtained with a culture of *Corynebacterium*. A breach occurs first at one terminal carbon atom and is followed by a second breach at the ω carbon atom, synthesizing first a primary alcohol and then an aliphatic acid. The second breach at the ω carbon gives an ω-hydroxymonoic acid, which is oxidized to the corresponding dioic acid. Certain species of *Pseudomonas* oxidize the fatty acids formed from heptane and hexane by classical β oxidation (Heringa, Huybregtse, and van der Linden, 1961). Decarboxylation of fatty acids does not occur in these particular systems.

Subterminal oxidation occurs with C_3 to C_6, and longer, alkanes with formation of a secondary alcohol and subsequent ketone; these may or may not be metabolized. However, this is probably not the primary metabolic pathway for most *n*-alkane-utilizing microorganisms.

At chain lengths greater than C_6, the degradability generally increases until about C_{11} to C_{12} (Hornick, Fisher, and Paolini, 1983). As the alkane chain length increases, the molecule becomes less soluble in water. However, at chain lengths of C_{11} to C_{12} and above, the liquid *n*-alkanes are "accommodated" in water at a higher concentration than would be extrapolated from the solubility of a series of lower alkanes. This may be due to a change in the structure of the water molecules surrounding the alkanes. During the microbial utilization of long-chain liquid alkanes, microbial attachment to droplets of alkane is seen with "transport" of these long-chain alkanes through the cell membrane. This is also thought to occur with solid, long-chain alkanes, although, in both cases, the actual mechanism is unknown. Entirely separate transport mechanisms may exist for gaseous alkanes, short-chain liquid alkanes, "accommodated" intermediate-chain liquid alkanes, and long-chain solid alkanes (Perry, 1968).

3.2.1.2 Degradation of Branched and Cyclic Alkanes

Isoalkanes (a group of branched alkanes) are degradable by a large number of microorganisms, although they are generally inferior to *n*-alkanes as growth substrates, especially if the branching is extensive or creates quaternary carbon atoms (Hornick, Fisher, and Paolini, 1983). A small number of methyl or ethyl side chains does not drastically decrease the degradability, but complex branched chains, and especially terminal branching, are harder to degrade. The position of the side chain also has an effect on the degradability. 1-Phenylalkanes are more degradable than interiorly substituted ones, and the farther the side group is into the molecule, the slower the degradation.

Cycloalkanes are more resistant to microbial attack than straight-chain alkanes and are highly toxic (Atlas, 1981). This is probably due to the absence of an exposed terminal methyl group for the initial oxidation (Hornick, Fisher, and Paolini, 1983). Complex alicyclic compounds, such as hopanes, are among the most persistent components of petroleum spillages in the environment. Up to six-membered, condensed ring structures have been reported to be subject to microbial degradation.

Bacterial degradation of aromatic hydrocarbons normally involves the formation of a phenolic diol followed by cleavage and formation of a *cis* diacid, while that of fungi forms a *trans*-diol, following ring cleavage (Hornick, Fisher, and Paolini, 1983). Light aromatic hydrocarbons are subject to microbial degradation in a dissolved state. Condensed ring aromatic structures are subject to microbial degradation by a similar metabolic pathway as monocyclic structures; however, these hydrocarbons are relatively resistant to enzymic attack. Structures with four or more condensed rings have been shown to be attacked by co-oxidation or as a result of commensalism.

A mixture of co-oxidation and commensalism by mixed microbial communities appears to be very important in the degradation of cycloalkanes, since only a few species of bacteria have been shown to use cyclohexane as a sole carbon source (Hornick, Fisher, and Paolini, 1983). Co-oxidation by one microbe changes the cycloalkane to a cycloalkanone, using a variety of oxygenases, such as peroxidase or polyphenoloxidase (Beam and Perry, 1974). A commensalistic symbiosis is then postulated to occur, with the cycloalkanones produced by co-oxidation being used as a sole carbon source by a wide range of microorganisms (Donoghue, Griffin, Norris, and Trudgill, 1976). Alkyl side chains on cycloalkanes are degraded by the normal alkane oxidation mechanism before degradation of the cycloalkane itself, depending upon the size of the side chain.

Some examples of microorganisms that can grow on branched alkanes include *Brevibacterium erythrogenes, Corynebacterium* sp., *Mycobacterium fortuitum, M. smegmatis, Nocardia* sp., and a *Pseudomonas aeruginosa*.

3.2.1.3 Degradation of Alkenes

Alkenes are unsaturated, meaning that, in addition to the carbon–hydrogen bonds, they possess one or more carbon–carbon double bonds.

Unsaturated 1-alkenes are generally oxidized at the saturated end of the molecule by the same mechanism as used for alkanes (Hornick, Fisher, and Paolini, 1983). Some microorganisms, such as the yeast, *Candida lipolytica*, attack at the double bond and convert the alkene into an alkane-1,2-diol. Other minor pathways have been shown to proceed via an epoxide, which eventually is converted into a fatty acid. For similar amounts of degradation, the chain length of 1-alkenes must be longer than the corresponding alkane. Many microorganisms will not grow on 1-alkenes less than C_{12}. Since 2-alkenes are more readily attacked than 1-alkenes, the presence of a terminal methyl group at each end of the molecule appears to make the molecule degradable for more organisms.

3.2.1.4 Degradation of Aromatic Compounds

Although aromatic hydrocarbons are not as readily biodegradable as are normal and branched alkanes, they are somewhat more easily degraded than are the alicyclic hydrocarbons (Perry, 1984). Initial enzymatic transformation of the parent compounds is referred to as primary biodegradation (Dawson and Chang, 1992). *Pseudomonas gladioli* BSU 45124 has a broad enzyme substrate specificity for the primary biodegradation of aromatic hydrocarbons.

Five phases can be distinguished for aerobic and anaerobic metabolism of aromatics (Evans, 1977).

1. Entry into the cell — this can be by free diffusion or with specific transport mechanisms;
2. Manipulations of the side chains and formation of substrates for ring cleavage;
3. Ring cleavage;
4. Conversion of the products of ring cleavage into amphibolic intermediates;
5. Utilization of the amphibolic intermediates.

The side groups of the ring are first modified by hydroxylation, demethylation, or decarboxylation (which are generally enzymatic reactions) to produce one or two basic molecules, which are then cleaved by the second group of enzymes and further degraded to molecules utilizable by the cell (Hornick, Fisher, and Paolini, 1983). The most common ring cleavage mechanism is the "ortho" pathway. This is followed by a series of enzymatic reactions, with the final products being low-molecular-weight organic acids and aldehydes that are readily incorporated into the tricarboxylic acid cycle. PAHs, such as anthracene and phenanthrene, are also degraded by the ortho cleavage pathway. The other major pathway is the "meta" cleavage mechanism, where the aromatic ring is cleaved by a dioxygenase to form a keto acid or an aldehydo-acid.

When an alkyl chain on a PAH is larger than an ethyl group, it is removed by β oxidation (Hornick, Fisher, and Paolini, 1983). Ring cleavage usually can occur when methyl side chains are present; however, if certain locations on the ring are substituted, the resulting compound is very resistant to degradation (McKenna and Heath, 1976; Gibson, 1976).

In one study, aromatic compounds were found to be susceptible to aerobic, but not anaerobic, biodegradation (Rittmann, Bouwer, Schreiner, and McCarty, 1980). However, halogenated aliphatic compounds evaluated were degradable only under anaerobic conditions and not aerobic conditions (Roberts, McCarty, Reinhard, and Schreiner, 1980). Degradation of chlorinated aromatics only under aerobic conditions suggests the need for mixed-function oxidase systems to bring about dehalogenation and ring cleavage of these compounds (Bitton and Gerba, 1985). Table 3.14 lists aromatic hydrocarbons known to be oxidizable by microorganisms (Gibson, 1977).

Some microorganisms can produce enzymes that can be used not only on the inducing compound but also on other compounds (Dawson and Chang, 1992). For instance, growth of *P. gladioli* BSU 45124 on phenol induces production of enzymes necessary for degradation of toluene. Induction of enzymes with broad specificity can thus make it possible for organisms to develop the capacity to utilize previously recalcitrant compounds.

3.2.1.5 Degradation of Specific Compounds

Some of the more common organic constituents of fuels will be discussed, including organisms that can degrade or transform them, intermediate and end products of the process, and factors affecting their

Table 3.14 Aromatic Hydrocarbons Known to Be Oxidized by Microorganisms

Monocyclic	*Tricyclic*
Benzene	Phenanthrene
Toluene	Anthracene
Xylenes	
Tri- and Tetramethylbenzenes	*Polycyclic*
Alkylbenzenes (linear and branched)	Pyrene
Cycloalkylbenzenes	Benz(a)pyrene
	Benzo(a)anthracene
Dicyclic	Dibenz(a)anthracene
Naphthalene	Benzperylene
Methylnaphthalenes (mono and di)	

Source: From Gibson, D.T. in *Fate and Effects of Petroleum Hydrocarbons in Marine Organisms and Ecosystems.* Wolfe, D.A., Ed. Pergamon, Oxford, 1977. With permission.

Table 3.15 Degradation of Anthropogenic Compounds by Different Groups of Microorganisms

Microorganism	Compound
Cyanobacteria (blue-green algae)	
Microcystis aeruginosa	Benzene, toluene, naphthalene, phenanthrene, pyrene
Algae	
Selanastrum capricornatum	Benzene, toluene, naphthalene, phenanthrene, pyrene
Actinomycetes	
Nocardia spp.	n-Paraffins: pentane, hexane, heptane, octane, 2-methylbutane, 2-methylpentane, 3-methylheptane, 2,2,4-trimethylpentane, ethylbenzene,[a] hexadecane, and kerosene (at 2% but not 4%)[b]
Yeasts	
Trichosporon, Pichia Rhodosporidium, Rhodotorula, Debaryomyces, Endomycopsis	Hexadecane and kerosene (at 2% but not 4%)[b]
Candida parapsilosis, C. tropicalis, C. guilliermondii, C. lipolytica, C. maltosa, Debaryomyces hansenii, Trichosporon sp., Rhodosporidium toruloides	(Naphthalene, biphenyl, benzo(a)pyrene)[c]

[a] Jamison, Raymond, and Hudson, 1976.
[b] Ahearn, Meyers, and Standard, 1971.
[c] Cerniglia and Crow, 1981.

Source: From Kobayashi, H. and Rittmann, B.E. *Environ. Sci. Technol.* 16:170A-183A. American Chemical Society. 1982. With permission.

biodegradation. It has been suggested, however, that growth on a pure hydrocarbon should not be used as evidence of oil-degrading activity (Davies and Westlake, 1979). Growth on complex substrates, such as oil, should be correlated with changes in the chemical composition of recovered substrate.

Table 3.15 presents some of the anthropogenic compounds that can be degraded by several groups of microorganisms (Kobayashi and Rittmann, 1982). In all, 32 organisms were isolated from groundwater contaminated with high-octane gasoline (Jamison, Raymond, and Hudson, 1976). These were used to study biodegradation of selected gasoline constituents (Table 3.16). Table 3.17 shows the biodegradation of gasoline components by mixed normal flora. The mixed population of natural flora in the groundwater biodegraded more constituents of the gasoline than the individual isolates, suggesting a form of mutualism

Table 3.16 Growth of Microorganisms on Components of Gasoline

Compound	Nocardia	Pseudomonas	Acinetobacter	Micrococcus	Flavobacterium	Unclassified
General Compound Classes					
n-Alkanes	+	−	−	+	−	−
Cyclic alkanes	−	−	−	−	−	−
Alkyl-substituted cyclicalkane	−	−	−	−	−	−
Monomethylalkanes	+	−	−	+	+	−
Dimethylalkanes	−	−	−	−	+	−
Trimethylalkanes	+	+	+	−	−	+
Aromatics	−	+	+	+	−	+
Specific Compounds					
n-Butane	−	−	−	−	−	−
n-Pentane	*+−	−+*	−	−+	−	−+
n-Hexane	*+	−+*	−+	−+	−	−
n-Heptane	*+−	−+*	−	−	−	−+
n-Octane	*+−	−+*	−	−	−	−+*
n-cis-Butene-2	−	−	−	−	−	−
n-Pentane-2	−	−	−	−	−	−
2,3-Dimethylbutene-1	−	−	−	−	−	−
Cyclopentane	−	−	−	−	−	−
Cyclohexane	−	−	−	−	−	−
Methylcyclopentane	−	−	−	−	−	−
2-Methylbutane	−+*	−+*	−	−	−	−
2-Methylpentane	−+*	−	−+	+	−	−+
3-Methylpentane	−	−	−	−	−	−
3-Methylhexane	−	−	−	−	−	−
2-Methylheptane	−	−	−	−	−	−
3-Methylheptane	*+−	−+	−	−	−	−
2,2-Dimethylbutane	−	−	−	−	−	−
2,3-Dimethylbutane	−	−	−	−	−	−
2,2-Dimethylpentane	−	−	−	−	−	−
2,4-Dimethylpentane	−	−	−	−	−	−
2,3-Dimethylpentane	−	−	−	−	−	−
2,3-Dimethylhexane	−	−	−+	−+	+	−+
1,2-Dimethylcyclohexane	−	−	−	−	−	−
2,2,4-Trimethylpentane (isooctane)	*+−	−+	*+−	−	−	*+−
2,3,4-Trimethylpentane (isooctane)	−	−	−	−	−	−
2,3,3-Trimethylpentane (isooctane)	−	−	−	−	−	−
Ethylcyclohexane	−	−	−	−	−	−
Benzene	−+*	−	+*	−	−	−+*
Ethylbenzene	+−*	−	+*	−+	−	−+
Toluene	−+	+*	+	+−	−	−+*
o-Xylene	−	−	−	−	−	−
m-Xylene	−	+−	−	−	−	−
p-Xylene	−	+−	−	−	−	−
Gasoline	+*	+*	+*	+	+−	+*

Note: Growth on the specific compounds: +− most isolates were +; −+ most isolates were −; * some isolates exhibited moderate to heavy growth.

Source: From Jamison, V.W. et al. in *Proc. 3rd Int. Biodegradation Symp.* Sharpley, J.M. and Kaplan, A.M., Eds. Elsevier, New York, 1976. With permission.

Table 3.17 Biodegradation of Gasoline Components by Mixed Normal Microflora

Component	Percent Biodegraded above Control	Component	Percent Biodegraded above Control
n-Propane	0	2,5-Dimethylhexane	20
n-Butane	0	2,4-Dimethylhexane	0
n-Pentane	70	2,3-Dimethylhexane	19
n-Hexane	46	3,4-Dimethylhexane	84
n-Heptane	49	2,2-Dimethylhexane	75
n-Octane	54	2,2-Dimethylheptane	62
Olefins — C4	0	1,1-Dimethylcyclopentane	25
Olefins — C5	16	1,2- and 1,3-Dimethylcyclopentane	78
Olefins — C6	18	1,3- and 1,4-Dimethylcyclohexane	0
Isobutane	0	1,2-Dimethylcyclohexane	26
Cyclopentane	0	2,2,3-Trimethylbutane	62
Cyclohexane	45	2,2,4-Trimethylpentane	13
Methylcyclopentane	10	2,2,3-Trimethylpentane	54
Methylcyclohexane	75	2,3,4-Trimethylpentane	13
2-Methylbutane	0	2,3,3-Trimethylpentane	16
2-Methylpentane	6	2,2,5-Trimethylpentane	23
3-Methylpentane	7	1,2,4-Trimethylcyclopentane	0
2-Methylhexane	23	Ethylpentane	0
3-Methylhexane	0	Ethylcyclopentane	31
2-Methylheptane	38	Ethylcyclohexane	95
3-Methylheptane	45	Benzene	100
4-Methylheptane	48	Ethylbenzene	100
2,2-Dimethylbutane	25	Toluene	100
2,3-Dimethylbutane	0	*o*-Xylene	100
2,2-Dimethylpentane	9	*m*-Xylene	100
2,4-Dimethylpentane	11	*p*-Xylene	100
3,3-Dimethylpentane	45	Heavy ends	87
2,3-Dimethylpentane	0		

Source: From Jamison, V.W. et al. in *Proc. 3rd Int. Biodegradation Symp.* Sharpley, J.M. and Kaplan, A.M., Eds. Elsevier, New York, 1976. With permission.

involved in the degradation of petroleum, where a variety of organisms are necessary for complete degradation. The percent of each constituent biodegraded by the mixture is given in the table. In Table 3.18 is a summary from the literature of many of the individual components of petroleum and related compounds and the organisms capable of degrading, mineralizing, or transforming them.

3.2.1.5.1 Mononuclear Aromatic Hydrocarbons and Derivatives
1. Phenol

Phenol, a substituted hydrocarbon, can be a breakdown product of other aromatic substances. It is rapidly degraded in aerobically incubated soil (Baker and Mayfield, 1980). Substituted phenols under aerobic conditions are not only sorbed irreversibly by clays and soils, but are also transformed into polymerized species (Sawhney and Kozloski, 1984). Phenol can be utilized as the sole carbon source via catechol by *Pseudomonas putida*, yeasts, etc. (Ghisalba, 1983). *Acinetobacter calcoaceticus* can degrade phenol and many aromatic compounds (Fewson, 1967). A *Pseudomonas* sp. mineralizes phenol rapidly in a medium with 0.2 mM phosphate at pH 5.2, but has little or no effect at pH 8.0. However, mineralization is greater at pH 8.0 than at pH 5.2, when the culture contains 10 mM phosphate.

Phenol and benzoic acid have been degraded after acclimatization to *p*-hydroxybenzoic acid (Healy and Young, 1979). A strain of *P. putida* has evolved a minimum number of genes for degrading phenol more efficiently via another pathway (Chakrabarty, 1994). The growth rate on this compound appears to be limited by the energy generation rate (Babel, 1994). For example, growth of *Alcaligenes eutrophus* JMP134 is much faster on phenol when formate is also present (R.J. Mueller, unpublished results). Fulvic acid can induce enzymes necessary for degradation of low levels of phenols (Boethling and Alexander, 1979a). Growth of *P. gladioli* BSU 45124 on phenol also induces production of enzymes necessary for degradation of toluene (Dawson and Chang, 1992).

BIODEGRADATION/MINERALIZATION/BIOTRANSFORMATION/BIOACCUMULATION

Table 3.18 Fuel Components/Hydrocarbons and Microorganisms Capable of Biodegrading/Biotransforming Them

Fuel Component/Hydrocarbon	Microorganisms
Acrylonitrile	(Mixed culture of yeast, mold, protozoa, bacteria; activated sludge)[e]
Alicyclics	*Pseudomonas anaerooleophila*[bg]
Aliphatics	*Pseudomonas anaerooleophila*[bg]
Alkanes	*Pseudomonas*[o,p] (*Arthrobacter, Acinetobacter* yeasts, *Penicillium* sp., *Cunninghamella blakesleeana, Absidia glauca, Mucor* sp.)[f]
n-Alkanes (C_1 to C_4) gaseous	*Mycobacterium ketoglutamicum*[f]
n-Alkanes (C_3 to C_{16})	*Mycobacterium rhodochrous*[g]
n-Alkanes (C_8 to C_{16})	(*Mycobacterium fortuitum, M. smegmatis*)[g]
n-Alkanes (C_{12} to C_{16})	(*Mycobacterium marinum, M. tuberculosis*)[g] *Corynebacterium*[f]
n-Alkanes (C_5 to C_{16})	(*Arthrobacter, Acinetobacter, Pseudomonas putida*, yeasts)[f]
n-Alkanes (C_{10} to C_{14})	*Corynebacterium*[g]
n-Alkanes (C_8 to C_{20})	*Acinetobacter*[q]
n-Alkanes (C_{11} to C_{19})	*Prototheca zopfii*[l], *Pseudomonas* spp.[k,m]
n-Alkanes (C_{16} to C_{20})	*Rhodotorula rubra*[cs]
Alkanes (straight chain)	*Pseudomonas putida*[f]
Alkenes (C_6 to C_{12})	*Pseudomonas oleovorans*[ab] *Candida lipolytica*[dw]
Alkylbenzenes	*Nocardia* sp.,[f] *Thauera selenatis* (anaerobe)[cr]
Anthracene	Stream bacteria,[e] *Flavobacterium*,[h,az,cn] Coryneform bacillus (*Aureobacterium*?),[ci] *Pseudomonas/Alcaligenes* sp., *Acinetobacter* sp.,[k,cl] pseudomonads,[cx] *Pseudomonas* spp.,[m,cl,cx] *Agrobacterium* spp.,[cl,cx] (*P. putida, Mycobacterium* sp., *M.* PYR-1)[da] *P. paucimobilis* EPA 505,[br] (*P. paucimobilis, P. cepacia, Rhodococcus* sp.)[cn] *Arthrobacter* sp.,[k,cn] *Nocardia*[cx] *Beijerinckia* sp.,[h,k,ax,cn,cx] *Cunninghamella elegans*,[h,ay,ba,cn] *Penicillium tordum*,[cl] *Bjerkandera* sp.,[cn] *B. adjusta*,[cf] *B.* sp. strain BOS55,[bn,bx,dj] *B. adusta* CBS 595.78,[ce] *Candida parapsilosis*,[bq] *Trametes versicolor* Paprican 52,[bx] *T. versicolor*,[u,ce,cf,cn] *Phanerochaete chrysosporium*,[ca,cf,cn] *P. chrysosporium* BKM-F-1767,[bx,u,ce] *Ramaria* sp.,[cn] *P. laevis* HHB-1625,[cw] (*Coriolopsis polyzona, Pleurotus ostreatus*)[u] (*Aspergillus terrus, A. flavus*)[cl] *Rhizoctonia solani*[cn]
Anthracene oil	*Flavobacterium* sp. in mixed culture[z]
Aromatics	*Pseudomonas* sp.,[j] *P. anaerooleophila*,[bg] *P. gladioli* BSU 45124[ck]
Benzene	*Pseudomonas putida*,[o,h,ae, aq] (sewage sludge, stabilization pond microbes)[e] (*P. rhodochrous, P. aeruginosa*)[f] *P. gladioli* BSU 45124,[ck] methanogens,[r,s] anaerobes,[t] *Acinetobacter* sp.,[ae] *Methylosinus trichosporium* OB3b,[ag] *Nocardia* sp.,[ah] *Flavobacterium* DS-711,[cm] (*Microcystis aeruginosa, Selanastrum capricornatum*)[e]
Benz(a)anthracene	*Alcaligenes denitrificans*,[cn] *Pseudomonas* sp.,[f] *P. putida*,[cn] *P. fluorescens*,[bm] *Cunninghamella elegans*,[e,f,u,bd,be,cn] *Beijerinckia* sp.,[c,g,cn] *B.* mutant,[bb] *Mycobacterium*[cz]
Benzo(a)pyrene	*Pseudomonas* sp.,[a] *Bacillus megaterium*,[b] *Mycobacterium* sp.,[cn,cz] *M.* PYR-1,[da] (*Candida parapsilosis, Trichosporon* sp., *Rhodosporidium toruloides, Candida lipolytica, C. guilliermondii*)[a] (*C. tropicalis, C. maltosa*)[a,cn] *Candida* spp.,[ct] *Debaryomyces hansenii*,[a] *Penicillum* sp.,[cn] *Beijerinckia* sp.,[c,g,cn] *Cunninghamella elegans*,[a,u,ap,bf,cn,cu] *C. banieri*,[bf] *Trametes versicolor*,[cf,cn] *T. versicolor* Paprican 52,[bx] *Aspergillus ochraceus*,[cn,bf] *Phanerochaete chrysosporium*,[bz,ca,cf,cn,cv] *P. chrysosporium* BKM-F1767,[bx] *P. laevis* HHB-1625,[cw] *Bjerkandera adusta*,[cf] *Bjerkandera* sp. BOS55,[bp,bx] (*Bjerkandera* sp., *Chrysosporium pannorum, Mortierella verrucosa, Ramaria* sp., *Trichoderma viride*)[cn] *Saccharomyces cerevisiae*,[cn,aj] *Selenastrum capricornutum*,[cn] *Irpex* sp.#232, *Fomitopsis* sp.#259, *Phanerochaete* spp. #326 and 404, *Heteroporus* sp.#501,[cu] Cyanobacteria, diatoms,[cx] *Neurospora cressa*[ai,cu]
Biphenyl	(*Pseudomonas* sp., *Flavobacterium* sp.)[ad] *P. putida*,[e,ae] (*Candida lipolytica, C. tropicalis, C. guilliermondii, C. maltosa, C. parapsilosis, Trichosporon* sp., *Rhodosporidium toruloides, Debaryomyces hansenii*)[a] *Beijerinckia* B8/36,[e,ae] *Cunninghamella elegans*,[h] (*Moraxella* sp., *Beijerinckia* sp.)[ae,aq] *Oscillatoria* sp.[ap,e,ae]
Branched hydrocarbons	*Pseudomonas* sp., *Acinetobacter*[dr,dq]
BTEX	*Phanerochaete chrysosporium*[by]
n-Butane	(*Arthrobacter, Brevibacterium*)[f] *Mycobacterium smegmatis*,[g] (*Pseudobacterium subluteum, Pseudomonas fluorescens, Actinomyces candidus*)[g,dp] *P. methanica*[g,do]

Table 3.18 (continued) Fuel Components/Hydrocarbons and Microorganisms Capable of Biodegrading/Biotransforming Them

Fuel Component/ Hydrocarbon	Microorganisms
Chlorobenzene	*Pseudomonas putida*[ae]
Chrysene	*Rhodococcus* sp.[cn,cp]
Cresols	*Methylosinum trichosporium* OB3b[ag]
o-Cresol	*Pseudomonas gladioli* BSU 45124[ck]
m-Cresol	*Pseudomonas gladioli* BSU 45124[ck]
p-Cresol	*Pseudomonas* sp.,[al] *P. gladioli* BSU 45124[ck]
Crude oils	(*Brevibacterium* sp., *Flavobacterium* sp., *Nocardia, Pseudomonas, Flavobacter, Vibrio, Achromobacter*)[aq] *Acinetobacter calcoaceticus*[bh]
Cyclohexane	(Cometabolism, *Xanthobacter* sp.)[f,h,d] *Nocardia* sp.,[f,h] *Pseudomonas anaerooleophila,*[bg] *Pseudomonas* sp.,[df] *Mycobacterium convolutum*[bs]
Cyclohexanol	*Xanthobacter autotrophicus,*[d] (*Acinetobacter, Nocardia globerula*)[h]
Cyclohexanone	*Xanthobacter autotrophicus*[d]
Decane	*Corynebacterium*[f]
Dibenzanthracene	Activated sludge[a]
Dibenzothiophene (DBT) derivitives	ARK strain[bk]
Dodecane	(*Arthrobacter, Acinetobacter, Pseudomonas putida,* yeasts)[f]
Ethane	*Methylosinus trichosporium,*[f] *Pseudomonas methanica,*[g,do] *P. putida*[h]
Ethylbenzene	*Pseudomonas putida,*[ae,o] *Nocardia* spp.,[dx] *Thauera selenatis* (anaerobe)[cr]
Fatty acids, straight chain (C$_2$ to C$_{16}$)	*Pseudomonas* spp.,[g] *Desulfonema magnum,*[dz] *P. putida*[dy]
Fluoranthene	Sewage sludge,[e] *Pseudomonas* spp.,[m,cl,cn] (*P. putida, P. cepacia*)[cn] *P. paucimobilis,*[bv,cn] *P. paucimobilis* EPA 505,[br,dv] *Alcaligenes denitrificans,*[ch,cn] *Rhodococcus* sp.,[cn,bv,ch] *Rhodococcus* sp.(*R. equi?*),[ci] *Mycobacterium* PYR-1,[da] *Mycobacterium* sp.,[cn] *Agrobacterium* spp.,[cl] Coryneform bacillus (*Aureobacterium?*),[ci] (*Aspergillus terrus, A. flavus, Penicillium tordum*)[cl] *Cunninghamella elegans,*[cn] *Phanerochaete chrysosporium*[ca]
Fluorene	*Phanerochaete chrysosporium,*[ca,cc] *Pseudomonas vesicularis,*[ch] *Rhodococcus* sp.(*R. equi?*)[ci] *Pseudomonas paucimobilis* EPA 505[bv]
Heating oil	*Acinetobacter calcoaceticus*[bh]
n-Heptane	*Pseudomonas aeruginosa,*[g] (*Arthrobacter, Acinetobacter, P. putida,* yeasts)[f] *Nocardia* sp.[e]
n-Hexane	*Mycobacterium smegmatis,*[g] *Nocardia* spp.,[e] (*Hyalodendron, Varicosporium, Paecilomyces, Cladosporium*)[cd]
Hexadecane	(*Pseudomonas putida,* yeasts, *Arthrobacter* sp.)[f] (*Micrococcus cerificans, C. parapsilosis, C. tropicalis, C. guilliermondii, C. lipolytica, Trichosporon* sp., *Rhodosporidium toruloides*)[i] *Prototheca zopfii,* (*alga*),[l] yeasts,[f] *Cladosporium resinae,*[ap,dk] *Nocardia* sp.,[i,ac] (*Pichia, Debaryomyces*)[i,an] (*Torulopsis, Candida,*[an] (*Rhodotorula, Endomycopsis*)[i] *Acinetobacter* sp.[aq,bc] (*Candida petrophilum, P. aeruginosa*)[f,aq]
Jet fuels	*Cladosporium, Hormodendrum*[aq]
Kerosene	(*Candida tropicalis, Torulopsis, Corynebacterium hydrocarboclastus*)[aq] (*Candida parapsilosis, C. guilliermondii, C. lipolytica, Trichosporon* sp., *Pichia, Rhodotorula, Debaryomyces, Endomycopsis, Rhohosporidium toruloides*)[i] *Cladosporium resinae*[ao]
Kerosene, jet fuel, paraffin wax	(*Aspergillus, Botrytis, Candida, Cladosporium, Debaromyces, Endomyces, Fusarium, Hansenula, Monilia, Penicillium, Actinomyces, Micromonospora, Nocardia, Proactinomyces, Streptomyces*)[aq]
Methane	*Pseudomonas methanica,*[g] (*Mycobacterium fortuitum, M. smegmatis*)[ak]
Methylcyclohexane	*Nocardia petroleophila*[bs]
2-Methylhexane	*Pseudomonas aeruginosa*[g]
Naphthalene	*Pseudomonas* sp.,[p,aq,cn] *P. aeruginosa* 19SJ,[bj] *P. stutzeri,*[aa,ci] (*Pseudomonas* NCIB 9816, *Pseudomonas* sp. 53/1 and 53/2, *P. desmolyticum*)[e,h] *P. rathonis,*[g] *P. fluorescens,*[h,as,cn] *P. oleovorans,*[g] *P. paucimobilis* EPA 505,[bv] (*P. paucimobilis, P. vesicularis, P. cepacia, P. testosteroni*)[cn] *P. putida,*[f,v,ar,as,cn] *P. putida* biotype B,[h,as] *Acinetobacter calcoaceticus,*[cn] (*Aeromonas,* stream bacteria)[e] (*Alcaligenes denitrificans, Bacillus cereus*)[cn] *Bacillus naphthalinicum nonliquifaciens,*[h] (*Flavobacterium, Alcaligenes, Corynebacterium*)[e] *Rhodococcus* sp. (*R. rhodochrous?*),[ci] (*Rhodococcus* sp., *Moraxella* sp., *Corynebacterium*

Table 3.18 (continued) Fuel Components/Hydrocarbons and Microorganisms Capable of Biodegrading/Biotransforming Them

Fuel Component/ Hydrocarbon	Microorganisms
	renale, Streptomyces sp. Mycobacterium sp.)[cn] Methylococcus trichosporium OB3b,[co] (Absidia glauca, Aspergillis niger, Basidiobolus ranarum, Circinella sp., Claviceps paspali, Cokeromyces poitrassi, Conidiobolus gonimodes, Emericellopsis sp.)[cn] Cunninghamella elegans,[e,c,e,h,aw,aa,ae,ci,cn] C. bainieri,[c,h,au,cn] C. japonica,[aa,ci,ae,cn] (C. echinulata, Syncephalastrum sp., S. racemosum, Mucor sp., M. hiemalis, Neurospora crassa)[aa,ci,ae,cn] (Claviceps paspali, Psilocybe strictipes, P. subaeruginascens, P. cubensis)[aa,ci,ae] (Candida parapsilosis, C. lipolytica, C. maltosa, C. tropicalis, C. guilliermondii, Debaryomyces hansenii)[a] C. utilis,[cn] (Rhodosporidium toruloides, Trichosporon)[a] (Nostoc sp., Nocardia)[e] (Nocardia strain R, Nocardia sp. NRRL 3385)[ae] (Epicoccum nigrum, Gilbertella persicaria, Gliocladium sp., Helicostylum piriforme, Hyphochytrium catenoides, Linderina pennispora, Panaeolus cambodginensis, P. subbalteatus, Penicillium chrysogenus, Pestalotia sp., Phytophthora cinnamomi, Psilocybe cubensis, Psilicybe strictipes, P. stuntzii, P. subaeruginascens, Rhizophlyctis harderi, R. rosea, Rhizopus oryzae, R. stolonifer, Saccharomyces cerevisiae, Saprolegnia parasitica, Smittium culicis, S. culisetae, S. simulii, Sordaria fimicola, Syncephalastrum racemosum, Thamnidium anomalum, Zygorhynchus moelleri, Phlyctochytrium reinboldtae, Phycomyces blakesleeanus, Penicillium chrysogenum, Choanephora campincta)[cn] Methylococcus trichosporium OB3b,[dl] Agmenellum quadruplicatum (strain PR-6),[av,cn] (Agmenellum, Chlorella sp., Dunaliella sp., Chlamydamonas sp., Cylindriotheca sp., Cyanobacteria)[ap] Oscillatoria,[e,cn] Oscillatoria sp. (strain JCM),[at,cn] Oscillatoria sp. (strain MEV),[cn] Microcoleus sp.,[e] (Nostoc sp., Microcoleus chthonoplastes)[cn] Anabaena,[e] Anabaena sp. (strain CA),[cn] Anabaena sp. (strain 1F),[cn] Aphanocapsa sp.,[e,cn] Coccochloris sp.,[e] (C. elabens, Chlorella sorokiniana, C. autotrophica, Dunaliella tertiolecta, Chlamydomonas angulosa, Ulva fasciata, Cylindrotheca sp., Nitzschia sp., Synedra sp., Navicula sp., Porphyridium cruentum)[cn] Amphora sp.[e,cn]
Octacosane	Rhodococcus sp. Q15[dt]
Octadecane	Micrococcus cerificans[g]
Octane	Pseudomonas,[g,dd] P. putida,[f,v] Corynebacterium sp. 7EIC,[f] Nocardia spp.[e]
Polycyclic aromatic hydrocarbons	Pseudomonas saccharophila P-15[bl]
Paraffins	Trichosporon pullulans,[aq] Nocardia sp.[j]
Pentadecane	Arthrobacter simplex[ds]
n-Pentane	Mycobacterium smegmatis,[g] Nocardia sp.[e]
Petroleum	Pseudomonas anaerooleophila[bg]
Petroleum derivitives	Acinetobacter calcoaceticus[bh]
Phenanthrene	(Pseudomonads, vibrios)[dc] Pseudomonas aeruginosa 19SJ,[bj,bj,dg] P. paucimobilis,[ch,cn] P. paucimobilis EPA 505,[bv] P. putida,[h,ax,cn] Pseudomonas spp.,[m,ci] Arthrobacter polychromogenes,[cj,cn] (Acinetobacter sp., Aeromonas sp., Alcaligenes faecalis, A. denitrificans)[cn] Flavobacterium,[h,w,az,cn] Mycobacterium PYR-1,[da] (Micrococcus sp., Mycobacterium sp., Rhodococcus sp., Streptomyces griseus, Streptomyces sp., Vibrio sp.)[cn] Beijerinckia sp.,[e,ax,cn] Bjerkandera BOS55,[db] Cunninghamella elegans,[h,ay,cn] Candida parapsilosis,[bq] Phanerochaete chrysosporium,[ca,cf,cn,de] P. laevis HHB-1625,[cw] Trametes versicolor,[cf,cn] Chrysosporium lignosum,[cf] Nocardia sp.[cn] (Sclerotium rolfsii, Trichoderma harzianum)[cl] Oscillatoria spp.,[dh,di] (Oscillatoria sp. strain JCM, Agmenellum quadruplicatum)[cn] (Microcystis aeruginosa, Selanastrum capricornatum)[e]
Phenol	(Pseudomonas, Vibrio, Spirillum, Bacillus, Flavobacterium, Chromobacter, Nocardia, Chlamydamonas ulvaensis, Phoridium, fuveolarum, Scenedesmus basiliensis, Euglena gracilis, Corynebacterium sp.)[e] Pseudomonas putida,[m,br] yeasts,[m] (Azotobacter sp., P. putida CB-173 (ATCC 31800))[ab] Acinetobacter calcoaceticus,[af,dn] P. gladioli BSU 45124,[ck] P. stutzeri (anaerobe)[cy]
Phenol cresols	Mutant strains of Pseudomonas putida, (strain U)[aq]
Pristane	Corynebacterium sp, Brevibacterium erythrogenes[f]
n-Propane	Mycobacterium smegmatis, M. rubrum, M. rubrum var. propanicum, M. carotenum, Pseudomonas, puntotropha, Pseudobacterium subluteum,[g] Pseudomonas methanica,[g,do] (Cunninghamella, elegans, Penicillium onatum)[f]
1-Propanol >2-propanol	(Nocardia paraffinica, Brevibacterium sp.)[f]

Table 3.18 (continued) Fuel Components/Hydrocarbons and Microorganisms Capable of Biodegrading/Biotransforming Them

Fuel Component/ Hydrocarbon	Microorganisms
Pyrene	Stabilization pond organisms,[e] *Pseudomonas saccharophila* P-15,[bl] (*P. Alcaligenes* sp., *Acinetobacter* sp., *Arthrobacter* sp.)[k] *Alcaligenes denitrificans*,[cn] *Bacillus subtilis*,[cl] Coryneform bacillus, (*Aureobacterium*?),[ci] *Mycobacterium* sp.,[bo,cn,cz] *Mycobacterium* PYR-1,[da] *Rhodococcus* sp.(*R. equi*?),[ci] *Rhodococcus* sp.,[cn,cp] *Phanerochaete chrysosporium*,[ca,cf,cg,cn] *Cunninghanella elegans*,[cn] *Bjerkandera* BOS55,[db] *Beijerinckia* sp.,[k] (*Microcystis aeruginosa*, *Selanastrum capricornatum*)[e]
Tetradecane	*Micrococcus cerificans*,[g] (*Arthrobacter*, *Acinetobacter*, *Pseudomonas*, *putida*, yeasts)[f] *P. anaerooleophila*,[bg] *Acinetobacter calcoaceticus*[bh,cq]
Toluene	*Bacillus* sp.,[e] *Pseudomonas putida*,[e,f,o,m,ae] *Cunninghamella elegans*,[h] *P. aeruginosa*,[f,am,bu] *P. mildenbergii*,[f] methanogens,[r,s] anaerobes,[s,y,t] *Methylosinus trichosporium* OB3b,[ag] (*Pseudomonas* sp., *Achromobacter* sp.)[f,ah] *P. anaerooleophila*,[bg] *Azoarcus tolulyticus* Tol-4[bi] (anaerobic), *Nocardia corallina*,[bt] *P. gladioli* BSU 45124,[ck] *P. putida*,[m,cm] psychrotrophic spp.,[dt] iron-reducing bacteria,[du] *Phanerochaete chrysosporium*,[by,dm] (*Microcystis aeruginosa*, *Selanastrum capricornatum*)[e] *Thauera selenatis* (anaerobe)[cr]
Total petroleum hydrocarbons	*Acremonium* sp.[bw]
n-Undecane	*Mycobacterium* sp.[g]
Xylenes	*Pseudomonas putida*,[f] *Phanerochaete chrysosporium*[by,dm]
o-Xylene	*Nocardia* sp.[bt]
p- and *m*-Xylene	*Pseudomonas putida*,[f,ae] *P. aeruginosa*,[bu] methanogens,[n] anaerobes[y,t]

References:

a = (Cerniglia and Crow, 1981)
b = (Poglazova, Fedoseeva, Khesina, Meissel, and Shabad, 1967)
c = (Gibson, Mahadevan, Jerina, Yagi, and Yeh, 1975)
d = (Magor, Warburton, Trower, and Griffin, 1986)
e = (Kobayashi and Rittmann, 1982)
f = (Hou, 1982)
g = (Zajic, 1964)
h = (Cerniglia and Gibson, 1977)
i = (Ahearn, Meyers, and Standard, 1971)
j = (Jamison, Raymond, and Hudson, 1975)
k = (Stetzenbach and Sinclair, 1986)
l = (Boehm and Pore, 1984)
m = (Ghisalba, 1983)
n = (Reinhard, Goodman, and Barker, 1984)
o = (Gibson, Koch, and Kallio, 1968)
p = (Solanas, Pares, Bayona, and Albaiges, 1984)
q = (Garvey, Stewart, and Yall, 1985)
r = (Grbic-Galic and Vogel, 1986)
s = (Grbic-Galic and Vogel, 1987)
t = (Battermann and Werner, 1984)
u = (Dodge and Gibson, 1980)
v = (Jain and Sayler, 1987)
w = (Foght and Westlake, 1985)
x = (Rees, Wilson, and Wilson, 1985)
y = (Zeyer, Kuhn, and Schwarzenback, 1986)
aa = (Cerniglia, Herbert, Szaniszlo, and Gibson, 1978)
ab = (Roberts, Koff, and Karr, 1988)
ac = (Mulkins-Phillips and Stewart, 1974b)
ad = (Stucki and Alexander, 1987)
ae = (Knox, Canter, Kincannon, Stover, and Ward, 1968)
af = (Fewson, 1981)
ag = (Higgins, Best, and Hammond, 1980)
ah = (Claus and Walker, 1964)
ai = (Lin and Kapoor, 1979)
aj = (Wiseman, Lim, and Woods, 1978)
ak = (Lukins, 1962)
al = (Dagley and Patel, 1957)
am = (Kitagawa, 1956)
an = (Scheda and Bos, 1966)
ao = (Atlas, 1977)
ap = (Atlas, 1981)
aq = (Savage, Diaz, and Golueke, 1985)
ar = (Jerina, Daly, Jeffrey, and Gibson, 1971)
as = (Jeffrey, Yeh, Herina, Patel, Davey, and Gibson, 1975)
at = (Cerniglia, Van Baalen, and Gibson, 1980)
au = (Ferris, MacDonald, Patrie, and Martin, 1976)
av = (Cerniglia, Gibson, and Van Baalen, 1979)
aw = (Cerniglia, Althaus, Evans, Freeman, Mitchum, and Yang, 1983)
ax = (Jerina, Selander, Yagi, Wells, Davey, Mahadevan, and Gibson, 1976)
ay = (Cerniglia and Yang, 1984)
az = (Colla, Fiecchi, and Treccani, 1959)
ba = (Cerniglia, 1982)
bb = (Gibson and Mahadevan, 1975)
bc = (Makula and Finnerty, 1972)
bd = (Cerniglia, Dodge, and Gibson, 1994)
be = (Fu, Cerniglia, Chou, and Yang, 1983)
bf = (Cerniglia and Gibson, 1979)
bg = (Imanaka and Morikawa, 1993)
bh = (Marin, Pedregosa, Ortiz, and Laborda, 1995)
bi = (Chee-Sanford, Frost, Fries, Zhou, and Tiedje, 1996)
bj = (Deziel, Paquette, Villemur, Lepine, and Bisaillon, 1996)
bk = (Kohata, Yamane, Hosomi, and Murakami, 1995)
bl = (Stringfellow, Chen, and Aitken, 1995)
bm = (Caldini, Cenci, Manenti, and Morozzi, 1995)
bn = (Field, Boelsma, Baten, and Rulkens, 1995)
bo = (Grosser, Warshawsky, and Kinkle, 1994)
bp = (Field, Feiken, Hage, and Kotterman, 1995)
bq = (Yong, Tousignant, Leduc, and Chan, 1991)
br = (Collins and Daugulis, 1996)
bs = (Perry, 1984)
bt = (Gibson and Subramanian, 1984)
bu = (Ribbons and Eaton, 1982)
bv = (Mueller, Chapman, Blattmann, and Pritchard, 1990)
bw = (McGugan, Lees, and Senior, 1995)
bx = (Field, de Jong, Feijoo-Costa, and de Bont, 1992)
by = (Yadav and Reddy, 1993)
bz = (Sanglard, Leisola, and Fiechter, 1986)
ca = (Bumpus, 1989)
cb = (Vyas, Bakowski, Sasek, and Matucha, 1994)
cc = (George and Neufeld, 1989)

Table 3.18 (continued) Fuel Components/Hydrocarbons and Microorganisms Capable of Biodegrading/Biotransforming Them

cd = (Rahman, Barooah, and Barthakur, 1995)
ce = (Hammel, Green, and Gai, 1991)
cf = (Cited by Field, de Jong, Feijoo-Costa, and de Bont, 1993)
cg = (Hammel, Kalyanaraman, and Kirk, 1986)
ch = (Weissenfels, Beyer, and Klein, 1990)
ci = (Bouchez, Blanchet, and Vandecasteele, 1995a)
cj = (Keuth and Rehm, 1991)
ck = (Dawson and Chang, 1992)
cl = (Mahmood and Rao, 1993)
cm = (Abe, Inoue, Usami, Moriya, and Horikoshi, 1995)
cn = (Cited by Cerniglia, 1992)
co = (Brusseau, Hsien-Chyang, Hanson, and Wackett, 1990)
cp = (Walter, Beyer, Klein, and Rehm, 1991)
cq = (Marin, Pedregosa, Rios, Ortiz, Laborda, 1981)
cr = (Rabus and Widdel, 1995)
cs = (Shailubhai, Rao, and Modi, 1984b)
ct = (Komagata, Nakase, and Katsu, 1964)
cu = (Lee, Fletcher, Avila, Callanan, Yunker, and Munnecke, 1995)
cv = (Bumpus, Tien, Wright, and Aust, 1985)
cw = (Bogan and Lamar, 1996)
cx = (Cerniglia, 1984)
cy = (Ehrlich, Godsy, Coerlitz, and Hult, 1983)
cz = (Schneider, Grosser, Jayasimhulu, and Warshawsky, 1994)
da = (Kelley and Cerniglia, 1995)
db = (Field, Baten, Boelsma, and Rulkens, 1996)
dc = (Kiyohara and Nagao, 1978)

de = (Sutherland, Selby, Freeman, Evans, and Cerniglia, 1991)
df = (de Klerk and van der Linden, 1974)
dg = (Hunt, Robinson, and Ghosh, 1994)
dh = (Phillips, Bender, Word, Niyogi, and Denovan, 1994)
di = (Bender, Vatcharapijarn, and Russell, 1989)
dj = (Field, Heessels, Wijngaarde, Kotterman, de Jong, and de Bont, 1994)
dk = (Cooney and Walker, 1973)
dl = (Colby, Sterling, and Dalton, 1977)
dm = (Paszczynski and Crawford, 1995)
dn = (Fewson, 1967)
do = (Foster, 1962)
dp = (Telegina, 1963)
dq = (Bausum and Taylor, 1986)
dr = (Fall, Brown, and Schaeffer, 1979)
ds = (Ratledge, 1978)
dt = (Whyte, Greer, and Inniss, 1996)
du = (Albrechtsen, 1994)
dv = (Siddiqi, Ye, Elmarakby, Kumar, and Sikka, 1994)
dw = (Hornick, Fisher, and Paolini, 1983)
dx = (Jamison, Raymond, and Hudson, 1976)
dy = (Williams, Cumins, Gardener, Palmier, and Rubidge, 1981)
dz = (Tiedje, Sexstone, Parkin, Revsbech, and Shelton, 1984)
bc = (Makula and Finnerty, 1972)
dd = (Chakrabarty, 1974)
z = (Walter, Beyer, Klein, and Rehm, 1990)

2. Benzene

The availability of dissolved oxygen is a dominant factor in the biodegradation of benzene (Barker and Patrick, 1985). A critical amount of microorganisms may also be required for benzene degradation to occur (Corseuil and Weber, 1994). Certain compounds, such as benzoate or phenylalanine, can be added to select preferentially for benzene degraders (Rotert, Cronkhite, and Alvarez, 1995).

It has been reported that only 0.5% of a large group of soil organisms could use benzene as the sole carbon source (Jones and Edington, 1968). There are organisms in water and soil that can degrade benzene; however, the unsubstituted aromatic nucleus appears relatively resistant to microbial attack.

A number of examples of benzene degraders have been reported. *Pseudomonas rhodochrous* and *P. aeruginosa* are able to metabolize benzene through catechol and *cis,cis*-muconic acid (Hou, 1982). Washed cell suspensions of *P. putida*, grown with toluene as sole source of carbon, are able to oxidize benzene (Gibson, Koch, and Kallio, 1968). The fungus, *Phanerochaete chrysosporium*, can degrade benzene found in gasoline and aviation fuels (Yadav and Reddy, 1993), under nonlignolytic culture conditions in a nitrogen-rich medium (Paszczynski and Crawford, 1995).

3. Toluene

Dissolved oxygen is important for biodegradation of toluene (Barker and Patrick, 1985). Turnover time for toluene is greater than 10,000 hours (Swindoll, Aelion, Dobbins, Jiang, Long, and Pfaender, 1988).

Toluene can undergo two types of attack: (1) immediate hydroxylation of the benzene nucleus, followed by ring cleavage or (2) oxidation of the methyl group, followed by hydroxylation and cleavage of the ring (Fewson, 1981).

Pseudomonas aeruginosa can oxidize toluene (Hou, 1982). Toluene can be converted to 3-methylcatechol by a *Pseudomonas* sp. and an *Achromobacter* sp. (Claus and Walker, 1964). Extracts of this organism grown on xylene can also oxidize toluene. The compound is also oxidized by other species of *Pseudomonas* (e.g., *P. mildenberger*) and *Achromobacter*. Several soil bacteria (*P. putida*, etc.) can utilize toluene as the sole carbon source (Ghisalba, 1983). More than 90% of toluene added to core samples from depths of 1.2, 3.0, and 5.0 m was degraded in 1 week (McNabb, Smith, and Wilson, 1981).

A number of psychrotrophic strains can mineralize toluene at both 23 and 5°C, indicating their potential for low-temperature bioremediation of petroleum hydrocarbon–contaminated sites (Whyte, Greer, and Inniss, 1996). Iron-reducing bacteria have been shown to be able to use organic matter from a landfill leachate as a carbon source and iron oxides in an aquifer as an electron acceptor to degrade toluene (Albrechtsen, 1994).

The fungus *Phanerochaete chrysosporium* can degrade benzene, toluene, ethylbenzene, and xylenes (BTEX) found in gasoline and aviation fuels (Yadav and Reddy, 1993), under nonlignolytic culture conditions in a nitrogen-rich medium (Paszczynski and Crawford, 1995).

The ability to use aromatic compounds can be an induced phenomenon in bacteria (Claus and Walker, 1964). Fulvic acid can be added to stimulate the degradation of toluene, if the toluene concentration is too low to induce the necessary enzymes (Boethling and Alexander, 1979a).

The bioavailability and microbial degradation of toluene can be enhanced by adding synthetic surfactants, such as Tween-80 (Strong-Gunderson and Palumbo, 1995). However, addition of Tween-80 has been seen to not only increase the bioavailability of toluene, but also that of natural organic matter, which introduces competition for nutrients and microbial metabolism.

There appear to be a critical number of microorganisms necessary for biodegradation to occur (Corseuil and Weber, 1994). The onset of microbial oxidation of readily degradable compounds like benzene is delayed if there are not enough microbes, even though nutrient and electron acceptor conditions are favorable.

4. Xylene (*o, m, p*)

An important factor in the biodegradation of xylenes is the availability of dissolved oxygen (Barker and Patrick, 1985). All three xylene isomers may be used by bacteria as sole carbon and energy sources under aerobic and anaerobic, denitrifying conditions (Kuhn, Colberg, Schnoor, Wanner, Zehnder, and Schwarzenbach, 1985). The *p*- and *m*-xylenes are degraded at equal rates but significantly faster than the *ortho* isomer.

Pseudomonas putida can directly oxidize the aromatic ring of *p*- and *m*-xylenes (Hou, 1982). This author was not able to isolate microorganisms that could grow on *o*-xylenes. The oxidation of *o*-xylene to *o*-toluene was demonstrated in a *Nocardia* sp. only with the co-oxidation technique. A *Pseudomonas* sp. was able to grow on *p*-xylene in the presence of toluene (Chang, Voice, and Criddle, 1993).

The fungus *Phanerochaete chrysosporium* can degrade xylenes under nonlignolytic culture conditions in a nitrogen-rich medium (Paszczynski and Crawford, 1995). A *Nocardia* sp. growing on hexadecane converts *p*-xylene to *p*-toluic acid and 2,3-dihydroxy-*p*-toluic acid (Ooyama and Foster, 1965). A *Nocardia* sp. co-oxidizes primarily *o*-xylene and *p*-xylene; *o*-xylene is oxidized to *o*-toluic acid, ethylcyclohexane to cyclohexane acid, *p*-xylene to *p*-toluic acid, and 2-methylheptane to a mixture of products, including ketones and aldehydes (Jamison, Raymond, and Hudson, 1976). When two components of gasoline are combined, one for growth and one as a co-oxidizable substrate, a *Pseudomonas* sp. is able to co-oxidize *o*-xylene with hexane as the growth substrate (Jamison, Raymond, and Hudson, 1976).

Certain compounds can be added to select preferentially for organisms with desired degradative capabilities, such as benzoate or phenylalanine, which can select for benzene, toluene, and xylene degraders, although addition of a nonaromatic substrate (i.e., acetate) does not stimulate these organisms (Rotert, Cronkhite, and Alvarez, 1994).

The onset of microbial oxidation of readily degradable compounds like xylenes can be delayed if there are not enough organisms, even though nutrient and electron acceptor conditions are highly favorable (Corseuil and Weber, 1994). Xylene has the longest critical population development period, which correlates with the comparatively low numbers of indigenous microbes capable of degrading this compound.

5. Alkylbenzenes

Long-chain alkylbenzenes are oxidized at the terminal methyl group (Hou, 1982). A *Nocardia* sp. grows on *n*-decylbenzene, *n*-dodecylbenzene, *n*-octadecylbenzene, and *n*-nonylbenzene. As the alkyl chain length grows and the substituent becomes the major part of the molecule, these compounds are more realistically regarded as substituted alkanes, rather than substituted benzenes. *n*-Alkylbenzenes are also oxidized by yeasts.

6. BTEX

BTX (benzene, toluene, and xylenes) is rapidly degraded under aerobic conditions, but persists in conditions of low dissolved oxygen (Barker and Patrick, 1985).

Phanerochaete chrysosporium can degrade BTEX, common pollutants derived from gasoline and aviation fuels (Yadav and Reddy, 1993), under nonlignolytic culture conditions in a nitrogen-rich medium (Paszczynski and Crawford, 1995). Washed cell suspensions of *Pseudomonas putida*, grown with toluene as the sole source of carbon, are able to oxidize benzene, toluene, and ethylbenzene at equal rates (Gibson, Koch, and Kallio, 1968).

Combining BTX substrates can result in competitive inhibition (Chang, Voice, and Criddle, 1993). Modeling and interpretation of BTX degradation by mixed cultures should consider that differences in transformation may be due to differences in the microorganisms and the complex enzymatic interactions within a single organism.

3.2.1.5.2 Polycyclic Aromatic Hydrocarbons (PAHs)

PAHs, also called polynuclear aromatics (PNAs), constitute a class of hazardous organic chemicals, made up of two or more fused benzene rings in linear, angular, or cluster arrangements, containing carbon and hydrogen (Cerniglia, 1992; Edwards, 1983). PAHs are on the U.S. Environmental Protection Agency (EPA) priority pollutant list, since some are known carcinogens and mutagens (Keith and Telliard, 1979). PAHs are hydrophobic and most are practically insoluble in water, which contributes to their persistence in the environment. Molecular weights range from 178 to 300. Their lipophilicity, environmental persistence, and genotoxicity increase as the molecular size increases up to four or five fused benzene rings (Jacob, Karcher, Belliardo, and Wagstaffe, 1986). A major source of PAH contamination in the environment is from petroleum products. Some of the PAHs of concern are anthracene, benzo(a)pyrene, benz(a)anthracene, fluoranthene, phenanthrene, perylene, pyrene, and fluorene. Biphenyl and naphthalene, although diaromatics not fitting Cerniglia's definitions of PAH, will nonetheless be included here.

Remediation of a PAH-contaminated soil showed that approximately 80% of the PAH disappearance was caused by physical or chemical factors following forced aeration (Yong, Tousignant, Leduc, and Chan, 1991). Bioremediation was effective, however, for anthracene and phenanthrene. PAHs are subject to chemical oxidation, photolysis, hydrolysis, volatilization, bioaccumulation, adsorption to soil particles, and leaching, but microbial degradation is generally the major process in the decontamination of PAHs in the environment (Callahan, Slimak, Gabel, May, Fowler, Freed, Jennings, Durfee, Whitmore, Maestri, Mabey, Holt, and Gould, 1979; Cerniglia, 1993).

Microorganisms can totally degrade (mineralize) or partially transform PAHs, through the action of individual microbes or interdependent communities (Gibson and Subramanian, 1984; Cerniglia and Heitkamp, 1989). However, complete mineralization of high-molecular-weight PAHs can be achieved by only a limited number of microorganisms (Cerniglia, 1992). Losses by abiotic processes may be important for two- and three-ring PAHs but not PAHs with more than three rings (Park, Sims, Doucette, and Matthews, 1988).

Bacteria, filamentous fungi, yeasts, cyanobacteria, diatoms, and eukaryotic algae have the enzymatic capacity to oxidize PAHs that range in size from naphthalene to benzo(a)pyrene (Cerniglia, 1984). Prokaryotic organisms, bacteria and cyanobacteria, use different biodegradation pathways than the eukaryotes, fungi and algae, but both involve molecular oxygen (Huddleston, Bleckmann, and Wolfe, 1986; Cerniglia, 1984). Bacteria employ dioxygenases to incorporate two oxygen atoms into the substrate to form dioxethanes, which are then oxidized to *cis*-dihydrodiols and then to dihydroxy products (Wilson and Jones, 1993). The rate-limiting step in the biodegradation of PAHs is the initial ring oxidation (Cerniglia and Heitkamp, 1989), the genes for which are localized on plasmids (Cerniglia, 1984; Huddleston, Bleckmann, and Wolfe, 1986). Degradation then proceeds rapidly with little or no accumulation of intermediates (Herbes and Schwall, 1978). The bacterial oxidation pathway of terminal ring cleavage appears to apply to all PAH compounds.

In contrast to bacteria, fungi produce cytochrome P-450 monooxygenases to incorporate one oxygen atom into the substrate to form arene oxides, which is followed by enzymatic addition of water to produce *trans*-dihydrodiols and phenols (Cerniglia, 1984; Wilson and Jones, 1993). Both ligninolytic and non-ligninolytic fungi have the ability to oxidize PAHs, and many have the enzymatic capacity to oxidize PAHs when grown on an alternative carbon source (Cerniglia, 1992). The products are both nontoxic metabolites, as well as compounds that have been implicated as biologically active forms of PAHs in higher organisms. Fungi seem to hydroxylate aromatic hydrocarbons as a prelude to detoxification, whereas bacteria oxidize aromatic hydrocarbons to dihydroxylated compounds, as a prelude to ring fission and assimilation (Dagley, 1981). Fungi also have the capacity to form glucuronide and sulfate conjugates of phenolic PAHs, which may be important in the detoxification and elimination of PAHs (Cerniglia, 1984). The fungal degradation of PAHs is of toxicological and environmental significance,

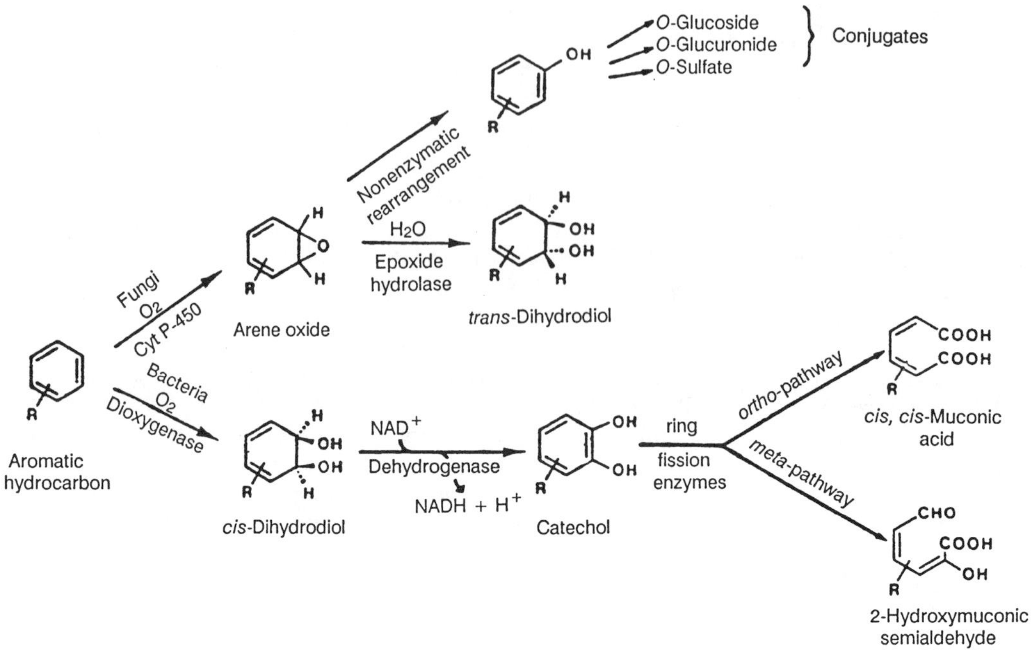

Figure 3.4 Pathways utilized by prokaryotic and eukaryotic microorganisms for the oxidation of PAHs. (From Cerniglia, C.E. *Adv. Appl. Microbiol.* 30:31–71, 1984. With permission.)

since some of the metabolic products have been implicated as carcinogenic, tumorigenic, or mutagenic in higher organisms. Figure 3.4 illustrates the differences in the pathways for oxidation of PAHs between the prokaryotes and eukaryotes. The role of fungi in biodegradation is further discussed in Section 5.2.1.4.

Microbial metabolism of PAHs has largely been studied using pure cultures and single-compound, laboratory-scale systems (Huddleston, Bleckmann, and Wolfe, 1986). There are few reports of PAH biodegradation under environmental field conditions and very few dealing with soil systems specifically. Little has been done to test extrapolation of laboratory results to the field (Cerniglia, 1984).

There is a rapid screening test for PAH degradation, which can be used to evaluate the metabolizing potential of a bacterial community isolated from contaminated soil (Maue and Dott, 1995). The test can be performed on a small scale within a few days using direct fluorometric quantitative analysis of selected PAHs. This enables a wide range of isolates and mixed cultures to be investigated under a variety of substrate conditions, requiring little time or materials. In one application, only a few isolates were found that could metabolize single PAHs as sole substrates; however, a mixed culture metabolized the PAHs rapidly regardless of precultivation. The mixed culture was also resistant to substrate changes that may occur during *in situ* bioremediation.

Varying results have been obtained with acclimated organisms. Microflora of soil contaminated with benzo(a)pyrene have been found to be more active in metabolizing benzo(a)pyrene than those in "clean" soil (Shabad, Cohan, Ilnitsky, Khesina, Shcherbak, and Smirnov, 1971). On the other hand, a PAH-degrading bacterial population added to various sediment systems did not significantly enhance PAH (naphthalene, phenanthrene, benzo(a)pyrene) mineralization rates (Sherrill and Sayler, 1982).

Concentration of PAHs does not appear to be an inhibitory factor, according to Bossert and Bartha (1986). A concentration of 5% of benzo(a)pyrene on a dry soil basis does not inhibit biodegradation of the compound in soil. In addition, repeated application of PAH-containing sludge does not inhibit hydrocarbon utilization (Bossert, Kachel, and Bartha, 1984).

According to Shabad, Cohan, Ilnitsky, Khesina, Shcherbak, and Smirnov (1971), the capacity of bacteria to degrade benzo(a)pyrene increases with benzo(a)pyrene content in the soil. Doubling the starting concentration of several PAHs (naphthalene, phenanthrene, anthracene, and pyrene) resulted in higher degradation rates (Wiesel, Wuebker, and Rehm, 1993). Kerr and Capone (1986) found a correlation between PAH contamination and the rate of mineralization of naphthalene. Rates of PAH mineralization in all environments investigated appeared to be primarily controlled by the amount of pollutant present. Other authors report an inverse relationship between ambient concentrations and mineralization/transformation rates (Shiaris and Jambard-Sweet, 1984).

Salinity of the environment may be an important factor in the biodegradation of PAHs (Shiaris, 1989). Biodegradation rates of these compounds are generally not affected by ambient salinities (Kerr and Capone, 1986). However, while Shiaris (1989) and Kerr and Capone (1988) reported a positive correlation between salinity and rates of mineralization of some PAHs, Ward and Brock (1978) obtained lower rates of mineralization in hypersaline environments, due probably to a reduction in microbial metabolic rates. Salinity may affect PAH-particle interactions and the solubility of the compounds. Four PAH-degrading bacteria have been found to have a high tolerance to salinity (to an NaCl concentration of up to 7.5%; Ashok, Saxena, and Musarrat, 1995). They are two *Micrococcus* isolates, a *Pseudomonas*, and an *Alcaligenes*, which all bear high-molecular-weight plasmid DNA, which probably aids in the metabolism of PAHs.

PAH mineralization is related to the length of incubation time, temperature, molecular weight of the hydrocarbon, and previous exposure to PAH or related contaminants (Sherrill and Sayler, 1982). In landtreatment, a period of 1 to 2 years might be needed to decompose PAHs (Overcash and Pal, 1979a). However, the time required for degradation can be reduced with a higher concentration of microorganisms (Nocentini, Tamburini, and Pasquali, 1995).

Loehr (1992) found that PAHs could be biodegraded in contaminated soils when the following environmental factors were controlled: aerobic conditions; moisture; nontoxic loading rates; indigenous, acclimated organisms; nutrients; and degradable organic matter. It was also shown that uncontaminated soil contained an indigenous microorganism population capable of quickly initiating PAH degradation.

The soil moisture requirement for microbial activity ranges from 25 to 85% of water-holding capacity, and 30 to 90% for optimum PAH degradation (Dibble and Bartha, 1979a). Microorganisms require a general soil pH of 5.5 to 8.5 and a pH of 7.0 (Weissenfels, Beyer, and Klein, 1990) or 7.5 to 7.8 for optimum PAH degradation. Aerobes and facultative anaerobes need a redox potential of >50 mV and anaerobes <50 mV for biodegradation to occur (Wilson and Jones, 1993). Aerobes require a minimum air-filled pore space of 10% and anaerobes <1% by volume oxygen content for activity and from 10 to 40% oxygen for optimum PAH degradation (Bauer and Capone, 1985). Whereas the N and P ratios for microbial growth are approximately C:N:P (120:10:1) (Wilson and Jones, 1993), for optimum PAH degradation they are C:N (60:1) and C:P (800:1) (Dibble and Bartha, 1979a). Microorganisms generally prefer a temperature in the range of 15 to 45°C (Wilson and Jones, 1993), but for optimum degradation of PAHs, the following temperatures have been reported: 30°C (Bauer and Capone, 1985; Weissenfels, Beyer, and Klein, 1990; Walter, Beyer, Klein, and Rehm, 1991); 24 to 30°C (Heitkamp, Franklin, and Cerniglia, 1988); 27°C (Song, Wang, and Bartha, 1990); or 20°C (Dibble and Bartha, 1979a). All of these factors will be discussed in detail in Section 5.

The soil type can greatly affect the degree of PAH biodegradation, even under the same optimum growth conditions (Weissenfels, Klewer, and Langhoff, 1992b). PAHs in some soils have been determined to be unbiodegradable, probably because the material is highly sorbed and is no longer available.

Since PAHs are hydrophobic, they are primarily attached to particles when they are present in the environment (Volkering, Breure, Sterkenburg, and van Andel, 1992). Organic carbon is the single most important factor determining the sorption of PAHs in soil (Weissenfels, Klewer, and Langhoff, 1992a). The soil organic carbon commonly represents a hydrophobic fraction with a strong binding affinity for hydrophobic compounds. Sorption of PAHs by soil organic matter may slow their biodegradation (Manilal and Alexander, 1991). Soilborne PAHs, if immobilized onto soil organic matter, are nonbiodegradable and are not released by leaching with rainwater (Weissenfels, Klewer, and Laughoff, 1992a).

Soil organic matter would slow biodegradation of PAHs that are otherwise readily metabolized (Manilal and Alexander, 1991). Phenanthrene sorbs to soil constituents, the extent of which is directly related to the percent organic matter in the soil. This may explain why mineralization of the compound occurs more slowly in soil than in liquid media. Fulvic acid has been shown to decrease mineralization of pyrene, apparently due to toxicity to the microbes and possible sorption of the compound making it less bioavailable (Grosser, Warshawsky, and Kinkle, 1994). Other studies found that the biodegradation of PAHs was increased significantly by the addition of compost, which stimulated the mineralization and fixation of PAHs in the soil (Mahro, Eschenbach, Kaestner, and Schaefer, 1994).

Hydrophobic PAHs may initially adsorb rapidly onto hydrophobic areas of the soil surface (Weissenfels, Klewer, and Langhoff, 1992b). This is believed to be followed by a slow absorption of the compounds into the soil matrix (Karickhoff, 1980; Robinson, Farmer, and Novak, 1990), until the soil organic capacity is filled and equilibrium reached. The portion trapped in the matrix may be the nonbioavailable, nonbiodegradable part of the contamination. The longer the exposure to the pollutants, the greater their incorporation into the matrix and the more recalcitrant they become.

After exposure to different soil types with high clay content, 100% of the anthracene and pyrene and up to 25% of the phenanthrene were not extractable (Karimi-Lotfabad, Pickard, and Gray, 1996). Anthracene was found to be oligomerized to higher-molecular-weight aromatic products. Water inhibited the reaction. The reacted anthracene was not available for biodegradation. Cutright and Lee (1994) recommend using a Hewlett-Packard HPLC 1050 to characterize PAH-contaminated soil quantitatively and qualitatively.

Erickson, Loehr, and Neuhausser (1993) conducted a laboratory study of the effect of temperature, pH, and nutrients on degradation of PAHs in contaminated soil. They concluded that PAHs in the soil resisted mineralization by microorganisms, that the test soils from a manufactured gas plant were nontoxic before and after testing, that significant populations of bacteria were present during the degradation studies, that addition of free naphthalene and phenanthrene resulted in rapid loss of these compounds but indigenous chemicals were unaffected, that PAHs were not soluble in a water extract of the soil, and that the PAHs in these soils were unavailable for microbial degradation.

In sandy soils, two-ring PAHs have half-lives of about 2 days (Sims, Doucette, McLean, Grenney, and Dupont, 1988). The three-ring PAHs, anthracene and phenanthrene, have half-lives of 16 and 134 days, respectively. In general, the four-, five-, and six-ring PAHs have half-lives of over 200 days. The half-lives determined under laboratory conditions may be considerably shorter than those in the field (Wild, Berrow, and Jones, 1991).

Biodegradation of high-molecular-weight PAHs sorbed to silt and clay particles can be enhanced by the presence of low-molecular-weight PAHs; for example, PAHs with four rings are degraded more readily in the presence of naphthalene (Ressler, Kaempf, and Winter, 1995). Sorption of PAHs to fine soil particles does not limit bacterial growth and activity.

Biodegradability and water solubility appear to be inversely related to the number of rings in the molecules (Huddleston, Bleckmann, and Wolfe, 1986). Higher-molecular-weight PAHs may appear to be more recalcitrant than lower-weight compounds, but this effect seems to be due to their low concentrations in the aqueous phase (Mihelcic, Lueking, Mitzell, and Stapleton, 1993). Tricyclic PAHs disappear rapidly from soil, tetracyclic PAHs more slowly, and pentacyclic PAHs only marginally, or not at all (Bossert and Bartha, 1986). The greater water solubility of phenanthrene, as compared with anthracene, may explain the faster biodegradation of phenanthrene. It has been suggested that naphthalene and phenanthrene are utilized only in the soluble form (aqueous solution), since the growth rates of bacteria are related to the solubilities of the hydrocarbons on which they are growing (Wodzinski and Coyle, 1974). Biodegradation of tetracyclic PAHs also corresponds to water solubilities, while pentacyclic PAHs are practically insoluble in water (Klevens, 1950). On the other hand, there is no general pattern of an inverse correlation of disappearance with the degree of ring condensation (clustered vs. linear arrangement of the same number of rings) (Bossert and Bartha, 1986). In addition, the thermodynamic stability and complexity of the tetracyclic compounds compared with the tricyclic make enzyme attack of these substances more difficult (Wiesel, Wuebker, and Rehm, 1993). Generally, PAHs with four or more rings can be degraded by microbes by metabolism via hydroxylation and ring fission (Ghisalba, 1983). When the soils (microbes) become acclimated to PAHs, their ability to degrade these compounds is enhanced. Degradation of some of the less-water soluble, such as benz(a)anthracene and benzo(a)pyrene, occurs only when the PAHs are mixed with soil, water, and a substance to stimulate growth of oxygenase-active organisms (Groenewegen and Stolp, 1981).

Another study reports that a mixed culture, isolated from PAH-contaminated soil, was able to degrade a range of PAHs, including fluoranthene, benzo(a)pyrene, anthracene, phenanthrene, acenaphthene, and fluorene (Trzesicka-Mlynarz and Ward, 1995). The predominant organisms, Gram-negative rods, were *Pseudomonas putida, P. aeruginosa*, and a *Flavobacterium* sp. The mixed culture and a synthetic mixture of the primary bacteria were more successful in degrading the compounds. The individual isolates could efficiently remove the more-water-soluble PAHs (acenaphthene, fluorene, phenanthrene, fluoranthene), but not the less-water-soluble compounds (anthracene and pyrene).

Di- and tricyclic aromatic hydrocarbons are very insoluble, but it appears that degradation of compounds, such as phenanthrene, may be more rapid in natural environments than was once imagined (Fewson, 1981). Wiesel, Wuebker, and Rehm (1993) report that the tetracyclic aromatics, fluoranthene and pyrene, can be metabolized but only after phenol, naphthalene, and most of the tricyclic compounds are degraded by a mixed bacterial culture. Increasing the concentration of the PAHs improves the degradation rate. Cell growth results from degradation of the tricyclic but not the tetracyclic compounds, possibly because the latter would not supply sufficient energy for cell growth. Bouchez, Blanchet, Besnainou, and Vandecasteele (1995) also report the formation of less biomass during degradation of four-ring than of two- or three-ring PAHs. Using strains of *Pseudomonas* and *Rhodococcus*, high mineralization rates of 56 to 77% and low production of soluble metabolites (7% to 23%) were obtained for a wide variety of PAHs. From 16 to 35% of the carbon was converted to biomass.

While greater solubility enhances biodegradation, inhibition is common when a PAH is more water soluble (Bouchez, Blanchet, and Vandecasteele, 1995a). As will be discussed in Section 4.1.9, the toxicity of PAHs to microorganisms is related to their water solubility (Sims and Overcash, 1983). Inhibition of biodegradation can occur with a mixture of PAHs. For instance, naphthalene is strongly toxic in a mixture of PAHs and can inhibit degradation of other components that would normally be biodegraded. This toxicity may be due to its high water solubility (about 30 ppm). Fluorene with a solubility of 2 ppm and phenanthrene at 1 ppm are also frequently inhibitory. On the other hand, inhibition by anthracene, which is poorly water soluble (around 50 ppb), can also occur.

It is suggested that the binding of PAHs within the soil matrix retards their solubilization and thus reduces their toxicity by elution with water (Weissenfels, Klewer, and Langhoff, 1992b). Volkering, Breure, Sterkenburg, and van Andel (1992) found that bacterial growth on crystalline or adsorbed PAHs can result in a linear increase in biomass concentration. Under these circumstances, mass transfer from the solid phase to the liquid phase is rate limiting for growth. This could explain the linear growth of bacteria and yeasts observed on slightly soluble substrates.

In the solution phase, sorption and desorption kinetics of PAHs by soils depends on the composition of the cosolvent/water mixture (Errett, Chin, Xu, and Yan, 1996). In general, the greater the proportion of cosolvent, the higher the PAH sorption/desorption.

After the degradation of PAHs in the soil reaches a certain amount, there is little change in PAH concentration. The problem of mass transfer will be a limiting factor in the field. However, it has been found that in addition to aqueous-phase substrate, sorbed-phase substrate that can be easily desorbed will also be available to microorganisms (Robinson, Farmer, and Novak, 1990).

Microorganisms could adhere to particles of insoluble substrates, and addition of chemical surfactants could promote solubilization of the PAH and increase its rate of utilization (Cox and Williams, 1980). Volkering, van de Wiel, Breure, van Andel, and Rulkens (1995) concluded that the bioavailability of PAHs could be enhanced by use of the surfactants Triton X-100 and Tergitol NPX at high concentrations. Biosurfactants produced by microorganisms growing on PAHs also have this capability (Deziel, Paquette, Villemur, Lepine, and Bisaillon, 1996). Fluorescence spectroscopy has been used to study the effect of surfactants on degradation of PAHs, since naphthalene and some degradation intermediates, such as salicylic acid, are fluorescent (Putcha and Domach, 1993). This revealed that the surfactant Triton-X-100 actually produces micelles that protect naphthalene from biodegradation. See Section 5.3.1 for a discussion of the use of chemical and biological surfactants to improve biodegradation of PAHs and other petroleum constituents.

The biotransformation process for PAHs with more than three rings appears to be cometabolism (Sims and Overcash, 1983; Keck, Sims, Coover, Park, and Symons, 1989). As discussed in Section 5.2.3, cometabolism has been observed to result in rapid rates of PAH degradation (Walter, Beyer, Klein, and Rehm, 1991; Weissenfels, Beyer, Klein, and Rehm, 1991). Fluorene and phenanthrene are cometabolized more easily than the higher-molecular-weight PAHs (Bouchez, Blanchet, Besnainou, and Vandecasteele, 1995), and benz(a)anthracene and pyrene are probably slow to be degraded (Ghisalba, 1983).

Microorganisms isolated for their degradative ability of a specific compound can show some diversity in their substrate specificity (Schneider, Grosser, Jayasimhulu, and Warshawsky, 1994). PAH-degrading bacteria may contain common enzymes for metabolizing other PAH substrates; i.e., inducers for degradation of low-molecular-weight PAHs may also induce that of higher-molecular-weight compounds (Stringfellow, Chen, and Aitken, 1995). For instance, salicylate can induce degradation of phenanthrene, and phenanthrene or salicylate can induce metabolism of pyrene and fluoranthene by *Pseudomonas saccharophila* P-15. While five- and six-ring PAHs are degraded along with two- through four-ring PAHs, levels of the former are only about half those of the latter (Castaldi, 1994). This may indicate a cometabolic enzyme induction.

When *P. paucimobilis* strain EPA 505 grows on fluoranthene as a sole source of carbon and energy, it is able to degrade benzo(a)pyrene, chrysene, and benz(a)anthracene (Siddiqi, Ye, Elmarakby, Kumar, and Sikka, 1994). The cells degrade 25.2% of benzo(a)pyrene to a major, highly polar metabolite. All the strains tested in another study were capable of cometabolizing PAHs but varied in the range of compounds attacked (Bouchez, Blanchet, and Vandecasteele, 1995a). There are even cases where non-growth-supporting analogs could induce enzymes needed for complete metabolism of growth-supporting substrates (Hegeman, 1966; van Eyk and Bartels, 1968; Rosenfeld and Feigelson, 1969). Such inductions could be utilized for compounds that are degraded slowly or are degraded only by nongrowth metabolism.

There can be inhibition even when cometabolism is taking place, because of the interactions of PAHs at several levels (Bouchez, Blanchet, and Vandecasteele, 1995a). This could involve competition at the active site of enzymes or at the level of enzyme induction. It could be due to accumulation of toxic end products. Addition of fluorene as a cosubstrate has been observed to be synergistic, increasing utilization of phenanthrene (Bouchez, Blanchet, Besnainou, and Vandecasteele, 1995). Normally, there is a lower degradation of the substrate PAH when a second PAH is added, whether the second PAH is cometabolized (as with fluorene) or not (as with anthracene). When the second PAH is also a growth substrate, the inhibition of the utilization of both PAHs is called substrate antagonism. Little is known about the biodegradation of complex mixtures of PAHs and the effect of one compound on the biodegradability of another.

Synergy among microorganisms also plays an important role. Degradation of a PAH mixture seems to be a cooperative process involving a number of organisms with complementary abilities (Bouchez, Blanchet, and Vandecasteele, 1995a). One or more organisms may be able to degrade partially transformed products of another microbe. For example, pyrene, 1,2-benzanthracene, 3,4-benzopyrene, and 1,2,5,6-dibenzanthracene can be degraded by a mixed culture of flavobacteria and pseudomonads (Cerniglia, 1984). The inhibition by a PAH in a mixture may be overcome by supplying an organism that is able to degrade that compound and thereby removing the inhibitory effect (Bouchez, Blanchet, and Vandecasteele, 1995a). Commensalism in biodegradation of PAHs and other petroleum constituents is treated in depth in Section 5.2.2.3.

Vegetation can enhance the rate and extent of degradation of PAHs in contaminated soil (Santharam, Erickson, and Fan, 1994). Plant roots release exudates capable of supplying carbon and energy to microflora for degrading PAHs. It has been established that the population of microorganisms in the rhizosphere is significantly greater than that in the nonvegetated soil and that these microorganisms are apparently responsible for enhanced biodegradation of PAHs.

A number of Gram-negative bacteria have been found to biodegrade PAHs in the soil. Anthracene, fluoranthene, and phenanthrene can be utilized by *Pseudomonas* spp. as the sole carbon source (Ghisalba, 1983). Mahmood and Rao (1993) also report degradation of anthracene and fluoranthene by *Pseudomonas* spp. Fluoranthene can be degraded by *P. paucimobilis* (Meuller, Chapman, Blattman, and Pritchard, 1990) and *Alcaligenes denitrificans* as the sole carbon source (Weissenfels, Beyer, and Klein, 1990). *Agrobacterium* spp. can degrade anthracene and fluoranthene (Mahmood and Rao, 1993).

Wiesel, Wuebker, and Rehm (1993) studied a mixed bacterial culture (MK1) which was able to degrade phenol completely in 1 day, naphthalene in 2 days, and phenanthrene after 15 days. Maximum degradation for the tricyclic compounds phenanthrene and anthracene in an anthracene oil occurred after 5 days. Significant metabolization of the tetracyclic aromatic hydrocarbons fluoranthene and pyrene was observed only after the degradation of phenol, naphthalene, and most of the tricyclic compounds were degraded. Results were the same for both immobilized and freely suspended cells.

While most studies have been conducted on the biodegradation of PAHs by Gram-negative bacteria, degradation by Gram-positive bacteria is also reported (Keuth and Rehm, 1991). Two examples are a *Mycobacterium* sp. (Guerin and Jones, 1988) and *Arthrobacter polychromogenes* (Keuth and Rehm, 1991). *Mycobacterium* sp. PYR-1 can mineralize fluoranthene, naphthalene, and pyrene (Kelley and Cerniglia, 1991; Kelley, Freeman, and Cerniglia, 1991; Kelley, Freeman, Evans, and Cerniglia, 1991). When provided with a complex carbon source, *Mycobacterium* PYR-1 was found to degrade simultaneously >74% of a mixture of phenanthrene, anthracene, fluoranthene, pyrene, and benzo(a)pyrene within

6 days (Kelley and Cerniglia, 1995). This compared quite favorably with an environmental microcosm test system with other organisms where there was a mineralization rate of about 50% of the mixture in 30 days, and mixtures of hydrocarbons were degraded more slowly than individual compounds. *Bacillus subtilis* has been shown to degrade pyrene (Mahmood and Rao, 1993). A *Rhodococcus* sp. can degrade pyrene and chrysene as the sole carbon source (Walter, Beyer, Klein, and Rehm, 1991).

Other microorganisms have also been studied for their effect on PAHs. Naphthalene, biphenyl, and benzo(a)pyrene can be metabolized by a number of different species of yeast, some of which are reported in high numbers in oil-polluted soils (Cerniglia and Crow, 1981). Cyanobacteria, diatoms, and some eukaryotic algae have the enzymatic capacity to oxidize PAHs that range in size from naphthalene to benzo(a)pyrene (Cerniglia, 1984). Multiple oxidative pathways may be involved in the cyanobacterial metabolism of PAHs. The photoautotrophs, cyanobacteria and green algae, produce both *cis*- and *trans*-dihydrodiols from PAHs (Cody, Radike, and Warshawsky, 1984; Narro, Cerniglia, Van Baalen, and Gibson, 1992; see Section 5.2.1.5).

A variety of fungi can attack PAHs. A filamentous fungus demonstrated the ability to degrade benzo(a)pyrene and benz(a)anthracene (Cerniglia and Gibson, 1979; Dodge and Gibson, 1980). The nonligninolytic fungus *Cunninghamella elegans* can oxidize numerous PAHs to *trans*-dihydrodiols, phenols, quinones, tetralones, and conjugates of these primary metabolites (Cerniglia, 1992). Most of the fungal transformation products formed by this fungus are less mutagenic than the parent compound. *Rhizoctonia solani* may also prove to be as useful as *C. elegans* in detoxifying PAHs. *Trametes versicolor* TV1 and *Chrysosporium lignorum* CL1 have also been shown to degrade PAHs (Field, de Jong, Feijoo-Costa, and de Bont, 1992). Anthracene and fluoranthene degradation was found to be dominated by *Aspergillus terrus, A. flavus,* and *Penicillium tordum* (Mahmood and Rao, 1993). *Sclerotium rolfsii* and *Trichoderma harzianum* were associated with degradation of phenanthrene.

The white-rot fungus *Phanerochaete chrysosporium* has been extensively studied for possible use in bioremediating contaminated soils. Its extracellular enzymes have demonstrated the ability to degrade numerous PAHs. *P. chrysosporium* has been observed to metabolize phenanthrene not only under ligninolytic conditions but also under nonligninolytic conditions (Dhawale, Dhawale, and Dean-Ross, 1992), similar to *C. elegans*. In the initial oxidative attack on PAHs, the fungus utilizes different enzymic mechanisms for the ligninolytic and nonligninolytic states (Hammel, Gai, Green, and Moen, 1992; Field, de Jong, Feijoo-Costa, and de Bont, 1993). It can initiate PAH degradation by employing free-radical mechanisms (Bogan, Schoenike, Lamar, and Cullen, 1996). PAHs with ionization potentials at or below around 7.55 eV are substrates for direct one-electron oxidation by lignin peroxidase; those with higher potentials may be acted upon by radical species formed during lipid peroxidation reactions dependent on manganese peroxidase (MnP). In a nutrient-limited medium, *P. chrysosporium* was able to remove 70 to 100% of 22 two-, three-, and four-ring PAHs, including phenanthrene, fluorene, fluroanthene, anthracene, and pyrene (Bumpus, 1989). Degradation of benzo(a)pyrene to CO_2 and water by this fungus has also been reported (Sanglard, Leisola, and Fiechter, 1986).

The white-rot fungus *Bjerkandera* sp. strain BOS55 is an outstanding degrader of PAHs (Field, Feiken, Hage, and Kotterman, 1995; Field, Baten, Boelsma, and Rulkens, 1996). The PAH-degrading ability of this strain is far superior to that of all other fungal strains tested, including *P. chrysosporium* (Field, de Jong, Feijoo-Costa, and de Bont, 1993). Cultures of this fungus can degrade 80% of the contaminants in PAH-contaminated soil in 2% ethanol. More-complex PAHs with more rings take longer to degrade. The 20% of benzo(a)pyrene found not available for the fungus to biodegrade within 22 days was rendered bioavailable by adding acetone (10% vol/vol of soil water) to the soil cultures. *Bjerkandera* sp. strain BOS55 is being studied for bioremediating PAH-polluted sites (Field, Boelsma, Baten, and Rulkens, 1996). Additional information on fungal degradation of PAHs and other petroleum constituents can be found in Section 5.2.1.4.

Preparation of a culture as a soil inoculum requires maintaining the specific degradative functions of the introduced organisms. Some microorganisms (e.g., an *Arthrobacter* sp.) can rapidly adapt to substrate (e.g., phenanthrene) degradation after having been grown in a substrate-free medium (Aamand, Bruntse, Jepsen, Jorgensen, and Jensen, 1995). Some bacteria, however, lose the ability to grow on a specific substrate, after growth on a medium without the compound. Selection of organisms that can retain this ability would allow larger-scale production of the microbes for soil inoculation.

When the specialized microorganisms are introduced into the subsurface, their survival must be assured (Lin, Lantz, Schultz, Mueller, and Pritchard, 1995). A technique called cell immobilization or encapsulation allows packaging of specific bacterial or fungal cells in a porous polymeric material (e.g., polyurethane foam and vermiculite) to reduce competition from indigenous microflora and yet permits expression of their particular metabolic capabilities. This technique has been used in the bioremediation of PAH-contaminated soil. See Section 5.2.2.7 for a description of cell encapsulation methods and their application in bioremediation.

For bioremediation to be successful, the proper environmental conditions should be determined and maintained for the specific set of contaminants and microorganisms (see Section 5). Wang, Yu, and Bartha (1990) reported almost complete elimination of PAHs from diesel oil–contaminated soil, by use of liming, fertilization, and tilling. Without bioremediation, 12.5 to 32.5% of the higher-molecular-weight PAHs were still present after the 12-week test period.

In a laboratory study to determine the best demonstrated available technology (BDAT) for biotreating PAHs in refinery API oil separator sludge, Huesemann, Moore, and Johnson (1993) found that naphthalene, anthracene, phenanthrene, and benzo(a)pyrene were completely biodegraded in the first 4 weeks in both a biotic, nutrient-amended, inoculated aerated slurry reactor and in a biotic, oxygen-sparged reactor, but not in a nitrogen-sparged control. Chrysene disappeared within 4 weeks in the aerated bioreactor, but required 16 weeks in the oxygen-sparged reactor. There was only 30% degradation of pyrene in the aerated bioreactor but none in the oxygen-sparged reactor. The nutrient-amended, aerated bioreactor proved more favorable. The reduced biodegradability of pyrene may have been due to the inherently low biodegradation rate of this compound or the limited bioavailability in the weathered oily sludge system.

Inoculation of soil heavily contaminated by coal tar with specific PAH-degrading organisms enhanced degradation of phenanthrene (Aamand, Bruntse, Jepsen, Jorgensen, and Jensen, 1995). However, such inoculations were ineffective in soil slightly polluted with coal tar, since the soil may have contained sufficient inherent microorganisms capable of the degradation.

1. Alkylnaphthalenes

Degradation of alkylnaphthalenes, by *Pseudomonas* spp. depends upon the position, number, and type of the substituents on the molecule (Solanas, Pares, Bayona, and Albaiges, 1984). Pure cultures of *Pseudomonas* spp. preferentially degrade the less-substituted aromatics in a light crude oil residue, providing indirect evidence that the organisms are attacking the ring system rather than the alkyl chains. Certain strains of bacteria, notably *Pseudomonas* spp., metabolize aromatic hydrocarbons with fused benzene rings through *cis*-dihydrodiol intermediates at different sites of the molecule (Jerina, Selander, Yagi, Wells, Davey, Mahadevan, and Gibson, 1976). It is, therefore, assumed that alkyl substituents may hinder the initial oxygenative attack of the molecule by the corresponding enzymatic system. However, in some cases, the oxidation rates seem to be enhanced, such as with *meta*-substituted naphthalenes (Solanas, Pares, Bayona, and Albaiges, 1984). Degradation of naphthalene by *Pseudomonas* involves the formation of 1,2-dihydroxynaphthalene as the first stable metabolite (Davies and Evans, 1976). Oxidation becomes difficult when a substituent is present in the *ortho* (1,2) positions, unless it is a methyl group adjacent to the sites of oxidative attack.

2. Naphthalene

Naphthalene and alkyl-substituted naphthalenes are among the most toxic components in the water-soluble fraction of crude and fuel oils (Winters, O'Donnell, Batterton, and Van Baalen, 1976). Naphthalene is strongly toxic in a mixture of PAHs and can inhibit degradation of other components that would normally be biodegraded (Bouchez, Blanchet, and Vandecasteele, 1995a). This toxicity may be due to its high water solubility (about 30 ppm). A way to overcome this toxicity is to introduce an organism that can degrade the naphthalene, which will allow biodegradation of the other PAHs to proceed. *Mycobacterium* sp. PYR-1 can mineralize naphthalene (Kelley and Cerniglia, 1991; Kelley, Freeman, and Cerniglia, 1991; Kelley, Freeman, Evans, and Cerniglia, 1991), even in a mixture of PAHs (Kelley and Cerniglia, 1995). Ressler, Kaempf, and Winter (1995) found that PAHs with four rings were degraded more readily in the presence of naphthalene.

Growth of pure cultures of bacteria is fastest on the solid substrates with the highest water solubilities (Wodzinski and Johnson, 1968). The rate of dissolution of naphthalene is directly related to its surface areas (Thomas, Yordy, Amador, and Alexander, 1986). Naphthalene may be utilized only in the soluble form (aqueous solution) (Wodzinski and Coyle, 1974).

The bioavailability and microbial degradation of naphthalene can be enhanced by adding synthetic surfactants (e.g., Tween-80), biosurfactants, or nutrients with surfactant-like properties (e.g., an oleophilic fertilizer) at or below the critical micelle concentration (Strong-Gunderson and Palumbo, 1995). Biodegradation of sorbed naphthalene can be stimulated by addition of the surfactants Triton X-100 and Tergitol NPX at high concentrations (Volkering, van de Wiel, Breure, van Andel, and Rulkens, 1995). Concentrations of Triton X-100 or Brij 30 above the critical micelle concentration are not toxic to naphthalene-degrading bacteria, and the presence of surfactant micelles does not inhibit mineralization of naphthalene (Liu, Jacobson, and Luthy, 1995). Volkering, Breure, van Andel, and Rulkens (1995) found that Triton X-100, Tergitol NPX, Brij 35, and Igepal CA-720 increased solubilities and rates of dissolution of naphthalene, with no toxic effects of the surfactants at concentrations up to 10 g/L. Naphthalene can be degraded rapidly in the absence of micelles; however, micelles of the nonionic surfactant Triton-X-100 can protect naphthalene against copper quenching and suppress biodegradation (Putcha and Domach, 1993). This sequestering effect may increase over time.

Disulfonates cause greater solubilization of naphthalene than monosulfonates and slightly lower solubilization than nonionics (Rouse, Sabatini, and Harwell, 1993). A petroleum sulfonate-oil (PSO) surfactant (commercial Petronate) appears to increase substantially the aqueous-phase concentrations of poorly water-soluble, nonionic organic contaminants, such as naphthalene (Sun and Boyd, 1993). Both aqueous and soil-sorbed PSO surfactants act as partition phases for these contaminants.

Biosurfactants can also aid in degradation of naphthalene. Ten cultures were found to produce a biosurfactant when grown on naphthalene or phenanthrene (Deziel, Paquette, Villemur, Lepine, and Bisaillon, 1996). Production of biosurfactant by *Pseudomonas aeruginosa* 19SJ is enhanced when the organism is grown on naphthalene, or with small amounts of biosurfactants and naphthalene degradation intermediates. Production of biosurfactants increases the aqueous concentration of naphthalene, indicating the bacterium is promoting the dispersion of the substrate.

Metabolism of naphthalene by the bacteria *Pseudomonas rathonis* and *P. oleovorans*, with optimal synthesis of salicylic acid, was accomplished by addition of 0.4% $Al(OH)_3$ (Zajic, 1964). Inorganic boron compounds also increase the yields of salicylic acid. *P. paucimobilis* seems to have the ability to metabolize naphthalene and other aromatic hydrocarbons, as well as substituted derivatives, by a single set of enzymes (Castaldi, 1994). Multiple plasmid transfer has been accomplished between *Pseudomonas* species (Chakrabarty, 1974) to construct a strain that can degrade several hydrocarbons, including naphthalene (Hornick, Fisher, and Paolini, 1983). A number of psychrotrophic strains have shown the ability to mineralize naphthalene (Whyte, Greer, and Inniss, 1996).

Naphthalene metabolism can be found in many genera of fungi. *Candida lipolytica, C. guilliermondii, C. tropicalis, C. maltosa,* and *Debaryomyces hansenii* are able to metabolize this compound (Cerniglia and Crow, 1981). *C. lipolytica* oxidizes naphthalene to 1-naphthol, 2-naphthol, 4-hydroxy-1-tetralone, and *trans*-1,2-dihydroxy-1,2-dihydronaphthalene. The primary metabolite is 1-naphthol. *Cunninghamella bainieri* oxidizes naphthalene through *trans*-1,2-dihydroxy-1,2-dihydronaphthalene (Gibson and Mahadevan, 1975). *C. elegans* oxidizes naphthalene by a sequence of reactions resulting in six metabolites: 1-naphthol, 4-hydroxy-1-tetralone, 1,4-naphthoquinone, 1,2-naphthoquinone, 2-naphthol, and *trans*-1,2-dihydroxy-1,2-dihydronaphthalene. A *Nocardia* sp. can biotransform naphthalene and substituted naphthalenes to their corresponding diols (Hou, 1982). Naphthalene is oxidized by species of *Cunninghamella, Syncephalastrum,* and *Mucor* to 2-naphthol, 4-hydroxy-1-tetralone, *trans*-naphthalene dihydrodiol, 1,2-naphthoquinone, 1,4-naphthoquinone, and predominantly 1-naphthol (Cerniglia, Hebert, Szaniszlo, and Gibson, 1978). *Neurospora crassa, Claviceps paspali,* and *Psilocybe* strains also show a similar degradative capacity.

Cunninghamella elegans oxidizes naphthalene, and several *Candida* spp. are able to metabolize naphthalene (Komagata, Nakase, and Katsu, 1964; Zajic and Gerson, 1977). This suggests that fungi metabolize aromatic hydrocarbons in a manner similar to mammalian systems, i.e., via a monooxygenase-catalyzed reaction (Cerniglia, Hebert, Szaniszlo, and Gibson, 1978). A cytochrome P-450-dependent reaction may be responsible for the initial oxygenation of naphthalene by these organisms.

Several brown- and white-rot fungi were able to degrade at least 40% of naphthalene in 2 weeks or less, producing more biomass at acidic to neutral pH, incubation at 30°C, 90% moisture saturation, with granulated corncobs or alfalfa pellets as a lignocellulosic substrate (Lee, Fletcher, Avila, Callanan, Yunker, and Munnecke, 1995). The isolates were *Irpex* sp. #232, *Fomitopsis* sp. #259, *Phanerochaete* spp. #326 and #404, *Heteroporus* sp. #501, and *Cunninghamella elegans*. Addition of Tween-80 to liquid PAH generally increased the rates of degradation of the compounds.

Cyanobacteria, diatoms, and some eukaryotic algae have the enzymatic capacity to oxidize naphthalene (Cerniglia, 1984; Naval Civil Engineering Laboratory, 1986; Phillips, Bender, Word, Niyogi, and Denovan, 1994). Prokaryotic and eukaryotic algae can oxidize naphthalene when grown photoautotrophically in the presence of naphthalene (Cerniglia, Gibson, and Van Baalen, 1980; 1982; Cerniglia, Van Baalen, and Gibson, 1980). An *Oscillatoria* sp. can oxidize naphthalene to a limited extent when growing photoautotrophically (Phillips, Bender, Word, Niyogi, and Denovan, 1994), and some *Oscillatoria* spp. can oxidize naphthalene in light and dark conditions (Bender, Vatcharapijarn, and Russell, 1989). There are green, brown, and red algae and diatoms that can partially oxidize naphthalene (Cerniglia, 1984). Species of *Oscillatoria, Microcoleus, Anabaena, Cocochloris, Nostoc, Chlorella, Dunaliella, Chlamydomonas, Ulva, Cylindretheca, Ampora*, and *Porphyridium* have been found to be capable of oxidizing naphthalene (Cerniglia, Gibson, and Van Baalen, 1980). The cyanobacteria *Agmenellum quadruplicatum* strain PR-6 and *Oscillatoria* sp. strain JCM oxidize naphthalene predominantly to 1-naphthol (Cerniglia, Van Baalen, and Gibson, 1980). The methyl-, dimethyl-, and trimethylnaphthalenes are more toxic than naphthalene to the freshwater alga *Selenastrum capricornutum* (Hsieh, Tomson, and Ward, 1980).

Naphthalene can be broken down by cell-free extracts, such as one containing naphthalene oxygenase purified from cells of *Corynebacterium renale* grown on naphthalene as the sole source of carbon and energy (Dua and Meera, 1981). The enzyme has a molecular weight of approximately 99,000 and forms *cis*-1,2-dihydroxy-1,2-dihydronaphthalene as the predominant metabolite from naphthalene.

Persistence of naphthalene in soil or water in nature may be due to oxygen limitation or to such low levels of the compound that microbial growth is not sustained (Alexander, 1985).

3. Biphenyl

Fungi appear to metabolize biphenyl to metabolites similar to those formed by mammalian systems, i.e., the *trans*-configuration (Hou, 1982). *Candida lipolytica, C. guilliermondii, C. tropicalis, C. maltosa,* and *Debaryomyces hansenii* are able to metabolize this compound (Cerniglia and Crow, 1981). *C. lipolytica* oxidizes biphenyl to 2-, 3-, and 4-hydroxybiphenyl, 4,4'-dihydroxybiphenyl, and 3-methoxy-4-hydroxybiphenyl, with 4-hydroxybiphenyl as the main metabolite. Yeasts primarily oxidize biphenyl in the *para* position to form 4-hydroxybiphenyl.

Strains of *Moraxella* sp., *Pseudomonas* sp., and *Flavobacterium* sp. able to grow on biphenyl have been isolated from sewage (Stucki and Alexander, 1987). Biphenyl can be metabolized by a number of different species of yeast (Cerniglia and Crow, 1981).

4. Benzo(a)pyrene (BaP)

Benzo(a)pyrene is a potent carcinogen (Cook, Hewett, and Hieger, 1933).

Fungi oxidize this compound by a mechanism similar to that observed in mammalian systems (Hou, 1982); e.g., it is oxidized by *Cunninghamella elegans* to polar products with *trans*-configuration. *C. elegans* degrades this compound (Zajic and Gerson, 1977), with formation of a highly mutagenic and tumorigenic product. Several *Candida* spp. are able to metabolize benzo(a)pyrene (Komagata, Nakase, and Katsu, 1964), e.g., *C. lipolytica, C. guilliermondii, C. tropicalis, C. maltosa,* and *Debaryomyces hansenii* are able to metabolize benzo(a)pyrene (Cerniglia and Crow, 1981). *C. lipolytica* oxidizes benzo(a)pyrene to 3-hydroxybenzo(a)pyrene and 9-hydroxybenzo(a)pyrene. *Neurospora crassa* and *Saccharomyces cerevisiae* have inducible hydroxylases that can attack this compound. *Cunninghamella bainieri, C. elegans, Aspergillus ochraceus* TS, and various yeast strains have also demonstrated hydroxylase activity (Cerniglia and Gibson, 1979).

Several brown- and white-rot fungi were able to degrade at least 40% of benzo(a)pyrene in 2 weeks or less (Lee, Fletcher, Avila, Callanan, Yunker, and Munnecke, 1995). They produced more biomass at acidic to neutral pH, incubation at 30°C, 90% moisture saturation, with granulated corncobs or alfalfa pellets as a lignocellulosic substrate. The isolates were *Irpex* sp. #232, *Fomitopsis* sp. #259, *Phanerochaete* spp. #326 and #404, *Heteroporus* sp. #501, and *Cunninghamella elegans*.

Isolates of the white-rot fungi *Phanerochaete chrysosporium* BKM-F-1767, *Trametes versicolor* Paprican 52, and *Bjerkandera adusta* CBS 595.78 are able to remove benzo(a)pyrene from liquid culture (Field, de Jong, Feijoo-Costa, and de Bont, 1992). There is no accumulation of quinones during degradation of benzo(a)pyrene. *Bjerkandera* sp. strain BOS55 can degrade benzo(a)pyrene at a rate of

8 to 14 mg/kg/day at 20 and 30°C (Field, Feiken, Hage, and Kotterman, 1995). Adequate aeration is essential. This strain was found to be the best degrader of both anthracene and benzo(a)pyrene (Field, de Jong, Feijoo-Costa, and de Bont, 1992).

P. chrysosporium produces powerful extracellular enzymes, peroxidases, and H_2O_2, which are involved in the initial attack of benzo(a)pyrene (Field, de Jong, Feojoo-Costa, and de Bont, 1993). *P. chrysosporium* has been shown to degrade benzo(a)pyrene in nutrient nitrogen-deficient cultures (Bumpus, Tien, Wright, and Aust, 1985). *P. laevis* HHB-1625 is able to transform benzo(a)pyrene by means of manganese peroxidase (MnP–Mn^{2+}) reactions or in MnP-based lipid peroxidation systems (Bogan and Lamar, 1996).

A number of bacteria have been found to biodegrade PAHs in the soil. A *Mycobacterium* sp. isolated from coal gasification site soils for its ability to mineralize pyrene can also degrade benzo(a)pyrene (Schneider, Grosser, Jayasimhulu, and Warshawsky, 1994). A *Mycobacterium* strain PYR-1 can simultaneously degrade >74% of a mixture of PAHs, including benzo(a)pyrene, in six days (Kelley and Cerniglia, 1995). A mixed culture was able to degrade a range of PAHs, including benzo(a)pyrene (Trzesicka-Mlynarz and Ward, 1995). The predominant organisms, Gram-negative rods, were *Pseudomonas putida, Pseudomonas aeruginosa*, and a *Flavobacterium* sp.

Bacillus megaterium strains accumulate benzo(a)pyrene, which is stored in the cytoplasm and the lipid inclusions of the cells but not in the form of benzo(a)pyrene–protein complexes (Poglazova, Fedoseeva, Khesina, Meissel, and Shabad, 1967). The chemical remains unaltered in some of the strains but gradually disappears or is reduced in others. When the organisms are grown on a medium that does not contain aromatic hydrocarbons, they eventually lose their ability to destroy benzo(a)pyrene. However, this ability can be restored and further enhanced by adding the compound to the medium.

Cyanobacteria, diatoms, and some eukaryotic algae can also oxidize benzo(a)pyrene (Cerniglia, 1984).

Often an organism that is adapted to metabolize a member of a homologous series of molecules may be capable of degrading the rest of the series. This has also been observed with related aromatics of fewer ring counts. For example, after growth with succinate plus biphenyl, a mutant strain of *Beijerinckia* could oxidize benzo(a)pyrene (Gibson and Mahadevan, 1975). Chemical analog adaptation can reduce the amount of time required for adaptation to a particular chemical. Analog enrichment can employ minimally mutagenic phenanthrene to increase the rate of degradation of highly mutagenic benzo(a)pyrene (Sims and Overcash, 1981). The rate of cometabolism of benzo(a)pyrene can be significantly increased with enrichment of the soil with phenanthrene as an analog (Sims and Overcash, 1981). When *P. paucimobilis* strain EPA 505 is grown on fluoranthene as a sole source of carbon and energy, it is able to degrade benzo(a)pyrene (Siddiqi, Ye, Elmarakby, Kumar, and Sikka, 1994). The cells degrade 25.2% of benzo(a)pyrene to a major, highly polar metabolite.

Concentration of PAHs does not appear to be an inhibitory factor, according to Bossert and Bartha (1986). A concentration of 5% of benzo(a)pyrene on a dry soil basis does not inhibit biodegradation of the compound in soil.

5. Benz(a)anthracene (BaA)

Benz(a)anthracene is considered to be a weak carcinogen and a weak tumor initiator (Wislocki, Kapitulnik, Levin, Lehr, Schaerer-Ridder, Karle, Jerina, and Conney, 1978).

A mutant strain of *Beijerinckia* is able to oxidize benz(a)anthracene after growing on succinate in the presence of biphenyl (Gibson and Mahadevan, 1975). This compound is metabolized to four dihydrodiols, primarily *cis*-1,2-dihydroxy-1,2-dihydrobenz(a)anthracene. Fungi, e.g., *Cunninghamella elegans*, oxidize this compound to polar products with *trans*-configuration (Hou, 1982). The product formed by *C. elegans* is highly mutagenic and tumorigenic (Thakker, Levin, Yagi, Ryan, Thomas, Karle, Lehr, Jerina, and Conney, 1979).

A strain of *Pseudomonas fluorescens* isolated from exhausted lubricating oil-polluted soil can grow and degrade chrysene and benz(a)anthracene as a sole carbon source (Caldini, Cenci, Manenti, and Morozzi, 1995).

A *Mycobacterium* sp. isolated from coal gasification site soils for its ability to mineralize pyrene can also degrade benzo(a)anthracene (Schneider, Grosser, Jayasimhulu, and Warshawsky, 1994).

6. Anthracene and Phenanthrene

Isomeric anthracene and phenanthrene and their metabolites are not acutely toxic, carcinogenic, or mutagenic; however, analogs of both structures are found in carcinogenic PAHs (Cerniglia, 1984).

Phenanthrene might be utilized only in the soluble form (aqueous solution), since the growth rates of bacteria are related to the solubilities of the hydrocarbons on which they are growing (Wodzinski and Johnson, 1968; Wodzinski and Coyle, 1974). Phenanthrene is about 20 times more soluble in water than anthracene (Leahy and Colwell, 1990), which may explain the faster biodegradation of phenanthrene (Bossert and Bartha, 1986). Phenanthrene may be more rapidly degraded in natural environments than predicted (Fewson, 1981). Other researchers found anthracene, an isomer of phenanthrene, to be degraded twice as fast as phenanthrene (Manilal and Alexander, 1991). They suggest that degradation of higher-ring PAHs may proceed through a phenanthrene intermediate or end point, producing artificially low degradation rates.

Induction and cometabolism can be utilized to improve degradation rates. Salicylate can induce degradation of phenanthrene, and phenanthrene or salicylate can induce metabolism of pyrene and fluoranthene by *Pseudomonas saccharophila* P-15 (Stringfellow, Chen, and Aitken, 1995). Phenanthrene may be a poor inducer of its own degradation, but addition of fluorene as a cosubstrate has been observed to be synergistic, increasing utilization of phenanthrene, possibly by a positive analog effect on enzyme induction (Bouchez, Blanchet, Besnainou, and Vandecasteele, 1995; Bouchez, Blanchet, and Vandecasteele, 1995a). An example of analog enrichment is the use of minimally mutagenic phenanthrene to increase the rate of degradation of highly mutagenic benzo(a)pyrene (Sims and Overcash, 1981). The rate of cometabolism of benzo(a)pyrene was significantly increased with enrichment of the soil with phenanthrene as an analog (Sims and Overcash, 1981). Phenanthrene is cometabolized more easily than the higher-molecular-weight PAHs (Bouchez, Blanchet, Besnainou, and Vandecasteele, 1995). Doubling the starting concentration of anthracene or phenanthrene also results in higher degradation rates (Wiesel, Wuebker, and Rehm, 1993).

Soil type affects biodegradation of these compounds. Manilal and Alexander (1991) report that mineralization of phenanthrene occurs more slowly in soil than in liquid media. This may be caused by sorption to soil constituents, such as organic matter. Mineralization of phenanthrene is enhanced by addition of phosphate but not potassium, and it is reduced by addition of nitrate. After exposure to different soil types with high clay content, 100% of the anthracene and up to 25% of the phenanthrene was not extractable (Karimi-Lotfabad, Pickard, and Gray, 1996). Anthracene had oligomerized to higher-molecular-weight aromatic products. Water inhibited the reaction. The reacted anthracene was not available for biodegradation. In sandy soils, the three-ring PAHs, anthracene and phenanthrene, have half-lives of 16 and 134 days, respectively (Sims, Doucette, McLean, Grenney, and Dupont, 1988).

Anthracene can be completely mineralized by soil pseudomonads (Cerniglia, 1984). Different organisms oxidize anthracene using different enzymes. These include strains of *Pseudomonas, Nocardia,* and *Beijerinckia.* Phenanthrene can be utilized by *Pseudomonas* spp. as the sole carbon source (Ghisalba, 1983). Strains of *Aeromonas*, various fluorescent and nonfluorescent pseudomonads, and vibrios utilize an alternative pathway for phenanthrene metabolism (Kiyohara and Nagao, 1978). The metabolic pathways for phenanthrene and anthracene are similar to the sequence for naphthalene (Hou, 1982). A *Pseudomonas/Alcaligenes* sp., an *Acinetobacter* sp., and an *Arthrobacter* sp. grow on anthracene or pyrene as the sole carbon source (Stetzenbach and Sinclair, 1986). Maximum rates of degradation for *P. paucimobilis* were found to be 1.0 mg phenanthrene/mL/day, near pH 7.0 and 30°C (Weissenfels, Beyer, and Klein, 1990).

A mixed culture was able to degrade a range of PAHs, including anthracene (Trzesicka-Mlynarz and Ward, 1995). The predominant organisms, Gram-negative rods, were *P. putida, P. aeruginosa,* and a *Flavobacterium* sp. The mixed culture and a synthetic mixture of the primary bacteria were more successful in degrading the compounds. The individual isolates could efficiently remove the more water-soluble PAHs, such as phenanthrene, but not the less water-soluble compounds, like anthracene. Anthracene degradation was found to be dominated by *Agrobacterium* spp., *Pseudomonas* spp., *Aspergillus terrus, A. flavus,* and *Penicillium tordum* (Mahmood and Rao, 1993). *Sclerotium rolfsii* and *Trichoderma harzianum* were associated with degradation of phenanthrene. Walter, Beyer, Klein, and Rehm (1990) also report degradation of anthracene oil by a mixed culture of 15 organisms, including species of *Pseudomonas, Achromobacter, Alcaligenes,* and *Flavobacterium.* The mixture was more efficient at degrading PAHs (including the less water-soluble PAHs) than the individual cultures.

Arthrobacter polychromogenes can degrade almost 50% of the phenanthrene in a solution containing 150 mg phenanthrene/L (Keuth and Rehm, 1991). Increased concentrations of phenanthrene results in increased growth and degradation rates. Adding low amounts of glucose (0.45 g/L) to the culture medium stimulate growth and phenanthrene degradation. High levels of glucose (3 g/L) inhibit phenanthrene degradation, because of preferential use of the glucose as an energy source. Raising the temperature from 30 to 35°C does not affect phenanthrene degradation but does inhibit microbial cell division.

White-rot fungi have shown the ability to degrade PAHs. Even if the organism is killed by 20% acetone or ethanol, there is partial bioconversion of the PAHs, probably due to the presence of extracellular peroxidases. PAH solubilization can be increased in water/solvent mixtures to enhance bioavailability (Field, Boelsma, Baten, and Rulkens, 1995). Acetone and ethanol at 5% are toxic to the white-rot fungus *Bjerkandera* sp. strain BOS55 at the time of inoculation. However, when 20% solvents were added to 9-day-old cultures, oxidation of anthracene to anthraquinone was observed. This may have been due to production of extracellular peroxidases. *Candida parapsilosis* is able to degrade phenanthrene and anthracene in soil (Yong, Tousignant, Leduc, and Chan, 1991). *Cunninghamella elegans* can oxidize anthracene with formation of a *trans*-dihydrodiol (Cerniglia, 1984).

All of the strains of the white-rot fungi *Phanerochaete chrysosporium* BKM-F-1767, *Trametes versicolor* Paprican 52, and *Bjerkandera adusta* CBS 595.78 that were tested were able to remove anthracene from liquid culture (Field, de Jong, Feijoo-Costa, and de Bont, 1992). *Bjerkandera* and *Phanerochaete* converted anthracene to anthraquinone, which was a dead-end metabolite. *Trametes* strains were able to degrade anthracene, as well as anthraquinone.

P. chrysosporium can also metabolize phenanthrene (Bumpus, 1989; Sutherland, Selby, Freeman, Evans, and Cerniglia, 1991). In the intact fungus, lipid peroxidation by MnP is the basis for the oxidation of phenanthrene (Moen and Hammel, 1994). The fungus can degrade phenanthrene in nitrogen-rich culture media (Sutherland, Selby, Freeman, Evans, and Cerniglia, 1991), although mineralization of phenanthrene and other structurally related compounds is greater at low nitrogen levels. In a nutrient-limited medium, *P. chrysosporium* was able to remove 70 to 100% of 22 two-, three-, and four-ring PAHs, including phenanthrene and anthracene (Bumpus, 1989). Degradation appears to occur under both ligninolytic and nonligninolytic conditions (Dhawale, Dhawale, and Dean-Ross, 1992) but involves different enzymes for each state (Field, de Jong, Feijoo-Costa, and de Bont, 1993). The fungus can also degrade anthracene (Vyas, Bakowski, Sasek, and Matucha, 1994) and PAHs present in anthracene oil (Bumpus, 1989). *Phanerochaete laevis* HHB-1625 is able to transform anthracene by means of MnP–Mn^{2+} reactions or in MnP-based lipid peroxidation systems (Bogan and Lamar, 1996).

The white-rot fungus *Bjerkandera* sp. BOS55 is an outstanding degrader of anthracene and phenanthrene (Field, Heessels, Wijngaarde, Kotterman, de Jong, and de Bont, 1994). This strain was found to be the best degrader of these compounds (Field, de Jong, Feijoo-Costa, and de Bont, 1992). The PAH-degrading ability of this strain is far superior to that of all other fungal strains tested, including *P. chrysosporium* (Field, de Jong, Feijoo-Costa, and de Bont, 1993). *Bjerkandera* sp. strain BOS55 requires a high level of dissolved oxygen or a high redox potential for rapid anthracene biodegradation.

Phenanthrene is very toxic to the freshwater alga *Selenastrum capricornutum* (Hsieh, Tomson, and Ward, 1980). However, *Oscillatoria* spp. are able to mineralize phenanthrene (Phillips, Bender, Word, Niyogi, and Denovan, 1994), whether incubated in light or dark conditions (Bender, Vatcharapijarn, and Russell, 1989).

Concentration of the surfactant is an important factor (Aronstein and Alexander, 1993). These authors report that the nonionic surfactant Novel II 1412-56 enhances the rate and extent of phenanthrene mineralization, when used at a concentration of 10 µg/mL in water pumped through soil. Stimulation is less if 100 µg surfactant/mL is employed. Microbial degradation of phenanthrene in soil–aqueous systems is inhibited by addition of alkyl ethoxylate, alkylphenyl ethoxylate, or sorbitan-type (Tween-type) nonionic surfactants at doses that result in micellar solubilization of phenanthrene from soil (Laha, Liu, Edwards, and Luthy, 1995).

Only a portion of phenanthrene partitioned into the micellar phase of nonionic surfactants is directly bioavailable (Guha and Jaffe, 1996b). The micellar-phase bioavailable fraction of phenanthrene decreases with an increasing concentration of Triton N101, Triton X-100, and Brij 30. For Brij 35, the micellar-phase fraction of phenanthrene is not directly bioavailable. Although micellized phenanthrene in the nonionic surfactant Tergitol NP-10 may not be directly available to the bacterium *Pseudomonas stutzeri*, the surfactant increases the phenanthrene dissolution rate, with an accompanying increase in bacterial growth (Grimberg, Stringfellow, and Aitken, 1996). Sun and Boyd (1993) report that sorbed PSO surfactants (commercial Petronates) are nearly as effective as aqueous-phase PSO emulsion as a partition phase for phenanthrene.

Biosurfactants can also aid in degradation of naphthalene. Several cultures were found to produce a biosurfactant when grown on naphthalene (Deziel, Paquette, Villemur, Lepine, and Bisaillon, 1996). *Pseudomonas aeruginosa* produces a biosurfactant when grown on phenanthrene (Deziel, Paquette, Villemur, Lepine, and Bisaillon, 1996). The biosurfactant excreted by *P. aeruginosa* PRP652 is a glycosylated, anionic, amphipathic compound, termed Rhamnolipid R1 (Hunt, Robinson, and Ghosh, 1994). The apparent solubility of phenanthrene increased from 1.2 mg/L with no surfactant to 34.4 mg/L (3500 mg/L biosurfactant). Rhamnolipid R1 can lower the surface tension of aqueous solutions to a minimum of 35.1 dyn/cm at a concentration of about 20 mg/L. Detectable amounts of glycolipids are also produced with growth on phenanthrene (Deziel, Paquette, Villemur, Lepine, and Bisaillon (1996).

7. Pyrene

Pyrene is probably slow to be degraded (Ghisalba, 1983), but doubling the starting concentration of the compound results in higher degradation rates (Wiesel, Wuebker, and Rehm, 1993). A sparing effect in the presence of other, more-biodegradable substrates has been proposed to account for the persistence of pyrene in one landtreatment study, while it had been quite biodegradable in another (Bossert and Bartha, 1986). Significant metabolization of the tetracyclic aromatic hydrocarbon pyrene was observed after the phenol, naphthalene, and most of the tricyclic compounds present were degraded (Wiesel, Wuebker, and Rehm, 1993).

A study of several surfactants — two sulfonated anionic surfactants (Dowfax C10L and Dowfax 8390), a phosphate-ester blend, weak-acid anionic surfactant (Rexophos 25/97), and two ethoxylated nonionic surfactants (Tergitol 15-S-9 and Triton X-100) — under acidic conditions determined that a decrease in pH from 8.3 to 1.0 would have no effect on pyrene solubilization (Van Benschoten, Ryan, Huang, Healy, and Brandl, 1995). Sorption of C10L and Triton X-100, however, increased significantly with a decrease in pH.

A *Pseudomonas/Alcaligenes* sp., an *Acinetobacter* sp., and an *Arthrobacter* sp. grow on pyrene as the sole carbon source (Stetzenbach and Sinclair, 1986). Pyrene degradation is correlated with the presence of oxygen with no decrease in concentration observed under anaerobic or microaerophilic conditions. A *Mycobacterium* strain PYR-1 can simultaneously degrade >74% of a mixture of PAHs, including anthracene and phenanthrene, within 6 days (Kelley and Cerniglia, 1995). Strains of *Mycobacterium* can degrade pyrene (Kelley and Cerniglia, 1991; Kelley, Freeman, and Cerniglia, 1991; Kelley, Freeman, Evans, and Cerniglia, 1991; Grosser, Warshawsky, and Kinkle, 1994). A *Mycobacterium* sp. isolated from coal gasification site soils for its ability to mineralize pyrene can also degrade benz(a)anthracine (Schneider, Grosser, Jayasimhulu, and Warshawsky, 1994). *Mycobacterium* strain PYR-1 can simultaneously degrade >74% of a mixture of PAHs, including pyrene, in 6 days (Kelley and Cerniglia, 1995). A *Rhodococcus* sp. can degrade pyrene and chrysene (Walter, Beyer, Klein, and Rehm, 1991). In another study, *Bacillus subtilis* was the primary degrader of pyrene (Mahmood and Rao, 1993).

Phanerochaete chrysosporium produces powerful extracellular enzymes, peroxidases, and H_2O_2, which are involved in the initial attack of pyrene (Bumpus, Milewski, Brock, Ashbaugh, and Aust, 1991; Hammel, 1992; Field, de Jong, Feijoo-Costa, and de Bont, 1993). The secondary metabolite, veratryl alcohol, enhances oxidation of pyrene (Cancel, Orth, and Tien, 1993). In a nutrient-limited medium, *P. chrysosporium* was able to remove a large percentage of PAHs, including pyrene (Bumpus, 1989). A *Beijerinckia* sp. can also oxidize pyrene (Stetzenbach and Sinclair, 1986). Pyrene can be degraded by 96.8 to 99.5% in 7 days by the white-rot fungus *Bjerkandera* sp. strain BOS55 (Field, Baten, Boelsma, and Rulkens, 1996).

The rate of cometabolism of pyrene is significantly increased with enrichment of the soil with phenanthrene or naphthalene (McKenna, 1977), and with a mixed culture of flavobacteria and pseudomonads in the presence of phenanthrene or naphthalene (Cerniglia, 1984). Naphthalene, and especially phenanthrene, can be used as growth substrates for the non-growth substrate pyrene (McKenna, 1977). Phenanthrene or salicylate can induce metabolism of pyrene by *Pseudomonas saccharophila* P-15 (Stringfellow, Chen, and Aitken, 1995).

Fulvic acid has been shown to decrease mineralization of pyrene, apparently due to toxicity to the microbes and possible sorption of the compound making it less bioavailable (Grosser, Warshawsky, and Kinkle, 1994). After exposure to soil with high clay content, 100% of the pyrene was not extractable (Karimi-Lotfabad, Pickard, and Gray, 1996).

8. Fluorene and Fluoranthene

Fluorene is very toxic to the freshwater alga *Selenastrum capricornutum* (Hsieh, Tomson, and Ward, 1980). A strain of *Pseudomonas paucimobilis* has the ability to use fluoranthene, a high-molecular-weight PAH that is slowly degraded, as the sole source of carbon and energy for growth (Mueller, Chapman, Blattmann, and Pritchard, 1990). Ghisalba (1983) reported that several *Pseudomonas* spp. can utilize fluoranthene as sole carbon source. *Alcaligenes denitrificans* can degrade 0.3 mg fluoranthene/mL/day (Weissenfels, Beyer, and Klein, 1990). Maximum rates of degradation for *A. denitrificans* occurred with 0.3 mg fluoranthene/mL/day as the sole carbon source around pH 7.0 and 30°C (Weissenfels, Beyer, and Klein, 1990). The primary degraders of fluoranthene in one study were *Agrobacterium* spp., *Pseudomonas* spp., *Aspergillus terrus*, *A. flavus*, and *Penicillium tordum* (Mahmood and Rao, 1993). *Mycobacterium* sp. PYR-1 can mineralize fluoranthene (Kelley and Cerniglia, 1991; Kelley, Freeman, and Cerniglia, 1991; Kelley, Freeman, Evans, and Cerniglia, 1991). When given a complex carbon source, a strain of *Mycobacterium* PYR-1 was found to degrade simultaneously >74% of a mixture of PAHs, including fluoranthene, within 6 days (Kelley and Cerniglia, 1995).

A mixed culture of predominantly Gram-negative rods — *Pseudomonas putida*, *P. aeruginosa*, and a *Flavobacterium* sp. — was able to degrade several PAHs, including fluorene and fluoranthene (Trzesicka-Mlynarz and Ward, 1995). The mixed culture and a synthetic mixture of the primary bacteria were more successful in degrading the compounds, while the individual isolates could efficiently remove the more water-soluble PAHs, including fluorene and fluoranthene. Wiesel, Wuebker, and Rehm (1993) studied a mixed bacterial culture (MK1) which was able to metabolize significantly the tetracyclic aromatic hydrocarbon fluoranthene, after the degradation of phenol, naphthalene, and most of the tricyclic compounds. Results were the same for both immobilized and freely suspended cells.

Addition of fluorene as a cosubstrate can increase utilization of phenanthrene (Bouchez, Blanchet, Besnainou, and Vandecasteele, 1995), and phenanthrene or salicylate can induce metabolism of fluoranthene by *P. saccharophila* P-15 (Stringfellow, Chen, and Aitken, 1995). Phenanthrene may be a poor inducer of its own degradation, but fluorene enhances phenanthrene biodegradation, possibly by a positive analog effect on enzyme induction (Bouchez, Blanchet, and Vandecasteele, 1995a).

Several brown- and white-rot fungi were able to degrade at least 40% of fluorene in 2 weeks or less (Lee, Fletcher, Avila, Callanan, Yunker, and Munnecke, 1995). They produced more biomass at acidic to neutral pH, incubation at 30°C, 90% moisture saturation, with granulated corncobs or alfalfa pellets as a lignocellulosic substrate. The isolates were *Irpex* sp. #232, *Fomitopsis* sp. #259, *Phanerochaete* spp. #326 and #404, *Heteroporus* sp. #501, and *Cunninghamella elegans*. Addition of Tween-80 to liquid PAH generally improved the degradation rate. In a nutrient-limited medium, *P. chrysosporium* was able to remove a large percentage of fluorene and fluroanthene (Bumpus, 1989).

3.2.1.5.3 Branched-Chain Aliphatics

1. Alkanes and Alkenes

Only a few microorganisms utilize these hydrocarbons, possibly because the oxidation enzymes are not capable of handling branched-chain substrates (Hou, 1982). The fatty acids resulting from oxidation of single-branched alkanes are incorporated into the cell lipid. Subsequent oxidation is usually then by the β oxidation pathway.

Size, position, and degree of branching all affect microbial alkane utilization: the larger the substituent, the nearer its position to the center of the alkane chain; and the more highly substituted a given carbon atom is in a chain, the more recalcitrant the alkane molecule is to utilization (McKenna, 1972). However, if the substituent is small (an ethyl or methyl group), the hydrocarbon can be utilized, unless the branches block β oxidation (or, in some cases, α oxidation) of the fatty acid intermediate. If properly placed, multiple methyl branches need not render an alkane unpalatable to the microbes. If a quaternary carbon atom (or neopentyl group) occurs at the end of an alkane chain, the result is a molecule quite resistant to microbial attack.

The accumulation of pristane in the biosphere would appear to be a consequence of the lack of proper conditions for oxidation rather than it having a refractory chemical structure (McKenna and Kallio, 1971). Microorganisms metabolize branched alkanes in much the same fashion as *normal* (*n*)alkanes. Information on the mechanism of *n*-alkane metabolism should then be relevant to the problem of branched alkane assimilation.

Multiple-branched alkanes, such as pristane, can be converted to succinyl-CoA (coenzyme A) by *Corynebacterium* sp. and *Brevibacterium erythrogenes*. In all, 11 out of 14 fungi and 4 out of 21 bacteria were able to degrade pristane (Cooney, 1980). Since 43% of the isolates tested could grow on pristane, which is relatively resistant to microbial attack, it would be expected that branched chain compounds would be readily degraded after an acute or chronic oil polluting event, if other conditions, such as nutrient requirements and absence of toxic materials, are met. Mixed populations may contribute to pristane degradation.

Hexadecane isomers with branches larger than ethyl can be used by *Nocardia corallina* for growth (McKenna, 1972). δ-Ethyltetradecane can be readily utilized by *N. corallina* and a species of *Moraxella*. Table 3.19 lists a number of microorganisms that can grow on some of the branched chain alkanes.

3.2.1.5.4 Straight-Chain Aliphatics

1. Alkanes

This series is oxidized by *Pseudomonas* spp. after 20 days at 20°C, the longer chains being converted into the isoprenoids pristane and phytane, the polycyclic sterane, and 17 α (H), 21 β (H)-hopane series, all of which are resistant to biodegradation (Gibson, Koch, and Kallio, 1968). The *n*-alkanes with shorter chains (from C_5 to C_9) are more easily used as a source of carbon and energy by microorganisms than those with longer chain lengths (from C_{10} to C_{14}) (Williams, Cumins, Gardener, Palmier, and Rubidge, 1981).

Oxidation of C_{10} to C_{14} alkanes is obtained with *Corynebacterium* (Zajic, 1964). Alkanes are metabolized by terminal oxidation, α oxidation, and diterminal oxidation. Fatty acids formed as by-products may be metabolized further by β oxidation. C_{12} to C_{16} are oxidized by *Corynebacterium* and further converted to the corresponding ketones or to esters of lower aliphatic acids (Hou, 1982). Alkanes (C_5 to C_{16}) have been utilized by selected strains of *Arthrobacter*, *Acinetobacter*, *Pseudomonas putida*, and yeasts. Heptane, dodecane, tridecane, tetradecane, and hexadecane have been reported as growth substrates. *n*-Alkane biodegradation is predominantly a bacterial activity (Song and Bartha, 1986). See Section 3.4 for the products formed from the microbial oxidation of alkanes.

Diesel oil consists mostly of linear and branched alkanes (Geerdink, van Loosdrecht, and Luyben, 1996). It can be degraded rapidly in batch culture with a mixture of microorganisms, at a growth rate of 0.55/h and a yield of 0.1 Cmol/Cmol. The branched alkanes are biodegraded after most of the linear alkanes. With a continuous stirred tank reactor, the maximum growth rate is 0.25/h, with a yield of 0.3 Cmol/Cmol, and much of the branched components is not degraded.

Free cells of *P. fluorescens* in diesel as a carbon source degraded 52.3% of C_{12} and 11.6% of C_{13}, but could not degrade C_{14} to C_{18}. Biofix-immobilized cells degraded 14.8% of C_{12} and an average of 53.5% of C_{13} to C_{18}. Drizit-immobilized cells degraded 24.5% of C_{12}, 52.4% of C_{13}, and an average of 91.2% of C_{14} to C_{18}. This shows that immobilized bacteria can be used to enhance the degradation of diesel in an aqueous system.

2. Methane

The ability to oxidize small amounts of methane is widely spread among soil microbes. It is oxidized by *Pseudomonas methanica* (Zajic, 1964). There is no absolute requirement for exogenous carbon dioxide. Methane is also oxidized by *Mycobacterium fortuitum* and *M. smegmatis* (Lukins, 1962).

Methylotrophs use single-carbon compounds as electron donors and carbon sources with oxygen serving as the electron acceptor (Woods, Williamson, Strand, Ryan, Polonsky, Ely, Gardner, and Defarges, 1987). They oxidize methane to carbon dioxide through the use of several enzymes in a series of reactions involving sequential transfers of two electrons. The initial oxidation of methane to methanol is catalyzed by the methane monooxygenase enzyme, which is thought to be largely responsible for the broad degradative competence of methylotrophs. Methanol is then further oxidized to form formaldehyde, formic acid, and carbon dioxide.

3. Ethane, Propane, Butane

Pseudomonas methanica cometabolically oxidizes ethane, propane, and butane while metabolizing its growth substrate, methane (Foster, 1962), with oxygen as a limiting factor (Zajic, 1964). Butane is utilized for growth by *Pseudobacterium subluteum*, *Pseudomonas fluorescens*, and *Actinomyces candidus* (Telegina, 1963).

BIODEGRADATION/MINERALIZATION/BIOTRANSFORMATION/BIOACCUMULATION

Table 3.19 Microbial Growth on Selected Branched Alkanes

Compound	Microorganism	Compound	Microorganism
Pristane	*Mycobacterium phlei* #451	3-Phenyleicosane	*Mycobacterium phlei* #451
	M. fortuitum #389		*M. fortuitum* #389
	M. rhodochrous #382		*M. smegmatis* #422
	M. smegmatis #422		*Nocardia rubra*
	Nocardia opaca	1-Phenylhendecane	*Micrococcus cerificans*
	N. rubra		*Mycobacterium phlei* #451
	N. erythropolis		*M. fortuitum* #389
	N. polychromogenes		*M. rhodochrous* #382
	N. corallina		*M. smegmatis* #422
	Micrococcus cerificans		*Nocardia opaca*
1-Phenyldecane	*Micrococcus cerificans*		*N. rubra*
	Pseudomonas aeruginosa		*N. erythropolis*
	Mycobacterium phlei #451		*N. polychromogenes*
	M. fortuitum #389		*N. corallina*
	M. rhodochrous #382	2-Phenylhendecane	*Nocardia opaca*
	M. smegmatis #422		*N. rubra*
	Nocardia opaca		*N. erythropolis*
	N. rubra		*N. polychromogenes*
	N. erythropolis		*N. corallina*
	N. polychromogenes	1-Phenyldodecane	*Micrococcus cerificans*
	N. corallina		*Pseudomonas aeruginosa*
1-Phenyl-4-methyldecane	*Mycobacterium phlei* #451		*P. fluorescens*
	M. fortuitum #389		*Mycobacterium phlei* #451
	M. rhodochrous #382		*M. fortuitum* #389
	M. smegmatis #422		*M. rhodochrous* #382
	Nocardia opaca		*M. smegmatis* #422
	N. rubra		*Nocardia opaca*
	N. erythropolis		*N. rubra*
	N. polychromogenes		*N. erythropolis*
	N. corallina		*N. polychromogenes*
1-Phenyleicosane	*Micrococcus cerificans*		*N. corallina*
	Pseudomonas aeruginosa	2-Phenyldodecane	*Mycobacterium phlei* #451
	P. fluorescens		*M. fortuitum* #389
	Mycobacterium phlei #451		*M. rhodochrous* #382
	M. fortuitum #389		*M.smegmatis* #422
	M. rhodochrous #382		*Nocardia opaca*
	M. smegmatis #422		*N. rubra*
	Nocardia opaca		*N. erythropolis*
	N. rubra		*N. polychromogenes*
	N. erythropolis		*N. corallina*
	N. polychromogenes	2-Phenyltridecane	*Nocardia opaca*
	N. corallina		*N. rubra*
2-Phenyleicosane	*Micrococcus cerificans*		*N. erythropolis*
	Pseudomonas aeruginosa		*N. polychromogenes*
	P. fluorescens		*N. corallina*
	Mycobacterium phlei #451	3-Phenyltetradecane	*Mycobacterium fortuitum* #389
	M. fortuitum #389		
	M. rhodochrous #382		
	M. smegmatis #422		
	Nocardia opaca		
	N. rubra		
	N. erythropolis		
	N. polychromogenes		
	N. corallina		

Source: From McKenna, E.J. in *Degradation of Synthetic Organic Molecules in the Biosphere, Proc. of Conf.*, San Francisco, National Academy of Sciences, Washington, D.C., 1972. With permission.

4. Alkenes

There are a number of reports of utilization of alkenes by microorganisms (Cooney, 1980). But many of them deal with 1-alkenes wherein growth could be at the expense of the saturated end of the molecule. Tetradecadiene could be used as sole carbon source by only 3 fungi out of 14 fungi and 21 bacteria tested. Mixed populations appear to be important in the degradation of this compound. The polyalkenes in refined oils should be relatively persistent.

Degradation of alkenes is not possible under methanogenic conditions (U.S. EPA, 1985a). Unsaturated 1-alkenes are generally oxidized at the saturated end of the molecule (Hornick, Fisher, and Paolini, 1983). Some microorganisms, such as the yeast *Candida lipolytica* attack at the double bond and convert the alkene into an alkane-1,2-diol. Other minor pathways proceed via an epoxide, which is converted into a fatty acid. Many microorganisms will not grow on 1-alkenes less than C_{12}. 2-Alkenes are more readily attacked than 1-alkenes. Terminal alkenes are readily degraded, being attacked at both ends and the double-bond region converted to an epoxide or diol (Ratledge, 1978). Utilization of internal alkenes can lead to both unsaturated and saturated fatty acids.

5. Tetradecane

Acinetobacter calcoaceticus MM5 is able to grow on tetradecane (as well as heating oil and crude oil), with increases of protein concentration and of caprilate-lipase and acetate-esterase enzymatic activities in the culture filtrate and a simultaneous pH drop (Marin, Pedregosa, Rios, Ortiz, and Laborda, 1995). A strong emulsification of petroleum by-products also occurs. Intracellular electron transparent inclusions were observed by transmission electron microscopy when grown on hydrocarbons.

Selected strains of *Arthrobacter, Acinetobacter, Pseudomonas putida*, and yeasts can use tetradecane as a growth substrate (Hou, 1982). A novel *P. anaerooleophila* strain HD-1 has the ability to grow in a medium containing *n*-tetradecane and petroleum and can be useful for treating environmental pollutants (Imanaka and Morikawa, 1993).

6. Hexane

The fungi *Hyalodendron, Varicosporium, Paecilomyces,* and *Cladosporium* are superior to bacterial isolates in degrading the *n*-hexane extractable fraction of crude oil, and *Penicillium* is superior to other fungi (Rahman, Barooah, and Barthakur, 1995).

7. Hexadecane

Hexadecane is suggested as an indicator of the microbial potential for degradation of hydrocarbons (Cooney, 1980). Fungi have been found to grow on hexadecane, but only some bacteria are able to utilize the material as sole carbon source. This suggests that use of hexadecane as a measure of hydrocarbon degradation potential may actually underestimate that potential.

Whyte, Greer, and Inniss (1996) tested 135 psychrotrophic microorganisms for ability to mineralize petroleum hydrocarbons. A number of strains mineralized toluene, naphthalene, dodecane, and hexadecane. Selected strains of *Arthrobacter, Acinetobacter, Pseudomonas putida*, and yeasts can use hexadecane as a growth substrate (Hou, 1982).

Cladosporium resinae has been found in soil (Walker, Austin, and Colwell, 1973) and repeatedly recovered as a contaminant of jet fuels (Atlas, 1981). This organism can grow on petroleum hydrocarbons and creates problems in the aircraft industry by clogging fuel lines. This fungus was found to oxidize 93% of hexadecane to CO_2 and to assimilate only 7% (Cooney and Walker, 1973).

Increasing the surface area of hexadecane increases its microbial destruction (Fogel, Lancione, Sewall, and Boethling, 1985). Extracellular free fatty acids proposed as emulsifiers are produced by *Acinetobacter* sp. in a medium containing *n*-hexadecane but not in nutrient broth–yeast extract medium (Makula and Finnerty, 1972).

Utilization of hexadecane is reduced by acetate (LePetit and Tagger, 1976), and addition of glucose to lake water represses hexadecane utilization by its microbial community in a diauxic manner (Bartha and Atlas, 1977).

When the nonmetabolizable compounds (cyclohexane, *p*-xylene, and toluene) were added to the metabolizable substrate *n*-hexadecane, the increased oxygen consumption was attributed to a stimulation of the oxidation of hexadecane by the nonoxidizable compounds (Cooney and Walker, 1973).

8. Heptane, Dodecane, Tridecane

Whyte, Greer, and Inniss (1996) isolated psychrotrophic microorganisms with the ability to mineralize dodecane. Selected strains of *Arthrobacter*, *Acinetobacter*, *Pseudomonas putida*, and yeasts can use these compounds as a growth substrate (Hou, 1982).

9. Octane

Multiple plasmid transfer has been accomplished between *Pseudomonas* species (Chakrabarty, 1974) to construct a strain that can degrade several hydrocarbons, including octane and naphthalene (Hornick, Fisher, and Paolini, 1983).

3.2.1.5.5 Fatty Acids and Carboxylic Acids

Since intermediary metabolites must also be removed for complete oil cleanup, non-hydrocarbon-utilizing microorganisms, such as fatty acid metabolizers, would also be required in the mixture (Atlas, 1977). During β-oxidation of an aliphatic carboxylic acid, fatty acids can form (Atlas, 1981; Hornick, Fisher, and Paolini, 1983). Some of these are toxic and can accumulate. Fatty acids produced from oxidation of single-branched alkanes may be directly incorporated into membrane lipids (Hou, 1982), instead of going through the β oxidation pathway (Dunlap and Perry, 1967). Certain species of *Pseudomonas* oxidize the fatty acids formed from heptane and hexane by classical β oxidation (Heringa, Huybregtse, and van der Linden, 1961). Decarboxylation of fatty acids does not occur in these particular systems. *P. putida* grown on aviation turbine fuel oxidizes straight-chain fatty acids from acetate (C_2) to palmitic acid (C_{16}) (Williams, Cumins, Gardener, Palmier, and Rubidge, 1981). No odd-numbered fatty acids are detected, indicating that β oxidation is the major pathway for fatty acid dissimilation.

Extracellular free fatty acids proposed as emulsifiers are produced by *Acinetobacter* sp. in a medium containing *n*-hexadecane (Makula and Finnerty, 1972).

3.2.1.5.6 Alcohols

Methanol and tertiary butyl alcohol (TBA) may each be present at 5% in some gasolines (Novak, Goldsmith, Benoit, and O'Brien, 1985). Methanol is readily biodegradable in both aerobic and anaerobic environments (Lettinga, DeZeeuw, and Ouborg, 1981; Novak, Goldsmith, Benoit, and O'Brien, 1985). It is degraded in soil samples to 31 m, especially in the saturated region (Novak, Goldsmith, Benoit, and O'Brien, 1985). TBA is more refractory. It is slowly degraded in anaerobic aquifers, with the degradation rate increasing with increasing concentration. At 1 mg/L TBA, it appears the microbial population receives insufficient energy to cause a population increase and utilization rates remain slow. Rates are faster at the highest concentration. Several species have been isolated that can degrade tertiary butyl alcohol under aerobic conditions (Benoit, Novak, Goldsmith, and Chadduck, 1985).

The first oxidation of methane produces methanol. Many bacteria and yeasts that can catabolize methanol under aerobic or anaerobic conditions and at low and high concentrations have been isolated from deep soil (to 100 ft) (Benoit, Novak, Goldsmith, and Chadduck, 1985).

3.2.1.5.7 Alicyclic Hydrocarbons

Cyclic aliphatics, or alicyclic hydrocarbons, are characterized by the presence of hydrocarbon rings. These compounds may be saturated or unsaturated and may possess alkyl side chains.

Pure strains do not grow on these compounds (Hou, 1982). Instead, these materials have been found to be co-oxidized by microorganisms growing on other substrates (see Section 5.2.3).

n-Alkyl-substituted alicyclic hydrocarbons are more susceptible to microbial attack than the unsubstituted parent compounds, e.g., methylcyclohexane and methylcyclopentane (Hou, 1982).

1. Cyclohexane and Oxygenates

Although some organisms can use alicyclic hydrocarbons as their sole source of carbon and energy (Sterling, Watkinson, and Higgins, 1977), several searches have not yielded pure cultures capable of degrading cyclohexane (Beam and Perry, 1974). Use of alicyclic compounds appears to be relatively rare. Co-oxidation by mixed microbial communities seems to be very important, since only a few species of bacteria have been shown to use cyclohexane as a sole carbon source (Hornick, Fisher, and Paolini, 1983).

In a process that uses both co-oxidation and commensalism, complete biodegradation of cyclohexane can be achieved in a two-step process using two organisms — an *n*-alkane oxidizer, such as a pseudomonad, that first converts cyclohexane into either cyclohexanol or cyclohexanone during growth on the *n*-alkane (e.g., *n*-heptane) (Ooyama and Foster, 1965) and another strain (e.g., a pseudomonad) that can grow on cyclohexanol or cyclohexanone (de Klerk and van der Linden, 1974; Donoghue, Griffin, Norris, and Trudgill, 1976). Microbial attack on cyclohexane could be initiated by hydroxylation. The same general results have been reported for the oxidation of other unsubstituted cycloparaffinic hydrocarbons (Beam and Perry, 1974).

The enzymes for cyclohexane oxidation are inducible by cyclohexane in a *Nocardia* sp. A *Nocardia* sp. has been identified as a primary cyclohexane utilizer, but growth occurs only in the presence of an unidentified pseudomonad that provides biotin and possibly other growth factors (Sterling, Watkinson, and Higgins, 1977; Hou, 1982). A *Nocardia* sp. was found to produce biotin, which is utilized by a *Pseudomonas* strain to grow on cyclohexane (Hou, 1982). A *Nocardia* sp. and a *Pseudomonas* sp. isolated on gasoline can co-oxidize a number of cyclic and aromatic compounds (Jamison, Raymond, and Hudson, 1976). *Acinetobacter* and *Nocardia globerula* grow rapidly with cyclohexanol as the sole source of carbon (de Klerk and van der Linden, 1974). *Mycobacterium convolutum* R-22 can completely mineralize cyclohexane only when an inducer, such as an *n*-alkane is present to induce the necessary hydroxylase (Perry, 1984). *Xanthobacter autotrophicus* cannot utilize cyclohexane but can grow with a limited range of substituted cycloalkanes, including cyclohexanol and cyclohexanone (Magor, Warburton, Trower, and Griffin, 1986). Another species of *Xanthobacter* can grow on cyclohexane. Both pathways produce adipic acid.

A novel *Pseudomonas anaerooleophila* strain can grow in a medium containing *n*-tetradecane, toluene, cyclohexane, and petroleum and can be useful for treating environmental pollutants (Imanaka and Morikawa, 1993).

Cyclohexane carboxylic acid and cyclohexane propionic acid are more suitable to microbial growth than cyclohexane acetic acid and cyclohexane butyric acid (Hou, 1982). Effective cleavage of an alicyclic ring occurs only when the side chain contains an odd number of carbon atoms.

3.2.1.5.8 Asphaltenes

Asphaltenes are that part of pure asphalt that is insoluble in petroleum naptha. It is hard and brittle and consists of oxidized hydrocarbons. Asphaltene components are relatively inert to microbial attack, since they consist of complex structures of sheets of aromatic and alicyclic ring structures with very short alkyl side chains (Ignasiak, Kemp-Jones, and Strausz, 1977). Lagoons may be employed for degradation of this material. Biodegradation of oil in lagoons may be effective for oil with a large proportion of asphaltenes (McLean, 1971).

3.2.1.6 Nonaqueous-Phase Liquids

Many pollutants are present in liquids that are immiscible with water rather than in the aqueous phase or sorbed to solids (Alexander, 1994). These nonaqueous-phase liquids (NAPLs), such as solvents and fuels, are common contaminants in soils and groundwater (Baker and Bierschenk, 1995; Anderson, 1994). They resist biodegradation because of their insolubility in water and can persist in the environment, unless removed by some special remediation technology (Alexander, 1994). Although NAPLs are considered immiscible in water, they are, in general, slightly soluble in water.

It is increasingly being recognized that nonaqueous phase material must be addressed for site assessment and remediation efforts to be cost effective and provide timely solutions to the risk of groundwater contamination (Reible, Malhiet, and Illangasekare, 1989).

Spills, leaking underground storage tanks, and improper disposal practices can result in the release and movement of NAPLs through soils. Movement of NAPLs through soil is considered to result from gravity- and/or capillarity-driven immiscible phase flow. Dispersive and convective transport of dissolved components, volatilization, sorption, and degradation are also important transport mechanisms. Thomson, Grahan, and Farquhar (1992) noted irregular immiscible liquid infiltration fronts in column experiments, indicating that very small-scale heterogeneities control the infiltration of immiscible liquids into soils.

These materials often exist as separate nonaqueous phases for long distances or times from the source due to their low aqueous solubilities (Reible, Malhiet, and Illangasekare, 1989). As NAPLs move through the subsurface as separate phase liquids, the light NAPLs (LNAPLs) can become trapped as residual saturation and lenses in the unsaturated and saturated zones and accumulate as free phase on the water table (Baker and Bierschenk, 1996; Seagren, Rittmann, and Valocchi, 1993; 1994). The dense NAPLs

(DNAPLs) pass through the soil and form a pool at the bottom of the aquifer (Alexander, 1994; Schwille, 1984). LNAPLs can also be in a nondrainable phase, where they are retained in the soil in the vadose zone, saturated zone, and capillary fringe (Baker and Bierschenk, 1996). And they can be present in the vapor phase in the vadose zone as soil gas.

The presence of a low permeability lens in a heterogeneous vadose zone can affect the dispersal of a NAPL (Smith, Reible, Koo, and Cheah, 1996). An essentially saturated layer of NAPL can be trapped at the base of a fine-grained lens, as a result of capillary forces. This would reduce the effectiveness of any *in situ* extraction methods, such as SVE, for removal of the NAPL. Mass transfer resistances in the existing vapor-phase control the release of compounds (e.g., toluene) in the NAPL from the fine lens at low air flowrate, while intralens diffusion resistances dominate at higher air flow rates. Liquid-phase resistances are important in the extraction of toluene from a NAPL mixture.

The composition of complex nonaqueous-phase liquids retained in the pore space of geologic material weathers until the residual NAPL no longer acts as a liquid and exists as discrete regions of hydrocarbon (residual hydrocarbons) in association with the geologic media (waterwet media), or as thin film coatings on the NAPL-wet media (Bouchard, Mravik, and Smith, 1990). These residuals from unleaded gasoline can resist separation from the soil solids.

NAPLs affect the concentration of contaminant available for organisms to degrade. Biodegradation rates are lower in multiphase systems than in aquatic systems (Gamerdinger, Achin, and Traxler, 1995). Reduced biodegradation rates in soil-NAPL-water systems can be explained by the lower aqueous phase concentration. This is not the case in a NAPL-water system, which suggests the NAPL has a direct effect on the bacteria. A NAPL may sequester a large fraction of a hydrophobic pollutant away from the aqueous phase and thus reduce the biodegradation of the hydrophobic compounds (Efroymson and Alexander, 1995; Efroymson and Alexander, 1991). This may result in unexpectedly slow biodegradation or in a contaminant concentration below the threshold for biodegradation.

Environmental contaminants may consist of a single NAPL, such as liquid alkanes, oils, or oil products, which represents 100 percent of the NAPL phase, or, more commonly, a mixture of compounds that represent a small fraction of the NAPL phase (Alexander, 1994). The equilibrium of each compound between the aqueous phase and the NAPL phase is critical. A chemical partitions between the two phases, based on its solubility in each. This relative partitioning can be expressed as an octanol-water partition coefficient, K_{ow}. A high value would indicate that a compound is hydrophobic and that there would be little present in water that is in equilibrium with NAPLs. A compound with a low value would exist at higher concentrations in water. When the aqueous concentrations of contaminants in NAPLs are high, rates of mineralization are higher than the rates of partitioning of a contaminant (e.g., phenanthrene) to water (Efroymson and Alexander, 1995). With low concentrations, biodegradation rates are slower than at higher concentrations and much slower than partitioning. Solubilization of the substrate can be utilized to improve biodegradation. Dissolution of PAHs from a nonaqueous-phase liquid to the aqueous phase renders these compounds bioavailable to microorganisms (Ghoshal, Ramaswami, and Luthy, 1995). Biodegradation of organic phase PAH then results in a depletion of the PAH from the NAPL.

The rate of mass transfer may control the overall rate of biotransformation of PAH compounds in certain systems (Ghoshal, Ramaswami, Luthy, 1995). This appears to occur during degradation of naphthalene from coal tar, a multicomponent, aromatic DNAPL (Ghoshal, Ramaswami, and Luthy, 1996). Coal tar partitions as a pseudocomponent in systems with appreciable solvent, but not in systems with only coal tar and water (Peters and Luthy, 1993). The rate of degradation of naphthalene is significantly influenced by the rate of external surface mass transfer from the coal tar in a mixed system, where coal tar exists as large globules (around 11 mm diameter) (Ghoshal, Ramaswami, and Luthy, 1996). Biodegradation depends upon relationships between NAPL composition and the equilibrium aqueous naphthalene concentration, the mass transfer rate of naphthalene from the NAPL to the aqueous phase, and the normal rate of naphthalene biodegradation. A coal tar with a predominance of PAHs and a bulk solubility in water of about 16 mg/L was studied (Peters and Luthy, 1993). Results found n-butylamine to be a good water-miscible solvent for coal tar dissolution. Mass transfer, on the other hand, does not limit biodegradation in slurry systems when coal tar is distributed in the micropores of a large number of small microporous silica particles (Ghoshal, Ramaswami, Luthy, 1995).

Bacterial activity is generally retained in the presence of a neat solvent and in biphasic systems where log K_{ow} of the solvent is >4 (Harrop, Woodley, and Lilly, 1992; Laane, Boeren, Hilhorst, and Veeger,

1987). The presence of aliphatic NAPLs will affect the biodegradation of PAHs (Gamerdinger, Achin, and Traxler, 1995). Naphthalene mineralization can occur in a multiphase system containing an aliphatic NAPL with log K_{ow} >5, and it is inhibited when the log K_{ow} of the NAPL is <4. It is suggested that, in petroleum-contaminated soil, the presence of aliphatic NAPLs may modify the distribution of PAHs, decreasing the PAH-sorbent interaction and reducing this barrier to biodegradation.

The extent of biodegradation suppression varies when different solvents are used as NAPLs (Labare and Alexander, 1995; Alexander, 1994). Partitioning and biodegradation of phenanthrene from NAPL to water varies with the NAPL and concentration of the test substrate (Efroymson and Alexander (1994a). Biodegradation is slow if the partitioning rate is slow. When the NAPL is hexadecane, dibutylphthalate, cyclohexane, commercial oils, crude oil, creosote, or kerosene, biodegradation is retarded. Although the rate of mass transfer of phenanthrene does not limit its mineralization by microorganisms in the soil, treatments that increase the rate of partitioning may enhance biodegradation, possibly because the treatment overcomes some other factor that limits degradation of the hydrocarbon (Orgega-Calvo, Birman, and Alexander, 1995).

Probably the most important factor in contamination by a NAPL is the amount released (Lyman, Noonan, and Reidy, 1990). With large releases, almost all of the petroleum product initially exists as NAPL; however, with time, a portion of it will transfer to the dissolved and sorbed phases. If the amount is very large, there will most likely be significant contamination of the groundwater. In small releases, all the NAPL may be trapped in the unsaturated zone as residual liquid.

Volatilization is the major physicochemical process affecting the behavior of NAPLs in inert porous media (Galin, Mcdowell, and Yaron, 1990). During volatilization, liquid kerosene gradually loses its light components (C9 to C13) and becomes more viscous, resulting in a decrease in the infiltration rate.

The rate and extent of biodegradation of hydrocarbons decreases with increasing viscosity of nontoxic NAPLs (Birman and Alexander, 1996b). When phenanthrene is present in kerosene, diesel fuel, fuel oil and mixtures of the NAPLs, it is degraded rapidly. However, if gasoline is present, the acclimation phase is prolonged and the rate of mineralization of phenanthrene is reduced. The slow mineralization of phenanthrene in 150 Bright stock or crude oils is enhanced by vigorous agitation of soil slurries, increasing the temperature, or addition of a nonionic surfactant with delayed inoculation. Thus, it is possible to improve bioremediation of pollutants present in viscous NAPLs.

Viscous and buoyancy forces are related to the capillary forces that retain organic liquids within a porous medium (Pennell, Pope, and Abriola, 1996). Buoyancy forces can contribute to NAPL mobilization and provide a novel approach for predicting NAPL displacement during surfactant flushing.

Soil type can affect the retention of NAPLs (Zytner, Biswas, and Bewtra, 1993). The large surface areas and exchange properties of soil organic material allow sorption of organic compounds, especially those that are hydrophobic (Devitt, Evans, Jury, Starks, Eklund, and Gholson, 1987; Alexander, 1994). Since hydrophobicity is related to sorption of compounds to soil organic material, the log K_{ow} values are good predictors of the extent of such binding (Oberbremer and Mueller-Hurtig, 1989). Retention of microbial or enzyme activity in the presence of organic solvents is related to solvent hydrophobicity (K_{ow}) (Chan, Kuo, Lin, and Mou, 1991). Retention capacity increases with an increase in the NAPL's density and the soil's porosity and a decrease in the soil bulk density (Zytner, Biswas, and Bewtra, 1993). These authors have developed a simple model to allow estimation of retention capacity values for different soils and NAPLs. Bouchard, Mravik, and Smith (1990) found that solute sorption was significantly higher for a low organic carbon soil contaminated with residual hydrocarbons than for natural soil organic carbon.

It has been shown that even hydrocarbons with very low water solubilities can be biodegraded and that a portion, if not all, of some chemicals that are not in the water phase can be utilized (Alexander, 1994). Alkanes up to at least C44 can be biodegraded (Haines and Alexander, 1974), even with a solubility lower than 1 ng/L. This amount of soluble aliphatic hydrocarbon would not be enough to support one bacterial cell per milliliter, indicating the microorganisms must be able to use the insoluble phase. Efroymson and Alexander (1994a) report that the maximum rate of biodegradation of pollutants in nonaqueous-phase liquids cannot always be predicted from the rates of their spontaneous partitioning to water.

Alexander (1994) proposes three mechanisms that could explain how microorganisms are able to degrade compounds in NAPLs or metabolize organic solvents having low water solubility. It is believed that the contaminant should be in contact with the surface of the microorganisms to allow its transport into the cell and subsequent enzymatic degradation. If microbes can utilize only those compounds that are water soluble, degradation would depend upon the rate of spontaneous partitioning into the water phase. Microorganisms might excrete material, such as surfactants or emulsifiers, that breaks down the compound to particles the size of 0.1 to 1 um for assimilation into the cell. The organisms might grow directly on the surfaces of the NAPL, absorbing the chemical at the area of contact.

Attachment of microorganisms to the organic solvent-water interface may be important in the transformation of organic contaminants (Efroymson and Alexander, 1991), although some organisms show stronger affinity for NAPLs than others (Rosenberg and Rosenberg, 1985). Ortega and Alexander (1994) found that naphthalene dissolved in diethylphthalate would exhibit first a slow then a rapid phase of degradation. Using Triton X-100 to prevent cell attachment to the interface, they concluded that the initial slow degradation was due to bacteria suspended in the aqueous phase and the rapid degradation was due to organisms present at the NAPL-H_2O interface. Colonization of the surface may not occur until after the compound in the aqueous phase is exhausted (Goswami, Singh, Bhagat, and Baruah, 1983). Some organisms, such as a strain of *Arthrobacter*, are able to attach to a NAPL and utilize a substrate (e.g., hexadecane) directly, without involving aqueous partitioning (Efroymson and Alexander, 1991). If extracellular enzymes are involved, they would probably function at the water/NAPL interface (Mattson and Volpenhein, 1966).

The greater the interfacial area between the water and the NAPL, the faster the degradation (Alexander, 1994). Soil flushing and *in situ* biodegradation can be employed to decrease the solute concentration, which increases the dissolution driving force (Seagren, Rittmann, and Valocchi, 1994; Seagren, Rittmann, and Valocchi, 1993). Biodegradation increases the dissolution flux from the pool only when Da-2 (ratio of biodegradation rate to the advection rate) is greater than 0.1. At that point, many different combinations of biodegradation and flushing can be used to improve the dissolution flux.

Inoculation of acclimated microbial cultures can enhance mineralization of some organic compounds in nonaqueous-phase liquids (Efroymson and Alexander (1994a). For instance, biodegradation of phenanthrene dissolved in the NAPL, dibutylphthalate, can be increased by addition of phenanthrene-degrading microorganisms, especially if the inoculum is previously grown in a medium containing the hydrocarbon and phthalate (Birman and Alexander, 1996a). Robertson and Alexander (1996) found that bacteria could mineralize phenanthrene that was initially present in di-*n*-butylphthalate but not toluene. However, bacteria sensitive to toluene were able to extensively degrade phenanthrene initially present in these toxic NAPLs if there was also a water phase and a separate nontoxic NAPL to lower the toxicant concentration in the aqueous phase. Efroymson and Alexander (1994b) report that transformation of phenanthrene is slow after an acclimation phase, when present in soil in some nonaqueous-phase liquids, while hexadecane and naphthalene are biodegraded rapidly.

If it is inconvenient, expensive, or too time-consuming to supply oxygen or other electron acceptors to organisms metabolizing oily contaminants *in situ*, it might be possible for the subsurface organisms to emulsify the hydrocarbons themselves (Wilson, Leach, Henson, and Jones, 1986). Many organisms excrete surface-active or emulsifying agents that are able to convert NAPLs, such as pure alkanes, various types of oil, and mixtures of hydrocarbons to small droplets that can then be assimilated and utilized by the microbes (Alexander, 1994; Rosenberg, Perry, Gibson, and Gutnick, 1979; Einsele, Schneider, and Fiechter, 1975). Several models have been proposed for the uptake of hydrocarbons by cells (Cooney, 1980). These include uptake of hydrocarbon dissolved in the aqueous phase, the necessity for direct contact between cells and large oil droplets, and an emphasis on formation of submicron-size oil droplets of micelles. Those models that include submicron droplets generally involve the production of surface active agents by the microorganisms.

Birman and Alexander (1996a) improved biodegradation of phenanthrene in dibutylphthalate, hexadecane, and diesel oil by adding organisms acclimated to the substrate dissolved in a NAPL and then intensely agitating the soil slurry. Growth of the microbes in the presence of a NAPL may produce an extracellular surfactant that promotes pseudosolubilization of the substrate (Singer and Finnerty, 1990; Hommel, 1990), or the cells may adhere to the NAPL, using the substrate more rapidly without the need for a surfactant (Efroymson and Alexander, 1991). Biodegradation of phenanthrene in 150 Bright stock oil and dibutylphthalate was also enhanced after preincubation with a nonionic surfactant (Alfonic 810-60), addition of acclimated organisms after the surfactant concentration fell to nontoxic levels, and vigorous agitation of the soil slurries (Birman and Alexander, 1996a). The agitation may have enhanced degradation by increasing diffusion of oxygen to the NAPL/water interface or by increasing the rate of partitioning of the phenanthrene from the NAPL to the water. Previous studies had found that the surfactant, Alfonic 810-60 increased partitioning but inhibited biodegradation (Ortega-Calvo, Birman, and Alexander, 1995). Use of surfactants to enhance biodegradation is covered in Section 5.3.1.

A wide range of wetting conditions can occur in the subsurface following spills of complex NAPL mixtures (Powers, Anckner, and Seacord, 1996). NAPLs with higher molecular weight constituents, such as creosote, or with added surfactants, such as gasoline, have a greater effect on the wettability of the subsurface system than lower molecular weight NAPLs. Neat solvents do not significantly impact quartz surface wettability, while many petroleum products form weakly water-wet surfaces. Under these conditions, NAPL recovery in the subsurface would be maximized, but recoveries for NAPL-wetting creosote and coal tar contaminants would be much lower.

There is a major obstacle, however, with the use of surfactants to clean up NAPLs from contaminated soil and groundwater (Ouyang, Mansell, and Rhue, 1995). Formation of macroemulsions with unfavorable flow characteristics in porous media can potentially result in low washing efficiency.

The typical *in situ* remediation technologies, such as pump-and-treat, soil vacuum extraction, soil flushing/washing, and bioremediation, for cleanup of contaminated sites are limited by flow channeling of chemical treatment agents (Bouillard, Enzien, Peters, Frank, Botto, and Cody, 1995). Argonne National Laboratory, the Gas Research Institute, and the Institute of Gas Technology have been experimenting with the use of foam to block pores and limit flow bypassing, to facilitate DNAPL remediation. Foam flushing is about 10 times greater than that of surfactant flushing alone (Enzien, Bouillard, Michelsen, Peters, Frank, Botto, and Cody, 1994).

Foams and oil-core aphrons (OCAs) have been tested for extracting, mobilizing, and dispersing nonaqueous-phase liquid contaminants for increased bioavailability (Enzien, Michelsen, Peters, Bouillard, and Frank, 1995). Foams caused the largest pressure drop in soil columns and had a >90% removal of NAPL pools. Surfactants removed <10%, while OCAs were ineffective.

Colloidal gas aphrons (CGAs) are foams generated *ex situ* using, for example, NaDBS (sodium dodecyl benzosulfonate) diluted in tap water (Bouillard, Enzien, Peters, Frank, Botto, and Cody, 1995). In a porous medium, foam exists as a gas phase dispersed in a continuous liquid phase (Bouillard, Enzien, Peters, Frank, Botto, and Cody, 1995). The diameter of the gas (air) bubble is about 60 um, and the gas volume fraction of the foams is about 65%. Foams flow through porous media as a front, instead of channeling as a surfactant would alone.

The surfactant in the foam fills the smallest pores and produces thin liquid lamellae (soap films) that stretch across pore spaces (Bouillard, Enzien, Peters, Frank, Botto, and Cody, 1995). Higher viscosity foams fill up larger channels (Lange, Bouillard, and Michelsen, 1995). The pressure drop builds up in the channel causing the foam to flow into less accessible spill zones. The stability of foams is affected by high capillary pressures. Unstable foams will rupture and become ineffective.

Application of colloidal gas aphrons (CGAs) was able to flush residual levels of a light nonaqueous-phase liquid (LNAPL), such as automatic transmission fluid (ATF), from a Superfund oil site (Roy, Kommalapati, Valsaraj, and Constant, 1995). Less pressure was required to pump the CGA suspension than surfactant solutions. CGA suspensions were more effective under downflow and upflow modes. Water floods seem to displace the ATF from the soil pores. The removal rate was not improved by increasing the surfactant concentration. In a laboratory study, about 80% of hexadecane in columns could be flushed with a 70% CGA quality (70% air, 30% liquid) (Lange, Bouillard, and Michelsen, 1995). A surfactant concentration of 1000 to 5000 ppm was needed to generate the microbubbles.

Application of magnetic resonance imaging (MRI) to visualize DNAPL interactions in the subsurface has shown that DNAPL ganglia are not uniformly distributed in the pore spaces (Bouillard, Enzien, Peters, Frank, Botto, and Cody, 1995). These new MRI tools should help visualize DNAPL mobilization in three dimensions, DNAPL interactions with foam, DNAPL emulsification, DNAPL scouring by the foam, and subsequent DNAPL mobilization/redeposition in the porous media.

Soil amendments can improve biodegradation of contaminating hydrocarbons. However, while addition of nitrogen and phosphorus can improve the degradation rate of phenanthrene, it did not affect the mineralization of DEHP di-2-ethylhexylphthalate dissolved in NAPLs (Efroymson and Alexander, 1994b). When NAPLs were added to a sandy, porous medium, air was entrapped, and NAPL and air were entrapped when water was added simulating a rising water table (Lenhard, Johnson, and Parker, 1993).

Using thermal gradients could prove to be effective for concentrating NAPLs in contaminated soil (Prunty, 1992). In columns packed with soil of initially uniform water and octane content with temperatures at opposite ends maintained at 40° and 6°C, octane moved toward the cool end within 1.5 hr. However, the water also, but more slowly, redistributed to the cool end and displaced the octanol, which then peaked on the warmer side of where the water content dropped sharply.

Free product recovery is often an essential prerequisite to further remedial action (Baker and Bierschenk, 1995). Vacuum-enhanced NAPL recovery (also called dual-phase extraction or bioslurping) increases LNAPL recovery rates compared with conventional methods (Leeson, Kittel, Hinchee, Miller, Haas, and Hoeppel, 1995; Baker and Bierschenk, 1995). It also accomplishes dewatering, while it facilitates vapor-based cleanup of the unsaturated zone (Baker and Bierschenk, 1995). Liquids may reside at negative gauge pressures and lack the potential energy to flow into a conventional recovery well. The subatmospheric pressures applied during vacuum-enhanced recovery can reduce the potential energy (i.e., entry suction) needed to allow extraction of liquid that otherwise would not flow into the well. It also induces pneumatic and hydraulic gradients toward the vacuum source. These increase the rate of water and NAPL recovery. A conceptual model has been successfully tested by these authors to take into account all of these crucial factors. Vacuum extraction can also be used to promote foam front advancement for enhancing DNAPL remediation (Enzien, Michelsen, Peters, Bouillard, and Frank, 1995).

Bioslurping combines the technologies of vacuum-enhanced recovery with bioventing to recover free-phase LNAPLs while simultaneously bioventing the vadose zone soil (Kittel, Hinchee, Hoeppel, and Miller, 1994). A single, aboveground vacuum pump can service several extraction wells to remove the LNAPL and soil gas in the same process stream. A slow extraction rate allows the soils to be aerated and minimizes volatilization. A bioslurper system at Fallon Naval Air Station, Nevada, employs a 10-hp liquid ring pump to extract 15 to 60 gpd LNAPL (JP-5), 50 cfm soil gas, and minimal groundwater (1 gpm). See Sections 2.2.2.2 and 2.2.2.3 for further discussion of the processes of bioventing and bioslurping, respectively.

An advanced microscopic technique, cryo-scanning electron microscopy (cryo-SEM) with X-ray analysis, was used to observe the location and form of a NAPL in a three-fluid-phase (air-NAPL-water) soil system (Hayden and Voice, 1993). Photomicrographs and X-ray dot maps confirmed the existence of continuous NAPL films on soil (with a residual water saturation) at high NAPL saturations. Photomicrographs revealed v-shaped wedges, pendular rings, and films on irregular-shaped sand grains. Low NAPL saturations probably exist as thin films or small isolated lenses or blobs.

Ryan and Dhir (1993) used a dual-beam gamma densiometer to measure residual light nonaqueous-phase liquid saturations after displacement by a dynamic water table. They determined the amount of entrapped contaminant was fairly consistent for portions of the column that were initially water-wet, producing residual saturations of 12%. The particle diameters studied had no effect. A second oscillation of the water table did not displace the entrapped contaminants, nor did additional contaminants become entrapped.

The design of vapor-extraction remedial systems and the analysis of their performance can be improved by using models that simulate processes affecting the multiple-compound, vapor-phase chemical mixtures and allow for natural (nonideal) conditions, such as inhomogeneous soil permeability, leakage of atmospheric air into the subsurface, and irregular contaminant distribution. (Benson, Huntley, and Johnson, 1993).

In recent years, various models have been developed to predict the fate of immiscible contaminants (NAPLs) in soils (Thomson, Grahan, and Farquhar, 1992). Hatfield and Stauffer (1992) demonstrate excellent predictions for breakthrough curves using a nonequilibrium model that can emulate hydrophobic solute transport and treat solute partitioning as a fully rate-limited process. The models prepared by Reible, Malhiet, and Illangasekare (1989) define the fate and transport processes of a gasoline phase from the point of spillage or leakage through the unsaturated zone. Physical processes considered include infiltration in the unsaturated zone with partitioning between the gasoline and soil, residual water and air phases, and subsequent contamination of aquifer recharge water by a residual gasoline phase.

3.2.2 ANAEROBIC DEGRADATION

Anaerobes make up about 10 to 15% of the bacterial population in soil, water, and sediment (Casella and Payne, 1996). They occur in almost every type of environmental niche and are likely to be available in a great variety of polluted ecosystems. Anaerobic zones containing polluting materials are commonly encountered, or entire contaminated systems may be oxygen poor or anoxic.

Anaerobic microbial transformations of organic compounds are important in anoxic environments (Young, 1984). Less cell material is formed under anaerobic conditions because of the lower growth yield, but organic fermentation products are likely to accumulate, unless they are converted into methane or other hydrocarbon gases. Anaerobiosis usually occurs in any habitat in which the oxygen consumption rate exceeds its supply rate and is a common phenomenon in many natural aquatic environments receiving organic materials (Berry, Francis, and Bollag, 1987). It also occurs in flooded soils and sediments.

An advantage to using anaerobes to degrade organic contaminants is the fact that, while their growth yield (i.e., mass) is lower than that of aerobes on a particular compound, the growth rate may be the same (Babel, 1994). Because of the inefficiency of anaerobiosis, the consumption rate of the compound can be many times higher than it is under aerobic conditions. Thus, the substrate disappears faster. If organisms are going to be supplied to a contaminant site, it may be advantageous to favor those that grow fast but inefficiently.

Many aromatic compounds can be cleaved under strict anaerobic conditions, producing carbon dioxide and methane (Healy and Young, 1979), although hydrocarbons that occur in a highly reduced state are slow to degrade anaerobically (Pettyjohn and Hounslow, 1983). Since anaerobic metabolism of hydrocarbons must begin with an oxidation step, an alternative electron acceptor is needed (Bausum and Taylor, 1986). Depending upon which of the electron acceptors (NO_3^-, SO_4^{-2}, or CO_2) is present in an anoxic environment, anaerobic respiration may be employed to degrade organic compounds (Berry, Francis, and Bollag, 1987). This may be performed by denitrifying bacteria, sulfate-reducing bacteria, or methanogens.

If nitrate, sulfate, and CO_2/bicarbonate are all present as electron acceptors, communities of anaerobes will utilize them in that order, ultimately producing methane from CO_2 (Alexander, 1994). Some organisms can use ferric iron as an electron acceptor (Lovley, Baedecker, Lonergan, Cozzarelli, Phillips, and Siegel, 1989). Nitrate and sulfate would probably have to be supplied to an anaerobic environment (Alexander, 1994).

A general reaction scheme for various electron acceptors is given by

$$\text{Substrate} + (NO_3^-, Mn^{4+}, Fe^{3+}, SO_4^{2-}, CO_2) \rightarrow \text{Biomass} + CO_2 + (N_2, Mn^{2+}, Fe_2 + S_2, CH_4)$$

The redox status of the system will determine which electron acceptor is used, i.e., the availability of the electron acceptor, the capability of the organisms, and the amount of carbon substrate (Turco and Sadowsky, 1995). The sequence of reduction, NO_3^-, Mn^{4+}, Fe^{3+}, SO_4^{2-}, and CO_2, is a reflection of the redox potential and oxidizing capacity of the chemical half-reaction (Zehnder and Strum, 1988). In an aqueous solution at pH 7, NO_3^- can be reduced at a standard reduction potential of +0.74 V (Stumm and Morgan, 1981). Denitrification can begin when the concentration of oxygen drops below 10 µmol/L (Roberston and Kuenen, 1984). After the NO_3^- is depleted, Mn^{4+} will be reduced at +0.52 V, Fe^{3+} at −0.05 V, SO_4^{2-} at −0.22 V and CO_2 to form CH_4 at −0.24 V (Stumm and Morgan, 1981).

It is now believed that three major groups of microorganisms are essential for complete mineralization of organic carbon to carbon dioxide and methane in anoxic sites that are without light and are low in electron acceptors other than carbon dioxide (Berry, Francis, and Bollag, 1987). These three groups are fermenters, proton reducers, and methanogens (Boone and Bryant, 1980; McInerney and Bryant, 1981; McInerney, Bryant, Hespell, and Costerton, 1981).

Anaerobic decomposition is performed mainly by bacteria utilizing either an anaerobic respiration or interactive fermentation/methanogenic type of metabolism (Parr, Sikora, and Burge, 1983). The decomposition of organic matter to carbon dioxide and methane involves interactions of heterotrophic bacteria and several fastidiously anaerobic bacteria: hydrolytic bacteria that catabolize major components of biomass, such as saccharides, proteins, and lipids; hydrogen-producing, acetogenic bacteria that catabolize products from the activity of the first group, such as fatty acids and neutral end products; homoacetogenic bacteria that catabolize multicarbon compounds to acetic acid; and methanogenic bacteria (Zeikus, 1980; Kobayashi and Rittmann, 1982). Both fastidious anaerobes, the organisms capable of living under anoxic, but not necessarily reduced, environmental conditions, and those capable of living facultatively (aerobically and anaerobically) are very important in anaerobic degradation.

Anaerobic degradation of xenobiotic compounds can be conducted by both single species of microbes and by microbial communities (Harder, 1981). Single species have been shown to degrade aromatic compounds in the presence of nitrate or light (Evans, 1977), whereas a number of microbial communities have been described that degrade these compounds in association with methane production (Balba and Evans, 1977) or in the presence of nitrate (Bakker, 1977). Interactions between different microbial populations are potentially important in the degradation of complex molecules (Berry, Francis, and Bollag, 1987; Hornick, Fisher, and Paolini, 1983), and cometabolism and synergism can occur anaerobically. Anaerobic degradation generally requires more than one species of anaerobes to mineralize a substrate completely (Alexander, 1994). Neither *Pelobacter acidigallici* nor *Acetobacterium woodii*, a demethylating microorganism, is able to degrade the aromatic ring when cultured separately (Bache and Pfennig, 1981). However, a coculture completely metabolizes syringic acid to acetate and carbon dioxide. In addition, if the isolates are cocultured with *Methanosarcina barkeri*, any metabolizable aromatic substrate can be completely mineralized to carbon dioxide and methane. The end products of anaerobic degradation are reduced compounds, some of which are toxic to microorganisms and plants (Parr, Sikora, and Burge, 1983). Organic fermentation products are likely to accumulate, unless they are converted into methane, or other hydrocarbon gases (Young, 1984). More than one species of anaerobes may be required to mineralize a substrate completely (Alexander, 1994).

Organic hydrocarbons vary in their susceptibility to different means of anaerobic biodegradation. Redox conditions can affect the transformation of a compound; i.e., a compound may be degraded under methanogenic conditions but not under denitrification conditions (Bouwer and McCarty, 1983a; 1983b). Petroleum hydrocarbons, straight-chain and branched alkanes, and alkenes are not degraded under methanogenic conditions (U.S. EPA, 1985a). Low-molecular-weight PAHs (e.g., naphthalene and acenaphthalene) can be completely degraded under denitrifying, nitrate-excess conditions, although more slowly than if degraded aerobically (Mihelcic and Luthy, 1988). Degradation depends on the desorption-adsorption kinetics of the PAH (Mihelcic and Luthy, 1991), the concentration of microorganisms, and the competition for nitrate with mineralization of soil organic carbon. When there is a limited amount of electron donor (petroleum hydrocarbons), sulfate-reducing bacteria can out-compete methanogens for available substrate at a lower concentration of sulfate (e.g., <1 mg/L) (Vroblesky, Bradley, and Chapelle, 1996). With abundant electron donor, methanogens may sequester part of the electron flow, even with enough sulfate to support sulfate reduction. Over a period of many years, aerobic nitrifiers may become more abundant in soils heavily contaminated with oil, while anaerobic nitrifiers may be more predominant in moderately contaminated soil (Amadi, Abbey, and Nima, 1996).

Normally, there are sufficient aerobic microorganisms present to biodegrade BTEX readily (Tiedje, 1995). However, if the soil oxygen is depleted, some anaerobic bacteria have the ability to degrade these substances (Lovley, Baedecker, Lonergan, Cozzarelli, Phillips, and Segal, 1989; Dolfing, Zeyer, Binder-Eicher, and Schwarzenbach, 1990; Evans, Mang, Kim, and Young, 1991). If conditions do not favor methanogens, certain classes of organic compounds, such as phenols, cresols, and xylenes, can be degraded by bacteria that respire nitrate or sulfate (Wilson, Leach, Henson, and Jones, 1986). Anaerobic degradation of aromatic hydrocarbons might also be facilitated in nature by the presence of other substrates and oxidants, including oxidized organics, metal–organic compounds, and perhaps even water (Kocki, Tang, and Bernath, 1972). If indigenous sulfate or ferric ions are prevalent at a contaminated site, they may effect removal of compounds, such as BTEX (Tiedje, 1995). Examples of compounds that can be degraded under anaerobic conditions include phenols, benzoates, toluene, ethylbenzene, *o*- and *m*-xylene, pyridine, quinoline, and *m*- and *p*-cresol (Alexander, 1994).

There are some compounds, most notably the lower-molecular-weight halogenated hydrocarbons, that will degrade only anaerobically (U.S. EPA, 1985a). Aromatic compounds consisting of either a homocyclic (e.g., benzoate, Figure 3.5a) or a heterocyclic (e.g., nicotinate, Figure 3.5b) aromatic nucleus can be metabolized by microorganisms under anaerobic conditions (Berry, Francis, and Bollag, 1987).

There are some compounds that are reported to be resistant to anaerobic degradation (Alexander, 1994). These include anthracene, naphthalene, benzene, aniline, 4-toluidine, 1- and 2-naphthol, pyridine, and saturated alkanes.

The chemistry of the anaerobic breakdown of aromatics involves an initial ring hydrogenation step (ring reduction followed by a ring hydration–ring cleavage reaction sequence) (Berry, Francis, and Bollag, 1987). This pathway is believed to be common to all microorganisms involved in benzenoid metabolism, including the denitrifiers, the sulfate reducers, and the fermenters.

The acclimation period required for anaerobic degradation can be immediate, or it can take up to 18 months (Sahm, Brunner, and Schoberth, 1986; Alexander, 1994).

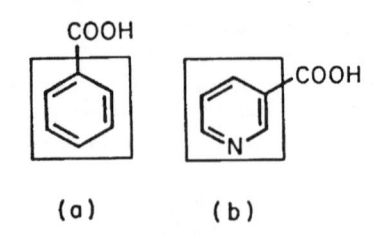

Figure 3.5 Homocyclic aromatic "benzenoid" nucleus (enclosed) of benzoate (a) and heterocyclic aromatic "pyridine" nucleus (enclosed) of nicotinate (b). (From Berry, D.F. et al. *Microbiol. Rev.* 51:43–59. American Society of Microbiology, Journals Division, 1987. With permission.)

Because of the complexity of natural ecosystems, laboratory results do not easily extrapolate to *in situ* environments (Berry, Francis, and Bollag, 1986; 1987). It can also be technically difficult to monitor *in situ* anaerobic microbial transformations of organic pollutants in subsurface environments. Methane production has been used to imply anaerobic biodegradation in some sites (Ehrlich, Goerlitz, Godsey, and Hult, 1982), while other sites with iron-sulfide precipitates have been designated as sulfate-reducing sites (Suflita and Gibson, 1985).

When alternative electron acceptors are being considered for bioremediation, among the variables to be evaluated are the size of the area to be treated; the demand for electron acceptor, monitoring requirements; incremental personnel costs; and remoteness of the site to the operating personnel (Norris, 1995).

3.2.2.1 Anaerobic Respiration

Without oxygen, oxygenases are inactive, and only those aromatic compounds with oxygen-containing functional groups (phenols and benzoates) are mineralized (Zeyer, Kuhn, and Schwarzenback, 1986). Under anaerobic conditions, the aromatic ring may be first reduced to a substituted cyclohexane before hydrolytic ring cleavage (Zeyer, Kuhn, and Schwarzenback, 1986). In some cases, removal or modification of a substituent must occur before reduction of the ring. The mechanisms used with aromatic hydrocarbons that have no activating groups to facilitate hydration of the ring are still unknown. However, it does appear that there is an oxidation of the ring in these instances. Apparently, attachment of an oxygen atom to the ring structure facilitates ring catabolism under anoxic conditions (Berry, Francis, and Bollag, 1987).

Since anaerobic metabolism of hydrocarbons must begin with an oxidation step, an alternate electron acceptor is needed (Bausum and Taylor, 1986). Depending upon which of the electron acceptors (NO_3^-, SO_4^{-2}, or CO_2) is present in an anoxic environment, anaerobic respiration may be employed to degrade organic compounds (Berry, Francis, and Bollag, 1987). This may be performed by denitrifying bacteria, sulfate-reducing bacteria, or methanogens.

Anaerobic respiration in soil involves biological oxidation-reduction reactions in which inorganic compounds (nitrate, sulfate, and carbonate ions; manganic and ferric ions) serve as the ultimate electron acceptor, instead of molecular oxygen. Nitrate may be used by nitrate reducers and sulfate by sulfate reducers as a terminal electron acceptor (U.S. EPA, 1985a). The organic waste serves as the electron donor or energy source. If oxygen is depleted during decomposition of a particular waste, the system will be dominated by facultative anaerobic bacteria, which are able to adapt from the aerobic to the anaerobic conditions, as necessary. Obligate anaerobes, on the other hand, cannot tolerate oxygen and are inhibited or killed by exposure to it.

In the reduction of sulfates and nitrates for the anaerobic degradation of petroleum, an alkane dehydrogenase is proposed to be the initial enzyme involved with production of an alkene, as the first intermediate compound (Shelton and Hunter, 1975). The fatty acids produced from the alkanes can be fermented under anaerobic conditions (Rosenfeld, 1947). Alkanes shorter than C_9 can be degraded anaerobically, while some of the longer-chain alkanes may be transformed into naphthalenes and other PAHs. For many anaerobic microorganisms, degradation of aromatic compounds can occur with both the aromatic ring and side chains of substituted aromatic compounds as carbon sources (Balba and Evans, 1977).

Different anaerobic bacteria can oxidize a variety of aromatic compounds completely to CO_2 via the intermediate, benzoyl-CoA, which becomes reduced to cyclohex-1-enecarboxyl-CoA (Koch and Fuchs, 1992). This reduction reaction is complex. The enzyme system involved is not found in aerobically grown cells. It is induced by aromatic compounds — probably by benzoyl-CoA rather than benzoate.

Sequencing anaerobic and aerobic biotreatment steps can be an alternative treatment approach (Field, Stams, Kato, and Schraa, 1995). The combined activity of anaerobic and aerobic organisms could also be obtained in a single treatment step, if the microbes are immobilized in particulate matrices, such as biofilms, or soil aggregates, within which anaerobic microniches can form.

Evidence strongly indicates that anaerobic biodegradation at the expense of sulfate, and perhaps nitrate, does occur, but only at very low rates (Bausum and Taylor, 1986). It may not be an important factor in natural ecosystems (Atlas, 1981), including soils (Blakebrough, 1978).

3.2.2.1.1 Denitrification

Catabolism of aromatic compounds can occur under anoxic conditions and in the presence of nitrate (Braun and Gibson, 1984). Many facultative bacteria, mainly heterotrophs, are capable of denitrification (Kaplan, Riley, Pierce, and Kaplan, 1984). Autotrophic denitrifying bacteria that have the capacity to oxidize inorganic energy sources, such as hydrogen and sodium sulfide, have been identified.

Biological denitrification may be defined as dissimilatory nitrate reduction, where nitrate serves as the terminal electron acceptor in the oxidation of an organic substance (Kaplan, Riley, Pierce, and Kaplan, 1984). Nitrogen gas is the final product in the case of dissimilatory denitrification for the production of energy via the respiratory transport chain.

Dissimilatory denitrification, which provides energy for the cell, is distinguished from assimilatory denitrification, which results in the reduction of nitrate to ammonia to serve in cellular synthesis (Kaplan, Riley, Pierce, and Kaplan, 1984). There is also a dissimilatory pathway that produces ammonia and does not produce energy. The function of this pathway may be detoxification or as an alternative electron sink.

Microorganisms that carry out nitrate respiratory metabolism (e.g., the denitrifiers) are facultative and appear to prefer oxygen as their electron acceptor (Gottschalk, 1979). Under aerobic conditions, this group of microorganisms uses a wide range of inorganic (Casella and Payne, 1996) and organic compounds as carbon and energy sources. In many instances, the same range of organic carbons is used under denitrifying conditions. Active nitrate-respiring microorganisms are found in a variety of anoxic environments, including soils, lakes, rivers, and oceans (Berry, Francis, and Bollag, 1987).

Denitrification can begin when the concentration of oxygen drops below 10 µmol/L (Roberston and Kuenen, 1984). Bacterial denitrification can supply the most effective type of biotransformation in polluted, anaerobic environments, especially if nitrate or other N-oxides are naturally available or added (Casella and Payne, 1996). Nitrate is more soluble and often less expensive than oxygen. It can be added to accelerate the degradation of monoaromatic hydrocarbons (except benzene) (Reinhard, 1994).

Under denitrifying, nitrate-excess conditions, low-molecular-weight PAHs, (e.g., naphthalene and acenaphthalene) can be completely degraded, although more slowly than in an aerobic system (Mihelcic and Luthy, 1988a). Degradation depends on the desorption–adsorption kinetics of the PAH (Mihelcic and Luthy, 1991), the concentration of microorganisms, and the competition for nitrate with mineralization of soil organic carbon. Nitrate-enhanced bioremediation of soil contaminated with JP-4 jet fuel resulted in an increase of the total number of denitrifiers by an order of magnitude (Thomas, Gordy, Bruce, Ward, Hutchins, and Sinclair, 1995).

Toluene is easily degraded under nitrate-reducing conditions; however, the toxic level of toluene is about 120 mg toluene/L and the reduction of high concentrations of NO_3^- may lead to accumulation of levels of NO_2^- that are toxic to the nitrate-reducing bacteria (Jorgensen, Mortensen, Jensen, and Arvin, 1991). Toluene degradation can be inhibited at 240 mg toluene/L and 13 NO_3^-–N/L, but not at 6 mg toluene/L and 520 mg NO_3^-–N/L, proving that toluene is the inhibiting factor, not nitrate. A concentration of 90 mg/L NO_2^- (the product of nitrate reduction) appears to inhibit degradation of toluene. NO_2^- has been reported to inhibit denitrifying bacteria at 28 to 47 mg N/L (Tiedje, 1988). When NO_3^- concentration is low relative to toluene concentration, the availability of the electron acceptor becomes limiting to degradation, and the bacteria switch to nitrite as the electron acceptor, which is then reduced to N_2 (Jorgensen, Mortensen, Jensen, and Arvin, 1991).

After acclimation under denitrifying conditions, organisms can rapidly metabolize about 90% of toluene with accompanying nitrate consumption, up to at least 93.5 mg/kg soil/day (Ramanand, Balba, and Duffy, 1995). Also after acclimation, soil microbes have been found to cross-adapt to degrade other aromatic substrates, such as ethylbenzene and some fluoro-chlorocompounds, but not *o*-xylene.

Degradation rates of mixtures of BTEX compounds tend to be sequential, with toluene most readily degraded under both denitrifying and sulfidogenic conditions (Reinhard, Hopkins, Orwin, Shang, and Lebron, 1995). Under denitrification conditions, transformation was complete within 8 days for toluene, ethylbenzene, and *m*-xylene and within 75 days for *o*-xylene. Benzene was not affected.

A mixed culture was found to be able to degrade a wide range of aromatic hydrocarbons and phenols, using nitrate as electron acceptor (Jensen and Arvin, 1994). Toluene, phenol, 2,4-dimethylphenol, *m*-cresol, and *p*-cresol could be degraded without a lag phase. Other compounds, such as *o*-xylene, butylbenzene, propylbenzene, and *o*-cresol, were dependent upon a primary substrate — toluene, ethylbenzene, or phenol — to be degraded. For *o*-xylene, the primary substrate, toluene, had to be degraded first, in a cometabolic relationship. These primary substrates might be inducing production of the necessary enzymes. This cometabolic relationship could be attributed to a mechanism linked to the initial oxidation of the methyl group (Jorgensen, Nielsen, Jensen, and Mortensen, 1995). Kuhn, Colberg, Schnoor, Wanner, Zehnder, and Schwarzenbach (1985) report that the three isomers of xylene can be degraded with NO_3^- as the electron acceptor. Zeyer, Kuhn, and Schwarzenbach (1986) found that toluene and *m*-xylene are degraded with NO_3^- as the electron acceptor. Toluene, ethylbenzene, the xylenes, and 1,2,3-trimethylbenzene, but not benzene, can be degraded by denitrification (Hutchins, Sewell, Kovacs, and Smith, 1991).

Another group, 3,5-dimethylphenol and 2,4,6-trimethylphenol, exhibited an initial slow degradation, and then seemed to become unbiodegradable. The following compounds were not degraded: benzene, 1,4-diethylbenzene, naphthalene, 1-methylnaphthalene, 1,4-methylnaphthalene, biphenyl, phenanthrene, 2,5-dimethylphenol, 2,6-dimethylphenol, and the nitrogen/sulfur/oxygen (NSO)-substituted compounds furane, fluorenone, and benzothiophene. Thus, nitrate limitation also might result in persistence of compounds that would be degraded under denitrifying conditions.

Both gasoline contamination and fertilizer application alter potential rates of denitrification under different conditions (Horowitz, Sexstone, and Atlas, 1978). Denitrification can be estimated as a potential activity only when nitrate (NO_3^-) and carbon sources are added. Denitrification activity, measured with NO_3^- alone added, can be interpreted as showing that prolonged exposure to hydrocarbons results in increased rates of denitrification activity. When glucose or peptones are added as a carbon source, in addition to NO_3^-, denitrification is greater than with NO_3^- alone.

Most high-nitrate industrial wastewaters do not contain sufficient electron donors to provide the energy for the reduction of all the nitrate to nitrogen gas (Kaplan, Riley, Pierce, and Kaplan, 1984). Therefore, an external source of carbon is necessary to provide the energy to promote this reduction. Methanol has long been used in biological denitrification as an external carbon source. However, large volumes of methanol are needed, and the expense has led to consideration of alternative carbon sources to provide the energy to drive the denitrification process.

When methanol, acetate, ethanol, acetone, and sugar were compared as carbon sources for reducing low concentrations of nitrates, all these compounds, except sugar, had approximately equal consumptive ratios; however, methanol was the least expensive of these (Kaplan, Riley, Pierce, and Kaplan, 1984). High nitrate removal (over 95%) was achieved, for the most part, while removal of soluble organic carbon was around 70%.

In another study, methanol, spent sulfite liquor, yeast, corn silage, and acid whey were compared, using COD-to-nitrate ratios to evaluate efficiency (Skrinde and Bhagat, 1982). Methanol was the most efficient with a ratio of 2.5:1. The ratios for the yeast and sulfite liquor were 2.8:1 and 2.9:1, respectively. Nitrate removal efficiencies never reached 95%. COD and BOD removal efficiencies ranged between 38 to 79% and 32 to 85%, respectively.

A denitrification system consisting of a single-stage, continuous-flow fermenter and receiving high nitrate loads (1259 mg/L nitrate or 285 mg/L nitrate-nitrogen) was used to compare alternative carbon sources (Kaplan, Riley, Pierce, and Kaplan, 1984). Methanol was the most effective, with $5/6$ mol of methanol being required to reduce 1 mol of nitrate completely to molecular nitrogen. However, an additional 30% over this amount is needed to satisfy the requirements for bacterial growth. There were several carbon sources that were close to the methanol results. Corn steep liquor and soluble potato solids, which are high in protein, can fuel the denitrification process by deamination of amino acids, yielding a variety of carboxylic acids that serve as electron donors. The sugars in acid whey and sweet whey are also used as an energy source by bacteria for denitrification. Nutrient broth, brewery spent grain, and sugar beet molasses failed to reach the 90% TOC (total organic carbon) removal. Sewage sludge digest contains insufficient available carbon to promote efficient denitrification.

Other authors determined that there was no apparent correlation between consumptive ratios and denitrification rates for different carbon sources and that carbon loads above those required for denitrification affected the consumptive ratio (Monteith, Bridle, and Sutton, 1980). Lactate and settled domestic sewage were found to be unsuitable as carbon sources for denitrification (Toit and Davies, 1973). Treatment systems dealing with both nitrates and hazardous organic compounds that are biodegraded only through cometabolism would have a dual benefit by supplementation of carbon sources (Kaplan, Riley, Pierce, and Kaplan, 1984).

When nitrate is the electron acceptor, 50.0 mg nitrate/L water would require 90,000 kg water/kg hydrocarbon (Hinchee and Miller, 1990). With 300.0 mg nitrate/L water, 15,000 kg water/kg hydrocarbon would be required for biodegradation.

A fluidized-bed technology has been developed for nitrate reduction by bacteriological denitrification on production scale (Patton, 1987). The system consists of four columns operating in series. In the fluidized-bed reactor, bacteria are allowed to grow and attach to 30- to 60-mesh anthracite coal particles to form "bioparticles." Wastewater is pumped through a bed of bioparticles at a velocity sufficient to fluidize the bed. As the water flows past the bioparticles, the nitrate degrades to N_2 and CO_2 gas, which is vented to the atmosphere. The anticipated denitrification load capacity is 2000 kg $NO_3^-(N)$/day.

Pseudomonas sp. strain PN-1 can use *p*-hydroxybenzoate, benzoate, and *m*-hydroxybenzoate, but not phenol, to grow under both aerobic and nitrate-reducing conditions (Braun and Gibson, 1984). This organism appears to have an oxygenase enzyme system that does not require oxygen as an inducer. Some facultative microorganisms retain low levels of oxygenase activity when grown in the presence of aromatic compounds, even under anaerobic conditions, while others do not (Taylor, Campbell, and Chinoy, 1970). A nitrate-respiring *P. stutzeri* is capable of using phenol as a substrate (Ehrlich, Godsy, Goerlitz, and Hult, 1983).

Two strains of Gram-negative, nitrate-reducing bacteria were isolated by Gorny, Wahl, Brune, and Schink (1992). One was a facultatively anaerobic organism, affiliated with *Alcaligenes denitrificans*. The other was strictly anaerobic, an apparent obligate nitrate reducer, and could not be grouped with any existing genus. Both utilized a direct, hydrolytic cleavage of resorcinol without initial reduction.

Four new denitrifying bacteria are closely related to *Thauera selenatis* (Rabus and Widdel, 1995). These strains exhibit different capacities for degrading the alkylbenzenes, specifically ethylbenzene, toluene, and propylbenzene. Polar aromatic compounds, such as benzoate, are utilized under both oxic and anoxic conditions, although none of the strains can grow on the alkylbenzenes with oxygen as electron acceptor. All grow anaerobically on crude oil, with depletion of toluene.

A sewage sludge isolate was able to degrade toluene anaerobically in the presence of nitrate, nitrite, nitric oxide, or nitrous oxide, converting 34% of the substrate into biomass, 53% into CO_2, and 6% into nonvolatile water-soluble products (Jorgensen, Flyvbjerg, Arvin, and Jensen, 1995).

Research on the effect of mixed oxygen/nitrate electron acceptor conditions on biodegradation of PAHs resulted in cultures able to degrade benzene, toluene, ethylbenzene, and naphthalene (2 mg/L) under conditions of 2 mg/L oxygen and high levels of nitrate (>150 mg/L NO_3^-) (Wilson, Durant, and Bouwer, 1995). Biodegradation was inhibited at high (8.6 mg/L) and low (<2 mg/L) oxygen levels, except for toluene, which was degraded under anaerobic denitrification conditions. Miller and Hutchins (1995) combined nitrate (27 to 28 mg(N)/L) with low levels of oxygen (0.8 to 1.2 mg O_2/L). They observed complete removal of toluene and partial removal of ethylbenzene, *m*-xylene, and *o*-xylene with nitrate as the only electron acceptor. Varying effects were produced with the nitrogen/oxygen combination. BTEX removal was unchanged in one soil, while it was reduced in another. The amount of nitrate utilized and nitrite production also varied among soil sources. Benzene removal was not observed under either condition.

While nitrate might be supplied to a contaminated site to promote denitrification, it is, however, itself a pollutant, limited to 10 ppm in drinking water. Consequently, it may be more difficult to obtain permits for use of nitrate than for oxygen or hydrogen peroxide. Also, degradation rates under anaerobic conditions are not as rapid, and the substrate range is more limited. Some researchers believe there is no reason nitrate respiration would be a better treatment approach, given the amount of success that has been already demonstrated with aerobic processes.

3.2.2.1.2 Sulfate Reduction

Those microorganisms that carry out dissimilatory sulfate reduction to obtain energy for growth are strict anaerobes (Berry, Francis, and Bollag, 1987). For these bacteria, organic carbon serves as a source of both carbon and energy. Reducible sulfur compounds (e.g., sulfate, thiosulfate) serve as terminal electron acceptors. Dissimilatory sulfate-reducing bacteria (sulfidogens) are most commonly associated with aquatic environments (i.e., marine and freshwater sediments) (Laanbroek and Pfennig, 1981; Lovley and Klug, 1983), although sulfate reducers can also be found in soil.

Organisms, such as *Desulfovibrio*, utilize sulfate as an electron acceptor, reducing it to sulfide, and by using organic acids as electron donors (U.S. EPA, 1985a). A sulfate-reducing organism, *Desulfonema magnum*, was isolated from marine sediment and is capable of mineralization of various fatty acids and benzoate (but not ethanol, cyclohexane carboxylate, or glucose) to carbon dioxide in the presence of reducible sulfur compounds (Tiedje, Sexstone, Parkin, Revsbech, and Shelton, 1984).

Sulfate can be added in conjunction with sodium benzoate and oxygen (Beeman, Howell, Shoemaker, Salazar, and Buttram, 1993). An interesting syntrophic association between *Pseudomonas aeruginosa* and *Desulfovibrio vulgaris* links sulfate respiration to the utilization of benzoate (Balba and Evans, 1980). *D. vulgaris* seems to produce organic acids, which are used by *P. aeruginosa* as electron acceptors, while it metabolizes the benzoate.

Under sulfidogenic conditions, toluene and *m*- and *o*-xylene can be transformed within 40 to 50 days (Reinhard, Hopkins, Orwin, Shang, and Lebron, 1995). Ethylbenzene removal accelerates after 30 days, while benzene removal is slow, but significant.

The availability of electron donor (petroleum hydrocarbons) affects the competition between sulfate-reducing bacteria and methanogenic bacteria (Vroblesky, Bradley, and Chapelle, 1996). When electron donor availability is limited, sulfate-reducing bacteria can outcompete methanogens for available substrate at a lower concentration of sulfate (e.g., <1 mg/L). In the presence of abundant electron donor, methanogens may be able to sequester part of the electron flow, even if there is sufficient sulfate to support sulfate reduction.

3.2.2.1.3 Methanogenesis

During biodegradation, certain anaerobic bacteria commonly produce short-chain organic acids that can be further broken down to methane, carbon dioxide, and inorganic substances by other bacterial forms (Freeze and Cherry, 1979). Methane bacteria are an example of obligate anaerobes that ferment organic acids to methane. This suggests that such chemicals that enter an anaerobic environment may not be refractory and can possibly be mineralized to carbon dioxide and methane (Healy and Young, 1979). However, organic compounds occurring in a highly reduced state, such as hydrocarbons, are degraded only slowly in an anaerobic environment (Pettyjohn and Hounslow, 1983). In addition, degradation of petroleum hydrocarbons, straight-chain and branched alkanes, and alkenes is not possible under methanogenic conditions (U.S. EPA, 1985a).

Bacterial methanogenesis is a process common to many anoxic environments (Berry, Francis, and Bollag, 1987). This strictly anaerobic process is frequently associated with the decomposition of organic matter in ecosystems, such as anoxic muds and sediments, the rumen and intestinal tract of animals, and anaerobic sewage sludge digesters (Stanier, Doudoroff, and Adelberg, 1970). Methane bacteria are able to use only a few simple compounds to support growth (Berry, Francis, and Bollag, 1987):

$$CO_2 + 4H_2 \rightarrow CH_4 + 2H_2O$$

$$4HCOOH \rightarrow CH_4 + 3CO_2 + 2H_2O$$

$$4CH_3OH \rightarrow 3CH_4 + CO_2 + 2H_2O$$

$$CH_3COOH \rightarrow CH_4 + CO_2$$

The importance of fastidious anaerobic consortia of organisms is illustrated by the types of detoxification reactions known to occur in the animal rumen, the best known of all anaerobic systems (Prins, 1978). These reactions include reductive dechlorination (or dehalogenation), possibly a limiting factor in degradation of certain compounds; nitrosamine degradation, a removal mechanism for a suspected carcinogen; reduction of epoxide groups in various compounds to olefins; reduction of nitro groups, as found in nitrophenol; and breakdown of aromatic structures (Kobayashi and Rittmann, 1982).

Methanogens are an essential component of anaerobic consortia degrading aromatics to methane (Ferry and Wolfe, 1976), but they appear to serve as electron sinks for other organisms rather than themselves attacking the primary substrates (Zeikus, 1977). Acetate and carbon dioxide plus hydrogen are probably the most important substrates for methane bacteria in natural ecosystems (Gottschalk, 1979). Since these organisms can use only simple compounds to support growth, they must rely on syntrophic associations with fermenters, which degrade complex organic compounds (i.e., aromatic compounds) into usable substrates. These associations are generally obligatory and may be similar from one methanogenic habitat to another (Suflita and Miller, 1985). It has been observed that the presence of –Cl or –NO$_2$ groups on phenol can inhibit methane production (Boyd, Shelton, Berry, and Tiedge, 1983).

Methanogens are very slow growing, especially the acetate-utilizing methanogens (Zehnder, Huser, Brock, and Wuhrmann, 1980; Switzenbaum, 1983). They are the rate-limiting factor in anaerobic reactors, which makes it necessary for solids to have long retention times. Methanogens require nickel, a component of factor F_{430}, for growth (Speece, Parkin, and Gallagher, 1983). They also require the cofactor, coenzyme M (McCarty, 1982) and factor F_{420}.

Many phenolic compounds in a creosote waste were degraded to carbon dioxide and methane by anaerobic bacteria in an aquifer (Ehrlich, Goerlitz, Godsey, and Hult, 1982).

Acetate and formate are important intermediate products in methanogenic fermentation and can indicate the presence of these organisms (McInerney and Bryant, 1981; McInerney, Bryant, Hespell, and Costerton, 1981).

Methane can be further broken down aerobically by methanotrophic bacteria and the reaction enhanced by the presence of iron (Boiesen, Arvin, and Broholm, 1993). Methanotrophs have been found to transform many xenobiotics (Colby, Stirling, and Dalton, 1977). For example, *Methylococcus trichosporium* OB3b can oxidize naphthalene.

3.2.2.2 Fermentation

Many microorganisms that inhabit anoxic environments obtain their energy for growth through fermentation of organic carbon (Schnitzer, 1982). Fermentation is a process that can be carried out in the absence of light by facultative or obligatory anaerobes. In fermentation reactions, the organic compound is both the electron donor and acceptor, since the substrate is both oxidized and reduced in the degradation process (Stanier, Adelberg, and Ingraham, 1976; Thibault and Elliott, 1979). This produces incompletely oxidized organic compounds, such as organic acids and alcohols.

3.2.2.3 Anaerobic Photometabolism

For the phototrophic purple nonsulfur bacteria (i.e., Rhodospiriaceae), organic compounds serve as the major source of electrons and of carbon for cellular components (Berry, Francis, and Bollag, 1986). Some species of purple nonsulfur bacteria are able to grow aerobically in the dark by respiratory metabolism of organic compounds. A few species, like *Rhodopseudomonas palustris* (Gottschalk, 1979) or *Rhodocyclus purpureus* (Pfennig, 1978b), can use thiosulfate or sulfide as an electron donor in addition to organic compounds. In general, however, purple nonsulfur bacteria are sensitive to sulfide and tend to grow only in those environments where sulfide concentration is low (Stanier, Doudoroff, and Adelberg, 1970). *Rhodopseudomonas palustris* can use benzoate as sole substrate under aerobic conditions via respiration or anaerobically by photometabolism (Proctor and Scher, 1960).

3.2.2.4 Specific Compounds
3.2.2.4.1 *Mononuclear Aromatic Hydrocarbons*
1. Phenol

Phenol can be degraded under anaerobic conditions (Alexander, 1994). Phenol, hydroquinone, and *p*-cresol are converted to CO_2 and CH_4 under different reducing conditions, and the rate improves with acclimation (Young and Bossert, 1984). Phenolic compounds have been anaerobically biodegraded in near-surface groundwater (Ehrlich, Goerlitz, Godsey, and Hult, 1982). Under anaerobic conditions, transformation of phenols into polymerized species can be inhibited, and these compounds may possibly leach through the soil more readily (Sawhney and Kozloski, 1984).

The phenol compounds degraded and the number of weeks required to degrade them in anaerobic sewage sludge diluted to 10% in a mineral salts medium were phenol (2 weeks), *o*-chlorophenol (3 weeks), *m*-cresol (7 weeks), *p*-cresol (3 weeks), *o*-, *m*-, and *p*-nitrophenol (1 week), *o*-methoxyphenol (2 weeks), *m*- and *p*-methoxyphenol (1 week) (Boyd, Shelton, Berry, and Tiedje, 1983). The presence of –Cl and –NO$_2$ groups on phenols inhibited methane production.

A nitrate-respiring *Pseudomonas stutzeri* is capable of using phenol as a substrate (Ehrlich, Godsy, Goerlitz, and Hult, 1983).

Cresol can be degraded in anoxic aquifers (Smolenski and Suflita, 1987). Both *m*- and *p*-cresol can be degraded under anaerobic conditions (Alexander, 1994). *p*-Cresol is degraded more easily than *m*-cresol, which in turn is more easily degraded than *o*-cresol. Biodegradation is favored under sulfate-reducing conditions compared with methanogenic conditions. A mixed culture was able to degrade *m*-cresol, and *p*-cresol without a lag phase, using nitrate as electron acceptor (Jensen and Arvin, 1994).

2. Benzene

Benzene has been reported by some researchers to be resistant to anaerobic degradation (Alexander, 1994), even by denitrification (Hutchins, Sewell, Kovacs, and Smith, 1991; Hutchins and Wilson, 1991; Reinhard, 1994). No anaerobic transformation of the compound was observed under methanogenic conditions by enrichment cultures from anaerobic sewage sludge, freshwater sediments, or marine sediments (Schink, 1985). Hutchins and Wilson (1991) note that conflicting results have been published on whether or not benzene can be degraded anaerobically, but suggest that oxygen may have intruded into those studies that reported rapid degradation (i.e., Batterman, 1986; Berry-Spark, Barker, Major, and Mayfield, 1986; Major, Mayfield, and Barker, 1988).

A number of authors indicate that benzene can be biodegraded without the presence of molecular oxygen (Reinhard, Goodman, and Barker, 1984; Grbic-Galic and Young, 1985; Kuhn, Colberg, Schnoor, Wanner, Zehnder, and Schwarzenbach, 1985). The anaerobic transformation of benzene might be a fermentation in which the substrates are partially reduced but also partially oxidized (Grbic-Galic and Vogel, 1987). Reduction results in the production of saturated alicyclic rings. Benzene degradation by mixed methanogenic cultures may lead to carbon dioxide and methane with intermediates of phenol, cyclohexane, and propanoic acid. The oxygen for ring-hydroxylation might be derived from water.

Benzene may be transformed by methanogenic cultures acclimated to lignin-derived aromatic acids under strictly anaerobic conditions with several intermediates, including demethylation products, aromatic alcohols, aldehydes and acids, cresols, phenol, alicyclic rings, and aliphatic acids, which are ultimately converted to carbon dioxide and methane (Grbic-Galic and Vogel, 1987).

Mixed oxygen/nitrate electron acceptor conditions enable cultures to degrade benzene under conditions of 2 mg/L oxygen and high levels of nitrate (>150 mg/L NO$_3^-$) (Wilson, Durant, and Bouwer, 1995). Biodegradation is inhibited at high (8.6 mg/L) and low (<2 mg/L) oxygen levels.

Under sulfidogenic conditions, benzene removal is slow, but significant (Reinhard, Hopkins, Orwin, Shang, and Lebron, 1995).

3. Toluene

Toluene can be biodegraded without the presence of molecular oxygen (Reinhard, Goodman, and Barker, 1984; Grbic-Galic and Young, 1985; Kuhn, Colberg, Schnoor, Wanner, Zehnder, and Schwarzenbach, 1985; Alexander, 1994). The anaerobic transformation of toluene may be a fermentation in which the substrates are partly oxidized and partly reduced (Grbic-Galic and Vogel, 1987). Reduction produces saturated alicyclic rings, while oxidation of the methyl group may give a primary product of benzyl alcohol, which, in turn, may be converted to benzaldehyde and benzoic acid.

No anaerobic transformation of the compound was observed under methanogenic conditions by enrichment cultures from anaerobic sewage sludge, freshwater sediments, or marine sediments by Schink (1985). Other authors, however, have observed methanogenic biodegradation. Toluene is transformed by methanogenic cultures acclimated to lignin-derived aromatic acids under strictly anaerobic conditions, with intermediates, including demethylation products, aromatic alcohols, aldehydes and acids, cresols, phenol, alicyclic rings, and aliphatic acids, which are ultimately converted to carbon dioxide and methane (Grbic-Galic and Vogel, 1987). Methanogenic alluvium from the floodplain of the South Canadian River, which receives leachate from a landfill, showed toluene degradation by an order of magnitude after 11 months (Rees, Wilson, and Wilson, 1985).

Toluene can be degraded by denitrification (Hutchins, Sewell, Kovacs, and Smith, 1991; Reinhard, 1994). After acclimation under denitrifying conditions, organisms can rapidly metabolize about 90% of toluene with accompanying nitrate consumption, up to at least 93.5 mg/kg soil/day (Ramanand, Balba, and Duffy, 1995).

Toluene is easily degraded under nitrate-reducing conditions; however, the toxic level of toluene is about 120 mg toluene/L and the reduction of high concentrations of NO_3^- may lead to accumulation of levels of NO_2^- that are toxic to the nitrate-reducing bacteria (Jorgensen, Mortensen, Jensen, and Arvin, 1991). Toluene degradation can be inhibited at 240 mg toluene/L and 13 NO_3^-–N/L, but not at 6 mg toluene/L and 520 mg NO_3^-–N/L, proving that toluene is the inhibiting factor, not nitrate. A concentration of 90 mg/L NO_2^- (the product of nitrate reduction) appears to inhibit degradation of toluene. NO_2^- has been reported to inhibit denitrifying bacteria at 28 to 47 mg N/L (Tiedje, 1988). When NO_3^- concentration is low, relative to toluene concentration, the availability of the electron acceptor becomes limiting to degradation, and the bacteria switch to nitrite as the electron acceptor, which is then reduced to N_2 (Jorgensen, Mortensen, Jensen, and Arvin, 1991).

A mixed culture was able to degrade toluene without a lag phase, with nitrate as electron acceptor (Jensen and Arvin, 1994). Four new denitrifying bacteria, which are closely related to *Thauera selenatis*, can degrade toluene (Rabus and Widdel, 1995). *Azoarcus tolulyticus* Tol-4 can grow anaerobically on toluene, with benzoate as an intermediate (Chee-Sanford, Frost, Fries, Zhou, and Tiedje, 1996). Two metabolites, ethylphenylitaconate and benzylsuccinic acid accumulated as a result.

A culture isolated from sewage sludge was able to degrade toluene under anaerobic conditions in the presence of nitrate, nitrite, nitric oxide, or nitrous oxide (Jorgensen, Flyvbjerg, Arvin, and Jensen, 1995). About 34% of toluene could be incorporated into biomass, 53% into CO_2, and 6% as nonvolatile water-soluble products. About 70% of the electrons donated during the oxidation of toluene were due to reduction of nitrate to nitrite. Reduction of nitrate to nitrogen gas accounted for 97% of the donated electrons.

Degradation of mixtures of BTEX compounds tends to be sequential, with toluene the most readily degraded under both denitrifying and sulfidogenic conditions (Reinhard, Hopkins, Orwin, Shang, and Lebron, 1995). Under sulfidogenic conditions, toluene can be transformed within 40 to 50 days.

Under denitrifying conditions, *o*-xylene removal relies on toluene degradation (Jorgensen, Nielsen, Jensen, and Mortensen, 1995). With toluene as primary carbon source and nitrate as an electron acceptor, *o*-xylene can be transformed cometabolically by a mixed culture of denitrifying bacteria (Arcangeli and Arvin, 1995). However, a toluene concentration of >1 to 3 mg/L reduces the *o*-xylene removal rate, and concentrations of *o*-xylene above 2 to 3 mg/L in turn inhibit toluene degradation. The two compounds may be competing for the same enzyme. Microorganisms adapted to growth on *m*-xylene with nitrate as an electron acceptor are also able to degrade toluene under denitrifying conditions (Zeyer, Kuhn, and Schwarzenback, 1986).

Mixed oxygen/nitrate electron acceptor conditions allow degradation of toluene under conditions of 2 mg/L oxygen and high levels of nitrate (>150 mg/L NO_3^-) (Wilson, Durant, and Bouwer, 1995). Biodegradation is inhibited at high (8.6 mg/L) and low (<2 mg/L) oxygen levels, except for toluene, which is degraded under anaerobic denitrification conditions. Miller and Hutchins (1995) combined nitrate (27 to 28 mg(N)/L) with low levels of oxygen (0.8 to 1.2 mg O_2/L). They observed complete removal of toluene with nitrate as the only electron acceptor. Varying effects were produced with the nitrogen/oxygen combination.

4. Xylene (*o, m, p*)

Xylenes can be biodegraded without the presence of oxygen (Reinhard, Goodman, and Barker, 1984; Grbic-Galic and Young, 1985; Kuhn, Colberg, Schnoor, Wanner, Zehnder, and Schwarzenback, 1985). All three xylene isomers may be used by bacteria as sole carbon and energy sources under aerobic and anaerobic, denitrifying conditions (Kuhn, Colberg, Schnoor, Wanner, Zehnder, and Schwarzenbach, 1985). The *p*- and *m*-xylenes are degraded at equal rates but significantly faster than the *ortho* isomer. Alexander (1994) reports anaerobic degradation of both *o*- and *m*-xylene.

m-Xylene can be mineralized in the absence of oxygen by reducing the redox potential, Eh, of the inflowing medium with sulfide to −0.11 V (Zeyer, Kuhn, and Schwarzenback, 1986). If anaerobes are acclimated to *o*-xylene, they can degrade the compound in anaerobic digesters at 41 and 62% for 250 mg/L and 500 mg/L, respectively (Lee, Melnyk, and Bishop, 1991).

Nitrate can support the degradation of xylenes in subsurface material (Kuhn, Colberg, Schnoor, Wanner, Zehnder, and Schwarzenbach, 1985; Hutchins, Sewell, Kovacs, and Smith, 1991). Up to 0.4 mM m-xylene was rapidly mineralized to carbon dioxide, in a laboratory aquifer column operated in the absence of molecular oxygen with nitrate as an electron acceptor (Zeyer, Kuhn, and Schwarzenback, 1986). Microorganisms adapted to growth on m-xylene in the absence of molecular oxygen and in the presence of nitrate, are also able to degrade toluene under these conditions (Zeyer, Kuhn, and Schwarzenback, 1986). Under denitrification conditions, transformation is complete within 8 days for m-xylene and within 75 days for o-xylene (Reinhard, Hopkins, Orwin, Shang, and Lebron, 1995).

With denitrification, o-xylene degradation occurs only cometabolically in combination with toluene (or ethylbenzene or phenol; Jensen and Arvin, 1994) degradation, which may be due to a mechanism linked to the initial oxidation of the methyl group (Jorgensen, Nielsen, Jensen, and Mortensen, 1995). The primary substrate, toluene, had to be degraded first. The primary substrate might be inducing production of the necessary enzymes (Jensen and Arvin, 1994). o-Xylene was also observed to be transformed cometabolically by a mixed culture of denitrifying bacteria with toluene as the primary carbon source (Arcangeli and Arvin, 1995). However, a toluene concentration of >1 to 3 mg/L reduces the o-xylene removal rate, and concentrations of o-xylene above 2 to 3 mg/L in turn inhibit toluene degradation. There may be competition between the two compounds for the same enzyme.

No anaerobic transformation of the xylenes was observed under methanogenic conditions by enrichment cultures from anaerobic sewage sludge, freshwater sediments, or marine sediments (Schink, 1985). However, the three isomers were preferentially removed over other petroleum products from a methanogenic landfill leachate (Reinhard, Goodman, and Barker, 1984).

Miller and Hutchins (1995) combined nitrate (27 to 28 mg(N)/L) with low levels of oxygen (0.8 to 1.2 mg O_2/L). They observed partial removal of m-xylene and o-xylene with nitrate as the only electron acceptor. Varying effects were produced with the nitrogen/oxygen combination.

Under sulfidogenic conditions, m- and o-xylene can be transformed within 40 to 50 days (Reinhard, Hopkins, Orwin, Shang, and Lebron, 1995).

5. Alkylbenzenes

Laboratory studies confirmed field tests that alkylbenzenes in groundwater could be transformed both aerobically and anaerobically (Wilson, Bledsoe, Armstrong, and Sammons, 1986). Anaerobic degradation may be a useful adjunct to the aerobic degradation in heavily contaminated areas of soil where the oxygen supply has been depleted.

Four new denitrifying bacteria, closely related to *Thauera selenatis*, exhibit different capacities for degrading the alkylbenzenes, ethylbenzene, toluene, and propylbenzene (Rabus and Widdel, 1995). None of the strains can grow on the alkylbenzenes with oxygen as electron acceptor.

6. BTEX

Normally, there are sufficient aerobic microorganisms present to readily biodegrade BTEX (Tiedje, 1995). However, if the soil oxygen is depleted, some anaerobic bacteria have the ability to degrade these substances (Lovley, Baedecker, Lonergan, Cozzarelli, Phillips, and Segal, 1989; Dolfing, Zeyer, Binder-Eicher, and Schwarzenbach, 1990; Evans, Mang, Kim, and Young, 1991). Anaerobic degradation of mixtures of BTEX compounds tends to be sequential, with toluene most readily degraded under both denitrifying and sulfidogenic conditions (Reinhard, Hopkins, Orwin, Shang, and Lebron, 1995). With denitrification, transformation is complete within 8 days for toluene, ethylbenzene, and m-xylene and within 75 days for o-xylene. Benzene is not affected.

Mixed oxygen/nitrate electron acceptor conditions produce cultures that are able to degrade benzene, toluene, and ethylbenzene under conditions of 2 mg/L oxygen and high levels of nitrate (>150 mg/L NO_3^-) (Wilson, Durant, and Bouwer, 1995). Biodegradation is inhibited at high (8.6 mg/L) and low (<2 mg/L) oxygen levels, except for toluene, which is degraded under anaerobic denitrification conditions. Miller and Hutchins (1995) combined nitrate (27 to 28 mg(N)/L) with low levels of oxygen (0.8 to 1.2 mg O_2/L). They observed complete removal of toluene and partial removal of ethylbenzene, m-xylene, and o-xylene with nitrate as the only electron acceptor. Varying effects were produced with the nitrogen/oxygen combination. BTEX removal was unchanged in one soil, while it was reduced in another. The amount of nitrate utilized and nitrite production also varied among soil sources. Benzene removal was not observed under either condition.

Under sulfidogenic conditions, toluene and *m-* and *o-*xylene can be transformed within 40 to 50 days (Reinhard, Hopkins, Orwin, Shang, and Lebron, 1995). Ethylbenzene removal accelerates after 30 days, while benzene is removed slowly.

7. Ethylbenzene

Ethylbenzene can be degraded under anaerobic conditions (Alexander, 1994), i.e., denitrification (Hutchins, Sewell, Kovacs, and Smith, 1991). After acclimation to toluene under denitrifying conditions, soil microbes can cross-adapt to degrade other aromatic substrates, such as ethylbenzene (Ramanand, Balba, and Duffy, 1995). With denitrification, transformation of ethylbenzene is complete within 8 days (Reinhard, Hopkins, Orwin, Shang, and Lebron, 1995). Four new denitrifying bacteria are closely related to *Thauera selenatis* (Rabus and Widdel, 1995). These strains exhibit different capacities for degrading the alkylbenzenes, ethylbenzene, toluene, and propylbenzene. Ethylbenzene removal increases rapidly after a month of exposure to sulfidogenic conditions (Reinhard, Hopkins, Orwin, Shang, and Lebron, 1995).

Under mixed oxygen/nitrate electron acceptor conditions, ethylbenzene can be degraded under conditions of 2 mg/L oxygen and high levels of nitrate (>150 mg/L NO_3^-) (Wilson, Durant, and Bouwer, 1995). Biodegradation is inhibited at high (8.6 mg/L) and low (<2 mg/L) oxygen levels. Miller and Hutchins (1995) combined nitrate (27 to 28 mg(N)/L) with low levels of oxygen and obtained partial removal of ethylbenzene (0.8 to 1.2 mg O_2/L). Varying effects were produced with the nitrogen/oxygen combination.

8. Benzoate

Rhodopseudomonas palustris photometabolizes benzoate anaerobically (Atlas and Schofield, 1975). *Pseudomonas* PN-1 and *P. stutzeri* grow anaerobically on benzoate–nitrate–mineral salts medium. *R. palustris* can use benzoate as sole substrate under aerobic conditions via respiration or anaerobically by photometabolism (Proctor and Scher, 1960). A sulfate-reducing organism, *Desulfonema magnum*, was isolated from marine sediment and is capable of mineralization of various fatty acids and benzoate to carbon dioxide in the presence of reducible sulfur compounds (Tiedje, Sexstonetone, Parkin, Revsbech, and Shelton, 1984). Four new denitrifying bacteria, closely related to *Thauera selenatis,* utilize polar aromatic compounds, such as benzoate, under both oxic and anoxic conditions (Rabus and Widdel, 1995). An interesting syntrophic association between *Pseudomonas aeruginosa* and *Desulfovibrio vulgaris* links sulfate respiration to the utilization of benzoate (Balba and Evans, 1980). *D. vulgaris* seems to produce organic acids, which are used by *P. aeruginosa* as electron acceptors, while it metabolizes the benzoate.

3.2.2.4.2 Polycyclic Aromatic Hydrocarbons

1. Naphthalene

No anaerobic transformation of this compound was observed under methanogenic conditions by enrichment cultures from anaerobic sewage sludge, freshwater sediments, or marine sediments (Schink, 1985). PAHs, such as naphthalene, were not degraded to carbon dioxide and methane by anaerobic bacteria in an aquifer (Ehrlich, Goerlitz, Godsey, and Hult, 1982). However, methanotrophs, such as *Methylococcus trichosporium* OB3b, have been found to be able to oxidize naphthalene (Colby, Stirling, and Dalton, 1977).

There was no evidence of anaerobic degradation of naphthalene in groundwater samples contaminated by wood creosoting products, although it disappeared at a faster rate in the aquifer than if only dilution were occurring (Erlich, Goerlitz, Godsey, and Hult, 1982).

Research on the effect of mixed oxygen/nitrate electron acceptor conditions on biodegradation of PAHs resulted in cultures able to degrade naphthalene (2 mg/L) under conditions of 2 mg/L oxygen and high levels of nitrate (>150 mg/L NO_3^-) (Wilson, Durant, and Bouwer, 1995). Biodegradation was inhibited at high (8.6 mg/L) and low (<2 mg/L) oxygen levels.

2. Anthracene and Pyrene

Pyrene degradation does not occur under anaerobic or microaerophilic conditions.

3.2.2.4.3 Straight-Chain Aliphatics

Degradation of straight-chain alkanes and alkenes is not possible under methanogenic conditions (U.S. EPA, 1985a). Alkanes shorter than C_9 can be degraded anaerobically, while some of the longer-chain alkanes may be transformed into naphthalenes and other PAHs (Rosenfeld, 1947).

3.2.2.4.4 Branched-Chain Aliphatics
Degradation of branched alkanes and alkenes is not possible under methanogenic conditions (U.S. EPA, 1985a).

3.2.2.4.5 Alcohols
Methanol is the first product resulting from oxidation of methane. Many bacteria and yeasts that can catabolize methanol under aerobic or anaerobic conditions and at low and high concentrations have been isolated from soil as deep as 100 ft (Benoit, Novak, Goldsmith, and Chadduck, 1985).

3.2.2.4.6 Alicyclic Hydrocarbons
1. Cyclohexane

A strain of *Acinetobacter anitratum* utilizes cyclohexane carboxylic acid through 2-oxocyclohexane carboxylic acid (Hou, 1982). Some anaerobic photosynthetic strains, as well as aerobic nonphotosynthetic strains, also metabolize cyclohexane carboxylic acid by this pathway. A *Nocardia* sp. and a *Pseudomonas* sp. grow using cyclohexane as the sole source of carbon and energy, via the same pathway. Cyclohexane induces the enzymes for cyclohexane oxidation in *Nocardia*. *Mycobacterium convolutum* R-22 requires an inducer, such as an *n*-alkane, to mineralize cyclohexane completely (Perry, 1984).

3.2.2.4.7 Fatty Acids
The fatty acids produced from the alkanes can be fermented under anaerobic conditions (Rosenfeld, 1947). A sulfate-reducing organism, *Desulfonema magnum*, was isolated from marine sediment and is capable of mineralization of various fatty acids and benzoate (but not ethanol, cyclohexane carboxylate, or glucose) to carbon dioxide in the presence of reducible sulfur compounds (Tiedje, Sexstone, Parkin, Revsbech, and Shelton, 1984).

3.3 HEAVY METALS

Sites polluted with organic compounds often contain metals, such as arsenic, mercury, lead, and zinc (Roane and Kellogg, 1996). About 50% of contaminated sites contain such inorganic pollutants, which may persist in the soil after biodegradation of the organic contaminants or which may inhibit the activity of native organisms or introduced organic compound degraders (Diels, Springael, Kreps, and Mergeay, 1991). Heavy metal contamination of soil decreases microbial diversity and causes bacterial communities to lose part of their degradative capability (Burkhardt, Insam, Hutchinson, and Reber, 1993). Unusual degradative capabilities become even rarer in bacterial communities affected by heavy metals. The annual combined toxicity of all human-induced mobilized trace metals is greater than the combined toxicity of all radioactive and organic waste generated each year (Wnorowski, 1991).

Metals can be classified as "hard" and "soft" acids and bases (Pearson, 1973) and as class A, class B, and borderline ions (Nieboer and Richardson, 1980). Some class A ions (e.g., Ca, Mg, Na) are essential for microorganisms, some borderline ions (e.g., Cu, Fe, Ni, Zn) are required as micronutrients, and some class B ions (e.g., Hg, Pb) are toxic and not necessary for biological functions (Collins and Stotzky, 1992).

In the last 200 years, microorganisms have been adapting to the changes in the distribution of elements at the surface of Earth, as a result of industrialization (Wood and Wang, 1983). Several strategies for resistance to metal ion toxicity have been identified in these organisms:

1. The development of energy-driven efflux pumps that keep toxic element levels low in the interior of the cell, e.g., for Cd(II) and As(V);
2. Oxidation (e.g., AsO_3^{-2} to AsO_4^{-3}) or reduction (e.g., Hg^{+2} to Hg^0), which can enzymatically and intracellularly convert a more toxic form of an element to a less toxic form;
3. Biosynthesis of intracellular polymers that serve as traps for the removal of metal ions from solution (e.g., for Cd, Ca, Ni, and Cu);
4. The binding of metal ions to cell surfaces;
5. The precipitation of insoluble metal complexes (e.g., metal sulfides and metal oxides) at cell surfaces. Precipitation, complexation, and crystallization can result in detoxification (Gadd, 1992);
6. Biomethylation (methylation and demethylation; Gadd, 1992) and transport through cell membranes by diffusion-controlled processes;
7. Volatilization (Roane and Kellogg, 1996);
8. Internal sequestration by metal-binding proteins or localization within certain organelles (Gadd, 1992).

Microorganisms that have short generation times, and, consequently, increased evolution rates, have adapted themselves to deal with high concentrations of metal ions (Wood and Wang, 1983). Microorganisms are evolving strategies to maintain low intracellular concentrations of toxic pollutants. Some resist high concentrations through their evolution under extreme environmental conditions (Brock, 1978). Others have achieved resistance to the recently polluted environment through acquisition of extrachromosomal molecules (plasmids) (Silver, 1983).

Metals can be removed by derived, induced, or excreted microbial products (Gadd, 1992). This includes cell wall constituents, pigments, polysaccharides, metallothioneins, phytochelatins, and other metal-binding proteins; microbial extracellular polymers (especially with bacteria; Lester, Sterritt, Rudd, and Brown, 1984), with a correlation existing between high anionic charge and metal-complexing capability (Kaplan, Christiaen, and Arad, 1987); siderophores, which can complex certain metals in addition to iron ((Macaskie and Dean, 1990); and fungal biomass, constituents, and related products, such as chitin (Macaskie and Dean, 1990), melanins (Bell and Wheeler, 1986), and phenolic polymers (Sakaguchi and Nakajima, 1982; 1987; Senesi, Sposito, and Martin, 1987). Metal-containing particulates can also be removed from solution by fungal biomass (Wainwright, Singleton, and Edyvean, 1990; Singleton, Wainwright, and Edyvean, 1990). Soil bacteria, such as *Arthrobacter globiformis*, are good producers of capsular polysaccharides (Grappelli, Hard, Pietrosanti, Tomati, Campanella, Cardarelli, and Cordatore, 1989). Purified, thiol-rich peptides or proteins, like metallothioneins (such as from cyanobacteria), are potentially more efficient in bioaccumulation than whole cells (Ron, Minz, Finkelstein, and Rosenberg, 1992). Biosorbents can achieve similar or higher uptake rates than ion exchangers (Roehricht, Weppen, and Deckwer, 1990). An exocellular phosphatidylethanolamine-containing residue appears to form when microbes are grown in minimal mineral medium in the presence of millimolar quantities of aluminum, iron, zinc, calcium, and gallium (Appanna, Finn, and St. Pierre, 1995). The metals are immobilized in this insoluble mass, as a response to multiple-metal stress.

There are two main mechanisms for microbiological metal accumulation — an active and a passive process (Hutchins, Davidson, Brierley, and Brierley, 1986). Passive accumulation occurs in nonviable or nonactive cells and is more tolerant of environmental conditions (Wnorowski, 1991). The term *bioaccumulation* usually refers to the active process of metal accumulation by living whole cells, while *biosorption* generally refers to the passive metal sequestration by cell components (Churchill, Walters, and Churchill, 1995). Metal sequestration at the cell surface could include complexation, chelation, coordination, ion exchange, adsorption, and inorganic microprecipitation processes (Volesky, 1990).

Accumulation by living organisms takes place in the range of pH, redox potential (Eh), temperature, and nutrient availability that will allow maintenance of their life functions (Hutchins, Davidson, Brierley, and Brierley, 1986). Heavy metals can be stored inside and on the surface of the cell. Metals accumulate on cell walls and extracellular slime, which serve as ion-exchange resins (Stary and Kratzer, 1984). Microorganisms may reduce metal ions adsorbed on themselves to an atomic form (Brunker and Bott, 1974; Hosea, Greene, McPherson, Henzl, Aleksander, and Darnall, 1986). Living biomass has an advantage of utilizing additional mechanisms for metal sequestration, such as active transport through cell membranes, intracellular traps, and intracellular precipitation (Gadd and Griffiths, 1978). However, yields of sequestered ions are usually higher with nonliving cells (Goddard and Bull, 1989). Nonactive cells seem to have several advantages in industrial applications.

Binding of metal cations on the surface of cells is influenced by the physicochemical characteristics of the environment (e.g., pH) and affects the electrokinetic properties of cells (Collins and Stotzky, 1992; Ivanov, Fomchenkov, Khasanova, Kuramshina, and Sadikov, 1992). Other environmental factors, such as Eh, inorganic anions and cations, particulate and soluble organic matter, clay minerals, and salinity, may also reduce or eliminate toxicity (Gadd and Griffiths, 1978).

Metal uptake by a microorganism is metabolism dependent (Hornick, Fisher, and Paolini, 1983). What the organism does not require can be precipitated intracellularly and stored. Microbes are also capable of producing organic compounds, such as citric acid and oxalic acid, which act as binding or chelating agents. Production of hydrogen sulfide by microbes is of great importance in that heavy metals form insoluble sulfides. Both bacteria and yeasts have exhibited hydrogen sulfide production and have created a more tolerant environment for more-sensitive organisms.

The availability of metal ions for transport into cells in restricted by their abundance and solubility in water (Wood and Wang, 1983). Solubility is greatly influenced by pH, temperature, standard reduction potential (E^0), the presence of competing anions and cations, and the presence of surface-active substances, such as particulates and macromolecules, including proteins, humic acids, and clays.

The fraction of a metal content that is available for uptake by organisms depends strongly on the chemical form in which the metals are present and where they are located in the soil system (Nederlof, Van Riemskijk, and De Haan, 1993). The distribution of the metals over different ion species and different phases in the soil will determine the toxicity. For instance, copper can be present in a soil solid phase, a soil solution, and a biotic phase (e.g., yeast cells). Most of the total metal content is bound to the soil solid phase, which controls the metal ion activity in the soil solution. Both metal binding to soils and to yeast cells as a function of solution concentration increase with pH. It is necessary to determine the pH effect on binding to soil and biota independently in order to predict the net pH effect on metal binding to biota in the soil environment. Ion exchange can be the most important process for mobilization of heavy metals from the solid phase (Andersen and Engelstad, 1993).

Bacterial resistance to heavy metals in the environment can result in bioaccumulation, biotransformation, changes in ecological diversity, and coselection of resistance factors for antibiotics (Sterritt and Lester, 1980). A simple, rapid method for the determination of bacterial resistance to a wide range of metals has been developed through modification of the antibiotic susceptibility test (Thompson and Watling, 1983). Microorganisms can be used for the removal of heavy metals from industrial effluents and as indicator organisms in bioassays (Anderson and Abdelghani, 1980). Bacteria make excellent biosorbents because of their high surface-to-volume ratios (Beveridge, 1989). Some microbial cells can concentrate metal ions up to several percent of their dry weight (Wnorowski, 1991). Bioleaching has been used successfully for recovery of a variety of metals, including zinc, manganese, and copper (Zagury, Narasiah, and Tyagi, 1994).

Many bacteria and fungi can detoxify or volatilize certain heavy metals by transforming them (Ron, Minz, Finkelstein, and Rosenberg, 1992). Metals thereby undergo changes in valency and/or conversion into organometallic compounds (Silver, 1991).

Both the pH and the E^0 can vary widely from outside the living cell to inside that cell (Wood and Wang, 1983). Most metal ions function as Lewis acids (electron acceptors), but, depending upon pH, oxidation state, and complexation, metal complexes also can function as bases. Living cells are not at equilibrium with the external environment, and, therefore, a kinetic approach to metal ion transport, binding, toxicity, and resistance to toxicity is much more meaningful than a thermodynamic approach.

Soil reactions can transform heavy metals to less toxic forms or make them unavailable to plants (Arthur D. Little, Inc., 1976). Modifications in soil conditions can promote inactivation of mercury and arsenic (e.g., increasing organic material, moisture, temperature, and pH of the soil for mercury and decreasing the oxygen level for arsenic).

Biosorption is stable over a wide range of pH and temperature conditions (Churchill, Walters, and Churchill, 1995). The pH is important for passive removal of heavy metals from aqueous solution (Wnorowski, 1991). A specific pH is required for every metal for successful biological removal, which makes it difficult to treat mixtures of metal ions. Optimal pH values for uptake of some metals are in the alkaline range, which may result in precipitation rather than biological removal. Optimal pHs will be discussed in the following sections with the specific metals. Removal with living organisms, on the other hand, would have to be performed at neutral pHs (Wnorowski, 1991).

Metal interactions in biology can be divided into three classes: ions in fast exchange with biological ligands (e.g., Na^+, K^+, Ca^{+2}, Mg^{+2}, and H^+), ions in intermediary exchange with biological ligands (e.g., Fe^{+2} and Mn^{+2}), and ions in slow exchange with biological ligands (e.g., Fe^{+3}, Zn^{+2}, Ni^{+2}, and Cu^{+2}) (Williams, 1983). Living cells have membranes that act as initial barriers to metal ion uptake (Wood and Wang, 1983). Prokaryotes select those ions in fast exchange. Eukaryotes use spatial partitioning of metals. Once the cell buffering capacity for essential metal ions is exceeded, toxicity becomes evident. Toxicity occurs at much lower concentrations for nonessential metals.

Insoluble metal complexes can be precipitated at the cell surface through the activities of membrane-associated sulfate reductases (Galun, Keller, Malki, Feldstein, Galun, Siegal, and Siegal, 1983) or through the biosynthesis of oxidizing agents, such as oxygen or hydrogen peroxide (Wood, 1983). The reduction of sulfate to sulfide and the diffusion of oxygen and hydrogen peroxide through the cell membrane provide highly reactive means by which metals can be complexed and precipitated. A green alga, *Cyanidium caldarium*, can grow in acidic conditions and at high temperatures (Lovelock, 1979). It can remove 68% of the iron, 50% of the copper, 41% of the nickel, 53% of the aluminum, and 76% of the chromium from solution. Anaerobic cultures growing in the dark produce hydrogen sulfide gas for sulfide precipitation of metals.

Most microorganisms have a high affinity for metal cations, since their cell walls have a net negative charge at physiological pH values (Beveridge and Doyle, 1989). Some heavy metals (e.g., Cu, Ni, Zn) cause the net negative charge of the cells and clays to change to a net positive charge (charge reversal) at pH values between 6.0 and 9.0 (Collins and Stotzky 1989; 1992). As the pH increases, metal ions in solution undergo hydrolysis. Charge reversal occurs in the pH range wherein the metal changes from the divalent to the monovalent hydroxylated cation. Some hydroxides (e.g., of Ca, Mg, Hg, Pb) precipitate at pH 9 or above, and some (e.g., of Al, Cu, Fe, Ni, Zn) at lower pH values (Collins and Stotzky, 1996). The different chemical species of a metal occurring at different pHs requires a critical, narrow pH range for adsorption. This is usually less than 1 pH unit, wherein the amount of metal adsorbed increases from almost 0 to 100%. The speciation form of the metal, rather than the bacterium, appears to determine adsorption. When the metal ion hydrolyzes and metal hydroxides are precipitated onto the negative surface of cells, the electrokinetic potential is reversed. With borderline metal ions (e.g., Ni, Cu, Zn), *Bacillus subtilis, Agrobacterium radiobacter,* and *Saccharomyces cerevisiae* cells remained viable when charge reversal occurred.

Different hydroxylated forms of the same metal have different toxicities to microorganisms (Babich and Stotzky, 1983b). Toxicity could occur as a result of the replacement of essential metals (e.g., Mg) by toxic metals (e.g., Ni) (Collins and Stotzky, 1992). The reversal in charge might affect physiological functions of the cell or its interactions with other cells and inanimate particulates in the soil.

Some microorganisms can use biomethylation to eliminate heavy metals, such as mercury, tin, and lead (Wong, Chao, Luxon, and Silverberg, 1975), and metalloids, such as arsenic and selenium (Wood and Wang, 1983). Products of methylation may be more toxic than the free metal, but they may also be volatile and be released to the atmosphere (Arthur D. Little, Inc., 1976; Ron, Minz, Finkelstein, and Rosenberg, 1992).

The synthesis of less polar organometallic compounds from polar inorganic ions has certain advantages for cellular elimination by diffusion-controlled processes (Wood, Cheh, Dizikes, Ridley, Rackow, and Lakowicz, 1978). Organometallic compounds can be degraded microbiologically and chemically and volatilized (Tonomura, Maeda, and Futai, 1968; Schottel, Mandal, Clark, and Silver, 1974). Mechanisms for B_{12}-dependent synthesis of metal alkyls (requiring the presence of vitamin B_{12}) have been discovered for the metals Hg, Pb, Tl, Pd, Pt, Au, Sn, and Cr and for the metalloids As and Se (Craig and Wood, 1981).

There is a possible sequence of activating measures to increase remediation of soils containing these metals (Figure 3.6) (Arthur D. Little, Inc., 1976). The first steps would be to add organic material,

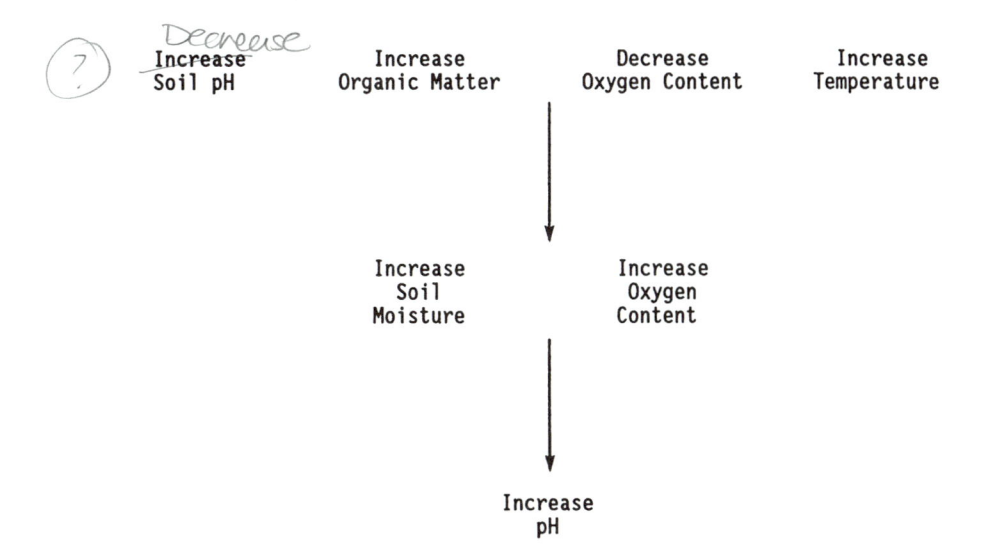

Increase Soil pH → Increase Organic Matter → Decrease Oxygen Content → Increase Temperature → Increase Soil Moisture → Increase Oxygen Content → Increase pH

Figure 3.6 Possible sequencing of soil manipulation. (From Arthur D. Little, Inc., Report No. EC-CR-76-76 on Contract No. DAAA 15-75-C-D188. Department of the Army, Aberdeen Proving Ground, MD, 1976.)

decrease oxygen, increase temperature, and decrease pH. Logistically, lowering the pH should come first, since ferrous sulfate should be added before irrigation begins. The organic matter should be incorporated into the soil next, immediately preceding irrigation; this will help produce anaerobic conditions. A PVC sheeting can then be applied. These measures should be kept in operation for about 2 to 3 years. Monitoring will indicate the actual time required. After the level of one toxicant is acceptable, conditions can be changed to transform and remove another. Finally, efforts must be made to decrease the availability of heavy metals and other undegradable toxicants left in the soil, e.g., increasing the pH of the soil with an application of limestone. Leaching of metals must be prevented throughout the treatment process.

Exposure to metals in the environment enhances microbial resistance (Baath, 1989). However, resistant bacteria have also been found in soils never exposed to high concentrations of metals, suggesting that metal-tolerant species already exist in nonpolluted habitats. Soils with significant soluble lead levels produce constant lead stress, resulting in a higher proportion of the microorganisms that are lead tolerant (Roane and Kellogg, 1996). Both heavy metal analyses and tolerance tests are necessary to understand the impact of heavy metal contamination on soil bacterial communities (Burkhardt, Insam, Hutchinson, and Reber, 1993).

The metal-leaching capability of indigenous bacteria can be increased by acclimating the organisms to the metal (Zagury, Narasiah, and Tyagi, 1994). Even strains that are not naturally resistant may become so by gradual adaptation (Wnorowski, 1991). Preselection and development of strains capable of growing in the presence of elevated levels of heavy metals can be an important step in building a collection of bioaccumulating strains for heavy metal removal studies.

Sources of inocula could be activated sludge, industrial cooling water, or streams with intensive industrial or mining activities in the vicinity, for instance. A study of 80 strains of bacteria, yeasts, and molds using such techniques identified 39 isolates that could bioaccumulate gold; 9, silver; 28, cadmium; and 22, nickel from diluted solutions (5 mg/L) to below 0.5 mg/L. Of these strains, 79% were resistant to 100 mg/L of $AuCl_4^-$, 42% to 100 mg/L Cd^{2+}, 34% to 100 mg/L Ag^+, and 31% to 200 mg/L of Ni^{2+}. Of these isolates, 14 belonged to the family Enterobacteriaceae, 15 to the genus *Aeromonas*, 11 to *Pseudomonas*, 7 to the genus *Bacillus*, 18 were other Gram-negative nonfermenting rods, 7 were molds and yeasts, and 8 were classified miscellaneous (Roane and Kellogg, 1996). Strains of *Aeromonas, Flavobacterium, Pseudomonas, Spirillium, Zoogloea, Arthrobacter*, and *Alcaligenes* have been shown to bind heavy metals (Ron, Minz, Finkelstein, and Rosenberg, 1992).

Eukaryotic organisms detoxify heavy metals mainly by binding to polythiols (Ron, Minz, Finkelstein, and Rosenberg, 1992). The prokaryotic bacteria have developed several efficient mechanisms for tolerating the metals. For instance, there are genes on bacterial plasmids that encode specific resistance for toxic heavy metal ions, such as Ag^+, AsO_2^-, AsO_4^{3-}, Cd^{2+}, Co^{2+}, CrO_4^{2-}, Cu^{2+}, Hg^{2+}, Ni^{2+}, Sb^{3+}, TeO_3^{2-}, and Zn^{2+} (Endo, Ji, and Silver, 1995). Bacteria with plasmids conferring resistance to heavy metals, gives the organisms an advantage over other microbes, when they are present (Ron, Minz, Finkelstein, and Rosenberg, 1992). Many of these bacteria are also resistant to antibiotics (Marques, Congregado, and Simon-Pujol, 1979; Misra, Brown, Haberstroh, Schmidt, Goddette, and Silver, 1985).

Alcaligenes spp. are known to be resistant to heavy metals (Diels and Mergeay, 1990). *A. eutrophus* var. *metallotolerans* strains harbor plasmids conferring multiple resistance to heavy metals (Diels, Springael, Kreps, and Mergeay, 1991). *Alcaligenes eutrophus* CH34 contains the plasmids pMOL28 (163 kb), specifying nickel, mercury, chromate, cobalt, and thallium resistance; and pMOL30 (240kb), specifying zinc, cadmium, cobalt, mercury, copper, lead, and thallium resistance. *A. eutrophus* DS185 contains pMOL85 (240 kb), specifying zinc, cadmium, cobalt, and copper resistance. These strains can remove cadmium, zinc, nickel, and copper from solution, under certain conditions (Diels, 1990). When provided with a carbon source, they can induce precipitation and crystallization of metal carbonates and hydroxides, due to the steep pH gradient at the cell surface.

Plasmid-determined heavy metal resistance can be due to efflux pumping of the toxic metal out of the bacterial cell, bioaccumulation in a physiologically inaccessible compound, and redox chemistry, in which a more toxic ion species is converted to a less toxic ion (Endo, Ji, and Silver, 1995). The most-promising system for bioremediation of toxic heavy metals is probably redox chemistry.

Heavy metal resistance genes, such as the Cd, Co, Zn genes (Nies, Mergeay, Schlegel, 1987), the Hg genes (Diels, Faelen, Mergeay, and Nies, 1985), and the Co, Ni, Cr genes (*cnr, chr*) (Nies, Nies, and Silver, 1989), can be cloned (Diels, Springael, Kreps, and Mergeay, 1991). Gene fusions have also been made with the Pb, Cu, and Tl resistance genes. Strains are being developed to combine the ability to degrade organics while being resistant to multiple heavy metals.

Algae (Laube, Ramamoorthy, and Kushner, 1979), bacteria (Strandberg, Shumate, and Parrott, 1981), and fungi (Marquis, Mayzel, and Carstensen, 1976) can complex relatively large quantities of metallic cations. Isolated Gram-positive and Gram-negative bacterial cell walls can also bind metal (Beveridge and Murray, 1980; Beveridge and Fyfe, 1985). Cell walls of the Gram-positive bacteria *Bacillus subtilis* and *B. licheniformis* can bind larger amounts of several metals than cell envelopes of the Gram-negative bacterium *Escherichia coli*.

Methane utilizers appear to have relatively discrete metal tolerance patterns with resistances to all metals depending upon the isolation site (Bowman, Sly, and Hayward, 1990). Methanotrophs are quite sensitive to mercury and cadmium but relatively resistant to copper, chromium, and zinc.

Higher plants can also be employed to accumulate metals in their roots and shoots (Kumar, Dushenkov, Motto, and Raskin, 1995). *Brassica juncea* (L.) Czern has a strong ability to accumulate Pb, Cr^{6+}, Cd, Ni, Zn, and Cu, with sulfates and phosphates as fertilizers.

There is no routine treatment for toxic metals in soils, other than extensive washing with acid or phosphates to remove and precipitate the metals from the soil matrix (Roane and Kellogg, 1996). Understanding the mechanisms of heavy metal resistance and the correlation between this resistance and population function within a microbial community may eventually lead to exploitation of these organisms for remediation of metal-contaminated soils. This could include enhanced sequestering methods, stimulated volatilization, or bacterial metal oxidation-reduction reactions.

There are several approaches for removing heavy metals from effluents (Gadd, 1992). One of these methods is sewage treatment, which involves a primary sedimentation, followed by an activated sludge or trickling filter system (Sterritt and Lester, 1986). Living cells are effective for binding heavy metals but require specific growth conditions, which may not allow optimum metal binding (Collins and Stotzky, 1989). Immobilized living cell systems also have potential, especially with continuous biomass replenishment (Macaskie and Dean, 1989). Living cell biofilms have the advantage of also being able to remove hydrocarbons. Living cell immobilization by entrapment within gels or other matrices can also be employed (Gadd, 1992).

Immobilized nonliving biomass has been used successfully for almost all microbial groups (Gadd, 1992). Dead biomass is frequently immobilized in particulate form, in processes similar to those used in ion exchange and activated carbon adsorption. Another method is to immobilize metal-binding compounds. Microorganisms can be inactivated by heating for 5 min in boiling steam (Churchill, Walters, and Churchill, 1995). The biomass is relatively stable. Nonliving biomass can also be prepared by acidic or alkaline conditions (Brierley, 1990; Rao, Iyengar, and Venkobachar, 1993). Alkali treatment may neutralize positively charged sites (Treen-Seers, 1986) or unmask metal-binding sites on cell-wall components (Brierley, 1990). An advantage of using nonliving biomass for passive biosorption is that optimum conditions for cell growth and biomass production and optimum conditions for metal sorption can be used separately (Volesky, 1990). Prepared biomass has been shown to absorb metals with equal or greater efficiency than viable cells (Brierley, Brierley, and Davidson, 1989). Biomass can be prepared from sphagnum peat moss, algae, yeast, and common duckweed (Bennett and Jeffers, 1990; Jeffers, Ferguson, and Bennett, 1991). Large-scale fermentation process by-products and fungal and freshwater algae culture can provide economical sources of biomass (Churchill, Walters, and Churchill, 1995).

Immobilized yeast cells are being investigated for application for metals removal. *Saccharomyces cerevisiae* yeast biomass was treated with hot alkali to increase its biosorption capacity for heavy metals, immobilized in alginate in gel, and biosorption capacity for Cu^{2+}, Cd^{2+}, and Zn^{2+} determined (Lu and Wilkins, 1995). Immobilized yeasts could be reactivated and reused in a manner similar to the ion-exchange resins. Immobilized caustic-treated yeast has high heavy metal biosorption capacity and high metal removal efficiency in a wide acidic pH range. The initial pH of polluted water affects the metal removal efficiency significantly, and the equilibrium biosorption capacity seems to be temperature independent at lower initial metal concentrations.

Also, there is the technique of growth-decoupled enzymic metal removal, which involves metal accumulation by growth-decoupled "resting cells" of a *Citrobacter* sp. (Gadd, 1992). The accumulation is catalyzed at the surface by an acid-type phosphatase enzyme, which releases HPO_4^{2-} from a supplied substrate and precipitates divalent cations (M^{2+}) as $MHPO_4$ at the surface (Macaskie and Dean, 1989; Macaskie, 1990; Plummer and Macaskie, 1990). The *Citrobacter* process has potential as a long-term

Table 3.20 Microbial Mechanisms for Metal Extracting/Concentrating/Recovery

Microorganism	Metal Removed	Method
Thiobacillus, Sulfolobus	Iron, sulfur	Oxidation
Sphaerotilus, Leptothrix, Hyphomicrobrium, Gallionella	Iron, manganese	Oxidation
Spirogyra, Oscillatoria, Rhizoclonium, Chara	Molybdenum, selenium, uranium, radium	Oxidation
Desulfovibrio spp.	Mercury, lead	Reduction
Scenedesmus, Synechococcus, Oscillatoria, Chlamydomonas, Euglena	Nickel	Surface ion-exchange
Saccharomyces cerevisiae, Rhizopus arrhizus	Uranium, cesium, radium	Surface ion-exchange
Penicillium digitatum	Uranium	Surface ion-exchange
Ustilago sphaerogena	Iron	Surface chelation
Aspergillus niger	Aluminum	Surface chelation
Cynanidium caldarium	Iron, copper, nickel, aluminum, chromium	Surface precipitation
Staphylococcus aureus, Escherichia coli	Cadmium, zinc, arsenate, arsenite, antimony	Chemosmotic efflux
Pseudomonas aeruginosa	Uranium, cesium, radium	Intracellular trap
Synechococcus	Nickel, copper, cadmium	Intracellular trap
Clostridium cochlearium	Mercury	Biomethylation
Pseudomonas spp.	Tin	Biomethylation

From Monroe, D. *Am. Biotechnol. Lab.* 3:10–19, International Scientific Communications, Inc. 1985. With permission.

biofilm-containing filter to achieve a high metal load and allow recovery and biomass regeneration (Macaskie and Dean, 1989).

Another method is to prepare heat-killed, oven-dried, freeze-thawed, and pulverized, semi-intact bacteria, for metal cation biosorption (Churchill, Walters, and Churchill, 1995). These porous cells have advantages over intact (viable or nonviable) cells in that metal binding can occur on both the cell barrier surface and the interior; this preparation is stable as a dry powder for prolonged periods at room temperature; and the method can be scaled up for industrial applications.

Disposal of large quantities of biomass, which has taken up the heavy metals and contains a high proportion of water, may be a problem (Brierley, Brierley, and Davidson, 1989). If metal recovery and biosorbent regeneration are desired, the mechanisms used will depend upon the element involved and the method of accumulation (Tsezos, 1984; 1990; Tsezos, McCready, and Bell, 1989; Volesky, 1990). Nondestructive methods similar to an ion-exchange process can often reverse biosorption that is not metabolism dependent. Metabolism-dependent accumulation and intracellular compartmentation or sequestration in organelles or binding to protein, etc. is often irreversible, which would require destructive recovery. Dilute mineral acids, carbonates, or bicarbonates can remove some heavy metals, or certain eluants could be selected for desorption of specific elements from a mixture of bound metals (Gadd, 1992). For stirred-batch and continuous-flow reactors, the loaded biosorbent can be separated by settling, flotation, centrifugation, or filtration and the biosorbent regenerated, ashed, or disposed of (Volesky, 1990). Other systems employ different methods. Kuyucak (1990) reports that sorbed metals can be recovered in an efficient manner, which minimizes waste.

Immobilized or pelleted preparations may be best for commercial use, with a cheap desorbing agent that can be recycled (Tsezos, 1990). Biosorptive treatments may be used as "polishing" systems for other processes that are not thorough (Gadd, 1992). There is a need to identify exopolymers selective toward Cu, Cd, Zn, Cr, and Ag, with little reactivity toward Na, K, Ca, or Mg (Jang, Geesey, Lopez, Eastman, and Wichlacz, 1990).

Table 3.20 lists bacteria, fungi, and algae; the metals they can remove; and the methods they use for removal (Monroe, 1985).

The following are elements that may be present in soils contaminated by the fuels addressed in this review and their potential for elimination by soil microbes (JRB Associates, Inc., 1984).

3.3.1 SPECIFIC ELEMENTS

A correlation between the metal content of soil samples and the degree of bacterial metal resistance was found in 38 isolates (Margesin and Schinner, 1995). Ultimately, it is planned to transfer the bacterial metal resistance of exceptional strains to bacterial decomposers of organic waste material polluted by heavy metals, thereby leading to an optimization of the process of bacterial biodegradation.

3.3.1.1 Arsenic (As)

Biotransformation can result in changes in valency of arsenic, with the production of less-toxic or volatile compounds, such as the oxidation of arsenite [As(III)] to arsenate [As(V)] (Ron, Minz, Finkelstein, and Rosenberg, 1992). Arsenic is apparently oxidized by aerobic heterotrophic organisms, and many heterotrophs can also reduce arsenate (Alexander, 1977). Methylation of arsenicals is an important process in soils, and trimethylarsine is an important gaseous product (Woolson, 1977). Soil contaminated with arsenic must be managed to minimize volatilization through microbial reduction. Addition of organic material and maintenance of aerobic conditions can help stimulate the oxidation of arsenite to arsenate (Sims and Bass, 1984). Further treatment with ferrous sulfate will form highly insoluble $FeAsSO_4$.

Resistance of microbes to arsenic occurs through the evolution of cellular exclusion mechanisms (Wood and Wang, 1983). Resistance in *Staphylococcus aureus* and *Escherichia coli* to arsenate and arsenite is induced by an operon-like system (Novick, Murphy, Gryczan, Baron, and Edellman, 1979). An operon is a DNA region that codes for several enzymes in a reaction pathway.

Arsenic behaves much like phosphorus in soils in that its adsorption increases as iron oxide content increases and that iron and aluminum hydrous oxides specifically adsorb the metal (Zitrides, 1978). Arsenate, however, can be reduced to arsenite, but arsenate is the most common form in soils. Generally, both forms are strongly retained in soils.

3.3.1.2 Cadmium (Cd)

Cadmium is one of the more toxic metals and is both carcinogenic and teratogenic (Ron, Minz, Finkelstein, and Rosenberg, 1992). It can be taken up by plants and enter the food chain. Detoxification of polluted water or soil involves concentration of the metal or binding it to make it biologically inert.

Growth of soil bacteria is retarded by cadmium, and the soil community structure is affected (JRB Associates, Inc., 1984). However, bacterial populations in contaminated sites have been found to be able to adapt to the heavy metal contamination (Tripp, Barkay, and Olson, 1983). Soil microbial biomass may contribute to the soil cadmium-binding capacity and affect cadmium availability, with dead cells sorbing more cadmium than live cells (Kurek, Czaban, and Bollag, 1982). Since cadmium exists in nature only in the valence state of +2, microbial oxidation or reduction of this element is unlikely.

Resistance to cadmium could be due to mechanisms and enzymes that make the bacterial cell wall impermeable to the metal, to efflux mechanisms, or to binding of the metal ions (Trevos, Oddie, and Belliveau, 1985).

Cellular exclusion mechanisms can be responsible for resistance of microorganisms to cadmium (Wood and Wang, 1983). This resistance is mediated by a plasmid. Resistant cells of *Staphylococcus aureus* have a very efficient chemosmotic efflux system specific for Cd^{2+} ions, as a result of two separate plasmid genes. Resistance to cadmium may be associated with resistance to other heavy metals (Ron, Minz, Finkelstein, and Rosenberg, 1992). The same gene could confer resistance to multiple metals or a plasmid may contain separate genes for the individual metals.

Cadmium can be bound by precipitation on the cell surface, the most important location for precipitation of heavy metals (Ron, Minz, Finkelstein, and Rosenberg, 1992). Capsular material of *Arthrobacter viscosus* and *Klebsiella aerogenes* can also bind cadmium (Scott and Palmer, 1988; 1990). Cadmium can also precipitate as $CdHPO_4$ (Ron, Minz, Finkelstein, and Rosenberg, 1992). Polythiol Cd^{2+}-binding peptides might also be involved in bacterial resistance to cadmium (Hamer, 1986; Sequin and Hamer, 1987).

Cadmium can be transported into cells of *Bacillus subtilis* via an energy-dependent manganese transport system (Laddaga, Bessen, and Silver, 1985). Intracellular traps can be biosynthesized as a temporary measure for organisms to remove metal ions, e.g., synthesis of metallothionen and removal of cadmium by this sulfhydryl-containing protein (Williams, 1981) to prevent metals from reaching toxic levels (Wood and Wang, 1983). Mutants with intracellular trapping mechanisms tend to bioconcentrate the toxic metal intracellularly to about 200 times over the external concentration. This strategy works well for some organisms, but is not as effective as the extracellular binding or precipitation of metals. On the other hand, *Arthrobacter* and *Pseudomonas* species appear to rely more heavily on detoxification systems that precipitate cadmium internally, rather than on their excreted polymers (Scott and Palmer, 1990).

In Gram-positive bacteria, Cd^{2+} can enter the cells as a toxic alternative substrate for the cellular Mn^{2+} transport system (Tynecka, Gos, and Zajac, 1981) and for the Zn^{2+} transport system in Gram-negative bacteria (Laddaga and Silver, 1985). A mutation in the membrane manganese transport system prevents a *B. subtilis* from taking up Cd^{2+} (Laddaga, Bessen, and Silver, 1985). There are several efflux mechanisms that can also confer resistance to cadmium (Silver and Walderhaug, 1992).

Cadmium is complexed by organic matter, oxides of iron and manganese, and chlorides (Chaney and Hornick, 1978). In alkaline, low-organic-matter, sandy soils, precipitation of cadmium compounds occurs. The soil pH is the most important factor governing cadmium solubility and resultant availability, with the solubility increasing as pH decreases. Several organocadmium compounds have been synthesized, and diorganocadmium compounds have been shown to be light sensitive and thermolabile (Ron, Minz, Finkelstein, and Rosenberg, 1992).

Cadmium-contaminated wastewaters are usually acidic (pH 1.0 to 4.0), while the optimum pH for cadmium uptake is pH 5.0, as observed with *Streptomyces pimprina* (Puranik, Chabukswar, and Paknikar, 1995). The sorption decrease under acidic conditions may be due to the increased competition of the hydrogen (H^+) and hydronium (H_3O^+) ions (Zhou and Kiff, 1991). The cadmium sorption process is thus governed by an ion-exchange mechanism.

In an investigation of PAH- and heavy metal–contaminated soil at a wood-preserving factory in Czechoslovakia, biological activities were low compared with unpolluted soil (Riha, Nymburska, Tichy, and Triska, 1993). However, several bacterial strains were found to be able potentially to survive a concentration of 2 mmol/L cadmium.

In another study, 28 strains were found to be able to remove cadmium from a diluted (6.5 mg/L) solution to below 0.5 mg/L (Wnorowski, 1991). The best sorption of cadmium was achieved at pH 11.0 for low metal concentrations. However, for more-concentrated solutions (20 mg/L), the highest pH obtainable without precipitation was 8.5, which was too low for maximum removal. Organisms showing this ability were *Pseudomonas* sp., *F. aquatile, Cytophage* sp., *Providentia* sp., *Enterobacter* sp., *Escherichia coli, Aeromonas hydrophila, Acinetobacter* sp., *Botritis* sp., *G. candidum, Alcaligenes* sp., *Klebsiella pneumoniae, Cytophaga* sp., *Providentia* sp., and *Pseudomonas rettgeri*. An *Enterobacter* sp. was able to accumulate silver, nickel, and cadmium. Both nickel and cadmium could be assimilated by an *Enterobacter* sp., a *Pseudomonas* sp., and an *Alcaligenes* sp., and silver and cadmium by an *Aeromonas hydrophila* sp. Gram-negative bacteria, especially *E. coli*, were found to remove more cadmium from solution at low concentrations than Gram-positive organisms (Mullen, Wolf, Ferris, Beveridge, Flemming, and Bailey, 1989).

Phormidium and *Myriophyllum spicatum* have high adsorption capabilities (mg metal/kg biomass) for cadmium removal (Wang, Weissman, Ramesh, Varadarajan, and Benemann, 1995). *Bacillus cereus*, another *Bacillus* sp., *Citrobacter* spp., and *Rhizopus arrhizus* are able to accumulate cadmium (Gadd, 1992).

Methanotrophs are quite sensitive to cadmium (Bowman, Sly, and Hayward, 1990). A few reference strains are somewhat resistant, in particular *Methylomonas methanica, Methylococcus whittenburyi,* and *Methylosinus trichosporium*.

Some natural, inexpensive materials, such as fungal biomass, have a higher cadmium-adsorption capacity than activated charcoals or ion-exchange resins, which are the alternative adsorbents (Azab and Peterson, 1989). Whole cells of the alga *Chlorella fusca* were able to take up 0.045 µmol Cd/mg dry mass (Gerhards and Weller, 1977). Walker, Flemming, Ferris, Beveridge, and Bailey (1989) report an uptake of 0.68 µmol Cd/mg dry cell wall for isolated cell walls of *Bacillus subtilis*.

Adsorption of cadmium (and other metals) is frequently associated with the secretion of exopolysaccharide or capsular material (Ron, Minz, Finkelstein, and Rosenberg, 1992). A polysaccharide-producing *Bacillus circulans* was grown in liquid medium with glucose as carbon source to produce biomass for metal accumulation (Sahoo, Kar, and Das, 1992). A biomass concentration of 1.48 to 1.52 g dry weight/L removed 44% of cadmium from a solution containing 492 ppm cadmium. Removal of cadmium at low concentrations by this organism was very efficient, but was affected by the pH. The exopolysaccharide from *Arthrobacter viscosus* accumulates 2.3 times more cadmium than the equivalent weight of whole cells and is 13.7 times more effective than the cells of *A. globiformis*, which does not produce exopolysaccharide (Scott and Palmer, 1988). The highest production of capsular polysaccharides in several *Arthrobacter* spp. was obtained with growth on 1% mannitol and correlated with greater adsorption of cadmium from solution (Grappelli, Hard, Pietrosanti, Tomati, Campanella, Cardarelli, and Cordatore, 1989).

A strain of *Pseudomonas* has the ability to secrete a massive amount of extracellular material that binds the cells together to form very large aggregates, which appears to also bind cadmium (Ron, Minz, Finkelstein, and Rosenberg, 1992). The bound cadmium can be released and separated from the bacteria at a low pH. An 0.1 M EDTA (ethylenediaminetetraacetic acid) solution can desorb cadmium loaded on *Streptomyces pimprina* biomass, but appears to be too strong a desorbing agent (Puranik, Chabukswar, and Paknikar, 1995).

Cadmium accumulation could occur in the cell envelope, as a result of thickening of the envelope and the proportion of peptidoglycan there (Hambuckers-Berhin and Remacle, 1990). *Proteus mirabilis* was able to grow in concentrations of 300 mg/L cadmium, with 80% accumulation in the envelope and 20% in the cytoplasm (Andreoni, Finoli, Manfrin, Pelosi, and Vecchio, 1991). Cadmium resistance is associated with a plasmid in *Alcaligenes eutrophus* CH34, and the metal is found mainly in the cell envelopes.

Couillard and Mercier (1993) have been able to solubilize metals biologically, including cadmium, from aerobically digested sewage sludge by using *Thiobacillus ferrooxidans* in a continuously stirred tank reactor. The metals did not reprecipitate during filtration. Added lime precipitated the metals, and selective precipitation was also possible. Sludge was also used to adsorb about 95% of total cadmium present at concentrations below 30 mg/L (Gourdon, Rus, Bhende, and Sofer, 1990).

There are several approaches for using cadmium-binding bacteria to detoxify contaminated water. One method is to use columns packed with immobilized cells of *Citrobacter* sp., growing as a biofilm on glass beads or incorporated into polyacrylamide gels that are then shredded (Macaskie, Wates, and Dean, 1987). The cadmium precipitates on the cell surface after release of phosphate. Another method immobilizes cells of *Zoogloea ramigera* 115 into calcium alginate beads for use in air-bubbled column reactors (Kuhn and Pfister, 1989). Adsorption efficiencies of 99% can be achieved with three bubbled columns in sequence. Granulated nonliving cellular mass from mixed microbial or algal cultures can be used in a similar manner (Silver, 1991).

A. eutrophus cells can be immobilized on a flat sheet reactor made of composite membranes of polysulfone with inorganic fillers, through which a nutrient solution is passed (Diels, 1990). About 90% of the cadmium could be removed from a solution of 320 ppm cadmium. The cadmium was bound to an exopolymer, with a binding ratio of 500 µg cadmium to 1 mg of polymer.

While there has been one report of possible biological methylation of cadmium (Huey, Brinckman, Iverson, and Grim, 1975), there is no conclusive evidence of microbial transformation of this heavy metal (Ron, Minz, Finkelstein, and Rosenberg, 1992).

3.3.1.3 Chromium (Cr)

Chromium must be removed or made immobile to reduce its toxicity in contaminated soil (Cifuentes, Lindemann, and Barton, 1996).

Both chromate reduction and chromate resistance occur in microorganisms, but plasmid resistance of Gram-negative bacteria seems to be unrelated to chromate reduction (Endo, Ji, and Silver, 1995). Chromate resistance is encoded by the *chr* determinant on plasmid pMOL28 and results from reduced cellular accumulation of CrO_4^{2-} (Dressler, Kues, Nies, and Friedrich, 1991). The *chr* determinant has been cloned and sequenced.

Chromium should be amenable to oxidation or reduction by microbes, since it commonly exists in several oxidation states, including Cr(III) (trivalent state) and Cr(VI) (hexavalent state) (Zajic, 1969). Cr(VI) is more toxic and mutagenic than Cr(III) (Ross, Sjogren, and Bartlett, 1981). Gram-negative bacteria are also more sensitive to Cr(VI) than Gram-positive organisms. It should not be assumed that Cr(III) is harmless to the soil microflora at high levels.

The most soluble, mobile, and toxic form of chromium in soils occurs in the hexavalent state as chromate or dichromate (Hornick, 1983). When aerobic conditions exist, the hexavalent form is rapidly reduced to trivalent chromium, which forms insoluble hydroxides and oxides and cannot leach. Liming soil to pH 6.5 with the presence of an alkaline oil waste in the soil will maintain the soil pH near neutral (Hornick, Fisher, and Paolini, 1983). A near neutral soil pH will also prohibit the formation of dichromates (Hornick, 1983).

Solid-phase chromium should not be readily mobilized during the retention of dredge slurries in disposal areas, possibly because of the slow oxidation of reduced chromium hydroxide (Lu and Chen, 1977). Little chromium is released from dispersed sediments under oxidizing conditions. Reduced chromium, Cr(III), is generally highly insoluble at pH values above 5.5, unless complexed with soluble organic compounds. Chromium has also not been noted to oxidize to more-soluble Cr(VI) forms under short-term oxidizing conditions.

Chromate reduction is potentially useful for bioremediation (Endo, Ji, and Silver, 1995). Bacteria from both Cr(VI)-contaminated and Cr(VI)-noncontaminated soils and sediments have been found to be capable of catalyzing the reduction of Cr(VI) to Cr(III) (Turick, Apel, and Carmiol, 1996). Efflux pumping mechanisms seem to be the most practical. There is also the possibility of building membranes with highly specific permeability to toxic metals. The green alga *Cyanidium caldarium* may be effective for selective removal of chromium from polluted wastewaters (Wood and Wang, 1983). Churchill, Walters, and Churchill (1995) report that semi-intact *Escherichia coli* cells are very efficient at binding chromium. Methanotrophs are relatively resistant to chromium (Bowman, Sly, and Hayward, 1990).

Sorption of Cr(III) is soil dependent, the maximum sorbed being 431 mmol/kg in a clay loam soil (Cifuentes, Lindemann, and Barton, 1996). Sorption of Cr(VI) is concentration dependent and independent of soil type, with a maximum of 63 mmol/kg soil sorbed. *Aspergillus niger* cells can sorb more Cr(VI) than Cr(III). Binding of Cr(III) by this organism depends on pH and is greatest at pH 9, but binding of Cr(VI) is not pH dependent. Soil amended with freeze-dried *Aspergillus* increases sorption of Cr(III). Organic amendment of Cr(VI)-contaminated soil causes direct sorption of Cr(VI) and reduction to Cr(III) by indigenous soil microorganisms. Inoculation of Cr-tolerant microorganisms does not appear necessary to immobilize Cr(VI).

A surfactant system containing Dowfax 8390 and diphenyl carbazide is quite effective in remediation of chromium-contaminated soil, by enhancing elution of the chromate (Nivas, Sabatini, Shiau, and Harwell, 1996).

3.3.1.4 Iron (Fe)

The concentration of free metal ions can be controlled by the biosynthesis of ligands in the form of small molecules with high stability constants, such as with the removal of iron (Wood and Wang, 1983). The cell may expend energy to pump the metal ion out of the cell, it may synthesize ligands that bind metals strongly at the cell surface, or it may use the activities of surface-bound enzymes to precipitate the metal extracellularly.

3.3.1.5 Lead (Pb)

Lead in the environment has typical turnover times of 220 to 5000 years, although about 30% of the lead input is mobile (Roane and Kellogg, 1996). Inorganic lead speciation is dominated by the +2 oxidation state and the +4 in the organolead compounds (Jarvie, Markall, and Potter, 1975). In a mixed microbial ecosystem, the following major lead compounds may be found: tetraethyl-lead, tetraphenyl-lead, hexamethyl-lead, hexaethyl-lead, and lead sulfide (Ibeanusi and Archibold, 1995). Inorganic lead is toxic to a broad range of microorganisms, including cyanobacteria, marine algae, fungi, and protozoa (JRB Associates, Inc., 1984). Lead and its compounds also affect microbial activities in soil, including inhibition of nitrogen mineralization, stimulation of nitrification, and the synthesis of soil enzymes. Species diversity is lower in lead-contaminated soils.

The pH affects removal of lead from the soil, by changing its solubility and motility. Biodegradation of petroleum hydrocarbons in soil has been found to reduce the pH conditions and, thus, increase the solubility of heavy metals and their concentrations in leachates (du Plessis, Phaal, and Senior, 1995). Lead solubility is determined by the amount of sulfate, phosphate, hydroxides, carbonates, and organic matter present in a soil system (Hornick, 1983). As the soil pH and available phosphorus in a soil decrease, soluble lead increases. The formation of lead sulfate at very low pH, lead phosphate and hydroxides at intermediate pH, and lead carbonates at a calcareous pH limits the mobility of lead in soils. The optimum pH (of the pHs tested) for lead removal is pH 5, with dextrose being a critical component (Vesper, Donovan-Brand, Paris, Al-Bed, Ryan, and Davis-Hoover, 1996). A low pH of 5 or 6 increases the toxicity of lead, while higher pH and other abiotic factors (phosphate and carbonate ions, clay minerals, particulate humic acid, and soluble organics) reduce toxicity (JRB Associates, Inc., 1984).

Fluctuating soluble lead concentrations *in situ* may allow lead-sensitive organisms to survive, while high lead levels may cause selection of lead-resistant strains (Roane and Kellogg, 1996). Lead-resistant isolates have been recovered from soil contaminated with lead, as well as from soil with no previous contamination, suggesting widespread capacity for lead resistance. Strains exhibiting lead resistance were found in the genera *Pseudomonas, Bacillus, Corynebacterium,* and *Enterobacter*. Plasmids for lead resistance have not been detected.

A *Pseudomonas aeruginosa* strain, designated CHL004, is able to remove lead from solidified media and soil (Vesper, Donovan-Brand, Paris, Al-Bed, Ryan, and Davis-Hoover, 1996). *Phormidium* and *Myriophyllum spicatum* have high adsorption capabilities (mg metal/kg biomass) for lead removal (Wang, Weissman, Ramesh, Varadarajan, and Benemann, 1995). A *Zoogloea* sp., *Bacillus* sp., and *Rhizopus arrhizus* can accumulate lead (Gadd, 1992). The alga *Chlorella fusca* showed a lead uptake of 0.077 μmol/mg for whole cells (Irmer, 1982).

A cell-free extract of a *Pseudomonas* sp. was found to biosorb lead ions more effectively than whole cells from aqueous solutions (Panchanadikar and Das, 1994). Presence of other cations and anions can inhibit lead uptake. Walker, Flemming, Ferris, Beveridge, and Bailey (1989) report uptake of 0.54 μmol Pb/mg dry cell wall for isolated cell walls of *Bacillus subtilis*.

Lead can be methylated by some microbes, being transformed into organometallic compounds (Ron, Minz, Finkelstein, and Rosenberg, 1992). Methylation can increase the volatility of the metal and its potential loss from the soil (JRB Associates, Inc., 1984). Hydrogen sulfide can effectively volatilize and precipitate lead compounds through disproportionation chemistry in the aqueous environment (Wood and Wang, 1983). Once in the atmosphere, the volatile compounds are unstable, since metal–carbon bonds are susceptible to homolytic cleavage by light.

3.3.1.6 Mercury (Hg)

Bioconversions of mercury can lead to changes in valency of the metal, resulting in production of volatile or less-toxic compounds, such as the reduction of mercury ions to metallic mercury (Ron, Minz, Finkelstein, and Rosenberg, 1992). Through chemical and microbial action, mercury can be volatilized or associated with clay particles and organic matter (CAST, 1976). Mercury reacts in soils with chlorides and sulfur to form insoluble HgS, $HgCl_3$, and $HgCl_4^{2-}$ (Hornick, 1983). Mercury can be chelated by organic matter as $HgCl_4^{2-}$ or can be absorbed by sesquioxide surfaces (CAST, 1976). Mercury is not very mobile in the soil profile due to its strong sorption reactions with soil constituents. Hydrogen sulfide is extremely effective at volatilization and precipitation of mercury through disproportionation chemistry in the aqueous environment (Wood and Wang, 1983).

Microbes oxidize, reduce, methylate, and demethylate mercury (JRB Associates, Inc., 1984). *Serratia marcescens* had the highest rate of transformation at pH 8 and *Enterobacter aerogenes* over the range tested (Mason, Anderson, and Shariat, 1979). Both aerobic and anaerobic heterotrophic bacteria were resistant to 14 ppm Hg^{2+} (Callister and Winfrey, 1983). Mercury-resistant populations of oil-degrading bacteria have been isolated (Walker and Colwell, 1974a). Petroleum biodegradation proceeded when mercury in the soil was present in the low-ppm-concentration range, but was absent at 85 ppm (Walker and Colwell, 1976b). A fungus that is able to accumulate mercury is *Rhizopus arrhizus* (Gadd, 1992).

Mercury methylation and demethylation are usually ascribed to different bacteria; however, the anaerobic *Clostridium cochlearium* T-2 was found to acquire demethylating capabilities, in addition to its methylating (Pan-Hou, Hosono, and Imura, 1980). This trait is probably on a plasmid. Biomethylation of Hg(II) salts to CH_3Hg^+ by a B_{12}-dependent strain of *C. cochlearium* allows detoxification and gives the organism an advantage in mercury-contaminated systems (Wood and Wang, 1983). Mercury methylation was highest in anaerobically incubated surface sediments (Callister and Winfrey, 1983). Sediment-bound mercury remained available for methylation over 7 days.

The products of methylation may be more toxic than the free metal, i.e., the methylated derivatives of mercury — methyl mercury (which is water- and lipid-soluble and more toxic than mercury) and dimethyl mercury (which is volatile) (Ron, Minz, Finkelstein, and Rosenberg, 1992). Dimethyl mercury compounds are light sensitive and thermolabile.

Mercuric and organomercurial strains of bacteria have been isolated from a variety of ecosystems, such as soil, water, and marine sediments (Vonk and Sjipesteijn, 1973; Friello and Chakrabarty, 1980). These organisms catalyze both the forward and reverse conversion of CH_3Hg^+ to Hg^{2+} and then to Hg^0 (Wood, 1983). A mercury-resistant strain of *Pseudomonas fluorescens* can break the carbon–mercury link of organomercurial compounds (Marcandella, Bicheron, and Bues, 1995). The resulting mercuric ion (Hg^{2+}) is reduced into volatile metallic mercury, Hg.

An enzyme capable of metabolizing phenylmercuric acetate has been isolated from soil microorganisms (Tonomura and Kanzaki, 1969). The enzyme(s) is capable of cleaving the carbon–mercury bond and requires both a sulfhydryl compound and the coenzyme, NADH, to carry out the reaction.

A variable volume extraction procedure and aqueous sodium boron tetraethyl derivatization were used to determine accurately and precisely methylmercury bioconcentration factors for fungi (Fischer, Rapsomanikis, Andreae, and Baldi, 1995). By *in vitro* experiments with axenic (i.e., free from other living organisms) cultures of *Coprinus comatus* and *C. radians*, it was shown that these saprophytic macromycetes are able to methylate mercury. The methylmercury content for fungi carpophores was 0.08 to 7.94 μg/g and total mercury was 6.2 to 144 μg/g. Bioconcentration factors are generally <1 for total mercury and 3 to 199 for methylmercury.

Mercury resistance is an inducible trait, and the genetic material coding for it is transposable (Silver and Kinscherf, 1982). The plasmids pMOL30 and pMOL28 encode narrow-range resistance to mercury (Dressler, Kues, Nies, and Friedrich, 1991). Resistance to inorganic mercury (Hg^{2+}) is widely found among eubacteria and is based on reduction of the mercuric ion to the volatile, less-toxic elemental mercury (Hg^0).

3.3.1.7 Nickel (Ni)

Nickel inhibits soil nitrification, carbon mineralization, and the activities of acid and alkaline phosphatase and arylsulfatase (JRB Associates, Inc., 1984). Toxicity to fungi is reduced by the presence of clay or an increase in pH to 7.0 (Babich and Stotzky, 1983a). Survival of certain bacteria and yeasts was improved by raising the pH from 4.9 to 7.7. The presence of ions such as Mg^{2+}, Zn^{2+}, S^{2-}, and PO_4^{-3}; alkaline pH; and type and amount of clay minerals greatly reduce the microbial toxicity of nickel. However, other ions (e.g., potassium, sodium, calcium, and iron), amino acids, tryptone, casamino acids, yeast extract, and chelating agents [citrate, EDTA, DPA (diphenylamine), NTA (nitroloacetic acid)] do not reduce Ni toxicity.

Both nickel binding and nickel toxicity are very pH dependent (Wood and Wang, 1983). The optimum pH for binding is between 8 and 8.5. Orientation of ligands at the cell surface must be important, since only surface-active substances, such as humic acids, could compete effectively for nickel binding (Galun, Keller, Malki, Feldstein, Galun, Siegal, and Siegal, 1983).

Nickel availability in soils is governed by iron and manganese hydrous oxides and by organic chelates that complex nickel less strongly than copper (CAST, 1976). Nickel differs from copper and zinc in that it is more available from organic sources than inorganic sources. Acid soils increase the solubility of nickel in the soil solution.

Green algae are much more resistant to high concentrations of Ni^{2+} than are the blue-green algae (Wood and Wang, 1983). Cyanobacteria and brown and green algae all bioconcentrate nickel (Fuge, 1973; Hirschberg, Skane, and Throsby, 1977; Karata, Yoichi, and Fumio, 1980; Ballester and Castellvi, 1980). The former are more sensitive to nickel toxicity than the green algae, indicating different transport mechanisms for prokaryotes and eukaryotes (Galun, Keller, Malki, Feldstein, Galun, Siegal, and Siegal, 1983). Nickel-tolerant mutants of the cyanobacterium *Synechoccus* can tolerate up to 20×10^{-5} M nickel sulfate by synthesizing an intracellular polymer that removes nickel from solution (Simon and Weathers, 1976). This intracellular trap prevents nickel toxicity. The green alga *Cyanidium caldarium* may be effective in treating waters polluted with nickel, so effluents can meet Federal standards (Wood and Wang, 1983).

Phormidium and *Myriophyllum spicatum* have high adsorption capabilities (mg metal/kg biomass) for nickel removal (Wang, Weissman, Ramesh, Varadarajan, and Benemann, 1995). Couillard and Mercier (1993) have been able to biologically solubilize metals, including nickel, from aerobically digested sewage sludge by using *Thiobacillus ferrooxidans* in a continuously stirred tank reactor. The metals did not reprecipitate during filtration. Added lime precipitates the metals, and selective precipitation is also possible.

An *Arthrobacter* sp. has been found to grow on up to 20 mM Ni^{2+} (Margesin and Schinner, 1995). This isolate has been well characterized, and its resistance will eventually be transferred to bacteria that can degrade organic waste material contaminated by heavy metals. The next year, Margesin and Schinner (1996) reported identification of four Gram-positive, plasmid-containing strains of *Arthrobacter* spp. that were also resistant to high concentrations (up to 20 mM) of nickel. Nickel resistance was induced in three strains and was constitutively expressed in one. They observed that nickel resistance was influenced by temperature.

It was found that 22 strains of bacteria were able to bioaccumulate nickel from dilute solutions (4 mg/L), with a saturation value of 22 to 57 mg/g dry weight (Wnorowski, 1991). The organisms were *Pseudomonas* sp., *Aeromonas hydrophila, Enterobacter* sp., *Achromobacter* sp., *A. lwoffi, Alcaligenes* sp., *Acinetobacter* sp., and *Klebsiella pneumoniae*. Optimum pH for Ni^{2+} was $\geqq 6.0$. Both nickel and cadmium could be assimilated by an *Enterobacter* sp., *Pseudomonas* spp., and an *Alcaligenes* sp. A *Zoogloea* sp. is able to accumulate nickel (Gadd, 1992). Churchill, Walters, and Churchill (1995) report that semi-intact *Escherichia coli* cells are very efficient at binding nickel.

Resistance to nickel and cobalt is mediated by an inducible energy-dependent cation efflux system encoded by the *cnr* resistance determinant adjoining the *chr* locus on the plasmid pMOL28 (Dressler, Kues, Nies, and Friedrich, 1991).

3.3.1.8 Selenium (Se)

Normal soils contain between 0.1 and 2.0 µg/g selenium (Girling, 1984). Selenium behaves similarly to sulfur in the soil solution, existing as selenates and selenite (Hornick, 1983). Selenate, the predominant form in alkaline soils, is quite soluble as $CaSeO_4$ and can move readily in these soils. Addition of lime will increase the availability of selenium (Gissel-Nielsen, 1971).

Toxic forms of selenium are found in oil refinery waste streams (Garbisu, Ishii, Smith, Yee, Carlson, Yee, Buchanan, and Leighton, 1995). A bacterial treatment system is being developed, using *Bacillus subtilis* and *Pseudomonas fluorescens* as model Gram-positive and Gram-negative soil bacteria, respectively. During growth, both organisms reduce selenite, a major soluble toxic form, to reduced elemental selenite, an insoluble product generally regarded as nontoxic. Reduction depended on growth substrate and was effected by an inducible system that removed selenite at concentrations typical of polluted sites (50 to 300 µg/L). The ability of *P. fluorescens* (but not *B. subtilis*) to remediate selenite was enhanced by growth in medium containing nitrate or sulfate.

Most of the selenium in the soil is biologically unavailable (Girling, 1984). Some bacteria and fungi reduce biologically available selenium to elemental insoluble forms, while others produce volatile organic forms of selenium that are lost to the atmosphere. Some bacteria are able to oxidize colloidal selenium to selenate or selenite so it becomes biologically available. Soil microflora are capable of several transformations of selenium, such as oxidation, reduction, and methylation (JRB Associates, Inc., 1984). The oxidized form, selenate, is very toxic (Alexander, 1977). Methylation is greatly accelerated when a readily available carbon source, such as glucose, is added to the soil.

Selenium salts can be biologically methylated to volatile organic products (Challenger and North, 1933). The fungus *Scopulariopsis brevicaulis* can produce dimethylselenide from inorganic selenite or selenate. In all, 11 strains of fungi have been found capable of producing dimethylselenide, including strains of *Penicillium, Fusarium, Cephalosporium*, and *Scopulariopsis* (Barkes and Fleming, 1974). Strains of *Bacillus* spp. are able to oxidize around 1.5% of the total selenium added to selenite and trace amounts of selenate (Shrift, 1973). Inorganic selenium can be reduced to elemental selenium. Isolates of bacteria, fungi, and actinomycetes from soils containing high selenium contain as much as 0.18% (dry weight) selenium inside their cells (Koval'skii, 1968). Their resistance to the high selenium levels depends upon their ability to reduce the soil selenium to the biologically inert elemental form.

Waste material containing significant amounts of selenium may present environmental problems if the selenium can be transformed into a form that is biologically available (Girling, 1984).

A method has been developed for bioremediation of selenium in agricultural drainage water (Cantafio, Hagen, Lewis, Bledsoe, Nunan, and Macy, 1996). A medium-packed, pilot-scale biological reactor was inoculated with the selanate-respiring bacterium, *Thauera selenatis*. From 91 to 96% of the total selenium recovered was elemental selenium, 97.9% of which could be removed with Nalmet 8072, a new, commercially available precipitant/coagulant.

3.3.1.9 Silver (Ag)

Silver is known to be bacteriocidal (Woodward, 1963); however, soil treated with 100 ppm Ag had about twice as many Ag^+-reducing bacteria as untreated soil (Klein and Molise, 1975; Sokol and Klein, 1975). Several bacteria have been found that precipitate silver as Ag_2S at the cell surface (Wood and Wang, 1983). Nine strains of bacteria (*Aeromonas hydrophila, Escherichia coli, Acinetobacter* sp., and *Enterobacter* sp.) were found to be capable of silver uptake from weak solution (0.5 mg/L or lower) (Wnorowski, 1991). Maximum uptake was high (35 to 207 mg/g dry weight). Optimum removal of 85% was achieved at pH 5.0. Both silver and cadmium could be taken up by *A. hydrophila*, a *Bacillus* sp., *Thiobacillus ferrooxidans*, *Rhizopus arrhizus*, and a *Phoma* sp. (fungus), while yeasts could accumulate silver (Gadd, 1992).

3.3.1.10 Other Metals

Riha, Nymburska, Tichy, and Triska (1993) found that while biological activity was low in soils contaminated with PAHs and heavy metals (mainly zinc, copper, and mercury), the bacteria were able to survive concentrations of 20 mmol/L zinc and 8.5 mmol/L copper. Actual concentrations of heavy metals in the soils were below levels where bacterial growth would be affected. *Phormidium* and *Myriophyllum spicatum* have high adsorption capabilities (mg metal/kg biomass) for zinc and copper removal (Wang, Weissman, Ramesh, Varadarajan, and Benemann, 1995).

Copper-resistant bacteria include *Escherichia coli, Pseudomonas syringae, Alcaligenes eutrophus* CH34, *Alcaligenes denitrificans*, and *P. paucimobilis* CD (Dressler, Kues, Nies, and Friedrich, 1991). *A. denitrificans* AH can tolerate a high copper concentration (MIC = 4 mM $CuSO_4$). Some of the strains also have various levels of resistance to other metal ions. Churchill, Walters, and Churchill (1995) report that semi-intact *E. coli* cells are very efficient at binding copper.

In a liquid medium with glucose as a carbon source, a polysaccharide-producing *Bacillus circulans* generated a biomass concentration of 1.48 to 1.52 g dry weight/L, which removed 80% of copper from a solution containing 495 ppm copper (Sahoo, Kar, and Das, 1992). The pH affected the uptake. A *Zoogloea* sp., a *Bacillus* sp., and *Rhizopus arrhizus* can accumulate copper, and a *Bacillus* sp. and the yeast *Saccharomyces cerevisiae* can accumulate zinc (Gadd, 1992). A maximum copper accumulation of 0.6 μmol/mg was obtained with dry cell walls of the alga *Vaucheria* (Christ, Oberholser, Shank, and Nguyen, 1980). Isolated cell walls of the fungus *Cunninghamella blakesleana* achieved a maximum copper adsorption of 0.86 μmol cell wall/mg (Venkateswerlu and Stotzky, 1989). Walker, Flemming, Ferris, Beveridge, and Bailey (1989) report uptake of 0.53 μmol Cu/mg dry cell wall for isolated cell walls of *Bacillus subtilis*.

At an equilibrium concentration of 1 μM, there was little difference between the most and least efficient bacteria, *B. subtilis* and *B. cereus*, for copper sorption (Mullen, Wolf, Ferris, Beveridge, Flemming, and Bailey, 1989). *Pseudomonas aeruginosa* was the most efficient at higher concentrations. While there was little difference between living cells of Gram-negative and Gram-positive bacteria, cell walls of *B. subtilis* and *B. licheniformis* bound 28 to 33 times more Cu^{2+} than did *E. coli* envelopes, and the cell walls were more efficient than the whole cells (Beveridge and Fyfe, 1985). Copper binding is basically controlled by interaction with negatively charged carboxyl and phosphoryl groups on the cell surface, with little or no precipitation.

Penicillium depends upon the parameters of pH, temperature, biomass, and initial metal concentration for biosorption of copper (Mishra and Chaudhury, 1996). Higher pH has a positive effect, while higher temperature has a negative effect on the process. Copper adsorption decreases in the presence of nickel.

Biosorption of Cu(II) by untreated and acid-treated fungal biomass has been studied (Huang and Huang, 1996). Acid washing of *Aspergillus oryzae* mycelia, but not that of *Rhizopus oryzae*, can strongly enhance adsorption capacity of Cu(II). Acid washing is both a pretreatment step and a regeneration step in the heavy metal removal. Cultivation of *A. oryzae* in pellet form is an effective means of immobilizing the mycelia. A high yield of uniformly sized particles (2 to 3 mm in diameter) for solid–liquid separation can be produced. A pellet column can completely remove metals before break point, after which there is still significant removal of Cu(II) over a long period of time. This may be due to intracellular uptake.

Methanotrophs are relatively resistant to copper and zinc (Bowman, Sly, and Hayward, 1990).

3.4 INTERMEDIATE METABOLITES AND END PRODUCTS OF BIODEGRADATION

In soil, organic chemicals are subject to alteration by biochemical reactions that are catalyzed by enzymes from a wide range of organisms (Kaufman and Doyle, 1978). In general, metabolites arising from these microbial reactions are usually nontoxic, polar molecules that exhibit little ability to accumulate in food chains. However, the breakdown products of many chemicals can be toxic; sometimes they are even more toxic than the parent compound. It is even possible, through microbial transformations, to convert a nontoxic parent compound into a toxic product (Alexander, 1994). This is called activation. Some partial degradation products might be more toxic than the parent compounds (Lee, Wilson, and Ward,

1987). Transformation of a toxic organic solute is no assurance that it has been converted to harmless or even less hazardous products (Mackay, Roberts, and Cherry, 1985).

Some of the mechanisms of activation are dehalogenation, nitrosamine formation, epoxidation, conversion of phosphorothionate to phosphate, metabolism of phenoxyalkanoic acids, oxidation of thioethers, and hydrolysis of esters (Alexander, 1994). The problem of toxic activation products may be avoided if the microorganisms employ another pathway to produce a different metabolite that is not toxic or cannot be activated. This is called defusing. Understanding the pathways and products of biodegradation can help develop approaches to prevent toxic materials from being released in the environment.

For efficient biodegradation, it is important to have a provision for the removal of toxic wastes and by-products (Texas Research Institute, 1982). This can be accomplished in two ways: (1) having a diverse microbial population so the by-products are consumed and (2) creating a flow through the system to remove toxins. Oxidizing agents can be used to degrade organic constituents in soil systems, although they may themselves be toxic to microorganisms or may cause the production of more-toxic or more-mobile oxidation products (Sims and Bass, 1984).

The end products of anaerobic degradation are reduced compounds, some of which are toxic to microorganisms and plants. As oxygen in the soil is depleted, microbial reactions produce malodorous compounds, such as amines, mercaptans, and H_2S, which can be phytotoxic. Acetate and formate are important intermediate products in methanogenic fermentation (McInerney and Bryant, 1981; McInerney, Bryant, Hespell, and Costerton, 1981). The fatty acids produced from anaerobic degradation of alkanes can be fermented under anaerobic conditions (Rosenfeld, 1947). Alkanes shorter than C_9 can be degraded anaerobically, while some of the longer-chain alkanes may be transformed into naphthalenes and other PAHs.

The anaerobic transformation of benzene might be a fermentation in which the substrates are partially reduced and partially oxidized, with reduction resulting in production of saturated alicyclic rings (Grbic-Galic and Vogel, 1987). Benzene degradation by mixed methanogenic cultures may lead to carbon dioxide and methane with intermediates of phenol, cyclohexane, and propanoic acid. Benzene and toluene may be transformed by methanogenic cultures acclimated to lignin-derived aromatic acids under strictly anaerobic conditions with several intermediates, including demethylation products, aromatic alcohols, aldehydes and acids, cresols, phenol, alicyclic rings, and aliphatic acids, which are ultimately converted to carbon dioxide and methane (Grbic-Galic and Vogel, 1987). *Azoarcus tolulyticus* Tol-4 can grow anaerobically on toluene, with benzoate as an intermediate and two metabolites, E-phenylitaconic acid and benzylsuccinic acid, accumulating as a result (Chee-Sanford, Frost, Fries, Zhou, and Tiedje, 1996).

Carbonic acid, organic acid intermediates, and nitrate and sulfate (most important for pH < 5), may accumulate during aerobic degradation of organic molecules (Zitrides, 1983). This can lower the soil pH and inhibit biological activity. The acid conditions can be controlled with reinoculation, use of chemical pH control agents, such as lime, or both. In fact, liming has been found to favor the biodegradation of oil (Dibble and Bartha, 1979a).

It was found that C_5 to C_9 alkanes were not toxic to a population of bacteria, but that the alcohols of these hydrocarbons were inhibitory (Bartha and Atlas, 1977). Toxic products formed during microbial hydrocarbon oxidation include the C_5 to C_9 primary alcohols (Liu, 1973); fatty acids, if β oxidation is lacking (Atlas and Bartha, 1973b); and oxidation products of aromatic hydrocarbons (Calder and Lader, 1976). The biodegradation of aromatic hydrocarbons yields phenolics and benzoic acid intermediates (Bartha and Atlas, 1977). Oxidation of alkanes and other aliphatics can generate fatty acids (Atlas, 1981; Hornick, Fisher, and Paolini, 1983). Some fatty acids are toxic and can accumulate during hydrocarbon biodegradation. Fatty acids produced from oxidation of single-branched alkanes may be directly incorporated into membrane lipids (Hou, 1982), instead of going through the β-oxidation pathway (Dunlap and Perry, 1967).

Since intermediary metabolites must also be removed for complete oil cleanup, nonhydrocarbon-utilizing microorganisms, such as fatty acid metabolizers, would also be required in the mixture (Atlas, 1977). Certain species of *Pseudomonas* oxidize the fatty acids formed from heptane and hexane by classical β oxidation (Heringa, Huybregtse, and van der Linden, 1961). Decarboxylation of fatty acids does not occur in these particular systems. *Arthrobacter simplex* is able to degrade some fatty acids by oxidative decarboxylation with the evolution of CO_2 (Ratledge, 1978).

Soil organic matter sorbs contaminants and can slow biodegradation of PAHs that are otherwise readily metabolized (Manilal and Alexander, 1991). Fulvic acid decreases mineralization of pyrene, apparently due to toxicity to the microbes and possible sorption of the compound making it less bioavailable (Grosser, Warshawsky, and Kinkle, 1994). Contaminants or their metabolites might also form complexes with the humic fraction of the soil or by attaching to reactive sites on the surfaces of organic colloids, creating new molecules that are not easily degraded (Alexander, 1994). The longer some compounds remain in soil, the more resistant they become to desorption and biodegradation (Alexander, 1994). Sorbed molecules may become available to microbes as biodegradation of soluble material causes the equilibrium to draw more of the sorbed form into solution. Microorganisms may also excrete metabolites that desorb the compound.

Fungi predominate under acidic conditions (pH < 7), which can result from contamination of soil by acidic hazardous wastes (Sims and Bass, 1984; JRB Associates, Inc., 1984). These organisms may transform aromatic hydrocarbons by means of oxygenases into arene oxides, the mutagenic forms of PAHs (Baver, Gardner, and Gardner, 1972). Fungi produce cytochrome P-450 monooxygenases to incorporate one oxygen atom into the substrate to form arene oxides, which is followed by enzymatic addition of water to produce *trans*-dihydrodiols and phenols (Cerniglia, 1984; Wilson and Jones, 1993). The products of fungal oxidation of PAHs are both nontoxic metabolites and metabolic products that are carcinogenic, tumorigenic, or mutagenic in higher organisms (Cerniglia, 1984; 1992). Higher organisms (above fungi) do not possess the oxygenases of aerobic bacteria and, like fungi, form *trans* aromatic diols, which tend to polymerize (Baver, Gardner, and Gardner, 1972).

The nonligninolytic fungus *Cunninghamella elegans* can oxidize numerous PAHs to *trans*-dihydrodiols, phenols, quinones, tetralones, and conjugates of these primary metabolites (Cerniglia, 1992). Most of the fungal transformation products formed by this fungus are less mutagenic than the parent compound. Several white-rot fungi (*Phanerochaete chrysosporium* BKM-F-1767, *Trametes versicolor* Paprican 52, and *Bjerkandera adusta* CBS 595.78) attack benz(a)pyrene, with no accumulation of quinones (Field, de Jong, Feijoo-Costa, and de Bont, 1992). *Bjerkandera* and *Phanerochaete* convert anthracene to anthraquinone, which is a dead-end metabolite. *Trametes* strains are able to degrade anthracene, as well as anthraquinone. *P. laevis* HHB-1625 is able to degrade the quinone intermediates of PAH metabolism even faster and more extensively than *P. chrysosporium* (Bogan and Lamar, 1996).

Candida lipolytica oxidizes naphthalene to 1-naphthol, 2-naphthol, 4-hydroxy-1-tetralone, and *trans*-1,2-dihydroxy-1,2-dihydronaphthalene (Cerniglia and Crow, 1981). The primary metabolite is 1-naphthol. *Cunninghamella bainieri* oxidizes naphthalene through *trans*-1,2-dihydroxy-1,2-dihydronaphthalene (Gibson and Mahadevan, 1975). *C. elegans* oxidizes naphthalene by a sequence of reactions resulting in six metabolites: 1-naphthol, 4-hydroxy-1-tetralone, 1,4-naphthoquinone, 1,2-naphthoquinone, 2-naphthol, and *trans*-1,2-dihydroxy-1,2-dihydronaphthalene. A *Nocardia* sp. can biotransform naphthalene and substituted naphthalenes to their corresponding diols (Hou, 1982). Naphthalene is oxidized by species of *Cunninghamella, Syncephalastrum*, and *Mucor* to 2-naphthol, 4-hydroxy-1-tetralone, *trans*-naphthalene dihydrodiol, 1,2-naphthoquinone, 1,4-naphthoquinone, and predominantly 1-naphthol (Cerniglia, Hebert, Szaniszlo, and Gibson, 1978). *Neurospora crassa, Claviceps paspali*, and *Psilocybe* strains also show a similar degradative capacity. 1-Naphthol is also produced by the filamentous fungus, *Cunninghamella elegans*, from oxidation of naphthalene (Dagley, 1984). *Candida lipolytica* oxidizes biphenyl to metabolites with the *trans*-configuration (Hou, 1982): 2-, 3-, and 4-hydroxybiphenyl, 4,4'-dihydroxybiphenyl, and 3-methoxy-4-hydroxybiphenyl, with 4-hydroxybiphenyl as the main metabolite (Cerniglia and Crow, 1981). *Candida guilliermondii, C. tropicalis, C. maltosa,* and *Debaryomyces hansenii* also metabolize this compound to *trans*-configuration metabolites. Yeasts primarily oxidize biphenyl in the *para* position to form 4-hydroxybiphenyl.

1-Naphthol is a major product of cyanobacterial metabolism of naphthalene, as well as being a fungal metabolite (Cerniglia, Gibson, and Van Baalen, 1980). The fate of 1-naphthol in natural environments is of particular interest because it appears to be very toxic (Fewson, 1981). This compound can be totally degraded by some microorganisms (Bollag, Czaplicki, and Minard, 1975), including a soil pseudomonad, which can grow on it as a sole source of carbon and energy (Walker, Janes, Spokes, and van Berkum, 1975). 1-Naphthol can also undergo other transformations, such as polymerization by the extracellular enzyme laccase (Fewson, 1981).

There are some beneficial products formed by fungi. These organisms have the capacity to form glucuronide and sulfate conjugates of phenolic PAHs, which may be important in the detoxification and elimination of PAHs (Cerniglia, 1984). The secondary metabolite, veratryl alcohol, produced by *P. chrysosporium* enhances oxidation of pyrene (Cancel, Orth, and Tien, 1993).

Whereas fungi convert aromatic hydrocarbons to *trans*-diols, with arene oxides (epoxides) as intermediates, bacteria oxidize aromatic hydrocarbons to *cis*-dihydrodiols (Dagley, 1984). Bacteria, which grow better at a neutral or slightly basic pH, would carry out the dioxygenation of the aromatic nucleus to form a *cis*-glycol as the first stable intermediate (Baver, Gardner, and Gardner, 1972). The products of fungal metabolism are often recognized carcinogens, a point that supports combining the fungi with bacteria for complete degradation. Selection of bacteria to dominate the bioremediation program may avoid the formation of mutagens, and pH may serve as an important engineering tool to direct the pathway of PAH degradation (Cerniglia, Hebert, Dodge, Szaniszlo, and Gibson, 1979).

Under aerobic conditions, microbial oxidation results in complete mineralization of organic substrates to end products of water, carbon dioxide, and biomass (Farmer, Chen, Kopchynski, and Maier, 1995). In addition, there are various hydroperoxides, alcohols, phenols, carbonyls, aldehydes, ketones, and esters that can result from incomplete oxidation (ZoBell, 1973). When optimum environmental conditions are maintained (appropriate oxygen, nutrients, moisture, pH, and temperature), there will be maximum treatment efficiency to help promote complete mineralization (Saberiyan, Wilson, Roe, Andrilenas, Esler, Kise, and Reith, 1994).

The biodegradation of higher-molecular-weight hydrocarbons involves many intermediates, some of which may accumulate to inhibitory levels (Bartha and Atlas, 1977). The higher PAHs produce higher mineralization and lower biomass, i.e., the four-ring as opposed to two- or three-ring PAHs. Growth of *Rhodococcus* and *Pseudomonas* strains on naphthalene, fluorene, phenanthrene, anthracene, fluoranthene, and pyrene yielded high mineralization (56 to 77% of the carbon) and good production of biomass (16 to 35% of carbon) and limited but significant accumulation of metabolites (5% to 23% of carbon) (Bouchez, Blanchet, and Vandecasteele, 1995; 1996). Guerin and Jones (1988) found that increasing the rate of uptake of phenanthrene by *Mycobacterium* sp. led to more biomass formation and less metabolite accumulation. Degradation of naphthalene by *Pseudomonas* involves the formation of 1,2-dihydroxynaphthalene as the first stable metabolite (Davies and Evans, 1976).

Dead-end metabolites from problem compounds can accumulate intracellularly (Leisinger, 1983). Even genetically constructing an organism by assembling degradative capabilities from different strains with different degradative pathways may not overcome this limitation (Leisinger, 1983). Cell-free extracts could be used instead of whole living cells. When naphthalene is broken down by cell-free extracts, such as one containing naphthalene oxygenase purified from cells of *Corynebacterium renale*, *cis*-1,2-dihydroxy-1,2-dihydronaphthalene is formed as the predominant metabolite from naphthalene (Dua and Meera, 1981).

Some microorganisms have the capability of cometabolically attacking certain contaminants by enzyme induction. When *P. paucimobilis* strain EPA 505 was grown on fluoranthene as a sole source of carbon and energy, it was able to degrade 25.2% of benz(a)pyrene to a major, highly polar metabolite (Siddiqi, Ye, Elmarakby, Kumar, and Sikka, 1994).

Since cometabolism rarely results in the complete oxidation of xenobiotics, it may allow accumulation of transformation products, which may be more or less toxic than the original substance (de Klerk and van der Linden, 1974; Perry, 1979). Incompletely degraded compounds include saturated hydrocarbons, halogenated hydrocarbons, many pesticides, and single-ringed PAHs (Horvath and Alexander, 1970; Horvath, 1972; Alexander, 1977). Although the compounds may not be completely mineralized, their toxicity might be removed by the partial degradation (Horvath, 1972).

Various microbes can be used in combination with chemical analogs of specific organic compounds to promote co-oxidation of the latter (Horvath, 1972). Pyrene, 3,4-benzopyrene, 1,2-benzanthracene, and 1,3,5,6-dibenzanthracene can be cometabolized by a mixed culture of flavobacteria and pseudomonads in the presence of either naphthalene or phenanthrene (Cerniglia, 1984). A *Nocardia* sp. has been identified as the primary cyclohexane utilizer, but growth occurs only in the presence of an unidentified pseudomonad that provides biotin and possibly other growth factors (Sterling, Watkinson, and Higgins, 1977).

By-products of the cometabolic process might be completely degraded through contribution of cometabolic reactions from the entire microbial community (Horvath, 1972; de Klerk and van der Linden, 1974; Perry, 1979; Alexander, 1994). For example, one organism may cometabolize a compound to an intermediate it can no longer utilize, but which another organism can then mineralize (Alexander, 1994). The first species might convert the compound to a toxic substance that is utilized or detoxified by another species (Pfennig and Biebl, 1976). Synergism may also allow one microbe to grow on the initial compound with incomplete degradation, which another organism can then continue to degrade (Johanides and Hrsak, 1976). Mutualistic relationships may involve relief of substrate inhibition (Osman, Bull, and Slater, 1976). Commensalism may be very widespread in nature with natural mixed populations employing each other's metabolic intermediates as growth substrates (Donoghue, Griffin, Norris, and Trudgill, 1976) (see Section 5.2.2.3). Many genera of bacteria, fungi, and actinomycetes can participate in the process (Horvath and Alexander, 1970; Alexander, 1977).

These interactions are potentially important in the degradation of complex molecules (Hornick, Fisher, and Paolini, 1983). Anaerobic degradation generally requires more than one species of anaerobes to mineralize a substrate completely and may even require a final aerobic transformation (Alexander, 1994). Mixed cultures are generally required for complete biodegradation of most surfactants and their metabolites, which can sometimes be toxic (van Ginkel, 1996). Complete oxidation is more likely when a diverse mixture of microbes is available (Texas Research Institute, 1982). Primary and intermediate oxidation products are more likely to accumulate for a pure culture than a mixed culture.

There can be inhibition even when cometabolism is taking place, because of the interactions of PAHs at several levels (Bouchez, Blanchet, and Vandecasteele, 1995a). This could involve competition at the active site of enzymes or at the level of enzyme induction. It could be due to accumulation of toxic end products.

Table 3.21 shows some of the end products formed from the microbial oxidation of a variety of hydrocarbons by specific microorganisms.

It is not completely known what the effect of microbial degradation of complex hydrocarbons may be on the environment (Texas Research Institute, 1982). It has been established that complete oxidation of many carcinogenic hydrocarbons is not required to render them noncarcinogenic. However, an EPA study reported that, with moderate temperatures and the presence of inorganic nitrogen and phosphorus, naturally occurring freshwater microorganisms are able to form mutagenic biodegradation products from crude oil that are bactericidal to *Escherichia coli* K-12 (Morrison and Cummings, 1982). The public health significance of this finding is undetermined.

Table 3.21 Products/Metabolites Formed from the Oxidation of Petroleum Hydrocarbons by Various Microorganisms

Substrate	Microbe	Product
Alkanes	*Pseudomonas* spp.[h]	Pristane, phytane, sterane, 17 α (H), 21 β (H)-hopane series
	Pseudomonas oleovorans[k]	Carboxylic acids
C_{12} to C_{16} (multiple-branched, e.g., pristane)	*Corynebacterium*[a] (*Brevibacterium erythrogenes, Corynebacterium* sp.)[a]	Ketones, esters of aliphatic acids Succinyl-CoA
C_{10} to C_{14}	*Corynebacterium*[i]	Fatty acids (metabolizable by β oxidation)
Alkenes (C_6 to C_{12})	*Pseudomonas oleovorans*[k]	1,2-Epoxides
Anthracene	Fungi[q]	Anthracene *trans*-1,2-dihydrodiol, 1-anthrol, anthraquinone, phthalate, glucuronide, sulfate and xyloside conjugates
	Rhizoctonia solani[t]	*trans*-1,2-dihydroxy-1,2-dihydroanthracene
	Bjerkandera, Phanerochaete[v]	Anthraquinone
Benzene	Anaerobes[u]	Saturated alicyclic rings
Benz(a)anthracene	*Cunninghamella elegans*[c]	*trans*-benzo(a)anthracene 3,4-, 8,9-, and 10,11-dihydrodiols
	Beijerinck (mutant, acclimated)[g]	Dihydrols, primarily *cis*-1,1-dihydroxy-1,2-dihydrobenzo(a)anthracene
	Fungi[q]	Benz(a)anthracene *trans*-3,4-dihydrodiol, benz(a)anthracene *trans*-8,9-dihydrodiol, benz(a)anthracene *trans*-10,11-dihydrodiol, phenolic and tetrahydroxy derivatives of benz(a)anthracene, glucuronide and sulfate conjugates

Table 3.21 (continued) Products/Metabolites Formed from the Oxidation of Petroleum Hydrocarbons by Various Microorganisms

Substrate	Microbe	Product
Benzo(a)pyrene	*Cunninghamella elegans*[b]	*trans*-7,8-Dihydroxy-7,8-dihydro-benzo(a)pyrene; *trans*-9,10-dihydroxy-9,10-dihydro-benzo(a)pyrene; benzo(a)pyrene-3- and 9-phenols; and benzo(a)pyrene-1,6- and 3,6-quinones. benzo(a)pyrene-7,8- and benzo(a)pyrene-9,10-diols are further oxidized to metabolites known to be carcinogenic, tumorigenic, and mutagenic to experimental animals
	Candida lipolytica[f]	3-Hydroxybenzo(a)pyrene, 9-hydroxybenzo(a)pyrene
	Beijerinckia (mutant, acclimated)[g]	Vicinal dihydrodiols, mainly *cis*-9,10-dihydroxy-9,10-dihydrobenzo(a)pyrene
	Neurospora crassa[o]	Mainly 3-hydroxybenzo(a)pyrene
	Saccharomyces cerevisiae[p]	7,8-dihydroxy-7,8-dihydrobenzo(a)pyrene, 9-hydroxybenzo(a)pyrene, 3-hydroxybenzo(a)pyrene
	Selenastrum capricornutum[r]	*cis*-11,12-Dihydroxy-11,12-dihydrobenzo(a)pyrene (benzo(a)pyrene *cis*-11,12-dihydrodiol)
	Fungi[q]	Benzo(a)pyrene *trans*-4,5-dihydrodiol, benzo(a)pyrene *trans*-7,8-dihydrodiol, benzo(a)pyrene *trans*-9,10-dihydrodiol, benzo(a)pyrene-1,6-quinone, benzo(a)pyrene-3,6-quinone, benzo(a)pyrene-6,12-quinone, 3-hydroxybenzo(a)pyrene, 9-hydroxybenzo(a)pyrene, 7β, 8α, 9α, and 10β-tetrahydrobenzo(a)pyrenes, 7β, 8α, 9β, and 10α-tetrahydroxy-7,8,9,10-tetrahydrobenzo(a)pyrenes, benzo(a)pyrene 7,8-dihydrodiol-9,10-epoxide, glucuronide and sulfate conjugates
Biphenyl	*Candida lipolytica*[a,f,i]	2-, 3-, and 4-Hydroxybiphenyl, 4,4′-dihydroxybiphenyl, 3-methoxy-4-hydroxybiphenyl, 4-hydroxybiphenyl
	(*Candida guilliermondii, C. tropicalis, C. maltosa, Debaryomyces hansenii*)[f]	*trans* Metabolites
	Yeasts[f]	4-Hydroxybiphenyl
	Cunninghamella elegans[m]	2-, 3-, 4-Hydroxybiphenyl and 4,4′-dihydroxybiphenyl, glucuronides, sulphates
n-Butane	*Mycobacterium smegmatis*[i]	2-Butanone
Decane	*Corynebacterium*[a]	1-Decanol, 1,10-decanediol
Ethane	*Methylosinus trichosporium*[i]	Acetone
Fluoranthene	Fungi[q]	Fluoranthene *trans*-2,3-dihydrodiol, 8- and 9-hydroxyfluoranthene, *trans*-2,3-dihydrodiols, glucoside conjugates
n-Heptane	*Pseudomonas aeruginosa*[i]	*n*-Hexanoic acid
Hexadecane	*Micrococcus cerificans*[a]	Cetyl palmitate
	Arthrobacter sp.[a]	Ketones
n-Hexane	*Mycobacterium smegmatis*[i]	2-hexanol, 2-hexanone
Kerosene, heavy naphthas, aromatic naphthas, petroleum pitches, tars, and asphalts	*Fusarium moniliforme*[i]	Giberellin
Kerosene	55 strains[i]	Amino acids
2-Methylhexane	*Pseudomonas aeruginosa*[i]	2-Methylhexanoic acid, 5-methylhexanoic acid
Naphthalene	*Nocardia* sp.[a,i]	Diols of naphthalene
	Cunninghamella elegans[i]	1-Naphthol, 2-napthtol *trans*-1,2-dihydroxy-1,2-dihydronaphthalene

Table 3.21 (continued) Products/Metabolites Formed from the Oxidation of Petroleum Hydrocarbons by Various Microorganisms

Substrate	Microbe	Product
	Oscillatoria sp. strain JCM[h]	1-Naphthol, *cis*-1,2-dihydroxy-1, 2-dihydronaphthalene, and 4-hydroxy-1-tetralone
	Candida lipolytica[b]	1-Naphthol, 2-naphthol, 4-hydroxy-1-tetralone, *trans*-1, 2-dihydroxy-1,2-dihydronaphthalene
	Candida elegans[c]	1-Naphthol, 4-hydroxy-1-tetralone, 1,4-naphthoquinone, 1,2-naphthoquinone, 2-naphthol, *trans*-1,2-dihydroxy-1,2-dihydronaphthalene
	Cunninghamella elegans[w]	1-Naphthol
	(*Cunninghamella elegans, C. echinulata, C. japonica, Syncephalastrum* sp., *S. racemosum, Mucor* sp., *M. hiemalis*)[j]	Mainly 1-naphthol; also 2-naphthol, 4-hydroxy-1-tetralone, *trans*-naphthalene dihydrodiol, 1,2-naphthoquinone, 1,4-naphthoquinone
	Methylococcus trichosporium OB3b[s]	1- and 2-Naphthol
	Cyanobacteria[x]	1-Naphthol
	Lake sediments[l]	Naph-*cis*-1,2-dihydroxy-1,2-dihydronaphthalene, 1-naphthol, salicylic acid, catechol
	Fungi[q]	1-Naphthol, 2-naphthol, naphthalene *trans*-1,2-dihydrodiol, 4-hydroxy-1-tetralone, 1,2-naphthoquinone, 1,4-naphthoquinone, sulfate and glucuronide conjugates
	Cell-free extracts of *Corynebacterium renale*[y]	*cis*-1,2-dihydroxy-1,2-dihydronaphthalene
Octadecane	*Micrococcus cerificans*[i]	Octadecyl stearate
n-Octane	*Pseudomonas*[i]	*n*-Octanol
n-Pentane	*Mycobacterium smegmatis*[i]	2-Pentanone
n-Propane	*Mycobacterium smegmatis*[i]	Acetone
Phenanthrene	(Not specified)[a]	1-hydroxy-2-naphthoic acid[a]
	Fungi[q]	Phenanthrene *trans*-1,2-dihydrodiol, phenanthrene *trans*-3,4-dihydrodiol, phenanthrene *trans*-9,10-dihydrodiol, glucoside conjugate of 1-phenanthrol, 3-, 4-, and 9-hydroxyphenanthrene, 2,2-diphenic acid
	Agmenellum quadruplicatum PR-6[q]	*trans*-9,10-Dihydroxy-9,10-dihydrophenanthrene, 1-methoxyphenanthrene
Pyrene	Fungi[q]	1-Hydroxypyrene, 1,6-pyrenequinone, 1,8-pyrenequinone, glucoside conjugates
Tetradecane	*Micrococcus cerificans*[i]	Myristyl palmitate
Toluene	(*Pseudomonas* sp., *Achromobacter* sp.)[n]	3-Methylcatechol
	Azoarcus tolulyticus Tol-4[z]	E-phenylitaconic acid, benzlsuccinic acid
Tridecane	*Pseudomonas aeruginosa*[a]	Tridecane-1-ol, undecan-1-ol
n-Undecane	*Mycobacterium* sp.[i]	Undecanoic acid, 1,11 undecandioic acid

References

a = (Hou, 1982)
b = (Cerniglia and Gibson, 1980)
c = (Dodge and Gibson, 1980)
d = (Cerniglia, Wyss, and Van Baalen, 1980)
e = (Lawlor, Shiaris, and Jambard-Sweet, 1986)
f = (Cerniglia and Crow, 1981)
g = (Gibson and Mahadevan, 1975)
h = (Gibson, Koch, and Kallio, 1968)
i = (Zajic, 1964)
j = (Cerniglia, Hebert, Szaniszlo, and Gibson, 1978)
k = (Roberts, Koff, and Karr, 1988)
l = (Heitkamp, Freeman, and Cerniglia, 1987)
m = (Dodge, Cerniglia, and Gibson, 1979)
n = (Claus and Walker, 1964)
o = (Lin and Kapoor, 1979)
p = (Wiseman, Lim, and Woods, 1978)
q = (As cited by Cerniglia, 1992)
r = (Warshawsky, Radike, Jayasimhulu, and Cody, 1988)
s = (Brusseau, Hsien-Chyang, Hanson, and Wackett, 1990)
t = (Sutherland, Selby, Freeman, Fu, Miller, and Cerniglia, 1992)
u = (Grbic-Galic and Vogel, 1987)
v = (Field, de Jong, Feijoo-Costa, and de Bont, 1982)
w = (Dagley, 1984)
x = (Cerniglia, Gibson, and Van Baalen, 1980)
y = (Dua and Meera, 1981)
z = (Chee-Sanford, Frost, Fries, Zhou, and Tiedje, 1996)

Section 4

Factors Affecting Biodegradation in Soil-Water Systems

The fate of a contaminant in the environment is determined by characteristics of the pollutants, the microorganisms, and the environment (Pfaender, Shimp, Palumbo, and Bartholomew, 1985; Madsen, 1991; Tiedje, 1993). The concentration, distribution, function, and structure of the contaminants; the physiology and genetics of the indigenous or introduced microorganisms; and the various soil factors, all result in complex interactions in the subsurface. If one of these critical components is suboptimal for conversion of organic contaminants, biodegradation will be slow or may not take place (Turco and Sadowsky, 1995).

The rate of biodegradation in soils is a function of the availability of the chemicals to the microorganisms that can degrade them, the quantity of these microorganisms, and the activity level of the organisms (Sepic, Leskovsek, and Trier, 1995). *n*-Alkane biodegradation rates may be affected by the fermentation conditions (agitation, aeration, etc.) and by the strain of microorganisms, while the behavior pattern of *n*-alkane degradation is essentially linked to the substrate characteristics (molecular structure, molecular weight, and density) (Setti, Pifferi, and Lanzarini, 1995). Contaminant properties and site characteristics can often provide a general indication of the applicability of the treatment technologies available for remediating the particular contamination incident (Ram, Bass, Falotico, and Leahy, 1993). Significant waste factors that will affect its biodegradation include (Parr, Sikora, and Burge, 1983):

Chemical composition of the waste
Its physical state (i.e., liquid, slurry, sludge)
Its carbon:nitrogen ratio
Water content and solubility
Chemical reactivity and dissolution effects on soil organic matter
Volatility
pH
Biochemical oxygen demand (BOD)
Chemical oxygen demand (COD)

If the COD at a site is more than an order of magnitude above the BOD, bioreclamation may not be the most cost-effective answer (Biosystems, Inc., 1986).

The behavior of toxic pollutants in the environment also depends upon a variety of processes (Ghisalba, 1983; Josephson, 1983; Pettyjohn and Hounslow, 1983; Bitton and Gerba, 1985):

1. Chemical processes, e.g.,
   Hydrolysis
   Photolysis
   Oxidation
   Reduction
   Hydration
2. Physical or transport processes, e.g.,
   Advection
   Dispersion and diffusion
   Sorption
   Volatilization
   Solubilization
   Viscosity
   Density
   Dilution

3. Biological processes, e.g.,
   Bioaccumulation
   Biotransformation
   Biodegradation
   Toxicity
4. Abiotic processes
5. Combined environmental factors

Crude petroleum and many of the products refined from petroleum contain thousands of hydrocarbons and related compounds (Cooney, Silver, and Beck, 1985). There are major compositional variations between different crude and refined oils (Atlas, 1978). Some oils contain toxic hydrocarbons, which may prevent or delay microbial attack; some refined oils have additives, such as lead, which can inhibit microbial degradation of polluting hydrocarbons. Oils contain varying proportions of paraffinic, aromatic, and asphaltic hydrocarbons. Both the rate and extent of biodegradation are dependent upon the relative proportions of these classes of hydrocarbons.

Under favorable conditions, microorganisms will degrade 30 to 50% of a crude oil residue, but degradation is never complete and does not affect the different hydrocarbon families in the same manner (Solanas, Pares, Bayona, and Albaiges, 1984). Many synthetic organic compounds (Pettyjohn and Hounslow, 1983) and many complex hydrocarbon structures in petroleum (Atlas, 1978) are not easily broken down by microbial action. However, with favorable conditions and the proper organisms, virtually all kinds of hydrocarbons — straight chain, branched chain, cyclic, simple aromatic, polynuclear aromatic, asphaltic — have been found to undergo oxidation (Texas Research Institute, Inc., 1982).

Each organic compound has unique characteristics that dictate which of the above mechanisms or combination of mechanisms controls its movement and degradability (Josephson, 1983). For a successful bioremediation program the natural heterogeneity of the soil systems must be overcome, the rate-limiting factors must be removed, and the microbial population promoted to remove the organic contaminants (Turco and Sadowsky, 1995).

The fate of toxic pollutants in soils is determined by a variety of chemical, physical, biological, and environmental processes that interact in a complex manner (Pfaender, Shimp, Palumbo, and Bartholomew, 1985). Polluted groundwater can also be a source of contamination of the unsaturated zone as water tables rise. It will, therefore, be included where appropriate in this discussion.

4.1 CHEMICAL AND PHYSICAL FACTORS

4.1.1 CHEMICAL SOLUBILITY

Organic compounds differ widely in their solubility, from infinitely miscible polar compounds, such as methanol, to extremely low solubility nonpolar compounds, such as polynuclear aromatic hydrocarbons (PAHs) (Horvath, 1982).

Many synthetic chemicals have low water solubilities (Stucki and Alexander, 1987). The availability of a compound to an organism will dictate its biodegradability (U.S. EPA, 1985a). Compounds with greater aqueous solubilities are generally more available to degradative enzymes. An example is *cis*-1,2-dichloroethylene, which is preferentially degraded relative to *trans*-1,2-dichloroethylene. This is probably due to "*cis*" being more polar than "*trans*" and, therefore, more water soluble. Surfactants can increase the solubility and, thus, the degradability of compounds (see Section 5.3.1).

Degradation of PAHs in general is limited because of their lower solubility (Wiesel, Wuebker, and Rehm, 1993). The order of degradation of PAHs is related to their water solubility, which is inversely related to ring condensation. Tetracyclic compounds are less available than di- and tricyclic compounds. Table 4.1 lists a number of common organic contaminants and relates their biodegradability to their solubility in water (Brubaker and O'Neill, 1982). The solubility represents the maximum concentration of the compound that will be dissolved in water under equilibrium conditions (Eckenfelder and Norris, 1993).

Mass transfer from the solid phase to the liquid phase is rate limiting for growth (Volkering, Breure, Sterkenburg, and van Andel, 1992). This could explain the linear growth of bacteria and yeasts observed on slightly soluble substrates. Higher-molecular-weight PAHs may appear to be more recalcitrant than lower-weight compounds, but this effect seems to be due to their low concentrations in the aqueous phase (Mihelcic, Lueking, Mitzell, and Stapleton, 1993). After the degradation of PAHs in the soil

Table 4.1 Solubility and Biodegradability of Some Common Organic Contaminants

Contaminant	Solubility in Water[a]	Biodegradability
Acetone	Miscible	++
Aniline	35 g/L	++
Anthracene	1 mg/L	+
Benzene	320 mg/L	++
	1791 mg/L[b]	
Benzo(a)pyrene[e]	3.8 g/L	
Butanol	77,000 mg/L[b]	
o-Cresol	31 g/L	++
Isopropanol	Miscible	++
Methanol	Miscible	++
Methylene chloride	20 g/L	(?)
Methylethylketone	370 g/L	++
Naphthalene	29 µg/mL[c]	
	31 µg/mL[b,d]	
Phenanthrene	0.9 mg/L[b]	
Phenol	82 g/L	+
	87,000 mg/L[b]	
Pyrene	0.2 mg/L	+
	0.1 mg/L[b]	
Toluene	470 mg/L	++

Note: ++ = Readily biodegradable; + = slow biodegradability; ? = materials of uncertain biodegradability.

[a] The actual solubility will be influenced greatly by other chemicals in the water.
[b] From Eckenfelder and Norris, 1993; at 25°C.
[c] From Thomas, Yordy, Amador, and Alexander, 1986.
[d] From Yalkowsky, Valvani, and Mackay, 1983.
[e] From Mackay and Shiu, 1977.

Source: From Brubaker, G.R. and O'Neill, E. in *Proc. 5th Natl. Symp. on Aquifer Restoration and Groundwater Monitoring,* Ground Water Pub. Co., Westerville, OH. 1982. With permission.

reaches a certain amount, there is little change in PAH concentration. The problem of mass transfer will be a limiting factor in the field.

It has been postulated that microorganisms might be able to take up a compound either through direct contact of the cells with hydrocarbon droplets larger than the cells, from the aqueous phase of droplets of substrate less than 1 µm in diameter, or from solubilized substrate (Gutnick and Rosenberg, 1977; Reddy, Singh, Roy, and Baruah, 1982; Singer and Finnerty, 1984). Most studies indicate that microorganisms require the compound to be available as a free solute, in aqueous phase (Mihelcic, Lueking, Mitzell, and Stapleton, 1993). Microbes have been found to grow on the soluble portion of water-insoluble compounds after dissolution from a solid phase (Stucki and Alexander, 1987).

In studies using a sensitive respirometric technique (Sapromat), Bouchez, Blanchet, and Vandecasteele (1995b) unambiguously demonstrated that a *Pseudomonas* strain required phenanthrene to be transferred to the aqueous phase for biodegradation to occur. There was close correlation between the rate of transfer and the rate of biodegradation during the second period of biodegradation following that of exponential growth. A solution of phenanthrene in a non-water-miscible solvent (silicone oil) allowed high transfer and biodegradation rates, but resulted in higher metabolite production and lower mineralization of the compound. At low cell densities, transfer of phenanthrene is not limited and exponential growth can occur. At higher cell densities, the substrate demand reaches the maximal phase-transfer capacities of the system, and biodegradation then proceeds at a limiting rate of transfer to the aqueous phase. Guerin and Jones (1988) found that increasing the rate of uptake of phenanthrene by *Mycobacterium* sp. led to more biomass formation and less metabolite accumulation.

The interface between the PAH and medium also affects the degradation rate (Wiesel, Wuebker, and Rehm, 1993). Some results suggest that biodegradation can occur if the microbes can attach to the surface of a separate phase. Bacteria have been observed to partition into an oil phase and orient themselves perpendicularly at the oil–water interface, because of the nonpolar nature of a portion of the bacterial surface (Marshall and Cruickshank, 1973). *Arthrobacter* cells have also attached to the surface of a solvent at the solvent–water interface to degrade naphthalene and *n*-hexadecane dissolved in the organic solvent (Efroysom and Alexander, 1991).

Ortega-Calvo and Alexander (1994) studied diphasic systems and found biodegradation rates higher than transfer rates and suggested that, after a partitioning-limited initial stage by bacteria in the aqueous phase, faster degradation could be carried out by bacteria present at the solvent–water interface. Early exponential growth may occur when the interface surface area is not saturated by bacteria (Dunn, 1968). At high cell numbers, the growth rate becomes independent of the total cell population and proportional to the interfacial area.

Microorganisms use various mechanisms to metabolize organic substrates present at concentrations that exceed their water solubility (Thomas, Yordy, Amador, and Alexander, 1986). The physical state of the insoluble phase of a compound, whether it is a liquid or a solid, may affect its degradation. Liquid hydrocarbons can be taken up and incorporated into the cell membrane (Johnson, 1964), whereas the mechanism of utilization of solid substrates is not fully understood. PAHs might be used only in the dissolved state. Growth of pure cultures of bacteria on naphthalene, phenanthrene, and anthracene is faster on the solid substrates having the highest water solubilities (Wodzinski and Johnson, 1968).

Emulsification can be employed to provide greater surface area (Liu, 1980), and some organisms produce their own emulsifiers (see Section 5.3.1.2) that increase the surface area of the substrate (Thomas, Yordy, Amador, and Alexander, 1986), or they may modify their cell surface to increase its affinity for hydrophobic substrates and, thus, facilitate their absorption (Kappeli, Walther, Mueller, and Fiechter, 1984). It has been suggested that microbial degradation of the insoluble phase of crystalline hydrocarbons is difficult because of the large amount of energy needed to disperse the solid (Zilber, Rosenberg, and Gutnick, 1980). If an organism cannot use the insoluble form of a chemical, it may be expected that the organism will first metabolize that portion of a chemical that is in solution and that the subsequent rate of transformation of the compound will be limited by the rate of dissolution.

It has been found that increasing the surface area of hexadecane increases the microbial destruction of the alkane (Fogel, Lancione, Sewall, and Boethling, 1985). Rates of dissolution of naphthalene have been found to be directly related to its surface area (Thomas, Yordy, Amador, and Alexander, 1986). Higher concentrations of PAH provide greater surface area, resulting in more-rapid degradation (Wiesel, Wuebker, and Rehm, 1993). See Section 5.3.1 for a discussion of the role of chemical and microbial surfactants in solubilizing organic compounds.

While growth of bacteria appears to be limited by the rate of dissolution of a hydrocarbon, exponential growth does not always continue in parallel with the available material (Stucki and Alexander, 1987). Strains of *Flavobacterium* and *Beijerinckia* in media containing 84 µM phenanthrene began to decline at densities of about 4×10^6/mL. The dissolution rate should have allowed exponential growth to a fivefold higher cell density. It is not clear why exponential growth ended so soon.

Many organic solutes with a low water solubility preferentially leave dilute water solutions and concentrate primarily in organic material of soil and sediments (Karickhoff, Brown, and Scott, 1979). This action is proportional to the partitioning between octanol-1 and water. The octanol–water partition coefficient (P) (also defined as K_{OW}) is defined as

$$P = C_o/C_w$$

where C_o and C_w are the concentrations of the solute in *n*-octanol and water, respectively. P values measured in the laboratory can be used to predict the environmental behavior of organic pollutants (Mallon and Harrison, 1984). The hydrophobic nature of a pollutant, as measured by the octanol/water partition coefficient or by the dielectric constant, is important in predicting its flow through clay soils (Green, Lee, and Jones, 1981).

Design of a bioremediation system requires understanding of the impact of transport of the contaminants (Eckenfelder and Norris, 1993). Transport in the aqueous phase is favored by high solubility and low (or more negative) log K_{ow} values. Table 4.2 lists the log K_{ow} values for several compounds. The subsurface transport of immiscible organic liquids is governed by a different set of factors than those for dissolved contaminants (Mackay, Roberts, and Cherry, 1985). The migration of an immiscible organic

Table 4.2 Log K_{ow} Values for Several Compounds

Compound	Log K_{ow}
Acetone	−0.24
Benzene	2.13
Butanol	0.88
Naphthalene	3.28
Phenanthrene	4.46
Phenol	1.46
Pyrene	5.20

Source: From Eckenfelder, W.W., Jr. and Norris, R.D. In *Emerging Technologies in Hazardous Waste Management* III. ACS. American Chemical Society. 1993. With permission.

liquid phase is governed largely by its density and viscosity. Some contaminants have moderately low solubilities (<1%). These may migrate as discrete nonaqueous phases, with some components dissolving into the surrounding water. Their concentration in the soil water may be limited by their very low solubilities; however, they may be toxic at very low concentrations (Freeze and Cherry, 1979). The presence of large quantities of high-density, low-solubility contaminants can provide a "hidden" source for long-term contamination. The more polar intermediary metabolites of biodegradation can also migrate downward through the subsoil (Lapinskas, 1989). See Section 3.2.1.6 for a discussion of nonaqueous-phase liquids (NAPLs) and Section 4.1.9 for information on the relationship of solubility of a chemical to its toxicity to microorganisms. The effect of concentration of chemicals on biodegradation is reviewed in Section 4.1.11.

4.1.2 ADVECTION

In sand and gravel aquifers, the dominant factor in the migration of a dissolved contaminant is advection, the process by which solutes are transported by the bulk motion of flowing groundwater (Mackay, Roberts, and Cherry, 1985). Groundwater generally flows from regions of the subsurface where water level is high to regions where water level is low, a process called hydraulic gradient. In most cases, the flow velocities under natural gradient conditions are probably between 10 and 100 m/year. In the zone of influence of a high-capacity well or well field, however, the artificially increased gradient substantially increases the local velocity.

4.1.3 DISPERSION AND DIFFUSION

The rate of movement of organic chemicals through air, water, and organic matter is directly proportional to the concentration of the toxicant and its diffusion coefficient (Kaufman and Plimmer, 1972). Percolating water is the principal means of movement of relatively nonvolatile chemicals, and diffusion in soil water is important only for transport over very small distances.

Dissolved contaminants spread as they move with groundwater (Mackay, Roberts, and Cherry, 1985). This dispersion results from molecular diffusion and mechanical mixing, and causes a net flux of the solutes from a zone of high concentration to a zone of lower concentration. This movement causes the concentrations to diminish with increasing distance from the source and the plume to become more uniform. Dispersion in the direction of flow is often much greater than dispersion in the directions transverse to the flow.

4.1.4 SORPTION

Solid surfaces can act by adsorption, which is the retention of solutes in solution by the surfaces of the solid material, or by absorption, which is the retention of the solute within the mass of the solid rather than on its surfaces (Alexander, 1994). The term *sorption* refers to both processes.

This is perhaps the most important single factor affecting the behavior of organic chemicals, including toxic and recalcitrant fractions of hazardous wastes (Sims and Bass, 1984), in the soil environment (Kaufman and Plimmer, 1972). Adsorption to soil constituents will affect the rate of volatilization, diffusion, or leaching, as well as the availability of chemicals to microbial or chemical degradation.

Adsorption may involve physical or van der Waals forces, hydrogen bonding, chemisorption, or ion exchange (Alexander, 1994). With the latter, cationic organic molecules may be sorbed to the cation-exchange sites of clay minerals or humic surfaces. Anionic compounds are moderately sorbed by organic surfaces but poorly retained by clay minerals. Nonionic organic compounds are primarily sorbed by organic matter. The main subsurface solids responsible for adsorption of organic chemicals are solid organic matter (especially good for hydrophobic organic compounds), clay minerals, and amorphous minerals (e.g., iron hydroxides) (Pettyjohn and Hounslow, 1983). Sorption of organic substrates influences the growth and activity of microorganisms (Alexander, 1994). It can affect biodegradation by removing the organic substrate from solution, binding extracellular enzymes, removing essential inorganic nutrients and growth factors, and lowering the pH immediately around negatively charged surfaces.

Before a compound can be biodegraded, it must be available, generally as a free solute (Mihelcic, Lueking, Mitzell, and Stapleton, 1993), and in aqueous phase (Ogram, Jessup, Ou, and Rao, 1985). However, it has been found that in addition to aqueous-phase substrate, sorbed-phase substrate that can be easily desorbed may also be available for biodegradation (Robinson, Farmer, and Novak, 1990).

Sorption of organic compounds can be influenced by the type and concentration of solutes, the type and quantity of clay minerals, the type of cation that is saturating the clay, the exchange capacity and specific surface area of clays, the amount of organic matter in the soil or sediment, pH, temperature, and the specific hydrocarbon contaminant (Alexander, 1994). The binding strength between a clay and a compound strongly affects the availability of the compound (Mihelcic, Lueking, Mitzell, and Stapleton, 1993). Montmorillonite adsorbs more strongly than kaolinite (Weber, Weed, and Ward, 1969). It has a larger cation exchange capacity and also swells to expose the inside to organic compounds, which may be sequestered there. Clay may also buffer pH changes at the clay–water–microbial interface (Marshall, 1976).

The surface properties of the soil determine where microorganisms will attach (Mihelcic, Lueking, Mitzell, and Stapleton, 1993). In soil, 60% of the total bacteria are located on particles coated with organic matter, which is 15% of the total particle surface area (Gray and Parkinson, 1968). On the other hand, bacteria colonize only 0.02% of the surface of sand grains.

Soil organic matter may slow biodegradation of bound PAHs that are otherwise readily metabolized (Manilal and Alexander, 1991). Phenanthrene is less degradable when sorbed to organic soil constituents than when in liquid culture. Fulvic acid can decrease mineralization of pyrene, partially due to possible sorption of the compound making it less bioavailable (Grosser, Warshawsky, and Kinkle, 1994).

Many compounds, especially those that are hydrophobic, sorb to the organic fraction of soils and sediments (Alexander, 1994). This includes many PAHs and other nonpolar contaminants, which are sorbed mainly by the native organic matter, rather than the clay constituents. The extent of the retention is directly related to the octanol–water partition coefficients.

Organic material, generally, has high surface areas and exchange properties ideal for adsorption of organic compounds (Devitt, Evans, Jury, Starks, Eklund, and Gholson, 1987). The K_{oc} value reflects the impact of this organic material to adsorb organic compounds out of solution. The K_{oc} for an organic compound is a coefficient that relates the partitioning of the organic compound between the adsorbed phase and the soil solution, relative to the organic carbon fraction. The K_{oc} reflects the affinity of an organic compound to adsorb out of solution onto soil organic material, when present. Compounds with low water solubility often have higher K_{oc} values (Wilson, Enfield, Dunlap, Cosby, Foster, and Baskin, 1981). When soils are fully hydrated, adsorption of the organic solutes by organic matter is more important than adsorption by soil minerals, which preferentially adsorb water (Chiou, 1985).

Sorption of a hydrophobic molecule may occur on the outer surface or within the micropores of humic material, where it is held by physical or chemical forces (Chiou, 1989; Alexander, 1994). Another explanation might be that the organic molecules partition into the solid portion of the organic matter (Alexander, 1994). This type of sorption is an "aging" process that renders the compounds inaccessible for biodegradation. A contaminant that becomes tightly bound after being in the soil for years could possibly have been easily desorbed and quickly metabolized, if treatment had been initiated immediately (Steinberg, Pignatello, and Sawhney, 1987). The binding of PAHs within the soil matrix may retard their solubilization, but this would also reduce their toxicity (Weissenfels, Klewer, and Langhoff, 1992b). Contaminants or their metabolites might also form complexes with the humic fraction of the soil or by attaching to reactive sites on the surfaces of organic colloids, creating new molecules that are not easily degraded (Alexander, 1994).

The role of phenol oxidases in determining the fate of xenobiotics may have been underestimated (Bollag and Liu, 1972). Polymerization reactions and enzymic oxidative coupling are probably important in the covalent bonding of phenolic compounds to soil organic polymers. The degree to which phenols become bound to soil humic molecules as a result of enzymatically mediated oxidative coupling reactions may be affected by substituent groups on the aromatic ring (Berry and Boyd, 1984). For instance, electron-donating groups, such as methoxy ($-OCH^3$) (common substituents on lignin-derived polyphenols), would facilitate this reactivity.

The pKa, or dissociation constant, of a compound indicates the degree of acidity or basicity that a compound will exhibit and, therefore, should be very important in determining both the extent of adsorption and the ease of desorption (Kaufman and Plimmer, 1972). Functional groups can affect the degree of sorption of a chemical (Brindley and Thompson, 1966). Chain molecules terminating in $-OH$, in $-COOH$, and in $-NH_2$ readily form complexes with montmorillonite, whereas similar molecules terminating in $-Cl$ and $-Br$ do not. All compounds are adsorbed strongly at low pH; anionic substances are adsorbed negatively at slightly basic conditions; and nonionic compounds are moderately adsorbed (Frissel, 1961). Adsorption processes are exothermic and desorption processes are endothermic, and an increase in temperature should reduce adsorption and favor desorption (Kaufman and Plimmer, 1972).

The sorption process is usually described by an adsorption isotherm, which expresses the relationship between the amount of constituent adsorbed onto a solid (soil, activated carbon, zeolite, organic matter, etc.) and the concentration of solute in solution at equilibrium (Sims and Bass, 1984). One frequently used relationship is the Freundlich isotherm, which is expressed as

$$S = KC^N$$

where S = amount of constituent adsorbed per unit dry weight of soil, K and N are constants, and C = solution-phase equilibrium concentration.

The soil adsorption constant (K) is a measure of the tendency of a pollutant to adsorb and stay on soil (Nash, 1987). Contaminants can be, thereby, grouped according to a K value and their removal efficiencies (RE) evaluated. Tables 4.3 and 4.4 list the soil adsorption constants and water quality criteria for a number of hydrophobic and hydrophilic compounds, respectively. The greater the value for K, the stronger the binding to the soil.

Table 4.3 Soil Adsorption Constants and Water Quality Criteria for Hydrophobic Organics

Substance	Soil Adsorption Constant K	EPA Water Quality Criteria (ppm)
Anthracene	700	—
Benz(a)anthracene	60,000	2.8×10^{-6} [a]
Benzo(a)pyrene	40,000	2.8×10^{-6} [a]
Pyrene	2,000	2.8×10^{-6} [a]
Naphthalene	600	—
Oil	(30,000)[b]	—
Grease	(5,000,000)[c]	—
Ethyl benzene	50	1.4
Cyclohexane	70	—
Benzo(b)pyrene	40,000	2.8×10^{-6} [a]

[a] Corresponds to an incremental increase in cancer risk of 10^{-6}
[b] Estimated based on n-C_{15}
[c] Estimated based on n-C_{25}

Source: Nash, J.H. Report No. EPA-600/2-87/110. U.S. EPA, Cincinnati, OH, 1987. PB 88146808.

Table 4.4 Soil Adsorption Constants and Water Quality Criteria for Hydrophilic Organics

Substance	Soil Adsorption Constant K	EPA Water Quality Criteria (ppm)
Xylene	30	—
Phenol	20	3.5
Toluene	30	14.3
Methylene chloride	5	—
Methyl isobutyl ketone	5	—
Benzene	10	6.6×10^{-4} [a]
Tetrahydropyran	4	—

[a] Corresponds to an incremental increase in cancer risk of 10^{-6}

Source: Nash, J.H. Report No. EPA-600/2-87/110. U.S. EPA, Cincinnati, OH, 1987. PB 88146808.

The octanol/water partition coefficient is a useful means of predicting soil adsorption, biological uptake, lipophilic storage, and biomagnification (Chiou, Freed, Schmedding, and Kohnert, 1977). The role of soil organic matter is similar to that of an organic solvent, and the partitioning of an organic solute between the soil material and water can be estimated by its tendency to partition between water and an immiscible organic solvent (Lambert, 1968). Octanol is believed to best imitate the fatty structures in plants (Karickhoff, Brown, and Scott, 1979) and the aqueous phases of living tissues (Hansch, Quinlan, and Lawrence, 1968).

The logarithm of the octanol/water partition coefficients (log P or log K_{ow}) is a measure of the tendency of a compound to dissolve in hydrocarbons, fats, or the organic component of soil rather than in water (Nash, 1987). Log K_{ow} is defined as the ratio of the concentration of the compound in octanol divided by the concentration of the same compound in water under equilibrium conditions (Eckenfelder and Norris, 1993). Many hydrophobics, some slightly hydrophilics, and no hydrophilics were detected in soil containing organic components that tend to adsorb other organics.

Log P can also be used to estimate the tendency of an organic compound to become (or remain) adsorbed in soil (Nash, 1987). The partitioning of a compound between organic soil components and a water solution is expressed as

$$K_{oc} = \frac{\text{g adsorbed/g organic carbon}}{\text{g/mL solution}}$$

The adsorption tendency is mainly dependent upon the weight of organic carbon (oc) in the soil. If the organic carbon content of a soil is known, the soil adsorption constant (K) can be derived from K_{oc} (Nash, 1987):

$$K = \frac{\%\text{ organic carbon}}{100} \left(K_{oc}\right)$$

$$K = \frac{\text{g adsorbed/g soil}}{\text{g/mL solution}}$$

Thus, K can be used to estimate what fraction of a compound will be adsorbed onto soil and what fraction will remain dissolved in water when the soil and water are in equilibrium with each other (Nash, 1987).

The linear partition coefficients (K_p) for sorption of hydrophobic pollutants on natural sediments are directly related to organic carbon content of the sediment biomass (Karickhoff, Brown, and Scott, 1979). Reasonable estimates of soil/water partition coefficients (K_{oc}) can be made from octanol/water partition coefficients. Association with soil or sediment is greatest for chemicals with low S and high K_{ow} values (Kenaga and Goring, 1980). Octanol/water partition coefficient is a much better predictor of extent of adsorption of microbial biomass than aqueous solubility. Reasonable estimation of the sorption behavior of organic pollutants can be made from a knowledge of organic carbon content of the biomass and octanol/water partition coefficient of the organic pollutants (Selvakumar and Hsieh, 1988).

There are occasions when the phenomenon of sorption could actually be beneficial for biodegradation. If a substrate is toxic, sorption can improve biodegradation by lowering its concentration in solution (van Loosdrecht, Lyklema, Norder, and Zehnder, 1990). Sorption may also store the substrate and later make it available for biodegradation by desorption.

Desorption of constituents is important in assessing treatment effectiveness through the extent of release of chemicals from soil into percolating water moving through the soil (Sims and Bass, 1984). Generally, the extent of desorption follows the Freundlich isotherm, but with constants different from the ones used for adsorption. Factors directly associated with desorption include the amount of leachate (soil:water ratio) and the amount of constituent contaminating the soil (soil:constituent ratio). The extent of desorption will decrease with an increase of these ratios. The longer some compounds remain in soil, the more resistant they become to desorption and biodegradation (Alexander, 1994). Sorbed molecules may become available to microbes as biodegradation of soluble material causes the equilibrium to draw more of the sorbed form into solution. Microbial metabolites may also desorb the compound. The rate of degradation is then governed by the rate of desorption.

The ability of microorganisms to attach to surfaces is variable and complex. Some, such as *Caulobacter*, require attachment to a solid surface for cell growth and reproduction (Shapiro, 1976). Attached bacteria may lose their flagella and become immotile. Attachment of microorganisms to a surface on which a contaminant is sorbed may facilitate biodegradation of that compound (Griffith and Fletcher, 1991). A high concentration of a desorbing substrate and a fast mass transfer of the substrate support rapid microbial growth on a solid surface, because of the shorter diffusion distance (van Loosdrecht, Lyklema, Norder, and Zehnder, 1990). Bacteria that are irreversibly adsorbed to a surface may be able to deplete a substrate in its vicinity more rapidly, but a reversibly adsorbed organism would be able to move to other locations when the initial material is depleted (Kefford, Kjelleberg, and Marshall, 1982). Mihelcic, Lueking, Mitzell, and Stapleton (1993) note that not all surface attachments are meaningful, and they should be interpreted carefully.

A mutant strain of a naphthalene-degrading *Pseudomonas fluorescens* TG-5 Nah⁻ possesses a higher sorptive capacity for naphthalene (partition coefficient of 380 cm^{-3}/g) than a soil with a 5.1% organic carbon content (Whitman, Mihelcic, and Lueking, 1995). If the soil system has a high organic carbon content, little of the naphthalene is bound to the cells. In a nonsorptive soil system, up to 10% of the compound can be associated with the cells.

A variety of models that combine sorption and biodegradation are described by Mihelcic, Lueking, Mitzell, and Stapleton (1993).

4.1.5 VOLATILITY

Volatile components of oils are toxic, and a temperature-dependent lag period of bacterial growth can be produced until these compounds evaporate (Spain, Pritchard, and Bourquin, 1980). The initial lag phase in biodegradation may result in part from the toxic or bacteriostatic effect of these low-molecular-weight hydrocarbons (Bausam and Taylor, 1986).

Volatility is not an important retardation mechanism for soil biodegradation after the organic compounds have migrated through the unsaturated zone either to the atmosphere or to the groundwater (Freeze and Cherry, 1979). As hydrocarbons are lost by biodegradation and volatilization, they are replaced to an extent by desorption from bound material (Carlson, 1981). See Section 6 for a discussion on volatile organic compounds in petroleum products.

4.1.6 VISCOSITY

Viscosity of polluting oils is an important property that determines, in part, the spreading and dispersion of the hydrocarbon mixture and, thus, the surface area available for microbial attack (Atlas, 1978). For example, about four times the volume of a light fuel oil in the high viscosity range would be retained by the average soil, compared with gasoline, a distillate with a lower viscosity (Noel, Benson, and Beam, 1983).

Viscosity and surface-wetting properties affect the transport of an organic liquid phase (Mackay, Roberts, and Cherry, 1985). Large quantities of immiscible liquid organic contaminants could be stored as droplets, even if the bulk of the migrating mass of liquid is removed. These droplets may then dissolve over time into water flowing past them.

Viscosity affects the migration of a chemical in groundwater (Noel, Benson, and Beam, 1983). Gasoline would spread over a wider area of an aquifer than a light fuel oil. Contaminants that are highly water soluble must be handled differently than those that float on the water table, like gasoline (Nielsen, 1983).

4.1.7 DENSITY

Migration of an immiscible organic liquid phase is governed largely by its density and viscosity. Density differences of about 1% influence fluid movement in the subsurface (Mackay, Roberts, and Cherry, 1985). The specific gravities of hydrocarbons (gasoline and other petroleum distillates) may be as low as 0.7, and halogenated hydrocarbons are almost without exception significantly more dense than water. The presence of large quantities of high-density, low-solubility contaminants can provide a "hidden" source for long-term contamination. See Section 3.2.1.6 for a discussion of nonaqueous-phase liquids (NAPLs).

When a compound reaches an aquifer, its density will determine where it will most likely be concentrated (Mackay, Roberts, and Cherry, 1985). Low-density hydrocarbons have a tendency to float on water and may be found in the upper portions of an aquifer. High-density hydrocarbons would sink to the lower portions of the aquifer, if they are heavier than water. It is important to recognize that the migration of dense organic liquids is largely uncoupled from the hydraulic gradient that drives advective transport and that the movement may have a dominant vertical component, even in horizontally flowing aquifers.

An organic liquid contaminant, such as gasoline, which is immiscible with and less dense than water, would migrate vertically through the soil to the water table and then float on the surface, spreading out in the downgradient direction (Mackay, Roberts, and Cherry, 1985). If the organic liquid contains a contaminant slightly soluble in water, e.g., benzene, a plume would form in the saturated zone. A complex pattern of overlapping plumes can develop when many contaminants are involved.

4.1.8 CHEMICAL STRUCTURE

The structure, concentration, and toxicity of a chemical are important in determining whether or not the material is accumulated in the environment and the environmental impact of the accumulation (Leisinger, 1983). A compound will accumulate if its structure prevents mineralization or biodegradation by organisms. This may be due to its insolubility or to a novel chemical structure to which microorganisms have not been exposed during evolutionary history. Such compounds are termed *xenobiotic*. This name refers to compounds of anthropogenic (man-made) origin, as well as to compounds that may occur naturally but exceed normal levels in the environment. Various laboratory culture techniques applied to samples from nature can be used to select or develop bacteria with the ability to biodegrade many of these chemicals (see Section 5.2.2.2).

The chemical structure of a contaminant can affect its biodegradation in two ways (Hutzinger and Veerkamp, 1981). First, the molecule may contain groups or substituents that cannot react with available or inducible enzymes (i.e., these chemical bonds cannot be broken). Secondly, the structure may be such that the compound is in a physical state (adsorbed, gas-phase) where microbial degradation does not easily occur. This seems to be a problem with many of the lipophilic compounds, which have very low solubilities in water.

Generally, the larger and more complex the structure of the hydrocarbon, the more slowly it is oxidized, but this depends largely upon the type of organism involved and the medium in which it was developed (Texas Research Institute, Inc., 1982). Some authors have proposed that aliphatic, long-chain molecules are attacked more readily than short chains, with hydrocarbons in the range of $C_{10}H_{22}$ to $C_{16}H_{34}$ being oxidized by soil bacteria more readily than those of smaller weight. Saturated fractions, e.g., *n*-alkanes, are highly degraded, while asphaltenes and aromatics are often resistant to microbial attack (Jobson, Cook, and Westlake, 1972). Other workers have suggested that although *n*-alkanes are probably metabolized more rapidly than naphthenes or other aromatics, the reactions appear to be slower with increasing chain length (possibly because of differences in water solubility), and that the *n*-alkanes with shorter chains (from C_5 to C_9) are more easily used as a source of carbon and energy by microorganisms than those with longer chain lengths (from C_{10} to C_{14}) (Williams, Cumins, Gardener, Palmier, and Rubidge, 1981). Aromatic compounds are the least degradable by microbes, and straight-chain paraffins are the easiest to degrade (Evans, Deuel, and Brown, 1980). The latter undergo oxidation to form alcohols, aldehydes, and acids. The most degradable alkanes or hydrocarbons are those with molecular weights in the C_6 to C_{28} range (Perry, 1968).

Huesemann and Moore (1993), studying biodegradation of crude oil–contaminated soil, noted that 93% of saturated and 79% of aromatic compounds were biodegraded. Some 96% of all compounds with carbon numbers from 10 to 20 were biodegraded, as were up to 85% with carbon numbers above 44. The heavy hydrocarbon fraction of this particular petroleum hydrocarbon waste may have consisted of biodegradable straight-chain or branched alkanes and naphthenes or aromatics with few saturated or aromatic rings.

Tetracyclic compounds are metabolized significantly only after phenol, naphthalene, and most of the tricyclic compounds are already degraded (Wiesel, Wuebker, and Rehm, 1993). The extreme thermodynamic stability and complexity of tetracyclic compared with tricyclic compounds make enzymatic attack more difficult. Cell growth occurs only with degradation of the tricyclic substances; biotransformation of the tetracyclics may not supply enough energy for cell growth. A species of *Mycobacterium* has been isolated, however, that shows greater metabolism of pyrene and fluoranthene than naphthalene or phenanthrene.

The degree of substitution affects the biodegradation. Compounds that possess amine, methoxy, and sulfonate groups, ether linkages, halogens, branched carbon chains, and substitutions at the *meta* position of benzene rings are generally persistent (Knox, Canter, Kincannon, Stover, and Ward, 1986). Addition of aliphatic side chains increases the susceptibility of cyclic hydrocarbons to microbial attack (Atlas, 1978). Linear nonbranched compounds are more easily biodegraded than are branched forms and rings (Pettyjohn and Hounslow, 1983). The side chains of the latter are generally attacked first. Changes in *n*-alkane to isoprenoid hydrocarbon ratios occur in oil spills (Haines, Pesek, Roubal, Bronner, and Atlas, 1981). Phenanthrenes and dibenzothiophenes with C_2 and greater substitution are relatively resistant to biodegradation, while unsubstituted and C_1 substituted two- and three-ring condensed aromatics are subject to abiotic and biotic losses. The number and locations of fused rings in polynuclear aromatics are important in determining the rates of their decomposition (Sims and Overcash, 1983). Hydrocarbons that are strong fat solvents may be less readily tolerated or assimilated than those that are less likely to dissolve cell lipids (ZoBell, 1946).

Table 4.5 summarizes organic groups subject to microbial metabolism by aerobic respiration, anaerobic respiration, and fermentation (U.S. EPA, 1985a). Oxidation indicates that the compound is used as a primary substrate, and co-oxidation indicates that the compound is cometabolized. These tables provide only a general indication of degradability of compounds, and treatability studies will usually be required to determine the degradability of specific waste components.

Table 4.5 Summary of Organic Groups Subject to Biodegradation

Substrate Compounds	Respiration Aerobic	Respiration Anaerobic	Fermentation	Oxidation	Co-oxidation
Straight-chain alkanes	+	+	+	+	+
Branched alkanes	+	+	+	+	+
Alcohols	+	+		+	
Aldehydes, ketones	+	+		+	
Carboxylic acids	+	+		+	
Cyclic alkanes	+		+	+	+
Unhalogenated aromatics	+	+		+	+
Phenols	+	+	+	+	+
Fused ring hydroxy compounds	+				
Phenols — dihydrides, polyhydrides	+			+	+
Two- and three-ring fused polycyclic hydrocarbons	+			+	

Source: From *Envirosphere,* 1985, reprinted in U.S. EPA Handbook No. EPA/625/6-85/006, 1985.

Table 4.6 Relative Persistence and Initial Degradative Reactions of Nine Major Organic Chemical Classes

Chemical Class	Persistence	Initial Degradative Process
Carbamates	2–8 weeks	Ester hydrolysis
Aliphatic acids	3–10 weeks	Dehalogenation
Nitriles	4 months	Reduction
Phenoxyalkanoates	1–5 months	Dealkylation, ring hydroxylation or oxidation
Toluidines	6 months	Dealkylation (aerobic) or reduction (anaerobic)
Amides	2–10 months	Dealkylation
Benzoic acids	3–12 months	Dehalogenation or decarboxylation
Ureas	4–10 months	Dealkylation
Triazines	3–18 months	Dealkylation or dehalogenation

Source: Kaufman, D.D. In *Land Treatment of Hazardous Wastes.* Parr, J.F. et al., Eds. Noyes, Park Ridge, NJ, 1983, 77–151. With permission.

Several linkages may be readily susceptible to biodegradation (Kearney and Plimmer, 1970):

1. $R-NH-CO_2R'$

2. $R-NH-COR'$

3. 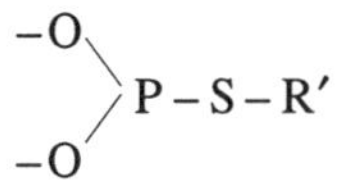 $(-O)_2 P-S-R'$

4. 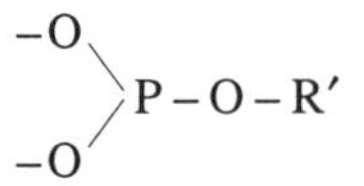 $(-O)_2 P-O-R'$

5. $R-CHCl-COO-$

6. $R-CHCl_2-COO-$

Aliphatic acids, anilides, carbamates, and phosphates are generally degraded within a short time in the soil (Kaufman and Plimmer, 1972). The rate at which linkages are hydrolyzed will depend upon the nature of R and R'. Chemicals that have an initial degradative reaction that is ester hydrolysis are relatively short-lived in soil, whereas those that initially undergo dealkylation generally tend to be somewhat more persistent. Chemicals that are initially dehalogenated are variable in their persistence. Halogenated aliphatic acids are readily degraded, whereas halogenated benzoic acids and *s*-triazines are intermediate in their persistence. In Table 4.6 are some of the major chemical classes and their relative persistence and initial degradation reactions.

An inverse relationship and a high correlation between microbial transformation rates and van der Waals radius of eight phenols have been found, suggesting these as useful for predicting degradability of xenobiotics (Paris, Wolfe, Steen, and Baughman, 1983).

Section 3 provides further information on the biodegradation of petroleum compounds in relation to their chemical structure.

4.1.9 TOXICITY

Organic compounds may not be readily degraded when the microbial population is low, the nutrient balance is inadequate, or because of contaminant overloading (toxicity) (Pettyjohn and Hounslow, 1983). Crude oils are mixtures of tremendous complexity, containing hundreds of hydrocarbon and nonhydrocarbon components, many of them still unidentified and some of these very toxic toward microorganisms (Atlas and Bartha, 1973a). Some compounds may be more toxic to microbes than others (Scholze, Wu, Smith, Bandy, and Basilico, 1986), and the presence of inhibitory substances in oil can delay or prevent the biodegradation of otherwise suitable hydrocarbon substrates (Bartha and Atlas, 1977).

When the structural features necessary for toxicity are compared with those features permitting degradation in the environment for target organisms, differences are found among the various chemical classes (Kaufman and Plimmer, 1972). In some classes, those structural features contributing to toxicity are coincident with those necessary for degradability; in other chemical classes they are diametrical. In all classes, however, the relationships are mediated by substituent type, number, and position.

The structure–toxicity and structure–degradability relationships of certain halogenated aliphatic acids are quite similar; i.e., the most phytotoxic structures are also the most readily degradable (Kaufman and Plimmer, 1972). *Meta*-substitution (*para* to a free *ortho* position) confers resistance to biodegradation and eliminates phytotoxic activity. Halogenation in the *para* position increases both phytotoxicity and biodegradability of phenoxyacetates. Increasing the length of the side chain affects both phytotoxicity and biodegradability.

Naphthalene and alkyl-substituted naphthalene are among the most toxic components in the water-soluble fraction of crude and fuel oils (Winters, O'Donnell, Batterton, and Van Baalen, 1976). Cycloalkanes are more resistant to microbial attack than straight-chain alkanes and are highly toxic (Atlas, 1981). This is probably due to the absence of an exposed terminal methyl group for the initial oxidation (Hornick, Fisher, and Paolini, 1983). Methyl-, dimethyl-, and trimethyl-naphthalenes are more toxic than naphthalene to the freshwater alga *Selenastrum capricornatum*, while dibenzofuran, fluorene, phenanthrene, and dibenzothiophene are the most toxic to this organism (Hsieh, Tomson, and Ward, 1980). In general, compounds with higher boiling points are more toxic.

The toxicity of PAHs to microorganisms is related to their water solubility (Sims and Overcash, 1983). Aromatic hydrocarbons in water-soluble fractions of petroleum products are toxic to aquatic organisms, but rapid volatilization of low-molecular-weight hydrocarbons reduces this toxicity (Coffey, Ward, and King, 1977). The vapor phase of short-chain alkanes is less toxic than the liquid phase. The toxicity of the short-chain alkanes is also related to temperature, since a higher temperature will increase the amount of alkane in the vapor phase and decrease the concentration of the liquid alkane. In cold water, however, these compounds may delay the onset of biodegradation for several weeks (Bartha and Atlas, 1977).

C_2 to C_6 alkanes are inhibitory to some microorganisms possibly because their size allows them to penetrate into cell membranes (Hornick, Fisher, and Paolini, 1983). This is also seen with cycloalkanes of similar size and could be the reason for the "toxicity" of short-chain alkanes seen with a few microorganisms. Long-chain normal alkanes from C_{10} to C_{22} are usually less toxic to microbes and more readily biodegradable than many highly water-soluble hydrocarbons (Bossert and Bartha, 1984).

Biodegradability and water solubility appear to be inversely related to the number of rings in the molecules (Huddleston, Bleckmann, and Wolfe, 1986). While low-molecular-weight aromatic hydrocarbons are quite toxic to microorganisms, they can be metabolized when present in low concentrations. Condensed polyaromatic hydrocarbons are less toxic to microorganisms but are metabolized only rarely and at slow rates. Cycloalkanes are highly toxic and serve as growth substrates for isolated organisms only in exceptional cases. Some are readily degraded, however, by the cometabolic attack of mixed microbial communities. In general, the biotransformation process for PAHs with more than three rings appears to be cometabolism (Sims and Overcash, 1983).

Fulvic acid in the soil has been shown to decrease mineralization of pyrene, apparently due to toxicity to the microbes and possible sorption of the compound making it less bioavailable (Grosser, Warshawsky, and Kinkle, 1994). It is suggested that the binding of PAHs within the soil matrix itself retards their solubilization and thus reduces their toxicity by elution with water (Weissenfels, Klewer, and Langhoff, 1992b).

Low-molecular-weight hydrocarbons solvate and, hence, destroy the lipid-containing pericellular and intracellular membrane structures of microorganisms (Bartha and Atlas, 1977). Liquid hydrocarbons of the *n*-alkane, *iso*-alkane, cycloalkane, and aromatic type with carbon numbers under 10 all share this property to varying degrees.

Floating oil is able to concentrate hydrophobic pollutants (Bartha and Atlas, 1977). This makes the material more toxic and interferes with microbial degradation. In certain situations, it may be possible to modify the chemical composition of oil, rendering it more susceptible to biodegradation (Atlas, 1977). The toxic components in the low molecular weight and low boiling range can be removed by temporary heating, ignition, and burning of the oil, and artificially increasing air movement over the oil (Atlas and Bartha, 1972; Atlas, 1975). Oil is particularly difficult to ignite in many aquatic environments (Fay, 1969). Burning would probably remove toxic components and many other hydrocarbons. However, burning of the substance is not without problems (Riser-Roberts, 1992) and, depending upon the chemical composition of the oil, may create a residual that is more resistant to biodegradation.

Inhibition of biodegradation can occur with a mixture of PAHs (Bouchez, Blanchet, and Vandecasteele, 1995a). For instance, naphthalene is strongly toxic in a mixture of PAHs and can inhibit degradation of other components that would normally be biodegraded. This toxicity may be due to its high water solubility (about 30 ppm). Fluorene, with a solubility of 2 ppm and phenanthrene at 1 ppm, are also frequently inhibitory. On the other hand, inhibition by anthracene, which is poorly water soluble (around 50 ppb), can also occur. The inhibition by a PAH in a mixture may be overcome by supplying an organism that is able to degrade that compound and thereby remove the inhibitory effect. However, there can be inhibition even during cometabolism, because of the interactions of PAHs at several levels (Bouchez, Blanchet, and Vandecasteele, 1995a). This could involve competition at the active site of enzymes or at the level of enzyme induction. It could be due to accumulation of toxic end products.

Toxic products formed during microbial hydrocarbon oxidation include the C_5 to C_9 primary alcohols (Liu, 1973); fatty acids, if β oxidation is lacking (Atlas and Bartha, 1973b); and oxidation products of aromatic hydrocarbons (Calder and Lader, 1976). The presence of a mixed microbial community, however, will often lead to further breakdown of toxic products (Bausam and Taylor, 1986).

The toxicity of an NAPL depends on the toxicity of the compounds in the solvent (Alexander, 1994). With some exceptions, organic solvents with high log K_{ow} values (4.0 or greater) do not inhibit the microorganisms; however, those with low values (2.0 or lower) are highly toxic (Laane, Boeren, Vos, and Veeger, 1987; Inoue and Horikoshi, 1991). Robertson and Alexander (1996) found that bacteria could mineralize phenanthrene that was initially present in di-n-butylphthalate but not toluene. However, bacteria sensitive to toluene were able to degrade extensively phenanthrene initially present in these toxic NAPLs if there was also a water phase and a separate nontoxic NAPL to lower the toxicant concentration in the aqueous phase.

Several processes may result in detoxification of a hydrocarbon:

Hydrolysis — cleavage of a bond by addition of water;
Hydroxylation — addition of OH to an aromatic or aliphatic molecule;
Dehalogenation — enzymic removal of a halogen;
Demethylation or other dealkylations;
Methylation — addition of a methyl group;
Nitro reduction;
Deamination;
Ether cleavage;
Conversion of nitrile to amide; or
Conjugation — reaction between a normal metabolite and a toxicant (Alexander, 1994).

These reactions would then be followed by mineralization of the compound.

A laboratory protocol has been developed to quantify the toxicity of chemicals encountered when setting up a bioremediation program (Arulgnanendran and Nirmalakhandan, 1995). Using the concepts of Toxic Units, Additivity Index, and Mixture Toxicity Index, laboratory results are tested for additive, synergistic, or antagonistic effects of the contaminants. In addition to the use of predictive models (Quantitative Structure Activity Relationship; QSAR) in evaluating cleanup levels for hazardous waste locations, these concepts can predict microbial toxicity in soils of new chemicals from a congeneric group acting by the same mode of toxicity. These models are applicable when the contaminants act singly or jointly in a mixture.

4.1.10 HYDROLYSIS AND OXIDATION

The effects of these processes on particular contaminants in the groundwater zone are unknown (Mackay, Roberts, and Cherry, 1985). It is believed, however, that most chemical reactions in the groundwater are likely to be slow in comparison with transformations mediated by microorganisms (Cherry, Gillham, and Barker, 1984).

See Section 2.2.1.3 for a discussion of the effect of hydrolysis and Section 2.2.1.2 for the effect of oxidation on soil contamination.

4.1.11 CONCENTRATION OF CONTAMINANTS

The concentration of a hydrocarbon can affect its biodegradability and toxicity to the degrading organisms (U.S. EPA, 1985a). High concentrations of a hydrocarbon can be inhibitory to microorganisms, and the concentration at which inhibition occurs will vary with the compound (Alexander, 1985). Concentrations of hydrocarbons in the range of 1 to 100 µg/mL of water or 1 to 100 µg/g of soil or sediment (on a dry weight basis) are not generally considered to be toxic to common heterotrophic bacteria and fungi.

The chemical concentration may affect the level of tolerance (Scholze, Wu, Smith, Bandy, and Basilico, 1986). The term *xenobiotic compound* refers not only to compounds with structural features foreign to life, but also to those compounds that are released in the environment by the action of humans and, thereby, occur in a concentration that is higher than natural (Leisinger, 1983). The concentration of hydrocarbons in the aqueous medium has two effects (Texas Research Institute, Inc., 1982). At low concentrations, all fractions are likely to be attacked, but, at high concentrations, only those fractions most susceptible to degradation will be attacked. Also, if the hydrocarbon mixture contains water-soluble toxic substances, their effect is intensified at high concentrations. Intermediate and end products of biological and chemical degradation could be more toxic or mobile than the original compound.

Microorganisms appear to require a contaminant to be available in aqueous phase for biodegradation to occur (Mihelcic, Lueking, Mitzell, and Stapleton, 1993). Higher-molecular-weight PAHs may appear to be more recalcitrant than lower-weight compounds, but this effect seems to be due to their low concentrations in the aqueous phase. For example, tetracyclic compounds are less available for degradation than the di- and tricyclic compounds (Wiesel, Wuebker, and Rehm, 1993). Microorganisms can grow on the soluble portion of water-insoluble compounds after dissolution from a solid phase (Stucki and Alexander, 1987). The dissolution rate of PAHs depends on the concentration in the aqueous phase and the interface between the PAH and the medium (Wiesel, Weubker, and Rehm, 1993). A higher concentration of the PAH provides a greater surface area and an improved degradation rate.

In some cases, relatively high concentrations of a pollutant (>100 µg/L) can stimulate multiplication of the microbes that metabolize the organic contaminant (Wilson and McNabb, 1983). However, concentrations of less than 10 µg/L of the material usually do not have this effect. It has been reported that if groundwater is taken from an anaerobic gasoline-contaminated or fuel oil–contaminated area with no significant degradation activity, and oxygen and nutrients are then added, rapid degradation of the hydrocarbons down to 1 µg/L or less starts with little lag time (Jensen, Arvin, and Gundersen, 1985). This requires an initial hydrocarbon concentration below about 10 mg/L. At higher concentrations, a considerable adaptation time may be expected. If the concentrations are greater than 1 to 10 mg/L, metabolism of the compound can entirely deplete the oxygen or other metabolic requirements available in the groundwater (Wilson and McNabb, 1983). See also Section 4.1.11.1 for the effect of low chemical concentrations on the rate of biodegradation.

The concentration of a contaminant will affect the number of organisms present. It has been noted that higher concentrations of gasoline in contaminated water were related to higher counts of microorganisms (McKee, Laverty, and Hertel, 1972; Litchfield and Clark, 1973). Waters containing less than 10 ppm of hydrocarbon had populations of bacteria less than 10^3, while concentrations of hydrocarbon in excess of 10 ppm sometimes supported growth of 10^6 bacteria/mL.

Table 4.7 lists concentrations at which certain compounds have been found to be toxic in industrial waste treatment (U.S. EPA, 1985a). Microorganisms present in the subsurface, however, may be more tolerant to high concentrations of these compounds. This must be determined on a case-by-case basis.

Tables 4.8 and 4.9 summarize from the literature concentrations of various organic compounds that have been tolerated and biodegraded in the field and in bioreactors or small-scale studies, respectively.

4.1.11.1 Low Concentrations

There appears to be a minimum concentration at which a single organic chemical can be decomposed under steady-state conditions, depending upon the decay of the bacteria, which, in turn, depends upon environmental factors (McCarty, Reinhard, and Rittmann, 1981). Trace organic materials may be biodegradable, but are

Table 4.7 Problem Concentrations of Selected Chemicals

Chemical	Problem Concentration (mg/L) Substrate Limiting[a]	Nonsubstrate Limiting[b]
Formaldehyde	—	50–100
Acetone	—	>1000
Phenol	>1000	300–1000
Ethyl benzene	>1000	—
Dodecane	>1000	—

[a] Substrate limiting represents the condition in which the subject compound is the sole carbon and energy source.
[b] Nonsubstrate limiting represents the condition in which other carbon and energy sources are present.

Source: U.S. EPA, Handbook No. EPA/625/6-85/006, 1985.

often below this minimum concentration. In general, biodegradation of such materials can occur only if they are used as secondary substrates, i.e., if there is an abundant primary organic substrate available and bacteria capable of decomposing both. An alternative is decomposition within a non-steady-state system that has a sufficiently large population of bacteria previously grown on a primary substrate. Biodecomposition is also possible, if several organic substrates are present in a sufficiently large total concentration.

Subthreshold concentrations may be characteristic of some pollutants that are found entirely in the aqueous phase of some environments (Alexander, 1994). Low levels, or trace concentrations, of some organic substrates could be less than 100 ng/mL, which is characteristic of the concentrations of pollutants in many fresh, estuarine, and marine waters (Alexander, 1986). This level of contamination is also important, since criteria and standards for water quality refer to maximum acceptable levels of many organic pollutants that are below 100 ng/mL (Patrick and Mahapatra, 1968) and since numerous toxicants are harmful at levels in the ppb range (Batterton, Winters, and van Baalen, 1978).

Microorganisms might not assimilate carbon from chemicals present in trace amounts (Alexander, 1985). They might not grow or produce the large, acclimated populations needed for enhanced biodegradation. It is possible to predict the minimum concentration of a chemical necessary to support microbial growth. However, erroneous conclusions may be reached, if data from laboratory studies of chemicals at high concentrations are extrapolated to environments in which the chemicals exist at low concentrations (Alexander, 1986). It is important to use concentrations characteristic of those in nature for laboratory investigations. It is also possible for a chemical to be mineralized at one concentration and cometabolized at another.

In one study, mineralization of glucose at ng/mL by *Salmonella* stopped and cell death occurred before the substrate was exhausted (Simkins, Schmidt, and Alexander, 1984). This suggests a significant role for maintenance energy in determining the kinetics of mineralization of organic chemicals at low concentrations.

Threshold concentrations of the carbon source below which multiplication does not occur have been identified for specific organisms (Alexander, 1994). The threshold is about 18 µg/L for *Escherichia coli* and *Pseudomonas* sp. growing on glucose (Shehata and Marr, 1971; Boethling and Alexander, 1979b), 180 µg/L for *Aeromonas hydrophila* growing on starch (van der Kooij, Visser, and Hijnen, 1980), 210 µg/L for a coryneform bacterium using glucose (Law and Button, 1997), about 300 µg/L for a *Pseudomonas* growing at the expense of 2,4-dichlorophenol (Goldstein, Mallory, and Alexander, 1985), about 5 µg/L for *Salmonella typhimurium* on glucose (Schmidt and Alexander, 1985), and about 2 µg/L for a bacterium mineralizing quinoline (Brockman, Denovan, Hicks, and Fredrickson, 1989). A model has been developed for theoretical estimation of the threshold concentration of organic compounds (Alexander, 1994).

The threshold can be lowered, if the microbes also have certain alternative carbon sources available (Alexander, 1994). Sometimes, microorganisms can mineralize contaminants at substrate levels below the threshold for replication, by utilizing other compounds for growth. On the other hand, when there are other sources of carbon available, the microorganisms may utilize these instead of contaminant hydrocarbons. Acclimation may also involve a threshold. In many cases, the rate of mineralization is directly related to the concentration.

Several anomalies have been detected in biodegradation of low concentrations of organic compounds (Alexander, 1986).

Table 4.8 Biodegradable Concentrations of Compounds in the Field

Compound	Time for Degradation	Initial Concentration	Final Concentration	Organism
South Louisiana crude oil and motor oil[a]	—	1.0%	5.0%	*Aeromonas, Alcaligenes, Pseudomonas, Vibrio*
Phenol[b]	40 d	31 ppm	<0.01 ppm	Mutant bacteria
o-Chlorophenol[c]	40 d	120 ppm	30 ppm	—
Mixed Fuels/Solvents[d]	2½ mo	22–45 ppm	<550 ppb	—
Gasoline[e]	—	100–500 ppm	2–5 ppm	—
Methylene chloride[f]	1 yr	91 ppm	<1 ppm	—
Acetone[f]	1 yr	54 ppm	<1 ppm	—
Organic chemicals[g]	—	<1000 ppm (soil)	<1 ppm	Indigenous and hydrocarbon-degrading bacteria
Acrylonitrile[h]	3 mo	1,000 ppm	1 ppm	Mutant bacteria
Acrylonitrile[i]	1 mo	1,000 ppm	lod	Mutant bacteria
p-Cresol[i]	—	8 ppm		
Ethylene glycol[j]	—	1,200 mg/L	<50 mg/L to lod	—
Propyl acetate[k]	—	500 mg/L	<50 mg/L to lod	—
Organic chemicals[l]	—	<1,000 ppm	<1 ppm	Indigenous and specific degrader
Methylene chloride[m]	2½ mo	2500 mg/L	<100 mg/L	Commercial hydrocarbon degrading bacteria
Dichlorobenzene[m]	2½ mo	800 mg/L	<50 mg/L	Commercial hydrocarbon degrading bacteria
Phenols[m]	7 h	1,500 ppm	>1 ppm	*Azotobacter*
Phenol[n]	—	32 ng/g soil	—	—
TPH[o]	<6 mo	100–4,030 ppm	<100 ppm	Indigenous
Phenol[p]	7 d	5 mg/L	(100%)	Domestic wastewater
		10 mg/L	(100%)	
Naphthalene[p]	7 d	5 mg/L	(100%)	Domestic wastewater
		10 mg/L	(100%)	
Benzene[p]	7 d	5 mg/L	(100%)	Domestic wastewater
		10 mg/L	(95%)	
Toluene[p]	7 d	5 mg/L	(100%)	Domestic wastewater
		10 mg/L	(100%)	
Anthracene[p]	7 d	5 mg/L	(92%)	Domestic wastewater
		10 mg/L	(51%)	
Phenanthrene[p]	7 days	5 mg/L	(100%)	Domestic wastewater
		10 mg/L	(100%)	
1,2-Benzanthracene[p]	7 d	5 mg/L	(0%)	Domestic wastewater
		10 mg/L	(0%)	
Pyrene[p]	7 d	5 mg/L	(100%)	Domestic wastewater
		10 mg/L	(0%)	
Tertiary butyl alcohol[q]	>1 mo	10 mg/L	<lod	Soil
	>1 yr	70 mg/L	<lod	
Methanol[q]	>30 d	100 mg/L	<lod	Soil (aerobic and anaerobic)
	>200 d	1,000 mg/L	<lod	
Hydrocarbon[r]	—	10 ppm	—	—
Gasoline[s]	10 mo	11,500 gal/75,000 ft²	<50 ppm	—
Petroleum distillate[t]	21 d	12,000 ppm	>1 ppm	BI-CHEM-SUS-8
Formaldehyde[t]	22 d	1,400 ppm	>1 ppm	PHENOBAC
TPH[u]	5.5 mo	12,200 mg/kg	1,400 mg/kg	Commercial organisms or indigenous
PAHs[v]+	—	—	75–83% reduction	—
BTEX[w]	—	—	94% reduction	—
TPH[x]	<8 mo	4,000–5,000 mg/kg	<1000 mg/kg	—
Diesel fuel, heavy bunker oil[y]++	few mo	—	>72% reduction, (i.e., 1,000 ppm)	—
BTX in gasoline[z]+++	50 wk	—	>67% reduction	—
Petroleum hydrocarbons[aa]	—	36,000 lbs	94% reduction	—
Stoddard Solvent	1 yr	3,500 mg/kg	<lod	Indigenous

Table 4.8 (continued) Biodegradable Concentrations of Compounds in the Field

Compound	Time for Degradation	Initial Concentration	Final Concentration	Organism
Mop oil	1 yr	—	10 mg/kg	
TPHCs[ab]	1 yr	9,700 mg/kg	<lod	
Heavy engine oil (TPH)[ac]	—	—	94% reduction	—
Heavy crude oil (field trial)[ad]	25 d	3.8 mg/g	84.5% reduction	—
Heavy crude oil[ad]	4 mo (winter)	100 tons	88% reduction	F-1 fertilizer and mixed culture

Note: lod = limits of detection = 50 ppb
 + = vacuum heaps
 ++ = biopiles
 +++ = in combination with pneumatic fracturing

References:
a = (Frieze and Oujesky, 1983)
b = (Brown, Loper, and McGarvey, 1985)
c = (Minugh, Patry, Keech, and Leek, 1983)
d = (Jhaveria and Mazzacca, 1982)
e = (Polybac Corporation, 1983)
f = (Walton and Dobbs, 1980)
g = (Ohneck and Gardner, 1982)
h = (Quince and Gardner, 1982b)
i = (Pritchard, Van Veld, and Cooper, 1981)
j = (Ehrlich, Schroeder, and Martin, 1985)
k = (Brown, Norris, and Brubaker, 1985)
l = (U.S. EPA, 1985b)
m = (Roberts, Koff, and Karr, 1988)
n = (Scow, Simkins, and Alexander, 1986)
o = (Loo, 1994)
p = (Tabak, Quave, Mashni, and Barth, 1981)
q = (Novak, Goldsmith, Benoit, and O'Brien, 1985)
r = (Zeyer, Kuhn, and Schwarzenbach, 1986)
s = (Brown, Longfield, Norris, and Wolfe, 1985)
t = (Sikes, 1984)
u = (Pearce, Snyman, Oellermann, and Gerber, 1995)
v = (Eiermann and Bolliger, 1995b)
w = (Meyer, Warrelmann, and von Reis, 1995)
x = (Graves, Chase, and Ray, 1995)
y = (Anenson, 1995)
z = (Venkatraman, Schuring, Boland, and Kosson, 1995)
aa = (Nelson, Hicks, and Andrews, 1994)
ab = (Schmitt, Lieberman, Caplan, Blaes, Keating, and Richards, 1991)
ac = (Ying, et al., 1990)
ad = (Rosenberg, Legmann, Kushmaro, Taube, Adler, and Ron, 1992)

Table 4.9 Biodegradable Concentrations of Compounds in Bioreactors or Small-Scale Studies

Compound	Time for Degradation	Initial Concentration	Final Concentration	Organism
Lubricating oil[a]+	5 mo	2,400 mg/kg	700 mg/kg	Indigenous
Diesel fuel[b]	—	—	91–95% reduction	—
Petroleum hydrocarbons[c]	7 d	50,000 ppm	50% reduction	Hydrocarbon degrading bacteria
	1 mo	—	85% reduction	
Oil and PAHs[d]	—	—	65% reduction	—
	—	—	92% reduction	—
Naphthalene acenaphthene[e]	10 d	7 mg/L	ndl	—
	10 d	1 mg/L	ndl	
Naphthalene acenaphthene[e]++	45 d	7 mg/L	ndl	—
	40 d	0.4 mg/L	ndl	

Note: ndl = nondetectable levels
 + = composting
 ++ = denitrification

References:
a = (Puustinen, Joergensen, Strandberg, and Suortti, 1995)
b = (Riis, Miethe, and Babel, 1995)
c = (Findlay and Fogel, 1994)
d = (Oostenbrink, Kleijntjens, Mijnbeek, Kerkhof, Vetter, and Luyben, 1995)
e = (Mihelcic and Luthy, 1988b)

1. The rate of mineralization may be less than anticipated if it is assumed that the rates are linearly related to concentration.
2. Chemicals mineralized at one concentration may not be converted to carbon dioxide at lower levels.
3. Organic compounds may not be mineralized at low and presumably nontoxic levels in water, but they may be metabolized to carbon dioxide at still lower concentrations.

4. Mineralization may not follow the commonly described kinetics but may proceed in a biphasic manner.
5. The extent of mineralization in samples from a single body of water may vary markedly.
6. Microbial communities may acclimate to mineralize a substrate even though the substrate concentration is below the threshold level to sustain growth.
7. Compounds may be mineralized in some but not all waters.

Some NAPLs may sequester hydrophobic compounds away from the aqueous phase to an extent that the concentration falls below the threshold for biodegradation or to a level that results in slower biodegradation than expected (Efroymson and Alexander, 1995). Excessive sorption of the compound may also make an insufficient amount available for biodegradation (Alexander, 1994). At concentrations of around 20 to 40 mg phenanthrene/kg soil, the degradation rate decreased, in spite of the presence of a high number of phenanthrene-degrading bacteria (Jorgensen, Aamand, Jensen, Nielsen, and Jacobsen, 1995). The bioavailable phenanthrene was probably limited by the slow rate of desorption/dissolution of the compound from soil particles.

4.2 BIOLOGICAL FACTORS

Increased persistence of chemicals may result from several types of biological interactions: (1) the biocidal properties of the chemicals to soil microorganisms may preclude their biodegradation, (2) direct inhibition of the adaptive enzymes of effective soil microorganisms, and (3) inhibition of the proliferation processes of effective microorganisms (Kaufman, 1983). Inhibition of microbial degradation may ultimately affect the mobility of a chemical in soil.

It should be realized that biodegradability of a petroleum compound is a result of the action of the mixed flora present at the location (Section 5.2.2.3), and that the material being degraded is actually a complex mixture of hydrocarbons, some of which contribute to the breakdown of others (Cooney, Silver, and Beck, 1985). There are cooperative and competitive effects between organisms *in situ*, as well as the potential for co-oxidation and cometabolism (Cooney, 1980) (Section 5.2.3). It is also possible that problems could arise involving the degradation, persistence, or toxicity of organic chemicals when several wastes or their residues are present in the soil together (Kaufman and Plimmer, 1972). These factors should be taken into account when assessing the microbiological potential for petroleum degradation.

It has been proposed that the observed recalcitrance of many compounds *in vitro* may be due to the lack of properly designed experiments under the appropriate conditions that are conducive to degradation (Hegeman, 1972). Recalcitrance could also be due to insufficient time to evolve enzymatic pathways to degrade certain chemicals. The acclimation or induction of enzymes that catalyze the necessary reactions in the microbial population is an important factor affecting biodegradability (Paris, Wolfe, Steen, and Baughman, 1983).

Hydrocarbon-degrading microorganisms could occur naturally in the contaminated soil. These will require adequate nutrients and aerobic or anaerobic soil conditions for biodegradation to occur. If the relevant microorganisms are not present, the soil can be augmented with acclimated microbes, mixtures of microorganisms, or genetically engineered organisms to target specific pollutants. Cometabolism and bioaccumulation play important roles. Enzymes produced by the soil flora can contribute to biodegradation, and cell-free enzymes could be added in place of live, whole cells. Many soil microbes produce antibiotics that can affect the viability of other microorganisms. Some microbes have also been found to produce biosurfactants, which enhance availability of the contaminants (Section 5.3.1.2).

Permeability of the soil can be affected by overgrowth of microorganisms (Olmsted, 1994). Both total biomass production and growth efficiency are inversely correlated with exopolysaccharide (EPS) production, which can lead to subsurface plugging, under aerobic or anaerobic conditions (Hanneman, Johnstone, Yonge, Petersen, Peyton, and Skeen, 1995).

The role of biological factors in degradation and how these factors can be optimized to enhance bioremediation are presented in Section 5.2.

4.3 SOIL/ENVIRONMENTAL FACTORS

Many microorganisms have specific ecological niches for proliferation and colonization (Daubaras and Chakrabarty, 1992). Metabolism by the indigenous microflora is influenced by soil and environmental factors, such as light, temperature (climate, daily and seasonal temperature fluctuations), pH, presence of cometabolites, reactive radicals, other organic and inorganic compounds, available oxygen and nutrients (nitrogen and phosphorus), as well as the physical state of the oil (Cooney, Silver, and Beck, 1985), moisture content, organic matter, oxidation-reduction potential, attenuation, and soil texture and structure. The environment influences biodegradation by regulating both the bioavailability of the compound and the activity of the degraders. Salinity, temperature, chlorophyll, nitrogen, and phosphorus concentrations have also been correlated with rates of biodegradation in surface environments. The numbers of hydrocarbon-using organisms may also be enhanced by prior pollution of a site.

Naturally occurring organic materials can influence the ability of microorganisms to degrade pollutants (Shimp and Pfaender, 1984). Amino acids, fatty acids, and carbohydrates stimulate biodegradation of monosubstituted phenols, while humics decrease biodegradation rates. Many aromatic compounds bind to particulate material and interfere with biodegradation (Horowitz and Tiedje, 1980). See Sections 4.1.4 and 5.1.6 for more information on the effect of organic matter in soil on bioremediation.

Aerobic and anaerobic soil conditions will influence the kind of microorganisms present and the type of biodegradation that will occur (see Sections 3 and 5.2.1). Predation can inhibit biodegradation (Section 5.2.1.6). Antibiotics present either as industrial wastes or in various human and livestock wastes are another area of concern (Kaufman and Plimmer, 1972). These chemicals are used extensively to control diseases and pests but, once excreted, are little understood with regard to their degradation and fate in the environment or their influence on various components of the environment.

The environment affects the evolutionary processes in microorganisms responsible for new degradative genes for degrading hazardous compounds (Daubaras and Chakrabarty, 1992). Environmental factors can determine the extent of microbial gene expression by activating or repressing genes through a sensory signal transduction process. For instance, the process of soil bioventing commonly increases the numbers of TOL (toluene) plasmid–containing bacteria by two to three orders of magnitude (Brockman, 1995). It is important to understand how the environment controls the expression of microbial genes and the evolution of new genes in bacteria.

Addition of nutrient supplements to enhance microbial growth and biodegradation can often lead to plugging near the area of injection (Olmsted, 1994). Overloading the soil with elements present in the waste or already in the soil can cause toxicity, leaching problems, and phosphate plugging (JRB Associates, Inc., 1984). Nutrient formulations often contain excessive orthophosphate to decrease the rate of peroxide decomposition (Aggarwal, Means, and Hinchee, 1991). Polyphosphates (e.g., pyrophosphate, tripolyphosphate, and trimetaphosphate) can serve as an alternative source of phosphate to avoid plugging problems. At low concentrations of oil contamination, medium levels of nutrients would have less impact on permeability over time (Olmsted, 1994). Low nutrient levels would reduce clogging effects even more.

Section 5.1 reviews in depth the environmental, or soil, factors of moisture, temperature, pH, oxygen levels, nutrients, organic matter, oxidation-reduction potential, attenuation, texture, and structure, and how these can be controlled to improve the process of bioremediation.

Section 5

Optimization of Bioremediation

There are three key issues that should be addressed to achieve a successful bioremediation (Tiedje, 1993). It must be determined whether the contaminant is biodegradable, whether the environment is habitable (presence of toxic chemicals or sufficient life-sustaining growth factors), and what the rate-limiting factor is and whether it can be modified. In an ecological approach to bioremediation, the important issue is to establish whether or not the conditions of natural selection can be expected to be met within the site vicinity. With this approach, more emphasis is placed on meeting requirements of the microorganisms, and less on measuring the actual pollutant.

Each contaminated site exhibits different characteristics and requires a site-specific remediation plan (Forsyth, Tsao, and Bleam, 1995). Decontaminating a site polluted with hazardous materials is a complex procedure involving systematic, step-by-step problem solving. The conditions necessary to optimize the efficiency of microbial systems in degrading environmental pollutants and the economics required must be assessed to select and implement cost-effective biotreatment. This requires understanding of the microorganisms and the conditions necessary for them to become established and maintained, and the scientific data must be translated into cost-effective, full-scale cleanup processes. Augmentation with proven contaminant-degrading microorganisms can save time and money over alternative approaches.

This section discusses how optimum conditions for bioremediation can be achieved through site manipulation, biological intervention, or chemical treatment. It is important to monitor the process to determine that biodegradation is occurring and to allow the conditions to be modified as necessary, to maintain optimum performance.

5.1 VARIATION OF SOIL FACTORS

The various chemical and physical properties of a soil determine the nature of the environment in which microorganisms are found (Parr, Sikora, and Burge, 1983). In turn, the soil environment affects the composition of the microbiological population both qualitatively and quantitatively. The rate of decomposition of an organic waste depends primarily upon its chemical composition and upon those factors that affect the soil environment. Factors having the greatest effect on microbial growth and activity will have the greatest potential for altering the rate of residue decomposition in soil.

The ability of the upper 6 in. of soil to absorb nutrients and hold water depends upon its physical and chemical properties of texture, infiltration and permeability, water-holding capacity, bulk density, organic matter content, cation exchange capacity, macronutrient content, salinity, and micronutrient content (Hornick, 1983). A typical mineral soil is composed of approximately 45% mineral material (varying proportions of sand, silt, and clay), 25% air and 25% water (i.e., 50% pore space, usually half saturated with water), and 5% organic matter, although this is highly variable. Any significant change in the balance of these components could affect the physical and chemical properties of the soil. This may alter the ability of the soil to support the chemical and biological reactions necessary to degrade, detoxify, inactivate, or immobilize toxic waste constituents.

Most soils have a tremendous capacity to detoxify organic chemical wastes by diluting the compounds, acting as a buffering system, and decomposing the material through microbial activity (JRB Associates, Inc., 1984). The most important soil characteristics for this detoxification are those that affect water movement and contaminant mobility, i.e., infiltration and permeability. Certain waste characteristics can also affect soil infiltration and permeability, and this interaction should be taken into account. Table 5.1 lists the site/soil properties that should be identified to be able to predict potential migration of the contaminating material and indicate what will be necessary for manipulating the soil characteristics for optimum results. Some of the soil factors, however, can be managed only near the surface for enhancing the soil treatment.

Unless all the proper conditions are met for a given compound, biodegradation is not likely to occur (Bitton and Gerba, 1985). Before *in situ* biological remedial actions can be initiated for treating hazardous

Table 5.1 Important Site and Soil Characteristics for *In Situ* Treatment

Site location/topography and slope	Hydraulic properties and conditions
Soil type, and extent	Soil/water characteristic curve
Soil profile properties	Field capacity/permanent wilting point
Boundary characteristics	Water-holding capacity[a]
Depth	Permeability (under saturated and range of unsaturated conditions)[a]
Texture[a]	Infiltration rates[b]
Amount and type of coarse fragments	Depth to impermeable layer or bedrock
Structure[a]	Depth to groundwater, including seasonal variations[b]
Color	Flooding frequency
Degree of mottling	Runoff potential[b]
Bulk density[a]	Geological and hydrogeological factors
Clay content	Subsurface geological features
Type of clay	Groundwater flow patterns and characteristics[b]
Cation exchange capacity[a]	Meteorological and climatological data
Organic matter content[b]	Wind velocity and direction
pH[b]	Temperature
Eh[b]	Precipitation
Aeration status[b]	Water budget

[a] Factors that may be managed to enhance soil treatment with shallow depth
[b] Factors that may be managed to enhance soil treatment

Source: JRB Associates, Inc. Report prepared for Municipal Environmental Research Laboratory, Cincinnati, OH, 1984. PB 85-124899.

waste–contaminated soils, both the site and waste characteristics must be evaluated (Solanas, Pares, Bayona, and Albaiges, 1984). These features will help determine whether or not a biological approach is the most feasible treatment option and, if selected, how biodegradation can be used most effectively with the prevailing conditions.

There are more than 1000 different soil types in the U.S. alone (Federle, Dobbins, Thornton-Manning, and Jones, 1986). The U.S. Soil Conservation Service has characterized certain chemical and physical parameters for many of them while preparing soil maps. These data are readily available. They would help in predicting biomass and activity in various profiles.

The most important soil factors that affect degradation are water; temperature; soil pH; aeration or oxygen supply; available nutrients, i.e., nitrogen (N), phosphorus (P), potassium (K), sulfur (S); oxidation/reduction potential; and soil texture and structure. Any treatments applied to the soil to enhance contaminant removal processes must not alter the physical or chemical environment to the extent that they would severely restrict microbial growth or biochemical activity (Sims and Bass, 1984). In general, this means that the soil water potential should be greater than –15 bar (Sommers, Gilmore, Wildung, and Beck, 1981); the pH should be between 5 and 9 (Atlas and Bartha, 1981; Sommers, Gilmore, Wildung, and Beck, 1981); and the oxidation-reduction (redox) potential should be between pe + pH of 17.5 to 2.7 (Baas Becking, Kaplan, and Moore, 1960). Soil pH and redox boundaries should be carefully monitored when chemical and biological treatments are combined.

Since the activity of microorganisms is so dependent upon soil conditions, modification of soil properties is a viable method of enhancing the microbial activity in the soil (Sims and Bass, 1984). To vary these factors for use as a treatment technology, the following information is required:

    Characterization and concentration of wastes, both organics and inorganics, at the site;
    Microorganisms present at site;
    Biodegradability of waste constituents (half-life, rate constant);
    Biodegradation products, particularly hazardous products;
    Depth, profile, and areal distribution of constituents;
    Soil moisture;
    Other soil properties for biological activity (pH, Eh, oxygen content, nutrient content, organic matter, temperature);
    Trafficability of soil and site.

The influence of soil factors, such as temperature and nutrient concentration, on phenol mineralization, for example, shows great variability as a function of soil type and horizon (Thornton-Manning, Jones, and Federle, 1987). Most of these factors do not function independently; i.e., a change in one may effect a change in others (Parr, Sikora, and Burge, 1983). While the soil factors play an important role in biodegradation, because of these interactions, it is not always easy to predict *a priori* how temperature or another environmental variable will affect biodegradation in a given soil environment (Thornton-Manning, Jones, and Federle, 1987). However, if any of the factors that affect degradation processes in soil are at less than an optimum level, microbial activity will be lowered accordingly and substrate decomposition decreased (Parr, Sikora, and Burge, 1983).

The inherent capacity of soil to degrade toxicants by chemical and biological mechanisms can be maximized by identification of the soil conditions that promote the degradation of each toxicant and manipulation of the soil environment to bring about these conditions (Arthur D. Little, Inc., 1976). Although each toxicant, in general, has a unique set of ideal soil conditions for degradation, for some compounds these ideal conditions overlap, and more than one toxic substance can be the focus of soil manipulation at one time. For other compounds, the ideal conditions do not overlap and are sometimes even contradictory; these materials must be treated in series.

Table 5.2 lists the soil factors that may have to be modified during the use of various treatment technologies (Sims and Bass, 1984).

5.1.1 SOIL MOISTURE

Biodegradation of waste chemicals in the soil requires water for microbial growth and for diffusion of nutrients and by-products during the breakdown process (JRB Associates, Inc., 1984). Extremes of very wet or very dry soil moisture markedly reduce waste biodegradation rates (Arora, Cantor, and Nemeth, 1982). Aerobic waste hydrocarbon decomposition is diminished under saturated soil moisture conditions because of low oxygen supply, while under very dry conditions, microbial activity is hindered due to insufficient moisture levels necessary for microbial metabolism (CONCAWE, 1980).

A typical soil is about 50% pore space and 50% solid matter (JRB Associates, Inc., 1984). Water entering the soil fills the pore spaces until they are full. The water then continues to move down into the subsoil, displacing air as it goes. The soil is saturated when it is at its maximum retentive capacity. When water then drains from the pores, the soil becomes unsaturated. Soils with large pores, such as sands, lose water rapidly. Larger pores are a less hospitable environment for microorganisms (Turco and Sadowsky, 1995), whereas the smaller pores inside the aggregate retain water (Papendick and Campbell, 1981).

If the soil is too impermeable, it will be difficult to circulate treatment agents or to withdraw the polluted water (Nielsen, 1983). Soils with a mixture of pore sizes, such as loamy soils, hold more water at saturation and lose water more slowly. The density and texture of the soil determine the water-holding capacity, which in turn affects the available oxygen, redox potential, and microbial activity (Parr, Sikora, and Burge, 1983). The actual microbial species composition of a soil is often dependent upon water availability. The migration of organisms in the soil can also be affected by pore size. Small bacteria are on the order of 0.5 to 1.0 μm in diameter (Bitton and Gerba, 1985). Larger bacteria tend to be immobilized in soils by physical straining or filtering.

The water content of soil typically ranges from 15 to 35 vol% (Huddleston, Bleckmann, and Wolfe, 1986). At 35%, most soils are water saturated. At the other extreme, the concentration can drop lower than 15% under unusually arid conditions. Soil water content is commonly addressed as percent of soil water-holding capacity. A soil water-holding capacity range of 25 to 100% is typically equivalent to a range of 7 to 28% volume percent. Dibble and Bartha (1979) report optimal biodegradation at a soil water-holding capacity of 30 to 90%.

Table 5.3 shows some of the conditions that can be selectively altered for removal of anthropogenic compounds by particular groups of microorganisms.

Field capacity refers to the percentage of water remaining in a soil after having been saturated and free gravitational drainage has ceased (JRB Associates, Inc., 1984). Gravitational water movement is important for mobilizing contaminants and nutrients, due to leaching. Slow drainage can reduce microbial activity as a result of poor aeration, change in oxidation-reduction potential, change in nutrient status, and increased concentration of natural minerals or contaminants to toxic levels in the pore water. The amount of water held in a soil between field capacity and the permanent wilting point for plants is known as available water. This is the water available for plants and for soil microbial and chemical reactions.

Table 5.2 Soil Modification Requirements for Treatment Technologies

Technology	Oxygen Content	Moisture Content	Nutrient Content	pH	Temperature
Extraction	—	—	—	X	X
Immobilization					
Sorption (heavy metals)					
Agricultural products	—	—	—	X	—
Activated carbon	—	—	—	X	—
Tetren	—	—	—	X	—
Sorption (organics)					
Soil moisture	—	X	—	—	—
Agricultural products	—	—	—	—	—
Activated carbon	—	—	—	—	—
Ion exchange					
Clay	—	—	—	X	—
Synthetic resins	—	—	—	X	—
Zeolites	—	—	—	X	—
Precipitation					
Sulfides	X	X	—	X	—
Carbonates, phosphates, and hydroxides	X	X	—	X	—
Degradation					
Oxidation					
Soil-catalyzed reactions	X	—	—	X	—
Oxidizing agents	X	—	—	X	—
Reduction					
Reducing agents	X	X	—	X	—
Chromium	X	—	—	X	—
Selenium	X	—	—	X	—
Polychlorinated biphenyls and dioxins	—	X	—	—	X
Polymerization	—	—	—	—	—
Modification of soil properties (for biodegradation)					
Soil moisture	—	X	—	—	—
Soil oxygen — aerobic	X	—	—	—	—
Soil oxygen — anaerobic	X	X	—	—	—
Soil pH	—	—	—	X	—
Nutrients	—	—	X	—	—
Nonspecific org. amendments	—	—	—	—	X
Analog enrichment for cometabolism	—	—	—	—	X
Exogenous acclimated or mutant microorganisms	—	—	X	—	X
Cell-free enzymes	—	—	—	—	X
Photolysis					
Proton donors	—	—	—	—	—
Enhance volatilization	—	X	—	—	—
Attenuation					
Metals	—	—	—	—	—
Organics	—	—	—	—	—
Reduction of Volatiles					
Soil vapor volume	—	X	—	—	—
Soil cooling	—	—	—	—	X

Source: Arthur D. Little, Inc., reprinted in Sims, R. and Bass, J. EPA Report No. EPA-540/2-84-003a, 1984.

Bacterial activity is highest in the presence of moisture (JRB Associates, Inc., 1984). Several authors have indicated ranges of moisture for optimum biodegradation (Bossert, Kachel, and Bartha, 1984; Ryan, Hanson, and Loehr, 1986; Huddleston, Bleckman, and Wolfe, 1986). Some indicate that 30 to 90% of field capacity is needed. Others, that 50 to 80% is a better range. Based on first-order regression relationships for O_2 uptake rates, moisture addition of 35 to 50% field capacity was found to accelerate *in situ* respiration in a JP-4-contaminated soil (Dupont, Doucette, and Hinchee, 1991). The aerobic

Table 5.3 Selective Use of Microorganisms for Removal of Different Anthropogenic Compounds

Microorganism	Selective[a] Characteristics	Significance
Fungi — Yeast, mold	pH < 5, ae-mae; high O_2 tension, pH < 5 moisture about 50%	Attacks and partially degrades compounds not readily metabolized by other organisms; wide range of nonspecific enzymes
Algae	ae-mae; light: 600 to 700 nm; low carbon flux	Self-sustaining population, light is primary energy source, partially degrades certain complex compounds, photochemical reactions, oxygenates effluent, no aeration needed, supports growth of other microbes, effective in bioaccumulation of hydrophobic substances
Cyanobacteria (blue-green algae)	ae-mae, an; light: 600 to 700 nm; low carbon flux	See algae
Bacteria		
Heterotrophs (aerobic)	ae; proper organic substrate, growth factors as required; Eh: 0.45 to 0.2 V	For many compounds degradation is more complete and faster than under anaerobic conditions, high sludge production
Anaerobic (fastidious)	an; Eh: <–0.2 to –0.4 V	Conditions for abiotic or biological reductive dechlorination, certain detoxification reactions not possible under aerobic conditions; no aeration, little sludge produced
Facultative anaerobes	ae, mae-an; Eh: <–0.2 V	No aeration, reductive dechlorination possible
Photosynthetic bacteria		
Purple sulfur	an (light), mae (dark); Eh: 0 to –0.2 V; S^{-2}: 2 to 8 mM, 0.4 to 1 mM; light: 800 to 890 nm at 1000 to 2000 lux, high intensities near limit; low C flux	Self-sustaining population able to use light energy, conditions right for reductive dechlorination, no aeration
Purple nonsulfur	an; Eh: 0 to –0.2 V; light: 800 to 890 nm; low C flux	See purple sulfur bacteria, also nonspecific enzymes
Actinomycetes	ae, moisture: 80 to 87%, temp.: 23 to 28° C, urea as nitrogen source	Universal scavengers with range of complex organic substrates often not used by other microbes
Oligotrophs (from almost any group above)	ae; carbon flux of <1 mg/L/d; favorable attachment sites	Removal of organic contaminants in trace concentrations, many inducible enzymes for multiple substrates

Abbreviations: ae = aerobic; mae = microaerophilic (<0.2 atm oxygen); an = anaerobic.

[a] Possible characteristics for selection, not growth range.

Source: From Kobayashi, H. and Rittmann, B.E. *Environ. Sci. Technol.* 16:170A–183A. American Chemical Society. Washington, D.C., 1982. With permission.

biodegradation of simple or complex organic material in soil is commonly greatest at 50 to 70% of the soil water-holding (field) capacity (Pramer and Bartha, 1972). Inhibition at levels below 30 to 40% is due to inadequate water activity, and high values interfere with soil aeration. The dependency on soil water content for biodegradation of petroleum constituents is compound specific and probably also soil specific (Holman and Tsang, 1995). Moisture is a critical parameter for degradation of two-, three-, and four-ring polycyclic aromatic hydrocarbons (PAHs), and it has been found that degradation is considerably greater at 80% than at 40% of field capacity (Loehr, 1992).

Holman and Tsang (1995) determined that a water content of 50 to 70% of field capacity was optimum for biodegradation of aromatic hydrocarbons to proceed at maximum rate. For simple monoaromatic and diaromatic hydrocarbons, such as toluene and naphthalene, a first-order kinetic model provides a good fit to mineralization data over a range of soil moisture content. However, for larger PAHs, such as phenanthrene and anthracene, the model provides a good fit only at soil water content below 50%. Since long-chain aliphatic hydrocarbons have such a low solubility, their mineralization is little affected by the soil water content.

There is a dramatic difference in characteristics of microbial communities as a result of different water content, which parallels mineralization measurements (Holman and Tsang, 1995). The greatest diversity and activity of microorganisms and the highest population densities are consistently observed in the sandy, water-bearing strata, whereas the dense, dry-clay layer zones have the least microbiological activity (Fredrickson and Hicks, 1987).

Soil gas humidities <30% cause considerable retardation of hydrocarbon vapors in all media (Batterman, Kulshrestha, and Cheng, 1995). Retardation coefficients decrease but remain large with increasing humidity in organic-rich soils. Based on soil–water isotherms, there may be competitive sorption between hydrocarbon and water vapors on soil surfaces, especially the mineral fraction.

Where it is necessary to predict and interpret the response of microorganisms in soils to organic wastes, both the water content and water potential should be reported (Parr, Sikora, and Burge, 1983). Water potential is useful for quantifying the energy status of water in soils containing waste chemicals. Generally, with decreasing water potentials, fewer organisms are able to grow and reproduce; and bacterial activity is usually greatest at high water potentials (wet conditions). Species composition of the soil microflora is regulated largely by water availability, which, in turn, is governed essentially by the energy of the water in contact with the soil or waste.

Some fungi can tolerate dry soils but do not grow well if the soil is wet (Clark, 1967). Bacteria may be antagonistic to fungi under moister conditions. At low potentials, bacteria are less active, allowing fungi to predominate (Cook and Papendick, 1970). Microbial decomposition of organic material in drier soils is probably due primarily to fungi (Gray, 1978; Harris, 1981). When soil becomes too dry, many microorganisms form spores, cysts, or other resistant forms, while many others are killed by desiccation (JRB Associates, Inc., 1984).

A well-drained soil (e.g., a loamy soil) is one in which water is removed readily but not rapidly (JRB Associates, Inc., 1984); a poorly drained soil (e.g., a poorly structured fine soil) remains waterlogged for extended periods of time, producing reducing conditions and insufficient oxygen for biological activity; and an excessively drained soil (e.g., a sandy soil) is one in which water can be removed readily to the point that drought conditions occur. For *in situ* treatment of hazardous waste–contaminated soils, the most desirable soil would be one in which permeability is only large enough to maximize soil attenuation processes (e.g., adequate aeration for aerobic microbial degradation) while still minimizing leaching. Although fine-textured soils may have the maximum total water-holding capacity, medium-textured soils have the maximum available water due to favorable pore size distribution.

Control of moisture content of soils at an *in situ* treatment site may be essential for control and optimization of some degradative and sorptive processes, as well as for suppression of volatilization of some hazardous constituents (Sims and Bass, 1984). The moisture content of soil may be controlled to immobilize constituents in contaminated soils and to allow additional time for accomplishing biological degradation. When contaminants are immobilized by this technique and anaerobic decomposition is desired, anaerobiosis must be achieved by a means other than flooding, such as soil compaction or organic matter addition. Control of soil moisture may be achieved through irrigation, drainage, or a combination of methods.

The need for moisture was demonstrated in the efforts to remediate oil-polluted Kuwaiti desert soil (Radwan, Sorkhoh, Fardoun, and Al-Hasan, 1995). The amount of alkanes in the untreated controls remained constant during the dry hot months then decreased during the rainy season. After 1 year, the desert had cleaned itself of half the contaminating extractable alkanes, but had required moisture to do so. Fertilized soils reduced these compounds to about a third in that time.

In another instance, a subsurface drip irrigation system was used to increase soil moisture during bioventing dry, sandy soils contaminated with gasoline, JP-5 jet fuel, and diesel fuel to a depth of 24 m (Zwick, Leeson, Hinchee, Hoeppel, and Bowling, 1995). *In situ* respiration rates increased significantly as a result.

Sometimes a site with shallow depth contamination may require soil mixing to dilute the wastes and incorporate nutrients and oxygen, as well as to enhance soil drying (Sims and Bass, 1984). It may be necessary to install a drainage system to reduce soil moisture. Increasing soil temperature will enhance surface soil drying. This can be achieved with landfarming. However, drying the soil may retard microbial activity, as well as increase volatilization of volatile waste components.

Excess moisture, extremely dry conditions, pooling, or flooding should be avoided (Zitrides, 1983). Biodecontamination programs should not be conducted during heavy rains or drought. However, an observed lack of inhibition at 30% of the field capacity suggests that the moisture requirement for maximum activity on hydrophobic petroleum may be different than the optimal moisture levels for the biodegradation of hydrophilic substrates (Dibble and Bartha, 1979a).

Rainfall dissolves contaminants and acts as a carrier as it percolates through the soil on its way to the groundwater, which can be useful to the bioremediation plan, if this is desired (Dietz, 1980). Rainwater also keeps the contaminated soil moist, and microorganisms will utilize the oxygen dissolved in interstitial water droplets (Thibault and Elliott, 1980).

Many organisms are capable of metabolic activity at water potentials lower than -15 bar (Soil Science Society of America, 1981). The lower limit for all bacterial activity is probably about -80 bar, but some organisms cease activities at -5 bar. Although many microbial functions continue in soils at -15 bar or drier, optimum biochemical activity is usually observed at soil water potentials of -0.1 to 1.0 bar (Sommers, Gilmore, Wildung, and Beck, 1981). The kinds of microorganisms that are metabolically active in the soil will be affected. Degradation rates are highest at soil water potential between 0 and -1 bar. When natural precipitation cannot maintain near optimal soil moisture for microbial activity, irrigation may be necessary (Sims and Bass, 1984).

Moisture control is widely practiced in agriculture; however, there is little information on its use to stimulate biological degradation of hazardous materials in soil (Sims and Bass, 1984). Most laboratory studies have been conducted at or near optimal soil moisture. The success of this technology depends upon the biodegradability of the waste constituents and the suitability of the site and soil for moisture control. Although degradation of hazardous organic compounds may be accelerated by soil moisture optimization, effectiveness of this treatment approach may be enhanced by combination with other techniques to increase biological activity. The technology is reliable in that it has been used in agriculture, but retreatment is necessary. There may be problems with leaching of soluble hazardous compounds and erosion.

Control of soil moisture content can be practiced to optimize degradative and sorptive processes and may be achieved by several means (Sims and Bass, 1984). Supplemental water may be added to the site (irrigation), excess water may be removed (drainage, well points), or these methods can be combined with other techniques, such as using soil additives, for greater moisture control.

5.1.1.1 Irrigation

Soil may be irrigated by subirrigation, surface irrigation, or overhead (sprinkler) irrigation (Fry and Grey, 1971). With subirrigation, water is applied below the ground surface and moves upward by capillary action. Water with high salinity may allow accumulation of salts in the surface soil, with an adverse effect on microbial activity. The site must be nearly level and smooth, with either a natural or perched water table, which can be maintained at a desired elevation. Check dams and gates in open ditches or jointed perforated pipe can be used to maintain the water level in the soil. These systems may be limited by the restrictive site criteria. A subirrigation system might be combined with a drainage system to optimize soil moisture content. At a hazardous waste site, though, raising the water table might produce undesirable groundwater contamination.

With trickle irrigation, filtered water is supplied directly on or below the soil surface through an extensive pipe network with low-flow-rate outlets only to areas that require irrigation (Fry and Grey, 1971). Coverage of an area will not be uniform, but with proper management, percolation and evaporation losses can be reduced. For most in-place treatment sites, this method would probably not be appropriate, but it may be applicable in an area where only "hot spots" of wastes are being treated.

Surface irrigation includes flood, furrow, or corrugation irrigation (Fry and Grey, 1971). Since off-site migration of hazardous constituents to groundwaters or surface waters should normally be prevented, surface irrigation should be considered with caution. Contaminated water may also be a hazard to on-site personnel.

In flood irrigation, water covers the surface of a soil in a continuous sheet (Fry and Grey, 1971). Theoretically, water should remain in place just long enough to apply the desired amount, but this is difficult or impossible to achieve under field conditions. Widrig and Manning (1995) determined that continuous saturation by flooding with nitrogen and phosphorus amendments was not as effective as periodic operation, consisting of flooding with nutrients, followed by draining and forced aeration. Monitoring CO_2 and O_2 levels *in situ* may allow optimization of the timing of flooding and aeration events to increase degradation rates.

In furrow irrigation, water is applied in narrow channels or furrows. As the water runs down the furrow, part of it infiltrates the soil (Fry and Grey, 1971). Irrigation of the soil between furrows requires considerable lateral water movement. Salts may accumulate between furrows. Furrow irrigation frequently requires extensive land preparation, which usually would not be possible or desirable at a hazardous waste site because of contamination and safety considerations.

In corrugation irrigation, as with furrow irrigation, water is applied in small furrows from a head ditch (Fry and Grey, 1971). The furrows are used in this case only to guide the water, and overflooding of the furrows can occur.

In general, control and uniform application of water is difficult with surface irrigation. Also, soils high in clay content tend to seal when water floods the surface, limiting water infiltration.

The basic sprinkler irrigation system consists of a pump to transfer water from the source to the site, a pipe or pipes leading from the pump to the sprinkler heads, and the spray nozzles (Fry and Grey, 1971). Sprinkler irrigation has many advantages. For instance, application rates can be adjusted for soils of different textures, even within the same area; water can be distributed more uniformly; and erosion and runoff of irrigation water can be controlled or eliminated. Sprinkler irrigation is also possible on steep, sloping land and irregular terrain. This method usually requires less water than surface flooding, and the amount of water applied can be controlled to meet the needs of the in-place treatment technique. Also, a larger soil surface area can be covered, which could facilitate soil washing.

There are several types of sprinkler irrigation systems (Fry and Grey, 1971):

1. Permanent installations with buried main and lateral lines;
2. Semipermanent systems with fixed main lines and portable laterals;
3. Fully portable systems with portable main lines and laterals, as well as a portable pumping plant.

The first two types (especially the first) would probably not be cost-effective or appropriate for a hazardous waste site because of the required land disturbance for installation and the limited time period for execution of the treatment.

There are fully portable systems available. These may have hand-moved or mechanically moved laterals (Fry and Grey, 1971). Portable systems are useful in difficult areas, such as forests, where they will not interfere with trees. Mechanically moved laterals may be side-roll/wheel-move, center-pivot systems, or traveling sprinklers. This equipment is more expensive but requires much less labor than the hand-moved systems. The health and safety of workers must be considered, as well as the cost, in the choice of an appropriate system.

5.1.1.2 Drainage

When irrigation is used, controls for erosion and proper drainage due to runoff are necessary (Sims and Bass, 1984). A properly designed drainage system removes excess water or lowers the groundwater level to prevent waterlogging (Fry and Grey, 1971). Open ditches and lateral drains are good for surface drainage, while a system of open ditches and buried tube drains into which water seeps by gravity is better for subsurface drainage. The collected water is conveyed to a suitable disposal point. Pumping from wells will also provide subsurface drainage by lowering the water table. The drainage water to be disposed of off-site must not be contaminated with hazardous substances, and must be collected, stored, treated, or recycled, if not acceptable for off-site release.

Subsurface drains can be used to lower the water table, while surface drains are used where subsurface drainage is impractical (e.g., impermeable soils, excavation difficult) to remove surface water or lower the water table (Donnan and Schwab, 1974). Construction materials for the drainage systems include clay or concrete tile, corrugated metal pipe, and plastic tubing. Selection of the materials depends upon strength requirements, chemical compatibility, and cost.

5.1.1.3 Additives

Various additives are available to enhance moisture control; e.g., the water-retaining capacity of the soil can be enhanced by adding water-storing substances (Nimah, Ryan, and Chaudhry, 1983). Evaporation retardants are available for retaining soil moisture. There are also water-repelling agents for diminishing water absorption by soils. Water-repelling soils can be treated with surface-active wetting agents to improve water infiltration and percolation. Surface-active agents also accelerate soil drainage, modify soil structure, disperse clays, and make soil more compactable.

5.1.2 TEMPERATURE

Soil temperature is one of the more important factors controlling microbiological activity and the rate of organic matter decomposition (Sims and Bass, 1984). Temperatures of both air and soil affect the rate of biological degradation processes in the soil, as well as the soil moisture content (JRB Associates, Inc., 1984). Temperature affects the physical nature and composition of the petroleum, the rate of microbial hydrocarbon metabolism, and the composition of the microbial communities (Atlas, 1994). There is an optimum temperature, beyond which biological activity often decreases rapidly, thus displaying a growth curve that is skewed to the right (JRB Associates, Inc., 1984).

Generally, raising the temperature increases the rate of degradation of organic compounds in soil (JRB Associates, Inc., 1982). Microbial growth usually doubles for every 10°C increase (Thibault and Elliott, 1979). There is a decrease in adsorption with rising temperature, which makes more organics available for the microorganisms to degrade (JRB Associates, Inc., 1984). On the other hand, higher temperatures increase evaporation of short-chain alkanes and other low-molecular-weight hydrocarbons, which usually cause solvent-type membrane toxicity to microorganisms (Atlas, 1994). They also decrease the viscosity of the petroleum hydrocarbons and their solubility in the soil aqueous phase. High temperatures, well above those normally experienced in soil, cause very rapid decreases in growth and metabolism and become lethal (Huddleston, Bleckmann, and Wolfe, 1986). If temperatures exceed 41 to 42°C, enzymes in the bacteria normally begin to break down, and life processes fail (Lapinskas, 1989).

Conversely, a lowering of the temperature is associated with a slowing of the microbial growth rate (Thibault and Elliott, 1979). Low temperatures can lengthen the acclimation period and delay onset of biodegradation (Zhou and Crawford, 1995). A microbial community will undergo an adaptation or selection process in the mineralization of a compound, which is reflected in a lag period that often increases with decreasing temperature (Thornton-Manning, Jones, and Federle, 1987). Low temperatures also can decrease microbial enzymatic activity — i.e., the "Q_{10}" effect (Zhou and Crawford, 1995). Low temperatures are not lethal to microorganisms, although repeated freezing and thawing will rupture some (Huddleston, Bleckmann, and Wolfe, 1986).

Microbial utilization of hydrocarbons can occur at temperatures ranging from −2 to 70°C (Texas Research Institute, Inc., 1982). Most soils, especially those in cold climates, contain psychrophilic microorganisms that grow best at temperatures below 20°C (JRB Associates, Inc., 1984) and are effective at temperatures below 0°C. Biodegradation can take place at a temperature of 5°C, but hydrocarbons are degraded more slowly at lower temperatures (Parr, Sikora, and Burge, 1983). Walworth and Reynolds (1995) report that bioremediation is effective for treating petroleum-contaminated soils in cold areas; however, diesel fuel loss is certainly greater in soil at 20 than at 10°C. At 10°C, the bioremediation rates are not affected by addition of phosphorus or nitrogen, but they are increased at 20°C by addition of phosphorus but not nitrogen. Dibble and Bartha (1979) found the optimum temperature for biodegradation to be 20°C or higher.

Whyte, Greer, and Inniss (1996) tested 135 psychrotrophic microorganisms for the ability to mineralize petroleum hydrocarbons. A number of strains mineralized toluene, naphthalene, dodecane, and hexadecane. *Rhodococcus* sp. Q15 was able to mineralize the C_{28} *n*-paraffin, octacosane. All the psychrotrophic biodegradative isolates were capable of mineralization activity at both 23 and 5°C, indicating their potential for low-temperature bioremediation of petroleum hydrocarbon–contaminated sites.

Soils in hot environments usually support many thermophilic microorganisms that are effective at temperatures above 60°C (Texas Research Institute, Inc., 1982). However, most soil microorganisms are mesophiles and exhibit maximum growth in the range of 20 to 35°C (Parr, Sikora, and Burge, 1983). The majority of hydrocarbon utilizers are most active in this range. Since many organisms multiply well at laboratory temperatures of 25 to 37°C but not at lower environmental temperatures, it would be beneficial to isolate appropriate organisms at temperatures and in media that correspond to the characteristics of the contaminated site (Alexander, 1994).

Temperatures in the thermophilic range (50 to 60°C) were shown to greatly accelerate decomposition of organic matter, in general (Parr, Sikora, and Burge, 1983). At these temperatures, actinomycetes will be naturally predominant over fungi and bacteria. Therefore, in certain situations, composting may offer potential for maximizing the biodegradation rate of waste industrial chemicals. It should be noted, however, that in another investigation in a test treatment facility, it was found that several aromatic hydrocarbons were not metabolized at 55°C, but were metabolized at 30°C (Phillips and Brown, 1975), while other researchers reported a leveling-off of the hydrocarbon biodegradation rate in soil above 20°C (Dibble and

Bartha, 1979a). Although elevated temperature has some advantage for potentially limiting the development of pathogenic microorganisms, too high a temperature would not be beneficial for stimulating petroleum biodegradation (Phillips and Brown, 1975). The increased availability of more-toxic hydrocarbons at higher temperatures may counteract the stimulation of metabolic processes (Dibble and Bartha, 1979a).

Mutant organisms are being developed to provide the optimal degradation at any given temperature. A commercially available mutant bacterial formulation (PETROBAC® Mutant Bacterial Hydrocarbon Degrader) provides degradation of crude oil over a range of temperatures from 5 to 35°C, with the greatest amount of degradation in the shortest amount of time at the higher temperatures (Thibault and Elliott, 1979).

A temperature gradient exists in the soil (Ahlert and Kosson, 1983). As a result of heat transfer phenomena, temperature responds less to daily weather fluctuations at increased depths. Microorganisms near the surface of the soil column must adapt more readily to temperature fluctuations than those at greater depth. Thus, the seasonal and geographic variations play a role in degradation rates (JRB Associates, Inc., 1984). Disposal sites for oil can be chosen in warm areas that receive direct sunlight to assure temperatures suitable for rapid metabolism by mesophilic microorganisms (Atlas, 1977). Even in near-Arctic environments, absorbance of solar energy raises temperatures into a range that allows for mesophilic microbial oil degradation (Atlas and Schofield, 1975).

Soil temperature is difficult to control in a field situation, but can be modified by regulating the incoming and outgoing radiation, or by changing the thermal properties of the soil (Baver, Gardner, and Gardner, 1972). Vegetation plays a significant role in soil temperature because of the insulating properties of plant cover (Sims and Bass, 1984). Bare soil unprotected from the direct rays from the sun becomes very warm during the hottest part of the day, but also loses its heat rapidly at night and during colder seasons. In the winter, vegetation acts as an insulator to reduce heat lost from the soil. Frost penetration is more rapid and deeper under bare soils than under a vegetative cover. On the other hand, during the summer months, a well-vegetated soil does not become as warm as a bare soil. These fluctuations in soil temperature decrease with increasing depth (Thornton-Manning, Jones, and Federle, 1987).

Soil temperature can be modified by soil moisture control and by the use of mulches of natural or artificial materials (JRB Associates, Inc., 1984). Mulches can affect soil temperature in several ways. In general, they reduce diurnal and seasonal fluctuations in soil temperature (Sims and Bass, 1984). In the middle of summer, there is little overall temperature difference between mulched and bare plots, but mulched soil is warmer in spring, winter, and fall, and warms up more slowly in the spring.

Mulches with low thermal conductivities decrease heat flow both into and out of the soil; thus, soil will be cooler during the day and warmer during the night. White paper, plastic, or other types of white mulch increase the reflection of incoming radiation, thereby reducing excessive heating during the day. A transparent plastic mulch transmits solar energy to the soil and produces a greenhouse effect. A black paper or plastic mulch absorbs radiant energy during the day and reduces heat loss at night. Placing a black covering over the soil to increase the soil temperature during the winter has been suggested as a means of overcoming the problem of slower biodegradation at the lower winter temperatures (Guidin and Syratt, 1975). Use of polyethylene sheeting as a landfarming cover during treatment of crude oil–contaminated soil does not appear to affect biodegradation kinetics adversely under laboratory conditions (Huesemann and Moore, 1993). Humic substances are dark, which increases the heat absorption of the surface soil (Sims and Bass, 1984). Use of film mulch as a means of stimulating waste oil biodegradation by increasing soil temperatures during the winter, however, would preclude tilling of the soil and, thus, decrease its aeration (Dibble and Bartha, 1979a). Some researchers believe this would not have an overall beneficial effect and may, in fact, be unnecessary, since the albedo decrease due to oil contamination can raise the temperature in the upper 10 to 20 cm of tundra soils as much as 5°C (Freedman and Hutchinson, 1976).

Mulches are also used to protect soil surfaces from erosion, reduce water and sediment runoff, conserve moisture, prevent surface compaction or crusting, and help establish plant cover (Soil Conservation Service, 1979). The type of mulch required determines the application method (Sims and Bass, 1984). Commercial machines for spraying mulches are available (Soil Conservation Service, 1979). Hydromulching is a process in which seed, fertilizer, and mulch are applied as a slurry. To apply plastic mulches, equipment is towed behind a tractor and mechanically applies plastic strips that are sealed at the edges with soil. For treatment of large areas, special machines that glue polyethylene strips together are available (Mulder, 1979). Table 5.4 describes the organic materials available for use as mulch and the situations when each would be most suitable.

Table 5.4 Mulch Materials

Organic Materials	Quality	Notes
Small-grain straw or tame hay	Undamaged, air-dried threshed straw, free of undesirable weed seed	Spread uniformly — at least ¼ of ground should be visible to avoid smothering seeding; anchor either during application or immediately after placement to avoid loss by wind or water; straw anchored in place is excellent on permanent seedings
Corn stalks chopped or shredded	Air-dried, shredded into 8 to 12 in. lengths	Relatively slow to decompose, resistant to wind blowing
Wood excelsior	Burred wood fibers approximately 4 in. long	A commercial product packaged in 80–90 lb bales; apply with power equipment; tie down, usually
Wood cellulose fiber	Air-dried, nontoxic with no growth-inhibiting factors	Must be applied with hydraulic seeder
Compost or manure	Shredded, free of clumps or excessive coarse material	Excellent around shrubs; may create problems with weeds
Wood chips and bark	Air-dried, free from objectionable coarse material	Most effective as mulch around ornamentals, etc.; resistant to wind blowing; may require anchoring with netting to prevent washing or floating off
Sawdust	Free from objectionable coarse material	More commonly used as a mulch around ornamentals, etc.; requires anchoring on slopes; tends to crust and shed water
Pine straw	Air-dried; free of coarse objectionable material	Excellent around plantings; resistant to wind blowing
Asphalt emulsion	Slow setting SS-1	Use as a film on soil surface for temporary protection without seeding; requires special equipment to apply; sheds water
Gravel or crushed stone	—	Apply as a mulch around woody plants; may be used on seeded areas subject to foot traffic (approx. weight — 1 ton/yd^3)
Wood excelsior mats	Blanket of excelsior fibers with a net backing on one side	Roll 36 in. × 30 yd covers 16½ yd^2; use without additional mulch; tie down as specified by manufacturer
Jute, mesh, or net	Woven jute yarn with ¾ in. openings	Roll 48 in. × 75 yd weighs 90 lb and covers 100 yd^2

Source: Soil Conservation Service. *Guide for Sediment Control on Construction Sites in North Carolina.* U.S. Department of Agriculture, Raleigh, NC, 1979.

Irrigation increases the heat capacity of the soil, lowers air temperature over the soil, raises the humidity of the air, and increases thermal conductivity, resulting in a reduction of daily soil temperature variations (Schweizer, 1976). Sprinkle irrigation, for example, has been used for temperature control, specifically frost protection in winter and cooling in summer and for reduction of soil erosion by wind (Schwab, Frevert, Edminster, and Barhes, 1981). Drainage decreases the heat capacity, which raises the soil temperature. Elimination of excess water in spring causes a more rapid temperature increase. The addition of humic substances improves soil structure, thus improving soil drainability, resulting indirectly in an increase in the insulative capacity of the soil.

Several physical characteristics of the soil surface can be modified to alter soil temperature (Schweizer, 1976). Compaction of the soil surface will increase the density and, thus, the thermal conductivity. Tillage, on the other hand, creates a surface mulch that, when dry, reduces heat flow from the surface to the subsurface. The diurnal temperature variation in a cultivated soil is often much greater than in an untilled soil. A loosened soil has more surface area exposed to the sun but is colder at night and more susceptible to frost.

Raising the temperature of a contaminated zone can also be achieved by pumping in heated water or recirculating groundwater through a surface heating unit (U.S. EPA, 1985a). Tentlike structures can be erected over treatment beds to elevate temperature, especially in winter (Ellis, Harold, and Kronberg, 1991).

5.1.3 SOIL pH

Soil pH contributes to the surface charge on many colloidal-sized soil particles (JRB Associates, Inc., 1984). Clays have a permanent negative charge, and it is primarily their coatings of organic and amorphous materials that change in charge. Thus, the pH of the groundwater and soil water in the vadose zone determines the degree of anion or cation adsorption by soil particles. In soils with pH-dependent charge, lowering the pH decreases the net negative charge and, thus, decreases anion repulsion or increases anion adsorption (Hornick, 1983). The soil pH may affect the solubility, mobility, and ionized forms of contaminants (JRB Associates, Inc., 1984).

Biological activity in the soil is greatly affected by the pH, through the availability of nutrients and toxicants and the tolerance of organisms to pH variations. Some microorganisms can survive within a wide pH range, while others can tolerate only small variations. The optimum pH for rapid decomposition of wastes and residues is usually in the range of 6.5 to 8.5. Bacteria and actinomycetes have pH optima near 7.0. A soil pH of 7.8 should be close to the optimum (Dibble and Bartha, 1979a). If the soil is acidic, these organisms often cannot compete effectively with soil fungi for available nutrients. The pH can influence the solubility or availability of macro- (especially phosphorus) and micronutrients, the mobility of potentially toxic materials, and the reactivity of minerals (e.g., iron or calcium) (Parr, Sikora, and Burge, 1983).

Hydrocarbon-contaminated soil could contain a number of heavy metals that are potentially toxic to the environment (Streebin, Robertson, Callender, Doty, and Bagawandoss, 1984). The pH range of 6.5 to 8.0 is also optimum for the formation of insoluble precipitates and, thus, results in the immobilization of certain heavy metals. pH is the most important aspect of the reaction between heavy metals and soils (Leeper, 1978). The leaching of metals will not be a problem at treatment sites employing proper pH control (Dibble and Bartha, 1979c).

Contamination by hydrocarbons can change the pH of the soil (Amadi, Abbey, and Nima, 1996). After exposure to oil spillage for 17 years, the pH of the soil varied from 4 to 6 in heavy and moderately impacted zones. Soil nutrients were similar in both areas. Petroleum hydrocarbon utilizers correlated positively with the distribution of oil. Aerobic nitrifiers, however, were more abundant in the heavy than the moderate zone, while anaerobic nitrifiers were higher in the moderate than in the heavy.

Carbonic acid, organic acid intermediates, and nitrate and sulfate (most important for pH < 5) may accumulate during aerobic degradation of organic molecules (Zitrides, 1983). This can lower the soil pH and inhibit biological activity. The acid conditions can be controlled with reinoculation or by addition of lime, which is favorable for the biodegradation of oil (Dibble and Bartha, 1979a).

When acidic wastes in hazardous waste–contaminated soil lower the soil pH, they change the microorganism distribution (Sims and Bass, 1984; JRB Associates, Inc., 1984). Fungi predominate under acidic conditions (pH < 7). These organisms may transform aromatic hydrocarbons by means of oxygenases into arene oxides, the mutagenic forms of PAHs (Baver, Gardner, and Gardner, 1972). Bacteria, on the other hand, growing better at a neutral or slightly basic pH, would carry out the dioxygenation of the aromatic nucleus to form a *cis*-glycol as the first stable intermediate, instead of the arene oxide. Higher organisms (above fungi) do not possess the necessary oxygenases and, thus, form *trans* aromatic diols, which tend to polymerize.

These differences in the mechanism of aromatic hydrocarbon metabolism by microorganisms have important implications concerning engineering techniques for controlling and possibly detoxifying simple aromatics and PAHs in contaminated soils (Cerniglia, Hebert, Dodge, Szaniszlo, and Gibson, 1979). It appears that selection for dominance of the microbial community by bacteria may avoid the formation of mutagens, and that pH may serve as an important engineering tool to direct the pathway of PAH degradation.

Control of soil pH at an in-place hazardous waste treatment site is a critical factor in several treatment techniques, including metal immobilization and optimum microbial activity (Sims and Bass, 1984). The pH of different soil types can vary. The goal of soil pH adjustment in agricultural application usually is to increase the pH to near neutral values, since most natural soils tend to be slightly acidic. Areas of the country in which the need for increasing soil pH is greatest are the humid regions of the East, South, Middle West, and Northwest States. In areas where rainfall is low and leaching is minimal, such as parts of the Great Plain States and the arid, irrigated saline soils of the Southwest, Intermountain, and Far West States, pH adjustment is usually not necessary but may require reduction.

A calcareous (containing calcium carbonate) soil can range from pH 7 to 8.3 (JRB Associates, Inc., 1984). A sodic (high in sodium carbonate) soil can go as high as pH 8.5 to 10. Saline soils tend to be around pH 7. The soil pH may need to be lowered by adjusting with sulfur or other acid-forming compounds, or raised by adding crushed limestone or lime products to bring it between pH 5.5 and 8.5 to encourage microbial activity. Phosphorus solubility is maximized at pH 6.5; this may be the ideal soil pH.

Since it is common to isolate microorganisms that grow well around pH 7, those selected for reintroduction into the environment may not survive or function as desired, if the pH of the contaminated soil is not in that range (Alexander, 1994). For growth of the appropriate organisms, the isolation medium should be maintained at pH values similar to those at the site of concern (Zaidi, Murakami, and Alexander, 1989).

5.1.3.1 Increasing Soil pH

Liming is a frequent agricultural practice and is the most common method of controlling pH, while acidification is much less common (Sims and Bass, 1984). Methods have been developed to determine the lime requirement of soils, taking into account the buffering capacity of the soil (McLean, 1982). A lime requirement test may be performed to find the loading rate to use for increasing soil pH. However, there are no readily available guidelines for reducing soil pH, and the acidification requirements for a particular soil have to be determined experimentally in the laboratory, taking into account the buffering capacity of the waste. Thorough mixing is required in the zone of contamination to change the pH. Runoff and minor controls are necessary to control drainages and erosion of the tilled soil. The achievable level of treatment is high, depending upon the wastes, site, and soil. It may be necessary to repeat the process during the treatment.

Liming is the addition to the soil of any calcium or calcium- and magnesium-containing compound capable of reducing acidity (i.e., raising pH) (Sims and Bass, 1984). Lime correctly refers only to calcium oxide, but is commonly used to refer to calcium hydroxide, calcium carbonate, calcium-magnesium carbonate, and calcium silicate slags.

There are several benefits of liming to biological activity (Sims and Bass, 1984). Manganese and aluminum are toxic to most plants but are less soluble at higher pH values. Phosphates and most microelements necessary for plant growth (except molybdenum) are more available at higher pH. Microbial activity is greater at or near neutral pH, which enhances degradation processes, mineralization, and nitrogen transformations (e.g., nitrogen fixation and nitrification).

The liming material to use depends upon several factors (Sims and Bass, 1984). Calcitic and dolomitic limestones are the most common. However, these must be ground in order to be effective quickly, since the velocity of reaction depends upon the surface in contact with the soil. The finer they are ground, the more rapidly they react with the soil. There will usually be a mixture of fine and coarse particles in a finely ground product. This material allows a rapid pH change, is relatively long lasting, and is reasonably priced. Many states require that 75 to 100% of the limestone pass an 8- to 10-mesh sieve and that 20 to 80% pass anywhere from an 8- to 100-mesh sieve. Calcium oxide and calcium hydroxide are manufactured as powders and react quickly. Other factors to consider in the selection of a limestone are neutralizing value, magnesium content, and cost per ton applied to the land.

Lime requirement for soil pH adjustment depends on soil factors, such as soil texture, type of clay, organic matter content, and exchangeable aluminum (Follett, Murphy, and Donahue, 1981). The buffering capacity reflects the soil cation exchange capacity and will directly affect the amount of lime needed to adjust soil pH. The amount of lime required is also a function of the volume of soil to be treated. The amount of lime necessary in a particular site/soil/waste system can be determined by a commercial soil testing laboratory in short-term treatability studies or soil-buffer tests (McLean, 1982). Lime requirements are also affected by acid-forming fertilizers. Commonly used liming materials are summarized in Table 5.5.

Limestone must be placed where needed, since it does not migrate easily in the soil and is only slightly soluble (Sims and Bass, 1984). Therefore, plowing or disking surface-applied lime into the soil may be required. The application of fluid lime is becoming more popular, especially when mixed with fluid nitrogen fertilizer. This combination results in fewer passes over the soil, and the lime is available to counteract acidity produced by the nitrogen. Also, limestone has been applied successfully to a pharmaceutical wastewater landtreatment facility through a spray irrigation system.

The addition of basic waste to acidic soil increases the pH of the surface layer (4 to 18 in.) but not the subsoil (Brown, 1975). The reaction neutralizes the buffer capacity of the soil. Basic waste can cause physical damage to the soil system; however, weak organic bases added to the soil may increase the soil buffer capacity and exchange capacity as the bases are degraded.

5.1.3.2 Decreasing Soil pH

Ferrous sulfate can be added to the soil to decrease alkalinity (Arthur D. Little, 1976). Under acidic conditions in soils, solubilities of complexed cations, such as those of copper and zinc, increase, and

Table 5.5 Liming Materials

Liming Material	Description	Calcium Carbonate Equivalent	Comments
Limestone, calcitic	$CaCO_3$, 100% purity	100	Neutralization value usually between 90 and 98% because of impurities; pulverized to desired fineness
Limestone, dolomitic	65% $CaCO_3$ + 20% $MgCO_3$, 87% purity	89	Pure dolomite (50% $MgCO_3$ and 50% $CaCO_3$) has neutralizing value of 109%; pulverized to desired fineness
Limestone, unslaked lime, burned lime, quick lime	CaO, 85% purity	151	Manufactured by roasting calcitic limestone; purity depends on purity of raw materials; white powder, difficult to handle — caustic; quick acting; must be mixed with soil or will harden and cake
Hydrated lime, slaked lime, builder's lime	$Ca(OH)_2$, 85% purity	85	Prepared by hydrating CaO; white powder, caustic, difficult to handle; quick acting
Marl	$CaCO_3$, 50% purity	50	Soft, unconsolidated deposits of $CaCO_3$, mixed with earth, and usually quite moist
Blast furnace slag	$CaSi_2O_3$	75–90	By-product in manufacture of pig iron; usually contains magnesium
Waste lime products	Extremely variable in composition	?	—

Source: Follett, R.H. et al. *Fertilizers and Soil Amendments.* Prentice-Hall, Englewood Cliffs, NJ, 1981. With permission.

those of simple ions of iron, manganese, and copper are easily reduced to more soluble forms (JRB Associates, Inc., 1984). Acidic wastes may also be used as a treatment process for saline-sodic soils.

5.1.4 OXYGEN SUPPLY

Although some xenobiotic organic compounds appear to require the slow anaerobic metabolism for decomposition, most of these compounds are susceptible to attack by aerobic organisms (Alexander, 1977; Brunner and Focht, 1983; Jain and Sayler, 1987). Therefore, assuring the aerobiosis of the soil will enhance the rate of biological decomposition for many compounds.

For degradation to occur, microorganisms must utilize an electron acceptor, such as oxygen (Turco and Sadowsky, 1995). Most biodegradation of petroleum hydrocarbons is aerobic, since hydrocarbon oxidation processes generally require oxygenases (Atlas and Bartha, 1987).

Molecular oxygen, which is soluble in oils, penetrates oil-contaminated soils and sediments to a degree that depends upon depth, the concentration of oil, and the presence of cracks and fissures in soils or of burrowing worms in sediments (Lee, 1977; Gordon, Dale, and Keizer, 1978). Typically, a gradient occurs in which biodegradation shows a strong negative relation to depth.

The degree to which the soil pore space is filled with water affects the exchange of gases through the soil (JRB Associates, Inc., 1984). Microbial respiration, plant root respiration, and the respiration of other organisms removes oxygen from the soil and replaces it with carbon dioxide. Gases diffuse into the soil from the air above, and gases in the soil diffuse into the air. However, the oxygen concentration in the surface, unsaturated soil may be only half that in air, while carbon dioxide concentrations may be many times that of air (Brady, 1974).

Air contains 79% nitrogen, which is essentially useless in bioremediation except in removing dissolved CO_2 to aid in pH control (Bergman, Greene, and Davis, 1994). However, the pH can be controlled chemically. As a simple diluent, nitrogen also reduces the amount of oxygen that can be dissolved in a body of waste by a factor of five, which decreases the oxygen dissolution rate. This in turn may limit the rate of biodegradation of the contaminant and make it more difficult to control dissolved oxygen concentration.

As soil becomes saturated, the diffusion of gases through the soil is severely restricted. In saturated soil, oxygen can be consumed faster than it can be replaced, and the soil becomes anaerobic (JRB Associates, Inc., 1984). This drastically alters the composition of the microflora. Facultative anaerobes, which use alternative electron acceptors, such as nitrate (denitrifiers) and strict anaerobic organisms become the dominant species. While many soil bacteria can grow under anaerobic conditions, though less actively, most fungi and actinomycetes do not grow at all (Parr, Sikora, and Burge, 1983). Microbial metabolism shifts from oxidative to fermentative and becomes less efficient in terms of biosynthetic energy production (JRB Associates, Inc., 1984). Soil structure and texture primarily determine the size of soil pores, and hence the water content at which gas diffusion is significantly limited in a given soil, and the rate at which anaerobiosis sets in.

Anaerobic reactions are accompanied by the production of malodorous compounds, such as amines, mercaptans, and H_2S (Parr, Sikora, and Burge, 1983). These can be phytotoxic, and, if the soil is heavily overloaded, it may remain anaerobic for some time. However, if the oxygen balance is maintained, relative to the amount of contaminants and the soil conditions, rapid aerobic decomposition will occur, and the end products will be inorganic carbon, nitrogen, and sulfur compounds.

The diffusion of air into soil is generally proportional to the square of the air-filled porosity (Turco and Sadowsky, 1995). Since, the active microorganisms appear to be located in areas with a pore neck diameter of 6 µm or less, the total exchange area for gases is limited by the pore neck, to an area of up to 28 µm².

If oil is also present, it can act as a diffusion barrier for oxygen moving into water (Downing and Truesdale, 1955). The diffusivity of oxygen moving in oil is about 2×10^{-3} cm²/s (Schwarzenbach, Gschwend, and Imboden, 1993). If the films are thicker than 100 µm, they can reduce the overall transfer velocity by as much as half. Thus, the ability to transfer oxygen will be limited in soils with large amounts of oil.

A great deal of oxygen-containing water is needed in fine-textured subsurface materials (Wilson, Leach, Henson, and Jones, 1986). Biodegradation of most organic contaminants requires approximately two parts of oxygen to completely metabolize one part of organic compound. The complete oxidation of 1 mg of hydrocarbon to carbon dioxide and water requires 3 to 4 mg of oxygen (Texas Research Institute, Inc., 1982). Less oxygen is needed when microbial biomass (new microorganisms) is generated or when oxidation is not complete. Lund and Gudehus (1990) report that 1.0 kg of hydrocarbons requires about 1.5 to 3.5 kg oxygen, depending upon the kind of hydrocarbons and the portion of which is converted into biomass. Typically, about half the carbon in hydrocarbons is converted into biomass (Green, Lee, and Jones, 1981). In general, the more oxygen, the faster the biodegradation (Zhou and Crawford, 1995). The concentration of the contaminants is an important factor (Wilson, Leach, Henson, and Jones, 1986). Dissolved oxygen should be maintained above the critical concentration for the promotion of aerobic activity, which ranges from 0.2 to 2.0 mg/L, with the most common being 0.5 mg/L (U.S. EPA, 1985a). Zhou and Crawford (1995) found that the optimal oxygen concentration for gasoline-degrading microorganisms was only about 10%, which was surprisingly low. This is about half of the atmospheric oxygen concentration. Other authors report that an oxygen concentration of 5% can be limiting to biodegradation (Wuerdemann, Wittmaier, Rinkel, and Hanert, 1994).

The problem is providing the necessary amount of oxygen to the site where it will be used. Oxygen can be provided to the subsurface through the use of air, pure oxygen, hydrogen peroxide, or ozone (U.S. EPA, 1985a). Oxygen levels can be increased about fivefold by sparging injection wells with oxygen instead of air (Wilson, Leach, Henson, and Jones, 1986). Fluid and semisolid systems can be aerated by means of pumps, propellers, stirrers, spargers, sprayers, and cascades (Texas Research Institute, Inc., 1982). The advantages and disadvantages of various oxygen supply alternatives are summarized in Table 5.6 (U.S. EPA, 1985a).

The flow of oxygen into the system is controlled by oxygen concentration in the carrier and the permeability of the geological material to that carrier (Wilson, Leach, Henson, and Jones, 1986). The amount of oxygen available for biodegradation of hydrocarbons varies considerably with the particular electron acceptor and its carrier medium (Hinchee and Miller, 1990). With water as a carrier, oxygen in air will supply 8.0 mg O_2/L water, and 400,000 kg water would be required to degrade 1 kg hydrocarbon. Pure oxygen delivers 40.0 mg O_2 mg/L water and requires 80,000 kg water/kg hydrocarbon. If the oxygen comes from H_2O_2, a level of 100.0 mg O_2/L H_2O_2 in water would require 65,000 kg carrier/kg hydrocarbon, or a level of 500.0 mg/L H_2O_2 in water would require 13,000 kg/kg hydrocarbon. With 20.9% oxygen in air, 13 kg/kg hydrocarbon would be needed. Dupont, Doucette, and Hinchee (1991) report that to obtain 1 g of O_2, it is necessary to supply 110,000 g of air-saturated water, 22,000 g of pure oxygen–saturated water, 2000 g of water containing 500 mg/L H_2O_2 (100% utilization), or 13 g of air (20.9% O_2).

Table 5.6 Oxygen Supply Alternatives

Substance	Application Method	Advantages	Disadvantages
Air	In-line	Most economical	Not practical except for trace contamination <10 mg/L COD
	In situ wells	Constant supply of oxygen possible	Wells subject to blow out
Oxygen-enriched air or pure oxygen	In-line	Provides much higher oxygen solubility than air	Not practical except for low levels of contamination <25 mg/L COD
	In situ wells	Constant supply of oxygen possible	Very expensive, wells subject to blow out
Hydrogen peroxide	In-line	Moderate cost, intimate mixing with groundwater, greater oxygen concentrations can be supplied to subsurface (100 mg/L), H_2O_2 provides 50 mg/L oxygen, helps to keep wells free of heavy growth	Chemical decomposes rapidly on contact with soil, and oxygen may bubble out prematurely unless properly stabilized
Ozone	In-line	Chemical oxidation will occur, rendering compounds more biodegradable	Ozone generation is expensive, toxic to microorganisms except at low concentrations, may require additional aeration

Source: U.S. EPA. Handbook No. EPA/625/6-85/006, 1985.

Table 5.7 Estimated Volumes of Water or Air Required to Completely Renovate Subsurface Material That Originally Contained Hydrocarbons at Residual Saturation

Texture	Hydrocarbons When Drained	Air When Drained	Water When Flooded	Air	Water
Stone to coarse gravel	0.005	0.4	0.4	250	5,000
Gravel to coarse sand	0.008	0.3	0.4	530	8,000
Coarse to medium sand	0.015	0.2	0.4	1,500	15,000
Medium to fine sand	0.025	0.2	0.4	2,500	25,000
Fine sand to silt	0.040	0.2	0.5	4,000	32,000

(Columns 2–4: Proportion of the Total Volume of the Subsurface Occupied by; Columns 5–6: Volumes Required to Meet the Oxygen Demand of the Hydrocarbons)

Source: De Pastrovich, T.L. et al. CONCAWE Report No. 3/79. The Oil Companies' International Study Group, The Hague, The Netherlands, 1979. With permission.

Whether the contaminant is above or below the water table, the rate of bioreclamation in hydrocarbon-contaminated zones is effectively controlled by the rate of supply of oxygen (Wilson, Leach, Henson, and Jones, 1986). Table 5.7 compares the number of times that water in contaminated material below the water table, or air in material above it, must be replaced to reclaim totally subsurface materials of various textures. The calculations assume typical values for the volume occupied by air, water, and hydrocarbons (De Pastrovich, Baradat, Barthal, Chiarelli, and Fussel, 1979). The actual values at a specific site will probably be different. It is also assumed that the oxygen content of the water is 10 mg/L, that of the air 200 mg/L, and that the hydrocarbons are completely metabolized to carbon dioxide.

After the oxygen in the air is consumed during the biological degradation of the contaminant, the remaining air should physically weather (remove volatiles by evaporation) the hydrocarbons (Wilson, Leach, Henson, and Jones, 1986). The extent of weathering depends upon the vapor pressure of the contaminant. Light hydrocarbons, such as gasoline, can be vaporized to a greater extent than they are metabolized with oxygen. The vapor pressure of gasolines varies from 100 to 1000 mm at 100°C. If the vapor pressure is reduced fourfold at typical groundwater temperatures of 10°C, and benzene is typical of the vapors, then the oxygen demand for complete metabolism of the gasoline vapors ranges from 2 to 20 times the oxygen content of air. The biological and physical weathering of the hydrocarbon should preferentially remove the more volatile and more water-soluble components (De Pastrovich, Baradat, Barthal, Chiarelli, and Fussel, 1979).

OPTIMIZATION OF BIOREMEDIATION

Oxygen content of the soil can be improved by the presence of sand or loam (heavy clay is undesirable), avoidance of unnecessary compaction (heavy trucks, etc.), and limited loading of rapidly biodegradable matter (Raymond, Hudson, and Jamison, 1976).

There are a multitude of chemical, photosynthetic, and electrochemical reactions that produce oxygen, either as a major or minor product (Texas Research Institute, Inc., 1982). The chemical reaction types most often encountered are

1. Decomposition of peroxides, superoxides (Shanley and Edwards, 1985)

$$\underset{\text{Hydrogen peroxide}}{H_2O_2} \rightarrow H_2O + \tfrac{1}{2}O_2$$

(a good, ecologically sound additive, used extensively in sewage treatment; March, 1968)

$$\underset{\text{Sodium peroxide}}{Na_2O_2} + H_2O \rightarrow NaOH + H_2O_2$$

$$\underset{\text{Sodium superoxide}}{NaO_2} + H_2O \rightarrow Na_2O_2 + O_2$$

Barium and strontium peroxides are used in the production of oxygenating cakes employed by fishermen for maintaining live bait (Texas Research Institute, Inc., 1982). A typical formulation would contain barium peroxide, manganese dioxide, calcium sulfate, and dental plaster, which releases oxygen slowly when in contact with water. However, use of materials such as these may not be advisable, because of the resulting heavy metal contamination of the water table. Barium peroxide is definitely highly poisonous.

There is also a urea–peroxide addition compound that has been used in conjunction with phosphate solutions to treat plants suffering from oxygen starvation in the root zone (U.S. Patent 3,912,490) (Texas Research Institute, Inc., 1982). The compound is available commercially from Western Europe. It is probably of the inclusion type, one in which H_2O_2 molecules are trapped within channels formed by the crystallization of urea (March, 1968). Since the molecules are held together only by van der Waals forces, when dissolved the solution will behave as a mixture of urea and hydrogen peroxide. By weight, 35% of the compound is H_2O_2.

2. Decomposition of Peroxyacids and Salts (*Austin American Statesman*, 1980)

Peroxy mono- and disulfuric acids, peroxy mono- and diphosphoric acid, and peroxyborates all produce acidic solutions, but the salts may be important for consideration. The exact mode of degradation of the salt $KHSO_5$ (potassium monoperoxysulfate) is uncertain, but it has been used as an aid in the degradation of atrazine (a pesticide) — presumably by virtue of its oxygen-producing ability. Degradation probably results in the formation of $KHSO_4$. Impure salts of peroxy monophosphoric acid (H_3PO_5) might prove useful.

3. Thermal decomposition of oxygen-bearing salts

$$2NaNO_3 \xrightarrow{\text{heat}} 2NaNO_2 + O_2$$

$$2KClO_3 \xrightarrow{\text{heat}} 2KCl + 3O_2$$

Generation of oxygen by this method has no particular advantage to treating underground contamination. Several alternative sources of oxygen have been suggested as a means to increase the degradative activity in contaminated aquifers (Texas Research Institute, Inc., 1982). Oxidizing agents can be used to degrade organic constituents in soil systems, although they may themselves be toxic to microorganisms or may cause the production of more-toxic or more-mobile oxidation products (Sims and Bass, 1984).

Two powerful oxidizing agents that have potential for in-place treatment are ozone and hydrogen peroxide.

5.1.4.1 Ozone

Ozone is an oxygen molecule containing three oxygen atoms. Ozone gas is a very strong oxidizing agent that is very unstable and extremely reactive (U.S. EPA, 1985a). It cannot be shipped or stored; therefore, it must be generated on-site prior to or during application.

Ozone can be employed as a pretreatment for wastes to break down refractory organics or to furnish a polishing step after biological or other treatment processes to oxidize untreated organics (Roberts, Koff, and Karr, 1988). It may be used to degrade recalcitrant compounds directly by creating an oxygenated compound without chemical degradation (Texas Research Institute, Inc., 1982). Ozonation is an oxidation process appropriate for aqueous streams that contain less than 1.0% oxidizable compounds (Roberts, Koff, and Karr, 1988). This chemical oxidation can be used on many organic compounds that cannot be easily broken down biologically, including chlorinated hydrocarbons, alcohols, chlorinated aromatics, pesticides, and cyanides (Lee and Ward, 1985, 1984; Lee, Wilson, and Ward, 1987). An added advantage of using ozone is its ability to react with PAHs (Hsu, Davies, and Masten, 1993). Ozone venting significantly shortens remediation time and lowers residual concentrations by an order of magnitude. The rate of ozone reaction can be controlled by adjusting the pH of the medium (Texas Research Institute, Inc., 1982). At high pH, hydroxyl free-radical reactions dominate over the more rapid direct ozone reactions.

Ozone increases the dissolved oxygen level in water for enhancing biological activity (Texas Research Institute, Inc., 1982). The most-effective and cost-effective uses of ozone in soil system decontamination appear to be in the treatment of contaminated water extracted from contaminated soil systems through recovery wells, and in the stimulation of biological activity in saturated soil (Nagel et al., 1982). Ozone treatment may be very effective for enhancing biological activity, if the organic contaminants are relatively biodegradable. However, if much of the material is relatively biorefractory, the amount of ozone required would greatly increase the cost of the treatment.

In commercially available ozone-from-air generators, ozone is produced at a concentration of 1 to 2% in air (U.S. EPA, 1985a). In bioreclamation, this ozone-in-air mixture could be contacted with pumped leachate using in-line injection and static mixing or using a bubble contact tank. A dosage of 1 to 3 mg/L of ozone can be used to attain chemical oxidation. However, the dosage should not be greater than 1 mg/L of ozone per mg/L total organic carbon (TOC); higher concentrations may be deleterious to the microorganisms. At many sites, this may limit the use of ozone as a pretreatment method to oxidize refractory organics, making them more amenable to biological oxidation.

Ozone has been used to treat groundwater contaminated with oil products to reduce dissolved organic carbon concentration (Nagel et al., 1982). Dosages of 1 g ozone/g dissolved organic carbon resulted in residual water ozone concentration of 0.1 to 0.2 ppm. The treated water was then infiltrated into the aquifer through injection wells. There was an increase in dissolved oxygen in the contaminated water. This increased microbial activity in the saturated soil zone, which stimulated microbial degradation of the organic contaminants.

Ozone was used in a petroleum-contamination incident in Karlsruhe, Germany that threatened a drinking water supply (Atlas and Bartha, 1973c). The polluted groundwater was withdrawn, treated with ozone, and infiltrated back into the system via three infiltration wells. About 1 g of ozone per gram of dissolved organic carbon (DOC) was added to the groundwater, with a contact time of 4 min in the aboveground reactor (Lee and Ward, 1985; U.S. EPA, 1985a). This increased the dissolved oxygen levels to 9 mg/L, with a residual of 0.1 to 0.2 g of ozone/m^3 in the treated water. The dissolved oxygen reached equilibrium at about 80% of the initial concentration injected. The oxygen consumption peaked at about 40 kg/day during the initial infiltration period. The microbial counts subsequently increased in the wells, with a decrease in dissolved organic carbon and mineral oil hydrocarbons. Total bacterial counts in the groundwater increased tenfold, but bacteria potentially harmful to humans did not increase. Levels of cyanide, a contaminant identified after the treatment began, also decreased, although biodegradation was not shown to be the cause. The ozone may have also reacted with the hydrocarbon for partial destruction of the organics. The drinking water from this aquifer contained no trace of contaminants after 1½ years of ozone treatment. It is conceivable that oxidizing the subsurface could result in the precipitation of iron and manganese oxides and hydroxides. If this is extensive, the delivery system and possibly even the aquifer could become clogged.

OPTIMIZATION OF BIOREMEDIATION

Saturated aliphatic compounds that do not contain easily oxidized functional groups are not readily reactive with ozone; for example, saturated aliphatic hydrocarbons, aldehydes, and alcohols (Sims and Bass, 1984). Reactivity of aromatic compounds with ozone is a function of the number and type of substituents. Substituents that withdraw electrons from the ring deactivate the ring toward ozone, for example, halogen, nitro, sulfonic acid, carbonyl, and carboxyl groups. Substituents that release electrons activate the ring toward ozone, for example, alkyl, methoxyl, and hydroxyl.

The following reactivity patterns with ozone are

>   phenol, xylene > toluene > benzene
>   pentachlorophenol < di-, tri-, and tetrachlorophenol

The relatively rapid decomposition rates of ozone in aqueous systems, especially in the presence of certain chemical contaminants or other agents that catalyze its decomposition to oxygen, preclude its effective application to subsurface waste deposits (Amdurer, Fellman, and Abdelhamid, 1985). The half-life of ozone in groundwater is less than ½ h (Ellis and Payne, 1984) (about 18 min; U.S. EPA, 1985a). Since the flow rates of water are likely to be in inches per hour or less, it is unlikely that effective doses of ozone could be delivered very far for chemical oxidation. However, it has been used successfully to supply oxygen for microbial biodegradation (Rice, 1984).

5.1.4.2 Hydrogen Peroxide (H_2O_2)

Amdurer, Fellman, and Abdelhamid (1985) state that hydrogen peroxide is a weaker oxidizing agent than ozone, but that it is considerably more stable in water. It decomposes to form water and oxygen, can supply improved oxygen levels (Lee and Ward, 1985, 1984; Lee, Wilson, and Ward, 1987), and has been used successfully to clean up several spill sites (U.S. EPA, 1985a). Advantages of hydrogen peroxide include

> Greater oxygen concentrations can be delivered to the subsurface. 100 mg/L H_2O_2 provides 50 mg/L oxygen.
> Less equipment is required to oxygenate the subsurface.
> Hydrogen peroxide can be added in-line along with the nutrient solution. Aeration wells are not necessary.
> Hydrogen peroxide keeps the well free of heavy biogrowth. Such growth and clogging can be a problem in air injection systems.

Hydrogen peroxide is used to degrade recalcitrant compounds and modify the mobility of some metals (Sims and Bass, 1984). It can be used to raise oxygen levels in the soil, which can increase microbial activity and degradation of organic contaminants (Nagel et al., 1982). Successful use of hydrogen peroxide requires careful control of the geochemistry and hydrology of the site (Wilson, Leach, Henson, and Jones, 1986).

Air sparging was able to maintain dissolved oxygen levels of only 1 to 2 ppm in a spill area (Nagel et al., 1982). However, addition of microbial nutrient (a specially formulated, hydrogen peroxide-based nutrient solution; FMC Aquifer Remediation Systems, Princeton, NJ) raised dissolved oxygen levels to over 15 ppm. This established the efficiency of hydrogen peroxide–based solutions for supplying increased oxygen levels to enhance the bioreclamation process. Hydrogen peroxide was selected as the source of oxygen for biodegration at the Kelly Air Force Base, TX because it could provide about five times more oxygen to the subsurface than aeration techniques (Wetzel, Davidson, Durst, and Sarno, 1986). The increase in microbial densities in stimulated underground spill sites is probably due to the increased oxygen from the hydrogen peroxide (Wilson, Leach, Henson, and Jones, 1986b).

Hydrogen peroxide is a strong oxidant and is nonselective (Sims and Bass, 1984). It will act with any oxidizable material present in the soil. It could thus lower the concentration of natural organic material in the soil, causing a reduced sorption capacity for some organics. The effectiveness of peroxide may be inhibited because it simultaneously increases mobility and decreases possible sorption sites, unless this result is desired, of course.

Hydrogen peroxide is effective for oxidizing cyanide, aldehydes, dialkyl sulfides, dithionates, nitrogen compounds, phenols, and sulfur compounds (FMC Corporation, 1979). The following chemical groups have incompatible reactions with peroxides (i.e., the reaction products are more mobile) (Sims and Bass, 1984):

Acid chlorides and anhydrides
Acids, mineral, nonoxidizing
Acids, mineral, oxidizing
Acids, organics
Alcohols and glycols
Alkyl halides
Azo, diazo compounds, hydrazine
Cyanides
Dithio carbamates
Aldehydes
Metals and metal compounds
Phenols and cresols
Sulfides, inorganic
Chlorinated aromatics/alicyclics

Hydrogen peroxide is more soluble in water than molecular oxygen and may provide more oxygen at specific sites of application (Britton, 1985). The enzymatic decomposition reactions are

$$2H_2O_2 \rightarrow 2H_2O + O_2$$

$$H_2O_2 + XH_2 \rightarrow 2H_2O + X$$

where X can be NADH, glutathione, or other biological reductants.

Hydrogen peroxide and ozone have been used in combination to degrade compounds that are refractory to either material individually (Nakayma et al., 1979). There is an ongoing debate as to what oxidant is the best. Some believe that hydrogen peroxide is the most-efficient way to move oxygen through a formation. Others find that air is the most cost-effective oxidizing agent. On the other hand, proponents of both hydrogen peroxide and ozone also use aeration in their bioreclamation systems.

Hydrogen peroxide is reasonably inexpensive and can be produced from a coproduct process, such as the initial conversion of glucose to gluconic acid with glucose-1-oxidase (Hou, 1982). It is nonpersistent and is not likely to represent a serious health hazard, if used properly (Texas Research Institute, Inc., 1982; Britton, 1985). However, it is cytotoxic (3% is commonly used as a general antiseptic) and may decompose (by enzymatic catalysis or nonenzymatically by *in situ* physicochemical processes) before reaching its targeted spill location.

This chemical can be toxic to the microorganisms it is intended to stimulate. However, the growth rate of hydrocarbon-utilizing bacteria is not necessarily inhibited by high hydrogen peroxide concentrations (Texas Research Institute, Inc., 1982). Even growth enhancement is sometimes observed. Whether or not a given hydrogen peroxide concentration will be toxic to bacteria depends upon the concentration of the organisms when the hydrogen peroxide is added. Large populations are more successful at surviving high hydrogen peroxide concentrations than are small populations.

Bacteria produce hydrogen peroxide themselves from respiratory processes, and almost all aerobic bacteria have enzymes (hydroperoxidases — catalase and peroxidase) to protect against the toxicity of the compound (Texas Research Institute, Inc., 1983; Britton, 1985). There appears to be a critical H_2O_2:organism ratio, above which the catalase-utilizing protective mechanisms of the organisms are overwhelmed. This ratio may be on the order of 2×10^{10}:1 in a given volume of solution, or it may be expressed as 1 ppm H_2O_2:8.9×10^5 bacteria.

Catalase buildup can result in too rapid H_2O_2 decomposition and wasteful off-gassing of oxygen (Spain, Milligan, Downey, and Slaughter, 1989). Britton (1985) noticed that nonviable cell material is just as capable as the enzyme in viable cells of catalyzing decomposition of hydrogen peroxide. It is impossible to distinguish between abiotic and biotic use of the oxygen produced by this reaction (Huling, Bledsoe, and White, 1991). Soils decompose H_2O_2 to the oxygen needed for *in situ* bioremediation by both biotic and abiotic catalysis, and they vary greatly in peroxide decomposition activity (Lawes, 1991). Autoclaving or treatment with acidic mercuric chloride should help estimate how much decomposition occurs by either mode. Phosphates cannot be used to estimate the extent of abiotic decomposition (Lawes, 1991).

Hydrogen peroxide has been shown to be toxic to fresh bacterial cultures at levels greater than 100 ppm, although mature cultures suffered less and could function at levels as high as 10,000 ppm. Subsequent experimentation with sand columns inoculated with gasoline and gasoline-degrading bacteria showed that 1.0% (10,000 ppm), 0.5% (5000 ppm), and 0.25% (2500 ppm) hydrogen peroxide solutions were toxic to the bacteria.

Other studies have shown that hydrocarbon-degrading bacteria can adapt to tolerate hydrogen peroxide equivalent to 200 mg/L oxygen, a 20-fold increase in oxygen over water sparged with air (Lee and Ward, 1985). In a mixed culture of gasoline degraders, the maximum concentration of H_2O_2 that could be tolerated was 0.05% (500 ppm), although by increasing the concentration gradually, the level of tolerance could be raised to 0.2% (2000 ppm) (Texas Research Institute, Inc., 1983; Britton, 1985). The remediation at Grange, IN, involved adding an initial concentration of 100 ppm, and increasing it to 500 ppm over the course of the treatment (U.S. EPA, 1985a). Most applications of H_2O_2 have used "safe" levels, generally less than 2000 mg/L (Brown and Norris, 1994). Approximately 2 kg of hydrogen peroxide are required to generate 1 kg of oxygen for biodegradation of 1 kg of hydrocarbon (Norris, Dowd, and Maudlin, 1994).

There is field evidence for enhanced degradation with the use of hydrogen peroxide (Yaniga and Smith, 1984). During air sparging with 100 ppm hydrogen peroxide, the dissolved oxygen concentrations in monitoring wells at a site contaminated by gasoline increased from 4 to 10 ppm. This was accompanied by an increase in the numbers of gasoline-utilizing organisms and a reduction in the size of the gasoline plume and a decrease from 4 to 2.5 ppm hydrocarbon. However, other restoration measures were concurrently being employed. Other authors report a greater benefit from using hydrogen peroxide in soils contaminated with JP-5 and diesel fuel than in soils contaminated with lubricating oil (Flathman, Carson, Whitehead, Khan, Barnes, and Evans, 1991). Ho, Shebl, and Watts (1995) have developed an injection system for *in situ* catalyzed peroxide remediation of contaminated soil.

Hydrogen peroxide could be used in sand columns but not in batch liquid cultures (Wilson, Leach, Henson, and Jones, 1986). The liquid cultures were extremely sensitive to the compound. This was probably due to the nature of the growth in the two environments: in sand, the organisms would grow as a film with multicellular depth; in liquid, they would be unicellular, unprotected by adjacent cells.

The rate of decomposition of hydrogen peroxide to oxygen must be controlled (Lee and Ward, 1985). Rapid decomposition of only 100 mg/L (100 ppm or 0.01%) hydrogen peroxide will exceed the solubility of oxygen in water, resulting in bubble formation, which could lead to gas blockage and loss of permeability. Nevertheless, it may be possible to overcome this limitation by stabilizing the hydrogen peroxide solution (Lee and Ward, 1985, 1984; Lee, Wilson, and Ward, 1987). Liberation of oxygen with resulting bubble formation should also occur in soil and groundwater with high concentrations of ferric iron (Wilson, Leach, Henson, and Jones, 1986). Addition of $FeCl_3$ to a pumping solution would be a way to form pockets of oxygen bubbles in a short time for bioreclamation of an underground gasoline spill.

The hydrogen peroxide might decompose before it reaches the depths required and cause precipitation of iron and manganese oxides and hydroxides (U.S. EPA, 1985a). Much of the decomposition of hydrogen peroxide in soil and groundwater will be due to reactions with iron salts (Haber and Weiss, 1934). Nonenzymatic decomposition can occur in a variety of reactions, including those in the presence of iron salts, known as Fenton chemistry (Fenton, 1894). The hydroxyl radical (OH$^-$) is known as Fenton's reagent (Ho, Shebl, and Watts, 1995). The following reactions show how different iron salts affect the decomposition of H_2O_2 (Haber and Weiss, 1934).

$$Fe^{++} + H_2O_2 \rightarrow Fe^{+++} + OH^- + OH^\cdot \text{ (hydroxyl radical)}$$

$$OH^\cdot + H_2O_2 \rightarrow H_2O + H^+ + O_2^{\cdot-} \text{ (superoxide radical)}$$

$$O^{\cdot-} + H_2O_2 \rightarrow O_2 + OH^- + OH^\cdot$$

$$Fe^{+++} + H_2O_2 \rightarrow Fe^{++} + 2H^+ + O_2^{\cdot-}$$

$$Fe^{+++} + O_2^{\cdot-} \rightarrow Fe^{++} + O_2$$

Most decreases of hydrogen peroxide occur rapidly in the top 5.5 cm of a sand column, with only slight decreases thereafter, which may be due to iron stimulation (Wilson, Leach, Henson, and Jones, 1986). On the one hand, the molecular oxygen produced from these reactions would help enhance gasoline biodegradation. On the other hand, iron can cause the hydrogen peroxide to decompose before it reaches the intended site. The decomposition rate of peroxide is also greatly accelerated in the presence of another heavy metal ion, Cu^{++} (Bambrick, 1985).

There are a number of ways to prevent the hydrogen peroxide from decomposing. Standard practice is to add enough phosphate to the recirculated water to precipitate the iron (Wilson, Leach, Henson, and Jones, 1986). High concentrations of phosphates (10 mg/L) (0.01 M monobasic potassium phosphate; Britton, 1985) can stabilize peroxide for prolonged periods of time in the presence of ferric chloride, an aggressive catalyst (U.S. EPA, 1985a). The stabilizing effect of phosphate is fortuitous, since it is a major nutrient for enhancement of underground biodegradation of gasoline. How phosphates are applied and tested are important factors in evaluating phosphates for stabilizing hydrogen peroxide (Lawes, 1991). Compounds used for *in situ* stabilization of peroxide in uranium mining might also improve stabilization by phosphate (Lawes and Watts, 1981).

There are problems associated with adding high phosphate concentrations, such as precipitation (U.S. EPA, 1985a). Also, phosphates only partially protect against abiotic decomposition (Lawes, 1991). Some suppliers add an organic inhibitor that will stabilize the peroxide at a rate appropriate to the rate of infiltration, so the oxygen demand of the bacteria attached to the solids is balanced by the oxygen supplied by decomposing peroxide in recirculated water (Wilson, Leach, Henson, and Jones, 1986). Dworkin Foster medium, or a similar medium containing the mineral components for growth (except for a source of carbon and energy), can also stabilize hydrogen peroxide and would be a suitable solution for pumping the material underground without premature decomposition (Britton, 1985).

Enhancement of the microbial population has also been reportedly used to reduce levels of iron and manganese in the groundwater (Hallberg and Martinelli, 1976). The process, known as the Vyrodex method, was developed in Finland and has been used in Sweden and other areas where high levels of the two elements are found in the groundwater. Iron bacteria and manganese bacteria oxidize the soluble forms of iron and manganese to insoluble forms; the bacteria use the electrons adsorbed from the oxidation process as sources of energy. Dissolved oxygen is added to the groundwater to stimulate the bacteria to first remove the iron and then later the manganese. As the iron bacteria population builds up and begins to die, it supplies the organic carbon necessary for the manganese bacteria. The efficiency of the process increases with the number of aerations.

Hydrogen peroxide can mobilize metals, such as lead and antimony, and, if the water is hard, magnesium and calcium phosphates can precipitate and plug the injection well or infiltration gallery (Wilson, Leach, Henson, and Jones, 1986). Heavy metal control procedures involve techniques that effectively prevent contact between the metals and the peroxide (Bambrick, 1985). This is accomplished by using a chelating agent and silicate. The most effective of the commercially available chelating agents is the pentasodium salt of diethylenetriaminepentaacetic acid (Na_5DTPA). This is a negatively charged compound that can form a ringed structure that alters the reactivity of a positively charged ion. The heavy metal ion is bound by covalent bonds off the nitrogens and ionic bonds off the acetate groups and, thus, is inhibited from entering into undesirable reactions, e.g., $Na_3MnDTPA$. The breakdown of peroxide can be decreased substantially using the chelate Na_5DTPA in combination with sodium silicate and $MgSO_4$. The real value of DTPA, even when the metal level is low, is in stabilizing the peroxide liquor solution. In the laboratory, this combination reduces the amount of peroxide decomposed to 55% after 2 h. Without DTPA, 95% of the peroxide is made useless after 1 h. These results have also been verified in field tests.

The pH does not strongly influence the rate of hydrogen peroxide decomposition by iron salts in aqueous media (Wilson, Leach, Henson, and Jones, 1986). The pathway of its decomposition depends upon the valence of the iron. Lawes (1991) found that an alkaline pH accelerates decomposition. Decreasing the ratio of hydrogen peroxide to soil (e.g., 1:4) also increases decomposition.

The plugging of interparticular spaces in the soil resulting from the growth of biomass can lead to formation of anaerobic conditions and, in some cases, formation of toxic or explosive gases and pollutants more toxic than the original ones (Kaufman, 1995). Prevention of biofouling during *in situ* bioremediation of hydrocarbon-impacted soil or groundwater involves appropriate engineering and hydrogeological considerations prior to process initiation, as well as the judicious use of hydrogen peroxide.

Although there are some compounds that will not react with hydrogen peroxide but will react with ozone or hypochlorite, hydrogen peroxide appears to be the most feasible for *in situ* treatment (U.S. EPA, 1985a).

5.1.4.3 Hypochlorite

Another potential oxidant is hypochlorite (Amdurer, Fellman, and Abdelhamid, 1985). It is generally available as potassium, calcium, or sodium hypochlorite (bleach) and is used in the treatment of drinking water, municipal wastewater, and industrial waste (U.S. EPA, 1985a). It reacts with organic compounds as both a chlorinating agent and an oxidizing agent. Hypochlorite additions may lead to production of undesirable chlorinated by-products (e.g., chloroform) rather than oxidative degradation products. Therefore, the use of hypochlorite for *in situ* treatment of organic wastes is not recommended.

5.1.4.4 Other Electron Acceptors

Oxygen can be supplemented with other electron acceptors, such as nitrate (Wilson, Leach, Henson, and Jones, 1986). Nitrate can support the degradation of xylenes in subsurface material (Kuhn, Colberg, Schnoor, Wanner, Zehnder, and Schwarzenbach, 1985). This approach is still experimental but offers considerable promise because nitrate is inexpensive, is very soluble, and is nontoxic to microorganisms, although it is of human health concern. Nitrate itself is a pollutant limited to 10 mg/L in drinking water (U.S. EPA, 1985b). Its use may also be limited by regulations and concerns for nitrite formation and potential for eutrophication (Brown, Mahaffey, and Norris, 1993). Section 3.2.2.1.1 describes the process of denitrification in depth and discusses applications for bioremediation.

While it has been investigated as an alternate electron acceptor for degradation of monoaromatic (except benzene) and polyaromatic compounds, nitrate does not result in degradation of aliphatic compounds (Brown, Mahaffey, and Norris, 1993). Another study found that neither nitrate nor sulfate as terminal electron acceptors in an anaerobic process is effective on the types of saturated hydrocarbons found in petroleum (Texas Research Institute, Inc., 1982).

When nitrate is the electron acceptor, 50.0 mg nitrate/L water would require 90,000 kg water/kg hydrocarbon (Hinchee and Miller, 1990). With 300.0 mg nitrate/L water, 15,000 kg water/kg hydrocarbon would be required for biodegradation.

Other options for electron acceptors include carbon dioxide, sulfate (see Section 3.2.2.1.2), and iron (Brown and Norris, 1994). Iron-reducing bacteria have been shown to be able to use organic matter from a landfill leachate as a carbon source and iron oxides in an aquifer as an electron acceptor to degrade toluene (Albrechtsen, 1994).

Cost may be the deciding factor when more than one electron acceptor would work, with hydrogen peroxide being the most expensive (Brown and Norris, 1994). There may be instances, however, where other methods cannot be used, may require too much time, or would be ineffective. Selection of the appropriate oxidizing agent depends, in part, upon the substance to be detoxified and also upon the feasibility of delivery and environmental safety (U.S. EPA, 1985a).

See Section 3.2.2 for a presentation of anaerobic biodegradation processes, biodegradable petroleum components, and the microorganisms capable of degrading petroleum compounds under anaerobic conditions.

5.1.4.5 Soil Oxygen Delivery Approaches

The contaminated soil should remain permeable to water and air during bioremediation (NcNabb, Johnson, and Guo, 1994). A common biological *in situ* remediation technique is to inject water saturated with air oxygen, pure oxygen, or enriched with hydrogen peroxide or ozone, and sometimes nutrients, into contaminated soil (Lund, Swinianski, Gudehus, and Maier, 1991). For technical, economic, and time reasons, the success of this approach is questionable (Lund and Gudehus, 1990).

1. Injecting Water

Oxygen must be dissolved in the interstitial water of the soil (Thibault and Elliott, 1980). Keeping the soil moist is, in itself, a simple and low-cost method of supplying some aeration, although it provides such a low amount of oxygen, it is not very effective.

2. Colloidal Gas Aphrons

A newly developed method that holds great promise for introducing oxygen to the subsurface is microdispersion of air in water using colloidal gas aphrons (CGA), which creates bubbles 25 to 50 µm in diameter (U.S. EPA, 1985a). With selected surfactants, dispersions of CGAs can be generated containing 65% air by volume. A surfactant concentration of 1000 to 5000 ppm is needed to generate the microbubbles (Lange, Bouillard, and Michelsen, 1995). Foam, in the form of microbubbles, consists of 60 to 70% dispersion of 55-μm microbubbles in water (CGA). Foam flows into less accessible spill areas, without channeling or poor sweep.

3. Injecting Air

Air is much less viscous than water (1.8×10^2 and 1.0×10^4 μP, respectively) (Wilson, Leach, Henson, and Jones, 1986). Air also has a 20-fold greater oxygen content on a volume basis. If the air- and water-filled porosity are about the same, and the pressure gradients are the same, then air should be about 1000 times more effective than water. Air should be particularly effective for oxygen supply to contaminated regions high in the unsaturated zone.

Air can be injected by use of pumps, propellers, stirrers, spargers, sprayers, and cascades (Texas Research Institute, Inc., 1982). A blower can be used to provide the flow rate and pressure for aeration, such as 5 psi pressure in a 10-ft aeration well, with an airflow of 5 ft^3/min (Texas Research Institute, Inc., 1982).

Air can also be added to extracted groundwater before reinjection, or it can be injected directly into an aquifer (U.S. EPA, 1985a). The first method, in-line aeration, involves adding air into the pipeline and mixing it with a static mixer to provide a maximum of 10 mg/L oxygen. This concentration will degrade about 5 mg/L hydrocarbons and would, therefore, provide an inadequate oxygen supply. A pressurized line can increase oxygen concentrations, as can the use of pure oxygen.

4. Oxygenation Systems

Oxygenation systems, either in-line or *in situ*, can also be installed to supply oxygen (U.S. EPA, 1985a). These can achieve higher oxygen solubilities and more-efficient oxygen transfer to the microorganisms than conventional aeration. Solubilities of oxygen in various liquids are four to five times higher under pure oxygen systems than with conventional aeration. In-line injection of pure oxygen can impart 40 to 50 ppm of dissolved oxygen to water (Brown, Norris, and Raymond, 1984), which will provide sufficient dissolved oxygen to degrade 20 to 30 mg/L of organic material, assuming 50% cell conversion. This oxygen will not be consumed immediately, as is the oxygen from aeration. Pure oxygen is expensive to use and the oxygen is likely to bubble out of solution (degas) before the microbes can utilize it.

Pure oxygen can be injected by use of pumps, propellers, stirrers, spargers, sprayers, and cascades (Texas Research Institute, Inc., 1982).

5. Injection of Liquified Gases

The injection of liquid oxygen or liquid air into the soil could utilize existing technology (Texas Research Institute, Inc., 1982). Intermittent injection of liquid oxygen would produce a high concentration of oxygen, which would slowly diffuse into the surrounding strata. Since oxygen is ten times more soluble in hydrocarbons than it is in water, the hydrocarbon phase could actually act as an oxygen reservoir to replace the oxygen being consumed in the aqueous phase (Faust and Hunter, 1971). Repeated injections would create a flow through the system, preventing buildup of carbon dioxide. Another technique would have to be used to add additional nutrients. This method would be best in an area where the soil contained abundant nutrients.

Liquid oxygen was trucked to a site contaminated with PAHs (Gupta, Djafari, and Zhang, 1995). A spacing of about 12 m was used between infiltration system laterals and for extension galleries. With the proposed infiltration trench and overlying collection gallery design for the waste matrix, fluid seepage velocities exceeded 2 m/day, with a minimum of 4 mg/L dissolved oxygen concentration conducive for bioremediation.

A technique for soil venting was used in Mont Belvieu, TX to flush leaked propane and ethane out of the ground (*Austin American Statesman,* 1980). Liquid nitrogen was pumped underground, and the large volumes of nitrogen gas generated swept the gases through the soil. Possibly, liquid oxygen or air could be utilized in this fashion to supply oxygen to the soil strata. It is not known what effect a stray spark or flame might have on a system such as this. The potential for an underground fire exists.

6. Injection of Oxygen-Releasing Compounds (with Nutrients)

The best material for implementing this approach is hydrogen peroxide (Texas Research Institute, Inc., 1982). Injection should ideally be made over the entire contaminated area, both into the water table and at points just above the water table into the gasoline-bearing soil. A large amount of the residual gasoline would be consumed by the bacteria in the soil. The organisms may produce emulsifiers that would help mobilize the gasoline into the water table where it could be collected at the producing wells. A recirculating system might be set up to treat the produced water by cleaning, fertilization, oxygenation, and reinjection into the water table. Oxygen is best provided through recirculated groundwater using hydrogen peroxide, if it has to be delivered to fractured bedrock, highly stratified aquifers, or where the saturated interval is 1 m or less (Brown, Mahaffey, and Norris, 1993).

A variation would be to use a physical oxygenation technique on the injection water instead of a chemical additive. See Section 5.1.4.2 for a discussion of the use of hydrogen peroxide for bioremediation.

7. Well Points

Site geology in most cases will determine the methods of aeration to be used in a given situation (Raymond, Jamison, and Hudson, 1976). For example, in a fractured dolomite and clay formation, lack of homogeneity makes well injection and distribution of oxygen difficult.

A well point injection system can be used to supply oxygen and nutrients to the subsurface. Air can be sparged into wells using diffusers. Diffusers attached to paint sprayer-type compressors can inject about 2.5 cfm air into a series of wells to enhance degradation (Raymond, Jamison, and Hudson, 1976; U.S. EPA, 1985a). The diffusers are positioned 5 ft from the bottom of the well and below the water table. Aeration through well points has been successfully used for saturated soils, but it is uncertain whether or not the technique would also work for unsaturated soils.

The solubility of oxygen is very low, approximately 8 mg/L at groundwater temperatures (Wilson and Rees, 1985). Diffusers that sparge compressed air into the groundwater cannot exceed the solubility of oxygen in water (Lee and Ward, 1986). Oxygen levels can be increased by sparging the injection wells with oxygen instead of air (Wilson, Leach, Henson, and Jones, 1986). This raises the oxygen levels about fivefold.

8. Shallow Injection Sites

Since a well or injection site is expensive to drill, relatively shallow soil injection sites that are just deep enough to get oxygen and nutrients past the plant growth zone may be adequate (Texas Research Institute, Inc., 1982).

9. Plowing/Tilling

Even though spills may contaminate just the upper layer of soil, an oxygen limitation can retard degradation (Thibault and Elliott, 1980). Active microflora have been observed in the top 15 cm of soil, and tilling is suggested as an effective means of promoting aeration (Raymond, Hudson, and Jamison, 1976). This provides aeration by turning over the soil layers with plows and exposing the soil to the atmosphere to a depth of 6 to 24 in. (Ju, Devinny, and Paspalof, 1993). Traditional machines for soil plowing are disk harrows and rototillers. Soils deeper than about 2 ft can be aerated by using construction equipment, such as a backhoe. Tilling equipment can aerate surface soils and mix wastes or reagents into the soil (Sims and Bass, 1984). Rototilling equipment promotes the aeration and mixing process more effectively than disks or bulldozers (Raymond, Hudson, and Jamison, 1976).

Plowing may be valuable for providing aeration to greater depths in the active layer of the soil, but this is only the case when soils are very active, and it is unlikely to maintain soils in aerobic conditions when this is the case (Devinny and Islander, 1989). Although plowing will ensure adequate mixing of water and fertilizer, plowed soil still remains highly heterogeneous (Ju, Devinny, and Paspalof, 1993). Plowing will break up large clods but may not affect small or hard aggregates, which prevent penetration of oxygen and protect entrapped petroleum hydrocarbons from degradation. Total petroleum hydrocarbon concentrations are higher in soils containing aggregates. Tillage is effective to only a depth of about 15 cm and destroys soil structure of wet soil, forming large clods and lowering the rate of bioremediation (NcNabb, Johnson, and Guo, 1994).

For a particular field experiment, rent for a disk harrow was $150/h (Ju, Devinny, and Paspalof, 1993). This equipment could plow about 20 ha/h but required two or three passes. Assuming two passes (10 ha/h), rental costs would be $15/ha for each treatment. With labor cost of $30/h, total treatment costs would be $18/ha. If the disked site were treated once a week for 6 weeks, the total cost would be $108/ha.

10. Pulverization

The time required for bioremediation can be reduced by substituting pulverization for the traditional plowing to aerate the contaminated soil (Ju, Devinny, and Paspalof, 1993). Experimentation determined that oil and grease elimination rates were higher for smaller particles, consistent with the greater availability of substrate in these particles. The numbers of microorganisms may not increase on pulverized particles, but the biodegradation of each organism will be more rapid because of the more easily degradable substrate exposed to them. It was unclear whether pulverization would affect total volatiles release. A 25% reduction in treatment time was observed in a field test of the pulverized system.

Costs of pulverization are modest (Ju, Devinny, and Paspalof, 1993). In a field test comparing pulverization with plowing, the pulverizer rented for about $260/h and could treat about 4 ha/h. Rental was about $65/ha, or $68/ha with labor. The pulverized site needed only four treatments, for a total of $272/ha.

11. Combined Air–Water Flushing

Oxygen can be better dispersed through the subsoil by creating nearly horizontal airflow through a network of positive (injection) and negative (suction) lances in the soil (Lund, Swinianski, Gudehus, and Maier, 1991). In addition to this, water percolating vertically downward from a sprinkling system on the surface detaches and solubilizes the hydrocarbons and provides moisture to the organisms for the biodegradation. Nutrients can also be dispersed with the water. Both air and water flow through the soil pores. During biodegradation, depleted oxygen in the water is replaced by diffusion from the air, and contaminants are removed in the air and water streams (see Section 2.2.2.7).

12. Hydraulic/Pneumatic Fracturing

Hydraulic fractures can be created in the subsurface and utilized for more even distribution of materials through the soil to enhance biodegradation (Davis-Hoover, Murdoch, Vesper, Pahren, Sprockel, Chang, Hussain, and Ritschel, 1991).

Fluid is injected into a well until the pressure nucleates a fracture. A proppant (a material that props the fracture) such as sand can be released into the fracture, at the time of formation. Granules of slow-release nutrients or oxygen-generating compounds in a viscous gel could then be pumped through the permeable sand conduits. Multiple layers of horizontal or vertical fractures can be created. Although an encapsulated solid peroxide proved to be toxic when dispersed with this approach, the method is still promising as a delivery system for appropriate materials. A test employing pneumatic fracturing at Tinker Air Force Base, Oklahoma City, OK successfully increased permeability of the unsaturated zone (Anderson, Peyton, Liskowitz, Fitzgerald, and Schuring, 1995). Postfracture airflows were 500 to 1700% higher after the treatment. See Section 2.2.2.5.

13. *In Situ* Electrobioreclamation/Electro-osmosis

Since contaminated soils may have a low permeability, it is difficult to deliver nutrients, molecular oxygen, other electron acceptors, or electron donors through the vadose zone (Lageman, Pool, van Vulpen, and Norris, 1995). Movement of these materials can be accelerated through the use of electro-kinetic transport.

The process of electrobioreclamation has been developed for low-permeability, unsaturated soils (Lindgren and Brady, 1995; Lageman, Pool, van Vulpen, and Norris, 1995). Alternating anode and cathode arrays induce an electrical current into the soil matrix and cause soil water to flow toward the cathode (electro-osmosis). The ionic pollutants dissolved in the soil–water solution migrate and collect around the oppositely charged electrode, where they are captured and removed. Production of electricity heats the soil, which increases hydraulic permeability and the solubility of organic constituents (Lageman, Pool, van Vulpen, and Norris, 1995).

The electrodes can be installed at any depth for *in situ* or *ex situ* electroreclamation (Lageman, Pool, van Vulpen, and Norris, 1995). Soil moisture can be controlled for optimum results. Because unsaturated conditions (typically <50% saturation) are maintained, there is a residual void space that could be utilized for dispersing gases (Hazen, Lombard, Looney, Enzien, Doughtery, Fliermans, Wear, and Eddy-Dilek, 1994). An enclosed system would permit control of temperature, moisture, pH, oxygen, nutrients, addition of surfactants, supplementation of highly efficient contaminant-degrading microorganisms, and monitoring of reactions and conditions, resulting in a more rapid and greater extent of biological degradation. This method is less dependent on favorable weather conditions, and seasonal fluctuations would not be a limiting factor. See Section 2.2.2.8 for a full description of this technology.

14. Lowering the Water Table

There is another treatment approach worth considering. If preliminary remediation has removed any hydrocarbons floating on the water table, and if the geology is favorable, then it might be possible to lower the water table to bring the entire contaminated soil into the unsaturated zone where it can be permeated by air (Wilson, Leach, Henson, and Jones, 1986). Dewatering the soil to eliminate saturated conditions and increase air/contaminant contact might improve biodegradation in these areas (Hinchee, Ong, Miller, Vogel, and Downey, 1992).

15. Air Sparging
Air sparging can be used in conjunction with a vacuum recovery system in a soil remediation strategy. Air sparging is a technology for injecting air below the water table (Brown, Mahaffey, and Norris, 1993). In the process of oxygenating the groundwater, volatile components can be transferred to the unsaturated zone, where they can then be removed by a vapor recovery system.

Air sparging alone could maintain dissolved oxygen levels of only 1 to 2 ppm in a spill area (Nagel et al., 1982). However, when it was combined with addition of a hydrogen peroxide–based nutrient solution (FMC Aquifer Remediation Systems, Princeton, NJ) dissolved oxygen levels rose to over 15 ppm.

16. Venting (Bioventing)
Bioventing is the process of aerating subsurface soils to stimulate *in situ* bioremediation, while simultaneously removing volatile compounds (Hinchee, Ong, Miller, Vogel, and Downey, 1992). This process uses soil vapor extraction (SVE) systems to transport oxygen to the subsurface, where microorganisms are stimulated to metabolize fuel components aerobically (Dupont, 1993). Bioventing systems are designed to optimize oxygen transfer and utilization and are operated at much lower flow rate and with configurations much different than those of conventional SVE systems. Lower flow rates allow longer vapor retention times in the soil. This promotes greater biodegradation (Miller, Vogel, and Hinchee, 1991) and reduces the amount of discharged vent gas requiring treatment (Dupont, Doucette, and Hinchee, 1991). Airflow rates can be adjusted to maintain oxygen levels between 2 and 4% for this purpose (Miller, Vogel, and Hinchee, 1991). An airflow rate of 0.5 air void volumes per day was found to be optimal at one test site. See Sections 2.2.1.11 and 2.2.2.2 for a full description of the processes of soil venting and bioventing.

Variations of the soil venting technique are being investigated (Downey, Frishmuth, Archabal, Pluhar, Blystone, and Miller, 1995). One alternative involves low rates of pulsed air injection, a period of high-rate SVE, and off-gas treatment followed by long-term air injection. An innovative remediation approach that combines regenerative resin *ex situ* vapor treatment with *in situ* bioventing to reduce overall costs of site remediation has been developed (Downey, Pluhar, Dudus, Blystone, Miller, Lane, and Taffinder, 1994). Cyclic, or surge, pumping vents air to the organisms as needed (Dupont, 1993). The retained vapors have a longer exposure to microbial degradation in the soil. Bioventing is being combined with bioslurping, a vacuum-enhanced, free-product recovery system for light nonaqueous-phase liquids (LNAPLs) (Kittel, Hinchee, Hoeppel, and Miller, 1994). In this process, LNAPLs are recovered while the vadose zone is biovented (see Section 3.2.1.6).

17. Forced Aeration with Biopiles
Biodegradation of organic contaminants by indigenous microorganisms can be stimulated by *ex situ* forced aeration of soil piles (Battaglia and Morgan, 1994). Slotted pipes extend through the pile and are attached to a vacuum blower, which draws air through the soil, aerating the microorganisms. Straw, sawdust, or manure can improve soil permeability. Without forced air, the piles would require a longer treatment time, as well as more space, although this would eliminate the need for treating off-gases (Benazon, Belanger, Scheurlen, and Lesky, 1995). This technology is fully described in Section 2.1.2.1.4.

18. Lagoons and Waste Stabilization Ponds
Lagoons will require aeration and additional nutrients for effective petroleum biodegradation (McLean, 1971). Aerating accelerates biodegradation (Johnson, 1978), but it may also be necessary to buffer the water and include algae to provide continuous oxygen.

19. Sludge Systems
More-rapid breakdown of chemicals can be accomplished by utilizing pure oxygen or oxygen-enriched air instead of air in activated sludge systems (Roberts, Koff, and Karr, 1988). Extended aeration involves longer detention times than conventional activated sludge. Addition of pure oxygen raised performance by 10% (Lopatowska, 1984).

20. Bioreactors
Bioreactors allow tighter control of oxygen levels during treatment of contaminated soil than can be achieved *in situ*. The range of 50 to 80% of the maximum water capacity provides high oxygen consumption in bioreactors, but may vary with the particular soil type (Stegmann, Lotter, and Heerenklage, 1991). There are many ways that air or oxygen can be added to the system, depending upon the type of reactor used. For instance, air can be added intermittently to soil-slurry reactors or by bubbleless oxygenation with a membrane gas-transfer system to prevent foaming (Stormo and Deobald, 1995). Check the individual processes in Section 2 for more specific information.

5.1.4.6 Commercial Soil Oxygen Delivery Approaches
1. Bio XL/Restore
The Aquifer Remediation System Bio XL process employs stabilized solutions of hydrogen peroxide (tradename Restore) to increase the amount of oxygen in the soil by more than 25 times, in comparison with air sparging (Chowdhury, Parkinson, and Rhein, 1986). Another of its products (Restore Microbial Nutrient) prevents precipitation of chemical nutrients. Bioreclamation can now be used in low-permeability formations, where the pumping rate from recovery wells is as low as 5 gal/min.

2. Enhanced Natural Degradation
Groundwater Technology has a similar *in situ* process, called END (Enhanced Natural Degradation). It is planning to introduce a new system that could cut the amount of hydrogen peroxide consumption by 75 to 90% by modifying the oxygen delivery system into a closed loop.

3. Tilling
There can be severe oxygen limitation to degradation within inches of the surface of soil (Zitrides, 1983). Polybac Corporation employs tilling of the soil to provide additional oxygen, as well as to better mix a microbial inoculum with the contaminant. Otherwise, the organisms will adhere to the top layers of soil and percolate only slowly to greater depths.

4. Biostim
A commercial product (Biostim) produced by Biosystems, Inc., can circulate 500 ppm oxygen in the soil, as opposed to the 10 ppm when air is used as the source of oxygen (Biosystems, Inc., 1986). This is achieved by using "Tysul" WW hydrogen peroxide from du Pont. It is environmentally safe and is a good source of dissolved oxygen, since the microbes can break down the peroxide into oxygen and water.

5. Mixflo Process
This process dissolves oxygen in two stages (Bergman, Greene, and Davis, 1994). First, slurry is pumped from a lagoon and pressurized to 2 to 4 atm. Oxygen is then injected into the pipeline as finely dispersed bubbles. This two-phase mixture flows turbulently through the pipeline with enough contact time to dissolve 60% of the injected oxygen. The solubility of the oxygen is high at the elevated pressure. In the second stage, the oxygen/slurry is returned to the lagoon with a liquid/liquid eductor, forming fine bubbles and mixing unoxygenated slurry with oxygenated and dispersing the mixture through the holding cell. The combined stages dissolve 90% of the injected oxygen.

Use of oxygen in the Mixflo system reduces the off-gas volume by over 99%, compared with other technologies using air (Bergman, Greene, and Davis, 1994). By reducing off-gas volumes, this system minimizes the problem of aqueous, oily organic wastes foaming during bioremediation and minimal emission being released to the atmosphere.

6. BioPurge[SM]/BioSparge[SM]
The low-volume airflow of bioventing is combined with a closed-loop concept to regulate soil moisture, nutrients, and oxygen (Burke and Rhodes, 1995). Vapors are extracted via wells and treated on-site. Oxygen, nutrients, heat, and moisture are added to the vapor stream, which is then reinjected into the contaminated soil above the water table, with no off-gases to treat. The system is called BioPurge[SM], when the vapor is injected above groundwater, and BioSparge[SM], when it is injected below groundwater level. It is primarily for use in permeable soils. See Section 2.2.2.4 for a full description of the process.

7. Deep Soil Fracture Bioinjection™
Deep Soil Fracture Bioinjection™ uses pressurized subsurface injection, in an overlapping grid pattern, of a slurry containing controlled-release nutrients, oxygen, and microbes (Burke and Rhodes, 1995).

OPTIMIZATION OF BIOREMEDIATION

Injected materials permeate all types of soil through desiccation cracks, fissures, and sand lenses to ensure excellent transmittance, even in silt and clay, to depths of 40 ft. It is an effective and economical technique. The process is further described in Section 2.2.2.6.

8. Vacuum Heap Biostimulation System
The Ebiox bioremediation system employs natural, contaminant-adapted microorganisms for large-scale treatment of excavated soil (Eiermann and Menke, 1993; Eiermann and Bolliger, 1995a). Contaminated soil is excavated and piled in several layers on a sealed biobed, with perforated plastic piping running between the layers and connected to a vacuum blower system. A vacuum draws outside air through the soil by bioventing to provide oxygen to the microorganisms. The vacuum heap is covered by a black plastic liner, and vapors are treated by biofilter. The leachate is collected, purified, enriched with oxygen and nutrients, and resprayed over the heap (see Section 2.1.2.1.5).

9. Detoxifier™
The Detoxifier™, by Toxic Treatments (U.S.A.), Inc., San Mateo, CA, is an adaptation of drilling technology that can be used to deliver treatment agents to the soil to a possible depth of 60 ft (25 ft has been successfully demonstrated) (Ghassemi, 1988). A process tower contains two cutter/mixer bits connected to separate, hollow Kelly bars, through which materials can be added to the soil in dry, liquid, slurry, or vapor form, with thorough mixing and homogenization of a vertical column of soil. Air, oxygen, or hydrogen peroxide could be easily added to the soil. A metal shroud captures off-gases, which are processed in a treatment train and recycled to the soil. On-line monitoring permits good process control and adjustment in treatment conditions. The system establishes a closed-loop operation, and it is completely mobile. The process can be operated *in situ*, using a treatment grid pattern, or *ex situ* in reactors. Sections 2.2.1.13 and 8.4 provide a full description of the process.

5.1.4.7 Creating Anaerobic Conditions
Both the compounds to be degraded and the specific microorganisms should be considered, when assessing whether anaerobic conditions are to be employed for bioremediation. Also, the size of the area, monitoring requirements, personnel costs, and remoteness of the site must be included in the evaluation (Norris, 1995).

Anaerobic degradation can occur immediately, or it can require an acclimation period of up to 18 months (Sahm, Brunner, and Schoberth, 1986; Alexander, 1994). When oxygen is consumed faster than it can be replaced (e.g., if oxygen is depleted during bioremediation), there will be a shift in the composition of the microflora (JRB Associates, Inc., 1984). The predominant organisms will then be facultative anaerobic bacteria, which use alternative electron acceptors (e.g., nitrate) and are able to adapt from aerobic to anaerobic conditions (U.S. EPA, 1985a). There will also be strict anaerobes that cannot tolerate oxygen and would be inhibited or killed by oxygen treatment or the presence of oxidized materials (Fulghum, 1983). Both fastidious anaerobes, the organisms capable of living under anoxic, but not necessarily reduced, environmental conditions, and facultative microbes are important in anaerobic degradation (Kobayashi and Rittmann, 1982). Many soil bacteria will grow under anaerobic conditions, but most fungi and actinomycetes will not (Parr, Sikora, and Burge, 1983).

Active nitrate-respiring microorganisms are found in a variety of anoxic environments, including soils, lakes, rivers, and oceans (Berry, Francis, and Bollag, 1987). When changing the soil from an aerobic to an anaerobic status, facultative microorganisms, such as the denitrifiers, can adapt to nitrate respiratory metabolism (Gottschalk, 1979), utilizing many of the same compounds as they did under aerobiosis (Casella and Payne, 1996). Bacterial denitrification is the most effective type of biotransformation in polluted, anaerobic environments, especially if nitrate or other N oxides are naturally available or added. Nitrate is more soluble and often less expensive than oxygen, although it is itself a pollutant, limited to 10 ppm in drinking water. When soil contaminated with JP-4 jet fuel was enhanced with nitrate, the total number of denitrifiers increased by an order of magnitude (Thomas, Gordy, Bruce, Ward, Hutchins, and Sinclair, 1995).

Denitrification can begin when the concentration of oxygen drops below 10 μmol/L (Roberston and Kuenen, 1984). Oxygen levels can be decreased by compacting the soil or by saturating the soil with water, which restricts the diffusion of gases (Sims and Bass, 1984). Soil structure and texture primarily determine the size of soil pores, and hence the water content at which gas diffusion is limited and the rate at which anaerobiosis occurs (JRB Associates, Inc., 1984). Reducing pore size and restricting aeration increases anaerobic microsite frequency in the soil. Compaction helps draw moisture to the soil surface. This lessens the problems of any leaching that may occur, if anaerobiosis is achieved by addition of water. Volatilization may also be suppressed by surface soil compaction. Water may still have to be added to reach the required degree of anaerobiosis; however, it would be less than for an uncompacted soil, also minimizing the leaching potential. Diking is a common agricultural practice that may be applicable to decreasing the soil oxygen content (Arthur D. Little, 1976). This would establish and maintain anaerobic conditions as long as the land is kept under water. If the soil is heavily overloaded with contaminants, it may become anaerobic and remain so for a long time.

If the soil contains montmorillonite, addition of moisture or water would cause the soil to swell and block any further water movement (Hornick, 1983). Runoff or flooding could result, inducing anaerobic conditions. If coarse soils of sand and gravel, with their large interconnecting pores, are excessively drained, nutrients in added material will not be sufficiently adsorbed on the soil. The groundwater could be contaminated, if there is no restrictive layer between the coarse layer and the water table.

Another possible method of rendering the site anaerobic would be to add excessive amounts of easily biodegradable organics so the oxygen will be depleted (U.S. EPA, 1985a). An example of this is addition of starchy potato-processing waste materials, which promote growth of the aerobic organisms until the available oxygen is depleted and anaerobiosis established (Stevens, Crawford, and Crawford, 1991; Kaake, Roberts, Stevens, Crawford, and Crawford, 1992). Contaminants may be immobilized in the soil to allow more time for degradation, by controlling soil moisture content. If anaerobic decomposition is desired, anaerobiosis must be achieved by a means other than flooding, such as soil compaction or organic matter addition. Control of soil moisture may be achieved through irrigation, drainage, or a combination of methods.

There can be economic advantages of using anaerobic degradative processes, since plugging and intensive management associated with hydrogen peroxide addition would be avoided. However, exopolysaccharide (EPS) can lead to subsurface plugging under denitrifying conditions.

The type of cometabolite (carbon source) used to stimulate biomass production under denitrifying conditions will influence the amount of EPS production, which could lead to subsurface plugging (Hanneman, Johnstone, Yonge, Petersen, Peyton, and Skeen, 1995). Both total biomass production and growth efficiency are inversely correlated with EPS production under aerobic or anaerobic conditions. The oxidation state of the carbon source can be used to estimate the potential for EPS production. Substrates more reduced than glucose yield more energy that can be used in synthesis of polysaccharide; substrates more oxidized than glucose allow energy to be used for cell synthesis rather than large quantities of EPS.

See Section 5.1.7 for a discussion of oxidation-reduction potential and how it relates to creating an anaerobic environment.

5.1.4.8 Combination Aerobic/Anaerobic Treatment

Some compounds require aerobic conditions for biodegradation, while others need anaerobic conditions. Some can be degraded under either condition, while others are not transformed at all. Compounds that appear to be resistant to anaerobic degradation include anthracene, naphthalene, benzene, aniline, 4-toluidine, 1- and 2-naphthol, pyridine, and alkanes (Alexander, 1994). While nitrate may serve as an alternative electron acceptor for degradation of monoaromatic (except benzene) and polyaromatic compounds, it does not promote the degradation of aliphatic compounds (Brown, Mahaffey, and Norris, 1993). Neither nitrate nor sulfate, as terminal electron acceptors, was effective on the types of saturated hydrocarbons found in petroleum (Texas Research Institute, Inc., 1982). Alkylbenzenes in groundwater can be transformed both aerobically and anaerobically (Wilson, Bledsoe, Armstrong, and Sammons, 1986). Anaerobic degradation may be a useful adjunct to the aerobic degradation in heavily contaminated areas of soil where the oxygen supply has been depleted.

In general, hydrocarbons are subject to both aerobic and anaerobic oxidation (Dietz, 1980). The first stage of biodegradation of insoluble hydrocarbons is predominantly aerobic, while the organic carbon content is then reduced by anaerobic action. Interactions between different microbial populations can be essential for the degradation of complex compounds (Hornick, Fisher, and Paolini, 1983). Sometimes, products from the anaerobic process require a final aerobic treatment (Alexander, 1994).

Many compounds can be transformed under anaerobic or alternating anaerobic/aerobic conditions (Wilson and Wilson, 1985), but not readily under strict aerobic conditions. Sequencing anaerobic and aerobic biotreatment steps can be a viable alternative treatment approach (Field, Stams, Kato, and Schraa,

1995). Anaerobic and aerobic organisms could be combined in a single treatment step, by immobilizing the microbes in biofilms or soil aggregates, within which anaerobic microniches can form.

The redox conditions can be controlled to achieve conditions under which specific compounds can be degraded, dehalogenated, or particular organisms or enzyme systems can be selected (Wilson and Wilson, 1985). Alteration of aerobic/anaerobic conditions by adjusting Eh through flooding or cultivation can, therefore, be a useful tool for engineering management to maximize detoxification and degradation of some compounds (Guenther, 1975).

Mixed oxygen/nitrate electron acceptor conditions showed that cultures could degrade benzene, toluene, ethylbenzene, and naphthalene (2 mg/L) under conditions of 2 mg/L oxygen and high levels of nitrate (>150 mg/L NO_3^-) (Wilson, Durant, and Bouwer, 1995). Biodegradation was inhibited at high (8.6 mg/L) and low (<2 mg/L) oxygen levels, except for toluene, which was degraded under anaerobic denitrification conditions. When Miller and Hutchins (1995) combined nitrate, 27 to 28 mg(N)/L, with low levels of oxygen, 0.8 to 1.2 mg O_2/L, they observed complete removal of toluene and partial removal of ethylbenzene, m-xylene, and o-xylene with nitrate as the only electron acceptor. The soil type, however, caused variable results.

The concentration of the contaminants affects the degradation (Wilson, Leach, Henson, and Jones, 1986). The critical concentration of dissolved oxygen for promotion of aerobic activity ranges from 0.2 to 2.0 mg/L and is usually 0.5 mg/L (U.S. EPA, 1985a). The solubility of benzene (1780 mg/L) is much greater than the capacity for its aerobic degradation in groundwater (Wilson, Leach, Henson, and Jones, 1986). However, anaerobic processes will often take over in these situations.

Most filamentous fungi are aerobic, and yeasts are often facultatively anaerobic. Some of the purple nonsulfur organisms can grow microaerobically and anaerobically as phototrophs, yet live as heterotrophs aerobically in the dark, by respiratory metabolism of organic compounds (Stanier, Doudoroff, and Adelberg, 1970; Pfennig, 1978a). Cyanobacteria and *Chlorella* tolerate low oxygen levels (Kobayashi and Rittmann, 1982). These authors developed a protocol for separating and treating refractory compounds by using the bioaccumulation capability of phototrophs (Sections 2.1.2.2.1.4 and 5.2.1.5). Aerobic/anaerobic degradation is used to treat easily degradable constituents. Refractory compounds are removed by sorption to microorganisms, concentrated, and disposed of by other means, such as burial or incineration. Relatively recalcitrant compounds can be degraded if they are first sorbed to photosynthetic organisms. The products of these reactions could then be further degraded by other microorganisms.

Aerobic or anaerobic conditions can be selected to encourage degradation by the most favorable process for the compounds of concern. A combination of the two may sometimes allow the most complete biodegradation to occur. A variety of methods is listed for modifying soil oxygen content to achieve aerobic (Sections 5.1.4.5 and 5.1.4.6) and anaerobic (Section 5.1.4.7) conditions.

BARR (bioanaerobic reduction and reoxidation) is a remedial technique for *in situ* degradation of organics in soil and groundwater employing broad genera, conditioned microorganisms (Dieterich, 1995). The process provides substrate and controlled aeration to create strong shifts in redox to destabilize contaminants and improve *in situ* biodegradation by means of direct metabolism, cometabolism, and surface catalysis.

If a surfactant is required, *Bacillus licheniformis* JF-2 synthesizes a surfactin-like lipopeptide that is the most effective biosurfactant known (Lin, Carswell, Georgiou, and Sharma, 1994). This peptide is produced under both aerobic and anaerobic conditions, but can be optimized by growing the cells under O_2-limiting conditions and a low dilution rate of 0.12/h.

5.1.5 NUTRIENTS

Microbial degradation of hazardous compounds requires the presence of nitrogen, phosphorus, and potassium, in addition to smaller levels of zinc, calcium, manganese, magnesium, iron, sodium, and sulfur for optimum biological growth (Arora, Cantor, and Nemeth, 1982). Nitrogen and phosphate are the nutrients most frequently present in limiting concentrations in soils (U.S. EPA, 1985a). The nutrients required by microorganisms are presented in Table 5.8 (Alexander, 1977).

Feeding nutrient solutions containing inorganic nutrients, such as nitrogen, phosphorus, and sulfur, to natural soil bacteria often enhances the ability of the microorganisms to degrade organic molecules into carbon dioxide and water (Stotzky and Norman, 1961a; 1961b). Even adding cane sugar molasses to Omani crude oil increased microbial respiration and n-alkane biodegradation (Al-Hadhrami, Lappin-Scott, and Fisher, 1996). At one site, decomposition of oil in the soil was shown to proceed at a rate of

Table 5.8 Nutrients Required by Microorganisms

1. Energy source	Organic compounds
	Inorganic compounds
	Sunlight
2. Electron acceptor	O_2
	Organic compounds
	NO_3^-, NO_2^-, N_2O, SO_4^{-2}, CO_2
3. Carbon source	CO_2, HCO_3^-
	Organic compounds
4. Minerals	N, P, K, Mg, S, Fe, Ca, Mn, Zn, Cu, Co, Mo
5. Growth factors[a]	
a. Amino acids	Alanine, aspartic acid, glutamic acid, etc.
b. Vitamins	Thiamine, biotin, pyridoxine, riboflavin, nicotinic acid, pantothenic acid, *p*-aminobenzoic acid, folic acid, lipoic acid, B_{12}, etc.
c. Others	Purine bases, pyrimidine bases, choline, inositol, peptides, etc.

[a] Where growth proceeds in the absence of growth factors, the compounds are presumably synthesized by the organism.

Source: Alexander, M. Introduction to Soil Microbiology. 2nd ed. John Wiley & Sons, New York. 1977. With permission.

0.5 lb/ft³/month without a nutrient source, and at 1.0 lb/ft³/month after the addition of fertilizer (Kincannon, 1972).

Normally, the aliphatics are more easily degraded than aromatics (Perry and Cerniglia, 1973). Without added nutrients, aromatic hydrocarbons are noted to be more readily attacked than saturated aliphatic hydrocarbons by the microbes (Atlas, 1981). Addition of nitrogen or phosphorus stimulates degradation of saturated hydrocarbons more than that of aromatic hydrocarbons. Aromatics might also exhibit more rapid degradation than aliphatics, if the soil contains more organisms with a preference for the former (Zhou and Crawford, 1995). However, addition of nutrients increases total biomass available for degradation of all compounds and reduces the difference noticed.

Some of these nutrients may be present in contaminating wastes, but may not be readily available or in the amount required (Sims and Bass, 1984). Their supplementation may be necessary. Various authors provide ratios for these nutrients. There is some disagreement on the exact ratios, but they are not too dissimilar and a range will be provided here. Three of the major nutrients, nitrogen, phosphorus, and potassium, can be supplied with common inorganic fertilizers (JRB Associates, Inc., 1984). The carbon, nitrogen, and phosphorus content of bacterial cells is generally in the ratio of 100 parts carbon to 15 parts nitrogen to 3 parts phosphorus (Zitrides, 1983). Theoretically, 150 mg of nitrogen and 30 mg of phosphorus are required to convert 1 g of hydrocarbon to cellular material (Rosenberg, Legmann, Kushmaro, Taube, Adler, and Ron, 1992). By knowing how much of the carbon in a spilled substance ends up as bacterial cells, it is possible to calculate the amount of nitrogen and phosphorus necessary to equal this ratio for optimum bacterial growth (Thibault and Elliott, 1980).

Measurement of soil organic carbon, organic nitrogen, and organic phosphorus allows the determination of its carbon-to-nitrogen-to-phosphorus (C:N:P) ratio and an evaluation of nutrient availability (Sims and Bass, 1984). If the ratio of organic C:N:P is wider than about 300:15:1, and available (extractable) inorganic forms of nitrogen and phosphorus do not narrow the ratio to within these limits, supplemental nitrogen and/or phosphorus should be added, such as by addition of commercial fertilizers (Kowalenko, 1978). One such product is POLYBAC® N Biodegradable Nutrients, which contains the proper balance of nitrogen and phosphorus required in a form readily available for microbial uptake (Thibault and Elliott, 1979). Sufficient nitrogen and phosphorus should be applied to ensure that these nutrients do not limit microbial activity (Alexander, 1981).

The C:N ratio might be the primary factor in determining the nutrient effect (Zhou and Crawford, 1995). The quantity of nitrogen and phosphorus required to convert 100% of the petroleum carbon to biomass may be calculated from the carbon-to-nitrogen (C:N) and carbon-to-phosphorus (C:P) ratios found in cellular material (Dibble and Bartha, 1979a). Accepted values for a mixed microbial population in the soil are C:N, 10:1 (Waksman, 1924); and C:P, 100:1 (Thompson, Black, and Zoellner 1954). In reality, a complete assimilation of petroleum carbon into biomass is not achievable under natural conditions. Some of the petroleum compounds are recalcitrant or are metabolized slowly over long periods. From petroleum compounds that are readily metabolized, some carbon will be mineralized to carbon dioxide. Thus, efficiency of conversion of substrate (petroleum) carbon to cellular material is less than 100%. The optimal C:N and C:P ratios are expected to be wider than the theoretical values.

It has been suggested that a C:N ratio of <25 leads to mineralization (excessive N present) and a C:N ratio of >38 leads to depletion of mineralized N (Routson and Wildung, 1970). The latter condition would limit biodegradation due to nitrogen starvation.

The ratio depends upon the rate and extent of degradation of the chemicals involved and may vary according to the particular contaminants present. Biodegradation of complex oily sludges in soil occurs most rapidly when nitrogen is added to reduce the C:N ratio to 9:1 (Brown, Donnelly, and Deuel, 1983b). That of petrochemical sludge is most rapid when nitrogen, phosphorus, and potassium are added at a rate of 124:1, C:NPK. The optimal ratios may be different for different soils (Zhou and Crawford, 1995). These authors found the optimal C:N ratio for a clay/loam soil to be around 50:1. Dibble and Bartha (1979) reported a C:N ratio of 60:1 and a C:P ratio of 800:1. On the other hand, excessive nitrogen (e.g., C:N = 1.8:1) can impair biodegradation, possibly due to ammonia toxicity, although a greater amount of fertilizer has not been shown to inhibit biodegradation (Zhou and Crawford, 1995).

During experiments on landfarming waste oil, it was determined that carbon-to-nitrogen and carbon-to-phosphorus ratios of 60:1 and 800:1, respectively, were optimal under the conditions used (Dibble and Bartha, 1979a). Addition of yeast extract or domestic sewage did not prove beneficial. Urea formaldehyde was found to be the most satisfactory nitrogen source tested, since it effectively stimulated biodegradation and did not leach nitrogen, which could contaminate the groundwater (Dibble and Bartha, 1979b). A problem with this technology is that runoff water from the site could contain high amounts of oil and fertilizer (Kincannon, 1972).

Under most growth conditions, about half of the carbon available from growth hydrocarbons eventually becomes cellular biomass (Texas Research Institute, Inc., 1982). This consists primarily of proteins, nucleic acids, amino acids, purines, pyrimidines, lipids, and polysaccharides. Huesemann and Moore (1993) observed an 11-fold increase in polar nonhydrocarbon compounds after 22 weeks of biodegradation of crude oil–contaminated soil. The initial amount of 430 mg/kg rose to 4725 mg/kg. The increase in polar compounds was probably due to biomass (cell mass) formation, citing, as above, constituents such as intracellular proteins, lipids, carbohydrates, and material found in cell walls and membranes. Growth of *Rhodococcus* and *Pseudomonas* strains on naphthalene, fluorene, phenanthrene, anthracene, fluoranthene, and pyrene yielded high mineralization (56 to 77% of the carbon) and good production of biomass (16 to 35% of carbon) and limited, but significant, accumulation of metabolites (5% to 23% of carbon) (Bouchez, Blanchet, and Vandecasteele, 1996). The higher PAHs produced higher mineralization and lower biomass.

The nitrogen requirement value (NRV) is the amount of nitrogen required by microorganisms to decompose/degrade a particular organic chemical waste (Parr, Sikora, and Burge, 1983). It depends mainly upon two factors: the chemical composition of the waste and the rate of decomposition. This value is also affected by the other soil factors.

After contact with an oily waste, microbial activity initially decreases (Hornick, Fisher, and Paolini, 1983). This may be due to the same initial decrease in mineral nitrogen resulting from nitrogen immobilization by hydrocarbon-metabolizing microbes using up all the available nitrogen. In time, the microorganisms will adapt to the high C:N ratio and increase the total microbial population (Overcash and Pal, 1979a).

Addition of sodium nitrate enhanced the oxidation of hydrocarbons and the ultimate decay of the resulting organic carbon compounds to inorganic carbon compounds (Dietz, 1980). Addition of potassium orthophosphates, KH_2PO_4 and K_2HPO_4, had no effect on biodegradation in this application. The phosphate precipitated in a very early stage due to the presence of calcium. In another case, mineralization of phenanthrene was found to be enhanced by addition of phosphate but not potassium, while it was reduced by addition of nitrate (Manilal and Alexander, 1991).

Calcium promotes flocculation, the clumping of tiny soil particulates, and may prevent thorough incorporation of phosphate into the soil (Brady, 1974). U.S. EPA (1985a) also reported that addition of phosphates can result in the precipitation of calcium and iron phosphates. At low concentrations in calcareous soils, phosphorus is adsorbed onto calcium carbonate; at high concentrations, calcium phosphate minerals are formed, and the phosphate is not available to the microorganisms (Mattingly, 1975). Phosphate can be added to sandy (quartz) soils, but not to calcareous soils (Aggarwal, Means, and Hinchee, 1991).

If calcium is present at 200 mg/L, it is likely that calcium supplementation is unnecessary (U.S. EPA, 1985a). Calcium deficiencies usually occur only in acid soils and can be corrected by liming (JRB Associates, Inc., 1984). If the soil is deficient in magnesium, the use of dolomitic lime is advised. It is desirable to have a high level of exchangeable bases (calcium, magnesium, sodium, and potassium) on the surface exchange sites of the soil for good microbial activity and for preventing excessively acid conditions. Sulfur levels in soils are usually sufficient; however, sulfur is also a constituent of most inorganic fertilizers.

The common form of phosphorus in soil is $H_2PO_4^{-1}$ in basic soil solutions (Mattingly, 1975). Phosphorus concentrations in the soil solution are usually low, ranging from 0.1 to 1 ppm, since this element is mostly associated with the solid phase in soils (Hornick, 1983). In acid soils, phosphorus reacts with iron and aluminum hydroxides to produce adsorbed forms of phosphorus that are in equilibrium with the soil solution or are precipitated and, thus, occluded by the minerals (Dietz, 1980). Robertson and Alexander (1992) report that the extent of growth of a *Corynebacterium* sp. was reduced with 2 or 10 mM phosphate in media containing high iron concentrations.

Phosphorus concentration affects the pH at which mineralization can occur (Robertson and Alexander, 1992). A *Pseudomonas* sp. mineralized phenol rapidly in medium with 0.2 mM phosphate at pH 5.2, but had little or no effect at pH 8.0. However, mineralization proved to be greater at pH 8.0 than at pH 5.2, when the culture contained 10 mM phosphate.

Microorganisms may be limited by phosphorus but not nitrogen (Thorn and Ventullo, 1986). Neither nitrogen nor phosphorus enrichment alone stimulated the biodegradation of phenol in topsoil (Atlas and Bartha, 1973d). However, in two different types of subsurface soils, addition of these nutrients significantly stimulated mineralization. Phosphorus enrichment had the greatest effects, and the effects of simultaneous nitrogen and phosphorus amendments were similar to those observed with phosphorus alone. Phosphorus limitation may be widespread in subsurface soils. If the input of phosphorus into the subsurface is disproportionate to that of organic compounds, phosphorus limitation could greatly reduce the ability of microbes in deeper soil to degrade pollutants as they migrate downward, assuming oxygen or another electron acceptor is not limiting.

Huesemann and Moore (1993) determined that nitrogen and phosphorus fertilizers were important for stimulating biodegradation of petroleum hydrocarbon–contaminated soils, while addition of bacteria from activated sludge solids alone was not sufficient. Degradation by addition of cow manure alone was equivalent to that of the fertilizer over the span of a year; however, the biodegradation rate was optimal with a mixture of the fertilizer, an activated sludge inoculum, and cow manure.

Ammonium phosphate generally provides the nitrogen and phosphorus required for maximum growth of hydrocarbon oxidizers (Rosenberg, Legmann, Kushmaro, Taube, Adler, and Ron, 1992). A mixture of other salts, such as ammonium sulfate, ammonium nitrate, ammonium chloride, sodium phosphate, potassium phosphate, and calcium phosphate, can also be used. However, these are all highly water soluble and can become too dilute in the environment to maintain their effectiveness. Oleophilic nitrogen and phosphorus compounds with low C:N and C:P ratios can overcome this problem.

Fixed nitrogen can initially be a limiting nutrient, but nitrogen limitation can sometimes be overcome by nitrogen fixation (Toccalino, Johnson, and Boone, 1993). This was observed in butane-amended soil but not in propane-amended soil.

The main danger at hazardous waste sites may be in overloading the soil with elements that may have been present in the waste (or already in the soil, e.g., phosphate plugging), causing toxicity and leaching problems (JRB Associates, Inc., 1984). Nutrient formulations often contain excessive orthophosphate to decrease the rate of peroxide decomposition (Aggarwal, Means, and Hinchee, 1991). The maximum orthophosphate concentration that may provide microbial nutrients and yet avoid significant precipitation in most geochemical environments is about 10 mg/L (Miller and Hinchee, 1990). In order to achieve this concentration at a greater distance, however, higher concentrations must be introduced at the injection point, where excessive precipitation may result. Thus, the plugging problem may be unavoidable with orthophosphate-based nutrient formulations. An available soil phosphate content of about 20 mg/L may be needed for microbial growth (Aggarwal, Means, Hinchee, Headington, Gavaskar, Scowden, Arthur, Evers, and Bigelow, 1990).

Polyphosphates (e.g., pyrophosphate, tripolyphosphate, and trimetaphosphate) can serve as an alternative source of phosphate to avoid plugging problems (Aggarwal, Means, and Hinchee, 1991). Restore 375 incorporates sodium tripolyphosphate, but in combination with orthophosphates. Polyphosphate hydrolysis is influenced by pH, ionic composition of the solution, microbial activity, concentration of the enzyme polyphosphatase, and chain length (Gilliam and Sample, 1968; Blanchar and Hossner, 1969; Blanchar and Riego, 1976; Busman and Tabatabai, 1985; Dick and Tabatabai, 1986; Hons, Stewart, and Hossner, 1986). Polyphosphates have a half-life of 1 to 10 days and can be used as *in situ*, slow-release sources of orthophosphate. Another polyphosphate that has been successfully applied at a fuel oil–contaminated site is sodium hexametaphosphate (Steiof and Dott, 1995). Poly- and metaphosphates both have complexing qualities for cations such as Ca^{2+} and Mg^{2+} and thus do not precipitate easily with these cations. They interact with soil, but their sorption is not strong, so they can be transported over longer distances, with metaphosphate being better for this purpose.

Influent substrate-loading rates directly affect the injection pressure and, thus, near-well biofouling (Jennings, Petersen, Skeen, Peyton, Hooker, Johnstone, and Yonge, 1995). Substrate-loading rates also determine biomass concentration in effluents. Continuous nutrient addition to the soil causes biomass to concentrate near the nutrient injection point.

Both total biomass production and growth efficiency are inversely correlated with EPS production, which can lead to subsurface plugging, under aerobic or anaerobic conditions (Hanneman, Johnstone, Yonge, Petersen, Peyton, and Skeen, 1995). The oxidation state of the carbon source can be used to estimate the potential for EPS production. Substrates more reduced than glucose yield more energy that can be used in synthesis of polysaccharide; substrates more oxidized than glucose allow energy to be used for cell synthesis rather than for producing large quantities of EPS. The type of cometabolite (carbon source) used to stimulate biomass production under denitrifying conditions will influence the amount of EPS production, which could lead to subsurface plugging.

Temperature also plays a role when nutrients are added (Walworth and Reynolds, 1995). At 10°C, bioremediation rates are not affected by addition of phosphorus or nitrogen; however, at 20°C, bioremediation is increased by addition of phosphorus but not nitrogen.

Key trace elements are essential to the stimulation of bacterial growth (Kincannon, 1972). These micronutrients are required in such small doses that most are already abundant in the soil. They include sulfur, sodium, calcium, magnesium, and iron. Copper, zinc, and lead are normally considered to exhibit harmful effects on biological growth. Addition of yeast cells can serve as a nutrient source (Lehtomakei and Niemela, 1975). Organic and inorganic nutrients in natural waters affect the rate of mineralization of organic compounds in trace concentrations (Kaufman and Doyle, 1978). Inorganic nutrients, arginine, or yeast extract often enhance, but glucose reduces, the rate of mineralization.

The concentration of nutrients and organics should be kept as uniform as possible to protect against shock loading (U.S. EPA, 1985a). Nitrogen must be applied with caution to avoid excessive application (Saxena and Bartha, 1983). Nitrate or other forms of nitrogen oxidized to nitrate in the soil may be leached to the groundwater (nitrate is itself a pollutant limited to 10 mg/L in drinking water) (U.S. EPA, 1985a). Some nitrogen fertilizers may also tend to lower the soil pH, necessitating a liming program to maintain the optimal pH for biological activity. Since the pH generally decreases with growth with the use of ammonium salts of strong acids, urea can serve as a nitrogen source (Rosenberg, Legmann, Kushmaro, Taube, Adler, and Ron, 1992). Low concentrations of readily metabolized organic compounds (peptone, calcium lactate, yeast extract, nicotinamide, riboflavin, pyridoxine, thiamine, ascorbic acid) often promote the growth of the oxidizer, but high concentrations will retard the degradation of the hydrocarbons (ZoBell, 1946; Morozov and Nikolayov, 1978). The quantity of organic material to add must be determined in treatability studies (Sims and Bass, 1984). Nutrient formulations should be devised with the help of experienced geochemists to minimize problems with precipitation and dispersion of clays (U.S. EPA, 1985a). Special soil preconditioners and nutrient formulations to reduce these problems and maximize nutrient mobility and solubility are being investigated.

Population turnover allows for the recycling of nutrients (Dibble and Bartha, 1979a). However, it is expected that fertilizer in the optimal ratios will have to be reapplied, as necessary. The best fertilizers for soil application are in a form of readily usable nitrogen and phosphorus and also in a slow-release form to provide a continuous supply of nutrients, which is beneficial in terms of fertilizer savings and minimized leaching from the oil–soil interface (Atlas, 1977). A liquid fertilizer containing: 3340 lb ammonium sulfate, 920 lb disodium phosphate, and 740 lb monosodium phosphate was injected into wells at a contaminated site in Marcus Hook, PA (Raymond, Jamison, and Hudson, 1976). Addition of nutrients in this form accelerated the removal of contaminating gasoline.

An additive has been developed to promote biodegradation of materials such as hydrocarbons and oil spills (Basseres, Eyraud, and Ladousse, 1994). The additive consists of a mixture of at least one assimilable nitrogen source, composed of at least one unsubstituted or substituted amino acid, and at least one phosphorus source, e.g., phosphate rocks, with the ratio of nitrogen to phosphorus ranging from 2 to 100. Meat or fish meal is acylated with lauryl acid chloride to render the additive oleophilic and mixed with an amino acid, e.g., lysine, methionine, cystine, threonine, tryptophan, hydroxylysine, or hydroxyproline. The composition is an effective nutrient and has been found to allow biodegradation of oil spills on ocean water.

Kopp-Holtwiesche, Weiss, and Boehme (1993) have patented an improved nutrient mixture for bioremediation of polluted soils and waters. The nutrient mixture to enhance biodegradation of pollutants, especially hydrocarbons, contains phosphoric acid ester emulsifiers as a phosphorus source and one or more water-soluble or dispersible nitrogen sources. It may also contain biodegradable surfactants, with the exception of glycerin esters. Nonionic surfactants are preferred. Suitable phosphoric acid esters include glycerophospholipids (Lipotin), other phospholipids, alkyl phosphates, and/or alkyl-ether phosphates. This composition is useful for biodegradation of oil spills on soils.

There are many substances that would be suitable as fertilizers, and their compositions and origins differ considerably (Sims and Bass, 1984). The choice of an appropriate fertilizer can be complicated, and an agronomist should be consulted to develop a fertilization plan at a hazardous waste site. A plan may include types and amounts of nutrients, timing and frequency of application, and method of application. The nutrient status of the soil and the nutrient content of the wastes must be determined to formulate an appropriate fertilization plan.

An optimum fertilization program has been proposed (Kincannon, 1972). Chemicals are added so as to attain a slight excess of nitrogen, phosphorus, and potassium in the contaminated area. After that, soil testing for ammonia and nitrates is conducted at regular monthly intervals. Small doses of ammonium nitrate are added, as needed, to maintain the ammonium or nitrate surplus. Urea is used as a nitrogen source to avoid the initial increase in soil salts, which may result from additions of other fertilizer stocks. Ammonium nitrate is subsequently applied, once urea is deemed no longer necessary. Potash is added as a potassium source.

An application method must also be selected. In agricultural application, fertilizers are either applied evenly over an area or concentrated at given points, such as banded along roots (Kincannon, 1972). However, at a hazardous waste site, fertilizer will likely be applied evenly over the whole contaminated area and incorporated by tilling, if necessary. Nutrients can also be injected through well points below the plow layer.

With broadcast fertilization, the fertilizer can be left on the surface or incorporated with a harrow (2 to 3 cm deep), a cultivator (4 to 6 cm deep), or a plow (a layer at bottom of furrow, e.g., 15 cm deep) (Kincannon, 1972). The depth depends upon the solubility of the fertilizer and the desired point of contact in the soil. In general, nitrate fertilizers move easily, while ammonia nitrogen is adsorbed by soil colloids and shows little movement until converted to nitrate. Potassium is also adsorbed and moves little except in sandy soils. Phosphorus does not move in most soils. Therefore, potassium and phosphorus need to be applied or incorporated to the desired point of use.

Vented percolation *in situ* can be used to stimulate naturally occurring organisms to degrade hydrocarbons by supplying nutrients and oxygen to the subsurface (Ram, Bass, Falotico, and Leahy, 1993). After oxygen is introduced by a vacuum-inducing airflow through the soil, nutrients can be percolated through the soil with the vent-system piping. Volatile hydrocarbons are removed by the venting, leaving adsorbed, heavier hydrocarbons to be biodegraded. Water (recovered, treated groundwater or freshwater) is amended with nutrients and injected under supplied or gravity pressure to the vent system with the blower off. The nutrients could otherwise be dispersed through horizontal slotted piping laid at intervals on the surface or in ditches just above the depth of contamination.

The BioPurge[SM] technology is based on the low-volume airflow of bioventing, but with a closed-loop system (Burke and Rhodes, 1995). Soil vapor is extracted from wells in the contaminated soil and treated in an *ex situ* unit, where the volatiles are absorbed and biodegraded. Oxygen, nutrients, heat, and moisture are added to the vapor stream, which is reinjected into the contaminated soil above the water table. The additives can be monitored, and there are no off-gases to treat. This approach has limited success in silts and clays. See Section 2.2.2.4 for a full description of the process.

Deep Soil Fracture Bioinjection™ employs pressurized subsurface injection to introduce controlled-release nutrients, oxygen, and microbes, which permeate all types of soil to ensure excellent transmittance

(Burke and Rhodes, 1995). This is an effective and economical technique, which is described in Section 2.2.2.6.

Nutrients can be supplied to the subsurface microorganisms by spraying onto or injecting the soil with a nutrient mixture immobilized on an organic carrier (Reichardt-Vorlaender, 1995). This results in an emulsion of oil and water in the soil. Organic nitrogen and phosphorus are fixed to the soil contaminants as nutrients for the microbes. Groundwater is not contaminated, and remediation of 2000 to 25,000 g oil in 1 t soil is possible.

Widrig and Manning (1995) determined that continuous saturation with nitrogen and phosphorus amendments was not as effective as periodic operation, consisting of flooding with nutrients, followed by draining and forced aeration. By monitoring CO_2 and O_2 levels *in situ*, it may be possible to optimize the timing of flooding and aeration events to maximize degradation rates.

Gaseous ammonia can be introduced with bioventing techniques to the subsurface at sites where surface application of liquid nutrients is not possible (Marshall, 1995). The ammonium ion is the preferred nutrient form of many microorganisms. Ammonia supplied to the subsurface dissolves readily in soil moisture and sorbs strongly to soil particles. Such ammonia applications should be conservative to enhance biodegradation. Ammonia-oxidizing organisms will convert some of the ammonia to nitrate, and excessive ammonia can promote formation of methane from anaerobic hydrocarbon degradation with nitrate as the electron acceptor.

Dineen, Slater, Hicks, Holland, and Clendening (1993) add anhydrous ammonia as a source of reduced nitrogen to an airstream through the unsaturated soil. This increases the soil oxygen and nitrogen levels, resulting in a 100-fold increase in microbial count. Maintaining viable cell counts at the level of 10^6 to 10^7 should result in a decrease of petroleum hydrocarbons *in situ* to a cleanup level of 100 ppm.

A microcapsule technique has been patented for degrading hydrocarbons (Schlaemus, Marshall, MacNaughton, Alexander, and Scott, 1994). It consists of a core material, a coating material, and at least one microorganism capable of degrading the hydrocarbon. The core is lipophilic material containing nutrients for the microorganisms. The coating material is water soluble. The capsule is such that the organisms are kept in close proximity to the hydrocarbon to be degraded.

Another encapsulation approach employs liposomes (Gatt, Bercovier, and Barenholz, 1991). The phospholipids in liposomes are naturally occurring membrane lipids (e.g., plant phospholipids), which can serve as a source of carbon, hydrogen, phosphorus, and nitrogen for microorganisms in the subsurface. The liposomes are in the shape of sealed microsacs containing water, which can provide a reservoir for nutrients, minerals, sugars, amino acids, vitamins, hormones, drugs, and growth factors. The membrane of the vesicle is hydrophobic on the inside and hydrophilic on the outside, which makes it compatible with a variety of neutral or charged, lipophilic or amphiphilic compounds. Continuous release of the nutrients could be controlled by the lipid composition. Liposomes rapidly induced a considerable enhancement (up to 7 logs) of growth of bacteria in soil contaminated with petroleum. Counts of over 10^{11} cells/mL were obtained. The liposomes could be used to clean up contaminated soils both by enhancing microbial growth and by modifying the physical properties of the contaminant (reducing the interfacial tension 10,000- to 50,000-fold), making it more biodegradable (See Section 5.3.1.2). There are numerous ways they could be dispersed, and they should have no toxic effects on the environment. Liposomes can be prepared on a small or large scale and be tailor-made to suit the requirements of the specific microorganisms and chemicals involved.

A similar patent involves a carrier for supporting microorganisms for soil remediation (Kozaki, Kato, Tanaka, Yano, Sakuranaga, and Imamura, 1994). This carrier contains pores, which hold a nutrient or are a nutrient for the organisms. A method for remediating the soil by administering the carrier has been developed.

In most cases, site geology will determine the method of fertilization to be used in a given situation (Raymond, Jamison, and Hudson, 1976). For example, in a fractured dolomite and clay formation, lack of homogeneity makes well injection and distribution of nutrients difficult. Use of diammonium phosphate could result in excessive precipitation, and nutrient solution containing sodium could cause dispersion of the clays, thereby reducing permeability (U.S. EPA, 1985a). High calcium could cause precipitation of added phosphate, rendering it unavailable to microbial metabolism. If a site is likely to encounter problems with precipitation, iron and manganese addition may not be desirable. If the total dissolved solids content in the water is extremely high, it may be desirable to add as little extra salts as possible.

Results of oil biodegradation in Marcus Hook, PA, and Corpus Cristi, TX, indicated that fertilizer was not a factor in biodegradation until approximately 50% of the oil had been degraded (Raymond, Hudson, and Jamison, 1976). This cannot be regarded as conclusive, since other environmental factors may have affected these studies.

A large kerosene spill (1.9 million L) in New Jersey was cleaned up by a combination of techniques (Dibble and Bartha, 1979c). Much of the kerosene was recovered by physical means and by removing 200 m³ of contaminated soil. Following stimulation of microbial degradation by liming, fertilization, and tillage, phytotoxicity was reduced.

Addition of nitrogen and phosphorus fertilizer at another site resulted in a doubling of the oil biodegradation rate of 70 bbl/acre/month to 1.0 lb/ft³/month (Kincannon, 1982). It is recommended that monthly determinations of nitrogen and phosphorus levels in the soil and periodic fertilizer application, when necessary, will optimize the fertilization process. The cost of soil disposal of oily wastes was estimated at $3.00/bbl. Degradation rates of up to 100 bbl/acre/month were reported, when the oil was applied to fertilized soils (Francke and Clark, 1974).

In the 1989 Exxon Valdez oil spill at Prince William Sound, AK, the rate of natural degradation was low and limited by environmental factors (Pritchard, Mueller, Rogers, Kremer, and Glaser, 1992; Atlas, 1995). This oil spill provided an opportunity for a major study of the effect of fertilizers on bioremediation. The efficiency of a fertilizer depends greatly on the environment and design of the treatment protocol. Hydrocarbon degraders are normally less than 1% of the total microbial community. This increases to about 10%, when oil pollutants are present. These are the organisms that need to be stimulated by providing the appropriate nutrients.

Fertilizers were able to enhance biodegradation of the indigenous hydrocarbon degraders in this cleanup effort (Pritchard, Mueller, Rogers, Kremer, and Glaser, 1992; Atlas, 1995). Three types of nutrient supplementation were used. One was a water-soluble (23:2 N:P) garden fertilizer formulation. Another was Customblen, a slow-release calcium phosphate, ammonium phosphate, and ammonium nitrate within a polymerized vegetable oil coating. The third was the oleophilic fertilizer Inipol EAP-22 (developed by Elf Aquitaine in France), an oil-in-water microemulsion with urea as a nitrogen source, tri(laureth-4)phosphate as a surfactant and a phosphorus source (an N:P ratio of 7.3:2.8; Glaser, 1991), 2-butoxyl-1-ethanol, and oleic acid as a carbon source. Multiple regression models showed nitrogen applications were effective in stimulating biodegradation rates. The failures and successes were discussed by these authors, as well as the necessary prerequisites that must be met for fertilizers to work.

A problem with Inipol EAP 22 is that it contains a large amount of oleic acid, which increases the C:N ratio and can serve as an alternative carbon source for the organisms (Rosenberg, Legmann, Kushmaro, Taube, Adler, and Ron, 1992). It also contains an emulsifier, and contact with water releases the urea to the water phase, where it is not available for the microorganisms.

A new, controlled-release, hydrophobic fertilizer, F-1, has been developed to overcome many of the problems associated with other sources of nitrogen and phosphorus (Rosenberg, Legmann, Kushmaro, Taube, Adler, and Ron, 1992). F-1 is a modified urea-formaldehyde polymer containing 18% N and 10% P as P_2O_5. It is insoluble and attaches to the oil/water interface with the microorganisms. Strains with a cell-bound, inducible enzyme for F-1 (e.g., *Pseudomonas* sp., *Gluconbacter* strain RT, *Pseudomonas* strain RL4, and *P. alcaligenes* strain RL3) can depolymerize the fertilizer to obtain nutrients at the site where they are needed. These microbes with high cell–surface hydrophobicity then desorb from the hydrocarbon when it is depleted, by releasing a hydrophilic capsular material, which repels them from the spent droplet and allows them to reattach to a fresh drop (Rosenberg, 1986; Rosenberg, Rosenberg, Shohan, Kaplan, and Sar, 1989). The fertilizer can continue to support growth and biodegradation, even after the aqueous phase is removed (Rosenberg, Legmann, Kushmaro, Taube, Adler, and Ron, 1992). A mixed culture of bacteria containing F-1-ase and growing on crude oil reached 1×10^8 cells/mL. F-1-ase activity is associated with the cells rather than the extracellular fluid, and it is influenced by the source of nitrogen and phosphorus.

Degradation of Nigerian light crude oil by *Bacillus* strains 28A and 61B was enhanced by addition of organic nitrogen sources (0.7% peptone, 0.14% urea, and 0.7% yeast extract), while inorganic nitrogen sources (0.46% KNO_3 and 0.3% $(NH_4)_2SO_4$) had a depressing effect on the degradation.

In the coral-derived sands of Kwajalein Island, in the Republic of the Marshall Islands, bioremediation of diesel fuel–contaminated soil by indigenous organisms was found to be feasible, but the degradation rates were very low (Siegrist, Phelps, Korte, and Pickering, 1994). The sand was alkaline (pH > 8) and deficient in nutrients (low nitrogen and phosphorus). Addition of nutrients enhanced the degradation somewhat.

Biodegradation is not always stimulated by addition of inorganic nutrients, because other factors may suppress microbial activity or interact with nutrient limitation to slow degradation (Steffensen and Alexander, 1995). The presence of other bacteria or other substrates may reduce the degradation of certain compounds, possibly as a result of competition for the nutrients.

5.1.6 ORGANIC MATTER

Organic material is very important in the soil matrix (Hornick, 1983). The presence of organic materials may have many effects on soil properties, including soil structure, water-holding capacity, bulk density, mobilization of nutrients (hindering degradation of organic wastes), reduction in soil erosion, and soil temperature (Atlas, 1978a).

Naturally occurring organic material can influence the ability of microorganisms to degrade pollutants (Shimp and Pfaender, 1984). Its role in metal reactions or sorption processes that occur in the soil determines the availability of metals and essential nutrients for plants and microorganisms (Hornick, 1983). Sorption of contaminants on soil particles can alter the molecular character and enzymatic attack of a given compound.

Soil generally contains 5 to 12% organic matter (Overcash and Pal, 1979a), and, of that, some 1 to 4% is mineralized annually (McGill, 1980). Assuming that 5% of the organic matter is nitrogen and 0.5% is phosphate, some 50 to 480 lb/acre nitrogen (60 to 550 kg/ha) and 5 to 48 lb/acre phosphorus (6 to 55 kg/ha) are available annually from soil for use by soil microflora to grow on waste added to the soil.

Soil contains organic material in varying stages of decomposition (JRB Associates, Inc., 1984). Organic matter is generally an amorphous organic residual in soils, which, when present in sufficient amounts, has a beneficial effect on the physical and chemical properties of the soil (Hornick, 1983). Around 65 to 75% of the organic material in soil (60 to 80% in most groundwaters and sediments; Khan, 1980) usually consists of humic substances, i.e., humic acid, fulvic acid, and humin (Schnitzer, 1978; Hornick, 1983). These humic substances have very large surface areas, large amounts of exchangeable bases, and high cation exchange capacities (the total amount of cations held exchangeably by a unit mass or weight of a soil). The crude humin consists of humic acid and hymatomelanic acids containing functional carboxyl and phenolic hydroxyl groups responsible for exchange and adsorption reactions. Both humates and fulvates show a high degree of reactivity due to their acidic functional groups. The reaction of these materials with cations in the soil solution is strongly pH dependent. These organics tend to be recalcitrant to degradation (Khan, 1980). The remainder of the organic material consists of polysaccharides and proteins, such as carbohydrates, proteins, peptides, amino acids, fats, waxes, alkanes, and low-molecular-weight organic acids, which are rapidly decomposed by the soil microorganisms (Schnitzer, 1982). This organic matter can also contribute nitrogen, phosphorus, sulfur, zinc, and boron, all of which add to the nutrient status of the soil (JRB Associates, Inc., 1984). Easily decomposed organic contaminants can become part of an important soil process and result in a substantial increase in beneficial organic materials (Hornick, 1983). It is likely that maintaining a supply of biodegradable organic matter in site soils would allow a higher population of diverse microbes capable of degrading many kinds of toxic organic compounds.

Organic matter is very important to the microbial ecology and activity of the soil (Sims and Bass, 1984; JRB Associates, Inc., 1984). Its high cation exchange capacity and high density of reactive functional groups help to bind both organic and inorganic compounds that may be added to the soil. These properties also help to retain the soil bacteria which can then attack the bound compounds. Thus, the sorbents may immobilize the organic constituents, as well as allow more time for biodegradation. Bacteria seem to be able to survive better in the presence of organic matter, especially in drier soils (Godbout, Comeau, and Greer, 1995). The presence of the solid organic matrix appears to be essential for enhanced degradation (Kaestner and Mahro, 1996). Humus increases the water-holding capacity of soil by swelling when wet to absorb two to three times its weight in water (Hornick, 1983). Because of its surface area, surface properties, and functional groups, humified soil can serve as a buffer, an ion exchanger, a surfactant, a chelating agent, and a general sorbent to help in the attenuation of hazardous compounds in soils (Ahlrichs, 1972).

On the one hand, humic polymers act as stabilizing agents, making contaminants less resistant to biodegradation (Verma, Martin, and Haider, 1975), while, on the other hand, compounds bound to humic material can become unavailable for biodegradation. Manilal and Alexander (1991) suggest that biodegradation of PAHs might be slowed by their sorption to soil organic matter. After adaptation of a microbial community to four types of compounds, it was found that amino acids, fatty acids, and carbohydrates stimulated biodegradation of monosubstituted phenols, while humics decreased biodegradation rates (Shimp and Pfaender, 1984). It appears that fulvic acids may be toxic to microbes and cause sorption of a contaminant, making it less bioavailable (Grosser, Warshawsky, and Kinkle, 1994).

Enzyme activities of soil organisms can be responsible for catalyzing the binding of xenobiotic compounds and their breakdown products to soil humic materials (de Klerk and van der Linden, 1974; Bollag, 1983). Bound hazardous organic compounds, including toxic metabolites, should be monitored. Humus-bound xenobiotic compounds may be slow to mineralize or be transformed to innocuous forms (Khan, 1982). In these cases, the humic content of the soil should probably not be increased. Hazardous constituents may be initially bound to organic materials, but later released as organic materials decompose (Sims and Bass, 1984). The released materials may be subject to leaching, volatilization, or reattachment to soil organic matter. This suggests that treatment is not complete until it can be demonstrated that these compounds are absent or at a safe level in the soil (Bollag, 1983).

In waste-amended soils, the addition of high amounts of organic matter ensures a predominance of organic matter reactions (Schnitzer and Khan, 1978). The mobility of heavy metals added by wastes is related to the organic matter content of soils, pH, hydrous oxide reactions, and the oxidation-reduction or redox potential of a soil. If the soil contains cracks and fractures that may increase the potential for mobilization and groundwater contamination, addition of an adsorbent can be useful (Sims and Bass, 1984; JRB Associates, Inc., 1984). It is especially important and effective in soils with low organic matter content, such as sandy and strip-mined soils. These sorbants include agricultural products and by-products, sewage sludges, other organic matter, and activated carbon.

Organic matter has a beneficial effect when bacteria are added immediately after soil contamination (Godbout, Comeau, and Greer, 1995). However, if the bacteria are added after 38 days pre-exposure to the chemicals, this beneficial effect is observed only in sand. The negative effect of soil texture on contaminant mineralization is more significant with time, possibly due to the formation of clay–humic acid complexes that increase the adsorption of substrate and nutrients on soil particles making them less bioavailable (Godbout, Comeau, and Greer, 1995).

Supplemental carbon and energy sources can be used to stimulate the metabolism of even recalcitrant xenobiotics, either through cometabolism (Alexander, 1981) or simply because of the presence of additional carbon and energy (Yagi and Sudo, 1980). However, if biodegradable organic materials are added to the soil in order to raise the C:N ratio higher than about 20:1, mineral nitrogen in the soil will be immobilized into microbial biomass, and the decomposition process will be slowed considerably (JRB Associates, Inc., 1984). Phosphorus is similarly immobilized when carbon is in excess (Alexander, 1977). If the soil must be managed to decompose organic matter during the treatment of hazardous waste–contaminated soils, nitrogen and phosphorus may be required to bring the C:N:P ratio close to that of the bacterial biomass. Terrestrial oil spillages will probably result in the death of plants, releasing large amounts of nonhydrocarbon organic matter into soil, which can serve as an alternative source of carbon for heterotrophic microorganisms, thereby interfering with the degradation of the contaminants (Atlas, 1977).

Natural organic matter can be added to the soil, such as in the form of synthetic commercial organics, cattle manure, sewage sludge, or crop residues (Arthur D. Little, Inc., 1976). Commercial synthetic organics are expensive and their suitability for microbial growth is uncertain.

Mixed results have been obtained by different researchers with using manure amendments to increase the rate of degradation of organic chemicals. While some workers reported that manure amendments increased the rate of degradation of ten organic chemicals tested, Doyle (1979) found that manure did not significantly reduce the degradation of any chemical examined. The breakdown of several compounds was positively correlated with the increased total microbial activity of manure-amended soil. Sewage sludge, however, enhanced the breakdown of only two compounds, while decreasing the rate of degradation of nine others. Some advantages of using municipal sludges in organic waste treatment are that they contain active indigenous populations of microorganisms with degradative potential, and they provide necessary nutrients for biodegradation (Sims and Bass, 1984).

Sewage sludge and cattle manure are the least-expensive supplements; however, their use is limited since they contain variable quantities of trace elements or heavy metals that may disturb the expected soil mechanisms for degradation (Arthur D. Little, Inc., 1976). They also contain populations of organisms, which, although they are usually enteric and do not survive long in the soil, may represent enough competition to slow the buildup of the desired soil microorganisms. Eight ton/acre of alfalfa meal has been shown to be as effective in stimulating microorganisms as 80 ton/acre of cattle manure. Considerable energy source is removed in the digestive tract of the cattle. The addition of organic wastes, such as animal manure and sewage sludge compost, decreases soil bulk density and increases infiltration and permeability, since organic wastes tend to increase soil aggregation and porosity (Hornick, Murray, and Chaney, 1979).

Nonspecific, readily biodegradable organic matter should be added and mixed into the soil as dry materials or as slurries (Sims and Bass, 1984). Straw has been added to soils to increase adsorption of *s*-triazine herbicides (Walker and Crawford, 1968). Fungal mycelium and baker's yeast also improve soil sorption, with nonliving cells exhibiting greater sorption capacity than living cells (Voerman and Tammes, 1969; Shin, Chodan, and Wolcott, 1970). The soil moisture level should be optimized when adding organic matter, and frequent mixing is required to maintain aerobic conditions (Sims and Bass, 1984). Controls to manage the run-on and runoff from the site, as a result of tillage, are necessary to prevent drainage and erosion problems (Kowalenko, 1978).

Amendments might be effective for a shorter length of time in clay soils than in sandy soils (Godbout, Comeau, and Greer, 1995). Slow-release nutrients or frequent small additions might then be more efficient and economical than a single large amendment at the beginning of a long remediation period. Retreatment may be necessary at intervals as nutrients are used up (Kowalenko, 1978). The potential achievable level of treatment ranges from low to high, depending upon the solubility, sorption, and biodegradability of the organic constituents.

It is generally accepted that subsurface microbes are oligotrophic (Wilson, McNabb, Wilson, and Noonan, 1983); however, in one study, carbon (cellulose) enrichment had little effect on mineralization of phenol in any soil examined (Thornton-Manning, Jones, and Federle, 1987). This response could have been due to the recalcitrance of the added carbon source or inorganic nutrient limitation. The most extensive mineralization occurred in a surface soil, which had the lowest content of organic matter.

It may be necessary to conduct laboratory experiments to determine the biochemical fate of given hazardous compounds in organically enriched soil or compost, as well as the environmental hazards associated with any residues (Kaplan and Kaplan, 1982). Residues may be more or less toxic than the parent compounds.

5.1.6.1 Addition of Products to Immobilize Heavy Metals

Many heavy metals have a strong affinity for organic matter (Sims and Bass, 1984). The retention of added metals is often well correlated with soil organic matter. Metals are readily chelated or complexed by functional groups in organic matter. These include carboxylic (–COOH), phenolic–OH, alcoholic–OH, and enolic–OH and carbonyl (C = O) structures of various types. The stability of these metal–organic complexes increases with pH because of the increased ionization of the functional groups. Sewage sludges from municipal areas often contain high concentrations of heavy metals themselves, and their use should be avoided (Sims and Bass, 1984). Waste materials may also contain soluble organic matter that chelates metals and increases their mobility.

The formation of organometal complexes through the organic matter chelation of metals is an important factor governing metal availability (Schnitzer and Khan, 1978). Other organic materials involved in metal reactions and complexation in soils are plant root exudates and various degradation products, which can serve as the base for the humic fraction of the soil (Hornick, 1983).

Theoretically, the addition of organic matter to a contaminated soil should remove metals from the soil solution, thus preventing their leaching to the groundwater (Sims and Bass, 1984). Organic materials most conducive for use with soils include agricultural products and by-products and activated carbon. Straw, sawdust, peanut hulls, and bark can be used for removal of heavy metals from wastewater solutions (Larsen and Schierup, 1981). One gram of barley straw can adsorb amounts of Zn, Cu, Pb, Ni, and Cd ranging from 4.3 to 15.2 mg, while pine sawdust removes 1.3 to 5.0 mg. The selective order of metal sorption for straw is Pb > Cu > Cd = Zn = Ni.

To obtain maximum sorption of metals by organic matter, soil pH must be adjusted or maintained at greater than 6.5. The addition of organic materials may result in a decreased pH, requiring continued pH adjustment (see Section 3.3 for specific pHs for individual heavy metals). Also, thorough mixing is required.

This is an effective method for removal of metals from wastes; however, agricultural products and by-products are highly susceptible to microbial activity. Degradation of the materials may result in the release of metals. Treatments would have to be repeated in the long term. There may also be competition of the metals with organics, which are also sorbable onto organic materials.

Addition of organic material and maintenance of aerobic conditions can result in the oxidation of arsenite to arsenate (Sims and Bass, 1984). Further treatment with ferrous sulfate will form highly insoluble $FeAsSO_4$. Anaerobic conditions must be avoided with this technology to prevent the reduction and methylation of arsenic to volatile forms, although anaerobic microsites can probably not be completely avoided even in carefully managed soils.

5.1.7 OXIDATION-REDUCTION POTENTIAL

The anaerobes require not only anoxic (oxygen-free) conditions, but also oxidation-reduction (OR) potentials of less than –0.2 V (Kobayashi and Rittmann, 1982) for initiation of growth (Fulghum, 1983).

The OR potential, or Eh, of the soil in question basically expresses the electron availability as it affects the oxidation states of hydrogen, carbon, nitrogen, oxygen, sulfur, manganese, iron, cobalt, copper, and other elements with multiple electron states in aqueous systems (Bohn, 1971). This indicates the electron density of a system. As a system becomes reduced, there is an increase in electron density and negative potential (Taylor, Parr, Sikora, and Willson, 1980). Eh decreases during flooding and increases during drying (Bouwer, 1984). The fastest changes occur within the top 2 cm.

The degradative pathways for some hazardous compounds may involve reductive steps (JRB Associates, Inc., 1984). This may occur as an initial reaction that requires anaerobiosis, or it may be expressed by more rapid degradation under anaerobic conditions. Many compounds can be transformed under anaerobic or alternating anaerobic–aerobic conditions (Wilson and Wilson, 1985), but not readily under strict aerobic conditions. Examples are chloroform, bromodichloromethane, dibromochloromethane, bromoform, and 1,1,1-trichloroethane (McCarty, Rittmann, and Bouwer, 1984). Different redox conditions may also affect the transformation of a compound. For example, chloroform and 1,1,1-trichloroethylene can be degraded under methanogenic conditions but not under denitrification conditions (Bouwer and McCarty, 1983a; 1983b). Some contaminated aquifers are anaerobic, and if the microbial population is capable of degrading the material, it may be possible to use anaerobic *in situ* techniques to treat certain compounds (Wilson and Wilson, 1985).

Different anaerobic bacteria can oxidize a variety of aromatic compounds completely to CO_2 via the intermediate, benzoyl-CoA (coenzyme A), which becomes reduced to cyclohex-1-enecarboxyl-CoA (Koch and Fuchs, 1992). This reduction reaction is complex. The enzyme system involved is not found in aerobically grown cells. It is induced by aromatic compounds — probably by benzoyl-CoA rather than benzoate.

Controlling the redox potential alters conditions to be more favorable for degradation of specific contaminants (Wilson and Wilson, 1985). It also allows for the selection of desired microorganisms. The Eh can be manipulated by cultivation or flooding to achieve aerobic or anaerobic conditions (Guenther, 1975). Regular cultivation of soil should maintain aerobic conditions, while anaerobic conditions can be maintained by keeping the soil saturated with water and limiting aeration. It may be feasible in some cases to enhance reducing conditions intentionally in the subsurface, thereby lowering the redox potential (U.S. EPA, 1985a). The pH can be adjusted with the addition of dilute acids or bases. Table 5.9 shows the succession of events that are related to the redox potential and can occur in poorly aerated soils receiving excessive loadings of organic material.

Oxygen levels in aquatic surface and subsurface environments can be expressed in terms of the logarithm of the electron concentration, pe (Bitton and Gerba, 1985). Values for Eh and pe for various microbiological processes are (at 25°C and pH 7):

Process	pe	Eh (in mV)
Aerobic respiration	+13.75	+810
Denitrification	+12.65	+750
Sulfate reduction	– 3.75	–220
Methane formation	– 4.13	–240

Table 5.9 Succession of Events Related to the Redox Potential, Which Can Occur in Waterlogged Soils or Poorly Drained Soils Receiving Excessive Loadings of Organic Chemical Wastes or Crop Residues

Period of Incubation	System	Redox Potential (mv)	Nature of Microbial Metabolism	Formation of Organic Acids
Early	Disappearance of O_2	+600 to +400	Aerobes	None
	Disappearance of NO_3^-	+500 to +300	Facultative anaerobes	Some accumulation after addition of organic matter
	Formation of Mn^{2+}	+400 to +200		
	Formation of Fe^{2+}	+300 to +100		
Later	Formation of S^{2-}	0 to −150	Obligate anaerobes	Rapid accumulation
	Formation of H_2	−150 to −220		Rapid decrease
	Formation of CH_4	−150 to −220		

Sources: Report for Municipal Environmental Research Laboratory. PB85-124899. JRB Assoc., Inc., Cincinnati, OH. 1984. Takai, Y. and Kamura, T. *Folia Microbiol.* 11:304–313. 1966.

The following classification of oxygen levels in soils, based upon their redox potential at pH 7, has been proposed (Patrick and Mahapatra, 1968):

Soil Type	Redox Potential, mV
Oxidized soil	>400
Moderately reduced soil	100 to 400
Reduced soils	−100 to 100
Highly reduced soils	−300 to −100

5.1.8 ATTENUATION

The basic principle of attenuation is the mixing of contaminated soil (or wastes) with clean soil to reduce the concentrations of hazardous compounds to acceptable levels (Sims and Bass, 1984). This is applicable to both inorganics and organics. The mixing of uncontaminated soil with the contaminated is a means of increasing the extent and effectiveness of immobilization of chemical contaminants at hazardous waste sites. It may also aid in decreasing toxicity of the contaminated soil to the soil microorganisms involved in biodegradation and bring the concentrations to levels that can be successfully biodegraded. In practice, attenuation systems have been designed, and acceptable concentration limits established, only for heavy metals. However, in principle, this technique should also apply to organic contaminants.

This treatment is applicable to all organic wastes (Sims and Bass, 1984). However, organics that are very soluble in water may be more effectively treated by other methods, since large amounts of soil may be required to reduce the mobility of the compound. If very toxic components are present in the waste, destructive treatment would be the preferable treatment alternative.

The indigenous soil profile is tilled to mix uncontaminated soil with the contaminated layers, importing soil or clay, if necessary (Sims and Bass, 1984). Clay soils have a greater capacity for physicochemical attenuation of contaminants than coarse sands or fissured rocks (Pye and Patrick, 1983). The ease of use of this method depends upon site/soil trafficability considerations for tillage and incorporation of added material (Sims and Bass, 1984). Tillage may cause erosion. The level of attenuation achievable is potentially high with suitable size, soil, and waste characteristics. The mixing of new material may alter the properties of the natural soil; therefore, the effectiveness of this approach may vary for different compounds and may not be as expected. However, this method should be reliable under most conditions. There is limited field experience in this technology.

5.1.9 TEXTURE AND STRUCTURE

Soil type may be an important determinant of whether or not pollutants are biodegraded as they pass through the unsaturated zone of a soil profile (Federle, Dobbins, Thornton-Manning, and Jones, 1986). The composition of soil influences infiltration rate and permeability, water-holding capacity, and adsorption capacity for waste components (Hornick, 1983). These, in turn, have an effect on the biodegradability of the contaminating wastes and the ability of microorganisms to metabolize the compounds.

Microorganisms have been shown to be present, often in large numbers, in the entire vertical profiles of sediments in wells several hundred feet deep (Federle, Dobbins, Thornton-Manning, and Jones, 1986). The vertical distribution of microorganisms in a soil profile differs greatly as a function of soil type. In four different soil types, biomass and activity declined with increasing depth; however, the magnitude and pattern of this decline differed for each soil type. The soil type also affects the types of microbial populations present. Horizon and soil type affect the time of transit of a contaminant, as well as the potential for biodegradation.

Soil texture, organic matter, and water content may influence xenobiotic degradation with time, possibly because of the formation of clay–humic acid complexes that would increase the adsorption of substrate and nutrients on soil particles making them less bioavailable (Godbout, Comeau, and Greer, 1995). A beneficial effect has been observed when bacteria are added immediately to contaminated soil that contains organic matter. If they are added later, this effect is seen in sand but not in sandy clay loam.

5.1.9.1 Texture
A predominance of clay and silt particles in finer-textured soils results in a very small pore size, with a slow infiltration rate of water (Hornick, 1983). Bacteria generally will not move large distances in fine-textured soil (less than a few meters, for example), but they can travel much larger distances in coarse-textured or fractured materials (Romero, 1970).

The presence of montmorillonite, with high shrink–swell tendencies, will cause swelling of the soil with added moisture or water and block any further water movement (Hornick, 1983). This can lead to runoff or flooding and inducement of anaerobic conditions. Coarse soils of sand and gravel have large interconnecting pores that allow rapid water movement. However, if such a site is excessively drained, nutrients in added material will move too rapidly to be sufficiently adsorbed on the soil. The groundwater can be contaminated, if there is no restrictive layer between the coarse layer and the water table.

When secondary sewage effluent was applied to sandy sites, trace organics were removed from the effluent, but the leachates contained traces of most of the organic pollutants present in the applied effluent (Bedient, Springer, Baca, Bouvette, Hutchins, and Tomson, 1983).

The pore tree model has been extended to describe the permeable pore structure characterizing the subsurface transport of gas and water in soil, the dispersion of contaminants, and *in situ* remediation (Simons, 1996). This extended model explains previous measurement errors in the permeability of soil due to the measurement scale size, and it predicts the bulk gaseous diffusivity in partially saturated soil as a function of a saturation scale size. It provides an analytic description of the port structure of soil upon which bulk transport, small-scale diffusion, and coupled chemical reactions may be added to accurately describe contaminant transport and *in situ* remediation.

5.1.9.2 Bulk Density
Bulk density is a measure of dry soil weight per unit volume and determines pore space through which water can move (Hornick, 1983). Pore size affects the rate of growth of organisms (McInerney, Weirick, Sharma, and Knapp, 1993). Growth of *Escherichia coli* is reduced in the presence of smaller pore sizes, possibly due to a restriction of bacterial cell division.

Frequent use of heavy machinery either to work the soil or to apply wastes will compact the soil and, thereby, increase the bulk density (Hornick, 1983). Clayey soil requires pretreatment to change its density and its transmissibility to improve bioremediation of petroleum contaminations (Elektorowicz, 1994). The addition of organic wastes, such as animal manure and sewage sludge compost, decreases soil bulk density and increases infiltration and permeability, since organic wastes tend to increase soil aggregation and porosity (Hornick, Murray, and Chaney, 1979).

5.1.9.3 Water-Holding Capacity
This capacity is directly related to the bulk density and texture of the soil (Hornick, 1983). Soils with very fine or very coarse textures or high bulk densities cannot maintain an adequate supply of water; the water content determines available oxygen, redox potential, and microbial activity of a soil system.

Table 5.10 Microbial Genera Degrading Hydrocarbons in Soil

Bacteria		Actinomycetes	Fungi	Yeasts
Achromobacter	*Escherichia*	*Actinomyces*	*Aspergillus*	*Candida*
Aerobacillus	*Flavobacterium*	*Endomyces*	*Cephalosporium*	*Rhodotorula*
Alcaligenes	*Gaffkya*	*Nocardia*	*Cunninghamella*	*Torula*
Arthrobacter	*Methanobacterium*		*Torulopsis*	
Bacillus	*Micrococcus*		*Trichoderma*	
Bacterium	*Micromonospora*		*Saccharomyces*	
Beijerinckia	*Mycobacterium*			
Botrytis	*Pseudomonas*			
Citrobacter	*Sarcina*			
Clostridium	*Serratia*			
Corynebacterium	*Spirillum*			
Desulfovibrio	*Thiobacillus*			
Enterobacter				

Source: Shailubhai, K. *Trends Biotechnol.* 4:202–206, Elsevier Trends Journals. 1986. With permission.

5.2 BIOLOGICAL ENHANCEMENT

5.2.1 MICROORGANISMS IN BIOREMEDIATION

Microorganisms are the principal agents responsible for the recycling of carbon in nature. In many ecosystems there is already an adequate indigenous hydrocarbonoclastic microbial community capable of extensive oil biodegradation, provided that environmental conditions are favorable for oil-degrading metabolic activity (Atlas, 1977). This has been shown for many soil and marine and freshwater environments (Mironov, 1970; Atlas and Bartha, 1973e; Litchfield and Clark, 1973; Cooney, 1974; Mulkins-Phillips and Stewart, 1974b; Cooney and Summers, 1976). It is suggested by some researchers (Atlas, 1977; McGill, 1977) that all soils, except those that are very acidic, contain the organisms capable of degrading oil products, that microbial seeding is not necessary, and that the problem is actually one of supplying the necessary nutrients at the site.

The ability to utilize hydrocarbons is widely distributed among diverse microbial populations (Atlas, 1981). Many species of bacteria, cyanobacteria, filamentous fungi, and yeasts coexist in natural ecosystems and may act independently or in combination to metabolize aromatic hydrocarbons (Gibson, 1982; Cerniglia, 1984; Fedorak, Semple, and Westlake, 1984). Table 5.10 provides some of the common microbial genera that can degrade hydrocarbons in soil (Shailubhai, 1986). See Table 3.18 for a more comprehensive list of microorganisms and the specific hydrocarbons they can degrade or transform.

In general, population levels of hydrocarbon utilizers and their proportions within the microbial community appear to be a sensitive index of environmental exposure to hydrocarbons (Atlas, 1981). In unpolluted ecosystems, hydrocarbon utilizers generally constitute less than 0.1% of the microbial community; in oil-polluted ecosystems, they can constitute up to 100% of the viable microorganisms. This difference seems to quantitatively reflect the degree or extent of exposure of an ecosystem to hydrocarbon contaminants.

Individual organisms are not restricted to one oil or a limited range of oil types (Bausum and Taylor, 1986). Each organism attacks many different oils with comparable facility (Stone, Fenske, and White, 1942). However, this is not the case when individual hydrocarbons or classes of hydrocarbons are considered (Bausum and Taylor, 1986). The overall effect of an organism on a complex substrate is limited by its capacity to attack only certain substances or to accumulate intermediates that it cannot further degrade. Extensive degradation of petroleum pollutants generally is accomplished by mixed microbial populations, rather than single microbial species (Atlas, 1978a). Combinations of bacteria, yeasts, and fungi provide about twice as much degradation of mixed hydrocarbon substrates as do bacterial or fungal strains individually (Walker and Colwell, 1974b). Mixed microbial populations are almost always encountered in natural systems.

There are advantages to relying on indigenous microorganisms rather than adding microorganisms to degrade wastes (U.S. EPA, 1985a). Through countless generations of evolution, natural populations have developed that are ideally suited for survival and proliferation in that environment. This is particularly true of uncontrolled hazardous waste sites where microorganisms have been exposed to the wastes for years or even decades.

Current evidence suggests that in aquatic and terrestrial environments, microorganisms are the chief agents of biodegradation of environmentally important molecules (Alexander, 1980). In 1946, ZoBell reported that nearly 100 species of bacteria, yeasts, and molds, representing 30 microbial genera had been discovered to have hydrocarbon-oxidizing properties (Texas Research Institute, Inc., 1982). Since that time, many other species and genera have been reported to have this ability and to be widely distributed in soils (Blakebrough, 1978; Atlas, 1981). Although many microorganisms appear limited to degradation of a specific group of chemicals, others have demonstrated a wide diversification of substrates they are capable of metabolizing.

The heterotrophic bacteria are the most important organisms in the transformation of organic hazardous compounds, and soil treatment schemes may be directed toward enhancing their activity (JRB Associates, Inc., 1984). Heterotrophs can use the organic contaminants as sources of both carbon and energy (Knox, Canter, Kincannon, Stover, and Ward, 1986). Some organic material is oxidized for energy while the rest is used as building blocks for cellular synthesis. There are three methods by which heterotrophic microorganisms can obtain energy: fermentation, aerobic respiration, and anaerobic respiration.

Bacteria are predominantly involved with degradation of those chemicals that have a higher degree of water solubility and are not strongly adsorbed (Kaufman and Plimmer, 1972). The binary fission-type reproductive methods of bacteria enable them to compete more successfully than fungi for readily available substrates.

The ability of certain microorganisms to oxidize simple aromatic hydrocarbons has been demonstrated. Most of our present knowledge of the microbial degradation of aromatic hydrocarbons has been obtained with single hydrocarbon substrates and pure cultures of different microorganisms (Poglazova, Fedoseeva, Khesina, Meissel, and Shabad, 1967). Little is known about how individual or communities of microorganisms interact with compounds when they are present in a sample of petroleum.

At a given density of oil-degrading microorganisms, their actual contribution to the elimination of oil depends upon their inherent metabolic capability, i.e., "heterotrophic potential," and the degree to which environmental conditions allow this potential to be expressed (Bartha and Atlas, 1977).

Different microorganisms utilize different metabolic processes to derive their energy (Davis, Dulbecco, Eisen, and Ginsberg, 1980). Heterotrophic aerobes employ respiration to oxidize organic compounds as a source of carbon and energy. Other organisms are autotrophic (use the reduction of carbon dioxide as a major source of the organic compounds needed for growth) and can derive energy from the absorption of visible light by photosynthesis or from the respiration of inorganic electron donors (lithotrophs). In their electron transport, some of these organisms use oxygen (Class I), while others use different electron acceptors (Class II; Table 5.11). Those organisms that can oxidize hydrogen can often similarly utilize carbon compounds. Table 5.11 presents examples of several forms of autotrophic metabolism.

Aerobic degradation in soil is dominated by a variety of organisms, including bacteria, actinomycetes, and fungi, which require oxygen during chemical degradation (Parr, Sikora, and Burge, 1983). This process involves OR reactions in which molecular oxygen serves as the ultimate electron acceptor, while an organic component of the contaminating substance functions as the electron donor or energy source. Most aerobic bacteria use oxygen to decompose organic compounds into carbon dioxide and other inorganic compounds (Freeze and Cherry, 1979). In soil, oxygen is supplied through diffusion. If the oxygen demand is greater than the supply, the soil becomes anaerobic. Maximum degradation rates are dependent upon the availability of molecular oxygen. Aerobic biodegradation occurs via the *ortho* pathway, a more efficient and rapid metabolic pathway than anaerobic reactions (Zitrides, 1983). Therefore, most site decontaminations are conducted under aerobic conditions.

Microbial population changes occur as oil composition changes during degradation, because the products formed by certain organisms serve as substrates for others (Bausum and Taylor, 1986).

5.2.1.1 Aerobic Bacteria

See Section 3.2.1 for an in-depth discussion of aerobically degradable petroleum hydrocarbons, including specific microorganisms capable of biodegrading individual constituents and mixtures of compounds found in petroleum.

OPTIMIZATION OF BIOREMEDIATION

Table 5.11 Autotrophic Modes of Metabolism

Organism or Group	Source of Energy	Remarks
I. Aerobic lithotrophs		Inorganic (litho-) electron donors
Hydrogen bacteria	$H_2 + \frac{1}{2} O_2 \rightarrow H_2O$	
Sulfur bacteria (colorless)	$H_2S + 1.2O_2 \rightarrow H_2O + S$	Can produce H_2SO_4 to pH as low as 0
	$S + 1.5O_2 + H_2O \rightarrow H_2SO_4$	
Iron bacteria	$2\ Fe^{2+} + \frac{1}{2} O_2 + H_2O \rightarrow 2\ Fe^{+3} + 2\ OH^-$	
Nitrifying bacteria		
Nitrosomonas	$NH_3 + 1.5\ O_2 \rightarrow HNO_2 + H_2O$	Convert soil N to nonvolatile form;
Nitrobacter	$HNO_2 + \frac{1}{2} O_2 \rightarrow HNO_3$	most can also use electron donors
II. Anaerobic respirers		
Denitrifiers[a]	$nH_2 + NO_3^- \rightarrow N_2O, N_2,$ or NH_3	Cause N loss from anaerobic soil
Desulfovibrio	$nH_2 + SO_4^{2-} \rightarrow S$ or H_2S	Odor of polluted streams, mud flat
Methane bacteria	$4H_2 + CO_2 \rightarrow CH_4 + 2H_2O$	Sewage disposal plants
Clostridium aceticum	$4H_2 + 2CO_2 \rightarrow CH_3COOH + 2H_2O$	
III. Photosynthesizers		
Purple sulfur bacteria	$4CO_2 + 2H_2S + 4H_2O \xrightarrow{light} 4(CH_2O) + 2H_2SO_4$	$H_2(A)$ = various electron donors
Nonsulfur purple	$CO_2 + 2H_2(A) \xrightarrow{light} (CH_2O) + H_2O + 2(A)$	$H_2(A)$ = various electron donors
Algae	$CO_2 + 2H_2O \xrightarrow{light} (CH_2O) + \frac{1}{2}O + 2$[b]	Plant photosynthesis

[a] Anaerobic respiration, with the use of nitrate instead of O_2, is also common for the oxidation of the usual organic substrates by heterotrophs. This metabolism bears no resemblance to autotrophy, as the energy is used for biosynthesis from organic compounds rather than from CO_2.

[b] The O_2 is derived directly from H_2O, and not from CO_2.

Source: Davis, B.D. et al. *Microbiology.* 3rd ed. Lippincott-Raven, Philadelphia, 1980. With permission.

Aerobic bacteria ultimately decompose most organic compounds into carbon dioxide, water, and mineral matter, such as sulfate, nitrate, and other inorganic compounds (Pettyjohn and Hounslow, 1983), and do not produce hydrogen sulfide or methane as reaction products (Amdurer, Fellman, and Abdelhamid, 1985).

The most commonly isolated organisms in areas of hydrocarbon contamination are heterotrophic bacteria of the genera *Pseudomonas, Achromobacter, Arthrobacter, Micrococcus, Vibrio, Acinetobacter, Brevibacterium, Corynebacterium, Flavobacterium* (Kobayashi and Rittmann, 1982), *Mycobacterium* (Gholson, Guire, and Friede, 1972; Soli, 1973), as well as the fungus, *Nocardia* (Canter and Knox, 1985). *Pseudomonas* species appear to be the most ubiquitous and the most adaptable to the different pollutants, while *Corynebacterium* species may be major agents for decomposing heterocyclic compounds and hydrocarbons in contaminated aquatic environments (Kobayashi and Rittmann, 1982).

5.2.1.2 Anaerobic Bacteria

See Section 3.2.2 for a review of the different anaerobic processes that are active in biodegradation of petroleum hydrocarbons, anaerobically degradable petroleum components, and specific microorganisms capable of biodegrading individual constituents and mixtures of compounds found in petroleum.

While many soil bacteria can grow under anaerobic conditions, most fungi and actinomycetes cannot grow at all (Parr, Sikora, and Burge, 1983). Obligately anaerobic bacteria are bacteria with no ability to synthesize an oxygen-linked respiratory chain (Fulghum, 1983). These strictly anaerobic bacteria cannot tolerate oxygen and are inhibited or killed by oxygen, oxidized components of media, etc. Anaerobic decomposition is performed mainly by bacteria utilizing either an anaerobic respiration or interactive fermentation/methanogenic type of metabolism (Parr, Sikora, and Burge, 1983).

The known mechanisms that cause bacteria to require anaerobic conditions include a requirement for a particular OR potential range for initiation of growth, sensitivity to oxidized products in media exposed to air, and an inability to cope with highly reactive molecular species or radicals produced during metabolism in the presence of oxygen (Fulghum, 1983). The anaerobes require not only anoxic (oxygen-free) conditions, but also OR potentials of less than -0.2 V (Kobayashi and Rittmann, 1982).

Anaerobic decomposition of organic matter to carbon dioxide and methane involves interactions within consortia of fastidiously anaerobic bacteria (Kobayashi and Rittmann, 1982), i.e., the fermenters, the proton reducers, and the methanogens (Boone and Bryant, 1980; McInerney and Bryant, 1981; McInerney, Bryant, Hespell, and Costerton, 1981). At least four interacting trophic groups of bacteria are involved (Kobayashi and Rittmann, 1982):

1. Hydrolytic bacteria that catabolize the major components of biomass, such as saccharides, proteins, and lipids;
2. Hydrogen-producing, acetogenic bacteria that catabolize products from the activity of the first group, such as fatty acids and neutral end products;
3. Homoacetogenic bacteria that catabolize multicarbon compounds to acetic acid;
4. Methanogenic bacteria (Zeikus, 1980).

The end products of anaerobic degradation are reduced compounds, some of which are toxic to microorganisms and plants. End products formed from degradation of specific organic hydrocarbons by different microorganisms are given in Table 3.21.

From the viewpoint of kinetic control of anaerobic reactors, the methanogens represent the rate-limiting step (Switzenbaum, 1983). In particular, the acetate-utilizing methanogens, which are important in anaerobic digesters, are quite slow growing. One species has been shown to have a doubling time of 9 days (Zehnder, Huser, Brock, and Wuhrmann, 1980). Because of this, long retention times are necessary for solids in anaerobic reactors.

Methanogens are dependent upon nickel — an unusual growth requirement (Speece, Parkin, and Gallagher, 1983). Nickel is a component of factor F_{430}, an oxygen-stable nonfluorescent chromophore. Failure of anaerobic treatment systems in the past may have been due to the lack of nickel in the medium. Coenzyme M has been uniquely associated with methanogens as a cofactor required for the reduction of methyl vitamin B_{12} to methanol (McCarty, 1982). Another coenzyme with important electron transfer functions not widely found elsewhere is factor F_{420}, which exhibits a blue-green fluorescence in ultraviolet (UV) light and has a strong absorption maximum at 420 nm. The fluorescence has been used as a technique for identifying methanogens (Doddema and Vogels, 1978) and for assessing their potential activities in reactors (Delafontaine, Naveau, and Nyns, 1979).

A marine coculture of blue-green algae has excellent methane-producing ability at seawater salinities, with 65% of the algal carbohydrate being converted to methane (Matsunaga and Izumida, 1984). Numerous anaerobic and facultative anaerobic bacteria and yeasts have been isolated from anaerobic reactors (Fannin, Conrad, Srivastava, Chynoweth, and Jerger, 1986). Halophilic methanogenic bacteria have also been found in salt environments (Mathrani and Boone, 1985; Paterek and Smith, 1985). A thermophilic, acetate-utilizing methanogenic organism had an optimum pH range of 7.3 to 7.5 and an optimum temperature of 60°C (Ahring and Westermann, 1985). A concentration of 10 μM phosphate was shown to be required by *Methanosarcina barkeri* for growth (Archer, 1985).

Section 5.2.2.3 discusses addition of anaerobes to enhance biodegradation under anoxic conditions.

5.2.1.3 Oligotrophs

Organisms living at organic concentrations <15 mg carbon/L are termed oligotrophs (Stetzenbach, Sinclair, and Kelley, 1983). They may be able to live under conditions of even lower carbon flux (<1 mg/L/day) (Poindexter, 1981). They do not constitute a special taxonomic grouping of organisms, but come from almost any group of bacteria or chemotrophs. They are generally adapted to life under low-nutrient conditions, but can readily be readapted to high-nutrient conditions. Reverse adaptation to the low-nutrient environment is, however, not readily achieved; therefore, oligotrophs are best obtained from low-nutrient environments.

Recent studies, however, indicate that bacteria can be resuscitated after an extended period of starvation (Bryers and Sanin, 1994). The creation of oligotrophic conditions can be used to reduce the size of specific hydrocarbon-degrading organisms to improve their transport in the subsurface and then revive them to resume their selective metabolic activity. This is further discussed in Section 5.2.2.5.

Oligotrophs generally have a high surface:volume ratio and high affinity for substrate (Kobayashi and Rittman, 1982). The minimum substrate concentration needed for measurable growth is lower than that required for eutrophic (high-nutrient) organisms, but the maximum growth rate is also lower. Oligotrophs degrade xenobiotics more slowly than natural compounds (Alexander, 1985); however, their capability of surviving on low concentrations makes them potentially useful for removal of trace concentrations of organic contaminants from water, or effluent from wastewater treatment processes

(Poindexter, 1981). Species of *Pseudomonas, Flavobacterium, Acinetobacter, Aeromonas, Moraxella, Alcaligenes,* and *Actinomyces* have been detected in water samples, with isolates surviving extended periods on low nutrient concentrations (Stetzenbach, Sinclair, and Kelley, 1983). These organisms are usually found living as biofilms (Poindexter, 1981). An important characteristic is that they often appear to have multiple inducible enzymes, are able to shift metabolic pathways, and can take up and use mixed substrates (e.g., a *Clostridium* sp.).

Some Actinomycetes (*Nocardia*), coryneforms, and mycobacteria have survived for 30 days under starvation conditions (Krulwick and Pelliccioni, 1979). Oligotrophic bacteria from surface water and those indigenous to the deeper subsurface fail to grow on complex media (Wilson, McNabb, Wilson, and Noonan, 1983). Apparently, many oligotrophic bacteria from surface water fail to use organic compounds that are used readily by eutrophic forms. There is an indication that the bacteria of the deeper subsurface will be active against a more limited range of organic compounds than are degraded in surface soil.

Microorganisms appear to have a threshold level below which some organic compounds cannot be converted to carbon dioxide (Alexander, 1985). This level is needed for growth, enzyme induction, and/or enzyme activity. At lower concentrations, substrate uptake by diffusion of the molecules will meet maintenance energy and survival requirements but will not support growth (Schmidt and Alexander, 1985).

It is not certain whether or not the trace amounts present in a contaminated site would be sufficient to allow microbial growth or propagation to numbers necessary for biodegradation of the material (Alexander, 1985). If not, this could explain the persistence of low levels of biodegradable organic substances, e.g., toluene, xylenes, naphthalene, and phthalate esters, in water or soil in nature (although this might also be due to oxygen limitation). It should be noted that if these organisms obtain energy and carbon for growth by using natural organic constituents of the environment, the threshold for a particular chemical contaminant may be below the level of detection possible using current analytical procedures. Some organic compounds are mineralized even at trace levels below 1.0 pg/ml. It is important to distinguish between compounds that can be transformed at low concentrations by large populations of nongrowing cells and substrates that must support growth for significant degradation to occur.

5.2.1.4 Fungi

Fungi are eukaryotic microorganisms that lack photosynthetic structures and depend upon heterotrophic metabolism (Solanas, Pares, Bayona, and Albaiges, 1984). They may be filamentous or unicellular (yeastlike or amoeboid), or they may aggregate to form large structures, such as a plasmodium or fruit bodies (mushrooms). Fungi are a large part of the microbial biomass in soil, especially in acidic conditions, and they contribute to most decomposition processes. When substrate or water availability is low, most of the fungal biomass is either dormant or dead. Fungal spores or other resistant structures can survive under adverse conditions for long periods of time and then quickly germinate and grow when environmental conditions become favorable. The species and diversity of fungi are affected by clay mineralogy, temperature, and other soil environmental conditions. Most filamentous fungi are aerobic, and yeasts are often facultatively anaerobic. The species of fungi that develop on plates are frequently those that produce spores in greatest abundance (Nannipieri, 1984). Additional colonies can, thus, grow from inactive spores or conidia and give an inaccurate estimate of population size.

Fungi play an important role in the hydrocarbon-oxidizing activities of the soil (Jones and Eddington, 1968). They seem to be at least as versatile as bacteria in metabolizing aromatics (Fewson, 1981). Their extracellular enzymes may help to provide substrates for bacteria, as well as for themselves, by hydrolyzing polymers. They are also important sources of secondary metabolites. Fungi are superior to bacteria in the amount of growth attained on various crude oils (Cerniglia, Hughes, and Perry, 1971). Some authors suggest that filamentous fungi might have greater potential than bacteria in cleaning the environment of spilled petroleum (Perry and Cerniglia, 1973). Several fungi (*Penicillium* and *Cunninghamella*) exhibit greater hydrocarbon biodegradation than bacteria (*Flavobacterium, Brevibacterium*, and *Arthrobacter*) (Cerniglia and Perry, 1973).

The consensus in the literature is that crude oil degradation is principally a result of bacterial activity (Davies and Westlake, 1979). This is, at least partly, due to the use of dilution plates to monitor the microbial biomass, which is biased in favor of bacteria and those fungi that produce large numbers or readily dispersed spores. However, many soil and water fungi produce few or no spores and would not be detected by this method. Many of these fungi in soil are bound up in the soil particles and would not be released during the dilution process (Warcup, 1955; 1957). While fungal numbers cannot be correlated with crude oil degradation, there is a change in the type of fungal colonies produced (Westlake, Jobson, and Cook, 1978).

Fungi appear to be predominantly involved in metabolizing those xenobiotics of lower water solubility and greater adsorptivity (Kaufman and Plimmer, 1972). The ability of fungi to degrade greater quantities of oil during growth may be due in part to their development as a mat on the surface of the oil (Perry and Cerniglia, 1973). The mycelial-type growth characteristics of fungi may enable them to encapsulate and penetrate soil particles to which xenobiotics may be adsorbed. Soil fungi are generally believed to play a more important role in the formation, metabolism, and interactions of soil–organic matter complexes than do bacteria. This may allow them to cope better with the various bonding mechanisms involved with adsorbed materials.

Fungi also have the ability to form spores that retain viability over extended periods of time without refrigeration (Perry and Cerniglia, 1973). Spores could be accumulated for use as seed inocula when and where they might be needed. Although bacteria have been found to initiate degradation of a synthetic petroleum mixture, twice as much is degraded when bacteria, fungi, and yeast are present (Davies and Westlake, 1979).

Some filamentous fungi, unlike other microorganisms that attack aromatic hydrocarbons, use hydroxylations as a prelude to detoxification rather than catabolism and assimilation (Dagley, 1984). These organisms do not degrade aromatic hydrocarbons as nutrients, but simply detoxify them. Fungi are able to form glucuronide and sulfate conjugates of phenolic PAHs, which may be important in the detoxification and elimination of PAHs (Cerniglia, 1984).

Fungi (yeasts and filamentous) have nonspecific enzyme systems for aromatic structures; however, fungal metabolism often results in incomplete degradation that necessitates bacterial association for complete mineralization (Gibson, 1978). Whereas bacteria oxidize aromatic hydrocarbons to *cis*-dihydrodiols, fungi convert them to *trans*-dihydrodiols, with arene oxides (epoxides) as intermediates (Dagley, 1984). Therefore, fungi may metabolize aromatic hydrocarbons in a manner similar to mammalian systems, i.e., via a monooxygenase-catalyzed reaction (Cerniglia, Hebert, Szaniszlo, and Gibson, 1978). It is probable that a cytochrome P-450-dependent reaction may be responsible for the initial oxygenation of PAHs by these organisms. Metabolism of benzo(a)pyrene, as measured using the aryl hydrocarbon hydroxylase assay, is P-450 dependent, and cytochrome P-450-mediated benzo(a)pyrene hydroxylation is present in *Aspergillus fumigatus* (Venkateswarlu, Marsh, Faber, and Kelly, 1996). A wide range of fungi have the enzymatic capacity to oxidize PAHs when grown on an alternative carbon source (Cerniglia, Hebert, Szaniszlo, and Gibson, 1978). Since the products of fungal metabolism are often carcinogenic, tumorigenic, or mutagenic in higher organisms, it may be best to use a combination of fungi and bacteria for complete biodegradation (Cerniglia, 1984).

The ability to utilize hydrocarbons occurs mainly in two orders, the Mucorales and the Moniliales (Nyns, Auquiere, and Wiaux, 1968). *Aspergillus* and *Penicillium* are rich in hydrocarbon-assimilating strains. Naphthalene oxidation predominates in the order Mucorales, which includes species of *Cunninghamella, Syncephalastrum, and Mucor* (Cerniglia, Hebert, Szaniszlo, and Gibson, 1978). *C. elegans* oxidizes naphthalene and benzo(a)pyrene (Zajic and Gerson, 1977).

The most effective strains of filamentous fungi in mineralizing hydrocarbons and crude oil have been found to be *C. elegans* (Cerniglia and Perry, 1973) and *P. zonatum* (Hodges and Perry, 1973). Other degradative fungi that grow well on hydrocarbons are *Neurospora crassa, Claviceps paspale*, and *Psilocybe* strains (Cerniglia, Hebert, Szaniszlo, and Gibson, 1978), as well as strains of *Aspergillus versicolor, Cephalosporium acremonium*, and *Penicillium ochrochlorens* (Perry and Cerniglia, 1973). Another fungus, *Rhizoctonia solani*, metabolizes anthracene and may be as useful as *Cunninghamella elegans* in detoxifying PAHs (Sutherland, Selby, Freeman, Fu, Miller, and Cerniglia, 1992). *Acremonium* sp. has shown high levels of total petroleum hydrocarbon removal (McGugan, Lees, and Senior, 1995). It has been concluded that the property of assimilating hydrocarbons is relatively rare and that it is a property of individual strains and not necessarily a characteristic of particular species or related taxa (Nyns, Auquiere, and Wiaux, 1968). Table 5.12 lists fungi capable of growth on a variety of crude oils (Davies and Westlake, 1979).

Table 5.12 Fungi Capable of Growth on a Variety of Crude Oils

Organism

Yeasts
    Candida sp. (H-33-F)[a]
    Torulopsis sp. (H-34-F)
    Rhodotorula sp. (H-35-F)
    Rhodotorula sp. (H-45-F)
    Saccharomycopsis lipolytica (H-6-F)

Fungi
    Acremonium sp. (H-21-F)
    Aspergillus ochraceus (UAMH 2666)
    A. versicolor (H-8-F)
    Beauveria bassiana (H-52-F)
    Cladosporium sp. (H-26-F)
    Cunninghamella elegans (ATCC 10028a)[b]
    C. blakesleeana (ATCC 8688a)
    Mortierella sp. (H-59-F)
    Paecilomyces sp. (H-49-F)
    Penicillium javanicum (UAMH 1747)
    Penicillium sp. (H-14-F)
    Penicillium sp. (H-16-F)
    Penicillium sp. (H-17-F)
    Penicillium sp. (H-19-F)
    Penicillium sp. (H-29-F)
    Penicillium sp. (H-31-F)
    Penicillium sp. (H-44-F)
    Penicillium sp. (H49-F)
    Penicillium sp. (H50-F)
    Penicillium sp. (H60-F)
    Phoma sp. (H-13-F)
    Phoma sp. (H-39-F)
    Scolecobasidium obovatum (H-37-F)
    S. obovatum (H-41-F)
    S. obovatum (H-42-F)
    S. obovatum (H-43-F)
    Tolypocladium inflatum (H-46-F)
    Trichoderma viride (H-36-F)
    Verticillium sp. (H-23-F)
    Verticillium sp. (H-28-F)
    Verticillium sp. (H-47-F)
    Verticillium sp. (H-51-F)
    Verticillium sp. (H-53-F)
    Verticillium sp. (H-54-F)
    Verticillium sp. (H-55-F)
    Verticillium sp. (H-58-F)

[a] Accession number.
[b] American Type Culture Collection.

Source: Davies, J.S. and Westlake, D.W.S. *Can. J. Microbiol.* 25:146–156, NRC of Canada. 1979. With permission.

The genera most frequently isolated from soils are those producing abundant small conidia, e.g., *Penicillium* and *Verticillium* spp. (Davies and Westlake, 1979). Oil-degrading strains of *Beauveria bassiana*, *Mortieriella* spp., *Phoma* spp., *Scolecobasidium obovatum*, and *Tolypocladium inflatum* have also been isolated. Of 500 yeasts studied, 56 were found to be able to degrade hydrocarbons; almost all of these were in the genus *Candida* (Komagata, Nakase, and Katsu, 1964). Several *Candida* spp. are able to metabolize naphthalene, biphenyl, and benzo(a)pyrene. Hydrocarbonoclastic strains of *Candida*, *Rhodosporidium*, *Rhodotorula*, *Saccharomyces*, *Sporobolomyces*, and *Trichosporon* have been identified from soil (Ahearn, Meyers, and Standard, 1971; Cook, Massey, and Ahearn, 1973).

Cladosporium resinae has been found in soil (Walker, Austin, and Colwell, 1973) and repeatedly recovered as a contaminant of jet fuels (Atlas, 1981). This organism can grow on petroleum hydrocarbons and creates problems in the aircraft industry by clogging fuel lines. This fungus was found to oxidize 93% of hexadecane to CO_2 and to assimilate only 7% (Cooney and Walker, 1973). If other substrates show this trend, the organism may be useful for seeding, since considerable hydrocarbon could be mineralized without accumulation of a large biomass. A prior period of adaptation is probably not necessary, and *C. resinae* would have a competitive advantage over many other hydrocarbon users in which hydrocarbon-oxidizing systems must be induced. This fungus appears to transport *n*-alkanes into the cell, with an initial intracellular oxidation. It accumulates hydrocarbons.

Bacteria and yeasts show decreasing ability to degrade alkanes with increasing chain length (Walker, Austin, and Colwell, 1975). However, *Rhodotorula rubra* is a yeast with demonstrated ability to rapidly degrade higher *n*-alkanes (*n*-C_{16} to *n*-C_{20}) as compared with the lower *n*-alkanes (*n*-C_7 to *n*-C_{12}) (Shailubhai, 1986). This organism can degrade 88% of the saturate fraction of an oil waste, 63% of the aromatic fraction, and 13% of the asphaltic fraction after 7 days treatment. During the treatment of oil sludge, the persistence of the asphaltic fraction along with the remaining aromatic fraction could be due to a sparing effect (Shailubhai, Rao, and Modi, 1984b). Filamentous fungi do not exhibit preferential degradation for particular chain lengths and appear to be better able to degrade or transform hydrocarbons of complex structure or long chain length (Walker, Austin, and Colwell, 1975). Nevertheless, isoprenoids are relatively resistant to fungal degradation (Davies and Westlake, 1979).

A strain of soil yeast, *Trichosporon cutaneum*, which uses phenol in preference to glucose, has been reported (Shoda and Udaka, 1980). This unusual metabolic response could be valuable in transforming xenobiotics. *T. cutaneum* has a broad specificity of enzyme induction and is, actually, as versatile in aromatic catabolism as any pseudomonad (Dagley, 1984).

The growth rate of fungi is affected by the temperature (Perry and Cerniglia, 1973). Most soil fungi are mesophiles with temperature optima between 25 and 35°C, but with an ability to grow from about 15 to 45°C (Cooke and Rayner, 1984). Even thermophilic fungi do not grow above around 65°C (JRB Associates, Inc., 1984). However, many fungi grow at temperatures below 10°C. The optimum temperature for *Cunninghamella elegans* is 30°C and for *Penicillium zonatum* is 37°C.

Fungi can grow under environmentally stressed conditions, such as low pH and poor nutrient status where bacterial growth might be limited (Davies and Westlake, 1979). The influence of soil acidity on fungal growth is difficult to assess because of the ability of fungi to alter radically the pH value of their environment (Cooke and Rayner, 1984). This capability indicates that acidifying soil will not improve growth of the organisms, but does reduce competition from other organisms (Kirk, Schultz, Connors, Lorenz, and Zeikus, 1978).

Lignin is a complex, biorecalcitrant, biogenic, irregular, nonhydrolyzable, and environmentally persistent wood polymer of phenol propane units (Chen and Chang, 1985; Kirk and Farrell, 1987). Because of the random nature of the structure of lignin, its degradation is nonspecific (Paszczynski and Crawford, 1995). As a result of the high degree of nonspecificity and the presence of similar structures, the organisms capable of degrading lignin may also have the ability to degrade aromatic compounds (Bumpus, Tien, Wright, and Aust, 1985; Kirk and Farrell, 1987). Fungal degradation of lignin by lignin peroxidases occurs at high rates only with nutrient limitation, and it is cometabolic, requiring cellulose or glucose as a primary growth substrate.

Phanerochaete chrysosporium, a white-rot Basidiomycete, has a superior ability to degrade lignin (Buswell and Odier, 1987; Kelley and Reddy, 1988). It initiates ligninolysis only after primary growth has ceased due to carbon, nitrogen, or sulfur limitation (Kirk and Farrell, 1987; Buswell, 1992). The fungus destroys the lignin matrix in order to access the hemicellulose and cellulose (Field, de Jong, Feijoo-Costa, and de Bont, 1993). The fungus generates a carbon-centered free-radical enzyme system that allows it to catalyze numerous nonspecific cleavage reactions of the lignin macrostructure (Holroyd and Caunt, 1995). It has demonstrated the ability to degrade chlorinated organics in pure liquid culture to carbon dioxide (Lamar, Larsen, Kirk, and Glaser, 1987; Holroyd and Caunt, 1995) and in contaminated soil (Holroyd and Caunt, 1995). The fungus can degrade benzene, toluene, ethylbenzene, and xylenes (BTEX), common pollutants derived from gasoline and aviation fuels (Yadav and Reddy, 1993), under nonlignolytic culture conditions in a nitrogen-rich medium (Paszczynski and Crawford, 1995). It can metabolize phenanthrene (Bumpus, 1989; Sutherland, Selby, Freeman, Evans, and Cerniglia, 1991). In the intact fungus, lipid peroxidation by manganese peroxidase is the basis for the oxidation of phenanthrene (Moen and Hammel, 1994). The fungus can degrade anthracene (Vyas, Bakowski, Sasek, and Matucha, 1994) and has demonstrated the ability to degrade PAHs present in anthracene oil (Bumpus, 1989). It has been shown to degrade benzo(a)pyrene in nutrient nitrogen–deficient cultures (Bumpus, Tien, Wright, and Aust, 1985) and phenanthrene in nitrogen-rich culture media (Sutherland, Selby, Freeman, Evans, and Cerniglia, 1991), although mineralization of phenanthrene and other structurally related compounds is greater at low nitrogen levels. Degradation appears to occur under both ligninolytic and nonligninolytic conditions (Dhawale, Dhawale, and Dean-Ross, 1992) and involves different enzymes for each of the two states (Field, de Jong, Feijoo-Costa, and de Bont, 1993). Peroxidase is not the only extracellular system utilized by white-rot fungi for PAH biodegradation.

The ability to degrade such a wide range of PAHs suggests that *P. chrysosporium* may have potential as an *in situ* hazardous waste degrader (Lamar, Larsen, Kirk, and Glaser, 1987) and that it could be used for degrading complex mixtures of aromatic hydrocarbons at petroleum-contaminated sites (Yadav and Reddy, 1993). A field test of this organism showed enhanced mineralization of oil tar–contaminated soil (Brodkorb and Legge, 1992). Degradation of mixtures of complex hydrocarbons by the fungus often proceeds faster than the rate of degradation of the pure chemicals (Bumpus, Fernando, Mileski, and Aust, 1987). Toxicity of chemicals to this fungus is rare but can be circumvented by using mature mycelia instead of fungal spores. Substrate concentration and oxygen requirements are codependent (Tabak, Glaser, Strohofer, Kupferle, Scarpino, and Tabor, 1991). The transformations by this fungus are, however, slow (Alexander, 1994). Degradation rates improve manyfold when the fungus is immobilized (Lewandowski, Armenate, and Pak, 1990).

P. laevis HHB-1625 is able to transform PAHs (e.g., anthracene, phenanthrene, benz(a)anthracene, and benzo(a)pyrene) by means of manganese peroxidase (MnP–Mn^{2+}) reactions or in MnP-based lipid peroxidation systems (Bogan and Lamar, 1996). This organism degraded the quinone intermediates of PAH metabolism even faster and more extensively than *P. chrysosporium*.

The white-rot fungus *Bjerkandera* sp. BOS55 is an outstanding degrader of PAHs, including benzo(a)pyrene (Field, Feiken, Hage, and Kotterman, 1995) and anthracene (Field, Heessels, Wijngaarde, Kotterman, de Jong, and de Bont, 1994). The PAH-degrading ability of this strain is far superior to that of all other fungal strains tested, including *P. chrysosporium* (Field, de Jong, Feijoo-Costa, and de Bont, 1993).

Unlike *P. chrysosporium*, the ligninolytic activity of *Bjerkandera* sp. BOS55 occurs during primary growth and is not repressed by high nitrogen (Field, Heessels, Wijngaarde, Kotterman, de Jong, and de Bont, 1994). This fungus produces a novel manganese-inhibited peroxidase (de Jong, Field, and de Bont, 1992). *Bjerkandera* sp. strain BOS55 requires a high level of dissolved oxygen or a high redox potential for rapid anthracene biodegradation. It can degrade benzo(a)pyrene at a rate of 8 to 14 mg/kg/d at 20 and 30°C (Field, Feiken, Hage, and Kotterman, 1995). Adequate aeration is essential. Aeration by addition of porous pumice stones to the soil contributes to the high degradation rates. Rice, annual plant stems, and wood are suitable cosubstrates for rapid benzo(a)pyrene biodegradation. Some 15 to 20% of the compound was found to resist biodegradation; aging increases the nonbioavailable fraction. Cosolvents, such as 10% (vol/vol) acetone, or surfactants may help render the refractory fraction bioavailable.

Several brown- and white-rot fungi were able to degrade at least 40% of naphthalene, fluorene, or benzo(a)pyrene in 2 weeks or less (Lee, Fletcher, Avila, Callanan, Yunker, and Munnecke, 1995). They produced more biomass at acidic to neutral pH, incubation at 30°C, 90% moisture saturation, with granulated corncobs or alfalfa pellets as a lignocellulosic substrate. The isolates were *Irpex* sp. #232, *Fomitopsis* sp. #259, *Phanerochaete* spp. #326 and #404, *Heteroporus* sp. #501, and *Cunninghamella elegans*. Addition of Tween-80 to liquid PAH generally increased the rates of degradation of the compounds. *Trametes versicolor* TV1 and *Chrysosporium lignorum* CL1 have also been shown to degrade PAHs (Field, de Jong, Feijoo-Costa, and de Bont, 1992).

All of the strains of the white-rot fungi *P. chrysosporium* BKM-F-1767, *T. versicolor* Paprican 52, and *B. adusta* CBS 595.78 that were tested were able to remove anthracene, and nine removed benzo(a)pyrene from liquid culture (Field, de Jong, Feijoo-Costa, and de Bont, 1992). The *Bjerkanera* sp. strain BOS55 was the best degrader of both anthracene and benzo(a)pyrene. *Bjerkandera* and *Phanerochaete* converted anthracene to anthraquinone, which was a dead-end metabolite. *Trametes* strains were able to degrade anthracene, as well as anthraquinone. There was no accumulation of quinones during degradation of benzo(a)pyrene.

Soil type has a significant effect on growth and growth habit of *P. chrysosporium* (Lamar, Larsen, Kirk, and Glaser, 1987). Nitrogen content appears to play a major role in mediating growth of the fungus in the soils. Raising the soil water potential from –1.5 to –0.03 MPa greatly increases growth of this organism. It might even benefit from soil water potentials above –0.03 MPa. Growth of the fungus is also significantly greater at 30 and 39°C than at 25°C.

The amount of nitrogen and phosphorus for maximum utilization of crude oil has been determined for fungi in general (Perry and Cerniglia, 1973). Growth is better on NH_4Cl than on $NaNO_3$, and 0.25 mg/mL N is sufficient for mineralization of crude oil. Organic nitrogen supplements (33 mM) as an amino acid mixture or peptone caused a 10- to 14-fold increase in the extracellular peroxidase titers compared with a basal nitrogen-limited (2.2 mM) medium (Kotterman, Wasseveld, and Field, 1995). These enzymes are involved in the initial attack on PAHs, but the peptone supplement increased the rate of anthracene elimination by only 2.5-fold. The absence of manganese decreased the manganese peroxidase titer, increased the lignin peroxidase titer, improved the degradation of anthracene, and increased the yield of anthraquinone, a product of peroxidase-mediated conversions of anthracene. With optimum conditions and peptone N supplementing Mn-free medium, the anthracene degradation rate was 31 mg/L/day.

The ability of supplemental glucose to increase the rate and extent of biodegradation of DDT suggests a dependency upon the availability of a carbon source that can serve as a growth substrate (Bumpus,

Fernando, Mileski, and Aust, 1987). The glucose may simply allow an increase in the overall rate of fungal metabolism or it may provide the substrate for fungal production of hydrogen peroxide, a required cosubstrate for ligninases that are partly responsible for oxidation of many organopollutants. Bulking agents, such as wood chips or corncobs, can also serve as a carbon source for the fungus.

5.2.1.4.1 Extracellular Enzymes

Metabolism of PAHs by some white-rot fungi have been correlated with the excretion of extracellular enzymes, such as lignin peroxidase (LiP) and manganese-dependent peroxidase (MnP) (Kirk and Farrell, 1987; Fritsche, Guenther, Hofrichter, and Sack, 1994). The enzymes may be able to penetrate soil fines more deeply than microbes. MnP is able to oxidize Mn^{2+} ions, which can initiate lignin oxidation at a distance from the enzyme (Lackner, Srebotnik, and Messner, 1991).

Phanerochaete chrysosporium produces powerful extracellular enzymes, peroxidases, and H_2O_2, which are involved in the initial attack of many aromatic xenobiotic compounds, including many PAHs, such as benz(a)pyrene, benzo(a)anthracene, and pyrene (Bumpus, Milewski, Brock, Ashbaugh, and Aust, 1991; Hammel, 1992; Field, de Jong, Feijoo-Costa, and de Bont, 1993). The enzyme is excreted as a secondary metabolite, typically under carbon or nitrogen limitation (Haapala and Linko, 1993). The ligninases have been characterized as glycoproteins, which in the presence of hydrogen peroxide, oxidize different aromatic substrates by a one-electron transfer mechanism (Harvey, Schoemaker, and Palmer, 1987; Kirk, 1987; Kirk and Farrell, 1987; Tien, 1987). During the ligninolytic stage, the fungus secretes multiple forms of ligninases, accumulates veratryl alcohol, and produces H_2O_2, which is essential for activation of the ligninases (Kern, 1989). Extracellular ligninases can be partially inactivated by H_2O_2, which reduces enzyme activity in culture fluid. Addition of solid manganese(IV)oxide to cultures of *P. chrysosporium* at the beginning of ligninolytic activity improves production, enzymatic activity, and stability of the ligninases. This may protect the ligninases against inactivation and damage by hydrogen peroxide. The secondary metabolite, veratryl alcohol, enhances oxidation of pyrene (Cancel, Orth, and Tien, 1993) and may function as a substrate or serve a protective role (Cancel, Orth, and Tien, 1993) for the ligninases. It actually stabilizes the lignin peroxidases against inactivation by H_2O_2 (Haemmerli, Leisola, Sanglard, and Fiechter, 1986).

Ligninolytic enzymes and mediators are active extracellularly. This indicates that white-rot fungi would be better for the bioremediation of highly apolar pollutants, compared with nonligninolytic microbes, which degrade the compounds intracellularly and require pollutant dissolution and diffusion into the cells (Field, de Jong, Feijoo-Costa, and de Bont, 1993). Since these enzymes have potential for application in bioremediation, much research has been directed toward optimizing enzyme production. The buffer system used for *P. chrysosporium* greatly affects LiP production (Haapala and Linko, 1993). With acetate as the buffer in shake-flask cultures, 20 to >100% more lignin peroxidase is produced than with tartrate-buffered systems. Increasing the concentration of Cu^{2+} and Zn^{2+} results in higher ligninase activities. Some Mn^{2+} must also be present. Highest LiP activities are obtained with lower phosphorus concentrations. Production of the enzyme is affected by pH and trace elements and cations (and their relative ratios), especially Ca^{2+}, Mg^{2+}, Mn^{2+}, Fe^{2+}, Mo^{6+}, and Cu^{2+}.

Purified LiP of *P. chrysosporium* oxidizes PAHs to PAH quinones (Haemmerli, Leisola, Sanglard, and Fiechter, 1986; Hammel, Kalyanaraman, and Kirk, 1986). Mn^{3+} ions in aqueous acetate oxidize PAH to acetoxy PAH derivatives and PAH quinones (Cremonesi, Hietbrink, Rogan, and Cavalieri, 1992). PAHs with a low ionization potential (<7.5 eV) can serve as substrates for direct one-electron oxidation by LiP (Hammel, Kalyanaraman, and Kirk, 1986). If the PAH has a high ionization potential ($\geqq 7.5$ eV), the LiP and Mn^{3+} will not be able to use it as a substrate (Cavalieri and Rogan, 1985; Hammel, Kalyanaraman, and Kirk, 1986), although such compounds may be degraded by the organism *in vivo*. These PAHs may be acted upon by radical species formed during manganese peroxidase-dependent lipid peroxidation reactions (Moen and Hammel, 1994; Bogan and Lamar, 1995; Bogan, Lamar, and Hammel, 1996). Transcription of the manganese peroxidase genes for these reactions may be regulated by controlling the growth conditions (Bogan, Schoenike, Lamar, and Cullen, 1996).

Fungal enzymes, such as laccase, peroxidase, and tyrosinase, all play an important role in catalyzing transformation of various aromatic compounds in the environment (Bollag and Dec, 1995). Contaminants are covalently bound to soil organic matter by enzyme-mediated oxidative coupling reactions. This is accompanied by a detoxification effect, which can be enhanced by optimizing reaction conditions and immobilizing the enzymes on solid materials.

5.2.1.4.2 Soil Inoculation

Remediation by direct inoculation with *Phanerochaete chrysosporium* has had limited success, mainly because soil is not its normal habitat (Glaser, 1990). Growth and survival are difficult because of competition from other soil microorganisms (Loske, Hutterman, Majcherczyk, Zadrazil, Lorsen, and Waldinger, 1990). However, Holyroyd and Caunt (1995) observed successful bioremediation when the fungus was pregrown on sterile, lignin-based substrate before adding to the contaminated soil. Toxic effects of the contaminants are more easily tolerated with the fungus growing on a support material. The fungus may be more effective if it is not disturbed, so soil and organic matter can be initially mixed and then aerated with a static-bed technology.

Bulking and tillage improve fungal biodegradation by supplying nutrients and improving aeration of the soil (McGugan, Lees, and Senior, 1995). When nitrogen levels are exceptionally low, nitrogen (urea) in poultry manure can enhance biodegradation. A lignocellulosic organic supplement provides an alternative carbon source for sustaining fungal biomass (and some bacteria) in the soil for long periods of time. There might be some loss of fungal viability as a result of culture maintenance on artificial media.

Soil contaminated with aromatic hydrocarbons, heterocyclic compounds, or monocyclic aromatic hydrocarbons or PAHs can be treated by biodegradation with exoenzymes from white-rot fungi (Kohlmeier, 1994). This patented process pumps an exoenzyme–water mixture over the wastes in a tank, allows the liquid to drain off for air exposure, then pumps it back in. Another patented process involves culturing the white-rot fungus until mycelia are produced (Stahl and Aust, 1994). The fungal mycelia are placed with the pollutant in an aqueous medium for aerobic mineralization to occur.

For successful application of fungi for soil remediation, an effective, low-cost means of applying the inoculum is needed (Lestan and Lamar, 1996). Lignin-degrading fungi have been inoculated into the soil using substrates, such as wheat straw, wood chips, corncobs, and mushroom spawn, which have yielded a low inoculum potential, are of inconsistent quality, and are expensive to produce. A novel fungal inoculum for bioaugmentation of soils contaminated with hazardous organic compounds has been developed. Pelleted solid substrates are coated with a sodium alginate suspension of fungal spores or mycelial fragments and incubated until overgrown with the mycelia of selected lignin-degrading fungi. These pellets resist competition from indigenous soil microbes, are lower in moisture content than previous methods, and are strong enough to withstand handling and injection into the soil. The technique has been successfully tried with *P. chrysosporium* (BKM F-1767, ATCC 42725), *P. sordida* (HHB-8922-Sp), *Irpex lacteus* (Mad-517, ATCC 11245), *Bjerkandera adusta* (FP-135160-Sp, ATCC 62023), and *Trametes versicolor* (MD-277).

5.2.1.4.3 Screening Strategies

Ligninolytic fungi have been observed to be able to decolorize a large number of dyes, which are usually resistant to biological treatment (Paszczynski and Crawford, 1995). Since the high-molecular-weight dyes are not taken up by the fungi, a reaction indicates extracellular activity (Field, de Jong, Feijoo-Costa, and de Bont, 1993).

Poly B-411, Poly R-481, and Poly Y-606 can serve as substrates for lignin-degrading enzymes (Glenn and Gold, 1983). Polymeric dyes could be a useful tool for mutant selection. Decoloration of these dyes has proved to be a good indicator of the initial transformation of xenobiotics mediated by the peroxidative activity of fungi (Field, de Jong, Feijoo-Costa, and de Bont, 1993). This technique provides a rapid, simple method for screening a large number of samples, when searching for new degradative strains.

The azo dyes are not readily degraded by microorganisms (Kulla, Krieg, Zimmermann, Leisinger, 1984) but can be attacked by *P. chrysosporium* (Cripps, Bumpus, and Aust, 1990). These dyes include Azure B, Tropaeolin O, Orange II, and Congo Red. This fungus can degrade azo dyes in concentrations up to 300 ppm (Paszczynski, Pasti, Goszczynski, Crawford, and Crawford, 1991), whereas strains of *Streptomyces* can degrade only about 500 ppm (Pasti, Hagen, Goszczynski, Paszczynski, Crawford, and Crawford, 1991). Degradability depends on the substitution of the aromatic ring, with lignin-like structures enhancing biodegradability. Veratryl alcohol is involved in the oxidation of some azo dyes by LiP (Paszczynski and Crawford, 1991). Polymeric, azo, heterocyclic, and triphenyl-methane dyes are decolorized by three major LiP isoenzymes (H2, H7, and H8) (Ollikka, Alhonmaki, Leppanen, Glumoff, Raijola, and Suominen, 1993). Purified enzymes are able to decolorize all these dyes, but some isoenzymes require veratryl alcohol for decolorization.

Several azo dyes have been found to be useful as substrates for assaying LiPs and manganese peroxidases of white-rot fungi (Pasti-Grigsby, Paszczysnski, Goszczynski, Crawford, and Crawford,

1994). These dyes are 3,5-dimethyl-4-hydroxyazobenzene-4-sulfonic acid and Orange I for assays of Mn(II)peroxidase and 3,5-difluoro-4-hydroxybenzene-4'-sulfonic acid and Orange II for lignin peroxidases. The rate of decolorization of the dye, Poly R-478, correlates well with PAH degradation and may be a useful screening method to identify potential PAH-degrading white-rot fungi (Field, de Jong, Feijoo-Costa, and de Bont, 1992). Poly R-478 decolorization probably indicates peroxidative activity (Field, de Jong, Feijoo-Costa, and de Bont, 1993). Decolorization of Poly R-478 proved not sensitive enough to show a relation between presence of manganese peroxidase or LiP and decolorization of the dyes (Freitag and Morrell, 1992). However, other authors have found good correlations, such as by using the capability of dye decolorization to find new fungal strains with significantly higher ability to decolorize Poly R than *P. chrysosporium* (de Jong, De Vries, Field, Van Der Zwan, and De Bont, 1992).

Many other dyes can be decolorized by white-rot fungi. Crystal violet degradation by *Phanerochaete chrysosporium* may involve a nonligninolytic enzyme system (Bumpus and Brock, 1988). Remazol brilliant blue R can also be decolorized (Ulmer, Leisola, and Fiechter, 1984). Another white-rot fungus, *Pycnoporus cinnabarinus,* also has the ability to decolorize pigments (Schiephake, Lonergan, Jones, and Mainwaring, 1993).

5.2.1.5 Phototrophs

The surface soil usually supports large populations of eukaryotic algae and cyanobacteria (blue-green algae) (JRB Associates, Inc., 1984). Since light cannot penetrate far into the soil, the algal biomass is usually low. These organisms may enhance photodecomposition of hazardous organic compounds at the soil surface. Although photosynthetic, cyanobacteria can use exogenous organic substrates under both lighted and dark conditions (heterotrophy) as a portion of the total carbon requirement for growth (Fogg, Stewart, Fay, and Walsby, 1973).

The ability to oxidize aromatic hydrocarbons is widely distributed among the cyanobacteria, algae, and photosynthetic bacteria (Kobayashi and Rittmann, 1982). These organisms are potentially important if there is a low concentration of nutrients, because they can obtain energy from sunlight and carbon by carbon dioxide fixation. This offers a great potential for exploitation in treatment processes because such organisms can be self-sustaining without the presence of organic matter in concentrations large enough to serve as carbon and electron donors (Kobayashi and Rittmann, 1982). Some of the cyanobacteria and photosynthetic bacteria are also able to fix nitrogen. Therefore, they can survive in situations in which the dissolved nitrogen concentration is inadequate to support bacterial growth.

Cyanobacteria are able to exist in inhospitable and caustic environments (Des Marais, 1990). Cyanobacteria, *Chlorella* (green algae), and especially the Chromatiaceae (photosynthetic bacteria) are pollution tolerant (Pfennig, 1978a; Kobayashi and Rittmann, 1982). Of the latter, the purple sulfur bacteria (Thiorhodaceae) are important. The purple nonsulfur (Rhodospirilliaceae) bacteria are also of interest. For these bacteria, organic compounds serve as the major source of electrons and carbon for cellular components (Berry, Francis, and Bollag, 1987).

The photosynthetic bacteria are known to be able to metabolize a wide variety of substances (e.g., simple sugars, alcohols, volatile fatty acids, tricarboxylic acid cycle intermediates, benzoates, aromatic compounds) (Stanier, Doudoroff, and Adelberg, 1970) and to have a wide range of inducible enzymes (Laskin and Lechevalier, 1974). *Rhodopseudomonas capsulata* is known to transform nitrosamines (carcinogens) to innocuous compounds (Kobayashi and Tchan, 1978). Some of the purple nonsulfur organisms can grow microaerobically and anaerobically as phototrophs, yet live as heterotrophs aerobically in the dark, by respiratory metabolism of organic compounds (Stanier, Doudoroff, and Adelberg, 1970; Pfennig, 1978a). Cyanobacteria and *Chlorella* are tolerant of low concentrations of dissolved oxygen (Kobayashi and Rittmann, 1982). *Dunaliella* can tolerate a wide salinity range. See Table 5.3 for a summary of the most favorable conditions for these organisms for removal of anthropogenic compounds.

Cyanobacteria can metabolize most gaseous and liquid pollutants fairly rapidly, reducing their levels in the atmosphere and effluents (Subramanian and Uma, 1996). Toxic compounds, such as phenolics, pesticides, and antibiotics, as well as recalcitrant chemicals, such as lignin, can be degraded and detoxified.

Cyanobacteria, diatoms, and some eukaryotic algae have the enzymatic capacity to oxidize PAHs that range in size from naphthalene to benzo(a)pyrene (Cerniglia, 1984). Prokaryotic and eukaryotic algae grown photoautotrophically in the presence of naphthalene have the ability to oxidize naphthalene

Table 5.13 Microorganisms That Can Bioaccumulate Anthropogenic Compounds

Microorganism	Compound
Cyanobacteria	
Microcystis aeruginosa	Benzene, toluene, naphthalene, phenanthrene, pyrene
Algae	
Selanastrum capricornatum	Benzene, toluene, naphthalene, phenanthrene, pyrene

Source: From Kobayashi, H. and Rittmann, B.E. *Environ. Sci. Technol.* 16:170A–183A. ©1982. American Chemical Society. With permission..

(Cerniglia, Gibson, and Van Baalen, 1980; 1982; Cerniglia, Van Baalen, and Gibson, 1980). Some cyanobacteria, e.g., an *Oscillatoria* sp., can oxidize hydrocarbons, such as naphthalene and biphenyl, to a limited extent when growing photoautotrophically (Cernilia, Gibson, and Van Baalen, 1980; Cerniglia, Van Baalen, and Gibson, 1980). Multiple oxidative pathways may be involved in the cyanobacterial metabolism of PAHs (Cerniglia, 1984). Some green, brown, and red algae and diatoms can partially oxidize hydrocarbons, such as naphthalene. Species of *Oscillatoria, Microcoleus, Anabaena, Cocochloris, Nostoc, Chlorella, Dunaliella, Chlamydomonas, Ulva, Cylindretheca, Ampora,* and *Porphyridium* have been found to be capable of oxidizing naphthalene (Cerniglia, Gibson, and Van Baalen, 1980). A marine cyanobacterium, *Agmenellum quadruplicatum* PR-6, can oxidize phenanthrene, and the freshwater green alga *Selenastrum capricornutum* can oxidize benzo(a)pyrene (Cerniglia, 1992).

A few species, such as *Rhodopseudomonas palustris* (Gottschalk, 1979) or *Rhodocyclus purpureus* (Pfennig, 1978b), can use thiosulfate or sulfide as an electron donor in addition to organic compounds. Generally, however, the sulfide concentration must be low (Schnitzer, 1982). *Rhodopseudomonas palustris* can use benzoate as the sole substrate under aerobic conditions via respiration or anaerobically by photometabolism (Proctor and Scher, 1960). The enzyme system used by this organism to photometabolize aromatic substrates is inducible and lacks substrate specificity (Dutton and Evans, 1969).

Phototrophic organisms have the ability to both transform and to bioaccumulate hydrophobic compounds (Table 5.13) (Kobayashi and Rittmann, 1982). Cyanobacteria can also bioaccumulate and biosorb heavy metals (Subramanian and Uma, 1996). Bioaccumulation of pollutants is also observed in other organisms (Finnerty, Kennedy, Lockwood, Spurlock, and Young, 1973). For example, *Acinetobacter* sp., yeasts, and filamentous fungi can accumulate hydrocarbons in the cytoplasm. An organic compound may be nontoxic in the amounts that exist free in the water or outside the microbial cell in soil, but if the chemical is subject to bioconcentration, species at higher trophic levels may be harmed (Alexander, 1986).

In general, phototrophs do not promote complete degradation, but only transformation (Kobayashi and Rittmann, 1982). The contribution of these microbes to complete degradation requires interactions with other organisms. The metabolic products they form stimulate growth of heterotrophic organisms. The proper balance between algae and bacteria can result in extensive biodegradation of anthropogenic compounds.

Facultative bacteria colonize the area under the cyanobacteria photozone (Phillips, Bender, Word, Niyogi, and Denovan, 1994). These cyanobacteria/bacteria biofilms form multilayered laminated mats in the sediment of shallow water. This type of mat was developed in the laboratory, inoculated with cyanobacteria (primarily *Oscillatoria* spp.) and incubated in light and dark conditions (Bender, Vatcharapijarn, and Russell, 1989). These organisms were able to mineralize the PAHs, naphthalene, phenanthrene, and chrysene, and <6% of hexadecane (Phillips, Bender, Word, Niyogi, and Denovan, 1994). Mineralization in the dark was probably due to bacteria and heterotrophic cyanobacteria. These mats have the capability to bioremediate petroleum-contaminated sites, as has been substantiated by the appearance of such mats along the contaminated Saudi Arabian coastline.

A scheme has been devised to separate and treat refractory compounds by use of the bioaccumulation capability of phototrophs (Section 2.1.2.2.1.4) (Kobayashi and Rittmann, 1982). Easily degradable compounds are treated directly by aerobic/anaerobic degradation. Refractory compounds are removed by sorption to microorganisms, concentrated, and may be disposed of by other means, such as burial or incineration. Relatively recalcitrant compounds can be degraded if they are first sorbed, for which photosynthetic organisms are potentially valuable. Other microorganisms could then further degrade the products of the initial reactions.

5.2.1.6 Higher Life Forms and Predation

The role of the soil macrofauna, such as insects, protozoa, earthworms, and slugs, in the decomposition of organic materials is significant, but predominantly indirect (Parr, Sikora, and Burge, 1983). It is minor compared with microorganisms, but it is still essential. Of the total respiration associated with soils amended with organic material, 10 to 20% could be from macrofauna. Because only a few of these organisms have the ability to produce their own enzymes for the degradation of substrate, their main mode of degradation is mechanical.

The gut of most soil macrofauna contains microorganisms that produce the necessary enzymes for the degradation of a substrate to the point where the macrofauna can absorb the nutrients (Parr, Sikora, and Burge, 1983). The remainder of the substrate passes into the soil where microorganisms complete the degradation.

Earthworms play a prominent role in the degradation of organic materials in soil (Parr, Sikora, and Burge, 1983). With their movement, the soil is aerated and nutrients are carried to deeper soil profiles, where these stimulate microbial growth and decomposition. Among the arthropods, the beetles and termites are most correlated with extensive degradation of organic material. Both animals often have a rich microflora in their guts, and these microorganisms produce the enzymes that degrade the cellulosic substrates. Earthworms have been found to bioaccumulate the metals cadmium, potassium, sodium, and zinc in applied oily waste; however, they do not bioaccumulate naphthalenes, alkanes, or specific aromatics (Loehr, Martin, Neuhauser, Norton, and Malecki, 1985). Macrofauna, in general, play an essential role in the decomposition of wastes, and the addition of materials to the soil that are toxic to these organisms can alter the rate of decomposition (Hornick, 1983).

Dry surface soil in the summer can cause worms to move deeper in the soil (Loehr, Neuhauser, and Martin, 1984). The most favorable conditions of moisture and temperature occur in the spring and fall. In landfarming, the earthworm biomass is reduced in the zone of incorporation by physical disturbance of the soil, such as by rototilling, and by large applications of waste. It is not known how long it takes for earthworms and microarthropods to recover from intermittent applications of oily waste (Loehr, Martin, Neuhauser, Norton, and Malecki, 1985). At least, land application of these wastes will not have an irreversible, adverse impact on these organisms.

Microbial predators also play a role in the degradation process (Texas Research Institute, Inc., 1982). There are higher eukaryotic (the nucleus is surrounded by a membrane) organisms in the soil, and these organisms graze on bacteria and fungi or feed on detrital matter and associated microflora (Sinclair and Ghiorse, 1985). Protozoa, nematodes, insects, and worms affect the decomposition process by controlling bacterial or fungal population size through grazing (Bryant, Woods, Coleman, Fairbanks, McClellan, and Cole, 1982), by harboring in their intestinal tract organisms that might decompose a compound of interest, by comminuting plant materials (by insects), or by mixing the soil and contributing to its aeration and homogeneity (JRB Associates, Inc., 1984). A cyst-forming amoeba was present at 111.1 organisms/g dry weight of soil and constituted 15% of the total biovolume of sediments in a groundwater interface zone (Sinclair and Ghiorse, 1985).

The environment can also contain bacteriophages, viruses that attack fungi, *Bdellovibrio*, mycobacteria, myxobacteria, slime molds, Acrasiales, and organisms that excrete enzymes that destroy cell walls of fungi and bacteria, causing them to lyse (Alexander, 1994). It has not been determined whether or not any of these organisms actually affect biodegradation, but protozoa have been shown to reduce the bacterial population substantially through grazing. For a single protozoan to divide, it requires 10^3 to 10^4 bacteria. A bacterial density of more than 10^6 to 10^7 cells/mL or cm is needed for grazing to occur, and the period of active grazing will end when the total population of susceptible bacteria has reached around 10^6 cells/mL. This becomes a limiting factor when there is a low density of a species of bacteria that is faced with a low concentration of contaminant to biodegrade. The numbers may not be great enough to initiate biodegradation, until the grazing rate declines and the bacteria are able to multiply, if they have not been totally suppressed or eliminated. Many species of hydrocarbon utilizers have been found to be ingested by a large number of ciliate and other cytophagic protozoans (Texas Research Institute, Inc., 1982). These higher organisms may reduce the microbial population from 10^7 to 10^2 bacteria/mL (ZoBell, 1973). Protozoan grazing is probably responsible for most of the acclimation period for the mineralization of organic compounds in sewage (Wiggins and Alexander, 1986; 1988). Even the large numbers of supplemented organisms would be greatly depleted (Acea, Moore, and Alexander, 1988). Small pore space probably offers a protective microhabitat for soil bacteria, and bacteria in larger pore spaces are subject to predation (Heijnen and van Veen, 1991). Protozoal predation of bacteria in pore spaces with neck diameters larger than 6 μm has a major influence on the survival of introduced strains. Sometimes it is necessary to inhibit protozoa and other eukaryotes with a compound such as cycloheximide, which will not affect the bacteria (Zaidi, Murakami, and Alexander, 1989).

Protozoa have been observed in some instances to be stimulatory for biodegradative bacteria and fungi by replenishing the soil with phosphorus and nitrogen from their grazing (Alexander, 1994) (see Section 5.2.5).

Vegetation can enhance the rate and extent of degradation of PAHs in contaminated soil (Santharam, Erickson, and Fan, 1994). Plant roots release exudates capable of supplying carbon and energy to microflora for degrading PAHs. The population of microorganisms in the rhizosphere is significantly greater than that in the nonvegetated soil. These microorganisms are apparently responsible for the enhanced biodegradation of PAHs. The plants themselves can also accumulate heavy metals in their roots and shoots (Kumar, Dushenkov, Motto, and Raskin, 1995); however, no uptake of anthracene or benzo(a)pyrene was observed with ryegrass (*Lolium multiflorum*), soybean, or cabbage (*Brassica oleracea* var. capitata L.) (Goodin and Webber, 1995). Knaebel and Vestal (1992) found that rhizosphere microbial communities enhance biodegradation of surfactants in soils by increasing the initial rates of mineralization by a factor 1.1 to 1.9.

5.2.2 BIOAUGMENTATION

It is unknown what is required for a microorganism to become "indigenous" (Turco and Sadowsky, 1995). However, such organisms have become stable members of a community and have a selective, competitive advantage in occupying available niches in that environment (Atlas and Bartha, 1993). A nonindigenous microbe may become a member of that population by prolonged and repeated application of soil inoculations (Dunigan, Bollich, Hutchinson, Hicks, Zaunbrecher, Scott, and Mowers, 1984; Yeung, Schell, and Hartel, 1989).

Biological treatment methods generally rely upon the stimulation and natural selection of indigenous microorganisms in the soil or groundwater (Sims and Bass, 1984). However, the natural soil flora may not have the metabolic capability to degrade certain compounds or classes of compounds or to emulsify the water-insoluble components. On the other hand, they may have the ability but not the biomass necessary to degrade the compounds rapidly enough to meet treatment criteria.

When a contamination incident occurs that requires immediate treatment or when indigenous bacteria at the site are insufficient in number or capability to degrade the pollutants involved, it may be necessary to employ bioaugmentation — the supplementation of microorganisms (Baud-Grasset and Vogel, 1995). One way to enhance biodegradation of organic compounds is to inoculate the environment with microorganisms that are known to metabolize these chemicals readily (Sepic, Leskovsek, and Trier, 1995).

Considering the diversity of enzymatic activities required, some have assumed that the natural microbial communities of many ecosystems would not possess all the needed enzymes, and, therefore, contaminating oil would not be extensively biodegraded unless hydrocarbon-degrading microorganisms were artificially added (Gholson, Guire, and Friede, 1972). In addition, many of the components would probably be recalcitrant to microbial attack.

There is considerable evidence that bioaugmentation enhances degradation of a variety of organic contaminants in soils and surface biological waste treatment processes. For this approach to be successful in subsurface environments (Goldstein, Mallory, and Alexander, 1985; Wilson, Leach, Henson, and Jones, 1986b):

1. The added microorganisms must be able to survive in what to them is a foreign, hostile environment and compete for nutrients with indigenous organisms.
2. The added microorganisms must be able to move from a point of injection to the location of the contaminant at what are very often low concentrations, in a medium where bacterial transport is normally very slight, especially in fine-grained materials.
3. The added microorganisms must be able to retain their selectivity for metabolizing compounds for which they were initially adapted.

Seeding microorganisms onto soil was first tried around 1968 (Gutnick and Rosenberg, 1979). An inoculum of *Cellumonas* sp. and nutrients was able to degrade the hydrocarbon contaminants more effectively than just fertilizer alone (Schwendinger, 1968). Since then, the seeding of microorganisms has been used in a number of different environments to degrade organics (U.S. EPA, 1985a), with varying success.

The addition of cultures of organisms could reduce the lag period required for indigenous populations to respond to petroleum pollutants (Atlas, 1978a). Some workers have suggested that the natural lag or response time is generally short, in which case, seeding would not usually reduce the impact of petroleum pollutants, except in a few open environments with limited microbial populations. Commercially available mutant bacteria generally require an adaptation period in order to adjust to site conditions. However, the time required is similar to the adjustment period for natural *in situ* bacteria. Therefore, additional factors, such as convenience and costs of application and labor must be considered in selecting the most efficient bacterial treatment.

Variations in chemical composition among different oils might require application of unique mixtures of microorganisms (Buckingham, 1981). Different mixtures of seed microorganisms would also be required for oil biodegradation in different soil environments (Westlake, Jobson, Phillippe, and Cook, 1974). Microbial inoculants covering a broad range of metabolic capabilities are available commercially, and they are being used increasingly in treating both contaminated soil and aquatic systems (Thibault and Elliott, 1980).

A special culture collection was begun as a depository for hydrocarbonoclastic (hydrocarbon-utilizing) microorganisms (Cobet, 1974). "Superbugs" are being sought to solve the world pollution problems (Gwynne and Bishop, 1975). Several commercial enterprises are marketing microorganism preparations for removing petroleum pollutants (Atlas and Bartha, 1973c; Azarowicz, 1973). A hydrocarbonoclastic superbug must be able to extensively degrade most of the components found in petroleum. It must be genetically stable, capable of being stored for long periods of time, and able to reproduce rapidly following storage. It must be capable of enzymatic activity and growth in the environment in which it is to be used, and it must be capable of competing with the naturally occurring microorganisms in that environment for a period sufficient to clean up the site. Finally, it must not produce adverse side effects; it must not be pathogenic or produce toxic metabolic end products.

Creation of a single superbug that combines the genetic information from many microorganisms into one could overcome the problem of interference of organisms with the metabolic activities of each other (Friello, Mylroie, and Chakrabarty, 1976). It would be difficult to find or engineer a single microbe with all these characteristics or even one capable of metabolizing most petroleum components (Atlas and Bartha, 1973c; Azarowicz, 1973). Therefore, it has been proposed that a mixture of microorganisms should be used as the seed culture for oil biodegradation (Kator, 1973). Each organism would be included for its ability to degrade particular petroleum components and would still have to meet the rest of the above criteria.

Inoculation of contaminated soil with microorganisms selected for their ability to degrade or transform specific hazardous materials is a very attractive treatment concept (JRB Associates, Inc., 1984). However, the treatment manager must realize that the soil environment is restrictive and the soil microbial system has a complex ecology. These factors may limit the ability of introduced organisms to become self-perpetuating and carry out their specialized functions for an extended period of time. It may be necessary to reinoculate the soil several times before satisfactory levels of treatment are achieved.

Even if microbial seeding is used, it will generally have to be accompanied by environmental modifications (Atlas, 1977). Seeding without simultaneous modification of environmental conditions is not likely to succeed. In most ecosystems, some environmental factor is going to limit the oil-degrading activity. The application of microbial amendments to the soils is frequently combined with other treatment techniques, such as soil moisture management, aeration, and fertilizer addition (Thibault and Elliott, 1980). This method may be most effective against one compound or closely related compounds. The effectiveness would be limited by toxicity of a contaminant or the inability of an organism to metabolize a wide range of substrates.

Seeding with hydrocarbonoclastic microorganisms appears to be a promising treatment for severely environmentally distressed soil (Cook and Westlake, 1974), such as in Arctic or near-Arctic climates, where the activity and growth rate of indigenous organisms may be limited (Hunt et al., 1973; Cook and Westlake, 1974). When an addition of 10^4 microorganisms/g dry soil was tested in such an environment (with addition of N and P and adjustment of the pH to 7), microbial activity over the controls was increased by at least a factor of four in 40 days.

The size of the inoculum required depends largely upon the size of the spill, how it is dispersed, and on the growth rate of the seed microorganisms. Some workers report that the size of the inoculum is of little importance, as long as a minimal inoculation of the oil occurs (Miget, 1973). Of greater importance is whether or not the oil will support the growth of the seed organism in the given environment.

Although highly favorable nutrient and electron acceptor conditions may exist at a site, onset of measurable microbial oxidation of readily degraded compounds can be delayed in subsurface systems containing small populations of microorganisms (Corseuil and Weber, 1994). There appears to be a need for some critical number of microbes. Organisms can be added to *in situ* treatment processes to enhance biodegradation by increasing biomass or by reducing the time necessary for acclimation to occur. Usually, natural soil microorganisms that have been acclimated to degrade the contaminants are used as "seed"; the microbes may have been selected by enrichment culturing, induced mutation, or genetic manipulation.

There have been varying degrees of success with inoculating acclimated organisms into a new environment (Goldstein, Mallory, and Alexander, 1985; Lee, Wilson, and Ward, 1987), and with seeding microbes onto petroleum-contaminated soils (Knox, Canter, Kincannon, Stover, and Ward, 1986). Some applications have not been satisfactory. Addition of 10^8 cells/g soil of two hydrocarbon-degrading isolates did not significantly influence oil concentrations (Lehtomakei and Niemela, 1975).

In one study, it was concluded that too low a level of application (10^6 bacterial cells/cm^2) allowed only a slight increase in the rate of oil utilization (C_{20} to C_{25}) obtained after seeding species of *Flavobacterium* and *Cytophaga* (41%), *Pseudomonas* (34%), *Xanthomonas* (10%), *Alcaligenes* (9%), and *Arthrobacter* (5%), rather than the inability of the added bacteria to survive under the natural field conditions (Jobson, McLaughlin, Cook, and Westlake, 1974). There is a threshold level of indigenous microorganisms that will prevent inoculation success with a similar or genetically related strain (McLoughlin, Hearn, and Alt, 1990; Thies, Singleton, and Bohlool, 1991). These authors estimate that number to be about 10 bacteria/g soil of a particular organism. When oil-degrading bacteria were added to soil of the boreal region of the Arctic, there was no increase in the changes of the recovered oil (Westlake, Jobson, and Cook, 1978). This was believed to be due to insufficient application.

Another possible solution to help bioaugmented organisms survive and compete with indigenous organisms would be to create a temporary niche for the introduced microbes (Dybas, Tatara, Knoll, Mayotte, and Criddle, 1995). For instance, this could be accomplished by making the environment alkaline, which reduces the bioavailability of essential trace metals, such as iron, thereby favoring organisms with efficient trace metal–scavenging systems. This was successfully applied with a *Pseudomonas* sp., which degraded the contaminant at the alkaline pH but then showed a rapid decline when the pH was lowered.

Bioaugmentation at a site contaminated by gasoline from an underground storage tank was unsuccessful (Maxwell and Baqai, 1995). It appears that the laboratory-cultured microorganisms (*Bacillus, Pseudomonas, Serratia,* and *Azotobacter*) did not survive in the soil. This may have been due to the lack of nutrients in the inoculum or contact with a toxic level of contaminants. The dead bacteria may have then clogged the well screens. Soil remediation did not occur within 15 months after inoculation.

The indigenous and applied microorganisms in soils may not be able to compete because of nutrient limitations and specificity, moisture requirements, pH soil texture and porosity, temperature, organic matter, bacteriocin and antibiotic production, solute types and concentrations, number and type of indigenous microorganisms, residence time in soils, mobility in soil, and time in laboratory culture before introduction into soils (Turco and Sadowsky, 1995).

Microorganisms able to degrade organic pollutants in culture sometimes may fail to function when inoculated into natural environments because the concentration of the hydrocarbon in nature is too low to support growth (Alexander, 1986). When oil-degrading bacteria were added to soil contaminated with an 0.5% concentration of light fuel oil or heavy waste oil, the bacteria had no significant effect on contaminant removal (Lehtomaki and Niemela, 1975). However, there was significant biodegradation with a 10% loading rate of an aromatic hydrocarbon mixture (Vecchioli, Del Panno, and Painceira, 1990); 1 and 5% loading rates were also favorable.

The organisms may be susceptible to toxins or predators such as protozoa, in the environment (Alexander, 1986; Turco and Sadowsky, 1995). Competition, predation, and parasitism are major obstacles in the soil (Vecchioli, Del Panno, and Painceira, 1990). A slow rate of growth may be slower than the rate of predation (Goldstein, Mallory, and Alexander, 1985).

The capability of microbes to use hydrocarbons for growth may be lost during cultivation on hydrocarbon-free media (Jenson, 1975). This ability can be preserved when the organisms are grown in hydrocarbon-free, nutrient-supplemented vermiculite (Vecchioli, Del Panno, and Painceira, 1990). Hydrocarbon-degrading strains retained their ability for at least a month when grown on this support and required no special storage.

Inoculant formulation is an important factor determining the success of bioaugmentation in bioremediation efforts (Pritchard, 1992). Two commercial bioaugmentation products were found to have no positive effects on oil degradation (Venosa, Haines, and Allen, 1992). The oil-degrading capacity of nine commercial bioaugmentation products was found to be inferior in liquid culture to that of activated sludge (Dott, Feidieker, Kaempfer, Schleibinger, and Strechel, 1989).

Added microorganisms may use other organic compounds in preference to the pollutant (Vecchioli, Del Panno, and Painceira, 1990). Most isolates used for bioaugmentation are Gram-negative, non-spore-forming bacteria, which are difficult to store in large quantities in a manner that preserves their viability (Bartha, 1986). A commercial, freeze-dried, bioaugmentation product containing bacteria and yeast (specially adapted to grow on various petroleum products, paraffins, naphthalenes, and lubricant oils) and small amounts of surfactants on a starch-based carrier was tested by Moller, Gaarn, Steckel, Wedebye, and Westermann (1995). Small amounts of the bioaugmentation product (1 mg/g soil) had no more effect on diesel oil degradation than addition of nutrients alone. Higher amounts of the product (10 mg/g soil or 100 mg/g soil) directly inhibited degradation. The excessive mineralization detected at the higher amounts of product appeared to be due to biodegradation of a readily degradable carbon source (starch) in the formulation of the product. This may also have depleted the soil of nutrients necessary for oil degradation. The introduced organisms were not able to express their oil-degrading capability. In addition, a marked CO_2 production occurred with the 100 mg/g soil inoculum, which could have led to partial anaerobiosis and suggests the need for a high rate of soil aeration in the field. The use of bioaugmentation products containing readily available carbon may not be appropriate for bioremediation of oil-polluted soil.

Addition of an allochthonous microbial population may not be necessary or effective in most cases (Vecchioli, Del Panno, and Painceira, 1990). Before microorganisms are added to a site, it should be determined if the existing population could be utilized instead (Maxwell and Baqai, 1995). Stimulation of indigenous, biodegrading microorganisms does show promise (Atlas and Bartha, 1993). In heavily polluted soil (e.g., with coal tar), a mixed population of added microorganisms was able to stimulate biodegradation (Aamand, Bruntse, Jepsen, Jorgensen, and Jensen, 1995). However, if the soil was only slightly contaminated with coal tar, supplementing the organisms was not as successful, since the indigenous microbes were capable of rapid, inherent mineralization. In another comparison study, it was found that biostimulation was significantly more effective than bioaugmentation, both in slurry reactor systems and in test plots containing weathered crude oil (Leavitt and Brown, 1994). These authors concluded that, while there is a place for bioaugmentation, it may not always be necessary.

There have been few studies that show survival of laboratory-cultured PAH-degrading microorganisms in the natural environment (Wilson and Jones, 1993). One such investigation was a field demonstration conducted on contaminated soil at the "Old Inger" CERCLA site (oil reclamation facility) to compare the effectiveness of a commercial microbial consortia of PAH-degrading bacteria with that of the indigenous microbes in degrading PAHs (Catallo and Portier, 1992). The sole advantage of the commercial inoculum was the lack of an extended acclimation period and the more rapid onset of PAH biodegradation. After a 14-day acclimation period with nutrient amendments, PAH degradation rates for the indigenous organisms was similar to that of the commercial preparation. If time is not a crucial limiting factor, it can be cost-effective and efficient to exploit the indigenous microflora.

A culture broth containing hydrocarbon-degrading microorganisms and the biosurfactants they produce can be used as the inoculum and introduced via a closed-loop system (Robertiello, Lucchese, Di Leo, Boni, and Carrera, 1994). Extracted groundwater is passed through three tanks. The first serves to volatilize the lighter components. The second tank is inoculated and acts as a fermenter for growing the microorganisms. The third tank further oxygenates the culture broth and is a settler for the recovery (by skimming) of residual hydrocarbons. The water is then reintroduced to the soil through well points. When this approach was applied to treat an accidental spill of gasoline and diesel fuel in sandy soil, the microflora became dominated by *Alcaligenes faecalis* and *Acinetobacter calcoaciticus*. The conclusion was that the inoculum should consist of those organisms that would naturally predominate in such an environment.

Although hydrocarbon-degrading bacteria have been found to be naturally present, microbial inoculation has been capable of substantial acceleration of biodegradation when appropriate conditions were provided (Vecchioli, Del Panno, and Painceira, 1990). The factors that could be limiting biodegradation by the supplemented microbes (e.g., oxygen, nutrients, etc.) should be evaluated and corrected (Maxwell and Baqai, 1995). If microorganisms are to be added, they must be hydrocarbon degraders and able to compete with the native population. The organisms may be unable to move through the soil to sites containing the chemical (Vecchioli, Del Panno, and Painceira, 1990). Appropriate methods must be used to ensure that the microbes can mobilize throughout the contaminated area (Maxwell and Baqai, 1995). Mixing may be necessary for contact of the organisms with the compounds to be destroyed (Vecchioli, Del Panno, and Painceira, 1990). Substantial monitoring should then be conducted during the project to evaluate site conditions and assess the effectiveness of the treatment. Repeated, high-dosage supplementations of organisms might be successful for *ex situ* and bioreactor applications, but may prove difficult and expensive *in situ* (Turco and Sadowsky, 1995).

5.2.2.1 Acclimated/Adapted Bacteria

The adaptation process during which microorganisms adjust to growing on a new substrate is defined functionally as an increase in rate of degradation with exposure to a compound (Swindoll, Aelion, Dobbins, Jiang, Long, and Pfaender, 1988). It may involve one or more of the following: (1) an induction or derepression of enzymes specific for degradation pathways of a particular compound, (2) an increase in the number of organisms in the degrading population, (3) a random mutation in which new metabolic pathways are produced that will allow degradation (Spain, Pritchard, and Bourquin, 1980), or (4) adaptation of existing catabolic enzymes (including associated processes, such as transport and regulatory mechanisms) to the degradation of novel compounds (Harder, 1981). In general, it is easier to obtain a stable and viable culture by adapting it first to specific compounds, then to a complex waste, rather than by trying to adapt it directly to the waste (Venkataramani, Ahlert, and Corbo, 1988).

Catallo and Portier (1992) differentiate between the terms *acclimation* and *adaptation*, which are often used interchangeably. While there is some definitional overlap, they point out that acclimation is the lag time during which indigenous organisms acquire the ability to degrade novel compounds, while adaptation is "the modification of characteristics of organisms that facilitates an enhanced ability to survive and reproduce in a particular environment" (Hochachka and Somero, 1984). Acclimated microorganisms are important to successful bioremediation (Loehr, 1992).

A rapid and simple screening technique for potential crude oil–degrading microorganisms was developed by Hanson, Desai, and Desai (1993). The method uses the redox indicator 2,6-dichlorophenol indophenol in Bushnell and Haas medium with crude oil and a microtiter plate. Bacteria possessing high crude oil–degrading potential turn the medium colorless after 24 h incubation, and, depending upon the time taken for the change in color, relative abilities of different cultures can be determined.

One approach to biodegradation is the addition to the system of microorganisms that have been especially acclimated to degrade a pollutant of concern (Wilson, Leach, Henson, and Jones, 1986). The organisms may be selected by enrichment culturing or genetic manipulation and can become acclimated to the degradation of compounds by repeated exposure to that substance (Lee and Ward, 1984). In this way, the metabolic activity and tolerance to different chemicals can be built up over time. The source of the samples, the media used, and growth conditions will determine the types of microbes isolated (Atlas, 1977). See Section 7.1.1.12 for a description of the enrichment culturing techniques. Frequently, microorganisms are isolated from a contaminated site itself (Wilson and Jones, 1993). Because of their exposure to the pollutants, these microbes have already initiated the adaptation process. Those with the fastest degradation rates are then reintroduced to the polluted site.

A short biologically active carbon adsorber can be an efficient reactor system for the growth, acclimation, and enrichment of indigenous microorganisms to be used for reinoculation (Weber and Corseuil, 1994). Activated carbon yields much higher levels of substrate biodegradation and higher specific growth rates when used as a biosupport medium than nonadsorbing or weakly adsorbing media.

To produce PAH-degrading bacteria on a large scale for soil inoculation, ideally, the organisms should be grown on a medium containing the intended substrate (Aamand, Bruntse, Jepsen, Joergensen, and Jensen, 1995). Some bacteria were found to lose the capability of degrading a PAH if they were grown on medium without the compound. Other bacteria did not lose this ability and readily adapted to the PAH degradation, following transfer to a PAH-containing medium. Therefore, selection of strains that are able to retain the capacity to degrade specific compounds when not grown in their presence could be utilized to produce a greater number of microbes in the substrate-free medium for soil inoculation than could be grown in substrate-specific media.

Selective pressure could be exerted by adding certain compounds to select preferentially for organisms with desired degradative capabilities. For instance, addition of aromatic compounds, such as benzoate or phenylalanine, can select for benzene, toluene, and xylene degraders, although addition of a nonaromatic substrate (i.e., acetate) does not stimulate these organisms (Rotert, Cronkhite, and Alvarez, 1995). A selective proliferation of target degraders would enhance the biodegradation kinetics, which should decrease the duration and cost of the bioremediation.

Adaptation affects biodegradation rates (Fournier, Codaccioni, and Soulas, 1981). The rate of transformation of a compound is often increased by prior exposure to the chemical. It can involve different mechanisms, such as gene transfer or mutation, enzyme induction, and population changes. There is a high degree of variability in the ability of a microbial community to adapt, depending upon many factors present at the site (Spain and Van Veld, 1983). Adaptation of aquatic microbial communities can last for several weeks after exposure to a xenobiotic compound.

Prior exposure and acclimatization of microorganisms to hydrocarbons in the soil can enhance degradation rates (Thomas, Lee, Scott, and Ward, 1989). In fact, PAH transformation rates have been recorded to be 3000 to 725,000 times greater in contaminated soil as in unacclimatized soil (Herbes and Schwall, 1978). A threshold PAH concentration or certain amount of time seems to be necessary for this to occur. Microbes may adapt to degrade only the compounds in the petroleum products at a given contaminated site (Lee and Hoeppel, 1991). For example, toluene was degraded slowly in soil contaminated with JP-5 (low in toluene content), while benzene, naphthalene, and methylnaphthalene were degraded rapidly in soil from this site.

Repeated exposure to a contaminant at a site will usually increase the adaptive capabilities of the microorganisms (Spain and Van Veld, 1983). Organisms from contaminated sites may be able to degrade a wider range of compounds, once they become acclimated to the organic compounds, a process that may require several months (Hamaker, 1966) or several years (Brubaker and O'Neill, 1982). It has been suggested that laboratory adaptation of the microorganism population is not a major factor in long-term degradation of petroleum in soils, because adaptation will eventually occur naturally. However, the addition of soil containing previously adapted microorganisms may improve the initial degradation rates (Hornick, Fisher, and Paolini, 1983). Addition of acclimated microbial cultures has also been shown to enhance markedly the rate and extent of mineralization of contaminants present in NAPLs (Labare and Alexander, 1995).

Often an organism that is adapted to metabolize a member of a homologous series of molecules may be capable of degrading the rest of the series. For example, phenol and benzoic acid have been degraded after acclimatization to p-hydroxybenzoic acid (Healy and Young, 1979). Microorganisms adapted to growth on m-xylene in the absence of molecular oxygen, with nitrate as an electron acceptor, are also able to degrade toluene under denitrifying conditions (Zeyer, Kuhn, and Schwarzenback, 1986). Sometimes the compounds are not similar in structure. For instance, after growth with succinate plus biphenyl, a mutant strain of *Beijerinckia* could oxidize benzo(a)pyrene and benz(a)anthracene (Gibson and Mahadevan, 1975). Chemical analog adaptation can reduce the amount of time required for adaptation to a particular chemical.

The enzymes are generally compound specific, and many microorganisms may be required to degrade all the PAHs in a contaminated site (Wilson and Jones, 1993). However, PAH-degrading bacteria may contain common enzymes for metabolizing other PAH substrates; i.e., inducers for low-molecular-weight PAH degradation may also induce that of higher-molecular-weight compounds (Stringfellow, Chen, and Aitken, 1995). For instance, salicylate can induce degradation of phenanthrene, and phenanthrene or salicylate can induce metabolism of pyrene and fluoranthene by *Pseudomonas saccharophila* P-15. Such inductions could be utilized for compounds that are degraded slowly or are degraded only by nongrowth metabolism.

Farbiszewska and Farbiszewska-Bajer (1993) applied acclimated organisms to treat soil contaminated from leaking airfield and gas station tanks. An enriched bacterial suspension of 10^9 cells/mL was added to soil containing 3.587 g petroleum products/kg plus 0.1 kg N–P–K fertilizer plus 0.1 kg NH_4NO_3 and bark, which acted as an air biofilter. After 8 to 12 weeks incubation, 50 to 84.75% of the original content of petroleum products vanished. The rate of oxygen uptake by microflora tripled, suggesting further decontamination was possible.

Enhanced mineralization of pyrene (55% enhancement in two days) was reported after bacteria isolated from the contaminated soil were reintroduced at 10^6 to 10^8 cells/g soil (Grosser, Warshewsky, and Robie Vestal, 1991). A 22% increase in hydrocarbon removal was achieved when specifically adapted strains from one site were added to another contaminated with petroleum (Vecchioli, Del Panno, and Painceira, 1990). A disadvantage to the latter approach is that the microbes from another source may not easily adapt to the specific conditions of the new location.

Acclimated microbes and nutrients can be employed to stimulate degradation (Quince and Gardner, 1982b). However, before a cleanup system employing acclimated bacteria can be implemented, a laboratory investigation of the kinetics of biodegradation for the acclimated bacteria, the potential for inhibition under various conditions, the oxygen and nutrient requirements, and the effects of temperature should be evaluated (Sommers, Gilmore, Wildung, and Beck, 1981). Emulsifiers may be necessary to increase the solubility of the contaminant. The hydrogeologic data that are needed are formation porosity, hydraulic gradient, depth to water, permeability, groundwater velocity and direction, and recharge/discharge information.

5.2.2.2 Mutant Microorganisms

Genetic engineering can be used to develop microorganisms with unique metabolic capabilities. The design and construction of microbial systems (cellular or enzymatic) to degrade specific compounds has been made possible by two developments (Thibault and Elliott, 1980). First, the ability of microorganisms to degrade some of these compounds has been shown to be encoded on extrachromosomal-DNA (plasmids), and, secondly, recent developments in recombinant DNA technology permit the "engineering" of DNA that codes for desired enzymatic capabilities.

For those compounds that are recalcitrant to attack by natural microbes, there is the theoretical possibility of constructing strains that can degrade them by the introduction into one bacterium of a number of enzyme activities from different bacteria, which under natural circumstances might have little or no chance to exchange genetic information (Williams, 1978). Expanding the ability to degrade a wide range of aliphatic and aromatic compounds should give strains a selective advantage over other bacteria in degrading a mixture of petroleum hydrocarbons (Hornick, Fisher, and Paolini, 1983).

Genetic engineering is trying to produce bacteria capable of complete degradation of xenobiotic compounds by using zymogenous organisms, such as *Pseudomonas* (Chakrabarty, 1982). Strains of *Arthrobacter*, more slowly growing autochthonous organisms, have also demonstrated xenobiotic-degrading capabilities, and these have also shown potential for treating hazardous waste–contaminated soils (Stanlake and Finn, 1982; Edgehill and Finn, 1983). A strain of *P. putida* has evolved a minimum number of genes for degrading phenol more efficiently via a different pathway (Chakrabarty, 1994). Understanding such phenomena may help researchers accelerate the evolutionary process.

Since genes for some enzymes involved in degrading alkanes and simple aromatic hydrocarbons can be found on plasmids (Chakrabarty, 1974; Alexander, 1981; Hou, 1982; Jain and Sayler, 1987), it may be possible to exchange enzymatic activity among closely related microorganisms through plasmid transfer.

By manipulating the exchange of this genetic material, it is possible to develop strains with extended degradative capability, i.e., organisms that can degrade more than one xenobiotic substrate or that can completely mineralize highly recalcitrant molecules (Kamp and Chakrabarty, 1979). Multiple plasmid transfer has been accomplished between *Pseudomonas* species (Chakrabarty, 1974) to construct a strain that can degrade several hydrocarbons, including octane and naphthalene (Hornick, Fisher, and Paolini, 1983). Friello, Mylroie, and Chakrabarty (1976) produced a multiplasmid-containing *Pseudomonas* strain capable of oxidizing aliphatic, aromatic, terpenic, and polyaromatic hydrocarbons.

Plasmids can be fused together to provide multiple degradative traits or to produce a novel or previously unexpressed degradative pathway (Pierce, 1982d). There is some indirect evidence this may occur in nature and that plasmid-borne genes may be transferred, to some extent, among indigenous bacteria in the soil (Pemberton, Corney, and Don, 1979). It is not known whether or not introduced microbes would also transfer these genes (Stotsky and Krasovsky, 1981).

Genetic engineering can be used to increase capacity of microbes by improving the stability of the enzyme systems (located on chromosomes instead of on plasmids), enhancing their activity, providing them with multiple degradative activities, and ensuring that they are safe, both to the environment and human health (Pierce, 1982d). Plasmids added to stable chromosomes should limit their potential for escape into the environment (Stotsky and Krasovsky, 1982).

The inability of a microorganism to degrade a certain substrate may just be related to its inability to induce certain enzymes or to transport the compound into the cell (Hornick, Fisher, and Paolini, 1983). Development of mutant strains through genetic alteration may increase growth rate or endow the organism with the desired biochemical capability (Thibault and Elliott, 1980).

Adaptation and mutation of microorganisms involve several steps (Zitrides, 1983). Wild strains known to degrade a specific organic chemical or functional group are exposed to successively increasing concentrations of that chemical. Those least inhibited by high concentrations will grow the fastest. These are then irradiated to induce genetic changes for increased growth rate.

Adaptation of the strains at high concentrations of a preferred substrate can induce increased production of the specific enzymes required to degrade that substrate (Thibault and Elliott, 1979). However, this is probably achieved at the expense of the ability to produce other enzymes. Therefore, an adapted mutant put back into the environment may be able to compete against indigenous organisms only when the preferred substrate is available.

Interstrain and interspecies genetic engineering promises to develop organisms with extraordinary abilities to degrade xenobiotic compounds (Chakrabarty, 1982). However, more information is needed on the ability of genetically engineered organisms to survive, grow, and function in the soil environment (Liang, Sinclair, Mallory, and Alexander, 1982). This technology has been demonstrated in the laboratory and used in several successful, full-scale soil decontamination operations, including cleanup of an oil spill (Thibault and Elliott, 1980). Other practitioners are skeptical because of the importance of the soil environment in determining the microbial activity and hence the success of applying exogenous organisms (Sims and Bass, 1984), especially since many compounds that can be degraded under laboratory conditions continue to persist in the environment (Pierce, 1982c).

Plasmids, such as TOL and NAH, mediate bacterial degradation of small aromatic compounds (Foght and Westlake, 1983). It was earlier believed that plasmids did not play a role in degradation of larger PAHs, such as phenanthrene, benzo(a)pyrene, or dibenzothiophene. However, plasmids have been implicated in phenanthrene degradation (Foght and Westlake, 1991). Almost all aromatic degraders have at least one plasmid, which can be divided into two groups: those with limited ability to degrade aromatics (e.g., only biphenyls) have small plasmids; those with extensive capability to degrade PAHs have larger plasmids (Foght and Westlake, 1983). Aromatic-degrading strains identified include members of the *Flavobacterium*, *Acinetobacter*, *Pseudomonas*, and *Aeromonas* genera. There is no evidence for any PAH-degrading isolates being capable of degrading saturates (e.g., *n*-alkanes) or *vice versa*. Growth on larger carbon chain-length *n*-alkanes (C_{11} to C_{20}) is chromosomally mediated and can be divided into three classes: C_{11} to C_{12}, C_{13} to C_{14}, and C_{15} to C_{20}.

Other authors report that by genetic and molecular techniques, it has been possible to transfer the plasmids NAH, TOL, and CAM-OCT to produce a multiplasmid *P. putida* (MPP), thereby constructing an organism that can transform, simultaneously, some linear alkanes and aromatic and polyaromatic hydrocarbons (Thibault and Elliott, 1980). Such an organism might be useful in the cleanup of oil spills or of wastes from industrial pollution. The creation of improved strains in this manner would accomplish the *in situ* treatment and removal of several environmental contaminants at the same time.

It may not be feasible to engineer strains that are resistant to heavy metals, since it is more than likely that the toxic conditions found in industrial wastewaters will cause plasmid losses from organisms, even though they function well in the laboratory (Wood and Wang, 1983). It looks more promising to employ organisms, such as the green alga *Cyanidium caldarium*, that naturally tolerate extreme conditions.

Greer, Masson, Comeau, Brousseau, and Samson (1993) have characterized genes from organic pollutant degradative pathways in bacteria and developed molecular techniques to study the microbial ecology of contaminated environments. The techniques include use of catabolic gene probes, nucleic acid hybridization, and the polymerase chain reaction to amplify specific target regions of nucleic acids. The *xylE* (toluene, xylene degradation), *ndoB* (naphthalene degradation), and *alkB* (C_6 to C_{12} *n*-paraffin degradation) gene probes have been used to isolate indigenous bacteria from hydrocarbon-contaminated soils and to enrich bacteria expressing these genotypes for subsequent bioaugmentation of soil. A strain was developed that could be unequivocally differentiated from indigenous bacteria and be detectable to ten viable cells per gram of soil. Foght and Westlake (1991) suggest that using TOL or NAH plasmids to probe an environmental population would underestimate the occurrence of PAH-degradative genes. They recommend employing a number of probes to evaluate the potential of a population.

Genetically engineered microorganisms may not be successful in the field (Pierce, 1982c). These microbes would have to compete with resident microflora. If there are already 10^8 to 10^9 bacteria/g soil, a very large number of the new organisms would have to be added for them to become established in the community. If environmental parameters, such as temperature, pH, substrate concentration, oxygen tension, are not optimal for the added organisms, there will be reduced activity (e.g., a bacterium that degrades *n*-alkanes optimally at 30°C in an aqueous medium would probably perform poorly in an oil spill in the North Sea). Also, most of the bacteria typically used in microbial genetic work are eutrophs of the families Enterobacteriaceae and Pseudomonadaceae, which may not be able to attack substrates in the ppb range that are often found in environmental samples (Johnston and Robinson, 1982).

The problem with the use of genetically selected microorganisms to degrade contaminants is that no conclusive evidence has been found that commercially available organisms are effective in establishing themselves or significantly enhancing biodegradation of pollutants in aeration basins or natural environments having an active native microbial population (Johnston and Robinson, 1982). The adapted mutant will probably be at a disadvantage in the competition with the native microbial population, and may only be able to proliferate on the substrate upon which it was isolated (Zitrides, 1978). Acclimated or genetically engineered organisms will not survive or offer significant advantage in treatment of hazardous wastes unless environmental parameters (oxygen, temperature, nutrients, etc.) can be controlled to promote survival of the added organisms (Lee, Wilson, and Ward, 1987).

Environmental stress, such as abiotic stress, starvation, biological antagonism (Liang, Sinclair, Mallory, and Alexander, 1982), unsuitable water availability, or the presence of toxicants may also affect an introduced microorganism (Johnston and Robinson, 1982). Microorganisms that are tolerant of multiple kinds of stresses have a higher potential for survival in the soil after genetic manipulation in the laboratory than organisms with less tolerance and versatility (Liang, Sinclair, Mallory, and Alexander, 1982).

To address these problems, recombinant *Escherichia coli* strains have been constructed to express the toluene monooxygenase (TMO) gene under control of the powerful GroEL stress promoter (Little, Fraley, McCann, and Matin, 1991). GroEL is a heat shock protein and an important biomolecule in a variety of cellular stress responses, including carbon, nitrogen, and phosphorus starvation, oxidative shock, and chemical toxicity (Matin, Auger, Blum, and Schultz, 1989). It appears to aid in renaturing damaged proteins in stressed cells.

Because of the large amount of information necessary before rational and reproducible experiments in strain construction can be performed, this approach has not yet been applied on a wide scale in biodegradation research (Leisinger, 1983). Assembling degradative pathways from different bacterial strains in one cell by genetic manipulation is a laborious technique by which to obtain organisms with novel degradative capacities. This method could be applied in cases where a precise strategy can be formulated and where obstacles to degradation, such as the intracellular accumulation of dead-end metabolites from problem compounds, cannot be overcome by cocultivation of strains with different degradative pathways.

The addition of adapted mutant microbes has not been completely successful, but it does have potential (Fox, 1985; Knox, Canter, Kincannon, Stover, and Ward, 1986). A number of companies are producing microbial strains to be used to treat abandoned hazardous waste sites and chemical spills (Anonymous, 1981). Mutant bacterial formulations have been developed by Polybac Corporation to degrade the most complex organic materials, including industrial surfactants, crude and refined petroleum products, pesticides and herbicides, and solvents (Thibault and Elliott, 1979). Polybac maintains an up-to-date library of information on the relative biodegradability of a wide range of organic chemicals by various mutant strains. This allows selection of the proper formulation. Researchers at government and industrial laboratories are developing artificially mutated bacteria that will be more active and selective for the chemicals they destroy (Chowdhury, Parkinson, and Rhein, 1986).

There have been concerns that genetically engineered organisms may continue to persist in the environment (Stotsky and Krasovsky, 1981). It is hard to predict the fate of these organisms in complex soils and aquatic ecosystems. Unless the new genetic material confers a wider substrate range, or increases the ability of the organism to detoxify its environment, it would seem energetically disadvantageous for the organism to maintain the genes. Eventually, the organism or its new genes would be selected against. Such an organism may decrease to insignificant numbers, if its target substrate is depleted (Chakrabarty, 1982). However, others argue that more stress-tolerant organisms could persist for extended periods (Stanlake and Finn, 1982; Edgehill and Finn, 1983).

It is important to be able to detect and monitor the fate of genes of an introduced organism (Jain and Sayler, 1987). Monitoring is required to predict whether the released organisms disappear quickly, whether they are able to compete with the indigenous population, whether they can proliferate and become temporarily or permanently established in the area, or whether they are being transported from the initial site of the release. Because of the concern over introduction of genetically engineered organisms into the environment, it may be some time before the issue is resolved and the use of these microbes to clean up hazardous waste sites is widespread (Fox, 1985). However, genetic engineering holds great promise, especially in treatment facilities where conditions can be controlled (Johnston and Robinson, 1982).

5.2.2.3 Microbial Consortia

Many studies of microorganisms capable of growing on oil have dealt with a single microorganism degrading a single hydrocarbon (Cooney, 1980). The effect of single microorganisms on multiple substrates and, conversely, the effect of multiple organisms on a single substrate are poorly understood. However, mixed population studies show that some compounds that are resistant to degradation by a single organism can be degraded by mixed populations (Beam and Perry, 1974; de Klerk and van der Linden, 1974).

Microbial interactions in the soil are very complex and undoubtedly play an important role in the transformation or decomposition of hazardous waste components (JRB Associates, Inc., 1984). The growth requirements among the organisms lead to intense competition for the available nutrients. As a result of metabolic specialization (e.g., autotrophic nitrification), some microbes are less dependent upon preformed organic substrates or growth factors. The majority, however, have acquired antagonistic abilities that help limit growth of competitors. Examples of these antagonistic agents are antibiotics, acids, bases, and other organic and inorganic compounds (Alexander, 1971; Atlas and Bartha, 1981).

True mutualistic or symbiotic relationships also exist among soil organisms (JRB Associates, Inc., 1984). It is common for degradation of a xenobiotic compound to involve sequential metabolism by two or more microorganisms (Beam and Perry, 1974) in a relationship that may benefit only one partner (commensalism) or both (protocooperation or synergism) (Atlas and Bartha, 1981). In such a commensalistic relationship, microbes cannot oxidize a given hydrocarbon individually, but collectively they are able to do so. Some members of a community might be able to provide important degradative enzymes, whereas others may supply surfactants or growth factors (Wiesel, Wuebker, and Rehm, 1993). This form of commensalism may be very widespread in nature with natural mixed populations employing each other's metabolic intermediates as growth substrates (Donoghue, Griffin, Norris, and Trudgill, 1976). Many genera of bacteria, fungi, and actinomycetes can participate in the process (Horvath and Alexander, 1970; Alexander, 1977).

When a particular chemical is not easily transformed by microorganisms using it as a sole source of carbon, it is sometimes possible to employ commensalism to encourage complete biodegradation (de Klerk and van der Linden, 1974; Donoghue, Griffin, Norris, and Trudgill, 1976). Bacterial mixtures, such as *Pseudomonas*, *Nocardia*, and *Arthrobacter* have been used as part of a "cocktail" that may consist of 12 to 20 strains of microbes for treating different types of wastes (Cooke and Bluestone, 1986). The mixture of organisms would depend upon the composition of the waste. Both selective and controlled mixed cultures are currently being used (Pfennig, 1978a). Specific organisms for target chemicals can be chosen. For example, purple nonsulfur bacteria are used to remove organic compounds and green sulfur bacteria to remove hydrogen sulfide. *Nocardia* cultures degrade the major n-paraffinic hydrocarbons, and *Pseudomonas* cultures degrade aromatics (Jamison, Raymond, and Hudson, 1976).

Mutualistic relationships (all members derive some benefit) are based not only upon growth-factor interdependence, but may also encompass removal of a (toxic) product of metabolism produced by one component of the mixed population and used by another (Pfennig and Biebl, 1976), combined metabolic attack (Johanides and Hrsak, 1976), or relief of substrate inhibition (Osman, Bull, and Slater, 1976). Some microorganisms thrive on metabolic products or products from lysis of other organisms, as a result of a commensalistic relationship (Harder, 1981). A *Nocardia* sp. has been identified as a cyclohexane utilizer, but growth occurs only in the presence of an unidentified pseudomonad that provides biotin and possibly other growth factors (Sterling, Watkinson, and Higgins, 1977).

More than 100 strains of bacteria were unable to use unsubstituted cycloparaffinic hydrocarbons as their sole source of carbon and energy; however, many could partially oxidize the hydrocarbons when a suitable energy source was present (e.g., an n-alkane) (Beam and Perry, 1973; 1974). The resulting cycloalkanones were readily oxidized and used as an energy source by other strains of bacteria. Unsubstituted cycloparaffinic hydrocarbons are readily mineralized in natural soil systems, presumably by a combination of cometabolism and commensalism.

An example of this is a two-step process using two microorganisms. A pseudomonad first co-oxidizes cyclohexane into either cyclohexanol or cyclohexanone while growing on an *n*-alkane, such as *n*-heptane; then another strain (e.g., a pseudomonad) grows on the cyclohexanol or cyclohexanone (de Klerk and van der Linden, 1974; Donoghue, Griffin, Norris, and Trudgill 1976). The same general results have been reported for the oxidation of other unsubstituted cycloparaffinic hydrocarbons (Beam and Perry, 1974).

A mixed culture was isolated from PAH-contaminated soil and was able to degrade a range of PAHs, including fluoranthene, benzo(a)pyrene, anthracene, phenanthrene, acenaphthene, and fluorene (Trzesicka-Mlynarz and Ward, 1995). Three of the four predominant isolates were the Gram-negative rods, *Pseudomonas putida, P. aeruginosa*, and a *Flavobacterium* sp. Better degradation was observed with the mixed culture. Cultures of the individual organisms showed efficient removal of the more water-soluble PAHs (acenaphthene, fluorene, phenanthrene, fluoranthene) but poor removal of the less water-soluble PAHs (pyrene and anthracene). A reconstituted mixture of the primary isolates performed in a manner similar to the original community.

Both free and immobilized cells of a stable community of a mixture of soil bacteria, which had the ability to degrade aliphatic and aromatic hydrocarbons, were able to use PAHs as their sole source of carbon and energy (Wiesel, Wuebker, and Rehm, 1993). Both communities showed the same degradation sequence of the PAHs — i.e., phenol, naphthalene, phenanthrene, then anthracene and pyrene.

A commercial microbial culture, Para-Bac (Micro-Bac Int., Inc., Austin, TX), can be added to anoxic environments with redox potentials of less than -200 mV to reduce viscosity and interfacial tension of a variety of crude petroleums, through biosurfactant-type activities (Schneider, 1993).

In the environment, mixed cultures utilize and detoxify hazardous organics; however, it has been difficult to construct a mixed population as a starter culture and multiply it, keeping the proportions constant (Babel, 1994). On the other hand, anthropogenic mixed cultures are certainly an alternative to genetically engineered superbugs.

5.2.2.4 Emulsifier Producers
Addition of microorganisms that produce surfactants to a site can be employed to help solubilize the organic contaminants. This would have the effect of making the compounds more accessible to microbial degradation. Section 5.3.1.2 describes the production of emulsifiers by microorganisms.

5.2.2.5 Microbial Transport
There is evidence that introduced organisms may not be successful in bioremediation because they cannot reach the contaminants (Goldstein, Mallory, and Alexander, 1985). Bacteria added to the soil surface for biodegradation may not be transported sufficiently to reach organic pollutants at sites distant from channels or macropores (Devare and Alexander, 1995).

The transport and adhesion properties of the added organisms would affect their delivery through the subsurface and their method of growing (Fletcher, 1994). Retention of microbes in the soil may be due to both adsorption (Hattori and Hattori, 1976; Marshall, 1980) and mechanical filtration (Pekdeger and Matthess, 1983; Smith, Thomas, White, and Ritonga, 1985). Some microbes adhere easily to soil particles near the injection site (Deflaun, Tanzer, McAteer, Marshall, and Levy, 1990; Fletcher, 1993). Microorganisms are resident on less than 0.17% of the surface of soil organic matter and on less than 0.02% of the soil mineral surfaces (Hissett and Gray, 1976). Gram-positive bacteria tend to attach to the outside of soil microaggregates, while Gram-negative bacteria are found within the aggregate (Kilbertus, Proth, and Vervier, 1980). The Gram-negatives, which do not have a protective cell wall, may prefer the more hospitable microenvironment inside the aggregate, as opposed to the less environmentally stable surfaces, where microbes can become desiccated and die or shift to a dormant phase. Most of the microbial activity in soil may occur within the aggregates (Martin and Foster, 1985; Foster, 1988; Killham, Amato, and Ladd, 1993).

The presence of microorganisms can limit the drainage efficiency of porous media and increase oil retention, although the effect is highly variable (Ducreux, Ballerini, and Bocard, 1994). If the organisms attach to surfaces in the soil, they might form biofilms — colonies of microorganisms embedded in a hydrated polysaccharide matrix (Fontes, Mills, Hornberger, and Herman, 1991). Syntrophic organisms might become localized together in this matrix. Strains that are nonadhesive may be chosen for better transport, by selection of variants or mutagenesis (Deflaun, Tanzer, McAteer, Marshall, and Levy, 1990).

Achieving the required dispersion and distances with inoculated organisms is difficult. Bacteria do not move appreciably through soil in the laboratory or in the field (Alexander, 1994). Few bacteria applied to a surface penetrate more than 5 cm (Edmonds, 1976). The extent of movement might be more pronounced through the larger pores of sandy soils, but bacterial transport with water is probably very limited in other soils (Alexander, 1994). For example, *Pseudomonas putida* would not move beyond 3 cm (Madsen and Alexander, 1982). On the other hand, certain bacteria (Gannon, Mingelgrin, Alexander, and Wagenet, 1991) and spores of some fungi (Hepple, 1960) may be more susceptible to transport with water.

Bacterial transport is strongly correlated with cell size (Gannon, Manilal, and Alexander, 1991). It should also be possible to transport smaller bacteria over greater distances than larger cells (Fontes, Mills, Hornberger, and Herman, 1991). Small soil bacteria are on the order of 0.5 to 1.0 μm in diameter (Bitton and Gerba, 1985), or even 0.3 μm (Bae, Cota-Robles, and Casida, 1972). Larger bacteria tend to be immobilized in soils by physical straining or filtering. The mean diameter of soil pores occupied by soil bacteria of 0.5 to 0.8 μm (Casida, 1971) is about 2 μm (Kilbertus, 1980). Organisms will occupy 6% of the inner volume of soil aggregates (Kilbertus, 1980).

Migration of microorganisms through the soil can occur via two processes: convective (or passive) transport and molecular "diffusion" (Ahlert and Kosson, 1983; Smith, Thomas, White, and Ritonga, 1985). Convective transport involves addition of significant quantities of an aqueous nutrient feed solution (or water) that causes movement of organisms with the feed and distribution throughout the soil column. Microbiological diffusion is analogous to molecular or surface diffusion. It occurs as a result of the life/death cycle and the natural movements of microorganisms. If the selective organisms added to *in situ* systems have to depend only on growth to diffuse through the subsurface, it will take too long for bioaugmentation to be cost-effective (Kobayashi and Rittmann, 1982). However, the organisms can be spread through the soil more rapidly by convective transport when the soil is flooded with nutrient solution and the water is continuously pumped from an aquifer, establishing a closed path that pulls the liquid and organisms through the soil. Active transport involves the expenditure of energy (Huysman and Verstraete, 1993). Passive movement carries bacteria greater distances than active movement (Madsen and Alexander, 1982).

Organic matter, clay content, pore-size distribution, physical structure, and water flow affect the transport of bacteria in soil (Devare and Alexander, 1995). Passive movement of bacteria in soil is slowed by their sorption onto soil particles or transport via small soil pores (Smith, Thomas, White, and Ritonga, 1985). It is actually uncertain whether microorganisms move through the soil matrix or through wider macropores or channels (Rahe, Hagedorn, McCoy, and Kling, 1978; Hagedorn, McCoy, and Rahe, 1981). While movement through macropores may be the major mechanism of bacterial transport in soils, movement through the soil matrix may be necessary for biodegradative microorganisms to reach soil contaminants (Gannon, Mingelgrin, Alexander, and Wagenet, 1991). Soil structure and the velocity of water flow also control bacterial movement (Smith, Thomas, White, and Ritonga, 1985; Harvey, George, Smith, and LeBlanc, 1989).

Microorganisms adhere to soil surfaces by electrostatic interactions, London–van der Waals forces, and hydrophobic interactions (van Loosdrecht, Lyklema, Norder, and Zehnder, 1990). The predominant factors affecting bacterial transport are the ionic strength of the suspending solution and the soil surface properties. Cell retention in soil is also a function of the type of bacterium (Huysman and Verstraete, 1993). A strain with hydrophilic properties is two to three times faster than those with hydrophobic properties, which has increased adhesion to soil particles. Transport of motile bacteria may be inhibited in soil with small pore size because of the large amount of flagella (Issa, Simmonds, and Woods, 1993a; 1993b). Flagella are also less useful in dry soil and may even enhance cell sorption onto soil surfaces. Selection of organisms less likely to adhere to soil particles near the injection site would help in their transport (DeFlaun, Tanzer, McAteer, Marshall, and Levy, 1990; Fletcher, 1993).

It has been suggested that using soil columns for studying bacterial transport may give misleading results (Bitton, Davidson, and Farrah, 1979). However, Gannon, Mingelgrin, Alexander, and Wagenet (1991) have since developed a soil column technique that overcomes many of the usual problems. The method makes it possible to isolate more-mobile bacteria that not only have the desired biodegradative capacity, but also a greater chance of reaching the subsurface contaminants. Resulting measurements also enable extrapolation of the potential penetrability of the bacteria to considerably greater depths. These authors concluded that adsorption of cells to soil particles that are the same size or smaller may not interfere with mobility as much as adsorption on particles that are larger. Cells with a low adsorption coefficient can move more easily through the soil. There was no consistent pattern in mobility noted among strains within genera of bacteria tested, but the organisms ranged from very mobile (*Enterobacter* IS1) to primarily bound (*Bacillus* CB3), and benzene degraders were more mobile than toluene degraders (Gannon, Mingelgrin, Alexander, and Wagenet, 1991).

Generally, the organisms may be applied in liquid suspension or with a solid carrier (Sims and Bass, 1984). Run-on and run-off controls may be necessary. The potential achievable level of treatment is high. Nevertheless, relatively long periods of time may be required to complete treatment, and excessive precipitation may wash out the inoculum, necessitating retreatment.

Movement of organisms through the soil can be enhanced by an *in situ* electrobioreclamation process. This technique applies an electric field (current density of 20 A/m^2) which has been shown to improve bioremediation in low-permeability soils (Lageman, Pool, van Vulpen, and Norris, 1995). Permeability of the subsurface is enhanced by the fracturing whereby microorganisms may be able to travel greater distances through the soil (see Section 2.2.2.5). Under neutral pH conditions, bacteria behave as anions, migrating toward the anode. This process is further discussed in Section 2.2.2.8.

The process of hydraulic or pneumatic fracturing can be used to distribute ionic forms of electron acceptors, electron donors, and nutrients in low-permeability areas. Recent studies have investigated a novel approach of selecting a species selectively enriched or genetically manipulated to metabolize a specific xenobiotic compound and then starving the organism of that compound (Bryers and Sanin, 1994). By creating oligotrophic conditions, the bacteria will exhibit responses that will affect their movement through porous media, such as a reduction in cell size, production of ultramicrobacteria (UMB), reduction in cell endogenous decay or maintenance of energy requirements, increase in ability to attach to surfaces, and a change in production of insoluble extracellular polysaccharides. The resulting UMB would provide wide distribution of a species throughout a contaminated subsurface area. The organism could then be resuscitated to resume its selective metabolic activity. This has been demonstrated with *Klebsiella pneumoniae*.

5.2.2.6 Microbial Preservation

Preserved microorganisms should have a high initial viability, retain the ability to degrade target contaminants, and be competitive with indigenous organisms for the necessary nutrients (Romich, Cameron, and Etzel, 1995). The form in which the organisms are preserved should be highly stable during storage, have a low specific weight and volume, be easy to use and distribute, and be inexpensive.

Seed microorganisms may have to be freeze-dried or frozen to maintain viability (Atlas, 1977). Growing cultures are metabolically active for immediate biodegradation, but would require large volumes and would be difficult to transport to a location. Frequent transferring also raises the danger of contamination or mutation. Mixing of organisms should not be done until just prior to application to prevent competition within the mixture. Freeze-dried or frozen cultures could be mixed well in advance without this problem. Freeze-dried cultures would occupy small volumes for easy transport. However, freeze-dried cultures are quite expensive to prepare and take some time to become metabolically active again, especially in the suboptimal environmental condition of the environments in which they would be used. An alternative method would be to store the culture in a freeze-dried state and then initiate growth in a laboratory just before actually seeding. Then the organisms can be added in the appropriate concentrations to the environment.

In a comparison of three preservation methods, freezing cultures proved more effective than freeze-drying or spray-drying for recovery of bacteria able to grow on the desired substrate (Romich, Cameron, and Etzel, 1995). Dehydration inactivation was detrimental for this purpose. It may not be necessary to grow these organisms on the target chemicals, since both biphenyl- and succinate-grown cells had the same ability to degrade biphenyl when recultivated; however, not all of the viable cells could grow on biphenyl.

Cells of *Alcaligenes eutrophus* H850 were initially grown to 10^9 to 10^{10} CFU/mL (10^{11} to 10^{13} CFU/g dry cell weight (DCW) (Romich, Cameron, and Etzel, 1995). Viable organisms recovered ranged from

    1×10^{12} to 3×10^{12} CFU/g DCW for freezing
    2×10^8 to 9×10^8 CFU/g DCW for freeze-drying
    1×10^8 to 5×10^9 CFU/g DCW for spray-drying

Biphenyl degraders recovered ranged from

1×10^{11} to 1×10^{12} CFU/g DCW for freezing
1×10^{8} to 3×10^{8} CFU/g DCW for freeze-drying
9×10^{2} to 5×10^{6} CFU/g DCW for spray-drying

Survival and preservation of the degradative capacities of aerobic, Gram-negative bacteria are higher after preservation by liquid drying in the presence of *myo*-inositol, skim milk, or activated charcoal as protective agents than after preservation by freeze-drying (Lang and Malik, 1996). Also, growing cells on complex medium renders them less sensitive to drying than growing them under selective pressure, such as on mineral medium with a sole carbon source. Loss of degradation capabilities may occur, even while viability levels remain high. Recovery of degraders can be improved by decreasing the storage temperature from 25 to 4°C, reactivating them in complex media, and then transferring them to selective media. Plasmids were not noted to be lost during this process; however, damage could not be ruled out. It is recommended that the most appropriate preservation method be selected, with the use of an effective protectant, to avoid genetic alteration and to maintain biodegradation capabilities during long-term preservation.

Mineralization of phenanthrene was doubled by the addition of skim milk to soil slurries (Weir, Dupuis, Providenti, Lee, and Trevors, 1995). The skim milk probably provides a nutrient source and enhances growth and activity of added and indigenous microorganisms, or it may protect mineralizing microbes from some of the toxic compounds in the creosote-contaminated soil. Addition of skim milk was more successful for promoting mineralization and survival of supplemented *Pseudomonas* cells than alginate encapsulation of the organisms.

5.2.2.7 Encapsulation/Immobilization

The technique of microencapsulation of nonindigenous degradative organisms can enhance the survival of these microorganisms in the subsurface (Stormo and Crawford, 1994). The transport characteristics of the encapsulated cells will affect the success of the bioremediation effort (Petrich, Stormo, Knaebel, Ralston, and Crawford, 1995).

The technique of cell immobilization or encapsulation can be used to envelop the bacterial or fungal cells in a porous polymeric material to improve storage and survival of inocula, to protect the introduced microbes from predators, and yet allow expression of their specific metabolic functions (Lin, Lantz, Schultz, Mueller, and Pritchard, 1995). Encapsulation of microbial cells may improve consistency in inoculum quality by providing a more uniform population and a slow release of cells from the matrix over time (Trevors, van Elsas, Lee, van Overbeek, 1992; Paul, Fages, Blane, Goma, and Pareilleux, 1993).

In a tracer experiment, agarose-encapsulated *Flavobacterium* microbeads ranging in diameter from about 2 to 80 μm appeared to be retarded with respect to the bromide and the 2-, 5-, and 15-μm-diameter polystyrene microsphere tracers (Petrich, Stormo, Knaebel, Ralston, and Crawford, 1995). Site heterogeneity was a primary factor in the transport of bromide and microsphere tracers. It was suggested that decreasing average encapsulated-cell microbead diameters, increasing uniformity in microbead size, and use of alternative materials for the encapsulated cell matrix might enhance encapsulated-cell transport.

Stormo and Crawford (1992) describe the method of encapsulating bacteria into microspheres of alginate, agarose, or polyurethane. Agarose, alginate, carrageenan, and polyacrylamide gels have been tried but were mechanically and chemically unstable (Alteriis, Scardi, Masi, and Parascandola, 1990). Porous polyurethane foam is a more durable and effective immobilization matrix (Lin, Lantz, Schultz, Mueller, and Pritchard, 1995). Peat, vermiculite, and activated carbon have also been employed for this purpose (Pritchard, 1992). Polyurethane foam and vermiculite have been used for encapsulation and immobilization for use in bioremediation of PAH-contaminated soil (Lin, Lantz, Schultz, Mueller, and Pritchard, 1995). Polyurethane-encapsulated cells are resistant to decomposition in the environment and would be more appropriate for use in reactors. The naturally derived vermiculite-immobilized cells can be tilled directly into soils. Coimmobilization of slow-release sources of nitrogen and phosphorus into the capsules is effective in promoting mineralization of PAHs and in reducing leaching of the nutrients into the environment. Also, the addition of adsorbents can moderate toxic concentrations of contaminants. Cells capable of degrading PAHs appear to have better survival in the immobilized form and are active even when mixed in creosote-contaminated soil.

A gentle and simple immobilization technique was developed, by using a syringe needle to add an ionic polysaccharide/immobilizant solution dropwise into a solution of a divalent cation (Kierstan and Bucke, 1977). The divalent ions cross-link the charged species on the polysaccharide, forming insoluble gel beads. A typical ionic polysaccharide is alginate, which may be used with calcium ions for cross-linking. This produces beads of uniform size and quality down to 1 mm in diameter. However, beads with a smaller dimension are difficult to produce.

Since smaller diameters would improve performance, a new method was developed for creating the beads (Poncelet, Lencki, Beaulieu, Halle, Neufeld, and Fournier, 1992). Small-diameter alginate microspheres are formed via internal gelation of alginate solution emulsified within vegetable oil. An oil-soluble acid initiates gelation and reduces the pH of the alginate solution, releasing soluble Ca^{2+} from the citrate complex. Diameters of the beads range from 200 to 1000 μm and are controlled by the reactor impeller design and rotational speed. This method can be used for large-scale and continuous application in immobilization.

Microorganisms encapsulated in alginate can be stored in a dry, uniform state and remain viable as long as a year (Cassidy, Leung, Lee, and Trevors, 1995). Encapsulation allows nutrients or protective agents to be incorporated into the matrix (Heijnen, Hok-A-Hin, and van Veen, 1992). It also protects *Pseudomonas* cells from freeze/thaw stress (Leung, Cassidy, Holmes, Lee, and Trevors, 1995).

Alginate encapsulation increased survival of an added *Pseudomonas* strain to creosote-contaminated soil slurries (Weir, Dupuis, Providenti, Lee, and Trevors, 1995). However, there was no significant difference between encapsulated cells or free cells with bead amendments. There was little protective effect due to encapsulation, and the noted increase in cell numbers was probably due to the presence of skim milk as a nutrient source.

Oxygen mass transfer is a major problem with immobilized cells, especially when high densities of aerobic cells are used in the gel beads (Omar, 1993; Speirs, Halling, and McNeil, 1995). The smaller the diameter of the particle size of the gel, the better the distribution of the cells in the interior of calcium alginate beads, but alginate concentration also hinders oxygen diffusion. The higher the gel concentration, the lower the porosity of the gel and thus the diffusivity and uptake rate of oxygen. Calcium alginate beads shrink rapidly in air, so size measurements should be made underwater (Speirs, Halling, and McNeil, 1995). The estimated diffusivity is not significantly affected by a distribution of bead sizes.

It has been found that a calcium–alginate gel can be chemically destabilized in many different ways by calcium chelators or by competition with nongelling cations (Hertzberg, Moen, Vogelsang, and Ostgaard, 1995). Over time, alginate-degrading bacteria can attack alginate beads — e.g., *Klebsiella aerogenes* produces alginate guluronate lyase (Ostgaard, Knutsen, Dyrset, and Aasen, 1993) and *Haliotis tuberculata* produces alginate mannuronate lyase (Boyen, Kloareg, Polne-Fuller, and Gibor, 1990). In the presence of Na^+, alginate gel beads swell dramatically in 1 mM Ca^{2+}, even if the high-guluronic alginate of the *Laminaria hyperborea* type, which is the most resistant of all the alginates, is used (Martinsen, Skjak-Braek, and Smidsrod, 1989). Pure alginate beads dissolve completely at 0.1 mM of Ca^{2+}.

More-stable and stronger gel beads can be made with photo-cross-linkable polyvinyl alcohol (PVA) bearing photosensitive stilbazolium (SbQ) groups (Hertzberg, Moen, Vogelsang, and Ostgaard, 1995). The purely synthetic PVA–SbQ network helps stabilize the beads against swelling and destruction. These beads have been successful carriers in a denitrification process and should have a longer life span than a natural biopolymer, such as alginate. However, a binary mixture with alginate is still useful for simplifying bead production. A 5% PVA–SbQ/2% alginate will maintain a high gel strength, even after loss of the alginate gel network. It would also avoid toxicity problems of the boric acid methods for making PVA beads (Kakiichi, Shibuya, Akita, Sugimoto, Oshida, Hayashi, Kamata, Komine, Otsuka, and Ushida, 1992).

A microcapsule technique has been patented for supplying microorganisms to degrade hydrocarbons (Schlaemus, Marshall, MacNaughton, Alexander, and Scott, 1994). The capsule consists of a core material, which is lipophilic and contains nutrients for the microbes; a coating, which is water soluble; and at least one microorganism capable of degrading the hydrocarbon. The organisms are kept close to the hydrocarbon to be degraded.

Nutrients, minerals, and growth factors can be encapsulated in liposomes, as a means of enhancing growth of microorganisms in the subsurface (Gatt, Bercovier, and Barenholz, 1991). Liposomes also modify the physical properties of oil or hydrophobic compounds, increasing their availability to microbial degradation. The liposomes could be custom-tailored for the appropriate microbes (Atlas, 1984) and specific contaminants (Racke and Coats, 1990). See Section 5.1.5 for a description of this technique and how it could be applied to bioremediation.

The fluorescein diacetate-hydrolyzing activity (FDA) assay can be used to determine the biological potential (fungal biomass produced per unit of substrate) of solid pelleted fungal inoculum intended for application in the bioaugmentation of soils with white-rot fungi (Lestan, Lestan, Chapelle, and Lamar, 1996). The FDA activity of *Phanerochaete chrysosporium* grown on pelleted substrates and on agar is proportional to quantities of fungal ergesterol and fungal dry matter, respectively. Formulation and structure of the substrate and temperature affect the biological potential. Higher manganese peroxidase activity can be obtained in soil inoculated with *Trametes versicolor* on pellets with high biological potential compared with pellets with a low potential.

5.2.3 COMETABOLISM AND ANALOG ENRICHMENT

Xenobiotic organic compounds are usually transformed or degraded by microorganisms in either a metabolic sequence that provides energy and nutrients (e.g., carbon, nitrogen, phosphorus) for growth or maintenance of the organism, or by a biochemically mediated reaction that provides neither energy nor nutrients to the cell (JRB Associates, Inc., 1984). The first process results in the complete biodegradation of the organic molecules to mineral products (e.g., carbon dioxide, methane, water, ammonia, phosphate) and is called mineralization. The second process usually results in only a minor transformation of the organic molecule and is called cometabolism or co-oxidation (Alexander, 1973; Alexander, 1977; Pierce, 1982b). This transformation product is still unusable to the organisms (Hornick, Fisher, and Paolini, 1983).

Often, two or more substrates will be required for cometabolism to occur; one is the nongrowth substrate that is neither essential for, nor sufficient to, support replication of the microorganism (Hulbert and Krawiec, 1977; Perry, 1979), while the other compound(s) does (do) support growth. The nongrowth substrate is only incidentally and incompletely transformed by the microorganism involved (de Klerk and van der Linden, 1974; Perry, 1979; Alexander, 1994). A nonspecific enzyme with a broad substrate specificity is attacking a recalcitrant molecule and metabolizing it (Horvath and Alexander, 1970; Alexander, 1981). The organism supplying the enzyme gains nothing from the metabolic transformation. The enzyme is made by the microorganism to metabolize some other organic compound for its energy (McKenna, 1977; de Klerk and van der Linden, 1974).

Oxygenases are often involved in cometabolism because they can be induced by, and can attack, a large set of substrates (de Klerk and van der Linden, 1974; McKenna, 1977). Cometabolic conversions that appear to involve a single enzyme include hydroxylations, oxidations, denitrations, deaminations, hydrolyses, acylations, or cleavages of ether linkages, although many conversions are complex and require several enzymes (Alexander, 1994).

Cometabolism probably occurs frequently in natural soil systems, since many genera of bacteria, fungi, and actinomycetes can participate in the process (Horvath and Alexander, 1970; Alexander, 1977). Cometabolism may be important in the biodegradation of complex organics in hazardous waste-contaminated soils. The chemical environment of petroleum permits a variety of microorganisms to attack compounds enzymatically that would not otherwise be degraded, as they grow on the multitude of potential primary substrates in the oil (Atlas, 1981). Many complex branched and cyclic hydrocarbons are removed as environmental contaminants after oil spills, as a result of co-oxidation. Structures with four or more condensed rings have been shown to be attacked by co-oxidation or as a result of commensalism. Cometabolism undoubtedly accounts for the disappearance of compounds in the saturate fraction (Westlake, Jobson, and Cook, 1978).

Soil microorganisms capable of cometabolically degrading a given xenobiotic may not necessarily follow the normal enrichment sequence of events, i.e., proliferation at the expense of the xenobiotic substrate (Horvath, 1972). Therefore, unique techniques may be necessary to isolate such organisms. Such techniques could possibly utilize as a primary substrate a chemical analog that would permit enrichment of microbial populations having the necessary cometabolic requirements to degrade the xenobiotic. If manometric experiments repeatedly show oxygen consumption to be well in excess of the endogenous respiratory rate, the occurrence of cometabolism is indicated. It should be noted, however, that experimental conditions may not truly represent what is occurring in nature, and caution should be exercised in extrapolating the results (Alexander, 1994).

Counts of bacteria capable of utilizing or co-oxidizing a compound can be obtained by spreading water and sediment samples on nutrient agar plates containing the compound as sole carbon source and on plates containing the compound and one alternative carbon source (Shiaris and Cooney, 1981). Colonies producing on the first plates are classified as utilizers, and those producing clear zones only on the second are considered co-oxidizers. Alternative substrates have been found to be effective in this order:

Yeast extract/peptone > glucose > benzoic acid > oil/kerosene

Laboratory evidence clearly indicates that under proper conditions, many potentially hazardous organic compounds can be biodegraded in the ground, even when present at very low concentrations (Bitton and Gerba, 1985). However, evidence also suggests that there may be a threshold concentration for a given contaminant, below which it will not be biodegraded. This would be expected, if the energy derived from low concentrations of the substrate was inadequate for maintenance of the bacterial cell or if higher concentrations were required to activate the transport and metabolic systems of the cell (Boethling and Alexander, 1979a). Actually, a chemical that is cometabolized at one concentration may be mineralized in the same environment at another concentration, or it may be mineralized in one environment and cometabolized in another (Alexander, 1994).

While the microbial community near a pollutant spill might be primed with small amounts of the pollutant itself to promote a more rapid microbial response (Spain and Van Veld, 1983), it is also possible to add a different compound at a relatively high concentration (the primary substrate) to the pollutant present at trace concentrations (the secondary substrate) with biotransformation of the latter, if the organisms are able to transform both substrates (Bouwer and McCarty, 1984). Thus, by adding *more* of the given contaminant (or other less- or nonhazardous similar substance) the threshold can be exceeded, since more inducible enzyme(s) is (are) produced (Boethling and Alexander, 1979a). An example is the addition of fulvic acid to stimulate degradation of benzene, toluene, or phenols. A small amount of an inducer chemical, to which an organism has become adapted, can also be added into a solution of a more persistent compound to facilitate a more rapid degradation of the compound (Kaufman and Plimmer, 1972). The growth rate on the xenobiotic compound appears to be limited by the energy generation rate (Babel, 1994). Adding a physiologically similar or functionally homologous but more energy-rich substrate (or adding a victim substrate) can improve efficiency by balancing the carbon/energy ratio.

Care must be used in selecting such analogs, since they or their degradation products could be hazardous (Sims and Bass, 1984). The analogs may be applied as solids, liquids, or slurries and mixed thoroughly with the contaminated soil, where feasible. Fertilization may be necessary to maintain microbial activity, and controls may have to be implemented to prevent drainage and erosion problems (Kaufman and Plimmer, 1972). They should be added in amounts large enough to stimulate microbial activity, but not enough to be toxic or to adversely affect public health or the environment (Sims and Bass, 1984). Consideration of the use of inducer molecules to enhance microbial degradation must be accompanied by an understanding of both the mechanisms involved and the ecological acceptability of adding the inducing substrate (Kaufman and Plimmer, 1972). Treatability studies are required to determine the feasibility, loading rate, and effectiveness of the analogs.

Many examples of cometabolism have been reported. There have been co-oxidative conversions of aromatic and cyclic compounds (Cooney, 1980). For instance, the enzymes for cyclohexane oxidation are inducible by cyclohexane in a *Nocardia* sp. A *Nocardia* sp. growing on hexadecane converted *p*-xylene to *p*-toluic acid and 2,3-dihydroxy-*p*-toluic acid (Ooyama and Foster, 1965). A *Nocardia* sp. oxidized *p*-xylene to *p*-toluic acid under co-oxidation conditions (Jamison, Raymond, and Hudson, 1976). *Mycobacterium convolutum* R-22 can completely mineralize cyclohexane only when an inducer, such as an *n*-alkane is present to induce the necessary hydroxylase (Perry, 1984). When two components of gasoline were combined, one for growth and one as a co-oxidizable substrate, a *Pseudomonas* sp. was able to co-oxidize 2-methylheptane, *o*-xylene, and ethylcyclohexane with hexane as the growth substrate (Jamison, Raymond, and Hudson, 1976). When the nonmetabolizable compounds (cyclohexane, *p*-xylene, and toluene) were added to the metabolizable substrate, *n*-hexadecane, the increased oxygen consumption was attributed to a stimulation of the oxidation of hexadecane by the nonoxidizable compounds (Cooney and Walker, 1973). Toluene-grown strains of *Pseudomonas* and *Achromobacter* oxidized, without lag, benzene, catechol, 3-methyl-catechol, benzyl alcohol, and, more slowly, *o*- and *m*-cresol, but not benzaldehyde or benzoic acid (Claus and Walker, 1964). The mutual adaptations to use benzene and toluene suggest that enzymes with similar activities may be involved in the metabolism of the two compounds.

An alkane-utilizing strain of *Mycobacterium vaccae* (JOB5) co-oxidizes a variety of alicyclic hydrocarbons to the corresponding ketones (Hou, 1982). A strain of *P. aeruginosa* grown on *n*-heptane co-oxidizes cycloalkanes to their corresponding alcohols. Methane-grown microorganisms will oxidize alicyclic hydrocarbons. An *n*-octadecane-grown *Nocardia* sp. was found to co-oxidize *n*-butylcyclohexane to cyclohexane acetic acid. A *Nocardia* sp. and a *Pseudomonas* sp. isolated on gasoline could co-oxidize a number of cyclic and aromatic compounds (Jamison, Raymond, and Hudson, 1976). A *Pseudomonas* sp. was able to grow on *p*-xylene in the presence of toluene (Chang, Voice, and Criddle, 1993). Competitive inhibition occurred when BTX substrates were combined. The complex enzymatic interactions within a single organism and among a mixture of microbes will all contribute to the transformations.

It was discovered that *P. methanica* cometabolically oxidizes ethane, propane, and butane while the organism metabolizes its growth substrate, methane (Foster, 1962). Soil aerobic organisms that can grow on aliphatic hydrocarbons, such as natural gas or propane, could possibly be added to the soil while methane or propane is pumped into the ground for degradation of a variety of chlorinated solvents, including TCE.

Cometabolism is an important process in the breakdown of PAHs (Texas Research Institute, Inc., 1982). It may play a major role in degradation of PAHs, such as phenanthrene, in polluted sediments (Shiaris and Cooney, 1981). Many compounds have been described as recalcitrant molecules because of repeated failures to isolate organisms capable of utilizing them as sole sources of carbon and energy for growth (Horvath, 1972). However, cometabolism of these materials has been shown to occur under laboratory conditions and may be an important process in their removal from the environment.

Keck, Sims, Coover, Park, and Symons (1989) report that four- and five-ring PAHs disappear more rapidly when the soil contains a mixture of contaminants, while three-ring compounds are degraded equally whether present as single constituents or in a complex mixture. This strongly suggests cooxidation is occurring and that this phenomenon could be used as a bioremediation tool for PAH-contaminated soils. The rate of cometabolism of benzo(a)pyrene is significantly increased with enrichment of the soil with phenanthrene as an analog (Sims and Overcash, 1981). Similar results are obtained by using naphthalene and, especially, phenanthrene as growth substrates for the nongrowth substrates of pyrene, 3,4-benzopyrene, 1,2-benzanthracene, 1,3,5,6-dibenzanthracene, with approximately 35% of the nongrowth substrate remaining after 4 weeks (McKenna, 1977). These compounds are also cometabolized by a mixed culture of flavobacteria and pseudomonads in the presence of either naphthalene or phenanthrene (Cerniglia, 1984). Minimally mutagenic phenanthrene can be used to increase the rate of degradation of highly mutagenic benzo(a)pyrene (Sims and Overcash, 1981). Biphenyl has also been used to stimulate cometabolic degradation of PCBs (Furukawa, 1982; Pierce, 1982d).

Since cometabolism rarely results in the complete oxidation of xenobiotics, it may allow accumulation of transformation products, which may be either more or less toxic than the original substance (de Klerk and van der Linden, 1974; Perry, 1979). Incompletely degraded compounds include saturated hydrocarbons, halogenated hydrocarbons, many pesticides, and single-ringed aromatic hydrocarbons (Horvath and Alexander, 1970; Horvath, 1972; Alexander, 1977). Although the compounds may not be completely mineralized, their toxicity may be removed by the partial degradation (Horvath, 1972).

Natural microbial populations present in the environment may be able to utilize by-products of the cometabolic process and completely degrade such substances through contribution of cometabolic reactions from the entire community (Horvath, 1972; de Klerk and van der Linden, 1974; Perry, 1979; Alexander, 1994) (see Section 5.2.2.3). Genetic engineering can construct organisms with enzymes capable of both the initial cometabolism and those that allow growth on the products of the initial sequence (Rubio, Engesser, and Knackmuss, 1986; Ramos, Wasserfallen, Rose, and Timmis, 1987).

Table 5.14 lists several species that have been shown to possess a cometabolic type of metabolism (Horvath, 1972). In Table 5.15 is a summary from the literature of a number of organic compounds that are subject to cometabolic action, the corresponding chemical analogs, organisms capable of cometabolizing them, and the resultant metabolic products. Products that can accumulate from incomplete metabolism of certain organic substances being cometabolized are given in Table 5.16 (Horvath, 1972).

Cometabolism can also occur anaerobically. Using nitrate as an electron acceptor, *o*-xylene can be transformed cometabolically by a mixed culture of denitrifying bacteria using toluene as the primary carbon source (Arcangeli and Arvin, 1995). However, a toluene concentration of >1 to 3 mg/L reduces the *o*-xylene removal rate, and concentrations of *o*-xylene above 2 to 3 mg/L in turn inhibits toluene degradation. The two compounds may be competing for the same enzyme.

Table 5.14 Microbial Species Exhibiting the Phenomenon of Cometabolism

Achromobacter sp.	*Micrococcus* sp.
Arthrobacter sp.	*Nocardia erythropolis*
Aspergillus niger	*Nocardia* sp.
Azotobacter chroococcum	*Pseudomonas fluorescens*
A. vinelandii	*P. methanica*
Bacillus megaterium	*P. putida*
Bacillus sp.	*Pseudomonas* sp.
Brevibacterium sp.	*Streptomyces aureofaciens*
Flavobacterium sp.	*Trichoderma viride*
Hydrogenomonas sp.	*Vibrio* sp.
Microbacterium sp.	*Xanthomonas* sp.
Micrococcus cerificans	

Source: From Horvath, R.S. *Bacteriol. Rev.* 36:147–155. American Society of Microbiology, Journals Division. 1972. With permission.

Table 5.15 Organic Compounds, Analogs/Growth Substrates, Microorganisms, and/or Products of Co-oxidation

Organic Compound	Analog/Cosubstrate	Organism	Products
n-Alkenes[b]	—	*Nocardia* sp.	—
1,2-Benzanthracene,[s] 1,3,5,6-dibenzanthracene	Naphthalene, phenanthrene	Flavobacteria or pseudomonads	—
Benzo(a)pyrene[d]	Phenanthrene	—	—
Benzo(a)pyrene[t]	Fluoranthene	*Pseudomonas paucimobilis* EPA 505	Polar metabolites
3,4-Benzopyrene[s]	Naphthalene, phenanthrene	Flavobacteria or pseudomonads	—
Biphenyl[j]	—	*Nocardia corallina*	—
2-Butanol[m]	—	—	2-Butanone
Cyclohexane[i]	n-Hexadecane	*Cladosporium resinae*	—
Cyclohexane[k]	—	*Pseudomonas* No.1, *Pseudomonas* No.2	Cyclohexanol
Cyclohexane[r]	Biotin	*Nocardia* + pseudomonad	—
Cycloparaffins from propane[b]	—	*Mycobacterium vaccae*	Cycloalkanones
Diethylbenzene[j]	—	*Nocardia corallina*	—
Dimethylnaphthalene[j]	—	*Nocardia corallina*	—
4,6-Dinitro-o-cresol[f]	Sucrose	—	—
Dodecylcyclohexane[p]	—	*Rhodococcus rhodochrous* + *Arthrobacter* sp.	—
Ethane, propane, butane[a]	Methane	*Pseudomonas methanica*	—
Ethylbenzene[h]	Hexadecane	*Nocardia* sp.	Phenylacetic acid
Ethylcyclohexane[c]	Hexane	*Pseudomonas* sp.	Cyclohexane acid
Ethylcyclohexane[c]	Hexadecane	*Nocardia* sp.	Cyclohexane acid
2-Methylheptane[c]	Hexane	*Pseudomonas* sp.	Ketones, aldehydes
Paraffins[g]	Diethoxymethane	*Pseudomonas aeruginosa*	—
Phenol[n]	—	—	cis,cis-Muconate
Propane[l]	—	—	Propionate, acetone
Pyrene[s]	Phenanthrene, naphthalene	Flavobacteria or pseudomonads	—
Pyrene, 3,4-benzopyrene, 1,2-benzanthracene, 1,3,5,6-dibenzanthracene[e]	Phenanthrene, naphthalene	—	—
Tetralin[j]	—	*Nocardia corallina*	—
Tetramethylbenzene[j]	—	*Nocardia corallina*	—
Toluene[i]	n-Hexadecane	*Cladosporium resinae*	—
Trimethylbenzene[j]	—	*Nocardia corallina*	—

Table 5.15 (continued) Organic Compounds, Analogs/Growth Substrates, Microorganisms, and/or Products of Co-oxidation

Organic Compound	Analog/Cosubstrate	Organism	Products
o-Xylene[c]	Hexane	*Pseudomonas* sp.	o-Toluic acid
	Hexadecane	*Nocardia* sp.	o-Toluic acid
o-Xylene[b]	—	*Nocardia* sp.	o-Toluene
o-Xylene[o]	—	—	o-Toluic acid
p-Xylene[c]	Hexadecane	*Nocardia* sp.	p-Toluic acid
p-Xylene[q]	Toluene	*Pseudomonas* sp.	—
p-Xylene[h]	Hexadecane	*Nocardia* sp.	a,a'-Dimethylmuconic acid
p-Xylene[i]	n-Hexadecane	*Cladosporium resinae*	—

References:
a = (Foster, 1962)
b = (Hou, 1982)
c = (Jamison, Raymond, and Hudson, 1976)
d = (Sims and Overcash, 1981)
e = (McKenna, 1977)
f = (Slonim, Lien, Eckenfelder, and Roth, 1985)
g = (Horvath, 1972)
h = (Gibson, 1977)
i = (Cooney and Walker, 1973)
j = (Jamison, Raymond, and Hudson, 1971)
k = (de Klerk and van der Linden, 1974)
l = (Leadbetter and Foster, 1959)
m = (Patel, Hou, Laskin, Derelanko, and Felix, 1979)
n = (Knackmuss and Hellwig, 1978)
o = (Raymond, Jamison, and Hudson, 1967)
p = (Feinberg, Ramage, and Trudgill, 1980)
q = (Chang, Voice, and Criddle, 1993)
r = (Sterling, Watkinson, and Higgins, 1977)
s = (Cerniglia, 1984)
t = (Siddiqi, Ye, Elmarakby, Kumar, and Sikka, 1994)

Table 5.16 Organic Substances Subject to Cometabolism and Accumulated Products

Substrate	Product
Ethane	Acetic acid
Propane	Propionic acid, acetone
Butane	Butanoic acid, methyl ethyl ketone
m-Chlorobenzoate	4-Chlorocatechol, 3-chlorocatechol
o-Fluorobenzoate	3-Fluorocatechol, fluoroacetate
2-Fluoro-4-nitrobenzoate	2-Fluoroprotocatechuic acid
4-Chlorocatechol	2-Hydroxy-4-chloro-muconic semialdehyde
3,5-Dichlorocatechol	2-Hydroxy-3,5-dichloro-muconic semialdehyde
3-Methylcatechol	2-Hydroxy-3-methyl-muconic semialdehyde
o-Xylene	o-Toluic acid
p-Xylene	p-Toluic acid, 2,3-dihydroxy-p-toluic acid
Pyrrolidone	Glutamic acid
Cinerone	Cinerolone
n-Butylbenzene	Phenylacetic acid
Ethylbenzene	Phenylacetic acid
n-Propylbenzene	Cynnamic acid
p-Isopropyltoluene	p-Isopropylbenzoate
n-Butyl-cyclohexane	Cyclohexaneacetic acid
2,3,6-Trichlorobenzoate	3,5-Dichlorocatechol
2,4,5-Trichlorophenoxy-acetic acid	3,5-Dichlorocatechol
p,p'-Dichlorodiphenyl methane	p-Chlorophenylacetate
1,1-Diphenyl-2,2,2-trichloroethane	2-Phenyl-3,3,3-trichloropropionic acid

Source: From Horvath, R.S. *Bacteriol. Rev.* 36:146–155. American Society of Microbiology, Journals Division. 1972. With permission.

Sucrose was used as a cometabolite in the degradation of 4,6-dinitro-o-cresol (DNOC), a phenolic priority pollutant, in wastewater, by use of an anaerobic recycle fluidized-bed reactor as a pretreatment stage, followed by an activated-sludge reactor as the aerobic treatment stage (Slonim, Lien, Eckenfelder, and Roth, 1985). There appeared to be a relationship between the sucrose concentration and degradation of the compound. A ratio of sucrose to DNOC of 2:1 or higher resulted in a 95 to 100% conversion of DNOC, while a lower ratio did not permit cometabolism or degradation of DNOC.

Co-oxidation and stationary transformation techniques are promising areas of hydrocarbon biotechnology for the future (Hou, 1982). These are systems actually operating in nature. When the transformation product of this process is not itself hazardous or is degradable by other organisms, analog enrichment may be an effective treatment for contaminated soil (Alexander, 1981).

5.2.3.1 Diauxie Effect

The opposite of cometabolism is a sparing, or diauxic, phenomenon, which occurs when a compound cannot be degraded in the presence of another compound (Atlas, 1981). The metabolic pathways of degradation are not altered, but the enzymes necessary for metabolic attack of a particular hydrocarbon may not be produced. This can lead to persistence of these hydrocarbons in a petroleum mixture.

Both processes of co-oxidation and sparing can occur within the context of a petroleum spillage (Atlas, 1981). When a microorganism with a broad substrate range is offered more than one type of organic substrate, it will not attack the substrates simultaneously, but rather in a definite sequence (Bartha and Atlas, 1977). The diauxie effect, where the presence of one compound will inhibit the degradation of another, may determine whether or not the hydrocarbon components of an oil spill are degraded and, if so, in what order.

A *Brevibacterium erythrogenes* strain was capable of utilizing pristane and other branched alkanes only in the absence of *n*-alkanes (Pirnik, Atlas, and Bartha, 1974). *B. erythrogenes* utilizes *n*-alkanes by a monoterminal β-oxidation sequence, but degrades isoalkanes by diterminal oxidation. *Rhodotorula rubra* preferentially utilizes higher *n*-alkanes to the lower *n*-alkanes and leaves residual aromatic and asphaltic fractions when degrading oil sludge (Shailubhai, Rao, and Modi, 1984a). The common phenomenon of the *n*-alkane components of an oil spill disappearing before the isoalkanes and other hydrocarbon classes show substantial biodegradation strongly suggests that such diauxic regulatory mechanisms can apply not only to pure cultures, but most likely also to a mixed microbial community (Bartha and Atlas, 1977).

The presence of nonhydrocarbon substrates may also repress the inductive synthesis of enzymes required for hydrocarbon oxidation (van Eyk and Bartels, 1968). Utilization of hexadecane is reduced by acetate (LePetit and Tagger, 1976). Addition of glucose to lake water repressed hexadecane utilization by its microbial community in a diauxic manner (Bartha and Atlas, 1977). In another study, the rate of mineralization of organic compounds in trace concentrations was found to be enhanced by the addition of inorganic nutrients, arginine, or yeast extract, but reduced by addition of glucose (Rubin and Alexander, 1983). Degradation of oil can be inhibited or delayed by the presence of peptone (Gunkel, 1967; LePetit and Barthelemy, 1968; van Eyk and Bartels, 1968). Sparing effects play a significant role in determining rates of oxidation of various components and, thus, affect the overall weathering process in complex mixtures (Atlas, 1981). This effect should be considered when selecting a substrate as a cometabolite.

Escherichia coli B was entrapped in κ-carrageenan and the diauxie phenomenon of the immobilized cells tested with lactose and glucose (Ariga, Saito, and Sano, 1995). Unlike with free cells, diauxie in immobilized cells depended on the cell concentration in the gel beads. At a low cell concentration, lactose in the medium containing both sugars was not utilized until the glucose was consumed. At a high cell concentration, both sugars were utilized at the same time, suggesting that the glucose concentration in the gel beads was low enough for the induction of lactose assimilation. These results are explained in terms of catabolite repression and the diffusion of the sugars.

A model for the diauxic growth of immobilized cells has been developed for estimating the dynamic behavior of cell growth and substrate consumption in the microbial degradation of mixed substrates by immobilized cells (Nakamura, Origasa, and Sawada, 1994). Microorganisms without inducible enzymes grow primarily on and near the surface of the gel beads, while those with inducible enzymes grow mainly in the inner part of the gel beads near the surface because of the exhaustion of the glucose substrate.

5.2.4 APPLICATION OF CELL-FREE ENZYMES

Low-molecular-weight compounds are generally believed to be degraded by intracellular rather than extracellular enzymes (Alexander, 1994). Organic matter in soil may bind enzymes or protect contaminants against hydrolysis by extracellular enzymes. However, the initial stages in the metabolism of some molecules, especially those of high molecular weight, are catalyzed by extracellular enzymes, which leads to transformation of these compounds.

There is evidence for the existence in soil of extracellular enzymes capable of degrading xenobiotics or their degradation products (Kaufman, 1983). Enzymatic degradation of organics with cell-free enzymes holds potential as a possible *in situ* treatment technique (U.S. EPA, 1985a). The sorption of enzymes by organic matter in soil may even furnish an approach to bioremediation using immobilized enzymes (Alexander, 1994). Industry commonly employs crude or purified enzyme extracts, in solution or immobilized on glass beads, resins, or fibers (or porous silica beads; Munnecke, 1981) to catalyze a variety of reactions, including the breakdown of carbohydrates and proteins (Sims and Bass, 1984; U.S. EPA, 1985a). Organic wastes are also amenable to this treatment.

Enzymatic methods show promise for removing aromatic compounds from high-strength industrial wastewater (Maloney, Manem, Mallevialle, and Fiessinger, 1985). Enzymatic oxidative coupling may be useful in eliminating aromatics that are not well removed in biological or physical water treatment. Wastewaters containing aromatic compounds are treated with horseradish peroxidase and hydrogen peroxide (Alberti and Klibanov, 1981). The resulting high-molecular-weight compounds are less soluble in water and can be removed by sedimentation or filtration.

Microbial enzymes that have the ability to transform hazardous compounds to nonhazardous or more-labile products could possibly be harvested from cells grown in mass culture and applied to contaminated soils (Sims and Bass, 1984). Some of the enzyme preparations are quite stable during storage and could be available for emergency responses to spills (Alexander, 1994). Theoretically, if enzymes remain active in the soil, they could quickly transform hazardous compounds, and the potential level of treatment is high. However, there is little information available on the use of this technique in soil or application in the field. Its reliability is unknown.

There are a number of advantages in using cell-free extracts. Enzyme activity can often be preserved in environments that would be inhospitable to the organisms, e.g., in soils with extremes of pH and temperature, high salinity, or high solvent concentrations. Increased mobility of a compound by extracellular enzymes in sites that inhibit microorganisms may render the material more susceptible to decomposition by the soil microflora present in less hostile areas (Munnecke, Johnson, Talbot, and Barik, 1982).

There are also limitations in using the extracts. For an enzyme to function outside the cell in the soil environment, it must not require cofactors or coenzymes. This would limit the application of many enzymes. Most important (especially *in situ*) is enzyme stability — they must be fairly stable in extracellular environments. Enzymes might be leached out of the treatment zone, and they might be inactive or have lower activity if they are bound to clay or humus in the soil. Outside of biochemical and environmental constraints, logistics and costs for producing enzymes in effective quantities may limit use of this concept.

One concern in the use of this technique for drinking water treatment is the nature of the products of the oxidative coupling (Maloney, Manem, Mallevialle, and Fiessinger, 1985). In one study, biphenyls accounted for 3% of the initial carbon concentration (Schwartz and Hutchinson, 1981). These may be the predominant incomplete polymerization products. They might not be removed by sedimentation or filtration and could pass through the treatment process in their altered (polymerized) form. Another problem in drinking water treatment is the presence of competing or interfering compounds (Maloney, Manem, Mallevialle, and Fiessinger, 1985). Raw water supplies usually have background organic carbon composed mainly of humic acids (McCarty, 1980). It has been suggested that humic acids may deactivate peroxidase (Pflug, 1980).

It is of interest that extracellular peroxidases have been found in soil (Kaufman, 1983). These could presumably be involved in a vast array of soil metabolic reactions affecting xenobiotic residues. The peroxidase–peroxide system is effective in eliminating chlorinated phenols from drinking water supplies, but does not remove their breakdown products from the water (Maloney, Manem, Mallevialle, and Fiessinger, 1985). Further work is necessary to determine if these by-products present a potential risk for human health and if they are removed in other unit processes.

An example of breakdown of a compound by a cell-free extract is the use of a naphthalene oxygenase purified from cells of *Corynebacterium renale* grown on naphthalene as the sole source of carbon and energy (Dua and Meera, 1981). The enzyme has a molecular weight of approximately 99,000 and forms *cis*-1,2-dihydroxy-1,2-dihydronaphthalene as the predominant metabolite from naphthalene.

The Cetus Company developed a novel multienzyme process for the oxidation of propylene (Hou, 1982). The first enzyme reaction converts the olefin to halohydrin in the presence of halide, hydrogen peroxide, and haloperoxidase. The latter can be obtained from horseradish, seaweed, or *Caldariomyces*. In the second reaction, propylene halohydrin is transformed to propylene oxide by halohydrin epoxidase or by whole cells of a *Flavobacterium* sp.

The soluble form of methane monooxygenase (sMMO) from the obligate methanotroph, *Methylosinus trichosporium* OB3b can oxidize a wide range of organic compounds, often not closely related to the natural substrate, methane (Sullivan and Chase, 1994). The soluble form of the enzyme can be induced in a low-concentration copper medium. The optimum pH for the transformation of naphthalene is 8.1 and the optimum temperature, 37°C.

Toluene-permeabilized cells having β-galactosidase activity as an enzyme can be encapsulated in PVA capsules (Ariga, Itoh, Sano, and Nagura, 1994). The long-term stability of the enzyme activity of the capsules was better than that of free cells and was improved at room temperature.

Fungal enzymes, such as laccase, peroxidase, and tyrosinase, are important in catalyzing the transformation of various aromatic compounds in the environment (Bollag and Dec, 1995). The enzyme-mediated oxidative coupling reaction results in covalent binding of chlorinated phenols and anilines to soil organic matter or polymerization of the substrates in aquatic systems, accompanied by a denitrification effect. The decontamination effect could be enhanced by optimization of the reaction conditions and immobilization of enzymes on solid materials.

White-rot fungi produce extracellular peroxidases that are involved in the initial attack of many PAHs (Field, Heessels, Wijngaarde, Kotterman, de Jong, and de Bont, 1994). *Phanerochaete chrysosporium* and *Bjerkandera* sp. strain BOS55 are excellent examples of PAH degraders, the latter of which has been shown to be active against anthracene and benzo(a)pyrene. See Section 3.2.1.5.2 for a discussion of the mechanism of this action.

FyreZyme™ is a proprietary bioremediation enhancing agent containing extracellular enzymes, microbial nutrients, and bioemulsifiers (Meaders, 1994). It is diluted from concentrate to a 4% aqueous solution. The enzymes and nutrients might initiate abiotic transformation of petroleum products, while stimulating exponential microbial growth. This may favor petroleum degraders. The bioemulsifiers increase bioavailability of soil-bound contaminants. The product appears to promote rapid biodegradation of petroleum products, including PAHs.

Cell-free enzymes for treating hazardous waste constituents are not currently in bulk production (Munnecke, Johnson, Talbot, and Barik, 1982). Only eight companies accounted for 90% of worldwide production of industrial enzymes in 1981, five of the firms in Western Europe. Only 16 enzymes (primarily amylases, proteases, oxidases, and isomerases) accounted for 99% of the 1981 market. This suggests that specialized enzyme production, even on a large scale, may be quite expensive. Prices for bulk enzyme materials range from $1.45 to 164/lb. If the enzyme could be produced through chemical synthesis, it would be much less expensive than if produced by microorganisms in fermenters.

5.2.5 ADDITION OF ANTIBIOTICS

Microbial predators also play a role in the degradation process (Texas Research Institute, Inc., 1982). There are eukaryotic organisms in the soil, and these organisms graze on bacteria and fungi or feed on detrital matter and associated microflora (JRB Associates, Inc., 1984; Sinclair and Ghiorse, 1985). Protozoa, insects, nematodes, and other worms affect the decomposition process by controlling bacterial or fungal population size through grazing (Bryant, Woods, Coleman, Fairbanks, McClellan, and Cole, 1982) (see Section 5.2.1.6).

Many species of hydrocarbon utilizers have been found to be ingested by a large number of ciliate and other cytophagic protozoans (Texas Research Institute, Inc., 1982). These higher organisms may reduce the microbial population from 10^7 to 10^2 bacteria/mL (ZoBell, 1973). Protozoa present in sewage can halt the bacterial action on specific pollutants in treatment plants by devouring the bacteria. Grazing of hydrocarbon-degrading microorganisms by protozoa can lengthen the acclimation period prior to biodegradation in wastewaters (Alexander, 1986). Protozoan grazing is responsible for most of the acclimation period for the mineralization of organic compounds in sewage (Wiggins and Alexander, 1986) and in wastewaters (Alexander, 1986).

This problem can be eliminated with the use of antibiotics that target the protozoa (Alexander, 1986). With eukaryotic inhibitors (cycloheximide and nystatin) added to untreated sewage, the counts of protozoa can be lowered to 60/mL and the acclimation period reduced from 11 to 1 day. With no inhibitors, protozoan counts can increase to 3.5×10^4/mL in 1 day (Wiggins and Alexander, 1986).

5.2.6 USE OF AEROBIC/ANAEROBIC CONDITIONS

Aerobic or anaerobic conditions can be selected to encourage degradation by the most favorable process for the compounds of concern. A combination of the two may sometimes allow the most complete biodegradation to occur. Sections 5.1.4.5 and 5.1.4.6 discuss how the soil oxygen content can be modified to achieve optimum aerobic conditions, while Section 5.1.4.7 covers methods for obtaining anaerobic conditions, and Section 5.1.4.8 deals with sequential or simultaneous application of both.

5.2.7 USE OF BIOSORPTION/BIOACCUMULATION/BIOCONCENTRATION

Microorganisms may ingest and utilize oil substances without obvious hydrocarbon degradation (Bausum and Taylor, 1986). The more lipophilic organisms, e.g., mycobacteria, may be preferentially attracted to the oil film, with the more hydrophilic remaining in the microlayer of organisms within the aqueous phase (Kjelleberg, Norkrans, Lofgren, and Larsson, 1976). Small microdroplets (micelles) may then be formed and ingested by the organisms in closer contact with the oil. Microorganisms utilizing hydrocarbons can develop cytoplasmic inclusions containing such substances. This has been reported in yeasts, filamentous fungi, and a strain of *Acinetobacter* (Finnerty, Kennedy, Lockwood, Spurlock, and Young, 1973). Hydrocarbon accumulation can even occur in protozoa (Andrews and Floodgate, 1974). Inside the inclusions, or microbodies, the oil molecules are in close contact with the membrane enzymes (Bausum and Taylor, 1986).

The extent of biosorption by the microorganisms depends upon the type of biomass (Selvakumar and Hsieh, 1988). Activated sludge biomass shows a relatively higher uptake capacity than nitrifying bacteria. The adsorption of organic compounds is related to the organic carbon content of the sorbent.

Bioaccumulation of petroleum substances may be accompanied by biotransformation (Bausum and Taylor, 1986). In a phenomenon similar to simple bioaccumulation, hydrocarbons are used to form normal microbial cell components with only very limited chemical alteration. This has been reported for a number of microorganisms with regard to their fatty acid composition. In a *Mycobacterium* growing on *n*-alkanes from C_{13} to C_{17}, the major fatty acid in the cells was of the same chain length as the substrate (Dunlap and Perry, 1967). The monoterminally oxidized substrates appeared to be utilized without further degradation. This has been observed with utilization by *Mycobacterium* of 1-alkanes C_{14} to C_{18}, 8-heptadecene, and the branched alkanes, 2- and 3-methyl octadecane (King and Perry, 1974). This was extended to the filamentous fungi, *Cunninghamella elegans* and *Penicillium zonatum* with regard to *n*-alkanes and 1-alkenes (Cerniglia and Perry, 1974). It has also been reported in *Acinetobacter* (Patrick and Dugan, 1974) and in the yeast *Candida* (Gill and Ratledge, 1973).

The capability of microorganisms to biosorb organic compounds has been combined with the use of aerobic or anaerobic reactors into a scheme for treating readily degradable and refractory compounds, the latter being adsorbed by photosynthetic organisms (Section 2.1.2.2.1.4).

5.2.8 USE OF VEGETATION

Bioremediation of petroleum-contaminated soil can be enhanced in the presence of vegetation (Lee and Banks, 1993). Microbial numbers are substantially higher in soil contaminated with polynuclear aromatic and aliphatic hydrocarbons in the presence of alfalfa plants, than in soil without these plants. This indicates that plant roots enhance microbial populations in contaminated soil. Higher plants accumulate metals in their roots and shoots and serve as insulation to reduce heat loss during winter. See also Sections 2.1.2.1.6, 2.2.2.11, 3.3, 5.1.2, 5.1.4, 5.1.6, and 5.2.1.6 for more information.

5.2.9 OTHER MICROBIAL APPLICATIONS

Microbially based enzymatic reactions that bind the pollutant into the soil matrix may be a good method to immobilize pollutants and prevent future environmental deterioration (Bollag, 1992).

5.3 CONTAMINANT ALTERATION

5.3.1 USE OF SURFACTANTS

Oil–water interfacial tension is one of the most important factors affecting the biodegradation of crude oil (Setti, Pifferi, and Lanzarini, 1995). A surfactant is a surface-active agent that lowers the surface tension of a liquid (Alexander, 1994). If it improves the stability of an emulsion by reducing the surface or interfacial tension, it is called an emulsifier. Most emulsifying agents are surfactants, but not all surfactants are emulsifying agents (Becher, 1965). Most microbial emulsifying agents are surfactants. Surfactants may be described as synthetic or natural (Nash, Traver, and Downey, 1987). Synthetic surfactants are those that have been made by humans by chemical processes and are available commercially. The natural surfactants are produced by microorganisms and are fatty acid and ester compounds that are by-products of the biological breakdown of organic material.

Natural and synthetic surface-active agents promote the wetting, solubilization, and emulsification of various types of organic chemicals (Amdurer, Fellman, and Abdelhamid, 1985). During wetting, surfactants decrease the interfacial tension between the aqueous and solid phases (soil), allowing the water to preferentially wet the soil, thereby displacing the contaminant (U.S. EPA, 1986). Surfactants can enhance the solubility of some contaminants. They may also enhance the emulsification, or dispersion of an insoluble organic phase within the aqueous phase, of a contaminant.

Surfactants are organic molecules, which can be either cationic, anionic, or nonionic (JRB Associates, 1984). The charge of the polar moiety specifies the type of the surfactant (van Ginkel, 1996). Cationic surfactants contain a quaternary ammonium group that has a positive charge. Anionic surfactants receive a negative charge from a sulfonate or sulfate constituent. The nonionic surfactants have no charge, and are primarily polymerization products of 1,2-epoxyethane. Amphoteric surfactants contain both positive and negative charges on the same molecule.

Surfactants accumulate at gas/liquid, liquid/liquid, and liquid/solid interfaces (van Ginkel, 1996). They are composed of a hydrocarbon part, called the hydrophobic or lipophilic group, and the hydrophilic, or water-soluble, group. The hydrophobic portion is generally a C_8 to C_{22} alkyl chain or an alkylaryl group, and the alkyl chain could be linear or branched.

Surfactants can promote the mobilization of hydrophobic contaminants from unsaturated soils (Ellis and Payne, 1984). Surfactants would be most effective in promoting the mobilization of organic compounds of relatively low water solubility and high lipid solubility (high K_{ow} values) (JRB Associates, 1984). The apparent water solubilities of nonionic organic contaminants can be substantially increased in the presence of conventional nonionic (e.g., Triton, Brij), cationic (e.g., cetyltrimethyl-ammonium bromide), and anionic (e.g., sodium dodecyl sulfate) surfactants (Sun and Boyd, 1993). The level of the rate of enhancement is specific for each surfactant, compound, and microorganism combination (Churchill and Griffin, 1992).

Both an oil and a water phase are present in petroleum products (Genner and Hill, 1981). Nutrients, inhibitors, and metabolic products are always partitioned between these phases. Microorganisms proliferate in the water phase and migrate into the hydrocarbon phase, where they usually die. Experiments with *Penicillium spiculisporum* showed that nearly 100% of the cells adhered to hydrophilic substrates, while most of the cells were suspended freely in culture broth when a hydrophobic substrate was used (Ban and Yamamoto, 1993).

There are many chemical oil dispersants that could be used as surfactants (Canevari, 1971). Successful chemical enhancement of oil biodegradation is dependent upon the particular emulsifying agent (Robichaux and Myrick, 1972). The problem with dispersing oil, even if it accelerates soil biodegradation, is that dispersed oil and many oil dispersants are more toxic than undispersed oil (Shelton, 1971). Thus, oil dispersion is not generally an ecologically acceptable treatment for stimulating oil biodegradation. On the other hand, some emulsifying agents may not produce these undesirable effects.

Surfactants are mixtures of homologues and isomers (Schoeberl, 1996). The biodegradation and the ecotoxicity of the different homologues and isomers are different. The surfactant residues and surfactant catabolites are significantly less toxic than the original surface-active substances. This fact is not taken into account by the official ecotoxicity test methods. If ecotoxicity data are used for an Environmental Risk Assessment, they must be determined by realistic test systems, as described by Schoeberl.

Several types of emulsifiers have been isolated and characterized to some degree (Cooney, 1980). They include glycolipids, fatty alcohols, fatty acids, polysaccharides, polysaccharide-proteins, peptide, peptidolipids, protein–lipid–carbohydrate complexes, and mixtures of lipids.

This led to the commercial development and application of synthetic biodegradable emulsifiers, which were then used for the cleanup of land-based spills of hydrophobic materials, such as crude and refined petroleum products (Thibault and Elliott, 1979). These emulsifiers act rapidly on the hydrocarbons and, therefore, assist in their rapid assimilation by the indigenous microorganisms. Use of mutant microorganisms that produce large quantities of biosurfactants has also been shown to be beneficial for improving petroleum hydrocarbon degradation (Thibault and Elliott, 1979).

Guymon (1993) patented a water/surfactant process for recovering hydrocarbons from soil contaminated by an oil spill, without emulsifying the oil. The surfactant is carefully selected from a group consisting of C_8 to C_{15} linear alcohols and oxylates having two to eight ethylene oxide units per chain. The surfactant concentration is held to about 0.5 vol% to minimize formation of an emulsion between the oil and the wash water. The process provides clean separation of oil from soil as well as from the water. The limited surfactant also minimizes dispersion of clay fines from the soil into the water. The water/surfactant can be heated for improved removal of oil from the soil.

It has been determined that the surfactant concentration at which the maximal spreading velocity occurs is independent of substrate surface energy (Lin, Hill, Davis, and Ward, 1994). It is dependent upon the microstructure of the surfactant dispersion rather than the energetics of the solution/substrate interface. The wetting velocity depends on surface energy in which the velocities are greatest on surfaces of moderate hydrophobicity. Significant water condensation on rough hydrophobic surfaces compared with smooth hydrophobic surfaces is caused by capillary condensation on the rough surfaces, which allows superspreading to occur almost exclusively on rough surfaces of high hydrophobicity.

There can be some drawbacks, however, to employing surfactants. For instance, the surfactants might be used as a preferential substrate by the microorganisms (Deschenes, Lafrance, Villeneuve, and Samson, 1995). A test of 13 priority PAHs in a creosote-contaminated soil was conducted using both biological (rhamnolipid produced by *Pseudomonas aeruginosa* UG2) and chemical (sodium dodecyl sulfate) surfactants at three concentrations. PAH biodegradation was not enhanced, and the surfactants (especially the chemical) caused the residual concentrations of the four-ring compounds to decrease more slowly than in untreated soil.

Adding surfactants to enhance contaminant degradation also introduces the risk of increasing the availability of natural organic matter, which would compete with the contaminants as a source of nutrients (Strong-Gunderson and Palumbo, 1995). Addition of Tween-80, a cell-free biosurfactant product, and an oleophilic fertilizer (nutrients with surfactant-like properties) at or below the critical micelle concentration enhanced bioavailability of toluene and naphthalene, while Tween-80 simultaneously increased the bioavailability of recalcitrant natural organic matter.

5.3.1.1 Chemical Surfactants

Synthetic biodegradable emulsifiers (e.g., POLYBAC® E Biodegradable Emulsifier) were developed for the cleanup of land-based spills of hydrophobic materials, such as crude and refined petroleum products (Thibault and Elliott, 1979). These emulsifiers act rapidly on the hydrocarbons and, therefore, assist in their rapid assimilation by the indigenous organisms or mutant microorganisms that exist in the aqueous phase of the Polybac mutant product, PHENOBAC® Mutant Bacterial Hydrocarbon Degrader.

With regard to relative detergencies of surfactants against different types of soil, it is generally true that alkyl-benzene sulfonates show optimum performance against particulate soil, whereas nonionics perform better on oily soils (laundry detergents) (Sutton, J.R., 1987). The optimum chain length for the alkyl group in the anionic detergent is dictated by solubility as C_{11} to C_{12}. Nonionic octyl and nonyl phenylethoxylates with from 9 to 12 ethoxylate units are the most effective for enhancing desorption of the PAHs anthracene, phenanthrene, and pyrene in soil/water suspensions (Liu, Laha, and Luthy, 1991). With soil:water ratios of 1:7 to 1:2, over 0.1% by volume of surfactant is required to initiate solubilization, and 1% by volume results in 70 to 90% solubilization.

Detergents are able to increase microbial membrane permeability (Gloxhuber, 1974), and substances from humic acids may have the same effect (Visser, 1982). The addition of humic products to a culture medium resulted in a 2000-fold increase in growth (Visser, 1985). These substances appear to induce a change in metabolism, allowing the organisms to proliferate on substrates they could not previously utilize. Tween-20-80 and Brij 35 increased microbial ATP levels, possibly as a result of an increased metabolic rate with the greater amount of nutrients; however, the mechanisms involved with the humic material have not yet been elucidated. It has been recommended that humic products be incorporated in media for determination of microbial activities in terrestrial and aquatic environments. Humic substances constitute the major part of the natural organic constituents of most waters and sediments, typically forming 60 to 80% of the total organic content (Khan, 1980).

Surfactant–substrate interactions, such as emulsification, solubilization, and partitioning of hydrocarbons between phases, can influence accessibility of substrates to microorganisms (Rouse, Sabatini, Suflita, and Harwell, 1994). An understanding of partitioning in aqueous surfactant systems is necessary for understanding mechanisms affecting the behavior of hydrophobic organic compounds in soil/water systems in which surfactants play a role in contaminant remediation or facilitated transport (Edwards, Luthy, and Liu, 1991).

Concentration of the surfactant is an important factor (Aronstein and Alexander, 1993). Surfactants have a hydrophobic and a hydrophilic portion and are soluble in water at low concentrations (Alexander, 1994). Below the critical micelle concentration (the lowest concentration at which micelles begin to form), surfactant monomers in the aqueous phase are relatively ineffective as a partitioning medium for nonionic organic compounds (NOCs), while the sorbed surfactant molecules increase the sorptive capacity of the solid phase (Sun, Inskeep, and Boyd, 1995). At higher concentrations, the surfactant molecules will emulsify oily material, creating increasingly fine droplets, and finally associating to form aggregates 10 to 100 Å in diameter called micelles (Higgins and Gilbert, 1978; van der Linden, 1978; Rosenberg, Zuckerberg, Rubinovitz, and Gutnick, 1979; Thomas, 1980; Alexander, 1994). In water, the hydrophilic ends of the surfactant molecules face toward the water, while the hydrophobic ends cluster in the center of the micelle, where they can isolate a portion of a hydrophobic substrate from an NAPL (Alexander, 1994). Micelles filled with the contaminant deliver it to the microbial cell, into which it diffuses and is then biodegraded (Guha and Jaffe, 1996a). The bioavailable fraction of the micellar-phase substrate is independent of the biomass concentration and is a function of the surfactant concentration, the polyoxyethylene chain length of the surfactant, and the biomass surface characteristics.

Addition of low concentrations (10 µg/g soil) of nonionic alcohol ethoxylate surfactants (Alfonic 810-60 and Novel II 1412-56) can enhance biodegradation of phenanthrene, even in the absence of surfactant-induced PAH desorption (Aronstein, Calvillo, and Alexander, 1991). These results may have been due to a nonspecific effect of the surfactants on the permeability of the indigenous organisms to the aromatics, or other nutrients whose rate of transport into the cell limited cell growth (Mihelcic, Lueking, Mitzell, and Stapleton, 1993). Novel II 1412-56 enhanced the rate and extent of phenanthrene mineralization and the extent but not initial rate of biphenyl mineralization when used at a concentration of 10 µg/mL in water pumped through soil (Aronstein and Alexander, 1993). Stimulation was less if 100 µg surfactant/mL was employed. The low concentrations of surfactants may be useful for *in situ* bioremediation of sites contaminated with hydrophobic contaminants without causing movement of the parent compounds to groundwaters.

A concentration of 100 ppm of most synthetic emulsifiers is not substantially deleterious to the microorganisms (Ellis and Payne, 1984). At concentrations over 100 mg/kg in soil, many surfactants adversely affect soil physical properties and may, at least temporarily, reduce microbial activity (Overcash and Pal, 1979a). Detergents and strong (especially synthetic) emulsifiers/surfactants may disrupt cell membranes (lipids) or cause toxicity by excessive uptake of toxic hydrophobic contaminants (Ellis and Payne, 1984). Due to the direct interaction of surfactants with microorganisms, the steric or conformational compatibility of the surfactants with cell membrane lipids and enzymes appears to be an important metabolic factor (Rouse, Sabatini, Suflita, and Harwell, 1994). Surfactant addition may enhance bioremediation of soil contaminated by polyaromatic hydrocarbons by increasing the contaminant bioavailability (Carriere and Mesania, 1995). The bioavailability and microbial degradation of toluene and naphthalene, for instance, have been improved by adding synthetic surfactants, such as Tween-80 (Strong-Gunderson and Pallumbo, 1995).

Solubilization of PAHs begins at the surfactant critical micelle concentration and is proportional to the concentration of surfactant in micelle form (Edwards, Luthy, and Liu, 1991; Guha and Jaffe, 1996a). Partitioning of organic compounds between surfactant micelles and aqueous solution is characterized by a mole fraction micelle-phase/aqueous-phase partition coefficient, Km.

In contaminated soil, PAHs desorb at different rates, with 90% of the three- and four-ring compounds desorbing in 24 h and 80% of the five- and six-ring PAHs desorbing in that time in the presence of a nonionic surfactant (Carriere and Mesania, 1995). Without surfactant, desorption was 30 and 15%, respectively. For surfactant-enhanced leaching of PAHs from soil, alkylphenol ethoxylate surfactants perform best, with the dominant removal mechanism being micellar solubilization, rather than mechanisms related to reducing surface tension (Ganeshalingham, Legge, and Anderson, 1994). These authors report that for adequate removal rates for PAHs, surfactant concentrations well above the critical micelle concentration are required. Contaminant removal is significantly reduced when more than one hydrocarbon is also present at high concentrations.

Other authors concluded that the bioavailability of PAHs could be increased by use of the surfactants, Triton X-100 and Tergitol NPX at high concentrations. Biodegradation of sorbed naphthalene was stimulated by addition of the surfactants (Volkering, van de Wiel, Breure, van Andel, and Rulkens, 1995). Surfactant concentrations of Triton X-100 or Brij 30 above the critical micelle concentration were not toxic to naphthalene-degrading bacteria, and the presence of surfactant micelles did not inhibit mineralization of naphthalene (Liu, Jacobson, and Luthy, 1995). Naphthalene solubilized by these surfactants was bioavailable and degradable by the bacteria. Breure, van Andel, and Rulkens (1995) found that Triton X-100, Tergitol NPX, Brij 35, and Igepal CA-720 increased apparent solubilities and maximal rates of dissolution of naphthalene and phenanthrene. Surfactants at concentrations higher than the critical micelle concentration affected the dissolution process, by increasing the dissolution rates. There were no toxic effects of the surfactants at concentrations up to 10 g/L.

Fluorescence spectroscopy and quenching experiments demonstrate that naphthalene can be degraded rapidly in the absence of micelles; however, when micelles of the nonionic surfactant, Triton X-100, are present, they protect naphthalene against copper quenching and also suppress biodegradation (Putcha and Domach, 1993). This sequestering effect may increase over time.

Formation of micelles does not necessarily result in degradation of the contaminant (Strong-Gunderson and Pallumbo, 1995). Substrate present in the micellar phase may not be readily available for degradation (Volkering, Breure, van Andel, and Rulkens, 1995). Phenanthrene in micelles of the nonionic surfactant, Tergitol NP-10, was not available to microorganisms; only phenanthrene in the aqueous phase could be degraded (Grimberg and Aitken, 1995). However, part of the phenanthrene partitioned into the micellar phase of some nonionic surfactants can be directly bioavailable to phenanthrene-degrading microorganisms, and this is described by a model (Guha and Jaffe, 1996a).

It has been shown that surfactant concentrations above the critical micelle concentration might inhibit hydrocarbon biodegradation (Rouse, Sabatini, Suflita, and Harwell, 1994). Microbial degradation of phenanthrene in soil/aqueous systems is inhibited by addition of alkyl ethoxylate, alkylphenyl ethoxylate, or sorbitan-type (Tween-type) nonionic surfactants at doses that result in micellar solubilization of phenanthrene from soil (Laha, Liu, Edwards, and Luthy, 1995). Above the critical micelle concentration, surfactant micelles in the aqueous phase begin to compete with the sorbed surfactant as an effective partitioning medium for the poorly water-soluble NOCs, resulting in a decrease in soil/water distribution coefficients (Sun, Inskeep, and Boyd, 1995).

The micellar-phase bioavailable fraction of phenanthrene decreases with an increasing concentration of Triton N101, Triton X-100, and Brij 30. For Brij 35, the micellar-phase fraction of phenanthrene is not directly bioavailable.

The surfactants, Triton X-114 and Corexit 0600, do not appear to enhance aqueous solubility of octadecane (Thai and Maier, 1993). The surfactants do, however, improve dispersion of the octadecane-surfactant complexes, thereby increasing the wetted surface area of octadecane, which raises the octadecane solubility rate. This is accompanied by an increase in the oxygen-uptake rate during biodegradation. Although micellized phenanthrene in the nonionic surfactant, Tergitol NP-10, may not be directly available to the bacterium, *Pseudomonas stutzeri*, the surfactant increases the phenanthrene dissolution rate, with an accompanying increase in bacterial growth (Grimberg, Stringfellow, and Aitken, 1996).

A petroleum sulfonate-oil (PSO) surfactant (commercial Petronate) appears to increase substantially the aqueous-phase concentrations of poorly water-soluble, nonionic organic contaminants, such as naphthalene and phenanthrene (Sun and Boyd, 1993). Both aqueous and soil-sorbed PSO surfactant act as partition phases for these contaminants. Sorbed PSO is nearly as effective as aqueous-phase PSO emulsion as a partition phase for phenanthrene, and sorbed PSO is about four times more effective as a sorptive phase for these contaminants than natural soil organic matter. These authors have developed a model for predicting the apparent soil/water distribution coefficient of a nonionic organic compound at different Petronate concentrations. A multiphase, multicompositional simulator (UTCHEM) has also been developed to model the migration and surfactant-enhanced remediation of an NAPL (Freeze, Fountain, Pope, and Jackson, 1995).

It has been found that gasoline and lead move much more effectively through soil during miscible flow of leaded-gasoline-in-water microemulsion than during immiscible flow of leaded gasoline (Ouyang, Mansell, and Rhue, 1995). In contrast to the adverse effect of macroemulsion on the transport of NAPLs, microemulsion enhances the transport of gasoline through water-saturated soil. A saline solution will displace only limited amounts of mobile gasoline and associated tetra-ethyl lead (TEL) components from soil (Ouyang, Mansell, and Rhue, 1996). However, 95% of the immobile or residual gasoline and 90% of the associated TEL entrapped in the soil pores can be removed by a surfactant/cosurfactant/water solution. Essentially, 1 g of surfactant (sodium lauryl sulfate) can remove 0.6 g immobile gasoline and 2 mg immobile lead from the soil. Pore clogging by gasoline droplets is greatly minimized with a leaded-gasoline-in-water microemulsion.

Using a blend of nonionic surfactants can avoid the formation of viscous emulsions, which can lead to reduced soil permeability or clogging, low surfactant losses, and low residual levels of oil (Sobisch, Kuhnemund, Hubner, Reinisch, and Kragel, 1994). The optimum surfactant composition will be quite different for weathered and unaltered diesel oil.

The removal of petroleum hydrocarbons from soil was improved by orders of magnitude by use of a 2% aqueous solution of Adsee 799 (Witco Chemical) and Hyonic NP-90 (Diamond Shamrock) rather than just water washing (Ellis, Payne, and McNabb, 1985). This combination has adequate solubility in water, minimal mobilization of clay-sized soil fines (to maintain soil permeability), good oil dispersion, and adequate biodegradability. It is potentially useful for *in situ* cleanup of hydrophobic and slightly hydrophilic organic contaminants in soil. Removal efficiency of the latter would be significantly improved by use of aqueous surfactants in soils with high TOC values.

A promising extraction procedure for large-scale treatment of oil-polluted soils is based on cloud point-phase separation of nonionic surfactants (Komaromy-Hiller and von Wandruszka, 1995). The clouding behavior of the detergent, Triton X-114, can be monitored by a fluorescence probe. Changes in the I_1/I_3 ratio indicate gradual dehydration of the detergent micelles upon heating. The rate of phase separation and the volume and water content of the micellar phase are determined. Washing for 15 min with 3 to 5% detergent caused 85 to 98% of the oil contamination to enter the micellar phase of the separated washing liquid. The extraction efficiency decreases with increasing carbon content of the soil.

An agent for emulsification of hydrocarbons has been developed for use in petroleum oil spills by Hosmer and LaRoche (1994). It comprises ethoxylated nonylphenol 59 to 69, soaps of tall oil, and an alkanolamine (ethanolamine) 17 to 21, and tripropylene glycol methanol ether 13 to 16 wt%. The agent can include sodium xylene sulfonate at less than 7.1 wt%, and water, perfume, and colorants.

Kopp-Holtwiesche, Weiss, and Boehme (1993) developed an improved nutrient mixture for bioremediation of soil and water polluted by hydrocarbons. It contains phosphoric acid ester emulsifiers as a phosphate source and water-soluble or dispersible nitrogen sources. The composition may also contain biodegradable surfactants, with the exception of glycerin esters, and preferably nonionic surfactants. Suitable phosphoric acid esters include glycerophospholipids (Lipotin), phospholipids, alkyl phosphates, and/or alkyl-ether phosphates.

The effectiveness of surfactants in enhancing carbon mineralization rates depends on the surface activity of the surfactants, the pH of both the oil and the surfactant, and the dispersibility of oil by the surfactant (Rasiah and Voroney, 1993). Nonionic surfactants may be more surface active than anionic surfactants of similar structure, since they could interact with both positive and negative charges of the oil. If the pH of the oil and surfactant are in the same range, the oil will disperse more easily. It was observed that the ion charge of the surfactant does not significantly enhance mineralization. Cometabolism of surfactant and oily waste, however, does contribute to enhanced waste mineralization.

The anionic surfactant mixture of mono- and diorganophosphate esters, Cedephos FA-600, had an alkaline pH similar to that of an oily waste and was very effective in enhancing carbon mineralization of the waste through emulsification (Rasiah and Voroney, 1993). The surfactant with the next highest enhancement of mineralization was Igepal CO-603, a nonionic ethoxylated alkylphenol, which had limited solubility under alkaline conditions. Surfactants should be evaluated for their potential to emulsify the specific wastes involved, and laboratory screening appears to be a cost-effective means of selecting the most appropriate surfactant.

When two sulfonated anionic surfactants (Dowfax C10L and Dowfax 8390), a phosphate ester blend weak-acid anionic surfactant (Rexophos 25/97), and two ethoxylated nonionic surfactants (Tergitol 15-S-9 and Triton X-100) were studied under acidic conditions to treat metal-polluted and hydrophobic organic compound–polluted soils determined that a decrease in pH from 8.3 to 1.0 would have no effect on naphthalene and pyrene solubilization (Van Benschoten, Ryan, Huang, Healy, and Brandl, 1995). Sorption of Dowfax C10L and Triton X-100, however, increased significantly with a decrease in pH.

At the same concentration, different types of surfactants reduce surface tension differently (Wanger and Poepel, 1995). Air bubble slip velocity is smaller in surfactant solutions than in clean water. Depending upon the type of surfactant, its concentration, and the type of water, the aeration coefficient decreases (down to 55%), the specific interfacial area increases (up to 350%), and the oxygen transfer coefficient decreases (down to 20%). Nonionic surfactants reduce oxygen transfer more strongly than anionic surfactants.

Experiments with foams and oil-core aphrons (OCA) for enhancing *in situ* bioremediation suggest that foams would be more effective for extracting, mobilizing, and dispersing NAPL contaminants than surfactant solutions (Enzien, Michelsen, Peters, Bouillard, and Frank, 1995; Lange, Bouillard, and Michelsen, 1995). Foams could remove >90% of NAPL pools, while surfactant solutions removed <10%. OCAs were ineffective. It was shown that foams flow through porous media as a front, whereas surfactant solutions tend to channel. Foams can encapsulate low-permeability lenses, and foam front advancement increases when vacuum extraction is applied.

Foam in the form of microbubbles consists of 60 to 70% dispersion of 55-µm microbubbles in water (colloidal gas aphrons, CGAs) (Lange, Bouillard, and Michelsen, 1995). Higher-viscosity foams flow forward and fill up larger channels. When the pressure drop builds up in the channel, the foam flows into less accessible spill areas. About 80% of hexadecane in columns could be flushed with a 70% CGA quality (70% air, 30% liquid). A surfactant concentration of 1000 to 5000 ppm is needed to generate the microbubbles. Microbubble scouring can be three to six times more effective than surfactant flushing. Channeling or poor sweep should not occur with the microbubble scouring as is seen with surfactant flushing. Application of this technique could be limited by the problem of pressure drop required to pump microbubbles into soil with low permeability. See Section 3.2.1.6 for additional information on the action of foams used for bioremediation.

Some bacteria (e.g., *Pseudomonas*) contain plasmids that encode metabolic pathways for metabolism of aromatic pollutants, through dispersion and accelerated cell barrier transport (Churchill and Griffin, 1992). Biosolve, Inipol EAP-22, and rhamnolipids accelerate mineralization of *p*-cresol, benzene, and toluene when the catabolic plasmid enzymes are induced, by reducing the lag phases and increasing the rate and extent of mineralization. Inipol EAP-22, the oleophilic fertilizer employed by the U.S.EPA in beach bioremediation of the Valdez oil spill in Prince William Sound, AK, accelerates biodegradation of aliphatic hydrocarbons by pure bacterial strains and mixed cultures with a concentration dependence similar to that of other synthetic surfactants, such as Bioversal and Biosolve. Aliphatic substrates are almost exclusively chromosome encoded. If the specific metabolic pathways are not induced, at least a decreased lag phase may be noticed.

Despite the potential of surfactant application as an aid to bioremediation, some limitations have been associated with the process. Both enhancements and inhibitions of biodegradation of organic compounds have been reported in the presence of surfactants (Rouse, Sabatini, Suflita, and Harwell, 1994). In some cases, chemical dispersants enhance hydrocarbon oxidation while others inhibit it (Cooney, 1984). A chemical surfactant was observed to inhibit a yeast strain growing on hydrocarbons (Mimura, Watanabe, and Takeda, 1971). The surfactant may have interfered with the direct interaction between cells and substrate. In the case of bacteria, Triton X-100 completely prevented degradation of hexadecane dissolved in heptamethylnonane (Efroymson and Alexander, 1991). The surfactant may have prevented adherence of the bacteria to the heptamethylnonane–water interface, thus preventing direct contact of the cells with the hexadecane. Triton X-100 may also have affected the cell membranes.

Another problem is that a chemical surfactant might be used as a growth substrate by microorganisms in preference to the contaminant, such as was observed in the application of sodium dodecyl sulfate to treat creosote-contaminated soil (Deschenes, Lafrance, Villeneuve, and Samson, 1995) and in the addition of Tween-80, which not only increased the bioavailability of toluene and naphthalene, but also that of natural organic matter (Strong-Gunderson and Palumbo, 1995). This would introduce additional nutrients for the microorganisms and interfere with bioremediation of the target chemicals. Ripper, Friedrich, and Ripper (1992) also determined that surfactants were ineffective, as a result of the bacteria utilizing the surfactant as a carbon source. Other researchers, however, tested a biodegradable surfactant and observed that the anionic surfactant was degraded along with kerosene but at a much slower rate, as the microbial communities preferred to degrade the kerosene (Sundaram, Sarwar, Bang, and Islam, 1994). At the same time, the presence of the surfactant enhanced the biodegradation of the petroleum contaminant.

The detergents, Serqua 710 and the Dobanols 91-5, 91-6, and 91-8, which contain C_9, C_{10}, and C_{11} alcohols and 5, 6, and 8 M of ethoxylate, respectively, were compared for their ability to solubilize residual oils in soil (Harmsen, 1991). Dobanol 91-6 gave the highest solubility. Use of these surfactants, though, may be of limited value in landfarming. Aerobic conditions are important for surfactant activity

(Ellis and Payne, 1984), and the Dobanols are easily biodegradable, which consumes oxygen, leaving less for degradation of the oil in the soil (Harmsen, 1991). High concentrations of the detergent are necessary, and residual concentrations still remain in the soil after one treatment.

Another drawback with surfactant-enhanced subsurface remediation is the loss of surfactants and their effectiveness, as a result of sorption, especially in clay soils, and precipitation (Falatko and Novak, 1992; Ducreux, Ballerini, and Bocard, 1994). Adsorption of surfactants onto solid matrices can lead to a decrease in micellar surfactant concentration, resulting in a smaller increase of the apparent solubility than expected from results of dissolution and desorption experiments (Breure, Volkering, Mulder, Rulkens, and van Andel, 1995). Anionic compounds may be precipitated in soil, thereby lowering the concentration of the surfactant in water (Falatko and Novak, 1992; Ducreux, Ballerini, and Bocard, 1994). Disulfonates (Na dodecylbenzenesulfonate and alkyl di-Ph oxide disulfonates) are less susceptible to sorption than monosulfonates and also less prone to sorption than nonionic surfactants (Rouse, Sabatini, and Harwell, 1993). Disulfonates also exhibit greater solubilization of naphthalene than monosulfonates and slightly lower solubilization than nonionics. Disulfonate surfactants are strong candidates for use in surfactant-enhanced subsurface remediation. Edwards, Liu, and Luthy (1994a) found that sorbed nonionic surfactant molecules tend to increase hydrophobic organic compound (HOC) sorption, while free surfactant monomers in solution tend to decrease HOC sorption by increasing the HOC aqueous solubility, modifying the HOC soil/water coefficient. Edwards, Liu, and Luthy (1994b) determined that treating soil with successive surfactant washings results in greater removal of HOC than a single washing with the same amount of surfactant.

In other studies testing three surfactants, surfactant plus soil venting failed to enhance biodegradation of JP-5 compared to soil venting alone (Arthur, O'Brien, Marsh, and Zwick, 1992). Soil venting appeared to be sufficient for promoting biodegradation and overcoming the oxygen limitations in the unsaturated soil of the particular site investigated. Nevertheless, use of surfactants alone to flush otherwise insoluble organics, or in combination with other treatments to solubilize waste materials (thereby, promoting biodegradation), is a promising avenue for further research (Ellis and Payne, 1984). Additional information on the nature of chemical surfactants, other evaluated compounds, and their application in soil washing techniques can be found in Section 2.2.1.7.

5.3.1.2 Microbial Surfactants

Synthetic surfactants might adversely affect the permeability of the microbial cell membrane, which would interfere with the capacity of a microorganism to biodegrade (Hunt, Robinson, and Ghosh, 1994). Surfactants produced by the organisms themselves are less likely to be detrimental to the microbes and this process. Microbial surfactants are generally much less toxic than chemical surfactants, at least as effective, and more readily biodegradable (Chakrabarty, 1995). Using microorganisms that produce their own biosurfactants could also lower treatment costs (Wilson and Jones, 1993).

Microorganisms must have contact with the hydrocarbon to degrade it (Rosenberg, Legmann, Kushmaro, Taube, Adler, and Ron, 1992). For this to occur, they have evolved specific adhesion mechanisms and the ability to emulsify the compounds. Microbes have been long known to produce surface-active agents when grown on specific substrates (Zajic and Panchal, 1976). Biologically produced surfactants (biosurfactants) are produced by microorganisms during growth on insoluble organic substrates in order to increase their solubility (Falatko and Novak, 1992). Microorganisms generally consume only soluble or solubilized (emulsified) organic molecules, and synthesis of an emulsifier may pseudosolubilize target hydrocarbons or enhance direct contact between microorganisms and a hydrocarbon substrate (Goma, Ani, and Pareilleus, 1976; Rosenberg, Perry, Gibson, and Gutnick, 1979; Rosenberg, Zuckerberg, Rubinovitz, and Gutnick, 1979; Thibault and Elliott, 1980).

Both membrane-bound and secreted biosurfactants enable transport of apolar, hydrophobic substances through the polar outer membrane barrier of the cells (Wiesel, Wuebker, and Rehm, 1993). By producing their own biosurfactants, they can decrease initial oil interface tension and increase hydrocarbon bioavailability (Setti, Pifferi, and Lanzarini, 1995). Many of these surfactants have excellent emulsifying properties (Panchal and Zajic, 1978). Microbial uptake of hydrocarbons can be facilitated by hydrophobization of the cell envelope or by hydrocarbon emulsifying with extracellular biosurfactants (Zajic and Seffens, 1984). These biosurfactants are composed of polysaccharides, polysaccharide–protein complexes, or glycolipids (Blanch and Einsele, 1973; Rosenberg, 1986; Bausum and Taylor, 1986). Emulsions formed are stabilized by the polysaccharide polymers secreted extracellularly by the microbes. These can then be absorbed through the lipophilic cell wall to be utilized by the microorganism as a carbon and energy source. These polymers may also act as flocculants (Zajic and Knettig, 1971).

Many oil-degrading microorganisms produce emulsifying agents (Reisfeld, Rosenberg, and Gutnick, 1972), and naturally occurring biosurfactants seem to be very important in the elimination of hydrocarbons from polluted biotopes (Rambeloarisoa, Rontani, Giusti, Duvnjak, and Bertrand, 1984). Surface-active agents are excreted into the aqueous medium when certain organisms are grown on liquid hydrocarbons, particularly n-alkanes (Zajic and Gerson, 1977). An example of the surfactants produced in the presence of hydrocarbon are the α, α-trehalose-6,6'dicornomycolates (glycolipids produced by n-alkanes in *Rhodococcus erythropolis*) (Kretschmer, Bock, and Wagner, 1982). When exposed to hydrocarbons, the lipid content of the cell wall increases significantly, which increases the affinity of the microbe for the hydrocarbon (Blanch and Einsele, 1973).

Spontaneous dissolution rates are important factors affecting the rates of biodegradation (Thomas, Yordy, Amador, and Alexander, 1986). Growth of pure cultures of bacteria on naphthalene, phenanthrene, and anthracene is fastest on the solid substrates with the highest water solubilities (Wodzinski and Johnson, 1968). The rate of dissolution of compounds, such as naphthalene, is directly related to their surface areas (Thomas, Yordy, Amador, and Alexander, 1986), and increasing the surface area of hexadecane increases its microbial destruction (Fogel, Lancione, Sewall, and Boethling, 1985). At the same time, the hydrocarbons come into better contact with available oxygen and inorganic nutrients (Atlas, 1981). In fact, the first step in hydrocarbon assimilation by *Candida lipolytica* is the microbial enhancement of the solubilization of the substrate (Goma, Pareilleux, and Durand, 1974).

Oil emulsification may be an integral part of the growth cycle of certain microorganisms (Horowitz, Gutnick, and Rosenberg, 1975). A bacterial strain, UP-2, is able to induce oil emulsification during the exponential growth phase. The cells (even at low concentrations) become tightly attached to the surface of fresh oil drops, where they multiply and form small colonies. In the exponential phase, they break up the oil drops into smaller droplets by producing an emulsifier. This provides new surface area for more growth. Production of these surface-active compounds is connected mainly with growth-limiting conditions in the late-logarithmic and the stationary-growth phases (Oberbremer and Mueller-Hurtig, 1989; Hommel, 1990).

Biosurfactants are synthesized and excreted into the environment by soil bacteria, such as *Pseudomonas, Rhodococcus,* and *Arthrobacter* (Itoh and Suzuki, 1972). Cooney (1980) found 129 bacterial cultures exhibited emulsifying activity. In every case where growth occurred in hydrocarbon (kerosene) medium, there was emulsification of the hydrocarbon. Emulsifiers were produced throughout the growth period. Production of emulsifiers during active cell growth on hydrocarbons suggests that emulsifiers are involved in growth on hydrocarbons. Biosurfactant production is associated with growth in four *Corynebacterium* species (Gerson and Zajic, 1979; Margaritis, Zajic, and Gerson, 1979) and in species of *Nocardia, Acinetobacter,* and *Arthrobacter* (Gerson and Zajic, 1979). A direct involvement of emulsifier in hydrocarbon use has been demonstrated in a *Pseudomonas*, which produces a rhamnolipid emulsifier. Although growth on hydrocarbons can occur without emulsification (Miura, Okazaki, Hamada, Murakawa, and Yugen, 1977), use of hydrocarbons by most freshwater bacteria involves production of emulsifiers (Cooney, 1980). The vast majority of the organisms form emulsifiers when they are exposed to hydrocarbons. Most isolates appear to produce emulsifiers, which are located at the cell surface and which can be released to the medium.

Oberbremer and Mueller-Hurtig (1989) determined there are two phases of biodegradation. During the first phase, the most water-soluble compounds are degraded. After the interfacial tension is lowered by production of biosurfactants, the more resistant compounds can be degraded in the second phase. This is accomplished by microorganisms with high cell-surface hydrophobicity, which allows them to adhere to high-molecular-weight hydrocarbons (Rosenberg, Legmann, Kushmaro, Taube, Adler, and Ron, 1992). This attachment could be due to hydrophobic fimbriae or fibrils, outer-membrane and other surface proteins, and lipids (Rosenberg, Hayer, Delaria, and Rosenberg, 1982). Bacterial capsules and other anionic exopolysaccharides seem to inhibit attachment (Rosenberg, Kaplan, Pines, Rosenberg, and Gutnick, 1983).

Hydrocarbon-degrading microorganisms are found mainly at the oil/water interface (Atlas, 1981). In soil, microbes break down oil at oil/water interfaces (Flowers, Pulford, and Duncan, 1984). These organisms can be seen growing over the entire surface of an oil droplet. Growth does not appear to occur within oil droplets in the absence of entrained water. The low surface area of paraffins normally reduces accessibility to microbes, which results in the retardation of the decomposition rate in the soil (Arora, Cantor, and Nemeth, 1982). Increasing the surface area improves the degradation rate. The material is then more readily available to microorganisms, and movement of emulsion droplets through the water column increases the availability of oxygen and nutrients. The high activity of xenobiotic-degrading microorganisms in a biphasic aqueous/organic system (with a hydrophobic organic solvent) depends mainly on the size of the interfacial area (Ascon-Cabrera and Lebeault, 1995). A large percentage of the biomass is attached to the interface as an interfacial biofilm. Growth and selection of the organisms should be based primarily on the formation of a biofilm at the interfacial area rather than on substrate transport to the aqueous phase.

Some organisms are encased in a matrix of extracellular polymeric substances (EPS) (Geesey, 1982) or algal slime, and this tends to protect them from the xenobiotic materials (Bausum and Taylor, 1986). Many fungi, algae, bacteria, and cyanobacteria produce this substance. The emulsifying action of these EPS-containing biofilms breaks up oily material into fine particles and then aggregate micelles (Higgins and Gilbert, 1978; van der Linden, 1978; Rosenberg, Zuckerberg, Rubinovitz, and Gutnick, 1979; Thomas, 1980).

It appears that, in some cases, emulsification may be the result of microorganisms detaching themselves from a hydrocarbon (Rosenberg, Legmann, Kushmaro, Taube, Adler, and Ron, 1992). *Acinetobacter calcoaceticus* RAG-1 adheres to petroleum droplets, where it grows on the utilizable hydrocarbons (Rosenberg, Rosenberg, Shohan, Kaplan, and Sar, 1989). When this source is exhausted, the bacterium releases its hydrophilic capsular material. This extracellular, amphipathic emulsan attaches to the hydrocarbon/water interface and displaces the cells from the hydrocarbon surface. This allows the capsule-free, hydrophobic organisms to reattach to fresh droplets.

Pseudomonas aeruginosa PRP652 produces and excretes a biosurfactant, which is a glycosylated, anionic, amphipathic compound, termed Rhamnolipid R1 (Hunt, Robinson, and Ghosh, 1994). The apparent solubility of phenanthrene increased from 1.2 mg/L with no surfactant to 34.4 mg/L (3500 mg/L biosurfactant). Rhamnolipid R1 can lower the surface tension of aqueous solutions to a constant minimum of 35.1 dyn/cm at a concentration of about 20 mg/L. Two other rhamnolipid biosurfactants were investigated by Zhang and Miller (1995). A methyl ester form of a dirhamnolipid reduced the interfacial tension between hexadecane and water to <0.1 dyn/cm, while an acid form decreased the interfacial tension to only 5 dyn/cm. Both forms (especially the methyl ester form) enhanced degradation of liquid hexadecane and solid octadecane by several strains of *Pseudomonas*. Degradation of octadecane required microorganisms to have a low cell surface hydrophobicity.

Other authors, however, studied the rhamnolipid biosurfactants produced by *P. aeruginosa* UG2 at three concentrations over a period of 45 weeks and observed no enhanced biodegradation of 13 EPA priority PAHs in a creosote-contaminated soil (Deschenes, Lafrance, Villeneuve, and Samson, 1995). In fact, for the four-ring PAHs, surfactant presence seemed harmful to the biodegradation process. This effect increased as a function of surfactant concentration. The surfactants may have been used as a preferential substrate by the indigenous microflora, rather than the PAHs.

A patented oil-emulsifying glycolipid from a strain of *P. aeruginosa* has the ability to mobilize oil from solid surfaces and disperse oil slicks (Chakrabarty, 1985). It has been used as a biosurfactant in the cleanup of oily sludge from an oil storage tank operated by Kuwait Oil Company, recovering >90% of the hydrocarbons trapped in the sludge (Banat, Samarah, Murad, Horne, and Banerjee, 1991). It was also employed to facilitate removal of oil from rocks during the Exxon Valdez cleanup (Harvey, Elashvili, Valdes, Kamely, and Chakrabarty, 1990). Addition of glycolipids at a concentration of 200 mg/L in excess of the critical micelle concentration shortened the adaptation phase and increased removal of C_{14} to C_{18} hydrocarbons and naphthalene from 81% to from 93 to 99% within 79 h (Oberbremer, Mueller-Hurtig, and Wagner, 1990).

An excessive pre-emulsification with 15 g (200 mg/L) biosurfactant/kg crude oil will maximize the degradation rate of a mixed population in continuous culture (Mattei and Bertrand, 1985). With this concentration, the initial interfacial tension can be reduced from 21 mN to a range of 2 to 16 mN (Oberbremer, Mueller-Hurtig, and Wagner, 1990). Glycolipids shortened the time for adaptation from around 35 h to from 17 to 23 h. After the depleted oxygen was restored and additional biosurfactant produced, the less water-soluble compounds were then degraded more rapidly (0 to 8 h instead of 21 h), and the metabolite, salicylic acid, was degraded. The degradation capacity of the soil under nonlimiting conditions was 25.7 g hydrocarbon/kg soil dry weight/day. Addition of biosurfactants increased this rate to 46.5 g hydrocarbon/kg soil/day. The degradation efficiency and velocity was increased by adding glycolipids, increasing biomass from 0.30 g/L protein to 0.70 to 2.03 g/L.

Acinetobacter calcoaceticus MM5 grows on heating oil, crude oil, and tetradecane, increasing protein concentration and caprilate-lipase and acetate-esterase enzymatic activities in the culture filtrate, with a simultaneous pH drop (Marin, Pedregosa, Rios, Ortiz, and Laborda, 1995). Strong emulsification of petroleum by-products was also noticed. The emulsifier is a high-molecular-weight product made of proteins, sugars, and fatty acids. It is resistant to high temperature.

Growth of microorganisms on gasoline produces a biosurfactant that acts like a commercial surfactant by increasing the solubility of gasoline compounds while not inhibiting biodegradation (Falatko and Novak, 1992). On the other hand, biosurfactants generated by growth on glucose and vegetable oil are effective at increasing the solubility of gasoline compounds but may inhibit biological degradation of these compounds.

Strains of *Pseudomonas* and *Corynebacterium* produce emulsifying agents (Zajic, Supplisson, and Volesky, 1974). *P. aeruginosa* UG2 and *A. calcoaceticus* RAG-1 produce high levels of extracellular biosurfactants (Van Dyke, Gulley, Lee, and Trevors, 1993). Culture filtrates containing the biosurfactants enhanced recovery of individual hydrocarbons and a mixture of hydrocarbons from soil. Biosurfactants from these organisms seemed to have the potential for remediation of hydrophobic pollutants in soil environments.

Extracellular free fatty acids proposed as emulsifiers are produced by *Acinetobacter* sp. in a medium containing *n*-hexadecane but not in nutrient broth–yeast extract medium (Makula and Finnerty, 1972). Addition of *n*-dodecane to bacteria growing on #6 fuel oil increased oil emulsification and yielded a greater dispersion of oil in the aqueous phase (Zajic, Supplisson, and Volesky, 1974). Emulsifier production is also inducible in *Endomycopsis lipolytica* (Roy, Singh, Bhagat, and Baruah, 1979) and in *Rhodococcus erythropolis* (Rapp, Bock, Wray, and Wagner, 1979). *E. lipolytica* may produce different emulsifying factors when grown on different hydrocarbons. When specific microorganisms are added to the soil to help promote biodegradation, the medium in which they are grown may influence their ability to develop a surfactant (du Plessis, Phaal, and Senior, 1995). Bacteria grown in a single-phase (liquid) medium might not be able to produce biosurfactants for solubilizing hydrocarbons from soil surfaces (two-phase system) (Oberbremer and Mueller-Hurtig, 1989; Jain, Lee, and Trevors, 1992).

A strain of *Corynebacterium* sp., isolated from sewage sludge, was grown in mineral salts medium with hexadecane (3.0% v/v) as a carbon and energy source (Panchal, Zajic, and Gerson, 1979). Both hydrophobic and hydrophilic emulsifiers were isolated from the same culture broth. The lipid extract was a very potent emulsifying agent, while the polysaccharide was very weak, unless used at a high concentration and in combination with Tween-20. Bioemulsifiers are highly substrate specific and are most effective with mixtures of compounds, which may not be emulsified individually (Rosenberg, Perry, Gibson, and Gutnick, 1979; Rosenberg, Zuckerberg, Rubinovitz, and Gutnick, 1979).

Surface-active exolipids called serrawettins and rubiwettins are produced by species of *Serratia* (Matsuyama, Murakami, Fujita, Fujita, and Yano, 1986; Matsuyama, Sogawa, and Yano, 1987). 3-Hydroxy fatty acids of C_{10} to C_{16} chain length are common structures in these exolipids and in other biosurfactants, such as the surfactin of *Bacillus subtilis* and the rhamnolipids of *P. aeruginosa* (Cooper and Zajic, 1980). The configuration of the 3-hydroxy fatty acids in biosurfactants can be determined by chiral high-pressure liquid chromatography (Nakagawa and Matsuyama, 1993). Serrawettin W2, a surface-active cyclodepsipeptide of *S. marcescens*, contains D-3-hydroxydecanoic acid. Rubiwettin R1 and RG1, surface active glycolipid and linked fatty acids of *S. rubidaea*, contain D-3-hydroxytetradecanoic acid and D-3-hydroxydecanoic acid.

The first study of biosurfactant production resulting from PAH metabolism was reported by Deziel, Paquette, Villemur, Lepine, and Bisaillon (1996). Ten cultures were found to produce a biosurfactant when grown on naphthalene or phenanthrene. Maximal productivity of biosurfactant by *P. aeruginosa* 19SJ grown on naphthalene was delayed compared with that in cultures grown on mannitol. However, if small amounts of biosurfactants and naphthalene degradation intermediates were present at the onset of the cultivation, the delay was substantially shortened. Production of biosurfactants was accompanied by an increase in the aqueous concentration of naphthalene, indicating the microorganism was promoting the solubility of the substrate. Detectable amounts of glycolipids were also produced on phenanthrene.

It has been suggested that PAH biodegradation in fine-grain soils will proceed only when the interfacial tension is lowered by the production of surface-active agents by the microorganisms involved (Castaldi, 1994). The bioavailability and microbial degradation of compounds like toluene and naphthalene can be enhanced by adding biosurfactants or nutrients with surfactant-like properties (e.g., an oleophilic fertilizer) at or below the critical micelle concentration (Strong-Gunderson and Palumbo, 1995). An alkylethoxylate (C12-E4) and alkylphenol ethoxylate (C8PE9.5 and C9PE10.5) surfactants are nonionic surfactants that solubilize PAHs in soil/water systems (Laha and Luthy, 1991). Mineralization of ^{14}C-phenanthrene is inhibited in the presence of these surfactants at concentrations that result in aqueous-phase critical micelle concentration or micelle formation. This is not due to toxicity of surfactant or micellized PAH, nor is the surfactant being utilized as a preferential substrate. Dilution of the surfactant to a concentration below that resulting in micellization reverses this inhibition. Subcritical micelle concentration levels of the surfactants in soil/water systems neither inhibit nor enhance the rate of degradation. The critical micelle concentration of biosurfactants ranges between 5 and 100 mg/L (Lang and Wagner, 1987).

Many antibiotics, which are natural extracellular products, act as surface active agents; i.e., they lower the surface tension (Zajic and Mahomedy, 1984). Surfactin, which is produced by *B. subtilis*, has been found to be the most active surfactant. Under both aerobic and anaerobic conditions, *B. licheniformis* JF-2 synthesizes a surfactin-like lipopeptide that is the most effective biosurfactant known (Lin, Carswell, Georgiou, and Sharma, 1994). The biosurfactant is produced by actively growing cells in the midlinear phase, but becomes internalized by the cells when they enter the stationary phase. This makes efficient production of the biosurfactant difficult. Synthesis of this peptide can be optimized by growing the cells under O_2-limiting conditions and a low dilution rate of 0.12/h. Maximum concentration of 110 mg/L lipopeptide is obtained in early stationary-phase culture, with 1% (w/v) glucose as the carbon source (Lin, Sharma, and Georgiou, 1993). Surfactant internalization can be reduced by optimizing the concentration of phosphate and magnesium. Production of lichenysin A is enhanced about twofold and fourfold by addition of L-glutamic acid and L-asparagine, respectively (Yakimov, Fredrickson, and Timmis, 1996). This surfactant has been purified and characterized (Lin, Minton, Sharma, and Georgiou, 1994). With this surfactant, the lowest interfacial tensions against octane occur when NaCl concentrations are ≥50 g/L, while calcium concentrations >25 g/L significantly increase the interfacial tension (McInerney, Javaheri, and Nagle, 1990).

B. licheniformis BAS50 produces a powerful lipopeptide surfactant when cultured aerobically or anaerobically on a variety of substrates at salinities of up to 13% NaCl, with an optimum at 5% NaCl and temperatures between 35 and 45°C (Yakimov, Timmis, Wray, and Fredrickson, 1995). The biosurfactant is termed *lichenysin A* and has been purified and chemically characterized. It decreases the surface tension of water from 72 to 28 mN/m and achieves the critical micelle concentration with as little as 12 mg/L.

A thermophilic *Bacillus* strain grows at up to 50°C and produces a biosurfactant with low surface and interfacial tension (27 to 29 and 1.5 mN/m, respectively) (Banat, 1993). It emulsified kerosene and other hydrocarbons efficiently and was able to recover >95% of the residual oil from sand pack columns. An *Arthrobacter* strain was found to emulsify oil extensively when growing on hydrocarbons (Reisfeld, Rosenberg, and Gutnick, 1972).

The organisms producing the emulsification may not be the ones responsible for hydrocarbon breakdown (Atlas, 1981). Where the hydrocarbonoclastic microbes do not have the emulsification ability, commercially produced, biodegradable emulsifiers can be added to accelerate the biodegradation process. Biodegradation rates can be enhanced by use of synthetic biodegradable emulsifiers or microbes that secrete large quantities of such surface-active agents (Zitrides, 1983).

Use of bioemulsification of oils for microbial enhancement of oil recovery from petroleum reservoirs (Cooper, 1982) should be directly applicable to petroleum product spills. Bacteria from a well contaminated by a spill of JP-5 jet fuel could emulsify the fuel, if the well water was supplemented with phosphate and nitrate (Ehrlich, Schroeder, and Martin, 1985). The surfactants not only emulsify the hydrocarbons, but also aid in mobilizing them through soil and water (Vanloocke, Verlinde, Verstraete, and DeBurger, 1979). In favorable geological situations, the mobile emulsions could be removed by pumping for treatment on the surface (Ehrlich, Schroeder, and Martin, 1985).

An interesting proposal is to use liposomes in bioremediation (Gatt, Bercovier, and Barenholz, 1991). These naturally occurring structures could not only be used to deliver nutrients to the subsurface, but also to modify the physical properties of oil or hydrophobic contaminants, which then become more available for biodegradation. The phospholipid liposomes can reduce interfacial tension 10,000- to 50,000-fold and increase oil wettability, allowing oil or waste molecules to break out of the small capillaries in the soil, coalesce, and be pushed by the water to the surface. Liposomes could be applied to the ground in water by spraying or flooding. For treating oil- or waste-contaminated soil, they could

Table 5.17 Biosurfactant-Producing Microorganisms

Pseudomonas sp.[a]	*Acinetobacter calcoaceticus* RAG-1[i]
P. aeruginosa 19SJ[b]	*A. calcoaceticus* MM5[j]
P. aeruginosa UG2[c]	*Bacillus* sp. (thermophilic)[k]
Candida lipolytica[d]	*Endomycopsis lipolytica*[l]
Corynebacterium sp.[e]	*Rhodococcus erythropolis*[m]
Nocardia[f]	*Bacillus subtilis*[n]
Acinetobacter sp.[g]	*Penicillium spiculisporum*[o]
Arthrobacter[h]	*Bacillus licheniformis* BOS50[p]

References:

a = (Zajic, Supplisson, and Volesky, 1974; Liu, 1980)
b = (Deziel, Paquette, Villemur, Lepine, and Bisaillon, 1996)
c = (Van Dyke, Gulley, Lee, and Trevors, 1993)
d = (Goma, Pareilleux, and Durand, 1974)
e = (Gerson and Zajic, 1979; Margaritis, Zajic, and Gerson, 1979; Panchal, Zajic, and Gerson, 1979)
f = (Gerson and Zajic, 1979)
g = (Makula and Finnerty, 1972; Gerson and Zajic, 1979)
h = (Reisfeld, Rosenberg, and Gutnick, 1972; Gerson and Zajic, 1979)
i = (Van Dyke, Gulley, Lee, and Trevors, 1993)
j = (Marin, Pedregosa, Rios, Ortiz, and Laborda, 1995)
k = (Banat, 1993)
l = (Roy, Singh, Bhagat, and Barnah, 1979; Rapp, Bock, Wray, and Wagner, 1979)
m = (Rapp, Bock, Wray, and Wagner, 1979; Kretschmer, Bock, and Wagner, 1982)
n = (Zajic and Mahomedy, 1984)
o = (Ban and Yamamoto, 1993)
p = (Yakimov, Timmis, Wray, and Fredrickson, 1995)

be dissolved in a light kerosene, such as hexane, and sprayed on the surface. This would deliver the phospholipids and help soften the weathered oil or hydrophobic waste. The kerosene would evaporate, and addition of water would hydrate the phospholipids, forming liposomes *in situ*. See Section 5.1.5 for a discussion of the application of liposomes for supplying nutrients to microorganisms in the soil.

While growing on hexadecane, *Acinetobacter* sp. HO1-N accumulates extracellular membrane vesicles of 20 to 50 nm in diameter (Kappeli and Finnerty, 1980). The vesicles may play a role in the uptake of alkanes by partitioning them in the form of a microemulsion. Unilamellar vesicles can aid in the transport of water-insoluble, solid hydrocarbons (Miller and Bartha, 1989). A *Pseudomonas* isolate was grown on octadecane with a K_s value of 2450 mg/L, compared with 60 mg/L when the hydrocarbon was presented in the form of liposomes.

Attempts have been made to isolate emulsifying agents for possible use in dispersing oil. Bioemulsifiers are generally easy and inexpensive to produce (Zajic and Gerson, 1977). Production of these compounds can be readily controlled by media manipulation, and emulsifiers of biological origin are being applied to industrial purposes (Zajic and Panchal, 1976). It is even possible to add or select for organisms that produce emulsifiers.

Table 5.17 lists several biosurfactant-producing microorganisms.

5.3.1.3 Biodegradation of Surfactants

Surfactants are susceptible to biodegradation (van Ginkel, 1996). This factor is important for limiting further environmental contamination, as a result of their application in bioremediation (Wilson and Jones, 1993). In natural ecosystems, surfactants may be completely biodegraded (van Ginkel, 1996). Ideally, a surfactant should be degradable by the soil microorganisms at a slow rate to maintain its effectiveness (Lyman, Noonan, and Reidy, 1990).

Surfactant biodegradation rates and extents are strongly related to molecular structure (Huddleston, Bleckmann, and Wolfe, 1986). Highly branched compounds are especially resistant, and some quaternary ammonium compounds may be so tightly bound to clays that biodegradation is difficult. Relative biodegradation rates and extents for principal classes of synthetic surfactants are

	Relative Rate	Extent (%)
Linear alcohol sulfates	100	100
Linear alkylbenzene sulfanates (LAS)	75	100
Linear alcohol ethoxylates	70	100
Quaternary ammonium compounds[a]	60	100
Alkyl (branched) phenol etholxylates	10	100
Alkyl (branched) benzene sulfonates	5	50

[a] Certain structures are more bioresistant.

Alkyl sulfates can be degraded by some microorganisms, with the release of sulfate (van Ginkel, 1996). Alkyl triethoxy sulfate-degrading organisms, such as *Pseudomonas* DES 1, initially attack one of the ether bonds. Alcohol ethoxylate metabolism can be initiated at the terminal alcohol group, terminal oxidation of the alkyl chain, or by a central fission of the ether bond next to the hydrophobic group (e.g., *Alcaligenes* sp.). Alkylphenol ethoxylates may be degraded by the same means employed for branched alcohol ethoxylates; for example, a *Pseudomonas* sp. was isolated with the ability to degrade the polyether moiety. The C_{alkyl}–N bond of ethoxylated fatty amines is attacked through a dehydrogenation reaction, forming diethanolamine as a breakdown product. Quaternary ammonium salts are first attacked with a hydroxylation of the far end of the alkyl chain (Dean-Raymond and Alexander, 1977), followed by a central fission of the C_{alkyl}–N bond (van Ginkel, van Dijk, and Kroon, 1992). Some microorganisms (e.g., *Pseudomonas* sp.; Kroon, Pomper, and van Ginkel, 1994) may access the alkyl chains of alkylamines and alkyltrimethylammonium salts by C_{alkyl}–N fission, with the production of ammonia, methylamine, and dimethylamine. Amphoteric surfactants, such as alkylbetaines, are mineralized by the presence of three microbes, the first of which cleaves the C_{alkyl}–N bond.

Low concentrations (<50 ng/g soil) of linear alcohol ethoxylate and linear alkyl benzene sulfonate surfactants can be degraded by natural soil microorganisms (Knaebel, Federle, and Vestal, 1990). However, only a few surfactants, such as alkane sulfonates, alkyl sulfates, and alkylamines, can be mineralized completely by one microorganism (van Ginkel, 1996). Because of the limited metabolic capabilities of individual microorganisms, however, mixed cultures are generally required for complete biodegradation of most surfactants and their metabolites, which can sometimes be toxic. A consortia of microbes may operate in a commensalistic or synergistic relationship, with some organisms utilizing the metabolic products of others.

Increasing anionic surfactant concentration from 1 to 4% does not inhibit microbial growth or biodegradation of the surfactant (Sundaram, Sarwar, Bang, and Islam, 1994). A 1% anionic surfactant solution in the presence of 5 and 10% kerosene showed that kerosene would degrade faster than the surfactant. Even though the surfactant was biodegraded, it enhanced the biodegradation of the petroleum contaminant, in a synergistic process.

The linear primary alcohol ethoxylates, fatty acid ethoxylates, and polyethylene glycol are all biodegradable (Adams, Spitzer, and Cowan, 1996). Biomineralization of the linear secondary alcohol ethoxylate, ethylene oxide/propylene oxide block copolymer, and alkylphenol ethoxylate is inhibited to varying degrees. Advanced oxidation pretreatment with hydrogen peroxide in combination with ozone is highly effective for enhancing biodegradability of both the linear secondary alcohol ethoxylate and the ethylene oxide/propylene oxide surfactants, but not that of alkylphenol ethoxylate.

Knaebel and Vestal (1992) determined that rhizosphere microbial communities could enhance the biodegradation of surfactants in soils by a factor of 1.1 to 1.9, by increasing the initial rate of mineralization.

5.3.2 PHOTOLYSIS

Photochemical reactions may be used for the enhancement of compound biodegradation at hazardous waste sites (Sims and Bass, 1984). Photolysis reactions are oxidative and should aid microbial degradation through the oxidation of resistant complex structures (Crosby, 1971; Sims and Overcash, 1981). Such reactions are limited to the surface of the soil or surface treatment of groundwater but, when coupled with soil mixing, may prove to be effective for treating relatively immobile chemicals.

Photolysis can be due either to direct light absorbed by the substrate molecule (direct) or to reactions mitigated by an energy-transferring sensitizer molecule (sensitized photo-oxidation) (Sims and Bass, 1984). Sensitized reactions result in substrate molecule oxidation rather than substrate isomerism, dehalogenation, or dissociation characteristic of direct photolysis. The reaction rates and breakdown

Table 5.18 Rate Constants for the Hydroxide Radical Reaction in Air with Various Organic Substances: $K_{OH°}$ in Units of $(mol\text{-}sec)^{-1}$

Substance	$\log_{air} K_{OH°}$	Substance	$\log_{air} K_{OH°}$
Acetaldehyde	9.98	Formaldehyde	9.78
Acrolein	10.42	Hexachlorocyclopentadiene	10.55
Acrylonitrile	9.08	Maleic anhydride	10.56
Allyl chloride	10.23	Methanol	8.78
Benzene	8.95	Methyl acetate	8.04
Benzyl chloride	9.26	Methyl chloroform	6.86
bis(Chloromethyl)ether	9.38	Methyl ethyl ketone	9.32
Carbon tetrachloride	<5.78	Methylene chloride	7.93
Chlorobenzene	8.38	Methyl propionate	8.23
Chloroform	7.78	Nitrobenzene	7.56
Chloromethylmethyl ether	9.26	Nitromethane	8.81
Chloroprene	10.44	2-Nitropropane	10.52
o-, m-, p-Cresol	10.52	n-Nitrosodiethylamine	10.19
p-Cresol	10.49	Nitrosoethylurea	9.89
Dichlorobromobenzene	8.26	n-Propylacetate	9.41
Diethyl ether	9.73	Perchloroethylene	8.01
Dimethyl nitrosamine	10.37	Phenol	10.01
Dioxane	9.26	Phosgene	Nonreactive
Epichlorohydrin	9.08	Polychlorinated biphenyls	<8.78
1,2-Epoxybutane	9.16	Propanol	9.51
Epoxypropane	8.89	Propylene oxide	8.89
Ethanol	9.28	Tetrahydrofuran	9.95
Ethyl acetate	9.06	Toluene	9.52, 9.56
Ethyl propionate	9.03	Trichloroethylene	9.12
Ethylene dibromide	8.18	Vinylidene chloride	9.38
Ethylene dichloride	8.12	o-, m-, p-Xylene	9.98
Ethylene oxide	9.08		

Source: Sims, R. and Bass, J. EPA Report No. EPA-540/2-84-003a, 1984.

products are only crudely understood. Photolysis can be affected by soil characteristics, such as soil organic content (Spencer, Adam, Shoup, and Spear, 1980), transition metal content (Nilles and Zabik, 1975), soil pigment content (Burkhard and Guth, 1979), and the soil moisture content (Burkhard and Guth, 1979).

The major photoreaction taking place with pesticides in the atmosphere is oxidation (Crosby, 1971) involving the OH radical or ozone, of which the OH radical is the species of greatest reactivity (Lemaire, Campbell, Hulpke, Guth, Merz, Philop, and Von Waldow, 1982). Based on a first-order rate of reaction of vapor-phase reactions with the OH radical, the half-life of a specific chemical species can be estimated, if its OH radical reaction rate constant is known, using:

$$t_{1/2} = 0.693 / K_{OH^0} [OH^0]$$

where $t_{1/2}$ = time to decrease component concentration by 50%
k_{OH^0} = OH radical reaction rate constant (cm³/molecule)
$[OH^0]$ = atmospheric OH radical concentration (4×10^5 molecules/cm³)
       = 6645×10^{-19} mol/cm³)

Table 5.18 presents the OH radical reaction rate constants for various organic compounds (Sims and Bass, 1984). The higher the number, the faster the oxidation of the compounds.

In order to assess the potential for use of photodegradation, the atmospheric reaction rate of the specific compound (log K_{OH^0}) and the anticipated reaction products must be known (Crosby, 1971). If a compound is poorly photoreactive (e.g., a $t_{1/2}$ in the atmosphere greater than 1 day; see equation above) volatilization suppression may be required to maintain safe ambient air concentrations at the site.

Photochemical reactivity depends upon the hydrocarbon family (olefinic > alkyl benzene > benzene > paraffinic hydrocarbons), the number of carbon atoms present, and the presence of oxygen, sulfur, chlorine, or nitrogen groups (Bell, Morrison, and Chonnard, 1987). Groups that typically do not undergo direct photolysis include saturated aliphatics, alcohols, ethers, and amines (Sims and Bass, 1984). Photodegradable organic wastes generally include compounds with moderate to strong absorption in the >290-nm wavelength range. Such compounds generally have an extended conjugated hydrocarbon system or a group with an unsaturated heteroatom (e.g., carbonyl, azo, nitro). Enhanced photodegradation of soil contaminants may be accomplished through the addition of various proton donor materials to the contaminated soils. Table 5.19 presents additional constants, with an estimation of the likelihood of a photolysis reaction occurring within the ambient atmosphere. Photolysis and photochemical oxidation are further discussed in Sections 2.1.1.2.6, 2.1.2.1.7, and 6.3.4.6.

Table 5.19 Atmospheric Reaction Rates and Residence Times of Selected Organic Chemicals

Compound	$K_{OH} \times 10^{12}$ (cm³/sec⁻¹)	Direct Photolysis Probability	Physical Removal Probability	Residence Time (days)	Anticipated Photoproducts
Acetaldehyde	16	Probable	Unlikely	0.03–0.7	H_2CO, CO_2
Acrolein	44	Probable	Unlikely	0.2	$OCH-CHO$, H_2CO, $HCOOH$, CO_2
Acrylonitrile	2	—	Unlikely	5.6	H_2CO, CN^O, $HC(O)CN$, $HCOOH$
Carbon Tetrachloride	<0.001	—	Unlikely	>11,000	Cl_2CO, Cl^O
o-, m-, p-Cresol	55	—	Unlikely	0.2	Hydroxynitrotoluenes, ring cleavage products
Formaldehyde	10	Probable	Unlikely	0.1–1.2	CO, CO_2
Nitrobenzene	0.06	Possible	Unlikely	190	Nitrophenols, ring cleavage products
2-Nitropropane	55	Possible	Unlikely	0.2	H_2CO, CH_3CHO
Phenol	17	—	Possible	0.6	Dihydroxybenzenes, nitrophenols, ring cleavage products
Toluene	6	—	Unlikely	1.9	Benzaldehyde, cresols, ring cleavage products, nitro compounds
o-, m-, p-Xylene	16	—	Unlikely	0.7	Substituted benzaldehydes, hydroxy xylenes, ring cleavage products, nitro compounds

Source: Sims, R. and Bass, J. EPA Report No. EPA-540/2-84-003a, 1984.

Section 6

Volatile Organic Compounds in Petroleum Products

6.1 EMISSIONS PRODUCED FROM SOIL CONTAMINATION

Management of hazardous wastes involves many operations that can result in air emissions (Allen and Blaney, 1985). For example, disposal of wastes in landfills may release volatile organic compounds (VOCs). Waste transfer and handling operations may also be a significant source of VOC emissions. Emissions from hazardous waste treatment storage and disposal facilities (TSDFs) have been estimated to be in excess of 1.5 Mt/year, and possibly over 5 Mt/year (Breton et al., 1983). It has been reported that at least one third of the total emissions of over 50 volatile, hazardous chemicals are from TSDFs (Springer, Valsaraj, and Thibodeaux, 1986). VOC emissions from industry, transportation, and other sources have been estimated to be 10.7, 7.7, and 3.9 Mt/year, respectively (Breton et al., 1983). Table 6.1 gives the estimated relative emissions from various types of TSDFs.

Landfarming can be used if volatilization of VOCs into the air is permitted; however, there is growing concern over VOC emissions and resulting air pollution (IT Corporation, 1987). Emissions from landtreatment facilities arise by volatilization from the wastes that have been spread on the soil prior to being incorporated within the top layers (Ehrenfeld, Ong, Farino, Spawn, Jasinski, Murphy, Dixon, and Rissmann, 1986). They later arise after the wastes have been mixed into the soil, as the materials volatilize and diffuse upward through the soil. The process of spreading and tilling of the contaminated soil results in volatilization of a significant fraction (up to 40 to 60%) of the volatile organics. By using the emission isolation flex chamber method to measure VOC emission rates, it was found that tilling caused a two- to tenfold increase in the emission rates, with peak emissions occurring within the first 4 h after tilling (Blaney, Eklund, Thorneloe, and Wetherold, 1986). Results from a landtreatment facility demonstrated that more than 90% of the organic compounds in hazardous oily wastes from a refinery were being biologically degraded, transformed, and volatilized in the soil (Fuller, Hinzel, Olsen, and Smith, 1986).

Gases may be generated by reactions in the subsurface (Ehrenfeld, Ong, Farino, Spawn, Jasinski, Murphy, Dixon, and Rissmann, 1986). Aerobic or anaerobic biological activity may decompose organics to produce methane, hydrogen sulfide, carbon dioxide, or other gases, which bubble up through the impoundment, carrying volatile materials to the surface. The addition of microorganisms or aeration of the soil for stimulation of biodegradation would also increase volatile emissions. Chemical reactions may also increase emissions, if gases are produced.

As VOCs pass through the soil, they can undergo a variety of transformations, such as biodegradation, adsorption onto the soil, dissolution in the soil water, and leaching into the groundwater (Valsaraj and Thibodeaux, 1988). Volatilization includes the loss of chemicals from surfaces in the vapor phase, indicating that it requires the vaporization and movement of chemicals from a surface into the atmosphere above the surface (Dupont and Reineman, 1986).

The abiotic process of evaporation can contribute significantly to the overall removal process of contaminants from soil (Kang and Oulman, 1996). The evaporation rate of VOCs can be predicted by a model, which indicates that the rate of evaporation for a particular volatile liquid is proportional to the square root of the product of diffusivity and partial pressure divided by the molecular weight of the liquid. This partially explains why evaporative losses from sand are so much higher for gasoline than for diesel fuel.

Volatile compounds are components in the soil and groundwater contamination at many, if not most, Superfund sites (Devitt, Evans, Jury, Starks, Eklund, and Gholson, 1987). Table 6.2 lists the 25 compounds most frequently reported at Superfund sites; 15 of these are volatile organic solvents. Table 6.3 shows more of the VOCs on the Hazardous Substance List (McDevitt, Noland, and Marks, 1987).

The predominant hazardous volatile organics in hazardous wastes from the petroleum refining industry are benzene and toluene (Overcash, Brown, and Evans, 1987). The total estimated amounts of benzene

Table 6.1 Emissions from Hazardous Waste Treatment, Storage, and Disposal Facilities

Facility Type	Estimated Annual Emissions[a] (10³ metric t/year)
Treatment tanks	530
Nonaerated surface impoundment — storage	420
Nonaerated surface impoundment — treatment	310
Landfill	190
Nonaerated surface impoundment — disposal	66
Aerated surface impoundment	66
Land applications	43
Storage tanks	10
Total	1635

[a] For 54 selected chemicals.

Source: Breton, M. et al. GCA Report No. GCA-TR-83-70-G, U.S. EPA. August 1983.

Table 6.2 Most Frequently Reported Substances at 546 MPL Sites

Substance	Percent of Sites
Trichloroethylene	33
Lead	30
Toluene	28
Benzene	26
Polychlorinated biphenyls (PCBs)	22
Chloroform	20
Tetrachloroethylene	16
Phenol	15
Arsenic	15
Cadmium	15
Chromium	15
1,1,1-Trichloroethane	14
Zinc and compounds	14
Ethylbenzene	13
Xylene	13
Methylene chloride	12
trans-1,2-Dichloroethylene	11
Mercury	10
Copper and compounds	9
Cyanides (soluble salts)	8
Vinyl chloride	8
1.2-Dichloroethane	8
Chlorobenzene	8
1,1-Dichloroethane	8
Carbon tetrachloride	7

Source: Devitt, D.A. et al. Report No. EPA-600/8-87/036, 1987.

and toluene in wastes treated by land annually by the U.S. petroleum industry are 150,000 and 950 lb, respectively. However, landtreatment of hazardous wastes is a minor source of VOCs, compared with other emissions sources.

The main focus of Federal and state regulations of TSDFs in the past has been to minimize contamination of surface and groundwater, to prevent air contamination by incineration, and to prevent accidental exposure (Springer, Valsaraj, and Thibodeaux, 1986). More recently, the emphasis has shifted to the emissions themselves and is focusing on specific hazardous constituents. The transfer of VOCs to ambient air is a concern of Michigan, for example (Love, Ruggiero, Feige, Carswell, Miltner, Clark, and Fronk,

Table 6.3 VOCs Included on the Hazardous Substance List (HSL)

	Detection Limits[a]	
VOCs	Low Water (μg/L)	Low Soil/Sediment (μg/kg)
Chloromethane	10	10
Bromomethane	10	10
Vinyl chloride	10	10
Chloroethane	10	10
Methylene chloride	5	5
Acetone	10	10
Carbon disulfide	5	5
1,1-Dichloroethene	5	5
1,1-Dichloroethane	5	5
trans-1,2-Dichloroethene	5	5
Chloroform	5	5
1,2-Dichloroethane	5	5
2-Butanone	10	10
1,1,1-Trichloroethane	5	5
Carbon tetrachloride	5	5
Vinyl acetate	10	10
Bromodichloromethane	5	5
1,1,2,2-Tetrachloroethane	5	5
1,2-Dichloropropane	5	5
trans-1,3-Dichloropropene	5	5
Trichloroethene	5	5
Dibromochloromethane	5	5
1,1,2-Trichloroethane	5	5
Benzene	5	5
cis-1,3-Dichloropropene	5	5
2-Chloroethyl vinyl ether	10	10
Bromoform	5	5
2-Hexanone	10	10
4-Methyl-2-pentanone	10	10
Tetrachloroethene	5	5
Toluene	5	5
Chlorobenzene	5	5
Ethyl benzene	5	5
Styrene	5	5
Total xylenes	5	5

[a] Medium water contract required detection limits (CRDL) for volatile HSL compounds are 100 times the individual low-water CRDL. Medium soil/sediment CRDL for volatile HSL compounds are 100 times the individual low soil/sediment CRDL. Detection limits listed for soil/sediment are based upon wet weight.

Source: McDevitt, N.P. et al. Report No. AMXTH-TE-CR-86092. ADA 178261. U.S. Army Toxic and Hazardous Materials Agency, Aberdeen, MD, 1987.

1983). Here, the "best available technology" must be applied to treatment trains discharging to the air, and carcinogens must not be transferred to ambient air.

The release rates of VOCs in a landtreatment system are characterized by a peak that occurs immediately upon waste application, followed by a rapid, exponential decline approaching steady conditions within minutes to hours (Figure 6.1; Overcash, Brown, and Evans, 1987). Additional tillage of the soil briefly releases another peak of volatiles of lesser magnitude, but the emission rates quickly return to those before the tillage.

Bolick and Wilson (1994) report that the extent to which the contaminant VOC has spread in the subsurface has a significant effect on the cleanup time required, indicating that very substantial savings in cleanup costs can result from rapid response after a spill has occurred. It is suggested that if preliminary pumping is started as soon after the incident as possible, even before negotiations are completed,

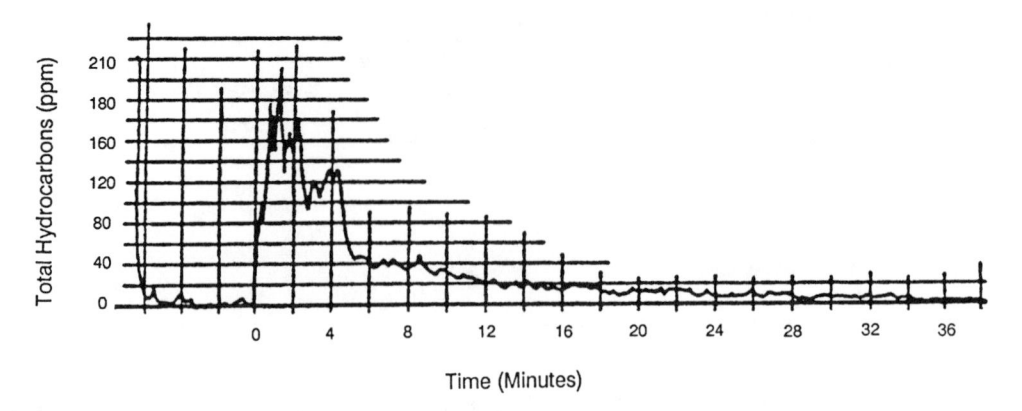

Figure 6.1 Typical chart trace from a total hydrocarbon monitor. (From Overcash, M. et al. Report No. ANL/EES-TM-340. DE88005571. Argonne National Laboratory, Argonne, IL, 1987.)

groundwater contamination may be reduced or avoided, and it would take considerably less time to remediate the site (Bolick and Wilson, 1994).

As volatile materials move through the hazardous waste management process, they must be destroyed, accumulated, emitted, or recycled to a prior step at each stage (Ehrenfeld, Ong, Farino, Spawn, Jasinski, Murphy, Dixon, and Rissmann, 1986). Design and operating practices, *in situ* treatment techniques, and pre- and posttreatment methods can help reduce emissions. It has been proposed that the choice of control method should be dictated by facility and environmental setting, rather than by waste properties, but waste properties can influence the effectiveness of the treatment. Permeability of the contaminated soil greatly affects cleanup times (Bolick and Wilson, 1994). The extent of the VOC contamination should be determined. However, greater emphasis should be placed on permeability measurements than on measurement of soil VOC concentrations, which have less effect, for calculating cleanup times by soil vapor extraction (SVE).

Effectiveness of a treatment is measured, primarily, by the degree of reduction of the *rate* of emissions, not by the reduction in total emissions over long periods of time (Ehrenfeld, Ong, Farino, Spawn, Jasinski, Murphy, Dixon, and Rissmann, 1986). Retarding the rate of loss by volatilization keeps the materials in the facility for longer periods. If other loss mechanisms, such as biodegradation, are also involved, then reducing the emission rate may change the absolute quantity emitted over time. An overall system for the treatment, storage, and disposal of VOCs is given in Figure 6.2.

Two methods of applying waste oily sludge were tested at a landtreatment site to compare the effects of application on the emissions (Eklund, Nelson, and Wetherold, 1987). Waste was applied either on the surface or injected 6 to 11 in. (0.15 to 0.28 m) below the surface and immediately disked. The plots were tilled two to three times a week for 5 weeks. The annual oil loading was about 8 to 27 kg/m^2 (300 to 1000 tons of wastes per acre at 12% oil). The volatile organics constituted about 0.8% of the sludge.

The average emission rate of volatile organics over the 5 weeks from surface application was 47.1 µg/m^2-s (RSD or relative standard deviation = ±14.9%), with a background of 6.16 µg/m^2-s (RSD = ±10.6%) (Eklund, Nelson, and Wetherold, 1987). The average emission from subsurface application was 53.9 µg/m^2-s (RSD = ±16.3%). The instantaneous emissions from the three were as high as 370.7, 38.5, and 324.9 µg/m^2-s, respectively. The emission rate decreased exponentially over time. The ratio of volatile organics released over 5 weeks to purgeable organics in the waste was 0.30 for the surface application and 0.36 for the subsurface. The ratios of volatile organics emitted over 5 weeks to the mass of applied oil were around 0.012 and 0.014 for the two plots, respectively. Table 6.4 shows the cumulative emissions for individual compounds and the percentage of the applied quantity for each. Individual compounds behaved in the same manner as the total volatile organics, with diurnal fluctuations.

Of 2896 tons of sludge applied to the landtreatment site over a year, approximately 43%, or 8900 kg/year, of the VOCs were emitted; 6,400 kg was released during the first 5 weeks after application. If emissions controls were 50% effective, a 4500 kg/year reduction in VOC emissions might be obtained. An emissions control of 90% efficiency would result in an 8000 kg/year reduction in emissions.

VOLATILE ORGANIC COMPOUNDS IN PETROLEUM PRODUCTS

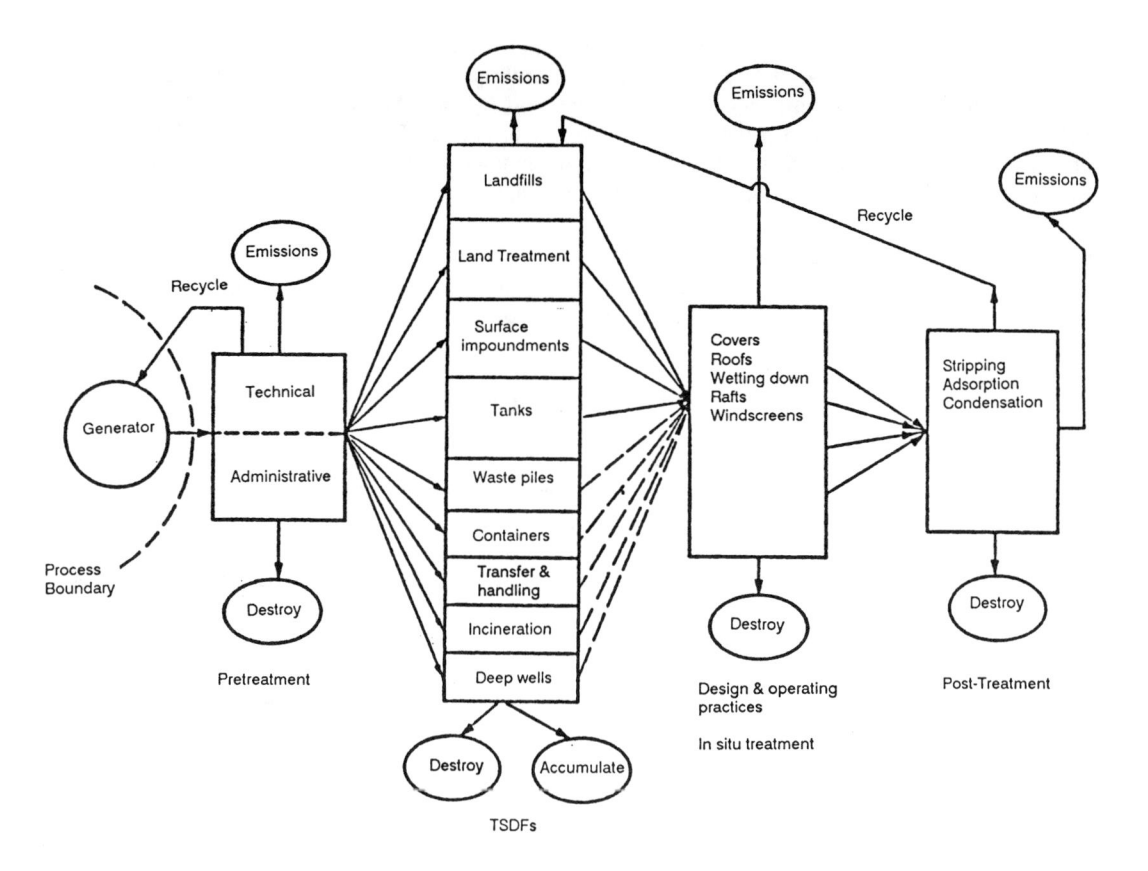

Figure 6.2 Overall system for treatment, storage, and disposal of VOCs. (With permission. Ehrenfeld, J.R. *Surface Impoundments*. Noyes, Park Ridge, NJ, 1986.)

Table 6.4 Cumulative Measured Emissions of Selected Individual Compounds

	Cumulative Emissions (g)			As Percent of Applied Control		
Compound	Plot A (Surface)	Plot B (Background)	Plot C (Subsurface)	Plot A (Surface)	Plot B (Background)	Plot C (Subsurface)
n-Heptane[a]	152	1.06	243	9.36	0.07	15.0
Methylcyclohexane	183	2.12	272	9.40	0.11	14.0
3-Methylheptane	161	3.05	249	8.04	0.15	12.4
n-Nonane[b]	131	4.20	190	8.83	0.28	12.8
1-Methylcyclohexane	32.0	0.802	56.2	6.97	0.18	12.2
1-Octene[a]	35.9	1.18	53.5	7.77	0.26	11.6
β-Pinene	58.5	3.43	75.5	2.61	0.15	3.37
Limonene[b]	39.3	2.73	45.8	3.47	0.24	4.04
Toluene[a]	271	2.18	403	6.24	0.05	9.29
p-, m-Xylene	219	3.22	311	5.59	0.08	7.93
1,3,5-Trimethylbenzene	74.1	3.85	97.8	3.23	0.17	4.27
o-Ethyltoluene[b]	88.9	4.1	122	4.89	0.23	6.70

[a] Relatively light compounds in each class.
[b] Heaviest compound in each class.

Source: Eklund, B.M. et al. EPA-600/2-87/086a, Cincinnati, OH, 1987.

A monoclonal antibody immunoassay for the rapid, on-site screening of gasoline- and diesel fuel–contaminated soil has been developed to detect volatile and semivolatile refined petroleum products (gasoline, diesel fuel, kerosene, and jet fuel) in the field (Mapes, McKenzie, Arrowood, Studabaker, Allen, Manning, and Friedman, 1993). The test involves use of PETRO RISc soil and ELISA (enzyme-linked immunosorbent assay), which detect these compounds at 100 and 75 ppm, respectively. It is simple to conduct, requires <20 min to perform, and is applicable to field testing.

6.1.1 GASOLINE VAPOR COMPOSITION

Gasoline is a clear, volatile liquid that is a complex mixture of paraffinic, olefinic, and aromatic hydrocarbons (Phillips and Jones, 1978). The liquid gasoline contains up to 250 constituents and the vapor phase, from 15 to 70 components.

The composition of gasoline varies greatly as a result of crude oil characteristics, processing techniques, and climate; therefore, there is no single threshold limit value (TLV) for all types of these materials (McDermott and Killiany, 1978). Light hydrocarbon compounds, such as butanes and pentanes, are blended into the gasoline to achieve the right volatility. To improve antiknock performance, branched chain aliphatic hydrocarbons or aromatic compounds resistant to detonation are blended into the gasoline. A gas chromatographic analytical technique can separate gasoline vapor into about 142 different components (Phillips and Jones, 1978). The aromatic hydrocarbon content generally determines the particular TLV. Consequently, the content of benzene, other aromatics, and additives should be determined to arrive at an appropriate TLV (Runion, 1975).

The vapor contains more lighter hydrocarbons when bulk liquid gasoline evaporates during loading or dispensing activities (McDermott and Killiany, 1978). About 92% of the total gasoline vapor by volume is represented by 21 hydrocarbon compounds (Table 6.5). About 43% of the vapor in an average sample consists of butanes. When benzene constitutes about 1% of the liquid gasoline, it contributes about 0.7% to the total gasoline vapor.

There is no Federal OSHA permissible exposure limit for gasoline, although there are separate limits for 14 individual hydrocarbon constituents and some additives (Phillips and Jones, 1978). The gasoline vapor component data were used to estimate a TLV for mixtures. Based on the toxicity of hydrocarbon compounds in gasoline vapor, 240 ppm TWA (time weighted average) exposure over 8 h and a 1000-ppm peak over 15 min have been suggested as reasonable criteria.

6.1.2 HUMAN HEALTH CRITERIA

The U.S. EPA is developing information to set standards, as necessary, to control emissions from hazardous waste TSDFs (Eklund, Nelson, and Wetherold, 1987). These regulations are intended to protect human health and the environment from emissions of volatile compounds and particulate matter.

The human health criteria provide estimates of ambient water concentrations, which, in the case of noncarcinogens, prevent adverse health effects in humans, and, in the case of suspected or proven carcinogens, represent various levels of incremental cancer risk (Bove, Lambert, Lin, Sullivan, and Marks, 1984).

There is no method to establish the presence of a threshold for carcinogenic effects. The EPA policy is that there is no scientific basis for estimating "safe" levels for carcinogens. Therefore, the criteria for carcinogens state that the recommended concentration for maximum protection of human health is zero. The EPA Water Quality Criteria for protection of human health are presented in Table 6.6 for a 10^{-5} risk level (i.e., one additional case of cancer in a population of 100,000).

The TLV listing, published by the American Conference of Governmental Industrial Hygienists (ACGIH), is a major source of guidelines for safe exposure to toxic compounds (American Conference of Governmental Industrial Hygienists, 1976).

Threshold limit values refer to airborne concentrations of substances and represent conditions under which it is believed nearly all workers may be repeatedly exposed day after day without adverse effect (American Conference of Governmental Industrial Hygienists, 1982). Threshold limits are based upon the best available information from industrial experience, from experimental human and animal studies, and, when possible, from a combination of the three. The TLVs, as issued by ACGIH, are recommendations to be used as guidelines for good practices. They do not have the force and effects of law. When two or more hazardous substances, which act upon the same organ system, are present, their combined

Table 6.5 Approximate Gasoline Vapor Components

Compound	Boiling Point (°C)	Airborne Gasoline Vapor Composition mean vol%
Alkanes		
Propane	−42.1	0.8
n-Butane	−0.5	38.1
Isobutane	−11.7	5.2
Isopentane	27.9	22.9
n-Pentane	36.1	7.0
Cyclopentane	49.3	0.7
2,3-Dimethylbutane	58.0	0.7
2-Methylpentane	60.3	2.1
3-Methylpentane	63.3	1.6
n-Hexane	68.7	1.5
Methyl cyclopentane	71.8	1.3
2,4-Dimethylpentane	80.3	0.4
2,3-Dimethylpentane	89.8	0.7
2,2,4-Trimethylpentane	99.2	0.5
Alkenes		
Isobutylene	−6.9	1.1
2-Methyl-1-butene	31.2	1.6
cis-2-Pentene	37.0	1.2
2-Methyl-2-butene	38.6	1.7
Aromatics		
Benzene	80.1	0.7
Toluene	110.6	1.8
Xylene (p, m, o)	142.0	0.5
Total Percent		92.1

Source: McDermott, H.J. and Killiany, S.E., Jr. *Am. Ind. Hyg. Assoc. J.* 39:110–117, 1978. With permission.

effect, rather than that of either individually, should be given primary consideration. If such information is not available, the effects of the different hazards should be considered as additive.

6.2 PARAMETERS AFFECTING VOLATILIZATION

For volatilization from soil to occur, organic compounds must move through a complex structure of solid particles and void spaces to the soil surface (Bell, Morrison, and Chonnard, 1987). At the surface, the pollutant must then traverse a relatively stagnant atmospheric film of air to escape into the atmosphere. An understanding of the mechanisms of transport of a pollutant through the soil is very important for predicting its volatilization from the soil. Several important mechanisms of pollutant transport are diffusion through the vapor and aqueous phases, flow of water-soluble pollutants to the surface due to capillary action, and evaporation of water from the soil surface.

The rate of contaminant volatilization is a complex function of the properties of the contaminant and its surrounding environment (Dupont and Reineman, 1986). For organics in soil systems, the factors that affect volatilization include (Spencer and Cliath, 1977; Ehrenfeld, Ong, Farino, Spawn, Jasinski, Murphy, Dixon, and Rissmann, 1986):

- Contaminant vapor pressure
- Contaminant concentration
- The Henry's law constant of the waste
- Soil/chemical adsorption reactions
- Contaminant solubility in soil water
- Contaminant solubility in soil organic matter
- Soil temperature, water content, organic content, porosity, and bulk density

Table 6.6 EPA Water Quality Criteria[a,b] for Protection of Human Health (10^{-5} Risk Level)

Contaminant	µg/L	Contaminant	µg/L
Volatile Organics			
Acrolein	320	1,2-Dichloropropylene	87
Acrylonitrile	0.58[d]	Ethylbenzene	1,400
Benzene	6.6[d]	Methylene chloride (dichloromethylene)	1.9[d]
Bis (chloromethyl) ether	0.000038[d]	1,1,2,2-Tetrachloroethane	1.7[d]
Carbon tetrachloride	4.0[d]	Tetrachloroethylene	8[d]
Chlorobenzene	488 (20)[e]	Toluene	14,300
Chlorodibromomethane	1.9[d]	1,1,1-Trichloroethane	18,400
Chloroform	1.9[d]	1,1,2-Trichloroethane	6.0[d]
Dichlorobromomethane	1.9[d]	Trichloroethylene	27[d]
Dichlorodifluoromethane	1.9[d]	Trichlorofluoromethane	1.9[d]
1,2-Dichloroethane	9.4[d]	Vinyl chloride	20[d]
1,1-Dichloroethylene	0.33[d]		
Acid Extractables			
Pentachlorophenol	1,010 (30)[e]	Phenol	3,500 (300)[e]
Base/Neutral Extractables			
Fluoranthene	42	Naphthalene	NDA[c]
Metals and Cyanide			
Antimony	146	Mercury	0.144
Arsenic	0.022[d]	Nickel	13.4
Beryllium	0.037[d]	Selenium	10
Cadmium	10	Silver	50
Chromium	50	Thallium	13
Copper	NDA	Zinc	NDA
Lead	50	Cyanide	200

[a] EPA water quality criteria documents (45 FR 79318, 28 November 1980).
[b] Values in micrograms per liter (µg/L).
[c] No definitive data available.
[d] 10^{-5} cancer risk criteria.
[e] Taste and odor (organoleptic) criteria.

Source: Bove, L.J. et al. Report to U.S. Army Toxic and Hazardous Materials Agency, Aberdeen Proving Ground, MD, on Contract No. DAAK11-82-C-0017. AD-A162 528/4. 1984.

Wind, humidity, and solar radiation
Adsorption to soil (Valsaraj and Thibodeaux, 1988)

The major contaminant property affecting volatilization is its vapor pressure, while the major environmental factors affecting contaminant mobility are the various soil/air, soil/water, and air/water partition coefficients for the various soil/water/air environments present within the soil system (Dupont and Reineman, 1986). It becomes more complex if the contaminant is added in a carrier fluid, such as oil in refinery wastes, where partitioning of the contaminant between the oil/soil, oil/water, and oil/air phases would also affect the volatilization of hazardous compounds in the waste.

The chemical and physical properties of organic contaminants and the properties of the unsaturated zone that affect emissions are discussed below.

6.2.1 TEMPERATURE

Soil temperature and the gradient that is established within the unsaturated zone can have an impact on the status of organic compounds (Devitt, Evans, Jury, Starks, Eklund, and Gholson, 1987). If there are large temperature gradients (surface layers), thermal diffusion will readily take place. Warming the soil lowers the suction and raises the vapor pressure of soil water (Hillel, 1971). Thus, a thermal gradient induces flow and distillation from warmer to cooler regions. Organic vapors migrating from the groundwater to the soil surface during warm months and during the daytime will have to move against a temperature gradient (i.e., movement by concentration gradient). During colder months, if the soil surface freezes, vapors may not be able to escape and would concentrate or move laterally. Organic compounds with boiling points lower than soil temperatures, such as the gaseous alkanes propane and isobutane, which boil at −42.1 and −11.17°C, will be highly volatile (Mackay and Shiu, 1981).

Increasing the temperature increases the vapor pressure of a compound (Ehrenfeld, Ong, Farino, Spawn, Jasinski, Murphy, Dixon, and Rissmann, 1986). There is a three- to fourfold increase in soil vapor pressure for every 10°C increase in temperature (Dibble and Bartha, 1979a). The temperature of water greatly affects the diffusivity, which increases approximately as the 1.8 power of the absolute temperature (Springer, Valsaraj, and Thibodeaux, 1986). Thus, a high liquid temperature favors more-rapid volatilization.

In an evaluation of a vapor-phase carbon adsorption system for the removal of toluene from a contaminated airstream, it was found that by increasing the relative humidity from 30 to 90%, at constant temperature, the carbon loading could be cut by about 50% (Foster, 1985). Increasing the operating temperature from 20°F, with constant relative humidity, also reduces the carbon loading by 50%.

There is no apparent correlation between soil bed temperature and VOC removal efficiency (McDevitt, Noland, and Marks, 1987). There does appear to be an inverse relationship between the inlet air temperature with air stripping and the VOC removal efficiency. Although decreasing inlet air temperature corresponds with increasing removal efficiency, it may not be the cause of it.

Section 5.1.2 further discusses the role of soil temperature in bioremediation and how it can be modified. Management of this factor can affect VOC emissions.

6.2.2 OPERATING SURFACE AREA

The rate of emissions is directly proportional to the operating surface area (Ehrenfeld, Ong, Farino, Spawn, Jasinski, Murphy, Dixon, and Rissmann, 1986). Minimizing surface area can help control emissions. This is essentially a linear relationship. The total quantity of materials volatilized would not be changed over the long run but would simply take longer to volatilize.

6.2.3 WIND/BAROMETRIC PRESSURE

The rate of emission into still air is slower than evaporation into the wind (Ehrenfeld, Ong, Farino, Spawn, Jasinski, Murphy, Dixon, and Rissmann, 1986). Surface turbulence by wind or mechanical agitation increases the rate of volatilization. Wind erosion of wastes depends upon waste type, wind velocity, moisture content, and surface geometry.

The effect of barometric pressure on soil gas transport is minor, being greatest at or near the soil surface (Buckingham, 1904). Periods of high wind and low barometric pressure may be optimal for maximum earth out-gassing (Reichmuth, 1984). However, this is a surface phenomenon.

6.2.4 SOIL MOISTURE/VOLUMETRIC WATER CONTENT

Soil moisture content provides an indication of VOC removal efficiency and possibly processed soil VOC residuals (McDevitt, Noland, and Marks, 1987). Soil moisture is important in determining the extent of adsorption of neutral, nonpolar molecules like most VOCs onto soil surfaces (Poe, Valsaraj, Thibodeaux, and Springer, 1988). Polar compounds show a greater degree of adsorption than nonpolar and slightly polar adsorbates. VOCs are strongly adsorbed to soils at low moisture contents. They are displaced from their adsorption sites as soil moisture increases, as a result of competition for adsorption sites on the polar mineral surface from polar water molecules.

The adsorption of VOCs by dry soils is considerable and is dominated by mineral adsorption (Poe, Valsaraj, Thibodeaux, and Springer, 1988). Most of the adsorption occurs on the external surface of the soils. The high degree of adsorption in dry soils retards the movement of volatile organics from hazardous waste landfills and from surface soil during land application of hazardous wastes.

The volumetric water content is the ratio of the volume of water in a porous medium to the total volume (Devitt, Evans, Jury, Starks, Eklund, and Gholson, 1987). When water fills the entire pore volume, the medium is saturated. Coarse soils have lower volumetric water contents at saturation than do medium-textured soils, and medium-textured soils have less than clayey soils. As the volumetric water content increases, the air-filled porosity decreases and the path for vapor flow becomes restricted.

It is common practice when wastes are being covered with soil to spray water over the soil as a dust control measure and to help compact the soil (Goring, 1962; Letey and Farmer, 1974). The amount of water added decreases the air-filled pore space available for vapor diffusion and, thus, affects the volatilization through the soil cover. By virtue of the solubility of benzene in water, increasing the soil water content will increase the capacity of the soil to retain benzene in the solution phase, reducing the quantity of benzene available for vapor-phase diffusion. The rate of volatilization and wicking of gasoline in soil is also reduced by an increase in soil water content (Smith, Stiver, and Zytner, 1995). Gasoline flux toward the soil surface is dependent upon wicking.

Organic contaminants in the unsaturated zone are susceptible to leaching, depending upon the frequency and amount of rainfall (Devitt, Evans, Jury, Starks, Eklund, and Gholson, 1987). Vertical concentration gradients are altered as vapors reaching the rainfall-saturated zone concentrate or move laterally or are resolubilized to some extent.

An equilibrium partitioning model has been developed to predict the effect of soil moisture content on vapor-phase sorption (Unger, Lam, Schaefer, and Kosson, 1996).

Soil moisture and how it can be controlled to improve biodegradation is further discussed in Section 5.1.1.

6.2.5 MASS TRANSFER COEFFICIENT/PARTITION COEFFICIENT

The VOC removal efficiency is directly related to the total VOC concentration in the feed soils; as the feed concentration increases, the VOC removal efficiency also increases (McDevitt, Noland, and Marks, 1987). The driving force for mass transfer is the difference between the VOC concentration in the airstream and the VOC concentration in the soil. An increase in the driving force causes an increase in mass transfer, with a corresponding increase in VOC removal efficiency.

The mass transfer process that governs the volatilization of almost all the chemicals of interest is the liquid-phase process (Springer, Valsaraj, and Thibodeaux, 1986). Therefore, differing rates from one chemical to another are largely a matter of liquid-phase diffusivity differences, and the rates do not vary greatly from one chemical to another, despite vapor pressure variations. It is assumed that the volatility is high enough that it is not limiting.

If the volatile materials are present as a dilute aqueous solution, the basic mass flow equation shows that the rate of emission depends upon the overall coefficient, the exposed area, and the concentration or mole fraction in the liquid (Ehrenfeld, Ong, Farino, Spawn, Jasinski, Murphy, Dixon, and Rissmann, 1986). The concentration in the gas phase in uncovered soil is essentially zero, with fresh air continually sweeping over the system.

Emissions can be controlled through the mass transfer coefficient (Ehrenfeld, Ong, Farino, Spawn, Jasinski, Murphy, Dixon, and Rissmann, 1986). The controllable parameters that determine the value of the mass transfer coefficient are the wind speed at the surface and the effective depth. Barriers and fences can reduce the wind speed. The dependence on depth is inverse. In theory, deeper impoundments have lower mass transfer coefficients, with reduced rates of volatilization.

In the case of a layer of lighter-than-water, immiscible organic compounds floating on the surface, the controlling mechanism will be diffusion in the gas phase, not the liquid phase (Ehrenfeld, Ong, Farino, Spawn, Jasinski, Murphy, Dixon, and Rissmann, 1986). The mass transfer coefficient then depends upon the operating parameters. Reducing the wind speed should inhibit the rate of volatilization. The rate of mass transfer depends upon the Henry's law constant, as well as the individual mass transfer coefficient. It is temperature dependent, increasing with increasing temperature. Controls that reduce the surface temperature would, therefore, inhibit the rate of volatilization.

Benzene may move by molecular diffusion in soil, in both the vapor phase and the solution phase. The relative importance of each phase is determined by the relative magnitude of the concentration in air (vapor density) and the concentration in solution (Goring, 1962; Letey and Farmer, 1974). Chemicals with partition coefficients between the soil water/soil air $\leqslant 10^4$ will diffuse mainly in the vapor phase, and those with higher coefficients will diffuse primarily in the solution phase. Since benzene has a partition coefficient of 4.6 at 25°C, it should diffuse primarily in the vapor phase.

6.2.6 EFFECTIVE DEPTH

Theoretically, the deeper the contaminant is in the subsurface, the lower the mass transfer coefficients and rates of volatilization (Ehrenfeld, Ong, Farino, Spawn, Jasinski, Murphy, Dixon, and Rissmann, 1986).

6.2.7 MOLE FRACTION OF DIFFUSING COMPONENT

The mole fraction, or concentration, refers to the amount of organic compound per unit amount of solvent (air/water) in such units as g/m^3 and ppb/v (Devitt, Evans, Jury, Starks, Eklund, and Gholson, 1987).

The concentration of the compound in the liquid affects the rate of emission (Ehrenfeld, Ong, Farino, Spawn, Jasinski, Murphy, Dixon, and Rissmann, 1986). The diffusion of a gas from areas of high concentration to low is the most important mechanism for gas transport in the unsaturated zone (Kreamer, 1982). In short test runs (230 to 285 min), aeration is not sufficient to promote volatilization when the driving force is low; i.e., low VOC concentrations (McDevitt, Noland, and Marks, 1987).

Pretreatment reduces the mole fraction in the liquid (Ehrenfeld, Ong, Farino, Spawn, Jasinski, Murphy, Dixon, and Rissmann, 1986). The effectiveness is linearly proportional to the degree of removal. If 50% of the volatile materials is removed, the rate of emission and the total quantity of wastes entering the atmosphere will be halved.

6.2.8 HUMIDITY

Humidity plays an important role in absorption and transport of vapors in soils (Batterman, Kulshrestha, and Cheng, 1995). Adsorption increases as the relative humidity drops below 90% (Chiou, 1985), and hydrocarbon vapors are considerably retarded in all media with soil gas humidities below 30%. This may significantly reduce the amount of organic vapors at or near the soil surface. Toluene has a retardation factor of 80. Methane is not retarded. As humidity increases in organic-rich soils, retardation coefficients decrease but remain large. Based on soil/water isotherms, there appears to be competitive sorption between hydrocarbon and water vapors on soil surfaces, especially the mineral fraction.

In an evaluation of a vapor-phase carbon adsorption system for the removal of toluene from a contaminated airstream, it was found that by increasing the relative humidity from 30 to 90%, at constant temperature, the carbon loading is cut by about 50% (Foster, 1985). Increasing the operating temperature from 20°F, with constant relative humidity, also reduces the carbon loading by 50%.

With air stripping, there may be an increase in removal efficiency with a decrease in moisture content of the inlet air (McDevitt, Noland, and Marks, 1987). The drier air could have a greater capacity to absorb moisture from the soil, and, as the moisture evaporates from the soil, the VOCs may also evaporate.

6.2.9 SOLAR RADIATION

If solar drying is used and volatile organics are present, emissions will be greatly increased (Ehrenfeld, Ong, Farino, Spawn, Jasinski, Murphy, Dixon, and Rissmann, 1986).

6.2.10 VAPOR PRESSURE

The pressure of the vapor of a liquid confined such that the vapor collects above it is referred to as the vapor pressure (Devitt, Evans, Jury, Starks, Eklund, and Gholson, 1987). It is a measure of the equilibrium between the liquid and vapor phases of a pure compound (Eckenfelder and Norris, 1993). Table 6.7 lists the vapor pressure values for several compounds. Thus, at spill sites, organic compounds with high vapor pressures would be expected to be present to some degree in the vapor phase of soil pores (Devitt, Evans, Jury, Starks, Eklund, and Gholson, 1987). Highly volatile fuels, such as gasoline, evaporate relatively rapidly, even in the subsoil, forming an envelope of hydrocarbon vapors around the core of the spill (Schwille, 1975).

The major contaminant property affecting volatilization is its vapor pressure in the soil airspace (Sims, Sorensen, Sims, McLean, Mahmood, and Dupont, 1985). The vapor pressure of the soil organic compound is the most important factor at low water content (presumably due to the vapor-phase diffusion), while with greater water content, aqueous-phase diffusion becomes most important (Ehlers, Letey, Spencer, and Farmer, 1969a; 1969b). The vapor pressure of an organic compound in the soil increases to an equilibrium value that corresponds to its vapor pressure (Bell, Morrison, and Chonnard, 1987). This is a result of increasing concentrations of the compound in the soil until there is saturation of adsorption sites on the soil mineral and organic fraction surfaces. Water has the ability to displace adsorbed organic molecules from the soil surface, due to preferential adsorption of water. Transport of a compound in the vapor phase is favored by high vapor pressures and Henry's law constants (Eckenfelder and Norris, 1993).

The aqueous vapor pressure measured in soil pores is considered to be vapor saturated (Devitt, Evans, Jury, Starks, Eklund, and Gholson, 1987). Vapor pressure increases with increasing temperature (Ehrenfeld,

Table 6.7 Vapor Pressure for Several Compounds

Compound	Vapor Pressure (mm Hg)
Acetone	231
Benzene	95
Butanol	7
Naphthalene	0.2
Phenanthrene	6.8×10^{-4}
Phenol	0.5
Pyrene	6.9×10^{-7}

Source: From Eckenfelder, W.W., Jr. and Norris, R.D. In *Emerging Technologies in Hazardous Waste Management* III. American Chemical Society, Washington, D.C. 1993. With permission.

Ong, Farino, Spawn, Jasinski, Murphy, Dixon, and Rissmann, 1986). Since vapors tend to move from warm to cold areas in a soil, they would tend to move downward during the day and upward during the night (Devitt, Evans, Jury, Starks, Eklund, and Gholson, 1987). At the soil surface and below for several inches, the vapor pressure can drop below saturation, because of higher gas mixing and exchange rates. The presence of electrolytes (often concentrated near the soil surface from evaporation) can also lower the vapor pressure. Reducing the vapor pressure in the soil pores will have a significant effect on the adsorption of organic vapors (Chiou, 1985). The mineral fraction of a dry or slightly hydrated soil is a powerful adsorbent for organic vapors at lower vapor pressures.

6.2.11 SOIL PROPERTIES

The texture of a soil refers to the proportions of various particle size groups in a soil mass, typically called sand, silt, and clay (Devitt, Evans, Jury, Starks, Eklund, and Gholson, 1987). Clayey soils have higher volumetric water content at saturation than medium-textured or coarse soils. As the clay content increases, the water-holding capacity and the exchange capacity increase, while the air-filled porosity and the rate of vapor diffusion decrease. A high clay content acts as a retarding layer to the vertical flux of VOCs. It is the rate of flux through the most-retarding layer that controls the vertical flux (Swallow and Gschwend, 1983). Soil monitoring will not be successful when the vadose zone contains high clay and water (Reid, Thompson, and Oberholtzer, 1985).

The steady-state vapor diffusion of benzene in soil under isothermal conditions and negligible water flow is directly related to soil air-filled porosity (Goring, 1962; Letey and Farmer, 1974). The volatilization flux of benzene through a soil cover is greatly reduced by an increase in soil bulk density and soil water content. Actual flux through the soil cover can be predicted from the soil porosity term, $P_a^{10/3}/P_T^2$, where P_a is the soil air-filled porosity and P_T is the total porosity. Adsorption of benzene by the soil matrix is not significant.

The markedly enhanced sorption of organic vapors at subsaturation humidities is attributed to adsorption on the mineral matter, which predominates over the simultaneous uptake by partition into the organic matter (Chiou and Shoup, 1985). Soil acts as a dual sorbent in which the mineral matter functions as a conventional solid adsorbent and organic matter as a partition medium. Also, interlayer swelling of montmorillonite can occur when polar molecules are adsorbed by dry soils, especially those with hydrogen-bonding potential, which causes an increase in adsorption (Poe, Valsaraj, Thibodeaux, and Springer, 1988).

Knowledge of the shape and size of pores or pore size distribution in soil is important for understanding the tortuous path vapors must travel to reach the soil surface (Devitt, Evans, Jury, Starks, Eklund, and Gholson, 1987). Some pores are totally blocked by interstitial water, reducing the rate of diffusion by orders of magnitude. The total porosity does not provide any indication of the pore size distribution. Clayey soils tend to have a more uniform pore size distribution than do coarser soils (Hillel, 1971), whereas the coarse soils tend to have larger mean pore sizes, which will transfer fluids faster under saturated conditions and vapors faster under unsaturated conditions. The diffusion coefficient must, therefore, compensate for this tortuous path for vapor flow. This is accomplished by replacing the diffusion coefficient with the effective diffusion coefficient.

See Section 5.1.9 for more information on soil texture and structure and how these factors affect biodegradation.

6.2.12 ADSORPTION ONTO SOIL

Toxic light aromatic hydrocarbons, such as toluene and naphthalene, from refined oil spillages do not evaporate when sorbed onto sediment particles (Horowitz, Sexstone, and Atlas, 1978). Adsorption onto the soil surfaces will greatly reduce the mobility of VOCs through soil (Poe, Valsaraj, Thibodeaux, and Springer, 1988). Adsorption can be both physical and chemical in nature; however, under natural environmental conditions and ambient temperatures, the predominant process is physical adsorption involving the London or van der Waals forces (Valsaraj and Thibodeaux, 1988). Chemisorption of VOCs is rarely seen, since it requires actual chemical bonding between the adsorbate and the adsorbent and is an energy-intensive process often occurring only at high temperatures (Adamson, 1982).

Soil matter consists mainly of mineral fractions, natural organic matter, and pore water (Valsaraj and Thibodeaux, 1988). Mineral matter, which is predominantly montmorillonite, illite, or kaolinite, and the organic matter provide large surface areas upon which physical adsorption or partitioning of molecules can occur. The amount of water in the soil affects adsorption by competing with the VOCs for these sites. Volatilization is retarded in soils with a higher organic content (du Plessis, Senior, and Hughes, 1994). Water is preferentially sorbed onto soil and displaces organic molecules (Bell, Morrison, and Chonnard, 1987). Over half the water in clay is bound tightly, in contrast with sand, which binds less than 5% (Fung, 1980). Gas movement is minimized in finely grained soil, while sand or gravel is the preferred soil for facilitating gas flow, as in venting systems.

The soil composition, in terms of mineral particle size and the fraction of organic matter, can affect the vapor pressure of a soil-applied organic compound (Spencer, 1970). The mineral fraction of dry soil at low vapor pressure sorbs organic vapors (Chiou, 1985). Organic compounds applied to the soil have a high affinity for the soil organic matter and preferentially adsorb at these sites, even under wet soil conditions.

In landfarming of oily wastes, the soils are often loaded with a relatively large amount of organic compounds (Bell, Morrison, and Chonnard, 1987), such as up to 5% wt oil/wt soil (Dibble and Bartha, 1979a). This is far in excess of the ability of the soil to adsorb it (0.001 to 0.01%). Thus, the vapor pressures and aqueous solubilities of the organic compound will be at saturation values during much of the volatilization.

See Sections 4.1.4 and 5.1.6 for more information on the effects of sorption on the biodegradation of organic compounds in soil and how these factors can be modified to enhance biodegradation.

6.2.13 EVAPORATION

Transport to the soil surface is controlled by diffusion and by mass flow of organic compounds during evaporation of water (Lyman, Rechl, and Rosenblatt, 1982). Initially, volatilization is controlled by the diffusion of the organic compound through the soil vapor and liquid phases and, only after a long time, does water evaporation become a dominant transport mechanism (Spencer and Claith, 1973; Jury, Grover, Spencer, and Farmer, 1980). This behavior is modified by the solubility of the organic compound in the aqueous phase. Evaporation of water increases volatilization from soils through the movement of highly water-soluble compounds to the soil surface by capillary action (Lyman, Rechl, and Rosenblatt, 1982). The organic compound then accumulates and volatilizes at the soil surface. The greatest VOC removal occurs during evaporation of moisture from the soil (Jury, Farmer, and Spencer, 1984). Evaporation most strongly influences volatilization for weakly adsorbed chemicals with nonnegligible vapor density.

6.2.14 WATER SOLUBILITY

The extent to which an organic compound (solute) dissolves in a solvent (water) is referred to as the water solubility of the compound (Devitt, Evans, Jury, Starks, Eklund, and Gholson, 1987). Organic compounds with high water solubility partition primarily into the liquid water phase. The rate at which these compounds move through the unsaturated zone is, therefore, controlled to a great extent by the unsaturated hydraulic conductivity of water in the porous medium. Compounds with high water solubility (from surface spills) would have shorter downward travel times. For oil spills, the hydrocarbon components with differing solubilities will dissolve out differentially and produce a simultaneous aging and leaching effect on the spill (Pfannkuch, 1985).

See Section 4.1.1 for a wider discussion of the effect of solubility on biodegradation of organic compounds in soil.

Table 6.8 Henry's Law Constants for Several Compounds

Compound	Henry's Law Constant (m^3/mol)
Acetone	3.7×10^{-5}
Benzene	5.4×10^{-3}
Butanol	5.6×10^{-6}
Naphthalene	4.6×10^{-4}
Phenanthrene	3.9×10^{-5}
Phenol	4×10^{-7}
Pyrene	1.1×10^{-5}

Source: From Eckenfelder, W.W., Jr. and Norris, R.D. In *Emerging Technologies in Hazardous Waste Management* III. American Chemical Society. Washington, D.C. 1993. With permission.

6.2.15 HENRY'S LAW CONSTANT

This constant is a ratio of partial pressure in the vapor to the concentration in the liquid (Mackay and Shiu, 1981). It is a coefficient that reflects the air/water partitioning. This is helpful in understanding in what phase an organic compound would most likely be found. Transport of a substance in the vapor phase is favored by high vapor pressures and Henry's law constants (Eckenfelder and Norris, 1993). Table 6.8 lists the Henry's law constants for several compounds.

6.2.16 DENSITY

The density of an organic compound refers to the amount of substance per unit volume (g/cm^3). Next to solubility, the difference in density between contaminant and groundwater is the next most important parameter in determining the contaminant migration relative to an aquifer (Schwille, 1984). Density differences of about 1% can significantly affect fluid movement, and the density differences between organic liquids and water are in excess of 1 and often 10% (Mackay, Roberts, and Cherry, 1985). Organic compounds with specific gravities of less than 1.0 associated with solubilities of less than 1% are referred to as floaters (New York State Department of Environmental Conservation, 1983).

Section 4.1.7 provides more information on the effect of density on biodegradation of organic compounds in soil.

6.2.17 VISCOSITY

The viscosity of a liquid organic compound is a measure of the degree to which it will resist flow under a given force measured in dyne-seconds per square centimeter (Devitt, Evans, Jury, Starks, Eklund, and Gholson, 1987). The viscosity of an organic fluid (such as oil) will affect the flow velocity (Schwille, 1984). The combination of density and viscosity will govern the migration of an immiscible organic liquid in the subsurface (Mackay, Roberts, and Cherry, 1985).

See Section 4.1.6 for background on the effect of viscosity on biodegradation of organic compounds in soil.

6.2.18 DIELECTRIC CONSTANT

The dielectric constant of a medium defines the relationship between two charges and the distance of separation of the two charges to the force of attraction (Devitt, Evans, Jury, Starks, Eklund, and Gholson, 1987). In a clay medium, this constant reflects the degree to which the clays will either shrink or swell. Liquids with a high dielectric constant (New York State Department of Environmental Conservation, 1983), such as water, would cause the clays to swell. Liquids with a low dielectric constant would cause the clays to shrink and, therefore, increase in permeability after exposure to concentrated organic liquids.

6.2.19 BOILING POINT

The boiling point of a compound is the temperature at which the external pressure of the liquid is in equilibrium with the saturation vapor pressure of the liquid (Devitt, Evans, Jury, Starks, Eklund, and Gholson, 1987). For higher boiling points, there is a general association with lower vapor pressures. If the boiling points of organic compounds are lower than the soil temperature (e.g., the gaseous alkanes, propane and isobutane, which boil at −42.1 and −11.17°C), the compounds will be highly volatile (Mackay and Shiu, 1981).

6.2.20 MOLECULAR WEIGHT
The molecular weight of an organic compound is the sum total of the weights of the atoms that compose it (Devitt, Evans, Jury, Starks, Eklund, and Gholson, 1987). For liquid alkanes, there is a tendency for the Henry's law constant to increase with increasing molecular weight, as the solubility falls more than the vapor pressure. High-molecular-weight hydrocarbons (especially aromatics) are decomposed through biodegradation at a much slower rate and, thus, would persist longer.

The effect of chemical structure on biodegradation is reviewed in Section 4.1.8.

6.2.21 AIR-FILLED POROSITY
The air-filled porosity of a porous medium, such as soil, is defined as the ratio of the volume of air in the soil pores to the total volume (volume of air, water, and soil combined) (Devitt, Evans, Jury, Starks, Eklund, and Gholson, 1987). It is the portion of the total soil volume not occupied by solid soil particles or by soil water. It is, thus, indicative of soil aeration and is inversely related to the degree of saturation. This is an important parameter for estimating the diffusion of gas in soil and unconsolidated material. The extent to which vapor-phase diffusion occurs in a soil will depend upon the air-filled porosity of that soil. Air-filled porosity, in turn, is determined in part by soil bulk density and soil water content. Generally, molecular diffusion in a soil is not related in a strict linear fashion to air-filled porosity but is modified by the tortuosity of the soil pores.

The diffusion coefficient of oxygen is approximately 10,000 times lower in water than in air (Letey and Stolzy, 1964). Thus, soil organic vapors migrating toward the soil surface would be restricted, if the water content increases and the air-filled porosity decreases. Vapors moving into low air-filled porosity zones could, potentially, be resolubilized. This parameter varies with rainfall and changes in soil texture (water-holding capacity).

The following equation can be used to reasonably estimate vapor diffusion at air-filled porosities >0.2 m³ m⁻¹ (Millington and Quirk, 1961):

$$D_s = D_o \, (P_a^{10/3}/P_T^2)$$

where D_o = vapor diffusion coefficient in air (m² s⁻¹)
P_a = air-filled porosity (m³ m⁻³)
P_T = total porosity (m³ m⁻³)

The soil porosity term, $P_a^{10/3}/P_T^2$, is applicable to predicting the diffusion in porous media of low-molecular-weight compounds like benzene.

The following equation can be used to assess the effect of altering the air-filled porosity of a soil cover on the volatilization flux of the compound from an industrial waste in a landfill:

$$J = -D_s \, (C_2 - C_s)/L$$

where J = vapor flux through the soil (kg m⁻² s⁻¹)
D_s = apparent steady-state vapor diffusion coefficient (m² s⁻¹)
C_2 = concentration in the air at the surface of the soil (kg m⁻³)
C_s = concentration in the air at depth L (kg m⁻³)
L = depth of the soil layer (m)

The following equation can be used to design a soil cover depth in order to reduce the flux to a specified value (Farmer, Yang, Letey, and Spencer, 1980):

$$J = D_s C_s /L$$

See Section 6.2.11, Soil Properties, which also discusses soil porosity.

Table 6.9 Time to Diffuse L = 1 m through a Soil[a]

	K_{OC} (mL/g)	K_H	T_D (day)
Benzene	8.3×10^1	2.2×10^{-1}	129
Biphenyl	1.4×10^3	6.6×10^{-2}	5100
Carbon tetrachloride	1.1×10^2	9.4×10^{-1}	45
Chlorobenzene	1.5×10^2	1.5×10^{-1}	292
Chloroform	2.9×10^1	1.2×10^{-1}	121
Chloromethane	3.9×10^1	2.2	16
DDT	2.4×10^5	2×10^{-3}	2.9×10^7
Dieldrin	1.2×10^4	6.7×10^{-4}	4.2×10^6
EPTC	2.8×10^2	5.9×10^{-4}	1.1×10^5
EDB	4.4×10^1	3.5×10^{-2}	500
Ethylene	8.5×10^1	3.6×10^2	9.3
Ethylene dichloride	2.9×10^1	4×10^{-2}	343
Lindane	1.3×10^3	1.3×10^{-4}	2.4×10^6
Methyl bromide	2.2×10^1	1.5	17
Naphthalene	1.3×10^3	5×10^{-2}	6.2×10^3
Nitrobenzene	7.1×10^1	1×10^{-3}	1.7×10^4
n-Octane	6.8×10^3	1.4×10^2	21
Phenol	2.7×10^1	7×10^{-6}	9.2×10^5
Trifluralin	7.3×10^3	6.7×10^{-3}	2.6×10^5
Trichloroethylene	9.8×10^1	4.4×10^{-1}	77
Toluene	1.4×10^2	3×10^{-1}	143
1,1,1-Trichloroethane	1×10^2	1.5	29
Vinyl chloride	4×10^2	9.7×10^1	10

[a] With $\phi = 0.5$, $a = 0.3$, $D_V^{air} = 4300$ cm²/day; foc = 0.005

Source: Devitt, D.A. et al. Report No. EPA-600/8-87/036, 1987.

6.2.22 RETENTION

Depending upon the solubility of the organic compound, the texture of the soil, and the pore size distribution, a portion of the liquid contaminant will be retained in the soil pores (Devitt, Evans, Jury, Starks, Eklund, and Gholson, 1987). The smaller capillaries in the soil are filled with gasoline only when the pressure drop overcomes capillary forces (Williams and Wilder, 1971). Thus, until the pressure drop $P_{gas} - P_{water}$ is greater than $P_{capillary}$, it is impossible to move the snapped-off gasoline bubbles through the throats of the pores. Water used for flushing the sand will, therefore, tend to flow through unblocked and continuous water-filled channels rather than through the gasoline-blocked channels.

6.2.23 DIFFUSION TRAVEL TIMES

For any spill with volatile components, a vapor phase will evolve above the dissolved phase as it migrates through groundwater (Devitt, Evans, Jury, Starks, Eklund, and Gholson, 1987). Vapor from the groundwater will move upward through the vadose zone by diffusion (reduced by dissolution and adsorption), by microbial degradation, and by chemical transformation. Table 6.9 gives the time for some compounds to diffuse 1 m through the soil.

A qualitative measure of diffusion for a chemical, called the characteristic diffusion time, t_D, is the time required for an organic chemical with an effective diffusion coefficient, D_E, to diffuse through a distance L (Jury, Spencer, and Farmer, 1983):

$$t_D = L^2/D_E$$

6.3 CONTROL OF VOC EMISSIONS

VOC emissions can be reduced by effective design and operating practices, *in situ* treatment techniques, and pre- and posttreatment methods (Ehrenfeld, Ong, Farino, Spawn, Jasinski, Murphy, Dixon, and Rissmann, 1986). In landtreatment systems, control can be achieved through the application of management techniques, along with the use of predictive models and laboratory analyses during the design and operation phases (Overcash, Brown, and Evans, 1987).

Several technologies have been recommended for removing volatile organics/solvents (Bove, Lambert, Lin, Sullivan, and Marks, 1984). These are *in situ* physical/chemical treatment (depending upon the treatment technology selected), *in situ* carbon, biological activated carbon, solvent refluxing (depending upon solubility of target contaminants in solvent), and aquaculture (for warm climatic conditions).

To evaluate a control scheme and to make a choice of control method, the purpose of the impoundment and the goals of the control strategy must be considered (Springer, Valsaraj, and Thibodeaux, 1986). To prevent volatile emissions entirely there must be some method of capture and destruction of the VOCs. If the aim is to reduce the rate of emission, which would lower the concentration of VOCs generated, then an absolute elimination might not be necessary.

Wastes containing VOCs that are treated and disposed of in surface impoundments and through land disposal have a high potential for emissions, because these operations have large, exposed surface areas that allow volatilization and are difficult to control (Allen and Blaney, 1985). If an impoundment is being used for aerobic digestion, then oxygen must be continually supplied. Devices, such as aerators, which are intended to increase oxygen transfer, will simultaneously increase the volatilization rate (Springer, Valsaraj, and Thibodeaux, 1986). On the other hand, methods that reduce volatilization will normally also reduce oxygen transfer.

There are several ways of controlling emissions from hazardous waste management operations (Allen and Blaney, 1985). Add-on (i.e., end-of-pipe) controls, such as flares or carbon canisters, can capture or destroy VOC emissions after they have migrated into the gas phase of the system. Capturing pollutants from emission sources of large areas can be expensive and may be complicated by factors, such as high humidity and mixtures of compounds, which can corrode flares or lead to adsorber breakthrough. The highly variable and complex composition of hazardous waste streams can also limit waste treatment for VOC removal.

The amount of organics in an aqueous or mixed stream will determine what treatment techniques to use for the physical separation or for the chemical or biological transformation of volatiles of concern (Allen and Blaney, 1985). Figure 6.3 shows the range of organic concentrations over which various treatment techniques are applicable. Treatment at the point of generation may use dedicated continuous or semicontinuous treatment processes, even though the unit may be small. Since biological systems are sensitive to some wastes, dilute aqueous waste streams may require pretreatment.

6.3.1 DESIGN AND OPERATING PRACTICES

If concerns about air emissions are incorporated into the initial decision-making process, surface impoundments can be designed and operated at each phase with the primary objective of reducing the emissions (Ehrenfeld, Ong, Farino, Spawn, Jasinski, Murphy, Dixon, and Rissmann, 1986). Potential design considerations to control air emissions are surface area minimization, freeboard depth, choice of cover materials, moisture control, and inflow/outflow drainage pipe locations. Operations during the active life involving temperature of influent, dredging frequency, draining frequency, cleaning frequency, handling of sediments and sludge from dredging, and types of wastes accepted at a facility can be designed to minimize emissions.

6.3.1.1 Surface Area Minimization

Minimization of surface area with respect to depth would decrease air emissions, as long as a large surface area (e.g., evaporation lagoon) is not required (Ehrenfeld, Ong, Farino, Spawn, Jasinski, Murphy, Dixon, and Rissmann, 1986).

Engineering difficulties would probably make costs higher with this approach. To reduce surface area, the side slopes of an impoundment could be increased — subject to erosion, the ability to hold a liner, and the ease of construction (trafficability). The efficiency of the treatment should not be significantly decreased, in the process. Aeration and mechanical mixing, however, would be reduced in efficiency because of increased depth.

6.3.1.2 Freeboard Depth

Increase in freeboard depth will decrease wind and wave action on the surface of the impoundment (Ehrenfeld, Ong, Farino, Spawn, Jasinski, Murphy, Dixon, and Rissmann, 1986). This will decrease

```
                                           Thin Film Evaporation
                                              |-------|__|
                                                   Fractional
                                                  Distillation
                                           |----------|_____|
                                              Steam Stripping
                                  |-----------------|_____|
                     Air Stripping
           _____|----------|
                  Carbon Adsorption
           _____|----------|
                 Ozone/UV Radiation
           _____|----------|
                  Wet Air Oxidation
           |_____|-------|
            |        |       |       |       |
           0.01     0.1     1.0     10      100
```

LEGEND

_____ Commercially applied

------- Potential extension

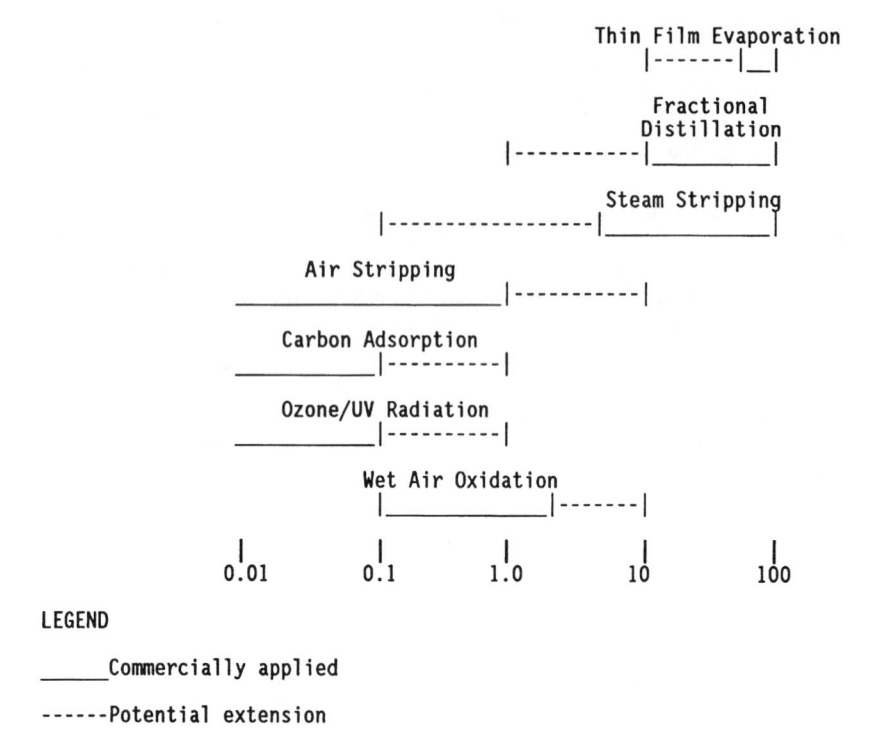

Figure 6.3 Approximate ranges of applicability of VOC removal techniques as a function of organic concentration in liquid waste stream. (From Allen, C.C. and Blaney, B.L. EPA-600/D-85/127, 1985. PB85218782.)

Volatilization
Turbulence on the impoundment surface
Spray formation
Erosion of dust and the dried surface of the impoundment

Field experiments with *in situ* windbreakers indicate that evaporation from reservoirs can be reduced significantly (Ehrenfeld, Ong, Farino, Spawn, Jasinski, Murphy, Dixon, and Rissmann, 1986). The effectiveness of the deeper freeboard is determined by the ratio of freeboard depth to diameter of the impoundment (distance from edge to edge). The larger this ratio, the more effective this control will be.

To increase freeboard, the height of the berm could be raised or the impoundment could be filled to a smaller depth (Ehrenfeld, Ong, Farino, Spawn, Jasinski, Murphy, Dixon, and Rissmann, 1986). The draft RCRA Guidance Document (July, 1982) suggests at least 60 cm (2 ft) of freeboard to prevent overtopping. Freeboard can be increased by minimizing run-on into the impoundment with a control system. Large lagoons may require other *in situ* controls to break the wind.

6.3.1.3 Inflow/Outflow Drainage Pipe Locations

This control minimizes disturbance of the surface (Ehrenfeld, Ong, Farino, Spawn, Jasinski, Murphy, Dixon, and Rissmann, 1986). Inflow pipes discharging above a liquid surface would create turbulence and spray formation and destroy the dry crust on the surface that forms a cover and reduces emissions. Inflow pipes should discharge as far as possible below the surface, and outflow systems should pump out liquid from the bulk of the impoundment.

6.3.1.4 Operating Practices
6.3.1.4.1 *Temperature of Influent*

The vapor pressure of a liquid increases with temperature (Ehrenfeld, Ong, Farino, Spawn, Jasinski, Murphy, Dixon, and Rissmann, 1986). Also, when two liquids of different temperatures are mixed, convective currents are induced which cause mixing and increased volatilization. Theoretically, to reduce air emissions, the influent should be discharged at as close a temperature to the bulk of the liquid as possible. This would not be appropriate for evaporation ponds, and treatment efficiencies will be reduced at lower temperatures. Oxidation ponds and aerobic and anaerobic lagoons may be affected by reduction of temperatures.

6.3.1.4.2 Dredging, Draining, and Cleaning Frequency
Emissions can be reduced by minimizing dredging, draining, and cleaning frequency (Ehrenfeld, Ong, Farino, Spawn, Jasinski, Murphy, Dixon, and Rissmann, 1986).

6.3.1.4.3 Handling of Sediments and Sludge
Some dewatering and disposal methods create more air emissions than others (Ehrenfeld, Ong, Farino, Spawn, Jasinski, Murphy, Dixon, and Rissmann, 1986). For example, solar drying causes more air emissions and fugitive dust than mechanical drying and filter presses.

6.3.1.4.4 Collecting Samples for Monitoring
Studies have shown that it is difficult to retain VOC concentrations while collecting and transferring intact soils for in-vial analysis (Hewitt and Lukash, 1996). This includes an intact soil sample held for <1 h in a metal core liner, held for days in a metal core liner sealed with tetrafluoroethylene sheets or aluminum foil, held for <2 min in a plastic bag after extruding from a sampling device, and immediately transferred to an empty vial to which a solvent is later added.

6.3.2 IN SITU CONTROLS

In situ technologies can be added on to a surface impoundment to control those parameters that influence emission rates, e.g., mass transfer coefficient, wind speed, and effective surface area (Ehrenfeld, Ong, Farino, Spawn, Jasinski, Murphy, Dixon, and Rissmann, 1986). These extend the efficiency and include covers, roofs, windscreens, rafts, barriers, shades, floating spheres, and surfactant layers. The objective is to reduce evaporation.

6.3.2.1 Air-Supported Structures and Synthetic Membranes
Huesemann and Moore (1993) determined that covering a bioremediation area with a polyethylene sheet does not limit biodegradation kinetics. The sheeting reduces release of volatile hydrocarbon emissions while allowing oxygen transfer for biodegradation. Complete enclosure of a surface impoundment by a domed, air-supported structure is a feasible control method, if a suitable method is available either to collect or dispose of generated vapors (Springer, Valsaraj, and Thibodeaux, 1986). This is the only feasible method if a surface aerator is to be used to improve oxygen transfer.

Synthetic covers reduce air emissions by reducing wind over the surface and by containing the emissions so they can be further treated (Ehrenfeld, Ong, Farino, Spawn, Jasinski, Murphy, Dixon, and Rissmann, 1986). Upjohn, Inc., in New Haven, CT, installed such a cover system over an aerated lagoon (425 × 150 ft and 8 ft deep), with two 75-hp aerators and 25 7.5-hp floating aerators. The air structure was made of a vinyl-coated polyester membrane coated on the inside with Teflon. It was fastened to the foundation around the impoundment by cables. The influent chemical oxygen demand (COD) was 5000 ppm and the effluent contained 700 ppm in solution. Bacteria were added every day and also recycled from the clarifier. The air was maintained at 19% oxygen. Exhaust was through either one of two carbon adsorbers installed at the vent outlet. The carbon adsorbers were steam regenerated every 2 days, and the condensate was recycled back to the lagoon to be further biodegraded.

A similar structure without a recovery system was installed over glauber salt storage ponds used by American Natural Gas in Beulah, ND, to keep out precipitation (Ehrenfeld, Ong, Farino, Spawn, Jasinski, Murphy, Dixon, and Rissmann, 1986). The air structure was movable from one pond to another. It covered around 2 acres and was made of vinyl-reinforced material over a concrete foundation. Two air blowers were used to keep the structure inflated. The facility was totally enclosed, except for one vent and a door for exit and entrance.

It has been suggested that a danger of using such membranes is that they may balloon and rupture (Fung, 1980). Air-supported structures are susceptible to wind damage, as well as weathering (Springer, Valsaraj, and Thibodeaux, 1986). The performance of a membrane as a vapor barrier cannot, generally, be predicted and may be disappointing. Little data on the permeability of various polymers to vapors are available. Some vapors may be harmful to the polymeric materials; however, the control effectiveness can approach 100%. Simple laboratory tests can be performed to measure the permeability of specific membrane materials to specific vapors. While vapors could be collected by controlled venting into an adsorption trap, or perhaps directly into an incinerator, they will be nearly saturated with water vapor, which can interfere with their adsorption.

Laboratory studies have been performed on various liner materials intended for use with hazardous and toxic wastes and municipal solid waste leachate (Haxo, Haxo, Nelson, Haxo, White, Dakessian, and Fong, 1985). Polymeric membrane liners proved to be the most promising for these applications. Flexible polymeric membranes have very low permeability to water and other fluids. The materials for the membranes vary considerably in polymer types, physical and chemical properties, interaction with various wastes, methods of installation, and costs. A given polymer type can vary from one manufacturer to another and even within the products of one manufacturer.

Polymers used in the production of lining materials include rubbers and plastics that differ in basic characteristics, e.g., chemical composition, polarity, chemical resistance, and crystallinity (Haxo, Haxo, Nelson, Haxo, White, Dakessian, and Fong, 1985). They can be classified into four types:

1. Rubbers (elastomers), which are vulcanized, i.e., cross-linked (XL);
2. Thermoplastic elastomers, which do not need to be vulcanized (TP);
3. Thermoplastics that are generally unvulcanized (TP), such as polyvinyl chloride (PVC);
4. Thermoplastics that have a relatively high crystalline (CX) content, such as the polyolefins.

The polymeric materials most frequently used in liners are (Haxo, Haxo, Nelson, Haxo, White, Dakessian, and Fong, 1985):

Polyvinyl chloride (PVC)
Chlorosulfonated polyethylene (CSPE)
Chlorinated polyethylene (CPE)
Butyl rubber (IIR)
Ethylene propylene rubber (EPDM)
Neoprene (CR)
High-density polyethylene (HPDE)

The thickness of polymeric membranes for liners ranges from 20 to 80 mils, with most in the 20- to 60-mil range (Haxo, Haxo, Nelson, Haxo, White, Dakessian, and Fong, 1985). Most of these liners are based upon single polymers and are usually compounded with fillers, plasticizers, antidegradants, and, if cross-linking is needed, curatives. There are blends of two or more polymers (e.g., plastic–rubber alloys) for this purpose. Membranes are supplied in pieces up to 100-ft wide to minimize the amount of field joining required (Fung, 1980). Adjacent sheets need to be sealed, not just overlapped, to produce watertightness. Noncrystalline, thermoplastic polymer compositions can be heat sealed or seamed with solvent or bodied solvent (generally, solutions of the liner compound) to increase the viscosity and reduce the rate of evaporation.

The most important liners are described below (Haxo, Haxo, Nelson, Haxo, White, Dakessian, and Fong, 1985). When these polymeric membrane lining materials were exposed to a typical municipal solid waste leachate for up to 56 months, there were only limited changes in properties. During the exposure, some of the materials swelled, with minor losses in tensile and other physical properties, but they did not become more permeable. Seaming of thermoplastic membrane liners by heat or by welding with solvents or bodied solvents appeared to yield seams in which the interface between the sheeting was almost eliminated. Adhesives usually differ in composition from the membrane and, thus, introduce additional interfaces in the seam assembly. The low-temperature vulcanizing adhesives required in the seaming of vulcanized sheetings are generally weaker on curing than the sheetings. These adhesives have fewer cross-links than the vulcanized sheeting and can swell considerably more and lose strength during long-term exposure. Some adhesives are initially weak but increase in strength over time as a result of additional cross-linking.

Among the polymeric linings, the partially crystalline thermoplastic materials produce the least amount of swell and the smallest change in properties on immersion at normal ambient temperatures (Haxo, Haxo, Nelson, Haxo, White, Dakessian, and Fong, 1985). These include polyethylenes, polybutylenes, polypropylenes, and thermoplastic elastomers. Often, the compatibility of a potential membrane must be tested against the wastes to which it will be exposed, and long exposures are probably necessary.

Butyl Rubber (BR)

Butyl rubber is a copolymer of isobutylene and a small amount of isoprene introduced in the polymer chain to furnish sites for vulcanization (cross-linking) (Haxo, Haxo, Nelson, Haxo, White, Dakessian, and Fong, 1985). Properties of these butyl rubber vulcanizates that relate to their use as a liner are

Thermal stability
Low gas and water vapor permeability
Ozone and weathering resistance
Chemical and moisture resistance
Resistance to animal and vegetable oils and fats

Some butyl compounds contain minor amounts of ethylene propylene diene monomer (EPDM) to improve ozone resistance (Haxo, Haxo, Nelson, Haxo, White, Dakessian, and Fong, 1985). Butyl vulcanizates are only mildly affected by oxygenated solvents and other polar liquids, but they swell considerably when exposed to hydrocarbon solvents and petroleum oils. They are highly resistant to mineral acids and extremes in temperature, and they remain flexible throughout their service lives. These sheetings are, generally, seamed with two-part, low-temperature curing adhesives, which may be less resistant to the service conditions than the liner itself, mainly because of their lower degree of ultimate cross-linking. These adhesives develop strength slowly (Fung, 1980).

A butyl rubber membrane was fabric reinforced with a nylon scrim and a vulcanized coating compound and exposed to acidic, alkaline, and lead wastes and pesticide water (Haxo, Haxo, Nelson, Haxo, White, Dakessian, and Fong, 1985). Basically, the butyl rubber showed good retention of its original properties on exposure to these wastes. It was not exposed to an oily waste, which would have caused softening and loss of tensile strength. Butyl rubber is supplied in folded rolls (Fung, 1980).

Chlorinated Polyethylene (CPE)

CPEs form a family of flexible, thermoplastic polymers produced by chlorinating high-density polyethylene and, as such, are saturated polymers with good aging and chemical resistance (Haxo, Haxo, Nelson, Haxo, White, Dakessian, and Fong, 1985). CPE can be cross-linked with peroxides, but in membrane liners uncross-linked thermoplastic compositions are usually used. Membranes of CPE are seamed thermally with solvent adhesives or by solvent welding. PVC or CSPE is now added to improve tensile and thermal properties.

Most CPE compositions withstand weathering, ozone, and ultraviolet (UV) light, and resist many corrosive chemicals, hydrocarbons (if cross-linked), microbiological attack, and burning (Haxo, Haxo, Nelson, Haxo, White, Dakessian, and Fong, 1985). Some compounds of CPE are also serviceable at low temperatures and are nonvolatile. A thermoplastic sheeting of CPE that was not fabric reinforced was exposed to acidic, alkaline, lead, pesticide, and oily wastes. Overall, the CPE membrane appeared to be satisfactory for the inorganic aqueous solutions but showed significant losses in properties in contact with oily wastes.

Chlorosulfonated Polyethylene (CSPE)

CSPEs form a family of saturated polymers prepared by treating polyethylene in solution with a mixture of chlorine and sulfur dioxide (Haxo, Haxo, Nelson, Haxo, White, Dakessian, and Fong, 1985). These polymers can be used in both thermoplastic (uncross-linked) and in vulcanized (cross-linked) compositions, and membranes are available in both forms.

Most CSPE compositions are resistant to ozone, light, heat, weathering, and deterioration by corrosive chemicals, such as acids and alkalies (Haxo, Haxo, Nelson, Haxo, White, Dakessian, and Fong, 1985). They have good resistance to mold, mildew, fungus, and bacteria, but only moderate resistance to oil. The thermoplastic versions cross-link slowly after exposure to moisture and the weather after placement. Usually, CSPE sheetings are reinforced with a polyester or nylon scrim. The fabric reinforcement gives needed tear strength and dimensional stability to the sheeting for use on slopes. It reduces distortion from shrinkage when placed on the base or exposed to the heat of the sun. A newly introduced grade of CSPE barrier compound is significantly more resistant to swelling caused by different liquids. Polyester is also replacing nylon as the reinforcing fabric.

CSPE membranes can be seamed by heat sealing, dielectric heat sealing, solvent welding, or with a bodied solvent adhesive. After PVC membranes, CSPE membranes are the most widely used of the polymeric flexible liner materials.

After exposure of a CSPE membrane with fabric-reinforced nylon to acidic, alkaline, lead, pesticide, and oily wastes, this sheeting tended to absorb water and wastes and some oil. Aging and exposure to wastes resulted in a modulus increase and decreases in elongation. Failures of seams were predominantly a combination of delamination and failure of the adhesive, indicating the loss in ply adhesion may be causing the loss in seam strength.

Elasticized Polyolefin (ELPO)

ELPO is a blend of rubbery and crystalline polyolefins (Haxo, Haxo, Nelson, Haxo, White, Dakessian, and Fong, 1985). It produces a black, unvulcanized, thermoplastic liner, which is heat sealable using a specially designed heat welder for use in the field or factory. ELPO has a low density (0.92) and is relatively resistant to weathering, acids, and alkalies. It has good aging, and moisture and chemical resistance. It can be fabricated into panels in the factory or shipped in rolls to a site for assembly in the field.

An ELPO membrane, which was a black thermoplastic sheeting based upon polyethylene, was exposed to acidic, alkaline, lead, pesticide, and oily wastes, as well as deionized water and well water. The ELPO lining material showed good retention of properties in the aqueous wastes. However, wastes containing significant amounts of oil were absorbed by the liner, accompanied by a loss of tensile strength, modulus, and tear strength. This membrane has the lowest permeability to water and to hydrogen ions.

Ethylene Propylene Rubber (EPDM)

EPDM is a terpolymer of ethylene, propylene, and a diene monomer with a few double bonds in the polymer molecule as sites for vulcanization of the rubber (Haxo, Haxo, Nelson, Haxo, White, Dakessian, and Fong, 1985). The unsaturation is in the side chains of the polymer and not in the main chain, as in the case of butyl rubber. This imparts good chemical, ozone, and aging resistance to the rubber. EPDM is chemically similar to, and can be covulcanized with, butyl rubber; consequently, it is now added to butyl rubber liner compounds to improve the resistance to oxidation, ozone, and weathering.

EPDM has excellent resistance to water absorption and permeation but relatively poor resistance to hydrocarbons. Sheeting is available in both unsupported and fabric-reinforced versions. Special care is needed when seaming cross-linked sheeting, because low-temperature vulcanizing adhesives are usually required. Sheeting with unvulcanized EPDM is thermoplastic and can be seamed with solvent adhesives and by thermal methods.

A cross-linked sheeting of EPDM rubber was exposed to acidic, alkaline, lead, and pesticide wastes. Oily wastes were not used because, as stated above, this type of rubber is sensitive to oil. The EPDM membrane was affected only moderately by these wastes, with the acidic waste being the most aggressive toward the EPDM compound. The seam strength was low before exposure and decreased with exposure, indicating a probable inadequacy of a Matrecon seaming method, employing a two-part vulcanizable adhesive and a gum tape.

Neoprene (CR)

Neoprene is the generic name of a family of synthetic rubbers based upon chloroprene (Haxo, Haxo, Nelson, Haxo, White, Dakessian, and Fong, 1985). These rubbers are vulcanizable, usually with metal oxides. They closely parallel natural rubber in mechanical properties, such as flexibility and strength. However, neoprene compositions are, generally, superior to those of natural rubber in their resistance to weathering, oils, ozone, and UV radiation, and have been used for containment of liquids containing traces of hydrocarbons. They also perform well with certain combinations of oils and acids. Most sheetings are relatively resistant to abrasion, puncture, and mechanical damage.

A vulcanized, unfabric-reinforced neoprene sheeting was exposed to alkaline, lead, pesticide, industrial, well water, and oily wastes. The neoprene was exposed to the oily wastes, since it is considered to be an oil-resistant rubber and since wastes of this type are aggressive to many of the lining materials. This membrane showed considerable absorption of both water and oily constituents. These tests did not demonstrate oil resistance.

Polybutylene (PB)

PB is a high-molecular-weight polymer synthesized from butene-1 (Haxo, Haxo, Nelson, Haxo, White, Dakessian, and Fong, 1985). Films of this material have good flexibility, heat sealability, low moisture vapor transmission, and good creep resistance. It is available in thin films but has not, as of yet, been manufactured for use as a liner. PB has the second lowest permeability to water of everything listed here.

Polyester Elastomer (PEL)
PELs form a family of polyether esters, which are semicrystalline and thermoplastic and feature oil, fuel, and chemical resistance (Haxo, Haxo, Nelson, Haxo, White, Dakessian, and Fong, 1985).

A polyester elastomer membrane was furnished by the supplier with a heat-sealed seam, since a high temperature was needed to prepare the seams. This material is reported to be resistant to hydrocarbons and other oily materials. It was exposed to acidic, alkaline, lead, pesticide, and oily wastes. This material failed by cracking and leaking on exposure to the acidic waste, which hydrolyzed the ester linkages. Its physical properties also degraded, when exposed to oily wastes. New improved versions of this type of material are now available.

Polyethylene (PE)
PEs are thermoplastic, crystalline polymers based upon ethylene (Haxo, Haxo, Nelson, Haxo, White, Dakessian, and Fong, 1985). They are currently made in three major types:

    Low-density polyethylene (LDPE)
    Linear low-density polyethylene (LLDPE)
    High-density polyethylene (HDPE)

The properties of a polyethylene depend upon density, molecular weight, and crystallinity (Haxo, Haxo, Nelson, Haxo, White, Dakessian, and Fong, 1985). Of the three types, HDPE polymers are the most resistant to oils, solvents, and permeation by water vapor and gases. Unpigmented, clear PE degrades readily on outdoor exposure, but addition of 2 to 3% carbon black confers UV protection. These membranes are normally free of additives, such as plasticizers and fillers. PE may have to be placed in the field at night, to prevent excessive softening and dimensional changes in hot climates (Fung, 1980).

LDPE and HDPE types of PE are used as liners (Haxo, Haxo, Nelson, Haxo, White, Dakessian, and Fong, 1985). Canals and ponds have been lined with non-fabric-reinforced membranes of LDPE. LDPE in thin sheeting tends to be difficult to handle and to field seam. It is also easily punctured under impact, such as when rocks are dropped on the lining; however, it has good puncture resistance when buried. LDPE films have been found to be deficient in puncture, crease, and fold resistances and unsatisfactory for impounding hazardous wastes, in spite of their good chemical resistance.

PE comes in folded rolls (Fung, 1980). Special seaming equipment has been developed for making seams of HDPE sheeting in the factory or the field (Haxo, Haxo, Nelson, Haxo, White, Dakessian, and Fong, 1985). This liner is stiff compared with most of the other membranes. Special adhesives that develop strength slowly can be used with PE, and panels can be joined with tape (Fung, 1980).

Polypropylene (PP)
PP is a partially crystalline thermoplastic polymer based upon propylene (Haxo, Haxo, Nelson, Haxo, White, Dakessian, and Fong, 1985). It is quite hard and stiff, has good chemical resistance, and has potential as a membrane liner.

Polyvinyl Chloride (PVC)
PVC is produced by any of several polymerization processes from vinyl chloride monomer (VCM) (Haxo, Haxo, Nelson, Haxo, White, Dakessian, and Fong, 1985). It is a versatile thermoplastic, which is compounded with plasticizers and other modifiers to produce sheeting with a wide range of physical properties. Polymeric membranes based upon PVC are the most widely used flexible liners.

PVC sheeting is manufactured in rolls in thicknesses of 20 to 30 mils and various widths up to 96 in. (Haxo, Haxo, Nelson, Haxo, White, Dakessian, and Fong, 1985). PVC is usually accordion-folded in both directions so the sheet can be opened lengthwise from the back of a truck (Fung, 1980). Most liners are used as unreinforced sheeting, but reinforced fabric is also available (Haxo, Haxo, Nelson, Haxo, White, Dakessian, and Fong, 1985). PVC compounds contain 25 to 35% plasticizers to make the sheeting flexible and rubberlike. They also contain 1 to 5% of a chemical stabilizer and various amounts of other additives but no water-soluble ingredients. PVC liners can deteriorate as a result of plasticizer loss caused by volatilization, extraction, and microbiological attack. The proper plasticizers and an effective biocide can virtually eliminate microbial attack and minimize extraction and volatility.

The PVC polymer is, generally, not affected under buried conditions, but it is affected by exposure to UV light (Haxo, Haxo, Nelson, Haxo, White, Dakessian, and Fong, 1985). PVC sheetings can deteriorate relatively quickly when weathered by wind, sunlight, and heat, which cause polymer degradation and loss of plasticizer; consequently, they are usually covered with soil. To prevent excessive softening and dimensional changes in hot climates, PVC may have to be installed at night (Fung, 1980).

Plasticized PVC sheeting is quite resistant to puncture and relatively easy to seam by "welding" with solvents or bodied solvents, adhesives, and heat (Haxo, Haxo, Nelson, Haxo, White, Dakessian, and Fong, 1985). PVC sheets can be joined and sealed by injecting a small amount of adhesive along a 4-in. overlap, then gently smoothing the adhesive-covered area to create a bond (Fung, 1980). The bonded sheets can be opened within a few minutes.

A polyvinyl liner membrane was exposed to acidic, alkaline, lead, pesticide, and oily wastes. This membrane showed considerable variation in its response to the different wastes. This was largely related to swelling and the loss of plasticizer from exposure.

Laboratory measurements of permeability of a 20-mil PVC membrane showed a high permeability for a number of volatile organic vapors and indicated that such a membrane would not provide significant protection against vapor flow out of a landfill (Springer, Valsaraj, and Thibodeaux, 1986). This membrane is the equivalent of only a few inches of porous soil covering.

6.3.2.2 Vapor Extraction Systems (VES)

Vacuum is a simple and effective means for *in situ* removal of spilled VOCs from soils to prevent contamination of groundwater (Bennedsen, 1987). *In situ* removal may be a cost-effective alternative to the more usual remedy of excavation and off-site disposal of the contaminated soil. Off-gas treatment may be required from SVE systems of air-stripping units to limit hydrocarbon discharge to the atmosphere.

Figure 6.4 is an illustration of a soil gas VES (Bennedsen, 1987). The vacuum is applied to the soil through extraction wells constructed with perforations above the water table. A conventional industrial blower provides the vacuum. In some cases, the extracted VOCs can be vented directly to the atmosphere. If they are too concentrated for direct discharge, they can be collected in a vapor-phase carbon adsorption system or piped to a boiler, if available, for mixing with the combustion air. The major system operating cost is for sampling and analysis of the extracted soil gas to monitor system performance. On a pound-for-pound basis of VOCs, it is less costly to remove volatiles from soil with a VES than to pump and treat contaminated groundwater. Several soil gas VESs have been installed and have proved to be effective.

The spilled materials must be volatile at usual ambient temperatures, the depth to groundwater must be great enough that there is a substantial cover of contaminated, unsaturated soil above the water table (at least 10 ft), and the contaminated soil must be pervious enough to permit a significant flow of air through the zone of contamination under a modest applied vacuum (Bennedsen, 1987). VESs operate at around 1×10^{-4} to 1×10^{-8} cm/s, and the blowers generally have a limit of about 8 in. mercury gauge vacuum to extract 100 ft^3/min.

A 55-gal air/water separator tank can be installed ahead of the blower to trap water extracted with the soil gas (Bennedsen, 1987). Air in soil will normally be near saturation with water vapor, which will cause condensation, if the temperature drops. The air/water separator protects the blower from the moisture in the extracted air. To measure the quantity of extracted soil gas, an orifice plate is connected to a U-tube manometer and installed on the blower discharge pipe. The concentrations of volatiles in the extracted soil gas are determined by laboratory analysis of samples drawn from a sampling port on the blower discharge line.

The design and operation of VES is an emerging technology. Selection of VES would depend upon the type and quantity of volatiles spilled, the concentrations of volatiles on the soil, the volume and depth of contaminated soil, the physical characteristics of the contaminated soil (particularly stratification and permeability), the depth to groundwater, and the surface of the contaminated area (i.e., paved, open, under a building).

Wilson (1994) developed a model for SVE in laboratory columns, which includes the effects of mass transport kinetics of VOCs between nonaqueous-phase liquid (NAPL) droplets and the aqueous phase and between the aqueous and vapor phases. The model provides a treatment of diffusion of VOCs through a stagnant aqueous boundary layer and permits time-dependent gas flow rates in the vapor extraction columns. Application of the model revealed high initial effluent soil gas VOC concentrations, which

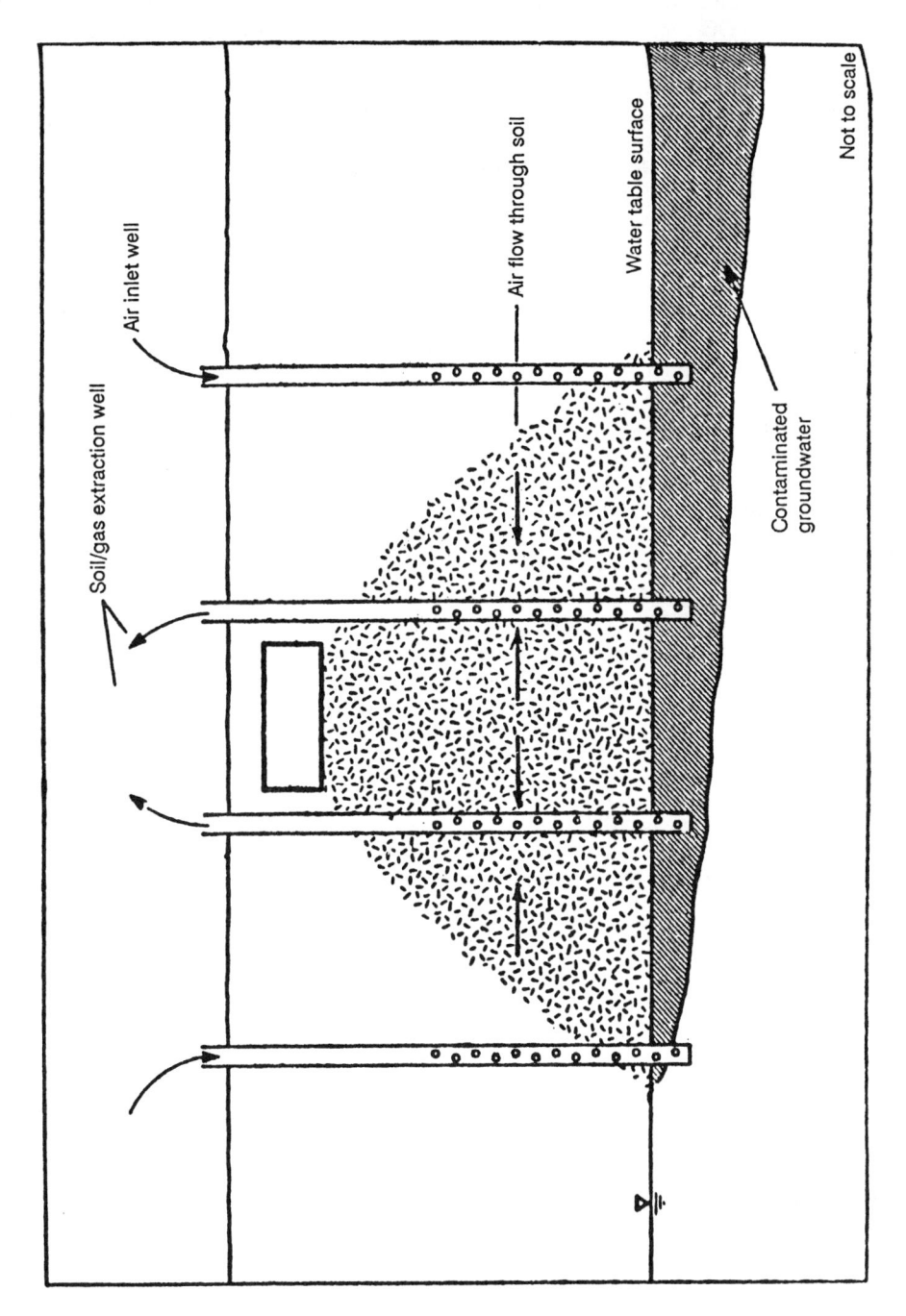

Figure 6.4 Schematic of soil gas VES. (From Bennedsen, M.B. *Pollut. Eng.* 19:66–68, 1987. With permission.)

typically decreased fairly rapidly. This was followed by a prolonged tailing region in which the effluent soil gas VOC concentrations slowly decreased until almost completely stripped from the system. The model shows it is useless to try to predict SVE cleanup times on the basis of short-term pilot-scale experiments, which give no idea as to the rate of VOC removal late in the remediation. The model permits gas flow to be varied with time. If the gas flow is shut off after partial cleanup, there can be rebounds in the soil gas VOC concentrations, which can be quite large, especially if some NAPL is still present.

The effects of variable airflow rates in diffusion-limited operations have been studied (Gomez-Lahoz, Rodriguez-Maroto, Wilson, and Tamamushi, 1994). The use of suitably selected airflow schedules in SVE can result in greatly reduced volumes of air to be treated for VOC removal with relatively little increase in time required to meet remediation standards. SVE using air as an oxygen carrier may help reduce costs by requiring less carrier medium and by lowering hydrocarbon concentration in the withdrawn soil gas (Urlings, Spuy, Coffa, and van Vree, 1991).

VESs are further discussed in Sections 2.2.1.11 and 2.2.2.2.

6.3.3 VOC PRETREATMENT TECHNIQUES

Pretreatment, as used in this book, refers to removing volatile components of a waste before it is put into a surface impoundment (Ehrenfeld, Ong, Farino, Spawn, Jasinski, Murphy, Dixon, and Rissmann, 1986). A cover and a vent can be used as a pretreatment control to collect emissions. There are several techniques for isolating the volatile components, provided they are in the form of a well-defined stream. For example, volatiles may be forced from the contaminated material via air stripping, then vented to a storage tank by refrigeration/condensation or adsorbed onto activated charcoal. The recovered volatile material could then be returned to storage or subjected to further treatment either to destroy the VOCs or to purify them for recycle.

Pretreatment controls are those administrative or technical procedures applied to wastes prior to their being sent to a TSDF, which will reduce their emissions in those treatment, disposal, and storage facilities (Ehrenfeld, Ong, Farino, Spawn, Jasinski, Murphy, Dixon, and Rissmann, 1986). Administrative controls include bans or restrictions on the disposal of volatile materials in landfills, surface impoundments, or landtreatment facilities. Technical controls include methods that separate the volatile materials from the wastes and either recycle them back to the generator or other potential users, or destroy the potentially volatile materials through subsequent treatment. Separation techniques include distillation, stripping, carbon adsorption, and solvent extraction.

Each of these methods will remove a fraction of the volatile materials from the waste stream (Ehrenfeld, Ong, Farino, Spawn, Jasinski, Murphy, Dixon, and Rissmann, 1986). The choice depends upon the composition of the waste. Once separated, the volatile fraction can be reused or destroyed by incineration with air oxidation or other method. The emissions occurring during the separation and treatment process will be less than those that would have occurred if the wastes had been deposited on the land or in treatment tanks. See Table 2.1 for the suitability of various treatment processes for volatile and nonvolatile organics and inorganics.

6.3.3.1 Pretreatment Processes for Organic Liquids

The solubility of the volatiles in the waste stream will determine ease of removal by thermal separation techniques, such as steam stripping or distillation, which are frequently used for mixed or high concentration aqueous streams (Allen and Blaney, 1985).

6.3.3.1.1 Distillation

Distillation can be used to remove VOCs from free-flowing liquid wastes (Allen and Blaney, 1985). It is more frequently used to separate components of organic streams but is also used by waste recyclers to remove organics from mixed streams. Distillation has the advantage over steam stripping in that additional water is not added to the system being treated. However, it cannot treat streams that are viscous or have a high degree of solids content, since these streams can foul the evaporation coils and the fractionation column plates. Distillation apparatus is more complex than batch steam strippers and requires more highly skilled personnel for operation (Spivey, Allen, Green, Wood, and Stallings, 1986). Like steam strippers, it is energy intensive, requiring 50 to 2500 kJ/kg (25 to 1200 Btu/lb) of waste feed.

6.3.3.1.2 Steam Stripping

Steam stripping is a distillation technique for removing organic compounds or dissolved gases from dilute aqueous solutions (IT Corporation, 1987). The degree of separation is governed by the volatility of the contaminants to be stripped relative to the volatility of water. The relative volatility is the ratio of vapor to liquid composition of the components (contaminants and water). Liquid and vapor phases in contact in a steam stripper are, essentially, at the same temperature and pressure.

Liquid stream treatment for organics removal is practiced commercially, and similar treatment processes will likely be applicable to hazardous waste streams (Allen and Blaney, 1985). Aqueous streams with low (<1000 ppm) organic contents can be treated with techniques used in purifying drinking water (Love, Ruggiero, Feige, Carswell, Miltner, Clark, and Fronk, 1983) or in municipal wastewater treatment (Shukla and Hicks, 1984), as long as air emissions are properly controlled. These techniques are being used to clean leachate at land disposal sites (O'Brian and Bright, 1983; Parmele and Allan, 1983). Highly concentrated liquid streams can be treated with techniques used for solvent recycling (Allen and Blaney, 1985). However, it is liquid streams of intermediate organic content (0.1 to 10%) and sludges that typically release much of the emissions at hazardous waste management facilities and are most difficult to treat.

Steam stripping has the same effects as elevated temperature air stripping (Knox, Canter, Kincannon, Stover, and Ward, 1984). Costs for air stripping have been estimated to be between 9 and 90¢/1000 gal of water treated for removal of 90% of trichloroethylene.

Steam stripping is normally carried out in a continuous, countercurrent flow stripping tower (IT Corporation, 1987). The major components in a stripping system consist of

- A heat exchanger to preheat the feed
- Plates, trays, or packing inside a stripping column to provide intimate contact of vapor and liquid phases
- A condenser
- A reboiler (if live steam injection is not practical)
- A decanter

Steam is injected at the bottom of the column to produce the proper amount of boil-up. The water to be treated or stripped is fed at the top of the column or part of the way down. The column overheads, containing steam and vaporized organics, are removed and condensed at the top, and the treated groundwater is removed at the bottom.

Steam stripping by direct injection of live steam can be used to treat aqueous and mixed wastes containing organic compounds that are at higher concentrations or that have lower volatility than those streams that are susceptible to air stripping (Allen and Blaney, 1985). Thus, this method is used by some waste recyclers for recovery of organics, including VOCs. This can provide the additional benefit of economic credits from the resale of off-gases collected during treatment for VOC emissions control.

There are disadvantages to this process, including an increased volume of treated aqueous waste to be managed in later process steps, because of the addition of steam to the waste (Allen and Blaney, 1985). The apparatus is energy intensive, and the cost of steam can account for a major portion of the operating costs for a unit (Spivey, Allen, Green, Wood, and Stallings, 1986). Steam stripping has generally been used commercially for removal and recovery of volatile organics, ammonia, and hydrogen sulfide from industrial wastewaters (IT Corporation, 1987). Application of steam stripping to VOC-contaminated groundwater treatment has been limited because of its high cost compared with air stripping. As with air stripping, the packing materials can become heavily coated with oxidized iron and manganese (Stover and Kincannon, 1983). However, metals can be removed from the water by pretreatment with lime at pH 10.0, with readjustment of the pH to 6.5 before the steam stripping.

Advantages include the ability of the method to handle waste streams of more variable composition than can a distillation unit (Allen and Blaney, 1985). The treatment temperature is, generally, lower than that for thin-film evaporators. This is useful for volatiles that react or decompose. Therefore, steam stripping may be useful as an initial VOC removal technique, which could then be followed by additional treatment to reduce the concentration of VOCs to an acceptable level. Activated carbon can effectively remove all the individual volatile organics from the off-gas (Stover and Kincannon, 1983). Thus, steam stripping can recover and concentrate organics with no resulting air emissions (IT Corporation, 1987).

This process can also be considered for water-soluble compounds, such as alcohols and ketones. Also, if batch treatment is required, steam stripping may be the preferred alternative (Allen and Blaney, 1985).

Vapor/liquid equilibria for toxic organic pollutants in dilute aqueous systems can help assess the relative ease or difficulty of applying steam stripping technology for the removal of organic pollutants in water (Goldstein, 1982). Steam stripping is also discussed in Section 2.2.1.14.2 as an *in situ* process.

6.3.3.1.3 Solvent Extraction

Solvent extraction is used in combination with steam stripping to remove materials, such as phenols, from waste streams (Allen and Blaney, 1985). The chosen solvent generally has low aqueous solubility and a strong affinity for the VOCs in the waste. The solvent in the treated waste is removed by steam stripping and can be regenerated by distillation. This process would probably be limited to on-site treatment.

6.3.3.1.4 Air Stripping

Aeration, as applied to water treatment, is a process allowing mass transfer between the liquid phase (water) and the gas phase (air) (Love, Ruggiero, Feige, Carswell, Miltner, Clark, and Fronk, 1983). Historically, it has been used in the absorption or release of gases to improve the palatability of water or to control objectionable inorganic substances, such as iron and manganese. Aeration has since been applied to remove VOCs from drinking water. Aeration of the water in a bioreactor can strip much of the VOCs, the amount of which depends upon the aeration rate and the competing removal mechanisms.

Air stripping units are inexpensive, employ equipment of simple design, and are easy to operate (Spivey, Allen, Green, Wood, and Stallings, 1986). Various configurations of equipment can be used in air stripping, including diffused aeration, cooling towers, aeration ponds, countercurrent packed columns, cross-flow towers, and coke tray aerators with countercurrent packed columns (Nielsen, 1983; IT Corporation, 1987). The latter is probably the most useful for treating waste streams contaminated with VOCs, since the countercurrent packed columns provide the most interfacial liquid area, low air pressure drop across the tower, and high air-to-water volumes, and they can be easily connected to vapor recovery equipment (Knox, Canter, Kincannon, Stover, and Ward, 1984; IT Corporation, 1987). Selection of an air stripping process will depend upon the volume to be treated and the concentration of organics. Figure 2.28 is an example of an air stripping unit (McDevitt, Noland, and Marks, 1987).

Air stripping is a cost-effective, mass transfer process that is now widely applied to removing many different volatile organic chemicals from contaminated groundwater (IT Corporation, 1987; American Petroleum Institute, 1983). The process consists of transferring the contaminants from solution in water to solution in air. The rate of mass transfer depends upon several factors according to the following equation (Canter and Knox, 1985; Nirmalakhandan, Lee, and Speece, 1987):

$$M = K_L a (C_L - C_g)$$

where M = mass of substance transferred per unit time and volume (g/h/m^3)
K_L = coefficient of mass transfer (m/h)
a = effective area (m^2/m^3)
$(C_L - C_g)$ = driving force (concentration difference between liquid phase and gas phase, g/m^3)

The rate of transfer from aqueous phase to gas phase is based not only upon the mass transfer rates of the components but also upon the equilibrium partitioning between the two phases and the actual conditions developed in the air stripping process (IT Corporation, 1987). The equilibrium partitioning of a contaminant in a water solution to air is a function of solubility, vapor pressure, and molecular weight. The driving force for mass transfer can be increased by increasing the volumetric air-to-water ratio, r (Treybal, 1980). Therefore, a high r value should increase the stripping rate and improve the performance. As r increases, the energy required to blow the air through the tower increases rapidly. An optimum air-to-water ratio has to be found to maximize the mass transfer rate and moderate the energy requirements.

Air stripping is effective on most hydrocarbon and chlorinated solvents. Aromatics and methyl *t*-butyl alcohol are effectively removed by air stripping. Alcohols and ketones are not amenable to cold air stripping. Heated air stripping can be used for these compounds, but costs are higher. Since many volatile organics in groundwater are not readily removed by air stripping, results can be improved by removal of iron and manganese prior to air stripping to minimize their coating of the packing material (Stover and Kincannon, 1983).

To determine whether or not stripping will give the desired results, equilibrium distribution data should be consulted (Hackman, 1978). These will give the concentration of the solute in the solvent and in the gas at given concentration, temperature, and pressure conditions, to indicate whether or not there is a strong tendency (large distribution coefficient favoring the vapor phase) for the solute to leave the liquid and be vaporized. For aqueous streams with low (<1%) concentrations of compounds of high volatility (Henry's law constant × 10^{-4} atm·m^3/g·mol), air stripping will likely be applicable (Spivey, Allen, Green, Wood, and Stallings, 1986).

Henry's law constant (H) is the most useful indicator of relative ease for air stripping VOCs (Love, Ruggiero, Feige, Carswell, Miltner, Clark, and Fronk, 1983). VOCs with Henry's law constants of 10^{-2} to 10^{-3} atm·m^3/mol are good candidates for air stripping (Stover and Kincannon, 1983). Extractable and volatile organics with Henry's law constants less than 10^{-3} to 10^{-4} atm·m^3/mol may require high-temperature air stripping or steam stripping for effective removal. When water containing VOCs is exposed to air, the air/water system tries to approach equilibrium, and aerating accelerates this natural phenomenon (Love, Ruggiero, Feige, Carswell, Miltner, Clark, and Fronk, 1983). By specifying Henry's law constant, or Henry's coefficient (H), as dimensionless, concentration in air divided by concentration in water, the reciprocal, $1/H$ (called the partition coefficient) (Day and Underwood, 1980), is the theoretical optimum air-to-water ratio (volume to volume) for removing a VOC by air stripping. At equilibrium, the volume of air can be determined for each volume of water to strip out a compound (Love, Ruggiero, Feige, Carswell, Miltner, Clark, and Fronk, 1983).

To express H as dimensionless at 20°C (293 K) and 1 atm,

Multiply "atm" by 0.000748, and
Multiply "atm·m^3/mol" by 41.6.

The "optimum" behavior of the process is useful, but the actual operating data deviate from the theoretical, especially at the higher air-to-water ratios, and a substantial range exists within the reported performance (Love, Ruggiero, Feige, Carswell, Miltner, Clark, and Fronk, 1983).

Films formed at the air/water interface influence the diffusion rate for volatile compounds and, thus, the time the system takes to approach equilibrium. The total resistance and impairment of transfer from the films can be reduced by agitation or by increasing the interfacial area for diffusion, such as by aeration.

Although air-to-water ratios as high as 3000:1 have been reported (McCarty, Sutherland, Graydon, and Reinhard, 1979), the ratios are, generally, less than 100:1, and when putting air through water, the practical upper limit is probably around 20:1 (Love, Ruggiero, Feige, Carswell, Miltner, Clark, and Fronk, 1983).

Table 6.10 summarizes estimated air-to-water ratios for stripping volatile organic chemicals from water (Love and Eilers, 1982). Both influent and effluent concentrations are given. For example, to reduce an influent concentration of trichloroethylene from 1000 to 0.1 µg/L, the estimated air-to-water ratio is 76:1. These ratios are used for both diffused air and packed tower aerators, but the practical upper limit for diffused air aeration is approximately 20:1.

The first step in optimizing the stripping process is selection of a packing material and then optimization of the parameters to achieve the treatment objectives for the predetermined site-specific worst conditions in the most economic manner (Nirmalakhandan, Lee, and Speece, 1987). Noninterlocking, symmetrical packings with uniform void distribution are preferable to avoid clogging, channeling, and compaction. The process will have a certain optimum region, for which the overall treatment cost will be the least. At higher air-to-water ratios, the pressure drop will be significantly higher, requiring a larger blower and higher operating costs. Increased temperatures can improve the removal efficiency of the stripping process for some compounds, such as aldehydes and alcohols (Law Engineering Testing Company, 1982). Air stripping is best run as a continuous process, since this minimizes the possibility of off-gas vapor concentrations drifting into the explosive range with batch processing (Allen and Blaney, 1985).

The data for VOCs in countercurrent flow packed columns provide a graphical indication of the process variability for design parameter changes (Speece, Nirmalakhandan, and Lee, 1987). A design-optimizing procedure for the removal of VOCs from water has been proposed (Nirmalakhandan, Lee, and Speece, 1987). Equations can be used to predict stripping losses of VOCs, which can be very

Table 6.10 Estimated Air-to-Water Ratios Necessary to Achieve Desired Water Quality

	Influent Conc. (μg/L)	Effluent Concentration (μg/L)				
		0.1	1	10	50	100
Trichloroethylene	1000	76:1	54:1	32:1	17:1	11:1
	100	54:1	32:1	11:1	1:1	—
	10	32:1	11:1	—	—	—
	1	11:1	—	—	—	—
Tetrachloroethylene	1000	96:1	72:1	45:1	26:1	18:1
	100	72:1	45:1	18:1	1:1	—
	10	45:1	18:1	—	—	—
	1	18:1	—	—	—	—
1,1,1-Trichloroethane	1000	198:1	90:1	35:1	14:1	8:1
	100	90:1	35:1	8:1	1:1	—
	10	35:1	8:1	—	—	—
	1	8:1	—	—	—	—
Carbon tetrachloride	1000	19:1	15:1	10:1	6:1	4:1
	100	15:1	10:1	6:1	1:1	—
	10	10:1	6:1	—	—	—
	1	6:1	—	—	—	—
cis-1,2-Dichloroethylene	1000	104:1	77:1	52:1	34:1	26:1
	100	77:1	52:1	26:1	4:1	—
	10	52:1	26:1	—	—	—
	1	26:1	—	—	—	—
1,2-Dichloroethane	1000	56:1	42:1	28:1	18:1	14:1
	100	42:1	28:1	14:1	—	—
	10	28:1	14:1	—	—	—
	1	14:1	—	—	—	—
1,1-Dichloroethylene	1000	10:1	8:1	5:1	3:1	3:1
	100	8:1	5:1	—	—	—
	10	5:1	3:1	—	—	—
	1	3:1	—	—	—	—

Source: From Love, O.T., Jr. and Eilers, R.G. *J. AWWA,* 74(8), 413–425. American Water Works Association. 1982. With permission.

significant and are frequently neglected (Truong and Blackburn, 1984). These losses can be determined and correlated into full-scale design criteria.

The contaminated air exhausted from air stripping systems may require treatment, such as with a vapor-phase activated carbon system or fume incineration, before being discharged to the atmosphere (IT Corporation, 1987). Air strippers have also been used in conjunction with other treatment technologies, such as carbon adsorption or UV/hydrogen peroxide oxidation. Treatment of the stripper off-gases to minimize air pollution adds significantly to the capital expenditure and operating costs (Spivey, Allen, Green, Wood, and Stallings, 1986). The air pollution control equipment is typically more expensive than the air stripping system. Some large waste management facilities are minimizing the additional cost of off-gas cleanup by using this airstream as part of the combustion air in the facility boiler or waste incinerator.

A technique similar to air stripping for the removal of some compounds from contaminated groundwater is that of dissolved air flotation, in which suspended fine particles or globules of oil and grease are floated to the surface by the action of pressurized air and then removed by skimming (Ehrenfeld and Bass, 1984). This technique has been used to remove up to 90% of the total suspended solids or oil and grease in wastewater containing 900 ppm of these substances.

Air stripping as an *in situ* soil treatment process is also covered in Section 2.2.1.10.

6.3.3.1.4.1 Aeration Devices
The following devices can be used for aeration (Love, Ruggiero, Feige, Carswell, Miltner, Clark, and Fronk, 1983):

Putting Air Through Water

1. *Diffused-Air Aeration* — Compressed air is injected into the water through perforated pipes or porous plates. These may be called injection or bubble aerators and are used for transferring oxygen into waste water. Such a system can be put into operation very quickly using existing facilities. Reservoirs, caisson wells, well casings, and well bores have served as temporary "aeration basins."

Operating at greater water depth results in improved performance at all air-to-water ratios tested (Singley and Williamson, 1982). The size and costs of operating the large compressors necessary for increased air-to-water ratios may negate any performance improvement of operating at the higher air-to-water ratio. Therefore, the most efficient performance possible at lower air-to-water ratios should be determined. Although this may be the best approach with diffused-air aeration basins, it is not necessarily the case with in-well diffused-air aeration, which also depends upon the location of the diffuser relative to the zones of contaminated strata (U.S. EPA Research Cooperative Agreement, 1985).

2. *Air Lift Pump* — This approach combines air stripping with pumping. These are simple devices with only two pipes in a well. One pipe introduces compressed air into the open bottom of the other pipe, called an eductor, in which air mixes with water. Since the mixture is less dense than the surrounding water, it will rise. The air and VOCs are then separated before the water is pumped into the distribution system. This method has poor pumping efficiency (35%) and very limited application (U.S. EPA Research Cooperative Agreement, 1985).

3. *Mechanical Surface Aeration* — There are several types commonly used in wastewater treatment, but they also offer many advantages for air stripping of VOCs (Roberts and Levy, 1983). They can be mounted on platforms or bridges, or be supported by columns or pontoons. The air and water are turbulently mixed by a motor-driven impeller-like turbine.

Putting Water Through Air

Figure 6.5 shows a number of aeration devices for putting water through air (Love, Ruggiero, Feige, Carswell, Miltner, Clark, and Fronk, 1983).

1. *Mechanical Surface Aeration* — Discussed above.

2. *Packed Tower Aerators* — Packed towers have been used for the transfer of material present in relatively large concentrations, for the transfer of highly volatile substances, and for absorption of air pollutants from various gas streams and the smaller concentrations involved in trace organic chemicals (Moore, 1986). Packed towers may reduce the concentrations of volatile trace organics and may be useful as the initial process in treating contaminated water. Packed towers have been employed in the chemical process and air pollution control industries. In the former, the concentrations are much greater than found in water treatment applications. In the latter, the towers serve as absorbers to transfer materials from the gas stream into the liquid.

The tower consists of a vessel filled with a low-weight medium that not only will provide a relatively large surface area to allow the mass transfer of the chemical from the gas to the liquid (absorbers) or from the liquid to the gas (strippers), but also has a large void ratio to provide a reasonably small headloss across the tower.

The particular organic chemicals to be removed and the site-specific considerations will determine whether the process is used alone or in conjunction with carbon (or other) absorption (Moore, 1986). The volatility of the substance is an important consideration. The Wisconsin Department of Natural Resources has established a 15 lb/day total volatile emission standard for release to the atmosphere (Nash, Traver, and Downey, 1987).

A packed tower consists of a column, which is 1 to 3 m in diameter, 5 to 10 m in height, and filled with "film" packing material (Love, Ruggiero, Feige, Carswell, Miltner, Clark, and Fronk, 1983). The packing can be glass, plastic, or ceramic, in numerous geometric shapes to create a water film for enhancing transfer of VOCs to the gas phase. Packing materials include raschig rings, pall rings, berl saddles, tellerite, and intalox saddles (Moore, 1986). On the inside wall of the aeration column are several "redistributors" that cause the water to flow over the packing rather than run down the walls (Love, Ruggiero, Feige, Carswell, Miltner, Clark, and Fronk, 1983). An "induced draft" packed tower employs a fan at the top. The "forced draft" packed tower has a blower at the bottom to force air up through the packing. These devices can remove 98% or more of the VOCs found in groundwater.

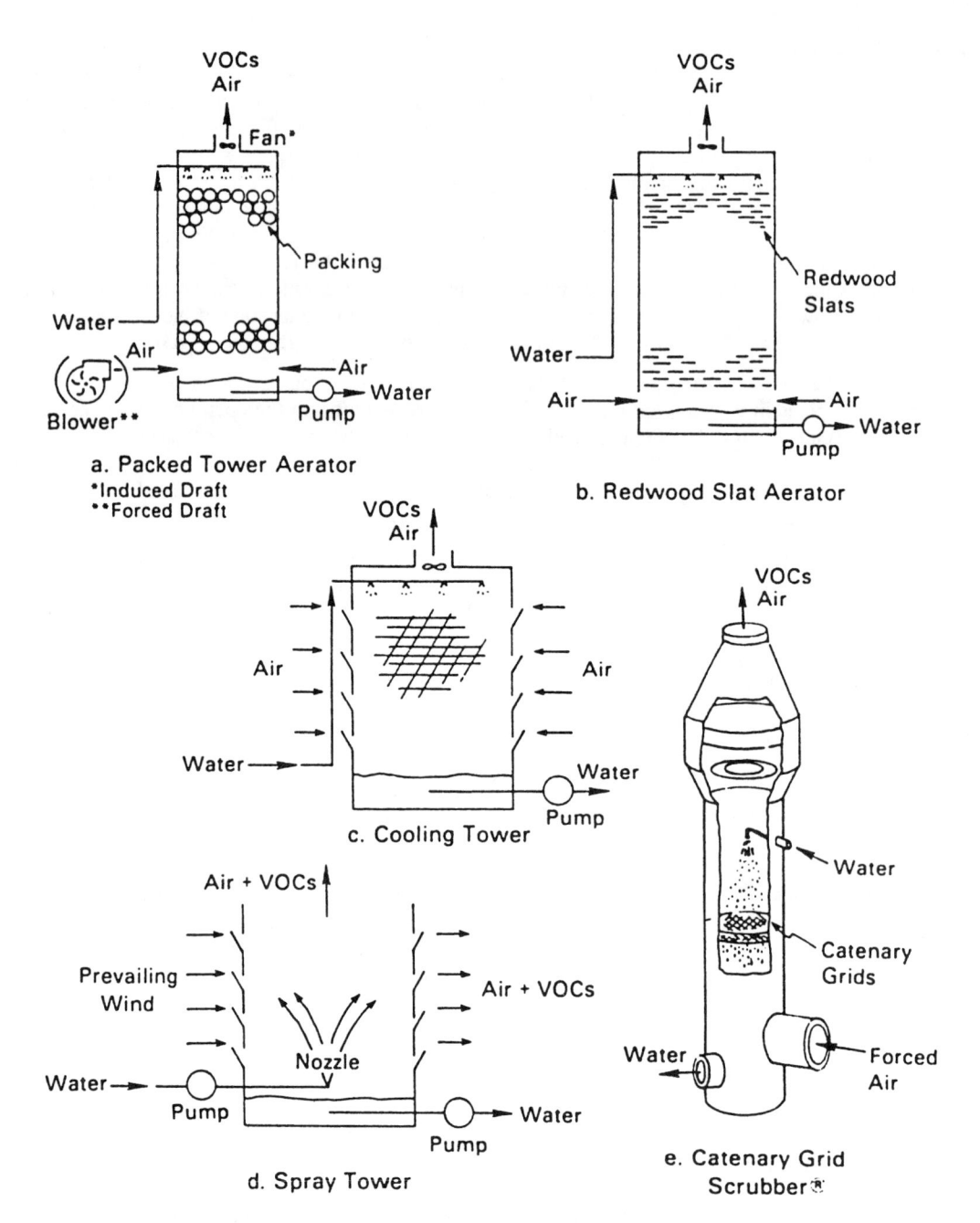

Figure 6.5 Aeration devices that put water through air. (From Love, O.T., Jr. et al. Report No. EPA-600/2-86/024. U.S. EPA, Cincinnati, OH, 1983. PB-84-130384.)

Packed towers have operated at air temperatures as low as −15°F (−26°C) with no freezing problems other than sample ports. Lime can be added to raise the pH to 7.0 to 7.5, if there is extensive precipitation of iron–organic floc with this technique (Love, Ruggiero, Feige, Carswell, Miltner, Clark, and Fronk, 1983). This produces the optimum oxidation and precipitation of the ferric hydroxide–organic complex.

A horizontal shaft Higee module air stripper (engineered by Glitsch) can be used in severe winter conditions that prevent installation of an ordinary air stripper to treat groundwater contaminated by leaking underground tanks. Higee units employ strong centrifugal forces, up to 1000 g, to increase permissible flow of both air and water phases in strippers and scrubbers. Formation of very thin liquid films, rapid renewal of wetted surfaces, and high vapor turbulence reduce the required thickness of packing from feet to inches. Stripper columns can thus be reduced to a fraction of their normal height, providing extremely high efficiencies in small space with low energy consumption. After 6 months of operation at 100 g/min, such a unit reduced benzene and toluene concentrations from 50 to 200 ppb and 150 to 500 ppb, respectively, to levels below those detectable by gas chromatography. VOC-laden air was catalytically incinerated before discharge to the atmosphere.

The design (and packing selection) of full-scale packed towers for removal of intermediate- and low-volatility organic chemicals should be based upon pilot-scale data.

3. *Wood Slat and Tray Aerators* — The spray, sparger, plate tower, and tray tower systems have been used mainly for either transfer of highly volatile substances, such as oxygen, hydrogen sulfide, ammonia, and methane, or for applications that do not require high transfer efficiencies (Moore, 1986). They are limited for stripping of moderate- and low-volatility substances, especially when high removal efficiencies are required. Plate and tray towers operate by creating thin streams of water and water droplets to achieve the necessary surface area for transfer of the chemical species desired.

Water is pumped into the top of the device, distributed over the cross-sectional area by a perforated plate, and allowed to trickle down and over "splash" packing, such as redwood slats, creating the air/water interface for mass transfer. Staggered slats prevent the water from short-circuiting. Air is pulled into the bottom of the tower by fans mounted on top or blown into the bottom. The air travels upward countercurrent to the water flow. This equipment is commonly used to control inorganic compounds (iron, manganese, carbon dioxide, hydrogen sulfide) in drinking water, and units have been modified (additional blower and slats) to contend with trace organic contamination (Silbovitz, 1982). Redwood slat aerators have been reported to be effective year round, with temperatures of the groundwater averaging 11°C (Fronk-Leist, Love, Miltner, and Eilers, 1983).

Tray aerators and cooling towers are also included in the water-through-air category, but cooling towers differ in that the air movement might be crosscurrent, rather than countercurrent.

4. *Spray Aerators* — Absorbing and stripping devices in the form of aerators are common (Moore, 1986). These include waterfall aerators, spray nozzles, cascade aerators, diffusion or bubble aerators, multiple tray and plate aerators, and mechanical aerators. Spray systems would be useful for stripping highly volatile materials, such as ammonia and methane, from water. Bubble aerators produce small bubbles to provide a large mass transfer surface area through which the contaminant can be reduced in concentration for stripping applications, or the oxygen concentration increased for absorption applications.

Spraying water through nozzles produces small droplets with large interfacial surface area (Love, Ruggiero, Feige, Carswell, Miltner, Clark, and Fronk, 1983). In a natural-draft spray tower, air movement depends upon the atmospheric conditions and the aspirating effect of the nozzles. Therefore, spray towers are usually placed side-by-side at right angles to the prevailing winds. Icing of the nozzles at low temperatures is a disadvantage of spray aeration (Kruithof, Graveland, van der Laan, and Reijnen, 1982).

5. *Catenary Grid Scrubber* — This is a proprietary device developed by Chem-Pro Corp. (17 Daniel Rd., P.O. Box 1248, Fairfield, NJ) as an air scrubber (Hesketh, Schifftner, and Hesketh, 1983), but which might also have application in removing VOCs from water (Love, Ruggiero, Feige, Carswell, Miltner, Clark, and Fronk, 1983). A fluidized gas–liquid contact is created by balancing the airflow velocity profile through catenary-shaped grids. The grids, generally in pairs, are stainless steel wire mesh with 0.6 to 2.5 cm openings. This device might be half the height of a packed tower and has a very low air pressure drop.

6.3.3.1.4.2 Secondary Effects of Aeration

Secondary effects may be beneficial or detrimental to water quality (Love, Ruggiero, Feige, Carswell, Miltner, Clark, and Fronk, 1983). Iron and manganese can precipitate as oxides of the metals and require sedimentation (detention) basins and sand filters. The precipitates may foul the aerator over time.

A shortcoming in the design of redwood slat aerators is the lack of inspection ports, making routine examination difficult (Fronk-Leist, Love, Miltner, and Eilers, 1983). This has prevented detection of accumulation of as much as 15 to 20 cm (6 to 8 in.) of iron sludge. Between 400 to 500 kg of iron could accumulate in such a unit, over the course of a year of continuous running. The iron sludge can be washed from the walls with fire hoses. If the concentration of iron in the raw water exceeds 0.5 mg/L, an iron control process should be considered. There should be easy access to the interior of all aerators for routine inspection and cleaning.

It has been suggested that aeration may introduce bacteria into the finished water (Love, Ruggiero, Feige, Carswell, Miltner, Clark, and Fronk, 1983). It is more likely that bacterial numbers would increase across an aerator because of the greater dissolved oxygen concentrations and an increase in nutrients deposited on packing. During continuous operation of a packed tower, a slime can accumulate on the packing material, decreasing the stripping performance and increasing the air pressure drop. The column could possibly be cleaned by flushing with a large volume of water, acid cleaning, or pretreatment to remove iron.

Aeration can decrease some forms of metal corrosion and can remove hydrogen sulfide and carbon dioxide that may be corrosive, but it could also increase corrosion as a result of an increase in dissolved oxygen and the resultant change in stability. For distribution systems made of iron pipe, dissolved oxygen at saturation could be detrimental.

Aeration does not destroy or alter VOCs; it simply transfers them to the ambient air where they are dispersed, diluted, and possibly photochemically degraded. Therefore, state or local ambient air quality standards must not be exceeded and must be considered in designing the aerator. Aerators are large air scrubbers, and the drinking water is in contact with 100,000 m^3 or more of air each day of operation. Without a particulate control, there could be an increase in particulate matter in the treated water.

6.3.3.1.5 Carbon Adsorption

Carbon adsorption can be used in two ways with a liquid waste stream containing organics for control of VOCs. One method is to bring the carbon into contact with the waste stream, thereby removing all organics from solution before any can volatilize. This is discussed in Sections 2.1.1.2.1 and 2.1.2.2.1.2. The other method is to drive off the VOCs, such as by use of air stripping or evaporation, and capture the emissions on the activated carbon.

Vapor-phase adsorption is the accumulation of a chemical from an off-gas stream onto the surface of a solid (Ram, Bass, Falotico, and Leahy, 1993). The contaminated off-gas may be treated by passing it sequentially through two vessels containing activated carbon until breakthrough is observed in the first carbon unit. Air stripping with carbon treatment of off-gas is often more cost-effective than liquid-phase carbon treatment alone, since vapor-phase carbon adsorbs 5 to 20 times more of a given contaminant per pound than liquid-phase carbon.

6.3.3.1.5.1 Gaseous Carbon Adsorption

Activated carbon is the most commonly used adsorbent for removal of volatile compounds from a gas stream, and other materials are available for specialized applications (Ehrenfeld, Ong, Farino, Spawn, Jasinski, Murphy, Dixon, and Rissmann, 1986). Activated carbon adsorption can achieve very low effluent concentrations, and can be used where removal of water-soluble organic contaminants is not required (IT Corporation, 1987). The components to be removed adhere to the surface of the grains and diffuse into and are trapped in the pores of the material (Ehrenfeld, Ong, Farino, Spawn, Jasinski, Murphy, Dixon, and Rissmann, 1986).

Because of the near-molecular dimensions of the pores of activated carbon, the toxic organics must be in the true vapor or molecular state before they can enter and accumulate in the pores (Hackman, 1978). Therefore, in air purification applications, activated carbon is generally considered only for true gases or vapors, and not for aerosols or particulate matter.

Physical adsorption is dependent upon the characteristics of the compounds to be adsorbed (adsorbate), temperature of the system, and concentration or partial pressure of the adsorbate in the gas stream (Hackman, 1978). Carbon, being generally a nonpolar material, prefers nonpolar adsorbates. Therefore, organic compounds are adsorbed in preference to polar inorganics and, in particular, water. Adsorption is also affected by the molecular weight and boiling point of a compound. Higher-molecular-weight compounds are usually more strongly adsorbed than lower ones. Most organic compounds can be successfully adsorbed on activated carbon. In cases where a variety of VOCs are being emitted, a carbon adsorption system may be at a disadvantage (Jennings, Krohn, Berry, Palazzolo, Parks, and Fidler, 1985). The adsorber may have difficulty with respect to desorbing capability and subsequent solvent separation. Thus, carbon adsorbers are particularly suitable for processes in which the waste stream is reasonably consistent with respect to VOC type.

Besides adsorbing molecules, the tremendous internal surface area can also provide a surface for chemical reactions to take place (Hackman, 1978). The carbon functions as a catalyst. The internal surface can also be impregnated, or coated with catalytic materials, with the carbon being a catalyst support. Reactive chemicals impregnated on the carbon can selectively react with, or "chemisorb," molecules from a gas stream.

For typical organic-adsorbing "Columbia" carbons supplied by Union Carbide Corp., the carbon tetrachloride activity is 60 to 65%, the bulk density is 0.45 to 0.48 g/cc, and the kindling point is about 350°C (Foster, 1985). A Pittsburgh Type BPL Activated Carbon from Calgon Corp. is designed for use in vapor-phase applications in several mesh sizes. A large portion of its micropore volume is in pores of 15 to 20 Å units in diameter. These small pores are accessible to all common gases and vapors and, therefore, provide the maximum surface area for adsorption. Type BPL is also permeated by a system of macropores (pores larger than 1000 Å in diameter), which serve as avenues for the rapid diffusion of gases to and from the micropore surfaces. This enhances both adsorption and reactivation characteristics.

Activated carbons and chars have been generally recognized as the most effective adsorbents for toxic organic vapors (Hackman, 1978). Gas masks containing activated carbon adsorb organic vapors, such as alcohol, aniline, benzene, ether, carbon disulfide, carbon tetrachloride, and toluene. Adsorption capacity onto activated carbon decreases with increasing TOC (total organic carbon), i.e., with increasing competition for adsorption sites (Miltner and Love, 1985). It is not well understood why compounds of higher adsorption capacity (tetrachloroethylene or trichloroethylene) are more affected by competitive matrices than compounds of lower capacity (trichloroethane 1,1,1- or cis-1,2-dichloroethylene).

Carbon adsorption presents a potential user with waste gas temperature constraints (Jennings, Krohn, Berry, Palazzolo, Parks, and Fidler, 1985). Adsorption is favored by lower temperatures. Adsorbers function best with inlet gas temperatures less than 100°F. Therefore, many installations include a gas cooler prior to the bed to reduce gas temperatures. However, this design may cause VOC condensation prior to adsorption in the carbon bed, which, in turn, gives rise to carbon-fouling problems. Filtering is one solution to this problem. In instances where cooling or filtering of waste gas is required, the costs of carbon adsorption vs. other technologies must be carefully weighed. In general, the solvent recovered must be of relatively high value in these instances for carbon adsorption to be the least costly alternative.

In an evaluation of a vapor-phase carbon adsorption system for the removal of toluene from a contaminated airstream, it was found that increasing the relative humidity from 30 to 90%, at constant temperature, would cut the carbon loading by about 50% (Foster, 1985). Increasing the operating temperature from 20°F, with constant relative humidity, also reduces the carbon loading by 50%. Reducing the steam regeneration ratio (steam-to-carbon) significantly reduces the loading of organics on the carbon.

Adsorption is favored by higher concentrations, although substantial amounts of odorous or toxic materials can be adsorbed even when present in the 1 to 50 ppm range, and important amounts of toxic organics can be adsorbed in the 1 to 100 ppb range (Hackman, 1978). Carbon adsorption systems tend to be constant outlet devices (the exit VOC concentration is relatively independent of the inlet VOC concentration). Thus, the efficiency tends to drop as VOC concentration in the inlet drops. Carbon adsorption systems are typically larger and heavier than recuperative thermal or catalytic incinerators. Although roof installation is sometimes feasible, it is not common.

Carbon systems require a great deal of data, such as knowledge of potential masking, fouling, and poisoning agents and types of VOC in the waste gas stream, in order to design separation systems (Jennings, Krohn, Berry, Palazzolo, Parks, and Fidler, 1985). These systems are batch-mode systems requiring large numbers of valves, flappers, dampers, controllers, etc. to switch airflows. There are two basic designs of carbon adsorbers: fixed-bed and fluidized-bed.

1. Fixed-Bed Carbon Adsorption

A typical granular activated carbon adsorption system consists of a fixed bed of carbon in a column (Jennings, Krohn, Berry, Palazzolo, Parks, and Fidler, 1985; IT Corporation, 1987). Fixed-bed carbon adsorption systems contain a minimum of two adsorbers. The adsorbers are cycled in and out of service resulting in each adsorber operating in a batch mode. The modes are characterized by an adsorption cycle and a desorption or regeneration cycle. An example of a fixed-bed carbon adsorption system is shown in Figure 6.6.

In the adsorption cycle, the VOC gas stream is directed to an adsorber containing freshly generated carbon (Jennings, Krohn, Berry, Palazzolo, Parks, and Fidler, 1985; IT Corporation, 1987). The VOCs are adsorbed onto the surface of the carbon, and the gas stream exits with a low VOC concentration. As the water flows continuously through the carbon bed and contaminants are adsorbed, the bed will approach saturation and the contaminants in the effluent from the column will begin to increase. At the

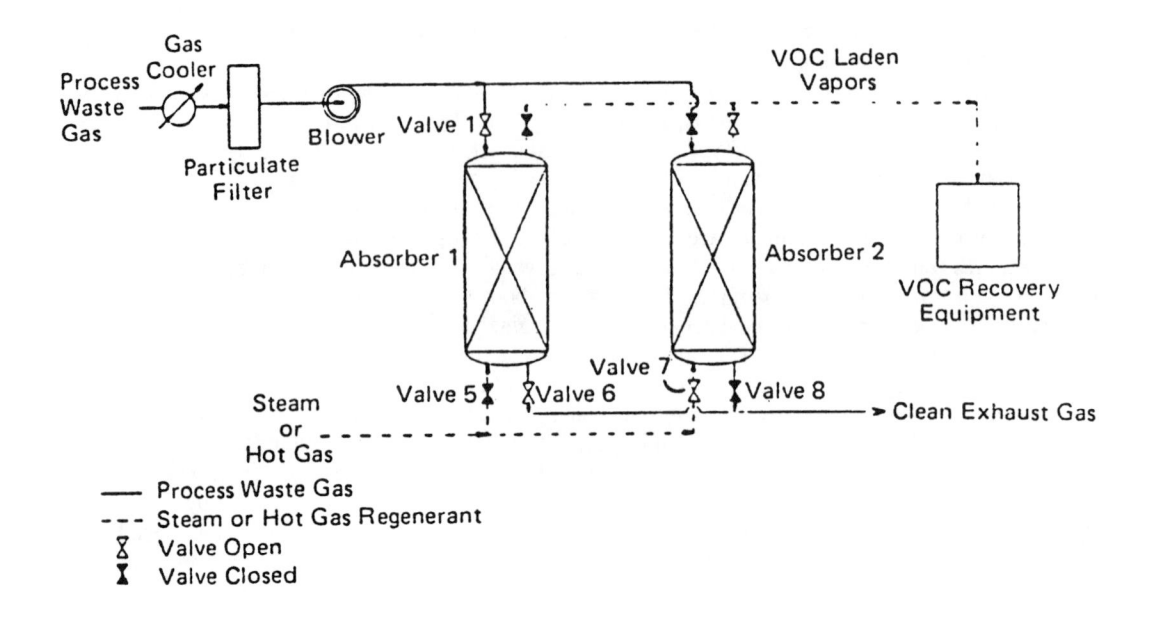

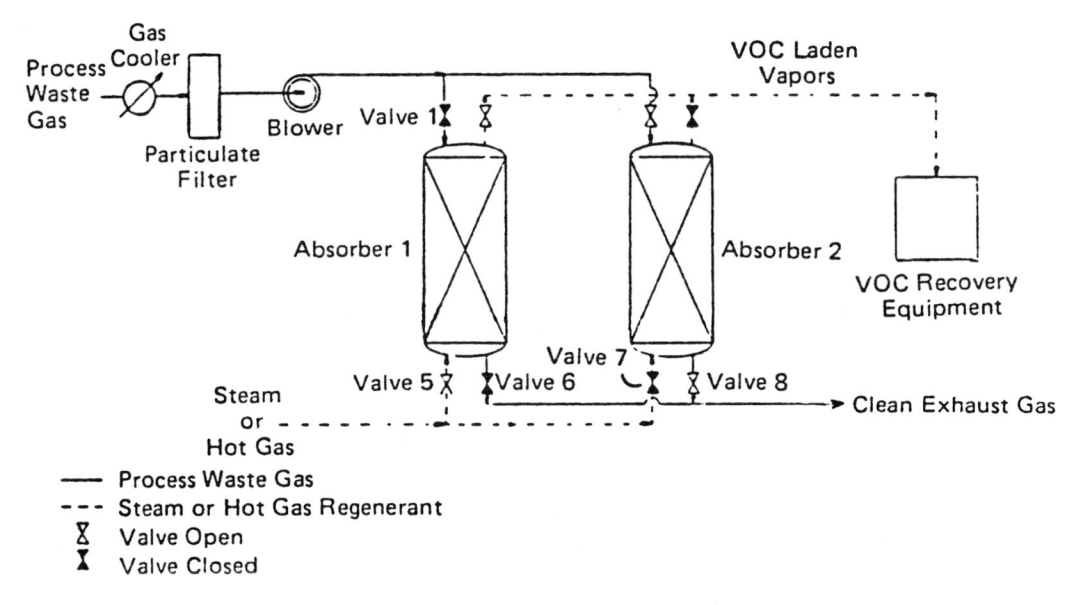

Figure 6.6 Fixed-bed carbon adsorption system. (From Jennings, M.S. et al. *Catalytic Incineration for Control of Volatile Organic Compound Emissions.* Noyes, Park Ridge, NJ, 1985. With permission.)

point where the effluent quality no longer meets the treatment requirements, the adsorptive capacity of the carbon is exhausted (or spent). In operating systems, the process is stopped before this point, called the breakthrough point (Ehrenfeld, Ong, Farino, Spawn, Jasinski, Murphy, Dixon, and Rissmann, 1986). The process can be reversed so that both the adsorbent material and the vapors that have been retained may be recovered. At this time, the flow of the VOC waste gas is transferred to a freshly regenerated bed and the used bed is regenerated (Jennings, Krohn, Berry, Palazzolo, Parks, and Fidler, 1985).

Most adsorptive forces and bonds holding pollutants, or the adsorbate, to an adsorbent are far less strong than chemical bond forces (Hackman, 1978). The most common techniques to remove the adsorbate involve raising the temperature, reducing the pressure, or using hot, inert gases. The molecular weight or boiling point of the adsorbate influences regeneration. Higher-molecular-weight materials, generally, require higher temperatures or lower pressures to release them from the carbon.

The spent carbon must be renewed by removal and replacement with virgin carbon or regeneration (IT Corporation, 1987). The regeneration cycle is designed to desorb most of the retained VOC and condition the carbon bed (Jennings, Krohn, Berry, Palazzolo, Parks, and Fidler, 1985). There are two basic types of bed regeneration: thermal and low pressure/vacuum regeneration.

Thermal Regeneration
The most common method of regeneration is to pass steam countercurrent to the direction of previous gas flow until the bed has reached a temperature sufficiently high to drive off the adsorbed vapors from the carbon (see Figure 6.6) (Hackman, 1978; Jennings, Krohn, Berry, Palazzolo, Parks, and Fidler, 1985). This temperature is usually below that which would cause a large loss of the carbon adsorbent by oxidation to carbon monoxide or carbon dioxide (Ehrenfeld, Ong, Farino, Spawn, Jasinski, Murphy, Dixon, and Rissmann, 1986). The steam/VOC mixture is then sent to a recovery system, where the mixture is condensed prior to separation (Jennings, Krohn, Berry, Palazzolo, Parks, and Fidler, 1985). Here, the water is separated from the VOCs so the VOCs can be reused. If the condensed VOC is miscible with water, a complex recovery system with one or more distillation columns or liquid–liquid extractors is needed. In Figure 6.7, the aqueous solution goes to a distillation column, where the VOCs are separated from the water and sent to a desiccator for final water removal. The water coming off the bottom of the distillation system is then either routed directly to the sewer or to an on-site wastewater treatment plant prior to disposal. The bed may then be dried, cooled, and returned to service (Hackman, 1978).

An immiscible water/VOC mixture requires less treatment (Jennings, Krohn, Berry, Palazzolo, Parks, and Fidler, 1985). In Figure 6.8, the immiscible mixture is sent to a decanter where the VOCs are removed through gravity separation. The wastewater either goes directly to a sewer or to an on-site wastewater treatment plant. In some systems, the wastewater is stripped of the remaining VOCs in stripping columns. The airstream from the stripping .column may be recycled to the carbon beds.

Steam regeneration is economically advantageous when large volumes (greater than 100 g/min for years) of water require treatment; otherwise, new carbon will probably be less expensive (Jennings, Krohn, Berry, Palazzolo, Parks, and Fidler, 1985). Cost savings from carbon regeneration must justify the added expense of the steam system. For some sorbates, continued regeneration of the sorbent by heat or by elution produces a carbon of diminished or even continually declining effectiveness (Robinson, 1979). Also, thermal desorption is risky for flammable sorbates. Thermal regeneration may use a hot gas other than steam, such as air (Jennings, Krohn, Berry, Palazzolo, Parks, and Fidler, 1985).

Vacuum Regeneration
The total regeneration cycle for vacuum regeneration is similar to that for steam regeneration (Jennings, Krohn, Berry, Palazzolo, Parks, and Fidler, 1985). In vacuum regeneration, the carbon adsorber containers are usually constructed with an external jacket. During the regeneration cycle, steam is directed through the jackets to heat the carbon indirectly to at least 50 to 60°F above the VOC adsorption temperature. A vacuum is then applied to the hot carbon to remove the VOCs. The vapors are condensed and cooled with no further recovery treatment required.

2. Fluidized-Bed Carbon Adsorption
Alternate approaches use moving bed adsorbers in which fresh carbon is continuously spent, then regenerated and returned to the adsorption cycle (Jennings, Krohn, Berry, Palazzolo, Parks, and Fidler, 1985; Ehrenfeld, Ong, Farino, Spawn, Jasinski, Murphy, Dixon, and Rissmann, 1986). Adsorption and desorption occur simultaneously in a single vertical container similar to a sieve tray distillation column (Figure 6.9). Spent sorbent is removed at one end of the system, passes through the regenerator, and is replaced at the gas inlet end. The activated carbon flows countercurrent to the exhaust gas. As the VOC waste gas flows upward in the adsorber, it contacts the activated carbon in the adsorption section and the VOCs are removed. The clean gas is released through the top of the vessel, while the carbon beads containing VOCs continue down to the regeneration section. Here, the carbon is indirectly heated with steam or other heating method. Then steam or other hot gas, such as nitrogen, strips the VOCs from the carbon on direct contact. A carrier gas transports the regenerated carbon back to the top of the container. The VOC/vapor mixture exits through the desorption section and is condensed. When nitrogen is used, no further VOC recovery treatment is needed. If steam is used, the VOCs are recovered by decantation or distillation.

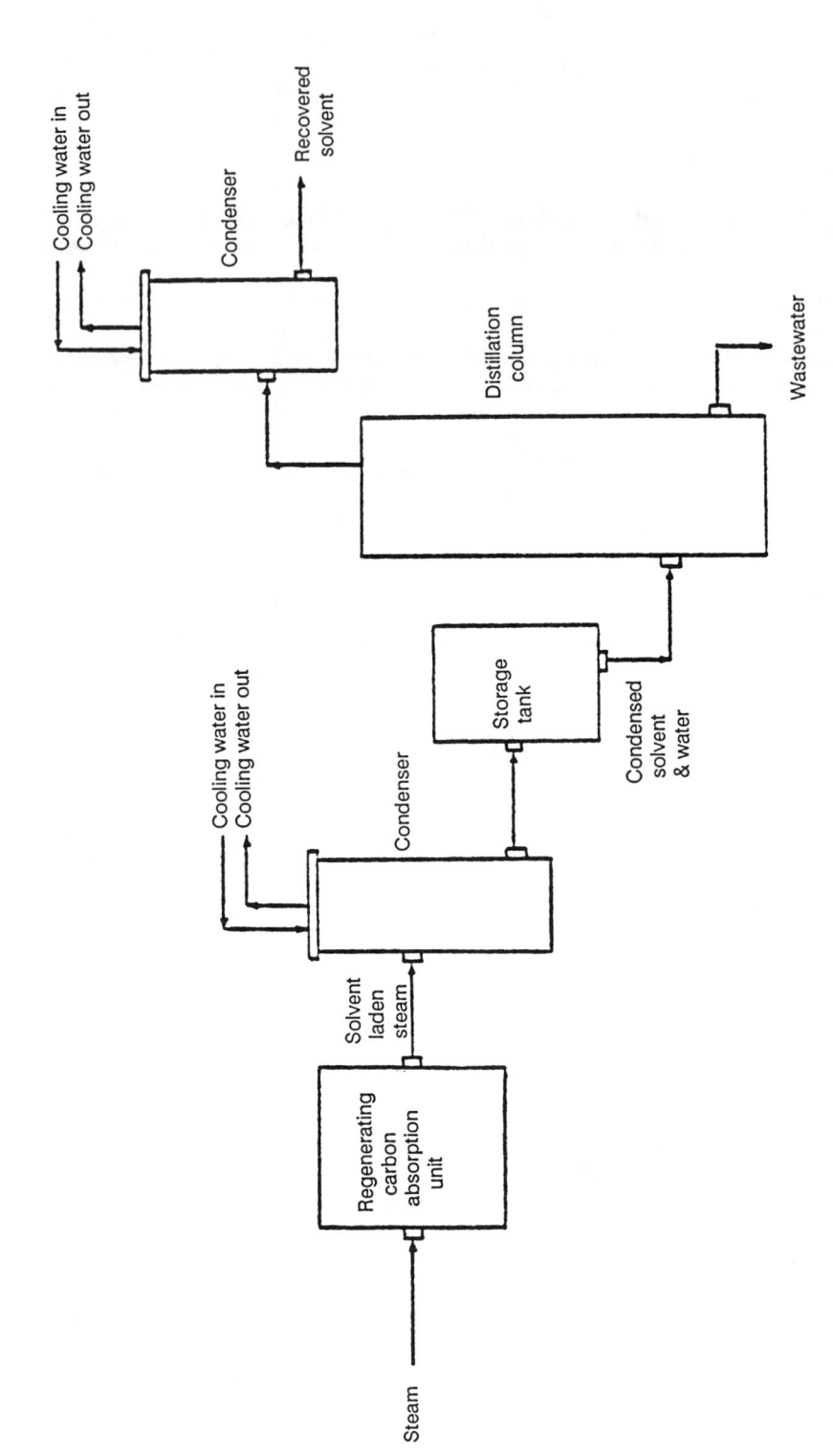

Figure 6.7 Schematic of solvent recovery by condensation and distillation. (From Jennings, M.S. et al. *Catalytic Incineration for Control of Volatile Organic Compound Emissions*. Noyes, Park Ridge, NJ, 1985. With permission.)

VOLATILE ORGANIC COMPOUNDS IN PETROLEUM PRODUCTS 355

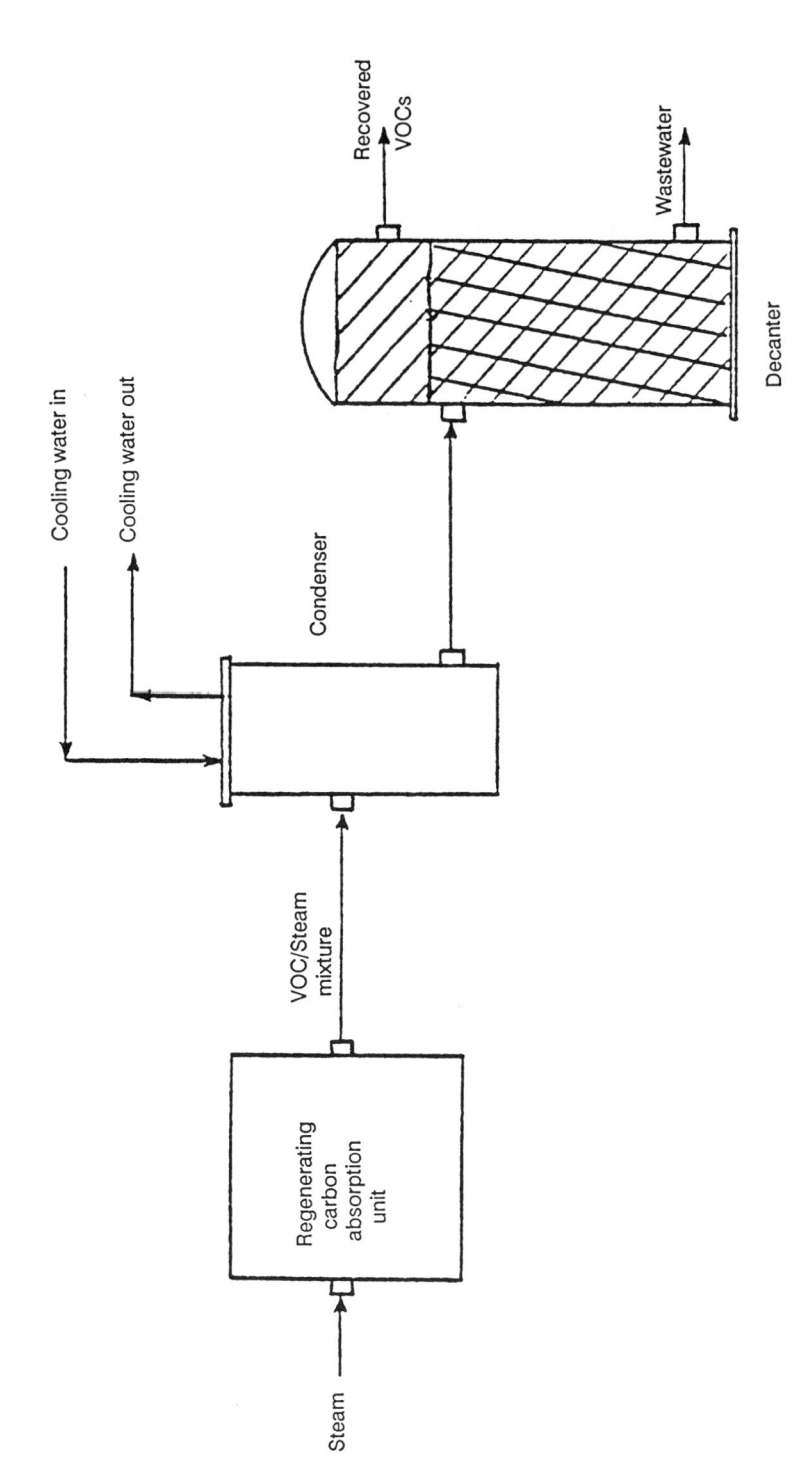

Figure 6.8 Schematic of VOC recovery by decantation. (From Jennings, M.S. et al. *Catalytic Incineration for Control of Volatile Organic Compound Emissions*. Noyes, Park Ridge, NJ, 1985. With permission.)

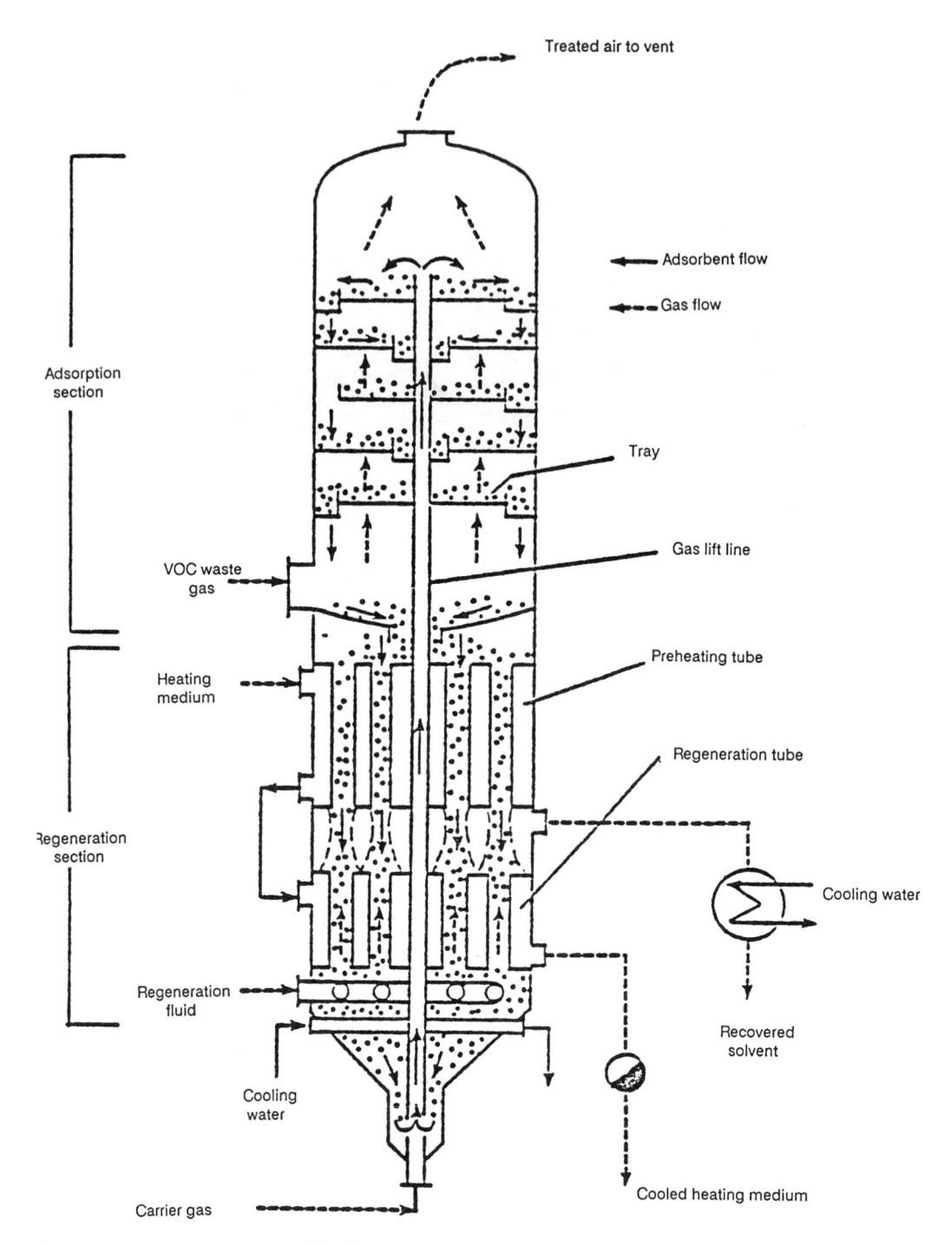

Figure 6.9 Schematic diagram of fluidized-bed carbon adsorption system. (From Jennings, M.S. et al. *Catalytic Incineration for Control of Volatile Organic Compound Emissions.* Noyes, Park Ridge, NJ, 1985. With permission.)

The main advantages of the fluidized-bed system arise from the single-vessel design (Jennings, Krohn, Berry, Palazzolo, Parks, and Fidler, 1985). The fluidized-bed process has a lower energy consumption for a given VOC waste gas flow than a fixed-bed design; however, the capital investment for fluidized-bed carbon adsorbers is higher (U.S. EPA, 1980a). Theoretically, this type of system is more efficient than a fixed-bed system using alternate units. Moving bed adsorbers are more expensive to construct and operate and more difficult to maintain. A VOC reduction efficiency of 90% has been reported with a fluidized-bed adsorber used on a stream with a low VOC concentration (100 ppm) (General Motors, 1980).

The loading capacities for the organics are normally determined by pilot studies for the water in question (IT Corporation, 1987). Isotherms from the literature can be used for feasibility studies and initial cost estimates. Costs also depend upon flow rate, concentration, adsorptivity of contaminants, carbon regeneration, and disposal cost of the carbon.

This technique is promising for emissions control and is being used in wastewater treatment (Spivey, Allen, Green, Wood, and Stallings, 1986). Carbon adsorption has been used extensively in hazardous wastewater treatment; however, it is not specific to volatiles removal from hazardous waste streams, and other nonvolatile or slightly volatile stream constituents may compete with the process dynamics. This method is of particularly limited applicability to a VOC removal program because there can be only limited suspended solids (<50 ppm) or oils and grease (<10 ppm) in the stream. However, some streams may be pretreated by filtering or centrifugation to minimize fouling of the carbon. Also, the carbon must be regenerated, adding to the expense of treatment.

Carbon adsorption systems are available as complete packages from several manufacturers (Ehrenfeld, Ong, Farino, Spawn, Jasinski, Murphy, Dixon, and Rissmann, 1986). Custom-designed systems are also available for larger or special-purpose applications.

6.3.3.1.6 Biological Treatment

Section 2.1.2.2 describes the aerobic and anaerobic systems that can be used for treatment of hazardous organic compounds in leachates and wastewater.

Limitations to the use of biological treatment of VOCs include formation of biodegradation products that may be more toxic or persistent than the original compounds, contaminant concentration that may be too high (toxic) or too low (inadequate energy source), and mixtures of organics that may contain compounds that are toxic or inhibitory to biological treatment.

Chemicals can be removed from contaminated water by adsorption onto biosludge. However, sorption of some VOCs, including benzene and toluene, is minimal and should not be used in design of a biological treatment system (Blackburn, 1987). The following removal efficiencies have been reported (Knox, Canter, Kincannon, Stover, and Ward, 1986):

	Percent Treatment Achieved		
	Air Stripping	**Sorption**	**Biodegradation**
Benzene	2.0	—	97.9
Toluene	5.1	—	94.9

Toluene has been removed with a laboratory-scale activated sludge treatment system at greater than 99% efficiency for an influent concentration of 195 mg/L, based upon aeration stripping in the reactor yielding less than 12% removal and negligible removal by sorption (Blackburn, Troxler, and Sayler, 1984).

6.3.3.1.7 Refrigeration/Condensation

The condensate containing volatile organics would require further treatment prior to discharge or disposal (Bove, Lambert, Lin, Sullivan, and Marks, 1984). See Section 6.3.4.2.

6.3.3.1.8 Evaporation

Agitated thin-film evaporation is widely used in the waste-recycling industry (Allen and Blaney, 1985). Its application is primarily for the separation of low boilers and solids from organic streams, but it can be used with mixed streams, if the relative volatility of VOC to other stream components is high. Treatment is possible to concentrations as low as 100 to 1000 ppm VOC. Thin-film evaporators are limited in the amount of solids or viscosity of the wastes they will accept. The size of the particulates in a waste stream is limited by the clearance of the agitating blades and vessel surface, typically 1 to 3 mm (0.03 to 0.1 in.). Streams with viscosities up to 100 P have been treated. Thus, thin-film evaporators appear to be strong candidates for VOC removal from waste streams that are between free-flowing liquids and viscous sludges. However, they are not appropriate for reactive organic streams that may polymerize at the high temperatures of the evaporator.

6.3.3.2 Pretreatment Processes for Sludge with Organics

Hazardous sludges, such as distillation bottoms, filter cakes, and various slurry and sludge by-products, are produced by many industrial production processes (Martin et al., 1982). Hazardous sludge streams require a different treatment process or different process operating conditions to remove volatile organics than is required for liquid streams (Allen and Blaney, 1985). This is mainly because the high solids mass can clog the reaction or adsorption interface in liquid processes, foul heating lines, or reduce mass transfer rates.

Of the techniques available to treating industrial sludges, only evaporation is applicable to the narrow problem of VOC removal, although steam stripping should also apply. With the exception of centrifugation, the other techniques (incineration, landtreatment, wet oxidation, and chemical fixation/stabilization/encapsulation) are for ultimate disposal or involve reactions that change more than just the VOC content of the waste.

6.3.3.2.1 Air Stripping with Carbon Adsorption

When volatile organics are released by the use of air stripping (described in Section 6.3.3.1.4), it may be necessary to treat the emissions before release to the atmosphere. Adsorption onto granular activated carbon (GAC) is a cost-effective treatment process for many hazardous volatile organic emissions (Love and Eilers, 1982). Section 6.3.3.1.5 covers the use of carbon adsorption for this purpose.

6.3.3.2.2 Evaporation with Carbon Adsorption

Evaporative removal of VOCs from sludges has the same advantages and disadvantages given above for air stripping, except that it will be more expensive, because of the need for thermal energy and waste mixing (Allen and Blaney, 1985). For example, a sludge treatment system might consist of a steam-jacketed mixer/evaporator and an activated carbon adsorber/regenerator. If the system can handle 24 Mg (tons) of sludge per day, the steam would account for about 35% of the annual operating costs, and the evaporator would account for 50% of the equipment capital costs (Spivey, Allen, Green, Wood, and Stallings, 1986). The additional water generated from many sludges during evaporation would require additional condenser capacity on the carbon adsorption system and would further increase costs. Controls may also be required to remove particulates emitted from the dry product.

6.3.3.2.3 Steam Stripping

The waste could be circulated through a pipe with a perforated auger, resulting in both sludge mixing and steam sparging (Allen and Blaney, 1985). The advantage of steam stripping of sludge over evaporation is the greater contact between the steam/air mixture. It also does not dewater the sludge, which leaves the treated material amenable to further processing. As with steam stripping of liquids (Section 6.3.3.1.2), this approach to VOC removal from sludges requires cleanup of the aqueous condensate and proper disposal of stripped organics.

6.3.3.3 Pretreatment Processes for Soils

Remediation technologies, such as landtreatment, soil washing/extraction, thermal desorption, soil venting, and *in situ* biodegradation can be used for treating VOCs in soils (IT Corporation, 1987). A general discussion of landtreatment and composting can be found in Sections 2.1.2.1.1 and 2.1.2.1.2, respectively. *In situ* bioremediation is described in Section 2.2.2.1. Biological techniques promote degradation of the organic hydrocarbons in the soil before they can volatilize, and retention of the volatiles in the soil further encourages their attack by microorganisms. Several techniques may be applicable to a given situation, in which case, selection may be based upon results of treatability studies, estimated capital and operating costs, historical applications, and regulatory acceptance.

6.3.3.3.1 Soil Washing/Extraction

This process is not appropriate if the soil is contaminated only by VOCs (IT Corporation, 1987). However, if both VOCs and heavier oils are present, it may be considered. This is discussed in Section 2.2.1.7.

With extraction, the extraction agent is mixed with the contaminated soil to dissolve or disperse the contaminants from the oil into the extracting agent and separate them from the soil particles (IT Corporation, 1987). The soil may be postwashed with clean extraction agent to remove residual contaminated extraction agent. The extraction agent is cleaned for reuse, and the chemical waste residue further treated or landfilled.

6.3.3.3.2 Thermal Desorption

This method can be used for VOC-contaminated soils that cannot be managed by other methods (IT Corporation, 1987). This approach involves heating the contaminated soil to between 300 and 1100°F to volatilize the hazardous chemicals adsorbed on the soil. The lower temperatures are adequate for VOC removal. The compounds in the vapor must then be further treated by vapor-phase carbon adsorption, fume incineration, or condensation followed by treatment or off-site disposal. A schematic illustration of a pilot system is shown in Figure 6.10 (Marks and Noland, 1986). The thermal processor is a commercially available indirect heat exchanger, which is commonly used to heat, cool, or dry bulk solids, slurries, pastes, or viscous liquids. A schematic process diagram of a full-scale, low-temperature thermal stripping system is shown in Figure 6.11.

Low-temperature thermal stripping is effective as a decontamination method for soils contaminated with VOCs (McDevitt, Marks, and Noland, 1986). Soil is treated in a thermal processor with a carrier gas (air) to enhance VOC removal. The soil is heated to evaporate volatile contaminants (Marks and Noland, 1986). Contaminants in the off-gases are thermally destroyed in an afterburner. The level of VOCs in the processed soil is a direct and predictable function of (McDevitt, Marks, and Noland, 1986)

- VOC concentration in the feed soil
- Processed soil temperature
- Soil residence time within the thermal processor
- Heat input rate to the thermal processor
- Moisture content of the feed soil and processed soil

A processed soil temperature of >233°C and a residence time of >90 min are required to obtain no detectable VOCs in the processed soils.

It has been suggested that an inert carrier gas, such as nitrogen or combustion gases from an oil heating unit, be used instead of air, not to improve VOC removal efficiency, but to improve the safety of the system (i.e., by avoiding the explosive limits associated with volatile hydrocarbons in air) (McDevitt, Noland, and Marks, 1987). The role of aeration in thermal stripping is minimal, which suggests that a minimal airflow rate should be used during thermal stripping.

Mobile thermal treatment systems include rotary kiln incinerators, thermal desorbers, fluidized-bed incinerators, infrared furnaces, fluid wall reactors, and plasma arc devices (DeCicco and Troxler, 1986). An indirect thermal heating unit with a design capacity of 13 ton/h and a low energy consumption of 10.6 gal of oil/ton has been developed (Van Hasselt, 1987).

Three different systems are recommended for low-temperature thermal stripping of VOCs from soil, depending upon the amount of soil to be treated (Table 6.11) (Marks and Noland, 1986). With or without flue gas scrubbing, System B is the most economical system evaluated for sites with 15,000 to 80,000 tons of soil to be processed. This is a single unit of the quad-screw design. The screws have a diameter of 24 in. and a length of 24 ft. System A is somewhat less expensive than System B for sites smaller than 10,000 tons of soil, but processing costs are in excess of $200/ton. This is a single unit of the double-screw design, with the same size screws. System C is somewhat less expensive than System B for sites larger than 85,000 tons of soil. This uses two quad-screw thermal processors arranged in series.

Thermal treatment as an on-site or *ex situ* soil treatment process is described in Section 2.1.1.1.1.

6.3.3.3.3 Soil Venting/in Situ Air Stripping

This treatment may be appropriate for VOCs in permeable soils (Koltuniak, 1986; IT Corporation, 1987). Generally, it requires loose, sandy soils with relatively unrestricted, uniform airflow and good soil/air interface. By mechanicallly injecting clean air into the soil, volatile solvents and hydrocarbons are stripped off the soil into the airstream. When emissions must be controlled, the contaminated air is treated in a vapor-phase carbon adsorption system or a fume incinerator before discharge to the atmosphere. Tightly packed soils, soils with high clay content, rock formations, debris, and a high water table restrict airflow and reduce the volatilization of VOCs and are, therefore, not good candidates for this technology.

Figure 2.28 (McDevitt, Noland, and Marks, 1987) is a schematic of a unit for aerating soil with a constant, low-pressure flow of diffused air to strip off the VOCs. Section 2.2.1.10 provides more information on soil air stripping.

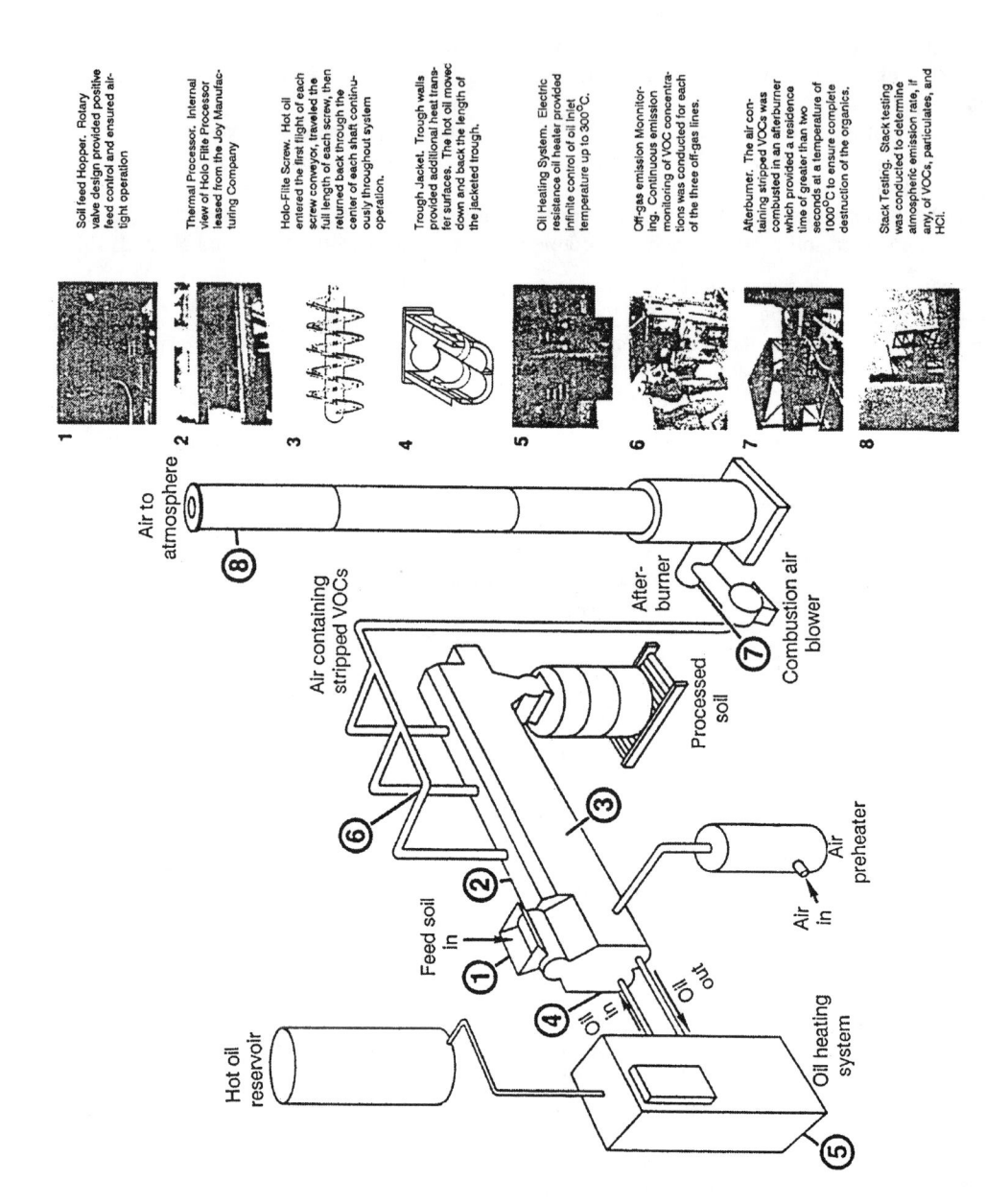

Figure 6.10 Schematic illustration of the low-temperature thermal stripping pilot system. (From Marks, P.J. and Noland, J.W. Report No. AMXTH-TE-CR 86085. ADA 171521. U.S. Army Toxic and Hazardous Materials Agency, 1986.)

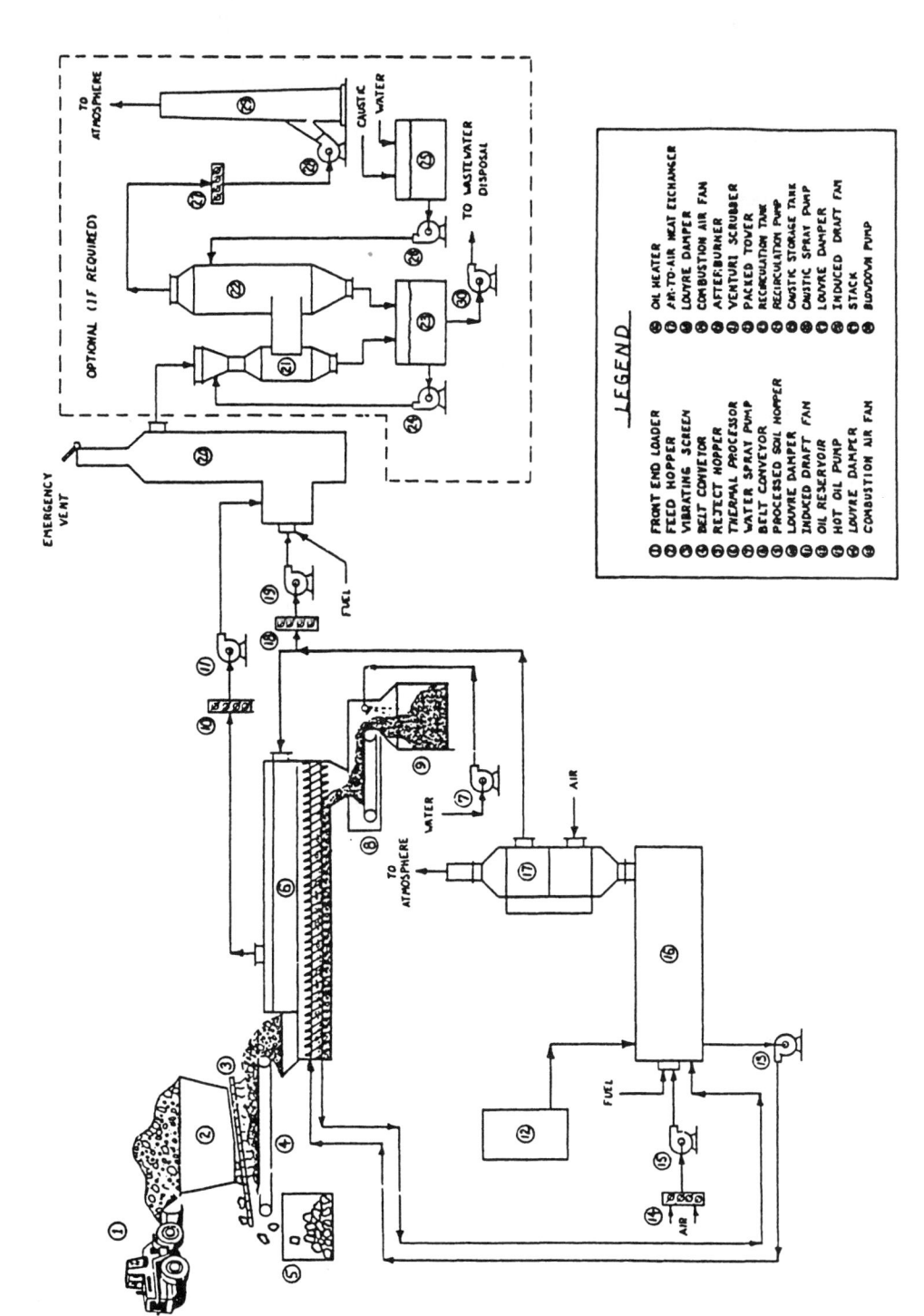

VOLATILE ORGANIC COMPOUNDS IN PETROLEUM PRODUCTS 361

Figure 6.11 Schematic process diagram of the full-scale low-temperature thermal stripping system. (From Marks, P.J. and Noland, J.W. Report No. AMXTH-TE-CR 86085. ADA 171521. U.S. Army Toxic and Hazardous Materials Agency, 1986.)

Table 6.11 Summary of LTTS System Data for Thermal Stripping Systems with and without Flue Gas Scrubbing

Description	System A	System B	System C
Thermal Processor Data			
Model number	D-2424-6	Q-2424-6	Q-2424-6
Number of units	1	1	2
Soil feed rate, lb/h	7,500	15,250	61,000
Soil moisture content, %	20	20	20
Soil discharge temperature, °F	400	400	400
Soil residence time, min	54	47	47
Fuel requirement, 10^6 Btu/h	2.75	5.6	11.21
Sweep airflow rate, ACFM	500	500	1,000
Sweep air temperature, °F	400	400	400
Total airflow to afterburner, ACFM	1,165	1,864	3,728
Air temperature to afterburner, °F	272	247	247
Relative humidity	4.9	24.3	24.3
Afterburner Data			
Number of units	1	1	1
Inside diameter, ft	5.0	6.0	7.0
Inside length, ft (each)	12.0	13.5	20.0
Refractory thickness, in.	9	9	9
Burner size, 10^6 Btu/h (each)	5.0	7.5	15.0
Fuel requirements, 10^6 Btu/h (each)	3.9	6.4	12.5
Exit gas temperature, °F	1,800	1,800	1,800
Exit gas flow rate, ACFM	6,978	11,523	22,786
Gas retention time, s	2.0	2.0	2.0
Scrubber Data			
Number of units	1	1	1
Exit gas temperature, °F	195	195	195
Exit gas flow rate, ACFM	3,780	6,160	12,300
Utilities			
Fuel type	Propane	Propane	Propane
Fuel burn rate, 10^6 Btu/h	6.65	12.0	47.4
Water consumption, gal/h (thermal processor and afterburner)	105	210	420
Water consumption, gal/h (scrubber)	805	1,380	2,760
Scrubber blowdown, gal/h	300	500	1,000

Source: Marks, P.J. and Noland, J.W. Report No. AMXTH-TE-CR 86085. ADA171521. U.S. Army Toxic and Hazardous Materials Agency, 1986.

6.3.3.3.4 Soil Vapor Extraction (SVE)

A vapor (or vacuum) extraction system has been employed to remove VOCs from contaminated soil by applying a vacuum to the soil through extraction wells constructed with perforations above the water table (Bennedsen, 1987). Air is taken in through the zone of contamination when vacuum is applied via air inlet wells. Figure 2.29 shows a schematic of a soil venting operation (Johnson, Kemblowski, and Colthart, 1990). See Section 2.2.1.11 for a full description of *in situ* SVE systems for treating contaminated soils.

6.3.3.3.5 Soil Vapor Extraction/Shallow Soil Mixing (SSM)

These two techniques were integrated to extract VOCs from soils at a Department of Energy facility in southern Ohio (Carey, Day, Pinewski, and Schroder, 1995). This combination has the advantages of a relatively rapid remediation compared with other *in situ* techniques at a lower cost, with less exposure of waste to the surface and elimination of off-site disposal. In implementing this integrated technology, it is important to have a proper site investigation, feasibility estimations, selection of appropriate materials and performance criteria, and bench-scale testing and construction. See Section 2.2.1.1 for a description of SSM.

6.3.3.3.6 Detoxifier™

A Detoxifier™ can be used to collect and treat extracted VOCs. See Section 2.2.1.13 for a description of this technology.

6.3.3.3.7 Photodegradation

Photodegradation of organic compounds may occur by two processes: (1) direct photodegradation and (2) sensitized photo-oxidation (Sims and Bass, 1984). The relative importance of photodegradation of chemicals on or within a soil will depend to a large extent upon its partitioning between the air/water/soil media within the soil system.

Enhancing volatilization of compounds that are susceptible to photodegradation may be a potential treatment technique (Sims and Bass, 1984). This method involves increasing the bulk density or drying of the soil system to increase soil vapor pore spaces and, subsequently, increase the vaporization rate of desired compounds, followed by photodegradation in air. The soil may be tilled to enhance vaporization. Drying of the soil to increase volatilization may be accomplished by tilling, or by installation of a drainage system.

The technique is applicable to compounds of low water solubility, with low K_D values, low K_W values, and those that are highly photoreactive and that, once within the lower atmosphere, would have a relatively short half-life (on the order of hours, or preferably, minutes). Generally, this includes compounds with moderate to strong absorption in the >290-nm wavelength range. Such compounds, generally, have an extended conjugated hydrocarbon system or a functional group with an unsaturated heteroatom (e.g., carbonyl, azo, nitro). Groups that, typically, do not undergo direct photolysis include saturated aliphatics, alcohols, ethers, and amines.

Hazardous products may result from photodegradation. Unless there is sufficient certainty that the constituents in a waste will not produce hazardous photodegradative products, this technology should not be used. To date, this technology is conceptual only. Photochemical oxidation is described in Section 2.1.1.2.6.

6.3.4 VOC POSTTREATMENT TECHNIQUES

Posttreatment techniques are defined here as procedures that can be used to deal with volatiles removed from waste streams after the imposition of *in situ* treatment techniques (Ehrenfeld, Ong, Farino, Spawn, Jasinski, Murphy, Dixon, and Rissmann, 1986). Basically, a posttreatment control involves treatment of emissions collected by a pretreatment control. This can be the use of a form of incineration on the emissions, or it could be the desorption of the VOCs from activated charcoal (see Section 6.3.3.1.5).

Off-gas treatment may be required from SVE systems or air stripping units to limit hydrocarbon discharge to the atmosphere (Ram, Bass, Falotico, and Leahy, 1993). This could be achieved by vapor-phase adsorption, catalytic oxidation, or thermal-oxidation treatment. Costs for these technologies have been reported by Kroopnick (1991).

6.3.4.1 Combustion/Incineration

Combustion is widely applicable for control of air emissions of combustible organic compounds (Cheremisinoff, 1988). The combustion device can be a thermal or catalytic incinerator, a boiler or process heater, or a flare. Combustion can destroy organic pollutants through oxidation, which forms water vapor and carbon dioxide. Any other elements in the organic compound will also be emitted as an oxide or acid gas; e.g., chlorine will be emitted as hydrogen chloride.

A number of hazardous waste facility managers dispose of their air emission residuals through thermal combustion (Allen and Blaney, 1985). At large facilities, a trunk line is used to route off-gases to the air intake of boilers or waste incinerators. At smaller facilities, off-gases are flared. Combustion of wastes in cement kilns, boilers, and incinerators results in >99.99% destruction of organics, as long as combustion temperatures are maintained above 1000°C (1800°F). Liquid injection incinerators, together with rotary kilns, multiple hearths, and fluidized beds, form the current basis of the hazardous waste incineration industry (Cheremisinoff, 1988).

A primary concern in thermal treatment of hazardous wastes, particularly by combustion and incineration, is potential sources for air pollution. Prior to discharge, flue gases must be cleaned by passage through one or more unit processes according to the pollutants present and stack gas quality requirements. Acids in the gas may be removed downstream by scrubbing. Particulate removal is achieved through venturis, electrostatic precipitators, baghouses, or ionizing wet scrubbers. In cases where a variety of

VOCs are being emitted, an incinerator can be adjusted to achieve the desired destruction of the mixture (Jennings, Krohn, Berry, Palazzolo, Parks, and Fidler, 1985).

The requirement of excess air for complete combustion adversely affects the cost of operation, because additional heat is needed to raise the air temperature to that of the exhaust gases (Cheremisinoff, 1988). For general applications in which fuel oil, methane, or sludge are used, the combustion requires 7.5 lb (3.4 kg) of air to release 10,000 Btu (10.55 MJ) from sludge or supplemental fuel.

Incineration is a two-step oxidation process involving first drying and then combustion (Cheremisinoff, 1988). Depending upon control parameters and temperature limitations, this can be performed in separate units or successively in the same unit. The drying step is not the same as preliminary dewatering, which is usually performed mechanically prior to incineration. In all furnaces, the drying and combustion processes follow the same phases: raising the temperature of the feed sludge to 212°F (100°C), evaporating water from the sludge, increasing the temperature of the water vapor and air, and increasing the temperature of the dried sludge volatiles to the ignition point.

The VOC waste gas must be heated to high temperatures (typically, 1000 to 1600°F) and held there long enough to ensure the completion of the oxidation (typically, 0.3 to 1 s) (Jennings, Krohn, Berry, Palazzolo, Parks, and Fidler, 1985). When incinerating solids, temperatures must be high enough to volatilize, partially oxidize, or convert waste organics to gases and to thermally degrade volatilized organics (Cheremisinoff, 1988). Residence time for solids in incineration equipment is longer than that for liquids. The residence time at temperatures less than 870°C is between 0.5 to 1.0 s (Ehrenfeld and Bass, 1983). Generally, a minimum heat content of 8000 to 10,000 Btu/lb (4400 to 5549 kcal/kg) is necessary to sustain combustion (Cheremisinoff, 1988).

In some cases, incineration may allow self-sustaining combustion, with little or no need for auxiliary fuel (Cheremisinoff, 1988). It may convert waste to energy to be used on-site or sold, significantly reducing waste disposal and fuel costs. Solids with a high fraction of combustible material, such as grease and scum, have high fuel values. Thermal oxidation units have lower initial costs than catalytic oxidation units but are usually more expensive to operate because of higher fuel requirements (Ram, Bass, Falotico, and Leahy, 1993).

Incinerators tend to exhibit less efficiency drop than carbon adsorption systems with decreases in VOC loading and may be adjusted over short time intervals for higher temperatures (Jennings, Krohn, Berry, Palazzolo, Parks, and Fidler, 1985). In instances where the VOC loading is highly variable and the VOC removal efficiency is of primary importance, incinerators would have an advantage over carbon adsorption systems.

While destroying one air pollutant, incineration may create other pollutants that require further treatment for removal from flue gases by flue gas scrubbing. See also Sections 2.1.1.1.1 and 2.1.1.1.2 for a description of these processes as they apply to on-site or *ex situ* soil treatment.

6.3.4.1.1 Thermal Incinerators

Most conventional thermal systems rely upon large amounts of supplemental fuel to heat the VOC waste gas (Jennings, Krohn, Berry, Palazzolo, Parks, and Fidler, 1985). To thoroughly heat the VOC waste gas, good mixing of the fuel combustion gas and the waste gas stream is accomplished by a baffle system (Rolke et al., 1972). A thermal incinerator system using regenerative heat exchange is available (Jennings, Krohn, Berry, Palazzolo, Parks, and Fidler, 1985). The system uses significantly less fuel than thermal incinerators with conventional recuperative heat exchangers. This could make thermal incinerators cost competitive at lower explosive limit (LEL) levels of less than 20%.

Thermal incineration is similar in emission control performance to catalytic incineration (Jennings, Krohn, Berry, Palazzolo, Parks, and Fidler, 1985). Thermal incinerators rely on high temperature, sufficient pollutant resident time, and turbulence to ensure high destruction efficiencies. Catalytic incinerators operate at somewhat lower temperatures, since a catalyst promotes the oxidation. There are three instances where thermal incineration may be preferred over catalytic incineration: (1) where exhaust streams contain significant amounts of catalyst poisons or fouling agents, (2) where extremely high destruction efficiencies of difficult to control VOCs are required, and (3) where the VOC waste gas streams are relatively rich (20 to 25% LEL and higher) and the potential for catalyst overheating may require the addition of dilution air to the waste gas stream. This makes catalytic incinerator control more expensive and thermal incineration cost-competitive. Thermal incinerators can potentially achieve higher VOC removal efficiencies than catalytic incinerators, since the latter are subject to operating temperature constraints of the catalyst. The higher temperatures achieved with the thermal incinerators allow greater oxidation of VOCs that are difficult to destroy.

Thermal incinerators require a general characterization of the emission stream to assist in the proper design of the unit. After start-up, the operation can be optimized by adjusting the operating temperature until the desired performance is achieved. This flexibility is important for emissions that are uncharacterized, subject to change, or variable over time.

Figure 6.12 is a schematic of a conventional thermal incineration system with an optional heat recovery system. Figure 6.13 illustrates a regenerative thermal incineration system (Jennings, Krohn, Berry, Palazzolo, Parks, and Fidler, 1985). This system can potentially achieve 85% heat recovery resulting in lower fuel usage. Recuperative thermal incinerators achieve about 30 to 60% recovery. The hot combustion gases can be used to preheat process gases entering the afterburner. Regenerative thermal incinerators are batch-mode systems containing a large number of valves, dampers, flappers, and controllers for switching airflows.

6.3.4.1.2 Afterburners

Afterburners are called vapor incinerators, since they are used to incinerate gases and vapors (Ehrenfeld, Ong, Farino, Spawn, Jasinski, Murphy, Dixon, and Rissmann, 1986). Dilute concentrations of organic vapors are burned along with additional fuel to generate a high temperature of up to 870°C. The fuels used include natural gas, liquified petroleum gas, and distillate and residual fuel oils. As incoming gases and vapors pass through the afterburner, they are decomposed and oxidized. This produces CO_2, water, and other combustion products, depending upon the composition of the incoming gases. Afterburners should be used only on those contaminants that will not produce undesirable oxidation products.

Afterburners can be operated with a heat recovery system (Ehrenfeld and Bass, 1983).

In designing an afterburner system, the following parameters must be specified (Jennings, Krohn, Berry, Palazzolo, Parks, and Fidler, 1985):

> Gas and vapor volume, both average and extremes, also variations due to changes in seasonal temperature;
> Identification of contaminants in gas;
> Concentration of contaminants in gas stream;
> Expected destruction efficiency of the afterburner, from bench or pilot tests.

The contaminants in the gas stream must be destructible to required efficiencies at the operating temperatures and residence time of the afterburner (Ehrenfeld and Bass, 1983). The efficiency of an afterburner system depends upon

> Residence time of the gases in the combustion process; the efficiency increases with residence time for times less than 1 s;
> Temperature of combustion; efficiency increases with temperature;
> Degree of mixing of chamber; efficiency increases with flame contact and oxygen concentration;
> Nature of waste gas;
> Concentration of contaminants in waste gas;
> Catalyst type, in the case of a catalytic process;
> Active surface area of catalyst, which depends upon how long the catalyst has been used.

An afterburner designed for use in a pilot study operated at a minimum temperature of 1000°C (1832°F) with a residence time of greater than 2 s (McDevitt, Noland, and Marks, 1987). It was propane fired, using a North American burner rated at 1.5 million Btu/h. The afterburner operated in conjunction with a refractory-lined stack that was 18 in. in diameter and 20 ft high.

Organics have been destroyed at efficiencies greater than 98% with well-designed and properly operated afterburners. The afterburner is a well-established method for destroying volatile organics. It is a conventional and well-demonstrated technology. In hazardous waste applications, changes in gas flow rate and composition may decrease the operating efficiency of the afterburner.

6.3.4.1.3 Catalytic Incinerators

A catalytic incinerator is, essentially, a thermal incinerator that employs a catalyst to promote the oxidation of VOCs to carbon dioxide and water at lower temperatures (Jennings, Krohn, Berry, Palazzolo, Parks, and Fidler, 1985). Without being altered, the catalyst accelerates the oxidation reaction by

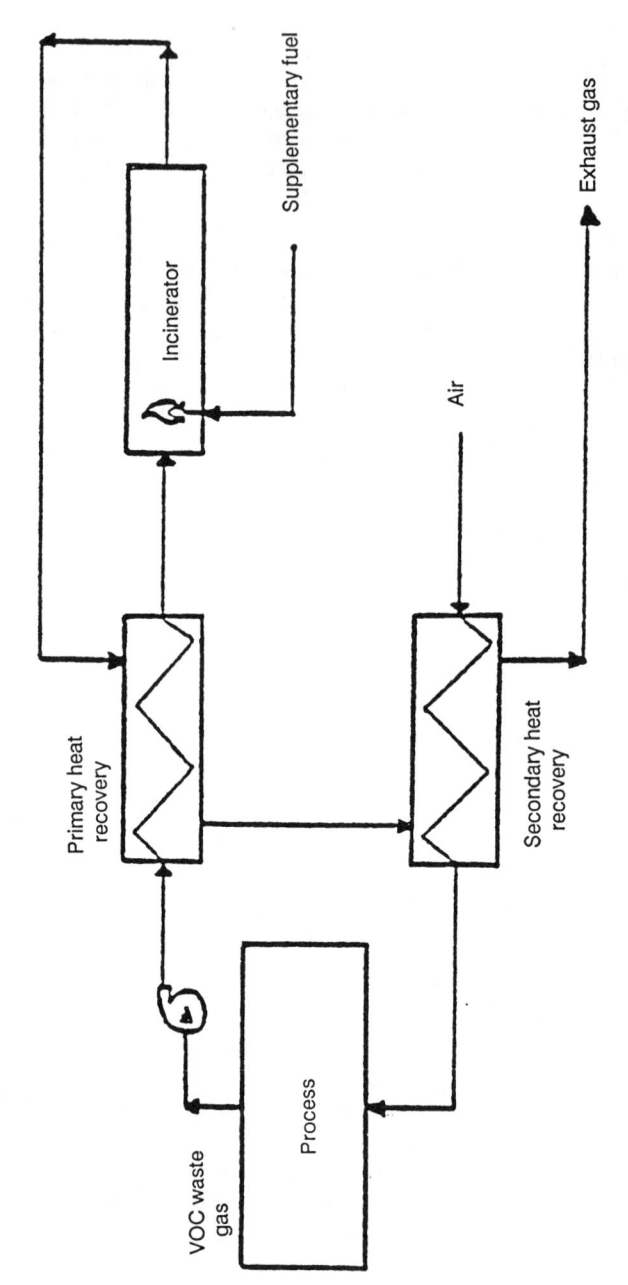

Figure 6.12 Thermal incinerator with primary and secondary heat recovery. (From Jennings, M.S. et al. *Catalytic Incineration for Control of Volatile Organic Compound Emissions*. Noyes, Park Ridge, NJ, 1985. With permission.)

concentrating the reactants (VOC and oxygen) at the active sites on the catalyst. The catalyst is heated electrically in units with a treatment capacity of less than 500 scfm and with propane or methane in larger units (Ram, Bass, Falotico, and Leahy, 1993). Catalytic incinerators have the *potential* to operate with less supplemental fuel (25% less energy) than thermal incinerators (Jennings, Krohn, Berry, Palazzolo, Parks, and Fidler, 1985). Supplemental fuel may be added to heat the incoming waste gas to a temperature suitable for catalytic oxidation.

The basic elements of a catalytic incinerator include

Preheat burner
Mixing chamber
Catalyst bed
Heat recovery equipment

An example of the layout of a typical catalytic incinerator is shown in Figure 6.14.

To oxidize the VOCs, the waste gas stream must be heated to a minimum temperature, which varies depending upon the VOC compound and concentration. The preheat burner heats the waste gas stream to the required temperature. The waste gas should be of uniform temperature before it enters the catalyst bed. The mixing chamber is downstream of the preheat burner and distributes the hot combustion products from the burner into the waste gas. The mixing produces an even temperature.

Next is the catalyst bed, which consists of finely divided platinum or other metal deposited on a ceramic or metal support structure (Jennings, Krohn, Berry, Palazzolo, Parks, and Fidler, 1985). This metal is the actual catalyst for the oxidation reaction (U.S. EPA, 1980a). The catalyst bed may consist of a fixed (single) bed, a modularized bed, or a fluidized bed (Jennings, Krohn, Berry, Palazzolo, Parks, and Fidler, 1985). A fixed bed (Figure 6.13) is assembled from precut pieces or pellets packed within a single container. A modularized bed consists of a number of cartridges of catalyst placed within a single vessel. A fluidized bed consists of spherical pellets supported by a screen in a single vessel (U.S. EPA, 1980b). Air movement through the bed causes the pellets to move randomly and behave as a fluid.

Catalytic afterburners operate at temperatures of between 540 and 870°C (800 and 1000°F; Jennings, Krohn, Berry, Palazzolo, Parks, and Fidler, 1985), although most combustion catalysts cannot be operated at temperatures greater than between 540 and 650°C (Ehrenfeld and Bass, 1983). Such catalysts include platinum, platinum alloys, copper chromite, copper oxide, chromium, manganese, nickel, and cobalt. The pollutants must not contaminate the catalysts, and the maximum concentration of VOC is limited to 25% of the lower flammability limit. Catalysts need to be replaced every 1 to 5 years.

Many catalytic incinerators operate with natural gas–fired burners without a forced draft air supply (the waste gas stream supplies air for combustion). These incinerators may also be oil fired.

Major factors to consider to optimize the performance of a catalytic incinerator include (1) operating temperature, (2) space velocity (reciprocal of residence time), (3) VOC species and concentration, (4) catalyst characteristics, (5) presence of waste contaminants, and (6) heat recovery used (Jennings, Krohn, Berry, Palazzolo, Parks, and Fidler, 1985). The potential for lower fuel costs is a major attraction of catalytic incinerators with a heat recovery system. Installing a recycle-type heat recovery catalytic incinerator on a formerly uncontrolled process could result in a net overall savings.

Both increased surface area and increased volume of the catalyst improve destruction efficiency, especially at low temperatures and high space velocities (Tichenor and Palazzolo, 1987). Catalytic incinerators commonly achieve destruction efficiencies in the range of 70 to 90% (Jennings, Krohn, Berry, Palazzolo, Parks, and Fidler, 1985). Efficiencies of 98% are possible, but might require large volumes of catalyst and high operating temperatures, depending upon the inlet concentration (U.S. EPA, 1976).

Using a test system with catalyst beds made of ceramic honeycombs coated with precious metal (platinum/palladium) catalyst, it was concluded that (1) VOC destruction efficiency increases with increasing temperature and concentration and with decreasing space velocity, (2) the destructibility of VOCs varies with the class of compound, (3) individual VOCs have different destruction efficiencies at a given temperature, concentration, and space velocity, and (4) different destruction efficiencies may result for a specific VOC incinerated in a mixture vs. burned alone (Table 6.12) (Tichenor and Palazzolo, 1987).

Except for clorinated hydrocarbons, all compound classes could be destroyed with 98 to 99% efficiency at sufficiently low space velocities and high temperature (Tichenor and Palazzolo, 1987).

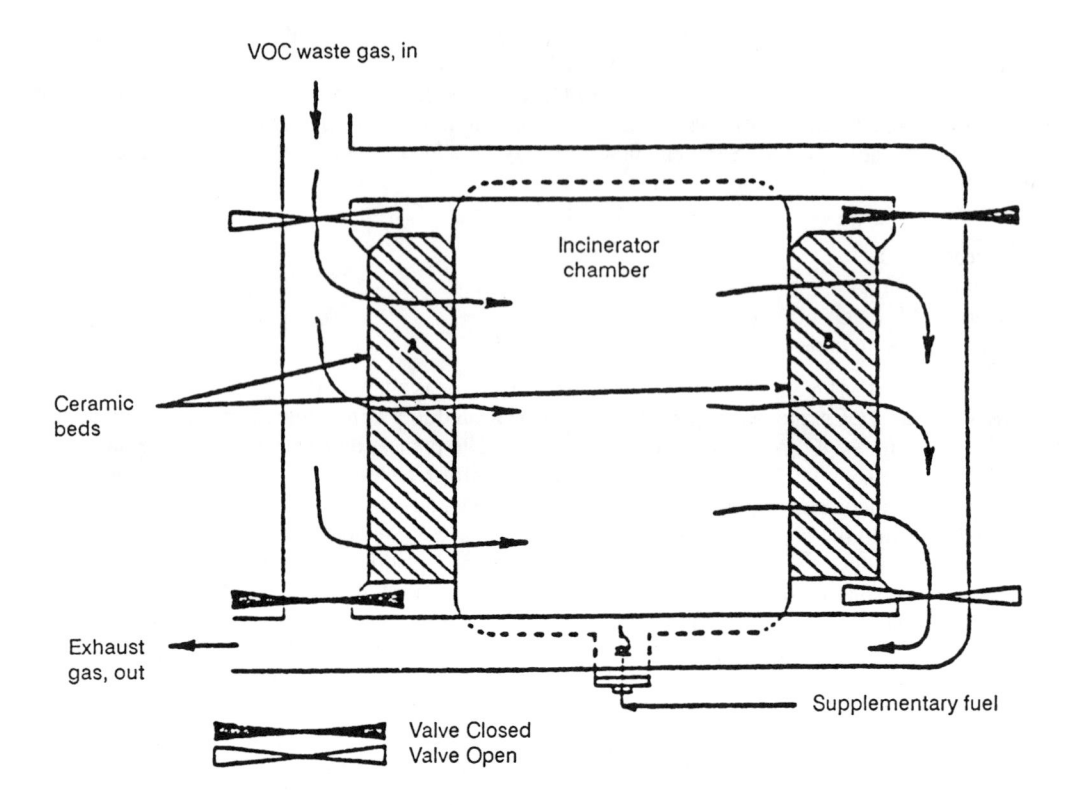

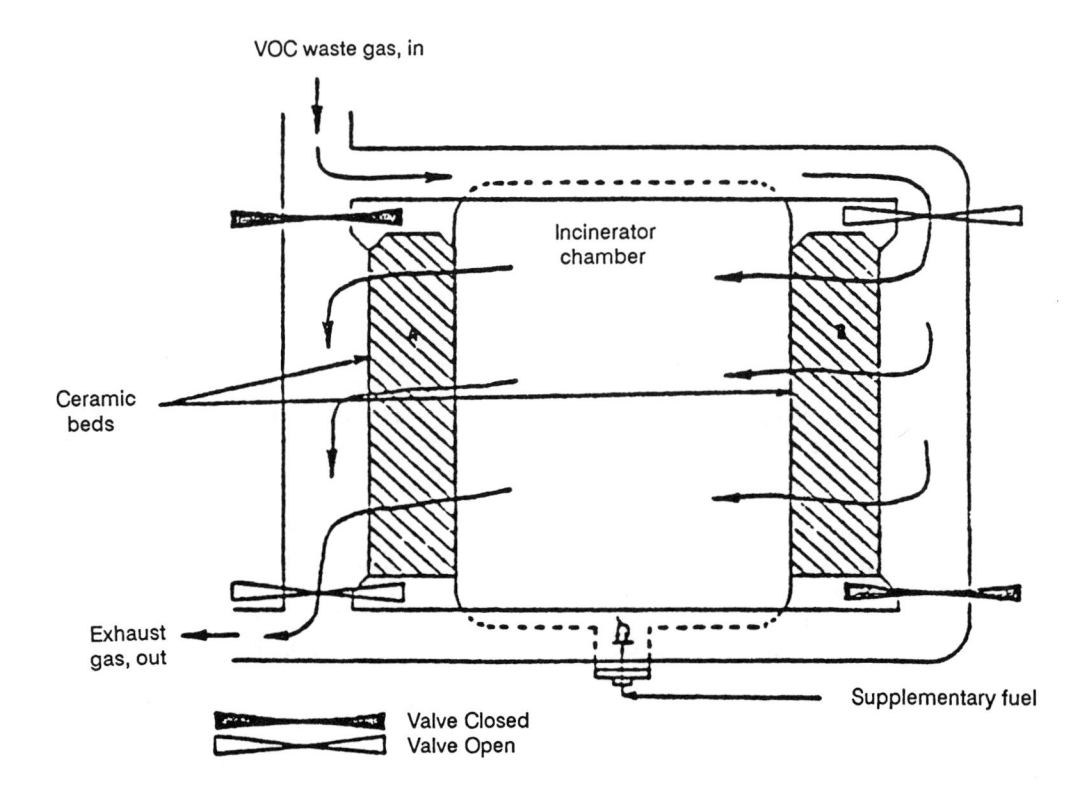

368 REMEDIATION OF PETROLEUM-CONTAMINATED SOILS

Specific compounds within each class may have different relative destructibilities. Some compounds, such as hexane, are more effectively destroyed when incinerated alone than in a mixture. Others, such as ethyl acetate, exhibit greater destruction when burned in a mixture than by themselves. Some compounds, such as benzene, show no difference in destruction efficiencies due to mixture.

Most VOCs are rapidly destroyed at temperatures over 1400°F, although some (e.g., halogenated hydrocarbons) require higher temperatures (Cheremisinoff, 1988). Generally, higher catalyst operating temperatures increase the VOC destruction efficiency; however, they also increase the energy usage (Jennings, Krohn, Berry, Palazzolo, Parks, and Fidler, 1985). Table 6.13 lists initiation temperatures that must be reached for several compounds when in contact with a precious metal catalyst.

Heat recovery equipment can be a part of the incineration system (Jennings, Krohn, Berry, Palazzolo, Parks, and Fidler, 1985). Heat produced from the combustion of VOCs in the catalyst bed, along with heat in the waste gas prior to entering the bed, can be recovered for use. Raising VOC concentration in the ovens increases waste gas VOC concentrations, which makes incinerators more energy efficient. However, most drying ovens maintain LEL levels at 25% or less to ensure the process does not become explosive. The released heat can be used in the following ways:

To heat the waste gas and reduce the energy use of the preheat burner;
To heat the process itself by recycling the exhaust back to the process;
To produce steam or hot water in a waste heat boiler.

Catalytic incinerator systems and components are available off-the-shelf from at least nine vendors (Jennings, Krohn, Berry, Palazzolo, Parks, and Fidler, 1985). The most common application of these systems is in control of VOC vapors emitted by driers and ovens used in surface-coating processes. Other applications include organic chemical manufacturing processes, fuel storage emissions, and refinery emissions.

Catalytic incinerators typically require 15 ft^2/1000 scfm of capacity (OxyCatalyst, Inc., 1980). Catalytic incinerators and recuperative thermal incinerators are smaller and lighter than regenerative thermal incinerators and carbon adsorption systems. Roof installation is possible but not common. Catalytic incinerators have an advantage for retrofit installations where land area is at a premium.

Marquis (1993) reports using an *in situ* VES with a combined thermal-catalytic oxidizer vapor treatment system for remediation of soils with low permeability.

6.3.4.1.4 Flares

When it is not economical to recover the heat value of the gases and the control process upsets vent gases, waste and purged gaseous organic compounds are commonly destroyed by flaring (Cheremisinoff, 1988).

Flares work effectively for organics destruction at petroleum refineries, but their efficiency in destroying complex mixtures of off-gases from hazardous waste facilities is unknown (Allen and Blaney, 1985). Their destruction efficiency for halogenated organics, which are not as combustible as other organics, is of particular concern.

Approximately 98% destruction efficiency can be achieved for flares, if they operate under proper conditions (Cheremisinoff, 1988). Air emissions having heating values of less than 300 Btu/scf (steam or air-assisted flares) or 200 Btu/scf (nonassisted flares) are not assured of achieving 98% destruction. This allows an estimate of the treatment efficiency in absence of other data for the compound.

6.3.4.1.5 Boiler/Process Heater

A combustion technique that may be used as a control device for toxic air pollutants is to inject the pollutants into process heaters or boilers (Cheremisinoff, 1988). Waste streams may provide supplemental fuel or even be the primary fuel in some operations.

6.3.4.2 Condensation

Condensation occurs when one or more volatile components of a vapor mixture are separated from the others by being converted to a liquid (Jennings, Krohn, Berry, Palazzolo, Parks, and Fidler, 1985). This

---

Figure 6.13 Regenerative thermal incinerator. (From Jennings, M.S. et al. *Catalytic Incineration for Control of Volatile Organic Compound Emissions.* Noyes, Park Ridge, NJ, 1985. With permission.)

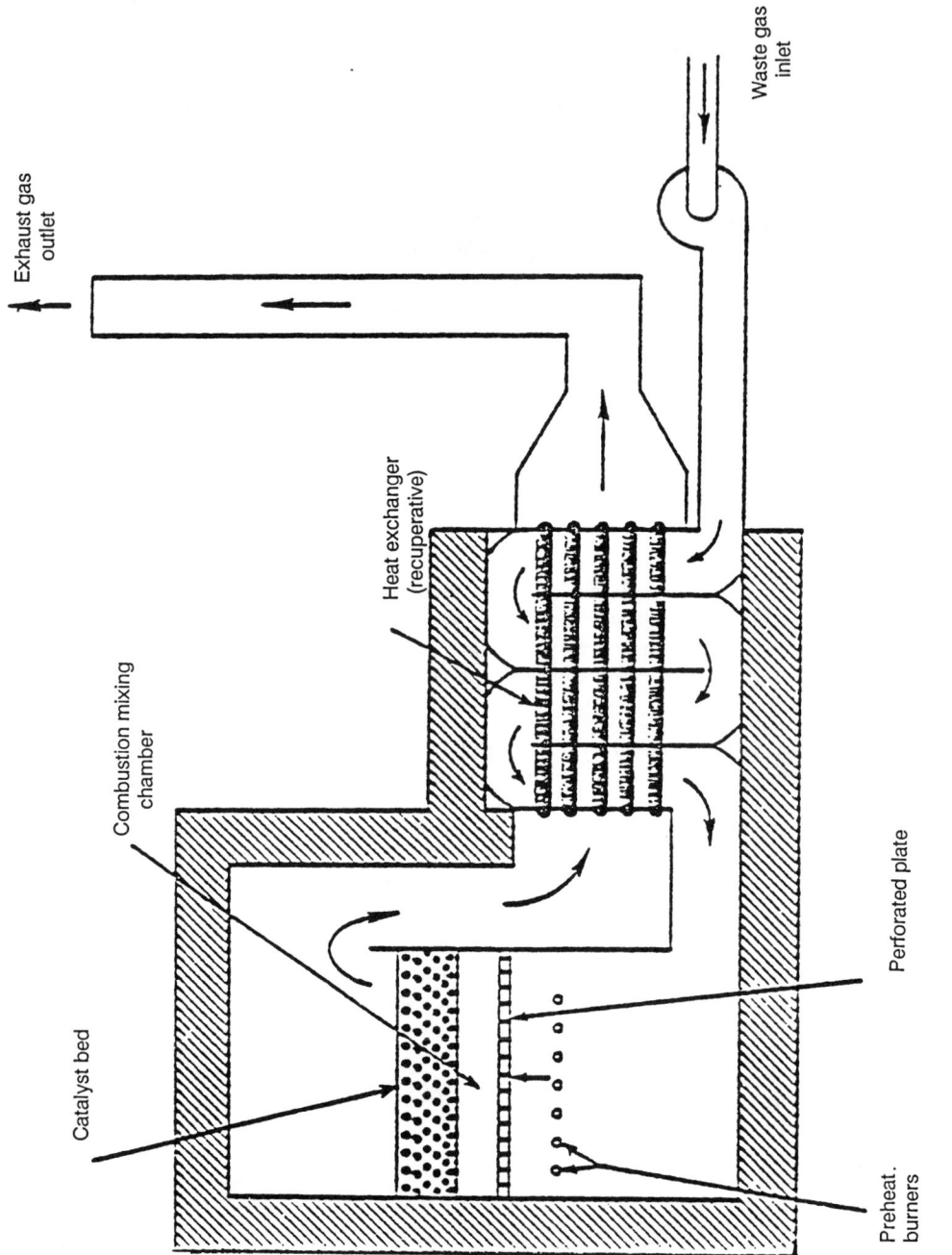

Figure 6.14 Catalytic incinerator with primary heat recovery. (From Jennings, M.S. et al. *Catalytic Incineration for Control of Volatile Organic Compound Emissions*. Noyes, Park Ridge, NJ, 1985. With permission.)

Table 6.12 The Destructibility of Different Compound Classes

Compound Class	Relative Destructibility
Alcohols	High
Cellosolves/dioxane	
Aldehydes	
Aromatics	
Ketones	
Acetates	
Alkanes	
Chlorinated hydrocarbons	Low

Source: Tichenor, B.A. and Palazzolo, M.A. Report No. EPA-600/J-87-182. PB 88159710. U.S. EPA, Research Triangle Park, NC, 1987.

Table 6.13 Temperatures for 90% Conversion in a Catalytic Incinerator

Component	Temperature (°F)
Hydrogen	105
Acetylene	200
Carbon monoxide	220
Propyne	240
Propadiene	250
Propylene	260
Ethylene	290
n-Heptane	300
Benzene	300
Toluene	300
Xylene	300
Ethanol	315
Methyl ethyl ketone	370
Methyl isobutyl ketone	370
Propane	410
Ethyl acetone	415
Dimethyl formamide	425
Ethane	430
Cyclopropane	455
Methane	490

Source: Jennings, M.S. et al. *Catalytic Incineration for Control of Volatile Organic Compound Emissions.* Noyes, Park Ridge, NJ, 1985. With permission.

is brought about by lowering the temperature of the vapor mixture until the vapor pressure of the condensable component equals its partial pressure. Lowering the temperature of the vapor mixture further causes the compound to condense from the vapor phase into the liquid phase. Gases that are noncondensable at the condenser temperature, such as the air in organic vapor, can severely reduce condenser capacity by blanketing the condensing surface (Hackman, 1978). Therefore, air should be excluded in steam systems. Condensation is also discussed in Section 6.3.3.1.7.

Condensation occurs when a saturated vapor comes in contact with a surface, which is at or below the saturation temperature (Hackman, 1978). A film of condensate will usually form on the surface, such as when toxic organic vapors are condensed by contact with cold metallic surfaces. When the wall is not uniformly wetted, the condensate collects in many small droplets at various points on the surface. If the contact angle of the droplet with the surface goes lower than 50°, the droplets spread unevenly and areas will be covered with a continuous film. If the contact angles are greater, individual droplets grow and coalesce with adjacent droplets, forming a small stream, which flows to the bottom, leaving dry surface in its wake. Film-type condensation is more common and more dependable.

During condensation, the partial pressure of the condensable component is always equal to its vapor pressure. Figure 6.15 shows a condensation system for controlling a solvent drying oven by employing a nitrogen inert gas atmosphere (Jennings, Krohn, Berry, Palazzolo, Parks, and Fidler, 1985).

All oxygen is displaced from the inert oven atmosphere (Jennings, Krohn, Berry, Palazzolo, Parks, and Fidler, 1985). This allows the oven VOC concentrations to be much higher than possible in a conventional drying oven, since there is no oxygen to support an explosion. Gaseous nitrogen provides inert oven atmospheres and liquid nitrogen as a refrigerant for condensation.

Heated, recirculated inert gases are used to dry the VOCs in the oven (Jennings, Krohn, Berry, Palazzolo, Parks, and Fidler, 1985). Upon drying, inert gas containing highly concentrated VOC vapors passes to a recovery vessel, where the VOCs are recovered by condensation in several stages of heat exchange. The stripped inert gas is reused in the oven. Nitrogen gas curtains seal the inert atmosphere from air leaks, which could cause an explosion. Only ovens that can be easily sealed can be used for condensation. This system is energy efficient as a result of not having to heat up large amounts of oven dilution air and draw them through the oven (U.S. EPA, 1981a).

Condensation is also technically limited to sources that emit waste gases with relatively high solvent concentrations (greater than 25% LEL) (U.S. EPA, 1976). The condensation temperature required at low concentrations is infeasibly low. This constraint limits the application of condensation control to very few of the common solvent evaporation processes.

A phase change takes place during condensation, and thermal energy must be removed (Hackman, 1978). The heat of vaporization is the minimum energy that must be removed, as well as any superheat above saturation temperature.

A humidity chart can be developed for an organic compound, plotting pounds of organic per pound of dry air vs. temperature. Saturation temperatures and percentages of saturation may then be plotted as curves. Then, knowing the organic vapor concentration for a given contaminated airstream, the temperature required for condensation may be read. By passing known concentrations of the vapor in air across the exterior surface of a polished metal tube, which is cooled and thermostatted on the interior, saturation temperatures can be determined. The saturation temperature is indicated when the metal surface starts to dull, as the tube slowly cools.

6.3.4.3 Distillation

If concentrations of dissolved or suspended solids are high in an incoming waste stream, fractional distillation units with heat exchange coils in their evaporators can foul (Allen and Blaney, 1985). These waste streams remove solids.

6.3.4.4 Absorption

Gas absorption occurs when one or more components of a vapor mixture are separated from the vapor by contact with an absorbent (Jennings, Krohn, Berry, Palazzolo, Parks, and Fidler, 1985).

6.3.4.4.1 Packed Columns

Packed columns are vertical structures containing manufactured packing elements, such as spiral rings, raschig rings, lessing rings, berl saddles, and intalox saddles (Jennings, Krohn, Berry, Palazzolo, Parks, and Fidler, 1985). The packing elements create large surface areas and promote turbulent mixing of gas and liquid. In a countercurrent absorber, the VOC waste gas stream enters the distributing space at the bottom of the column and flows upward through the packing crevices. A scrubbing liquid is simultaneously introduced at the top of the column and flows down over the packing. When the gas and liquid streams contact, the gaseous pollutants are absorbed from the gas into the liquid (see Figure 6.16).

6.3.4.4.2 Plate Columns

In contrast to packed columns, where gas and liquid absorbent are in continuous contact throughout the packed bed, plate columns employ stepwise contact (Jennings, Krohn, Berry, Palazzolo, Parks, and Fidler, 1985). A number of trays or plates are arranged so the gas passes through a layer of liquid on each plate. Figure 6.17 illustrates a bubble-cap tray tower. There are openings (vapor risers) on each plate covered by bubble caps. The gas flows through the column and the openings in each plate. It

VOLATILE ORGANIC COMPOUNDS IN PETROLEUM PRODUCTS

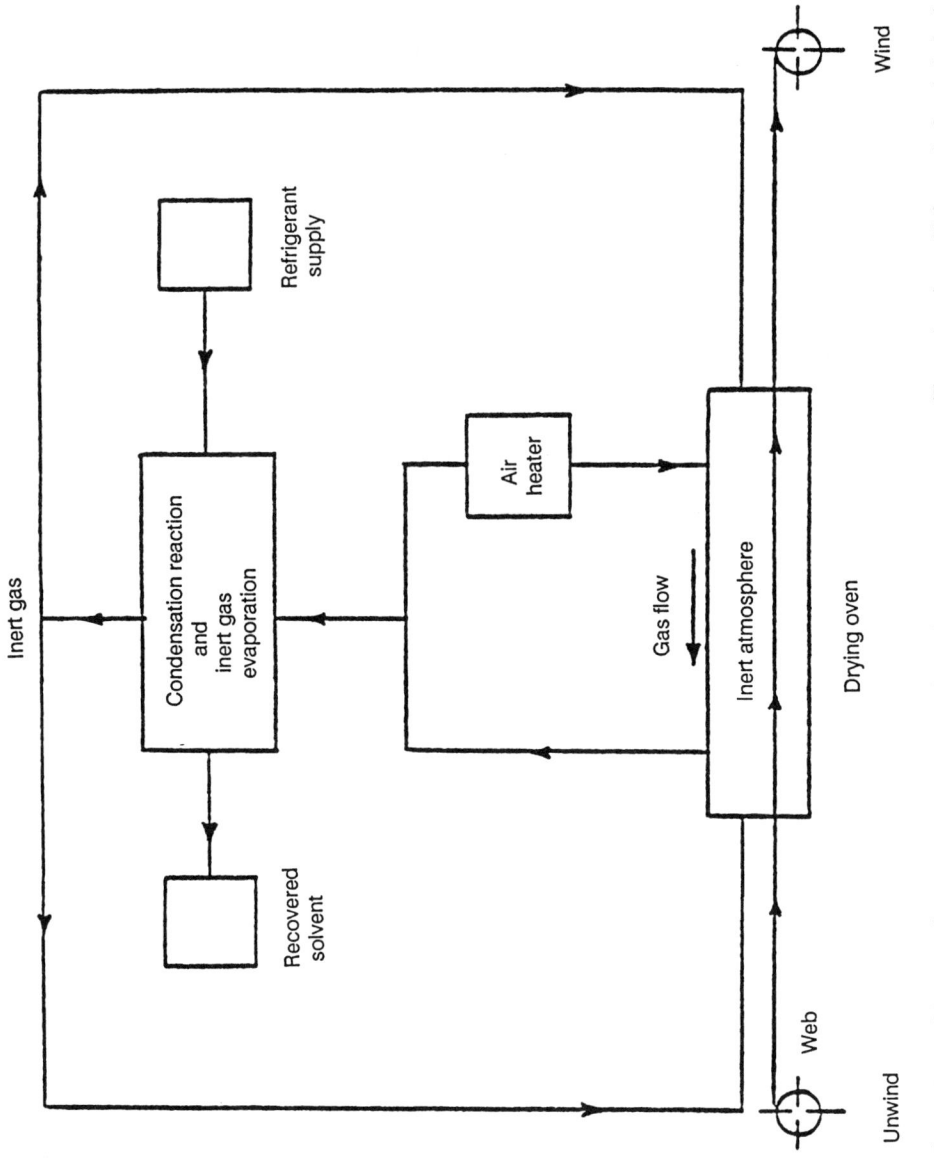

Figure 6.15 Diagram of an inert gas condensation solvent recovery system. (From Jennings, M.S. et al. *Catalytic Incineration for Control of Volatile Organic Compound Emissions.* Noyes, Park Ridge, NJ, 1985. With permission.)

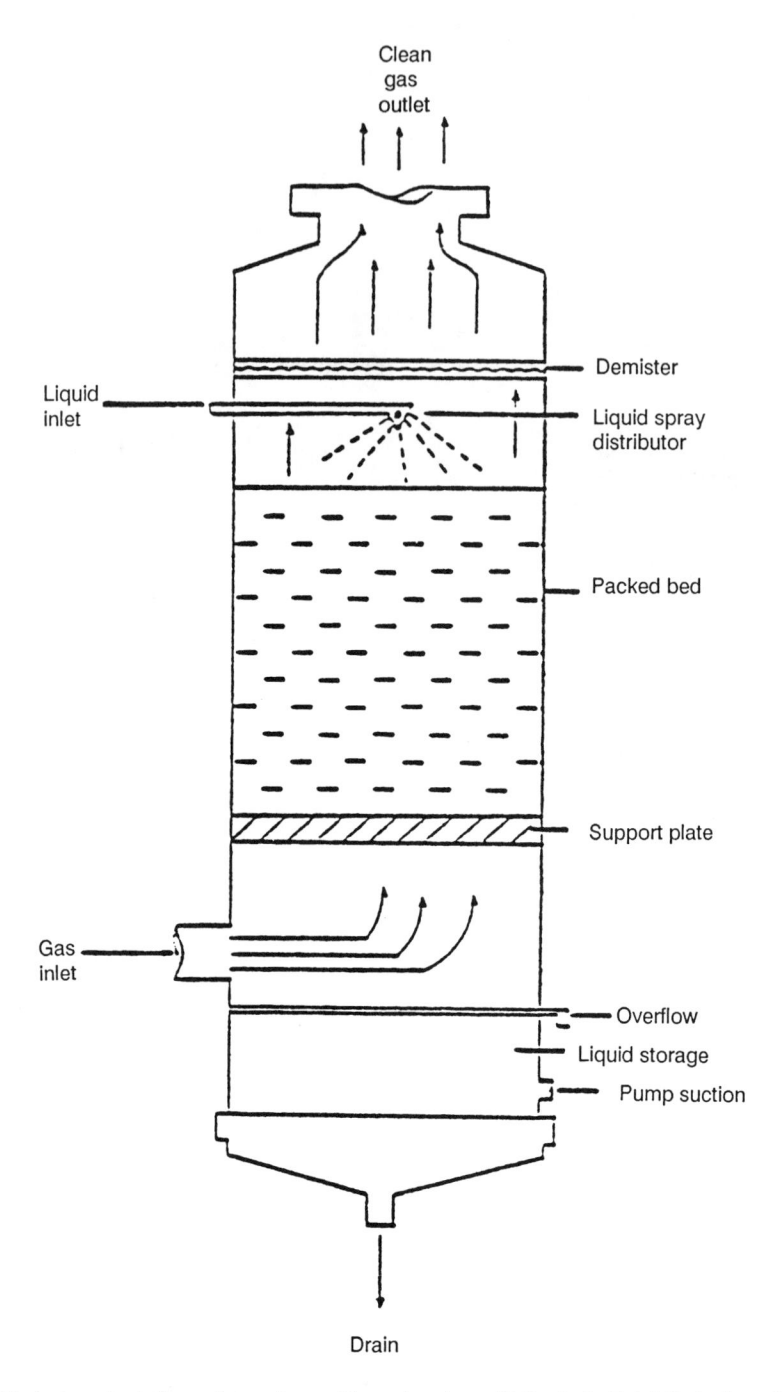

Figure 6.16 Typical packed absorption column. (From Jennings, M.S. et al. *Catalytic Incineration for Control of Volatile Organic Compound Emissions.* Noyes, Park Ridge, NJ, 1985. With permission.)

bubbles through the liquid absorbent, which enters at the top of the column and flows downward across each plate and through the down spouts. As the liquid passes down the column, the gas bubbles through it. The bubble caps permit good contact between the gas and the liquid. The depth of liquid on the plates is maintained at a design depth that allows adequate mixing without excessive pressure loss. There are advantages and disadvantages to both types of columns. Selection of one depends upon many factors. These are summarized in Table 6.14.

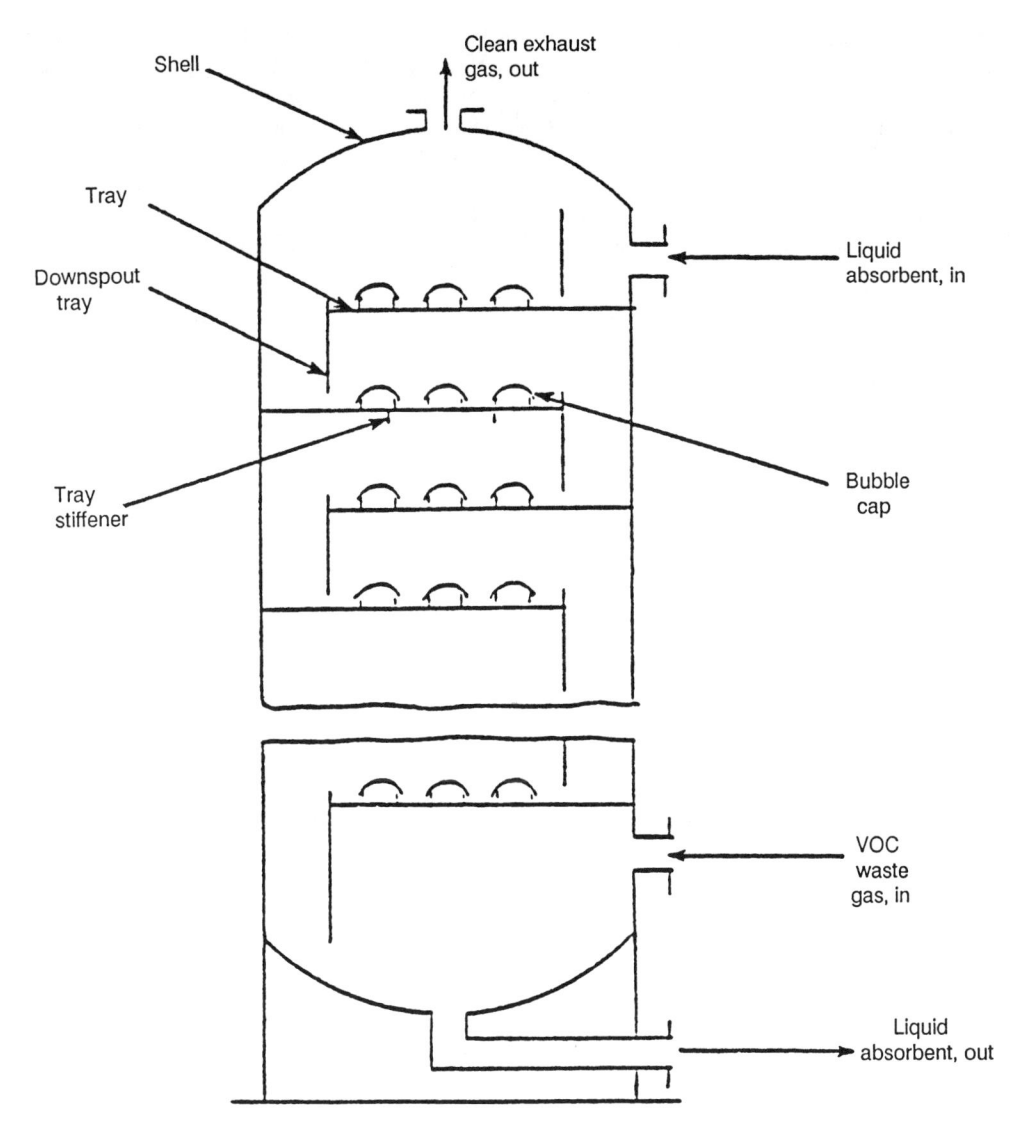

Figure 6.17 Bubble cap absorption column. (From Jennings, M.S. et al. *Catalytic Incineration for Control of Volatile Organic Compound Emissions.* Noyes, Park Ridge, NJ, 1985. With permission.)

Table 6.14 Comparison of Packed and Plate Columns

Packed Towers
• Packed towers are less expensive than plate towers when materials of construction must be corrosion resistant.
• Packed towers have smaller pressure drops than plate towers designed for the same throughput.
• Packed towers are preferred in sizes up to 2 ft in diameter, if other conditions are nearly equal.
Plate Towers
• Plate towers are preferred when the liquid contains suspended solids, since plate towers are more easily cleaned.
• Plate towers are more suitable when the process involves appreciable temperature variation, since expansions and contractions due to temperature changes may crush the tower packing.

Source: Jennings, M.S. et al. *Catalytic Incineration for Control of Volatile Organic Compound Emissions.* Noyes, Park Ridge, NJ, 1985. With permission.

For common surface-coating VOC emission sources, absorbers are limited to applications in which the used liquid absorbent can be reused directly without removing the absorbed VOCs (Jennings, Krohn, Berry, Palazzolo, Parks, and Fidler, 1985). Since the residual organic concentration in the liquid must be extremely low for it to be suitable for reuse, desorption requirements can be prohibitively expensive. Absorption has been used to control VOCs in surface-coating operations, degreasing operations, waste-handling and treatment plants, asphalt batch plants, ceramic tile manufacturing plants, petroleum coker units, chromium-plating units, and varnish and resin cookers (U.S. EPA, 1978). The typical removal efficiencies for absorption run from 90 to 99% (Jennings, Krohn, Berry, Palazzolo, Parks, and Fidler, 1985).

6.3.4.4.3 Polymer-Based Adsorbent

Polyad FB is a novel polymer-based Bonopore adsorbent developed by Nobel Chematur in Sweden (O'Sullivan, 1988). It consists of highly porous plastic microspheres, about 0.5 mm in diameter, that are both hydrophobic and polar. Each gram provides a surface area of about 800 m^2.

Contaminated air enters at the base of a tray containing the beads and moves upward (O'Sullivan, 1988). The particles become a fluidized bed and adsorb the chemicals. The loaded beads migrate to the top of a desorption column, while being replaced by regenerated beads. Loaded beads move down through the stripper and are heated to release the adsorbed chemicals, which are carried by circulating air to a condenser and stored in tanks.

This system can recover a wide variety of chemicals from air, such as aromatic and chlorinated solvents, ketones, and alcohols, and is ideal when solvent concentration of air is in the range of 0.1 to 10 g/m^3 (O'Sullivan, 1988). It can handle hourly flow rates from a few hundred cubic meters to several hundred thousand. The beads can be customized, and contaminated air with a high moisture content does not impair the efficiency.

6.3.4.5 Biofiltration

Biofiltration is an emerging technology for the control of VOCs emitted in off-gases.

The main disadvantage of nonbiological methods is the creation or persistence of undesirable end products (Douglass, Armstrong, and Korreck, 1991). Contaminants may be transferred from one medium to another, or ultimately produce potentially toxic emissions. Such off-gas may be released to the atmosphere without treatment or controlled by incineration or activated carbon, perpetuating the waste stream cycle. Treating withdrawn soil vapors is quite expensive, usually more than half the total remediation costs (Miller, Hinchee, Vogel, Dupont, and Downey, 1990).

SVE with air-based biodegradation may help reduce these costs. Recently, biological cleaning of off-gases has been explored as an alternative treatment because of its low cost, technical simplicity, and production of little or no secondary waste (Lei, Lord, Arneberg, Rho, Greer, and Cyr, 1995). This process is common in several countries, including Germany, The Netherlands, and Japan. Vapor phase biofilters can decompose VOCs from secondary airstreams (Saberiyan, Wilson, Roe, Andrilenas, Esler, Kise, and Reith, 1994; Paca, 1994). Biological treatment of off-gas is inexpensive and effective for converting the contaminant to benign products (Douglass, Armstrong, and Korreck, 1991), by biodegradation or transformation (Ottengraf, 1986). Biochemical oxidation results in complete mineralization of organic substrates to end products of water and carbon dioxide (Farmer, Chen, Kopchynski, and Maier, 1995). Biofiltration allows on-site treatment without creation of hazardous residues or the need for expensive carbon regeneration or high fuel costs for combustion (King, Long, and Sheldon, 1992; Devinny, Medina, and Hodge, 1994).

Biofiltration involves passing a stream of gas containing VOCs through a solid support on which microorganisms are growing, with destruction of the contaminants (Alexander, 1994). An air biofilter consists of one or more beds of packing material, such as compost, peat, wood chips, sphagnum moss, or soil, which provide a high-surface-area support for the microorganisms (Saberiyan, Wilson, Roe, Andrilenas, Esler, Kise, and Reith, 1994). Sawdust, activated carbon, clay particles, or porous glass can also be used for the solid support (Alexander, 1994). Peat buffered with CaCO$_3$ would be the lightest, least-expensive, and among the best biological support media for a mobile treatment system (Douglass, Armstrong, and Korreck, 1991). Compost has been found to support higher concentrations of microorganisms than peat or carbon (Evans, Bourbonais, Peterson, Lee, and Laakso, 1995). A mixture of compost and perlite (80:20 vol/vol) is an optimal compromise based on pressure drop and volumetric surface area. However, while soil and compost are commonly used in commercial applications, they may be too unhomogeneous and may provide organic material that would compete for degradation by the organisms (Medina, Devinny, and Ramaratnam, 1995). Biofilters with pellets or structured media allow better gas distribution, improved pH control by employing buffers in the nutrient solution trickling through the bed, and the ability to remove excess biomass from the media (Bishop and Govind, 1995).

The appropriate reactor packing material depends on the chemical composition, contaminant solubility, contaminant concentration, and vapor stream flow rate (Douglass, Armstrong, and Korreck, 1991). The packing must have good pneumatic conductivity and high surface area, good wetting characteristics, and good sorptive capacity. Retention and distribution of a thin water layer and biofilm over the packing is necessary to colonize the microbes. The support media and size of reactor are chosen to provide the proper contact and retention time, solubility, diffusion, and packing sorptive capacity. An ideal packing has high porosity, high specific surface area, low bulk density, and good moisture retention (Fischer and Bardtke, 1984). Porous media improves nutrient retention, which maintains the level of essential minerals in a moist biofilm (Bishop and Govind, 1995). The geometry of the medium controls the biofilm surface area per unit volume. The biofilm thickness depends on adhesion between the cells and the surface of the medium (Utgikar, 1993). Adsorptive media could allow the adsorbed substrate to back-diffuse into the biofilm (Bishop and Govind, 1995). The filters can be customized with specific carriers, nutrients, or cultures (King, Long, and Sheldon, 1992).

A comparison of several types of media showed that adsorbing media, such as activated carbon, was superior to smooth, nonadsorbing media, such as ceramic or resin (Bishop and Govind, 1995). The success of the activated-carbon-coated biofilter was probably due to improved adsorption of contaminants on the surface of the medium, which increased the length of time for biodegradation, and to better biofilm attachment and thicker biofilms, as a result of back-diffusion of adsorbed contaminants into the biofilm.

The elimination capacity (grams of contaminant removed per cubic meter of biofilter per hour) is largely a function of the characteristics of the packing medium (Devinny, Medina, and Hodge, 1994). Biofilters have been built to treat up to 90,000 ft^3/min of airflow with filters up to 20,000 ft in wetted area (King, Long, and Sheldon, 1992).

Humidifying the off-gas prior to passing it through the filter prevents the filter material from drying out (Leson, Tabatabai, and Winer, 1992). As a result of interparticle porosity, the air permeates the spaces between the support particles (Devinny, Medina, and Hodge, 1994). Intraparticle porosity, within the particles, is part of the solids/water phase (where biodegradation occurs) and allows for the sorption capacity of the medium. Highly sorptive support media that produce high mass partition coefficients may be beneficial, while those with lower porosities may be limited by pressure drop across the biofilter and biological clogging. The particles in the packing material have a diameter of <10 mm, which gives a high specific surface area (300 to 1000/mL) (Diks and Ottengraf, 1991). Biofilters can handle rapid airflow rates and VOC concentrations over 1000 ppm (King, Long, and Sheldon, 1992).

The packing is inoculated with hydrocarbon-degrading bacteria (e.g., *Pseudomonas* sp.) (Saberiyan, Wilson, Roe, Andrilenas, Esler, Kise, and Reith, 1994), mixed cultures from petroleum-contaminated soils (Devinny, Medina, and Hodge, 1994), or other heterotrophic microorganisms capable of degrading the organic contaminants of concern (Miller, Saberiyan, DeSantis, Andrilenas, and Esler, 1995). Waste gases and oxygen pass through the inoculated packing material. The VOC diffuses into the film and is sorbed by the moist solid phase, where it is metabolized by the attached film of microorganisms, releasing CO_2 and water. Biodegradation continuously removes contaminants, maintaining concentrations below the saturation value, which transfers more from the air phase (Medina, Devinny, and Ramaratnam, 1995).

White-rot fungi have been used successfully as organisms in biofilters (Braun-Luellemann, Johannes, Majcherczyk, and Huettermann, 1995). *Trametes versicolor, Pleurotus ostreatus, Bjerkandera adusta*, and *Phanerochaete chrysosporium* can remove organic compounds from the gas phase. Filters with these fungi have a very high biologically active surface area. This provides good retention, low pressure drop, and high physical stability. When the fungus is grown on lignocellulosic substrates, the extracellular enzymes produced have an unspecific degradation capacity for a wide variety of similar compounds. Styrene, ethylbenzene, xylenes, and toluene can be degraded with this method.

When optimum environmental conditions are maintained (appropriate oxygen, nutrients, moisture, pH, and temperature), there will be maximum treatment efficiency (Saberiyan, Wilson, Roe, Andrilenas, Esler, Kise, and Reith, 1994). The end products are CO_2, water, and biomass. Acidification of the filter bed can be prevented by the addition of neutralizers (Diks and Ottengraf, 1991). To prevent the accumulation of neutralization products, which can inhibit biological activity, a biological trickling filter can be used. A model has been developed to describe the biological processes inside the biofilter (Ottengraf and van den Oever, 1983). With the appropriate size and design of the filter, control efficiencies of more than 90% can be achieved (Leson, Tabatabai, and Winer, 1992).

These treatment systems are relatively inexpensive and allow destruction of compounds at low concentrations (i.e., <800 ppmv) (Bishop and Govind, 1995). It is suggested, however, that a minimum, as well as a maximum, concentration of influent is required for biodegradation, and this will vary with the packing material (Saberiyan, Wilson, Roe, Andrilenas, Esler, Kise, and Reith, 1994). High pollutant concentrations could be toxic to the microorganisms in the filter (Leson, Tabatabai, and Winer, 1992). In general, total VOC concentrations of 3000 to 5000 mg/m^3 should not be exceeded, while individual compounds may require a lower value. VOC concentrations >1 g/m^3 begin to be more expensive than energy-efficient incineration technologies. For compounds of medium degradability, an off-gas concentration <500 ppm as carbon would indicate that biofiltration is economically viable.

A trickle-bed reactor can be designed for VOCs, by having the vapors pass through bacteria fixed on a column and then be dissolved in a solution (Alexander, 1994). Bioscrubbers are similar, but the gases and oxygen usually pass first into a unit where the VOCs dissolve in water, then this solution may go into an activated sludge system where the organic compounds are degraded by microbes in an aqueous phase. Biofilters can be open, single-bed systems, or they can be fully enclosed, usually containing stacked beds, for use where space is limited or control over the filter moisture content is desired (Leson, Tabatabai, and Winer, 1992).

The elimination rate for a given contaminant depends mainly on its water solubility, its biodegradability, and its concentration in the off-gas (Leson, Tabatabi, and Winer, 1992). Elimination rates typically range from 10 to 100 g/m^3/h, but can vary considerably depending upon the concentration in the off-gas and the particular filter material. The rate of contaminant reduction increases as the retention time increases, leveling off when the organisms are operating at their maximum removal efficiency at a given loading rate (Saberiyan, Wilson, Roe, Andrilenas, Esler, Kise, and Reith, 1994). Greater than 90% removal efficiency is possible with a less than maximum loading rate and a 500-ppmv influent concentration, with a removal efficiency of 16 g/h of gasoline vapors per 1 ft^3 (0.028 m^3) of packing material. Microbial counts greater than 10^7 to 10^8 CFU/mL have been achieved after equilibration of the system. Biofilters are generally sized to handle combinations of contaminant levels and flow rates (Yudelson and Tinari, 1995). Higher contaminant loadings can be tolerated, but overall removal rates (g/m^3/h) will decrease.

Naphthalene, acetone, toluene, benzene, propane and n- and isobutane (Kampbell, Wilson, Read, and Stocksdale, 1987), fumes from aviation gasoline (Kampbell and Wilson, 1991), alcohols, ketones, aldehydes, phenols, carboxylic acids, carboxylic acid esters, cresols, hexane, octane, butadiene, sulfur- and nitrogen-containing compounds, hydrogen sulfide, ammonia (Leson, Tabatabai, and Winer, 1992), and aromatic components of benzene, toluene, ethylene, and xylenes (BTEX) are examples of VOCs that can be degraded in biofilters (Alexander, 1994). Foght and Westlake (1991) report development of strains capable of growing on naphthalene vapors as a sole carbon source. Cometabolism can also take place in biofilters or bioscrubbers (Alexander, 1994).

Organic carbon removed by biofilters may be oxidized or incorporated into biomass (Medina, Devinny, and Ramaratnam, 1995). For example, in a treatment of toluene vapors, about a third of toluene removed was oxidized, while about two thirds was synthesized into biomass. Such rapid growth could lead to clogging in the biofilters. This particular test of a biofilter with a packed bed of activated carbon pellets removed 70% of the toluene in the vapors, for a total removal of 64 g/m^3/h, at an input concentration of 2700 µg/L, with an empty bed detention time of 1.8 min.

A bench-scale biofiltration system consists of three ceramic-bed bioreactors connected in series, which allows for changing the gas flow sequence (Farmer, Chen, Kopchynski, and Maier, 1995). Important process advantages can be realized by switching the sequence of flow. As a result of better control and distribution of biomass throughout the filter depths of the forward reactors, there can be longer operation, minimal pore plugging, and no increase in pressure drop across the reactors.

A 50-L laboratory-scale biofilter consisting of composted peat moss and chicken manure blend was used to degrade benzene, toluene, and xylenes (BTX) in contaminated airstreams (Tahraoui, Samson, and Rho, 1995). High performance was achieved with this biofilter, which responded rapidly to a BTX load variation, with no need to adapt to the load shock. The dynamic behavior of the biofilter was influenced by moisture content, oxygen consumption, and temperature, all of which could seriously affect the aerobic degradation and biofilter performance. Since biodegradation is an exothermic reaction, there was a rise in temperature during the active phase. Moisture content was a critical operating parameter and was difficult to maintain at a constant level throughout the filter bed. Water accumulation at any portion of the bed would result in a reduction of aerobic degradation at that level.

A pilot-scale study was conducted using biofilters for purification of waste air containing the aromatic hydrocarbons, ethylacetate, ethanol, butanol, and butane-2-one (Eitner, 1996). Long-term solvent concentration should not exceed 1.5 g/m^3, although peak loads of ≥4.5 g/m^3 can be treated. The specific load of the filter ranged from 40 to 65 g/m^3/h solvent. A special compost filter material gave better results than root wood chips.

Biofilters can be employed to treat contaminated vapors removed during bioventing. A field test of an air/water separator, trickling filter, and biofilter in series was conducted to treat volatile hydrocarbon-contaminated gases generated during *in situ* bioventing and air sparging of subsurfaces contaminated with gasoline (Lei, Lord, Arneberg, Rho, Greer, and Cyr, 1995). An average removal of 90% of BTX and 72% of the total hydrocarbons was achieved. In another instance, a biofilter was installed in conjunction with a bioventing system to remove gaseous hydrocarbons from air extracted from soil and to clean air from a stripping tower at a contaminated retail gasoline station (van Eyk and Vreeken, 1991). Both air extracted from the soil and air from the air stripper were passed into the biofilter up to a rate of 38 m^3/h. Air transport calculations predicted a value of 68 m^3/h. An average removal rate of BTX was calculated at 0.3 kg/m^3/day. A sludge or carrier type of bioreactor can be used to mineralize the vapors in withdrawn soil vapor and the dissolved contaminants in pumped groundwater (Urlings, Spuy, Coffa, and van Vree, 1991). A very simple, but effective, biofilter can consist of pipes running through a pile of soil, through which the VOCs can pass (Alexander, 1994).

Gasoline vapors typically contain both aromatic and aliphatic hydrocarbons (Evans, Bourbonais, Peterson, Lee, and Laakso, 1995). Application of biofilters to petroleum hydrocarbons is complicated by the different mass transfer characteristics of aliphatics and aromatics. Earlier researchers reported that while biofilters had been shown to be effective for removing contaminants following air stripping of groundwater at fuel spill sites, they exhibited poor removal of total gasoline vapor, suggesting that they may not be feasible for treatment of other vapor sources, such as soil venting (Douglass, Armstrong, and Korreck, 1991).

Since then, a full-scale system for treating gasoline vapors generated by a vapor-extraction and groundwater-treatment system has demonstrated that an air biofiltration unit can have high organic-vapor-removal efficiency (Miller, Saberiyan, DeSantis, Andrilenas, and Esler, 1995). The system employed was composed of two cylindrical reactors with a total packing volume of 3 m^3. They were packed with sphagnum moss and inoculated with species of *Pseudomonas* and *Arthrobacter*. They were connected in series for airflow passage, while parallel lines allowed injection of nutrients, water, and buffer. Another example was the development of bioreactors to treat SVE off-gas containing light-end gasoline components (mainly C_4 to C_8 aliphatics and some aromatics) (Evans, Bourbonais, Peterson, Lee, and Laakso, 1995). The elimination capacity (EC) of the smaller of the two was 7.2 g/m^3/h, while that of the larger (70 x) was 70 to 80% of the smaller. Low EC values may have been due to a combination of mass transfer and kinetic limitations.

Biofilters are generally not 100% efficient at removing VOCs and can be backed up with GAC filters, which would then have to be transported elsewhere for contaminant destruction (Yudelson and Tinari, 1995). Biofilters with backup GAC units cost much less than conventional GAC filters and catalytic/thermal oxidation (Catox) units for controlling VOC emissions. Biofilters can reduce the cost of remediation of gas stations by 24% ($10,000) per facility in 24-month projects and by 32% ($16,000) per facility in 36-month projects. Savings would be greater in longer-term projects and in those with higher contaminant loadings (to about 1000 ppmv), at which point Catox units start to provide better economics than biofilters.

Several types of biofilters are commercially available in the U.S. (Leson, Tabatabai, and Winer, 1992). The size and capital cost will depend mainly on the pollutant load in the off-gas and its rate of elimination in the biofilter. For treatment of multiple compounds, the sizing of a full-scale unit may require pilot testing in a small biofilter, containing around 2 yd^3 of filter material.

Inhibitory components, dry filter bed material, nutrition depletion zones, and dead zones are detrimental to the distribution of an active microbial population in the filter bed (Rho, Mercier, Jette, Samson, Lei, and Cyr, 1995). Biologically active sections of the biofilter can be identified. One method entails measuring the specific oxygen consumption rate (qO_2) of the filter bed, which is highly correlated with the contaminant EC. At low organic load, the radiorespirometric test, which determines the degradation rate of a target pollutant, is more sensitive and specific for measurements of field-scale biofilters than the respirometric test, which determines oxygen consumption of the whole microflora. Laboratory and field results cannot be compared, if there are differences in filter material, pollutants, and microbial populations of each filter bed, all of which impact the qO_2 parameter. It may be necessary to remove particulates and to cool off hot off-gases (Leson, Tabatabai, and Winer, 1992). Off-gas temperature above 110°F and high particulate loads require pretreatment, with additional capital and operating costs. After 3 to 5 years, the biological activity in the filter material will lead to compaction and require replacement. Spent filters can serve as fertilizer.

Bioreactor design is further described by Eitner and Gethke, 1987; Ottengraf, 1987; Prokop and Bohn, 1985. Leson and Winer (1991) also provide a summary of biofiltration technology and a detailed discussion of the relevant technical aspects.

6.3.4.5.1 BIOPUR®
This is a patented, aerated, packed-bed, fixed-film reactor, which uses reticulated polyurethane (PUR) as a carrier material for microorganisms for the simultaneous treatment of soil vapor and groundwater contaminated with VOCs and nonvolatile organic compounds (Oosting, Urlings, van Riel, and van Driel, 1992; Marsman, Appelman, Urlings, and Bult, 1994). See Section 2.1.2.2.1.2 for a full description of the bioreactor.

6.3.4.6 Photo-Oxidation
UV light-induced and other radical oxidation processes have been developed for on-site destruction of organic contaminants in water (Blystone, Johnson, Haag, and Daley, 1993). However, there are limitations with this method. An alternative approach is to strip the VOCs and then carry out the UV photolysis in the resulting airstream (AWWA Research Foundation, 1989; Cobiella, 1989). The problems with this approach include difficulty in obtaining gas-phase hydrogen peroxide and permits when using gaseous ozone, low rates of direct photolysis with available light sources, and competitive absorption by oxygen when using a mercury lamp (Blystone, Johnson, Haag, and Daley, 1993). A new UV light source is a xenon plasma flashlamp, which generates greater light intensity at 200 to 250 nm than mercury lamps, with more-rapid direct photolysis of VOCs. See also Sections 2.1.1.2.6, 2.1.2.1.7, 5.3.2, and 6.3.4.6.

6.3.5 RECYCLE
Volatile constituents can be removed early in the management cycle, when they are most concentrated and where the potential for recycling is greatest (Allen and Blaney, 1985). Wastes that are recycled or incinerated should have limited emissions, if proper storage and transfer practices are employed. Generally, waste streams of high volume can be treated on-site more cost-effectively than by shipping the waste to a waste management facility. There are also additional incentives, if the recovered volatiles can be recycled on-site. Distillation can be used to remove VOCs from free-flowing liquid wastes. It is more frequently used to separate components of organic streams but is also used by waste recyclers to remove organics from mixed streams. Mixing of various types of waste before shipping to a commercial treatment facility, however, can cause treatment problems, since additional multiple distillation columns could be required to separate the various components. The cost of simple distillation is approximately $9.21/L, but multiple steps can increase the cost (Allen and Brant, 1984).

The extent to which recyclers treat a waste tends to be dictated by the economic value of the recovered organics, which usually results in the bottoms being left with several percent organics (Allen and Blaney, 1985). Maximum temperatures could be set in thin-film evaporation to avoid decomposition or reactions in the waste. An acceptable Btu value of thin-film evaporator bottoms will sometimes limit the degree of low boilers, if the bottoms are to be sold as fuel.

Recyclers use both stripping and distillation for much lower organic concentrations than would be dictated by market considerations, and local regulations for waste disposal would dictate the lower organic content (Allen and Blaney, 1985). Steam stripping can clean aqueous waste to acceptable levels for discharge to the local wastewater treatment facility. The organic content must be reduced to 0.1%, which could also be met by using fractional distillation. One waste management facility strips wastes initially containing 1 to 3% organics for disposal in an evaporation pond. A local air pollution permit requires the concentration of organics in the vapor phase above this pond to be less than 300 ppm.

6.3.6 TREATMENT RESIDUALS

Treatment of wastes for VOC removal can itself result in air emissions and other residuals that must be disposed of properly (Allen and Blaney, 1985). Most waste treatment processes generate much less air emissions than the amount of VOCs recovered. Few solid or liquid residual streams are produced from the treatment processes discussed here. Table 6.15 lists such streams for a number of potential VOC removal processes.

Table 6.15 Treatment Process Residuals

Process	Air[a]	Water[b]	Solids[b]
Agitated thin-film evaporation	Condenser vents, vacuum pumps	Contaminated cooling water	—
Batch steam stripper	Condenser vents	Contaminated cooling water	—
Steam stripping	Process air	—	—
Fractional distillation	Condenser vents, vent pumps	Contaminated cooling water	—
Carbon adsorption/regeneration	Regeneration furnace	Reacted carbon transport water	Spent carbon
Wet air oxidation	Excess process vapors	Contaminated process cooling water	Catalyst residue
Ozonation/UV radiation	Process off-gas	—	—

[a] Will include storage tank vents, as well as transfer line flanges, valves, pumps, etc. for all processes.
[b] May include waste solids and wastewaters from filtering to remove solids prior to treatment for VOC removal.

Source: Allen, C.C. and Blaney, B.L. EPA-600/D-85/127, 1985. PB 85218782.

Section 7

Monitoring Bioremediation

In order to demonstrate that biodegradation is taking place in the field, the chemistry or microbial population must be shown to change in ways that would be predicted if bioremediation were occurring (National Research Council, 1993). Measurements of field samples, experiments run in the field, and modeling experiments can all improve our understanding of the fate of the contaminants.

A bench-scale biotreatability methodology has been designed to assess bioremediation of contaminated soil in the field (Saberiyan, MacPherson, Andrilenas, Moore, and Pruess, 1995). The first phase involves characterization of the physical, chemical, and biological aspects of the contaminated soil, where soil parameters, contaminant type, presence of indigenous contaminant-degrading bacteria, and bacterial population size are defined. The second phase is experimentation, consisting of a respirometry test to measure the growth of microbes indirectly (via generation of CO_2) and the consumption of their food source directly (via contaminant loss). The half-life of a contaminant can be calculated by a Monod kinetic analysis. Abiotic losses are accounted for based on a control test. The contaminant molecular structure is used to generate a stoichiometric equation, which yields a theoretical ratio for milligrams of contaminant degraded per milligrams of CO_2 produced. Data collected from the respirometry test are compared with theoretical values to evaluate bioremediation feasibility.

A field-portable instrument is being tested to utilize infrared transmitting optical fibers and Fourier transform infrared spectroscopy (FTIR) to perform a quick and accurate chemical analysis of unknown waste materials at a contaminated site without removing a sample for analysis (Druy, Glatkowski, Bolduc, Stevenson, and Thomas, 1995).

There should be the use of chemical analytical data in mass balance calculations, and there should be laboratory microcosm studies using samples collected from the site as evidence to support the remediation proposal. An important element of the bioremediation effort is establishing a field control for comparison (Atlas, 1991). Without a control, the effectiveness of the bioremediation treatment is unknown, and an opportunity to add the information gained from each experience toward a better understanding and refinement of the technology is lost.

The general strategy for demonstrating that *in situ* bioremediation is working should include documented loss of contaminants from the site, laboratory assays showing that microbes in site samples have the potential to transform the contaminants under expected site conditions, and evidence showing that the biodegradation potential is actually realized in the field (National Research Council, 1993). Since biorestoration can fail, it is important to collect and analyze samples of the soil and microbial populations to ascertain that the desired reactions are occurring and to be able to maintain optimum conditions for these reactions to continue. Methods selected for this purpose should allow distinction between biotic and abiotic processes (Madsen, 1991).

7.1 MICROBIAL COUNTS

Microorganisms are widely distributed in nature, but reports of the actual numbers present are confusing because of the methodological differences used to enumerate the microbes (Atlas, 1981). No place has been found in the U.S. or Canada — at depths to 400 ft — where sufficient organisms are not present to be brought up in 72 h to a significant population (Rich, Bluestone, and Cannon, 1986). The extent of the modification of organic contaminants depends upon biological reactions (Webster, Hampton, Wilson, Ghiorse, and Leach, 1985). In order to be able to predict the fate of pollutants, it is essential to be able to measure the biological activity present in subsurface material. The bacteria are present; the problem is establishing the right conditions for their growth, in the laboratory, as well as in the field.

Microbial counts are often used to monitor the bioremediation process. In general, the more microbes, the more quickly the contaminants will be degraded. Correlating an increase in the number of contaminant-degrading bacteria above normal field conditions is one indicator that bioremediation is taking place.

Enumeration of microorganisms can be difficult, since most subsurface bacteria exist in an ecosystem low in organic carbon and do not grow well, if at all, in conventional growth media with high organic carbon concentrations (Wilson, Leach, Henson, and Jones, 1986).

Counting colonies growing on culture media is not directly applicable to subsurface microbes that may have unknown growth requirements (Wilson, Leach, Henson, and Jones, 1986). It is difficult to cultivate all of the heterotrophic bacteria present in a soil or water sample on a single medium. Nutritional requirements for individual bacteria vary. Even complex nutrient media may not provide essential growth factors for fastidious organisms, resulting in unrealistically low plate counts. In addition, many organisms attach firmly to particles (Federle, Dobbins, Thornton-Manning, and Jones, 1986). Because of aggregation and formation of microcolonies in the environment, the colonies that form on plates may not represent a single viable cell in the sample, which would also lower the count.

It is important to be able to distinguish between viable and nonviable cells. However, it is believed that many organisms in the subsurface will be in a dormant state until stimulated by an appropriate concentration of a suitable substrate (Alexander, 1977). The deeper the soil, the more oligotrophic the organisms will become and, hence, the more fastidious their requirement for low nutrient concentrations.

It appears that different soil types vary in the distribution of biomass and enzymatic activity through their vertical profile (Federle, Dobbins, Thornton-Manning, and Jones, 1986). Biomass and activity are significantly correlated with each other and negatively correlated with depth. While biomass and activity decrease with increasing soil depth, the magnitude of decline differs for different soils. It is difficult to generalize on the level of biomass or activity to expect in a soil based on depth or horizon alone. Soil type is also important in determining the types of microbial populations present. Depth may be responsible for as much as 75% of the variation in biomass, but an additional 11% of the variation can be explained by pH and silt, clay, and organic contents. Depth also explains 78% of the variation in microbial activity; silt content explains another 4.5%.

Soil is extremely heterogeneous. Microorganisms seem to be distributed in patches in the subsurface, depending upon the quality of the soil and the effect of usage (Turco and Sadowsky, 1995). Where the contamination is located in the soil matrix will affect its subsequent turnover (Killham, Amato, and Ladd, 1993).

Variable results have been reported from attempts to calculate the number of viable organisms in a sample. Typically, more than 25% of the microorganisms isolated will fail to grow on subculture on an artificial medium (Stetzenbach, Kelley, Stetzenbach, and Sinclair, 1985). Dilution plating techniques with artificial media may yield only 1 to 10% of the number of cells determined by microscopic direct counting (Alexander, 1977; Nannipieri, 1984). Not all organisms capable of degrading petroleum hydrocarbons will grow on culture media. On the other hand, less than 30% of the organisms that form colonies on oil agar may actually be capable of metabolizing hydrocarbons (Atlas, 1991). Counts in soil samples taken a few centimeters from each other and even among subsamples have been found to vary by orders of magnitude (Federle, Dobbins, Thornton-Manning, and Jones, 1986). The huge variation has been attributed to the inadequacies of the enumeration procedures, as well as heterogeneity of the soils. The difference between total and viable cell counts usually obtained may be due to many of the bacteria in the subsurface being dormant (Larson and Ventullo, 1983). It should also be recognized that prolonged storage of some core samples may decrease biological activity (Thomas, Lee, and Ward, 1985).

The proportion of hydrocarbon-degrading organisms to total heterotrophs is now considered to be a more significant indicator of the biological activity in the subsurface, rather than total numbers of petroleum-degraders per se (Walker and Colwell, 1975; Alexander, 1977). Normalizing the data, by comparing the percentage of petroleum-degrading bacteria in the total viable, heterotrophic count with the percentage of specific hydrocarbon-extractable material, provides a better estimate of degrading activity. However, there appears to be a "threshold" concentration of oil in the environment or percentage of petroleum-degrading microorganisms in the microbial population of the environment below which there is little correlation between the two. Incubation temperature and presence of oil were found to influence the numbers of petroleum-degrading microorganisms recovered from a given sampling site.

Collecting subsurface samples by removing cylindrical cores from below ground is expensive and time-consuming, and every effort should be made to prevent contamination of the samples (National Research Council, 1993).

7.1.1 METHODS FOR ENUMERATING SUBSURFACE MICROORGANISMS

There are a variety of methods available for obtaining microbial counts. These range from simple observation of the microorganisms on a slide to the more-sophisticated and precise nucleic acid–based techniques. A wide selection is presented here for application to soil or water samples.

7.1.1.1 Direct Microscopic Counts

Direct microscopic counting is a traditional method of enumerating bacteria and may employ stains to distinguish microbes from debris on a slide (National Research Council, 1993). It does not distinguish between living and dead cells. An acid dye, such as rose bengal or erythrosin in 5% phenol, will stain the organisms and not the soil colloids (Thimann, 1963).

Specialized microscope slides have been developed for counting cells. A Helber counting chamber is a slide with a central platform surrounded by a ditch (Collins, Lyne, and Grange, 1990). A cover slip is placed over the slide and sample, creating a uniform depth. A 1-mm^2 area on the platform is ruled with 400 squares, each 0.0025 mm^2, giving a volume over each square of 0.00005 mL. The suspension should be diluted until there are five to ten organisms per square, and the cells are counted in 50 to 100 squares. Then, with the volume and dilution factors, the total number of bacteria per milliliter can be calculated.

A rough but useful technique is to employ the Breed slide, on which is marked an area of 1 cm^2 (Collins, Lyne, and Grange, 1990). Then, 0.01 mL of sample is placed on the square, dried, stained with methylene blue, examined with the oil immersion lens, and the number of organisms in several fields entered into an equation to derive the count per milliliter.

7.1.1.2 Direct Counts with Acridine Orange

The difficulty of applying standard enumeration techniques to environmental samples has led to the use of other methods, including the direct microscopic examination of samples with acridine orange counting (AODC) of the organisms (Alexander, 1977; Ghiorse and Balkwill, 1983; 1985). This dye binds to nucleic acids, especially DNA, and is excited with blue light. The method allows bacteria to be distinguished from abiotic particles. AODC provides total bacterial numbers (Heitzer and Sayler, 1993). Monoclonal antibodies can be combined with AODC, creating very good specificity for target bacterial groups.

7.1.1.3 Direct Viable Counts by Cell Enlargement

In this assay, cells are enlarged by preincubation in yeast extract medium containing nalidixic acid (Roszak and Colwell, 1987; Desmonts, Minet, Colwell, and Cormier, 1992). Nalidixic acid inhibits DNA replication, but not an increase in volume.

7.1.1.4 Direct Viable Counts from Cell Division

Viability of bacteria can be confirmed by microscopically observing the first initial cell divisions on a slide (Postgate, Crumpton, and Hunter, 1961; Torrella and Morita, 1981). This method has a good correlation with the number of macrocolonies formed on agar plates (Bakken and Olsen, 1987), although growth may not continue beyond the first division (Rodrigues and Kroll, 1988).

7.1.1.5 Dip Slides

Plastic slides are attached to caps of screw-capped bottles (Collins, Lyne, and Grange, 1989). These can be either a single- or double-sided tray containing agar culture media or a membrane filter bonded to an absorbent pad with dehydrated culture media. Both contain a grid. The slides are dipped into the sample, drained, returned to the bottles, incubated, and the colonies counted.

7.1.1.6 INT Activity Test

When another dye, 2-(p-iodophenyl)-3-(p-nitrophenyl)-5-phenyl-tetrazolium chloride (INT), is used, bacteria with active respiratory enzymes will reduce the INT and deposit red-purple INT-formazan granules in their cells, which can also be counted. The proportion of respiring cells then reflects the metabolic activity of a population. Sometimes the intensity of color is difficult to assess; however, if the weakly positive cells are even marginally metabolically active, they would be significant in decomposition of a pollutant (Webster, Hampton, Wilson, Ghiorse, and Leach, 1985). The INT activity test identifies only those bacteria that are active in electron transport, the main force behind all metabolism (National Research Council, 1993).

7.1.1.7 ATP Content

Another counting method uses a biochemical indicator, such as adenosine-5′-triphosphate (ATP), to determine the biomass, or amount of living material present (Hampton, Webster, and Leach, 1983; Webster, Hampton, Wilson, Ghiorse, and Leach, 1985). This technique is involved and requires extraction of the chemical with a mixture composed of H_3PO_4, EDTA, adenosine, urea, DMSO, and Zwittergent 3,10, followed by sensitive and specific analysis. A recovery of 98% of the ATP has been obtained with the method. The amount of ATP in bacteria during exponential growth is fairly constant. However, when bacteria are exposed to extreme environmental conditions, there can be a wide variation in ATP content (as much as 30-fold). This can affect the cell count.

7.1.1.8 Direct Epifluorescence Filtration Technique (DEFT)

This is a rapid, sensitive, and economical counting method (Collins, Lyne, and Grange, 1990). About 2 mL of the sample is passed through a 24-mm polycarbonate membrane, stained with acridine orange, and examined with an epifluorescence microscope.

7.1.1.9 Microcolony Epifluorescence Technique

The filter count technique of Rodrigues and Kroll (1988) was modified by combining a microcolony assay with epifluorescence microscopy to detect subpopulations of viable, nonculturable bacteria in soil (Binnerup, Jensen, Thordal-Christensen, and Sorensen, 1993). Soil bacteria are sonicated and filtered onto an 0.2 μm Nuclepore filter, which is placed on the surface of Kings B agar, citrate minimal medium, or soil extract medium for 3 to 4 days. Careful washing and staining of kanamycin-resistant cells with acridine orange does not disrupt the microcolonies resulting from two to three cell divisions growing on media supplemented with kanamycin. The method yields about 20% recovery of the initial inoculum and correlates well with the number of macrocolonies on agar. It may be useful for monitoring specific bacteria in soils.

There are limitations with this approach. The technique requires that cell aggregates from soil samples be adequately disrupted, low numbers of viable but nonculturable cells may not always be detected, and high numbers may cause overgrowth of the filters. However, there are possible means of circumventing these problems.

7.1.1.10 Immunofluorescence Microscopy

This is a sensitive, accurate, and highly specific detection technique, which can contribute to quantification of the persistence of specific microbes, including genetically engineered microorganisms (Jain and Sayler, 1987). Immunofluorescence microscopy, which is based upon an interaction between an antibody and its corresponding antigen, has still not been widely used for environmental samples. However, the technique has been employed to determine survival of *Escherichia coli* cells suspended in seawater and showed the greater sensitivity of this method over plate counts (Grimes and Colwell, 1986).

7.1.1.11 Plate Counts

This technique quantifies the number of bacteria capable of growing on a selected solid medium, by counting the colonies formed (National Research Council, 1993).

Tubes of 10 mL of melted medium are cooled to 45°C; 1 mL of each dilution of the sample is pipetted into two or more petri dishes, one tube of medium added to each, and the plates swirled to evenly distribute the mixture. The plates are allowed to set, then are inverted and incubated. Only those plates with 30 to 300 colonies are counted. The colonies are reported as colony forming units (CFUs). Semi- or fully automatic counters are available for large-scale operations.

Plate counts for total heterotrophs provide a moderate representation of *in situ* conditions, with moderate specificity, providing counts of all viable microorganisms on the medium used (Heitzer and Sayler, 1993). Selective plate counts are more specific and yield counts of specific catabolic phenotypes. Plate count techniques can be used for field demonstrations. Dyes can be incorporated to demonstrate metabolism of aromatic hydrocarbons by organisms on agar plates or in liquid culture in microtiter plates (Shiaris and Cooney, 1983).

Viable heterotrophs can be enumerated by plating samples on a medium designated TGA (0.75% trypticase peptone, 0.25% phytone peptone, 0.25% NaCl, 0.1% unleaded gasoline, 1.5% agar) (Horowitz, Sexstone, and Atlas, 1978). Counts of gasoline-utilizing microorganisms can be determined with medium GA (Bushnell Haas agar with 0.5% emulsified leaded MOGAS) (Horowitz and Atlas, 1977). Presumptive heterotrophic denitrifiers can be enumerated on Difco nitrate agar incubated at 15°C for 1 week under an atmosphere of helium (Horowitz, Sexstone, and Atlas, 1978).

Silica gel–oil medium and a yeast medium are recommended for enumeration of petroleum-degrading bacteria, and yeasts and fungi, respectively (Walker and Colwell, 1975). The use of silica gel as a solidifying agent has been shown to improve the reliability of procedures for counting hydrocarbon utilizers (Seki, 1976). Addition of Amphotericin B permits selective isolation of hydrocarbon-utilizing bacteria (Walker and Colwell, 1976a). The medium found to be best by these authors for counting petroleum-degrading microorganisms contains 0.5% (vol/vol) oil and 0.003% phenol red, with Fungizone added for isolating bacteria, and streptomycin and tetracycline added for isolating yeasts and fungi (Walker and Colwell, 1976a). Addition of Fungizone to oil agar no. 2 is selective for actinomycetes (Walker and Colwell, 1975). Washing the inoculum does not improve recovery of petroleum degraders.

Other researchers report that plate counts, using either agar or silica gel solidifying agents, are unsuitable for enumerating hydrocarbon-utilizing microorganisms (Higashihara, Sato, and Simidu, 1978). They based this conclusion on the observation that many marine bacteria can grow and produce microcolonies on small amounts of organic matter.

Bogardt and Hemmingsen (1992) present an agar plate overlay technique specifically for enumeration of bacteria that degrade polycyclic aromatic hydrocarbons (PAH) in soil samples. Greer, Masson, Comeau, Brousseau, and Samson (1993) describe a spread-plate technique employing glass beads and minimal salts medium containing yeast extract, tryptone, and starch.

7.1.1.12 Enrichment Techniques

One of the procedures for enumerating specific bacterial populations in environmental samples is the use of selective enrichment techniques (Jain and Sayler, 1987). This method is based upon the assumption that organisms capable of growth on liquid or agar media containing a pollutant or recalcitrant compound as a sole carbon source must be capable of catabolism of that substrate. This assumption has some serious flaws that affect the utility and reliability of the approach. Selective media prepared for such isolations have usually incorporated the xenobiotic as a primary energy or nutrient source. In theory, this approach encourages the isolation of all those organisms capable of metabolizing the xenobiotic. In fact, however, it isolates only those microorganisms that are capable of utilizing the xenobiotic as a primary or supplemental source of nutrients and of proliferating at the expense of the xenobiotic.

Nevertheless, while these techniques may not be feasible for determining accurate counts, they can be employed for isolating target microbes, including potential hydrocarbonoclastic seed organisms (ZoBell, 1973). The types of organisms that are isolated depend upon the source of the inoculum, the conditions used for the enrichment, and the substrate (Westlake, Jobson, Phillippe, and Cook, 1974; Atlas, 1977). Microorganisms selected by enrichment culturing can have their metabolic activity and tolerance to a particular substance built up over time. This repeated exposure acclimates the microorganisms to certain components or related compounds, enabling them to degrade these materials (Zajic and Daugulis, 1975).

Dworkin Foster is a mineral medium that is commonly used in studies with hydrocarbon-degrading bacteria and contains the minimal components for growth, except for a source of carbon and energy, such as gasoline (Horowitz and Atlas, 1977). A low-nutrient medium, R2A, has also been employed for the primary isolation and enumeration of bacteria from well water (Stetzenbach, Sinclair, and Kelley, 1983). Soil suspensions are plated onto R2A medium (Reasoner and Geldreich, 1985) and incubated at the average *in situ* soil temperature of 11°C for at least 7 days (Cerniglia, Gibson, and Van Baalen, 1980). Representative colonies are restreaked onto R2A agar for isolation of pure cultures. Enrichment of well water with low concentrations (100 μg carbon/L or 1000 μg carbon/L) of glucose, acetate, succinate, or pyruvate was able to enhance the growth of *Acinetobacter* isolates and an unidentified, oxidase negative, pigmented bacterium (Jobson, Cook, and Westlake, 1972).

Various hydrocarbons have been tested as the sole carbon source for enrichment cultures (Gibson, 1971; Walker, Austin, and Colwell, 1975). Organisms have, thus, been isolated that can degrade various branched paraffins, as well as aromatic and alicyclic hydrocarbon petroleum components (Gibson, 1971; Dean-Raymond and Bartha, 1975). Many investigators have used n-paraffins for these enrichments (Atlas and Bartha, 1972; Miget, 1973). However, the n-paraffins rarely constitute the major percentage of the compounds found in an oil, and the organisms isolated often do not possess the enzymatic capability to degrade the other classes of hydrocarbon components in petroleum (Kallio, 1975). Use of a crude or refined oil as the substrate is an improvement, but the initial organisms isolated are often those that metabolize the n-paraffins. An important consideration is that any isolation and enrichment culturing should try to simulate the environment into which the organisms will be released (Alexander, 1994). This includes adjusting the medium, pH, and temperature to approximate those of the contaminated site to help ensure success of the reinoculated organisms.

Cyclodextrins can be incorporated into agar to produce a homogeneous mixture of water-immiscible lipophilic organic liquids and solids as substrates for surface microbial growth (Bar, 1990). Otherwise, there will be a phase separation of the hydrophobic hydrocarbon source from the agar gel. Cyclodextrins are produced enzymatically from starch and are biocompatible with enzymes and microorganisms. The cyclodextrins complex water-insoluble chemicals inside their hydrophobic cavities and form molecular inclusion compounds.

A technique using solid agar was developed to allow rapid analysis of a large number of individual strains or mixtures of fungi for those that grow well on a given hydrocarbon (Nyns, Auquiere, and Wiaux, 1968). It can also be used to increase the ability of a wild strain to assimilate a hydrocarbon by subculturing of resistant colonies. This method has been varied slightly to determine the ability of fungi to grow on crude oils and single hydrocarbons by substituting another medium (Davies and Westlake, 1979). Slants are inoculated with spores. When mycelia appear, crude oil or n-tetradecane is pipetted halfway up the agar slope. Naphthalene, sterilized by ultraviolet (UV) irradiation, is sprinkled over inoculated plates, which are then incubated in air. Toluene is supplied in the vapor phase by incubating inoculated plates in a closed system containing air and toluene.

Oil-utilizing fungi can be isolated by adding soil to a liquid medium, washing mold colonies that develop on the surface of the enrichment medium, and transferring them to plates of Cooke's aureomycin–rose bengal medium (Cooke, 1973). Yeast colonies are then streaked on 2% malt agar. Molds are maintained on slants of mixed cereal agar (Carmichael, 1962) and yeasts on yeast-malt agar (Wickerham, 1951). Another method for isolating hydrocarbonoclastic yeasts is to spread oil-impregnated waters directly onto an isolation agar medium containing 0.7% yeast–nitrogen base and 0.5% chloramphenicol (Ahearn, Meyers, and Standard, 1971). The defined yeast–nitrogen base medium of Wickerham (Wickerham, 1951) has been employed in assimilation studies.

Sequential enrichment techniques are a modification of enrichment culturing and can be used to isolate microorganisms capable of degrading most of the components of petroleum (Horowitz, Gutnick, and Rosenberg, 1975; U.S. EPA, 1985a). A crude or refined oil or a hydrocarbon mixture is used as the initial substrate and inoculated with a microbial population. The organisms that can degrade it are isolated. The undegraded, residual hydrocarbons left after the first enrichment usually do not contain n-paraffins. The former are recovered and used for a second enrichment from which other microorganisms are isolated. This presumably recovers microbes that can attack petroleum components that are progressively more difficult to degrade. This continues until none of the substrate remains or no new isolates are recovered. A combination of these organisms then will have the enzymatic capability of degrading many different petroleum components. The mixture is more effective and has demonstrated better crude oil degradation than any of the single isolates. Different combinations of organisms may be obtained from soil samples, if the enrichments are carried out at 4 rather than 20°C (Jobson, Cook, and Westlake, 1972).

This process may allow isolation of various microorganisms that could degrade the low-solubility, high-molecular-weight compounds, as well as the more soluble, toxic hydrocarbons and intermediates of hydrocarbon metabolism (Zajic and Daugulis, 1975). Such selective continuous enrichments may be occurring in nature in areas subjected to constant input of petroleum hydrocarbons. Since intermediary metabolites must also be removed for complete oil cleanup, non-hydrocarbon-utilizing microorganisms, such as fatty acid metabolizers, would also be required in the mixture (Atlas, 1977). However, organisms isolated individually in the sequential enrichments may not be able to degrade the oil simultaneously, since one organism in the mixture may interfere with another.

A technique has been developed by Weber and Corseuil (1994) to increase a mixture of subsurface populations of specific microorganisms rapidly. A short biologically active carbon adsorber is used as an efficient reactor system for the growth, acclimation, and enrichment of indigenous microorganisms for reinoculation. The technique was tested in laboratory soil columns using benzene, toluene, and xylene as organic target compounds and a natural aquifer sand as a subsurface medium. Empty-bed reactor contact times of about 40 s were sufficient for continuous production of effluent streams of enriched indigenous microbes for reinoculation. The number of organisms rapidly rose to more than 10^5 cells/g dry solids. This resulted in increased rates of *in situ* degradation of the target hydrocarbons over the range of 25 to 9000 μg/L.

7.1.1.13 Fume Plate Method

The fume plate method has been tried for enumerating colonies capable of growing on mineral medium in the presence of specific hydrocarbon fumes (Randall and Hemmingsen, 1994a). This procedure was evaluated and found to give erroneous results if colony formation was the sole criterion for hydrocarbon utilization. Counts developing from exposure to fumes or from colony formation on mineral agar plates containing hydrocarbons are much higher than those from the MPN (most probable number) method or TOL (toluene) plasmid estimation (Randall and Hemmingsen, 1994b). Many environmental bacteria, which are not hydrocarbon degraders, can form colonies on mineral agar plates in the presence of hydrocarbons. Thus, use of this type of medium may yield counts that are too high.

To determine counts of JP-5-utilizing bacteria, 0.1 mL of well water, or a dilution thereof, is spread over the surface of a sterile plate of mineral salts agar, which is then inverted over a piece of JP-5-saturated filter paper in the petri dish lid and incubated at ambient conditions (18 to 22°C) for 7 days (Ehrlich, Schroeder, and Martin, 1985). Gasoline hydrocarbon–utilizing microorganisms can be enumerated on medium BA-G (Bushnell Haas agar exposed to volatile gasoline hydrocarbons) incubated at 15°C for 1 week (Horowitz and Atlas, 1977).

7.1.1.14 Drop Count Method

In the Miles and Misra method, pipettes with a standard dropper size of 0.02 mg (50 drops/mL) or unground 19-gauge hypodermic needles are used to place five drops of the sample onto agar plates (Collins, Lyne, and Grange, 1990). After incubation, the colonies are counted and total counts calculated.

7.1.1.15 Droplette Method

This accurate and rapid method involves making serial, replicate dilutions of the sample in agar medium in 0.1-mL amounts, and 0.1-mL drops are automatically placed in petri dishes (Collins, Lyne, and Grange, 1990). The viewer with a grid screen and the electromechanical counter offer great savings in time and labor.

7.1.1.16 Broth Cultures

Liquid cultures can be used to measure actual hydrocarbon disappearance to establish that particular organisms are, in fact, hydrocarbon degraders (Atlas, 1991).

Bacteria have been the predominant organisms isolated from enrichment experiments in which the soil perfusion technique has been employed. Soil fungi capable of degrading xenobiotics have been more frequently isolated from enrichment experiments that have used shake-culture techniques. The cultural techniques employed seem ultimately to affect those microorganisms isolated.

Metabolic adaptation can be documented by comparing laboratory flask biodegradation assays of samples from contaminated and uncontaminated areas (Madsen, 1991). Adaptation can indicate *in situ* biodegradation only if combined with other evidence, such as enhanced numbers of protozoan predators.

7.1.1.17 Most-Probable-Number (MPN) Method

The MPN technique is based on the assumption that microorganisms are equally distributed in liquid media and that repeated samples from one source will contain the same average number of organisms (Collins, Lyne, and Grange, 1990). The average number is termed the *most probable number*. The technique can be used for most organisms (e.g., aerobes, anaerobes, yeasts, molds), as long as growth is observable, such as by turbidity or acid production. The sample is shaken and 10-mL amounts pipetted into each of three (or five) tubes of 10 mL of double-strength medium, 1-mL amounts (or 1 mL of a 1:10 dilution) into each of three (or five) tubes of 5 mL of single-strength medium, and 0.1-mL amounts into each of three (or five) tubes of 5 mL of single-strength medium. If testing water, 50 mL of water is also added to 50 mL of double-strength broth. Incubate and observe growth or acid and gas. Record the numbers of positive tubes in each set of three (or five) and consult the MPN tables provided in a book on microbiological methods to determine the approximate number of viable organisms.

The method is most accurate when the mean number of cells is 1.59/tube (Gerhardt, Murray, Costilow, Nester, Wood, Krieg, and Phillips, 1981). Outside of the range of 1 to 2.5 cells/tube, the accuracy falls rapidly. Since the method is simple, but wasteful, statistical methods have been developed to give goodness-of-fit. Programmable calculators can replace the classical MPN tables for more accurate determinations.

MPN with a selected substrate is more specific, and can provide total specific catabolic phenotypes (Heitzer and Sayler, 1993). For accurate enumerations of microbial populations that degrade hydrocarbons in marine environments, an MPN procedure is recommended, using hydrocarbons as the source of carbon and trace amounts of yeast extract for necessary growth factors. The MPN method can also be used for counts of protozoa (National Research Council, 1993).

Methanogenic bacteria can be determined by multiple-tube procedures, according to the method of Godsy (Godsy, 1980). Sulfate-reducing bacteria can be determined by multiple-tube procedures using American Petroleum Institute (API) broth (Difco, Detroit) (Ehrlich, Schroeder, and Martin, 1985). Heterotrophic anaerobic bacteria can be determined by multiple-tube techniques using prereduced, anaerobically sterilized, peptone-yeast extract glucose broth (Holdeman and Moore, 1972).

The method can be automated with machines that fill the wells of plastic trays with up to 144 depressions (Gerhardt, Murray, Costilow, Nester, Wood, Krieg, and Phillips, 1981). Scanning devices distinguish wells with and without growth. Automatic and semiautomatic pipettes can be used to fill test tubes. However, since many more cultures can be examined with the rapid automation, the standard table of fixed numbers of tubes and dilutions series is no longer appropriate.

A miniaturized MPN method has also been developed to determine the number of total heterotrophic, aliphatic hydrocarbon-degrading, and PAH-degrading microorganisms (Heitkamp and Cerniglia, 1986). An MPN procedure can now separately enumerate aliphatic and aromatic hydrocarbon–degrading bacteria, which were previously undistinguishable (Wrenn and Venosa, 1996). The size of the two populations are estimated using separate 96-well microtiter plates. The alkane-degrader MPN method uses hexadecane as the selective growth substrate and positive wells are detected by reduction of iodonitrotetrazolium violet, which is added after incubation for 2 weeks at 20°C. PAH degraders are grown on a mixture of PAHs in another plate. Positive wells turn yellow to greenish brown from accumulation of the partial oxidation products of the aromatic substrates after 3 weeks incubation. Heterotrophic plate counts on a nonselective medium and the appropriate MPN procedure also provide estimates of pure culture densities. This method is simple enough for use in the field and provides reliable estimates for the density and composition of hydrocarbon-degrading populations.

The MPN method is statistically inefficient, which requires use of a large number of tubes, or it will give a very approximate cell count (Gerhardt, Murray, Costilow, Nester, Wood, Krieg, and Phillips, 1981). Preparation of nonliquid samples, both in the extraction of microorganisms and in the even distribution of the material in the diluent used are potential sources of error with the method (O'Leary, 1990). Although relatively inaccurate, it can allow detection of very low concentrations of microorganisms. Another advantage is that it does not require growing the organisms on solid media (Gerhardt, Murray, Costilow, Nester, Wood, Krieg, and Phillips, 1981). It is also useful if the growth kinetics of the different organisms are highly variable.

7.1.1.18 Membrane Filter Counts

Liquid containing bacteria is passed through a porous, 120-μm-thick, cellulose ester filter disk (Collins, Lyne, and Grange, 1990). The bacteria are trapped in the 0.5- to 1.0-μm pores in the upper layers of the filter. Culture medium is able to rise from below through the 3- to 5-μm pores in the lower layers to reach the cells above. The upper surface of the filters contains a grid to facilitate counting the colonies that develop after incubation. The colonies can be stained.

7.1.1.19 Rapid Automated Methods

Rapid automated methods may have greater initial and running costs, but this could offset the time and labor costs of conventional methods (Collins, Lyne, and Grange, 1990). The techniques include electronic particle counting, changes in pH and Eh by bacterial growth, changes in optical properties, bioluminescence (as measured by bacterial ATP), detection of ^{14}C in CO_2 evolved from a substrate, changes in impedance or conductivity, and microcalorimetry.

7.1.1.20 Fatty Acid Analysis/Lipid Biomarkers

An alternative approach is to determine biomass by analyzing the phospholipids extractable from soil (Nannipieri, 1984). Fatty acid analysis makes use of the characteristic "signature" of fatty acids present in the membranes of cells (National Research Council, 1993). Determination of biomass through analysis of the extractable lipids avoids many of the problems associated with some of the other quantification methods (Federle, Dobbins, Thornton-Manning, and Jones, 1986). Estimates of biomass are not dependent upon growth of the organisms and are not biased by the germination of inactive forms of the microbes, such as spores. They are made on a large sample and are not hindered by the problem of differentiating living and dead cells. This method has been used to estimate microbial biomass in estuarine and marine environments (Gillan, 1983; White, 1983) and in subsurface soils (Federle, Dobbins, Thornton-Manning, and Jones, 1986). Very low levels of microbial biomass can be determined from the glycerol content of phospholipids from environmental samples (Gehron and White, 1983). Analysis of the acid labile glycerol can indicate a community composition.

A signature microbial lipid biomarker (SLB) specifically related to viable biomass and to both prokaryotic and eukaryotic biosynthetic pathways can be used to monitor the effectiveness of *in situ* bioremediation (Pinkart, Ringelberg, Stair, Sutton, Pfiffner, and White, 1995). An application of this technique at one site detected an increase in monoenoic fatty acids, which suggested an increase in Gram-negative bacteria during the treatment. Ratios of specific phospholipid fatty acids indicative of nutritional stress decreased with a nutrient amendment.

A phospholipid ester–linked fatty acid analysis can be combined with a test of sole carbon source utilization to distinguish communities from disparate origins (Lehman, Colwell, Ringelberg, and White, 1995). Since these community-level characterization methods simultaneously provide specific information about individual community members and about community-level function, they can help monitor controlled bioprocesses and environmental remediation.

7.1.1.21 Dehydrogenase-Coupled Respiratory Activity

This technique has been proposed for determination of viable, metabolically active bacteria in environmental samples (Zimmermann, Iturriaga, and Becker-Birk, 1978; Rodriguez, Phipps, Ishiguro, and Ridgway, 1992).

7.1.1.22 Microautoradiography

This method can be used to enumerate viable bacteria in environmental samples (Meyer-Reil, 1978).

7.1.1.23 Protozoan Counts

Since protozoans prey on bacteria, an increase in their number suggests a major increase in the number of bacteria (National Research Council, 1993). The MPN method can be used for protozoan counts.

7.1.1.24 Fungal Counts

Fungi can be stained with Calcofluor W® to determine total hyphal length and number of fungal spores and yeast cells (Zvyagintsev, 1994). See also Sections 7.1.1.11, 7.1.1.12, 7.1.1.16, and 7.1.1.20.

7.1.1.25 Opacity Tube Method

International Reference Opacity Tubes are tubes containing glass powder of increasing opacity that are correlated with a table relating opacity to counts (Collins, Lyne, and Grange, 1990). The opacity of the sample is matched against that of the standards.

7.1.1.26 Turbidimetric Measurement

Growth in a liquid nutrient medium produces turbidity, which can be correlated with cell number (O'Leary, 1990). Standard curves can be constructed to estimate the counts from the observed turbidity values. There are filter photometers, spectrophotometers, and direct-reading turbidimeters (nephelometers) that can be used for this purpose.

Table 7.1 Distribution of Microorganisms in Various Horizons of a Soil Profile

Depth (cm)	Aerobic Bacteria	Anaerobic Bacteria	Actinomycetes	Fungi	Algae
3–8	7,800,000	1,950,000	2,080,000	119,000	25,000
20–25	1,800,000	379,000	245,000	50,000	5,000
35–40	472,000	98,000	49,000	14,000	500
65–75	10,000	1,000	5,000	6,000	100
135–145	1,000	400	—	3,000	—

Organisms/g of Soil

Source: Alexander, M. *Introduction to Soil Microbiology.* 2nd ed. John Wiley & Sons, New York. 1977. With permission.

Table 7.2 Distribution of Aerobic and Anaerobic Heterotrophic Bacteria and Fungi with Depth in a Retorted Shale Lysimeter

Depth (cm)	Aerobic/Facultative Aerobic (A)	Anaerobic/Facultative Anaerobic (B)	Ratio A/B[a]	Fungi[a]
0–30	1×10^6	2×10^4	50	5×10^4
30–60	3×10^5	7×10^3	42	8×10^2
60–90	1×10^4	$<10^2$	<40	3×10^2
90–120	4×10^3	$<10^2$	<40	10^2
20–150 (near saturated zone)	9×10^5	2×10^5	<5	$<10^2$

Bacteria[a]

[a] CFU/g soil.

Source: Wildung, R.E. and Garland, T.R. In *Soil Reclamation Processes — Microbiological Analyses and Applications.* Tate, R.L. III and Klein, D.A., Eds. Chapter 4. p. 117. Marcel Dekker, New York. 1985. With permission. Adapted from Rogers et al. (1981).

Light-scattering methods are generally employed to monitor the growth of pure cultures (Gerhardt, Murray, Costilow, Nester, Wood, Krieg, and Phillips, 1981). They can be powerful, useful, and rapid, but may provide information about a quantity not of interest. Primarily, they give information about macromolecular content (dry weight) and not about the number of organisms.

7.1.2 COUNTS IN UNCONTAMINATED SOIL

Hydrocarbon-utilizing organisms typically constitute a small percentage of the total heterotrophic population in uncontaminated ecosystems (Bausum and Taylor, 1986). Direct counts of bacteria in uncontaminated soil ranged from 10^6 to 10^7 organisms/g in the literature, while viable counts were reported from 0 to 10^8 CFU/g. On a gram dry weight basis, bacteria often exceed 10^8; actinomycetes, 10^6; and fungi, 10^5 (Turco and Sadowsky, 1995). Over 10,000 different species of bacteria have been found per gram of soil (Torsvik, Goksoy, and Daae, 1990; Torsvik, Salte, Sorheim, and Goksoyr, 1990). Microorganisms can exceed 500 mg biomass C/kg soil (Jenkinson and Ladd, 1981). In spite of these numbers, microorganisms make up only about 3% of the soil organic carbon (Sparling, 1985).

Microbial numbers decrease with depth from the soil surface (Hissett and Gray, 1976). The distribution is nonuniform and reflects soil structure and available nutrients (Richaume, Steinberg, and Jocteru-Monrozier, 1993). Table 7.1 shows the distribution of various microorganisms at different depths (Alexander, 1977). Table 7.2 compares aerobic and anaerobic bacterial counts and fungal counts at different soil depths (Wildung and Garland, 1985). All organisms and the ratio of aerobes to anaerobes decreased with depth, reflecting reduced oxygen levels. An increase in total numbers near the saturated zone was probably due to the presence of nutrient-rich water in the pore spaces, with a selection for the facultative anaerobes.

Other counts taken by Federle, Dobbins, Thornton-Manning, and Jones (1986) assumed that there are 50 µmol phospholipid/g dry weight of bacteria and that there are 10^{12} bacteria/g (Gehron and White,

Table 7.3 Bacterial Populations in Subsurface Soils

Depth (m)	Soil Extract (CFU/g)	AO Direct Count (organisms/g)
	Site 1	
0	$9.7 \times 10^6 \pm 5.7 \times 10^5$	$7.6 \times 10^6 \pm 3.0 \times 10^6$
3	$<10^3$	$5.4 \times 10^6 \pm 2.7 \times 10^6$
4.5	$3.3 \times 10^6 \pm 4.0 \times 10^5$	$2.3 \times 10^6 \pm 1.6 \times 10^6$
9	$5.6 \times 10^5 \pm 7.1 \times 10^3$	$2.9 \times 10^6 \pm 2.5 \times 10^6$
	Site 2	
0	$3.0 \pm 0.3 \times 10^7$	$5.6 \pm 1.9 \times 10^7$
3–4	$3.5 \pm 2.1 \times 10^3$	$3.9 \pm 1.4 \times 10^7$

Source: From Novak, J.T., Goldsmith, C.D., Benoit, R.E., and O'Brien, J.H. *Water Sci. Technol.* 17:71–85. Pergamon Press, London. 1985.

1983). However, these estimates may be low since subsurface bacteria are smaller than surface bacteria, due to severe nutrient limitation (Webster, Hampton, Wilson, Ghiorse, and Leach, 1985), and a gram of bacteria may contain many more than 10^{12} organisms/g of soil. Many of the bacteria in soil environments exist as cells around 0.5 to 1.0 µm in diameter (Bitton and Gerba, 1985). In fact, nonrhizosphere soil bacteria can measure less than 0.3 µm in diameter (Bae, Cota-Robles, and Casida, 1972).

Other investigations used two methods for measuring bacterial populations in soil at different depths (Novak, Goldsmith, Benoit, and O'Brien, 1985). These detected considerable differences among viable counts but little variation in direct counts with depth. These counts are presented in Table 7.3 for samples taken from different sites. Table 7.4 summarizes the results of a number of studies from the literature that are also presented below.

7.1.3 COUNTS IN CONTAMINATED SOIL

There appears to be a critical number of microorganisms necessary for biodegradation to occur (Corseuil and Weber, 1994). These investigators found that the onset of microbial oxidation of readily degradable compounds (benzene, toluene, and xylene) was delayed in systems with small populations of microorganisms, even though nutrient and electron acceptor conditions were highly favorable. Xylene had the longest critical population development period, which correlated with the comparatively low numbers of indigenous microbes capable of degrading this compound. Sometimes contaminant levels are so low or biodegradable compounds are so inaccessible that bacterial counts may not be significantly greater than the background counts (National Research Council, 1993). This does not mean that bioremediation is unsuccessful.

The presence of gasoline results in changes in microbial populations and metabolic activities (Horowitz, Sexstone, and Atlas, 1978). Microbial numbers and activity are initially depressed by even light hydrocarbon contamination (Odu, 1972). However, this is followed by a stimulation of activity. The number of hydrocarbon-utilizing organisms in a soil reflects the past exposure of the soil to hydrocarbons (Atlas, 1981). These organisms are most abundant in places that have been chronically exposed to hydrocarbon pollution (Texas Research Institute, Inc., 1982). Few or none is found in unpolluted groundwater or petroleum directly from wells. Substantial adapted populations exist in contaminated zones, with the bacterial biomass increasing as the organic contaminants are metabolized (U.S. EPA, 1985a). Numbers of hydrocarbon-utilizing microorganisms have been high in sediment 1 year after spillage (Horowitz, Sexstone, and Atlas, 1978).

The total numbers of microbes increase greatly after a petroleum spill. An increase was noted from 10^6 to 10^8 organisms/g after an oil well blowout (Odu, 1972). Bacterial counts were 100 to 1000 times higher inside than outside a zone of contamination of an aquifer containing JP-5 jet fuel (Ehrlich, Schroeder, and Martin, 1985). Hydrocarbon-using fungi in soil increased from 60 to 82% and hydrocarbon-using bacteria from 3 to 50%, following a fuel oil spill (Pinholt, Struwe, and Kjoller, 1979).

Ratios of hydrocarbon utilizers to viable heterotrophs show dominance of hydrocarbon utilizers in gasoline-contaminated sediment (Horowitz, Sexstone, and Atlas, 1978). It appears that only a few species of specialized bacteria, presumably those able to assimilate the hydrocarbons, are preferentially selected

Table 7.4 Summary of Viable and Direct Counts in Uncontaminated Soils from Several Studies

Viable Counts (CFU/g)	Direct Counts (organisms/g)	Unspecified Counts (organisms/g)
[a]8×10^3 to 3.4×10^6	[a]3.4×10^6 to 9.8×10^6	
[b]1.5×10^5 to 8×10^5	[b]4×10^6 to 9×10^6	
	[b]1.2×10^7 to 1.6×10^7	
[c]5×10^5	[c]10^6	
	[d]2.1×10^7	
3.5×10^3	2.9×10^6	
		[e]10^5

[a] The total number of bacteria in the B and C horizons of an undefined soil series was determined by use of fluorescent microscopy (Wilson, McNabb, Balkwill, and Ghiorse, 1983). These numbers did not decline with depth and ranged from 3.4 to 9.8×10^6 bacteria/g dry weight soil. The viable cell counts were more variable, ranging from 8×10^3 to 3.4×10^6.

[b] Fluorescent microscopy on uncontaminated soil samples determined total counts ranging from 4×10^6 to 9×10^6 bacteria/g at one site and 1.2×10^7 to 1.6×10^7 bacteria/g at another (Webster, Hampton, Wilson, Ghiorse, and Leach, 1985). Only a small percentage (<5%) of these cells were actively respiring, as measured by their ability to reduce INT. Based on this, the active bacteria ranged from 1.5×10^5 to 8×10^5 bacteria/g.

[c] The microflora of saturated and unsaturated subsurface samples (depths of 4 to 16 ft) were examined (Balkwill and Ghiorse, 1982). Total cells, determined by epifluorescence light microscopy (EF) counts of acridine orange–stained preparations, numbered 10^6/g dry weight in all samples. The population appeared to be entirely bacterial. The predominant cell types were small, coccoid rods, mainly Gram-positive. Plating on soil extract agar showed that at least 50% of the cells counted by EF were viable. Counts on a nutritionally rich medium were three to five orders of magnitude lower.

[d] High-permeability subsurface soils in a pristine area contained 2.1×10^7 cells/g dry soil using AODC (Thomas, Lee, Scott, and Ward, 1986).

[e] The number of microorganisms in soil before application of waste oil was 1×10^5 (Raymond, Hudson, and Jamison, 1980).

for in a contaminated zone (Ehrlich, Schroeder, and Martin, 1985). The relative occurrence of hydrocarbon utilizers in the microbial community can be used to monitor contamination of the environment by hydrocarbons (Horowitz, Sexstone, and Atlas, 1978).

Direct counts in contaminated soil were found to range from 10^3 to 10^8 organisms/g, while viable counts were recorded from less than 100 to 10^6 CFU/g. Hydrocarbon degraders have been measured at naturally occurring levels of 10^2 to 10^5 organisms/g. These results were summarized from the studies below and are listed in Table 7.5.

Catallo and Portier (1992) reported a decrease in bacterial counts in soil contaminated with PAHs and trace metals. The numbers gradually declined from 5×10^3 CFUs at trace PAH to 7×10^2 bacterial CFUs at 49,207 mg PAH/kg soil/sediment dry weight. Fungal counts went from 58 to 0.25 at the same PAH concentrations. Protozoa counts fell from 5731 to 31. The greatest drop for all microbes occurred between 33,820 and 49,207 mg PAH.

7.1.4 EFFECT OF BIOSTIMULATION ON COUNTS

Addition of stimulants, such as electron acceptors, electron donors, and nutrients, should increase biodegradation but not abiotic contaminant removal processes (National Research Council, 1993).

Growth rates of bacteria in the subsurface soil have been found to range between 0.51 and 1.94×10^5 cells/g/day (Thorn and Ventullo, 1986). Application of fertilizer stimulates greater microbial growth and utilization of some components of oil, while other components of oil are not easily attacked by the microbes and may persist in the soil (Westlake, Jobson, and Cook, 1978). Saturated fractions are highly degraded, while asphaltenes and aromatics are often resistant to microbial attack (Jobson, Cook, and

Table 7.5 Summary of Viable and Direct Counts in Contaminated Soils from Several Studies

Viable Counts (CFU/g or mL)	Direct Counts (organisms/g)	Hydrocarbon-Degraders (organisms/g or mL)	Unspecified Counts (organisms/g)
[a]10^3 to 10^5	[a]7.8×10^6	[a]8.5×10^5	
		[a]1.2×10^5	
	[b]10^6		
[c]<100 to 7×10^6	[c]7.6×10^6 to 1.7×10^8		
		10^2 to 10^5	
	[f]10^3		
		[d]10^3	
			[e]10^7
2.5×10^6		8.4×10^5	
[g]1.2×10^7 to 1.2×10^8		$\leq 9.4 \times 10^7$	
[h]10^8 to 10^{11}		10^8 to 10^9	
		[i]10^{10} to 10^{11}	

Note: The ratios of gasoline-utilizing organisms (enumerated on media GA) to viable heterotrophs (enumerated on medium TGA) indicated that hydrocarbon-degrading bacterial populations had developed in lake sediments in response to the presence of gasoline hydrocarbons (Horowitz, Sexstone, and Atlas, 1978). All ratios were greater than 0.3. Ratios for uncontaminated regions of this lake were found to be less than 0.002 (Horowitz and Atlas, 1977). The presumptive counts of denitrifiers showed no differences between any sites. The mean probable number of denitrifiers was 3×10^6 CFU/g dry weight sediment.

[a] High-permeability subsurface soils in an area contaminated with jet fuel contained 7.8×10^6 cells/g dry soil using AODCs (Thomas, Lee, Scott, and Ward, 1986). Viable counts were one to three orders of magnitude lower, but were higher in contaminated than in uncontaminated soil. Additions of 1000 ppb of benzene and 1000, 100, 10, and 1 ppb toluene could not be detected after 4 weeks. The MPN of benzene and toluene degraders in contaminated soil was 8.5×10^5 and 1.2×10^5 cells/g dry soil, while none of these organisms were detected in uncontaminated soil. This indicates the microflora exposed to jet fuel adapted and multiplied to degrade these compounds.

[b] Core samples collected from petroleum-contaminated and uncontaminated soil revealed an even distribution of bacteria for both soil conditions from 0.3 to 2.0 m (10^6 bacteria/g dry weight of soil) (Stetzenbach, Kelley, Stetzenbach, and Sinclair, 1985). However, bacteria isolated from the contaminated soil were able to degrade naphthalene more quickly in the laboratory than the isolates from the uncontaminated soil. Some PAHs (fluorene, anthracene, pyrene, and naphthalene) were used as sole carbon sources, indicating utilization by the indigenous population.

[c] In contaminated soil from Kelly Air Force Base, direct counts of organisms from subsurface samples ranged from 7.6×10^6 to 1.7×10^8 cells/g; viable cells counts ranged from less than 100 to 7×10^6 cells/g (Wetzel, Davidson, Durst, and Sarno, 1986). Similar yields of cells for seven different substrate media indicated the presence of highly adaptive bacteria.

[d] A natural flora of gasoline-utilizing organisms were present at levels of 10^3/mL (Jamison, Raymond, and Hudson, 1975) in an area contaminated with over 3000 barrels of high-octane gasoline. This population was increased 1000-fold by supplementing the groundwater with air, inorganic nitrogen, and phosphate salts.

[e] Soil microbes increased due to oil application from 1×10^5 to 1×10^7 microorganisms/g of soil (Arora, Cantor, and Nemeth, 1982).

[f] At a site contaminated with over 3000 barrels of high-octane gasoline, the natural flora of gasoline-utilizing organisms were present at levels of 10^3/mL (Jamison, Raymond, and Hudson, 1975).

[g] Five batches of 200 to 300 m^3 of contaminated soil from a refinery were treated with indigenous or specially selected microorganisms (Bosecker, Hollerbach, Kassner, Teschner, and Wehner, 1993). The beds were irrigated, nutrients added, or the test sites heated. After 2 years of bioremediation, the total amount of hydrocarbons decreased from a concentration of 15,000 to 35,000 mg/kg to a level of 3750 to 9400 mg/kg dry weight. Saturated hydrocarbons were reduced by 20 to 60%. However, heterocompounds and asphaltenes increased. PAHs measured 16 to 31 mg/kg dry weight; phenols, 130 to 170 ug/kg dry weight. Heterotrophic aerobes were present at 1.2×10^7 to 1.2×10^8 CFU/mL and oil-degrading bacteria at about 9.4×10^7 cells/mL, showing high potential for degradation of saturated hydrocarbons.

[h] Total heterotrophs were predominantly hydrocarbon degraders (Huesemann and Moore, 1993), except in uncontaminated soil. Counts of hydrocarbon degraders were higher in soil with addition of nitrogen and phosphorus.

[i] Oil-polluted Kuwaiti desert samples showed high counts of 10^{10} to 10^{11} oil-utilizing bacteria/g soil (Radwan, Sorkhoh, Fardoun, and Al-Hasan, 1995). They were predominantly *Bacillus, Pseudomonas, Rhodococcus,* and *Streptomyces.* Oil-utilizing fungi were much less frequent and were predominantly *Aspergillus* and *Penicillium.*

Westlake, 1972). Low concentrations of readily metabolized organic compounds (peptone, calcium lactate, yeast extract, nicotinamide, riboflavin, pyridoxine, thiamine, or ascorbic acid) often promote the growth of the oxidizer, but high concentrations can retard degradation of the hydrocarbons (ZoBell, 1946; Morozov and Nikolayov, 1978).

Dave, Ramakrishna, Bhatt, and Desai (1994) studied biodegradation of slop oil from a petrochemical industry. Slop oil contains at least 240 hydrocarbon components of which 54% are from C_5 to C_{11} and the rest from C_{12} to C_{23}. Of 22 bacterial cultures able to degrade slop oil, 7 could each degrade about 40%, and a mixture of all 7 could degrade $\leqq 50\%$ in liquid medium. Bioaugmentation of soil contaminated with slop oil with mixed cultures led to degradation of 70% of the slop oil in more than 30 days, compared with 40% degradation without augmentation. Wheat sown on bioaugmented soil grew better than on nonaugmented soil and led to increased degradation of up to 80% of the oil. These results show the value of adding nutrients and may illustrate a commensalism between mixed cultures and mixed plant forms.

Stimulation of pleomorphs (bacteria having multiple forms) in response to adding fertilizer suggests that such organisms adapt to oil degradation more easily than others under improved growth conditions (Lode, 1986). The very high stimulation of non-spore-forming, rod-shaped bacteria after sludge and fertilizer application supports the assumption that many of these bacteria (particularly the pigmented types) live on degradation products of hydrocarbons.

When antarctic mineral soils were tested by addition of nitrogen, phosphorus, and potassium, the estimated number of metabolically active bacteria were in the range of 10^7 to 10^8/g dry weight soil with a biomass of 0.03 to 0.26 mg/g soil. Amoebae numbered around 10^6 to 10^7/g soil, with a biomass of 2 to 4 mg/g soil. The highest populations were found in fertilized, contaminated soils, which were the only soils where petroleum degraders were demonstrated (Kerry, 1993).

The following accounts indicate the favorable influence biostimulation can have on the total counts of hydrocarbon-degrading organisms at actual field locations. These are summarized in Table 7.6.

7.2 OTHER MONITORING METHODS

Although there are selective isolation procedures for many microorganisms, most components of the natural bacterial communities are nonculturable and their identity remains unknown (Brock, 1978; Torsvik, Goksoyr, and Daae, 1990).

7.2.1 BIOMOLECULAR/NUCLEIC ACID-BASED METHODS

Biomolecular methods are now being used to characterize the nucleic acids or cell membranes of organisms in the environmental sample and to monitor *in situ* bioremediation (Brockman, 1995a; 1995b). An advantage is that the analyses are direct and preserve the *in situ* metabolic status and microbial community composition. Direct extraction of nucleic acids or cell membranes can account for the very large proportion of microorganisms (90 to 99.9%) that are not easily culturable but may be responsible for most of the biodegradation in the field. This approach includes methods based on nucleic acids (DNA and RNA) and on cell membranes. In theory, these methods enable a more comprehensive perspective and a more defensible interpretation of the response of the microbial community to intrinsic and engineered bioremediation processes.

Randomly amplified polymorphic DNA (RAPD) can be used to characterize the bacterial flora in biodegradation (Persson, Quednau, and Ahrne, 1995).

Nucleic acid–based methods allow sensitive, direct detection and determination of the levels of catabolic genes in environmental samples (Brockman, 1995b). These methods analyze nucleic acids extracted from samples taken during *in situ* bioremediation to demonstrate that contaminant loss in the field is due to biological processes. They are more accurate than culture-based enumerations. Analyses are performed on material frozen immediately after sampling, and the nucleic acids are extracted from most of the microorganisms in the sample, including those that cannot be cultured on media.

Seven nucleic acid–based methods can be applied to environmental samples:

1. Hybridization to colony DNA (Sayler, Shields, Tedford, Breen, Hooper, Sirotkin, and Davis, 1985);
2. Hybridization to DNA from enrichments (Fredrickson, Bezdicek, Brockman, and Li, 1988);
3. Hybridization to community DNA (Ogram, Sayler, and Barkay, 1987);

Table 7.6 Summary of Effect of Biostimulation on Counts in Contaminated Soils from Several Studies

Before (organisms/mL)			After (organisms/mL)		
Viable	Total	Hydrocarbon Degraders	Viable	Total	Hydrocarbon Degraders
				[a]6×10^6 times more organisms	
		[b]10^2 to 10^5			>[b]10^6
	[c]1.8×10^3			[c]1.6×10^6	
		[d]10^3			[d]10^6
[e]10^3 to 10^4			[e]4×10^3 to 4×10^4		

Note: After the biostimulation program ended at Ambler, PA, the numbers of gasoline-utilizing bacteria declined, suggesting a depletion of nutrients and gasoline (Raymond, Jamison, and Hudson, 1976).

[a] After biostimulation at a LaGrange, OR, site contaminated with gasoline, bacterial levels increased up to 6 million times the initial levels (Minugh, Patry, Keech, and Leek, 1983).

[b] At a contaminated site in Millville, NJ, a microbial population of 10^2 to 10^5 gasoline-utilizing organisms/mL in contaminated groundwater responded to the addition of nutrients and oxygen with a ten- to 1000-fold increase in the numbers of gasoline-utilizing and total bacteria in the vicinity of the spill. There were levels of hydrocarbon utilizers in excess of 10^6/mL in several wells. The microbial response was an order of magnitude greater in the sand than the groundwater.

[c] Aeration of the groundwater contaminated with methylene chloride, n-butyl alcohol, dimethyl aniline, and acetone (temperature 12 to 14°C) in a monitoring well with a small sparger and the subsequent addition of nutrients resulted in an increase of bacteria from 1.8×10^3/mL to 1.6×10^6/mL in a 7-day period (Jhaveri and Mazzacca, 1985).

[d] A natural flora of gasoline-utilizing organisms were present at levels of 10^3/mL (Jamison, Raymond, and Hudson, 1975) in an area contaminated with over 3000 barrels of high-octane gasoline. This population could be increased 1000-fold by supplementing the groundwater with air, inorganic nitrogen, and phosphate salts.

[e] In the solvent contamination at the Biocraft Laboratories, Waldwick, NJ, the wells had populations of 10^3 to 10^4 colonies/mL prior to biostimulation; addition of nitrogen and phosphorus increased the numbers of resident organisms as high as four times that of the control level (Lee and Ward, 1985).

4. Polymerase chain reaction (PCR) performed on community DNA (Steffan and Atlas, 1988);
5. Hybridization to community RNA (Pichard and Paul, 1991);
6. Ribonuclease protection assay using community RNA (Fleming, Sanseverino, and Sayler, 1993); and
7. PCR performed on DNA synthesized from community RNA (Ogram, Sun, Brockman, and Fredrickson, 1995).

Colony hybridization provides very good specificity for catabolic genotypes; direct DNA extraction, for catabolic genes; and direct mRNA extraction, for catabolic activity (Heitzer and Sayler, 1993). Direct gene probe detection (direct detection of specific DNA in the organism) can be employed to determine the presence and persistence of genetically engineered microorganisms without culturing (Jain and Sayler, 1987; Atlas, 1992). DNA and RNA gene probes can be used to assess distribution of potential catabolic expression (Wong and Crosby, 1978; Olson, 1991).

Oligonucleotide probes are small pieces of DNA that can identify bacteria by the unique sequence of molecules encoded in their genes. When the small DNA probe bonds with a complementary region of the genetic material of the target cell, the amount of bound probe can be quantified and correlated with the number of cells. This method identifies which types of bacteria are present and can also show whether or not the gene for a particular biodegradation reaction is present. It requires knowing the DNA sequence in the degradative gene.

Colony hybridization procedures can positively identify the colony-forming units with the genetic capability for degrading specific aromatic hydrocarbons (Sayler, Shields, Tedford, Breen, Hooper, Sirotkin, and Davis, 1985). An example is using gene probes to identify the naphthalene catabolic genes in a colony. In colony hybridization, bacteria are grown on agar media (Atlas, 1992). Gene probes and nucleic acid hybridization can detect colonies with specific, targeted nucleic acid sequences. These are transferred to hybridization filters, lysed, and hybridized. This technique is useful for detecting, enumerating, and isolating bacteria with specific genotypes or phenotypes, and for developing gene probes (Ford and Olson, 1988). Specific gene sequences can be amplified using PCR, a method for the *in vitro* replication of defined sequences of DNA, whereby gene segments can be amplified exponentially (Mullis and Faloona, 1987).

Nucleic acid hybridization or DNA probe techniques are already proving useful and sensitive for detecting and monitoring the critical populations recovered from the environment; for example, in the enumeration of toluene- and naphthalene-degradative populations in environmental microcosms contaminated with synthetic oils (Pettigrew and Sayler, 1986). A number of probes should be employed to evaluate the PAH-degrading potential of a mixed population, since using TOL or NAH (naphthalene) plasmids would underestimate the presence of PAH-degradative genes (Foght and Westlake, 1991).

Probe technology can even detect a single colony containing target genes among 10^6 colonies from an environmental community (Sayler, Shields, Tedford, Breen, Hooper, Sirotkin, and Davis, 1985). Use of specific chromosomal or plasmid DNA probes to monitor the maintenance of ABS10, AHS24, AOS23, and *Pseudomonas putida* (TOL and RK2) inoculated into a groundwater microcosm showed that regardless of the presence of chemical pollutants or selective pressure (by toluene, chlorobenzene, or styrene), these organisms were maintained at approximately 1×10^5 positive hybrid colonies/g of aquifer microcosm material throughout an 8-week incubation period (Jain and Sayler, 1987). Use of specific naphthalene-degrading DNA probes to determine the naphthalene-degrading population in a complex biological wastewater system (a completely mixed aerobic reactor) demonstrated the significance and sensitivity of this technology.

Limitations of this technique can include inefficient extraction of cells, DNA, and RNA from environmental samples and divergence between nucleotide sequences obtained from laboratory and naturally occurring microorganisms (Madsen, 1991). These methods also rarely indicate whether the microorganisms are viable and active (Edwards, Diaper, Porter, Deere, and Pickup, 1994). However, when performed in conjunction with other field and laboratory measurements, gene probing can help explain how biodegradation is controlled and expressed (Madsen, 1991).

7.2.1.1 Reporter Genes

Most genetically engineered microorganisms that have been released into the environment contain marker genes for their detection (Atlas, 1992). A genetically engineered microorganism can be fitted with a reporter gene that is expressed only when a degradative gene of interest is also expressed (National Research Council, 1993). For instance, the protein product of the reporter gene could signal by emitting light to indicate that the degradative gene is present and is being expressed in the *in situ* population. Activity from the bioreporter gene would indicate successful bioremediation (Burlage, Kuo, and Palumbo, 1994). Strains with bioreporter genes can be used to study expression of the catabolic genes with a variety of substrates and to help optimize bioremediation.

A direct system for monitoring bacteria in the environment has been developed (Greer, Masson, Comeau, Brousseau, and Samson, 1994). The genes for lactose utilization (*lacYZ*) and for bioluminescence (*luxAB*) are integrated into a single site in the chromosome of the desired organism. This produces a genetically stable, nontransferable marker system (Masson, Comeau, Brousseau, Samson, and Greer, 1993). Marked bacteria can be differentiated from indigenous bacteria and detected on solid media as blue, light-emitting colonies, at a level of sensitivity below 10 viable cells/g soil. The *lacYZ* system has been used as a marker or reporter of recombinant *Pseudomonas* for determining survival and movement of these organisms in soil (Atlas, 1992).

The bioluminescent *lux* genes of *Vibrio fischeri* were fused to the promoter of the upper pathway for toluene degradation from the TOL plasmid to produce a bioreporter strain, *P. putida* mt-2 (Burlage, Kuo, and Palumbo, 1994). *o*-Xylene acted as a gratuitous inducer of the catabolic genes and produced strong bioluminescence. Results suggested that conditions for optimal expression of the catabolic operon might not be the same as those for optimum growth, which questions the appropriate operating conditions for efficient biodegradation. The *luxAB* genes were integrated into the chromosome of a *Pseudomonas* strain, which allowed the organism to be recovered from contaminated soil and unambiguously enumerated by bioluminescence of its colony-forming units in the presence of *n*-decanal vapor (Weir, Dupuis, Providenti, Lee, and Trevors, 1995).

A 2,4-D degrading strain of *P. cepacia* was marked and shown to be effective in mineralizing the substrate in test soils (Greer, Masson, Comeau, Brousseau, and Samson, 1994). The bacterium could survive in soil for more than 4 months but was substrate and contaminant dependent. Temperature and competition by indigenous bacteria for available nutrients affected the numbers that could be established and maintained.

Dyes and aerosol sprays can be used to identify recombinant colonies, such as those with the *xylE* gene of the TOL plasmid, which has been cloned as a transcriptional fusion reporter gene (Atlas, 1992).

7.2.1.2 mRNA Extraction

Since levels of mnp mRNA correlate with manganese peroxidase production and the disappearance of high ionization potential PAHs during soil remediation with *Phanerochaete chrysosporium*, extraction of mRNA and RT-PCR (reverse transcription-coupled-PCR) analysis should provide a useful tool for monitoring the physiological state of the fungus and the progress of the bioremediation (Bogan, Schoenike, Lamar, and Cullen, 1996).

There are advantages to using mRNA determination over other approaches, such as quantitation of ergosterol (Davis and Lamar, 1992) or PCR-based quantitation of fungal DNA (Johnston and Aust, 1994). Tailoring of PCR primers allows species specificity, and mRNA quantitation data reveal the physiological status of the organism (Bogan, Schoenike, Lamar, and Cullen, 1996).

This approach has also been applied to bacteria, such as the expression of dmpN during phenol degradation by *P. putida* (Selvaratnam, Schoedel, McFarland, and Kulpa, 1995) and nahA during mineralization of naphthalene in contaminated soils (Fleming, Sanseverino, and Sayler, 1993).

A method has been developed to measure soluble methane monooxygenase mRNA directly in bulk soil samples to confirm that the gene for soluble methane monooxygenase is present and actively producing mRNA (Hart, 1996).

7.2.1.3 Chromosomal Painting

This technique allows for microscopic localization of genetic material (Lanoil and Giovannoni, 1997). It can be applied at the subcellular level to identify regions of eukaryotic chromosomes. Bacteria can be identified by bacterial chromosomal painting (BCP) by purifying the genomic DNA and labeling by nick translation with the fluorochrome Fluor-X, Cy3, or Cy5. The probes are hybridized to formaldehyde-fixed bacterial cells attached to slides and viewed by fluorescence microscopy. Additional treatment may be required to identify some species.

7.2.1.4 rRNA Methods

Ribosomal RNA (rRNA) molecules are ubiquitous, with a natural amplification up to several thousands of copies per cell, a composition of conserved and variable regions, and a single-stranded nature (Olsen, Lane, Giovannoni, Pace, and Stahl, 1986). The sequences of the rRNAs (5S, 16S, and 23S) have proved to be useful tools for studying the phylogeny of microorganisms (Woese, 1987). rRNA-directed oligonucleotide probes can detect or identify microbes with or without prior isolation (Stahl, Flesher, Mansfield, and Montgomery, 1988; Giovannoni, Britschgi, Moyer, and Field, 1990; Hahn, Starrenburg, and Akkermans, 1990; Hahn, Kester, Starrenburg, and Akkermans, 1990) and monitor population dynamics (Stahl, Devereux, Amann, Flesher, Lin, and Stromley, 1989).

Hybridization can be detected using probes labeled with radioisotopes and subsequent autoradiography (Yang, Horvath, Hontelex, van Kammen, and Bisseling, 1991); fluorescent-labeled rRNA-targeted oligonucleotide probes (Amann, Stromley, Devereux, Key, and Stahl, 1992); enzymes (Amann, Zarda, Stahl, and Schleifer, 1992); and nonradioactive reporter molecules, such as biotin or digoxigenin (Hahn, Amann, and Zeyer, 1993).

In vitro transcription of rRNA is a simple procedure, which permits the large-scale quantification of uncultured microorganisms in the environment, without the need for total nucleic acids extraction from pure cultures (Polz and Cavanaugh, 1997). *In vitro* transcribed 16S rRNA can serve as a template for midpoint dissociation temperature determinations of specific oligonuleotide probes and as a standard in quantitative probing.

rRNA-targeted oligonucleotides can be applied for *in situ* detection of bacterial communities in heterogeneous systems (Hahn and Zeyer, 1994). The amount of rRNA per cell and the permeability of the cells for probes are limiting factors. Nutrient addition may be necessary for adequate *in situ* hybridization with rRNA-targeted probes in soil (Hahn, Amann, Ludwig, Akkermans, and Schleifer, 1992).

7.2.1.5 Polymerase Chain Reaction

A rapid freeze-and-thaw direct DNA extraction procedure can be part of a screening method for monitoring the presence and activity of xenobiotic catabolic genes in soils (Berthelet and Greer, 1995). It would be appropriate for assessing the survival of introduced hydrocarbon-degrading microorganisms.

7.2.2 BIOMARKERS

A biomarker should be a conservative (not degraded), source-specific reference compound in a contaminant (Douglas, Prince, Butler, and Steinhauer, 1994). The contamination should preferably be from a single source. This chemical should not be formed from weathering or biodegradation, and it should not be degraded during the weathering process (Prince, Hinton, Bragg, Elmendorf, Lute, Grossman, Robbins, Hsu, Douglas, Bare, Haith, Senius, Minak-Bernero, McMillen, Roffall, and Chianelli, 1993). The extraction efficiency of the marker should be the same as the rest of the oil (Douglas, Prince, Butler, and Steinhauer, 1994).

7.2.2.1 Carboxylic/Hopanoic Acids

Crude oil goes through a maturing process where easily biodegradable alkanes and aromatics may be removed, while more-recalcitrant "biomarker" molecules, such as hopanes and steranes, are left behind (Connan, 1983). Hopanes (C_{27} alicyclic hydrocarbons) are more resistant to biodegradation than isoprenoid hydrocarbons (Atlas and Cerniglia, 1995). Hopanes are biodegradable, but slowly relative to aromatic and aliphatic hydrocarbons (Butler, Douglas, Steinhauer, Prince, Aczel, Hsu, Bronson, Clark, and Lindstrom, 1991). These compounds can be easily identified by means of their monocarboxylic acid fingerprints (Jaffe and Gallardo, 1993).

Pentacyclic (hopanoic) monocarboxylic acids have been found to be more resistant to biodegradation than their alkane counterparts or steranes and can be used as indicators of biodegradation and oil migration (Jaffe and Gallardo, 1993). Total acidity (TA) also increases with the degree of biodegradation. A significant proportion of the TA is due to the presence of complex, high-molecular-weight compounds with acidic functional groups, such as asphaltenes and resins. The TA and the ratio of tricyclic to pentacyclic acids are good parameters of biodegradation in crude oils. When used together, they differentiate "mixed" oils (biodegraded with nonbiodegraded) and early generated oils from others. The isomeric distribution of these acids, with the carbon preference index (CPI) and the C_{28}/C_{18} ratio of the *n*-acids are useful in determining migration patterns of the oils.

The ratio of biodegradable petroleum components to hopanes, using source crude oil as an initial standard, can be an indicator of *in situ* biodegradation (Madsen, 1991). However, most of the biodegradable components of crude oil have different solubility, transport, and volatility characteristics than those of hopanes, which makes it difficult to select appropriate ones for comparison.

The triterpane C_{30} 17α (H) 21β (H) hopane is neither generated nor biodegraded during biodegradation of crude oil fractions and has appropriate characteristics to serve as an internal standard for monitoring biodegradation of both specific petroleum compounds and total oil in crude oil in the environment (Prince, Elmendorf, Lute, Hsu, Haith, Senius, Dechert, Douglas, and Butler, 1994; Douglas, Prince, Butler, and Steinhauer, 1994). By using this hopane as an internal reference, it has been possible to calculate the amount and percent lost for aliphatics, aromatics, and total oil in surface and subsurface field sediment samples (Butler, Douglas, Steinhauer, Prince, Aczel, Hsu, Bronson, Clark, and Lindstrom, 1991). An unidentified, tricyclic acid (C_{26}) also appears to be a useful indicator of biodegradation (Jaffe and Gallardo, 1993).

The use of hopane as an internal, undegraded reference compound demonstrated that bioremediation was a successful component of the treatment of the Exxon Valdex oil spill in Prince William Sound, AK (Bragg, Prince, and Atlas, 1994).

7.2.2.2 Bicyclic Alkanes, Pentacyclic Terpanes, and Steranes

The study of bicyclic alkanes and pentacyclic terpanes by GC/MS/MS (GC = gas chromatography and MS = mass spectroscopy) and of steranes by GC/MS/MS revealed good stability of these biomarkers, which represent several classes of compounds (Jacquot, Doumenq, Guiliano, Munoz, Guichard, and Mille, 1996).

7.2.2.3 Phenanthrenes/Anthracenes

For refined petroleum products, such as diesel fuel and fuel oil #2, that do not contain the hopane, C_4-phenanthrenes/anthracenes may be substituted (Douglas, Prince, Butler, and Steinhauer, 1994). These compounds are degraded, but very slowly.

7.2.2.4 Pristane and Phytane

The ratio of biodegradable to nonbiodegradable substances in a mixture of contaminants will decrease over time, if biodegradation is occurring (National Research Council, 1993). This approach can be used to compare early ratios with those from samples taken at stages during the remediation process. It can compare different compounds that display different degrees of biodegradability, or it can compare single contaminants having different forms, one of which may be biodegradable and one not.

Assuming that pristane and phytane are undegraded, the ratio of straight-chain alkanes to these highly branched alkanes will give an estimate of the hydrocarbon degradation taking place (Atlas, Boehn, and Calder, 1981). Traditional weathering indicators of biodegradation are C_{18}/phytane, C_{17}/pristane, and the aromatic weathering ratio (Butler, Douglas, Steinhauer, Prince, Aczel, Hsu, Bronson, Clark, and Lindstrom, 1991). This technique can be applied to full-scale field demonstrations (Heitzer and Sayler, 1993).

Changes in the n-alkane:isoprenoid ratio are an early indication that biodegradation has begun, but isoprenoids are themselves quite biodegradable (Kennicutt, 1988; Tabak, Haines, Venosa, Glasser, Desai, and Nisamaneepong, 1991), so later readings may substantially underestimate the extent of biodegradation (Butler, Douglas, Steinhauer, Prince, Aczel, Hsu, Bronson, Clark, and Lindstrom, 1991).

The accumulation of pristane in the biosphere may be due to the lack of proper conditions for oxidation rather than it being a refractory chemical (McKenna and Kallio, 1971). Microorganisms metabolize branched alkanes in much the same fashion as *normal* alkanes. Multiple-branched alkanes, such as pristane, can be converted to succinyl-CoA (coenzyme A) by *Corynebacterium* sp. and *Brevibacterium erythrogenes*. Cooney (1980) found that 11 out of 14 fungi and 4 out of 21 bacteria were able to degrade pristane. Since 43% of the isolates tested could grow on pristane, which is relatively resistant to microbial attack, it would be expected that it could be easily degraded after an acute or chronic oil-polluting event, if other conditions, such as nutrient requirements and absence of toxic materials, are met. Mixed populations may contribute to pristane degradation. During the Exxon Valdez oil spill, pristane and phytane were found to be degraded rapidly and could be used as an internal marker for only a few weeks to months (Atlas and Cerniglia, 1995). This may have been due to hydrocarbon degraders that had evolved to attack terpenes from pine trees, after long exposure to these compounds. Terpenes are isoprenoids with chemical structures similar to those of pristane and phytane.

7.2.3 HYDROCARBON CONCENTRATION

Reduction in soil hydrocarbon concentration during treatment is the typical method of measuring petroleum hydrocarbon biodegradation in field-scale soil remediations (Hater, Green, Solsrud, Bower, and Barbush, 1995). Interpretation of such data, however, depends on the analytical procedures used, which are subject to error (Atlas, 1991).

Total petroleum hydrocarbon (TPH) measurements can be used as an indicator of *in situ* biodegradation of gasoline, diesel fuel, and jet fuel, if pollutant losses due to abiotic reactions can be excluded (Heitzer and Sayler, 1993). This generic approach is accepted in many U.S. states as a regulatory tool for setting cleanup standards for underground storage tank leaks and other petroleum-related contaminations (Michelsen and Boyce, 1993). However, standards and procedures vary among the states.

The technique measures the total concentration of a wide range of petroleum hydrocarbon constituents detectable by the analysis (Michelsen and Boyce, 1993). It is simple, easy, and inexpensive. However, using a range of molecular weights as the basis of the measurements does not identify the compounds or the risks associated with the actual contaminants, including their mobility and toxicity. TPH standards are usually more concerned with drinking water aesthetics than with health risk assessments or state-of-the-art fate and transport evaluations.

TPH analysis employs infrared spectroscopy or GC with a flame ionization detector (FID) or a photoionization detector (PID) to measure bulk concentration of petroleum hydrocarbons in a sample (Michelsen and Boyce, 1993). Results are reported simply as the concentration of "gasoline" or "diesel" TPH present, even if these are not the actual compounds.

7.2.4 OTHER ORGANIC INDICATORS

Levels of total and dissolved organic carbon (TOC, DOC) or chemical oxygen demand (COD) can indicate *in situ* biodegradation of all organic pollutants, if abiotic losses can be excluded (Heitzer and Sayler, 1993). Proof of complete mineralization using TOC, DOC, or COD measurements requires establishment of complete mass balances and analytical measurement of the pollutants in all relevant process areas.

Oil and grease soil concentrations should be used only as a general indicator for measuring the success of bioremediation (Huesemann and Moore, 1993). Since polar compounds are sometimes formed during the biodegradation process, TPH is a more reliable indicator of contaminant removal.

7.2.5 ELECTRON ACCEPTOR CONCENTRATION

Since bacteria consume electron acceptors (e.g., O_2, NO_3^-, or SO_4^{2-}) during biodegradation, a simultaneous reduction in the electron acceptor concentration and contaminant level is evidence that bioremediation is occurring (National Research Council, 1993). O_2 would be monitored for aerobically degradable pollutants and NO_3^- for pollutants degradable under denitrifying conditions (Heitzer and Sayler, 1993). It would be applicable for full-scale field demonstrations, but only if electron acceptor consumption is directly related to pollutant biodegradation.

7.2.6 SOIL GAS MONITORING

Soil gas monitoring involves sampling and analyzing soil gas for volatile constituents or trace gases at shallow depths (i.e., around 1.5 to 2 m below the surface) to infer subsurface contamination (Davis, Barber, Buselli, and Sheehy, 1991). Methods based on changes in soil gas composition may not be accurate, since these changes may be caused by processes other than hydrocarbon degradation (Aggarwal and Hinchee, 1991). For example, during remediation of a JP-4 jet fuel spill employing *in situ* air injection, soil vacuum extraction (SVE), and enhanced biodegradation, the biodegradation rate of non-aqueous-phase liquids (NAPLs) was apparently overestimated based on oxygen consumption in the soil gas (Cho, DiGiulio, Wilson, and Vardy, 1997).

7.2.6.1 Carbon Dioxide and Oxygen

Hydrocarbon degradation by aerobic bacteria is commonly determined by monitoring soil gas CO_2 and oxygen (Hater, Green, Solsrud, Bower, and Barbush, 1995). GC is the method of choice for determining gaseous CO_2 concentrations (National Research Council, 1993). A decrease in oxygen consumption and an increase in CO_2 production over time have been used as indicators of bioremediation (Hinchee and Ong, 1992; van Eyk, 1994). An *in situ* respiration test employs bioventing initially to aerate contaminated soil of the unsaturated zone and then periodically monitors the depletion of oxygen and production of carbon dioxide over 4 to 5 days to measure aerobic biodegradation rates (Hinchee and Ong, 1992).

However, when soil gas measurements are used to calculate biodegradation rates, they significantly underestimate the amount of petroleum removal as calculated from soil analysis (Hater, Green, Solsrud, Bower, and Barbush, 1995).

Soil gas monitoring can be affected by the soil environment. *In situ* field respiration studies determined that O_2 uptake was a better indicator of biological activity than CO_2 production, which gave inconsistent results (Dupont, Doucette, and Hinchee, 1991). Soil water content and other variable soil conditions can affect the sensitivity of CO_2 measurements. Sometimes CO_2 production can be difficult to measure because of the balance of lime and carbon (Wuerdemann, Wittmaier, Rinkel, and Hanert, 1994). It can be difficult to distinguish CO_2 levels resulting from the breakdown of hydrocarbons from those due to dissolved calcium carbonate (Hart, 1996), and under alkaline conditions, the CO_2 resulting from microbial respiration may be converted to carbonate (Hinchee and Ong, 1992). The consumption of oxygen can produce an incorrect indication of hydrocarbon degradation, since other substrates may be furnishing the carbon for growth (Atlas, 1991). Oxygen content of the soil might be influenced by several biological processes (Wuerdemann, Wittmaier, Rinkel, and Hanert, 1994). Rittmann et al. (1994) compare the number of molecules of oxygen lost and the number of molecules of CO_2 gained.

During *in situ* remediation employing biodegradation, vapor extraction, and air sparging in low-permeability soils, O_2 values were useful for vadose zone soil vapor sampling but were not sensitive enough as a biological indicator for exhaust gases (Aelion, Widdowson, Ray, Reeves, and Shaw, 1995). It was difficult to know to what extent O_2 and CO_2 levels in the soil were augmented or diminished, respectively, by the addition of air from the SVE operations or from atmospheric air.

Electromagnetic techniques and/or shallow soil gas sampling offers a remote, cost-effective means of determining the distribution of contaminants in groundwater (Davis, Barber, Buselli, and Sheehy, 1991). However, concentrations of volatiles in soil gas are sensitive to changes in soil porosity; thus, shallow soil gas measurements are not accurate for detecting changes with time in subsurface contamination.

Zeng (1995) presents a simple equation to account for the effect of CO_2 absorption and the dissociation of carbonic acid in liquid on the measurement of the CO_2 evolution rate of both anaerobic and aerobic continuous cultures. Deviation of calculated gas-phase measurements from true values are assessed with two parameters: one accounts for the influence of pH, resulting from dissociation of carbonic acid, the other for the influence of operating conditions. The resulting plots may be used as a guideline for choosing proper operating conditions for the reliable measurement of CO_2.

The Sapromat, which is manufactured by Voith in Germany, was designed to measure oxygen consumption in water; however, it has also been applied to measuring oxygen consumption in soil (Harmsen, 1991).

An infrared gas analyzer can be used for measuring the CO_2 evolution rate in the exit gas of a sequencing batch reactor system (Buitron, Koefoed, and Capdeville, 1993). This permits monitoring and control of the biodegradation process. Since starvation reduces the biodegradation rate, maximal activity can be maintained by monitoring this parameter.

Radiolabeled carbon dioxide produced from radiolabeled hydrocarbon substrates can demonstrate hydrocarbon utilization (Caparello and LaRock, 1975). When used in conjunction with MPN calculations, an accurate number of hydrocarbon degraders can be determined (Atlas, 1979). Respiration measurements, however, require a closed system, which is impractical for the field.

The microbial respiration quotient is a suitable indicator for easily biodegradable carbon sources in the soil (Hund and Schenk, 1994). It is an adequate, rapid, and easy indicator of the efficiency of the bioremediation process.

7.2.6.2 Nitrous Oxide

Formation of N_2O is a useful indicator under anaerobic conditions; however, denitrification may or may not be linked to contaminant loss (Madsen, 1991). Also, a trace amount of N_2O may be produced during nitrification.

Denitrification in sediments can be examined using the acetylene blockage of N_2O reduction technique (Balderston, Sherr, and Payne, 1976). Nitrogen fixation can be estimated by the acetylene reduction method (Hardy, Burns, and Holsten, 1973).

An initiation of anaerobic processes can be used to monitor oxygen levels in the soil during bioremediation (Wuerdemann, Wittmaier, Rinkel, and Hanert, 1994). Measurements of N_2O, an intermediate of denitrification, can be used to indicate oxygen deficiency. Gas sampling can be carried out with a vacuum pump and soil–air probes.

7.2.6.3 Methane

Methane production has been used to imply anaerobic biodegradation (Ehrlich, Goerlitz, Godsey, and Hult, 1982). Methane can be measured to determine biodegradation of organic pollutants that are degradable under methanogenic conditions (Heitzer and Sayler, 1993). The origin of the contaminants must be known, and proof of complete mineralization requires complete mass balances and analytical measurement of the contaminants. This is a reliable method, if other methane sources are absent; however, methanogenesis may or may not be tied to contaminant loss (Madsen, 1991).

7.2.7 ANAEROBIC BY-PRODUCTS

In situ monitoring of anaerobic microbial transformations of organic pollutants in subsurface environments is important in order to assess their fate and transport, but it is technically difficult (Berry, Francis, and Bollag, 1986; 1987). Because of the complexity of natural ecosystems, it can be hard to extrapolate laboratory results to *in situ* environments.

An increase in methane, sulfides, reduced forms of iron and manganese, and nitrogen gas can suggest an increase in anaerobic activity (National Research Council, 1993). If this is accompanied by evidence of an anaerobic environment, loss of electron acceptors other than oxygen, and consumption of electron donors responsible for the loss of the electron acceptors, it is good evidence of successful anaerobic bioremediation. Sites with iron-sulfide precipitates have been designated as sulfate-reducing sites (Suflita and Gibson, 1985).

7.2.8 INORGANIC INDICATORS

Analysis of inorganic compounds in lysimeter samples can be used to evaluate geochemical changes during bioremediation of the vadose zone and can be used to improve the remediation design (Capuano and Johnson, 1996). P-CO$_2$ is a measure of the effectiveness of bioremediation and vapor extraction. If vapor extraction occurs alone removing CO$_2$ from solution without coupling of a process to buffer the pH, large amounts of carbonate minerals could precipitate, reducing sediment permeability. High background bicarbonate concentrations or dissolution of calcareous minerals mask respiratory production of inorganic carbon (National Research Council, 1993). Stable isotope analysis of the carbon might be able to distinguish bacterially produced inorganic carbon from mineral carbon.

7.2.9 STABLE ISOTOPE ANALYSIS

Analysis of stable carbon isotope ratios in soil gas CO$_2$ can be used to monitor and evaluate *in situ* bioremediation (Aggarwal and Hinchee, 1991; Van de Velde, Wagner, and Marley, 1994). The variation in stable carbon isotope ratio due to fractionation during physical, chemical, and biological processes allows evaluation of biological systems where carbon compounds are mineralized, resulting in liberation of CO$_2$ (Van de Velde, Marley, Studer, and Wagner, 1995). Proof of complete mineralization requires establishment of complete mass balances and analytical measurement of pollutants in all relevant process compartments (Heitzer and Sayler, 1993).

It may be possible to determine whether CO$_2$ or other inorganic carbon in a sample is an end product of contaminant biodegradation or is from some other source, by the carbon isotope ratios and isotope fractionation (National Research Council, 1993). Stable isotope analysis can be an inexpensive and effective tool for monitoring and evaluating biological systems, and the service is commercially available at many laboratories. It has been successfully applied to identify carbon dioxide produced from aerobic biodegradation of hydrocarbons at jet fuel–contaminated sites (Aggarwal and Hinchee, 1991).

If the contaminant source can be located and identified, selected contaminants could be labeled by synthesizing versions containing a known amount of a stable isotope, usually ^{13}C or deuterium, and be introduced into the site (National Research Council, 1993). The metabolic byproducts would be monitored to confirm biodegradation.

Some metabolic processes, such as methanogenesis are selective between light and heavy isotopes (Trudell, Gillham, and Cherry, 1986). Aggarwal and Hinchee (1991) report that if methanogenesis is occurring, the typical decrease in ^{13}C/^{12}C carbon isotope ratios will not be observed. The ratios can be determined with a mass spectrometer; however, the procedures are elaborate and expensive. This method has been useful for identifying microbiological sources of CH$_4$ and CO$_2$ in groundwaters but may not yet be reliable for monitoring biodegradation of recent contaminations (Madsen, 1991).

7.2.10 LABELED CONTAMINANTS

Biodegradation can be measured by using either specific analytical or radiochemical techniques (Larson and Ventullo, 1983). Both of these methods allow biodegradation to be measured in complex environmental matrices at realistic (µg/L) environmental concentrations. Radiochemical procedures are especially gaining increased acceptance for use in environmental fate studies. The use of certain isotopes, such as ^{14}C allows the actual metabolism of the compound to be measured. Whereas analytical procedures follow disappearance of the parent compound, studies with ^{14}C-labeled materials allow the complete (ultimate) biodegradation to ^{14}CO$_2$ to be measured. If cumulative ^{14}CO$_2$ production is measured as a function of time, both the rate and extent of ultimate biodegradation can also be quantitated. Use of ^{14}C uniformly labeled compounds and the turnover time-tracer approach will permit measurement of heterotrophic activity (Azam and Holm-Hansen, 1973; Gocke, 1977).

Formation of radiolabeled carbon dioxide from radiolabeled hydrocarbon substrates indicates hydrocarbon utilization (Caparello and LaRock, 1975). This reaction can form the basis of a ^{14}C-radiorespirometric MPN technique (Atlas, 1978b; Lehmicke, Williams, and Crawford, 1979). The method uses the conversion of the radiolabeled compound to radiolabeled carbon dioxide to establish positive results in the MPN procedure. This will provide an accurate number of hydrocarbon degraders (Atlas, 1979). However, measuring only mineralization to estimate the amount of substrate metabolized may produce erroneous conclusions, as seen in the differences in ^{14}CO$_2$ evolved for different compounds (Swindoll, Aelion, Dobbins, Jiang, Long, and Pfaender, 1988).

A technique was developed to measure both mineralization and uptake of radiolabeled substrates into biomass, as well as a mass balance of added label (Long, Dobbins, Aelion, and Pfaender, 1986; Dobbins, Aelion, Long, and Pfaender, 1986). ^{14}C-hydrocarbons can be utilized as a means of estimating the hydrocarbon-degrading potential of bacteria. This technique allows assessment of the metabolic activity of subsurface microbial communities, and use of several concentrations permits calculation of metabolic kinetics. The amount of mineralization of ^{14}C-hexadecane, for instance, can be equated with the total number of petroleum-degrading bacteria and the percentage of the total heterotrophic population, that they represent. A community can be adapted to metabolize at concentrations significantly higher than have been suggested to exist in its environment. The technique works well for substrates of various polarity and volatility.

Thus, contaminants can be isotopically labeled (stable or radioactive), released into the soil, and the labeled metabolites monitored to test for biodegradation (Madsen, 1991). ^{3}H thymidine incorporation can also be used to measure growth rates of bacteria in subsurface soils (Thorn and Ventullo, 1986).

7.2.11 ENZYME ASSAYS

Bacterial communities can be described by total community enzyme activity (Kanazawa and Filip, 1986).

Bacterial numbers are indicative of a stimulated biodegradation process, but they do not represent an accurate measurement of the actual biodegradation (Van Der Waarde, Dijkhuis, Henssen, and Keuning, 1995a). Dehydrogenase activity, measured with the reduction of triphenyltetrazolium chloride, appears to have a good relation with CO_2 production in several soils. Esterase activity, measured with the hydrolysis of fluorescein diacetate, is indicative of the onset of biodegradation, but is less accurate in measuring the decline of activity.

When enzyme assays were used as a monitoring instrument for a full-scale soil bioremediation process, the dehydrogenase activity had the best correlation with hydrocarbon removal and CO_2 production (Van der Waarde, Dijkhuis, Henssen, and Keuning, 1995b).

7.2.12 INTERMEDIARY METABOLITE FORMATION

Detection of unique intermediary metabolites in site-derived samples is accepted as evidence for the occurrence of *in situ* contaminant biodegradation (Wilson and Madsen, 1996). Intermediary metabolites are formed during the process of biodegradation. Presence of indicative metabolites can demonstrate that biodegradation is progressing (National Research Council, 1993). Some of these metabolites are degraded too rapidly to detect. Their absence does not mean bioremediation is not occurring. They can be determined with GC, high-performance liquid chromatography, or one of these methods combined with MS. Some products of microbial metabolic transformations are more stable but do not indicate when the reactions occurred, but others can give an indication of what has taken place at the time of the sampling (Wilson and Madsen, 1996).

A transient, but reliable, intermediary metabolite, 1,2-dihydroxy-1,2-dihydronaphthalene, of the oxidation of naphthalene can be used to indicate real-time *in situ* naphthalene biodegradation (Wilson and Madsen, 1996). Release of isotopically labeled (stable or radioactive) contaminants can lead to production of labeled metabolites unique to metabolic pathways (Madsen, 1991).

Acetate and formate are important intermediate products in methanogenic fermentation and can indicate the presence of these organisms (McInerney and Bryant, 1981; McInerney, Bryant, Hespell, and Costerton, 1981).

7.2.13 MONITORING CONSERVATIVE TRACERS

Conservative tracers have chemical and transport properties similar to those of microbiologically reactive chemicals but are not microbiologically reactive themselves (National Research Council, 1993). They can distinguish abiotic chemical changes, such as volatilization, sorption, and dilution, from chemical changes caused by microorganisms. For instance, helium gas could be used as a conservative tracer to determine how much sparged oxygen is being consumed by microorganisms and how much is disappearing through abiotic routes, such as dilution. Bromide could be used as a tracer by adding it to water used to supply a dissolved electron acceptor, such as NO_3^-, SO_4^{2-}, or dissolved O_2. Such conservative tracers may also be present in the contaminants. A decreasing NO_3^-/Br^- ratio has been used to indicate *in situ* denitrification, which may or may not be due to contaminant loss (Madsen, 1991).

7.2.14 GAS CHROMATOGRAPHY AND MASS SPECTROMETRY (GC/MS)

GC analysis of the aliphatic fraction, often following column chromatographic separation of this fraction, is a definitive analytical procedure. Degradation of individual hydrocarbons and classes of hydrocarbons can be determined. The aromatic hydrocarbon fraction can be analyzed using capillary column GC.

These analyses are often combined with MS to track aromatic hydrocarbons and classes of aromatic hydrocarbons. These methods, however, do not resolve all the compounds in an oil mixture (Atlas, 1991). High-molecular-weight PAHs are very difficult to analyze. High-pressure liquid chromatography has been used for such compounds (Heitkamp and Cerniglia, 1989).

The GC traces from a TPH analysis can be used as a "fingerprint" to provide more information about the contaminants (Michelsen and Boyce, 1993). They can determine the type of compounds, degree of weathering, and, often, the specific origin of the hydrocarbons, which can be useful in some situations to help identify the source of contamination.

The EPA Method 8270 (high-resolution GC/MS) can be used for semivolatile organic compounds, such as PAHs and methyl-*tert*-butyl ether (MTBE, a gasoline additive) (Michelsen and Boyce, 1993). EPA Methods 8020 and 8240 are appropriate for other volatile organic compounds and ICP/MS (inductively coupled plasma/mass spectrometry), for organic lead additives (Baugh and Lovegreen, 1990). Sources of PAH contamination can be traced with Method 8270 alone or in combination with measurement of total aliphatic hydrocarbons (TAH). The ratios of low-molecular-weight PAH/high-molecular-weight PAH (LPAH/HPAH) and TAH/total PAH can determine the origin of contamination, even in soils far from the source (Nestler, 1974; Clark and Brown, 1977).

7.2.15 THIN-LAYER CHROMATOGRAPHY–FLAME IONIZATION DETECTION

Thin-layer chromatography–flame ionization detection (TLC-FID) has been used to determine biodegradation of tetradecane and phenanthrene by mixed and pure bacterial cultures (Cavanagh, Juhasz, Nichols, Franzmann, and McMeekin, 1995). This method for monitoring hydrocarbon biodegradation allows a number of samples to be replicated and run simultaneously in a short period of time (e.g., 45 min). This system can detect concentrations as low as 0.1 µg/µL, as applied on the chromarods, with a reproducibility of plus or minus 10%. Low-molecular-weight hydrocarbons can be qualitatively assessed with this system.

TLC can be used in the field and can provide chemical fingerprinting to identify a wide range of petroleum products, including higher-molecular-weight mixtures, such as creosote or asphalt (Michelsen and Boyce, 1993).

7.2.16 ANTIBIOTIC-RESISTANT MICROORGANISMS

This test monitors hydrocarbon-degrading bacteria that have been released into the environment containing resistance to a particular antibiotic. A problem with this method is that there are already organisms in the soil resistant to antibiotics (Atlas, 1992). It is also undesirable to release into the environment drug-resistance genes that could be transferred to pathogenic microorganisms. Only antibiotic resistance markers for antibiotics that have limited therapeutic use should be used.

7.2.17 ELISA

Species-specific, enzyme-linked immunosorbent assay (ELISA) techniques can determine the amount and distribution of different species of methanogens in complex bacterial populations in anaerobic digesters (Bryniok and Troesch, 1989). These data reflect the functional status of the degradation process and can aid in the control of methanogenic processes, especially if the methods are used for acidogenic and acetogenic bacteria.

There are two ELISA approaches using polyclonal antisera (Bryniok and Troesch, 1989). If the cell titer is $\geqq 10^8$, the indirect ELISA method can be a rapid and inexpensive method to determine the composition of methanogenic populations or quantitative determination of methanogenic bacteria in pure culture. If the counts are too low, the Sandwich-ELISA technique must be used. Kemp, Archer, and Morgan (1988) present three assays: a Sandwich- and a competitive ELISA for the determination of *Methanosarcina mazei* and a Sandwich-ELISA for *Methanobacterium bryantii*.

7.2.18 BIOLOG® SYSTEM

The Biolog system (Biolog, Inc.) consists of a 96-well microtiter plate with 95 different carbon sources and a control well with no carbon source (Winding, 1994). The redox dye, tetrazolium violet, is reduced during respiration, and insoluble formazan (violet) accumulates inside the cells (Bochner and Savageau, 1977; Bochner, 1978). Formazan formation in individual wells is measured by a video image analyzing system, CREAM® (Kem-En-Tec A/S), designed to read ELISA plates.

Incubation of whole bacterial communities results in fingerprints based on the metabolic potentials, and a subsequent cluster analysis shows the relationship between the communities (Winding, 1994). Bacterial communities from different soils can be discriminated, as well as communities associated with different size fraction of soils. The bacteria are grown for only a few generations in the microtiter plates and are tested during growth. When the samples are tested immediately after collection, the metabolic potential of the *in situ* community is more reliable. This reduces the effect of media on community composition normally obtained as a result of the isolation procedures. The method is fast and easy to perform, allowing description of a large number of bacterial communities within a reasonable time.

7.2.19 RESPIROMETRY/RADIORESPIROMETRY

In situ respirometry has been widely used in bench-scale bioremediation studies and in the field (Heitzer and Sayler, 1993). It is a moderate representation of *in situ* conditions with good specificity for general catabolic activity, but low specificity for biodegradation of selected pollutants. Radiorespirometry used in biotreatability studies is moderately representative of *in situ* conditions but has very good specificity for selected pollutant degradation.

7.2.20 MICROCALORIMETRY

A microcalorimeter was developed by the Department of Biological and Agricultural Engineering to offer a convenient and relatively rapid way of determining whether or not a contaminant is metabolizable and what the maximum level of concentration of the contaminant can be (Scholze, Wu, Smith, Bandy, and Basilico, 1986). The microbial activity in terms of heat output for a 1-g sample is a "thermogram." A normalization test performed on a sample that is the organic medium used to sustain the microbiological community will reflect the indigenous activity of the microorganisms present. The heat output is proportional to the numerical density. A baseline thermogram shows background activity in the soil. The area between the baseline and a normalization thermogram is a quantitative measure of the heat contributed by the organisms. The effect of adding a contaminant to the organisms demonstrates whether the compound is toxic (lower heat output) or can be metabolized (greater heat flux than for the normalized thermogram). This is the degradation response. The results indicate the feasibility of using the microbiological community to degrade an organic contaminant.

7.2.21 FLOW CYTOMETRY

Flow Cytometry combines microscopy and biochemical techniques to analyze thousands of cells per second (Edwards, Diaper, Porter, Deere, and Pickup, 1994). Cells are injected in a pressurized stream of fluid and pass through an excitation beam. Light scattered at the cell surface, or narrow angle light scatter (NALS), reflects the size. Light scattered after passage through the cells, or wide angle light scatter (WALS), reflects internal structure as refractility. Computer manipulation allows study of subpopulations. Cells can be sorted by size or fluorescence. More information on this procedure can be found in reviews by Kell, Ryder, Kaprelyants, and Westerhoff, 1991; Edwards, Porter, Saunders, Diaper, Morgan, and Pickup, 1992; and Edwards, 1993.

For application to soil samples, the organisms must be extracted and most of the particulates removed (Edwards, Diaper, Porter, Deere, and Pickup, 1994). The cells could be fixed in soil before extraction to help preserve them during the purification. Light scatter is basically useful for counting total cells in samples that give distinctive profiles.

Addition of fluorescent dyes enhances the technique (Edwards, Diaper, Porter, Deere, and Pickup, 1994). Hoechst 33342 for DNA allows assessment of macromolecular composition, and mithramycin/ethidium bromide staining shows chromosome number per cell. Fluorescein isothiocyanate (FITC) permits monitoring of total cell protein (Allman, Hann, Phillips, Martin, and Lloyd, 1990). Propidium iodide is useful for estimating the nucleic acid (primarily rRNA) content of bacteria (Edwards, Diaper, Porter, Deere, and Pickup, 1994). Different dyes and conjugates can be employed for enumeration and measuring a variety of bacterial properties, even with soil samples that have no meaningful light scatter response. Fluorescent conjugated molecules, such as the fluorescent-antibody and d16S ribosomal RNA (rRNA)–fluorochrome conjugates are also available, but not without limitations. Some microorganisms (e.g., methanogens) produce their own fluorescent molecules, which could be used for their identification and enumeration.

Flow cytometry can be employed to enumerate viable bacteria that cannot be cultured (Edwards, Diaper, Porter, Deere, and Pickup, 1994). It can be used with Rhodamine 123, which is especially sensitive for demonstrating viability (Kaprelyants and Kell, 1992). This approach is only moderately successful with bacteria from soil, since soil components produce a background fluorescence. However, a colorless dye conjugate, such as carboxyfluorescein diacetate, would be taken only into viable cells, where it would be cleaved to release fluorescein, with less nonspecific binding of the background. Flow cytometric cell sorting may be usable for recovery of viable cells and isolation of a target subpopulation, based on the intensity of the fluorescence.

7.2.22 BIOCHEMICAL TESTING

Bacterial isolates are transferred from culture plates to a variety of media to test for biochemical reactions, which will help identify the organisms. The procedure is rather time-consuming, and only a limited number of isolates can be thus processed.

7.2.23 MODELING

Models can be used to determine whether bioremediation is progressing as predicted, and they can serve as a valuable aid for the management of contaminated field sites (National Research Council, 1993). They can range from conceptual models to complex mathematical models, as more data are acquired. Models link concepts of the biodegradation process with field observations and may be powerful tools for assessing bioremediation.

Models are limited in that there are no "off-the-shelf" models, and each must be validated for a given site (National Research Council, 1993). Determining all of the modeling parameters needed may be as demanding and expensive as using other verification methods.

7.3 RATE OF BIODEGRADATION

It is important to have information on the rates of pollutant biodegradation under environmental conditions to be able to assess the potential fate of the compounds (Pfaender and Klump, 1981), to evaluate the efficacy of the *in situ* biodegradation treatment, and to assign appropriate approaches to enhance the degradation rates (Fu, Pfanstiel, Gao, Yan, Govind, and Tabak, 1996). Demonstrating that bacteria are capable of performing the desired reactions at significant rates helps to provide evidence that the bioremediation is successful (National Research Council, 1993).

Biodegradation rates can help provide a rough estimate of the remediation time (Eckenfelder and Norris, 1993). *In situ* estimates of remediation time are difficult to predict, because of the heterogeneous soils and uncertain contaminant distribution. Degradation rates from the literature are useful but do not take into account the soil-specific factors and bacterial acclimatization. With optimum conditions, the remediation time depends upon the degradation half-life, initial concentration of the contaminants, and the final objective concentration.

There are two types of parameters influencing the rate of biodegradation:

1. Those that determine the availability and concentration of the compound to be degraded or that affect the microbial population site and activity and
2. Those that control the reaction rate.

The kinetics of biodegradation are zero-order, if the concentration is high relative to microbes that can degrade it, or first order, if the concentration is not high enough to saturate the ability of the microbes (Kaufman and Plimmer, 1972). First-order kinetics apply where the concentration of the chemical being degraded is low relative to the biological activity in the soil; i.e., at any one time, the rate of chemical loss is proportional to its concentration in the soil (Hill, McGrahen, Baker, Finnerty, and Bingeman, 1955; Kearney, Nash, and Isensee, 1969). The concentration at a given time can be predicted from the biodegradation reaction half-life, i.e., the time it takes for the concentration to decrease by one half (Eckenfelder and Norris, 1993). Michaelis-Menten kinetics seem to apply when the chemical concentration increases and the rate of decomposition changes from being proportional to being independent of concentration (Hamaker, 1966).

The reaction term, r, is often difficult to identify in a given situation (Devitt, Evans, Jury, Starks, Eklund, and Gholson, 1987). Therefore, simple, idealized forms are often used to give approximate estimates. The most common form is the first-order degradation model

$$r = uC_T$$

where u (day^{-1}) is a first-order degradation rate coefficient. It is related to the effective half-life $T_{1/2}$ (day) of the compound by the equation

$$T_{1/2} = 0.693/u$$

In most natural ecosystems, the numbers of hydrocarbon-utilizing microorganisms present will initially limit the rate of hydrocarbon degradation (Atlas, 1978). But after a short period of exposure to petroleum pollutants, the numbers of hydrocarbon utilizers increase and will no longer be the principal rate-limiting factor.

Also in natural ecosystems, a variety of factors probably alter the shapes of substrate disappearance curves (Alexander, 1986). These factors may include predation by protozoa, the time for induction of the active organisms, the accumulation of toxins produced by other microorganisms, depletion of inorganic nutrients or growth factors, the presence of other substrates that may repress utilization of the compound of interest, and binding of the compound to colloidal matter. The impacts or interactions of such potentially important factors may make it difficult to predict the kinetics of mineralization or disappearance of a particular substrate.

Rates of biodegradation vary enormously between the various classes of substances present in petroleum (Bausum and Taylor, 1986). The rates of degradation of long-chain alkanes will depend upon the availability of the hydrocarbon to microorganisms (Atlas, 1978). Availability will be greatly restricted by very low solubility and low surface area of long-chain alkanes, which are solid at normal environmental temperatures. Extensive branching also tends to reduce rates of hydrocarbon degradation.

The following generalizations are possible (Van der Linden and Thijsse, 1965; Cooney and Summers, 1976; Bartha and Atlas, 1977; Ratledge, 1978; Atlas, 1981). The *n*-alkanes from C_{10} to C_{25} are the most readily degraded and utilized. Shorter alkanes are quite volatile. As chain length increases above C_9, the yield of oxidized material increases, while the rate of oxidation decreases. Saturated compounds are degraded more readily than unsaturated ones and straight-chain compounds more readily than branched ones, especially where branching is extensive or creates quaternary carbons. If the alkanes are branched at the β-position, β oxidation is usually blocked (Atlas, 1981). Branched-chain alkanes and alkenes and cycloalkanes are attacked by a very limited range of organisms. Aromatic compounds are partially oxidized by many but are assimilated by few organisms. Polynuclear aromatics, while less toxic than simpler aromatics, are metabolized by fewer organisms and at low rates. Cycloalkanes are fairly toxic, and the initial degradative attack is, generally, accomplished through cometabolism. Degradation rates are low. The most resistant classes seem to be (1) polynuclear aromatics (PAHs), (2) alicyclic substances, such as the tripentacyclics (hopanes), and (3) very long chain aliphatics that seem to be largely a product of the environmental alteration (union) of shorter-chain molecules. The biodegradability of organic compounds found in petroleum products is further discussed in Section 3.

The persistence of a compound increases as the initial concentration increases (Hamaker, 1972). The reduced rate is explained either by the limited active sites available (Hance and McKone, 1971) or by a toxic effect on microorganisms or enzyme inhibition (Hurle and Walker, 1980). With PAHs, there is an increasing trend of initial rate of degradation as the initial concentration increases (Sims and Overcash, 1983). Table 7.7 shows the effect of initial concentration on the rate of degradation.

In many instances, it is possible that the minimum set of factors or variables (at least for substrates that are mineralized) that need to be considered in assessing the rate of degradation of a compound are the concentration of the compound and the abundance of active organisms (Alexander, 1986). At 1 mg/L *t*-butyl alcohol, it appears that the microbial population receives insufficient energy to cause a population increase and utilization rates remain slow (Novak, Goldsmith, Benoit, and O'Brien, 1985). Rates are faster at the highest concentration. Other studies indicate that the rate of mineralization of naphthalene

Table 7.7 Kinetic Parameters Describing Rates of Degradation of Aromatic Compounds in Soil Systems

PAH	Initial Concentration (μg/g soil)	k (day^{-1})	Rate of Transformation (μg/g-day)	$t_{1/2}$[a] (days)
Pyrocatechol	500	3.47	1,735	0.2 m
Phenol	500	0.693	364.5	1.0 m
	500	0.315	157.5	2.2 l
Fluorene	0.9	0.018	0.016	39 m
	500	0.347	173.3	2 m
Indole	500	0.693	364.5	1.0 m
	500	0.315	157.5	2.2 l
Naphthol	500	0.770	385	0.9 m
Naphthalene	7.0	5.78	40.4	0.12 m
	7.0	0.005	0.035	125 l
	25,000	0.173	4,331	4 h
1,4-Naphthoquinone	500	0.578	288.8	1.2 m
Acenaphthene	500	0.173	86.6	4 m
	5	2.81	22.6	0.3 m
Anthracene	3.4	0.21	0.714	3.3 l
	13.7	0.004	0.054	175 m
	10.3	0.005	0.050	143 m
	11.4	0.006	0.073	108 m
	40.0	0.005	0.208	138 m
	36.4	0.005	0.196	129 m
	25,000	0.198	4,950	3.5 h
Phenanthrene	2.1	0.027	0.056	26 m
	25,000	0.277	6,930	2.5 h
Carbazole	500	0.067	33	10.5 m
	5	0.231	1.16	3 m
Benz(a)anthracene	0.12	0.046	0.005	15.2 l
	0.12	0.0001	0.00001	6,250 m
	3.5	0.007	0.024	102 m
	20.8	0.003	0.062	231 m
	25.8	0.005	0.134	133 m
	17.2	0.008	0.060	199 m
	22.1	0.006	0.130	118 m
	42.6	0.003	0.118	252 m
	72.8	0.004	0.257	196 m
	25,000	0.173	4,331	4 h
Fluoranthene	3.9	0.016	0.061	44 m
	18.8	0.004	0.072	182 m
	23.0	0.007	0.152	105 m
	16.5	0.005	0.080	143 m
	20.9	0.006	0.125	109 m
	44.5	0.004	0.176	175 m
	72.8	0.005	0.379	133 m
Pyrene	3.1	0.020	0.061	35 m
	500	0.067	33	10.5 m
	5	0.231	1.16	3 m
Chrysene	4.4	0	0	—
	500	0.067	33	10.5 m
	5	0.126	0.63	5.5 m
Benzo(a)pyrene	0.048	0.014	0.007	50 l
	0.01	0.001	0.00001	694 l
	3.4	0.012	0.041	57 m
	9.5	0.002	0.022	294 m
	12.3	0.005	0.058	147 m
	7.6	0.003	0.020	264 m
	18.5	0.023	0.312	30 m

Table 7.7 (continued) Kinetic Parameters Describing Rates of Degradation of Aromatic Compounds in Soil Systems

PAH	Initial Concentration (μg/g soil)	k (day⁻¹)	Rate of Transformation (μg/g-day)	$t_{1/2}$ [a] (days)
	17.0	0.002	0.028	420 m
	32.6	0.004	0.129	175 m
	1.0	0.347	0.347	2 h
	0.515	0.347	0.179	2 h
	0.00135	0.139	0.0002	5 h
	0.0094	0.002	0.00002	406 l
	0.545	0.011	0.006	66 l
	28.5	0.019	0.533	37 l
	29.2	0	0	—
	9,100	0.018	161.7	39 h
	19.5	0.099	1.93	7 h
	19.5	0.139	2.70	5 h
	19.5	0.231	4.50	3 h
	130.6	0.173	22.63	4 h
	130.6	0.116	15.08	6 h
Dibenz(a,h)anthracene	9,700	0.033	320.1	21 h
	25,000	0.039	962.5	18 h

[a] l = low temperature range (<15°C); m = medium temperature range (15 to 25°C); h = high temperature range (>25°C).

Source: Sims, R.C. and Overcash, M.R. *Res. Rev.* 88:1–68, 1983. With permission.

is determined primarily by the presence of elevated hydrocarbon-degrading microbial populations and may not be directly related to elevated populations of heterotrophic bacteria or sediment organic carbon content. Mineralization activity has been found to be related to the activity of the bacterial populations, with the following rates of mineralization (Walker and Colwell, 1976a).

[^{14}C]hexadecane > [^{14}C]naphthalene > [^{14}C]toluene > [^{14}C]cyclohexane

Other factors that appear to be a major influence on rates of biodegradation include temperature and whether the site has fresh or saline water (Palumbo, Pfaender, Paerl, Bland, Boyd, and Cooper, 1983). Salinity affects biodegradation of PAHs, possibly by affecting PAH–particle interactions, as well as the PAH solubility (Shiaris, 1989). However different authors have reported opposite results on the correlation between salinity and rates of mineralization (Ashok, Saxena, and Musarrat, 1995).

Degradation rates are also affected by seasonal fluctuations (Kerr and Capone, 1986) — high in spring and low in winter (Bartha and Atlas, 1977). Total particle concentrations and chlorophyll *a* concentration do not appear to significantly influence biodegradation rates (Palumbo, Pfaender, Paerl, Bland, Boyd, and Cooper, 1983). The action of a *Nocardia* sp. on hexadecane suggests that the rate of natural biodegradation of oil in marine environments is limited by low temperatures and phosphorus concentration, but not by the concentrations of naturally occurring nitrogen (Mulkins-Phillips and Stewart, 1974a).

A continuous oxygen demand and low diffusion rate in soil reduces oxygen concentrations and thus lowers decomposition rates (Freijer, 1996). A kinetic second-order model, compared with experimental results, determined that O_2 concentration and biomass concentration are important rate-controlling variables. The availability of oxygen must be monitored in closed-system experiments to avoid erroneous interpretation of the results.

When the biodegradability or composition of a waste constituent is unknown, it is necessary to undertake a laboratory investigation of the kinetics of biodegradation of the material by the various available bacterial products (Thibault and Elliott, 1980). Modern, automated respirometric techniques (such as those developed by Polybac Corporation) have been developed to establish biodegradation rates (kinetics), the potential for inhibition of these rates at various concentrations, oxygen and nutrient requirements, and temperature effects. A 1-gal sample of the "pure" organic or 1 ft^3 sample of contaminated soil can be analyzed in 24 h. A respirometric technique showed that simultaneous substrate biodegradation has a relatively small effect on kinetics, so that single substrate extant kinetic tests should be adequate to describe the capabilities of a biomass for degrading a particular substrate (Ellis, Smets, and Grady, 1995).

Rates of biodegradation under optimal laboratory conditions have been reported to be as high as 2500 to 100,000 g/m^3/day (Bartha and Atlas, 1987). Under *in situ* conditions petroleum biodegradation rates are orders of magnitude lower. *In situ* natural rates have been reported in the range of 0.001 to 60 g/m^3/day.

In soils, rates of 0.9 to 15 g/m^2/month have been reported for crude oils (ZoBell, 1950). Rates up to 500 g/m^2/month for a #2 fuel oil have also been noted (Raymond, Hudson, and Jamison, 1976). Huddleston, Bleckmann, and Wolfe (1986) presented a literature review on oil and grease biodegradation kinetics in soil. Biodegradation rates were found to span a wide range from 3 to 454 g oil and grease/kg soil/year of treatment. Despite these substantial degradation rates, complete destruction is not readily achieved (Bausum and Taylor, 1986). In one study, oil loss was 46 to 90% at the end of 1 year (Raymond, Hudson, and Jamison, 1976). In a study with laboratory soil columns, 30% of a light fuel oil remained after 65 weeks in the upper 10 cm of the column and 70% at lower levels (Blakebrough, 1978). Turnover times in oil-contaminated sediment have been reported to be 7 h for naphthalene, 400 h for anthracene, and 30,000 h for benzo(a)pyrene.

Very rapid mineralization rates (e.g., for naphthalene) have been reported for some sediments that are chronically exposed to very high concentrations of degradable hydrocarbons. The mineralization rate and half-life calculated for naphthalene was about 2.9%/day and 2.4 weeks (17 days), respectively, for such a source; while the half-life was 4.4 weeks with sediment from a pristine environment.

Microorganisms have been found to be able to degrade one- and two-ringed aromatic hydrocarbons with high reaction rates down to extremely low concentration levels (i.e., <1 µg/L), given sufficient oxygen and nutrients (Gray, 1978). Degradation rates for hexadecane have been measured to be 0.050 g (Knetting and Zajic, 1972) and 0.015 g/m^3/day (Knetting and Zajic, 1972; Seki, 1976) at summer temperatures, 0.001 g/m^3/day in the colder waters of Alaska, and a rate even lower than this in the open waters of the Arctic Ocean (Robertson, Arhelger, Kinney, and Button, 1973).

The rate of solubilization may not be the sole factor determining the degradation of lipophilic compounds (Thomas and Alexander, 1983). Naphthalene had the highest degradation rate of the relatively nonvolatile hydrocarbons tested, followed in decreasing order by methylnaphthalene, heptadecane, hexadecane, octadecane, fluorene, and benzopyrene. The calculated turnover times (time required to convert all hydrocarbon to carbon dioxide) are presented in Table 7.8 (Science Applications International Corporation, 1985).

Relatively little is known of the kinetics of degradation of mixed substrates at low concentrations, or the possible interactions among primary and secondary substrates and bacteria (McCarty, Reinhard, and Rittmann, 1981). The presence of additional substrates in the soil may also alter the kinetics of mineralization of low concentrations of organic pollutants (Schmidt, Scow, and Alexander, 1985). With a pure culture of *Pseudomonas acidovorans*, acetate and phenol disappeared at an equal rate, when they were at low concentrations. However, phenol mineralization was repressed at high acetate concentrations.

When comparing slurry, wafer, and compacted soil tube bioreactors, it has been found that biodegradation rates in intact soil systems are slower than in soil slurry reactors and that soil tube reactors can be used in conjunction with respirometry to assess bioremediation rates in intact soil systems (Fu, Pfanstiel, Gao, Yan, Govind, and Tabak, 1996). In the soil slurry reactor, biodegradation occurs in the aqueous phase by suspended and soil-immobilized microorganisms. In the soil wafer reactor, diffusivity of pollutant in the soil matrix controls the biodegradation rate. In the porous tube reactor, oxygen limitations occur inside the tube due to diffusional resistances, and oxygen consumption occurs due to biodegradation.

Table 7.9 gives rates of biodegradation observed with various whole petroleum products in soil (Huddleston, Bleckmann, and Wolfe, 1986). The petroleum biodegradation rate in soil appears to be similar to that for vegetable oils. These data should not be closely compared, since significant environmental differences undoubtedly existed for the studies.

Laboratory microcosms duplicating the field conditions can furnish a direct means for estimating biodegradation rates (National Research Council, 1993). Field samples, microbes, substrate concentration, and environmental conditions can be controlled in the microcosms and allow for more accurate measurement of contaminant loss or other biodegradation markers than could be obtained in the field. Microcosms are well suited for assessing microbial adaptation to the contaminants by allowing measurements of increasing degradation rates. Laboratory experiments, however, can impose artifacts or might not allow for the delicate balance of chemical, physical, and biological relationships occurring in the field. Microbes also may behave differently in the laboratory and the contaminated site.

Table 7.8 Turnover Times for Microbial Hydrocarbon Degradation in Coastal Waters

Compound	Concentration (ppb)	Date — Locality	Turnover Time (days)
Benzopyrene	5	Feb — Sk	0
	5	May — Sk	3500
	5	May — O	0
Fluorene	30	Feb — Sk	0
	30	Feb — O	0
	30	May — Sk	0
	30	June — Sk	1000
Heptadecane	8	May — Sk	80
	15	May — Sk	60
	30	May — Sk	54
	30	May — Sk	170
Hexadecane	25	Feb — Sk	500
	25	April — Sk	210
	25	April — O	1000
Naphthalene	40	Feb — Sk	500
	40	May — Sk	46
	40	May — Sk	79
	130	May — Sk	30
	130	May — O	330
Methylnaphthalene	40	Feb — Sk	500
	40	May — Sk	50
Octadecane	16	May — Sk	100
Toluene	20	May — Sk	17
	20	May — Sk	17
	20	May — O	40

Note: Sk = Skidaway River (3m); O = offshore water (10m).

Source: Lee, R.F. and Ryan, C. In *Proc. 3rd Int. Biodegradation Symp.,* Aug. 17–23, 1975. Kingston, R.I., Sharply, J.M., and Kaplan, A.M., Eds. Applied Science Publishers, London, 1976. pp. 119–125. With permission.

A multistep protocol was developed to determine the important kinetics parameters for *in situ* biodegradation of toxic compounds in soils (Tabak, Govind, Fu, Yan, Gao, and Pfanstiel, 1997). The procedure involves abiotic evaluation of adsorption/desorption rates and equilibria followed by the use of respirometry to determine biodegradation rates.

7.4 DIFFERENTIATING BIOTIC AND ABIOTIC PROCESSES

Abiotic reactions include inorganic, organic, photolytic, surface-catalyzed, sorptive, and transport processes (Sepic, Leskovsek, and Trier, 1995). Abiotic losses play an important role in the disappearance of organic soil contaminants, especially those with lower molecular weight, such as PAHs with fewer than four benzene rings and straight-chain alkanes with fewer than 20 carbon atoms.

In studies of selected aromatic and aliphatic compounds, abiotic losses amounted to 78% of the starting concentration within 17 days of incubation at room temperature (Sepic, Leskovsek, and Trier, 1995). These losses were substantially lower at around 5°C, indicating that evaporation was a major contributor for single compounds at low concentrations. Aromatic compounds showed lower biotic degradation, again depending upon the molecular weight. For instance, phenol demonstrated only around 3% biotic degradation and about 92% abiotic losses. PAHs with higher molecular weights are affected only slightly by both biotic and abiotic losses. Abiotic losses are, however, much lower in mixtures of compounds and in actual contaminated field samples, where the biological degradation is more efficient. Also, higher concentrations of hydrocarbons reduce the abiotic losses.

Controlled-release experiments, which determine an enhanced loss of contaminants in fertilized or inoculated field plots, may help to distinguish biotic and abiotic activity (Madsen, 1991). Detection of

Table 7.9 Reported Oil and Grease Biodegradation Rates in Soil

Type — Location	Rate (Oil) g/kg soil/year	lb/ft³ soil/year
PETROLEUM		
Crude oil		
— TX, OK, PA, Avg.	15	1.3
— OK	10	0.9
— Canada	55–237[a]	4.9–20.9
(Paraffins) — OK	3	0.3
(Aromatics) — OK	3.5	0.3
(Asphaltics) — OK	4	0.4
Raffinate — TX	139	10.9
Heating oil — TX, OK, PA, Avg.	13	1.1
#6 Fuel oil		
— TX, OK, PA, Avg.	16	1.4
— TX	97	12.8
Refinery waste		
— TX	69	6.2
— OK	75	6.6
— OK	12.5	1.1
— OK	28	4.5
— OK	12	2.7
— TX	14	1.2
— TX	11	1.0
— TX	29	2.5
— OH	8	3.1
— OH	10	3.2
(Paraffin) — OK	184	16.2
(Aromatics) — OK	49	4.3
(Asphaltenes) — OK	37	3.2
Tank sludge — Canada	16	1.5
— TX	24	7.6
Oil wastes — TN	171	15.0
Lube oil waste		
— TN	171	15.0
— TX, OK, PA, Avg.	18	1.6
— New Zealand	17–454	2.3–40
MINERAL OIL	38–77	3.3–6.8
VEGETABLE OIL		
Palm oil	14–91	1.8–8.0
Soybean oil	17–96	1.5–8.5
Waste cooking oil	59–190	5.2–16.8

[a] Range = low to high loading rate.

Source: Huddleston, R.L. et al. In *Land Treatment: A Hazardous Waste Management Alternative. Water Resources Symposium Number 3*. Loehr, R.C. and Malina, J.F., Jr., Eds. Center for Research in Water Resources, University of Texas, Austin, 1986. pp 41–62. With permission.

unique microbial metabolites, stimulation of protozoa, or decreasing ratios of biodegradable to nonbiodegradable contaminants may also be used to indicate that biological processes are occurring. Addition of nutrients or other key metabolic stimulants to a portion of the site or as a pulse in time can result in production of unique biotic responses (Raymond, Hudson, and Jamison, 1976; Westlake, Jobson, and Cook, 1978; Lamar and Dietrich, 1990). An isotopically labeled (stable or radioactive) contaminant can be released at the site, followed by monitoring metabolite production (Cheng and Lehmann, 1985; Fuehr, 1985). This approach is reliable if labeled metabolites are unique to metabolic pathways.

Section 8

Treatment Trains

8.1 LIMITATIONS OF SOIL TREATMENT SYSTEMS

8.1.1 PHYSICAL/CHEMICAL TREATMENT SYSTEMS

Many of the chemical treatments listed in Section 2 may have limited application for organic wastes. Physical methods, such as stripping or sorption, are not as effective as biological methods for treating hazardous organic compounds (Knox, Canter, Kincannon, Stover, and Ward, 1986). Air stripping and vapor extraction are limited to volatile compounds in porous, homogeneous soil, and channeling is a problem. These are also difficult to monitor. Chemical methods may have to be used to remove heavy metals. The physical and chemical treatment systems are most useful in combination with biological methods, as dictated by the site and specific waste requirements, many serving as a pretreatment prior to biodegradation.

Incineration is an effective treatment process for destroying organic contaminants in soil; however, it is considerably more expensive than biodegradation. Soil flushing is not feasible with complex wastes, and when the subsurface is not homogeneous, channeling through the soil prevents even distribution of the eluant. In addition, the use of treatment agents on the soil can change the pH or other soil properties.

8.1.2 LAND TREATMENT

As with landfilling, this technology depends upon the availability of land. It is subject to weather conditions, which may interfere with application schedules. The heavy components of petroleum oils are not easily degradable and may accumulate in the soil. Volatilization of the lighter compounds can result in uncontrolled hazardous emissions. Waste constituents that are sufficiently volatile, mobile, or might bioaccumulate may be difficult to treat by this method.

8.1.3 *IN SITU* BIODEGRADATION

In situ bioreclamation is a versatile tool for treating contaminated groundwater and soil; however, it is not the answer to all contamination problems (Brown, Loper, and McGarvey, 1986). Its applicability must be determined for each site and depends upon local site microbiology, hydrogeology, and chemistry.

There are several important limitations for application of conventional biological treatment methods (Brown, Loper, and McGarvey, 1986). These include:

1. Environmental parameters must be appropriate for support of microbial growth (pH, temperature, redox state, and available nutrients);
2. Some chemicals are nonbiodegradable, according to current knowledge;
3. By-products of biodegradation may be more toxic or persistent than the original compound;
4. Substrate concentration may be too high (toxic) or too low (inadequate energy source); and
5. Complex mixtures of organics may include inhibitory compounds.

It appears that microorganisms that are able to degrade chemicals in culture sometimes may not do so when introduced into natural environments because of pH, inability to survive, or preferential use of other substrates (Zaidi, Stucki, and Alexander, 1986). Some *Pseudomonas* strains were able to mineralize biphenyl or *p*-nitrophenol in lake water at the natural pH of 8.0, while another strain required the pH to be adjusted to 7.0, and yet another did not mineralize the substrate, although its population density rose.

Biodegradation in the field may also be hindered by protozoa grazing (Alexander, 1994). Also, added microorganisms do not easily infiltrate the soil beyond the first 5 cm (Edmonds, 1976). A method of dispersing them uniformly throughout the contaminated site would have to be devised (Alexander, 1994).

Introduction of nutrients into the environment and the residues generated by the organisms can adversely affect water quality (Lee, Wilson, and Ward, 1987). A field demonstration in a very gravelly clay loam was not very successful, due partly to the low permeability (1×10^{-6} cm/s), which made it difficult to inject nutrients and produce water (Science Applications International Corporation, 1985).

Other factors contributing to the poor success of this demonstration were the complexities of the site, possible mobilization of lead and antimony by the hydrogen peroxide treatment, and reductions in the permeability of the soil due to precipitation of the nutrients.

Some partial degradation products might be more toxic than the parent compounds (Lee, Wilson, and Ward, 1987). Transformation of a toxic organic solute is no assurance that it has been converted to harmless or even less hazardous products (Mackay, Roberts, and Cherry, 1985). Given our limited understanding of transformation processes and the factors influencing them, hazardous contaminants must be assumed, in the absence of site-specific evidence to the contrary, to persist indefinitely. Microorganisms can mobilize hydrocarbons by transforming them to polar compounds, such as alcohols, ketones, and phenols, or to organic acids, such as formic, acetic, proprionic, and benzoic, when contaminated with JP-5 (Perry, 1979; Ehrlich, Schroeder, and Martin, 1985). If it is impossible to verify "no migration" of hazardous constituents, landtreatment may be prohibited as a management alternative for RCRA wastes, as well as the use of *in situ* biological treatment, at Superfund sites (Scholze, Wu, Smith, Bandy, and Basilico, 1986). Also, because of the complexity of waste streams and Agency time constraints, it will be difficult for the EPA to establish a "dilute" concentration level by which biological treatment performances can be evaluated.

Bacterial growth can plug the soil and reduce circulation of the groundwater (Lee, Wilson, and Ward, 1987). The plugging of well screens and the neighboring interstitial zones of an aquifer can be a direct result of biofilm generation (Cullimore, 1983). This can result in reduced flow from the wells (sometimes with complete shutdown of the system), reduced quality of the water (through the generation of turbidity, taste, odor, and color), and, eventually, the generation of serious anaerobic corrosion problems. The resulting degeneration in well productivity has, on occasion, been expensive, with an estimated annual cost between $10 and 12 million (Canadian dollars).

Organisms found to be responsible for this plugging have been *Gallionella*, other bacteria able to deposit iron or manganese oxides or hydroxides in or around the cell (e.g., *Leptothrix, Crenothrix,* or *Sphaerotilus*), and heterotrophic bacteria able to grow in a biofilm. The extensive growth at the aquifer/well interface is probably due to the increase in oxygen concentration at the site of injection. However, disinfectants and physical techniques have been reported for controlling this problem.

In order to be a useful pollution abatement method, biodegradation of petroleum pollutants would have to occur rapidly (Atlas, 1977). This is generally not the case; natural biodegradation of petroleum hydrocarbons occurs relatively slowly. It is likely that the treatment will be time-consuming and expensive, with costs ranging from tens of thousands of dollars for simple treatment programs up to the tens of millions of dollars for complex, large sites (Lee, Wilson, and Ward, 1987).

There is the possibility that potentially pathogenic noncoliform microorganisms capable of oligotrophic life could be introduced to the subsurface during well drilling and establish resident populations, which would go undetected with the standard methods of bacterial analysis of water (Stetzenbach, Sinclair, and Kelley, 1983). Species of *Pseudomonas, Flavobacterium, Acinetobacter, Aeromonas, Moraxella, Alcaligenes,* and *Actinomyces* capable of surviving for extended periods in low nutrient concentrations have been isolated from water samples. These organisms could become established in the subsurface with a rapid and significant impact. Therefore, the identification and characterization of either autochthonous (native microorganisms) or transient noncoliform bacteria in well water is essential in understanding the overall quality of the water. Usually, oil oxidizers do not cause infections in higher organisms, although a few species of human pathogens have been induced to metabolize hydrocarbons; e.g., *Mucor* sp. (Texas Research Institute, Inc., 1982).

There are few additional safety hazards associated with *in situ* bioreclamation aside from those hazards normally associated with being on a hazardous waste site or a drill site (U.S. EPA, 1985a). Since wastes are treated in the ground, the danger of exposure to contaminants is minimal during a bioreclamation operation relative to excavation and removal. The only treatment reagent that could pose a hazard, if used, is the concentrated hydrogen peroxide solution. The complexity of *in situ* treatment and the difficulty of obtaining data required for the permitting process, can cause delays in feasibility demonstrations.

Although bioremediation may seem the method of choice, there are a variety of factors to consider before it is selected over another treatment method or disposal (Amdur and Clark-Clough, 1994). This includes evaluation of contaminants, soil type, space limitations, time of year, critical path, and local requirements. Managers must be realistic in their goals regarding treatment time and treatment levels, and understand the limits of the bioremediation process. These authors discuss these considerations and present a framework for selecting and contracting with a bioremediation firm.

8.1.4 ON-SITE/*EX SITU* BIOLOGICAL SYSTEMS

Since the organisms used in the various on-site biological treatment systems may grow slowly, population retention is important (Kobayashi and Rittmann, 1982). Microbes could be washed out (total loss of the organism from a reactor) or taken over by other microorganisms. Fixed-film processes may be the best mechanism to assure population retention, when appropriate, to avoid total loss of slow-growing components. The cell concentrations in these processes are higher than those found in suspended growth systems. Efficient removal of the organic compounds is possible only when the biomass concentration is large (Matter-Muller, Gujer, Giger, and Strumm, 1980). Sometimes it is necessary to maintain a series of microorganisms selectively in order to achieve complete degradation (Pfennig, 1978a).

8.2 REMEDIATION GUIDELINES

The decision to remediate fuel hydrocarbon contamination can be linked to risk-based corrective actions (RBCAs) (Benson, Frishmuth, and Downey, 1995). It is too resource intensive and unjustifiable to try to remediate all fuel hydrocarbon–contaminated sites to non-site-specific cleanup goals. The RBCA approach provides a site-specific framework for defining the level of remediation necessary to protect human health and the environment. Often, natural attenuation processes (e.g., intrinsic biodegradation) and land management controls may be sufficient, without engineered remediation. Sometimes, low-cost, source reduction technologies adequately supplement the natural processes. Remedial requirements depend upon which exposure pathways may reasonably be expected to be completed at a particular site.

No two contamination incidents are exactly alike (Bartha and Atlas, 1977). Consequently, control responses should be flexible and tailored to the situation. A thorough understanding of the hydrogeologic and geochemical characteristics of the area will permit full optimization of all selected remedial actions, maximum predictability of remediation effectiveness, minimum remediation costs, and more reliable cost estimates (Wilson, Leach, Henson, and Jones, 1986). The design of remediation strategies depends upon contaminant properties and distribution, infrastructure, lithology, regulatory requirements, site usage, and time restrictions. A limiting factor is delivering the contaminated subsurface material to the treatment unit, or the treatment process to the contaminated material, in the case of *in situ* processes (Wilson, Leach, Henson, and Jones, 1986).

The total petroleum hydrocarbon (TPH) measurement is a common tool for establishing cleanup standards for underground storage tank sites and other petroleum-contaminated areas (Michelsen and Boyce, 1993). There are, however, alternative techniques for developing site-specific cleanup standards, such as chemical fingerprinting, constituent analysis, and risk assessment methods.

There are advantages and disadvantages associated with all the available treatment options (Eckenfelder and Norris, 1993). However, adequate assessment of each contamination incident will allow selection of the appropriate process or combination of technologies to achieve the required remediation. Figure 8.1 presents a decision framework for remediation technologies to use at sites contaminated with petroleum hydrocarbons (Ram, Bass, Falotico, and Leahy, 1993). It provides a structured progression of decision points consisting of technology or site applicability criteria. This will help the user to select appropriate technologies for the specific remediation requirements. In decreasing order of importance, liquid-phase hydrocarbon (LPH) removal processes are considered first, then *in situ* vadose and saturated zone technologies, then groundwater pump-and-treat approaches. Site characterization and closure goals are necessary for assessing technology applicability and remediation criteria. Table 2.2 summarizes several technologies for their applicability in different situations. The interaction between technologies often used at sites contaminated by petroleum hydrocarbons is illustrated in Figure 8.2.

8.3 COMBINED TECHNOLOGIES

8.3.1 ON SITE/*EX SITU*

Contaminated waste streams are normally composed of a complex mixture of compounds of variable concentration (Wilson, Leach, Henson, and Jones, 1986; Sutton, P.M., 1987). The compounds may be

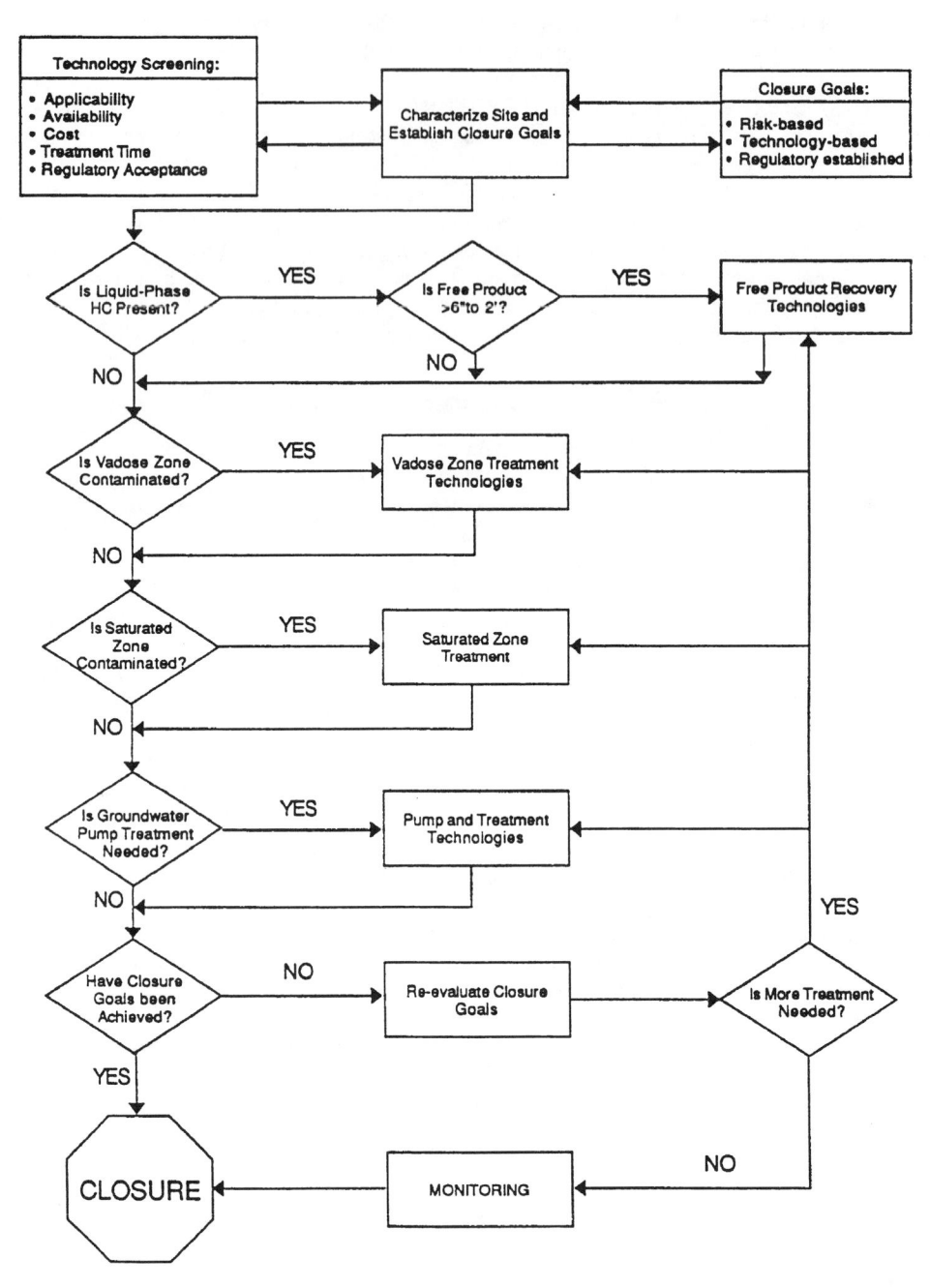

Figure 8.1 Decision framework for remediation technologies. (From Ram, N.M., Bass, D.H., Falotico, R., and Leahy, M. *J. Soil Contam.* 2(2):167–189. Lewis Pubishers, Boca Raton, FL. 1993.)

degradable, inhibitory, or recalcitrant to various degrees. Soils may be so heavily contaminated that they have to be removed or attenuated (Brubaker and O'Neill, 1982).

Excavation of contaminated soils can be costly; however, it allows a more rapid treatment of the material *ex situ* (Eckenfelder and Norris, 1993). There are a number of approaches whereby contaminated soil can be excavated and subjected to treatment in a reactor on- or off-site. By treating soil in a reactor, a near-perfect environment for biodegradation can be created (King, Long, and Sheldon, 1992). This controlled environment allows a combination of biological, physical, and chemical processes to be applied

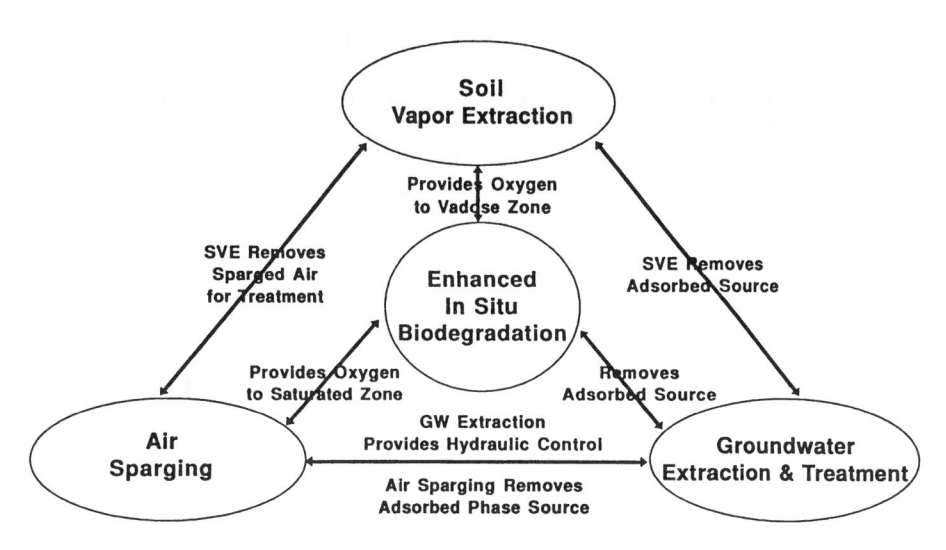

Figure 8.2 System integration. (From Ram, N.M., Bass, D.H., Falotico, R., and Leahy, M. *J. Soil Contam.* 2(2):167–189. ©1993 Lewis Pubishers, Boca Raton, FL)

in treatment trains tailored to the specific pollutants. It permits optimization of many of the parameters necessary for biodegradation, while controlling the release of volatile organic compounds (VOCs) and leachate produced during the process. This is one of the most cost-effective and exemplary approaches available for resolution of the problem of environmental contamination (Madsen, 1991). Bioreactors can increase the rate of polycyclic aromatic hydrocarbon (PAH) degradation in contaminated soil (Wilson and Jones, 1993). However, running costs are generally higher than *in situ* and other on-site treatments.

Volatile organics, extractable organics, and inorganics (heavy metals) of concern in contaminated waste streams can be treated (removed) successfully by two alternative processes (Stover and Kincannon, 1983). One process consists of chemical precipitation to remove metals and steam stripping, followed by activated carbon adsorption of organics. The alternative consists of combined physical–chemical and biological treatment. The physical–chemical techniques may remove nonbiodegradable constituents and may render the contaminated material less inhibitory to microbial treatment. Metals treatment would be a safety measure against possibly higher concentrations than anticipated; it would also be required for removing high levels of iron and manganese. Chemical detoxification techniques include injection of neutralizing agents for acid or caustic leachates, addition of oxidizing agents to destroy organics or precipitate inorganic compounds, addition of agents that promote photodegradation or other natural degradation processes, extraction of contaminants, immobilization, or reaction in treatment beds (Lee and Ward, 1985, 1984; Lee, Wilson, and Ward, 1987). Physical–chemical treatment will normally be provided in conjunction with the biological step (Wilson, Leach, Henson, and Jones, 1986). Combining the unit processes of chemical precipitation, steam stripping, and biological treatment is the more feasible alternative of these two, otherwise concentrations of residual organics, measured as TOC, would still be too high (Stover and Kincannon, 1983). Various treatment trains for treating leachates to remove organic compounds are described by Enzminger, Robertson, Ahlert, and Kosson (1987).

Bioreclamation with Innovative On-Site Controlled Environment Landtreatment Systems (BIOCELS) can be accomplished by integrating the information presented in this book on the variety of processes available for remediating petroleum-contaminated soils and waste streams with the methods for optimizing biodegradation of the contaminants. This will allow development of incident-specific bioreactors or other customized *ex situ* treatment trains.

8.3.2 *IN SITU*

In most polluted hydrogeologic systems, a remediation process is so complex in terms of contaminant behavior and site characteristics that no one system or unit will usually meet all requirements (Wilson, Leach, Henson, and Jones, 1986; Sutton, P.M., 1987). It is often necessary to combine several unit operations, in series or in parallel, into one treatment process train to bring the contamination to an acceptable level.

In situ biodegradation has sometimes been applied as a treatment for spill management after partial recovery of a contaminant by physical means, such as by excavation, free-product recovery, pumping, air sparging, air stripping, or vapor extraction (Raymond, Jamison, and Hudson, 1976; Walton and Dobbs, 1980; Brown, Mahaffey, and Norris, 1993). These auxiliary physical treatments can also be employed *during* bioreclamation (Brubaker and O'Neill, 1982). Biodegradation is an alternative to physical recovery processes once they become nonproductive in terms of cost and effectiveness. Integration of other removal mechanisms with the biological step should be cost-effective, if technically feasible. One case study initiated stimulation of soil organisms, after estimating that continued physical recovery methods would require 100 years of operation and maintenance to make the contaminated water potable (Raymond, Jamison, and Hudson, 1976). Biological treatment, itself, is the least-expensive method of organic destruction (U.S. EPA, 1985a). About 99% of all organic compounds can be destroyed by biological reactions. When used with other treatment technologies, essentially all the organic contaminants can be removed and destroyed. Enhanced bioremediation may not necessarily replace other control measures, but it should rather add further flexibility to integrated control programs (Bartha and Atlas, 1977).

While physical and chemical procedures may help promote biodegradation, bioremediation might also render a site more amenable to treatment with nonbiological methods (Brown, Mahaffey, and Norris, 1993). The synergistic effects of the different techniques when employed in a given incident should help maximize contaminant removal. Bioremediation is currently utilized as part of an integrated system (e.g., with soil vapor extraction or air sparging) for treating highly mobile (volatile or soluble) or degradable substrates, such as gasoline or diesel fuel. It is also employed as a primary system for treating recalcitrant or nonmobile substrates, such as heavier petroleum products.

Multiple technologies are being used at many sites (Brown, Mahaffey, and Norris, 1993). Since pump-and-treat methods remove only those contaminants with water solubilities greater than 10,000 mg/L, pockets of free-phase liquids and adsorbed-phase organics will remain, requiring other means of removal. A combination of *in situ* bioremediation, air sparging, and/or vapor extraction may be the best approach for dealing with VOCs. For example, using intermittent or low airflow rates for vapor extraction will reduce off-gas treatment and encourage biodegradation. Also, combined vapor recovery and bioremediation would probably be good for unsaturated soils contaminated with biodegradable compounds with vapor pressures exceeding about 1.0 mmHg. If the soil is polluted with biodegradable contaminants that are minimally volatile, such as PAHs and heavy fuels, bioremediation may be a stand-alone technology.

Treatment trains employing one or more treatment processes may be required for complex waste streams (Lee and Ward, 1986). Bioreclamation can be preceded by, or otherwise used in combination with, other on-site or *in situ* treatment techniques that could destroy, degrade, or by other means reduce the toxicity of contaminants (U.S. EPA, 1985a).

The information presented in this book on the variety of processes available for remediating petroleum-contaminated soils and waste streams can be combined or used in conjunction with the methods for optimizing biodegradation of the contaminants to develop site-specific, *in situ* treatment trains. Additional information on bioremediation of groundwater, freshwater, estuarine, and marine environments contaminated with petroleum products is available to supplement the soil, leachate/wastewater, and VOC treatment processes presented in this book (Riser-Roberts, 1992).

8.3.3 PROCESSES FOR TREATMENT TRAINS

Options for *in situ* or on-site/*ex situ* treatment of soils, leachates, and emissions that have been contaminated by petroleum products are listed in the Table of Contents and described throughout the text of this book. Depending upon the specific site and contaminant requirements, a combination or sequence of processes may be developed to achieve acceptable contaminant levels.

8.4 EXAMPLES OF THE USE OF TREATMENT TRAINS

1. A laboratory-scale treatment train was used on a Department of Energy site soil contaminated by mixed wastes, including petroleum hydrocarbons (Portier (1994). It consisted of a liquids–solids contact soil slurry reactor and an immobilized microbe bioreactor for treating aqueous wastes in the process waters from the slurry reactor. Soils were treated in a defined sequence of roughing for 3 days, biological

treatment for 27 days, and continuous polishing involving metals chelation. The immobilized microbe bioreactor was a modified fixed-film reactor with a controlled pore surface.

2. A pilot-scale bioremediation system was integrated with pneumatic fracturing to increase subsurface permeability and establish a broader bioremediation zone with aerobic, denitrifying, and methanogenic populations (Venkatraman, Schuring, Boland, and Kosson, 1995). Phosphate and nitrogen were added to the subsurface over 50 weeks, resulting in >67% reduction in BTX.

3. Radio-frequency heating was combined with soil vapor extraction to enhance recovery of #2 fuel oil in silty soil at a depth of 20 ft (Price, Kasevich, and Marley, 1994).

4. Shallow soil mixing with bentonite, cement, or other compounds, and soil vacuum extraction were integrated and enhanced to extract VOCs from soils at a Department of Energy facility in Ohio (Carey, Day, Pinewski, and Schroder, 1995). The advantages were a relatively rapid remediation, lower cost, less exposure of waste to the surface, and elimination of off-site disposal, with this *in situ* approach. (See Section 2.2.1.1.)

5. In a study of a site polluted by hydrocarbons, chlorinated hydrocarbons, and organochloride pesticides, it was found that no single technology could remove or destroy all of the contaminants (Rickabaugh, Clement, Martin, and Sunderhaus, 1986). However, when microbial degradation, surfactant scrubbing, photolysis, and reverse osmosis were combined, nearly total destruction of these compounds could be attained on-site.

6. A combined technology approach was also employed at a site where 130,000 gal of several organics had been spilled (Lee and Ward, 1985). Treatment of the site was by clarification, adsorption onto granular activated carbon, air stripping, then reinjection. After levels of the contaminants had fallen below 1000 mg/L, a biodegradation program employing facultative hydrocarbon-degrading bacteria, nutrients, and oxygen was begun. Biodegradation by both the indigenous microbes and the added organisms reduced the levels of the contaminants in soil cores from 25,000 to 2000 mg/L within 2 months. The monitoring wells showed no levels above 1 mg/L at the end of the program.

7. Another example involving the use of multiple treatment processes was a bench-scale study on a site in Muskegon, MI, contaminated by several priority pollutants and at least 70 other organics at levels in the hundreds of ppm (Lee and Ward, 1985). Acclimation of an activated sludge culture to the contaminated groundwater was unsuccessful, and a commercial microbial culture was ineffective at degrading the contaminants. However, coupling an activated sludge process to granular activated carbon treatment proved beneficial, as the organisms were able to degrade the organics that passed through the carbon system. This treatment train was able to remove up to 95% of the total organic carbon in the wastewater, as long as the activated carbon continued to function properly.

8. The release of phenol and chlorinated derivates in the soil in the Midwest was corrected by installing a recovery system and using activated carbon filters on the groundwater (Walton and Dobbs, 1980). Surface waters were contained in a pond. Mutant bacteria were injected into the pond and into the contaminated soil. After an incubation and adaptation period, the phenol was completely degraded in 40 days, while the *o*-chlorophenol was reduced from 120 to 30 ppm.

9. The Thermatrix flameless oxidation process combined with other contaminant separation and removal technologies can result in effective integrated systems. Thermal desorption can be combined with the Thermatrix flameless oxidation process for near zero emissions when treating contaminated soil *in situ* (see Section 2.2.1.2).

10. Combined biological–carbon systems can be used for leachate treatment (see Section 2.1.1.2.1).

11. Vapor phase biofilters can be used to decompose VOCs, in combination with SVE and air-based biodegradation (see Section 6.3.4.5). This will help lower costs of treating off-gases. For instance, activated carbon treatment of leachate can be used with biological pretreatment of effluent in a sequencing-batch reactor.

12. Wet air oxidation can be used for hazardous waste leachate treatment for treating concentrated organic streams generated by other processes, e.g., steam stripping, ultrafiltration, reverse osmosis, still bottoms, biological treatment process waste sludges, and regeneration of powdered activated carbon (see Section 2.1.1.2.4).

13. Chemical oxidation removes most organics poorly from wastewaters but could facilitate treatment by other processes. Chemical oxidants can be used as a pretreatment to oxidize partially refractory, toxic, or inhibitory organic compounds, e.g., ozonation alone or in combination with ultraviolet irradiation as a pretreatment for biological treatment (see Section 2.1.1.2.6).

14. Ozonation and granular activated carbon combination depends upon the composition of the wastewater (see Section 2.1.1.2.6).

15. Hydrogen peroxide or ozone plus ultraviolet light degrade or destroy VOCs in water (see Section 2.1.1.2.6).

16. Sedimentation must be used with another technique, such as chemical precipitation, or as a pretreatment prior to another process, such as carbon or resin adsorption (see Section 2.1.1.2.10.1).

17. Flocculation must be used with a solid/liquid separation process, e.g., sedimentation, as a pretreatment for carbon adsorption. It is often preceded by precipitation (see Section 2.1.1.2.11).

18. Air stripping is a useful pretreatment for adsorption, and steam stripping is a good pretreatment to reduce VOC levels for following treatments (see Section 2.1.1.2.13).

19. Filtration can be used as a polishing step subsequent to precipitation and sedimentation or as a dewatering process for sludges generated by other processes (see Section 2.1.1.2.15).

20. Ion exchange could serve as a polishing step to remove ionic constituents that could not be reduced by other methods (see Section 2.1.1.2.18).

21. A chemical coagulation/flocculation process removed up to 100% of suspended solids and 98% of BOD_5 (see Section 2.1.2.2.1.2).

22. Biological activated carbon and wet air oxidation are not feasible alone, but in combination can treat dilute contaminated groundwater (see Section 2.1.2.2.1.2).

23. The ability of microorganisms to biosorb organic compounds can be combined with the use of reactors for both readily degradable or more refractory contaminants. The degradable are removed by biodegradation in the reactors, and the refractory, by microbial absorption. The reactors can be aerobic, anaerobic, chemotrophic, phototrophic, or a series of several types, depending upon the nature of the contaminants (see Section 2.1.2.2.1.4).

24. A reactor combining an upflow anaerobic sludge blanket with a fixed film gives better performance than a sludge blanket alone (see Sections 2.1.2.2.2.2 and 6.3.3.3.5).

25. An air/water separator, trickling filter, and biofilter in series can be used to treat VOCs generated during *in situ* bioventing and air sparging (see Section 6.3.4.5).

26. Combined oxidation, i.e., combined use of ozone and other chemical or physical treatment, can improve destruction of biodegradation-resistant organic compounds (see Section 2.2.1.2).

27. Soil containing organics and inorganics can be pretreated with land application to remove or reduce metals by precipitation followed by land application of the elutriate (see Soil Flushing, Section 2.2.1.7.)

28. The pump-and-treat method can be combined with bioventing, if the groundwater has become contaminated (see Section 2.2.1.7).

29. Soil vapor extraction removes VOCs adsorbed to unsaturated soils, but fluctuations in groundwater level affect the rate of removal. When combined with air sparging, VOCs are removed from saturated soils and groundwater, as well (see Section 2.2.1.11).

30. A soil vapor extraction system can be used *in situ* with a combined thermal-catalytic oxidizer vapor treatment system on petroleum-contaminated soils with low permeability (see Section 2.2.1.11).

31. Cyclic steam injection can be combined with vacuum extraction to lower residual hydrocarbon content of JP-5 jet fuel–contaminated soil. Additional cycles should remove even more (see Section 2.2.1.14.2).

32. Wet or dry steam can strip and vaporize contaminants from fuel-contaminated soil, while a vacuum extracts and condenses the vapors. This can employ portable steam systems and a packed-bed thermal oxidizer for the vapors. Extracted liquids can go to an oil/water separator to collect fuel for recycling.

TREATMENT TRAINS

Contaminated groundwater and condensate are treated with filters and carbon adsorption (see Section 2.2.1.14.2).

33. Radio-frequency heating was combined with soil vapor extraction to improve recovery of #2 fuel oil in silty soil at a depth of 20 ft. Radio-frequency heating vaporizes pollutants or decomposes or pyrolyzes them to more-volatile compounds, which are removed by soil vapor extraction. The method is appropriate for higher-boiling-point soil contaminants (see Section 2.2.1.14.3).

34. Active warming of contaminated soil in cold climates can be combined with bioventing to improve biodegradation (see Section 2.2.2.2).

35. An alternative bioventing approach includes low rates of pulsed air injection, a period of high-rate soil venting extraction, and off-gas treatment followed by long-term air injection (see Section 2.2.2.2).

36. Regenerative resin for *ex situ* vapor treatment can be combined with *in situ* bioventing to reduce bioremediation costs (see Section 2.2.2.2).

37. Bioslurping combines vacuum-enhanced recovery of free product with *in situ* bioventing to aerate the vadose zone for improved biodegradation of low-volatility hydrocarbons, while promoting vapor extraction of the more volatile fractions (see Section 2.2.2.3).

38. Hydraulic/pneumatic soil fracturing might be combined with other processes to enhance bioremediation (see Section 2.2.2.5). For instance, it can be used to facilitate transport of materials in the subsurface.

39. Aerobic and anaerobic biotreatment may be used in sequence (see Sections 3.2.2.1, 5.1.4.5, 5.1.4.6, 5.1.4.7, and 5.1.4.8).

40. Aerobic and anaerobic treatment can be employed in a single step, if the microorganisms are immobilized in matrices where anaerobiosis can occur (see Section 3.2.2.1).

41. Fungi, yeasts, and bacteria can be combined for more complete biodegradation. Bacteria should be present to break down mutagens produced by the fungi. Mixed microbial populations are more effective (see Sections 3.4, 5.2.1, 5.2.1.4, and 5.2.2.3).

42. Microorganisms can be combined with chemical analogs of organic compounds to promote co-oxidation of the latter (see Sections 3.4 and 5.2.3).

43. When chemical and biological treatments are combined, the soil pH and redox boundaries should be carefully monitored (see Section 5.1).

44. Soil moisture can be controlled through irrigation, drainage, soil additives, or combinations of these. Moisture optimization may be enhanced when used with other techniques to increase biological activity (see Section 5.1.1).

45. Combined air–water flushing disperses oxygen through soil and detaches solubilized hydrocarbons, while providing moisture for biodegradation. Contaminants are removed in the airstreams and water streams (see Sections 2.2.2.7 and 5.1.4.5).

46. The low-volume airflow of bioventing is combined with a closed-loop concept to regulate soil moisture, nutrients, and oxygen with Biopurge^{SM}, when the vapor is injected above groundwater, and with Biosparge^{SM}, when the vapor is injected below groundwater level. The vapors are extracted in wells and treated on-site (see Sections 2.2.2.4 and 5.1.4.6).

47. Seeding of microorganisms should be combined with soil moisture management, aeration, and fertilization (see Section 5.2.2).

48. Surfactants can be used in conjunction with other treatments to solubilize contaminants (see Sections 5.3.1.1 and 5.3.1.2).

49. Solvent extraction can be used with steam stripping to remove contaminants from waste streams. Solvents are chosen with low aqueous solubility and strong affinity for the VOCs in the waste. The solvent is removed by steam stripping and regenerated by distillation (see Section 6.3.3.1.3).

50. A biofilter can be used to clean gasoline-contaminated air from a stripping tower (see Section 6.3.4.5).

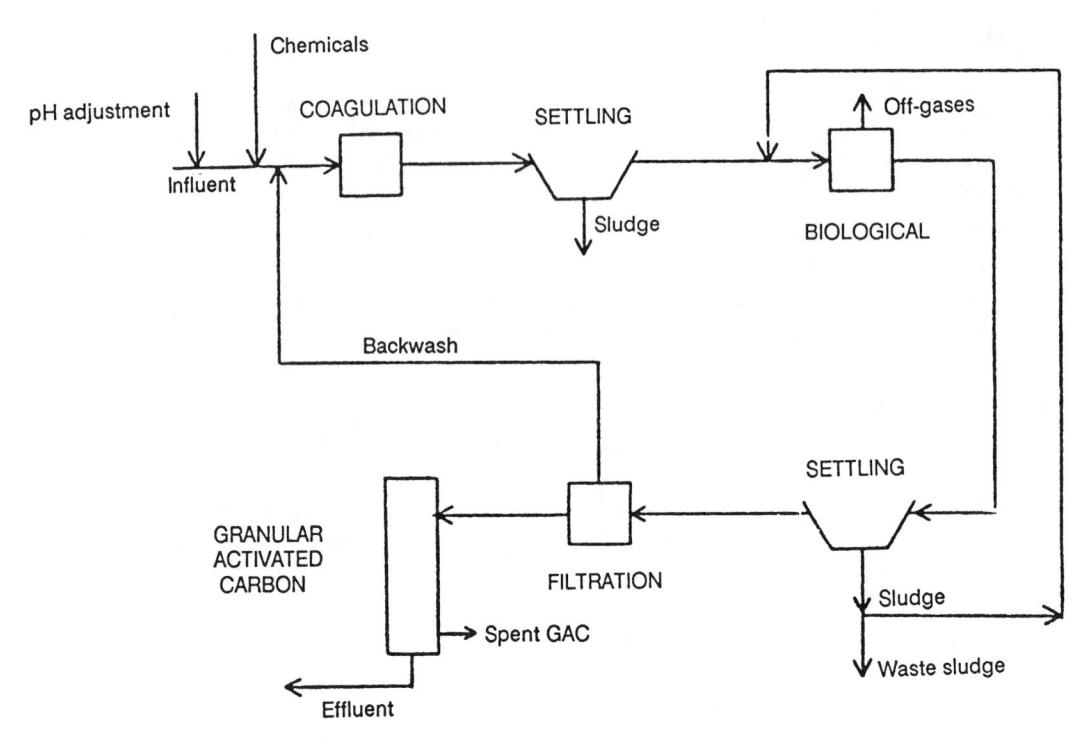

Figure 8.3 Schematic of biological/carbon sorption process train. (From Shuckrow, A. J. et al. *Hazardous Waste Leachate Management Manual.* Noyes Data Corp., Park Ridge, NJ, 1982. With permission.)

51. Biofilters can be backed up with GAC filters for more efficient VOC removal. Such units are much less expensive than conventional GAC filters and catalytic/thermal oxidation (see Section 6.3.4.5).

52. Bioventing can be combined with bioslurping to recover LNAPLs while bioventing the vadose zone (see Sections 2.2.1.11, 2.2.2.2, 3.2.1.6, and 5.1.4.5).

53. Bioventing can also be applied to excavated soil, such as with the Ebiox vacuum heap bioremediation system (see Sections 2.1.2.1.5 and 5.1.4.5).

54. Bioventing can be used to introduce gaseous ammonia as a form of nutrients to the subsurface (see Section 5.1.5).

55. A vacuum-inducing airflow can supply oxygen through the soil, then nutrients percolated through the soil with the vent-system piping. VOCs are removed by the venting (see Section 5.1.5).

56. Photochemical reactions combined with soil mixing can be effective for treating relatively immobile contaminants (see Sections 2.1.1.2.6, 2.1.2.1.7, 5.3.2, and 6.3.4.6).

57. Figure 8.3 shows how biological and carbon sorption processes can be combined in a treatment train (Shuckrow, Pajak, and Touhill, 1982b). Figure 8.4 illustrates a process train that could be used for a leachate that contains metals. In Figure 8.5, the processes of air stripping, carbon adsorption, and ion exchange are combined in a treatment train (Bove, Lambert, Lin, Sullivan, and Marks, 1984). Figures 8.6 and 8.7 show the *in situ* use of a Detoxifier™ system in a treatment train on a site contaminated by hydrocarbons (Ghassemi, 1988) (see also Sections 2.2.1.13 and 6.3.3.3.6). The system is composed of

- The process tower, including the drill bit assemblies, tower shroud, and the rotary and hydraulic motors that control the up-and-down and rotating motions of the drill assemblies;
- The control room containing the on-line monitoring equipment;
- The crawler tractor, which moves the drilling rig, the control room, and a diesel engine power generator;
- Gas treatment and power feed systems, mounted on two trailers, and consisting of
    Suction blowers
    Cooling coil

TREATMENT TRAINS

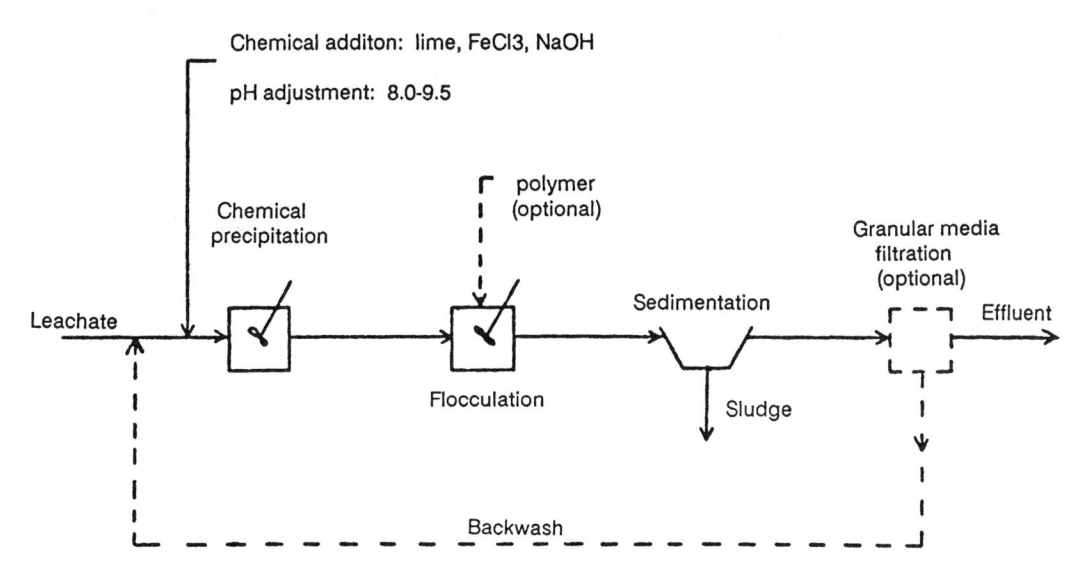

Figure 8.4 Process train for leachate containing metals. (From Shuckrow, A. J. et al. *Hazardous Waste Leachate Management Manual*. Noyes Data Corp., Park Ridge, NJ, 1982. With permission.)

    Demisters
    Refrigeration and heating coils
    Activated carbon adsorption unit
    Powder storage bins and feeding system
    Primary and auxiliary compressors;
- Mixing and pumping systems (trucks) for preparing the treatment agent solution;
- Steam production boiler, mounted on a separate trailer.

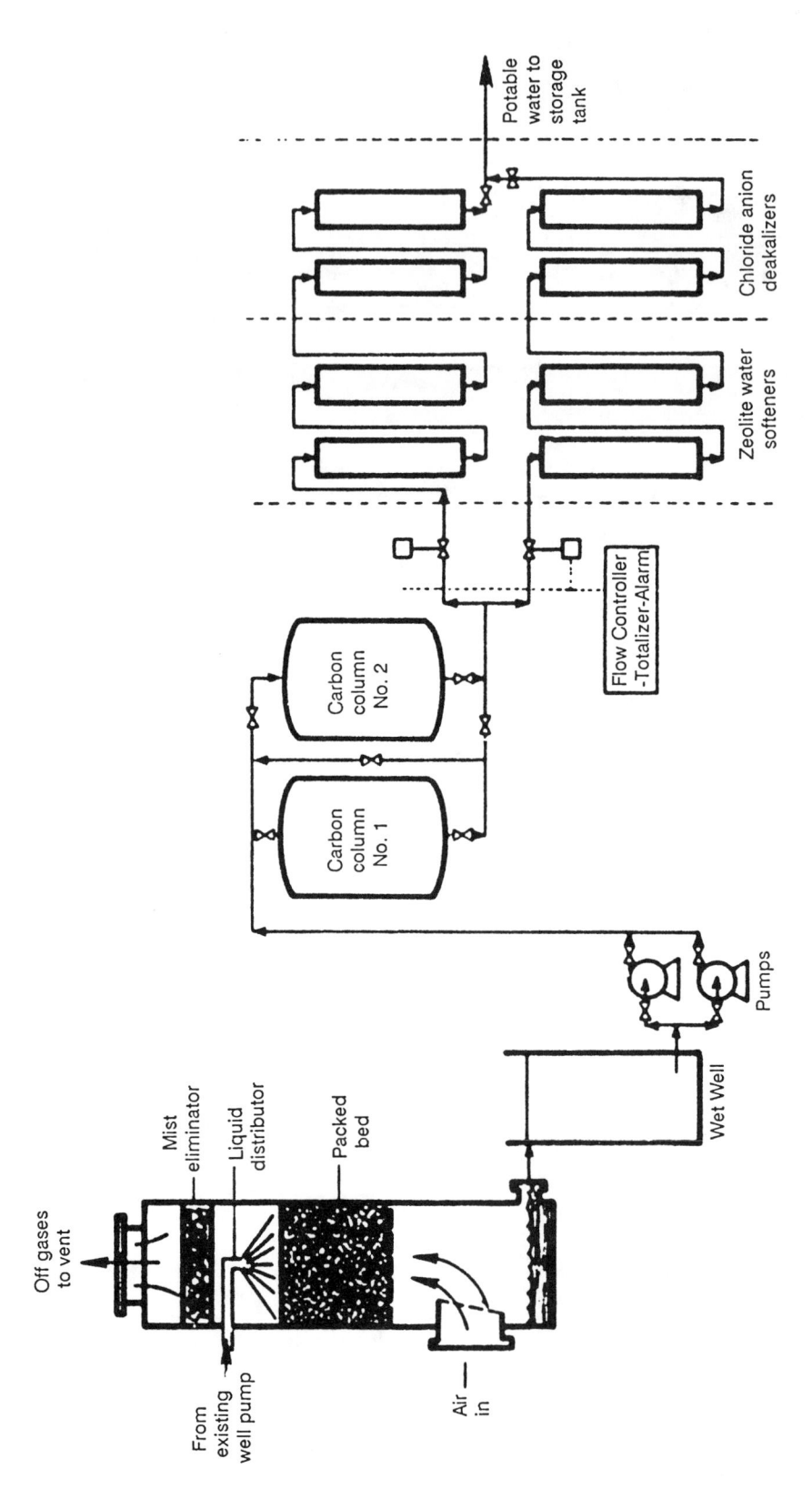

Figure 8.5 Air stripping, carbon adsorption, and ion exchange process flow diagram (typical). (From Bove, L.J. et al. Report to U.S. Army Toxic and Hazardous Materials Agency, Aberdeen Proving Ground, MD, on Contract No. DAAK11-82-C-0017, 1984. AD-A162 528/4.)

TREATMENT TRAINS

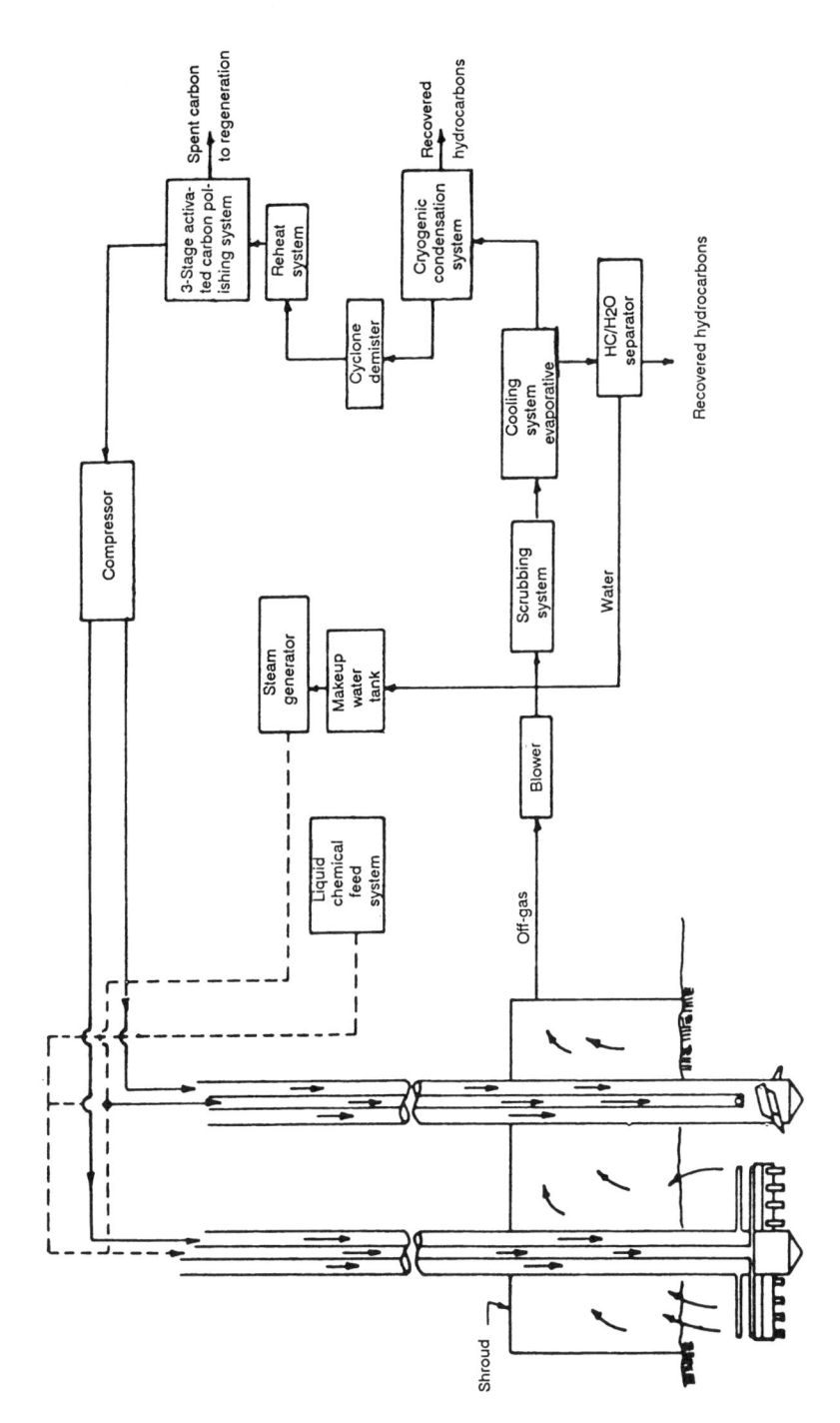

Figure 8.6 Process diagram for the Detoxifier™ II Treatment Train. (From Ghassemi, M. J. *Haz. Mat.* 17:189–206, Elsevier Science Publishers, Academic Division. 1988. With permission.)

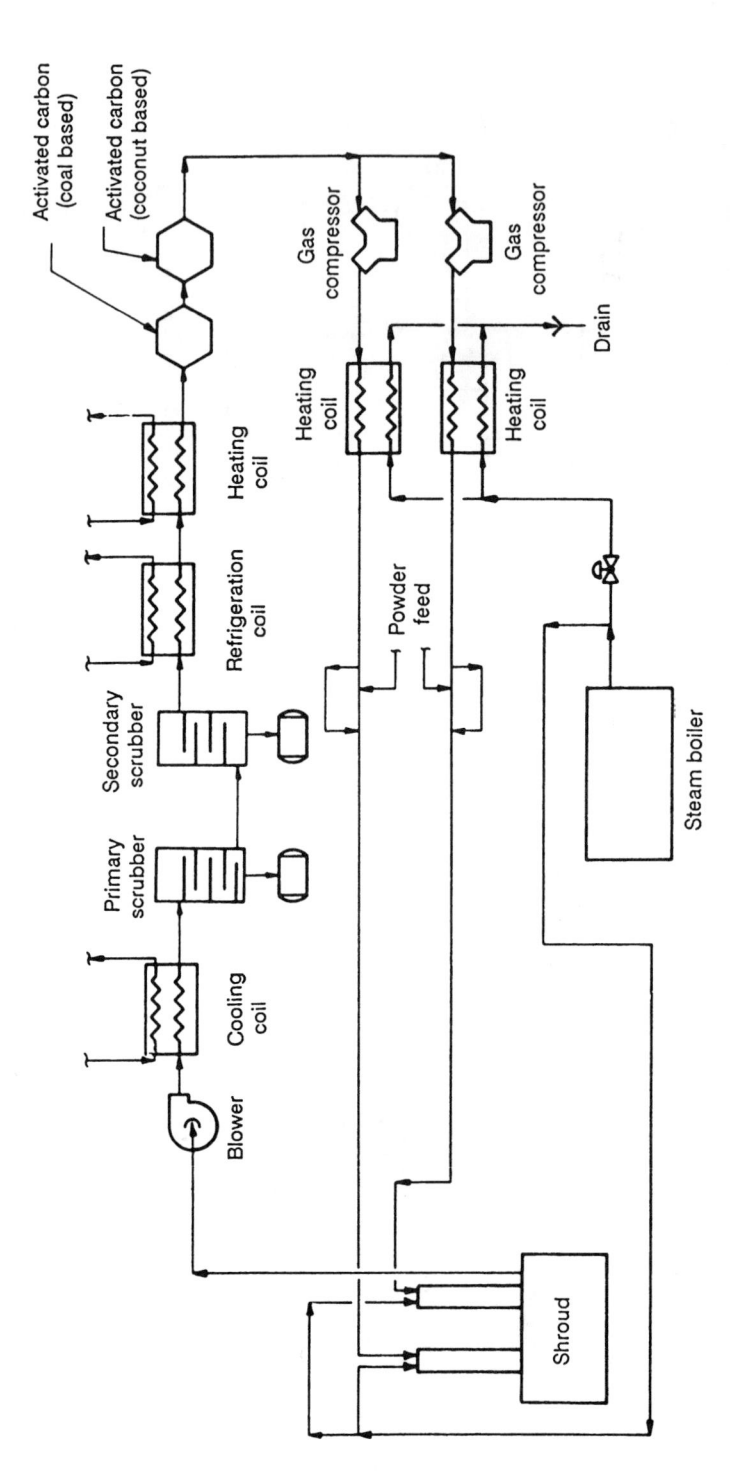

Figure 8.7 Treatment train used at a southern California site. (From Ghassemi, M. J. *Haz. Mat.* 17:189–206, Elsevier Science Publishers, Academic Division. 1988. With permission.)

REFERENCES

A

Aamand, J., Bruntse, G., Jepsen, M., Jorgensen, C., and Jensen, B.K. 1995. Degradation of PAHs in soil by indigenous and inoculated bacteria. In *Bioaugmentation for Site Remediation. Pap. 3rd Int.* In Situ *On-Site Bioreclam. Symp.* Hinchee, R.E., Fredrickson, J., and Alleman, B.C., Eds. Battelle Press, Columbus, OH. pp. 121–127.

Abdul, A.S., Gibson, T.L., Ang, C.C., Smith, J.C., and Sobczynski, R.E. 1992. *In situ* surfactant washing of polychlorinated biphenyls and oils from a contaminated site. *Ground Water.* 30(2):219–231.

Abe, A., Inoue, A., Usami, R., Moriya, K., and Horikoshi, K. 1995. Degradation of polyaromatic hydrocarbons by organic solvent-tolerant bacteria from deep sea. *Biosci. Biotech. Biochem.* 59(6):1154–1156.

Abeliovich, A. 1985. Nitrification of ammonia in wastewater. Field observations and laboratory studies. *Water Res.* (G.B.). 19:1097–1099.

Absalon, J.R. and Hockenbury, M.R. 1983. Treatment alternatives evaluation for aquifer restoration. In *Proc. Aquifer Restoration and Ground Water Monitoring, 3rd Natl. Symp.,* Columbus, OH, 1983. Nielsen, D.M., Ed. pp. 98–104. National Water Well Assoc. (NWWA), Worthington, OH.

Acea, M., Moore, C.R., and Alexander, M. 1988. *Soil Biol. Biochem.* 20:509–515.

Adams, C.D., Spitzer, S., and Cowan, R.M. 1996. Biodegradation of nonionic surfactants and effects of oxidative pretreatment. *J. Environ. Eng.* 122(6):477–483.

Adams, V.D., Watts, R.J., and Pitts, M.E. 1986. Organics. *J. Water Pollut. Control Fed.* 58:449–471.

Adamson, A.W. 1982. *Physical Chemistry of Surfaces.* 4th ed. John Wiley & Sons, New York.

Ademoroti, C.M. 1985. Integrated biological/chemical wastewater treatment. *Effluent Water Treat. J.* 25:237–241.

Aelion, C.M., Widdowson, M.A., Ray, R.P., Reeves, H.W., and Shaw, J.N. 1995. Biodegradation, vapor extraction, and air sparging in low-permeability soils. In In situ *Aeration: Air Sparging, Bioventing, Relat. Rem. Processes (Pap. 3rd Int.* in Situ *On-Site Bioreclam. Symp.* Hinchee, R.E., Miller, R.N., and Johnson, P.C., Eds. Battelle Press, Columbus, OH. pp. 127–134.

AeroVironment, Inc. 1988. Report prepared for Naval Civil Engineering Laboratory, Port Hueneme, CA, on Contract No. N62474-87-C-3062.

Aggarwal, P.K. and Hinchee, R.E. 1991. Monitoring *in situ* biodegradation of hydrocarbons by using stable carbon isotopes. *Environ. Sci. Technol.* 25(6):1178–1180.

Aggarwal, P.K., Means, J.L., and Hinchee, R.E. 1991. Formulation of nutrient solutions for *in situ* bioremediation. In *In Situ Bioreclamation. Applications and Investigations for Hydrocarbon and Contaminated Site Remediation.* Hinchee, R.E. and Olfenbuttel, R.F., Eds. Butterworth-Heinemann, Stoneham, MA. pp. 51–66.

Aggarwal, P.K., Means, J.L., Hinchee, R.E., Headington, G.L., Gavaskar, A.R., Scowden, C.M., Arthur, M.F., Evers, D.P., and Bigelow, T.L. 1990. Methods to select chemicals for *in-situ* biodegradation of fuel hydrocarbons. Final report to the Air Force Engineering and Services Center on Contract FO8635-85-C-0122, subtask 3.05.

Ahearn, D.G., Meyers, S.P., and Standard, P.G. 1971. The role of yeasts in the decomposition of oils in marine environments. *Dev. Ind. Microbiol.* 12:126–134.

Ahlert, R.C., Black, W., Bruger, J., Kosson, D., and Suenter, J. 1990. In *Remedial Action, Treatment, and Disposal of Hazardous Waste. Proc. 16th Annu. RREL Haz. Waste Res. Symp.* U.S. EPA, Washington, D.C. pp. 536–547.

Ahlert, R.C. and Kosson, D.S. 1983. *In Situ* and On-Site Biodegradation of Industrial Landfill Leachate. Report to Dept. of the Interior, Washington, D.C., on Contract No. 14-34-0001-1132. PB84-136787.

Ahlrichs, J.L. 1972. The soil environment. In Goring, C.A.I. and Hamaker, J.W., Eds. *Organic Chemicals in the Soil Environment.* Vol. 1. Marcel Dekker, New York.

Ahring, B.K. and Westerman, P. 1985. Methanogenesis from acetate: physiology of a thermophilic, acetate-utilizing methanogenic bacterium. *FEMS Microbiol. Lett.* 28:15–19.

Alberti, B.N. and Klibanov, A.M. 1981. Enzymatic removal of dissolved aromatics from industrial aqueous effluents. In *Biotech. Bioeng. Symp.,* No. 11. John Wiley & Sons, New York. pp. 373–379.

Albrechtsen, H.-J. 1994. Bacterial degradation under iron-reducing conditions. In *Hydrocarbon Bioremediation.* Hinchee, R.E., Alleman, B.C., Hoeppel, R.E., and Miller, R.N., Eds. CRC Press, Boca Raton, FL. pp. 418–423.

Alexander, M. 1971. *Microbial Ecology.* John Wiley & Sons, New York. 511 pp.

Alexander, M. 1973. Nonbiodegradable and other recalcitrant molecules. *Biotech. Bioeng.* 15:611–647.

Alexander, M. 1977. *Introduction to Soil Microbiology.* 2nd ed. John Wiley & Sons, New York. 467 pp.

Alexander, M. 1979. In *Microbial Degradation of Pollutants in Marine Environments.* Bourquin, A.W. and Pritchard, P.H., Eds. U.S. EPA, Gulf Breeze, FL. pp. 67–75.

Alexander, M. 1980a. Biodegradation of chemicals of environmental concern. *Science.* 211:132–138.
Alexander, M. 1980b. Biodegradation of toxic chemicals in water and soil. In *Dynamics, Exposure and Hazard Assessment of Toxic Chemicals.* Haque, R., Ed. Ann Arbor Science, Ann Arbor, MI. pp. 179–190.
Alexander, M. 1981. Biodegradation of chemicals of environmental concern. *Science.* 211:132–138.
Alexander, M. 1985. Biodegradation of organic chemicals. *Environ. Sci. Technol.* 19:106–111.
Alexander, M. 1986. Biodegradation of Chemicals at Trace Concentrations. Report to Army Research Office on Grant No. DAAG29-83-K-0068.
Alexander, M. 1994. *Biodegradation and Bioremediation.* Academic Press, San Diego, CA. 302 pp.
Al-Hadhrami, M.N., Lappin-Scott, H.M., and Fisher, P.J. 1996. Effects of the addition of organic carbon sources on bacterial respiration and *n*-alkane biodegradation of Omani crude oil. *Mar. Pollut. Bull.* 32(4):351–357.
Alleman, B.C., Hinchee, R.E., Brenner, R.C., and McCauley, P.T. 1995. Bioventing PAH contamination at the Reilly tar site. In In Situ *Aeration: Air Sparging, Bioventing, Relat. Rem. Processes. Pap. 3rd Int.* In Situ *On-Site Bioreclam. Symp.* Hinchee, R.E., Miller, R.N., Johnson, P.C., Eds. Battelle Press, Columbus, OH. pp. 473–482.
Allen, C.C. and Blaney, B.L. 1985. Techniques for Treating Hazardous Wastes to Remove Volatile Organic Constituents. EPA-600/D-85/127. PB 85218782.
Allen, C.C. and Brant, G. 1984. EPA Contract No. 68-03-3149. Task 25–1.
Allman, R., Hann, A.C., Phillips, A.P., Martin, K.L., and Lloyd, D. 1990. Growth of *Azotobacter vinelandii* with correlation of coulter cell size, flow cytometric parameters and ultrastructure. *Cytometry.* 11:822–831.
Allred, B. and Brown, G.O. 1996. Anionic surfactant transport characteristics in unsaturated soil. *Soil Sci.* 161(7):415–425.
Alteriis, E.D., Scardi, V., Masi, P., and Parascandola, P. 1990. Mechanical stability and diffusional resistance of a polymeric gel used for biocatalysts immobilization. *Enzyme Microbiol. Technol.* 12:539–545.
Alvares, A.P. 1981. Cytochrome P-450s: research highlights of the last two decades. *Drug Metab. Rev.* 12:431–436.
Amadi, A., Abbey, S.D., and Nima, A. 1996. Chronic effects of oil spill on soil properties and microflora of a rainforest ecosystem in Nigeria. *Water Air Soil Pollut.* 86(1–4):1–11.
Amann, R.I., Stromley, J., Devereux, R., Key, R., and Stahl, D.A. 1992. Molecular and microscopic identification of sulfate-reducing bacteria in multispecies biofilms. *Appl. Environ. Microbiol.* 58:614–623.
Amann, R.I., Zarda, B., Stahl, D.A., and Schleifer, K.-H. 1992. Identification of individual prokaryotic cells by using enzyme-labeled, rRNA-targeted oligonucleotide probes. *Appl. Environ. Microbiol.* 58:3007–3011.
Amdur, J. and Clark-Clough, A. 1994. Guidelines for selecting the bioremediation option. In *Proc. HAZMACON 94, 11th Haz. Mat. Mgmt. Conf. Exhib.* pp. 190–197.
Amdurer, M., Fellman, R., and Abdelhamid, S. 1985. *In situ* treatment technologies and superfund. In *Proc. Int. Conf. on New Frontiers for Haz. Waste Mgmt.,* Sept. 1985. EPA Report No. EPA-600/9-85/025. Hazardous Waste Engineering Research Laboratory, Cincinnati, OH.
Amend, L.J. and Lederman, P.B. 1991. Critical evaluation of PCB remediation technologies. Presented at Amer. Inst. Chem Eng. 1991 Summer Natl. Mtg., Aug. 1991, Pittsburgh.
American Conference of Governmental Industrial Hygienists. 1976. *Threshold Limit Values for 1976.* ACGIH, Cincinnati, OH.
American Conference of Governmental Industrial Hygienists. 1982. *Industrial Ventilation: A Manual of Recommended Practice.* 17th ed. Committee on Industrial Ventilation, Lansing, MI.
American Petroleum Institute. May, 1983 Report: Treatment Technology for Removal of Gasoline Components from Ground Water. Vols. 1 and 2. Washington, D.C.
American Petroleum Institute. 1990. *Petroleum Release Decision Framework.*
Amiran, M.C. and Wilde, C.L. 1994. Remediation of contaminated soil and sediment using the BioGenesis washing process. *Hydrocarbon Contam. Soils.* 4:425–437.
Anastos, G., Corbin, M.H., and Coia, M.F. 1986. In *7th Natl. Conf. Mgmt. Uncontrolled Haz. Waste Sites,* Dec. 1–3, 1986, Washington, D.C.
Andersen, S. and Engelstad, F. 1993. Application of factorial designs: estimating the pollution potential inferred from changes in soil water chemistry. *Soil Environ.* (Integrated Soil and Sediment Research: A Basis for Proper Protection). 1:227–130.
Anderson, A.C. and Abdelghani, A.A. 1980. Toxicity of selected arsenical compounds in short-term bacterial bioassays. *Bull. Environ. Contam. Toxicol.* 24:124–127.
Anderson, D.B., Peyton, B.M., Liskowitz, J.J., Fitzgerald, C., and Schuring, J.R. 1995. Enhancing *in situ* bioremediation with pneumatic fracturing. In *Applied Bioremediation of Petroleum Hydrocarbons.* Hinchee, R.E., Kittel, J.A., and Reisinger, H.J., Eds. Battelle Press, Columbus, OH. pp. 467–473.
Anderson, M.A. 1994. Interfacial tension-induced transport of nonaqueous phase liquids in model aquifer systems. *Water Air Soil Pollut.* 75(1–2):51–60.
Andreoni, V., Finoli, C., Manfrin, P., Pelosi, M., and Vecchio, A. 1991. Studies on the accumulation of cadmium by a strain of *Proteus mirabilis. FEMS Microbiol. Ecol.* 85:183–192.
Andrews, A.R. and Floodgate, G.D. 1974. Some observations on the interactions of marine protozoa and crude oil residues. *Mar. Biol.* 25:7–12.

REFERENCES

Anenson, T.B. 1995. Biological forced air soil treatment (BIOFAST) system: a field application. In *Proc. HAZMACON 95, 12th Haz. Mat. Mgmt. Conf. Exhib.* pp. 344–353.

Ang, C.C. and Abdul, A.S. 1994. Evaluation of an ultrafiltration method for surfactant recovery and reuse during *in situ* washing of contaminated sites: laboratory and field studies. *Ground Water Monit. Rem.* 14(3):160–171.

Annokkee, G.J. 1990. MT-TNO research into the biodegradation of soils and sediments contaminated with oils and polyaromatic hydrocarbons (PAHs). In *Contaminated Soil 1990*. Wolf, K., Van de Brink, J., and Colon, F.J., Eds. Kluwer, Amsterdam. pp. 941–945.

Anonymous. 1981. Pyrolysis bids for hazardous-waste jobs. *Chem. Week.* 128:40–41.

Appanna, V.D., Finn, H., and St. Pierre, M. 1995. Microbial response to multiple-metal stress. In *Environmental Biotechnology: Principles and Applications*. Moo-Young, M., Anderson, W.A., and Chakrabarty, A.M., Eds. Kluwer, Amsterdam. pp. 105–113.

Aprill, W. and Sims, R.C. 1990. Evaluation of the use of prairie grasses for stimulating polycyclic aromatic hydrocarbon treatment in soil. *Chemosphere.* 20:253–265.

Arcangeli, J.-P. and Arvin, E. 1995. Cometabolic transformation of *o*-xylene in a biofilm system under nitrate reducing conditions. *Biodegradation.* 6:19–27.

Archer, D.B. 1985. Uncoupling of methanogenesis from growth of *Methanosarcina barkeri* by phosphate limitation. *Appl. Environ. Microbiol.* 50:1233–1237.

Ariga, O., Itoh, K., Sano, Y., and Nagura, M. 1994. Encapsulation of biocatalyst with PVA capsules. *J. Ferment. Bioeng.* 78(1):74–78.

Ariga, O., Saito, M., and Sano, Y. 1995. Diauxie in immobilized microorganisms. *J. Ferment. Bioeng.* 79(5):519–521.

Aronstein, B.N. and Alexander, M. 1993. Effect of a non-ionic surfactant added to the soil surface on the biodegradation of aromatic hydrocarbons within the soil. *Appl. Microbiol. Biotechnol.* 39(3):386–390.

Aronstein, B.N., Calvillo, Y.M., and Alexander, M. 1991. Effect of surfactants at low concentrations on the desorption and biodegradation of sorbed aromatic compounds in soil. *Environ. Sci. Technol.* 25(10):1728–1731.

Arora, H.S., Cantor, R.R., and Nemeth, J.C. 1982. Land treatment: a viable and successful method of treating petroleum industry wastes. *Environ. Int.* 7:285–291.

Arthur D. Little, Inc. 1976. State-of-the-Art Survey of Land Reclamation Technology. Report No. EC-CR-76-76 on Contract No. DAAA 15-75-C-0188. Dept. of the Army, Edgewood Arsenal. Aberdeen Proving Ground, MD. 105 pp.

Arthur, M.F., O'Brien, J.G.K., Marsh, S.S., and Zwick, T.C. 1992. Evaluation of Innovative Approaches to Stimulate Degradation of Jet Fuels in Subsoils and Groundwater. Report NCEL-CR92-004; Order No. AD-A252359; Available NTIS. From *Gov. Rep. Announce. Index* (U.S.). 92(20), Abstr. No. 256,903.

Arulgnanendran, V.R.J. and Nirmalakhandan, N. 1995. Determining cleanup levels in bioremediation: quantitative structure activity relationship techniques. *Monit. Verif. Biorem. 3rd Int.* In Situ *On-Site Bioreclam. Symp.* Hinchee, R.E., Douglas, G.S., and Ong, S.K., Eds. Battelle Press, Columbus, OH. pp. 165–174.

Arvin, E., Godsy, E.M., Grbic-Galic, D., and Jensen, B.K. 1988. Microbial degradation of oil and creosote related aromatic compounds under aerobic and anaerobic conditions. *Int. Conf. Physicochem. Biol. Detoxification Haz. Wastes,* May 3–5, 1988. Atlantic City, NJ.

Ascon-Cabrera, M.A. and Lebeault, J.-M. 1995. Interfacial area effects of a biphasic aqueous/organic system on growth kinetic of xenobiotic-degrading microorganisms. *Appl. Microbiol. Biotechnol.* 43:1136–1141.

Ascon-Cabrera, M.A., Thomas, D., and Lebeault, J.-M. 1995. Activity of synchronized cells of a steady-state biofilm recirculated reactor during xenobiotic biodegradation. *Appl. Environ. Microbiol.* 61(3):920–925.

Ashok, B.T., Saxena, S., and Musarrat, J. 1995. Isolation and characterization of four polycyclic aromatic hydrocarbon degrading bacteria from soil near an oil refinery. *Lett. Appl. Microbiol.* 21:246–248.

Assink, J.W. and Rulkens, W.W. 1984. In *5th Natl. Conf. on Mgmt. of Uncontrolled Haz. Waste Sites Conf.,* Nov. 7–9, 1984, Washington, D.C. pp. 576–583.

Atlas, R.M. 1975. Effects of temperature and crude oil composition on petroleum biodegradation. *Appl. Microbiol.* 30:396.

Atlas, R.M. 1977. Stimulated petroleum biodegradation. *CRC Crit. Rev. Microbiol.* 5:371–386.

Atlas, R.M. 1978a. Microorganisms and petroleum pollutants. *BioScience.* 28:387–391.

Atlas, R.M. 1978b. In *Native Aquatic Bacteria: Enumeration, Activity and Ecology.* Costerton, J.W. and Colwell, R.R., Eds. ASTM-STP 695. American Association of Testing and Materials, Philadelphia.

Atlas, R.M. 1981. Microbial degradation of petroleum hydrocarbons: an environmental perspective. *Microbiol. Rev.* 45:180–209.

Atlas, R.M. 1984. *Petroleum Microbiology.* Macmillan, New York.

Atlas, R.M. 1991. Bioremediation of fossil fuel contaminated soils. In In Situ *Bioreclamation. Applications and Investigations for Hydrocarbon and Contaminated Site Remediation.* Hinchee, R.E. and Olfenbuttel, R.F., Eds. Butterworth-Heinemann, Stoneham, MA. pp. 14–33.

Atlas, R.M. 1992. Molecular methods for environmental monitoring and containment of genetically engineered microorganisms. *Biodegradation.* 3:137–146.

Atlas, R.M. 1994. Microbial hydrocarbon degradation-bioremediation of oil spills. *J. Chem. Tech. Biotechnol.* 52:149–156.

Atlas, R.M. 1995. Bioremediation of petroleum pollutants. *Int. Biodeterior. Biodeg.* 35(1–3):317–327.

Atlas, R.M. and Bartha, R. 1972. Biodegradation of petroleum in seawater at low temperatures. *Can. J. Microbiol.* 18:1851–1855.
Atlas, R.M. and Bartha, R. 1973a. Fate and effects of polluting petroleum on the marine environment. *Residue Rev.* 49:49–85.
Atlas, R.M. and Bartha, R. 1973b. Inhibition by fatty acids of the biodegradation of petroleum. *Antonie van Leeuwenhoek J. Microbiol. Serol.* 39:257–271.
Atlas, R.M. and Bartha, R. 1973c. Effects of some commercial oil herders, dispersants and bacterial inocula on biodegradation of oil in seawater. In *The Microbial Degradation of Oil Pollutants, Workshop,* Dec. 4–6, 1972, Atlanta, GA. Ahearn, D.G. and Meyers, S.P., Eds. Pub. No. LSU-SG-73-01. Louisiana State University, Center for Wetland Resources, Baton Rouge, LA. pp. 283–289.
Atlas, R.M. and Bartha, R. 1973d. Stimulated biodegradation of oil slicks using oleophilic fertilizers. *Environ. Sci. Technol.* 13:538–541.
Atlas, R.M. and Bartha, R. 1973e. Abundance, distribution and oil-biodegrading potential of microorganisms in Raritan Bay. *Environ. Pollut.* 4:291–300.
Atlas, R.M. and Bartha, R. 1981. *Microbial Ecology; Fundamentals and Applications.* Addison-Wesley, Reading, MA. 560 pp.
Atlas, R.M. and Bartha, R. 1987. *Microbial Ecology — Fundamentals and Applications.* 2nd ed. Benjamin Cummings, Menlo Park, CA.
Atlas, R.M. and Bartha, R. 1993. *Microbial Ecology: Fundamentals and Applications.* Benjamin/Cummings, Redwood City, CA.
Atlas, R.M., Boehn, P.D., and Calder, J.A. 1981. *Est. Coastal Shelf Sci.* 12:598–608.
Atlas, R.M. and Cerniglia, C.E. 1995. Bioremediation of petroleum pollutants. *BioScience.* 45(5):332–338.
Atlas, R.M. and Schofield, E.A. 1975. Petroleum biodegradation in the Arctic. In *Proc. Impact of the Use of Microorganisms on the Aquatic Environment.* Bourquin, Q.W., Ahearn, D.G., and Meyers, S.P., Eds. EPA Report No. EPA 660-3-75-001. Environmental Protection Agency, Corvallis, OR. p. 185.
Augustijn, D.C.M., Jessup, R.E., Rao, P.S.C., and Wood, A.L. 1994. Remediation of contaminated soils by solvent flushing. *J. Environ. Eng.* (N.Y.). 120(1):42–57.
Auria, R., Christen, P., Favela, E., Gutierrez, M., Guyot, J.P., Monroy, O., Revah, S., Roussos, S., Saucedo-Castaneda, G., and Viniegra-Gonzalez, G. 1995. Biotreatment of liquid, solid or gas residues: an integrated approach. In *Environmental Biotechnology: Principles and Applications.* Moo-Young, M., Anderson, W.A., and Chakrabarty, A.M., Eds. Kluwer, Amsterdam. pp. 221–236.
Austin American Statesman. Nov. 19, 1980.
AWWA Research Foundation. 1989. *Advanced Oxidation Processes for Control of Off-Gas Emissions from VOC Stripping.* The Foundation, Denver, CO.
Azab, M.S. and Peterson, P.J. 1989. The removal of cadmium from water by the use of biological sorbents. *Water Sci. Technol.* 21(12):1705–1706.
Azam, F. and Holm-Hansen, O. 1973. Use of tritiated substrates in the study of heterotrophy in seawater. *Mar. Biol.* 23:191–196.
Azarowicz, E.M. 1973. Microbial degradation of petroleum. *Off. Gaz. U.S. Patent Office.* 915, 1835.

B

Baas Becking, L.G.M., Kaplan, I.R., and Moore, D. 1960. Limits of the natural environment in terms of pH and oxidation-reduction potentials. *J. Geol.* 68:243–284.
Baath, E. 1989. Effects of heavy metals in soil on microbial processes and populations (a review). *Water Air Soil Pollut.* 47:335–379.
Babel, W. 1994. Bioremediation of ecosystems by micro-organisms. Approaches for exploiting upper limits and widening bottlenecks. In *Bioremediation: the Tokyo '94 Workshop.* Organization of Economic Co-operation and Development, Paris. pp. 101–115.
Babich, H. and Stotzky, G. 1983a. Nickel toxicity to microbes and a bacteriophage in soil and aquatic ecosystems: mediation by environmental characteristics. *Abstr. Annu. Mtg. Am. Soc. Microbiol.* p. 261.
Babich, H. and Stotzky, G. 1983b. Influence of chemical speciation on the toxicity of heavy metals to the microbiota. In *Aquatic Toxicology.* Nriagu, J.O., Ed. John Wiley & Sons, New York. pp. 1–46.
Bache, R. and Pfennig, N. 1981. Selective isolation of *Acetobacterium woodii* on methoxylated aromatic acids and determination of growth yields. *Arch. Microbiol.* 130:255–261.
Bae, H.C., Cota-Robles, E.H., and Casida, L.E. 1972. Microflora of soil as viewed by transmission electron microscopy. *Appl. Microbiol.* 23:637–648.
Bair, T.I. and Camp, C.E. 1995. Adsorbent biocatalyst porous beads. PCT Int. ppl. WO 95 23,768 (Cl.C02F3/10), 8 Sept. 1995. U.S. Appl. 205,689. 3 Mar. 1994. 23 pp.
Baker, J.M. 1970. The effect of oils on plants. *Environ. Pollut.* 1:27–44.
Baker, M.D. and Mayfield, C.I. 1980. Microbial and nonbiological decomposition of chlorophenols and phenols in soil. *Water Air Soil Pollut.* 13:411–424.

REFERENCES

Baker, R.S. and Bierschenk, J. 1995. Vacuum-enhanced recovery of water and NAPL: concept and field test. *J. Soil Contam.* 4(1):57–76.

Baker, R.S. and Bierschenk, J. 1996. *Contam. Pollut. Eng.* March 1996.

Bakken, L.R. and Olsen, R.A. 1987. The relationship between cell size and viability of soil bacteria. *Microb. Ecol.* 103–114.

Bakker, G. 1977. Anaerobic degradation of aromatic compounds in the presence of nitrate. *FEMS Microbiol. Lett.* 1:103–108.

Balba, M.T. and Evans, W.C. 1977. The methanogenic fermentation of aromatic substrates. *Biochem. Soc. Trans.* 5:302–304.

Balba, M.T. and Evans, W.C. 1980. The anaerobic dissimilation of benzoate by *Pseudomonas aeruginosa* coupled with *Desulfovibrio vulgaris* with sulphate as terminal electron acceptor. *Biochem. Soc. Trans.* 8:624–625.

Balderston, W.L., Sherr, B., and Payne, W.J. 1976. Blockage by acetylene of nitrous oxide reduction in *Pseudomonas perfectomarinus*. *Appl. Environ. Microbiol.* 31:504–508.

Balkwill, D.L. and Ghiorse, W.C. 1982. Characterization of subsurface microorganisms. *Abstr. Annu. Mtg. Am. Soc. Microbiol.* p. 192.

Ballester, A. and Castellvi, J. 1980. Bioaccumulation of V and Ni by marine organisms and sediments. *J. Invest. Pesquera.* 44:1–12.

Bambrick, D.R. 1985. The effect of DTPA on reducing peroxide composition. *Tappi J.* 68:96–100.

Ban, T. and Yamamoto, S. 1993. Adhesion of microbial cells to porous hydrophilic and hydrophobic solid substrata. *Dev. Petrol. Sci.* (Microbial Enhancement of Oil Recovery: Recent Advances). 39:159–169.

Banat, I.M. 1993. The isolation of a thermophilic biosurfactant-producing *Bacillus* sp. *Biotechnol. Lett.* 15(6):591–594.

Banat, I.M., Samarah, N., Murad, M., Horne, R., and Banerjee, S. 1991. Biosurfactant production and use in oil tank cleanup. *World J. Microbiol. Biotechnol.* 7:80–88.

Bar, R. 1990. A new cyclodextrin-agar medium for surface cultivation of microbes on lipophilic substrates. *Appl. Microbiol. Biotechnol.* 32:470–472.

Barker, J.F. and Patrick, G.C. 1985. Natural attenuation of aromatic hydrocarbons in a shallow sand aquifer. In *Conf. Petrol. Hydro. and Organ. Chem. in Groundwater.* pp. 160–177.

Barkes, L. and Fleming, R.W. 1974. Production of dimethylselenide gas from inorganic selenium by eleven soil fungi. *Bull. Environ. Contam. Toxicol.* 12:308–311.

Bartha, R. 1986. Biotechnology of petroleum pollutant biodegradation. *Microb. Ecol.* 12:155–172.

Bartha, R. and Atlas, R.M. 1977. The microbiology of aquatic oil spills. *Adv. Appl. Microbiol.* 22:225–266.

Bartha, R. and Atlas, R.M. 1987. In *Long-Term Environmental Effects of Offshore Oil and Gas Developments.* Boesch, D.F. and Rabalais, N.N., Eds. Elsevier, New York. pp. 287–341.

Basseres, A., Eyraud, P., and Ladousse, A. 1994. Nutrient additive for biodegradation of materials such as hydrocarbons and oil spills. PCT Int. Appl. WO 9405,773. 17 Mar 1994.

Battaglia, A. and Morgan, D.J. 1994. *Ex situ* forced aeration of soil piles: a physical model. *Environ. Prog.* 13(3):178–187.

Battelle. 1995. Test Plan and Technical Protocol for Bioslurping. Report prepared by Battelle Columbus Operations for the U.S. Air Force Center for Environmental Excellence, Brooks Air Force Base, TX.

Batterman, G. 1986. In *1985 Int. TNO Cong. Contam. Soil.* Assink, J.W. and van den Brink, W.J., Eds. Nijhoff, Dordrecht. pp. 711–722.

Batterman, S., Kulshrestha, A., and Cheng, H.Y. 1995. Hydrocarbon vapor transport in low moisture soils. *Environ. Sci. Technol.* 29(1):171–180.

Batterman, G. and Werner, P. 1984. Beseitigung einer Untergrundkontamination mit Kohlenwasserstoffen durch Mikorbiellen Abbau. *Grundwasserforschung-Wasser/Abwasser.* 125:366–373.

Batterton, J., Winters, K., and van Baalen, C. 1978. Anilines: selective toxicity to blue-green algae. *Science.* 199:1068–1070.

Baud-Grasset, F. and Vogel, T.M. 1995. Bioaugmentation: biotreatment of contaminated soil by adding adapted bacteria. In *Bioaugmentation for Site Remediation. Pap. 3rd Int. In Situ On-Site Bioreclam. Symp.*. Hinchee, R.E., Fredrickson, J., and Alleman, B.C., Eds. Battelle Press, Columbus, OH. pp.39–48.

Bauer, J.E. and Capone, D.G. 1985. Degradation and mineralization of the polycyclic aromatic hydrocarbons anthracene and naphthalene in intertidal marine sediments. *Appl. Environ. Microbiol.* 50:81–90.

Baugh, A.L. and Lovegreen, J.R. 1990. Differentiation of crude oil and refined petroleum products in soil. In *Petroleum Contaminated Soils.* Vol. 3. Calabrese, E.J., Kostecki, P.T., and Bell, C.E., Eds. Lewis Publishers, Chelsea, MI. pp. 141–149.

Baum, R. 1988. Low-cost cleanup of toxic petrochemicals. *Chem. Eng. News.* 66:24–25.

Bausum, H.T. and Taylor, G.W. 1986. Literature Survey and Data Base Assessment: Microbial Fate of Diesel Fuels and Fog Oils. U.S. Army Medical Bioengineering Research and Development Laboratory, Fort Detrick, Frederick, MD. Technical Report 8408. ADA167799.

Baver, L.D., Gardner, W.H., and Gardner, W.R. 1972. *Soil Physics.* 4th ed. John Wiley & Sons, New York. 498 pp.

Beam, H.W. and Perry, J.J. 1973. Co-metabolism as a factor in microbial degradation of cycloparaffinic hydrocarbons. *Arch. Mikrobiol.* 91:87–90.

Beam, H.W. and Perry, J.J. 1974. Microbial degradation of cycloparaffinic hydrocarbons via cometabolism and commensalism. *J. Gen. Microbiol.* 82:163–169.

Becher, P. 1965. *Emulsions: Theory and Practice.* Reinhold, New York.

Becker, S. and Garzonetti, G.A. 1985. Primary effluent filtration: the Warminster experience. *Public Works.* 116:54–56.

Bedient, P.B., Springer, N.K., Baca, E., Bouvette, T.C., Hutchins, S.R., and Tomson, M.B. 1983. Groundwater transport from wastewater infiltration. *J. Environ. Eng.* 109:485–501.

Beeman, R.E., Howell, J.E., Shoemaker, S.H., Salazar, E.A., and Buttram, J.R. 1993. A field evaluation of *in situ* microbial reductive dehalogenation by the biotransformation of chlorinated ethenes. In *Bioremediation of Chlorinated and Polycyclic Aromatic Compounds.* Hinchee, R.E., Leeson, A., Semprini, L., and Ong., S.K., Eds. Lewis Publishers, Ann Arbor, MI.

Bell, A.A. and Wheeler, M.H. 1986. Biosynthesis and functions of fungal melanins. *Annu. Rev. Phytopathol.* 24:411–451.

Bell, R.L., Morrison, B.J., and Chonnard, D.R. 1987. Research and Development of Methods for the Engineering Evaluation and Control of Toxic Airborne Effluents. California Air Resources Board Report No. ARB/R-87/312. Sacramento, CA.

Belyaev, S.S., Borzenkov, I.A., Milekhina, E.I., Zvyagintseva, I.S., and Ivanov, M.V. 1993. *Dev. Petrol. Sci.* 39:79–88.

Benazon, N., Belanger, D.W., Scheurlen, D.B., and Lesky, M.J. 1995. Bioremediation of ethylbenzene- and styrene-contaminated soil using biopiles. In *Biological Unit Processes for Hazardous Waste Treatment.* Hinchee, R.E., Sayles, G.D., and Skeen, R.S., Eds. Battelle Press, Columbus, OH. pp. 179–190.

Bender, J., Vatcharapijarn, Y., and Russell, A. 1989. Fish feeds from grass clippings. *Aquacult. Eng.* 8:407–419.

Bennedsen, M.B. 1987. Vacuum VOCs from soil. *Pollut. Eng.* 19:66–68.

Bennett, P.G. and Jeffers, T.H. 1990. Removal of metal contaminants from a waste stream using Bio-Fix beads containing sphagnum moss. In *Min. Mineral Process Wastes Proc. Waste Reg. Symp.* Doyle, F.M., Ed. Society for Mineral and Metallurgical Exploration, Littleton, CO. pp. 279–286.

Benoit, R., Novak, J., Goldsmith, C., and Chadduck, J. 1985. Alcohol biodegradation in groundwater microcosms and pure culture systems. In *Abstr. Annu. Mtg. Am. Soc. Microbiol.* p. 258.

Benson, D.A., Huntley, D., and Johnson, P.C. 1993. Modeling vapor extraction and general transport in the presence of NAPL mixtures and nonideal conditions. *Ground Water.* 31(3):437–445.

Benson, L.A., Frishmuth, R.A., and Downey, D.C. 1995. Methodology to develop risk-based corrective actions at petroleum–hydrocarbon contaminated sites. *Proc. 65th Water Environ. Fed. Ann. Conf. Expo.* 2:377-387.

Bergman, T.J., Jr., Greene, J.M., and Davis, T.R. 1994. Use of pure oxygen dissolution system enhances *in situ* slurry-phase bioremediation. In *Applied Biotechnology for Site Remediation.* Hinchee, R.E., Anderson, D.B., Metting, F.B., Jr., and Sayles, G.D., Eds. Lewis Publishers, Boca Raton, FL. pp. 379–383.

Berry, D.F. and Boyd, S.A. 1984. Division S-3 — soil microbiology and biochemistry. Oxidative coupling of phenols and anilines by peroxidase: structure–activity relationships. *Soil Sci. Soc. Am. J.* 48:565–569.

Berry, D.F. and Boyd, S.A. 1985. Decontamination of soil through enhanced formation of bound residues. *Environ. Sci. Technol.* 19:1132–1133.

Berry, D.F., Francis, A.J., and Bollag, J.-M. 1986. Microbial Metabolism of Aromatic Compounds under Anaerobic Conditions. DOE Report DOE/ER-0289. DE86014211.

Berry, D.F., Francis, A.J., and Bollag, J.-M. 1987. Microbial metabolism of homocyclic and heterocyclic aromatic compounds under anaerobic conditions. *Microbiol. Rev.* 51:43–59.

Berry-Spark, K.L., Barker, J.F., Major, D., and Mayfield, C.I. 1986. In *Proc. Pet. Hydrocarbons Org. Chem. Ground Water: Prevention, Detection, Restoration.* NWWA/API. Water Well Journal Publishing Co., Dublin, OH. pp. 613–623.

Berthelet, M. and Greer, C.W. 1995. Detection of catabolic genes in soil using the polymerase chain reaction. In *Environmental Biotechnology: Principles and Applications.* Moo-Young, M., Anderson, W.A., and Chakrabarty, A.M., Eds. Kluwer, Amsterdam. pp. 635–644.

Beveridge, T.J. 1989. The role of cellular design in bacterial metal accumulation and mineralization. *Annu. Rev. Microbiol.* 43:147–171.

Beveridge, T.J. and Doyle, R.J., Eds. 1989. Metal Ions and Bacteria. John Wiley & Sons, New York.

Beveridge, T.J. and Fyfe, W.S. 1985. Metal fixation by bacterial cell walls. *Can. J. Earth Sci.* 22:1893–1898.

Beveridge, T.J. and Murray, R.G.E. 1980. Sites of metal deposition in the cell wall of *Bacillus subtilis. J. Bacteriol.* 141:876–887.

Bhandari, A., Dove, D.C., and Novak, J.T. 1994. Soil washing and biotreatment of petroleum-contaminated soils. *J. Environ. Eng.* 120(5):1151–1169.

Bhargava, D.S. and Datar, M.T. 1984. Nitrification during aerobic digestion of activated sludge. *Effluent Water Treat. J.* (G.B.). 24:352–355.

Binnerup, S.J., Jensen, D.F., Thordal-Christensen, H., and Sorensen, J. 1993. Detection of viable, but non-culturable *Pseudomonas fluorescens* DF57 in soil using a microcolony epifluorescence technique. *FEMS Microbiol. Ecol.* 12:97–105.

Biosystems, Inc. 1986. Giving nature a boost. *Dupont Mag.* May/June.

Birman, I. and Alexander, M. 1996a. Optimizing biodegradation of phenanthrene dissolved in nonaqueous-phase liquids. *Appl. Microbiol. Biotechnol.* 45(1–2):267–272.

Birman, I. and Alexander, M. 1996b. Effect of viscosity of nonaqueous-phase liquids (NAPLs) on biodegradation of NAPL constituents. *Environ. Toxicol. Chem.* 15(10):1683–1686.

Bishop, D.F. and Govind, R. 1995. Development of novel biofilters for treatment of volatile organic compounds. In *Biological Unit Processes for Hazardous Waste Treatment. Pap. 3rd Int. In Situ On-Site Bioreclam. Symp.* Hinchee, R.E., Skeen, R.S., and Sayles, G.D., Eds. Battelle Press, Columbus, OH. pp. 219–226.

REFERENCES

Bitton, G., Davidson, J.M., and Farrah, S.R. 1979. On the value of soil columns for assessing the transport pattern of viruses through soils: a critical outlook. *Water Air Soil Pollut.* 12:449–457.

Bitton, G. and Gerba, C.P., Eds. 1985. *Groundwater Pollution Microbiology*. John Wiley & Sons, New York.

Black, W.V., Ahlert, R.C., Kosson, D.S., and Brugger, J.E. 1991. Slurry-based biotreatment of contaminants sorbed onto soil constituents. In *On-Site Bioreclamation: Processes for Xenobiotic and Hydrocarbon Treatment*. Hinchee, R.E. and Olfenbuttel, R.F., Eds. Butterworth-Heinemann, Stoneham, MA. pp. 408–422.

Blackburn, J.W. 1985. EPA/68-02-6919 (Draft). U.S. EPA, Cincinnati, OH.

Blackburn, J.W. 1987. Prediction of organic chemical fates in biological treatment systems. *Environ. Prog.* 6 (4):217–223.

Blackburn, J.W., Lee, M.K., and Horn, W.C. 1995. Enhanced treatment of refinery soils with "open system" slurry reactors. In *Biological Unit Processes for Hazardous Waste Treatment. Pap. 3rd. Int.* In Situ On-Site Bioreclam. Symp. Hinchee, R.E., Skeen, R.S., and Sayles, G.D., Eds. Battelle Press, Columbus, OH. pp. 137–144.

Blackburn, J.W., Troxler, W.L., and Sayler, G.S. 1984. Prediction of the fates of organic chemicals in a biological treatment process — an overview. *Environ. Prog.* 3:163–176.

Blakebrough, N. 1978. Interactions of oil and microorganisms in soil. In *The Oil Industry and Microbial Ecosystems*. Charter, K.W.A. and Somerville, H.J., Eds. Heyden and Sons, Institute of Petroleum, London. pp. 28–40.

Blanch, H.W. and Einsele, A. 1973. The kinetics of yeast growth on pure hydrocarbons. *Biotechnol. Bioeng.* 15:861.

Blanchar, R.W. and Hossner, L.R. 1969. Hydrolysis and sorption of ortho-, pyro-, tripoly-, and trimetaphosphate in 32 midwestern soils. *Soil Sci. Soc. Am. Proc.* 33:622–625.

Blanchar, R.W. and Riego, D.C. 1976. Tripolyphosphate and pyrophosphate hydrolysis in sediments. *Soil Sci. Soc. Am. J.* 225–229.

Blaney, B.L., Eklund, B.M., Thorneloe, S.A., and Wetherold, R.G. 1986. Assessment of Volatile Organic Emissions from a Petroleum Refinery Land Treatment Site. EPA-600/D-86/074. PB 86184603.

Block, R.N., Clark, T.P., and Bishop, R. 1990. In *Petroleum Contaminated Soils*. Kostecki, P.T. and Calabrese, E.J., Eds. Lewis Publishers, Chelsea, MI. 3:167–175.

Blystone, P.G., Johnson, M.D., Haag, W.R., and Daley, P.F. 1993. Advanced ultraviolet flash lamps for the destruction of organic contaminants in air. In *Emerging Technologies in Hazardous Waste Management*, III. Tedder, D.W. and Pohland, F.G., Eds. ACS Symposium Series. ACS, Washington, D.C. pp. 380–392.

Bochner, B.R. 1978. Device, Composition and Method for Identifying Microorganisms. U.S. Patent No. 4 129 483.

Bochner, B.R. and Savageau, M. 1977. Generalised indicator plate for genetic, metabolic, and taxonomic studies with microorganisms. *Appl. Environ. Microbiol.* 33:434–444.

Boehm, D.F. and Pore, R.S. 1984. Studies on hexadecane utilization by *Prototheca zopfii*. In *Abstr. Annu. Mtg. Am. Soc. Microbiol.* p. 213.

Boehm, K. 1992. A thermal method for cleaning contaminated soil. In *Contam. Land Treat. Technol. Pap. Int. Conf.*, 1992. pp. 195–219.

Boethling, R.S. and Alexander, M. 1979a. Effect of concentration of organic chemicals on their biodegradation by natural microbial communities. *Appl. Environ. Microbiol.* 37:1211–1216.

Boethling, R.S. and Alexander, M. 1979b. *Environ. Sci. Technol.* 13:989–991.

Bogan, B.W. and Lamar, R.T. 1995. One-electron oxidation in the degradation of creosote polycyclic aromatic hydrocarbons by *Phanerochaete chrysosporium*. *Appl. Environ. Microbiol.* 61:2631–2635.

Bogan, B.W. and Lamar, R.T. 1996. Polycyclic aromatic hydrocarbon-degrading capabilities of *Phanerochaete laevis* HHB-1625 and its extracellular ligninolytic enzymes. *Appl. Environ. Microbiol.* 62(5):1597–1603.

Bogan, B.W., Lamar, R.T., and Hammel, K.E. 1996. Fluorene oxidation *in vivo* by *Phanerochaete chrysosporium* and *in vitro* during managenese peroxidase-dependent lipid peroxidation. *Appl. Environ. Microbiol.* 62:1788–1792.

Bogan, B.W., Schoenike, B., Lamar, R.T., and Cullen, D. 1996. Manganese peroxidase mRNA and enzyme activity levels during bioremediation of polycyclic *Phanerochaete chrysosporium*. *Appl. Environ. Microbiol.* 62(7):2381–2386.

Bogardt, A.H. and Hemmingsen, B.B. 1992. Enumeration of phenanthrene-degrading bacteria by an overlayer technique and its use in evaluation of petroleum-contaminated sites. *Appl. Environ. Microbiol.* 58(8):2579–2582.

Bohn, H.L. 1971. Redox potentials. *Soil Sci.* 112:39–45.

Boiesen, A., Arvin, E., and Broholm, K. 1993. Effect of mineral nutrients on the kinetics of methane utilization by methanotrophs. *Biodegradation.* 4:163–170.

Bolick, J.J., Jr. and Wilson, D.J. 1994. Soil clean up by *in situ* aeration. XIV. Effects of random permeability variations on soil vapor extraction cleanup times. *Sep. Sci. Technol.* 29(6):701–725.

Bollag, J.-M. 1983. Cross-coupling of humus constituents and xenobiotic substances. In *Aquatic and Terrestrial Humic Materials*. Christman, R.F. and Gjessing, E.T., Eds. Ann Arbor Science Publ., Ann Arbor, MI. pp. 127–141.

Bollag, J.-M., Czaplicki, E.J., and Minard, R.D. 1975. Bacterial metabolism of 1-naphthol. *J. Agric. Food Chem.* 23:85–90.

Bollag, J.-M. and Dec, J. 1995. Detoxification of aromatic pollutants by fungal enzymes. In *Microbial Processes for Bioremediation. Pap. 3rd Int.* In Situ On-Site Bioreclam. Symp. Hinchee, R.E., Vogel, C.M., and Brockman, F.J., Eds. Battelle Press, Columbus, OH. pp. 67–73.

Bollag, J.-M. and Liu, S.-Y. 1972. Fungal degradation of 1-naphthol. *Can. J. Microbiol.* 18:1113–1117.

Boone, D.R. and Bryant, M.P. 1980. Propionate-degrading bacterium, *Syntrophobacter wolinii* sp. nov. gen. nov. from methanogenic ecosystems. *Appl. Environ. Microbiol.* 40:626–632.

Booth, M.G. and Tramontini, E. 1984. In *Sewage Sludge Stabilization and Disinfection. Water Res. Cent. Conf. 1983.* Bruce, A.M., Ed. Horwood, Chichester, U.K. 293 pp.
Booz, Allen and Hamilton, Inc. 1983. Hazardous Waste Materials: Hazardous Effects and Disposal Methods. Vol. 2. Prepared for U.S. EPA.
Bosecker, K., Hollerbach, A., Kassner, H., Teschner, M., and Wehner, H. 1993. *Biohydrometal. Technol., Proc. Int. Biohydrometal. Symp.* 2:365–737.
Bossert, I. and Bartha, R. 1984. The fate of petroleum in soil ecosystems. In *Petroleum Microbiology.* Atlas, R.M., Ed. Macmillan, New York. pp 435–473.
Bossert, I.D. and Bartha R. 1986. Structure-biodegradability relationships of polycyclic aromatic hydrocarbons in soil. *Bull. Environ. Contam. Toxicol.* 37:490–495.
Bossert, I., Kachel, W.M., and Bartha, R. 1984. Fate of hydrocarbons during oily sludge disposal in soil. *Appl. Environ. Microbiol.* 47:763–767.
Bouchard, D.C., Mravik, S.C., and Smith, G.B. 1990. Benzene and naphthalene sorption on soil contaminated with high molecular weight residual hydrocarbons from unleaded gasoline. *Chemosphere.* 21(8):975–990.
Bouchez, M., Blanchet, D., Besnainou, B., and Vandecasteele, J.-P. 1995. Diversity of metabolic capacities among strains degrading polycyclic aromatic hydrocarbons. In *Microbial Processes for Bioremediation.* Hinchee, R.E., Vogel, C.M., and Brockman, F.J., Eds. Battelle Press, Colombus, OH. pp. 153–159.
Bouchez, M., Blanchet, D., and Vandecasteele, J.-P. 1995a. Degradation of polycyclic aromatic hydrocarbons by pure strains and by defined strain associations: inhibition phenomena and cometabolism. *Appl. Microbiol. Biotechnol.* 43:156–164.
Bouchez, M., Blanchet, D., and Vandecasteele, J.-P. 1995b. Substrate availability in phenanthrene biodegradation: transfer mechanism and influence on metabolism. *Appl. Microbiol. Biotechnol.* 43:952–960.
Bouchez, M., Blanchet, D., and Vandecasteele, J.-P. 1996. The microbiological fate of polycyclic aromatic hydrocarbons: carbon and oxygen balances for bacterial degradation of model compounds. *Appl. Microbiol. Biotechnol.* 45:556–561.
Bouillard, J.X., Enzien, M., Peters, R.W., Frank, J., Botto, R.E., and Cody, G. 1995. Hydrodynamics of foam flows for *in situ* bioremediation of DNAPL-contaminated subsurface. In *Applied Bioremediation of Petroleum Hydrocarbons.* Hinchee, R.E., Kittel, J.A., and Reisinger, H.J., Eds. Battelle Press, Columbus, OH. pp. 311–317.
Bouwer, E.J. and McCarty, P.L. 1983a. Transformations of 1- and 2-carbon halogenated aliphatic organic compounds under methanogenic conditions. *Appl. Environ. Microbiol.* 45:1286–1294.
Bouwer, E.J. and McCarty, P.L. 1983b. Transformation of halogenated organic compounds under denitrification conditions. *Appl. Environ. Microbiol.* 45:1295–1299.
Bouwer, E.J. and McCarty, P.L. 1984. Modeling of trace organics biotransformation in the subsurface. *Ground Water.* 22:433–440.
Bouwer, H. 1984. Elements of soil science and groundwater hydrology. In *Groundwater Pollution Microbiology.* Bitton, G. and Gerba, C.P., Eds. John Wiley & Sons, New York.
Bove, L.J., Lambert, W.P., Lin, L.Y.H., Sullivan, D.E., and Marks, P.J. 1984. Installation Restoration General Environmental Technology Development. Report to U.S. Army Toxic and Haz. Mat. Agency, Aberdeen Proving Ground, MD, on Contract No. DAAK11-82-C-0017. AD-A162 528/4.
Bowman, J.P., Sly, L.I., and Hayward, A.C. 1990. Patterns of tolerance to heavy metals among methane-utilizing bacteria. *Lett. Appl. Microbiol.* 10:85–87.
Boyd, S.A., Shelton, D.R., Berry, D., and Tiedje, J.M. 1983. Anaerobic biodegradation of phenolic compounds in digested sludge. *Appl. Environ. Microbiol.* 46:50–54.
Boyen, C., Kloareg, B., Polne-Fuller, M., and Gibor, A. 1990. Preparation of alginate lyases from marine molluscs for protoplast isolation in brown algae. *Phycologia.* 29:173–181.
Boynton, W.P. and Brattain, W.H. 1929. Interdiffusion of gases and vapors. In *International Critical Tables of Numerical Data, Physics, Chemistry, and Technology.* Vol. 5. Washburn, E., Ed. McGraw-Hill, New York. pp. 62–63.
Bracker, G.P. 1993. Sealed systems for the manufacture of compost from biowastes. *Umvelt-Technol. Aktuell.* (Germany). 3(6):372.
Brady, N.C. 1974. *The Nature and Properties of Soils.* 8th ed. Macmillan, New York.
Bragg, J.R., Prince, R.C., and Atlas, R.M. 1994. Effectiveness of bioremediation for oiled intertidal shorelines. *Nature.* 366:413–418.
Braun, K. and Gibson, D.T. 1984. Anaerobic degradation of 2-aminobenzoate (anthranilic acid) by denitrifying bacteria. *Appl. Environ. Microbiol.* 48:1187–1192.
Braun-Luellemann, A., Johannes, C., Majcherczyk, A., and Huettermann, A. 1995. The use of white-rot fungi as active biofilters. *Biol. Unit. Processes Hazard. Waste Tmt. Pap. 3rd Int. In Situ On-Site Bioreclam. Symp.* Hinchee, R.E., Skeen, R.S., and Sayles, G.D., Eds. Battelle Press, Columbus, OH.
Breedveld, G.D., Olstad, G., Briseid, T., and Hauge, A. 1995. Nutrient demand in bioventing of fuel oil pollution. In *In Situ Aeration: Air Sparging, Bioventing, Relat. Rem. Processes. Pap. 3rd Int. In Situ On-Site Bioreclam. Symp.* Hinchee, R.E., Miller, R.N., Johnson, P.C., Eds. Battelle Press, Columbus, OH. pp. 391–399.
Breton, M. et al. 1983. GCA Corporation for the U.S. EPA. GCA Report No. GCA-TR-83-70-G, August.
Breure, A.M., Volkering, F., Mulder, H., Rulkens, W.H., and van Andel, J.G. 1995. Enhancement of bioavailability by surfactants. *Soil Environ.* 5 (Contaminated Soil 95, Vol. 2). pp. 939–948.

REFERENCES

Brierley, C.L., Brierley, J.A., and Davidson, M.S. 1989a. Applied microbial processes for metals recovery and removal from wastewater. In *Metal Ions and Bacteria*. Beveridge, T.J. and Doyle, R.J., Eds. John Wiley & Sons, New York. pp. 303–323.

Brierley, C.L., Brierley, J.A., and Davidson, M.S. 1989b. Applied microbial processes for metals recovery and removal from wastewater. In *Metal Ions and Bacteria*. Beveridge, T.J. and Doyle, R.J., Eds. John Wiley & Sons, New York. pp. 359–382.

Brierley, J.A. 1990. Production and application of a *Bacillus*-based product for use in metals biosorption. In *Biosorption of Heavy Metals*. Volesky, B., Ed. CRC Press, Boca Raton, FL. pp. 305–311.

Brindley, G.W. and Thompson, T.D. 1966. Clay organic studies. XI. Complexes of benzene, pyridine, piperidine and 1,3-substituted propanes with a synthetic Ca-fluorhectorite. *Clay Miner.* 6:345.

Britton, L.N. 1985. *Feasibility Studies on the Use of Hydrogen Peroxide to Enhance Microbial Degradation of Gasoline*. API Publication No. 4389. American Petroleum Institute, Washington, D.C.

Brock, T.D. 1978. *Thermophilic Microorganisms and Life at High Temperatures*. Springer-Verlag, Heidelburg.

Brockman, F.J. 1995a. Overview of biomolecular methods for monitoring bioremediation performance. In *Monitoring and Verification of Bioremediation*. Hinchee, R.E., Douglas, G.S., and Ong, S.K., Eds. Battelle Press, Columbus, OH.

Brockman, F.J. 1995b. Nucleic-acid-based methods for monitoring the performance of *in situ* bioremediation. *Mol. Ecol.* 4(5):567–578.

Brockman, F.J., Denovan, B.A., Hicks, R.J., and Fredrickson, J.F. 1989. *Appl. Environ. Microbiol.* 55:1029–1032.

Brodkorb, T.S. and Legge, R.L. 1992. Enhanced biodegradation of phenanthrene in oil tar-contaminated soil supplemented with *Phanerochaete chrysosporium*. *Appl. Environ. Microbiol.* 58:3117–3121.

Brombach, M., Schwabe, S., and Theissen, H. 1996. Improved biological reclamation of abandoned polluted areas by high-frequency fields. *Chem. Ind.* (Duesseldorf). 119(3):37–38.

Brooks, K. and McGinty, R. 1987. Groundwater treatment know-how comes of age. *Chem. Week.* pp. 50–52.

Brown, K.L., Davila, B., and Sanseverino, J. 1995. Combined chemical and biological oxidation of slurry phase polycyclic aromatic hydrocarbons. In *Proc. HAZMACON 95, 12th Haz. Mater. Mgmt. Conf. Exhib.* pp. 363–369.

Brown, K.L., Davila, B., Sanseverino, J., Thomas, M., Lang, C., Hague, K., and Smith, T. 1995. Chemical and biological oxidation of slurry-phase polycyclic aromatic hydrocarbons. In *Biological Unit Processes Hazardous Waste Treatment. Pap. 3rd. Int. In Situ On-Site Bioreclam. Symp.* Hinchee, R.E., Skeen, R.S., and Sayles, G.D., Eds. Battelle Press, Columbus, OH. pp. 113–127.

Brown, K.W. 1975. An Investigation of the Feasibility of Soil Disposal of Waste Water for the Jefferson Chemical Co., Monroe, Texas. Final Report Phase I. Texas A&M Research Foundation, P.O. Box H, College Station, TX.

Brown, K.W. 1981. In *Proc. Symp. and Workshop on Haz. Waste Mgmt. Protection of Water Resources*, Nov. 16–20, Worm, B.G., Dantin, E.J., and Seals, R.K., Eds. Louisiana State University, Baton Rouge, LA.

Brown, K.W. and Associates, Inc. 1981. U.S. EPA Report under Contract 68-03-2943. U.S. EPA, Cincinnati, OH.

Brown, K.W., Deuel, L.E., Jr., and Thomas, J.C. 1983. Land treatability of refinery and petrochemical sludges. EPA Report No. EPA-600/S2-074. Municipal Environmental Research Laboratory, Cincinnati, OH.

Brown, K.W., Donnelly, K.C., and Deuel, L.E., Jr. 1983. Effects of mineral nutrients, sludge application rate, and application frequency on biodegradation of two oily sludges. *Microb. Ecol.* 9:363–373.

Brown, R.A. and Bass, D. 1991. Use of aeration in environmental clean-ups. Paper presented at Mid-Atlantic Environ. Expo, April 9–11, 1991, Baltimore, MD.

Brown, R.A., Longfield, J.Y., Norris, R.D., and Wolfe, F.G. 1985. Enhanced bioreclamation: designing complete solution to ground water problems. *Proc. 2nd Wastes Symp. Water Poll. Control Fed.* Kansas City, KS.

Brown, R.A., Loper, J.R., and McGarvey, D.C. 1986. In *In Situ Treatment of Ground Water: Issues and Answers. Proc. 1986 Hazar. Mat. Spills Conf.*, May 5–8, 1986, St. Louis, MO. pp. 261–264.

Brown, R.A., Mahaffey, W., and Norris, R.D. 1993. *In situ* bioremediation: The state of the practice. In *In Situ Bioremediation. When Does it Work?* National Research Council. Water Science and Technology Board, Com. Eng. Tech. Systems. National Academy of Sciences. National Academy Press, Washington, D.C. pp. 121–135.

Brown, R.A. and Norris, R.D. 1986. U.S. Patent Number 4,591,443.

Brown, R.A. and Norris, R.D. 1994. The evolution of a technology: hydrogen peroxide in *in situ* bioremediation. In *Hydrocarbon Bioremediation*. Hinchee, R.E., Alleman, B.C., Hoeppel, R.E., and Miller, R.N., Eds. CRC Press, Boca Raton, FL. pp. 148–162.

Brown, R.A., Norris, R.D., and Brubaker, G.R. 1985. Aquifer restoration with enhanced bioreclamation. *Pollut. Eng.* 17:25–28.

Brown, R.A., Norris, R.D., and Raymond, R.L. 1984. Oxygen transport in contaminated aquifers. In *Proc. NWWA/API Conf. on Petroleum Hydrocarbons and Organic Chemicals in Groundwater — Prevention, Detection, and Restoration*, Houston, TX. National Water Well Association, Worthington, OH. pp. 441–450.

Brubaker, G.R. and O'Neill, E. 1982. Remediation strategies using enhanced bioreclamation. In *Proc. 5th Natl. Symp. on Aquifer Restoration and Groundwater Monitoring*. Ground Water Pub. Co., Westerville, OH. pp. 1–5.

Brunker, R.L. and Bott, T.L. 1974. Reduction of mercury to the elemental state by a yeast. *Appl. Microbiol.* 27:870–873.

Brunner, G.H. 1994. Extraction and destruction of waste with supercritical water. *NATO ASI Ser.*, Ser. E. 273 (Supercritical Fluids). pp. 697–705.

Brunner, W. and Focht, D.D. 1983. Persistence of polychlorinated biphenyls (PCB) in soils under aerobic and anaerobic conditions. In *Abstr. Annu. Mtg. Am. Soc. Microbiol.* p. 266.

Brusseau, G.A., Hsien-Chyang, T., Hanson, R.S., and Wackett, L.P. 1990. Optimization of trichloroethylene oxidation by methanotrophs and the use of a colorimetric assay to detect soluble methane monooxygenase activity. *Biodegradation.* 1:19–29.

Brusseau, M.L., Wang, X., and Hu, Q. 1994. Enhanced transport of low-polarity organic compounds through soil by cyclodextrin. *Environ. Sci. Technol.* 28(5):952–956.

Bryant, C.W. 1986. Lagoons, ponds, and aerobic digestion. *J. Water Pollut. Control Fed.* 58:501–504.

Bryant, R.J., Woods, L.E., Coleman, D.C., Fairbanks, B.C., McClellan, J.F., and Cole, C.V. 1982. Interactions of bacterial and amoebal populations in soil microcosms with fluctuating moisture content. *Appl. Environ. Microbiol.* 43:747–752.

Bryers, J.D. and Sanin, S. 1994. Resuscitation of starved ultramicrobacteria to improve *in situ* bioremediation. *Ann. N.Y. Acad. Sci.* 745:61-76.

Bryniok, D. and Troesch, W. 1989. ELISA techniques for the determination of methanogenic bacteria. *Appl. Microbiol. Biotechnol.* 32:235–242.

Buckingham, E. 1981. Contributions to Our Knowledge of the Aeration of Soils. USDA, Bureau of Soils Bulletin No. 25.

Buitron, G., Koefoed, A., and Capdeville, B. 1993. Control of phenol biodegradation by using CO_2 evolution rate as an activity indicator. *Environ. Technol.* 14:227–236.

Bulman, T.L., Newland, M., and Wester, A. 1993. *In situ* bioventing of a diesel fuel spill. *Hydrol. Sci. J.* 38(4):297–308.

Bumpus, J.A. 1989. Biodegradation of polycyclic aromatic hydrocarbons by *Phanerochaete chrysosporium*. *Appl. Environ. Microbiol.* 55:154–158.

Bumpus, J.A. and Brock, B.J. 1988. Biodegradation of crystal violet by the white-rot fungus *Phanerochaete chrysosporium*. *Appl. Environ. Microbiol.* 54:1140–1150.

Bumpus, J.A., Fernando, T., Mileski, G.J., and Aust, S.D. 1987. Biodegradation of organopollutants by *Phanerochaete chrysosporium*: practical considerations. In *Proc. 13th Annu. Res. Symp. on Land Disposal, Remedial Action, Incineration and Treatment of Haz. Waste,* May 6–8, Cincinnati, OH.

Bumpus, J.A., Milewski, G., Brock, B., Ashbaugh, W., and Aust, S.D. 1991. In *Innovative Hazardous Waste Treatment Technology Series.* Freeman, H.M. and Sferra, P.R., Eds. Technomic Pub. Corp., Lancaster, PA. 3:47–54.

Bumpus, J.A., Tien, M., Wright, D., and Aust, S.D. 1985. Oxidation of persistent environmental pollutants by a white rot fungus. *Science.* 228:1434–1436.

Burke, G.K. and Rhodes, D.K. 1995. Alternative systems for *in situ* bioremediation: enhanced control and contact. In *In Situ Aeration: Air Sparging, Bioventing, and Related Remediation Processes.* Hinchee, R.E., Miller, R.N., and Johnson, P.C., Eds. Battelle Press, Columbus, OH. pp. 527–534.

Burkhard, N. and Guth, J.A. 1979. Photolysis of organophosphorus insecticides on soil surfaces. *Pest. Sci.* 10:313.

Burkhardt, C., Insam, H., Hutchinson, T.C., and Reber, H.H. 1993. Impact of heavy metals on the degradative capabilities of soil bacterial communities. *Biol. Fertil. Soils.* 16:154–156.

Burlage, R.S., Kuo, D., and Palumbo, A.V. 1994. Monitoring catabolic gene expression by bioluminescence in bioreactor studies. In *Hydrocarbon Bioremediation.* Hinchee, R.E., Alleman, B.C., Hoeppel, R.E., and Miller, R.N., Eds. CRC Press, Boca Raton, FL.

Busman, L.M. and Tabatabai, M.A. 1985. Hydrolysis of trimetaphosphate in soils. *Soil Sci. Soc. Am. J.* 45:630–636.

Buswell, J.A. 1992. In *Handbook of Applied Mycology.* Vol. 1: *Soil and Plants.* Arora, D.K., Rai, B., Mukerji, K.G., and Kundsen, G., Eds. Marcel Dekker, New York. pp. 425–480.

Buswell, J.A. and Odier, E. 1987. Lignin biodegradation. *Crit. Rev. Biotechnol.* 6:1–60.

Butler, E.L., Douglas, G.S., Steinhauer, W.G., Prince, R.C., Aczel, T., Hsu, C.S., Bronson, M.T., Clark, J.R., and Lindstrom J.E. 1991. Hopane, a new chemical tool for measuring oil biodegradation. In *On-Site Bioreclamation: Processes for Xenobiotic and Hydrocarbon Treatment.* Hinchee, R.E. and Olfenbuttel, R.F., Eds. Butterworth-Heinemann, Stoneham, MA. pp. 515–521.

C

Cail, R.G. and Barford, T.P. 1985. The development of granulation in an upflow floc digester and an upflow anaerobic sludge blanket digester treating cane juice stillage. *Biotechnol. Lett.* 7:493–498.

Calder, J.A. and Lader, J.H. 1976. Effect of dissolved aromatic hydrocarbons on the growth of marine bacteria in batch culture. *Appl. Environ. Microbiol.* 32:95–101.

Caldini, G., Cenci, G., Manenti, R., and Morozzi, G. 1995. The ability of an environmental isolate of *Pseudomonas fluorescens* to utililze chrysene and other four-ring polynuclear aromatic hydrocarbons. *Appl. Microbiol. Biotechnol.* 44(1–2):225–229.

Callahan, M.A., Slimak, M.W., Gabel, N.W., May, I.P., Fowler, C.F., Freed, J.R., Jennings, P., Durfee, R.C., Whitmore, F.C., Maestri, B., Mabey, W.R., Holt, B.R., and Gould, C. 1979. Water Related Environmental Fate of 129 Priority Pollutants, Vol II. Halogenated Aliphatic Hydrocarbons, Halogenated Ethers, Monocyclic Aromatics, Phthalate Esters, Polycyclic Aromatic Hydrocarbons, Nitrosoamines and Miscellaneous Compounds. EPA-440/4-79-029 b.

Callister, S.M. and Winfrey, M.R. 1983. Microbial mercury resistance and methylation rates in the upper Wisconsin River. *Abstr. Annu. Mtg. Am. Soc. Microbiol.* p. 261.

REFERENCES

Cancel, A.M., Orth, A.B., and Tien, M. 1993. Lignin and veratryl alcohol are not inducers of ligninolytic system of *Phanerochaete chrysosporium. Appl. Environ. Microbiol.* 59:2909–2913.

Canevari, G.P. 1971. Oil spill dispersants: current status and future outlook. In *Proc. Joint Conf. on Prevention and Control of Oil Spills.* American Petroleum Institute, Washington, D.C. p. 263.

Cantafio, A.W., Hagen, K.D., Lewis, G.E., Bledsoe, T.L., Nunan, K.M., and Macy, J.M. 1996. Pilot-scale selenium bioremediation of San Joaquin drainage water with *Thauera selenatis. Appl. Environ. Microbiol.* 62(9):3298–3303.

Canter, L.W. and Knox, R.C. 1985. *Ground Water Pollution Control.* Lewis Publishers, Chelsea, MI. 526 pp.

Caparello, D.M. and LaRock, P.A. 1975. *Microb. Ecol.* 2:28–42.

Capuano, R.M. and Johnson, M.A. 1996. Geochemical reactions during biodegradation-vapor-extraction remediation of petroleum contamination in the vadose zone. *Ground Water.* 34(1):31–40.

Carey, M.J., Day, S.R., Pinewski, R., and Schroder, D. 1995. Case study of shallow soil mixing and soil vacuum extraction remediation project. In *Innov. Technol. Site Rem. Haz. Waste Mgmt., Proc. Natl. Conf.* pp. 21–29.

Carlson, R.E. 1981. Engineering and Services Laboratory, Air Force Engineering and Services Center, Tyndall Air Force Base, FL. Report No. ESL-TR-81-50.

Carmichael, J.W. 1962. *Chrysosporium* and some other aleuriosporic hyphomycetes. *Can. J. Bot.* 40:1137–1173.

Carriere, P.P.E. and Mesania, F.A. 1995. Enhanced biodegradation of creosote contaminated soil using a nonionic surfactant. *Haz. Ind. Wastes.* 27th. pp. 294–303.

Casella, S. and Payne, W.J. 1996. Potential of denitrifiers for soil environment protection. *FEMS Microbiol. Lett.* 140(1):1–8.

Casida, L.E. 1971. Microorganisms in unamended soil as observed by various forms of microscopy and staining. *Appl. Environ. Microbiol.* 21:1040–1045.

Cassidy, M.B., Leung, K., Lee, H., and Trevors, J.T. 1995. Survival of *lac-lux* larked *Pseudomonas aeruginosa* UG2Lr cells encapsulated in *k*-carrageenan and alginate. *J. Microbiol. Methods.* 23(3):281–290.

CAST. 1976. Application of sewage sludge to cropland. EPA Report No. 64, EPA-430/9-76-013. U.S. EPA, Washington, D.C.

Castaldi, F.J. 1994. Slurry bioremediation of polycyclic aromatic hydrocarbons in soil wash concentrates. In *Applied Biotechnology for Site Remediation.* Hinchee, R.E., Anderson, D.B., Metting, F.B., Jr., and Sayles, G.D., Eds. Lewis Publishers, Boca Raton, FL. pp. 99–108.

Catallo, W.J. and Portier, R.J. 1992. Use of indigenous and adapted microbial assemblages in the removal of organic chemicals from soils and sediments. *Water Sci. Technol.* 25(3):229–237.

Catalytic, Inc. 1984. Evaluation of thickening and dewatering character of SRC-1 wastewater treatment sludges. NTIS DE84013560. DOE/OR/03054-101.

Cavalieri, E. and Rogan, E. 1985. Role of radical cations in aromatic hydrocarbon carcinogenesis. *Environ. Health Perspect.* 64:69–84.

Cavanagh, J.-A., Juhasz, A.L., Nichols, P.D., Franzmann, P.D., and McMeekin, T.A. 1995. Analysis of microbial hydrocarbon degradation using TLC-FID. *J. Microbiol. Methods.* 22:119–130.

Cerniglia, C.E. 1981. Aromatic hydrocarbons: metabolism by bacteria, fungi, and algae. *Rev. Biochem. Toxicol.* 3:321–361.

Cerniglia, C.E. 1982. Initial reactions in the oxidation of anthracene by *Cunninghamella elegans. J. Gen. Microbiol.* 128:2055–2061.

Cerniglia, C.E. 1984. Microbial metabolism of polycyclic aromatic hydrocarbons. *Adv. Appl. Microbiol.* 30:31–71.

Cerniglia, C.E. 1992. Biodegradation of polycyclic aromatic hydrocarbons. *Biodegradation.* 3:351–368.

Cerniglia, C.E., Althaus, J.A., Evans, F.E., Freeman, J.P., Mitchum, R.K., and Yang, S.K. 1983. Stereochemistry and evidence for an arene oxide-NIH shift pathway in the fungal metabolism of naphthalene. *Chem. Biol. Interact.* 44:119–132.

Cerniglia, C.E. and Crow, S.A. 1981. Metabolism of aromatic hydrocarbons by yeasts. *Arch. Microbiol.* 129:9–13.

Cerniglia, C.E., Dodge, R.H., and Gibson, D.T. 1994. Metabolism of benz(a)anthracene by the filamentous fungus *Cunninghamella elegans. Appl. Environ. Microbiol.* 60:3931–3938.

Cerniglia, C.E. and Gibson, D.T. 1977. Metabolism of naphthalene by *Cunninghamella elegans. Appl. Environ. Microbiol.* 34:363–370.

Cerniglia, C.E. and Gibson, D.T. 1979. Oxidation of benzo(a)pyrene by the filamentous fungus *Cunninghamella elegans. J. Biol. Chem.* 254:12174–12180.

Cerniglia, C.E. and Gibson, D.T. 1980. Fungal oxidation of benzo(a)pyrene: evidence for the formation of a BP-7,8-diol 9,10-epoxide and BP-9,10-diol 7,8-epoxides. *Abstr. Annu. Mtg. Am. Soc. Microbiol.* p. 138.

Cerniglia, C.E., Gibson, D.T., and Van Baalen, C. 1979. Algal oxidation of aromatic hydrocarbons: formation of 1-naphthol from naphthalene by *Agmenellum quadruplicatum*, strain PR-6. *Biochem. Biophys. Res. Commun.* 88:50–58.

Cerniglia, C.E., Gibson, D.T., and Van Baalen, C. 1980. Oxidation of naphthalene by cyanobacteria and microalgae. *J. Gen. Microbiol.* 116:495–500.

Cerniglia, C.E., Gibson, D.T., and Van Baalen, C. 1982. Naphthalene metabolism by diatoms isolated from the Kachemak Bay Region of Alaska. *J. Gen. Microbiol.* 128:987–990.

Cerniglia, C.E., Hebert, R.L., Dodge, R.H., Szaniszlo, P.J., and Gibson, D.T. 1979. Some approaches to studies on the degradation of aromatic hydrocarbons by fungi. In *Microbial Degradation of Pollutants in Marine Environments.* Bourquin, A.L. and Pritshard, H., Eds. EPA Report No. EPA-600/9-79-012.

Cerniglia, C.E., Hebert, R.L., Szaniszlo, P.J., and Gibson, D.T. 1978. Fungal transformation of naphthalene. *Arch. Microbiol.* 117:135–143.

Cerniglia, C.E. and Heitkamp, M.A. 1989. Microbial degradation of polycyclic aromatic hydrocarbons in the aquatic environment. In *Metabolism of Polycyclic Aromatic Hydrocarbons in the Aquatic Environment.* Varanasi, U., Ed. CRC Press, Boca Raton, FL. pp. 41–68.

Cerniglia, C.E. Hughes, T.J., and Perry, J.J. 1971. Microbial degradation of hydrocarbons in marine environments. *Bacteriol. Proc.* p. 56.

Cerniglia, C.E. and Perry, J.J. 1973. Oil degradation by microorganisms isolated from the marine environment. *Z. Allg. Mikrobiol.* 13:299–306.

Cerniglia, C.E. and Perry, J.J. 1974. Effect of substrate on the fatty acid composition of hydrocarbon-utilizing filamentous fungi. *J. Bacteriol.* 118:844–847.

Cerniglia, C.E., Van Baalen, C., and Gibson, D.T. 1980. Metabolism of naphthalene by the cyanobacterium *Oscillatoria* sp., strain JCM. *J. Gen. Microbiol.* 116:485–495.

Cerniglia, C.E., Wyss, O., and Van Baalen, C. 1980. Recent studies on the algal oxidation of naphthalene. *Abst. Annu. Mtg. Am. Soc. Microbiol.* p. 197.

Cerniglia, C.E. and Yang, S.K. 1984. Stereoselective metabolism of anthracene and phenanthrene by the fungus *Cunninghamella elegans. Appl. Environ. Microbiol.* 47:119–124.

Chakrabarty, A.M. 1974. Dissociation of a degradative of plasmid aggregate in *Pseudomonas. J. Bacteriol.* 118(2):815-820.

Chakrabarty, A.M. 1982. Genetic mechanisms in the dissimilation of chlorinated compounds. In *Biodegradation and Detoxification of Environmental Pollutants.* Chakrabarty, A.M., Ed. CRC Press, Inc., Boca Raton, FL. pp. 127–139.

Chakrabarty, A.M. 1985. Genetically manipulated microorganisms and their products in the oil service industries. *Trends Biotechnol.* 3:32–38.

Chakrabarty, A.M. 1995. Microbial degradation of hazardous chemicals: evolutionary consideration and bioremediative developments. In *Bioremediation: the Tokyo '94 Workshop.* Org. Econ. Co-operation Devel. Paris. pp. 71–80.

Challenger, F. and North, H.J. 1933. Production of organo-metalloidal compounds by micro-organisms. II. Dimethyl selenide. *J. Chem. Soc.* 18:68–71.

Champagne, A.T. and Bienkowski, P.R. 1995. The supercritical fluid extraction of anthracene and pyrene from a model soil: an equilibrium study. *Sep. Sci. Technol.* 30(7–9):1289–1307.

Chan, D.B., Yeh, S.L., and Bialecki, A. 1994. Demonstration of the steam injection and vacuum extraction (SIVE) technology for removal of JP-5 jet fuel in soil. *Hydrocarbon Contam. Soils.* 4:17-42.

Chan, E.C., Kuo, J., Lin, H.P., and Mou, D.G. 1991. Stimulation of *n*-alkane conversion to dicarboxylic acid by organic-solvent- and detergent-treated microbes. *Appl. Microbiol. Biotechnol.* 34:772–777.

Chaney, R.L. and Hornick, S.B. 1978. Accumulation and effects of cadmium on crops. In *Proc. 1st Int. Cadmium Conf. Metals Bulletin,* Ltd., San Francisco, London.

Chang, M.-K., Voice, T.C., and Criddle, C.S. 1993. Kinetics of competitive inhibition and cometabolism in the biodegradation of benzene, toluene, and *p*-xylene by two *Pseudomonas* isolates. *Biotechnol. Bioeng.* 41:1057–1065.

Chao, A., Chang, Y.-T., Bricka, M., and Neale, C.N. 1995. Transient response of a multi-stage counter-current washing system for removal of metals from contaminated soils. In *Proc. 49th Ind. Waste Conf.,* 1994. pp. 99–110.

Chee-Sanford, J.C., Frost, J.W., Fries, M.R., Zhou, J., and Tiedje, J.M. 1996. Evidence for acetyl coenzyme A and cinnamoyl coenzyme A in the anaerobic toluene mineralization pathway in *Azoarcus tolulyticus. App. Environ. Microbiol.* 62(3):964–973.

Chen, C.L. and Chang, H.M. 1985. Chemistry of lignin biodegradation. In *Biosynthesis and Biodegradation of Wood Components.* Higuchi, T., Ed. Academic Press, New York. pp. 535–556.

Chen, J.S. and Maier, W.J. 1993. Sorption, desorption, and biodegradation of phenanthrene in soil. In *Proc. 47th Ind. Waste Conf.,* 1992. pp. 159–166.

Cheng, J.J. and Lehmann, R.G. 1985. *Weed Sci.* 33(Suppl. 2):7–10.

Cheremisinoff, P.N. 1988. Thermal treatment technologies for hazardous wastes. *Pollut. Eng.* 20:50–55.

Chern, H.-T. and Bozzelli, J.W. 1994. Thermal desorption of organic contaminants from sand and soil using a continuous feed rotary kiln. *Haz. Ind. Wastes.* 26th. pp. 417–424.

Cherry, J.A., Gillham, R.W., and Barker, J.F. 1984. Chapter 3. *Groundwater Contamination* (Studies in Geophysics). National Academy Press, Washington, D.C. pp. 46–64.

Chevalier, P. and de la Noue, J. 1985. Wastewater nutrient removal with microalgae immobilized in carrageenan. *Enzyme Microb. Technol.* 7:621–624.

Chian, E.S.K. and deWalle, F.B. 1977. Evaluation of Leachate Treatment. Vol. 1: Characterization of Leachate. EPA-600/2-77-186a. U.S. EPA, Cincinnati, OH. 226 pp.

Chiou, C.T. 1985. Partition coefficients of organic compounds in lipid–water systems and correlations with fish bioconcentration factors. *Environ. Sci. Technol.* 19:57–62.

Chiou, C.T. 1989. In *Reactions and Movement of Organic Chemicals in Soils.* Sawhney, B.L. and Brown, K., Eds. Soil Science Society of America, Madison, WI. pp. 1–29.

Chiou, C.T., Freed, V.H., Schmedding, D.W., and Kohnert, R.L. 1977. Partition coefficient and bioaccumulation of selected organic chemicals. *Environ. Sci. Technol.* 11:475–478.

REFERENCES

Chiou, C.T. and Shoup, T.D. 1985. Soil sorption of organic vapors and effects of humidity on sorptive mechanism and capacity. *Environ. Sci. Technol.* 19:1196–1200.

Cho, J.S., DiGiulio, D.C., Wilson, J.T., and Vardy, J.A. 1997. In-situ air injection, soil vacuum extraction and enhanced biodegradation: a case study in a JP-4 jet fuel contaminated site. *Environ. Prog.* 16(1):35–42.

Cho, S.H. and Bowers, A.R. 1991. Treatment of toxic or refractory aromatic compounds by chemical oxidants. In *On-Site Bioreclamation: Processes for Xenobiotic and Hydrocarbon Treatment.* Hinchee, R.E. and Olfenbuttel, R.F., Eds. Butterworth-Heinemann, Stoneham, MA. pp. 273–292.

Choi, Y.-B., Lee, J.-Y., and Kim, H.-S. 1992. A novel bioreactor for the biodegradation of inhibitory aromatic solvents: experimental results and mathematical analysis. *Biotechnol. Bioeng.* 40:1403–1411.

Chowdhury, J., Parkinson, G., and Rhein, R. 1986. CPI go below to remove groundwater pollutants. *Chem. Eng.* 93:14–19.

Christ, R.H., Oberholser, K., Shank, N., and Nguyen, M. 1980. Nature of bonding between metallic ions and algal cell walls. *Environ. Sci. Technol.* 15:1212–1217.

Chudoba, J. 1985. Quantitative estimation in COD units of refractory organic compounds produced by activated sludge microorganisms. *Water Res.* (G.B.). 19:37–43.

Churchill, P.F. and Griffin, R.A. 1992. Biosurfactant enhanced bioremediation of hazardous substances. *Poster Abstr./J. Haz. Mater.* 32:323–382.

Churchill, S.A., Walters, J.V., and Churchill, P.F. 1995. Sorption of heavy metals by prepared bacterial cell surfaces. *J. Environ. Eng.* 121:706–711.

Cifuentes, F.R., Lindemann, W.C., and Barton, L.L. 1996. Chromium sorption and reduction in soil with implications to bioremediation. *Soil Sci.* 161(4):233.

Circeo, L.J., Camacho, S.L., Jacobs, G.K., and Tixier, J.S. 1994. Plasma remediation of *in-situ* materials – the PRISM concept. *In-situ Remediation: Scientific Basis for Current and Future Technology. 33rd Hanford Symp. Health Environ.* 2:707–719.

Clark, F.E. 1967. Bacteria in soil. In *Soil Biology.* Burges, A. and Raw, F., Eds. Academic Press, New York. pp. 15–49.

Clark, R.C. and Brown, D.W. 1977. Petroleum: properties and analyses in biotic and abiotic systems. In *Effects of Petroleum on Arctic and Subarctic Marine Environments and Organisms.* Vol. 1. Malins, D.C., Ed. Academic Press, New York. pp. 15–16.

Claus, D. and Walker, N. 1964. The decomposition of toluene by soil bacteria. *J. Gen. Microbiol.* 36:107–122.

Clemons, G.P., Fields, S., and Hazaga, D. 1984. In *5th Natl. Conf. Mgmt. Uncontrolled Haz. Waste Sites,* November 7–9, Washington, D.C. pp. 404–406.

Clewell, H.J., III. 1981. The Effect of Fuel Composition on Groundfall from Aircraft Fuel Jettisoning. Report to Eng. and Services Lab., Air Force Engineering and Services Center, Tyndall Air Force Base, FL.

Cobet, A.B. 1974. Hydrocarbonoclastic repository. In *Progress Report Abstracts.* Microbiology Program, Office of Naval Research, Arlington, VA. p. 131.

Cobiella, R. 1989. A method for removing volatile substances from water, using flash volatilization. Forum on Innovative Haz. Waste Treat. Technol., Atlanta, GA.

Cody, T.E., Radike, M.J., and Warshawsky, D. 1984. The phytotoxicity of benzo(a)pyrene in the green alga, *Selenastrum capricornutum. Environ. Res.* 35:122–131.

Coffey, J.C., Ward, C.H., and King, J.M. 1977. Effects of petroleum hydrocarbons on growth of fresh-water algae. *Dev. Ind. Microbiol.* 18:661–672.

Cohen, A., Breure, A.M., Schmedding, D.J.M., Zoetemeyer, R.J., and van Andel, J.G. 1985. Significance of partial pre-acidification of glucose for methanogenesis in an anaerobic digestion process. *Appl. Microbiol. Biotechnol.* 21:404–408.

Colby, J., Stirling, D.I., and Dalton, H. 1977. The soluble methane monooxygenase of *Methylococcus capsulatus* (Baath): its ability to oxygenate n-alkanes, n-alkenes, ethers and alicyclic, aromatic and heterocyclic compounds. *Biochem. J.* 165:395–402.

Colla, A., Fiecchi, A., and Treccani, V. 1959. *Ann. Microbiol.* 9:87.

Collins, C.H., Lyne, P.M., and Grange, J.M. 1990. *Collins and Lyne's Microbiological Methods.* 6th ed. Butterworths, London.

Collins, L.D. and Daugulis, A.J. 1996. Use of a two phase partitioning bioreactor for the biodegradation of phenol. *Biotechnol. Tech.* 10(9):643–648.

Collins, Y.E. and Stotzky, G. 1989. Factors affecting the toxicity of heavy metals to microbes. In *Metal Ions and Bacteria.* Beveridge, T.J. and Doyle, R.J., Eds. John Wiley & Sons, New York.

Collins, Y.E. and Stotzky, G. 1992. Heavy metals alter the electrokinetic properties of bacteria, yeasts, and clay minerals. *Appl. Environ. Microbiol.* 58(5):1592–1600.

Collins, Y.E. and Stotzky, G. 1996. Changes in the surface charge of bacteria caused by heavy metals do not affect survival. *Can. J. Microbiol.* 42:621–627.

Compeau, G.C., Mahaffey, W.D., and Patras, L. 1991. In *Environmental Biotechnology for Waste Treatment.* Sayler, G.S., Fox, R., and Blackburn, J.W., Eds. Plenum Press, New York. pp. 91–109.

CONCAWE. 1980. Report No. 3/80. The Hague, Netherlands.

Connan, J. 1983. Biodegradation of crude oils in reservoirs. In *Advances in Petroleum Geochemistry.* Vol. 1. Brooks, J. and Welte, D.H., Eds. Academic Press, New York. PP. 299–335.

Conner, J.R. 1995. Recent findings on immobilization of organics as measured by total constituent analysis. *Waste Manage.* 15(5–6):359–369.

Cook, F.D. and Westlake, D.W.S. 1974. Microbiological Degradation of Northern Crude Oils. Environmental-Social Committee, Northern Pipelines, Task Force on Northern Oil Development. Report No. 74-1. Catalog No. R72-12774. Information Canada, Ottawa.

Cook, J.W., Hewett, C.L., and Hieger, I. 1933. *Proc. R. Soc. London,* Ser. B. III:395.

Cook, R.J. and Papendick, R.I. 1970. Soil water potential as a factor in the ecology of *Fusarium roseum* f. sp. cerealis "colmorum." *Plant Soil.* 32:131–145.

Cook, W.L., Massey, J.K., and Ahearn, D.G. 1973. The degradation of crude oil by yeasts and its effect on *Lesbistes reticulatus.* In *The Microbial Degradation of Oil Pollutants,* Workshop, Dec. 4–6, 1972, Atlanta, GA. Ahearn, D.G. and Meyers, S.P., Eds. Pub. No. LSU-SG-73-01. Center for Wetland Resources, Louisiana State University, Baton Rouge, LA. p. 279.

Cooke, R.C. and Rayner, A.D.M., Eds. 1984. *Ecology of Saprophytic Fungi.* Longman, New York. 415 pp.

Cooke, S. and Bluestone, M. 1986. Invasion from Ireland: grease-eating microbes. *Chem. Week.* p. 20.

Cooke, W.B. 1973. U.S. Dept. of Health, Education, and Welfare, Publication No. 999-WP-L.

Cooney, J.J. 1974. Microorganisms Capable of Degrading Refractory Hydrocarbons in Ohio Waters. NTIS PB-237293/6ST. National Technical Information Service, Springfield, VA.

Cooney, J.J. 1980. Microorganisms Capable of Degrading Refractory Hydrocarbons in Ohio Waters. Ohio Water Resources Center, Columbus, OH for the Office of Water Research and Technology, Washington, D.C. Report No. 493X. PB83-108290.

Cooney, J.J., Silver, S.A., and Beck, E.A. 1985. Factors influencing hydrocarbon degradation in three freshwater lakes. *Microb. Ecol.* 11:127–137.

Cooney, J.J. and Summers, R.J. 1976. Hydrocarbon-using microorganisms in three fresh-water ecosystems. In *Proc. 3rd Int. Biodegradation Symp.,* Aug. 17–23, 1975, Kingston, RI. Sharply, J.M. and Kaplan, A.M., Eds. Applied Science Publishers, London. pp. 141–156.

Cooney, J.J. and Walker, J.D. 1973. Hydrocarbon utilization by *Cladosporium resinae.* In *Proc. Workshop,* Dec. 4–6, 1972, Georgia State University, LSU-SG-73-01. Center for Wetland Resources, Louisiana State University, Baton Rouge, LA. pp. 25-32.

Cooper, D.G. 1982. Biosurfactants and enhanced oil recovery. *Proc. 1982 Int. Conf. on Microbiological Enhancement of Oil Recovery,* May 16–21, ShangriLa, Afton, OK. pp. 112–113.

Cooper, D.G. and Zajic, J.E. 1980. Surface-active compounds from microorganisms. *Adv. Appl. Microbiol.* 26:229-253.

Corseuil, H.X. and Weber, W.J., Jr. 1994. Potential biomass limitations on rates of degradation of monoaromatic hydrocarbons by indigenous microbes in subsurface soils. *Water Res.* 28(6):1415–1423.

Couillard, D. and Mercier, G. 1993. Removal of metals and fate of nitrogen and phosphorus in the bacterial leaching of aerobically digested sewage sludge. *Water Res.* 27(7):1227–1235.

Cox, D.P. and Williams, A.L. 1980. Biological process for converting naphthalene to *cis*-1,2-dihydroxy-1,2-dihydronaphthalene. *Appl. Environ. Microbiol.* 39:320–326.

Craig, P.J. and Wood, J.M. 1981. *Environmental Lead.* Lynam, D.R., Piantanida, L.E., and Cole, J.F., Eds. Academic Press, New York. p. 333.

Cremonesi, P., Hietbrink, B., Rogan, E.G., and Cavalieri, E.L. 1992. One-electron oxidation of dibenzo(a)pyrenes by manganic acetate. *J. Org. Chem.* 57:3309–3312.

Cresswell, L.W. 1977. Proc. 1977 Oil Spill Conf.: Prevention, Behavior, Control, Cleanup, March 8–10, 1977, Washington, D.C. EPA API/UCSG Joint Oil Spill Conf.

Cripps, C., Bumpus, J.A., and Aust, S.D. 1990. Biodegradation of azo and heterocyclic dyes by *Phanerochaete chrysosporium. Appl. Environ. Microbiol.* 56:1114–1118.

Cripps, R.E. and Watkinson, R.J. 1978. Polycyclic aromatic hydrocarbons: metabolism and environmental aspects. In *Developments in Biodegradation of Hydrocarbons* — 1. Watkinson, R.J., Ed. Applied Science, London. pp. 113–134.

Crosby, D.G. 1971. Environmental photooxidation of pesticides. In *Degradation of Synthetic Organic Molecules in the Biosphere.* National Academy of Sciences, Washington, D.C. pp. 260–290.

Crosby, R.A. 1996. Thermal desorption unit for removing chemical contaminants from soil. U.S. Patent 5,514,286 (Cl. 210-742; B01D35/01), 7 May 1996, Appl. 145,486, 29 Oct 1993. 12 pp.

Cudahy, J.J., DeCicco, S.G., and Troxler, W.L. 1987. Paper presented at Int. Conf. Haz. Mat. Mgmt., Chattanooga, TN.

Cullimore, D.R. 1983. Microbiological parameters controlling plugging of wells and groundwater systems. In *Land Treatment of Hazardous Wastes.* Parr, J.F., Marsh, P.B., and Kla, J.M., Eds. Noyes Data Corp., Park Ridge, NJ. p. 12.

Cupitt, L.T. 1980. Fate of toxic and hazardous materials in the air environment. EPA Report No. EPA-600/53-80-084. U.S. EPA, Athens, GA.

Cutright, T.J. and Lee, S. 1994. Quantitative and qualitative analysis of PAH contaminated soil. *Fresenius Environ. Bull.* 3(1):42–48.

REFERENCES

D

Dablow, J.F., III. 1992. Steam injection to enhance removal of diesel fuel from soil and groundwater. In *Hydrocarbon Contaminated Soils and Groundwater. Proc. 1st Annu. West Coast Conf. Hydrocarbon Contam. Soils Groundwater,* February 1990, Newport Beach, CA. Vol. 1. Calabrese, E.J. and Kostechi, P.T., Eds. Lewis Publishers, Chelsea, MI. pp. 410–407.

Dablow, J. Hicks, R., and Cacciatore, D. 1995. Steam injection and enhanced bioremediation of heavy fuel oil contamination. In *Applied Bioremediation of Petroleum Hydrocarbons.* Hinchee, R.E., Kittel, J.A., and Reisinger, H.J., Eds. Battelle Press, Columbus, OH. pp. 115–121.

Dablow, J., Hicks, R., Cacciatore, D., and van de Meene, C. 1995. Steam injection and enhanced bioremediation of heavy fuel oil contamination. *Soil Environ.* 5(Contaminated Soil 95, Vol. 2). pp. 1183–1184.

Daglet, S. and Patel, M.D. 1957. Oxidation of p–creosol and related compounds by a *Pseudomonas. Biochem. J.* 66:227–233.

Dagley, S. 1977. *Degradation of Synthetic Organic Molecules in the Biosphere.* National Academy of Science, Washington, D.C. ERAC in Fates of Pollutants.

Dagley, S. 1981. New perspectives in aromatic catabolism. In *FEMS Symposium* No. 12. *Microbial Degradation of Xenobiotics and Recalcitrant Compounds.* Leisinger, T., Hutter, R., Cook, A.M., and Nuesch, J., Eds. Academic Press, New York. 12:141:179.

Dagley, S. 1984. Microbial degradation of aromatic compounds. *Dev. Ind. Microbiol.* 25:53–65.

Datar, M.T. and Bhargava, D.S. 1984. Effect of temperature on BOD and COD kinetics during aerobic digestion. *Indian J. Environ. Health.* 26:285–297.

Datar, M.T. and Bhargava, D.S. 1985. Effect of initial solids on kinetics of solids reduction during aerobic digestion. *Indian J. Environ. Health.* 27:31–45.

Daubaras, D. and Chakrabarty, A.M. 1992. The environment, microbes and bioremediation: microbial activities modulated by the environment. *Biodegradation.* 3:125–135.

Dave, H., Ramakrishna, C., Bhatt, B.D., and Desai, J.D. 1994. Biodegradation of slop oil from a petrochemical industry and bioreclamation of slop oil contaminated soil. *World J. Microbiol. Biotechnol.* 10(6):653–656.

Davies, J.I. and Evans, W.C. 1976. Oxidative metabolism of naphthalene by soil pseudomonads. The ring-fission mechanism. *Biochem. J.* 91:5988–5996.

Davies, J.S. and Westlake, D.W.S. 1979. Crude oil utilization by fungi. *Can. J. Microbiol.* 25:146–156.

Davis, G.B., Barber, C., Buselli, G., and Sheehy, A. 1991. Potential applications for monitoring remediations in Australia using geoelectric and soil gas techniques. In In situ *Bioreclamation: Applications and Investigations for Hydrocarbon and Contaminated Site Remediation.* Hinchee, R.E. and Olfenbuttel, R.F., Eds. Butterworth-Heinemann, Stoneham, MA. pp. 337–350.

Davis, J.B. 1967. *Petroleum Microbiology.* Elsevier-North Holland. New York.

Davis, M.W. and Lamar, R.T. 1992. Evaluation of methods to extract ergosterol for quantitation of fungal biomass. *Soil Biol. Biochem.* 24:189–198.

Davis-Hoover, W.J., Murdoch, L.C., Vesper, S.J., Pahren, H.R., Sprockel, O.L., Chang, C.L., Hussain, A., and Ritschel, W.A. 1991. Hydraulic fracturing to improve nutrient and oxygen delivery for *in situ* bioreclamation. In In Situ *Bioreclamation. Applications and Investigations for Hydrocarbon and Contaminated Site Remediation.* Hinchee, R.E. and Olfenbuttel, R.F., Eds. Butterworth-Heinemann, Stoneham, MA. pp. 67–82.

Dawson, T.D. and Chang, F.-H. 1992. Screening test of the biodegradative capability of a new strain of *Pseudomonas gladioli* (BSU 45124) on some xenobiotic organics. *Bull. Environ. Contam. Toxicol.* 49:10–17.

Day, R.A. and Underwood, A.L. 1980. *Quantitative Analysis.* 4th ed. Prentice-Hall, Englewood Cliffs, NJ. p. 462.

Daymani, M.A., Forster, K., Ahlfeld, D.P., Hoag, G.E., and Carley, R.J. 1992. Evaluation of TCLP for gasoline-contaminated soils. In *Proc. Petrol. Hydrocarbons Org. Chem. in Ground Water: Prevention, Detection, and Restoration,* Nov. 4–6, Houston, TX. pp. 19–33.

Dean-Raymond, D. and Alexander, M. 1977. Bacterial metabolism of quaternary ammonium compounds. *Appl. Environ. Microbiol.* 33:1037–1041.

Dean-Raymond, D. and Bartha, R. 1975. Biodegradation of Some Polynuclear Aromatic Petroleum Components by Marine Bacteria. NTIS No. AD/A-006 346/1st, National Technical Information Service, Springfield, VA.

DeCicco, S.G. and Troxler, W.L. 1986. *Standard Handbook for Hazardous Waste Treatment and Disposal.* Freeman, H.M., Ed. McGraw-Hill, New York.

Deflaun, M.F., Tanzer, A.S., McAteer, A.L., Marshall, B., and Levy, S.B. 1990. Development of an adhesion assay and characterization of an adhesion-deficient mutant of *Pseudomonas fluorescens. Appl. Environ. Microbiol.* 56:1112–1119.

Deis, G.C. et al. 1985. Modular plastic media for shallow bed trickling filter applications. In *Environmental Engineering — Proc. of the 1985 Specialty Conf.,* July 1–5, 1985, Boston, MA. O'Shaughnessy, J.C., Ed. American Society of Civil Engineers Press, New York. pp. 254–260.

de Jong, E., De Vries, F.P., Field, J.A., Van Der Zwan, R.P., and de Bont, J.A.M. 1992. Isolation and screening of basidiomycetes with high peroxidase activity. *Mycol. Res.* 96:1098–1104.

de Jong, E., Field, J.A., and de Bont, J.A.M. 1992. *FEBS Lett.* 299:107–110.

de Klerk, H. and van der Linden, A.C. 1974. Bacterial degradation of cyclohexane. Participation of a co-oxidation reaction. *Antonie van Leeuwenhoek* 40:7–15.

De Laat, J., Bouanga, F., Dore, M., and Mallevialle, J. 1985. Influence of bacterial growth in granular activated carbon filters on the removal of biodegradable and of non-biodegradable organic compounds. *Water Res.* 19(12):1565–1578.

Delafontaine, M.J., Naveau, H.P., and Nyns, E.J. 1979. *Biotechnol. Lett.* 1:71–74.

Del Borghi, M., Palazzi, E., Parisi, F., and Ferrajolo, G. 1985. Influence of process variables on the modelling and design of a rotating biological surface. *Water Res.* 19:573–580.

DePaoli, D.W., Wilson, J.H., and Thomas, C.O. 1996. Conceptual design of soil venting systems. *J. Environ. Eng.* 122(5):399–406.

De Pastrovich, T.L., Baradat, Y., Barthal, R., Chiarelli, A., and Fussel, D.R. 1979. Protection of ground water from oil pollution. CONCAWE Report No. 3/79. The Oil Companies' Intl. Study Group, The Hague, The Netherlands.

de Percin, P.R. 1991. Demonstration of *in situ* steam and hot-air stripping technology. *J. Air Waste Manage. Assoc.* 41(6):873–877.

De Rome, L. and Gadd, G.M. 1991. Use of pelleted and immobilized yeast and fungal biomass for heavy metal and radionuclide recovery. *J. Ind. Microbiol.* 7:97–104.

Deschenes, L., Lafrance, P., Villeneuve, J.-P., and Samson, R. 1995. Surfactant influence on PAH biodegradation in a creosote-contaminated soil. In *Microbial Processes for Bioremediation. Pap. 3rd Int. In Situ On-Site Bioreclam. Symp.* Hinchee, R.E., Vogel, C.M., and Brockman, F.J., Eds. Battelle Press, Columbus, OH. pp. 51–56.

Des Marais, D.J. 1990. Microbial mats and the early evolution of life. *Trends Ecol. Evol.* 5(5):140–144.

Desmonts, C., Minet, J., Colwell, R., and Cormier, M. 1992. An improved filter method for direct viable count of *Salmonella* in seawater. *J. Microbiol. Methods.* 16:195–201.

Dev, H., Bridges, J.E., and Sresty, G.C. 1984. Decontamination for volatile organics: a case history. In *Proc. 1984 Haz. Mater. Spills Conf.,* Nashville, TN, Ludwigson, J., Ed. Government Institutes, Inc., Rockville, MD. pp. 57–64.

Devare, M. and Alexander, M. 1995. Bacterial transport and phenanthrene biodegradation in soil and aquifer sand. *Soil Sci. Soc. Am. J.* 59:1316–1320.

Devine, K. 1994. Bioremediation: The state of usage. In *Applied Biotechnology for Site Remediation.* Hinchee, R.E., Anderson, D.B., Metting, F.B., Jr., and Sayles, G.D., Eds. Lewis Publishers, Boca Raton, FL.

Devinny, J.S. and Islander, R.L. 1989. Oxygen limitation in land treatment of concentrated wastes. *Haz. Waste Haz. Mater.* 6(4):421–433.

Devinny, J.S., Medina, V.F., and Hodge, D.S. 1994. Biofiltration for treatment of gasoline vapors. In *Hydrocarbon Bioremediation.* Hinchee, R.E., Alleman, B.C., Hoeppel, R.E., and Miller, R.N., Eds. CRC Press, Boca Raton, FL. pp. 12–19.

Devitt, D.A., Evans, R.B., Jury, W.A., Starks, T.H., Eklund, B., and Gholson, A. 1987. Soil gas sensing for detection and mapping of volatile organics. EPA-600/8-87/036.

De Wit, J.C.M., Urlings, L.G.C.M., and Alphenaar, P.A. 1995. Application of mechanistic models for bioventing. In *In situ Aeration: Air Sparging, Bioventing, Relat. Rem. Processes. Pap. 3rd Int. In Situ On-Site Bioreclam. Symp.* Hinchee, R.E., Miller, R.N., and Johnson, P.C., Eds. Battelle Press, Columbus, OH. pp. 463–471.

Deziel, E., Paquette, G., Villemur, R., Lepine, F., and Bisaillon, J.-G. 1996. Biosurfactant production by a soil *Pseudomonas aeruginosa* strain growing on polycyclic aromatic hydrocarbons. *Appl. Environ. Microbiol.* 62(6):1908–1912.

Dhawale, S.W., Dhawale, S.S., and Dean-Ross, D. 1992. Degradation of phenanthrene by *Phanerochaete chrysosporium* occurs under ligninolytic as well as nonligninolytic conditions. *Appl. Environ. Microbiol.* 58:3000–3006.

Dibble, J.T. and Bartha, R. 1979a. Effect of environmental parameters on the biodegradation of oil sludge. *Appl. Environ. Microbiol.* 37:729–739.

Dibble, J.T. and Bartha, R. 1979b. Leaching aspects of oil sludge biodegradation in soil. *Soil Sci.* 127:365–370.

Dibble, J.T. and Bartha, R. 1979c. Rehabilitation of oil-inundated agricultural land: a case history. *Soil Sci.* 128:56–60.

Dick, R.P. and Tabatabai, M.A. 1986. Hydrolysis of polyphosphate in soils. *Soil Sci.* 142:132–140.

Diels, L. 1990. Accumulation and Precipitation of Cd and Zn ions by *Alcaligenes eutrophus* strains. *Biohydrometallurgy.* pp. 369–377.

Diels, L., Faelen, M., Mergeay, M., and Nies, D. 1985. Mercury transposons from plasmids governing multiple resistance to heavy metals in *Alcaligenes eutrophus* CH34. *Arch. Int. Physiol. Biochim.* 93:B27–B28.

Diels, L. and Mergeay, M. 1990. DNA probe mediated detection of resistant bacteria from soils highly polluted by heavy metals. *Appl. Environ. Microbiol.* 56:1485–1491.

Diels, L., Springael, D., Kreps, S., and Mergeay, M. 1991. Construction and characterization of heavy metal resistant, PCB-degrading *Alcaligenes* sp. strains. In *On-Site Bioreclamation: Processes for Xenobiotic and Hydrocarbon Treatment.* Hinchee, R.E. and Olfenbuttel, R.F., Eds. Butterworth-Heinemann, Stoneham, MA. pp. 483–493.

Dieterich, L. 1995. BARR: bio anaerobic reduction and re-oxidation. *Soil Environ.* 5(Contaminated Soil 95, Vol. 2). pp. 1185–1186.

Dietz, D.M. 1980. The intrusion of polluted water into a groundwater body and the biodegradation of a pollutant. In *Proc. 1980 Conf. Control of Haz. Mater. Spills.* Sponsored by U.S. EPA et al. pp. 236–244.

Diks, R.M.M. and Ottengraf, S.P.P. 1991. A biological treatment system for the purification of waste gases containing xenobiotic compounds. In *On-Site Bioreclamation: Processes for Xenobiotic and Hydrocarbon Treatment.* Hinchee, R.E. and Olfenbuttel, R.F., Eds. Butterworth-Heinemann, Stoneham, MA. pp. 452–463.

REFERENCES

Dineen, D., Slater, J.P., Hicks, P., Holland, J. and Clendening, L.D. 1993. *In situ* biological remediation of petroleum hydrocarbons in unsaturated soils. In *Princ. Pract. Petrol. Contam. Soils*. Calabrese, E.J., Ed. pp. 453–463.

Dobbins, D.C., Aelion, M.C., Long, S.C., and Pfaender, F.K. 1986. Methodology for assessing the metabolism of subsurface microbial communities. *Abstr. Annu. Mtg. Am. Soc. Microbiol.* p. 300.

Dobbs, R.A., and Cohen, J.M. 1980. Carbon adsorption isotherms for toxic organics. EPA-600/8-80-023. Research Reporting Series 8, "Special Reports." Municipal Environmental Research Laboratory, Cincinnati, OH.

Dockstader, A.M. 1994. Success and difficulties of two remediation technologies. In *Proc. HAZMACON 94, 11th Haz. Mater. Mgmt. Conf. Exhib.* pp. 418–427.

Doddema, H.J. and Vogels, G.D. 1978. Improved identification of methanogenic bacteria by fluorescence microscopy. *Appl. Environ. Microbiol.* 36:752–754.

Dodge, R.H., Cerniglia, C.E., and Gibson, D.T. 1979. Fungal metabolism of biphenyl. *Biochem. J.* 178:223–230.

Dodge, R.H. and Gibson, D.T. 1980. Fungal metabolism of benzo(a)anthracene. *Abstr. Annu. Mtg. Am. Soc. Microbiol.* p. 138.

Dohanyos, M., Kosova, B., Zabranska, J., and Grau, P. 1985. Production and utilization of volatile fatty acids in various types of anaerobic reactors. *Water Sci. Technol.* (G.B.). 17:191–205.

Doi, Y. 1990. *Microbial Polyesters*. VCH Publishers, New York.

Dolfing, J., Zeyer, J., Binder-Eicher, P., and Schwarzenbach, R.P. 1990. Isolation and characterization of a bacterium that mineralizes toluene in the absence of molecular oxygen. *Arch. Microbiol.* 154:336–341.

Donnan, W.W. and Schwab, G.O. 1974. Current drainage methods in the U.S.A. In *Proc. 1st Int. Symp. on Acid Precipitation and the Forest Ecosystem*. U.S. Forest Service General Technical Report NE-23, USDA — Forest Service, Upper Darby, PA.

Donoghue, N.A., Griffin, M., Norris, D.B., and Trudgill, P.W. 1976. The microbial metabolism of cyclohexane and related compounds. In *Proc. 3rd Int. Biodegradation Symp.*, Aug. 17–23, 1975, Kingston, RI. Sharply, J.M., and Kaplan, A.M., Eds. Applied Science Publishers, London. pp. 43–56.

Dosoretz, C.G., Chen, H.C., and Grethlein, H.E. 1990. Effect of environmental conditions on extracellular protease activity in ligninolytic cultures of *Phanerochaete chrysosporium*. *Appl. Environ. Microbiol.* 56:395–400.

Dott, W., Feidieker, D., Kaempfer, P., Schleibinger, H., and Strechel, S. 1989. Comparison of autochthonous bacteria and commercially available cultures with respect to their effectiveness in fuel oil degradation. *J. Ind. Microbiol.* 4:365–374.

Douglas, G.S., Prince, R.C., Butler, E.L., and Steinhauer, W.G. 1994. The use of internal chemical indicators in petroleum and refined products to evaluate the extent of biodegradation. In *Hydrocarbon Bioremediation*. Hinchee, R.E., Alleman, B.C., Hoeppel, R.E., and Miller, R.N., Eds. CRC Press, Boca Raton, FL. pp. 219–236.

Douglass, R.H., Armstrong, J.M., and Korreck, W.M. 1991. Design of a packed column bioreactor for on-site treatment of air stripper off gas. In *On-Site Bioreclamation: Processes for Xenobiotic and Hydrocarbon Treatment*. Hinchee, R.E. and Olfenbuttel, R.F., Eds. Butterworth-Heinemann, Stoneham, MA. pp. 209–225.

Downey, D.C., Frishmuth, R.A., Archabal, S.R., Pluhar, C.J., Blystone, P.G., and Miller, R.N. 1995. Using *in situ* bioventing to minimize soil vapor extraction costs. In *In Situ Aeration: Air Sparging, Bioventing, Relat. Rem. Processes. Pap. 3rd Int. In Situ On-Site Bioreclam. Symp.* Hinchee, R.E., Miller, R.N., and Johnson, P.C., Eds. Battelle Press, Columbus, OH. pp. 247–266.

Downey, D.C., Pluhar, C.J., Dudus, L.A., Blystone, P.G., Miller, R.N., Lane, G.L., and Taffinder, S. 1994. Remediation of gasoline-contaminated soils using regenerative resin vapor treatment and *in situ* bioventing. *Proc. Pet. Hydrocarbons Org. Chem. Ground Water: Prev., Detect., Rem. Conf.* pp. 239–254.

Downing, A.L. and Truesdale, G.A. 1955. Some factors affecting the rate of solution of oxygen in water. *J. Appl. Chem.* 5:570–581.

Doyle, R.C. 1979. The effect of dairy manure and sewage sludge on pesticide degradation in soil. Ph.D. Dissertation. University of Maryland. College Park, MD.

Dressler, C., Kues, U., Nies, D.H., and Friedrich, B. 1991. Determinants encoding resistance to several heavy metals in newly isolated copper-resistant bacteria. *Appl. Environ. Microbiol.* 57(11):3079–3085.

Druy, M.A., Glatkowski, P.J., Bolduc, R., Stevenson, W.A., and Thomas, T.C. 1995. Hazardous waste identification (HWI) system based on infrared transmitting optical fibers and Fourier transform infrared (FTIR) spectroscopy. *Proc. SPIE Int. Soc. Opt. Eng.* 2367(Optical Sensors for Environ. and Chem. Process Monitoring). pp. 24–32.

Dua, R.D. and Meera, S. 1981. Purification and characterization of naphthalene oxygenase from *Corynebacterium renale*. *Eur. J. Biochem.* 120:461–465.

Ducreux, J., Ballerini, D., and Bocard, C. 1994. The role of surfactants in enhanced *in situ* bioremediation. In *Hydrocarbon Bioremediation*. Hinchee, R.E., Alleman, B.C., Hoeppel, R.E., and Miller, R.N., Eds. CRC Press, Boca Raton, FL. pp. 237–242.

Ducreux, J., Baviere, M., Seabra, P., Razakarisoa, O., Shaefer, G., and Arnaud, C. 1995. Surfactant-aided recovery/*in situ* bioremediation for oil-contaminated sites. In *Applied Bioremediation of Petroleum Hydrocarbons*. Hinchee, R.E., Kittel, J.A., and Reisinger, H.J., Eds. Battelle Press, Columbus, OH. pp. 435–443.

Dunigan, E.P., Bollich, P.K., Hutchinson, R.L., Hicks, P.M., Zaunbrecher, F.C., Scott, S.G., and Mowers, R.P. 1984. Introduction and survival of an inoculant strain of *Rhizobium japonicum* in soil. *Agron. J.* 76:463–466.

Dunlap, K.R. and Perry, J.J. 1967. Effect of substrate on the fatty acid composition of hydrocarbon-utilizing microorganisms. *J. Bacteriol.* 94:1919–1923.

Dunn, I.J. 1968. An interfacial kinetics model for hydrocarbon oxidation. *Biotechnol. Bioeng.* 10:891–894.

du Plessis, C.A., Phaal, C.B., and Senior, E. 1995. Bioremediation of soil contaminated with hydrocarbons and heavy metals. In *Applied Bioremediation of Petroleum Hydrocarbons.* Hinchee, R.E., Kittel, J.A., and Reisinger, H.J., Eds. Battelle Press, Columbus, OH. pp. 107–113.

du Plessis, C.A., Senior, E., and Hughes, J.C. 1994. The physical-chemical approach to organic pollutant attenuation in soil. In *Applied Biotechnology for Site Remediation.* Hinchee, R.E., Anderson, D.B., Metting, F.B., Jr., and Sayles, G.D., Eds. Lewis Publishers, Boca Raton, FL.

Dupont, R.R. 1992. Application of bioremediation fundamentals to the design and evaluation of *in situ* soil bioventing systems. In *Proc. 85th Annu. Mtg. Exhib. Air Waste Management Association,* Kansas City, MO. Paper 92-30.03.

Dupont, R.R. 1993. Fundamentals of bioventing applied to fuel contaminated sites. *Environ. Prog.* 12(1):45–53.

Dupont, R.R., Doucette, W.J., and Hinchee, R.E. 1991. Assessment of *in situ* bioremediation potential and the application of bioventing at a fuel-contaminated site. In In Situ *Bioreclamation. Applications and Investigations for Hydrocarbon and Contaminated Site Remediation.* Hinchee, R.E. and Olfenbuttel, R.F., Eds. Butterworth-Heinemann, Stoneham, MA. pp. 262–282.

Dupont, R.R. and Reineman, J.A. 1986. Evaluation of volatilization of hazardous constituents at hazardous waste land treatment sites. EPA-600/2-86/071. U.S. EPA, Ada, OK.

Dutton, P.L. and Evans, W.C. 1969. The metabolism of aromatic compounds by *rhodopseudomonas palustris. Biochem. J.* 113:525–536.

Dybas, M.J., Tatara, G.M., Knoll, W.H., Mayotte, T.J., and Criddle, C.S. 1995. Niche adjustment for bioaugmentation with *Pseudomonas* sp. strain KC. In *Bioaugmentation for Site Remediation. Pap. 3rd Int. In Situ On-Site Bioreclam. Symp.* Hinchee, R.E., Fredrickson, J., and Alleman, B.C., Eds. Battelle Press, Columbus, OH. p. 77.

E

Eckenfelder, W.W., Jr. and Norris, R.D. 1993. Applicability of biological processes for treatment of soils. In *Emerging Technologies in Hazardous Waste Management,* III. Tedder, D.W. and Pohland, F.G., Eds. American Chemical Society, Washington, D.C. pp. 138–158.

Edelstein, W.A., Iben, I.E.T., Mueller, O.M., Uzgiris, E.E., Philipp, H.R., and Roemer, P.B. 1994. Radiofrequency ground heating for soil remediation: science and engineering. *Environ. Prog.* 13(4):247–252.

Edgehill, R.U. and Finn, R.K. 1983. Microbial treatment of soil to remove pentachlorophenol. *Appl. Environ. Microbiol.* 45:1122–1125.

Edmonds, R.L. 1976. *Appl. Environ. Microbiol.* 32:537–546.

Edwards, C. 1993. The significance of *in situ* activity on the efficiency of monitoring methods. In *Monitoring Genetically Manipulated Microorganisms in the Environment.* Edwards, C., Ed. John Wiley & Sons, Chichester, U.K.

Edwards, C., Diaper, J., Porter, J., Deere, D., and Pickup, R. 1994. Analysis of microbial communities by flow cytometry and molecular probes: identification, culturability and viability. In *Beyond the Biomass.* Ritz, K., Dighton, J., and Giller, K.E., Eds. British Society of Soil Science. Wiley-Sayce Publication. John Wiley & Sons, Chichester, U.K.

Edwards, C., Porter, J., Saunders, J.R., Diaper, J., Morgan, J.A.W., and Pickup, R.W. 1992. Flow cytometry and microbiology. *SGM Q.* 19:105–108.

Edwards, D.A., Liu, Z., and Luthy, R.G. 1994a. Surfactant solubilization of organic compounds in soil/aqueous systems. *J. Environ. Eng.* (N.Y.) 120(1):5–22.

Edwards, D.A., Liu, Z., and Luthy, R.G. 1994b. Experimental data and modeling for surfactant micelles, HOCs, and soil. *J. Environ. Eng.* (N.Y.). 120(1):23–41.

Edwards, D.A., Luthy, R.G., and Liu, Z. 1991. Solubilization of polycyclic aromatic hydrocarbons in micellar nonionic surfactant solutions. *Environ. Sci. Technol.* 25(1):127–142.

Edwards, N.T. 1983. Reviews and analyses. Polycyclic aromatic hydrocarbons (PAHs) in the terrestrial environment — a Review. *J. Environ. Qual.* 12:427–441.

Efroymson, R.A. and Alexander, M. 1991. Biodegradation by an *Arthrobacter* sp of hydrocarbons partitioned into an organic solvent. *Appl. Environ. Microbiol.* 57(5):1441–1447.

Efroymson, R.A. and Alexander, M. 1994a. Role of partitioning in biodegradation of phenanthrene dissolved in nonaqueous-phase liquids. *Environ. Sci. Technol.* 28(6):1172–1179.

Efroymson, R.A. and Alexander, M. 1994b. Biodegradation in soil of hydrophobic pollutants in nonaqueous-phase liquids (NAPSs). *Environ. Toxicol. Chem.* 13(3):405–411.

Efroymson, R.A. and Alexander, M. 1995. Reduced mineralization of low concentrations of phenanthrene because of sequestering in non aqueous-phase liquids. *Environ. Sci. Technol.* 29(2):515-521.

Ehlers, W., Letey, J., Spencer, W.F., and Farmer, W.J. 1969a. Lindan diffusion in soils: I. Theoretical considerations and mechanism of movement. *Soil Sci. Soc. Am. Proc.* 33:501–504.

Ehlers, W., Letey, J., Spencer, W.F., and Farmer, W.J. 1969b. Lindan diffusion in soils: II. Water content, bulk density, and temperature effects. *Soil Sci. Soc. Am. Proc.* 33:505–508.

REFERENCES

Ehrenfeld, J. and Bass, J. 1983. Handbook for Evaluating Remedial Action Technology Plans. EPA Report No. EPA-600/2-83-076. Municipal Environmental Research Laboratory, Cincinnati, OH.

Ehrenfeld, J. and Bass, J. 1984. Evaluation of Remedial Action Unit Operations at Hazardous Waste Disposal Sites. Noyes Publications, Park Ridge, NJ. 435 pp.

Ehrenfeld, J.R., Ong, J.H., Farino, W., Spawn, P., Jasinski, M., Murphy, B., Dixon, D., and Rissmann, E. 1986. *Surface Impoundments*. Noyes Publications, Park Ridge, NJ.

Ehrhardt, J.M. and Rehm, H.J. 1985. Phenol degradation by microorganisms adsosrbed on activated carbon. *Appl. Microbiol. Biotechnol.* 21:32.

Ehrhardt, H.M. and Rehm. H.J. 1989. *Appl. Microbiol. Biotechnol.* 30:312–317.

Ehrlich, G.G., Godsy, E.M., Goerlitz, D.F., and Hult, M.F. 1983. Microbial ecology of a creosote-contaminated aquifer at St. Louis Park, MN. *Dev. Ind. Microbiol.* 24:235–245.

Ehrlich, G.G., Goerlitz, D.F., Godsey, E.M., and Hult, M.F. 1982. Degradation of phenolic contaminants in ground water by anaerobic bacteria: St. Louis Park, MN. *Ground Water.* 20:703–710.

Ehrlich, G.G., Schroeder, R.A., and Martin, P. 1985. Microbial populations in a jet-fuel-contaminated shallow aquifer at Tustin, California. U.S. Geological Survey. Open File Report 85-335. Prepared in cooperation with the U.S. Marine Corps. pp. 85–335.

Eiermann, D.R. and Bolliger, R. 1995a. Bioremediation of a PAH-contaminated gasworks site with the Ebiox Vacuum Heap System. In *Applied Bioremediation of Petroleum Hydrocarbons*. Hinchee, R.E., Kittel, J.A., and Reisinger, H.J., Eds. Battelle Press, Columbus, OH. pp. 241–248.

Eiermann, D.R. and Bolliger, R. 1995b. Vacuum heap bioremediation of a PAH-contaminated gasworks site. *Soil Environ.* 5(Contaminated Soil 95, Vol. 2). pp. 1189–1190.

Eiermann, D.R. and Menke, R. 1993. No-bug biostimulation new remediation standard. *Tech. News Int.* 10:19–20.

Einsele, A., Schneider, H., and Fiechter, A. 1975. *J. Ferment. Technol.* 53:241–243.

Eitner, D. 1996. Potentials and limits of biofilters for degradation of organic solvents in waste air streams. *Wasser Luft Boden* (Germany). 40(7–8):44–47.

Eitner, D. and Gethke, H.G. 1987. Paper presented at the 80th Annu. Mtg. APCA, June 21–26, 1987. New York.

Eklund, B.M., Nelson, T.P., and Wetherold, R.G. 1987. Field assessment of air emissions and then control at a refinery land treatment facility. EPA-600/2-87/086a. U.S. EPA, Cincinnati, OH.

Elektorowicz, M. (1994). Bioremediation of petroleum-contaminated clayey soil with pretreatment. *Environ. Technol.* 15(4):373–380.

El-Haggar, S.M., Hamoda, M.F., and Elbieh, M.S. 1996. Mobile composting unit for organic waste suitable for severe hot weather. *Int. J. Environ. Pollut.* 6(2–3):322–327.

Elias, H.H. and Pfrommer, C., Jr. 1983. Paper presented at NJ Water Pollut. Control Assoc. Conf. on Control Tech. at Hazardous Waste Sites, Tinton Falls, NJ, Nov. 3.

Ellis, B., Harold, P., and Kronberg, H. 1991. Bioremediation of a creosote contaminated site. *Environ. Technol.* 12:447–459.

Ellis, T.G., Smets, B.F. and Grady, C.P.L, Jr. 1995. Influence of simultaneous multiple substrate biodegradation on the kinetic parameters for individual substrates. *Proc. Water Environ. Fed. 68th Annu. Cong. Expo.* 1:167–178.

Ellis, W. and Payne, J.R. 1984. Application of MHF Technology to a Tight Gas Sand in the Fort Worth Basin. EPA Report No. EPA-600/D-84-148840. NVO-684-1.

Ellis, W.D. and Payne, J.R. 1985. Treatment of Contaminated Soils with Aqueous Surfactants. Interim Report to EPA Releases Control Branch. Sept. 6.

Ellis, W.D., Payne, J.R., and McNabb, G.D. 1985. Treatment of contaminated soils with aqueous surfactants. EPA Report No. EPA-600/2-85/129. Haz. Waste Eng. Res. Lab., Cincinnati, OH.

Ellis, W.D., Payne, J.R., Tafuri, A.N., and Freestone, F.J. 1984. The development of chemical countermeasures for hazardous waste contaminated soil. In *Proc. 1984 Haz. Mater. Spills Conf.*, Nashville, TN. Ludwigson, J., Ed. Government Institutes, Inc., Rockville, MD. pp. 116–124.

El-Shoubary, Y.M. and Woodmansee, D.E. 1996. Soil washing enhancement with solid sorbents. *Environ. Prog.* 15(3):173–178.

Endo, G., Ji, G., and Silver, S. 1995. Heavy metal resistance plasmids and use in bioremediation. In *Environmental Biotechnology: Principles and Applications*. Moo-Young, M., Anderson, W.A., and Chakrabarty, A.M., Eds. Kluwer Academic Publishers, Amsterdam. pp. 47–62.

Enfield, C.G., Wilson, J.T., Piwoni, M.D., and Walters, D.M. 1985. Toxic Organic Volatilization from Land Treatment Systems. EPA/600/D-85/031. PB 85164523.

Enzien, M.J., Bouillard, J.X., Michelsen, D.L., Peters, R.W., Frank, J., Botto, R.E., and Cody, G. 1994. NAPL-Contaminated Soil/Groundwater Remediation Using Foams. Technical Report, Argonne National Laboratory.

Enzien, M.V., Michelsen, D.L., Peters, R.W., Bouillard, J.X., and Frank, J.R. 1995. Enhanced *in situ* bioremediation using foams and oil aphrons. In In situ *Aeration: Air Sparging, Bioventing, Relat. Rem. Processes. Pap. 3rd Int.* In situ *On-Site Bioreclam. Symp.* Hinchee, R.E., Miller, R.N., and Johnson, P.C., Eds. Battelle Press, Columbus, OH. pp. 503–509.

Enzminger, J.D., Robertson, D., Ahlert, R.C., and Kosson, D.S. 1987. Treatment of landfill leachates. *J. Haz. Mater.* 14:83–101.

Epstein, E. and Alpert, J.E. 1980. Composting of industrial wastes. In *Toxic and Hazardous Waste Disposal*. Vol. 4. Ann Arbor Sciences Publishers/The Butterworth Group, Ann Arbor, MI. pp. 243–252.

Erickson, D.C., Loehr, R.C., and Neuhauser, E.F. 1993. PAH loss during bioremediation of manufactured gas plant site soils. *Water Res.* 27(5):911–919.

Ermisch, O. and Rehm, H.-J. 1989. *DECHEMA Biotechnol. Conf.* 3:905–908.

Errett, D.H., Chin, Y.-P., Xu, Y., and Yan, Y. 1996. The sorption and desorption kinetics of polycyclic aromatic hydrocarbons in methanol-water mixtures. *Haz. Waste Haz. Mater.* 13(2):177–195.

ERT, Inc. 1984. The Land Treatability of Appendix VIII Constituents Present in Petroleum Industry Wastes. API Publ. 4379.

Evans, D., Elder, R., and Hoffman, R. 1992. Bioremediation of diesel contamination associated with oil and gas operations. In *Gas, Oil, Environ. Biotechnol.* IV. 4th Pap. Int. Symp., 1991. Akin, C., Markuszewski, R., and Smith, J.R.W., Eds. IGT, Chicago.

Evans, G.B., Jr., Deuel, L.E., Jr., and Brown, R.W. 1980. Mobility of Water Soluble Organic Constituents of API Separator Waste in Soils. Report prepared on EPA Grant No. R805474010.

Evans, P.J., Bourbonais, K.A., Peterson, L.E., Lee, J.H., and Laakso, G.L. 1995. Vapor-phase biofiltration: laboratory and field experience. In *Biological Unit Processes for Hazardous Waste Treatment. Pap. 3rd. Int. In Situ On-Site Bioreclam. Symp.* Hinchee, R.E., Skeen, R.S., and Sayles, G.D., Eds. Battelle Press, Columbus, OH. pp. 249–255.

Evans, P.J., Mang, D.T., Kim, K.S., and Young, L.Y. 1991. Anaerobic degradation of toluene by a denitrifying bacterium. *Appl. Environ. Microbiol.* 57:1139–1145.

Evans, W.C. 1977. Biochemistry of the bacterial catabolism of aromatic compounds in anaerobic environments. *Nature (London).* 270:17–22.

F

Falatko, D.M. and Novak, J.T. 1992. Effects of biologically produced surfactants on the mobility and biodegradation of petroleum hydrocarbons. *Water Environ. Res.* 64(2):163–169.

Fall, R.R., Brown, J.L., and Schaeffer, T.L. 1979. Enzyme recruitment allows the biodegradation of recalcitrant branched hydrocarbons by *Pseudomonas citronellolis*. *Appl. Environ. Microbiol.* 38:715–722.

Fannin, K.F., Conrad, J.R., Srivastava, V.J., Chynoweth, D.P., and Jerger, D.E. 1986. *J. Water Pollut. Control Fed.* 58:504–510.

Farbiszewska, T. and Farbiszewska-Bajor, J. 1993. *Fizykochem. Probl. Mineralurgii.* 27:219–224.

Farmer, R.W., Chen, J.-S., Kopchynski, D.M., and Maier, W.J. 1995. Reactor switching: proposed biomass control strategy for the biofiltration process. In *Biological Unit Processes for Hazardous Waste Treatment. Pap. 3rd. Int. In Situ On-Site Bioreclam. Symp.* Hinchee, R.E., Skeen, R.S., and Sayles, G.D., Eds. Battelle Press, Columbus, OH. p. 243.

Farmer, W.J., Igue, K., and Spencer, W.F. 1973. Effect of bulk density on the diffusion and volatilization of dieldrin from soil. *J. Environ. Qual.* 2:107–109.

Farmer, W.J., Yang, M.S., Letey, J., and Spencer, W.F. 1980. Land disposal of hexachlorobenzene wastes: controlling vapor movement in soil. EPA-600/2-80-119. Environmental Protection Technology Series. U.S. EPA, Cincinnati, OH.

Farr, J.M., McMillan, J., and Shibberu, D. 1995. Optimizing soil vapor extraction system design and operations. In *Proc. HAZMACON 95, 12th Haz. Mater. Mgmt. Conf. Exhib.* pp. 287–295.

Faust, S.J. and Hunter, J.V. 1971. *Organic Compounds in Aquatic Environments.* Marcel Dekker, New York.

Fay, J.A. 1969. The spread of oil slicks on a calm sea. In *Oil on the Sea.* Hoult, D.P., Ed. Plenum Press, New York. p. 53.

Federle, T.W., Dobbins, D.C., Thornton-Manning, J.R., and Jones, D.D. 1986. Microbial biomass, activity, and community structure in subsurface soils. *Ground Water.* 24:365–374.

Fedorak, P.M., Semple, K.M., and Westlake, D.W.S. 1984. Oil degrading capabilities of yeasts and fungi isolated from coastal marine environments. *Can. J. Microbiol.* 30:565–571.

Feinberg, E.L., Ramage, P.I.N., and Trudgill, P.W. 1980. *J. Gen. Microbiol.* 121:507–511.

Felgener, G., Janitza, J., Koscielski, S. 1993. Treatment of highly polluted landfill leachate using brown coal coke. *Braunkohle* (Duesseldorf). 45(3):17–23.

Felten, D.W., Leahy, M.C., Bealer, L.J., and Kline, B.A. 1992. Case study: site remediation using air sparging and soil vapor extraction. In *Proc. Petrol. Hydrocarbons Org. Chem. in Ground Water: Prevention, Detection, and Restoration,* Nov. 4–6, 1992, Houston, TX. pp. 395–406.

Fenton, H.J. 1894. Oxidation of tartaric acid in presence of iron. *J. Chem. Soc.* (London). 65:899.

Ferris, J.P., MacDonald, L.H., Patrie, M.A., and Martin, M.A. 1976. Aryl hydrocarbon hydroxylase activity in the fungus *Cunninghamella bainieri*: evidence for the presence of cytochrome P-450. *Arch. Biochem. Biophys.* 175:443–452.

Ferry, J.G. and Wolfe, R.S. 1976. Anaerobic degradation of benzoate to methane by a microbial consortium. *Arch. Microbiol.* 107:33–40.

Ferry, J.G. and Wolfe, R.S. 1977. Nutritional and biochemical characterization of *Methanospirillum hungatii*. *Appl. Environ. Microbiol.* 34:371–376.

Fewson, C.A. 1967. The growth and metabolic versatility of the gram-negative bacterium NCIB 8250 ('Vibrio 01'). *J. Gen. Microbiol.* 46:255–266.

Fewson, C.A. 1981. Biodegradation of aromatics with industrial relevance. In *FEMS Symposium No. 12. Microbial Degradation of Xenobiotics and Recalcitrant Compounds.* Leisinger, T., Hutter, R., Cook, A.M., and Nuesch, J., Eds. Academic Press, New York. 12:141–179.

REFERENCES

Fiebig, R. and Dellweg, H. 1985. Comparison between the process performance of an UASB-reactor and an UASB-fixed film-combination with an acetic acid enrichment culture. *Biotechnol. Lett.* 7:487–492.
Field, J.A., Baten, H., Boelsma, F., and Rulkens, W.H. 1996. Biological elimination of polycyclic aromatic hydrocarbons in solvent extracts of polluted soil by the white rot fungus, *Bjerkandera* sp. strain BOS55. *Environ. Technol.* 17(3):317–323.
Field, J.A., Boelsma, F., Baten, H., and Rulkens, W.H. 1995. Oxidation of anthracene in water/solvent mixtures by the white-rot fungus, *Bjerkandera* sp. strain BOS55. *Appl. Microbiol. Biotechnol.* 44(1–2):234–240.
Field, J.A., de Jong, E., Feijoo-Costa, G.F., and de Bont, J.A.M. 1992. Biodegradation of polycyclic aromatic hydrocarbons by new isolates of white rot fungi. *Appl. Environ. Microbiol.* 58(7):2219–2226.
Field, J.A., de Jong, E., Feijoo-Costa, G.F., and de Bont, J.A.M. 1993. Screening for ligninolytic fungi applicable to the biodegradation of xenobiotics. *Trends Biotechnol.* 11(2):44–49.
Field, J.A., Feiken, H., Hage, A., and Kotterman, M.J.J. 1995. Application of a white-rot fungus to biodegrade benzo(a)pyrene in soil. In *Bioaugmentation for Site Remediation. Pap. 3rd Int. In Situ On-Site Bioreclam. Symp.* Hinchee, R.E., Fredrickson, J., and Alleman, B.C., Eds. Battelle Press, Columbus, OH. pp. 165–171.
Field, J.A., Heessels, E., Wijngaarde, R., Kotterman, M., de Jong, E., and de Bont, J.A.M. 1994. The physiology of polycyclic aromatic hydrocarbon biodegradation by the white-rot fungus, *Bjerkandera* sp. strain BOS55. In *Applied Biotechnology for Site Remediation.* Hinchee, R.E., Anderson, D.B., Metting, F.B., Jr., and Sayles, G.D., Eds. Lewis Publishers, Boca Raton, FL. pp. 143–151.
Field, J.A., Stams, A.J.M., Kato, M., and Schraa, G. 1995. Enhanced biodegradation of aromatic pollutants in cocultures of anaerobic and aerobic bacterial consortia. *Antonie van Leeuwenhoek.* 67:47–77.
Finnerty, W.R., Kennedy, R.S., Lockwood, P., Spurlock, B.O., and Young, R.A. 1973. Microbes and petroleum: perspectives and implications. In *Microbial Degradation of Oil Pollutants,* Workshop, Georgia State University, Atlanta, GA, Dec. 4–6, 1972. Ahearn, D.G. and Meyers, S.P., Eds. Pub. No. LSU-SG-73-01. Center for Wetland Resources, Louisiana State University, Baton Rouge, LA. pp. 105–126.
Fischer, K. and Bardtke, D. 1984. Presented at the Intl. Symp. Characterization and Control of Odiferous Pollutants in Process Industries. Society Belge de Filtration, Louvain-La-Neuve, Belgium. April 25–27.
Fischer, R.G., Rapsomanikis, S., Andreae, M.O., and Baldi, F. 1995. Bioaccumulation of methylmercury and transformation of inorganic mercury by macrofungi. *Environ. Sci. Technol.* 29(4):993–999.
Flathman, P.E. and Caplan, J.A. 1985. Biological cleanup of chemical spills. In *Proc. Hazmacon 85,* April, 23–25, Oakland, CA. Preprint.
Flathman, P.E., Carson, J.H., Whitehead, S.J., Khan, K.A., Barnes, D.M., and Evans, J.S. 1991. Laboratory evaluation of the utilization of hydrogen peroxide for enhanced biological treatment of petroleum hydrocarbon contaminants in soil. In In Situ *Bioreclamation. Applications and Investigations for Hydrocarbon and Contaminated Site Remediation.* Hinchee, R.E. and Olfenbuttel, R.F., Eds. Butterworth-Heinemann, Stoneham, MA. pp. 125–142.
Fleming, J.T., Sanseverino, J., and Sayler, G.S. 1993. Quantitative relationship between naphthalene catabolic gene frequency and expression in predicting PAH degradation in soils at town gas manufacturing plants. *Environ. Sci. Technol.* 27:1068–1074.
Fletcher, M. 1994. Microbial ecology and bioremediation. In *Bioremediation: the Tokyo '94 Workshop.* Org. Econ. Co-operation Devel., Paris, pp. 109–115.
Flowers, T.H., Pulford, I.D., and Duncan, H.J. 1984. Studies on the breakdown of oil in soil. *Environ. Pol. Ser. B.* 8:71–82.
FMC Aquifer Remediation Systems. 1986. *In Situ* Treatment of Ground Water: Principles, Methods, and Choices. Report to U.S. Marine Corps. Jan. 7.
FMC Corporation. 1979. *Industrial Waste Treatment with Hydrogen Peroxide.* Industrial Chemicals Group, Philadelphia, PA.
Fogel, S., Lancione, R., Sewall, A., and Boethling, R.S. 1985. Application of biodegradability screening tests to insoluble chemicals: hexadacane. *Chemosphere.* 14:375–382.
Fogg, G.E., Stewart, W.D.P., Fay, P., and Walsby, A.E. 1973. *The Blue-Green Algae.* Academic Press, New York.
Foght, J.M. and Westlake, D.W.S. 1983. Evidence for plasmid involvement in bacterial degradation of polycyclic aromatic hydrocarbons. *Abstr. Annu. Mtg. Am. Soc. Microbiol.* p. 275.
Foght, J.M. and Westlake, D.W.S. 1985. Degradation of some polycyclic aromatic hydrocarbons by a plasmid-containing *Flavobacterium* species. *Abstr. Annu. Mtg. Am. Soc. Microbiol.* p. 261.
Foght, J.M. and Westlake, D.W.S. 1991. Cross hybridization of plasmid and genomic DNA from aromatic and polycyclic aromatic hydrocarbon degrading bacteria. *Can. J. Microbiol.* 37:924–932.
Foght, J.M. et al. 1985. AOSTRA (Alberta Oil Sands Technology & Research Authority) *J. Res.* 1:139.
Follett, R.H., Murphy, L.S., and Donahue, R.L. 1981. *Fertilizers and Soil Amendments.* Prentice-Hall, Englewood Cliffs, NJ.
Fontes, D.E., Mills, A.L., Hornberger, G.M., and Herman, J.S. 1991. Physical and chemical factors influencing transport of microorganisms through porous media. *Appl. Environ. Microbiol.* 57:2473–2481.
Ford, S.F. and Olson, B. 1988. Methods for detecting genetically engineered microorganisms in the environment. *Adv. Microbial Ecol.* 10:45–79.
Forsyth, J.V., Bleam, R., and Wrubel, N. 1995. Bioremediation: a systematic, tiered approach. III. *Haz. Ind. Wastes.* 27th. pp. 194–202.

Forsyth, J.V., Tsao, Y.M., and Bleam, R.D. 1995. Bioremediation: when is augmentation needed? *Bioaugmentation for Site Remediation. Pap. 3rd Int.* In Situ *On-Site Bioreclam. Symp.* Hinchee, R.E., Fredrickson, J., and Alleman, B.C., Eds. Battelle Press, Columbus, OH. pp. 1–14.

Foster, J.W. 1962. Bacterial oxidation of hydrocarbons. In *Oxygenases.* Hayaishi, O., Ed. Academic Press, New York. 241 pp.

Foster, M.L. 1985. 78th Annu. Mtg. of the Air Pollut. Control Assoc. *Air Pollut. Control Assoc. J.* Detroit, MI.

Foster, R.C. 1988. Microenvironments of soil microorganisms. *Biol. Fertil. Soils.* 6:189–203.

Fouhy, K. and Shanley, A. 1992. Mighty microbes. *Environ. Eng.* (Chem. Eng. Special Suppl., Nov. 1992). pp. 20–22.

Fournier, J.C., Codaccioni, P., and Soulas, G. 1981. Soil adaptation of 2,4-D degradation in relation to the application rates and the metabolic behaviour of the degrading microflora. *Chemosphere.* 10:977–984.

Fowler, G.D., Sollars, C.J., Ouki, S.K., and Perry, R. 1994. Thermal conversion of gasworks contaminated soil into carbonaceous adsorbents. *J. Haz. Mater.* 39(3):281–300.

Fox, J.L. 1985. Fixed up in Philadelphia: genetic engineers meet with ecologists. *ASM News.* 51:382–386.

Fox, R.D. et al. 1991. Thermal treatment for the removal of PCBs and other organics for soil. *Environ. Prog.* 10(1).

Francke, H.C. and Clark, R.E. 1974. Disposal of Oil Wastes by Microbial Assimilation. U.S. Atomic Energy Report Y-1934.

Frandsen, C.F. 1980. Landfarming disposes of organic wastes. *Pollut. Eng.* 12:55–57.

Fredericks, K.M. 1966. Adaptation of bacteria from one hydrocarbon to another. *Nature* (London). 209:1047–1048.

Fredrickson, J.K., Bezdicek, D.F., Brockman, F.J., and Li, S.W. 1988. Enumeration of Tn5 mutant bacteria in soil by using a most-probable-number-DNA hybridization procedure and antibiotic resistance. *Appl. Environ. Microbiol.* 54:446–453.

Fredrickson, J.K. and Hicks, R.J. 1987. Probing reveals many microbes beneath earth's surface. *ASM News.* 53:78–79.

Freedman, W. and Hutchinson, T.C. 1976. Physical and biological effects of experimental crude oil spills on low Arctic tundra in the vicinity of Tuktoyaktut. *Can. J. Bot.* 54:2219–2230.

Freeze, G.A., Fountain, J.C., Pope, G.A., and Jackson, R.E. 1995. Numerical simulation of surfactant-enhanced remediation using UTCHEM. AIChE Symp. Ser. 306(Heat Transfer, Portland, 1995). pp. 68–73.

Freeze, R.A. and Cherry, J.A., Eds. 1979. *Goundwater.* Prentice-Hall, Englewood Cliffs, NJ. pp. 26–30.

Freijer, J.I. 1996. Mineralization of hydrocarbons in soils under decreasing oxygen availability. *J. Environ. Qual.* 25:296–304.

Freitag, M. and Morrell, J.J. 1992. Decolorization of the polymeric dye Poly R-487 by wood-inhabiting fungi. *Can. J. Microbiol.* 38:811–822.

Freitas dos Santos, L.M. and Livingston, A.G. 1995. Novel membrane bioreactor for detoxification of VOC wastewaters: biodegradation of 1,2-dichloroethane. *Water Res.* 29(1):179–194.

Friello, D.A. and Chakrabarty, A.M. 1980. *Transposable Mercury Resistance in* Pseudomonas putida *Plasmids and Transposons.* Suttard, C. and Rozec, K.R., Eds. Academic Press, New York. p. 249.

Friello, D.A., Mylroie, J.R., and Chakrabarty, A.M. 1976. Use of genetically engineered multi-plasmid microorganisms for rapid degradation of fuel hydrocarbons. In *Proc. 3rd Int. Biodegradation Symp.,* Aug. 17–23, 1975, Kingston, RI. Sharply, J.M. and Kaplan, A.M., Eds. Applied Science Publishers, London. p. 205.

Frieze, M.P. and Oujesky, H. 1983. Oil degrading bacterial populations in municipal wastewater. *Abstr. Annu. Mtg. Am. Soc. Microbiol.* p. 269.

Frishmuth, R.A., Ratz, J.W., Blicker, B.R., Hall, J.F., and Downey, D.C. 1995. *In situ* bioventing in deep soils at arid sites. In *Innov. Technol. Site Rem. Haz. Waste Mgmt. Proc. Natl. Conf.* pp. 157–164.

Frissel, M.J. 1961. *Versl. Landbouwk. Onderzoek.* 76:3.

Fritsche, W., Guenther, T., Hofrichter, M., and Sack, U. 1994. Metabolisms of polycyclic aromatic hydrocarbons by fungi of different ecological groups. *Biol. Abwasserreinig.* (Germany). 4:167–182.

Fronk-Leist, C.A., Love, O.T., Jr., Miltner, R.J., and Eilers, R.G. 1983. Treatment of VOCs in drinking water. In *Proc. 11th AWWA Water Quality Tech. Conf.,* Norfolk, VA. EPA/600/8-83-019.

Frostell, B. 1982. In *Proc. 36th Industrial Waste Conf.,* Purdue University. pp. 269–291. Ann Arbor Science Publishers, Inc., Ann Arbor, MI.

Fry, A.W. and Grey, A.S. 1971. *Sprinkler Irrigation Handbook.* Rain Bird Sprinkler Mfg. Corp., Glendora, CA.

Fu, C., Pfanstiel, S., Gao, C., Yan, X., Govind, R., and Tabak, H.H. 1996. Studies on contaminant biodegradation in slurry, wafer, and compacted soil tube reactors. *Environ. Sci. Technol.* 30(3):743–750.

Fu, P.P., Cerniglia, C.E., Chou, M.W., and Yang, S.K. 1983. In *Polynuclear Aromatic Hydrocarbons: Formation, Metabolism, and Measurement.* Cooke, M. and Dennis, A.J., Eds. Battelle Columbus, OH. p. 531.

Fuehr, F. 1985. *Weed Sci.* 33(Suppl. 2):11–17.

Fuge, R. 1973. Trace metal concentrations in brown seaweeds, Cardigan Bay, Wales. *Mar. Chem.* 1:281–293.

Fulghum, R.S. 1983. *Am. Soc. Microbiol. News.* 49:432–434.

Fuller, P.R., Hinzel, E.J., Olsen, R.L., and Smith, P. 1986. In *Mgmt. of Uncontrolled Haz. Waste Sites Conf.* Dec. 1–3. pp. 313–317.

Fuller, W.H. 1977. Movement of selected metals, asbestos, and cyanide in soil: applications to waste disposal problems. EPA 600/2-77-020. U.S. EPA.

Fung, R., Ed. 1980. *Protective Barriers for Containment of Toxic Materials.* Noyes Data Corp., Park Ridge, NJ.

Furukawa, K. 1982. Microbial degradation of polychlorinated biphenyls (PCBs). In *Biodegradation and Detoxification of Environmental Pollutants.* Chakrabarty, A.M., Ed. CRC Press, Boca Raton, FL.

REFERENCES

G

Gadd, G.M. 1988. Accumulation of metals by microorganisms and algae. In *Biotechnology — A Comprehensive Treatise.* Vol. 6b. *Special Microbial Processes.* Rehm, H.-J., Ed. VCH Verlagsgesellschaft, Weinheim. pp. 401–433.

Gadd, G.M. 1992. Microbial control of heavy metal pollution. In *Microbial Control of Pollution.* Fry, J.C., Gadd, G.M., Herbert, R.A., Jones, C.W., and Watson-Craik, I.A., Eds. 48th Symp. Soc. Gen. Microbiol., March, 1992, University Cardiff. Cambridge University Press, Cambridge, U.K.

Gadd, G.M. and Griffiths, A.J. 1978. Microorganisms and heavy metal toxicity. *Microb. Ecol.* 4:303–317.

Galin, T., Mcdowell, C., and Yaron, B. 1990. The effect of volatilization on the mass flow of a nonaqueous pollutant liquid mixture in an inert porous medium: experiments with kerosene. *J. Soil Sci.* 41(4):631–642.

Galun, M., Keller, P., Malki, D., Feldstein, H., Galun, E., Siegal, S.M., and Siegal, B.Z. 1983. *Science.* 219:285–286.

Gamerdinger, A.P., Achin, R.S., and Traxler, R.W. 1995. Biodegradation and bioremediation: effect of aliphatic nonaqueous phase liquids on naphthalene biodegradation in multiphase systems. *J. Environ. Qual.* 24:1150–1156.

Ganeshalingham, S., Legge, R.L., and Anderson, W.A. 1994. Surfactant-enhanced leaching of polyaromatic hydrocarbons from soil. *Process Saf. Environ. Prot.* 72(B4):247–251.

Gannon, J.T., Manilal, V.B., and Alexander, M. 1991. Relationship between cell surface properties and transport of bacteria through soil. *Appl. Environ. Microbiol.* 57:190–193.

Gannon, J.T., Mingelgrin, U., Alexander, M., and Wagenet, R.J. 1991. Bacterial transport through homogeneous soil. *Soil Biol. Biochem.* 23:1155–1160.

Garbisu, C., Ishii, T., Smith, N.R., Yee, B.C., Carlson, D.E., Yee, A., Buchanan, B.B., and Leighton, T. 1995. Mechanisms regulating the reduction of selenite by aerobic Gram (+) and (–) bacteria. *Biorem. Inorg. Pap. 3rd Int.* In Situ *On-Site Bioreclam. Symp.* Hinchee, R.E., Means, J.L., and Burris, D.R., Eds. Battelle Press, Columbus, OH. pp. 125–131.

Garcia-Herruzo, F., Gomez-Lahoz, C., Rodriguez-Jimenez, J.J., Wilson, D.J., Garcia-Delgado, R.A., and Rodriguez-Maroto, J.M. 1994. Influence of water evaporation on soil vapor extraction (SVE). *Water Sci. Technol.* 30(7):115–118.

Garrett, B.C., McKown, M.M., Miller, M.P., Riggin, R.M., and Warner, J.S. 1981. Development of a solid waste leaching procedure and manual. In *Proc. 7th Annu. Res. Symp.*, Philadelphia, March 16–18, 1981. PB81-173882.

Garvey, K.J., Stewart, M.H., and Yall, I. 1985. A genetic characterization of hydrocarbon growth substrate utilization by *Acinetobacter* (Strain P-7): evidence for plasmid mediated decane utilization. *Abstr. Annu. Mtg. Am. Soc. Microbiol.* p. 261.

Gatchett, A. and Banerjee, P. 1995. Evaluation of the BioGenesis^{SM} soil washing technology. *J. Haz. Mater.* 40(2):165–173.

Gatt, S., Bercovier, H., and Barenholz, Y. 1991. Use of liposomes for combating oil spills and their potential application to bioreclamation. In *On-Site Bioreclamation: Processes for Xenobiotic and Hydrocarbon Treatment.* Hinchee, R.E. and Olfenbuttel, R.F., Eds. Butterworth-Heinemann, Stoneham, MA. pp. 293–312.

Gaydardjiev, S., Hadjihristova, M., and Tichy, R. 1996. Opportunities for using two low-cost methods for treatment of metal bearing aqueous streams. *Miner. Eng.* 9(9):947–964.

Geerdink, M.J., van Loosdrecht, M.C.M., and Luyben, K.Ch.A.M. 1996. Biodegradability of diesel oil. *Biodegradation.* 7:73–81.

Geerdink, M.J., Kleijintjensm, R.H., van Loosdrecht, M.C.M., and Luyben, K.C.A.M. 1996. Microbial decontamination of polluted soil in a slurry process. *J. Environ. Eng.* 122(11):975–982.

Geesey, G.G. 1982. *Am. Soc. Microbiol. News.* 48:9–14.

Gehron, M.J. and White, D.C. 1983. Sensitive measurements of phospholipid glycerol in environmental samples. *J. Microbiol. Methods.* 1:23–32.

General Motors. 1980. Report on Phase I Testing and Phase II Continuance Pursuant to a Delayed Compliance Order Issued on February 15, 1980 by the U.S. EPA. Prepared for U.S. EPA, Region IX.

Genner, C. and Hill, E.C. 1981. Fuels and oils. In *Microbial Biodeterioration: Economic Microbiology.* Rose, A.H., Ed. Academic Press, New York. 6:260–306.

George, E.J. and Neufeld, R.D. 1989. *Biotechnol. Bioeng.* 33:1306–1310.

Gerhards, U. and Weller, H. 1977. Die Aufnahme von Quecksilber, Cadmium and Nickel durch *Chlorella pyrenoidosa. Z. Pflanzenphysiol.* 82:292–300.

Gerhardt, P., Murray, R.G.E., Costilow, R.N., Nester, E.W., Wood, W.A., Krieg, N.R., and Phillips, G.B. 1981. *Manual of Methods for General Bacteriology.* ASM, Washington, D.C.

Gerson, D.F. and Zajic, J.E. 1979. Comparison of surfactant production from kerosene by four species of *Corynebacterium. Antonie van Leeuwenhoek J. Microbiol. Serol.* 45:81–94.

Ghassemi, M. 1988. Innovative *in situ* treatment technologies for cleanup of contaminated sites. *J. Haz. Mat.* 17:189–206.

Ghassemi, M., Panahloo, A., and Quinlivan, S. 1984a. Comparison of physical and chemical characteristics of shale oil fuels and analogous petroleum products. *Environ. Toxicol. Chem.* 3:511–535.

Ghassemi, M., Panahloo, A., and Quinlivan, S. 1984b. Physical and chemical characteristics of some widely used petroleum fuels: a reference data base for assessing similarities and differences between synfuel and petrofuel products. *Energ. Sourc.* 7:377–401.

Ghayeni, S.B.S., Madaeni, S.S., Fane, A.G., and Schneider, R.P. 1996. Aspects of microfiltration and reverse osmosis in municipal wastewater reuse. *Desalination.* 106(1–3, Desalination and Water Reuse). pp. 25–29.

Ghiorse, W.C. and Balkwill, D.L. 1983. Enumeration and morphological characterization of bacteria indigenous to subsurface environments. *Dev. Ind. Microbiol.* 24:213–224.

Ghiorse, W.C. and Balkwill, D.L. 1985. Microbiological characterization of subsurface environments. In *Ground Water Quality.* Ward, C.H., Giger, W., and McCarty, P.L., Eds. John Wiley & Sons, New York.

Ghisalba, O. 1983. Microbial degradation of chemical waste, an alternative to physical methods of waste disposal. *Experientia.* 39:1247–1257.

Gholson, R.K., Guire, P., and Friede, J. 1972. Assessment of Biodegradation Potential for Controlling Oil Spills. NTIS Report No. AD-759 848. National Technical Information Service, Springfield, VA.

Ghosh, S., Ombregt, J.P., and Pipyn, P. 1985. Methane production from industrial wastes by two-phase anaerobic digestion. *Water Res.* (G.B.). 19:1083–1088.

Ghoshal, S., Ramaswami, A., and Luthy, R.G. 1995. Biodegradation of naphthalene from nonaqueous-phase liquids. In *Microbial Processes for Bioremediation. Pap. 3rd Int. In Situ On-Site Bioreclam. Symp.* Hinchee, R.E., Vogel, C.M., and Brockman, F.J., Eds. Battelle Press, Columbus, OH. pp. 75–82.

Ghoshal, S., Ramaswami, A., and Luthy, R.G. 1996. Biodegradation of naphthalene from coal tar and heptamethylnonane in mixed batch systems. *Environ. Sci. Technol.* 30(4):1282–1291.

Ghuman, A.S. 1995. TESVE model for design of soil vapor extraction systems with thermal enhancement. In *Proc. HAZMACON 95, 12th Haz. Mater. Mgmt. Conf. Exhib.* pp. 235–243.

Gibson, D.T. 1971. The microbial oxidation of aromatic hydrocarbons. *CRC Crit. Rev. Microbiol.* 1:199.

Gibson, D.T. 1976. Microbial degradation of polycyclic aromatic hydrocarbons. In *Proc. 3rd Int. Biodegradation Symp.,* Aug. 17–23, 1975, Kingston, RI., Sharply, J.M., and Kaplan, A.M., Eds. Applied Science Publishers, London. pp. 57–66.

Gibson, D.T. 1977. Biodegradation of aromatic petroleum hydrocarbons. In *Fate and Effects of Petroleum Hydrocarbons in Marine Organisms and Ecosystems.* Wolfe, D.A., Ed. Pergamon Press, Oxford. pp. 36–46.

Gibson, D.T. 1978. Microbial transformations of aromatic pollutants. In *Aquatic Pollutants.* Hutzinger, O., Van Lelyveld, L.H., and Zoeteman, B.C.J., Eds. Pergamon Press, New York.

Gibson, D.T. 1982. Microbial degradation of hydrocarbons. *Environ. Toxicol. Chem.* 5:237–250.

Gibson, D.T., Koch, J.R., and Kallio, R.E. 1968. Oxidative degradation of aromatic hydrocarbons by microorganisms. I. Enzymatic formation of catechol from benzene. *Biochemistry.* 7:2653–2662.

Gibson, D.T. and Mahadevan, V. 1975. Oxidation of the carcinogens benzo[a]pyrene and benzo[a]anthracene to dihydrols by a bacterium. *Science.* 189:295–297.

Gibson, D.T., Mahadevan, V., Jerina, D.M., Yagi, H., and Yeh, H.J.C. 1975. *Science.* 189:295.

Gibson, D.T. and Subramanian, V. 1984. Microbial degradation of aromatic hydrocarbons. In *Microbial Degradation of Organic Compounds.* Gibson, D.T., Ed. Marcel Dekker, New York. pp. 181–252.

Giddens, J. 1976. Spent motor oil effects on soil and crops. *J. Environ. Qual.* 5:179–181.

Giger, W. and Roberts, P.V. 1978. Characterization of persistent organic carbon. Vol. 2. In *Water Pollution Microbiology.* Mitchell, R., Ed. Wiley-Interscience, New York. pp. 135–176.

Gill, C.O. and Ratledge, C. 1973. Regulation of *de novo* fatty acid biosynthesis in the *n*-alkane-utilizing yeast, *Candida* 107. *J. Gen. Microbiol.* 78:337–347.

Gillan, F.T. 1983. Analysis of complex fatty acid methyl ester mixtures on non-polar capillary GC columns. *J. Chromatgr. Sci.* 21:293–297.

Gilliam, J.W. and Sample, E.C. 1968. Hydrolysis of pyrophosphate in soils: pH and biological effects. *Soil Sci.* 106:352–357.

Giovannoni, S.J., Britschgi, T.B., Moyer, C.L., and Field, K.G. 1990. Genetic diversity in Sargasso Sea bacterioplankton. *Nature.* 345:60–63.

Girling, C.A. 1984. Selenium in agriculture and the environment. *Agric. Ecosyst. Environ.* 11:37–65.

Gissel-Nielsen, G. 1971. Selenium content of some fertilisers and their influence on uptake of selenium in plants. *J. Agric. Food Chem.* 19:565–566.

Glaser, J.A. 1990. Hazardous waste degradation by wood-degrading fungi. In *Biotechnology and Biodegradation, Advances in Applied Biotechnology Series.* Vol. 4. Kamely, D., Chakrabarty, A., and Ommen, G.S., Eds. Gulf, New York.

Glaser, J.A. 1991. Nutrient-enhanced bioremediation of oil-contaminated shoreline: the Valdez experience. In *On-Site Bioreclamation.* Hinchee R.E. and Olfenbuttel, R.F., Eds. Butterworth-Heinemann, Stoneham, MA. pp. 336–384.

Glaser, J.A., Tzeng, J.-W., and McCauley, P.T. 1995. Slurry biotreatment of organic-contaminated solids. In *Biological Unit Processes Hazardous Waste Treatment. Pap. 3rd. Int. In Situ On-Site Bioreclam. Symp.* Hinchee, R.E., Skeen, R.S., and Sayles, G.D., Eds. Battelle Press, Columbus, OH. pp. 145–152.

Gleim, D., Milch, H., and Kracht, M. 1995. Database on microbial degradation of soil pollutants. In *Biological Unit Processes for Hazardous Waste Treatment. Pap. 3rd. Int. In Situ On-Site Bioreclam. Symp.* Hinchee, R.E., Skeen, R.S., and Sayles, G.D., Eds. Battelle Press, Columbus, OH. pp. 331–332.

Glenn, J.K. and Gold, M.H. 1983. Decolorization of several polymeric dyes by the lignin degrading basidiomycete Phanerochaete chrysosporium. *Appl. Environ. Microbiol.* 45:1741–1747.

Gloxhuber, C. 1974. Toxicological properties of surfactants. *Arch. Toxicol.* 32:245–270.

Gocke, K. 1977. Comparisons of methods for determining the turnover time of dissolved organic compounds. *Mar. Biol.* 42:131–141.

REFERENCES

Godbout, J.G., Comeau, Y., and Greer, C.W. 1995. Soil characteristics effects on introduced bacterial survival and activity. In *Bioaugmentation for Site Remediation. Pap. 3rd Int. In Situ On-Site Bioreclam. Symp.* Hinchee, R.E., Fredrickson, J., and Alleman, B.C., Eds. Battelle Press, Columbus, OH. pp. 115–120.

Goddard, P.A. and Bull, A.T. 1989. Accumulation of silver by growing and non-growing populations of *Citrobacter intermedius* B6. *Appl. Microbiol. Biotechnol.* 31:314–319.

Godsy, E.M. 1980. Isolation of *Methanobacterium bryantii* from a deep aquifer by using a novel broth-antibiotic dish method. *Appl. Environ. Microbiol.* 39:1074–1075.

Goldstein, D.J. 1982. Air and Steam Stripping of Toxic Pollutants. Vol. 1, No. 68-03-002. U.S. EPA.

Goldstein, R.M., Mallory, L.M., and Alexander, M. 1985. Reasons for possible failure of inoculation to enhance biodegradation. *Appl. Environ. Microbiol.* 50:977–983.

Goma, G., Al Ani, D., and Pareilleus, A. 1976. Hydrocarbon uptake by microorganisms. Paper presented at 5th Int. Fermentation Symp. Berlin.

Goma, G., Pareilleux, A., and Durand, G. 1974. Aspects physicochemiques de l'assimilation des hydrocarbures par *Candida lipolytica. Agric. Biol. Chem.* 38:1273–1280.

Gomez-Lahoz, C., Rodriguez, J.J., Rodriguez-Maroto, J.M., and Wilson, D.J. 1994. Biodegradation phenomena during soil vapor extraction. III. Sensitivity studies for two substrates. *Sep. Sci. Technol.* 29(10):1275–1291.

Gomez-Lahoz, C., Rodriguez-Maroto, J.M., and Wilson, D.J. 1994. Soil cleanup by *in-situ* aeration. XVII. Field-scale model with distributed diffusion. *Sep. Sci. Technol.* 29(10):1251–1274.

Gomez-Lahoz, C., Rodriguez-Maroto, J.M., and Wilson, D.J. 1995. Soil cleanup by *in-situ* aeration. XXII. Impact of natural soil organic matter on cleanup rates. *Sep. Sci. Technol.* 30(5):659–682.

Gomez-Lahoz, C., Rodriguez-Maroto, J.M., Wilson, D.J., and Tamamushi, K. 1994. Soil clean up by in-situ aeration. XV. Effects of variable air flow rates in diffusion-limited operation. *Sep. Sci. Technol.* 29(8):943–969.

Gonenc, I.E. and Harremoes, P. 1985. Nitrification in rotating disc systems — I. Criteria for transition from oxygen to ammonia rate limitation. *Water Res.* 19:1119–1127.

Goodin, J.D. and Webber, M.D. 1995. Persistence and fate of anthracene and benzo(a)pyrene in municipal sludge treated soil. *J. Environ. Qual.* 24(2):271–178.

Gordon, D.C., Dale, J., and Keizer, P.D. 1978. Importance of sediment working by the deposit-feeding lychaete *Arenicola marina* on the weathering rate of sediment-bound oil. *J. Can. Fisheries Res. Board.* 35:591–603.

Goring, C.A.I. 1962. Theory and principles of soil fumigation. *Adv. Pest Control Res.* 5:47–84.

Gorny, N., Wahl, G., Brune, A., and Schink, B. 1992. A strictly anaerobic nitrate-reducing bacterium growing with resorcinol and other aromatic compounds. *Arch. Microbiol.* 158:48–53.

Gossett, J.M. and Lincoff, A.H. 1984. In *Gas Transfer at Water Surfaces.* Brutsaert, W. and Jirka, G.H., Eds. D. Reidel, Boston. pp. 17–25.

Goswami, P.C., Singh, H.D., Bhagat, S.D., and Baruah, J.N. 1983. *Biotechnol. Bioeng.* 25:2929–2943.

Gottschalk, G. 1979. *Bacterial Metabolism.* Springer-Verlag, New York.

Gourdon, R., Rus, E., Bhende, S., and Sofer, S.S. 1990. Mechanism of cadmium uptake by activated sludge. *Appl. Microbiol. Biotechnol.* 34:274–278.

Grabowski, T.M. and Raymond, A.J. 1984. Disposal and transportation of refinery wastes. In *Hazardous and Toxic Wastes: Technology, Management, and Health Effects.* Majumdar, S.K. and Miller E.W., Eds. The Pennsylvania Academy of Sciences. Easton, PA. pp. 188–203.

Grappelli, A., Hard, J.S., Pietrosanti, W., Tomati, U., Campanella, L., Cardarelli, E., and Cordatore, M. 1989. Cadmium decontamination of liquid streams by *Arthrobacter* species. In *Water Pollution Research and Control,* Brighton, Part 5, Vol. 21. Lijklema, L. et al., Eds. *Water Sci. and Technol.* pp. 1759–1762.

Graves, D., Chase, L., and Ray, J. 1995. Analytical and statistical approaches to validate biological treatment of petroleum hydrocarbons in soil. In *Proc. HAZMACON 95, 12th Haz. Mater. Mgmt. Conf. Exhib.* pp. 323–332.

Graves, D., Dillon, T., Jr., Hague, K., Klein, J., McLaughlin, J., Wilson, B., and Olson, G. 1995. Application and performance of remote bioventing systems powered by wind. In *In Situ Aeration: Air Sparging, Bioventing, Relat. Rem. Processes. Pap. 3rd Int. In Situ On-Site Bioreclam. Symp.* Hinchee, R.E., Miller, R.N., and Johnson, P.C., Eds. Battelle Press, Columbus, OH. pp. 401–407.

Gray, T.R.G. 1978. Microbiological aspects of the soil, plant, aquatic, air, and animal environments: the soil and plant environments. In *Pesticide Microbiology.* Hill, I.R. and Wright, S.J.L., Eds. Academic Press, New York. pp. 19–38.

Gray, T.R.G. and Parkinson, D. 1968. *The Ecology of Soil Bacteria.* Liverpool University Press, Liverpool, U.K.

Grbic-Galic, D. and Vogel, T.M. 1986. Toluene and benzene trnsformation by ferulate-acclimated methanogenic consortia. *Abstr. Annu. Mtg. Am. Soc. Microbiol.* p. 303.

Grbic-Galic, D. and Vogel, T.M. 1987. Transformation of toluene and benzene by mixed methanogenic cultures. *Appl. Environ. Microbiol.* 53:254–260.

Grbic-Galic, D. and Young, L.Y. 1985. Methane fermentation of ferulate and benzoate: anaerobic degradation pathways. *Appl. Environ. Microbiol.* 50:292–297.

Green, M.K. and Hardy, P.J. 1985. *Water Pollut. Control.* 84:44.

Green, W.J., Lee, G.F., and Jones, R.A. 1981. Clay-soils permeability and hazardous waste storage. *J. Water Pollut. Control Fed.* 53:1347–1354.

Greer, C., Masson, L., Comeau, Y., Brousseau, R., and Samson, R. 1993. Application of molecular biology techniques for isolating and monitoring pollutant-degrading bacteria. *Water Pollut. Res. J. Can.* 28(2):275–287.

Greer, C.W., Masson, L., Comeau, Y., Brousseau, R., and Samson, R. 1994. Monitoring the fate of bacteria released into the environment using chromosomally integrated reporter genes. In *Applied Biotechnology for Site Remediation.* Hinchee, R.E., Anderson, D.B., Metting, F.B., Jr., and Sayles, G.D., Eds. Lewis Publishers, Boca Raton, FL. pp. 405–409.

Griffin, C.J., Kampbell, D., and Blaha, F.A. 1993. Biosparging an aviation gasoline spill. *Hydrocarbon Contam. Soils.* 3:351–361.

Griffith, P.C. and Fletcher, M. 1991. Hydrolysis of protein and model dipeptide substrates by attached and nonattached marine *Pseudomonas* sp. strain NCIMB 2021. *Appl. Environ. Microbiol.* 57:2186–2191.

Grimberg, S.J. and Aitken, M.D. 1995. Biodegradation kinetics of phenanthrene solubilized in surfactant micelles. In *Microbial Processes for Bioremediation. Pap. 3rd Int. In Situ On-Site Bioreclam. Symp.* Hinchee, R.E., Vogel, C.M., and Brockman, F.J., Eds. Battelle Press, Columbus, OH. pp. 59–66.

Grimberg, S.J., Stringfellow, W.T., and Aitken, M.D. 1996. Quantifying the biodegradation of phenanthrene by *Pseudomonas stutzeri* P16 in the presence of a nonionic surfactant. *Appl. Environ. Microbiol.* 62(7):2387–2392.

Grimes, D.J. and Colwell, R.R. 1986. Viability and virulence of *Escherichia coli* suspended by membrane chamber in semitropical ocean water. *FEMS Microbiol. Lett.* 34:161–165.

Groenewegen, D. and Stolp, H. 1981. Microbial breakdown of polycyclic aromatic hydrocarbons. In *Decomposition of Toxic and Nontoxic Organic Compounds in Soils.* Overcash, M.R., Ed. Ann Arbor Science Publishers, Ann Arbor, MI. pp. 233–240.

Gromicko, G.J., Smock, M., Wong, A.D., and Sheriday, B. 1995. Pilot study: fixed-film bioreactor to enhance carbon adsorption. In *Biological Unit Processes for Hazardous Waste Treatment.* Hinchee, R.E., Sayles, G.D., and Skeen, R.S., Eds. Battelle Press, Columbus, OH. pp. 63–69.

Grosser, R.J., Warshawsky, D., and Kinkle, B.K. 1994. The effects of fulvic acids extracted from soils on the mineralization of pyrene by an isolated *Mycobacterium* sp. *Proc. 9th Annu. Conf. Haz. Waste Rem.* Erickson, L.E., Ed. pp. 309–321.

Grosser, R.J., Warshawsky, D., and Robie Vestal, J. 1991. Indigenous and enhanced mineralisation of pyrene, benzo(a)pyrene and carbazole in soils. *Appl. Environ. Microbiol.* 57:3462–3469.

Grove, G.W. 1980. In *Disposal of Industrial and Oily Sludges by Land Cultivation.* Shilesky, D.M., Ed. Resource Systems and Management Association, Northfield, NJ. pp. 25–32.

Gruiz, K. and Kriston, E. 1995. *In situ* bioremediation of hydrocarbon in soil. *J. Soil Contam.* 4(2):163–173.

Guenther, W.B. 1975. *Chemical Equilibrium: A Practical Introduction for the Physical and Life Sciences.* Plenum Press, New York.

Guerin, W.F. and Jones, G.E. 1988. Mineralization of phenanthrene by a *Mycobacterium* sp. *Appl. Environ. Microbiol.* 54:937–944.

Guha, S. and Jaffe, P.R. 1996a. Bioavailability of hydrophobic compounds partitioned into the micellar phase of nonionic surfactants. *Environ. Sci. Technol.* 30(4):1382–1392.

Guha, S. and Jaffe, P.R. 1996b. Biodegradation kinetics of phenanthrene partitioned into the micellar phase of nonionic surfactants. *Environ. Sci. Technol.* 30(2):605–611.

Guidin, C. and Syratt, W.J. 1975. Biological aspects of land rehabilitation following hydrocarbon contamination. *Environ. Pollut.* 8:107–112.

Guiot, S.R. and Van den Berg, L. 1985. Dynamic performance of an anaerobic reactor combining an upflow sludge blanket and a filter for the treatment of sugar waste. In *Proc. 39th Ind. Waste Conf.,* May 8–10, 1984, Purdue University. Butterworth, Boston. pp. 705–717.

Gunkel, W. 1967. *Helgol. Wiss. Meeresunters.* 15:210–224.

Gupta, S.K., Djafari, S.H., and Zhang, J. 1995. Oxygen transport in an *in situ* bioremediation application. In *Innov. Technol. Site Rem. Haz. Waste Mgmt. Proc. Natl. Conf.* pp. 165–172.

Gutnick, D.L. and Rosenberg, E. 1977. Oil tankers and pollution: a microbiological approach. *Annu. Rev. Microbiol.* 31:379–396.

Gutnick, D.L. and Rosenberg, E. 1979. Oil tankers and pollution: a microbiological approach. *Annu. Rev. Microbiol.* 31:379–396.

Guymon, E. 1993. Water/surfactant process for recovering hydrocarbons from soil in the absence of emulsifying the oil. U.S. Patent 5,252,138. 12 Oct. 1993. 9 pp.

Gvozdyak, P.I., et al. 1985. *Sov. J. Water Chem. Technol.* 7:93.

Gwynne, P. and Bishop, J., Jr. 1975. Bring on the bugs. *Newsweek.* 86:116.

H

Haapala, R. and Linko, S. 1993. Production of *Phanerochaete chrysosporium* lignin peroxidase under various culture conditions. *Appl. Microbiol. Biotechnol.* 40:494–498.

Haber, F. and Weiss, J. 1934. *Proc. R. Soc. Ser. A.* 147:332.

REFERENCES

Hackman, E.E., III. 1978. *Toxic Organic Chemicals: Destruction and Waste Treatment.* Noyes Data Corp., Park Ridge, NJ.
Haemmerli, S.D., Leisola, M.S.A., Sanglard, D., and Fiechter, A. 1986. Oxidation of benzo(a)pyrene by extracellular ligninase of *Phanerochaete chrysosporium. J. Biol. Chem.* 261:6900–6903.
Hagedorn, C., McCoy, E.L., and Rahe, T.M. 1981. The potential for groundwater contamination from septic effluents. *J. Environ. Qual.* 10:1–8.
Hager, D.G., Loven, C.G., and Giggy, C.L. 1987. In *Haz. Mater. Control Res. Inst. Natl. Conf. Exhib.,* Washington, D.C.
Hager, D.G. and Smith, C.E. 1985. In *Proc. 2nd Annu. Haz. Mater. Mgmt. Conf. West,* Long Beach, CA.
Hahn, D., Amann, R.I., Ludwig, W., Akkermans, A.D.L., and Schleifer, K.-H. 1992. Detection of microorganisms in soil after *in situ* hybridization with rRNA targeted fluorescently labeled oligonucleotides. *J. Gen. Microbiol.* 138:879–887.
Hahn, D., Amann, R.I., and Zeyer, J. 1993. Whole cell hybridisation of *Frankia* strains with fluorescence- or digoxigenin-labeled 16S rRNA targeted oligonucleotide probes. *Appl. Environ. Microbiol.* 59:1709–1716.
Hahn, D., Kester, R., Starrenburg, M.J.C., and Akkermans, A.D.L. 1990. Extraction of ribosomal RNA from soil for detection of *Frankia* with oligonucleotide probes. *Arch. Microbiol.* 154:329–335.
Hahn, D., Starrenburg, M.J.C., and Akkermans, A.D.L. 1990. Oligonucleotide probes that hybridise with rRNA as a tool to study *Frankia* strains in root nodules. *Appl. Environ. Microbiol.* 56:1342–1346.
Hahne, H.C.H. and Kroontje, W. 1973. The simultaneous effect of pH and chloride concentrations upon mercury (II) as a pollutant. *Proc. Soil Sci. Soc. Am.* 37:838–843.
Haider, K.M. and Martin, J.P. 1988. Mineralization of ^{14}C-labeled humic acids and of humic-acid bound ^{14}C-xenobiotics by *Phanerochaete chrysosporium. Soil Biol. Biochem.* 20:425–429.
Haines, J.R. and Alexander, M. 1974. Microbial degradation of high-molecular-weight alkanes. *Appl. Microbiol.* 28:1084–1085.
Haines, J.R., Pesek, E., Roubal, G., Bronner, A., and Atlas, R. 1981. Microbially mediated chemical evolution of crude oils spilled into differing ecosystems. *Abstr. Annu. Mtg. Am. Soc. Microbiol.* p. 213.
Hallberg, R.O. and Martinelli, R. 1976. *In-situ* purification of ground water. *Ground Water.* 14:88–93.
Ham, R.K., Anderson, M.A., Stegmann, R., and Stanforth, R. 1979. Background Study on the Development of a Standard Leaching Test. EPA-600/2-79-109. 274 pp. U.S. EPA, Cincinnati, OH.
Hamaker, J.W. 1966. Mathematical prediction of cumulative levels of pesticides in soil. *Adv. Chem. Ser.* 60:122–131.
Hamaker, J.W. 1972. Decomposition: quantitative aspects. In *Organic Chemicals in Soil Environment.* Goring, C.A.I. and Hamaker, J.W., Eds. Marcel Dekker, New York.
Hambuckers-Berhin, F. and Remacle, J. 1990. Cadmium sequestration in cells of two strains of *Alcaligenes eutrophus. FEMS Microbiol. Ecol.* 73:309–316.
Hamer, D.H. 1986. Metallothioneins. *Annu. Rev. Biochem.* 55:913–951.
Hammel, K.E. 1992. Oxidation of aromatic pollutants by lignin-degrading fungi and their extracellular enzymes. In *Metal Ions in Biological Systems.* Sigel, H. and Sigel, A., Eds. Vol. 28. Marcel Dekker, New York. pp. 41–60.
Hammel, K.E., Gai, Z.G., Green, B., and Moen, M.A. 1992. Oxidative degradation of phenanthrene by the ligninolytic fungus *Phanerochaete chrysosporium. Appl. Environ. Microbiol.* 58:1831–1838.
Hammel, K.E., Green, B., and Gai, W.Z. 1991. *Proc. Natl. Acad. Sci. U.S.A.* 88:10605–10608.
Hammel, K.E., Kalyanaraman, B., and Kirk, T.K. 1986. Oxidation of polycyclic aromatic hydrocarbons and dibenzo(p)dioxins by *Phanerochaete chrysosporium* ligninase. *J. Biol. Chem.* 261:16948–16952.
Hampton, G.J., Webster, J.J., and Leach, F.R. 1983. The extraction and determination of ATP from soil and subsurface material. In *Land Treatment of Hazardous Wastes.* Parr, J.F., Marsh, P.B., and Kla, J.M., Eds. Noyes Data Corp., Park Ridge, NJ. p. 15.
Hampton, G.J., Wilson, J.T., Ghiorse, W.C., and Leach, F.R. 1985. Determination of microbial cell numbers in subsurface samples. *Ground Water.* 23:17–25.
Hance, R.J. and McKone, C.E. 1971. Effect of concentration on the decomposition rates in soil of atrizine, linuron and picloram. *Pest. Sci.* 3:31–34.
Hand, D.W., Crittendan, J.C., Gehin, J.C., and Lykins, Jr., B.W. 1986. Design evaluation of an air-stripping tower for removing VOCs from groundwater. *J. Am. Water Works Assoc.* 78:87–97.
Hanneman, T.F., Johnstone, D.L., Yonge, D.R., Petersen, J.N., Peyton, B.M., and Skeen, R.S. 1995. Control of bacterial exopolysaccharide production. In *Microbial Processes for Bioremediation.* Hinchee, R.E., Vogel, C.M., and Brockman, F.J., Eds. Battelle Press, Columbus, OH. pp. 323–328.
Hansch, C., Quinlan, J.E., and Lawrence, G.L. 1968. The linear free-energy relationship between partition coefficients and the aqueous solubility of organic liquids. *J. Org. Chem.* 33:347–350.
Hanson, K.G., Desai, J.D., and Desai, A.J. 1993. A rapid and simple screening technique for potential crude oil degrading microorganisms. *Biotechnol. Tech.* 7(10):745–748.
Harder, W. 1981. Enrichment and characterization of degrading organisms. In *FEMS Symposium No. 12. Microbial Degradation of Xenobiotics and Recalcitrant Compounds.* Leisinger, T., Hutter, R., Cook, A.M., and Nuesch, J., Eds. Academic Press, New York. 12:77–96.
Hardy, R.W., Burns, R.C., and Holsten, R.D. 1973. Applications of the acetylene-ethylene assay for measurement of nitrogen fixation. *Soil Biol. Biochem.* 5:47–81.

Harmsen, J. 1991. Possibilities and limitations of landfarming for cleaning contaminated soils. In *On-Site Bioreclamation: Processes for Xenobiotic and Hydrocarbon Treatment.* Hinchee, R.E. and Olfenbuttel, R.F., Eds. Butterworth-Heinemann, Stoneham, MA. pp. 255–272.

Harmsen, J., Velthorst, H.J., and Bennehey, I.P.A.M. 1994. Cleaning of residual concentrations with an extensive form of landfarming. In *Applied Biotechnology for Site Remediation.* Hinchee, R.E., Anderson, D.B., Metting, F.B., Jr., and Sayles, G.D., Eds. Lewis Publishers, Boca Raton, FL. pp. 84–91.

Harris, P.A. and Ramelow, G.J. 1990. Binding of metal ions by particulate biomass derived from *Chlorella vulgaris* and *Scenedesmus quadricauda. Environ. Sci. Technol.* 24:220–228.

Harris, R.F. 1981. Effect of water potential on microbial growth and activity. In *Water Potential Relations in Soil Microbiology.* SSSA Special Publication No. 9. Soil Science Society of America, Madison, WI. pp. 23–95.

Harrop, A.J., Woodley, J.M., and Lilly, M.D. 1992. Production of naphthalene-*cis*-glycol by *Pseudomonas putida* in the presence of organic solvents. *Enzyme Microb. Technol.* 14:725–730.

Hart, S. 1996. *In situ* bioremediation: defining the limits. *Environ. Sci. Technol.* 30(9):398–401.

Hartley, G.S. 1969. Evaporation of pesticides. *Adv. Chem. Ser.* 86:115–134.

Harvey, P.J., Schoemaker, H.E., and Palmer, J.M. 1987. Lignin degration by white rot fungi. *Plant Cell Environ.* 10:709–714.

Harvey, R.G. 1982. Polycyclic hydrocarbons and cancer. *Am. Sci.* 70:386–393.

Harvey, R.W., George, L.H., Smith, R.L., and LeBlanc, D.R. 1989. Transport of microspheres and indigenous bacteria through a sandy aquifer: results of natural- and forced-gradient tracer experiments. *Environ. Sci. Technol.* 23:51–56.

Harvey, S., Elashvili, I., Valdes, J.J., Kamely, D., and Chakrabarty, A.M. 1990. Enhanced removal of Exxon Valdez spilled oil from Alaskan gravel by a microbial surfactant. *Bio/Technol.* 8:228–230.

Hater, G.R., Green, R.B., Solsrud, T., Bower, D.A., Jr., and Barbush, J.A. 1995. Prediction of bioremediation rates at multiple fixed-site soil centers. In *Biological Unit Processes for Hazardous Waste Treatment. Pap. 3rd. Int.* In Situ *On-Site Bioreclam. Symp.* Hinchee, R.E., Skeen, R.S., and Sayles, G.D., Eds. Battelle Press, Columbus, OH. pp. 171–178.

Hatfield, K. and Stauffer, T.B. 1992. Nonequilibrium modeling of transport in fine sand containing residual decane. *Haz. Waste Haz. Mater.* 9(4):369–382.

Hattori, T. and Hattori, R. 1976. The physical environment in soil microbiology: an attempt to extend principles of microbiology to soil microorganisms. *Crit. Rev. Microbiol.* 4:423–461.

Hatzinger, P.B. and Alexander, M. 1995. Effect of aging of chemicals in soils on their biodegradability and extractability. *Environ. Sci. Technol.* 29(2)537–554.

Haxo, H.E., Jr., Haxo, R.S., Nelson, N.A., Haxo, P.D., White, R.M., Dakessian, S., and Fong, M.A. 1985. *Liner Materials for Hazardous and Toxic Wastes and Municipal Solid Waste Leachate.* Noyes Publications, Park Ridge, NJ.

Hayden, N.J. and Voice, T.C. 1993. Microscopic observation of a NAPL in a three-fluid-phase soil system. *J. Contam. Hydrol.* 12(3):217–226.

Hayes, K.W., Meyers, J.D., and Huddleston, R.L. 1995. Biopile treatability, bioavailability, and toxicity evaluation of a hydrocarbon-impacted soil. In *Applied Bioremediation of Petroleum Hydrocarbons.* Hinchee, R.E., Kittel, J.A., and Reisinger, H.J., Eds. Battelle Press, Columbus, OH. pp. 249–256.

Hazen, T.C., Lombard, K.H., Looney, B.B., Enzien, M.V., Doughtery, J.M., Fliermans, C.B., Wear, J., and Eddy-Dilek, C.A. 1994. Summary of *in-situ* bioremediation demonstration (methane biostimulation) via horizontal wells at the Savannah River Site Integrated Demonstration Project. In *33rd Hanford Symp. on Health and the Environment — In-Situ Remediation: Scientific Basis for Current and Future Technologies.* Gee, G.W. and Wing, N.R., Eds. Nov. 7–11, Pasco, WA. p. 137.

HazTech News. 1992. 7(10):78.

Healy, J.B., Jr. and Young, L.Y. 1979. Anaerobic biodegradation of eleven aromatic compounds to methane. *Appl. Environ. Microbiol.* 38:84–89.

Heath, W.O., Caley, S.M., Peurrung, L.M., Lerner, B.D., and Moss, R.W. 1994. Feasibillity of *in situ* electrical corona for soil detoxification. In *In-Situ Rem.: Sci. Basis Curr. Future Technol., Hanford Symp. Health Environ.* 33rd. 2:799–818.

Hegeman, G.D. 1966. Synthesis of the enzymes of the mandelate pathway by *Pseudomonas putida.* I. Synthesis of enzymes by the wild type. *J. Bacteriol.* 91:1140–1154.

Hegeman, G.D. 1972. The evolution of metabolic pathways in bacteria. In *Degradation of Synthetic Organic Molecules in the Biosphere. Natural, Pesticidal, and Various Other Man-Made Compounds.* National Academy of Sciences, Washington, D.C. pp. 56–72.

Heijnen, C.E., Hok-A-Hin, C.H., and van Veen, J.A. 1992. Improvements to the use of bentonite clay as a protective agent, increasing survival levels of bacteria introduced into soil. *Soil Biol. Biochem.* 24:533–538.

Heijnen, C.E. and van Veen, J.A. 1991. A determination of protective microhabitats for bacteria introduced into soil. *FEMS Microb. Ecol.* 85:73–80.

Heitkamp, M.A. and Cerniglia, C.E. 1986. Microbial degradation of *tert*-butylphenyl diphenyl phosphate: a comparative microcosm study among five diverse ecosystms. *Toxicity Assessment Bull.* 1:103–122.

Heitkamp, M.A. and Cerniglia, C.E. 1989. *Appl. Environ. Microbiol.* 55:1968–1973.

Heitkamp, M.A., Franklin, W., and Cerniglia, C.E. 1988. Microbial metabolism of polycyclic aromatic hydrocarbons: isolation and characterisation of a pyrene degrading bacterium. *Appl. Environ. Microbiol.* 54:2549–2555.

REFERENCES

Heitkamp, M.A., Freeman, J.P., and Cerniglia, C.E. 1987. Naphthalene biodegradation in environmental microcosms: estimates of degradation rates and characterization of metabolites. 53:129–136.

Heitzer, A. and Sayler, G.S. 1993. Monitoring the efficacy of bioremediation. *Tibtech.* 11:334.

Henley, E. and Seader, J.D. 1981. *Equilibrium-Stage Separation Operation in Chemical Engineering.* John Wiley & Sons, New York.

Henriksen, N. 1996. Method and equipment for the purification of a liquid such as oil removal from waters. PCT Int. Appl. WO 96 12,678 (Cl. C01F1/52), 2 May 1996, NO Appl. 94/3,956, 19 Oct. 1994. 17 pp.

Hepple, S. 1960. *Trans. Br. Mycol. Cos.* 43:73–79.

Herbes, S.E. and Schwall, L.R. 1978. Microbial transformation of polycyclic aromatic hydrocarbons in pristine and petroleum contaminated sediments. *Appl. Environ. Microbiol.* 35:306–316.

Heringa, J.W., Huybregtse, R., and van der Linden, A.C. 1961. n-Alkane formation by a *Pseudomonas.* Formation and beta-oxidation of intermediate fatty acids. *Antonie van Leeuwenhoek J. Microbiol. Serol.* 27:51–58.

Hertzberg, S., Moen, E., Vogelsang, C., and Ostgaard, K. 1995. Mixed photo-cross-linked polyvinyl alcohol and calcium-alginate gels for cell entrapment. *Appl. Microbiol. Biotechnol.* 43:10–17.

Hesketh, H.E., Schifftner, K.C., and Hesketh, R.P. 1983. Paper No. 38-55.1, 76th Annu. Air Pollut. Control Assoc. Mtg., Atlanta, GA, June 19–24.

Heuckeroth, D.M., Eberle, M.F., and Rykaczewski, M.J. 1995. *In situ* vacuum extraction/bioventing of a hazardous waste landfill. In In Situ *Aeration: Air Sparging, Bioventing, Relat. Rem. Processes. Pap. 3rd Int.* In Situ *On-Site Bioreclam. Symp.* Hinchee, R.E., Miller, R.N., and Johnson, P.C., Eds. Battelle Press, Columbus, OH. pp. 341–349.

Hewitt, A.D. and Lukash, N.J. 1996. Obtaining and Transferring Soils for In-Vial Analysis of Volatile Organic Compounds. Report (CRREL-SR-96-5, SFIM-AEC-ET-Cr-96002; Order No. AD-A306918). Avail. NTIS. From Gov. Rep. Ann. Index. (U.S.). 96(18). Abstr. No. 18-01,204. 16 pp.

Hicks, R.J. and Brown, R.A. 1990. *In-situ* bioremediation of petroleum hydrocarbons. Paper presented at Water Pollut. Control Fed. Annu. Conf., October 7–11, 1990, Washington, D.C.

Higashihara, T., Sato, A., and Simidu, U. 1978. An MPN method for the enumeration of marine hydrocarbon degrading bacteria. *Bull. Jpn. Soc. Sci. Fish.* 44:1127–1134.

Higgins, I.J., Best, D.J., and Hammond, R.C. 1980. New findings in methane-utilizing bacter highlight their importance in the biosphere and their commercial potential. *Nature (London)* Z86(5773):561–564.

Higgins, I.J. and Gilbert, P.D. 1978. The biodegradation of hydrocarbons. In *The Oil Industry and Microbial Ecosystems.* Clater, K.W.A. and Sommerville, H.J., Eds. Heyden and Sons, for Institute of Petroleum, London. pp. 80–117.

Higgins, I.J., Hammond, R.C., Sariaslani, F.S., Best, D., Davies, M.M., Tryhorn, S.E., and Taylor, F. 1979. Biotransformation of hydrocarbons and related compounds by whole organism suspensions of methane-grown *Methylosinus trichosporium* OB 3b. *Biochem. Biophys. Res. Commun.* 89:671–677.

Hildebrandt, W.W. and Wilson, S.B. 1991. JPT, *J. Petrol. Technol.* 43:18–22.

Hill, G.D., McGrahen, J.W., Baker, H.M., Finnerty, D.W., and Bingeman, C.W. 1955. The fate of substituted urea herbicides in agricultural soils. *Agron. J.* 47:93–104.

Hillel, D. 1971. *Soil and Water, Physical Principles and Processes.* Academic Press, New York. 288 pp.

Hinchee, R.E., Downey, D.C., Dupont, R.R., Aggarwal, P., and Miller, R.E. 1991. Enhancing biodegradation of petroleum hydrocarbons through soil venting. *J. Haz. Mater.* 28(3).

Hinchee, R.E. and Miller, R.N. 1990. Bioreclamation of hydrocarbons in the unsaturated zone. In *Hazardous Waste Management of Contaminated Sites and Industrial Risk Assessment.* Pillman, W. and Zirm, K., Eds. Vienna. pp. 641–650.

Hinchee, R.E. and Ong, S.K. 1992. A rapid *in situ* respiration test for measuring aerobic biodegradation rates of hydrocarbons in soil. *J. Air Waste Manage. Assoc.* 42:1305–1312.

Hinchee, R.E., Ong, S.K. et al. 1991. A field treatability test for bioventing. In *Proc. 84th Annu. Mtg. Exhib. Air Waste Mgmt. Assoc.,* Vancouver, British Columbia, Reprint #91-19.4. p. 13.

Hinchee, R.E., Ong, S.K., Miller, R.N., Vogel, C., and Downey, D.C. 1992. Vacuum extraction induced biodegradation of petroleum hydrocarbons. In *Proc. Int. Conf. Subsurf. Contam. Immiscible Fluids.* 1992. Weyer, K.U., Ed. pp. 361–369.

Hirschberg, H., Skane, H., and Throsby, E. 1977. *Plant Cell Physiol.* 16:1167.

Hissett, R. and Gray, T.R.G. 1976. Microsites and time changes in soil microbial ecology. In *The Role of Terrestrial and Aquatic Organisms in Decomposition Processes.* Anderson, J.M. and MacFadyen, A., Eds. Blackwell, Oxford, U.K. pp. 23–39.

Ho, C.L., Shebl, M. A.-A., and Watts, R.J. 1995. Development of an injection system for *in situ* catalyzed peroxide remediation of contaminated soil. *Haz. Waste Haz. Mater.* 12(1):15–25.

Hoag, G.E., Nadim, F., and Dahmani, A.M. 1996. Reaching soil cleanup levels by vapor extraction: laboratory approach. In *Proc. 50th Ind. Waste Conf.* May 8–10, Purdue University, West Lafayette, IN. Ann Arbor Press, Chelsea, MI. 1995. pp. 23–30.

Hobby, M.M. Ltd. 1993. *BioSparge^{SM} Kinetics.* Las Vegas, NV. pp. 10–13.

Hochcachka, P.W. and Somero, G.N. 1984. *Biochemical Adaptation.* Princeton University Press, Princeton, NJ. p. 3.

Hodges, C.S. and Perry, J.J. 1973. A new species of *Eupenicillium* from soil. *Mycologia.* 65:697–702.

Hoeppel, R.E., Kittel, J.A., Goetz, F.E., Hinchee, R.E., and Abbott, J.E. 1995. Bioslurping technology applications at Naval middle distillate fuel remediation sites. In *Applied Bioremediation of Petroleum Hydrocarbons.* Hinchee, R.E., Kittel, J.A., and Reisinger, H.J., Eds. Battelle Press, Columbus, OH. pp. 389–399.

Holdeman, L.V. and Moore, W.E.C. 1972. *Anaerobe Laboratory Manual.* 2nd ed. Virginia Polytechnic Institute and State University. Blacksburg, VA, 130 pp.

Holdren, M.W., Smith, D.L., and Smith, R.N. 1986. Comparison of Ambient Air Sampling Techniques for Volatile Organic Compounds. EPA/600/4-85/067.

Holland, C.D. 1975. *Fundamentals and Modeling of Separation Processes.* Prentice-Hall, Englewood Cliffs, NJ.

Holman, H.-Y. and Tsang, Y.W. 1995. Effects of soil moisture on biodegradation of petroleum hydrocarbons. In In Situ *Aeration: Air Sparging, Bioventing, Relat. Rem. Processes. Pap. 3rd Int.* In Situ *On-Site Bioreclam. Symp.* Hinchee, R.E., Miller, R.N., and Johnson, P.C., Eds. Battelle Press, Columbus, OH. pp. 323–332.

Holroyd, M.L. and Caunt, P., 1995. Large-scale soil bioremediation using white-rot fungi. In *Bioaugmentation for Site Remediation. Pap. 3rd Int.* In Situ *On-Site Bioreclam. Symp.* Hinchee, R.E., Fredrickson, J., and Alleman, B.C., Eds. Battelle Press, Columbus, OH. pp. 181–187.

Hommel, R.K. 1990. Formation and physiological role of biosurfactants produced by hydrocarbon-utilizing microorganisms. *Biodegradation.* 1:107–119.

Hons, F.M., Stewart, W.M., and Hossner, L.R. 1986. Factor interactions and their influence on hydrolysis of condensed phosphates in soils. *Soil Sci.* 141:408–416.

Hornick, S.B. 1983. The interaction of soils with waste constituents. In *Land Treatment of Hazardous Wastes.* Parr, J.F., Marsh, P.B., and Kla, J.M., Eds. Noyes Data Corp., Park Ridge, NJ. pp. 4–19.

Hornick, S.B., Fisher, R.H., and Paolini, P.A. 1983. Petroleum wastes. In *Land Treatment of Hazardous Wastes.* Parr, J.F., Marsh, P.B., and Kla, J.M., Eds. Noyes Data Corp., Park Ridge, NJ. pp. 321–337.

Hornick, S.B., Murray, J.J., and Chaney, R.L. 1979. Overview on utilization of composted municipal sludges. In *Proc. Natl. Conf. Municipal and Industrial Sludge Composting.* Information Transfer, Inc., Silver Springs, MD.

Hornsby, R.G. and Jensen, D.J. 1992. Operating results from an integrated soil and groundwater remediation system: selected for the Superfund Innovative Technology Evaluation Program winner of the Air and Waste Mgmt. Assoc. J. Deane Sensenbaugh award for waste material engineering projects. 1992. *Ground Water Manage.* (Proc. 1992 Petrol. Hydrocarbons and Organic Chem. in Ground Water: Prevention, Detection, and Restoration). 14:205–220.

Horowitz, A. and Atlas, R.M. 1977. Response of microorganisms to an accidental gasoline spillage in an Arctic freshwater ecosystem. *Appl. Environ. Microbiol.* 33:1252–1258.

Horowitz, A.D., Gutnick, D., and Rosenberg, E. 1975. Sequential growth of bacteria on crude oil. *Appl. Microbiol.* 30:10–19.

Horowitz, A., Sexstone, A., and Atlas, R.M. 1978. *Artic.* 31:180–191.

Horowitz, A. and Tiedje, J.M. 1980. Anaerobic degradation of substituted monoaromatic compounds. *Abstr. Annu. Mtg. Am. Soc. Microbiol.* p. 196.

Horvath, A.L. 1982. *Halogenated Hydrocarbons.* Marcel Dekker, New York, NY.

Horvath, R.S. 1972. Microbial co-metabolism and the degradation of organic compounds in nature. *Bacteriol. Rev.* 36:146–155.

Horvath, R.S. and Alexander, M. 1970. Cometabolism of m-chlorobenzoate by an *Arthrobacter. Appl. Microbiol.* 20:254–258.

Hosea, M., Greene, B., McPherson, R., Henzl, M., Aleksander, M.D., and Darnall, D.W. 1986. Accumulation of elemental gold on the alga *Chlorella vulgaris. Inorg. Chim.* 123:161–165.

Hou, C.T. 1982. Microbial transformation of important industrial hydrocarbons. In *CRC Microbial Transformations of Bioactive Compounds.* Vol. I. Rosazza, J.P., Ed. CRC Press, Boca Raton, FL. pp. 81–107.

Hsieh, Y.-P., Tomson, M.B., and Ward, C.H. 1980. Toxicity of water-soluble extracts of No. 2 fuel oil to the freshwater alga *Selenastrum capricornutum. Dev. Ind. Microbiol.* 21:401–409.

Hsu, M.I., Davies, S.H.R., and Masten, S.J. 1993. The use of ozone for the removal of residual trichloroethylene from unsaturated soils. In *Proc. 48th Ind. Waste Conf.,* 1993 (Pub. 1994). pp. 215–225.

Hsu, T.S. and Bartha, R.J. 1976. Hydrolysable and nonhydrolysable 3,4-dichloroaniline humus complexes and their respective rate of biodegradation. *J. Agric. Food Chem.* 24:118–122.

Hu, Z.-C., Korus, R.A., and Stormo, K.E. 1993. Characterization of immobilized enzymes in polyurethane foams in a dynamic bed reactor. *Appl. Microbiol. Biotechnol.* 39:289–295.

Huang, C. and Huang, C.P. 1996. Application of *Aspergillus oryzae* and *Rhizopus oryzae* for Cu(II) removal. *Water Res.* 30(9):1985–1990.

Huang, C.-P. and Morehart, A.L. 1990. The removal of Cu(II) from dilute aqueous solutions by *Saccharomyces cerevisiae. Water Res.* 24:433–439.

Huang, J.-C., Chang, S.-Y., Liu, Y.-C., and Jiang, Z. 1985. Biofilm growths with sucrose as substrate. *J. Environ. Eng.* 111:353–363.

Huddleston, R.L. 1979. Solid waste disposal: landfarming. *Chem. Eng.* Feb. 26, 1979. pp. 119–124.

Huddleston, R.L., Bleckmann, C.A., and Wolfe, J.R. 1986. Land treatment — biological degradation processes. In *Land Treatment: A Hazardous Waste Management Alternative.* Water Resources Symposium Number 13. Loehr, R.C. and Malina, J.F., Jr., Eds. Center for Research in Water Resources, University of Texas, Austin. pp. 41–62.

REFERENCES

Huddleston, R.L. and Cresswell, L.W. 1976. In *Proc. of the Mgmt. of Petroleum Refinery Wastewaters Forum.*

Hudel, K., Forge, F., Klein, M., Schroeder, H.-Fr., and Dohmann, M. 1995. Steam extraction of organically contaminated soil and residue — process development and implementation on an industrial scale. *Soil Environ.* 5 (Contaminated Soil 95, Vol. 2). pp. 1103–1112.

Huesemann, M.H. and Moore, K.O. 1993. Compositional changes during landfarming of weathered Michigan crude oil–contaminated soil. *J. Soil Contam.* 2(3):245–264.

Huesemann, M.H., Moore, K.O., and Johnson, R.N. 1993. The fate of BDAT polynuclear aromatic compounds during biotreatment of refinery API oil separator sludge. *Environ. Prog.* 12(1):30–38.

Huey, C.W., Brinckman, F.E., Iverson, W.P., and Grim, S.O. 1975. Bacterial volatilization of cadmium. In *Int. Conf. Heavy Metals in the Environ.*, Toronto. C214–C216.

Huismann, S.S., Peterson, M.A., and Jardine, R.J. 1995. Bioremediation of chlorinated solvents and diesel soils. In *Innov. Technol. Site Rem. Haz. Waste Mgmt. Proc. Natl. Conf.* pp. 173–180.

Hulbert, M.H. and Krawiec, S. 1977. Cometabolism: a critique. *J. Theor. Biol.* 69:287–291.

Huling, S.G., Bledsoe, B.E., and White, M.V. 1991. The feasibility of utilizing hydrogen peroxide as a source of oxygen in bioremediation. In *In Situ Bioreclamation. Applications and Investigations for Hydrocarbon and Contaminated Site Remediation.* Hinchee, R.E. and Olfenbuttel, R.F., Eds. Butterworth-Heinemann, MA. pp. 83–102.

Hund, K. and Schenk, B. 1994. The microbial respiration quotient as indicator for bioremediation processes. *Chemosphere.* 28(3):477–490.

Hunt, W.P., Robinson, K.G., and Ghosh, M.M. 1994. The role of biosurfactants in biotic degradation of hydrophobic organic compounds. In *Hydrocarbon Bioremediation.* Hinchee, R.E., Alleman, B.C., Hoeppel, R.E., and Miller, R.N., Eds. Lewis Publishers, Boca Raton, FL.

Hunt, P.G. et al. 1973. In *Proc. Joint Conf. on Prevention and Control of Oil Slicks.* American Petroleum Institute, Washington, D.C. pp. 773–740.

Hunter, L.L. and Engleman, V.S. 1992. VOC screening model: choosing VOC emission control technologies. In *Proc. 85th Annu. Mtg. Air Waste Manage. Assoc.* Vol. 4. Paper No. 92/114.07. 14 pp.

Hupe, K., Lueth, J.-C., Heerenklage, J., and Stegmann, R. 1995. Blade-mixing reactors in the biological treatment of contaminated soils. In *Biological Unit Processes for Hazardous Waste Treatment. Pap. 3rd. Int. In Situ On-Site Bioreclam. Symp.* Hinchee, R.E., Skeen, R.S., and Sayles, G.D., Eds. Battelle Press, Columbus, OH. pp. 153–159.

Hupka, J. and Wawrzacz, B. 1996. Effectiveness of release of oil from soil under stagnant conditions. *Fizykochem. Probl. Mineralurgii.* (Poland). 30:177–186.

Hurle, K. and Walker, A. 1980. Persistence and its prediction. In *Interaction between Herbicides and the Soil.* Hance, R.J., Ed. Academic Press, London.

Hutchins, S.R., Davidson, M.S., Brierley, J.A., and Brierley, C.L. 1986. Microorganisms in reclamation of metals. *Annu. Rev. Microbiol.* 40:311–336.

Hutchins, S.R., Sewell, G.W., Kovacs, D.A., and Smith, G.A. 1991. *Environ. Sci. Technol.* 25:68–76.

Hutchins, S.R. and Wilson, J.T. 1991. Laboratory and field studies on BTEX biodegradation in a fuel-contaminated aquifer under denitrifying conditions. In *In Situ Bioreclamation: Applications and Investigations for Hydrocarbon and Contaminated Site Remediation.* Hinchee, R.E. and Olfenbuttel, R.F., Eds. Butterworth-Heinemann, Stoneham, MA. pp. 157–172.

Hutzinger, O. and Veerkamp, W. 1981. Xenobiotic chemicals with pollution potential. In *FEMS Symposium No. 12. Microbial Degradation of Xenobiotics and Recalcitrant Compounds.* Leisinger, T., Hutter, R., Cook, A.M., and Nuesch, J., Eds. Academic Press, New York. 12:3–45.

Hutzler, N.J., Murphy, B.E., and Gierke, J.S. 1988. State of technology review: Soil Vapor Extraction Systems. Final report to the U.S. EPA, Hazardous Waste Engineering Research Laboratory, Cincinnati, OH.

Huysman, F. and Verstraete, W. 1993. Water-facilitated transport of bacteria in unsaturated soil columns: influence of cell surface hydrophobicity and soil properties. *Soil Biol. Biochem.* 25:83–90.

Hwang, S.T. 1982. Toxic emissions from land disposal facilities. *Environ. Prog.* 1:46–52.

I

Ibeanusi, V.M. and Archibold, E.A. 1995. Mechanisms of heavy metal uptake in a mixed microbial ecosystem. In *Bioremediation of Pollutants in Soil and Water.* Schepart, B.S., Ed. ASTM, Philadelphia. pp. 191–203.

Ignasiak, T., Kemp-Jones, A.V., and Strausz, O.P. 1977. The molecular structure of athabasca asphaltene cleavage of the carbon-sulfur bonds by radical ion electron transfer reactions. *J. Org. Chem.* 43:312–320.

Ijah, U.J.J. and Ukpe, L.I. 1992. Biodegradation of crude oil by *Bacillus* strains 28A and 61B isolated from oil spilled soil. *Waste Manage.* 12(1):55–60.

Imanaka, T. and Morikawa, M. 1993. Novel *Pseudomonas anaerooleophila* resistant to carbohydrates. Jpn. Kokai Tokkyo Koho JP 05,276,933 [93,276,933], 26 Oct. 1993.

Inoue, A. and Horikoshi, K. 1991. *J. Ferment. Bioeng.* 71:194–196.

Irmer, U. 1982. Die Wirkung der Schwermetalle Blei, Cadmium und Mangan auf die Suesswassergruenalgen *Chlamydomonas reinhardii* Dangeard und *Chlorella fusca* Shihira et Krauss. Dissertation, University Hamburg, Germany.

Irvine, R.L. and Cassidy, D.P. 1995. Periodically operated bioreactors for the treatment of soils and leachates. In *Biological Unit Processes for Hazardous Waste Treatment. Pap. 3rd. Int.* In Situ *On-Site Bioreclam. Symp.* Hinchee, R.E., Skeen, R.S., and Sayles, G.D., Eds. Battelle Press, Columbus, OH. pp. 289–298.

Issa, S., Simmonds, L.P., and Woods, M. 1993a. Passive movement of chickpea and bean rhizobium through soils. *Soil Biol. Biochem.* 25:959–965.

Issa, S., Simmonds, L.P., and Woods, M. 1993b. Active movement of chickpea and bean rhizobium in dry soils. *Soil Biol. Biochem.* 25:951–958.

IT Corporation. 1987. Report for Naval Civil Engineering Laboratory, Port Hueneme, CA, on Contract N 62474-87B-C-3063.

Itoh, M., Ohguchi, M., and Doi, S. 1968. Studies on hydrocarbon-utilizing microorganisms: microbial treatment of petroleum waste. *J. Ferment. Technol.* 46:34.

Itoh, S. and Suzuki, T. 1972. Effect of rhamnolipids on growth of *Pseudomonas aeruginosa* mutant deficient in *n*-paraffin-utilizing ability. *Agr. Biol. Chem.* 36:2233–2235.

Ivanov, A.Y., Fomchenkov, V.M., Khasanova, L.A., Kuramshina, Z.M., and Sadikov, M.M. 1992. Effect of heavy metal ions on the electro-physical properties of *Anacystis nidulans* and *Escherichia coli*. *Microbiology.* 61:319–326.

IWACO Consultancy. 1989. Report 20.358. IWACO, Groeingen, The Netherlands.

J

Jacob, J., Karcher, W., Belliardo, J.J., and Wagstaffe, P.J. 1986. Polycyclic aromatic hydrocarbons of environmental and occupational importance. *Fresenius. Z. Anal. Chem.* 323:1–10.

Jacobs, R.A., Sengun, M.Z., Hicks, R.E., and Probstein, R.F. 1994. Model and experiments on soil remediation by electric fields. *J. Environ. Sci. Health,* Part A. A29(9):1933–1955.

Jacquot, F., Doumenq, P., Guiliano, M., Munoz, D., Guichard, J.R., and Mille, G. 1996. Biodegradation of the (aliphatic + aromatic) fraction of Oural crude oil. Biomarker identification using GC/MS SIM and GC/MS/MS. *Talanta.* 43(3):319–330.

Jaffe, R. and Gallardo, M.T. 1993. Application of carboxylic acid biomarkers as indicators of biodegradation and migration of crude oils from the Maracaibo Basin, Western Venezuela. *Org. Geochem.* 20(7):973–984.

Jain, D.K., Lee, H., and Trevors, J.T. 1992. Effect of addition of *Pseudomonas aeruginosa* UG2 inocula or biosurfactants on biodegradation of selected hydrocarbons in soil. *J. Ind. Microbiol.* 10:87–93.

Jain, R.K. and Sayler, G.S. 1987. Problems and potential for *in situ* treatment of environmental pollutants by engineered microorganisms. *Microbiol. Sci.* 4:59–63.

Jamison, V.W., Raymond, R.L., and Hudson, J.O. 1971. *Dev. Ind. Microbiol.* 12:99–105.

Jamison, V.W., Raymond, R.L., and Hudson, J.O., Jr. 1975. Biodegradation of high-octane gasoline in groundwater. In *Developments in Industrial Microbiology.* Underkofler, L.A. and Murray, E.D., Eds. American Institute of Biological Sciences, Washington, D.C. 16:305–312.

Jamison, V.W., Raymond, R.L., and Hudson, J.O., Jr. 1976. Biodegradation of high-octane gasoline. In *Proc. 3rd Int. Biodegradation Symp.* Sharpley, J.M. and Kaplan, A.M., Eds. Elsevier, New York. pp. 187–196.

Jang, L.K., Geesey, G.G., Lopez, D.L., Eastman, S.L., and Wichlacz, 1990. Sorption equilibrium of copper by partially-coagulated calcium alginate gel. *Cehm. Eng. Commun.* 94:63–77.

Jarvie, A.W., Markall, R.N., and Potter, H.R. 1975. Chemical alkylation of lead. *Nature.* 255:217–218.

Jeffers, T.H., Ferguson, C.B., and Bennett, P.G. 1991. Biosorption of metal contaminants from acidic mine waters. In *Proc. Conf. Mineral Bioprocessing.* Smith, R.W. and Mistra, M., Eds. Minerals, Metals, and Materials Society, Warrendale, PA. pp. 289–298.

Jeffrey, A.M., Yeh, H.J.C., Jerina, D.M., Patel, T.R., Davey, J.F., and Gibson, D.T. 1975. Initial reactions in the oxidation of naphthalene by *Pseudomonas putida*. *Biochemistry.* 14:575–583.

Jenkinson, D.S. and Ladd, J.M. 1981. Microbial biomass in soil: movement and turnover. In *Soil Biochemistry.* Paul, E.A. and Ladd, J.M., Eds. Marcel Dekker, New York. pp. 415–471.

Jennings, D.A., Petersen, J.N., Skeen, R.S., Peyton, B.M., Hooker, B.S., Johnstone, D.L., and Yonge, D.R. 1995. An experimental study of microbial transport in porous media. In *Bioaugmentation for Site Remediation. Pap. 3rd Int.* In Situ *On-Site Bioreclam. Symp.* Hinchee, R.E., Fredrickson, J., and Alleman, B.C., Eds. Battelle Press, Columbus, OH. pp. 97–103.

Jennings, M.S., Krohn, N.E., Berry, R.S., Palazzolo, M.A., Parks, R.M., and Fidler, K.K. 1985. *Catalytic Incineration for Control of Volatile Organic Compound Emissions.* Noyes Publications, Park Ridge, NJ. 251 pp.

Jensen, B.K. and Arvin, E. 1994. Aromatic hydrocarbon degradation specificity of an enriched denitrifying mixed culture. In *Hydrocarbon Bioremediation.* Hinchee, R.E., Alleman, B.C., Hoeppel, R.E., and Miller, R.N., Eds. CRC Press, Boca Raton, FL. pp. 411–417.

Jensen, B., Arvin, E., and Gundersen, A.T. 1985. The degradation of aromatic hydrocarbons with bacteria from oil contaminated aquifers. In *Proc. Petrol. Hydro. and Organ. Chem. in Groundwater Conf.* pp. 421–435.

Jensen, R.E. and Miller, J.A. 1994. Field demonstrations of bioremediation and low-temperature thermal treatment technologies for petroleum-contaminated soil. *Hydrocarbon Contam. Soils Groundwater.* 4:227–261.

Jenson, V. 1975. Bacterial flora of soil after application of oily waste. *Oikos.* 26:152–158.

REFERENCES

Jerina, D.M., Daly, J.W., Jeffrey, A.M., and Gibson, D.T. 1971. Communication. *cis*-1,2-dihydroxy-1,2-dihydronaphthalene: a bacterial metabolite from naphthalene. *Arch. Biochem. Biophys.* 142:394–396.

Jerina, D.M., Selander, H., Yagi, H., Wells, M.C., Davey, J.F., Mahadevan, V., and Gibson, D.T. 1976. Dihydrodiols from anthracene and phenanthrene. *Biochem. J.* 91:5988–5996.

Jhaveri, V. and Mazzacca, A.J. 1982. Bioreclamation of Ground and Groundwater by the GDS Process. Groundwater Decontamination Systems, Inc., Waldwick, NJ.

Jhaveri, V. and Mazzacca, A.J. 1983. Paper presented at 4th Natl. Conf. on Mgt. of Uncontrolled Haz. Waste Sites, Nov. 1. Washington, D.C.

Jhaveri, V. and Mazzacca, A.J. 1985. In Case History. Groundwater Decontamination Systems, Inc. Waldwick, NJ.

Jobson, A., Cook, F.D., and Westlake, D.W.S. 1972. Microbial utilization of crude oil. *Appl. Microbiol.* 23:1082–1089.

Jobson, A., McLaughlin, M., Cook, F.D., and Westlake, D.W.S. 1974. Effect of amendments on the microbial utilization of oil applied to soil. *Appl. Microbiol.* 27:166–171.

Johanides, V. and Hrsak, D. 1976. In *Abstr. 5th Int. Fermentation Symp.*, Berlin, 1976. Dellweg, H., Ed. Westkreuz, Berlin. p. 124.

Johns, F.J., II and Nyer, E.K. 1996. Miscellaneous *in situ* treatment technologies. In In Situ *Treatment Technology.* Nyer, E.K., Kidd, D.F., Palmer, P.L., Crossman, T.L., Fam, S., Johns, F.J., II, Boettcher, G., and Suthersan, S.S., Eds. CRC Press, Boca Raton, FL. pp. 289–319.

Johnson, L., Kauffman, J., and Krupka, M. 1982. Paper presented at Summer Natl. Mtg. Am. Inst. Chem. Eng., Aug. 29–Sept 1, 1982, Cleveland, OH.

Johnson, L.A. and Leuschner, A.P. 1992. The CROW process and bioremediation for *in situ* treatment of hazardous waste sites. In *Hydrocarbon Contaminated Soils and Groundwater. Proc. 1st Annu. West Coast Conf. Hydrocarbon Contam. Soils Groundwater,* February 1990, Newport Beach, CA. Vol. 1. Calabrese, E.J. and Kostechi, P.T., Eds. Lewis Publishers, Chelsea, MI. pp. 343–357.

Johnson, M.J. 1964. Utilization of hydrocarbons by microorganisms. *Chem. Ind.* (London). 1964:1532–1537.

Johnson, P.C., Kemblowski, M.W., and Colthart, J.D. 1990. Quantitative analysis. Cleanup of hydrocarbon-contaminated soils *in situ* soil venting. *Ground Water.* 28(3):413–429.

Johnson, P.C., Stanley, C.C., Byers, D.L., Benson, D.A., and Acton, M.A. 1991. Soil venting at a California site: field data reconciled with theory. In *Hydrocarbon Contaminated Soils and Groundwater: Analysis, Fate, Environmental and Public Health Effects, and Remediation.* Vol.1 Kostecki, P.T. and Calabrese, E.J., Eds. Lewis Publishers, Chelsea, MI. pp. 253–281.

Johnson, P.C., Stanley, C.C., Kemblowski, M.W., Byers, D.L., and Colthart, J.D. 1990. A practical approach to the design, operation, and monitoring of *in situ* soil venting systems. *Groundwater Monit. Rev.,* Spring.

Johnson, S.L. 1978. Biological treatment. In *Unit Operations for Treatment of Hazardous Industrial Wastes.* Berkowitz, J.B., Funkhouser, J.T., and Stevens, J.I., Eds. Noyes Data Corp., Park Ridge, NJ. pp. 168–268.

Johnston, C.G. and Aust, S.D. 1994. Detection of *Phanerochaete chrysosporium* in soil by PCR and restriction enzyme analysis. *Appl. Environ. Microbiol.* 60:2350–2354.

Johnston, J.B. and Robinson, S.G. 1982. Opportunities for development of new detoxification processes through genetic engineering. In *Detoxification of Hazardous Waste.* Exner, J.H., Ed. Ann Arbor Science Publishers, Ann Arbor, MI. pp. 301–314.

Jones, J.G. and Edington, M.A. 1968. An ecological survey of hydrocarbon-oxidizing microorganisms. *J. Gen. Microbiol.* 52:381–390.

Jorgensen, C., Aamand, J., Jensen, B.K., Nielsen, S.D., and Jacobsen, C.S., 1995. Microbial properties governing the microbial degradation of polycyclic aromatic hydrocarbons. In *Environmental Biotechnology: Principles and Applications.* Moo-Young, M., Anderson, W.A., and Chakrabarty, A.M., Eds. Kluwer, Amsterdam. pp. 178–192.

Jorgensen, C., Flyvbjerg, J., Arvin, E., and Jensen, B. 1995. Stoichiometry and kinetics of microbial toluene degradation under denitrifying conditions. *Biodegradation.* 6:147–156.

Jorgensen, C., Mortensen, E., Jensen, B.K., and Arvin, E. 1991. Biodegradation of toluene by a denitrifying enrichment culture. In In Situ *Bioreclamation. Applications and Investigations for Hydrocarbon and Contaminated Site Remediation.* Hinchee, R.E. and Olfenbuttel, R.F., Eds. Butterworth-Heinemann, Stoneham, MA. pp. 480–486.

Jorgensen, C., Nielsen, B., Jensen, B.K., and Mortensen, E. 1995. Transformation of *o*-xylene to *o*-methyl benzoic acid by a denitrifying enrichment culture using toluene as the primary substrate. *Biodegradation.* 6:141–146.

Josephson, J. 1983. Restoration of aquifers. *Environ. Sci. Technol.* 17:347A–350A.

Joshi, M.M. and Lee, S. 1996. Optimization of surfactant-aided remediation of industrially contaminated soils. *Energ. Sourc.* 18(3):291–301.

JRB Associates, Inc. 1982. Handbook for Remedial Action at Waste Disposal Sites. EPA Report No. EPA-625/6-82-006. U.S. EPA, Cincinnati, OH.

JRB Associates, Inc. 1984. Summary report: Remedial Response at Hazardous Waste Sites. Prepared for Municipal Environmental Research Laboratory, Cincinnati, OH. PB 85-124899.

Ju, M., Devinny, J.S., and Paspalof, 1993. Effects of pulverization on soil bioremediation. *Haz. Waste Haz. Mater.* 10(3):357–364.

Jury, W.A., Farmer, W.J., and Spencer, W.F. 1984. Behavior assessment model for trace organics in soil. II. Chemical classification and parameter sensitivity. *J. Environ. Qual.* 13:567–572.

Jury, W.A., Grover, R., Spencer, W.F., and Farmer, W.F. 1980. Modeling vapor losses of soil-incorporated triallate. *Soil Sci. Soc. Am. J.* 44:445-450.

Jury, W.A., Spencer, W.F., and Farmer, W.J. 1983. Behavior assessment model for trace organics in soil: I. Model descriptions. *J. Environ. Qual.* 12:558–564.

K

Kaake, R.H., Roberts, D.J., Stevens, T.O., Crawford, R.L., and Crawford, D.L. 1992. Bioremediation of soils contaminated with the herbicide 2-sec-butyl-4,6-dinitrophenol (dinoseb). *Appl. Environ. Microbiol.* 58:1683–1689.

Kaestner, M., Breuer-Jammali, M., and Mahro, B. 1994. Enumeration and characterization of the soil microflora from hydrocarbon-contaminated soil sites able to mineralize polycyclic aromatic hydrocarbons. *Appl. Microbiol. Biotechnol.* 41:267–273.

Kaestner, M., Lotter, S., Heerenklage, J., Breuer-Jammali, M., Stegmann, R., and Mahro, B. 1995. Fate of ^{14}C-labeled anthracene and hexadecane in compost-manured soil. *Appl. Microbiol. Biotechnol.* 43:1128–1135.

Kaestner, M. and Mahro, B. 1996. Microbial degradation of polycyclic aromatic hydrocarbons in soils affected by the organic matrix of compost. *Appl. Microbiol. Biotechnol.* 44(5):668–675.

Kakiichi, N., Shibuya, S., Akita, K., Sugimoto, Y., Oshida, T., Hayashi, M., Kamata, S., Komine, K., Otsuka, H., and Ushida, K. 1992. Immobilization of denitrifying bacteria using poly(vinyl alcohol) and denitrification by immobilized cells. *Anim. Sci. Technol.* 63:47–53.

Kallio, R.E. 1975. Microbial degradation of petroleum. In *Proc. 3rd Int. Biodegradation Symp.*, Aug. 17–23, Kingston, RI. Sharply, J.M. and Kaplan, A.M., Eds. Applied Science Publishers, London. p. I-1.

Kamp, P.F. and Chakrabarty, A.M. 1979. Plasmids specifying *p*-cholorbiphenyl degradation in enteric bacteria. In *Plasmids of Medical, Environmental and Commercial Importance.* Timmis, K.N. and Puhler, A., Eds. Elsevier/North Holland Biomedical Press, Amsterdam, The Netherlands. pp. 275–285.

Kampbell, D.H. and Wilson, J.T. 1991. *J. Haz. Mater.* 28:75–80.

Kampbell, D.H., Wilson, J.T., Read, H.W., and Stocksdale, T.T. 1987. *J. Air Pollut. Control Assoc.* 37:1236–1240.

Kanazawa, S. and Filip, Z. 1986. Distribution of microorganisms, total biomass, and enzyme activities in different particles of brown soil. *Microb. Ecol.* 12:205–215.

Kang, S.-H. and Oulman, C.S. 1996. Evaporation of petroleum products from contaminated soils. *J. Environ. Eng.* 122(5):384–387.

Kaplan, D., Christiaen, D., and Arad, S.M. 1987. Chelating properties of extracellular polysaccharides from *Chlorella* spp. *Appl. Environ. Microbiol.* 53:2953–2956.

Kaplan, D.L. and Kaplan, A.M. 1982. Composting industrial wastes — biochemical consideration. *Biocycle.* 23:42–44.

Kaplan, D.L., Riley, P.A., Pierce, J., and Kaplan, A.M. 1984. U.S. Army Natick R&D Center. Report No. NATICK/TR-85/003.

Kappeli, O. and Finnerty, W.R. 1980. Characteristics of hexadecane partition by the growth medium of *Acinetobacter* sp. *Biotechnol. Bioeng.* 22:495–503.

Kappeli, O. Walther, P., Mueller, M., and Fiechter, A. 1984. Structure of the cell surface of the yeast *Candida tropicalis* and its relation to hydrocarbon transport. *Arch. Microbiol.* 138:279–282.

Kaprelyants, A.D. and Kell, D.B. 1992. Rapid assessment of bacterial viability and vitality using rhodamine 123 and flow cytometry. *J. Appl. Bacteriol.* 72:410–422.

Karan, K., Chakma, A., and Mehrotra, A.K. 1995. Air stripping of hydrocarbon-contaminated soils: investigation of mass transfer effects. *Can. J. Chem. Eng.* 73(2):196–203.

Karata, A., Yoichi, Y., and Fumio, T. 1980. *Jpn. J. Mar. Chem.* 18:1.

Karickhoff, S.W. 1980. Sorption kinetics of hydrophobic pollutants in natural sediments. In *Contaminants and Sediments.* Vol. 2. Baker, R.A., Ed. Ann Arbor Science Publishers, Ann Arbor, MI. pp. 193–205.

Karickhoff, S.W. 1984. Organic pollutant sorption in aquatic systems. *J. Hydraul. Eng.* 110:707–735.

Karickhoff, W.W., Brown, D.S., and Scott, T.A. 1979. Sorption of hydrophobic pollutants on natural sediments. *Water Res.* 13:241–248.

Karimi-Lotfabad, S., Pickard, M.S., and Gray, M.R. 1996. Reactions of polynuclear aromatic hydrocarbons on soil. *Environ. Sci. Technol.* 30(4):1145–1151.

Katin, R.A. 1995. Operation and maintenance of remediation systems. In *Proc. HAZMACON 95, 12th Haz. Mater. Mgmt. Conf. Exhib.* pp. 224–232.

Kator, H. 1973. Utilization of crude oil hydrocarbons by mixed cultures of marine bacteria. In *The Microbial Degradation of Oil Pollutants, Workshop,* Atlanta, GA, Dec. 4–6, 1972. Ahearn, D.G. and Meyers, S.P., Eds. Pub. No. LSU-SG-73-01. Center for Wetland Resources, Louisiana State University, Baton Rouge, LA. p. 47.

Kaufman, A.K. 1995. Biofouling: a critical consideration for evaluating *in situ* biotreatment feasibility. In *Proc. HAZMACON 95, 12th Haz. Mater. Mgmt. Conf. Exhib.* pp. 358–362.

REFERENCES

Kaufman, D.D. 1983. Fate of toxic organic compounds in land-applied waste. In *Land Treatment of Hazardous Wastes*. Parr, J.F., Marsh, P.B., and Kla, J.M., Eds. Noyes, Park Ridge, NJ. pp. 77–151.

Kaufman, D.D. and Doyle, R.D. 1978. Biodegradation of organics. In *Proc. Natl. Conf. Composting of Municipal Residues and Sludges,* Aug. 23–25, 1977. Information Transfer, Inc., Rockville, MD. pp. 75–80.

Kaufman, D.D. and Plimmer, J.R. 1972. Approaches to the synthesis of soft pesticides. In *Water Pollution Microbiology.* Mitchell, R., Ed. Wiley Interscience, New York. pp. 173–203.

Kaufman, N. 1984. Solvent waste management in a high technology corporation. *Haz. Waste and Haz. Mater.* 1:83–92.

Kawai, H., Grieco, V.M., and Jureidini, P. 1984. A study of the treatability of pollutants in high rate photosynthetic ponds and the utilization of the proteic potential of algae which proliferate in the ponds. *Environ. Technol. Lett.* 5:505–515.

Kearney, P.C., Nash, R.G., and Isensee, A.R. 1969. Persistence of pesticide residues in soils. In *Chemical Fallout: Current Research on Persistent Pesticides.* Miller, M.W. and Berg, G.G., Eds. Charles C Thomas, Springfield, IL. pp. 54–67.

Kearney, P.C. and Plimmer, J.R. 1970. Relation of structure to pesticide decomposition. In *Int. Symp. Pesticides in the Soil: Ecology, Degradation and Movement.* Michigan State University, East Lansing, MI. pp. 65–172.

Keck, J., Sims, R.C., Coover, M., Park, K., and Symons, B. 1989. Evidence for cooxidation of polynuclear aromatic hydrocarbons in soil. *Water Res.* 23(12):1467–1476.

Keet, B.A. 1995. Bioslurping state of the art. In *Applied Bioremediation of Petroleum Hydrocarbons.* Hinchee, R.E., Kittel, J.A., and Reisinger, H.J., Eds. Battelle Press, Columbus, OH. pp. 329–334.

Kefford, B., Kjelleberg, S., and Marshall, K.C. 1982. Bacterial scavenging: utilization of fatty acids localized at a solid–liquid interface. *Arch. Microbiol.* 133:257–260.

Keith, L.H. and Telliard, W.A. 1979. Priority pollutants. I. A perspective view. *Environ. Sci. Technol.* 13:416–423.

Kell, D.B., Ryder, M.M., Kaprelyants, A.S., and Westerhoff, H.V. 1991. Quantifying heterogeneity: flow cytometry of bacterial culture. *Antonie van Leeuwenhoek.* 60:145–158.

Kelley, I. and Cerniglia, C.E. 1991. The metabolism of fluoranthene by a species of *Mycobacterium. J. Ind. Microbiol.* 7:19–26.

Kelley, I. and Cerniglia, C.E. 1995. Degradation of a mixture of high-molecular-weight polycyclic aromatic hydrocarbons by a *Mycobacterium* strain PYR-1. *J. Soil Contam.* 4(1):77–91.

Kelley, I., Freeman, J.P., and Cerniglia, C.E. 1991. Identification of metabolites from the degradation of naphthalene by a *Mycobacterium* sp. *Biodegradation.* 1:283–290.

Kelley, I., Freeman, J.P., Evans, F.E., and Cerniglia, C.E. 1991. Identification of a carboxylic acid metabolite from the catabolism of fluoranthene by a *Mycobacterium* sp. *Appl. Environ. Microbiol.* 57:636–641.

Kelley, R.L. and Reddy, C.A. 1988. Glucose oxidase of *Phanerochaete chrysosporium*. In *Methods in Enzymology.* Vol. 161. *BIOMASS Part B, Lignin, Pectin, and Chitin.* Wood, W.A. and Kellog, S.T., Eds. Academic Press, New York. pp. 307–315.

Kelly, A., Pennock, S.J., Bohn, and White, M.K. 1992. Expert software that matches remediation site and strategy. Remediation. U.S. Department of Energy Pacific Northwest Laboratories. pp.183–198.

Kemp, H.A. Archer, D.B., and Morgan, M.R.A. 1988. Enzyme linked immunosorbent assays for the specific and sensitive equantification of *Methanosarcina mazei* and *Methanobacterium bryantii. Appl. Environ. Microbiol.* 54:1003–1080.

Kenaga, E.E. and Goring, C.A.I. 1980. In *Aquatic Toxicology.* Eaton, J.G., Parrish, P.R., and Hendricks, A.C., Eds. ASTM, Philadelphia, pp. 78–115.

Kennedy, K.T. 1985. *Biotechnol. Bioeng.* 27:1152.

Kennedy, K.T., Muzar, M., and Copp, G.H. 1985. Stability and performance of mesophilic anaerobic fixed-film reactors during organic overloading. *Biotechnol. Bioeng.* 27:86–93.

Kennicutt, M.C., II. 1988. The effect of biodegration on crude oil bulk and molecular composition. *Oil Chem. Pollut.* 4:89–112.

Kern, H.W. 1989. Improvement in the production of extracellular lignin peroxidases by *Phanerochaete chrysosporium*: effect of solid manganese(IV)oxide. *Appl. Microbiol. Biotechnol.* 32:223–234.

Kerr, R.P. and Capone, D.G. 1986. Salinity effects on naphthalene and anthracene mineralization by sediment microbes in coastal environments. *Abstr. Annu. Mtg. Am. Soc. Microbiol.* p. 303.

Kerr, R.P. and Capone, D.G. 1988. The effect of salinity on the microbial mineralization of two polycyclic aromatic hydrocarbons in estuarine sediments. *Mar. Environ. Res.* 26:181–198.

Kerry, E. 1993. Bioremediation of experimental petroleum spills on mineral soils in the Vestfold Hills Antarctica. *Polar Biol.* 13(3):163–170.

Kester, A.S. and Foster, J.W. 1963. Biterminal oxidation of long-chain alkanes by bacteria. *J. Bacteriol.* 85:859–869.

Keuth, S. and Rehm, H.-J. 1991. Biodegradation of phenanthrene by *Arthrobacter polychromogenes* isolated from a contaminated soil. *Appl. Microbiol. Biotechnol.* 34:804–808.

Khan, S.U. 1980. Role of humic substances in predicting fate and transport of pollutants in the environment. In *Dynamics, Exposure and Hazard Assessment of Toxic Chemicals.* Haque, R., Ed. Ann Arbor Science, Ann Arbor, MI. pp. 215–230.

Khan, S.U. 1982. Studies on bound ^{14}C-prometryn residues in soil and plants. *Chemosphere.* 11:771–795.

Kidd, D.F. 1996. Fracturing. In In Situ *Treatment Technology.* Nyer, E.K., Kidd, D.F., Palmer, P.L., Crossman, T.L., Fam, S., Johns, F.J., II, Boettcher, G., and Suthersan, S.S., Eds. CRC Press, Boca Raton, FL. pp. 245–269.

Kiehne, M., Berghof, K., Mueller-Kuhrt, L., and Buchholz, R. 1995. Mobile revolving tubular reactor for continuous microbial soil decontamination. *Soil Environ.* 5 (Contaminated Soil 95, Vol. 2). pp. 873–882.

Kierstan, M. and Bucke, C. 1977. The immobilization of microbial cells, subcellular organelles, and enzymes in calcium alginate gels. *Biotechnol. Bioeng.* 19:387–397.

Kilbertus, G. 1980. Etude des microhabitats contenus dans les aggregats du soil. Leur relation avec la biomasse bacterienne et la taille des procaryotes presents. *Rev. Ecol. Biol. Soil.* 17:543–557.

Kilbertus, G., Proth, J., and Vervier, B. 1980. Effects de la dessiccation sur les bacteria gram-negatives d'un sol. *Soil Biol. Biochem.* 11:109–114.

Killham, K., Amato, M., and Ladd, J.N. 1993. Effect of substrate location in soil and soil pore-water regime on carbon turnover. *Soil Biol. Biochem.* 25:57–62.

Kincannon, C.B. 1972. Oily waste disposal by soil cultivation process. EPA Report No. EPA-P2-72-110. Office of Research and Monitoring, U.S. EPA, Washington, D.C.

Kincannon, D.F. and Lin, Y.S. 1985. Microbial degradation of hazardous waste by land treatment. In *Proc. of the 40th Industrial Waste Conf.* May 14–16, Purdue University Butterworth, Boston. pp. 607–619.

King, D.H. and Perry, J.J. 1974. The origin of fatty acids in the hydrocarbon-utilizing microorganism, *Mycobacterium vaccae*. *Can. J. Microbiol.* 21:85–89.

King, R.B., Long, G.M., and Sheldon, J.K. 1992. Bioreactors: the technology of total control. In *Practical Environmental Bioremediation.* Lewis Publishers, Boca Raton, FL. p. 83.

Kinner, N.E. and Eighmy, T.T. 1986. Biological fixed-film systems. *J. WPCF.* 58:498–501.

Kinner, N.E., Maratea, D., and Bishop, P.L. 1985. An electron microscopic evaluation of bacteria inhabiting rotating biological contactor biofilms during various loading conditions. *Environ. Technol. Lett.* 6:455–466.

Kirk, T.K. 1987. Lignin-degrading enzymes. *Phil. Trans. Royal Soc.* London. A. 321:561–474.

Kirk, T.K. and Farrell, R.L. 1987. Enzymatic "combustion": the microbial degradation of lignin. *Annu. Rev. Microbiol.* 41:465–505.

Kirk, T.K., Schultz, E., Connors, W.J., Lorenz, L.F., and Zeikus, J.G. 1978. Influence of culture parameters on lignin metabolism by *Phanerochaete chrysosporium*. *Arch. Microbiol.* 117:277–285.

Kirts, R.E. 1995. Review of Advanced Treatment Technologies Applicable to Navy Hazardous Wastes. TM-2150-ENV. Naval Facilities Engineering Services Center, Port Hueneme, CA. 64 pp.

Kitagawa, M. 1956. Studies on the oxidation mechanisms of methyl group. *Jour. Biochem* [Tokyo]. 43(4):553–563.

Kittel, J.A., Hinchee, R.E., Hoeppel, R., and Miller, R. 1994. Bioslurping — vacuum-enhanced free-product recovery coupled with bioventing: a case study. In *Proc. Petrol. Hydrocarbons Org. Chem. Ground Water: Prev., Detect., Rem. Conf.* pp. 255–270.

Kiyohara, H. and Nagao, K. 1978. The catabolism of phenanthrene and naphthalene by bacteria. *J. Gen. Microbiol.* 105:69–75.

Kjelleberg, S., Norkrans, B., Lofgren, H., and Larsson, K. 1976. Surface balance study of the interaction between microorganisms and lipid monolayer at the air/water interface. *Appl. Environ. Microbiol.* 31:609–611.

Klein, D.A. and Molise, E.M. 1975. Ecological ramifications of silver iodide nucleating agent accumulation in a semi-arid grassland environment. *J. Appl. Meteorol.* 14:673–680.

Klevens, H.B. 1950. *J. Phys. Chem.* 54:283–298.

Knackmuss, H.-J. 1981. Degradation of halogenated and sulfonated hydrocarbons. In *FEMS Symposium No. 12. Microbial Degradation of Xenobiotics and Recalcitrant Compounds.* Leisinger, T., Hutter, R., Cook, A.M., and Nuesch, J., Eds. Academic Press, New York. 12:189–212.

Knackmuss, H.-J. and Hellwig, M. 1978. *Arch. Microbiol.* 117:1–7.

Knaebel, D.B., Federle, T.W., Vestal, J.R. 1990. Mineralization of linear alkylbenzene sulfonate (LAS) and linear alcohol ethoxylate (LAE) in 11 contrasting soils. *Environ. Tox. Chem.* 9(8):981–988.

Knaebel, D.B. and Vestal, J.R. 1992. Effects of intact rhizosphere microbial communities on the mineralization of surfactants in surface soils. *Can. J. Microbiol.* 38(7):643-653.

Knetting, E. and Zajic, J.E. 1972. Flocculant production from kerosene. *Biotechnol. Bioeng.* 14:379–390.

Kniebusch, M.M., Hildebrandt-Moeller, A., and Wilderer, P.A. 1994. Biological degradation of polycyclic aromatic hydrocarbons with a membrane biofilm system. *Biol. Abwasserreinig.* (Germany). 4:123–134.

Knox, R.C., Canter, L.W., Kincannon, D.F., Stover, E.L., and Ward, C.H. 1984. State-of-the-Art of Aquifer Restoration. National Center for Ground Water Research.

Knox, R.C., Canter, L.W., Kincannon, D.F., Stover, E.L., and Ward, C.H., Eds. 1986. *Aquifer Restoration: State of the Art.* Noyes Publications, Park Ridge, NJ. 750 pp.

Kobayashi, H. and Rittmann, B.E. 1982. Microbial removal of hazardous organic compounds. *Environ. Sci. Technol.* 16:170A-183A.

Kobayashi, M. and Tchan, Y.T. 1978. Formation of demethylnitrosamine in polluted environment and the role of photosynthetic bacteria. *Water Res.* 12:199–201.

Koch, J. and Fuchs, G. 1992. Enzymatic reduction of benzoyl-CoA to alicyclic compounds, a key reaction in anaerobic aromatic metabolism. *Eur. J. Biochem.* 205:195–202.

REFERENCES

Kocher, B.S., Azzam, F.O., and Lee, S. 1995. Single-stage remediation of contaminated soil-sludge. *Energ. Sourc.* 17(5):553–563.

Kochi, J.D., Tang, R., and Bernath, T. 1972. Mechanism of aromatic substitution: role of cation-radicals in the oxidative substitution of arenes by cobalt (III). *J. Am. Chem. Soc.* 95:7114–7123.

Koe, L.C.C. and Ang, F.G. 1992. Bioaugmentation of anaerobic digestion with a biocatalytic addition: the bacterial nature of the biocatalytic addition. *Water Res.* 26(3):389–392.

Kohata, N., Yamane, A., Hosomi, M., and Murakami, A. 1995. Selective degradation of dibenzothiophene in *n*-alkane by microorganisms. *Seibutsu-Kogaku Kaishi.* 73(6):489–495.

Kohlmeier, E. 1994. Degradation of ring-containing hydrocarbon compounds using white rot fungi exoenzymes. Ger. Offen. DE 4,314,352 (Cl. A62D3/00), Nov. 3, 1994.

Koltuniak, D.L. 1986. *In situ* air stripping cleans contaminated soil. *Chem. Eng.* 93:30–31.

Komagata, K., Nakase, T., and Katsu, N. 1964. Assimilation of hydrocarbons by yeast. I. Preliminary screening. *J. Gen. Appl. Microbiol.* 10:313–321.

Komaromy-Hiller, G. and von Wandruszka, R. 1995. Decontamination of oil-polluted soil by cloud point extraction. *Talanta.* 42(1):83–88.

Kommalapati, R.R. and Roy, D. 1996. Bioenhancement of soil microorganisms in natural surfactant solutions: I — aerobic. *J. Environ. Sci. Health,* Part A: Environ. Sci. Eng. Toxic Haz. Sub. Cont. A31(8):1951–1964.

Kopp-Holtwiesche, B., Weiss, A., and Boehme, A. 1993. Improved nutrient mixtures for bioremediation of polluted soils and waters. Ger. Offen. DE 4,218,243. December 9, 1993. 7 pp.

Kotterman, M.J.J., Wasseveld, R., and Field, J.A. 1995. Influence of nitrogen sufficiency and manganese deficiency on PAH degradation by *Bjerkandera* sp. In *Bioaugmentation for Site Remediation. Pap. 3rd Int. In Situ On-Site Bioreclam. Symp.* Hinchee, R.E., Frederickson, J., and Alleman, B.C., Eds. Battelle Press, Columbus, OH. pp. 189–195.

Koval'skii, S.V. 1968. Geochemical ecology of microorganisms under conditions of different selenium content in soils. *Mikrobiologiya.* 37:122–130 [in Russian].

Kowalenko, C.G. 1978. Organic nitrogen, phosphorus and sulfur in soils. In *Soil Organic Matter.* Schnitzer, M. and Khan, S.U., Eds. Elsevier Scientific, New York. pp. 95–135.

Kozaki, S., Kato, K., Tanaka, K., Yano, T., Skuranaga, M., and Imamura, T. 1994. Carrier for supporting microorganisms, soil remediating agent using carrier, and method for remediating soil. European Patent Apl. EP 594,125. April 27, 1994.

Kreamer, D.K. 1982. Ph.D. Dissertation. University of Arizona, Tucson.

Kremser, A. 1930. *Natl. Petrol. News.* 22:48.

Kretschmer, A., Bock, H., and Wagner, F. 1982. Chemical and physical characterization of interfacial-active lipids from *Rhodococcus erythropolis* grown on *n*-alkanes. *Appl. Environ. Microbiol.* 44:864–870.

Kroon, A.G.M., Pomper, M.A., and van Ginkel, C.G. 1994. Metabolism of dodecyldimethylamine by *Pseudomonas* MA3. *Appl. Microbiol. Biotechnol.* 42:134–139.

Kroopnick, P.M. 1991. Life-cycle costs for the treatment of hydrocarbon vapor extracted during soil venting. In *Proc. Petrol. Hydrocarbons and Org. Chem. in Groundwater,* NWWA, November, Houston, TX.

Krueger, M., Harrison, A.B., and Betts, W.B. 1995. Treatment of hydrocarbon contaminated soil using a rotating drum bioreactor. *Microbios.* 83:243–247.

Kruithof, J.C., Graveland, A., van der Laan, J., and Reijnen, G.K. 1982. Removal of trichloroethene by combination of intensive aeration and activated carbon filtration. In *Proc. Atlantic Workshop on Org. Chem. Contamination of Ground Water.* AWWA/IWSA, Nashville, TN. pp. 219–237.

Krulwick, T.A. and Pelliccioni, N.J. 1979. Catabolic pathways of Coryneforms, Nocardias, and Mycobacteria. *Annu. Rev. Microbiol.* 33:95–111.

Kuhn, E.P., Colberg, P.J., Schnoor, J.L., Wanner, O., Zehnder, A.J.B., and Schwarzenbach, R.P. 1985. Microbial transformation of substituted benzenes during infiltration of river water to groundwater: laboratory column studies. *Environ. Sci. Technol.* 19:961–968.

Kuhn, S.P. and Pfister, R.M. 1989. Adsorption of mixed metals and cadmium by calcium-alginate immobilized *Zoogloea ramigera*. *Appl. Microbiol. Biotechnol.* 31:613–618.

Kuhn, S.P. and Pfister, R.M. 1990. Accumulation of cadmium by immobilized *Zoogloea ramigera* 115. *J. Ind. Microbiol.* 6:123–128.

Kulla, H.G., Krieg, R., Zimmermann, T., and Leisinger, T. 1984. Biodegradation of xenobiotics. Experimental evaluation of azo dye-degrading bacteria. In *Current Perspectives in Microbial Ecology. Proc. 3rd Int. Symp. Microb. Ecology,* Aug. 1983, Michigan Stage University. Klug, M.J. and Reddy, C.A., Eds. American Society of Microbiology. Washington, D.C. pp. 663–667.

Kumar, P.B.A.N., Dushenkov, V., Motto, H., and Raskin, I. 1995. Phytoextraction: the use of plants to remove heavy metals from soils. *Environ. Sci. Technol.* 29(5):1232–1238.

Kurek, E., Czaban, J., and Bollag, J.-M. 1982. Sorption of cadmium by microorganisms in competition with other soil constituents. *Appl. Environ. Microbiol.* 43:1011–1015.

Kuyucak, N. 1990. Feasibility of biosorbents application. In *Biosorption of Heavy Metals.* Volesky, B., Ed. CRC Press, Boca Raton, FL. pp. 317–378.

L

Laanbroek, H.J. and Pfennig, N. 1981. Oxidation of short-chain fatty acids by sulfate-reducing bacteria in freshwater and marine sediments. *Arch. Microbiol.* 128:330–335.

Laane, C., Boeren, S., Hilhorst, R., and Veeger, C. 1987. Optimization of biocatalysis in organic media. In *Biocatalysis in Organic Media.* Laane, C. et al., Eds. Elsevier Science Publishers, Amsterdam. pp. 65–84.

Laane, C., Boeren, S., Vos, K., and Veeger, C. 1987. *Biotechnol. Bioeng.* 30:81–87.

Labare, M.P. and Alexander, M. 1995. Enhanced mineralization of organic compounds in nonaqueous-phase liquids. *Environ. Toxicol. Chem.* 14(2):257–265.

Lackner, R., Srebotnik, E., and Messner, K. 1991. Oxidative degradation of high molecular weight chlorolignin by manganese peroxidase of *Phanerochaete chrysosporium. Biochem. Biophys. Res. Commun.* 178:1092–1098.

Laddaga, R.A. and Silver, S. 1985. Cadmium uptake in *Escherichia coli* K-12. *J. Bacteriol.* 162:1100–1105.

Laddaga, R.A., Bessen, R., and Silver, S. 1985. Cadmium-resistant mutant of *Bacilllus subtilis* 168 with reduced cadmium transport. *J. Bacteriol.* 162:1106–1110.

Lageman, Pool, van Vulpen, and Norris, 1995. *In situ* electrobioreclamation in low-permeability soils. In *Applied Bioremediation of Petroleum Hydrocarbons.* Hinchee, R.E., Kittel, J.A., and Reisinger, H.J., Eds. Battelle Press, Columbus, OH. pp. 287–292.

Laha, S., Liu, Z., Edwards, D.A., and Luthy, R.G. 1995. Surfactant solubilization of phenanthrene in soil-aqueous systems and its effects on biomineralization. *Adv. Chem. Ser.* 244 (Aquatic Chem.). pp. 339–361.

Lamar, R.T. and Dietrich, D.M. 1990. *Appl. Environ. Microbiol.* 56:3093–3100.

Lamar, R.T., Larsen, M.J., Kirk, T.K., and Glaser, J.A. 1987. Growth of the white-rot fungus *Phanerochaete chrysosporium* in soil. In *Proc. U.S. EPA 13th Annu. Res. Symp. on Land Disposal, Remedial Action, Incineration, and Treatment of Haz. Waste,* May 6–8, Cincinnati, OH.

Lambert, S.M. 1968. Omega, a useful index of soil sorption equilibria. *J. Agric. Food Chem.* 16:340–343.

La Mori, P.N. 1994. Site closure using *in-situ* hot air/steam stripping (HASS) of hydrocarbons in soils. *Hydrocarbon Contam. Soils.* 4:335–357.

Lang, E. and Malik, K.A. 1996. Maintenance of biodegradation capacities of aerobic bacteria during long-term preservation. *Biodegradation.* 7:65–71.

Lang, S. and Wagner, F. 1987. Structure and properties of biosurfactants. In *Biosurfactants and Biotechnology.* Kosaric, N., Cairns, W.L., and Gray, N.C.C., Eds. Marcel Dekker, New York. pp. 21–45.

Lange, T.A., Bouillard, J.X., and Michelsen, D.L. 1995. Use of microbubble dispersion for soil scouring. In *In Situ Aeration: Air Sparging, Bioventing, Relat. Rem. Proceses. Pap. 3rd Int. In Situ On-Site Bioreclam. Symp.* Hinchee, R.E., Miller, R.N., and Johnson, P.C., Eds. pp. 511–518.

Lanoil, B.D. and Giovannoni, S.J. 1997. Identification of bacterial cells by chromosomal painting. *Appl. Environ. Microbiol.* 63(3):1118–1123.

Lapinskas, J. 1989. Soil Biotreatment: bacterial degradation of hydrocarbon contamination in soil and groundwater. *Chem. Ind.* 4 Dec. pp. 784–789.

Larsen, F.S., Silcox, G.D., and Keyes, B.R. 1994. The development of a thermal treatment assessment procedure for soils contaminated with hydrocarbons. *Combust. Sci. Technol.* 101(1–6):443–459.

Larsen, V.J. and Schierup, H.-H. 1981. The use of straw for removal of heavy metals from waste water. *J. Environ. Qual.* 10:188–193.

Larson, R.J. and Ventullo, R.M. 1983. Biodegradation potential of groundwater bacteria. In *Proc. 3rd Natl. Symp. Aquifer Restoration and Groundwater Monitoring,* May 25–27, Columbus, OH.

Laskin, A. and Lechevalier, H.A., Eds. 1974. *Handbook of Microbiology.* CRC Press, Cleveland, OH.

Laube, V., Ramamoorthy, S., and Kushner, D.J. 1979. Mobilization and accumulation of sediment bound heavy metals by algae. *Bull. Environ. Contam. Toxicol.* 21:763–770.

Law, A.T. and Button, D.K. 1977. *J. Bacteriol.* 129:115–123.

Law Engineering Testing Company. 1982. *Literature Inventory: Treatment Techniques Applicable to Gasoline Contaminated Ground Water.* American Petroleum Institute, Washington, D.C. 60 pp.

Lawes, B.C., 1991. Soil-induced decomposition of hydrogen peroxide. In *In Situ Bioreclamation. Applications and Investigations for Hydrocarbon and Contaminated Site Remediation.* Hinchee, R.E. and Olfenbuttel, R.F., Eds. Butterworth-Heinemann, Stoneham, MA. pp. 143–156.

Lawlor, G.F., Jr., Shiaris, M.P., and Jambard-Sweet, D. 1986. Phenanthrene metabolites by pure and mixed bacterial cultures as examined by HPLC. *Abstr. Annu. Mtg. Am. Soc. Microbiol.* p. 302.

Leadbetter, E.R. and Foster, J.W. 1959. *Arch. Biochem. Biophys.* 82:491–492.

Leahy, J. and Colwell, R. 1990. *Microb. Rev.* 54:305–315.

Leavitt, M.E. and Brown, K.L. 1994. Biostimulation vs. bioaugmentation — three case studies. In *Hydrocarbon Bioremediation.* Hinchee, R.E., Alleman, B.C., Hoeppel, R.E., and Miller, R.N., Eds. CRC Press, Boca Raton, FL. pp. 72–79.

Lee, C.J.B., Fletcher, M.A., Avila, O.I., Callanan, J., Yunker, S., and Munnecke, D.M. 1995. In *Bioaugmentation for Site Remediation. Pap. 3rd Int. In Situ On-Site Bioreclam. Symp.* Hinchee, R.E., Fredrickson, J., and Alleman, B.C., Eds. Battelle Press, Columbus, OH. pp. 195–202.

REFERENCES

Lee, E. and Banks, M.K. 1993. Bioremediation of petroleum contaminated soil using vegetation: a microbial study. *J. Environ. Sci. Health,* Part A. A28(10):2187–2198.

Lee, J., Choi, Y., and Kim, H. 1993. Simultaneous biodegradation of toluene and *p*-xylene in a novel bioreactor: experimental results and mathematical analysis. *Biotechnol. Prog.* 9:46-53.

Lee, K.M., Melnyk, I.R., and Bishop, D.F. 1991. Anaerobic treatment of *o*-xylene. In *On-Site Bioreclamation: Processes for Xenobiotic and Hydrocarbon Treatment.* Hinchee, R.E. and Olfenbuttel, R.F., Eds. Butterworth-Heinemann, Stoneham, MA. pp. 226–238.

Lee, M.D. and Ward, C.H. 1984. Reclamation of contaminated aquifers: biological techniques. In *Proc. 1984 Hazardous Mater. Spills Conf.,* Apr. 9–12, Nashville, TN. Ludwigson, J., Ed. Government Institutes, Inc., Rockville, MD. pp. 98–103.

Lee, M.D. and Ward, C.H. 1985. Environmental and biological methods for the restoration of contaminated aquifers. *Environ. Toxicol. Chem.* 4:743–750.

Lee, M.D., Wilson, J.T., and Ward, C.H. 1984. Microbial degradation of selected aromatics in a hazardous waste site. *Dev. Ind. Microbiol.* 25:557–565.

Lee, M.D., Wilson, J.T., and Ward, C.H. 1987. In situ restoration techniques for aquifers contaminated with hazardous wastes. *J. Haz. Mater.* 14:71–82.

Lee, R.F. 1977. In *Fate and Effects of Petroleum Hydrocarbons in Marine Organisms and Ecosystems: Proc. Symp.,* Nov. 10–12, 1976, Seattle, WA. Wolfe, D.A., Ed. Pergamon Press, New York. pp. 60–70.

Lee, R.F. and Hoeppel, R. 1991. Hydrocarbon degradation potential in reference soils and soils contaminated with jet fuel. In In Situ *Bioreclamation. Applications and Investigations for Hydrocarbon and Contaminated Site Remediation.* Hinchee, R.E. and Olfenbuttel, R.F., Eds. Butterworth-Heinemann, Stoneham, MA. p. 570.

Lee, R.F. and Ryan, C. 1976. Biodegradation of petroleum hydrocarbons by marine microbes. In *Proc. 3rd Intl. Biodegradation Symp.,* Aug. 17–23, 1975, Kingston, RI, Sharply, J.M., and Kaplan, A.M., Eds. Applied Science Publishers, London. pp. 119–125.

Leeper, G.W. 1978. *Managing the Heavy Metals on the Land.* Marcel Dekker, New York.

Leeson, A., Hinchee, R.E., Kittel, J., Sayles, G., Vogel, C.M., and Miller, R.N. 1993. Optimizing bioventing in shallow vadose zones and cold climates. *Hydrol. Sci. J.* 38(4):283–295.

Leeson, A., Kittel, J.A., Hinchee, R.E., Miller, R.N., Haas, P.E., and Hoeppel, R.E. 1995. Test plan and technical protocol for bioslsurping. In *Applied Bioremediation of Petroleum Hydrocarbons.* Hinchee, R.E., Kittel, J.A., and Reisinger, H.J., Eds. Battelle Press, Columbus, OH. pp. 335–347.

Leeson, A., Kumar, P., Hinchee, R.E., Downey, D., Vogel, C.M., Sayles, G.D., and Miller, R.N. 1995. Statistical analyses of the U.S. Air Force bioventing initiative results. In In Situ *Aeration: Air Sparging, Bioventing, Relat. Rem. Processes. Pap. 3rd Int.* In Situ *On-Site Bioreclam. Symp.* Hinchee, R.E., Miller, R.N., and Johnson, P.C., Eds. Battelle Press, Columbus, OH. pp. 223–225.

Leffrang, U., Ebert, K., Flory, K., Galla, U., and Schnieder, H. 1995. Organic waste destruction by indirect electrooxidation. *Sep. Sci. Technol.* 30(7–9):1883–1899.

Lehman, R.M., Colwell, F.S., Ringelberg, D.B., and White, D.C. 1995. Combined microbial community-level analyses for quality assurance of terrestrial subsurface cores. *J. Microbiol. Methods.* 22:263–281.

Lehmicke, L.G., Williams, R.T., and Crawford, R.L. 1979. ^{14}C-most-probable-number method for enumeration of active heterotrophic microorganisms in natural waters. *Appl. Environ. Microbiol.* 38:644–649.

Lehtomakei, M. and Niemela, S. 1975. Improving microbial degradation of oil in soil. *Ambio.* 4:126–129.

Lei, J., Lord, D., Arneberg, R., Rho, D., Greer, C., and Cyr, B. 1995. Biological treatment of waste gas containing volatile hydrocarbons. In *Biological Unit Processes for Hazardous Waste Treatment. Pap. 3rd Int.* In Situ *On-Site Bioreclam. Symp.* Hinchee, R.E., Skeen, R.S., and Sayles, G.D., Eds. Battelle Press, Columbus, OH. p. 275.

Leisinger, T. 1983. I. General aspects. Microorganisms and xenobiotic compounds. *Experientia.* 39:1183–1191.

Lemaire, J., Campbell, I., Hulpke, H., Guth, J.A., Merz, W., Philop, L., and Von Waldow, C. 1982. An assessment of test methods for photodegradation of chemicals in the environment. *Chemosphere.* 11:119–164.

Lendvay, L. 1992. Safe disposal of hazardous wastes by vitrification. *Epitoanyag.* 44(6):230–231.

Lenhard, R.J., Johnson, T.G., and Parker, J.C. 1993. Experimental observations of nonaqueous-phase liquid subsurface movement. *J. Contam. Hydrol.* 12(1–2):79–101.

LePetit, J. and Barthelemy, M.H. 1968. *Ann. Inst. Pasteur Paris.* 114:149–158.

LePetit, J. and Tagger, S. 1976. Degradation des hydrocarbures en presence d'autres substances organiques par des bacteries isolees de l'eau de mer. *Can. J. Microbiol.* 22:1654–1657.

Leson, G., Tabatabai, F., and Winer, A.M. 1992. Control of hazardous and toxic air emissions by biofiltration. 92-116.03. 85th Annu. Mtg. Exhib. Air Waste Mgmt. Assoc., June 21–26, 1992, Kansas City, MO.

Leson, G. and Winer, A. 1991. Biofiltration: an innovative air pollution control technology for VOC emissions. *J. Air Waste Manage. Assoc.* 41:1045.

Lestan, D. and Lamar, R.T. 1996. Development of fungal inocula for bioaugmentation of contaminated soils. *Appl. Environ. Microbiol.* 62(6):2045–2052.

Lestan, D., Lestan, M., Chapelle, J.A., and Lamar, R.T. 1996. Biological potential of fungal inocula for bioaugmentation of contaminated soils. *J. Ind. Microbiol.* 16(5):286–294.

Lester, J.N., Sterritt, R.M., Rudd, T., and Brown, M.J. 1984. Assessment of the role of bacterial extracellular polymers in controlling metal removal inbiological waste water treatment. In *Microbiological Methods for Environmental Biotechnology.* Grainger, J.M. and Lynch, J.M., Eds. Academic Press, London. pp. 197–217.

Letey, J. and Farmer, W.J. 1974. Movement of pesticides in soil. In *Pesticides in Soil and Water.* Guenzi, W.D., Ed. Soil Science Society of America, Madison, WI. pp. 67–97.

Letey, J. and Stolzy, L.H. 1964. Measurement of oxygen diffusion rates with the platinum microelectrode theory and equipment. *Hilgardia.* 35(2):545–554.

Lettinga, G., DeZeeuw, W., and Ouborg, E. 1981. Anaerobic treatment of wastes containing methanol and higher alcohols. *Water Res.* 15:171–182.

Leung, K., Cassidy, M.B., Holmes, S.B., Lee, H., and Trevors, J.T. 1995. Survival of κ-carrageenan-encapsulated and unencapsulated *Pseudomonas aeruginosa* UG2Lr cells in forest soil monitored by polymerase chain reaction and spread plating. *FEMS Microb. Ecol.* 16(1):71–82.

Levine, H., MacDonald, G., Rothaus, O., Ruderman, M., and Treiman, S. 1995. Subsurface science. Report JSR-94-330. Order No. AD-A298759. Avail. NTIS. From Gov. Rep. Announce. Index, 1996. 96(5). Abstr. No. 05-01,597. 107 pp.

Lewandowski, G., Armenate, P.A., and Pak, D. 1990. Reactor design for hazardous waste treatment using a white rot fungus. *Water Res.* 24:75–82.

Lewis, R.F. 1993. SITE demonstration of slurry-phase biodegradation of PAH contaminated soil. *Air Waste.* 43 (April):503.

Li, Y., Yang, I.C.-Y., Lee, K.-I., and Yen, T.F. 1994. *In situ* biological encapsulation: biopolymer shields. In *Applied Biotechnology for Site Remediation.* Hinchee, R.E., Anderson, D.B., Metting, F.B., Jr., and Sayles, G.D., Eds. Lewis Publishers, Boca Raton, FL. pp. 275–286.

Liang, L.N., Sinclair, J.L., Mallory, L.M., and Alexander, M. 1982. Fate in model ecosystems of microbial species of potential use in genetic engineering. *Appl. Environ. Microbiol.* 44:708–714.

Liao, P.H. and Lo, K.V. 1985. Methane production using whole and screened dairy manure in conventional and fixed-film reactors. *Biotechnol. Bioeng.* 27:266.

Lieberman, S.H., Apitz, S.E., Borbridge, L.M., and Theriault, G.A. 1993. Subsurface screening of petroleum hydrocarbons in soils via laser induced fluorometry over optical fibers with a cone penetrometer system. In *Proc. SPIE-Int. Soc. Opt. Eng.* 1716 (Int. Conf. Monitoring Toxic Chem. and Biomarkers, 1992). pp. 392–402.

Lin, J.-E., Lantz, S., Schultz, W.W., Mueller, J.G., and Pritchard, P.H. 1995. Use of microbial encapsulation/immobilization for biodegradation of PAHs. In *Bioaugmentation for Site Remediation. Pap. 3rd Int. In Situ On-Site Bioreclam. Symp.* Hinchee, R.E., Fredrickson, J., and Alleman, B.C., Eds. Battelle Press, Columbus, OH. pp. 211–220.

Lin, S.-C., Carswell, K.S., Georgiou, G., and Sharma, M.M. 1994. Continuous production of the lipopeptide biosurfactant of *Bacillus licheniformis* JF-2. *Appl. Microbiol. Biotechnol.* 41(3):281–285.

Lin, S.-C., Minton, M.A., Sharma, M.M., and Georgiou, G. 1994. Structural and immunological characterization of a biosurfactant produced by *Bacillus licheniformis* JF-2. *Appl. Environ. Microbiol.* 60(1):31–38.

Lin, S.-C., Sharma, M.M., and Georgiou, G. 1993. Production and deactivation of biosurfactant by *Bacillus licheniformis* JF-2. *Biotechnol. Prog.* 9(2):138–145.

Lin, W.S. and Kapoor, M. 1979. Induction of aryl hydrocarbon hydroxylase in *Neurospora crassa* by benzo(a)pyrene. *Curr. Microbiol.* 3:177–180.

Lin, Z., Hill, R.M., Davis, H.T., and Ward, M.D. 1994. Determination of wetting velocities of surfactant superspreaders with the quartz crystal microbalance. *Langmuir.* 10:4060–4068.

Lindgren, E.R. and Brady, P.V. 1995. Electrokinetic control of moisture and nutrients in unsaturated soils. In *Applied Bioremediation of Petroleum Hydrocarbons.* Hinchee, R.E., Kittel, J.A., and Reisinger, H.J., Eds. Battelle Press, Columbus, OH. pp. 475–481.

Litchfield, J.H. and Clark, L.C. 1973. Bacterial Activity in Ground Waters containing Petroleum Products. Project OS 21.1. API Publication No. 4211. American Petroleum Institute Committee on Environmental Affairs, Washington, D.C.

Little, C.D., Fraley, C.D., McCann, M.P., and Matin, A. 1991. Use of bacterial stress promoters to induce biodegradation under conditions of environmental stress. In *On-Site Bioreclamation: Processes for Xenobiotic and Hydrocarbon Treatment.* Hinchee, R.E. and Olfenbuttel, R.F., Eds. Butterworth-Heinemann, Stoneham, MA. pp. 493–498.

Liu, D.L.S. 1973. Microbial Degradation of Oil Pollutants, Workshop, Louisiana State University Publ. No. LSU-Sg-73-01. pp. 95–104.

Liu, D. 1980. Enhancement of PCBs biodegradation by sodium ligninsulfonate. *Water Res.* 14:1467–1475.

Liu, J, Crittenden, J.C., Hand, D.W., and Perram, D.L. 1996. Regeneration of adsorbents using heterogeneous photocatalytic oxidation. *J. Environ. Eng.* 122(8):707–713.

Liu, M. and Roy, D. 1995. Surfactact-induced interactions and hydraulic conductivity changes in soil. *Waste Management.* 15(7):463–470.

Liu, Z., Laha, S., and Luthy, R.G. 1991. Surfactant solubilization of polycyclic aromatic hydrocarbon compounds in soil-water suspensions. *Water Sci. Technol.* 23:475–485.

Liu, Z., Jacobson, A.M., and Luthy, R.G. 1995. Biodegradation of naphthalene in aqueous nonionic surfactant systems. *Appl. Environ. Microbiol.* 61(1):145–151.

REFERENCES

Lode, A. 1986. Changes in the bacterial community after application of oily sludge to soil. *Appl. Microbiol. Biotechnol.* 25:295–299.
Loehr, R.C. 1986. In *Land Treatment: A Hazardous Waste Management Alternative. Water Resources Symposium Number 13.* Loehr, R.C. and Malina, J.F., Jr., Eds. University of Texas, Austin.
Loehr, R.C. 1992. Bioremediation of PAH compounds in contaminated soil. In *Hydrocarbon Contaminated Soils and Groundwater. Proc. 1st Annu. West Coast Conf. Hydrocarbon Contam. Soils Groundwater,* Feb. 1990, Newport Beach, CA. Vol. 1. Calabrese, E.J. and Kostechi, P.T., Eds. Lewis Publishers, Chelsea, MI. pp. 213–222.
Loehr, R.C., Martin, J.H., Neuhauser, E.F., Norton, R.A., and Malecki, M.R. 1985. Land Treatment of an Oily Waste-Degradation, Immobilization, and Bioaccumulation. EPA-600/S2-85/009.
Loehr, R.C., Neuhauser, E.F., and Martin, J.H., Jr. 1984. *16th Mid-Atlantic Ind. Waste Conf.* pp. 568–579.
Loehr, R.C., Tewell, W.J., Novak, J.D., Clarkson, W.W., and Freidman, G.S. 1979. *Land Application of Wastes.* Vol. 1. Van Nostrand Reinhold, New York.
Long, G. 1992. Bioventing and vapor extraction: innovative technologies for contaminated site remediation. *J. Air Waste Manage. Assoc.* 42(3):345–348.
Long, S.C., Dobbins, D.C., Aelion, M.C., and Pfaender, F.K. 1986. Metabolism of naturally occurring and xenobiotic compounds by subsurface microbial communities. *Abstr. Annu. Mtg. Am. Soc. Microbiol.* p. 285.
Loo, W.W. 1994. Electrokinetic enhanced passive in-situ bioremediation of soil and groundwater containing gasoline, diesel and kerosene. *Proc. 11th HAZMACON 94, Haz. Mater. Mgmt. Conf. Exhib.* pp. 254–264.
Lopatowska, D. 1984. *Acta Microbiol. Pol.* 33:263.
Loske, D., Hutterman, A., Majcherczyk, A., Zadrazil, F., Lorsen, H., and Waldinger, P. 1990. In *Advances in Biological Treatment of Lignocellulosic Materials.* Coughlan, M.P., Eds. Elsevier Science, London.
Love, O.T., Jr. and Eilers, R.G. 1982. Treatment of drinking water containing trichloroethylene and related industrial solvents. *J. Am. Water Works Assoc.* 74:413–425.
Love, O.T., Jr., Miltner, R.J., Eilers, R.G., and Fronk-Leist, C.A. 1983. Treatment of Volatile Organic Compounds in Drinking Water. EPA-600/8-83-019. Municipal Environmental Research Laboratory, Cincinnati, OH.
Love, O.T., Jr., Ruggiero, D.D., Feige, W.A., Carswell, J.K., Miltner, R.J., Clark, R.M., and Fronk, C.A. 1983. Field Evaluation of Aeration Processes for Organic Contaminant Removal from Groundwater. EPA-600/2-86/024. U.S. EPA, Cincinnati, OH. PB-84-130384.
Lovelock, J.E. 1979. *Gaea: A New Look at Life on Earth.* Oxford University Press, Oxford, U.K.
Lovley, D.R. 1985. Minimum threshold for hydrogen metabolism in methanogenic bacteria. *Appl. Environ. Microbiol.* 49:1530–1531.
Lovley, D.R., Baedecker, M.J., Lonergan, D.J., Cozzarelli, I.M., Phillips, E.J.P., and Segal, D.I. 1989. Oxidation of aromatic contaminants coupled to microbial iron reduction. *Nature.* 339:297–300.
Lovley, D.R. and Klug, M.J. 1983. Sulfate reducers can outcompete methanogens at freshwater sulfate concentrations. *Appl. Environ. Microbiol.* 45:187–192.
Lovley, D.R., Woodward, J.C., and Chapelle, F.H. 1994. Stimulated anoxic biodegradation of aromatic hydrocarbons using Fe(III) liquids. *Nature.* 370:128–136.
Lowe, P. and Williamson, S.M. 1984. Accelerated cold anaerobic digestion. In *Sewage Sludge Stabilization and Disinfection.* 1983. (G.B.). Bruce, A.M., Ed. Published for the Water Research Centre, E. Horwood, NY, p. 224–238.
Lowenbach, W. 1978. Compilation and Evaluation of Leaching Test Methods. EPA-600/2-78-095. U.S. EPA Cincinnati, OH. 111 pp.
Lu, J.C.S. and Chen, K.Y. 1977. Migration of trace metals in interfaces of seawater and polluted surficial sediments. *Environ. Sci. Technol.* 11:174–182.
Lu, Y. and Wilkins, E. 1995. Heavy metal removal by caustic-treated yeast immobilized in alginate. In *Biorem. Inorg. Pap. 3rd Int. In Situ On-Site Bioreclam. Symp.* Hinchee, R.E., Means, J.L., and Burris, D.R., Eds. Battelle Press, Columbus, OH. pp. 117–124.
Ludzack, F.L. and Kinkead, D. 1956. Persistence of oily wastes in polluted water under aerobic conditions: motor oil class of hydrocarbons. *Ind. Eng. Chem.* 48:263.
Lukins, H.B. 1962. On the Utilization of Hydrocarbons, Methyl Ketones, and Hydrogen by Mycobacteria. Thesis. The University of Texas.
Lund, N.-Ch. and Gudehus, G. 1990. Biologische *in situ*-Sanierung kohlenwasserstoffbelasteter koerniger Boeden. Ph.D. Dissertation, Report No. 119. Inst. Soil Mech. Rock Mech., University Karlsruhe, FRG.
Lund, N.-Ch., Swinianski, J., Gudehus, G., and Maier, D. 1991. Laboratory and field tests for a bibological *in situ* remediation of a coke oven plant. In In Situ *Bioreclamation. Applications and Investigations for Hydrocarbon and Contaminated Site Remediation.* Hinchee, R.E. and Olfenbuttel, R.F., Eds. Butterworth-Heinemann, Stoneham, MA. pp. 396–412.
Lyman, W.J., Noonan, D.C., and Reidy, P.J. 1990. Cleanup of petroleum contaminated soils at underground storage tanks. *Pollut. Technol. Rev.* No. 195. Noyes Data Corp., Park Ridge, NJ.
Lyman, W.J., Rechl, W.F., and Rosenblatt, D.H. 1982. In *Handbook of Chemical Properties Estimation Methods: Environmental Behavior of Organic Chemicals.* Chap. 16. McGraw-Hill, New York.
Lynch, J.M. and Poole, N.J., Eds. 1979. *Microbial Ecology, a Conceptual Approach.* John Wiley & Sons, New York.

M

Macaskie, L.E. 1990. An immobilized cell bioprocess for the removal of heavy metals from aqueous flows. *J. Chem. Technol. Biotechnol.* 49:357–379.

Macaskie, L.E. and Dean, A.C.R. 1987. Use of immobilized biofilm of *Citrobacter* sp. for the removal of uranium and lead from aqueous flows. *Enzyme Microbial Technol.* 9:2–4.

Macaskie, L.E. and Dean, A.C.R. 1989. Microbial metabolism, desolubilization, and deposition of heavy metals: metal uptake by immobilized cells and application to the treatment of liquid wastes. In *Biological Waste Treatment*. Mizrahi, A., Ed. Alan R. Liss, New York. pp. 150–201.

Macaskie, L.E. and Dean, A.C.R. 1990. Metal-sequestering biochemicals. In *Biosorption of Heavy Metals*. Volesky, B., Ed. CRC Press, Boca Raton, FL. pp. 199–248.

Macaskie, L.E., Wates, J.M., and Dean, A.C.R. 1987. Cadmium accumulation by a *Citrobacter* sp. immobilized on gel and solid supports: applicability to the treatment of liquid wastes containing heavy metal cations. *Biotechnol. Bioeng.* 38:66–73.

Mackay, D.M., Roberts, P.V., and Cherry, J.A. 1985. Transport of organic contaminants in groundwater. *Environ. Sci. Technol.* 19:384–392.

Mackay, D. and Shiu, W.Y. 1977. Aqueous solubility of polynuclear aromatic hydrocarbons. *J. Chem. Eng. Data.* 22:399–402.

Mackay, D. and Shiu, W.Y. 1981. A critical review of Henry's law constants for chemicals of environmental interest. *J. Phys. Chem. Ref. Data.* 10(4):1175–1199.

Madsen, E.L. 1991. Determining *in situ* biodegradation: facts and challenges. *Environ. Sci. Technol.* 25(10):1663–1673.

Madsen, E.L. and Alexander, M. 1982. Transport of *Rhizobium* and *Pseudomonas* through soil. *Soil Sci. Soc. Am. J.* 46:557–560.

Magee, L.A. and Colmer, A.R. 1960a. Factors affecting the formation of gum by *Alcaligenes faecalis*. *J. Bacteriol.* 80:477–483.

Magee, L.A. and Colmer, A.R. 1960b. Properties of gums formed by *Alcaligenes faecalis*. *J. Bacteriol.* 81:800–802.

Magor, A.M., Warburton, J., Trower, M.K., and Griffin, M. 1986. Comparative study of the ability of three *Xanthobacter* species to metabolize cycloalkanes. *Appl. Environ. Microbiol.* 52:665–671.

Mah, R.A. 1981. In *Trends in the Biology of Fermentations for Fuels and Chemicals*. Hollaender, A., Rabson, R., Rogers, P., San Pietro, A., Valentine, R., and Wolfe, R., Eds. Plenum Press, New York. pp. 357–373.

Mah, R.A. 1982. Methanogenesis and methanogenic partnerships. *Philos. Trans. R. Soc. London.* Ser. B. 297:599–616.

Mahmood, S.K. and Rao, P.R. 1993. Microbial abundance and degradation of polycyclic aromatic hydrocarbons in soil. *Bull. Environ. Contam. Toxicol.* 50:486–491.

Mahro, B., Eschenbach, A., Kaestner, M., and Schaefer, G. 1994. Investigation of possibilities for targeted stimulation of biogenic mineralization and humification of PAK in soils. *Biol. Abwasserreinig.* 4:55–68.

Major, D.W., Mayfield, C.I., and Barker, J.F. 1988. *Ground Water.* 26:8–14.

Makula, R.A. and Finnerty, W.R. 1972. Microbial assimilation of hydrocarbons: cellular distribution of fatty acids. *J. Bacteriol.* 112:398–407.

Mallon, B.J. and Harrison, F.L. 1984. Octanol-water partition coefficient of benzo(a)pyrene: measurement, calculation, and environmental implications. *Bull. Environ. Contam. Toxicol.* 32:316–323.

Maloney, S.W., Manem, J., Mallevialle, J., and Fiessinger, F. 1985. The potential use of enzymes for removal of aromatic compounds from water. *Water Sci. Technol.* 17:273–278.

Manilal, V.B. and Alexander, M. 1991. Factors affecting the microbial degradation of phenanthrene in soil. *Appl. Microbiol. Biol.* 35(3):401–405.

Mapes, J.P., McKenzie, K.D., Arrowood, S.P., Studabaker, W.B., Allen, R.L., Manning, W.B., and Friedman, S.B. 1993. PETRO RISc soil: a monoclonal antibody immunoassay for the rapid, onsite screening of gasoline and diesel fuel contaminated soil. *Hydrocarbon Contam. Soils Groundwater.* 3:47–56.

Marcandella, E., Bicheron, C., and Bues, M.A. 1995. Bacterial degradation of an organomercurial micropollutant in natural sediments. In *Microbial Processes for Bioremediation*. Hinchee, R.E., Vogel, C.M., and Brockman, F.J., Eds. *Bioremediation* 3 (8). Battelle Press, Columbus, OH. p. 281.

March, J. 1968. *Advanced Organic Chemistry: Reactions, Mechanisms and Structure*. McGraw Hill, New York.

Margaritis, A., Zajic, J.E., and Gerson, D.F. 1979. Production and surface-active properties of microbial surfactants. *Biotechnol. Bioeng.* 21:1151–1162.

Margesin, R. and Schinner, F. 1995. Metal resistance of bacteria isolated from contaminated soils. In *Biodeterior. Biodegrad.* 9. Proc. 9th Int. Biodeterior. Biodegrad. Symp., 1993. Bousher, A., Chandra, M., and Edyvean, R., Eds. Institute of Chemical Engineering, Rugby, U.K. pp. 395–399.

Margesin, R. and Schinner, F. 1996. Bacterial heavy metal-tolerance: extreme resistance to nickel in *Arthrobacter* spp. strains. *J. Basic Microbiol.* 36(4):269–282.

Marin, M.M., Pedregosa, A.M., Ortiz, M.L., and Laborda, F. 1995. Study of a hydrocarbon-utilizing and emulsifier-producing *Acinetobacter calcoaceticus* strain isolated from heating oil. *Microbiologia* (Madrid). 11(4):447–454.

REFERENCES

Marin, M., Pedregosa, A., Rios, S., Ortiz, M.L., and Laborda, F. 1995. Biodegradation of diesel and heating oil by *Acinetobacter calcoaceticus* MM5: its possible applications on bioremediation. *Int. Biodeterior. Biodeg.* 35(1–3):269–285.

Markovetz, A.J. 1971. Subterminal oxidation of liphatic hydrocarbons by microorganisms. *Crit. Rev. Microbiol.* 1:225–237.

Marks, P.J. and Noland, J.W. 1986. Evaluation of Low Temperature Thermal Stripping of VOCs from Soil. Report No. AMXTH-TE-CR 86085. ADA 171521. U.S. Army Toxic and Haz. Mater. Agency.

Marques, A.M., Congregado, F., and Simon-Pujol, D.M. 1979. Antibiotic and heavy metal resistance of *Pseudomonas aeruginosa* isolated from soils. *J. Appl. Bacteriol.* 47:347–350.

Marquis, R.E., Mayzel, K., and Carstensen, E.L. 1976. Cation exchange in cell walls of gram-positive bacteria. *Can. J. Microbiol.* 22:975–982.

Marquis, S.A., Jr. 1993. Soil remediation of a lower-permeability unit using an *in situ* vapor extraction system with a combined thermal-cataliticus oxidizer vapor treatment system. *Hydrocarbon Contam. Soils Groundwater.* 3:351–370.

Marshall, K.C. 1976. *Interfaces in Microbial Ecology.* Harvard University Press, Cambridge, MA.

Marshall, K.C. 1980. Adsorption of microorganisms to soils and sediments. In *Adsorption of Microorganisms to Surfaces.* Bitton, G. and Marshall, K.C., Eds. Wiley, New York. pp. 317–329.

Marshall, K.C. and Cruickshank, R.H. 1973. Cell surface hydrophobicity and the orientation of certain bacteria at interfaces. *Arch. Mikrobiol.* 91:29–40.

Marshall, T.R. 1995. Nitrogen fate model for gas-phase ammonia-enhanced *in situ* bioventing. In *In Situ Aeration: Air Sparging, Bioventing, Relat. Rem. Processes. Pap. 3rd Int. In Situ On-Site Bioreclam. Symp.* Hinchee, R.E., Miller, R.N., and Johnson, P.C., Eds. Battelle Press, Columbus, OH. pp. 351–359.

Marsman, E.H., Appelman, J.M.M., Urlings, L.G.C.M., and Bult, B.A. 1994. BIOPUR, an innovative bioreactor for the treatment of groundwater and soil vapor contaminated with xenobiotics. In *Applied Biotechnology for Site Remediation.* Hinchee, R.E., Anderson, D.B., Metting, F.B., Jr., and Sayles, G.D., Eds. Lewis Publishers, Boca Raton, FL. pp. 391–399.

Martin, E.J. et al. 1982. Paper presented at 5th U.S.–Japan Governmental Conf. on Solid Waste Mgmt. Tokyo, Japan, Sept., 1979.

Martin, J.K. and Foster, R.C. 1985. A model system for studying the biochemistry and biology of the root–soil interface. *Soil Biol. Biochem.* 17:261–269.

Martin, J.P. and Sims, R.C. 1986. *Haz. Waste Haz. Mater.* 3:261–280.

Martin, J.P., Sims, R.C., and Mathews, J. 1986. Review and Evaluation of Current Design and Management Practices for Land Treatment Units Receiving Petroleum Wastes. EPA-600/J-86/264. PB 87166339.

Martinsen, A., Skjak-Braek, G., and Smidsrod, O. 1989. Alginate as immobilization material. I. Correlation between chemical and physical properties of alginate gel beads. *Biotechnol. Bioeng.* 33:79–89.

Mason, J.W., Anderson, A.C., and Shariat, M. 1979. Rate of demethylation of methylmercuric chloride by *Enterobacter aerogenes* and *Serratia marcescens. Bull. Environ. Contam. Toxicol.* 21:262–268.

Masson, L., Comeau, Y., Brousseau, R., Samson, R., and Greer, C.W. 1993. Construction and application of chromosomally integrated *lac-lux* gene markers to monitor the fate of a 2,4-dichlorophenoxyacetic acid-degrading bacterium in contaminated soils. *Microbial Releases.* 1:209–216.

Mathrani, I.M. and Boone, D.R. 1985. Isolation and characterization of a moderately halophilic methanogen from a solar saltern. *Appl. Environ. Microbiol.* 50:140–143.

Matin, A., Auger, E.A., Blum, P.B., and Schultz, J.E. 1989. *Annu. Rev. Microbiol.* 43:293–316.

Matsunaga, T. and Izumida, H. 1984. *Biotechnol. Bioeng. Symp.* 14:407.

Matsuyama, T., Murakami, T., Fujita, M., Fujita, S., and Yano, I. 1986. Extracellular vesicle formation and bio-surfactant production by *Serratia marcescens. J. Gen. Microbiol.* 132:865-875.

Mattei, G. and Bertrand, J.-C. 1985. Production of biosurfactants by a mixed bacteria population grown in continuous culture on crude oil. *Biotechnol. Lett.* 7:217–222.

Matter-Muller, C., Gujer, W., Giger, W., and Strumm, W. 1980. Non-biological elimination mechanisms in a biological sewage treatment plant. *Prog. Water Technol.* 12:299–314.

Mattingly, G.E.G. 1975. Labile phosphates in soils. *Soil Sci.* 119:369–375.

Mattson, F.H. and Volpenhein, R.A. 1966. *J. Am. Oil Chem. Soc.* 43:286–289.

Maue, G. and Dott, W. 1995. Degradation tests with PAH-metabolizing soil *in situ* bioremediation. In *Monit. Verif. Biorem. Pap. 3rd Int. In Situ On-Site Bioreclam. Symp.* Hinchee, R.E., Douglas, G.S., and Ong, S.K., Eds. Battelle Press, Columbus, OH. pp. 127–133.

Maxwell, C.R. and Baqai, H.A. 1995. Remediation of petroleum hydrocarbons by inoculation with laboratory-cultured microorganisms. In *Bioaugmentation for Site Remediation. Pap. 3rd Int. In Situ On-Site Bioreclam. Symp.* Hinchee, R.E., Fredrickson, J., and Alleman, B.C., Eds. Battelle Press, Columbus, OH. pp. 129–138.

Mayer, R., Letey, J., and Farmer, W.J. 1974. Models for predicting volatilization of soil-incorporated pesticides. *Proc. Soil Sci. Soc. Am.* 38:563.

McBean, E.A. and Anderson, W.A. 1996. A two-stage process for the remediation of semi-volatile organic compounds. In *Environmental Biotechnology: Principles and Applications.* Moo-Young, M., Anderson, W.A., and Chakrabarty, A.M., Eds. Kluwer Academic Publishers, Amsterdam. pp. 269–277.

McCabe, W.L. and Smith, J.C. 1979. *Unit Operations of Chemical Engineering.* 3rd ed. McGraw-Hill, New York.

McCarty, P.L. 1980. Organics in water...an engineering challenge. *J. Environ. Eng. Div. Am. Soc. Civil Eng.* 106:1.

McCarty, P.L. 1982. One hundred years of anaerobic treatment. In *Anaerobic Digestion 1981: Proc. 2nd Int. Symp. on Anaerobic Digestion,* Sept. 6–11, 1981, Travemunde, FRG., Hughes, D.E., Stafford, D.A., Wheatley, B.I., Baader, W., Lettinga, G., Nyns, E.J., Verstraete, W., and Wentworth, R.L., Eds. Elsevier Biomedical Press, Amsterdam. pp. 3–22.

McCarty, P.L., Reinhard, M., and Rittmann, B.E. 1981. Trace organics in groundwater. *Environ. Sci. Technol.* 15:40–51.

McCarty, P.L., Rittmann, B.E., and Bouwer, E.J. 1984. Microbiological processes affecting chemical transformations in groundwater. In *Groundwater Pollution Microbiology.* Bitton, G. and Gerba, C.P., Eds. John Wiley & Sons, New York. pp. 89–115.

McCarty, P.L., Sutherland, K.H., Graydon, J., and Reinhard, M. 1979. In *Proc. AWWA Seminar, Controlling Organics in Drinking Water. 99th Annu. Natl. AWWA Conf.,* June, 1979. San Francisco, CA.

McDermott, H.J. and Killiany, S.E., Jr. 1978. Quest for a gasoline TLV. *Am. Ind. Hyg. Assoc. J.* 39:110–117.

McDevitt, N.P., Marks, P.J., and Noland, J.W. 1986. Installation Restoration General Environmental Technology Development. Task 1. Pilot investigation of low temperature thermal stripping of volatile organic compounds from soil. Report No. AMXTH-TE-CR-86074. U.S. Army Toxic and Haz. Mater. Agency, Aberdeen, MD. ADA 169439.

McDevitt, N.P., Noland, J.W., and Marks, P.J. 1987. Installation Restoration General Environmental Technology Development. Task 4. Bench-scale investigation of air stripping of volatile organic compounds from soil. Report No. AMXTH-TE-CR-86092. ADA 178261. U.S. Army Toxic and Haz. Mater. Agency, Aberdeen, MD.

McDowell, T. 1992. Microencapsulation of hydrocarbons in soil using reactive silicate technology. In *Hydrocarbon Contaminated Soils and Groundwater. Proc. 1st Annu. West Coast Conf. Hydrocarbon Contam. Soils Groundwater,* Feb. 1990, Newport Beach, CA. Vol. 1. Calabrese, E.J. and Kostechi, P.T., Eds. Lewis Publishers, Chelsea, MI. pp. 327–341.

McFarland, M.J., Qiu, X.J., Sims, J.L., Randolph, M.E., and Sims, R.C. 1992. Remediation of petroleum impacted soils in fungal compost bioreactors. *Water Sci. Technol.* 25(3):197–206.

McGill, W.B. 1976. Alberta Inst. of Pedology, API Pub. No. G-76-1, July.

McGill, W.B. 1977. Soil restoration following oil spills — a review. *J. Can. Petrol. Technol.* 16:60–67.

McGill, W.B. 1980. In *Disposal of Industrial and Oily Sludges by Land Cultivation.* Shilesky, D.M., Ed. Resource Systems and Management Association, Northfield, NJ. pp. 103–122.

McGugan, B.R., Lees, Z.M., and Senior, E. 1995. Bioremediation of an oil-contaminated soil by fungal intervention. In *Bioaugmentation for Site Remediation. Pap. 3rd Int. In Situ On-Site Bioreclam. Symp.* Hinchee, R.E., Fredrickson, J., and Alleman, B.C., Eds. Battelle Press, Columbus, OH. pp. 149–156.

Mcguinness, T.G. 1996. Supercritical oxidation reactor for treating hazardous wastes. U.S. US 5,558,783 (Cl. 210-761; C02F1/72), 24 Sept. 1996, US Appl. 14,345, 5 Feb. 1993. Cont.-in-part of U.S. 5,384,051. 24 pp.

McInerney, M.J. and Bryant, M.P. 1981. Anaerobic degradation of lactate by syntrophic associations of *Methanosarcina barkeri* and *Desulfovibrio* species and effect of H_2 on acetate degradation. *Appl. Environ. Microbiol.* 41:346–354.

McInerney, M.J., Bryant, M.P., Hespell, R.B., and Costerton, J.W. 1981. *Syntrophomonas wolfei* gen. nov. sp. nov., an anaerobic, syntrophic, fatty acid-oxidizing bacterium. *Appl. Environ. Microbiol.* 41:1029–1039.

McInerney, M.J., Bryant, M.P., and Pfenning, N. 1979. Anaerobic bacterium that degrades fatty acids in syntrophic association with methanogens. *Arch. Microbiol.* 122:129–135.

McInerney, M.J., Javaheri, M., and Nagle, D.P., Jr. 1990. Properties of the biosurfactant produced by *Bacillus licheniformis* strain JF-2. *J. End. Microbiol.* 5(2–3):95–102.

McInerney, M.J., Weirick, E.W., Sharma, P.K., and Knapp, R.M. 1993. Noninvasive methodology to study the kinetics of microbial growth and metabolism in subsurface porous materials. *Dev. Petrol. Sci.* (Microbial Enhancement of Oil Recovery: Recent Advances) 39:151–157.

McKee, J.R., Laverty, F.B., and Hertel, R.M. 1972. Gasoline in groundwater. *J. Water Pollut. Control Fed.* 44:293–302.

McKenna, E.J. 1972. Microbial metabolism of *normal* and branched chain alkanes. In *Degradation of Synthetic Organic Molecules in the Biosphere. Proc. of Conf.,* San Francisco, June 12–13, 1971. National Academy of Sciences, Washington, D.C. pp. 73–97.

McKenna, E.J. 1977. In *Biodegradation of Polynuclear Aromatic Hydrocarbon Pollutants by Soil and Water Microorganisms.* 70th Annu. Mtg. Am. Inst. Chem. Eng., New York.

McKenna, E.J. and Heath, R.D. 1976. Biodegradation of Polynuclear Aromatic Hydrocarbon Pollutants by Soil and Water Microorganisms. UILU-WRC-76-0113, University of Illinois at Urbana Water Resources Center, Urbana, IL. 25 pp.

McKenna, E.J. and Kallio, R.E. 1971. Microbial metabolism of the isoprenoid alkane pristane. *Proc. Natl. Acad. Sci. U.S.A.* 68:1552–1554.

McLean, E.O. 1982. Soil pH and lime requirement. In *Methods of Soil Analysis.* Part 2 — *Chemical and Microbiological Properties.* 2nd ed. Page, A.L., Ed. American Society of Agronomy, Inc., Madison, WI.

McLean, T.C. 1971. Sump oil digested by bacteria at Santa Barbara. *Petrol. Eng.* (Los Angeles). 43:68.

McLoughlin, T.J., Hearn, S., and Alt, S.G. 1990. Competition for nodule occupancy of introduced *Bradyrhizobium japonicum* strains in a Wisconsin soil with low numbers of indigenous bradyrhizobia population. *Can. J. Microbiol.* 36:839–845.

McNabb, D.H., Johnson, R.L., and Guo, I. 1994. Aggregation of oil- and brine-contaminated soil to enhance bioremediation. In *Hydrocarbon Bioremediation.* Hinchee, R.E., Alleman, B.C., Hoeppel, R.E., and Miller, R.N., Eds. CRC Press, Boca Raton, FL. pp. 296–302.

REFERENCES

McNabb, J.F., Smith, B.H., and Wilson, J.T. 1981. Biodegradation of toluene and chlorobenzene in soil and ground water. *Abstr. Annu. Mtg. Am. Soc. Microbiol.* p. 213.

Meaders, R.H. 1994. Enzyme-enhanced bioremediation. In *Applied Biotechnology for Site Remediation.* Hinchee, R.E., Anderson, D.B., Metting, F.B., Jr., and Sayles, G.D., Eds. Lewis Publishers, Boca Raton, FL. pp. 410–416.

Means, J.L., Kucak, T., and Crerar, D.A. 1980. Relative degradation rates of NTA, EDTA, and DTPA and environmental implications. *Environ. Pollut.,* Ser. B. 1:45–60.

Medina, V.F., Devinny, J.S., and Ramaratnam, M. 1995. Biofiltration of toluene vapors in a carbon-medium biofilter. In *Biological Unit Processes for Hazardous Waste Treatment. Pap. 3rd Int. In Situ On-Site Bioreclam. Symp.* Hinchee, R.E., Skeen, R.S., and Sayles, G.D., Eds. Battelle Press, Columbus, OH. p. 257.

Meegoda, J., Ho, W., Bhattacharjee, M., Wei, C.F., Cohen, D.M., Magee, R.S., and Frederick, R.M. 1995. Ultrasound enhanced soil washing. In *27th Mid-Atlantic Industrial Waste Conf. Haz. Ind. Wastes.* pp. 733–742.

Mehrotra, A.K., Karan, K., and Chakma, A. 1996. Model for *in situ* air stripping of contaminated soils: effects of hydrocarbon adsorption. *Energ. Sourc.* 18(1):21–36.

Meunier, A.D. and Williamson, K.J. 1981. Packed bed biofilm reactors: simplified model. *Am. Soc. Civ. Eng. Environ. Eng. Div. J.* 107:307–317.

Meyer, O., Warrelmann, J., and von Reis, H. 1995. Pilot plant stage bioremediation of CKW- and BTEX-contaminated soil by *in situ* infiltration in combination with on-site water and air treatment at the model site Eppelheim. *Soil Environ.* 5 (Contaminated Soil 95, Vol. 2). pp. 843–852.

Meyer-Reil, L. 1978. Autoradiography and epifluorescence microscopy combined for the determination of number and spectrum of actively metabolizing bacteria in natural waters. *Appl. Environ. Microbiol.* 36:506–512.

Meyers, J.D. and Huddleston, R.L. 1979. Treatment of oily refinery wastes by landfarming. In *Proc. 34th Annu. Ind. Waste Conf.,* May 8–10, Purdue University, Lafayette, IN. Ann Arbor Science Publishers, Ann Arbor, MI. 1980. pp. 686–698.

Michelsen, T.C. and Boyce, C.P. 1993. Cleanup standards for petroleum hydrocarbons. Part 1. Review of methods and recent developments. *J. Soil Contam.* 2(2):109–124.

Miget, R.J. 1973. Bacterial seeding to enhance biodegradation of oil slicks. In *The Microbial Degradation of Oil Pollutants, Workshop,* Atlanta, GA, Dec. 4–6, 1972. Ahearn, D.G. and Meyers, S.P., Eds. Pub. No. LSU-SG-73-01. Center for Wetland Resources, Louisiana State University, Baton Rouge, LA. p. 291.

Mihelcic, J.R., Lueking, D.R., Mitzell, R.J., and Stapleton, J.M. 1993. Bioavailability of sorbed- and separate-phase chemicals. *Biodegradation.* 4:141–153.

Mihelcic, J.R. and Luthy, R.G. 1988a. Microbial degradation of acenaphthene and naphthalene under dennitrification conditions in soil-water systems. *Appl. Environ. Microbiol.* 54:1188–1198.

Mihelcic, J.R. and Luthy, R.G. 1988b. Degradation of polycyclic aromatic hydrocarbon compounds under various redox conditions in soil-water systems. *Appl. Environ. Microbiol.* 54(5):1182–1187.

Mihelcic, J.R. and Luthy, R.G. 1991. Sorption and microbial degradation of naphthalene in soil-water suspensions under denitrification conditions. *Environ. Sci. Technol.* 25:169–177.

Miller, D.E. and Hutchins, S.R. 1995. Petroleum hydrocarbon biodegradation under mixed denitrifying/microaerophilic conditions. In *Microbial Processes for Bioremediation. Pap. 3rd Int. In Situ On-Site Bioreclam. Symp.* Hinchee, R.E., Vogel, C.M., and Brockman, F.J., Eds. Battelle Press, Columbus, OH. pp. 129–136.

Miller, R.M. and Bartha, R. 1989. Evidence from liposome encapsulation for transport-limited microbial metabolism of solid alkanes. *Appl. Environ. Microbiol.* 55:269–274.

Miller, R.M., Singer, G.M., Rosen, J.D., and Bartha, R. 1988. Photolysis primes biodegradation of benzo(a)pyrene. *Appl. Environ. Microbiol.* 54:1724–1730.

Miller, R.N. 1990. Ph.D. Dissertation, Utah State University.

Miller, R.N. and Hinchee, R.E. 1990. Enhanced Biodegradation through Soil Venting. U.S. Air Force AFESC, Final Report Contract FO8635-85-L-0122, Tyndall AFB, FL.

Miller, R.N., Hinchee, R.E., Vogel, C.M., Dupont, R.R., and Downey, D.C. 1990. A field scale investigation of enhanced petrol hydrocarbon biodegradation in the vadose zone at Tyndall AFB, Florida. In *Proc. NATO/CCMS Mtg.,* Dec. 1990, France.

Miller, R.N., Vogel, C.C., and Hinchee, R.E. 1991. A field-scale investigation of petroleum hydrocarbon biodegradation in the vadose zone enhanced by soil venting at Tyndall AFB, Florida. In *In Situ Bioreclamation. Applications and Investigations for Hydrocarbon and Contaminated Site Remediation.* Hinchee, R.E. and Olfenbuttel, R.F., Eds. Butterworth-Heinemann, Stoneham, MA. pp. 283–302.

Miller, R.S., Saberiyan, A.G., DeSantis, P., Andrilenas, J.S., and Esler, C.T. 1995. Removal of gasoline volatile organic compounds via air biofiltration. In *Biological Unit Processes for Hazardous Waste Treatment. Pap. 3rd Int. In Situ On-Site Bioreclam. Symp.* Hinchee, R.E., Skeen, R.S., and Sayles, G.D., Eds. Battelle Press, Columbus, OH. p. 265.

Millington, R.J. and Quirk, J.P. 1961. Permeability of porous solids. *Trans. Faraday Soc.* 57:1200–1207.

Miltner, R.J. and Love, O.T., Jr. 1985. Comparison of Procedures to Determine Adsorption Capacity of Volatile Organic Carbons on Activated Carbon. EPA-600/D-85/030. PB 85161552.

Mimura, A., Watanabe, S., and Takeda, I. 1971. Biochemical engineering analysis of hydrocarbon fermentation III. Analysis of emulsification phenomena. *J. Fermet. Technol.* 49:255–262.

Minugh, E.M., Patry, J.J., Keech, D.A., and Leek, W.R. 1983. A case history: cleanup of a subsurface leak of refined product. In *Proc. 1983 Oil Spill Conf.: Prevention, Behavior, Control and Cleanup,* Feb. 28–Mar. 3, 1983, San Antonio, TX. pp. 397–403.

Mironov, O.G. 1970. Role of microorganisms growing on oil in self-purification and indication of oil pollution in sea. *Oceanol. Rev.* 10:650.

Mishra, S.P. and Chaudhury, G.R. 1996. Biosorption of heavy metal ions by *Penicillium* sp. — A case study. *Trans. Indian Inst. Met.* 49(1–2):85–87.

Misra, T.K., Brown, N.L., Haberstroh, L., Schmidt, A., Goddette, D., and Silver, S. 1985. Mercuric reductase structural genes from plasmid R100 and transposon Tn501: functional domains of the enzyme. *Gene.* 34:253–262.

Miura, Y., Okazaki, M., Hamada, S.-I., Murakawa, S.-I., and Yugen, R. 1977. Assimilation of liquid hydrocarbon by microorganisms. I. Mechanism of hydrocarbon uptake. *Biotechnol. Bioeng.* 19:701–714.

Moen, M.A. and Hammel, K.E. 1994. Lipid peroxidation by the manganese peroxidase of *Phanerochaete chrysosporium* is the basis for phenanthrene oxidation by the intact fungus. *Appl. Environ. Microbiol.* 60:1956–1961.

Moller, J., Gaarn, H., Steckel, T., Wedebye, E.G., and Westermann, P. 1995. Inhibitory effects on degradation of diesel oil in soil-microcosms by a commercial bioaugmentation product. *Bull. Environ. Contam. Toxicol.* 54:913–918.

Molleron, H.J. 1994. Thermal treatment of petroleum contaminated soils: waste tracking and performance analysis in an off-site facility. In *Therm. Treat. Radioact., Haz. Chem., Mixed, Munitions, Pharm. Wastes, Proc. 13th Int. Incineration Conf.,* University of California, Irvine. pp. 47–53.

Monroe, D. 1985. *Am. Biotechnol. Lab.* 3:10–19.

Monteith, H.D., Bridle, T.R., and Sutton, P.M. 1980. Industrial waste carbon sources for biological denitrification. *Prog. Water Technol.* 12(6):127–141.

Moore, J.W. 1986. Evaluation of Packed Towers for Removing Volatile Organic Compounds from Surface Waters. Arkansas Water Resources Research Center, Fayetteville. PUB-124, June 1986. U.S. Dept. of the Interior, Geological Survey, Reston, VA. PB 87-149332.

Morales, M., Perez, F., Auria, R., and Revah, S. 1994. Toluene removal from air stream by biofiltration. Presented at 1st Int. Symp. Bioprocess Eng. Cuernavaca, Mor. UNAM. Kluwer, Amsterdam.

Mori, P.L. and Mori, J.L. 1992. Development of post-treatment cleanup criteria for *in situ* soils remediation. In *Hydrocarbon Contaminated Soils and Groundwater. Proc. 1st Annu. West Coast Conf. Hydrocarbon Contam. Soils Groundwater,* Feb. 1990, Newport Beach, CA. Vol. 1. Calabrese, E.J. and Kostechi, P.T., Eds. Lewis Publishers, Chelsea, MI. pp. 311–324.

Morin, T.C. 1997. Enhanced intrinsic bioremediation speeds site cleanup. *Pollut. Eng.* 29(2):44–47.

Morozov, N.V. and Nikolayov, V.N. 1978. Environmental influences on the development of oil-decomposing microorganisms. *Hydrobiol. J.* 14:47–53.

Morrison, S.M. and Cummings, B.A. 1982. Microbiologically Mediated Mutagenic Activity of Crude Oil. EPA Report No. EPA-600/S3-81-053. Environmental Research Laboratory, Corvallis, OR.

Mueller, J.G., Chapman, P.J., Blattmann, B.O., and Pritchard, P.H. 1990. Isolation and characterization of a fluoranthene-utilizing strain of *Pseudomonas paucimobilis. Appl. Environ. Microbiol.* 56:1079–1086.

Mueller, J.G., Lantz, S.E., Blattmann, B.O., and Chapman, P.J. 1991. *Environ. Sci. Technol.* 25:1045–1055.

Mulder, D., Ed. 1979. *Soil Disinfection.* Elsevier Scientific, Amsterdam.

Mulkins-Phillips, G.J. and Stewart, J.E. 1974a. Effect of environmental parameters on bacterial degradation of bunker C oil, crude oils, and hydrocarbons. *Appl. Microbiol.* 28:915–922.

Mulkins-Phillips, G.J. and Stewart, J.E. 1974b. Distribution of hydrocarbon-utilizing bacteria in northwestern Atlantic waters and coastal sediments. *Can. J. Microbiol.* 20:955–962.

Mullen, M.D., Wolf, D.C., Ferris, F.G., Beveridge, T.J., Flemming, C.A., and Bailey, G.W. 1989. Bacterial sorption of heavy metals. *Appl. Environ. Microbiol.* 55(12):3143–3149.

Mullis, K.B. and Faloona, F.A. 1987. Specific synthesis of DNA in vitro via a polymerase-catalyzed chain reaction. *Methods Enzymol.* 155:335–351.

Munnecke, D.M. 1981. In *Microbial Degradation of Xenobiotics and Recalcitrant Compounds.* Leisinger, T., Cook, A.M., Huetter, R., and Nuesch, J., Eds. Academic Press, London. pp. 251–269.

Munnecke, D.M., Johnson, L.M., Talbot, H.W., and Barik, S. 1982. Microbial metabolism and enzymology of selected pesticides. In *Biodegradation and Detoxification of Environmental Pollutants.* Chakrabarty, A.M., Ed. CRC Press, Boca Raton, FL. pp. 1–32.

Murdoch, L. 1991. Feasibility of Hydraulic Fracturing of Soil to Improve Remedial Actions. EPA/600/S2-91/012. U.S. EPA.

Muszynski, A., Karwowska, E., and Kaliszewski, M. 1996. Bioremediation of soil with removal of petroleum products using microorganisms immobilized on solid supports. *Gaz. Woda Tech. Sanit.* 70(8):299–301.

N

Nagel, G. et al. 1982. Sanitation of groundwater by infiltration of ozone treated water. *GWF-Wasser/Abwasser.* 123:399–407.

Nakagawa, Y. and Matsuyama, T. 1993. Chromatographic determination of optical configuration of 3-hydroxy fatty acids composing microbial surfactants. *FEMS Microbiol. Lett.* 108:99–102.

REFERENCES

Nakajima, A. and Sakaguchi, T. 1986. Selective accumulation of metals by microorganisms. *Appl. Microbiol. Biotechnol.* 24:59–64.

Nakamura, Y., Origasa, H., and Sawada, T. 1994. Mathematical modeling for diauxic growth in immobilized cell culture. *J. Ferment. Bioeng.* 78(5):361–367.

Nakayma, S. et al. 1979. Ozone. *Sci. Eng.* 1:119–132.

Nannipieri, P. 1984. Microbial biomass and activity measurements in soil: ecological significance. In *Current Perspectives in Microbial Ecology*. Klug, M.J. and Reddy, C.A., Eds. American Society of Microbiology, Washington, D.C. pp. 515–521.

Naphtali, L.M. and Sandholm, D.P. 1971. Multicomponent separation calculations by linearization. *AIChE J.* 17:148–153.

Narro, M.L., Cerniglia, C.E., Van Baalen, C., and Gibson, D.T. 1992. Metabolism of phenanthrene by the marine cyanobacterium *Agmenellum quadruplicatum*, strain PR-6. *Appl. Environ. Microbiol.* 58:1351–1359.

Nash, J.H. 1987. Field Studies of *in Situ* Soil Washing. EPA-600/2-87/110. U.S. EPA. Cincinnati, OH. PB 88146808.

Nash, J., Traver, R.P., and Downey, D.C. 1987. Surfactant-Enhanced *in Situ* Soils Washing. U.S. EPA/HWERL Report No. AFE SL-ESL-TR-87-18. ADA 188066. Eng. and Services Lab., Air Force Eng. and Services Center, Tyndall Air Force Base, FL.

National Research Council. 1993. *In Situ Bioremediation. When Does It Work?* Water Science and Technology Board, Com. Eng. Tech. Systems. National Academy of Sciences. National Academy Press, Washington, D.C.

Naval Civil Engineering Laboratory. 1986. Plan of Action and Cost Estimate. Demonstration project for on-site remediation of JP-5 and other fuel oil pollution from NAS Patuxent River, Maryland. Report to Chesapeake Division, Naval Facilities Engineering Command.

Nazarian, D. 1996. Enhanced vapor extraction system and method of *in-situ* remediation of contaminated soil zones. U.S. Patent 5,553,974 (Cl. 405-128; B09B3/00), 10 Sept. 1996, Appl. 348,608, 2 Dec. 1994. 7 pp.

Neal, D.M., Glover, R.L., and Moe, P.G. 1977. In *Land as a Waste Management Alternative. Proc. 1976 Cornell Agricultural Waste Mgmt. Conf.* Loehr, R.C., Ed. Ann Arbor Science Publishers, Ann Arbor, MI. pp. 757–767.

Nederlof, M.M., Van Riemsdijk, W.H., and De Haan, F.A.M. 1993. Effect of pH on the bioavailability of metals in soils. *Soil Environ.* (Integrated Soil and Sediment Research: A Basis for Proper Protection). 1:215–219.

Neely, N.S., Walsh, J.S., Gillespie, D.P., and Schauf, F.J. 1981. Remedial actions at uncontrolled waste sites. In *Land Disposal: Hazardous Waste, Proc. 7th Annu. Res. Symp.*, Philadelphia. Sheltz, D.W., Ed. EPA Report No. EPA-600/9-81-002b. Environmental Protection Agency, Cincinnati, OH. pp. 312–319.

Nelson, C.H., Hicks, R.J., and Andrews, S.D. 1994. *In situ* bioremediation: an integrated system approach. In *Hydrocarbon Bioremediation*. Hinchee, R.E., Alleman, B.C., Hoeppel, R.E., and Miller, R.N., Eds. CRC Press, Boca Raton, FL. p. 125.

Nestler, F.H.M. 1974. The Characterization of Wood-Preserving Creosote by Physical and Chemical Methods of Analysis. Research Paper FPL 195, U.S. Government Printing Office 1974-754-556/82. U.S. Dept. of Agriculture Forest Service, Forest Prod. Lab., Madison, WI. 31 pp.

New York State Department of Environmental Conservation. 1983. *Technology for the Storage of Hazardous Liquids, a State of the Art Review*. Albany, NY.

Nicholas, R.B. 1987. Biotechnology in hazardous-waste disposal: an unfulfilled promise. *Am. Soc. Microbiol. News.* 53:138–142.

Nieboer, E. and Richardson, D.H.S. 1980. The replacement of the nondescript term "heavy metals" by a biologically and chemically significant classification of metal ions. *Environ. Pollut. Ser. B.* 1:3–26.

Nielsen, D.M. 1983. Remedial methods available in areas of ground water contamination. In *Ground Water Quality, Proc. 6th Natl. Symp.*, Atlanta, GA. Nielsen, D.M. and Aller, L., Eds. National Water Well Association, Worthington, OH. pp. 219–227.

Nies, A., Nies, D., and Silver, S. 1989. Cloning and expression of plasmid genes encoding resistance to chromate and cobalt in *Alcaligenes eutrophus*. *J. Bacteriol.* 171:5065–5070.

Nies, D., Mergeay, M., Friedrich, B., and Schlegel, H.G. 1987. Cloning of plasmid genes encoding resistance to cadmium, zinc and cobalt in *Alcaligenes eutrophus* CH34. *J. Bacteriol.* 169:4865–4868.

Nilles, G.P. and Zabik, M.J. 1975. Photochemistry of bioactive compounds. Multiphase photodegradation and mass spectral analysis of basagran. *J. Agric. Food Chem.* 23:410.

Nimah, M.H., Ryan, J., and Chaudhry, M.A. 1983. Effect of synthetic conditioners on soil water retention, hydraulic conductivity, porosity, and aggregation. *Soil Sci. Soc. Am. J.* 47:742–745.

Nirmalakhandan, N., Lee, Y.H., and Speece, R.E. 1987. Designing a cost-efficient air-stripping process. *J. Am. Water Works Assoc.* 79:56–63.

Nivas, B.T., Sabatini, D.A., Shiau, B.-J., and Harwell, J.H. 1996. Surfactant enhanced remediation of subsurface chromium contamination. *Water Res.* 30(3):511–520.

Nocentini, M., Tamburini, D., and Pasquali, G. 1995. Biodegradation of polycyclic aromatic hydrocarbons in semisold-phase reactors. *Ing. Ambientale.* 24(9):475–481.

Noel, M.R., Benson, R.C., and Beam, P.M. 1983. Advances in mapping organic contamination: alternative solutions to a complex problem. In *Proc. Natl. Conf. on Mgmt. of Uncontrolled Haz. Waste Sites*. Hazardous Material Control Research Institute, Silver Spring, MD. pp. 71–75.

Noordkamp, E.R., Grotenhuis, J.T.C., and Rulkens, W.H. 1995. Optimal process conditions for extraction of pyrene and benzo(a)pyrene from soil with organic solvents. *Soil Environ.* 5 (Contaminated Soil 95, Vol. 2). pp. 1317–1318.

Norouzian, M.Y. and Gonzalez-Martinez, S. 1984 (in 1985 vol.). A performance evaluation of a full scale rotating biological contactor (9RBC) system. *Environ. Technol. Lett.* 6:79–86.

Norouzian, M., Gonzalez-Martinez, S., Pedroza-de-Brenes, R., and Duran-de-Bazua, C. 1985. Treatment of wastewater from the alkaline cooking of maize in an RBC system. In *Environmental Engineering — Proc. of the 1985 Specialty Conf.* O'Shaughnessy, J.C. Ed. American Society of Civil Engineers Press, New York. pp. 606–613.

Norris, R.D. 1995. Selection of electron acceptors and strategies for *in situ* bioremediation. In *Applied Bioremediation of Petroleum Hydrocarbons.* Hinchee, R.E., Kittel, J.A., and Reisinger, H.J., Eds. Battelle Press, Columbus, OH. pp. 483–487.

Norris, R.D., Dowd, K., and Maudlin, C. 1994. The use of multiple oxygen sources and nutrient delivery systems to effect *in situ* bioremediation of saturated and unsaturated soils. In *Hydrocarbon Bioremediation.* Hinchee, R.E., Alleman, B.C., Hoeppel, R.E., and Miller, R.N., Eds. CRC Press, Boca Raton, FL. pp. 405–410.

Novak, J.T., Goldsmith, C.D., Benoit, R.E., and O'Brien, J.H. 1985. Biodegradation of methanol and tertiary butyl alcohol in subsurface systems. *Water Sci. Technol.* 17:71–85.

Novak, J.T., Schuman, D., and Burgos, W. 1995. Characterization of petroleum biodegradation patterns in weathered contaminated soils. In *Monit. Verif. Biorem. Pap. 3rd Int. In Situ On-Site Bioreclam. Symp.* Hinchee, R.E., Douglas, G.S., and Ong, S.K., Eds. Battelle Press, Columbus, OH. pp. 29–38.

Novick, R.P., Murphy, E., Gryczan, T.J., Baron, E., and Edellman, I. 1979. Penicillinase plasmids of *Stapahylococcus aureus*: restriction-deletion maps. *Plasmid.* 2:109–129.

Novotny, J.F., Jr. 1992. Characterization of thermophilic microorganisms that utilize recalcitrant substrates. Avail. University Microfilms Int., Order No. DA9214426. *Diss. Abstr. Int. B.* 53(2):685.

Nowatzki, E.A., Lang, R.J., Medellin, M.C., and Sellers, S.M. 1994. Electroosmotic bioremediation of hydrocarbon-contaminated soils *in situ*. In *Applied Biotechnology for Site Remediation.* Hinchee, R.E., Anderson, D.B., Metting, F.B., Jr., and Sayles, G.D., Eds. Lewis Publishers, Boca Raton, FL. pp. 295–299.

Nyns, E.J., Auquiere, I.P., and Wiaux, A.L. 1968. Taxonomic value of property of fungi to assimilate hydrocarbons. *Antonie van Leeuwenhoek J. Microbiol. Serol.* 34:441–457.

O

Oberbremer, A. and Mueller-Hurtig, R. 1989. Aerobic stepwise hydrocarbon degradation and formation of biosurfactants by an original soil population in a stirred reactor. *Appl. Microbiol. Biotechnol.* 31:582–586.

Oberbremer, A., Mueller-Hurtig, R., and Wagner, F. 1990. Effect of the addition of microbial surfactants on hydrocarbon degradation in a soil population in a stirred reactor. *Appl. Microbiol. Biotechnol.* 32:485–489.

O'Brian, R.P. and Bright, R.L. 1983. In *Proc. 1st Annu. Haz. Mater. Mgmt. Conf.* July 1983. Wheaton, IL.

Ockeloen, H., Overcamp, T.J., and Grady, C.P.L., Jr. 1992. A biological fixed-film simulation model for the removal of volatile organic air pollutants. In *Proc. 85th Annu. Mtg. Air Waste Mgmt. Assoc.* Vol. 4. Paper No. 92/116.05. 15 pp.

Odu, C.T.I. 1972. Microbiology of soils contaminated with petroleum hydrocarbons. I. Extent of contamination and some soil and microbial properties after contamination. *J. Inst. Petrol.* 58:201–208.

Oedegaard, H., Rusten, B., and Westrum, T. 1994. A new moving bed biofilm reactor — applications and results. *Water Sci. Technol.* 29(10–11, Biofilm Reactors). pp. 157–165.

Ogram, A., Sayler, G.S., and Barkay, T. 1987. The extraction and purification of microbial DNA from sediments. *J. Microbiol. Methods.* 7:57–66.

Ogram, A., Sun, W., Brockman, F.J., and Fredrickson, J.K. 1995. Isolation and characterization of RNA from low-biomass deep subsurface sediments. *Appl. Environ. Microbiol.* 61:763–768.

Ogram, A.V., Jessup, R.E., Ou, L.T., and Rao, P.S.C. 1985. Effects of sorption on biochemical degradation rates of (2,4-dichlorophenoxy)acetic acid in soils. *Appl. Environ. Microbiol.* 49:582–587.

Ogunseitan, O.A. 1996. Analytical prerequisites for environmental bioremediation. *Environ. Test. Anal.* 5(1):36–40.

Ohneck, R.J. and Gardner, G.L. 1982. Restoration of an aquifer contaminated by an accidental spill of organic chemicals. *Ground Water Monit. Rev.* 2(4):50–53.

O'Leary, W.M. 1990. *Practical Handbook of Microbiology.* CRC Press, Boca Raton, FL.

Ollikka, P., Alhonmaki, K., Leppanen, V.-M., Glumoff, T., Raijola, T., and Suominen, I. 1993. Decolorization of azo, triphenyl methane, heterocyclic, and polymeric dyes by lignin peroxidase isoenzymes from *Phanerochaete chrysosporium. Appl. Environ. Microbiol.* 59:4010–4016.

Olmsted, L.P. 1994. A bench-scale assessment of nutrient concentration required to optimize hydrocarbon degradation and prevent clogging in fully saturated aerobic sands. In *Applied Biotechnology for Site Remediation.* Hinchee, R.E., Anderson, D.B., Metting, F.B., Jr., and Sayles, G.D., Eds. Lewis Publishers, Boca Raton, FL.

Olsen, G.J., Lane, D.J., Giovannoni, S.J., Pace, N.R., and Stahl, D.A. 1986. Microbial ecology and evolution: a ribosomal RNA approach. *Annu. Rev. Microbiol.* 40:337–365.

Olson, B.H. 1991. *Environ. Sci. Technol.* 25:604–611.

Omar, S.H. 1993. Oxygen diffusion through gels employed for immobilization. 2. In the presence of microorganisms. *Appl. Microbiol. Biotechnol.* 40:173–181.

REFERENCES

Oostenbrink, I.M., Kleijntjens, R.H., Mijnbeek, G., Kerkhof, I., Vetter, P., and Luyben, K.Ch.A.M. 1995. Biotechnological decontamination of oil and PAH polluted soils and sediments using the 4 M^3 pilot plant of the "slurry decontamination process." *Soil Environ.* 5 (Contaminated Soil 95, Vol. 2). pp. 863–872.

Oosting, R., Urlings, L.G.C.M., van Riel, P.H., and van Driel, C. 1992. BIOPUR: alternative packaging for biological systems. In *Biotechniques for Air Pollution and Odour Control Policies*. Draft, A.J. and van Ham, J., Eds. Elsevier Science Publishers, New York. pp. 63–70.

Ooyama, S. and Foster, J.W. 1965. Bacterial oxydation of cycloparaffinic hydrocarbons. *Antonie van Leeuwenhoek J. Microbiol. Serol.* 31:45–65.

Ortega-Calvo, J.J. and Alexander, M. 1994. Roles of bacterial attachment and spontaneous partitioning in the biodegradation of napthalene initially present in nonaqueous-phase liquids. *Appl. Environ. Microbiol.* 60(7):2643–2646.

Ortega-Calvo, J.J., Birman, I., and Alexander, M. 1995. Effect of varying the rate of partitioning of phenanthrene in nonaqueous-phase liquids on biodegradation in soil slurries. *Environ. Sci. Technol.* 29:2222–2225.

Osman, A., Bull, A.T., and Slater, J.H. 1976. In *Abstr. 5th Int. Fermentation Symp.* Dellweg, H., Ed. Westkreuz, Berlin. p. 124.

Ostgaard, K., Knutsen, S.H., Dyrset, N., and Aasen, I.M. 1993. Production and characterization of guluronate lyase from *Klebsiella pneumoniae* for applications in seaweed biotechnology. *Enzyme Microb. Technol.* 15:756–763.

O'Sullivan D. 1988. Polymer-based adsorbent cleans plant emissions. *Chem. Eng. News.* July 25, p.24.

Otten, A.M., Spuij, F., Lubbers, R.G.M., Okx, J.P., Schoen, A., and de Wit, J.C.M. 1995. Soil vapor extraction as a containment technique. *Soil Environ.* 5 (Contaminated Soil 95, Vol. 2). pp. 789–798.

Ottengraf, S.P.P. 1986. Exhaust gas purification. In *Biotechnology.* Rehm, H.J. and Reed, G., Eds. VCH, Weinheim, Germany. 8:301–332.

Ottengraf, S.P.P. 1987. Biological systems for waste gas elimination. *Trends in Biotech.* 5:132–136.

Ottengraf, S.P.P. and van den Oever, A.H.C. 1983. *Biotechnology and Bioengineering.* Wiley, New York. pp. 3089–3102.

Ouyang, Y., Mansell, R.S., and Rhue, R.D. 1995. Flow of gasoline-in-water microemulsion through water-saturated soil columns. *Ground Water.* 33(3):399–406.

Ouyang, Y., Mansell, R.S., and Rhue, R.D. 1996. A microemulsification approach for removing organolead and gasoline from contaminated soil. *J. Haz. Mater.* 46(1):23–35.

Overcash, M., Brown, K.W., and Evans, G.B., Jr. 1987. Hazardoous Waste Land Treatment: A Technology and Regulatory Assessment. ANL/EES-TM-340. DE88005571. Argonne National Laboratory, Argonne, IL.

Overcash, M.R., Nutter, W.L., Kendall, R.L., and Wallace, J.R. 1985. Field and laboratory evaluation of petroleum landtreatment system closure. EPA-600/2-85/134. PB 86130564.

Overcash, M.R. and Pal, D. 1979a. *Design of Land Treatment Systems for Industrial Wastes — Theory and Practices.* Ann Arbor Science Publishers, Ann Arbor, MI. pp. 246–255.

Overcash, M.R. and Pal, D. 1979b. *J. Am. Inst. Chem. Eng.* 75:357–361.

OxyCatalyst, Inc. 1980. Oxycat Catalytic Abatement Systems; Gas Fired. Sales Information.

P

Paca, J. 1994. Performance characteristics of biofilter during xylene and toluene degradation. In *Proc. 8th Forum Appl. Biotechnol. Bruges.* 2:2175–2184.

Palumbo, A.V., Pfaender, F.K., Paerl, H.W., Bland, P.T., Boyd, P.E., and Cooper, D. 1983. Biodegradation rates in coastal exosystem: influence of environmental factors. *Abstr. Annu. Mtg. Am. Soc. Microbiol.* p. 265.

Panchal, C.J. and Zajic, J.E. 1978. Isolation of emulsifying agents from a species of *Corynebacterium. Dev. Ind. Microbiol.* 19:569–576.

Panchal, C.J., Zajic, J.E., and Gerson, D.F. 1979. Multiple-phase emulsions using microbial emulsifyers. *J. Colloid Interface Sci.* 68:295–307.

Panchanadikar, V.V. and Das, R.P. 1994. Biosorption process for removing lead(II) ions from aqueous effluents using *Pseudomonas* sp. *Int. J. Environ. Stud.* 46(4):243–250.

Pan-Hou, H.S.K., Hosono, M., and Imura, N. 1980. Plasmid-controlled mercury biotransformation by *Clostridium cochlearium* T-2. *Appl. Environ. Microbiol.* 40:1007–1011.

Papendick, R.I. and Campbell, G.S. 1981. Theory and measurement of water potential. In *Water Potential Relations in Soil Microbiology:* Proc. of a Symposium sponsored by Divisions S-1 and S-3 of the Soil Science Society of America. Elliott, L.F., Papendick, R.I., and Wildung, R.E. (Organizing Committee) Parr, J.F., Gardner, W.R., and Elliott, L.F., Eds. Soil Science Society of America, Madison, WI. pp. 1–22.

Paris, D.F., Wolfe, N.L., Steen, W.C., and Baughman, G.L. 1983. Effect of phenol molecular structure on bacterial transformation rate constants in pond and river samples. *Appl. Environ. Microbiol.* 45:1153–1155.

Park, K.S., Sims, R.C., Doucette, W.J., and Matthews, J.E. 1988. Biological transformation and detoxification of 7,12-dimethylbenz(a)anthracene in soil systems. *J. Water Pollut. Control Fed.* 60:1822–1825.

Parker, N.H. 1993. Survey of internal combustion engines used in soil vapor extraction systems. *Hydrocarbon Contam. Soils Groundwater.* 3:309–340.

Parmele, C.S. and Allan, R.D. 1983. Treatment technique for removal of dissolved gasoline components from ground water. In *Proc. 3rd Natl. Symp. Aquifer Restoration and Ground-water Monitoring,* July 12–14, Wheaton, IL. pp. 51–59.

Parr, J.F., Sikora, L.J., and Burge, W.D. 1983. Factors affecting the degradation and inactivation of waste constituents in soils. In *Land Treatment of Hazardous Wastes.* Parr, J.F., Marsh, P.B., and Kla, J.M., Eds. Noyes Data Corp., Park Ridge, NJ. pp. 20–49, 321–337.

Parsa, J., Munson-McGee, S.H., and Steiner, R. 1996. Stabilization/solidification of hazardous wastes. *J. Environ. Eng.* 122(10):935–940.

Parthen, J. 1992. Untersuchungen zur mikrobiologischen Sanierung von mit PAK verunreinigten feinkoernigen Boeden im Drehtrommelreaktor. Ph.D. Thesis, University Hannover, Germany.

Pasti, M.B., Hagen, S.R., Goszczynski, S., Paszczynski, A., Crawford, R.L., and Crawford, D.L. 1991. The influence of guaiacol and syringyl groups in azo dyes on their degradation by lignocellulolytic *Streptomyces* spp. In *Abstr. Int. Symp. Appl. Biotechnol. Tree Culture, Protection, Utilization.* Columbus, OH. pp. 119–120.

Pasti-Grigsby, M.B., Paszczynski, A., Goszczynski, S., Crawford, D.L., and Crawford, R.L. 1994. Use of dyes in assaying *Phanerochaete chrysosporium* Mn(II)-peroxidase and ligninase. *Proc. Inst. Mol. Agric. Gen. Eng.* (IMAGE). 1:1–12.

Paszczynski, A. and Crawford, R.L. 1991. Degradation of azo compounds by ligninase from *Phanerochaete chrysosporium*: involvement of veratryl alcohol. *Biochem. Biophys. Res. Commun.* 178:1056–1063.

Paszczynski, A. and Crawford, R.L. 1995. Potential for bioremediation of xenobiotic compounds by the white-rot fungus *Phanerachaete chrysosporium. Biotechnol. Prog.* 11:368–379.

Paszczynski, A., Pasti, M.B., Goszczynski, S., Crawford, D.L., and Crawford, R.L. 1991. Designing biodegradability: lessons from lignin. In *Abstr. Int. Symp. Appl. Biotechnol. Tree Culture, Protection, Utilization.* Columbus, OH. pp. 73–78.

Patel, R.N., Hou, C.T., Laskin, A.I., Derelanko, P., and Felix, A. 1979. *Appl. Environ. Microbiol.* 38:219–223.

Paterek, T.R. and Smith, P.H. 1985. Isolation and characterization of a halophilic methanogen from Great Salt Lake. *Appl. Environ. Microbiol.* 50:877–881.

Patrick, M.A. and Dugan, P.R. 1974. Influence of hydrocarbons and derivatives on the polar lipid fatty acids of an *Acinetobacter* isolate. *J. Bacteriol.* 119:76–81.

Patrick, W.H., Jr. and Mahapatra, I.C. 1968. Transformation and availability to rice of nitrogen and phosphorus in waterlogged soils. *Adv. Agron.* 20:323–359.

Patton, J.B. 1987. Abstr. Paper FMPC-2078. Chemical Processing Table Top Seminar, Kingsport, TN, Feb. 25.

Paul, E., Fages, J., Blane, P., Goma, G., and Pareilleux, A. 1993. Survival of alginate-entrapped cells of *Axospirillum lipoferum* during dehydration and storage in relation to water properties. *Appl. Microbiol. Biotechnol.* 40:34–39.

Paulsrud, B. et al. 1984. Commission European Communities, EUR 9129 (G.B.). p. 107.

Pearce, K., Snyman, H.G., Oellermann, R.A., and Gerber, A. 1995. Bioremediation of petroleum-contaminated soil. In *Bioaugmentation for Site Remediation. Pap. 3rd Int. In Situ On-Site Bioreclam. Symp.* Hinchee, R.E., Fredrickson, J., and Alleman, B.C., Eds. Battelle Press, Columbus, OH. pp. 71–76.

Pearson, R.G. 1973. *Hard and Soft Acids and Bases.* John Wiley & Sons, New York.

Pedersen, A.R. and Arvin, E. 1995. Removal of toluene in waste gases using a biological trickling filter. *Biodegradation.* 6:109–118.

Pekdeger, A. and Matthess, G. 1983. Factors of bacteria and virus transport in groundwater. *Environ. Geol.* 5:49–52.

Pemberton, J.M., Corney, B., and Don, R.H. 1979. Evaluation and spread of pesticide degrading ability among soil microorganisms. In *Plasmids of Medical, Environmental, and Commercial Importance.* Timmis, K.N. and Puhler, A., Eds. Elsevier/North Holland Biomedical Press, Amsterdam.

Pennell, K.D., Pope, G.A., and Abriola, L.M. 1996. Influence of viscous and buoyancy forces on the mobilization of residual tetrachloroethylene during surfactant flushing. *Environ. Sci. Technol.* 39(4):1328–1335.

Perry, J.J. 1968. Substrate specificity in hydrocarbon utilizing microorganisms. *Antonie van Leeuwenhoek.* 34:27–36.

Perry, J.J. 1979. Microbial cooxidations involving hydrocarbons. *Microbiol. Rev.* 43:59–72.

Perry, J.J. 1984. Microbial metabolism of cyclic alkanes. In *Petroleum Microbiology.* Macmillan, New York. pp. 61–97.

Perry, J.J. and Cerniglia, C.E. 1973. Studies on the degradation of petroleum by filamentous fungi. In *The Microbial Degradation of Oil Pollutants.* Workshop held December 1972, Georgia State University, Atlanta, GA. Ahearn, D.G. and Meyers, S.P., Eds. Baton Rouge. Center for Wetland Resources, Louisiana State University pp. 89–94.

Persson, A., Quednau, M., and Ahrne, S. 1995. Composting oily sludges: characterizing microflora using randomly amplified polymorphic DNA. In *Bioremediation 3: Monit. Verif. Biorem. Pap. 3rd Int. In Situ On-Site Bioreclam. Symp.* Hinchee, R.E., Douglas, G.S., and Ong, S.K., Eds. Battelle Press, Columbus, OH. pp. 147–155.

Persson, N.A. and Welander, T.G. 1994. Biotreatment of petroleum hydrocarbon-containing sludges by land application: a case history and prospects for future treatment. In *Hydrocarbon Bioremediation.* Hinchee, R.E., Alleman, B.C., Hoeppel, R.E., and Miller, R.N., Eds. CRC Press, Boca Raton, FL. pp. 334–342.

Peters, C.A. and Luthy, R.G. 1993. Coal tar dissolution in water-miscible solvents: experimental evaluation. *Environ. Sci. Technol.* 27(13):2831–2843.

Petrich, C.R., Stormo, K.E., Knaebel, D.B., Ralston, D.R., and Crawford, R.L. 1995. A preliminary assessment of field transport experiments using encapsulated cells. In *Bioaugmentation for Site Remediation. Pap. 3rd Int. In Situ On-Site Bioreclam. Symp.* Hinchee, R.E., Fredrickson, J., and Alleman, B.C., Eds. Battelle Press, Columbus, OH. pp. 237–244.

Pettigrew, C. and Sayler, G.S. 1986. The use of DNA-DNA colony hybridization in the rapid isolation of 4-chlorobiphenyl degradative bacteria phenotypes. *J. Microbiol. Methods.* 5:205–213.

REFERENCES

Pettyjohn, W.A. and Hounslow, A.W. 1983. Organic compounds and ground-water pollution. *Ground Water Monit. Rev.* 3:41–47.

Pfaender, F.K. 1992. Biodegradation of hydrocarbon contaminants by Pataxent River soil microbial communities. Order No. AD-A253944, 26 pp. Avail NTIS from Govt. Rep. Annound Indes 1992, 92(23) Abstr. No. 266,812.

Pfaender, F.K. and Klump, J.V. 1981. Assessment of environmental biodegradation of organic pollutants. *Abstr. Annu. Mtg. Am. Soc. Microbiol.* p. 214.

Pfaender, F.K., Shimp, R.J., Palumbo, A.V., and Bartholomew, G.W. 1985. Comparison of environmental influences on pollutant biodegradation: estuaries, rivers, lakes. *Abstr. Annu. Mtg. Am. Soc. Microbiol.* p. 266.

Pfannkuch, H. 1985. In *Proc. NWWA/API Conf. on Petroleum Hydrocarbons and Organic Chemicals in Ground Water — Prevention, Detection, and Restoration,* Nov. 5–7, 1984, Houston, TX. NWWA, Worthington, OH.

Pfennig, N. 1978a. General physiology and ecology of photosynthetic bacteria. In *The Photosynthetic Bacteria.* Clayton, P.K. and Sistrom, W.R., Eds. Plenum Press, New York. pp. 3–17.

Pfennig, N. 1978b. *Rhodocyclus purpureus* gen. nov. and sp. nov., a ring-shaped vitamin B_{12}-requiring member of the family Rhodospirillaceae. *Int. J. Syst. Bacteriol.* 28:283–288.

Pfennig, N. and Biebl, H. 1976. *Desulfuromonas acetoxidans* gen. nov. and sp. nov., a new anaerobic, sulfur-reducing, acetate-oxidizing bacterium. *Arch. Microbiol.* 110:3–12.

Pflug, Z. 1980. Effect of humic acids on the activity of two peroxidases. *Z. Pflannernechr. Bodenkd.* pp. 430–440.

Phelan, J.M. and Webb, S.W. 1994. Thermal enhanced vapor extraction systems — design, application, performance prediction, including contaminant behavior. *In-Situ Rem; Sci. Basis Curr. Future Technol., Hanford Symp. Health Environ.* 33rd. 2:737–762.

Phelps, T.J., Conrad, R., and Zeikus, J.G. 1985. Sulfate-dependent interspecies H_2 transfer between *Methanosarcina barkeri* and *Desulfovibrio vulgaris* during coculture metabolism of acetate or methanol. *Appl. Environ. Microbiol.* 50:589–594.

Phillips, C.F. and Jones, R.K. 1978. Gasolene vapor exposure during bulk handling operations. *Am. Ind. Hyg. Assoc. J.* 39:118–128.

Phillips, P., Bender, J., Word, J., Niyogi, D., and Denovan, B. 1994. Mineralization of naphthalene, phenanthrene, chrysene, and hexadecane with a constructed silage microbial mat. In *Applied Biotechnology for Site Remediation.* Hinchee, R.E., Anderson, D.B., Metting, F.B., Jr., and Sayles, G.D., Eds. Lewis Publishers, Boca Raton, FL. pp. 305–309.

Phillips, W.E., Jr. and Brown, L.R. 1975. The effect of elevated temperature on the aerobic microbiological treatment of a petroleum refinery wastewater. In *Developments in Industrial Microbiology.* Underkofler, L.A., Cooney, J.J., and Walker, J.D., Eds. Amer. Inst. Biol. Sci., Washington, D.C. 16:296–304.

Pichard, S.L. and Paul, J.H. 1991. Detection of gene expression in genetically engineered microorganisms and natural phytoplankton populations in the marine environment by mRNA analysis. *Appl. Environ. Microbiol.* 57:1721–11727.

Pierce, G.E. 1982a. Paper presented at Summer Natl. Mtg. of the AIChE, Aug. 29–Sept. 1.

Pierce, G.E. 1982b. Diversity of microbial degradation and its implications in genetic engineering. In *Symp. on Impact of Appl. Genetics in Pollut. Cont.,* May 24–26, University Notre Dame. Kulpa, C.F., Irvine, R.L., and Sojka, S.A., Eds.

Pierce, G.E. 1982c. Development of genetically engineered microorganisms to degrade hazardous organic compounds. In *Hazardous Waste Management for the 1980s.* Ann Sweeney, T.L., Bhat, H.G., Sykes, R.N., and Sproul, O.J., Eds. Arbor Science Publishers, Ann Arbor, MI. pp. 431–439.

Pierce, G.E. 1982d. Potential role of genetically engineered microorganisms to degrade toxic chlorinated hydrocarbons. In *Detoxication of Hazardous Waste.* Exner, J.H., Ed. Ann Arbor Science Publishers, Ann Arbor, MI. pp. 315–322.

Pinholt, Y., Struwe, S., and Kjoller, A. 1979. *Holarctic Ecol.* 2:195–200.

Pinkart, H.C., Ringelberg, D.B., Stair, J.O., Sutton, S.D., Pfiffner, S.M., and White, D.C. 1995. Phospholipid analysis of extant microbiota for monitoring *in situ* bioremediation effectiveness. In *Monit. Verif. Biorem. Pap. 3rd Int. In Situ On-Site Bioreclam. Symp.* Hinchee, R.E., Douglas, G.S., and Ong, S.K., Eds. Battelle Press, Columbus, OH. pp. 49–57.

Piotrowski, M.R. 1991. Bioremediation of hydrocarbon contaminated surface water, groundwater, and soils: the microbial ecology approach. In *Hydrocarbon Contaminated Soils and Groundwater: Analysis, Fate, Environmental and Public Health Effects, Remediation.* Vol. 1. Kostecki, P.T. and Calabrese, E.J., Eds. Lewis Publishers, Chelsea, MI. pp. 203–238.

Pirnik, M.P. 1977. Microbial oxidation of methyl branched alkanes. *Crit. Rev. Microbiol.* 5:413–422.

Pirnik, M.P., Atlas, R.M., and Bartha, R. 1974. Hydrocarbon metabolism by *Brevibacterium erythrogenes*: normal and branched alkenes. *J. Bacteriol.* 119:868–878.

Pizarro, D.R. 1985. Alternative sludge handling and disposal at Kent County, Delaware. *J. Water Pollut. Control Fed.* 57:278–284.

Plehn, S.W. 1979. Draft Economic Analysis (regulatory analysis supplement) for Subtitle C, Resource Conservation and Recovery Act of 1976. Office of Solid Waste U.S. EPA, Washington, D.C.

Plice, M.J. 1948. Some effects of petroleum on soil fertility. *Soil Sci. Am. Proc.* 13:413–416.

Plumb, R.H., Jr. 1985. Disposal site monitoring data: observations and strategy implications. In *Proc. 2nd Canadian/American Conf. Hydrogeology,* June 25–29, Banff, Alberta, Canada. Hitchon, B. and Trudell, M., Eds.

Plummer, E.J. and Macaskie, L.E. 1990. Actinide and lanthanum toxicity towards a *Citrobacter* sp.: uptake of lanthanum and a strategy for the biological treatment of liquid wastes containing plutonium. *Bull. Environ. Contam. Toxicol.* 44:173–180.

Poe, S.H., Valsaraj, K.T., Thibodeaux, L.J., and Springer, C. 1988. Equilibrium vapor phase adsorption of volatile organic chemicals on dry soils. *J. Haz. Mater.* 19:17–32.

Poels, J., Van Assche, P., and Verstraete, W. 1985. Influence of H_2 stripping o methane production in conventional digesters. *Biotechnol. Bioeng.* 27:1692–1698.

Poglazova, M.M., Fedoseeva, G.E., Khesina, A.J., Meissel, M.N., and Shabad, L.M. 1967. Destruction of benzo(a)pyrene by soil bacteria. *Life Sci.* 6:1053–1062.

Poindexter, J. 1981. Workshop on Trace Organic Contaminant Removal. July 27–28, 1981. Advanced Environmental Control Technology Res. Center, Univ. of Illinois, Urbana, IL.

Polybac Corporation. 1983. Product Information Packet. Allentown, PA.

Polz, M.F. and Cavanaugh, C.M. 1997. A simple method for quantification of uncultured microorganisms in the environment based on *in vitro* transcription of 16S rRNA. *Appl. Environ. Microbiol.* 63(3):1028–1033.

Poncelet, D., Lencki, R., Beaulieu, C., Halle, J.P., Neufeld, R.J., and Fournier, A. 1992. Production of alginate beads by emulsification/internal gelation. I. Methodology. *Appl. Microbiol. Biotechnol.* 38:39–45.

Porta, A. 1991. A review of European bioreclamation practice. In *In Situ Bioreclamation: Applications and Investigations for Hydrocarbon and Contaminated Site Remediation.* Hinchee, R.E. and Olfenbuttel, Eds. Battelle Memorial Inst., Columbus, OH. pp. 1–13.

Portier, R.J. 1994. Remediation of mixed wastes in soils by combined biological and chelation technologies. In *Proc. 87th Annu. Mgt. Air Waste Mgmt. Assoc.,* Vol. 13. 16 pp. 94-WP84B.03.

Postgate, J.R., Crumpton, J.E., and Hunter, J.R. 1961. The measurement of bacterial viabilities by slide culture. *J. Gen. Microbiol.* 24:15–24.

Powers, S.E., Anckner, W.H., and Seacord, T.F. 1996. Wettability of NAPL-contaminated sands. *J. Environ. Eng.* 122(10):889–896.

Pramer, D. and Bartha, R. 1972. Preparation and processing of soil samples for biodegradation studies. *Environ. Lett.* 2:217–224.

Pramer, D. and Schmidt, E.L. 1964. *Experimental Soil Microbiology.* Burgess, Minneapolis, MN.

Premuzic, E.T., Lin, M.S., Racaniello, L.K., and Manowitz, B. 1993. Chemical markers of induced microbial transformations in crude oils. *Dev. Petrol. Sci.* (Microbial Enhancement of Oil Recovery: Recent Advances). 39:37–54.

Price, S.L., Kasevich, R.S., Marley, M.C., 1994. Enhancing site remediation through radio frequency heating. *Hydrocarbon Contam. Soils.* 4:399–411.

Prill, R.C., Oaksford, E.T., and Potorti, J.E. 1979. A Facility Designed to Monitor the Unsaturated Zone during Infiltration of Tertiary-Treated Sewage. U.S. Geological Survey, Water — Resources Investigation 79-48. Washington, D.C. PB 80-102700.

Prince, R.C., Elmendorf, D.L., Lute, J.R., Hsu, C.S., Haith, C.E., Senius, J.D., Dechert, G.J., Douglas, G.S., and Butler, E.L. 1994. 17 alpha(H)-21 beta(H)-hopane as a conserved internal marker for estimating the biodegradation of crude oil. *Environ. Sci. Technol.* 28(1):142–145.

Prince, R.C., Hinton, S.M., Bragg, J.R., Elmendorf, D.L., Lute, J.R., Grossman, M.J., Robbins, W.K., Hsu, C.S., Douglas, G.S., Bare, R.E., Haith, C.E., Senius, J.D., Minak-Bernero, V., McMillen, S.J., Roffall, J.C., and Chianelli, R.R. 1993. Laboratory studies of oil spill bioremediation; toward understanding field behavior. In *Proc. ACS Mtg.,* April, Denver, CO.

Prins, R.A. 1978. Chapter 7. Nutritional impact of intestinal drug-microbe interations. In *Int. Symp. on Nutrition and Drug Interrelations.* Hathcock, J. and Coon, J., Eds. Academic Press, New York. pp. 189–251.

Pritchard, P.H. 1992. Use of inoculation in bioremediation. *Curr. Opin. Biotechnol.* 3:232–243.

Pritchard, P.H., Mueller, J.G., Rogers, J.C., and Kremer, F.V. 1992. Oil spill bioremediation: experiences, lessons and results from the Exxon Valdez oil spill in Alaska. *Biodegradation.* 3(2–3):315–335.

Pritchard, P.H., Van Veld, P.A., and Cooper, W.P. 1981. Biodegradation of *p*-cresol in artificial stream channels. *Abstr. Annu. Mtg. Am. Soc. Microbiol.* p. 210.

Proctor, M.H. and Scher, S. 1960. Decomposition of benzoate by a photosynthetic bacterium. *Biochem. J.* 76:33.

Prokop, W.H. and Bohn, H.L. 1985. *J. Air Pollut. Control Assoc.* 35:1332–1338.

Prunty, L. 1992. Thermally driven water and octane redistribution in unsaturated, closed soil cells. *Soil Sci. Soc. Am. J.* 56(3):707–714.

Punch, J.D. 1966. The Production and Composition of the Slime of *Alcaligenes viscolactis*. Ph.D. Dissertation, University of Minneapolis, MN.

Puranik, P.R., Chabukswar, N.S., and Paknikar, K.M. 1995. Cadmium biosorption by *Streptomyces pimprina* waste biomass. *Appl. Microbiol. Biotechnol.* 43:1118–1121.

Putcha, R.V. and Domach, M.M. 1993. Fluorescence monitoring of polycyclic aromatic hydrocarbon biodegradation and effect of surfactants. *Environ. Prog.* 12(2):81.

Puustinen, J., Joergensen, K.S., Strandberg, T., and Suortti, A.-M. 1995. Bioremediation of oil-contaminated soil from service stations: evaluation of biological treatment. 5 (Contaminated Soil 95, Vol. 2). pp. 1325–1326.

Pye, V.I. and Patrick, R. 1983. Ground water contamination in the United States. *Science.* 221:713–718.

Pye, V.I., Patrick, R., and Quarles, J. 1983. *Groundwater Contamination in the United States.* University of Pennsylvania Press, Philadelphia.

REFERENCES

Q

Quince, J.R. and Gardner, G.L. 1982a. Recovery and treatment of contaminated ground water: Part 1. *Ground Water Monit. Rev.* 2:(Summer):18–22.

Quince, J.R. and Gardner, G.L. 1982b. Recovery and treatment of contaminated ground water: Part 2. *Ground Water Monit. Rev.* 2:(Fall):18–25.

R

Rabus, R. and Widdel, F. 1995. Anaerobic degradation of ethylbenzene and other aromatic hydrocarbons by new denitrifying bacteria. *Arch. Microbiol.* 163:96–103.

Racke, K.D. and Coats, J.R., Eds. 1990. *Enhanced Biodegradation of Pesticides in the Environment.* American Chemical Society.

Radwan, S.S., Sorkhoh, N.A., Fardoun, F., and Al-Hasan. 1995. Soil management enhancing hydrocarbon biodegradation in the polluted Kuwaiti desert. *Appl. Microbiol. Biotechnol.* 44(1–2):265–270.

Rahe, T.M., Hagedorn, C., McCoy, E.L., and Kling, G.F. 1978. Transport of antibiotic-resistant *Escherichia coli* through western Oregon hillslope soils under conditions of saturated flow. *J. Environ. Qual.* 7:487–494.

Rahman, S.A.S., Barooah, M., and Barthakur, H.P. 1995. Microbial degradation of n-hexane extractable crude fraction under different conditions. *J. Assam Sci. Soc.* 37(3):129–134.

Rai, D., Serne, R.J., and Swanson, J.L. 1980. Solution species of plutonium in the environment. *J. Environ. Qual.* 9:417–420.

Ram, N.M., Bass, D.H., Falotico, R., and Leahy, M. 1993. A decision framework for selecting remediation technologies at hydrocarbon-contaminated sites. *J. Soil Contam.* 2(2):167–189.

Ramanand, K., Balba, M.T., and Duffy, J. 1995. Biodegradation of select organic pollutants in soil columns under denitrifying conditions. *Haz. Waste Haz. Mater.* 12(1):27–36.

Rambeloarisoa, E., Rontani, J.F., Giusti, G., Duvnjak, Z., and Bertrand, J.C. 1984. Degradation of crude oil by a mixed population of bacteria isolated from sea-surface foams. *Mar. Biol.* 83:69–81.

Ramos, J.L., Wasserfallen, A., Rose, K., and Timmis, K.N. 1987. *Science.* 235:593–596.

Rand, M.C. et al. 1975. Standard methods for the examination of water and wastewater. APHA–AWWA–WPCF. Washington, D.C.

Randall, J.D. and Hemmingsen, B.B. 1994a. A critical evaluation of the fume plate method for the enumeration of bacteria capable of growth on volatile hydrocarbons. In *Applied Biotechnology for Site Remediation.* Hinchee, R.E., Anderson, D.B., Metting, F.B., Jr., and Sayles, G.D., Eds. Lewis Publishers, Boca Raton, FL. pp. 400–404.

Randall, J.D. and Hemmingsen, B.B. 1994b. Evaluation of mineral agar plates for the enumeration of hydrocarbon-degrading bacteria. *J. Microbiol. Methods.* 20:103–113.

Rao, C.R.N., Iyengar, L., and Venkobachar, C. 1993. Sorption of copper(II) from aqueous phase by waste biomass. *J. Environ. Eng.* 2:369–377.

Rapp, P., Bock, H., Wray, V., and Wagner, F. 1979. Formation, isolation and characterization of trehalose dimycolates from *Rhodococcus erythropolis* grown on n-alkanes. *J. Gen. Microbiol.* 115:491–503.

Rasiah, V. and Voroney, R.P. 1993. Assessment of selected surfactants for enhancing C mineralization of an oily waste. *Water Air Soil Pollut.* 71(3–4):347–355.

Ratledge, C. 1978. Degradation of aliphatic hydrocarbons. In *Developments in Biodegradation of Hydrocarbons — 1.* Watkinson, R.J., Ed. Applied Science Publishers, London. pp. 1–46.

Ratz, J.W., Pierson, G.D., Caskey, K.K., and Barry, W.L. 1995. *In situ* bioventing: results of three pilot tests performed in Hawaii. In In Situ *Aeration: Air Sparging, Bioventing, Relat. Rem. Processes. Pap. 3rd Int. In Situ On-Site Bioreclam. Symp.* Hinchee, R.E., Miller, R.N., and Johnson, P.C., Eds. Battelle Press, Columbus, OH. pp. 383–389.

Raymond, R.L. 1974. U.S. Patent 3,846,290. Patented 5 Nov. 1974.

Raymond, R.L., Hudson, J.O., and Jamison, V.M. 1976. Oil degradation in soil. *Appl. Environ. Microbiol.* 31:522–535.

Raymond, R.L., Hudson, J.O., and Jamison, V.M. 1980. *AIChE Symp.* 75:340–356.

Raymond, R.L., Jamison, V.W., and Hudson, J.O. 1967. *Appl. Microbiol.* 15:857–865.

Raymond, R.L., Jamison, V.W., and Hudson, J.O. 1976. Beneficial stimulation of bacterial activity in groundwaters containing petroleum products. *AIChE Symp. Series. Water — 1976.* 73:390–404.

Reasoner, D.J. and Geldreich, E.E. 1985. A new medium for the enumeration and subculture of bacteria from potable water. *Appl. Environ. Microbiol.* 49:1–7.

Reddy, P.G., Singh, H.D., Roy, P.K., and Baruah, J.N. 1982. Predominant role of hydrocarbon solubilization in the microbial uptake of hydrocarbons. *Biotechnol. Bioeng.* 224:1241–1269.

Reed, B.E. and Thomas, B. 1995. Treatment of heavy metal-bearing wastewaters using activated carbon. In *Proc. 12th Annu. Int. Pittsburgh Coal Conf.* pp. 358–363.

Rees, J.F., Wilson, B.H., and Wilson, J.T. 1985. Biotransformation of toluene in methanogenic subsurface material. *Abstr. Annu. Mtg. Am. Soc. Microbiol.* p. 258.

Reible, D.D., Malhiet, M.E., and Illangasekare, T.H. 1989. Modeling gasoline fate and transport in the unsaturated zone. *J. Haz. Mater.* 22(3):359–376.

Reichardt-Vorlaender, C. 1995. Biological reclamation of petroleum-contaminated soil. *Energie.* 47(5):12–116.

Reichmuth, D.R. 1984. In *Proc. NWWA/API Conf. on Petroleum Hydrocarbons and Organic Chemicals in Ground Water — Prevention, Detection, and Restoration.* Worthington, OH. NWWA.

Reid, G.R., Thompson, G., and Oberholtzer, C. 1985. Soil vapor monitoring as a cost-effective method for assessing ground water degradation from volatile chlorinated hydrocarbons in an alluvial environment. In *2nd Annu. Can./Am. Conf. on Hydrogeology — Haz. Waste in Ground Water: A Soluble Dilemma.* National Water Works Association. pp. 133–140.

Reinhard, M. 1994. *In-situ* bioremediation technologies for petroleum-derived hydrocarbons based on alternate electron acceptors (other than molecular oxygen). In *Handbook of Bioremediation.* Norris, R.D. et al., Eds. Lewis Publishers, Boca Raton, FL. p. 134.

Reinhard, M., Goodman, N.L., and Barker, J.F. 1984. Occurrence and distribution of organic chemicals in two landfill leachate plumes. *Environ. Sci. Technol.* 18:953–961.

Reinhard, M., Hopkins, G.D., Orwin, E., Shang, S., and Lebron, C.A. 1995. *In situ* demonstration of anaerobic BTEX biodegradation through controlled-release experiments. In *Applied Bioremediation of Petroleum Hydrocarbons.* Hinchee, R.E., Kittel, J.A., and Reisinger, H.J., Eds. Battelle Press, Columbus, OH. pp. 263–270.

Reisfeld, A., Rosenberg, E., and Gutnick, D. 1972. Microbial degradation of crude oil: factors affecting oil dispersion in sea water by mixed and pure cultures. *Appl. Microbiol.* 24:363–368.

Ressler, B.P., Kaempf, C., and Winter, J. 1995. Effect of sorption and substrate pattern on PAH degradability. In *Biological Unit Processes for Hazardous Waste Treatment. Pap. 3rd. Int. In Situ On-Site Bioreclam. Symp.* Hinchee, R.E., Skeen, R.S., and Sayles, G.D., Eds. Battelle Press, Columbus, OH. pp. 333–338.

Rho, D., Mercier, P., Jette, J.-F., Samson, R., Lei, J., and Cyr, B. 1995. Respirometric oxygen demand detrminations of laboratory- and field-scale biofilters. In *Biological Unit Processes for Hazardous Waste Treatment. Pap. 3rd Int. In Situ On-Site Bioreclam. Symp.* Hinchee, R.E., Skeen, R.S., and Sayles, G.D., Eds. Battelle Press, Columbus, OH. pp. 211–218.

Ribbons, D.W. and Eaton, R.W. 1982. Chemical transformations of aromatic hydrocarbons that support the growth of microorganisms. *Biodegrad. Detoxif. Environ. Pollut.* Chakrabarty, A.M., Ed. CRC Press, Boca Raton, FL.

Rice, R.G. 1984. Paper presented at 5th Conf. on Mgmt. of Uncontrolled Haz. Waste Sites, Nov. 7–9. Washington, D.C.

Rich, L.A., Bluestone, M., and Cannon, D.R. 1986. *In situ* bioreclamation. *Chem. Week.* 139(8):62–64.

Richaume, A., Steinberg, C., and Jocteru-Monrozier, L. 1993. Differences between direct and indirect enumeration of soil bacteria: the influence of soil structure and cell location. *Soil Biol. Biochem.* 25:641–643.

Riha, V., Nymburska, K., Tichy, R., and Triska, J. 1993. Microbiological, chemical and toxicological characterization of contaminated sites in Czechoslovakia. *Sci. Total Environ.* (Suppl. Pt. 1):185–193.

Riis, V., Miethe, D., and Babel, W. 1995. Comparison of the microbial degradability of refinery products and oils from polluted sites. 5 (Contaminated Soil 95, Vol. 2). pp. 1331–1332.

Rijnaarts, H.H.M., Hesselink, P.G.M., and Doddema, H.J. 1995. Activated *in-situ* bioscreens. *Soil Environ.* 5 (Contaminated Soil 95, Vol. 2). pp. 929–937.

Ripper, J., Friedrich, L., and Ripper, P. 1992. Laboratory experiments for technological process optimization and planning of microbiological cleanup of Pintsch site, Hanau. *Mikrobiol. Reinig. Boeden Beitr. Dechema-Fachgespraechs Umweltschutz,* 9th, 1991. pp. 423–432.

Rippon, G.M. 1983. The bioenergy process — an overview. *Water Pollut. Control.* 82:29–36.

Riser-Roberts, E. 1992. *Bioremediation of Petroleum Contaminated Sites.* CRC Press, Boca Raton, FL. 461 pp.

Rittmann, B.E. et al. 1994. In Situ *Bioremediation.* 2nd ed. Noyes, Park Ridge, NJ.

Rittmann, B.E., Bouwer, E.J., Schreiner, J.E., and McCarty, P.L. 1980. Technical Report No. 255. Stanford University, Dept. of Civil Eng., Palo Alto, CA.

Roane, T.M. and Kellogg, S.T. 1996. Characterization of bacterial communities in heavy metal contaminated soils. *Can. J. Microbiol.* 42:593–603.

Roberston, L.A. and Kuenene, J.G. 1984. Aerobic denitrification: a controversy revived. *Arch. Microbiol.* 139:351–354.

Robertiello, A., Lucchese, G., Di Leo, C., Boni, R., and Carrera, P. 1994. *In situ* bioremediation of a gasoline and diesel fuel contaminated site with integrated laboratory simulation experiments. In *Hydrocarbon Bioremediation.* Hinchee, R.E., Alleman, B.C., Hoeppel, R.E., and Miller, R.N., Eds. CRC Press, Boca Raton, FL. pp. 133–140.

Roberts, P.V. and Levy, J.A. 1983. AWWA Seminar, Controlling Trihalomethanes. Annu. Natl. AWWA Conf., Las Vegas, NV.

Roberts, P.V., McCarty, P.L., Reinhard, M., and Schreiner, J. 1980. Organic contaminant behavior during groundwater recharge. *J. Water Pollut. Control. Fed.* 52:161–172.

Roberts, P.V., Reinhard, M., and Valocchi, A.J. 1982. Movement of organic contaminants in groundwater: implications for water supply. *J. Am. Water Works Assoc.* 74:408–413.

Roberts, R.M., Koff, J.L., and Karr, L.A. 1988. Enzyme and microbe immobilization, toxic and hazardous waste handling. Hazardous Waste Minimization Initiation Decision Report. TN-1787. pp. J-1–J-8.

Robertson, L.A. and Kuenen, J.G. 1984. Aerobic denitrification: a controversy revived. *Arch. Microbiol.* 139:351–354.

Robertson, B., Arhelger, S., Kinney, P.J., and Button, D.K. 1973. Hydrocarbon biodegradation in Alaskan waters. In *The Microbial Degradation of Oil Pollutants, Workshop,* Atlanta, Dec. 4–6, 1972. Ahearn, D.G. and Meyers, S.P., Eds. Pub. No. LSU-SG-73–01. Center for Wetland Resources, Louisiana State University, Baton Rouge. pp. 171–184.

REFERENCES

Robertson, B.K. and Alexander, M. 1992. Influence of calcium, iron, and pH on phosphate availability for microbial mineralization of organic chemicals. *Appl. Environ. Microbiol.* 58(1):38–41.

Robertson, B.K. and Alexander, M. 1996. Mitigating toxicity to permit bioremediation of constituents of nonaqueous-phase liquids. *Environ. Sci. Technol.* 30(6):2066–2070.

Robichaux, T.J. and Myrick, H.N. 1972. Chemical enhancement of the biodegradation of crude-oil pollutants. *J. Petrol. Technol.* 24:16.

Robinson, J.S., Ed. 1979. Hazardous chemical spill cleanup. *Pollut. Technol. Rev.* No. 59. Noyes Data Corp., Park Ridge, NJ.

Robinson, K.G., Farmer, W.S., and Novak, J.T. 1990. Availability of sorbed toluene in soils for biodegradation by acclimated bacteria. *Water Res.* 24:345–350.

Rodriguez, G.G., Phipps, D., Ishiguro, K., and Ridgway, H.F. 1992. Use of a fluorescent redox probe for direct visualization of actively respiring bacteria. *Appl. Environ. Microbiol.* 58:1801–1808.

Rodriguez, U.M. and Kroll, R.G. 1988. Rapid selective enumeration of bacteria in foods using a microcolony epifluorescence microscopy technique. *J. Appl. Bacteriol.* 64:65–78.

Roehricht, M., Weppen, P., and Deckwer, W.D. 1990. Abrennung von Schwermetallen aus Abwasserstroemen — Biosorption im Vergleich zu herkoemmlichen Verfahren. *Chem. Ing. Tech.* 62:582–583.

Rogers, J.E., Riley, R.G., Li, S.W., Mann, D.C., and Wildung, R.E. 1981. Microbiological degradation of organic components in oil shale retort water: organic acids. *Appl. Environ. Microbiol.* 42(5):830–837.

Rolke, R.W. et al. 1972. EPA-R272062. U.S. EPA, Research Triangle Park, NC.

Romero, J.C. 1970. The movement of bacteria and viruses through porous media. *Ground Water.* 8:37–48.

Romich, M.S., Cameron, D.C., and Etzel, M.R. 1995. Three methods for large-scale preservation of a microbial inoculum for bioremediation. In *Bioaugmentation for Site Remediation. Pap. 3rd Int. In Situ On-Site Bioreclam. Symp.* Hinchee, R.E., Fredrickson, J., and Alleman, B.C., Eds. Battelle Press, Columbus, OH. pp. 229–235.

Ron, E.Z., Minz, D., Finkelstein, N.P., and Rosenberg, E. 1992. Interactions of bacteria with cadmium. *Biodegradation.* 3:161–170.

Rose, W.W. and Mercer, W.A. 1968. Progress Report, Part I, National Canners Association, Berkeley, CA, July.

Rosenberg, D.G. et al. 1976. Assessment of hazardous waste practices in the petroleum refining industry. EPA Report FLD/GP 7A 13B. PB-259-097.

Rosenberg, E. 1986. Microbial surfactants. *CRC Crit. Rev. Biotechnol.* 3:109–132.

Rosenberg, E., Kaplan, N., Pines, O., Rosenberg, M., and Gutnick, D. 1983. Capsular polysaccharides interfere with adherence of *Acinetobacter. FEMS Microbiol. Lett.* 17:157–161.

Rosenberg, E., Legmann, R., Kushmaro, A., Taube, R., Adler, E., and Ron, E.Z. 1992. Petroleum bioremediation — a multiphase problem. *Biodegradation.* 3:337–350.

Rosenberg, E., Perry, A., Gibson, D.F., and Gutnick, D.L. 1979. Emulsifier of *Arthrobacter* RAG-1: specificity of hydrocarbon substrate. *Appl. Environ. Microbiol.* 37:409–413.

Rosenberg, E., Rosenberg, M., Shoham, Y., Kaplan, N., and Sar, N. 1989. Adhesion and desorption during growth of *Acinetobacter calcoaceticus* on hydrocarbons. In *Microbial Mats.* Cohen, Y. and Rosenberg, E., Eds. ASM Publ., Washington, D.C.

Rosenberg, E., Zuckerberg, A., Rubinovitz, C., and Gutnick, D.L. 1979. Emulsifier of *Arthrobacter* RAG-1: isolation and emulsifying properties. *Appl. Environ. Microbiol.* 37:402–408.

Rosenberg, M., Hayer, E.A., Delaria, J., and Rosenberg, E. 1982. Role of thin fimbriae in adherence and growth of *Acinetobacter calcoaceticus* RAG-1 on hexadecane. *Appl. Environ. Microbiol.* 44:929–937.

Rosenberg, M. and Rosenberg, E. 1985. *Oil Petrochem. Pollut.* 2:155–162.

Rosenfeld, H. and Feigelson, P. 1969. Synergistic and product induction of the enzymes of tryptophan metabolism in *Pseudomonas acidovorans. J. Bacteriol.* 97:697–704.

Rosenfeld, W.D. 1947. Anaerobic oxidation of hydrocarbons by sulfate-reducing bacteria. *J. Bacteriol.* 54:664–665.

Ross, A., Tremblay, C., and Boulanger, C. 1995. *In situ* remediation of hydrocarbon contamination using an injection-extraction process. In *Applied Bioremediation of Petroleum Hydrocarbons.* Hinchee, R.E., Kittel, J.A., and Reisinger, H.J., Eds. Battelle Press, Columbus, OH. pp. 235–239.

Ross, D.S., Sjogren, R.E., and Bartlett, R.J. 1981. Behavior of chromium in soils: IV. Toxicity to microorganisms. *J. Environ. Qual.* 10:145–148.

Roszak, D.B. and Colwell, R.R. 1987. Metabolic activity of bacterial cells enumerated by direct viable count. *Appl. Environ. Microbiol.* 53:2889–2983.

Rotert, K.H., Cronkhite, L.A., and Alvarez, P.J.J. 1995. Enhancement of BTX biodegradation by benzoate. In *Microbial Processes for Bioremediation. Pap. 3rd Int. In Situ On-Site Bioreclam. Symp.* Hinchee, R.E., Vogel, C.M., and Brockman, F.J., Eds. Battelle Press, Columbus, OH. pp. 161–168.

Rouse, J.D., Sabatini, D.A., and Harwell, J.H. 1993. Minimizing surfactant losses using twin-head anionic surfactants in subsurface remediation. *Environ. Sci. Technol.* 27(10):2072–2078.

Rouse, J.D., Sabatini, D.A., Suflita, J.M., and Harwell, J.H. 1994. Influence of surfactants on microbial degradation of organic compounds. *Crit. Rev. Environ. Sci. Technol.* 24(4):325–370.

Routson, R.C. and Wildung, R.E. 1970. What happens in soil-disposal of wastes? *Ind. Water Eng.* 7(10):25–27.

Roy, D., Kommalapati, R.R., Valsaraj, K.T., and Constant, W.D. 1995. Soil flushing of residual transmission fluid: application of colloidal gas apron suspensions and conventional surfactant solutions. *Water Res.* 29(2):589–595.

Roy, P.K., Singh, H.D., Bhagat, S.E., and Baruah, J.N. 1979. Characterization of hydrocarbon emulsification and solubilization occurring during the growth of *Endomycopsis lipolytica* on hydrocarbons. *Biotechnol. Bioeng.* 21:955–974.

Rubin, D.L. and Mon, G.J. 1994. Extensive review of treatment technologies for the removal of hydrocarbons from soil. *Hydrocarbon Contam. Soils.* 4:413–424.

Rubin, H.E. and Alexander, M. 1983. Effect of nutrients on the rates of mineralization of trace concentrations of phenol and *p*-nitrophenol. *Environ. Sci. Technol.* 17:104–107.

Rubin, J. 1983. Transport of reacting solutes in porous media: relation between mathematical nature of problem formulation and chemical nature of reactions. *J. Water Resour. Res.* 19:1231–1252.

Rubio, M.A., Engesser, K.-H., and Knackmuss, J.-J. 1986. *Arch. Microbiol.* 145:116–122.

Runion. 1975. *Am. Ind. Hyg. Assoc. J.* 24:99.

Ryan, J.R., Hanson, M.L., and Loehr, R.C. 1987. Land treatment practices in the petroleum industry. In *Land Treatment, A Hazardous Waste Management Alternative.* Loehr, R.C. and Malina, J.F., Jr., Eds. Center for Research in Water Resosurces. University of Texas, Austin. pp. 319–346.

Ryan, R.G. and Dhir, V.K. 1993. The effect of soil particle size on hydrocarbon entrapment near a dynamic water table. *Hydrocarbon Contam. Soils Groundwater.* 3:213–235.

S

Saberiyan, A.G., MacPherson, J.R., Jr., Andrilenas, J.S., Moore, R., and Pruess, A.J. 1995. A bench-scale biotreatability methodology to evaluate field bioremediation. In *Monit. Verif. Biorem. Pap. 3rd Int. In Situ On-Site Bioreclam. Symp.* Hinchee, R.E., Douglas, G.S., and Ong, S.K., Eds. pp. 185–192.

Saberiyan, A.G., Wilson, M.A., Roe, Andrilenas, J.S., Esler, C.T., Kise, G.H., and Reith. 1994. Removal of gasoline volatile organic compounds via air biofiltration: a technique for treating secondary air emissions from vapor-extraction and air-stripping systems. In *Hydrocarbon Bioremediation.* Hinchee, R.E., Alleman, B.C., Hoeppel, R.E., and Miller, R.N., Eds. CRC Press, Boca Raton, FL. pp. 1–11.

Sahm, H., Brunner, M., and Schoberth, S.M. 1986. *Microb. Ecol.* 12:147–153.

Sahoo, D.K., Kar, R.N., and Das, R.P. 1992. Bioaccumulation of heavy metal ions by *Bacillus circulans*. *Bioresour. Tech.* 41:177–179.

Sakaguchi, T. and Nakajima, A. 1982. Recovery of uranium by chitin phosphate and chitosan phosphate. In *Chitin and Chitosan.* Mirano, S. and Tokura, S., Eds. Japan Soc. Chitin and Chitosan, Tottori. pp. 177–182.

Sakaguchi, T. and Nakajima, A. 1987. Accumulation of uranium by biopigments. *J. Chem. Technol. Biotechnol.* 40:133–141.

Samson, R., van den Berg, L., and Kennedy, K.J. 1985a. Mixing characteristics and startup of anaerobic downflow stationary fixed film (DSFF) reactors. *Biotechnol. Bioeng.* 27:10–19.

Samson, R., van den Berg, L., and Kennedy, K.J. 1985. Influence of continuous vs. channels on mixing characteristics and performance of anaerobic downflow stationary fixed film (DSFF) reactors before and during waste treatment. *Proc. 39th Ind. Waste Conf.*, May 8–10, 1984, Purdue University, West Lafayette, IN. Ann Arbor Science Publishers, Ann Arbor, MI. p. 677–685.

Saner, M., Bollier, D., Schneider, K., and Bachofen, R. 1996. Mass transfer improvement of contaminants and nutrients in soil in a new type of closed soil bioreactor. *J. Biotechnol.* 48(1–2):25–35.

Sanglard, D., Leisola, M.S.A.A., and Fiechter, M.S.A. 1986. Role of extracellular ligninases in biodegradation of benzo(a)pyrene by *Phanerochaete chrysosporium*. *Enzyme Microb. Technol.* 8:209–212.

Sanning, D.E. and Olfenbuttel, R. 1987. NATO/CCMS pilot study on demonstraton of remedial action technologies for contaminated land and groundwater. In *Proc. 13th Annu. Res. Symp. on Land Disposal, Remedial Action, Incineration and Treatment of Haz. Waste,* May 6–8, 1987, Cincinnati, OH.

Santharam, S.K., Erickson, L.E., and Fan, L.T. 1994. Modeling the fate of polynuclear aromatic hydrocarbons in the rhizosphere. In *Proc. 9th Annu. Conf. Haz. Waste Rem.* Erickson, L.E., Ed. pp. 333–350.

Sarkis, B.E. and Cooper, D.G. 1994. Biodegradation of aromatic compounds in a self-cycling fermenter (SCF). *Can. J. Chem. Eng.* 72:874–880.

Savage, G.M., Diaz, L.F., and Golueke, C.G. 1985. Disposing of organic hazardous wastes by composting. *BioCycle.* 26:31–34.

Sawhney, B.L. and Kozloski, R.P. 1984. Organic pollutants in lechates from landfill sites. *J. Environ. Qual.* 13:349–352.

Saxena, A. and Bartha, R. 1983. Microbial mineralization of humic acid — 3,4-dichloroaniline complexes. *Soil Biol. Biochem.* 15:59–62.

Sayler, G.S., Shields, M.S., Tedford, E.T., Breen, A., Hooper, S.W., Sirotkin, K.M., and Davis, J.W. 1985. Application of DNA-DNA colony hybridization to the detection of catabolic genotypes in environmental samples. *Appl. Environ. Microbiol.* 49:1295–1303.

Schaeffer, T.L., Cantwell, S.G., Brown, J.L., Watt, D.S., and Fall, R.R. 1979. Microbial growth on hydrocarbons: terminal branching inhibits biodegradation. *Appl. Environ. Microbiol.* 38:742–746.

Scheda, R. and Bos, P. 1966. Hydrocarbons as substrates for yeasts. *Nature.* 211(5049):660.

Schiephake, K., Lonergan, G.T., Jones, C.L., and Mainwaring, D.E. 1993. Decolorization of pigment plant effluent by *Pycnoporus cinnabarinus* in packed-bed reactor. *Biotechnol. Lett.* 15:1185–1188.

Schink, B. 1985. Degradation of unsaturated hydrocarbons by methanogenic enrichment cultures. *FEMS Microbiol. Ecol.* 31:69–77.
Schink, B. and Pfennig, N. 1982. Fermentation of trihydroxybenzenes by *Pelobacter acidigallici* gen. nov. sp. nov., a new strictly anaerobic nonsporeforming bacterium. *Arch. Microbiol.* 133:195–201.
Schlaemus, H.W., Marshall, M.C., MacNaughton, M.G., Alexander, M.L., and Scott, J.R. 1994. Microcapsules and Method for Degrading Hydrocarbons. U.S. Patent 5,348,803. Sept. 20.
Schlauch, M.B. and Clark, D.C. 1992. Biodegradation studies of diesel-contaminated soils and high-chloride sediments. *Proc. 85th Annu. Mtg. Air Waste Mgmt. Assoc.* Vol. 8. Paper No. 92/27.05. 21 pp.
Schleussinger, A., Ohlmeier, B., Reiss, I., and Schulz, S. 1996. Moisture effects on the cleanup of PAH-contaminated soil with dense carbon dioxide. *Environ. Sci. Technol.* 30(11):3199–3204.
Schmid, K. and Hahn, H.H. 1995. Reclamation of the fine-particle fraction in hydrocarbon-contaminated soils. In *Biological Unit Processes for Hazardous Waste Treatment. Pap. 3rd Int. In Situ On-Site Bioreclam. Symp.* Hinchee, R.E., Skeen, R.S., and Sayles, G.D., Eds. Battelle Press, Columbus, OH. pp. 161–169.
Schmidt, S.K. and Alexander, M. 1985. Effects of dissolved organic carbon and second substrates on the biodegradation of organic compounds at low concentrations. *Appl. Environ. Microbiol.* 49:822–827.
Schmidt, S.K., Scow, K.M., and Alexander, M. 1985. The kinetics of simultaneous mineralization of two substrates in soil, lake water, and pure cultures of bacteria. *Abstr. Annu. Mtg. Am. Soc. Microbiol.* p. 263.
Schmitt, E.K., Lieberman, M.T., Caplan, J.A., Blaes, D., Keating, P., and Richards, W. 1991. Bioremediation of soil and groundwater contaminated with Stoddard solvent and mop oil using the PetroClean® Bioremediation system. In *In Situ Bioreclamation. Applications and Investigations for Hydrocarbon and Contaminated Site Remediation.* Hinchee, R.E. and Olfenbuttel, R.F., Eds. Butterworth-Heinemann, Stoneham, MA. pp. 581–599.
Schneider, D.R. 1993. Behavior of microbial culture product (Para-Bac) isolates in anaerobic environments. *Dev. Petrol. Sci.* (Microbial Enhancement of Oil Recovery: Recent Advances). 39:107–113.
Schneider, J., Grosser, R.J., Jayasimhulu, K., and Warshawsky, D. 1994. Biodegradation of PAHs by two bacterial species isolated from coal gasification sites. In *Proc. 9th Annu. Conf. Haz. Waste Rem.* Erickson, L.E., Ed. pp. 294–308.
Schnitzer, M. 1978. Humic substances: chemistry and reactions. In *Soil Organic Matter.* Schnitzer, M. and Khan, S.U., Eds. Elsevier North-Holland, New York. pp. 1–64.
Schnitzer, M. 1982. Organic matter characterization. In *Methods of Soil Analysis.* Part 2. *Chemical and Microbiological Properties.* 2nd ed. Agronomy Monograph No. 9. Page, A.L., Miller, R.H., and Keeney, D.R., Eds. American Society of Agronomy, Madison, WI. pp. 581–594.
Schnitzer, M. and Khan, S.U. 1978. *Soil Organic Matter. Developments in Soil Science 8.* Elsevier Scientific, Amsterdam.
Schoeberl, P. 1996. Ecological evaluation of surfactants. *Tenside Surfactants Deterg.* 33(2):120–124.
Scholz, R. and Milanowski, J. 1982. Mobile system for extracting spilled hazardous materials from excavated soils. In *Proc. 1982 Conf. Control Haz. Mater. Spills.* pp. 111–115.
Scholze, R.J., Jr., Wu, Y.C., Smith, E.D., Bandy, J.T., and Basilico, J.V., Eds. 1986.In *Proc. Int. Conf. Innovative Biological Treatment of Toxic Wastewaters,* June 24–26, Arlington, VA.
Schottel, J., Mandal, A., Clark, D., and Silver, S. 1974. Volatilization of mercury and organomercurials determined by inducible R-factor systems in enteric bacteria. *Nature.* 251:335–357.
Schroepfer, G.J., Fullen, W.J., Johnson, A.S., Ziemke, N.R., and Anderson, J.J. 1955. Industrial wastes, the anaerobic contact process as applied to packing house wastes. *Sewage Ind. Wastes.* 27:460–486.
Schulz, S., Reiss, I., and Schleussinger, A. 1995. Supercritical fluid extraction of contaminants from soil with a pilot plant. 5 (Contaminated Soil 95, Vol. 2). pp. 1353–1354.
Schuring, J.R. and Chan, P.C. 1993. Pneumatic fracturing of low permeability formations — technology status paper. Unpublished paper.
Schuring, J.R., Jurka, V., and Chan, P.C. 1991/1992. Pneumatic fracturing to remove VOCs. *Remediation.* Winter.
Schwab, G.O., Frevert, R.K., Edminster, T.W., and Barhes, K.K. 1981. *Soil and Water Conservation Engineering.* 3rd ed. John Wiley & Sons, New York.
Schwartz, R.D. and Hutchinson, D.B. 1981. Microbial and enzymatic production of 4,4′ dihydroxybiphenyl via phenol coupling. *Enzyme Microbiol. Technol.* 3:361.
Schwarzenbach, R.P., Gschwend, P.M., and Imboden, D.M. 1993. *Environmental Organic Chemistry.* John Wiley & Sons, New York.
Schweizer, E.E. 1976. Persistence and movement of ethofumesate in soil. *Weed Sci.* 14:22–26.
Schwendinger, R.B. 1968. Reclamation of soil contaminated with oil. *J. Inst. Petrol.* (London). 54:182–197.
Schwille, F. 1975. Groundwater pollution by mineral oil products. In *Proc. Ground Water Pollut. Symp.,* Aug. 1971, Moscow. AISH Pub. No. 103. pp. 226–240.
Schwille, F. 1984. Migration of organic fluids immiscible with water in the unsaturated zone. In *Pollutants in Porous Media: The Unsaturated Zone between Soil Surface and Groundwater.* Yaron, B., Dagan, G., and Goldshmid, S., Eds. Springer-Verlag, New York. pp. 27–48.
Science Applications International Corporation. 1985. Company Literature.
Scott, J.A. and Palmer, S.J. 1988. Cadmium bio-sorption by bacterial exopolysaccharide. *Biotechnol. Lett.* 10:21–24.

Scott, J.A. and Palmer, S.J. 1990. Sites of cadmium uptake in bacteria used for biosorption. *Appl. Microbiol. Biotechnol.* 33:221–225.

Scow, K.M., Simkins, S., and Alexander, M. 1986. Kinetics of mineralization of organic compounds at low concentrations in soil. *Appl. Environ. Microbiol.* 51:1028–1035.

Seagren, E.A., Rittmann, B.E., and Valocchi, A.J. 1993. Quantitative evaluation of flushing and biodegradation for enhancing *in-situ* dissolution of nonaqueous-phase liquids. *J. Contam. Hydrol.* 12(1–2):103–132.

Seagren, E.A., Rittmann, B.E., and Valocchi, A.J. 1994. Quantitative evaluation of the enhancement of NAPL-pool dissolution by flushing and biodegradation. *Environ. Sci. Technol.* 28(5):833–839.

Seigal, B.Z. 1983. *Science.* 219:255.

Seki, H. 1976. Method for estimating the decomposition of hexadecane in the marine environment. *Appl. Environ. Microbiol.* 31:439–441.

Selvakumar, A. and Hsieh, H.-N. 1988. Correlation of compound properties with biosorption of organic compounds. *J. Environ. Sci. Health.* A23:543–557.

Selvaratnam, S., Schoedel, B.A., McFarland, B.L., and Kulpa, C.F. 1995. Application of reverse transcriptase PCR for monitoring expression of the catabolic dmpN gene in a phenol-degrading sequencing batch reactor. *Appl. Environ. Microbiol.* 61:3981–3985.

Senesi, N., Sposito, G., and Martin, J.P. 1987. Copper(II) and iron(III) complexation by humic acid-like polymers (melanins) from soil fungi. *Sci. Total Environ.* 62:241–252.

Sepehr, M. and Samani, Z.A. 1993. *In-situ* soil remediation using vapor extraction wells development and testing of a three-dimensional finite-difference model. *Ground Water.* 31(3):425–436.

Sepic, E., Leskovsek, H., and Trier, C. 1995. Aerobic bacterial degradation of selected polyaromatic compounds and *n*-alkanes found in petroleum. *J. Chromatogr. A.* 697:515–523.

Sequin, C. and Hamer, D.H. 1987. Regulation *in vitro* of metallothionein gene binging factors. *Science.* 235:1383–1387.

Setti, L., Pifferi, P.G., and Lanzarini, G. 1995. Surface tension as a limiting factor for aerobic *n*-alkane biodegradation. *J. Chem. Technol. Biotechnol.* 64:41–48.

Shabad, L.M., Cohan, Y.L., Ilnitsky, A.P., Khesina, A. Ya., Shcherbak, N.P., and Smirnov., G.A. 1971. The carcinogenic hydrocarbon benzo(a)pyrene in the soil. *J. Natl. Cancer Inst.* 47:1179–1191.

Shah, F.H., Hadim H.A., and Korfiatis, G.P. 1995. Laboratory studies of air stripping of VOC-contaminated soils. *J. Soil Contam.* 4(1):93–109.

Shah, M.M. and Stevens, D.K. 1995. Residence time distribution studies and design of aerobic bioreactors. In *Biological Unit Processes for Hazardous Waste Treatment.* Hinchee, R.E., Sayles, G.D., and Skeen, R.S., Eds. Battelle Press, Columbus, OH. pp. 7–12.

Shaikh, A.U., Hawk, R.M., Sims, R.A., and Scott, H.D. 1985. Redox potential and oxygen diffusion rate as parameters for monitoring biodegradation of some organic wastes in soil. *Nucl. Chem. Waste Manage.* 5:337–343.

Shailubhai, K. 1980. In *Management of Environment.* Patel, B. Ed. John Wiley & Sons, New Delhi. pp. 172–180.

Shailubhai, K. 1986. Treatment of petroleum oil sludge in soil. *Trends Biotechnol.* 4:202–206.

Shailubhai, K., Rao, N.N., and Modi, V.V. 1984a. Degradation of petroleum industry oil sludge by *Rhodotorula rubra* and *Pseudomonas aeruginosa. Oil Petrochem. Pollut.* 2:133–136.

Shailubhai, K., Rao, N.N., and Modi, V.V. 1984b. Treatment of petroleum industry oil sludge by *Rhodotorula* sp. *Appl. Microbiol. Biotechnol.* 19:437–438.

Shanley, E.S. and Edwards, J.O. 1985. Peroxides and peroxy compounds, in *Kirk-Othner Encyclopedia of Chemical Technology,* 2nd ed. Vol. 14. Wiley, New York.

Shapiro, L. 1976. Differentiation in the *Caulobacter* cell cycle. *Annu. Rev. Microbiol.* 30:377–402.

Sharpley, J.M. 1964. TDR No. ASD-TDR-63-752. Air Force Systems Command, Wright-Patterson Air Force Base, OH.

Shehata, T.E. and Marr, A.G. 1971. *J. Bacteriol.* 107-210–216.

Shelton, R.G.J. 1971. Effects of oil and oil dispersants on the marine environment. *Proc. R. Soc. London. Ser. B.* 177:411.

Shelton, T.B. and Hunter, J.V. 1975. Anaerobic decomposition of oil in bottom sediments. *J. Water Pollut. Control Fed.* 47:2256–2270.

Sherrard, J.H. and Schroeder, E.D. 1975. Stoichiometry of industrial biological wastewater treatment. In *Proc. 30th Ind. Waste Conf.,* May 6–8, Purdue University.

Sherrill, T.W. and Sayler, G.S. 1982. Enhancement of polyaromatic hydrocarbon mineralization rates by polyaromatic hydrocarbon and synthetic oil contamination of freshwater sediments. *Abstr. Annu. Mtg. Am. Soc. Microbiol.* p. 216.

Shiaris, M.P. 1989. Seasonal biotransformation of naphthalene, phenanthrene and benzo(a)pyrene in surficial estuaring sediments. *Appl. Environ. Microbiol.* 55:1391–1399.

Shiaris, M.P. and Cooney, J.J. 1981. Phenanthrene-degrading and cooxidizing microorganisms in estuarine sediments. *Abstr. Annu. Mtg. Am. Soc. Microbiol.* p. 213.

Shiaris, M.P. and Cooney, J.J. 1983. *Appl. Environ. Microbiol.* 45:706–710.

Shiaris, M.P. and Jambard-Sweet, D. 1984. Potential transformation rates for polynuclear aromatic hydrocarbons (PNAHs) in surficial estuarine sediments. *Abstr. Annu. Mtg. Am. Soc. Microbiol.* p. 217.

Shimp, R.J. and Pfaender, F.K. 1984. Influence of naturally occurring carbon substrates on the biodegradation of mono-substituted phenols by aquatic bacteria. *Abstr. Annu. Mtg. Am. Soc. Microbiol.* p. 212.

REFERENCES

Shin, H.M. and Huang, C.P. 1994. The feasibility study of lead removal by soil washing. *Haz. Ind. Wastes.* 26th. pp. 506–513.
Shin, Y., Chodan, J.J., and Wolcott, A.R. 1970. Adsorption of DDT by soils, soil fractions, and biological materials. *J. Agric. Food Chem.* 18:1129–1133.
Shoda, M. and Udaka, S. 1980. Preferential utilization of phenol rather than glucose by *Trichosporon cutaneum* possessing a partially constitutive catechol 1,2-oxygenase. *Appl. Environ. Microbiol.* 39:1129–1133.
Short, H. and Parkinson, G. 1983. *Chem. Eng.* 90:26.
Shrift, A. 1973. Selenium in biomedicine. In *Organic Selenium Compounds: Their Chemistry and Biology.* Klaymann, D.L. and Gunther, W.H.H., Eds. Wiley, London. pp. 764–814.
Shuckrow, A.J. and Pajak, A.P. 1981. In *Proc. Land Disposal Haz. Waste,* Philadelphia, Mar. 16–18. EPA Report No. EPA-600/9-81-992b. U.S. EPA, Cincinnati, OH. pp. 341–351.
Shuckrow, A.J., Pajak, A.P., and Osheka, J.W. 1981. Concentration Technologies for Hazardous Aqueous Waste Treatment. EPA-600/S2-81-019. PB-81-150583. U.S. EPA, Cincinnati, OH.
Shuckrow, A.J., Pajak, A.P., and Touhill, C.J. 1982a. Management of Hazardous Waste Leachate. Report No. SW-871. EPA 530/SW-871, (NTIS No. PB81-166354). U.S. EPA, Washington, D.C.
Shuckrow, A.J., Pajak, A.P., and Touhill, C.J. 1982b. *Hazardous Waste Leachate Management Manual.* Noyes Data Corp., Park Ridge, NJ.
Shukla, H.M. and Hicks, R.E. 1984. *Processes Design Manual for Stripping of Organics.* U.S. EPA, Cincinnati, OH.
Siddiqi, M.A., Ye, D., Elmarakby, S.A., Kumar, S., and Sikka, H.C. 1994. Microbial metabolism of polycyclic aromatic hydrocarbons (PAH and aza-PAH). *Polycyclic Aromat. Compd.* 7(1–3):115–122.
Siegrist, R.L., Phelps, T.J., Korte, N.E., and Pickering, D.A. 1994. Characterization and biotreatability of petroleum contaminated soils in a coral atoll in the Pacific Ocean. *Appl. Biochem. Biotech.* 45–46:757–773.
Sikes, D.J. 1984. The containment and mitigation of a formaldehyde rail car spill; Naval chemical and biological *in situ* treatment techniques. Hazardous Materials Spills Conference, Nashville, TN.
Silbovitz, A.M. 1982. In *Proc. Am. Soc. Civil Eng. Natl. Conf.* New Orleans, LA.
Silver, S. 1983. Bacterial interactions with mineral cations and anions: good ions and bad. In *Biomineralization and Biological Metal Accumulation.* Westbroek, P. and deJong, E.W., Eds. D. Reidel, Amsterdam. pp. 439–457.
Silver, S. 1991. Bacterial heavy metal resistance systems and possibility of bioremediation. In *Biotechnology: Bridging Research and Applications.* Kamely, D. et al., Eds. Kluwer Academic Publishers, Boston. pp. 264–287.
Silver, S. and Kinscherf, T.G. 1982. Genetic and biochemical basis for microbial transformations and detoxification of mercury and mercurial compounds. In *Biodegradation and Detoxification of Environmental Pollutants.* Chakrabarty, A.M., Ed. CRC Press, Boca Raton, FL. pp. 85–103.
Silver, S. and Walderhaug, M. 1992. Gene regulation of plasmid- and chromosome-determined inorganic ion transport in bacteria. *Microbiol. Rev.* 56:195–228.
Simkins, S., Schmidt, S.K., and Alexander, M. 1984. Kinetics of mineralization by bacteria and in environmental samples fit models incorporating only the variables of substrate concentration and cell density. *Abstr. Annu. Mtg. Am. Soc. Microbiol.* p. 212.
Simon, R.D. and Weathers, P. 1976. Determination of the structure of the novel polypeptide containing aspartic acid and arginine which is found in cyanobacteria. *Biochim. Biophys. Acta.* 420:165–176.
Simons, G.A. 1996. Extension of the "pore tree" model to describe transport in soil. *Ground Water.* 34(4):683–690.
Sims, R.C. 1982. Land Treatment of Polynuclear Aromatic Compounds. Dissertation. North Carolina State University at Raleigh. Available from: University Microfilms, Ann Arbor, MI. 391 pp.
Sims, R.C. 1986. Loading rates and frequencies for land treatment systems. In *Land Treatment, A Hazardous Waste Management Alternative.* R.C. Loehr and J.F. Malina, Eds. Water Resources Symposium Number 13. Center for Research in Water Resources, College of Engineering, The University of Texas, Austin. pp. 151–170.
Sims, R. and Bass, J. 1984. Review of In-Place Treatment Techniques for Contaminated Surface Soils. Volume 1: Technical Evaluation. EPA Report No. EPA-540/2-84-003a.
Sims, R.C., Doucette, W.J., McLean, J.E., Grenney, W.J., and Dupont, R.R. 1988. Treatment potential for 56 EPA-listed hazardous chemicals in soil. EPA/600/6-88/001. Robert Kerr Environmental Research Laboratory, Ada, OK.
Sims, R.C. and Overcash, M.R. 1981. Land treatment of coal conversion wastewaters. In Environmental Aspects of Coal Conversion Technology VI. A Symposium on Coal-based Synfuels. EPA Report No. EPA-600-9-82-017. U.S. EPA, Washington, D.C. pp. 218–230.
Sims, R.C. and Overcash, M.R. 1983. Fate of polynuclear aromatic compounds (PNAs) in soil-plant systems. *Res. Rev.* 88:1–68.
Sims, R.C., Sorensen, D.L., Sims, J.L., McLean, J.E., Mahmood, R., and Dupont, R.R. 1985. Review of In-Place Treatment Techniques for Contaminated Surface Soils, Vols. 1 and 2. EPA-540/S2-84-003 a and b. Office of Solid Waste and Emergency Response. U.S. EPA, Cincinnati, OH.
Sinclair, J.L. and Ghiorse, W.C. 1985. Isolation and characterization of a subsurface amoeba. *Abst. Annu. Mtg. Am. Soc. Microbiol.* p. 258.
Singer, M.E. and Finnerty, W.R. 1984. Microbial metabolism of straight-chain and branched alkanes. In *Petroleum Microbiology.* Atlas, R.M., Ed. Macmillan, New York. pp. 1–59.
Singer, M.E.V. and Finnerty, W.R. 1990. Physiology of biosurfactant synthesis by *Rhodococcus* special H13-A. *Can. J. Microbiol.* 36(11):741–745.

Singleton, I., Wainwright, M., and Edyvean, R.G.J. 1990. Some factors influencing the absorption of particulates by fungal mycelium. *Biorecovery.* 1:271–289.

Singley, J.E. and Williamson, D. 1982. Aeration for the removal of volatile synthetic organic chemicals. In *Proc. Atlantic Workshop on Organic Chemical Contamination of Ground Water.* AWWA/IWSA, Nashville, TN. pp. 199–218.

Skrinde, J.R. and Bhagat, S.K. 1982. Industrial waste as carbon sources in biological denitrification. *J. Water Pollut. Control Fed.* 54:370–377.

Sleat, R. and Robinson, J.P. 1984. The bacteriology of anaerobic degradation of aromatic compounds. *J. Appl. Bacteriol.* 57:381–394.

Slonim, Z., Lien, L.-T., Eckenfelder, W.W., and Roth, J.A. 1985. Anaerobic-Aerobic Treatment Process for the Removal of Priority Pollutants. EPA Report No. EPA-600/S2-85/077.

Smith, J.H., Harper, J.C., and Jaber, H. 1981. SRI Int., Menlo Park, CA. Report No. ESL-TR-81-54. Engineering and Services Laboratory, Tyndall Air Force Base, FL.

Smith, J.L., Reible, D.D., Koo, Y.S., and Cheah, E.P.S. 1996. Vacuum extraction of a nonaqueous phase residual in a heterogeneous vadose zone. *J. Haz. Mater.* 49(2–3):247–265.

Smith, M., Stiver, W.H., and Zytner, R.G. 1995. The effect of varying water content on passive volatilization of gasoline from soil. In *Proc. 49th Ind. Waste Conf.,* 1994. pp. 111–115.

Smith, M.S., Thomas, G.W., White, R.E., and Ritonga, D. 1985. Transport of *Escherichia coli* through intact and disturbed soil columns. *J. Environ. Qual.* 14:87–91.

Smolenski, W. and Suflita, J.M. 1987. Biodegradation of cresol isomers in anoxic aquifers. *Appl. Environ. Microbiol.* 53:710–716.

Sobish, T., Kuhnemund, L., Hubner, H., Reinisch, G., and Kragel, J. 1994. Physicochemical aspects of *in situ* surfactant washing of oil contaminated soils. *Prog. Colloid Polym. Sci.* 95 (Surfactants and Colloids in the Environ.). pp. 125–129.

Sobisch, T., Kuehnemund, L., Huebner, H., Reinisch, G., and Olesch, T. 1995. Investigations on enhanced washing of tar contaminated soils. *Soil Environ.* 5 (Contaminated Soil 95, Vol. 2). pp. 1357–1358.

Soil Conservation Service. 1979. *Guide for Sediment Control on Construction Sites in North Carolina.* U.S. Dept. of Agriculture. Soil Conservation Service, Raleigh, NC.

Soil Science Society of America. 1981. Water Potential Relations in Soil Microbiology. SSSA Special Publication No. 9. Soil Science Society of America, Madison, WI. 151 pp.

Sojka, S.A. 1984. *BIOTECH '84 USA.* Online Publications, Pinner, U.K.

Sokol, R.A. and Klein, D.A. 1975. The responses of soils and soil microorganisms to silver iodide weather modification agents. *J. Environ. Qual.* 4:211–214.

Solanas, A.M., Pares, R., Bayona, J.M., and Albaiges, J. 1984. Degradation of aromatic petroleum hydrocarbons by pure microbial cultures. *Chemosphere.* 13:593–601.

Soli, G. 1973. Marine hydrocarbonoclastic bacteria: types and range of oil degradation. In *The Microbial Degradation of Oil Pollutants,* Workshop, Atlanta, GA, Dec. 4–6, 1972. Ahearn, D.G. and Meyers, S.P., Eds. Pub. No. LSU-SG-73-01. Center for Wetland Resources, Louisiana State University, Baton Rouge, LA. p. 141.

Sommers, L.E., Gilmore, C.M., Wildung, R.E., and Beck, S.M. 1981. The effect of water potential on decomposition processes in soils. In *Water Potential Relations in Soil Microbiology.* SSSA Special Publication No. 9. Soil Science Society of America, Madison, WI. pp. 97–117.

Song, H.-G. and Bartha, R. 1986. Bacterial and fungal contributions to hydrocarbon mineralization in soil. *Abstr. Annu. Mtg. Am. Soc. Microbiol.* p. 302.

Song, J.-G., Wang, X., and Bartha, R. 1990. Bioremediation potential of terrestrial fuel spills. *Appl. Environ. Microbiol.* 56:652–656.

Sonnleitner, B. and Fiechter, A. 1985. Microbial flora studies in thermophilic aerobic sludge treatment. *Conserv. Recycl.* 8:303–313.

Spain, J.C., Milligan, J.D., Downey, D.C., and Slaughter, J.K. 1989. Excessive bacterial decomposition of hydrogen peroxide during enhanced biodegration. *Ground Water.* 27(2):163–167.

Spain, J.C., Pritchard, P.H., and Bourquin, A.W. 1980. Effects of adaptation on biodegradation rates in sediment/water cores from estuarine and freshwater environments. *Appl. Environ. Microbiol.* 40:726–734.

Spain, J.C. and Van Veld, P.A. 1983. Adaptation of natural microbial communities to degradation of xenobiotic compounds: effects of concentration, exposure, time, inoculum, and chemical structure. *Appl. Environ. Microbiol.* 45:428–435.

Sparling, G.P. 1985. The soil biomass. In *Soil Organic Matter and Biological Activity.* Vaughan, D. and Malcolm, R.E., Eds. Martinus Nijhoff Junk, Dordrecht. pp. 223–262.

Speece, R.E. 1983. Anaerobic biotechnology for industrial wastewater treatment. *Environ. Sci. Technol.* 17:416A–427A.

Speece, R.E., Nirmalakhandan, N., and Lee, Y.H. 1987. Nomograph for air stripping of VOC from water. *J. Environ. Eng.* 113:434–443.

Speece, R.E., Parkin, G.F., and Gallagher, D. 1983. Nickel stimulation of anaerobic digestion. *Water Res.* 17:677–683.

Speirs, E.D., Halling, P.J., and McNeil, B. 1995. The importance of bead size measurement in mass-transfer modelling with immobilised cells. *Appl. Microbiol. Biotechnol.* 43:440–444.

Spencer, W.F. 1970. In *Pesticides in the Soil: Ecology, Degradation, and Movement. Int. Symp. on Pesticides in the Soil,* Michigan State University pp. 120–128.

REFERENCES

Spencer, W.F., Adam, J.D., Shoup, T.D., and Spear, R.C. 1980. Conversion of parathion to paraoxon on soil dusts and clay minerals as affected by ozone and UV light. *J. Agric. Food Chem.* 28:369.
Spencer, W.F. and Claith, M.M. 1973. Pesticide volatilization as related to water loss from soil. *J. Environ. Qual.* 2:284–289.
Spencer, W.F. and Cliath, M.M. 1977. The solid-air interface: transfer of organic pollutants between the solid-air interface. In *Fate of Pollutants in the Air and Water Environments,* Vol. 8, Part 1. *Advances in Environmental Science and Technology.* Suffet, I.H., Ed. John Wiley & Sons, New York. pp. 107–126.
Spivey, J.J., Allen, C.C., Green, D.A., Wood, J.P., and Stallings, R.L. 1986. Preliminary assessment of hazardous waste pretreatment as an air pollution control technique. Research Triangle Inst., Research Triangle Park, NC. Contract No. 68-03-3149. Report No. EPA-600-28-6028. PB 86172095.
Sprehe, T.G., Streebin, L.E., Robertson, J.M., and Bowen, P.T. 1985. Process considerations in land treatment of refinery sludges. In *Proc. 40th Ind. Waste Conf.* May 14–16, Purdue University, West Lafayette, IN. Butterworth, Boston. pp. 529–534.
Springer, C., Valsaraj, K.T., and Thibodeaux, L.J. 1986. In-Situ Methods to Control Emissions from Surface Impoundments and Landfills. EPA-600/S2-85/124. U.S. EPA, Cincinnati, OH.
Spuij, F., Lubbers, R.G.M., Okx, J.P., Schoen, A., and de Wit, J.C.M. 1995. Soil vapor extraction as a containment technique. *Soil Environ. 5 (Contaminated Soils 95, Vol. 2)* pp. 789–798.
Stahl, D.A., Devereux, R., Amann, R.I., Flesher, B., Lin, C., and Stromley, J. 1989. Ribosomal RNA based studies of natural microbial diversity and ecology. In *Recent Advances in Microbial Ecology.* Hattori, T., Ishida, Y., Maruyama, Y., Morita, R., and Uchida, A., Eds. Japan Scientific Society Press, Tokyo.
Stahl, D.A., Flesher, B., Mansfield, H.R., and Montgomery, L. 1988. Use of phylogenetically based hybridisation probes for studies of ruminal microbial ecology. *Appl. Environ. Microbiol.* 54:1079–1084.
Stahl, J.D. and Aust, S.D. 1994. Biodegradation of environmental pollutants using white rot fungus. PCT Int. App. WO 94 21,394 (Cl. B09B3/00), Sept. 29, 1994. US Appl. 36,113, Mar. 22, 1993.
Stamer, J.R. 1963. A Study of Slime Formation by *Alcaligenes viscolactis.* Ph.D. Dissertation, Cornell University, Ithaca, NY.
Standard Methods for the Examination of Water and Wastewater. 1975. Amer. Pub. Health Assn., Water Works Assn. Water Poll. Control Fed. 14th ed. APHA, Washington, D.C.
Stanier, R.Y., Adelberg, E.A., and Ingraham, J. 1976. *The Microbial World.* Prentice-Hall, Englewood Cliffs, NJ.
Stanier, R.Y., Doudoroff, M., and Adelberg, E.A. 1970. *The Microbial World,* 3rd ed. Prentice-Hall, Englewoods Cliffs, NJ. pp. 566–570.
Stanlake, G.J. and Finn, R.K. 1982. Isolation and characterization of a pentachlorophenol degrading bacterium. *Appl. Environ. Microbiol.* 44:1421–1427.
Stary, J. and Kratzer, K. 1984. Mechanism of the uptake of metal cations by algal cell walls. *Toxicol. Environ. Chem.* 9:115–125.
State of California, 1987. Leaking Underground Fuel Tank Field Manual: Guidelines for Site Assessment, Cleanup, and Underground Storage Tank Closure. Leaking Underground Fuel Tank Task Force. Academic Press, Orlando, FL.
Steelman, B.L. and Ecker, R.M. 1984. Organics contamination of groundwater — an open literature review. 5th DOE Environmental Protection Mtg., Nov. 5, 1984, Albuquerque, NM.
Steffan, R. and Atlas, R. 1988. DNA amplification to enhance detection of genetically engineered bacteria in environmental samples. *Appl. Environ. Microbiol.* 54:2185–2191.
Steffensen, W.S. and Alexander, M. 1995. Role of competition for inorganic nutrients in the biodegradation of mixtures of substrates. *Appl. Environ. Microbiol.* 61(8):2859–2862.
Stegmann, R., Lotter, S., and Heerenklage, J. 1991. Biological treatment of oil-contaminated soils in bioreactors. In *On-Site Bioreclamation: Processes for Xenobiotic and Hydrocarbon Treatment.* Hinchee, R.E. and Olfenbuttel, R.F., Eds. Butterworth-Heinemann, Stoneham, MA. pp. 188–208.
Steinberg, S.M., Pignatello, J.J., and Sawhney, B.L. 1987. *Environ. Sci. Technol.* 21:1201–1208.
Steiof, M. and Dott, W. 1995. Application of hexametaphosphate as a nutrient for *in situ* bioreclamation. In *Applied Bioremediation of Petroleum Hydrocarbons.* Hinchee, R.E., Kittel, J.A., and Reisinger, H.J., Eds. Battelle Press, Columbus, OH. pp. 301–310.
Sterling, L.A., Watkinson, R.J., and Higgins, I.J. 1977. Microbial metabolism of alicyclic hydrocarbons: isolation and properties of a cyclohexane-degrading bacterium. *J. Gen. Microbiol.* 99:119–125.
Sterritt, R.M. and Lester, J.N. 1980. Interactions of heavy metals with bacteria. *Sci. Total Environ.* 14:5–17.
Sterritt, R.M. and Lester, J.N. 1986. Heavy metal immobilisation by bacterial extracellular polymers. In *Immobilisation of Ions by Bio-sorption.* Eccles, H. and Hunt, S., Eds. Ellis Horwood, Chichester. pp. 201–218.
Stetter, J.R., Zaromb, S., Penrose, W.R., Findlay, M.W., Jr., Otagawa, T., and Sincali, A.J. 1984. Portable device for detecting and identifying hazardous vapors. In *Proc. 1984 Haz. Mater. Spills Conf.,* April 9–12, 1984, Nashville, TN. pp. 183–190.
Stetzenbach, L.D., Kelley, L.M., Stetzenbach, K.J., and Sinclair, N.A. 1985. Decreases in hydrocarbons by soil bacteria. In *Proc. Symp. on Groundwater Contamination and Reclamation.* American Water Research Association. pp. 55–60.
Stetzenbach, L.D. and Sinclair, N.A. 1986. Degradation of anthracene and pyrene by soil bacteria. *Abstr. Annu. Mtg. Am. Soc. Microbiol.* p. 302.
Stetzenbach, L.D., Sinclair, N.A., and Kelley, L.M. 1983. Isolation and growth of "naturally occurring" bacteria from ground water. In *Land Treatment of Hazardous Wastes.* Parr, J.F., Marsh, P.B., and Kla, J.M., Eds. Noyes Data Corp., Park Ridge, NJ. p. 28.

Stevens, T.O., Crawford, R.L., and Crawford, D.L. 1991. Selection and isolation of bacteria capable of degrading dinoseb (2-sec-butyl-4,6-dinitrophenol). *Biodegradations.* 2:1–13.

Stone, R.W., Fenske, M.R., and White, A. 1942. *J. Bacteriol.* 44:169–178.

Stormo, K.E. and Crawford. 1992. Preparation of encapsulated microbial cells for environmental applications. *Appl. Environ. Microbiol.* 58(2):727–730.

Stormo, K.E. and Crawford, R.L. 1994. Pentachlorophenol degradation by micro-encapsulated *Flavobacterium* and their enhanced survival for *in situ* aquifer bioremediation. In *Applied Biotechnology for Site Remediation.* Hinchee, R.E., Anderson, D.B., Metting, F.B., Jr., and Sayles, G.D., Eds. Lewis Publishers, Boca Raton, FL. pp. 422–427.

Stormo, K.E. and Deobald, L.A. 1995. Novel slurry bioreactor with efficient operation and intermittent mixing capabilities. In *Biological Unit Processes for Hazardous Waste Treatment.* Hinchee, R.E., Sayles, G.D., and Skeen, R.S., Eds. Battelle Press, Columbus, OH. pp. 129–135.

Stotsky, G. and Krasovsky, V.N. 1981. Ecological factors that affect the survival, establishment, growth, and genetic recombination of microbes in natural habitats. In *Molecular Biology, Pathogenicity, and Ecology of Bacterial Plasmids.* Levy, S.B., Clowes, R.C., and Koenig, E.L., Eds. Plenum Press, New York. pp. 31–42.

Stotzky, G. and Norman, A.G. 1961a. *Arch. Mikrobiol.* 40:341–369.

Stotzky, G. and Norman, A.G. 1961b. *Arch. Mikrobiol.* 40:370–382.

Stover, E.L. and Kincannon, D.F. 1983. Contaminated groundwater treatability — a case study. *Res. Technol.* pp. 292–298.

Strand, S.E. and McDonnell, A.J. 1985. Mathematical analysis of oxygen and nitrate consumption in deep microbial films. *Water Res.* 19:345–352.

Strand, S.E., McDonnell, A.J., and Unz, R.F. 1985. Concurrent denitrification and oxygen uptake in microbial films. *Water Res.* 19:335–344.

Strandberg, G.W., Shumate, S.E., II, and Parrott, J.R., Jr. 1981. Microbial cells as biosorbents for heavy metals: accumulation of uranium by *Saccharomyces cerevisiae* and *Pseudomonas aeruginosa. Appl. Environ. Microbiol.* 41:237–245.

Stratton, R.B., Jr. 1981. M.S. Thesis, Department of Civil Engineering, University of Illinois, Urbana, IL.

Streebin, L.E., Robertson, J.M., Callender, A.B., Doty, L., and Bagawandoss, K. 1984. Closure Evaluation for Petroleum Residue Land Treatment. EPA-600/2-84-162. U.S. EPA, Ada, OK.

Stringfellow, W.T., Chen, S.-H., and Aitken, M.D. 1995. Induction of PAH degradation in a phenanthrene-degrading pseudomonad. In *Microbial Processes for Bioremediation. Pap. 3rd Int. In Situ On-Site Bioreclam. Symp.* Hinchee, R.E., Vogel, C.M., and Brockman, F.J., Eds. Battelle Press, Columbus, OH. pp. 83–89.

Strong-Gunderson, J.M. and Palumbo, A.V. 1995. Bioavailability enhancement by addition of surfactant and surfactant-like compounds. In *Microbial Processes for Bioremediation. Pap. 3rd Int. In Situ On-Site Bioreclam. Symp.* Hinchee, R.E., Vogel, C.M., and Brockman, F.J., Eds. Battelle Press, Columbus, OH. p. 33.

Stronguilo, M.L., Vaquero, M.T., Comellas, L., and Broto-Puig, F. 1994. The fate of petroleum aliphatic hydrocarbons in sewage sludge-amended soils. *Chemosphere.* 29(2):273–281.

Stucki, G. and Alexander, M. 1987. Role of dissolution rate and solubility in biodegradation of aromatic compounds. *Appl. Environ. Microbiol.* 53:292–297.

Stumm, W. and Morgan, J.J. 1981. *Aquatic Chemistry.* 2nd ed. John Wiley & Sons, New York.

Subramanian, G. and Uma, L. 1996. Cyanobacteria in pollution control. *J. Sci. Ind. Res.* 55(8–9):685–692.

Suflita, J.M. and Gibson, S.A. 1985. In *Proc. of the 2nd Int. Conf. on Ground Water Quality Research.* Durham, N.N. and Redelfs, A.E., Eds. Stillwater, OK. pp. 30–32.

Suflita, J.M., Horowitz, A., Shelton, D.R., and Tiedje, J.M. 1982. Dehalogenation: a novel pathway for the anaerobic biodegradation of haloaromatic compounds. *Science.* 218:1115–1116.

Suflita, J.M. and Miller, G.D. 1985. The microbial metabolism of chlorophenolic compounds in ground water aquifers. *Environ. Toxicol. Chem.* 4:751–858.

Sujata, A.D. 1961. *Hydrocarbon Process.* Dec.:137.

Sullivan, J.P. and Chase, H.A. 1994. The development of a multi-stage system for the aerobic treatment of organic pollutants. *IChemE Res. Event, Two-Day Symp.*, Rugby, U.K. 1:413–415.

Sun, S. and Boyd, S.A. 1993. Sorption of nonionic organic compounds in soil-water systems containing petroleum sulfonate-oil surfactants. *Environ. Sci. Technol.* 27(7):1340–1346.

Sun, S., Inskeep, W.P., and Boyd, S.A. 1995. Sorption of nonionic organic compounds in soil-water systems containing a micelle-forming surfactant. *Environ. Sci. Technol.* 29(4):903–913.

Sundaram, N.S., Sarwar, M., Bang, S.S., and Islam, M.R. 1994. Biodegradation of anionic surfactants in the presence of petroleum contaminants. *Chemosphere.* 29(6):1253–1261.

Surampalli, R.Y. and Baumann, E.R. 1985. Role of air in improving first-stage RBC performance. In *Proc. of the 1985 Specialty Conference-Environmental Engineering,* July 1–5, Northeastern University, Boston. O'Shaughnessy, J.C., Ed. ASCE Press, New York.

Sutherland, J.B., Selby, A.L., Freeman, J.P., Evans, F.E., and Cerniglia, C.E. 1991. Metabolism of phenanthrene by *Phanerochaete chrysosporium. Appl. Environ. Microbiol.* 57:3310–3316.

Sutherland, J.B., Selby, A.L., Freeman, J.P., Fu, P.P., Miller, D.W., and Cerniglia, C.E. 1992. Identification of xyloside conjugates formed from anthracene by *Rhizoctonia solani. Mycol. Res.* 96:509–517.

REFERENCES

Sutton, J.R. 1987. Formulation of industrial hygiene products: art or science. In *Industrial Applications of Surfactants. Proc. Symp.* organized by the NW region of the Ind. Div. Royal Soc. Chem., University Salford, Apr. 15-17, 1986. Karsa, D.R., Ed. Special Pub. No. 59. Royal Society of Chemistry, Burlington House, London. pp. 208-234.

Sutton, P.M. 1987. Biological treatment of surface and groundwater. *Pollut. Eng.* 19:86-89.

Swallow, J.A. and Gschwend, P.M. 1983. Volatilization of organic compounds from unconfined aquifers. Volatilization of organic compounds from unconfined aquifers. In *Proc. of the 3rd Natl. Symp. on Aquifer Restoration and Ground Water Monitoring.* NWWA. pp. 327-333.

Swindoll, C.M., Aelion, C.M., Dobbins, D.C., Jiang, O.U., Long, S.C., and Pfaender, F.K. 1988. Aerobic biodegradation of natural and xenobiotic organic compounds by subsurface microbial communities. *Environ. Toxicol. Chem.* 7:291-299.

Switzenbaum, M.S. 1983. *ASM News.* 49:532-536.

Switzenbaum, M.S. and Jewell, W.J. 1980. Anaerobic attached-film expanded-bed reactor treatment. *J. Water Pollut. Control Fed.* 52:1953-1965.

T

Tabak, H.H., Glaser, J.A., Strohofer, S., Kupferle, M.J., Scarpino, P., and Tabor, M.W. 1991. Characterization and optimization of treatment of organic wastes and toxic organic compounds by a lignolytic white rot fungus in bench-scale bioreactors. In *On-Site Bioreclamation: Processes for Xenobiotic and Hydrocarbon Treatment.* Hinchee, R.E. and Olfenbuttel, R.F., Eds. Butterworth-Heinemann, Stoneham, MA. pp. 341-365.

Tabak, H.H., Govind, R., Fu, C., Yan, X., Gao, C., and Pfanstiel, S. 1997. Development of bioavailability and biokinetics determination methods for organic pollutants in soil to enhance *in-situ* and on-site bioremediation. *Biotechnol. Prog.* 13:43-52.

Tabak, H.H., Govind, R., Pfanstiel, S., Fu, C., Yan, X., and Gao, C. 1995. Protocol development for determining kinetics of *in situ* bioremediation. In *Monit. Verif. Biorem. Pap. 3rd Int.* In Situ *On-Site Bioreclam. Symp.* Hinchee, R.E., Douglas, G.S., and Ong, S.K., Eds. pp. 203-209.

Tabak, H.H., Haines, J.R., Venosa, A.D., Glasser, J.A., Desai, S., and Nisamaneepong, W. 1991. Enhancement of biodegradation of Alaskan weathered crude oil by indigenous microbiota with the use of fertiizers and nutrients. In *Proc. 1991 Int. Oil Spill Conf.* American Petroleum Institute, Washington, D.C. pp. 583-590.

Tabak, H.H., Quave, S.A., Mashni, C.I., and Barth, E.F. 1981. Biodegradability studies with priority pollutant compounds. *J. Water Pollut. Control Fed.* 53:1503-1518.

Tahraoui, K., Samson, R., and Rho, D. 1995. BTX degradation and dynamic parameters interaction in a 50-L biofilter. In *Applied Bioremediation of Petroleum Hydrocarbons.* Hinchee, R.E., Kittel, J.A., and Reisinger, H.J., Eds. Battelle Press, Columbus, OH. pp. 257-262.

Takahashi, N. 1994. Degradation of biorefractory organic compounds using ozone. *Shigen to Kankyo* (Japan). 9(3):177-185.

Takai, Y. and Kamura, T. 1966. The mechanism of reduction in waterlogged paddy soil. *Folia Microbiol.* (Prague). 11:304-313.

Tariq, M.M. and Ahmad, K. 1985. *Asian Environ.* (Philippines). 6:15.

Taylor, B.F., Campbell, W.L., and Chinoy, I. 1970. Anaerobic degradation of the benzene nucleus by a facultatively anaerobic microorganism. *J. Bacteriol.* 102:430-437.

Taylor, J.M., Parr, J.F., Sikora, L.J., and Willson, G.B. 1980. Considerations in the land treatment of hazardous wastes: principles and practices. In *Proc. 2nd Oil and Haz. Mater. Spills Conf. and Exhib.* Hazardous Material Control Research Institute, Silver Springs, MD.

Telegina, Z.P. 1963. A study of the capacity of individual saprophytic microflora species to adapt themselves to gaseous hydrocarbon oxidation. *Mikrobiologiya.* 32:398-402.

Ten Brummeler, E., Hulshoff Pol, L.W., Dolfing, J., Lettinga, G., and Zehnder, A.J.B. 1985. Methanogenesis in an upflow anaerobic sludge blanket reactor at pH 6 on an acetate-proprionate mixture. *Appl. Environ. Microbiol.* 49:1472-1477.

Texas Research Institute, Inc. 1982. Enhancing the Microbial Degradation of Underground Gasoline by Increasing Available Oxygen. Report to the American Petroleum Institute, Washington, D.C.

Texas Research Institute, Inc. 1983. Progress Report: Biostimulation Study. Feb.

Thai, L.T. and Maier, W.J. 1993. Solubilization and biodegradation of octadecane in the presence of two commercial surfactants. In *Proc. 47th Ind. Waste Conf.,* 1992. pp. 167-175.

Thakker, D.R., Levin, W., Yagi, H., Ryan, D., Thomas, P.E., Karle, J.M., Lehr, R.E., Jerina, D.M., and Conney, A.H. 1979. Metabolism of benzo(a)anthracene to its tumorigenic 3,4-dihydrodiol. *Mol. Pharmacol.* 15:138.

Thibault, G.T. and Elliott, N.W. 1979. Accelerating the biological clean-up of hazardous materials spills. In *Proc. Oil and Haz. Mater. Spills: Prevention — Control — Cleanup — Recovery — Disposal,* Dec. 3-5. Sponsored by Haz. Mater. Control Res. Inst. and Info. Transfer, Inc. pp. 115-120.

Thibault, G.T. and Elliot, N.W. 1980. Biological detoxification of hazardous organic chemical spills. In *Proc. 1980 Conf. Haz. Mat. Spills,* May 13-15, Louisville, KY. Sponsored by Environmental Protection Agency. Vanderbilt Univ., Nashville, TN. pp. 398-402.

Thibodeaux, L.J. 1979. *Chemodynamics, Environmental Movement of Chemicals in Air, Water and Soil.* John Wiley & Sons, New York.

Thibodeaux, L.J. and Hwang, S.T. 1982. Landfarming of petroleum wastes — modeling the air emission problems. *Environ. Prog.* 1:42–46.

Thies, J.E., Singleton, P.W., and Bohlool, B.B. 1991. Influence of size of indigenous rhizobial populations on establishment and symbiotic performance of introduced rhizobia on field-grown legumes. *Appl. Environ. Microbiol.* 57:19–28.

Thimann, K.V. 1963. *The Life of Bacteria, Their Growth, Metabolism and Relationships.* 2nd Ed. Macmillan, NY.

Thomas, J.K. 1980. Radiation-induced reactions in organized assemblies. *Chem. Rev.* 80:283–299.

Thomas, J.M. and Alexander, M. 1983. Solubilization and mineralization rates of water-insoluble compounds. *Abstr. Annu. Mtg. Am. Soc. Microbiol.* p. 274.

Thomas, J.M., Gordy, V.R., Bruce, C.L., Ward, C.H., Hutchins, S.R., and Sinclair, J.L. 1995. Microbial activity in subsurface samples before and during nitrate-enhanced bioremediation. EPA/600/A-95/109. Order No. PB95-274239GAR. Avail. NTIS. From Gov. Rep. Ann. Index. 95(24), Abstr. No. 24-01,224. 17 pp.

Thomas, J.M., Lee, M.D., Scott, M.J., and Ward, C.H. 1986. Microbial adaptation to a jet fuel spill. *Abstr. Annu. Mtg. Am. Soc. Microbiol.* p. 303.

Thomas, J.M., Lee, M.D., Scott, M.J., and Ward, C.H. 1989. Microbial ecology of the subsurface at an abandoned creosote waste site. *J. Ind. Microbiol.* 4:109–120.

Thomas, J.M., Lee, M.D., and Ward, C.H. 1985. Microbial numbers and activity in the subsurface at a creosote waste site. *Abstr. Annu. Mtg. Am. Soc. Microbiol.* p. 258.

Thomas, J.M., Gordy, V.R., Amador, J.A., and Alexander, M. 1986. Rates of dissolution and biodegradation of water-insoluble organic compounds. *Appl. Environ. Microbiol.* 52:290–296.

Thompson, G.A. and Watling, R.J. 1983. A simple method for the determination of bacterial resistance to metals. *Bull. Environ. Contam. Toxicol.* 31:705–711.

Thompson, J.C., Hatfield, J.H., and Reed, B.E. 1994. Electrokinetic (EK) soil flushing on a saturated silt clay loam. *Haz. Ind. Wastes.* 26th. pp. 436–445.

Thompson, L.M., Black, C.A., and Zoellner, J.A. 1954. Occurrence and mineralization of organic phosphorus in soils, with particular reference to associations with nitrogen, carbon, and pH. *Soil Sci.* 77:185–196.

Thomson, B.M., Morris, C.E., Stormont, J.C., and Ankeny, M.D. 1996. Development of tensiometric barriers for containment and remediation at waste sites. *Radioact. Waste Manage. Environ. Restor.* 20(2–3):167–189.

Thomson, N.R., Graham, D.N., and Farquhar, G.J. 1992. One-dimensional immiscible displacement experiments. *J. Contam. Hydrol.,* 10(3):197–223.

Thorn, P.M. and Ventullo, R.M. 1986. Growth of bacteria in aquifer soil as measured by ^{3}H thymidine incorporation. *Abstr. Annu. Mtg. Am. Soc. Microbiol.* p. 300.

Thornton-Manning, J.R., Jones, D.D., and Federle, T.W. 1987. Effects of experimental manipulation of environmental factors on phenol mineralization in soil. *Environ. Toxicol. Chem.* 6:615–621.

Tichenor, B.A. and Palazzolo, M.A. 1987. Destruction of Volatile Organic Compounds via Catalytic Incineration. EPA-600/J-87-182. PB 88159710. U.S. EPA, Research Triangle Park, NC.

Tiedje, J.M. 1988. Chapter 4. In *Biology of Anaerobic Microorganisms.* Zehnder, A.J.B., Ed. Wiley, New York.

Tiedje, J.M. 1993. Bioremediation from an ecological perspective. In In Situ *Bioremediation. When Does It Work?* National Research Council. Water Science and Technology Board, Com. Eng. Tech. Systems. National Academy of Sciences. National Academy Press, Washington, D.C. pp. 110–120.

Tiedje, J.M. 1995. New advances and challenges for bioremediation. In *Bioremediation: the Tokyo '94 Workshop.* Org. Econ. Co-operation Devel., Paris, France. pp. 81–88.

Tiedje, J.M., Sexstone, A.J., Parkin, T.B., Revsbech, M.P., and Shelton, D.R. 1984. Anaerobic processes in soil. *Plant Soil.* 76:197–212.

Tien, M. 1987. Properties of ligninase from *Phanerochaete chrysosporium* and their possible applications. *CRC Crit. Rev. Microbio.* 15:141–168.

Tierney, J.W. and Bruno, J.A. 1967. Equilibrium stage calculations. *AIChE J.* 13:556–563.

Tierney, J.W. and Yanosik, J.L. 1969. Simultaneous flow and temperature correction in the equilibrium stage problem. *Am. Inst. Chem. Eng. J.* 15:897–901.

Toccalino, P.L., Johnson, R.L., and Boone, D.R. 1993. Nitrogen limitation and nitrogen fixation during alkane biodegradation in a sandy soil. *Appl. Environ. Microbiol.* 59(9):2977–2983.

Toda, K. and Ohtake, H. 1985. Comparative study on performance of biofilm reactors for waste treatment. *J. Gen. Appl. Microbiol.* 31:177–186.

du Toit, P.J. and Davies, T.R. 1973. Denitrification. Studies with laboratory-scale continuous-flow units. *Water Res.* 7:489–500.

Tonelli, F.A. and Behmann, H. 1996. Aerated Hot Membrane Bioreactor Process for Treating Recalcitrant Compounds. U.S. Patent 5,558,774 (Cl. 210-612; C02F3/12), 24 Sep. 1996, U.S. Appl. 773,226, 9 Oct. 1991. 19 pp.

Tonomura, K. and Kanzaki, F. 1969. The reductive decomposition of organic mercurials by cell-free extract of a mercury-resistant pseudomonad. *Biochim. Biophys. Acta.* 184:227–229.

REFERENCES

Tonomura, K., Maeda, K., and Futai, F. 1968. Studies on the action of mercury resistant microorganisms on mercurials. II. The vaporization of mercurials stimulated by mercury-resistant bacterium. *J. Ferment. Technol.* 46:685–692.

Torrella, F. and Morita, R.Y. 1981. Microcultural study of bacterial size changes and microcolony and ultramicrocolony formation by heterotrophic bacteria in seawater. *Appl. Environ. Microbiol.* 41:518–527.

Torsvik, V., Goksoy, J., and Daae, F.L. 1990. High diversity in DNA of soil bacteria. *Appl. Environ. Microbiol.* 56:782–787.

Torsvik, V., Salte, K., Sorheim, R., and Goksoyr, J. 1990. Comparison of phenotypic diversity and DNA heterogeneity in a population of soil bacteria. *Appl. Environ. Microbiol.* 56:776–781.

Treen-Seers, M.E. 1986. Propagation and characterization of *Rhizopus* biosorbents. Ph.D. dissertation, McGill University, Montreal, Canada.

Trevors, J.T., van Elsas, J.D., Lee, H., and van Overbeek, L.S. 1992. Use of alginate and other carriers for encapsulation of microbial cells for use in soil. *Microb. Releases.* 1:61–69.

Trevos, J.T., Oddie, K.M., and Belliveau, B.H. 1985. Metal resistance in bacteria. *FEMS Microbiol. Rev.* 32:39–54.

Treybal, R.E. 1963. *Liquid Extraction.* 2nd ed. McGraw-Hill, New York.

Treybal, R. 1980. *Mass Transfer Operations.* 2nd ed. McGraw-Hill, New York.

Trim, B.C. and McGlashan, J.E. 1985. Sludge stabilization and disinfection by means of autothermal aerobic digestion with oxygen. *Water Sci. Technol.* 17:563–573.

Tripp, S., Barkay, T., and Olson, B.H. 1983. The effect of cadmium on the community structure of soil bacteria. *Abstr. Annu. Mtg. Am. Soc. Microbiol.* p. 260.

Trudell, M.R., Gillham, R.W., and Cherry, J.A. 1986. *J. Hydrol.* 83:251–268.

Truong, K.N. and Blackburn, J.W. 1984. The stripping of organic chemicals in biological treatment processes. *Environ. Prog.* 3(3):143–152.

Trzesicka-Mlynarz, D. and Ward, O.P. 1995. Degradation of polycyclic aromatic hydrocarbons (PAHs) by a mixed culture and its component pure cultures, obtained from PAH-contaminated soil. *Can. J. Microbiol.* 41:470–476.

Tsezos, M. 1984. Recovery of uranium from biological adsorbents — desorption equilibrium. *Biotechnol. Bioeng.* 26:973–981.

Tsezos, M. 1990. Engineering aspects of metal binding by biomass. In *Microbial Mineral Recovery.* Ehrlich, H.L. and Brierley, C.L., Eds. McGraw-Hill, New York. pp. 323–339.

Tsezos, M. and Benedek, A. 1980. Removal of organic substances by biologically activated carbon in a fluidized-bed reactor. *J. Water Pollut. Control Fed.* 52(3):578–586.

Tsezos, M., McCready, R.G.L., and Bell, J.P. 1989. The continuous recovery of uranium from biologically leached solutions using immobilized biomass. *Biotechnol. Bioeng.* 34:10–17.

Tsuda, S. and Kuwabara, M. 1996. Process for treating organic wastewater by evaporation and bio-film filtration. Jpn. Kokai Tokyo Koho JP 08,192,188 [96,192,188] (Cl. C02F9/00), 30 Jul. 1996, Appl. 95/23,280, 19 Jan. 1995. 5 pp.

Turco, R.F. and Sadowsky, M. 1995. The microflora of bioremediation. In *Bioremediation: Science and Applications.* SSSA Special Pub. 43. Soil Science Society of America, American Society of Agronomy and Crop Science Society, Madison, WI. pp. 87–102.

Turick, C.E., Apel, W.A., and Carmiol, N.S. 1996. Isolation of hexavalent chromium-reducing anaerobes from hexavalent-chromium-contaminated and noncontaminated environments. *Appl. Microbiol. Biotechnol.* 44(5):683–686.

Turney, G.R., Aten, T.P., and Zikopoulos, J.N. 1992. Multiphase biological remediation. In *Gas, Oil, Environ. Biotechnol.* IV. *4th Pap. Int. Symp.,* 1991. pp. 29–39.

Tyagi, R.D., Tran, F.T., and Chowdhury, A.K.M.M. 1993. Biodegradation of petroleum refinery wastewater in a modified rotating biological contactor with polyurethane foam attached to the disks. *Water Res.* 27(1):91–99.

Tynecka, Z., Gos, Z., and Zajac, J. 1981. Reduced cadmium transport determined by a resistance plasmid in *Staphylococcus aureus. J. Bacteriol.* 147:305–312.

U

Ulmer, D.C., Leisola, M.S.A., and Fiechter, A. 1984. Possible induction of the ligninolytic system of *Phanerochaete chrysosporium. J. Biotechnol.* 1:13–24.

Unger, S.R., Lam, R.R., Schaefer, C.E., and Kosson, D.S. 1996. Predicting the effect of moisture on vapor-phase sorption of volatile organic compounds to soils. *Environ. Sci. Technol.* 30(4):1081–1091.

Urlings, L.C.C.M., Spuy, F., Coffa, S., and van Vree, H.B.R.J. 1991. Soil vapour extraction of hydrocarbons: *in situ* and on-site biological treatment. In *In Situ Bioreclamation. Applications and Investigations for Hydrocarbon and Contaminated Site Remediation.* Hinchee, R.E. and Olfenbuttel, R.F., Eds. Butterworth-Heinemann, Stoneham, MA. pp. 321–336.

U.S. EPA. 1971. Water Pollution Control Research Series, 12050 DSH 03/71. EPA, Washington, D.C.

U.S. EPA. 1974. 625/6-74-003a. Environmental Research Center, Cincinnati, OH.

U.S. EPA. 1976a. EPA-450/2-76-028. Research Triangle Park, NC.

U.S. EPA. 1976b. 600/4-76-023.

U.S. EPA. 1978. EPA-450/2-78-022. Research Triangle Park, NC. p. 75.
U.S. EPA. 1980a. EPA-450/3-80-031a. Chapter 4.
U.S. EPA. 1980b. *Federal Resister,* Vol. 45: 71538-71556, Part IV. Washington, D.C. U.S. Government Printing Office.
U.S. EPA. 1981a. EPA Contract No. 68-02-3058. Research Triangle Park, NC.
U.S. EPA. 1981b. U.S. EPA background Document 3 (MS 1941.36).
U.S. EPA. 1982. EPA-625/6-82-006. *Handbook for Remedial Action at Waste Disposal Sites.* Office of Emergency and Remedial Response, Washington, D.C.
U.S. EPA. 1983. Innovative and Alternative Technology Assessment Manual. Draft report.
U.S. EPA. 1985a. *Handbook for Remedial Action at Waste Disposal Sites.* Revised. EPA/625/6-85/006.
U.S. EPA. 1985b. No. 600/S2-85-009. M.E.R.L., Ada, OK.
U.S. EPA. 1986. Systems to Accelerate *in Situ* Stabilization of Waste Deposits. EPA/540/2-86/002. Hazardous Waste Engineering Research Laboratory. Office Res. Devel., Cincinnati, OH.
U.S. EPA. 1988. Technology screening Guide for Treatment of CERCLA Soils and Sludges. EPA/540/2-88/004. Sept.
U.S. EPA. 1989. Estimating Air Emissions from Petroleum UST Cleanups. U.S. EPA, Office of Underground Storage Tanks, Washington, D.C.
U.S. EPA. 1991. Research and Development (RD-681). EPA/600/M-91/049. Alternative treatment technology information center (ATTIC). Nov.
U.S. EPA. 1994a. Soil vapor extraction (SVE) treatment technology resource guide and soil vapor extraction treatment technology resource matrix. Report 1994. EPA/542/B-94/007. Order No. PB95-138681. Avail. NTIS. From Gov. Rep. Announce Index. 1995. 95(5), Abstract No. 511,613. 81 pp.
U.S. EPA. 1994b. Demonstration Program of Hughes Environmental Systems, Inc. (Steam Enhanced Recovery Process), in SITE: Technology Profiles, EPA/540/R-94/526. Nov.
U.S. EPA. 1994c. Demonstration Program of NOVATERRA, Inc. (In Situ Steam and Air Stripping), in SITE: Technology Profiles, EPA/540/R-94/526. Nov.
U.S. EPA. 1994d. Demonstration Program of Hrubetz Environmental Services, Inc. (HRUBOUT Process), in SITE: Technology Profiles, EPA/540/R-94/526. Nov.
U.S. EPA. 1994e. Emerging Technology Program of Electrokinetics, Inc. (Electrokinetic Remediation), in SITE: Technology Profiles, EPA/540/R-94/526. Nov.
U.S. EPA. 1994f. Emerging Technology Program of Battelle Memorial Institute (*In Situ* Electroacoustic Soil Decontamination), in SITE: Technology Profiles, EPA/540/R-94/526. Nov.
U.S. EPA. 1994g. SITE Demonstration Bulletin: In Situ Vitrification, Geosafe Corp. EPA/540/MR-94/520.
U.S. EPA Research Cooperative Agreement CR809758. 1985. U.S. EPA, Cincinnati, OH.
U.S. Patent No. 3,912,490.
Utgikar, V. 1993. Fundamental studies on the biodegradation of volatile organic chemicals in a biofilter. Ph.D. Thesis, University of Cincinnati, Cincinnati, OH.

V

Valsaraj, K.T. and Thibodeaux, L.J. 1988. Equilibrium adsorption of chemical vapors on surface soils, landfills, and landfarms — a review. *J. Haz. Mater.* 19:79–99.
Van Benschoten, J.E., Ryan, M.E., Huang, C., Healy, T.C., and Brandl, P.J. 1995. Remediation of metal/organic contaminated soils by combined acid extraction and surfactant washing. *Haz. Ind. Wastes.* 27th. pp. 551–560.
Van den Berg, L., Kennedy, K.J., and Samson, R. 1985. Anaerobic downflow stationary fixed film reactors: performance under steady-state and non-steady state conditions. *Water Sci. Technol.* (G.B.). 17:89–102.
van der Kooij, D., Visser, A., and Hijnen, W.A.M. 1980. *Appl. Environ. Microbiol.* 39:1198–1204.
Van der Linden, A.C. 1978. In *Degradation of Oil in the Marine Environment.* Chapter 6. Applied Science, Barking, Essex, U.K..
Van der Linden, A.C. and Thijsse, G.J.E. 1965. *Adv. Enzymol.* 27:469–546.
Van der Waarde, J.J., Dijkhuis, E.J., Henssen, M.J.C., and Keuning, S. 1995a. Enzyme assays as indicators for biodegradation. In *Monit. Verif. Biorem. Pap. 3rd Int.* In Situ *On-Site Bioreclam. Symp.* Hinchee, R.E., Douglas, G.S., and Ong, S.K., Eds. Battelle Press, Columbus, OH. pp. 59–63.
Van der Waarde, J.J., Dijkhuis, E.J., Henssen, M.J.C., and Keuning, S. 1995b. Enzyme assays as indicators for biodegradation. *Soil Environ.* 5 (Contaminated Soil 94, Vol. 2). pp. 1377–1378.
Van de Velde, K.D., Marley, M.C., Studer, J., and Wagner, D.M. 1995. Stable carbon isotope analysis to verify bioremediation and bioattenuation. In *Monit. Verif. Biorem. Pap. 3rd Int.* In Situ *On-Site Bioreclam. Symp.* Hinchee, R.E., Douglas, G.S., and Ong, S.K., Eds. Battelle Press, Columbus, OH. pp. 241–257.
Van de Velde, K.D., Wagner, D.M., and Marley, M.C. 1994. Application of carbon isotope analysis in quantifying biological degradation at bioventing/biosparging sites. In *Proc. Petrol. Hydrocarbons Org. Chem. Ground Water: Prev., Detect., Rem. Conf.* pp. 271–285.
Van Dyke, M.I., Gulley, S.L., Lee, H., and Trevors, J.T. 1993. Evaluation of microbial surfactants for recovery of hydrophobic pollutants from soil. *J. Ind. Microbiol.* 11(3):163–170.

REFERENCES

van Eyk, J. 1994. Venting and bioventing for *in situ* removal of petroleum from soil. In *Hydrocarbon Bioremediation.* Hinchee, R.E., Alleman, B.C., Hoeppel, R.E., and Miller, R.N., Eds. Lewis Publishers, Boca Raton, FL. pp. 243–251.

van Eyk, J. and Bartels, T.J. 1968. Paraffin oxidation in *Pseudomonas aeruginosa.* I. Induction of paraffin oxidation. *J. Bacteriol.* 96:706–712.

van Eyk, J. and Vreeken, C. 1991. *In situ* and on-site subsoil and aquifer restoration at a retail gasoline station. In *In Situ Bioreclamation: Applications and Investigations for Hydrocarbon and Contaminated Site Remediation.* Hinchee, R.E. and Olfenbuttel, R.F., Eds. Butterworth-Heinemann, Stoneham, MA. pp. 303–320.

van Eyk, J. and Vreeken, C. 1992. The application of (bio)venting for *in situ* subsoil restoration following a hydrocarbon spill at a retail gasoline station: design of the system. In *Proc. Int. Conf. Subsurf. Contam. Immiscible Fluids.* Weyer, K.U., Ed. pp. 511–518.

van Ginkel, C.G. 1996. Complete degradation of xenobiotic surfactants by consortia of aerobic microorganisms. *Biodegradation.* 7:151–164.

van Ginkel, C.G., Dijk, J.B., and Kroon, A.G.M. 1992. Metabolism of hexadecyltrimethylammonium chloride in *Pseudomonas* strain B1. *Appl. Environ. Microbiol.* 58:3083–3087.

Van Hasselt, H.J. 1987. NBM Bodemsanering, Zonweg 23, 2516 AS 's-Gravenhage.

Vanloocke, R., Verlinde, A.M., Verstraete, W., and DeBurger, R. 1979. Microbial release of oil from soil columns. *Environ. Sci. Technol.* 13:346–348.

van Loosdrecht, M.C.M., Lyklema, J., Norder, W., and Zehnder, A.J.B. 1990. Influences of interfaces on microbial activity. *Microbiol. Rev.* 54:75–87.

Vanneck, P., Beeckman, M., Saeyer, N.De., D'Haene, S., and Verstraete, W. 1994. Biodegradation of polynuclear aromatic hydrocarbons in a biphasic (aqueous-organic) system. *Meded — Fac. Landbouwkd. Toegepaste Biol. Wet.* 59(4A):1867–1876. [In English.]

Varga, G.M., Jr., Lieberman, M., and Avella, A.J., Jr. 1985. The Effects of Crude Oil and Processing on JP-5 Composition and Properties. Report to Naval Air Propulsion Center under Contract N00140-81-C-9601. NAPC-PE-121C.

Vaughn, J.M., Landry, E.F., Beckwith, C.A., and Thomas, M.Z. 1981. Virus removal during groundwater recharge: effects of infiltration rate on adsorption of poliovirus to soil. *Appl. Environ. Microbiol.* 41:139–147.

Vecchioli, G.I., Del Panno, M.T., and Painceira, M.T. 1990. Use of selected autochthonous soil bacteria to enhance degradation of hydrocarbons in soil. *Environ. Pollut.* 67:249–258.

Velazquez, L.A. and Noland, J.W. 1993. Low temperature stripping of volatile compounds. *Principles and Practices for Petroleum Contaminated Soils.* Calabrese, E.J. and Kostecki, P.T., Eds. Lewis Publishers, Boca Raton, FL, pp. 423–431.

Venkataramani, E.S., Ahlert, R.C., and Corbo, P. 1988. Aerobic and anaerobic treatment of high-strength hazardous liquid wastes. *J. Haz. Mater.* 17:169–188.

Venkateswarlu, K., Marsh, R.M., Faber, B., and Kelly, S.L. 1996. Investigation of cytochrome P450 mediated benzo(a)pyrene hydroxylation in *Aspergillus fumigatus. J. Chem. Technol. Biotechnol.* 66(2):139–144.

Venkateswerlu, G. and Stotzky, G. 1989. Binding of heavy metals by cell walls of *Cunninghamella blakesleana* grown in the presence of copper and cobalt. *Appl. Microbiol. Biotechnol.* 31:619–625.

Venkatraman, S.N., Schuring, J.R., Boland, T.M., and Kosson, D.S. 1995. Integration of pneumatic fracturing with bioremediation for the enhanced removal of BTX from low permeability gasoline-contaminated soils. In *Innov. Technol. Site Rem. Haz. Waste Mgmt. Proc. Natl. Conf.* pp. 189–196.

Venosa, D.V., Haines, J.R., and Allen, D.M. 1992. Efficacy of commercial inocula in enhancing biodegradation of weathered crude oil contaminating a Prince William Sound beach. *J. Ind. Microbiol.* 10:1–11.

Verma, L., Martin, J.P., and Haider, K. 1975. Decomposition of carbon-14-labeled proteins, peptides, and amino acids; free and complexed with humic polymers. *Soil Sci. Soc. Am. Proc.* 39:279–284.

Vesper, S.J., Donovan-Brand, R., Paris, K.P., Al-Bed, S.R., Ryan, J.A., and Davis-Hoover, W.J. 1996. Microbial removal of lead from solid media and soil. *Water Air Soil Pollut.* 86(1–4):207–219.

Vesper, S.J., Murdoch, L.C., Hayes, S., and Davis-Hooper, W.J. 1993. Solid oxygen source for bioremediation in subsurface soils. *J. Haz. Mater.* 36(3):265–274.

Viraraghavan, T. et al. 1985a. *Public Health Eng.* 13:42.

Viraraghavan, T., Landine, R.C., Winchester, E.L., and Wasson, G.P. 1985b. Activated biofilter process for wastewater treatment. *Effluent. Water Treat. J.* 25:129–134.

Vismara, R. 1985. A model for autothermic aerobic digestion. *Water Res.* (G.B.). 19:441–447.

Visser, S.A. 1982. Surface active phenomena by humic substances of aquatic origin. *Rev. Francaise Sci. Eau.* 1:285–295.

Visser, S.A. 1985. Physiological action of humic substances on microbial cells. *Soil Biol. Biochem.* 17:457–462.

Voerman, S. and Tammes, P.M. 1969. Adsorption and desorption of lindane and dieldrin by yeast. *Bull. Environ. Contam. Toxicol.* 4:271–277.

Voice, T.C., Zhao, X., Shi, J., and Hickey, R.F. 1995. Biological activated carbon fluidized-bed system to treat gasoline-contaminated groundwater. In *Biological Unit Processes for Hazardous Waste Treatment.* Hinchee, R.E., Sayles, G.D., and Skeen, R.S., Eds. Battelle Press, Columbus, OH. pp. 29–36.

Volesky, B., Ed. 1990. *Biosorption of Heavy Metals.* CRC Press, Boca Raton, FL.

Volkering, F., Breure, A.M., Sterkenburg, A., and van Andel, J.G. 1992. Microbial degradation of polycyclic aromatic hydrocarbons: effect of substrate availability on bacterial growth kinetics. *Appl. Microbiol. Biotechnol.* 36:548–552.

Volkering, F., Breure, A.M., van Andel, J.G., and Rulkens, W.H. 1995. Influence of nonionic surfactants on bioavailability and biodegradation of polycyclilc aromatic hydrocarbons. *Appl. Environ. Microbiol.* 61(5):1699–1705.

Volkering, F., van de Wiel, R., Breure, A.M., van Andel, J.G., and Rulkens, W.H. 1995. Biodegradation of polycyclic aromatic hydrocarbons in the presence of nonionic surfactants. In *Microbial Processes for Bioremediation. Pap. 3rd Int. In Situ On-Site Bioreclam. Symp.* Hinchee, R.E., Vogel, C.M., and Brockman, F.J., Eds. Battelle Press, Columbus, OH. pp. 145–151.

Vonk, J.W. and Sjipesteijn, A.K. 1973. Studies on the methylation of mercuric chloride by pure cultures of bacteria and fungi. *Antonie van Leeuwenhoek.* 39:505–513.

Vroblesky, D.A., Bradley, P.M., and Chapelle, F.H. 1996. Influence of electron donor on the minimum sulfate concentration required for sulfate reduction in a petroleum hydrocarbon-contaminated aquifer. *Environ. Sci. Technol.* 30:1377–1381.

Vyas, B.R.M., Bakowski, S., Sasek, V., and Matucha, M. 1994. Degradation of anthracene by selected white rot fungi. *FEMS Microbiol. Ecol.* 14:65–70.

W

Wainwright, M., Singleton, I., and Edyvean, R.G.J. 1990. Magnetite adsorption as a means of making fungal biomass susceptible to a magnetic field. *Biorecovery.* 2:37–53.

Waksman, S.A. 1924. Influence of microorganisms upon the carbon:nitrogen ratio in the soil. *J. Agric. Sci.* 14:555–562.

Walker, A. and Crawford, D.V. 1968. The role of organic matter in adsorption of the triazine herbicides by soil. In *Isotopes and Radiation in Soil Organic Matter Studies. Proc. 2nd Symp.* International Atomic Energy Agency, Vienna, Austria.

Walker, J.D., Austin, H.F., and Colwell, R.R. 1975. Utilization of mixed hydrocarbon substrate by petroleum-degrading microorganisms. *J. Gen. Appl. Microbiol.* 21:27–39.

Walker, J.D., Cofone, L., Jr., and Cooney, J.J. 1973. Microbial petroleum degradation: the role of *Cladosporium resinae*. In *Proc. Joint Conf. Prevention and Control of Oil Spills.* American Petroleum Institute, Washington, D.C. pp. 821–825.

Walker, J.D. and Colwell, R.R. 1974a. Mercury-resistant bacteria and petroleum degradation. *Appl. Microbiol.* 27:285–287.

Walker, J.D. and Colwell, R.R. 1974b. Microbial degradation of model petroleum at low temperatures. *Microbiol. Ecol.* 1:63–95.

Walker, J.D. and Colwell, R.R. 1975. Factors affecting enumeration and isolation of actinomycetes from Chesapeake Bay and Southeastern Atlantic Ocean sediments. *Mar. Biol.* 30:193–210.

Walker, J.D. and Colwell, R.R. 1976a. Measuring the potential activity of hydrocarbon-degrading bacteria. *Appl. Environ. Microbiol.* 31:189–197.

Walker, J.D. and Colwell, R.R. 1976b. Oil, mercury, and bacterial interactions. *Environ. Sci. Technol.* 10:1145–1147.

Walker, N., Janes, N.F., Spokes, J.R., and van Berkum, P. 1975. Degradation of 1-naphthol by a soil pseudomonad. *J. Appl. Bacteriol.* 39:281–286.

Walker, N. and Wiltshire, G.H. 1953. The breakdown of naphthalene by a soil bacterium. *J. Gen. Microbiol.* 8:273–276.

Walker, S.G., Flemming, C.A., Ferris, F.G., Beveridge, T.J., and Bailey, G.W. 1989. Physicochemical interaction of *Escherichia coli* cell envelopes and *Bacillus subtilis* cell walls with two clays and ability of the composite to immobilize heavy metals from solution. *Appl. Environ. Microbiol.* 55:2976–2984.

Walter, U., Beyer, M., Klein, J., and Rehm, H.-J. 1990. Biodegradation of polycyclic aromatic hydrocarbons by bacterial mixed culture. In *Lectures Held at the 8th Dechema Annu. Mtg. Biotechnol.,* May 28–30, 1990, Frankfurt am Main, Germany. Behrens, D. and Kramer, P., Eds. Dechema Deutsche Gesellschaft fuer Chemisches Apparatewesen, Chemische Technik und Biotechnologie, Frankfurt am Main, Germany. pp. 489–492.

Walter, U., Beyer, M., Klein, J., and Rehm, H.J. 1991. Degradation of pyrene by *Rhodococcus* sp. UW 1. *Appl. Microbiol. Biotechnol.* 34:671–676.

Walton, G.C. and Dobbs, D. 1980. Biodegradation of hazardous materials in spill situations. In *Proc. 1980 Conf. on Control of Haz. Mater. Spills,* Louisville, KY. Sponsored by Environmental Protection Agency et al. pp. 23–29.

Walworth, J.L. and Reynolds, C.M. 1995. Bioremediation of a petroleum-contaminated cryic soil: effects of phosphorus, nitrogen, and temperature. *J. Soil Contam.* 4(3):299–310.

Wang, T.C., Weissman, J.C., Ramesh, G., Varadarajan, R., and Benemann, J.R. 1995. Bioremoval of toxic elements with aquatic plants and algae. In *Biorem. Inorg. Pap. 3rd Int. In Situ On-Site Bioreclam. Symp.* Hinchee, R.E., Means, J.L., and Burris, D.R., Eds. pp. 65–69.

Wang, X. and Bartha, R. 1990. *Soil Biol. Biochem.* 22(4):501–506.

Wang, X., Yu, X., and Bartha, R. 1990. Effect of bioremediation of polycyclic aromatic hydrocarbon residues in soil. *Environ. Sci. Technol.* 24(7):1086–1089.

Wang, Y.-T., Suidan, M.T., and Rittmann, B.E. 1985. Performance of expanded-bed methanogenic reactor. *J. Environ. Eng., Amer. Soc. Civ. Eng. Div.* 111:460–471.

Wanger, M.R. and Poepel, H.J. 1995. Influence of surfactants on oxygen transfer. *Proc. 68th Water Environ. Fed. Annu. Conf. Expo.* 1:297–306.

Warcup, J.H. 1955. Isolation of fungi from hyphae present in soil. *Nature* (London). 175:953–954.

Warcup, J.H. 1957. Studies on the occurrence and activity of fungi in a wheat-field soil. *Trans. Br. Mycol. Soc.* 40:237–262.
Ward, C.H., Tomson, M.B., Bedient, P.B., and Lee, M.D. 1986. Transport and fate processes in the subsurface. In *Land Treatment: A Hazardous Waste Management Alternative.* Water Resources Symp. Number 13. Loehr, R.C. and Malina, J.F., Jr., Eds. University of Texas, Austin, TX. pp. 19–39.
Ward, D., Atlas, R.M., Boehm, P.D., and Calder, J.A. 1980. Microbial biodegradation and chemical evolution of oil from the Amoco spill. *Ambio.* 9:277–283.
Ward, D.M. and Brock, T.D. 1978. Hydrocarbon biodegration in hypersaline environments. *Appl. Environ. Microbiol.* 35(2):353–359.
Warshawsky, E., Radike, M., Jayasimhulu, K., and Cody, T. 1988. Metabolism of benzo(a)pyrene by a dioxygenase enzyme system of the freshwater green alga *Selenastrum capricornutum. Biochem. Biophys. Res. Commun.* 152:540–544.
Watson, J.S., Scott, C.D., and Faison, B.D. 1990. Evaluation of a cell-biopolymer sorbent for uptake of strontium from dilute solution. *Am. Chem. Soc. Symp. Ser.* 422:173–186.
Watts, J.R. and Corey, J.C. 1981. D.E.-AC09-SR00001, U.S. Department of Energy.
Weber, J.B., Weed, S.B., and Best, J.A. 1969. Displacement of diquat from clay and its phototoxicity. *J. Agric. Food Chem.* 17:1075–1076.
Weber, W.J., Jr. and Corseuil, H.X. 1994. Inoculation of contaminated subsurface soils with enriched indigenous microbes to enhance bioremediation rates. *Water Res.* 28(6):1407–1414.
Webster, J.J., Hampton, G.J., Wilson, J.T., Ghiorse, W.C., and Leach, F.R. 1985. Determination of microbial cell numbers in subsurface samples. *Ground Water.* 23:17–25.
Wehrheim, B. and Wettern, M. 1994. Biosorption of cadmium, copper and lead by isolated mother cell walls and whole cells of *Chlorella fusca. Appl. Microbiol. Biotechnol.* 41:725–728.
Weir, B.A., Sundstrom, D.W., and Klei, H.E. 1987. Destruction of benzene by ultraviolet light-catalyzed oxidation with hydrogen peroxide. *Haz. Waste Haz. Mater.* 4:165–176.
Weir, S.C., Dupuis, S.P., Providenti, M.A., Lee, H., and Trevors, J.T. 1995. Nutrient-enhanced survival of and phenanthrene mineralization by alginate-encapsulated and free *Pseudomonas* sp. UG14Lr cells in creosote-contaminated soil slurries. *Appl. Microbiol. Biotechnol.* 43:946–951.
Weisman, R.J., Falatko, S.M., Kuo, B.P., and Eby, E. 1994. Effectiveness of innovative technologies for treatment of hazardous soil. 94-TP62.04. In *Proc. 87th Annu. Mtg. Exhib. Air Waste Mgmt. Assoc.,* June 19–24, Cincinnati, OH.
Weissenfels, W.D., Beyer, M., and Klein J. 1990. Degradation of phenanthrene, fluorene and fluoranthene by pure bacterial cultures. *Appl. Microbiol. Biotechnol.* 32:479–484.
Weissenfels, W.D., Beyer, M., Klein, J., and Rehm, H.J. 1991. Microbial metabolism of fluoranthene: isolation and identification of ring fission products. *Appl. Microbiol. Biotechnol.* 34:528–535.
Weissenfels, W.D., Klewer, H.-J., and Langhoff, J. 1992a. Sorption of organic pollutants by oil particles: influence on microbial PAH-degradation and potential risk of contaminated sites. *DECHEMA Biotechnol. Conf.* 5(B):1023–1028.
Weissenfels, W.D., Klewer, H.-J., and Langhoff, J. 1992b. Adsorption of polycyclic aromatic hydrocarbons (PAHs) by soil particles: influence on biodegradability and biotoxicity. *Appl. Microbiol. Biotechnol.* 36:689–696.
Welch, J.F., Bateman, B.R., Perkins, R.A., and Roberts, R.M. 1987. Report No. TM 71-87-20. Naval Civil Engineering Laboratory, Port Hueneme, CA.
Wentsel, R.S. et al. 1981. Restoring Hazardous Spill-Damaged Areas: Technique Identification, Assessment. EPA Report No. EPA-600/2-7-81-1-208. U.S. EPA, Cincinnati, OH.
Werner, P. 1982. *Vom Wasser.* 57:157.
Westlake, D.W.S., Jobson, A.M., and Cook, F.D. 1978. *In situ* degradation of oil in a soil of the boreal region of the Northwest Territories. *Can. J. Microbiol.* 24:254–260.
Westlake, D.W.S., Jobson, A., Phillippe, R., and Cook, F.D. 1974. Biodegradability and crude oil composition. *Can. J. Microbiol.* 20:915–928.
Wetzel, D.M. and Reible, D.D. 1982. Report to LSU Hazardous Waste Research Center, Louisiana State University, Baton Rouge, LA.
Wetzel, R.S., Davidson, D.H., Durst, C.M., and Sarno, D.J. 1986. Field demonstration of *in situ* biological treatment of contaminated groundwater and soils. In *Proc. 12th Annu. Res. Symp. on Land Disposal, Remedial Action, Incineration, and Treatment of Haz. Waste,* Apr. 21–23. Sponsored by Environmental Protection Agency, Hazardous Waste Engineering Research Laboratory.
White, D.C. 1983. Analysis of microorganisms in terms of quantity and activity in natural environments. In *Microbes and Their Natural Environment.* Slater, J.H., Whittenbury, R., and Wimpenny, J.W.T., Eds. Cambridge University Press, New York. pp. 37–66.
Whitman, B.E., Mihelcic, J.R., and Lueking, D.R. 1995. Naphthalene biosorption in soil-water systems of low or high sorptive capacity. *Appl. Microbiol. Biotechnol.* 43(3):539–544.
Whitman, W.G. 1923. *Chem. Met. Eng.* 24:147.
Whyte, L.G., Greer, C.W., and Inniss, W.E. 1996. Assessment of the biodegradation potential of psychrotrophic microorganisms. *Can. J. Microbiol.* 42(2):99–106.
Wickerham, L.J. 1951. Taxonomy of yeasts. *U.S. Dept. Agric. Tech. Bull.* 1029:1–55., USDA, Washington, D.C.

Widdowson, M.A., Aelion, C.M., Ray, R.P., and Reeves, H.W. 1995. Soil vapor extraction pilot study at a Piedmont UST site. In In situ *Aeration: Air Sparging, Bioventing, Relat. Rem. Processes. Pap. 3rd Int.* In situ *On-Site Bioreclam. Symp.* Hinchee, R.E., Miller, R.N., and Johnson, P.C., Eds. Battelle Press, Columbus, OH. pp. 455–461.

Widrig, D.L. and Manning, J.F., Jr. 1995. Biodegradation of No. 2 diesel fuel in the vadose zone: a soil column study. *Environ. Toxicol. Chem.* 14(11):1813–1822.

Wiesel, I., Wuebker, S.M., and Rehm, H.J. 1993. Degradation of polycyclic aromatic hydrocarbons by an immobilized mixed bacterial culture. *Appl. Microbiol. Biotechnol.* 39:110–116.

Wiggins, B.A. and Alexander, M. 1986. Role of protozoa in microbial acclimation at low concentrations of organic chemicals. *Abstr. Annu. Mtg. Am. Soc. Microbiol.* p. 300.

Wiggins, B.A. and Alexander, M. 1988. *Can. J. Microbiol.* 34:661–666.

Wilbourn, R.G., Newburn, J.A., and Schofield, J.T. 1994. Treatment of hazardous wastes using the Thermatrix treatment system. In *Therm. Treat. Radioact., Haz. Chem., Mixed, Munitions, Pharm. Wastes, Proc. 13th Int. Incineration Conf.*, University of California, Irvine. pp. 221–223.

Wild, S.R., Berrow, M.L., and Jones, K.C. 1991. The persistence of polynuclear aromatic hydrocarbons (PAHs) in sewage sludge-amended agricultural soils. *Environ. Pollut.* 72:141–157.

Wildung, R.E. and Garland, T.R. 1985. Chapter 4. Microbial development on oil shale wastes: influence on geochemistry. In *Soil Reclamation Processes — Microbiological Analyses and Applications.* Tate, R.L., III and Klein, D.A., Eds. Marcel Dekker, New York. pp. 107–139.

Wilkinson, R.R., Kelso, G.L., and Hopkins, F.C. 1978. State of the Art Report: Pesticide Disposal Research. EPA Report No. EPA-600/2-78-183. U.S. EPA, Cincinnati, OH.

Williams, D.E. and Wilder, D.G. 1971. Gasoline pollution of a ground-water reservoir — a case history. *Ground Water.* 9:50–56.

Williams, G.R., Cumins, E., Gardener, A.C., Palmier, M., and Rubidge, T. 1981. The growth of *Pseudomonas putida* in AVTUR aviation turbine fuel. *J. Appl. Bactiol.* 50:551–557.

Williams, P.A. 1978. *Dev. Biodegradation Hydrocarbons.* 1:85–112.

Williams, R.J.P. 1981. Physicochemical aspects of inorganic element transfer through membranes. *Philos. Trans. R. Soc. Lond. Ser. B Biol. Sci.* 294:57–74.

Williams, R.J.P. 1983. In *Dahlem Konferenzen on Changing Biogeochemical Cycles of Metals and Human Health,* Berlin, F.R.G., March 20–25.

Willson, G.B., Sikora, L.J., and Parr, J.F. 1983. Composting of chemical industrial wastes prior to land application. In *Land Treatment of Hazardous Wastes.* Parr, J.F., Marsh, P.B., and Kla, J.M., Eds. Noyes Data Corp., Park Ridge, NJ. pp. 263–273.

Wilson, B.H., Bledsoe, B.E., Armstrong, J.M., and Sammons, J.H. 1986. Biological fate of hydrocarbons at an aviation gasoline spill site. In *Proc. Petroleum Hydrocarbons and Organic Chemicals in Ground Water: Prevention, Detection and Restoration Conf. and Expo.,* Nov. 12–14, Houston, TX. Presented by National Water Well Association and American Petroleum Institute.

Wilson, B.H. and Rees, J.F. 1985. In *Proc. Biotransformation of Gasoline Hydrocarbons in Methanogenic Aquifer Material. Proc. NWWA/API Conf. on Petroleum Hydrocarbons and Organic Chemicals in Groundwater,* Nov. 13–15. Houston, TX. pp. 128–141.

Wilson, D.J. 1994. Soil clean up by in-situ aeration. XIII. Effects of solution rates and diffusion in mass-transport-limited operation. *Sep. Sci. Technol.* 29(5):579–600.

Wilson, J.T., Enfield, C.G., Dunlap, W.J., Cosby, R.L., Foster, D.A., and Baskin, L.B. 1981. Transport and fate of selected organic pollutants in a sandy soil. *J. Environ. Qual.* 10:501–506.

Wilson, J.T., Leach, L.E., Henson, M., and Jones, J.N. 1986. *In situ* biorestoration as a groundwater remediation technique. *Ground Water Monitor. Rev.* 56–64.

Wilson, J.T. and McNabb, J.F. 1983. Biological transformation of organic pollutants in groundwater. *Eos.* 64:505–507.

Wilson, J.T., McNabb, J.F., Balkwill, D.L., and Ghiorse, W.C. 1983. Enumeration and characterization of bacteria indigenous to a shallow water table aquifer. *Ground Water.* 21:134–142.

Wilson, J.T., McNabb, J.F., Wilson, B.H., and Noonan, M.J. 1983. Biotransformation of selected organic pollutants in ground water. *Dev. Ind. Microbiol.* 24:225–233.

Wilson, J.T. and Wilson, B.H. 1985. Biotransformation of trichloroethylene in soil. *Appl. Environ. Microbiol.* 49:242–243.

Wilson, L.P., Durant, N.D., and Bouwer, E.J. 1995. Aromatic hydrocarbon biotransformation under mixed oxygen/nitrate electron acceptor conditions. In *Microbial Processes for Bioremediation. Pap. 3rd Int.* In Situ *On-Site Bioreclam. Symp.* Hinchee, R.E., Vogel, C.M., and Brockman, F.J., Eds. Battelle Press, Columbus, OH. pp. 137–143.

Wilson, M.A., Saberiyan, A.G., Andrilenas, J.S., Miller, R.S., Esler, C.T., Kise, G.H., and DeSantis, P. 1994. Bioremediation of waste oil-contaminated gravels via slurry reactor technology. In *Hydrocarbon Bioremediation.* Hinchee, R.E., Alleman, B.C., Hoeppel, R.E., and Miller, R.N., Eds. CRC Press, Boca Raton, FL. pp. 383–387.

Wilson, M.S. and Madsen, E.L. 1996. Field extraction of a transient intermediary metabolite indicative of real time *in situ* naphthalene biodegradation. *Environ. Sci. Technol.* 30(6):2099–2103.

Wilson, N.G. and Bradley, G. 1996. Enhanced degradation of petrol (Slovene diesel) in an aqueous system by immobilized *Pseudomonas fluorescens. J. Appl. Bacteriol.* 80(1):99–104.

Wilson, S.C. and Jones, K.C. 1993. Bioremediation of soil contaminated with polynuclear aromatic hydrocarbons (PAH)s: a review. *Environ. Pollut.* 81:229–249.

Winding, A. 1994. Fingerprinting bacterial soil communities using Biolog microtitre plates. In *Beyond the Biomass*. Ritz, K., Dighton, J., and Giller, K.E., Eds. British Society of Soil Science. Wiley-Sayce Publication.

Winters, K., O'Donnell, R., Batterton, J.C., and Van Baalen, C. 1976. Water-soluble components of four fuel oils: chemical characterization and effects on growth of microalgae. *Mar. Biol.* 36:269.

Wiseman, A., Lim, T.-K., and Woods, L.F.J. 1978. Regulation of the biosynthesis of cytochrome P-450 in Brewer's yeast. *Biochim. Biophys. Acta.* 544:615–623.

Wislocki, P.G., Kapitulnik, J., Levin, W., Lehr, R.F., Schaefer-Ridder, M., Karle, J.M., Jerina, D.M., and Conney, A.H. 1978. Exceptional carcinogenic activity of benz(a)anthracene 3,4-dihydrodiol in the newborn mouse and the Bay Region Theory. *Cancer Res.* 38:693–696.

Wnorowski, A.U. 1991. Selection of bacterial and fungal strains for bioaccumulation of heavy metals from aqueous solutions. *Water Sci. Technol.* 23:309–318.

Wodzinski, R.S. and Coyle, J.E. 1974. Physical state of phenanthrene for utilization by bacteria. *Appl. Microbiol.* 27:1081–1084.

Wodzinski, R.S. and Johnson, M.J. 1968. Yields of bacterial cells from hydrocarbons [sic]. *Appl. Microbiol.* 16:1886–1891.

Woese, C.R. 1987. Bacterial evolution. *Microbiol. Rev.* 51:221–271.

Wolinski, W.K. and Bruce, A.M. 1984a. Doc. Eur. Abwassar Abfallsymp. (Ger.). p. 385.

Wolinski, W. and Bruce, A.M. 1984b. Comm. Eur. Communities EUR9129 (G.B.). p. 110.

Wong, A.S. and Crosby, D.G. 1978. In *Pentachlorophenol*. Ranga Rao, K., Ed. Plenum Press, New York. pp. 19–25.

Wong, P.T.S., Chao, Y.K., Luxon, P.L., and Silverberg, B. 1975. Methylation of lead and selenium in the environment. Int. Conf. Heavy Metals in the Environ., Toronto. C220–C221.

Wood, J.M. 1982. Chlorinated hydrocarbons: oxidation in the biosphere. *Environ. Sci. Technol.* 16:291A–297A.

Wood, J.M. 1983. Selected biochemical reactions of environmental significance. *Chem. Sci.* 21:155–160.

Wood, J.M., Cheh, A., Dizikes, L.J., Ridley, W.P., Rackow, S., and Lakowicz, J.R. 1978. Mechanisms for the biomethylation of metals and metalloids. *Fed. Proc.* 37:16–21.

Wood, J.M. and Wang, H.-K. 1983. Microbial resistance to heavy metals. *Environ. Sci. Technol.* 17:582A–590A.

Woods, S., Williamson, K., Strand, S., Ryan, K., Polonsky, J., Ely, R., Gardner, K. and Defarges, P. 1987. Use of submerged aerobic biofilm processes for degradation of halogenated organic compounds. Preprints of papers presented at the 194th ACS Natl. Mtg. Dept. Civil Eng., Oregon State University, Corvallis, and College of Forest Res., University of Washington, Seattle. pp. 41–44.

Woodward, R.L. 1963. Review of the bactericidal effectiveness of silver. *J. Am. Water Works Assoc.* 55:881–888.

Woolson, E.A. 1977. Fate of arsenicals in different environmental substrates. *Environ. Health Perspect.* 19:73–81.

Wrenn, B.A. and Venosa, A.D. 1996. Selective enumeration of aromatic and aliphatic hydrocarbon degrading bacteria by a most-probable-number procedure. *Can. J. Microbiol.* 42:252–258.

Wuerdemann, H., Wittmaier, M., Rinkel, U., and Hanert, H.H. 1994. A simple method for determining deficiency of oxygen during soil remediation. In *Hydrocarbon Bioremediation*. Hinchee, R.E., Alleman, B.C., Hoeppel, R.E., and Miller, R.N., Eds. CRC Press, Boca Raton, FL. pp. 454–485.

Y

Yadav, J.S. and Reddy, C.A. 1993. Mineralization of 2,4-dichlorphenoxyacetic acid (2,4-D) and mixtures of 2,4-D and 2,4,5-trichlorophenoxyacetic acid by *Phanerochaete chrysosporium*. *Appl. Environ. Microbiol.* 59:2904–2908.

Yagi, O. and Sudo, R. 1980. Degradation of polychlorinated biphenyls by microorganisms. *J. Water Pollut. Control Fed.* 52:1035–1043.

Yakimchuck, B.M., Zhurba, M.G., Prikhod'ko, V.P., and Shevchuk, B.I. 1985. Tertiary treatment of waste water in a hydroautomatic biofilter-filter with a floating media. *Sov. J. Water Chem. Technol.* 7:99–104.

Yakimov, M.M., Fredrickson, H.L., and Timmis, K.N. 1996. Effect of heterogeneity of hydrophobic moieties on surface activity of lichenysin A, a lipopeptide biosurfactant from *Bacillus licheniformis* BAS50. *Biotechnol. Appl. Biochem.* 23(1):13–18.

Yakimov, M.M., Timmis, K.N., Wray, V., and Fredrickson, H.L. 1995. Characterization of a new lipopeptide surfactant produced by thermotolerant and halotolerant subsurface *Bacillus licheniformis* BAS50. *Appl. Environ. Microbiol.* 61(5):1706–1713.

Yalkowsky, S.H., Valvani, S.C., and Mackay, D. 1983. Estimation of the aqueous solubility of some aromatic compounds. *Residue Rev.* 85:43–55.

Yang, G.C.C. and Ku, Y.-C. 1994. Remediation of a spiked, oil-contaminated soil by a thermal process. *Haz. Ind. Wastes.* 26th. pp. 446–453.

Yang, I. C.-Y., Li, Y., Park, J.K., and Yen, T.F. 1994. Subsurface application of slime-forming bacteria in soil matrices. In *Applied Biotechnology for Site Remediation*. Hinchee, R.E., Anderson, D.B., Metting, F.B., Jr., and Sayles, G.D., Eds. Lewis Publishers, Boca Raton, FL. pp. 268–274.

Yang, W.-C., Horvath, B., Hontelez, J., van Kammen, A., and Bisseling, T. 1991. *In situ* localisation of *Rhizobium* mRNAs in pea root nodules: *nif* A and *nif* H localisation. *Mol. Plant Microb. Interact.* 4:464–468.

Yaniga, P.M. and Smith, W. 1984. Aquifer restoraton via accelerated *in situ* biodegradation of organic contaminants. In *Proc. NWWA/API Conf. on Petroleum Hydrocarbons and Organic Chemicals in Ground Water — Prevention, Detection and Restoration,* Houston, TX. National Water Well Association, Worthington, OH. pp. 451–472.

Yare, B.S. 1991. A comparison of soil-phase and slurry-phase bioremediation of PNA-containing soils. In *On-Site Bioreclamation: Processes for Xenobiotic and Hydrocarbon Treatment.* Hinchee, R.E. and Olfenbuttel, R.F., Eds. Butterworth-Heinemann, Stoneham, MA. pp. 173–187.

Yau, C.H. 1985. *Water Pollut. Control.* 84:93.

Yeung, K.-H., Schell, A.M., and Hartel, P.G. 1989. Growth of genetically engineered *Pseudomonas aeruginosa* and *Pseudomonas putida* in soil and rhizosphere. *Appl. Environ. Microbiol.* 55:3243–3246.

Yezzi, J.J., Jr., Brugger, J.E., Wilder, I., Freestone, F., Miller, R.A., Pfrommer, C., Jr., and Lovell, R. 1984. The EPA-ORD mobile incineration system trial burn. In *Proc. 1984 Haz. Mater. Spills Conf.,* April 9–12, Nashville, TN. Ludwigson, J., Ed. Gov. Inst., Inc., Rockville, MD. pp. 80–91.

Ying, A. et al. 1990. Bioremediation of heavy petroleum oil in soil at railroad maintenance yard. In *Haz. Waste Treatment: Treatment of Contam. Soils. A&WMA Int. Symp. Proc.,* Cincinnati, OH. pp. 134–147.

Yong, R.N., Tousignant, L.P., Leduc, R., and Chan, E.C.S. 1991. Disappearance of PAHs in a contaminated soil from Mascouche, Quebec. In *In Situ Bioreclamation. Applications and Investigations for Hydrocarbon and Contaminated Site Remediation.* Hinchee, R.E. and Olfenbuttel, R.F., Eds. Butterworth-Heinemann, Stoneham, MA. pp. 377–395.

Young, D.A. 1985. Biodegradation of Waste Coolant Fluid. Final Report, Bendix Corp. BDX-613–3174. NTIS DE85005793.

Young, L.Y. 1984. Anaerobic degradation of aromatic compounds. In *Microbial Degradation of Organic Compounds.* Gibson, D.T., Ed. Marcel Dekker, New York. pp. 487–523.

Young, L.Y. and Bossert, I. 1984. Degradation of chlorinated and nonchlorinated phenols and aromatic acids by anaerobic microbial communities. *Abstr. Annu. Mtg. Am. Soc. Microbiol.* p. 217.

Yudelson, J.M. and Tinari, P.D. 1995. Economics of biofiltration for remediation projects. In *Biological Unit Processes for Hazardous Waste Treatment. Pap. 3rd Int. In Situ On-Site Bioreclam. Symp.* Hinchee, R.E., Skeen, R.S., and Sayles, G.D., Eds. Battelle Press, Columbus, OH. pp. 205–241.

Z

Zagury, G.J., Narasiah, K.S., and Tyagi, R.D. 1994. Adaptation of indigenous iron-oxidizing bacteria for bioleaching of heavy metals in contaminated soils. *Environ. Technol.* 15:517–530.

Zaidi, B.R., Murakami, Y., and Alexander, M. 1989. *Environ. Sci. Technol.* 23:859–863.

Zaidi, B.R., Stucki, G., and Alexander, M. 1986. Inoculation of lake water to promote biodegradation. *Abstr. Annu. Mtg. Am. Soc. Microbiol.* p. 286.

Zajic, J.E. 1964. Chapter 2. Biochemical reactions in hydrocarbon metabolism. *Dev. Ind. Microbiol.* 6:16–27.

Zajic, J.E. 1969. *Microbial Biogeochemistry.* Academic Press, New York. 345 pp.

Zajic, J.E. and Daugulis, A.J. 1975. Selective enrichment processes in resolving hydrocarbon pollution problems. In *Impact of the Use of Microorganisms on the Aquatic Environment.* Bourquin, A.W., Ahearn, D.G., and Meyers, S.P., Eds. EPA Report No. EPA-600/3-75-001. Environmental Protection Agency, Corvallis, OR. p. 169.

Zajic, J.E. and Gerson, D.F. 1977. *Am. Chem. Soc. Symp. Oil Sands and Oil Shale.* 22:195.

Zajic, J.E. and Knettig, E. 1971. Flocculants from paraffinic hydrocarbons. *Dev. Ind. Microbiol.* 12:87–98.

Zajic, J.E. and Mahomedy, A.Y. 1984. Biosynthesis of surface active agents. In *Petroleum Microbiology.* Atlas, R.M., Ed. Macmillan, New York. pp. 221–297.

Zajic, J.E. and Panchal, C.J. 1976 (in 1977 vol). Bio-emulsifiers. *CRC Crit. Rev. Microbiol.* 5:39–66.

Zajic, J.E. and Seffens, W. 1984. Biosurfactants. *CRC Crit. Rev. Microbiol.* 5:87–107.

Zajic, J.E., Supplisson, B., and Volesky, B. 1974. Bacteria degradation and emulsification of No. 6 fuel oil. *Environ. Sci. Technol.* 8:664–668.

Zehnder, A.J.B., Huser, B.A., Brock. T.D., and Wuhrmann, K. 1980. Characterization of an acetate-decarboxylating, non-hydrogen-oxidizing methane bacterium. *Arch. Microbiol.* 124:1–11.

Zehnder, A.J.B. and Strumm, W. 1988. Geochemistry and biogeochemistry of anaerobic habitats. In *Biology of Anaerobic Microorganisms.* Zehnder, A.J.B., Ed. John Wiley & Sons, New York.

Zeikus, J.G. 1977. The biology of methanogenic bacteria. *Bacteriol. Rev.* 41:514–541.

Zeikus, J.G. 1980. Chemical and fuel production by anaerobic bacteria. *Annu. Rev. Microbiol.* 34:423–464.

Zeng, A.-P. 1995. Effect of CO_2 absorption on the measurement of CO_2 evolution rate in aerobic and anaerobic continuous cultures. *Appl. Microbiol. Biotechnol.* 42:688–691.

Zeyer, J., Kuhn, E.P., and Schwarzenback, R.P. 1986. Rapid microbial mineralization of toluene and 1,3–dimethylbenzene in the absence of molecular oxygen. *Appl. Environ. Microbiol.* 52:944–947.

Zhang, Y. and Miller, R.M. 1995. Effect of rhamnolipid (biosurfactant) structure on solubilization and biodegradation of n-alkanes. *Appl. Environ. Microbiol.* 61(6):2247–2251.

REFERENCES

Zhao, Q. and Wang, B. 1996. Evaluation on a pilot-scale attached-growth pond system treating domestic wastewater. *Water Res.* 30(1):242–245.

Zhou, E. and Crawford, R.L. 1995. Effects of oxygen, nitrogen, and temperature on gasoline biodegradation in soil. *Biodegradation.* 6:127–140.

Zhou, J.L. and Kiff, R.J. 1991. The uptake of copper from aqueous solution by immobilized fungal biomass. *J. Chem. Technol. Biotechnol.* 52:317–330.

Zilber, I.K., Rosenberg, E., and Gutnick, D. 1980. Incorporation of ^{32}P and growth of pseudomonad UP-2 on *n*-tetracosane. *Appl. Environ. Microbiol.* 40:1086–1093.

Zimmermann, R., Iturriaga, R., and Becker-Birk, J. 1978. Simultaneous determination of the total number of aquatic bacteria and the number thereof involved in respiration. *Appl. Environ. Microbiol.* 36:926–935.

Zitrides, T.G. 1978. *Pesticide Disposal Research and Development Symposium,* Sept. 6–7, Reston, VA. 23:1082–1089.

Zitrides, T. 1983. Biodecontamination of spill sites. *Pollut. Eng.* 15:25–27.

ZoBell, C.E. 1946. Action of microorganisms on hydrocarbons. *Bacteriol. Rev.* 10:1–49.

ZoBell, C.E. 1950. *Adv. Enzymol.* 10:443–486.

ZoBell, C.E. 1973. Microbial-facilitated degradation of oil: a prospectus. In *The Microbial Degradation of Oil Pollutants,* Workshop, Atlanta, GA, Dec. 4–6, 1972. Ahearn, D.G. and Meyers, S.P., Eds. Pub. No. LSU-SG-73-01. Center for Wetland Resources, Louisiana State University, Baton Rouge, LA.

Zvyagintsev, D.G. 1994. Vertical distribution of microbial communities in soils. In *Beyond the Biomass.* Ritz, K., Dighton, J., and Giller, K.E., Eds. British Soil Science Society. Wiley-Sayce Publication.

Zweifelhofer, H.P. 1985. Aerobic-thermophilic/anaerobic-mesophilic two-stage sewage sludge treatment: practical experiences in Switzerland. *Conserv. Recycl.* 8:285–301.

Zwick, T.C., Leeson, A., Hinchee, R.E., Hoeppel, R.E., and Bowling, L. 1995. Soil moisture effects during bioventing in fuel-contamnated arid soils. In In Situ *Aeration: Air Sparging, Bioventing, Relat. Rem. Processes. Pap. 3rd Int. In Situ On-Site Bioreclam. Symp.* Hinchee, R.E., Miller, R.N., and Johnson, P.C., Eds. Battelle Press, Columbus, OH. pp. 333–340.

Zytner, R.G., Bhat, N., Rahme, Z., Secker, L., and Stiver, W.H. 1995. The use of supercritical CO_2 to remediate soil. In *Innov. Technol. Site Rem. Haz. Waste Mgmt. Proc. Natl. Conf.* pp. 197–203.

Zytner, R.G., Biswas, N., and Bewtra, J.K. 1993. Retention capacity of dry soils for NAPLs. Environ. Tech. 14(11):1073–1080.

INDEX

A

ABFs (activated biofilters), see Fixed-film systems, activated biofilters
Abiotic degradation, 405, 413
　of PAHs, 53, 141
Abiotic losses, 383
　of PAHs in landtreatment, 40
Abiotic processes, 1
　evaporation, 317
　vs. biotic, 413–414
Abiotic stress, 285
Absorption, 203–207; see also Adsorption; Sorption
　of VOCs, 372–376
Absorption column, 374, 375
Acclimated bacteria, 281–283
　addition to aquifer, 85
　degradation of homologous series of molecules, 282
　evaluation of applicability of, 283
Acclimated microorganisms, 281–283
　in bioslurry reactors, 49
　metal leaching by, 182
　modification of environment for, 285
　for NAPLs, 163
　PAH degradation by, 142
　in return activated sludge (RAS), 56
Acclimation
　causes of, 281
　　under denitrifying conditions, 169, 170
　effect on degradation rates, 282
　effect on lag time, 281
　effect of trace concentrations on, 214
　by enrichment culturing, 387
　to PAHs, 144
　site specificity of, 283
　to subthreshold levels, 217
　threshold concentration for, 214
　for PAH degradation, 282
Acclimation period, 214
　anaerobic, 167, 247
　in BAC systems, 64
　effect of protozoan grazing on, 276, 277
　effect of temperature on, 227
　in NAPLs, 162
Accumulation of heavy metals, microbial, see Heavy metal binding/bioaccumulation/biosorption
Accumulation
　of metabolites, 195, 285
　of refractory compounds, landtreatment, 37
　of transformation products, 294
Acetate-utilizing methanogens, 173
Acetogens, 70
Acetone, degradation in anaerobic reactor, 74
ACGIH (American Conference of Governmental Industrial Hygienists), 322
Acinetobacter, 2
　bioaccumulation by, 275
　cytoplasmic inclusions in, 300
Acinetobacter lwoffi, in fixed-film systems, 58
Acinetobacter spp., citronellol degradation pathway for branched hydrocarbons in, 126
Acrasiales, 276

Acridine orange direct counts (AODC), 385
Acrylonitrile-contaminated soil, treatment of by hydrolysis, 80
Actinomycetes, 2, 233
　in composting, 44
　as oligotrophs, 267
　optimum pH for, 230
　in thermophilic range, 227
Activated biofilters (ABFs), 58, 65
Activated carbon
　acidic/basic, 19
　batch reactors, 20
　biofilter packing material, 376
　for cell encapsulation, 290
　column reactors, 20
　effluent concentrations with, 21
　for enrichment culturing, 389
　for fluidized-bed reactors, 20, 60
　granular, 19
　hydrophilic/hydrophobic, 19
　for immobilized cells in fixed-film systems, 65
　influent concentrations with, 21
　for leachate polishing, 83
　oleophilic/liphobic, 19
　for organics/heavy metal removal, 19, 20
　powdered, 19
　selectivity of, 19
　thermal regeneration of, 19
　for VOC removal, 343, 350
Activated carbon adsorption capacity, 351
Activated carbon filters/bioremediation (mutant bacteria), 421
Activated carbon reactors/systems, 20
　isotherms for adsorption capacity of, 20
　loading capacity and clogging with, 20
Activated carbon leachate treatment/biological pretreatment, 421
Activated charcoal for preservation, 290
Activated sludge/air stripping, 56
Activated sludge processes, 17, 56
　anaerobic, 72
　biaugmentation of, 54
　capability of microorganisms in, 54
　food-to-microorganism ratio, 56
　with granular activated carbon, 421
　with immobilized bacteria, 58
　for leachate treatment, 54
Activated sludge semicontinuous reactors, 56
Activation, 193
Active soil vapor extraction, 89
Adaptation, 281
　effect on biodegradation rates, 282
　to heavy metals, 178, 182
　site specificity of, 283
Adaptation period, 278
　effect of concentration on, 213
Adapted bacteria, 281–283
Additives for soil moisture control, 226
　with blade-mixing reactors, 50
Additivity Index, 212
Adsee, 799, 305
Adsorbents for chemical recovery, 376
Adsorber breakthrough, 333
Adsorption

503

with brown coal, 21
  coefficient, 209
  with resin, 20
Adsorption/air stripping, 28, 422
Adsorption by microorganisms in reactors, 67
Adsorption capacity, 20, 351
Adsorption isotherm, 205
Adsorption sites, soil, 327
Adsorption, soil, 203–207, 329; see also Absorption; Sorption
  effect on diffusion, 203
  in landtreatment, 30
  of microbes to soil, 287, 288
  of PAHs, 141
  van der Waals forces in, 204
  of VOCs, 329
Adsorption tendency, 206
Advection of contaminants, 203
Aeration
  devices, 346–349
  enclosed mechanical, 12
  forced, 245
  lagoons, 245
Aeration coefficient with surfactants, 305
Aeration ponds/air stripping, 344
Aeration, soil, 331
  by bioventing, 245
  in bioreactors, 246
  commercial oxygen delivery approaches, 246–247
  in composting, 44
  by deep soil fracture bioinjection, 246
  by Detoxifier™, 247
  effect on volatilization rate, 333
  by electrobioreclamation, 244
  by hydraulic/pneumatic fracturing, 244
  pretreatment with blade-mixing reactors for, 50
  secondary effects of, 349
Aerobic/anaerobic
  biofilm reactor, 76
  conditions, 299
  treatment, 67–68, 76–78, 169, 248
Aerobic bacteria, 2, 264–265
Aerobic degradation, 123–165
  microorganisms, 264
  oxidation reduction reactions in, 264
Aerobic polishing, 76
Aerobic reactors/microbial adsorption, 67
  by photosynthetic organisms, 68
  for recalcitrant compounds, 67
Aerobiosis, 232
Aeromonas hydrophila, in electrobioreclamation, 111
AFCEE (U.S. Air Force Center for Excellence Technology Transfer Division Bioslurper Initiative), 106
Afterburners, 365
Agar for cell immobilization, 58
Agarose for cell encapsulation, 290
Agar plate overlay technique, 387
Aggregated soil
  in anaerobic reactors, 72
  stability with amendments, 47
  water retention characteristics, 47
Aggregation of microorganisms
  bacterial adhesion to soil, 287
  effect on colony counts, 384
Aging process, 204
Air biofilter, 376
Air bubble slip velocity, 305

Air-filled porosity, 233, 325, 328, 331
Airflow rates, bioventing, 104
Air injection, 242
Air injection/soil venting/off-gas treatment, 423
Air lift bioreactors, 57
Air lift pump, 347
Air pollution control equipment for air stripping, 346
Air pollution control technologies for leachate treatment, 15
Air pollution permit, 380
Air sparging, 93
  dissolved oxygen levels with, 237, 245
  for soil aeration, 245
Air sparging/soil vapor extraction, 91, 422
Air sparging/vacuum recovery system, 93
Air/steam stripping in Detoxifier™, 93
Air stripping, 87–88
  air-to-water ratios with 345, 346
  costs, 343
  improving stripping rate, 344
  of leachates, 28
  of leachate treatment by-products, 18
  limitations, 88, 415
  mass transfer equation, 344
  metals removal prior to, 344–345
  moisture content for, 327
  off-gas treatment from, 346
  optimization of, 345
  packing material, 345
  preheating air, 88
  pretreatment for adsorption, 28
  pretreatment for organic liquids with VOCs, 344
  temperature, 345
  units, 344
  with UV/hydrogen peroxide, 346
Air stripping/adsorption, 422
Air stripping airstream treatment
  by fume incineration, 87
  by vapor-phase carbon adsorption, 87
Air stripping/carbon adsorption, 346, 358
Air stripping/carbon adsorption/ion exchange, 426
Air stripping/clarification/granular activated carbon/reinjection/bioremediation, 421
Air stripping evaluation, 345
Air stripping of groundwater
  applicability, 8
  biofilter treatment of vapors from, 379
Air stripping process equipment, 88
Air stripping process, selection of, 344
Air stripping/steam stripping, 422
Air stripping of VOCs, 88, 344–350, 359
  countercurrent flow packed columns for, 345
  equations for predicting, 345
  Henry's law constant for, 345
Air-supported structures, 335–340
Air-to-water ratios with air stripping, 345, 346
Air-water flushing combined, 109–110
  for soil aeration, 244
Air/water partitioning
  Henry's law constant, 330
Air-water separator/trickling filter/biofilter, 422
Albedo effect on temperature, 228
Alcaligenes, for biopolymer shields, 112
  eutrophus (ATCC 17699)
  faecalis (ATCC 49677)
  viscolactis (ATCC 21698)
Alcohol ethoxylate surfactant, biodegradation of, 313

INDEX

Alcohols
    aerobic degradation of, 159
    anaerobic degradation of, 178
    resin adsorption of, 20
Alconox for soil washing, 11
Aldehydes, 20
Alfalfa
    meal, 259
    for PAH degradation, 150, 155
    pellets, 149
Alfonic 810–60, 303
    for NAPLs, 163, 164
Algae, 2, 190
    cadmium adsorption by, 186
    for heavy metal removal, 180, 183
    for lagoons, 57
    lead toxicity to, 188
    PAH degradation by, 147, 150, 151
    phototrophs, 274
Algae/bacteria degradation, 275
Algal/bacterial mats for PAH degradation, 275
Algal resistance to nickel
    transport systems for, 190
Alginates, for cell encapsulation/immobilization, 290, 291
    in fixed-film systems, 58, 65
Alicyclic hydrocarbons
    anaerobic degradation of, 178
    biodegradability of, 131
    cometabolism of, 159–160
Aliphatics
    diterminal oxidation of, 129
    NAPLs, 162
Alkaline chlorination for leachate treatment, 16
Alkane degradation, 128–130
    branched and cyclic, 130
    effect of branching on, 155
    by landfarming, 37
    microorganisms in, 127
Alkane dehydrogenase, 168
n-Alkane:isoprenoid ratio, 401
Alkanes, branched, biodegradation of, 130, 155–156
    chain length accommodation, 130
    cyclic, biodegradation of, 130
    diterminal oxidation of, 129
    half-life in soil, 39
    straight-chain, biodegradation of, 156, 158
    transport through cell membrane, 130
$alk\,B$ (C_6 to C_{12} n-paraffin degradation) gene probe, 284
Alkenes, biodegradation of, 131, 158
Alkenes, branched-chain, biodegradation of, 155–156
Alkylbenzenes
    aerobic degradation of, 140
    anaerobic degradation of, 176
    denitrification of, 176
Alkylbenzene sulfonates
    for soil flushing, 83, 301
Alkylbetaine surfactants, biodegradation of, 313
Alkyl ether sulfate
    for soil flushing, 83
Alkylnaphthalenes, biodegradation of, 148
Alkyl sulfate surfactants, biodegradation of, 313
Alkyl triethoxy sulfate biosurfactants, biodegradation of, 313
Allied Signal Immobilized Cell Bioreactor (ICB), 65
Allochthonous microbial population, 280
Alternate electron acceptors, 166, 168, 241
Alternating aerobic/anaerobic conditions, 248

Alternative carbon sources for anaerobiosis, 214, 248
    for denitrification, 170–171
Alternative Treatment Technology Information Center (ATTIC), EPA, 5
Alum, 27
Amberlite beads in tubular reactors, 49
Ambersorb 563 regeneration, 19
Amendments, inorganic, in bioreactors, 47
Amendments, organic, 258, 259
    in bioreactors, 47
American Conference of Governmental Industrial Hygienists (ACGIH), 322
Amines for resin and carbon adsorption, 20
Ammonia, gaseous, 255, 424
Amoeba, 276, 396
Amphibolic intermediates, 131
Amphotericin B, 387
Amphoteric surfactants, 301
    biodegradation of, 313
Anaerobic activated carbon filter, 77
Anaerobic activated sludge process, 72
Anaerobic/aerobic biofilm reactor, 76
Anaerobic/aerobic treatment, 67–68, 76–78, 169, 248–249
Anaerobic bacteria, 2
    aggregate formation in reactors, 72
    anaerobic consortia, 166, 196, 265–266; see also Microbial consortia; Mixed cultures
    biofilm formation in reactors, 72
    denitrifiers, 166
    effect of contaminant concentration on, 167
    methanogens, 166, 167
    proportion of bacterial population, 165
    redox potential range for, 265
    sulfidogens, 166
Anaerobic baffled reactor, 74
Anaerobic bioconversion process, 70–71
Anaerobic biodegradation, 165–178
    acclimation period for, 167, 247
    advantages of, 166, 248
    alternate electron acceptors for, 166, 168
    anaerobe competition in, 167
    in bioreactors, 47
    by-products as indicator of, 403
    cometabolism in, 167, 294
    degradable compounds by, 167
    effect of redox potential on, 166
    electron acceptor for, 166
    end products of, 167
    of fatty acids, 168, 178, 193
    $in\ situ$ monitoring, 168
    iron-sulfide precipitation in, 168
    methane as indicator of, 168
    microorganisms in, 166, 167, 196, 247, 249, 265–266
    monitoring, 403; see also Bioremediation monitoring methods
    of mononuclear aromatic hydrocarbons, 173–177
    oxygen level for denitrification, 166
    of PAHs, 167
    Para-Bac enhancement of, 287
    of petroleum, 168
    of resistant compounds, 167
    sequence of reduction in, 166
    synergism in, 167
    of xenobiotic compounds, 166
Anaerobic conditions, see Anaerobiosis
Anaerobic consortia, 166, 167, 196, 247, 265–266

interaction of trophic groups, 266
  for methanogenesis, 172–173
Anaerobic corrosion, 416
Anaerobic decomposition, 166
Anaerobic degradation pathway for aromatics, 167
Anaerobic digester, conventional, 71–72
Anaerobic downflow stationary fixed-film reactor, 74
Anaerobic filter system, 74–75
Anaerobic filtration, 54, 74
  for leachate by-products, 17
Anaerobic fixed-film systems, 74–76
  baffled reactors, 74
  Desulfovibrio vulgaris in, 74
  expanded bed, 74
  expanded-bed anaerobic GAC reactor, 77
  expanded/fluidized bed, 75–76
  fluidized-bed, 74, 75, 61
  upflow fixed-bed systems, 74
Anaerobic lagoons, 54, 73
Anaerobic microsites, 248, 260
Anaerobic photometabolism, by phototrophic purple nonsulfur bacteria, 173
Anaerobic reactors, 71
  bioaugmentation in, 72
  biocatalysts in, 72
  rate-limiting factor in, 173
  retention times in, 70
Anaerobic respiration, 166, 168–173
Anaerobic systems, 70–78
Anerobic transformations, 166
Anaerobic upflow sludge blanket (UASB), 70, 72–73
Anaerobic wastewater treatment systems, 70
Anaerobiosis, 2, 166, 232, 260
  alternate carbon sources for, 248
  in bioslurry reactors, 47, 48
  in composting, 44
  creating, 247–248
  in landfarming, 34, 36, 38
  malodorous compounds, 233
  with montmorillonite, 262
Analog adaptation, 282
Analog enrichment, 292–297
  inducer molecules, 293
  PAH degradation by, 151, 152, 154, 155
  treatability studies, 293
  victim substrate, 293
Analog enrichment/microorganisms, 423
Anionic detergent, optimum chain length, 302
Anionic sulfonated alkyl ester, 85
Anionic surfactants, 301
  oxygen transfer with, 305
  precipitation of, 307
Anoxic conditions, 166
Antagonism, 285
Antagonistic effects of contamination, 212
Antarctic mineral soils, microbial counts in, 396
Anthracene biodegradation, 151–154
  composting of, 43
  degradation pathway for, 131
  effect of starting concentration on, 152
  emissions in bioslurry reactors, 49
  enzymes in, 152
  in fixed-film systems, 65
  half-life, 152
  by mixed culture, 152
  by *Trichoderma harzianum*, 65
  turnover time for, 412
Anthracene biomarker, 400

Anthracene-contaminated soil, treatment
  by composting, 45
  by supercritical fluid extraction, 13
Anthracene soil transport with cyclodextrin, 84
Anthropogenic compounds, 126, 208
  degradation by different microorganisms, 132; see also Aromatic hydrocarbons, biodegradation of; Bioaugmentation microorganisms; Hydrocarbonoclastic microorganisms
Antibiotic-resistant microorganisms, 406
Antibiotics, 299
  Amphotericin B, 387
  as antagonistic agent, 286
  Cooke's aureomycin–rose bengal medium for fungi with, 388
  Fungizone for isolating actinomycetes, 387
  against protozoa, 299
  as surfactants, 311
AODC (acridine orange direct counts), 385
Application frequency in landtreatment, 38
Application rates in landtreatment, 38–39; see also Loading rate
  for aromatics, asphaltics, and saturated hydrocarbons, 38
  effect on biodegradation rates, 39
  pretreatment for higher rates, 39
Aquaculture for VOC removal, 333
Aquastore hydrogel for composting, 45
Aqueous-phase diffusion, 327
Aqueous streams, pretreatment, 342–359
Aqueous vapor pressure in soil, 327
Arctic climates
  bioaugmentation for, 278
Arene oxides
  formation of, 127
  from fungi, 124, 194, 230
Aromatic hydrocarbons, biodegradation of, see also Anthropogenic hydrocarbons; Bioaugmentation microorganisms; Hydocarbonoclastic microorganisms
  anaerobic, 70
  anaerobic, resistance to, 131
  bacterial, 124, 130, 131
  eukaryotic and prokaryotic, 125
  enzyme induction for, 140
  fungal oxidation, 124–125, 130
  with landfarming, 34
  oxidizable compounds, 132
  oxygenases for, 125
  in sequencing batch reactor (SBR), 69
Aromatic metabolism, 131
Aromatics-contaminated soil, treatment of, by landfarming, 34
Aromatic solvents, bioreactor for, 68
Aromatics, resin adsorption of, 20
Aromatics, unsubstituted, degradation by landfarming, 37
Arsenic
  biotransformation of, 185
  microbial resistance to, 185
  oxidation by potassium permanganate, 80
Arsenic-contaminated aqueous stream, treatment of by chemical precipitation, 27
Arsenite oxidation by organic material, 260
Arthrobacter, 2
  attachment to NAPL, 163
  in biofilters, 379
Arthrobacter viscosus, capsular material for cadmium binding, 186
Arthropods, 276
Artifacts in laboratory experiments, 412
Artificial fibrous carriers in waste stabilization ponds, 57
Aspergillus oryzae, immobilized with PUF, 66
Asphaltenes, 400

INDEX

Asphaltenes, biodegradation of
   in lagoons, 57, 160
   by landfarming, 34
Asphaltics, landtreatment of, 38
Asphalt, soil recycle, 13
Assimilatory denitrification, 169
ATA MBR (autothermal membrane reactor), 70
ATF (automatic transmission fluid), 164
Atmospheric reaction rate of specific compounds (log $K_{OH}°$), 314
ATP content of cells, 386
Attached growth ponds (AGP), 57
Attached growth systems, 65
Attachment of microorganisms
   for facilitating biodegradation, 207
   to interface, 201–202
   on NAPLs, 163
   to soil, 207, 287
Attenuation, 261, 257
Attrition scrubber, 11
Attrition scrubbing in bioslurry reactors, 48
Autochthonous microorganisms, 283, 416
Automated counting methods, rapid, 390–391
Automatic transmission fluid (ATF), 164
Autoradiography, 399
Autothermal aerobic membrane reactor (ATA MBR), 70
Autotrophic denitrifying bacteria, 169
Autotrophic modes of metabolism, 265
Autotrophic nitrification, 286
Aviation gasoline-contaminated soil, treatment of, 34
Azo dyes, 273

B

BAC (Biological activated carbon) systems, see Fixed-film systems
Bacillus circulans, for copper removal, 192
Bacillus licheniformis BAS50, lipopeptide surfactant from, 311
Bacillus licheniformis JF-2, surfactin-like lipopeptide from, 249
Bacillus megaterium, accumulation of benzo(a)pyrene in, 151
Bacillus stearothermophilus, in TAD reactor, 56
Bacillus strain, biosurfactant-producing thermophile, 311
Bacteria, see also Aromatic hydrocarons, degradation of; Bioaugmentation microorganisms; Hydrocarbonoclastic microorganisms
   aerobic, 264–265
   anaerobic, 265–266
   as biosorbants, 180
   cell size, 108, 266, 288, 289, 407
   in composting, 44
   for degrading/transforming specific fuel components/hydrocarbons, 132–139, 263–277
   metal-binding, 183
   oligotrophs, 266
Bacteria/algae degradation, 275
Bacterial counts with cometabolism, 293
Bacteriophages, 276
Bacteriostatic effect of volatiles, 207
Baker's yeast for soil sorption, 259
Barium-contaminated soil, treatment of, by stabilization processes, 12
Bark for heavy metal removal, 255
Barometric pressure, 325
BARR (bioanaerobic reduction/reoxidation), 249
Barriers, 99, 326
Basidiomycete, 270
BDAT (best demonstrated available technology), 148
Beads as biocatalyst, 70
   for immobilization, 291
   for microbes with inducible enzymes, 297
   polymer-based chemical adsorbent, 376
Beetles, 276
Beggiatoa in RBCs, 60
Beijerinckia degradation of PAHs, 151, 154
Belt filter press in bioslurry reactors, 48
Beltsville system for composting, 44
Beneficial reuse of contaminated soil, 13
Benz(a)anthracene biodegradation, 151
Benzene, anaerobic degradation of, 174
   biodegradation of, 139
   emissions in bioslurry reactors, 49
Benzene-contaminated aqueous stream, treatment of by UV light-catalyzed hydrogen peroxide, 27
Benzene-contaminated soil, treatment of, by dual vacuum extraction system, 92
Benzo(a)pyrene (BaP)
   accumulation in *Bacillus megaterium*, 151
   covalent binding to soil humus by fungi, 51
Benzo(a)pyrene-contaminated soil, treatment
   by chemical extraction, 11
   by supercritical fluid extraction, 13
Benzo(a)pyrene degradation, 150–151
   turnover time for, 412
   by UV light/hydrogen peroxide, 27
Benzo(a)pyrene oxidation, by cyanobacteria, diatoms, and algae, 151
Benzoate anaerobic degradation, 177
Best available technology for VOC control, 319
Best demonstrated available technolgy (BDAT), 148
BFAB, biological interactions, 217
Bicyclic alkane biomarkers, 400
Binary fission reproduction, 264
Bioaccumulation, 300
   by *Acinetobacter* sp., 275
   by filamentous fungi, 275
   in landtreatment, 37
   of metals, 66–67, 179, 180, 182, 183, 184, 192; see also Heavy metal binding/bioaccumulation/biosorption
   by microbial products, 179
   of organic compounds, 115–178
   of PAHs, 141
   of petroleum constituents, 115
   by photoautotrophs, 68
   by phototrophs, 275
   by plants, 53, 183, 277
   by protozoa, 300
   for refractory compounds, 275
   by specific microorganisms, 276
   by yeasts, 275
Bioanaerobic reduction/reoxidation (BARR), 249
Bioaugmentation, 2, 277–292; see also Inoculant formulations, Seeding microorganisms; Soil inoculation of activated sludge, 54
   adaptation period, 278
   adsorption of microorganisms, 287
   with aerobic or anaerobic conditions, 300
   in anaerobic reactors, 72
   with antibiotics, 299
   application methods for, 289
   for Arctic climates, 278
   biologically active carbon adsorber for, 281
   biosorption, bioaccumulation, bioconcentration for, 300; se also Biosorption, Bioaccumulation, Bioconcentration
   with cell-free enzymes, 297–299; see also Cell-free enzymes; Extracellular enzymes; Fungi, extracellular enzymes
   closed-loop system for, 280

commercial products for, 278, 280
competition with, 148
control of in bioreactors, 46
encapsulation for, 147–148
environmental modifications for, 278, 285
foam for, 148
immobilization for, 147–148
indigenous microorganisms and, 279
inoculant formulations for, 279, 280, 287
of lagoons, 57
lag period with, 278
level of application for, 279
limiting factors for, 281
monitoring, 281
for NAPLs, 163
with nutrients, 249–257
with oxygen, 232–249; see also Soil oxygen
for PAHs, 48, 148
PHENOBAC® mutant bacterial hydrocarbon degrader, 302
polyurethane foam for, 148
protozoan grazing in, 276–277, 279
reduction of competition with, 148
requirements for successful, 277
in rotating drum reactor, 48
soil inoculum for, 147, 148; see also Soil inoculum
with surfactants, 300–313
transport of microorganisms in, 287
vermiculite for, 148
Bioaugmentation/deep soil fracture bioinjection, 109
Bioaugmentation/hydraulic/pneumatic fracturing, 108
Bioaugmentation microorganisms, see also Aromatic hydrocarbons, biodegradation of; Anthropogenic hydrocarbons; Hydrocarbonoclastic microorganisms; Seeding microorganisms
  acclimated/adapted, 281–283
  biosurfactant producers, 287, 307–312
  cell size, 108, 266, 288, 289, 407
  consortia, 286–287
  counts in bioslurry reactors, 48
  critical number of microorganisms, 279
  detection and monitoring of, 286
  fungi as, 273
  indigenous microorganisms as, 280
  inoculant formulations for, 279, 280
  inoculum preparation for, 147, 279
  mutants, 283–286
  starter culture, 287
  superbugs, 101, 103, 278, 283
  temporary niche for, 279
Bioavailability of PAHs with surfactants, 145
Biocatalysts
  beads, 70
  in anaerobic reactors, 72
  immobilized enzymes in fixed-film systems, 66
Biochemical oxygen demand (BOD), 116
Biochemical removal of contaminants from water, 67
Biochemical testing, 408
Bioconcentration, 186, 275, 300
Biodegradability database, 117, 122–123
Biodegradability of hydrocarbons, relative, 127
Biodegradable surfactants, 313
Biodegradation, 115–198; see also Bioremediation
  biological enhancement of, 262–300
  in biopiles, 52
  of branched alkanes, 130
  of cyclic alkanes, 130
  description/definition, 1–2
  effect of biological factors on, 217
  factors affecting, 199–218
  of heavy metals, 178–198
  of insoluble phase of contaminants, 162
  in landtreatment/landfarming, 30
  limitations of, 415
  microorganisms for specific fuel components/hydrocarbons, 132–139, 263–277
  of NAPLs, 163
  of organic compounds, 115–178
  of PAHs, 141
  pH for, 36, 229–232
  rates of, see Rates of biodegradation; Remediation time
  of surfactants, 312–313
  as treatment alternative, 1–3
  treatment criteria, 3
Biodegradation monitoring methods, 396–408; see also Bioremediation monitoring methods
Biodegradation of fuel components/hydrocarbons, by specific microorganisms, 132–139, 263–277
Biodegradation of petroleum constituents, 115
Biodegradation of specific compounds, 131–141, 173–178
Biodegradation of surfactants, 312–313
Biodegradation potential, hexadecane indicator, 158
Biodegradation rate, 408–413, 414; see also Remediation time
  anaerobic, 170
  of biphasic systems, 202
  effect
    of acclimation on, 282
    of application rate on, 39
    of half-lives on, 38, 39
    of seasons on, 411
    of soil factors on, 220
    of soil water potential on, 225
  for estimating remediation time, 408
  factors affecting, 219
  in landtreatment, 38
  for oily sludges, 38
  of PAHs, 142–143
  in porous tube reactors, 412
  of petroleum, 412
  of phenols, 41
Bioemulsification for oil recovery, 311
Bioemulsification with NAPLs, 163
Bioemulsifiers, 159, 202, 307–312
BIOFAST (biological forced-air soil treatment), 52
Biofilms, 287–288
  in anaerobic reactors, 72
  cyanobacteria/bacteria as, 275
  in fixed-film systems, 57, 58
  heavy metal removal by, 59
  oligotrophs as, 267
  in RBCs, 60
Biofilters, see also Biofiltration
  with bioventing, 379
  clogging in, 378
  composition of, 376
  factors affecting, 379
  with GAC filters, 379, 424
  with injection–extraction process, 87
  interpartical/intrapartical porosity, 377
  microorganisms in, 377, 379
  packing material in, 50, 376–377
  trickle-bed reactor for, 378
  VOC concentration with, 377
  for VOCs from SVE, 89
  white-rot fungi in, 377
Biofilter/stripping tower, 423
Biofiltration, 376–380; see also Biofilters

INDEX **509**

with air stripping, 379
bench scale system, 378
biomass production with, 378
bioreactor design, 380
bioventing vapors treatment by, 379
concentration of influent for, 378
costs of, 379
degradable VOCs with, 378
effect of moisture on, 378, 379
effect of oxygen on, 378
effect of temperature on, 378, 379
effect of water accumulation on, 379
elimination capacity with, 377
elimination rate with, 378
gasoline vapors treatment by, 379
loading rate for, 378
microbial counts in, 378
off-gas prehumidification for, 377
optimization of, 377
removal efficiency of, 378
SVE off-gas treatment by, 379
trickle-bed for, 378
for VOC treatment, 376–380
Biofiltration/air-water separator/trickling filter, 422
Biofiltration model for processes, 378
Biofix-immobilized cells, 156
Biofouling, *in situ* for bioscreens, 112
Bioinjection, deep soil, 246–247
Bioleaching, 180
Biological activated carbon (BAC), for production of bioaugmentation inocula, 281
Biological activated carbon (BAC) systems, 64–65; see also Fixed-film systems
heavy metal removal by, 65
with immobilized bacteria, 58
VOC removal by, 333
Biological activated carbon/wet air oxidation, 24, 422
Biological antagonism, 285
Biological/carbon systems
for leachate treatment, 19, 421
sorption train, 424
Biological enhancement, 263–300
Biological factors affecting biodegradation, 217
Biological forced-air soil treatment (BIOFAST), 52
Biological interactions, 217
Biological processes, 29–78
in bioreactors, 46
in leachate/wastewater treatment systems, 54–78
in soil treatment systems, 29–53, 100–113
Biological towers, 57, 59
Biological treatment, see also Bioremediation
costs, 37
processes
leachate/wastewater, 16, 54–78
soil, 29–53, 100–113
for VOCs in aqueous streams, 357
BIOLOG® system, 406–407
Bioluminescence, 398
Biomarkers, 400–401
Biomass
for biofiltration, 378
effect of PAHs on, 251
effect of soil depth on, 384
determination methods for, 391
for heavy metal sorption, 183
increase in landtreatment efficiency with, 30
recycle in lagoons/waste stabilization ponds, 57
sources for heavy metal sorption, 183

Biomethylation of heavy metals, 178, 181
Biomethylation of mercury, 189
Biomolecular/nucleic acid-based monitoring methods, 396–400
Bioparticles, in denitrification fluidized-bed reactor, 171
Biopiles, 51–52, 245
Biopolymer shields, 112
BIOPUR®, 66, 380
BioPurge^SM, 106–107
for nutrient addition, 254
for soil aeration, 246
BioPurge^SM/BioSparge^SM schematic, 107
Bioreactors, 46–51
anaerobic degradation in, 47
bioaugmentation in, 46
combined technologies in, 417–419
groundwater treatment applicability, 8
leachate control in, 47
soil aeration in, 245, 246
Bioreclamation, see Bioremediation
Bioremediation, 100–113; see also Biodegradation
bioaugmentation for, 2, 277–292
BOD levels for feasibility of, 199
cell-free enzymes for, 100
database for, 117, 122–123
degradable compounds, 124
environmental modifications for, 219, 221
key issues to address in, 219
limitations of, 415
microorganisms in, 132–139, 263–277
monitoring of, 383–414; see also Monitoring bioremediation
optimization of, 219–315
by biological enhancement, 263–300
by contaminant alteration, 300–315
by variation of soil/environmental factors, 219–262
superbugs for, 101, 103, 278, 283
with surfactants, 300–313
system design, 202
Bioremediation/clarification/granular activated carbon/air stripping/reinjection, 421
Bioremediation, *in situ*
applicability of, 7
limitations of, 415–417
processes for, 100–113
site and soil characteristics for, 220
Bioremediation monitoring for anaerobic conditions, 403, 406
Bioremediation monitoring methods, 383–414; see also Counts of microorganisms
anaerobic, 403
autoradiography, 399
for bioaugmentation, 281, 286
biochemical testing, 408
in bioreactors, 46–47
bioluminescence gene (*luxAB*), 398
biomarkers, 400–401
biomolecular/nucleic acid-based methods, 396–400
bioreporter genes, 398
carbon dioxide and oxygen, 402
carboxylic acids, 400
chromosomal painting, 399
COD, 401
colony hybridization, 397
conservative tracers, 404, 405
dehydrogenase activity, 405
DNA gene probes, 397, 398
effect of clay on, 328
electron acceptor concentration, 402
enzyme assays, 405
field control, 383

flame ionization detector (FID), 401
flow cytometry, 408
flow cytometry with FITC, 407
GC/MS, 405
gene fusion, 398
gene probe detection, direct, 397
hopanoic acids, 400
hybridization to colony DNA, 396
hybridization to community DNA, 396
hybridization to community RNA, 397
hybridization to DNA from enrichments, 396
hydrocarbon concentration, 401
hydrocarbon utilizers, 394
infrared gas analyzer, 403
infrared spectroscopy, 401
inorganic indicators, 403
intermediary metabolite formation, 405
labeled contaminants, 404
lacYZ (lactose utilization gene), 398
in landtreatment, 39
limitations of gene probes, 398
marker genes, 398
methane, 403
microbial counting methods, 383–392
microbial respiration quotient, 403
mineralization, 404
modeling, 408
MPN with radiolabeled carbon, 403
nitrous oxide, 403
nonradioactive reporter genes, 399
nucleic acid-based methods, 396–397
nucleic acid hybridization, 398
oligonucleotide probes, 397
other organic indicators, 401
PCR, 397
pentacyclic terpanes, 400
petroleum-degraders:to total heterotrophic bacteria ratio, 405
phenanthrenes/anthracenes, 400
photoionization detector (PID), 401
polymerase chain reaction, 400
radiolabeled CO_2, 403
radiolabeled CO_2 with MPN, 403
RAPD, 396
rate of biodegradation, 408
reporter genes, 398
respirometry/radiorespirometry, 407
ribonuclease protection assay with community RNA, 397
mRNA extraction, 399
rRNA transcription, 399
RT-PCR analysis, 399
sapromat, 403
soil gas monitoring, 402
stable carbon isotope analysis, 404
steranes, 400
TLC-FID, 406
TOC, 401
tricyclic to pentacyclic ratio, 400
WALS and NALS, 407
Bioremediation (mutant bacteria)/activated carbon filter, 421
Bioremediation/physical treatment processes, 420
Bioremediation/pneumatic fracturing, 421
Bioremediation program set up, 212
Bioremediation/steam stripping/chemical precipitation, 419
Bioremediation/surfactant scrubbing/photolysis/reverse osmosis, 421
Bioremediation/SVE/vapor-phase biofilters, 421
Bioremediation/UV/ozonation, 422
Bioreporter genes, 398

Biorestoration, see Bioremediation
Bioscreens, 112
Bioscrubbers
    in trickle-bed reactor, 378
    in tubular bioreactors, 50
Bioslurping, 105–106, 165
    AFCEE Technology Transfer Division Bioslurper Initiative, 106
    for JP-5-contaminated soil, 106
Bioslurping/bioventing, 424
Bioslurry reactors, 47–49
Biosolve, 306
Biosorbent, heavy metal, 179
    bacteria as, 180
    regeneration of, 184
Biosorption, 300
    effect of pH on, 180
    heavy metal removal by, 179, 180, 183, 184
    as polishing treatment, 184
BioSparge^{SM}, 106–107
    for groundwater aeration, 246
BioSparge^{SM}/BioPurge^{SM} schematic, 107
Biostim, 246
Biostimulation
    effect on microbial counts, 394, 396, 397
    for soil aeration, 246
    vacuum heap system, 247
Biosurfactant-producing microorganisms, 149, 154, 307–312
Biosurfactants, 307–312; see also Microbial surfactants
    from *Bacillus* strains, 249, 311
    for naphthalene degradation, 149, 154
    from *Pseudomonas* strains, 149, 154, 302
Biotic vs. abiotic processes, 413–114
Biotin
    for cometabolism, 195
    for cyclohexane degradation, 160
Biotransformation
    examples of, 300
    of heavy metals, 178–198
    of organic compounds, 115–178
    of petroleum constituents, 115
Biotreatability studies, 407
Bioventing, 103–105
    airflow rates in, 104
    alternative approaches for, 104
    with biofilters, 379
    with gaseous ammonia, 255
    for soil aeration, 245
    for soil moisture, 224
    for VOC-, SVOC-, and PAH-contaminated soil, 104
    wind-powered systems, 105
Bioventing/bioslurping, 165, 424
Bioventing/cyclic, or surge, pumping, 104
Bioventing/electrobioremediation, 112
Bioventing/extraction, 104
Bioventing/gaseous nutrient application, 424
Bioventing/heating soil, 103, 423
Bioventing/injection–extraction process, 87
Bioventing installation schematic, 103
Bioventing parameters, U.S. Air Force Bioventing Initiative, 104
Bioventing/pump-and-treat, 422
Bioventing/regenerative resin, 104, 423
Bioventing/sparging, 104
Bioversal, 306
Bio XL/Restore, 246
Biphasic mineralization, 217
Biphasic systems, 161
Biphenyl biodegradation, 150

Bjerkandera adjusta in biofilters, 377
Bjerkandera adjusta CBS 595.78, 271
    PAH biodegradation by, 150, 153
    metabolites of, 194
Bjerkandera sp. strain BOS55
    extracellular peroxidases, 153
    manganese-inhibited peroxidase in, 271
    PAH degradation by, 147, 151, 153, 154, 271
Blade-mixing reactors, 50
Blue-green algae, methane production of, 266
BOD (biochemical oxygen demand), 116, 199
BOD/COD ratios, 116
BOD_5/COD ratios for various organic compounds, 124
Boilers, 363, 369
Boiling point, 325, 330–331
Bonopore adsorbent for chemical recovery, 376
Bound residue formation, 51
Boyancy forces, NAPL mobilization, 162
Branched alkanes, see Branched-chain alkanes
Branched-chain aliphatics, biodegradation of, 155–156
    conversion to coenzyme A, 156
Branched-chain alkanes, biodegradation of, 130, 155, 157
    microorganisms for, 130
Branched-chain alkenes, biodegradation of, 155
Branched hydrocarbons, cometabolism of, 292
Branching, effect on alkane degradation, 127
Breakdown products, 2
Breakthrough point with carbon adsorption, 352
Breed slide, 385
Brij 30, 304, 149, 153
Brij 35, 149, 302, 304
    micellar phase of phenanthrene unavailable with, 153
Broadcast fertilization, 254
Broth cultures, 389
Brown algae, 275
    bioconcentration of nickel by, 190
    naphthalene degradation by, 150
Brown coal adsorption, 21
Brown coal coke adsorption, 21
Brown-rot fungi, 149, 150, 271
BTEX biodegradation
    aerobic, 140–141
    aerobic/anaerobic sequential, 167, 176
    in Allied Signal immobilized cell bioreactor (ICB), 65
    anaerobic, 176–177
    anaerobic degradation rate, 170
    denitrification of, 175, 176
    effect of sulfate or ferric ions on, 167
    fungal, 270
    sulfate reduction, 175, 176, 177
BTEX-contaminated aqueous stream, treatment of
    with BIOPUR®, 66
    by landfarming, 34
BTEX-contaminated soil, treatment of, by prepared-bed reactors, 50
    by SVE, 91
    by quartz furnace, 9
BTX-contaminated soil, treatment of, by pneumatic fracturing, 108
B_{12}-dependence, 181
Bubble aerators, 347
Bubble cap absorption column, 375
Bubble caps in plate columns, 372, 374
Bubble interface, 58
Buffering capacity of soil, 231
Bulk density, 328
    decrease with amendments, 47
    effect on pore size, 262
    for enhancement of fungal degradation, 273
    increase in, 262
Bulking agents
    as carbon source for fungi, 271
    for composting, 44
Bunker C fuel oil, accumulation in landtreatment, 37
Bunker C fuel oil-contaminated soil, treatment of
    by BIOFAST, 52
    by prepared bed reactors, 50
Burning of oil, 212
Bushnell Haas medium, 281, 387, 389
Butane cometabolism, 156
Butyl rubber, 337
By-products, see also End products; Metabolites
    of anaerobic degradation, 167, 403
    cometabolism of, 195, 294
    of leachate treatment, 14, 15, 17–18
    of mercury biomethylation, 189
    for metals removal, 179
    catalysis of leachate, 17

C

Cadmium, 185–187
    binding by capsular material, 186
    bioaccumulation of, 66, 187
    bioconcentration of, 186
    biomethylation of, 187
    detoxification of, 185, 186, 187
    effect on landtreatment, 39
    microbial resistance to, 185
    microorganisms for biosorptions of, 186
Cadmium complexation, 186
Cadmium-contaminated aqueous stream, treatment of, by chemical precipitation, 27
Cadmium-contaminated soil, treatment of, by Chapman soil-washing process, 11
Cadmium resistance
    by cellular exclusion mechanisms, 185
    by efflux mechanisms, 186
    by intracellular traps, 185
    plasmids in, 187
Calcareous soil pH, 230
Calcium hydroxide, 11
Caldoactive microorganisms, 70
CAM-OCT plasmid, 284
Candida, adsorption with BAC, 64
Capillary condensation, 302
Capillary forces, 332
Capsular polysaccharides for heavy metal removal, 186–187
Carbonaceous resins, 21
Carbon adsorbers for VOCs, 351
Carbon adsorption, 19–20, 350–357
    in BAC systems, 64
    breakthrough point with, 352
    candidates for, 20
    carbon-fouling problems, 351
    effect of contaminant concentration on, 351
    effect of temperature on, 351
    efficiency of, 20
    factors affecting adsorption, 19
    fixed-bed, 351–353
    fluidized-bed, 353, 356–357
    inhibition of, 20
    for leachate treatment, 16, 17
    for VOCs, 342, 350–353, 356–357
    for VOCs, SVOCs, and NVOCs, 357
Carbon adsorption/air stripping, 346, 358

Carbon adsorption/air stripping/ion exchange process, 426
Carbon adsorption/evaporation for sludges, 358
Carbon adsorption/sedimentation, 28, 422
Carbon adsorption systems, 351; see also Vapor-phase carbon adsorption systems
Carbon adsorption/vacuum extraction/oil–water separator/steam stripping/filters, 422
Carbon adsorption/wet air oxidation, 22
Carbon canisters, 333
Carbon dioxide
  as alternate electron acceptor, 241
  for monitoring bioremediation, 402–403
Carbon isotope ratios, 404
Carbon tetrachloride, 9
Carbon-to-nitrogen (C:N) ratio, 35, 250
Carbon-to-nitrogen-to-phosphorus (C:N:P) ratio, 250
Carboxyfluorescein diacetate, 408
Carboxylic acids
  biodegradation of, 159
  as biomarker, 400
Carcinogens, 127, 141, 142, 150, 151
  cadmium, 185
  fungal products, 142
  safe levels of, 322
Carcinogenic products of fungal metabolism, 194
Carrageenan, 290
Catabolic activity, 407
Catabolic enzyme systems, 124
Catabolic genes, 397, 398, 400
Catabolic gene probes, 284
Catabolic genotypes, 397
Catabolic pathways, 127
Catalysis of leachate by-products, 17
Catalytic afterburners, 367
Catalytic combustion, 89
Catalytic incinerators, 367
  with primary heat recovery, 370
  temperatures for different compounds in, 371
  for VOC destruction, 363, 365, 367, 369, 370, 371
Catalytic oxidation of off-gas, 8, 363
Catenary grid scrubber, 349
Cedephos FA-600, 305
Cell analysis, 407
Cell biomass, composition, 251
Cell concentration in fixed-film systems, 57
Cell detachment in fixed-film systems, 58
Cell division counting method, 385
Cell enlargement counting method, 385
Cell-free enzymes, 195, 297–299
  in bioremediation, 100
  naphthalene degradation by, 150
Cell-free extract, 150, 189
Cell immobilization, see Encapsulation/immobilization, 290
Cell lipid, 155, 159
Cell membrane, 130, 202
Cell permeability with surfactants, 303
Cell retention, 288
Cell retention time in fixed-film systems, 57
Cell size, 108, 266, 288, 289, 407
Cell structure analysis, WALS, 407
Cellular exclusion mechanisms, 185
Cellulose for immobilization, 58
Cell walls
  charge reversal on, 181
  heavy metal adsorption by, 186, 192
  net negative charge, 181
Cement kilns, 363
Cement for stabilization, 80

Centrifugation, 83
CERCLA (Comprehensive Environmental Response, Compensation, and Liability Act), 89
Cereal agar, 388
CGA (colloidal gas aphrons), 241
Channeling, 415
  with air stripping, 88
  with bioremediation, 102
  foams for, 164, 242
  in landtreatment, 36
  overcoming problem of, 10, 164, 242
  with soil flushing, 86
  of surfactants, 306
Chapman soil-washing process for heavy metals, 11
Characteristic diffusion time, 332
Charge reversal of cell walls, 181
Charge, soil surface, see Soil surface charge
Chelating agents
  for soil flushing/washing, 82
  for solubilizing, 14
Chelation of mercury, 189
Chemical analog adaptation, 282
Chemical catalysts, 24, 27
Chemical composition of fuel oils, 115–124
Chemical extraction procedures, 11
Chemical factors affecting biodegradation, 200–217
Chemical fingerprinting, 406
Chemical fixation, 37
Chemical oxidation, 23–27
  for leachate treatment, 16, 17
  with ozone, 24
  as pretreatment, 24
  for PAHs, 141
  for VOCs, 15
Chemical oxygen demand (COD), 116
Chemical/physical treatment, costs, 37
  limitations of, 415
  processes
    leachate/wastewater, 13–29
    soil, 6–13, 79–99
Chemical precipitation, 27–28
  for heavy metals, 27
  for leachate treatment, 16, 17
  process design, 27
Chemical precipitation/steam stripping/bioremediation, 419
Chemical recovery, 376
Chemical reduction for leachates, 16, 17
Chemical soil treatment, see Chemical/physical treatment processes
Chemical solubility, 200–203
Chemical structure
  cometabolism of organic groups, 209
  effect on biodegradation, 208–210
  modification to reduce toxicity, 212
Chemical surfactants, 302–307
Chemisorption of VOCs, 329
Chemoheterotrophic bacteria, anaerobic bioconversion by, 70
Chemosmotic efflux system, 185
Chemotrophs, 266
Chlorella, 274
Chlorinated hydrocarbons, 9, 20
Chlorinated polyethylene membrane liner, 337
Chlorosulfonated polyethylene membrane liner, 337
Chromate reduction, 188
Chromatiaceae, 274
Chromium, 187–188
  chromate reduction, 188
  precipitation of, 27

reducing agents for, 79
removal by green algae, 181
resistant microorganisms, 188
Chromium-contaminated aqueous streams, treatment of
 by algae, 188
 by Chapman soil-washing process, 11
 by chemical precipitation, 27
 by reduction/precipitation, 27
Chromium-contaminated soil, treatment of
 by diphenyl carbazide, 86
 by Dowfax 8390
 by stabilization, 12
 by surfactants, 188
Chromium resistance, 187
Chromosomal painting, 399
Chromosomes, hydrocarbonoclastic, 283, 284, 398
Chromosomes, resistance, 306
Ciliate protozoans, 299
Cis-dihydrodiols, 141, 147, 195, 268
Cis-diols, 126
Cis-glycol, bacterial, 230
Citrobacter sp.
 citronellol pathway, 126
 heavy metals removal by, 183
Citronellol pathway, 126
Cladosporium resinae, 269
Clarification/granular activated carbon, air stripping, reinjection, bioremediation, 421
Clay
 for bioslurry reactors, 48
 bound water in, 329
 effect on microbial transport, 288
 effect on sorption, 204
 as packing material for biofilters, 376
 support in fixed-film systems, 58
Clay-humic acid complexes, 262
Clay minerals
 reduction of nickel toxicity by, 90
 sorption by, 204
Clay soils, 415
 adsorption of pyrene by, 154
 air-filled porosity of, 328
 attenuation by, 261
 BioPurge[SM] for, 107
 deep soil fracture bioinjection for, 109
 dual vacuum extraction system for, 92
 effect of air stripping on, 87
  of bioventing on, 105
  of dielectric constant on, 330
  on irrigation, 226
  on monitoring, 328
  on soil flushing, 86
  on soil washing, 85
  on variation in biomass, 384
  on VOCs, 328
 electrobioreclamation for, 110, 111
 organic amendments for, 259
 PAH sorption on, 144
 pneumatic fracturing for, 109
 pore size of, 328
 pretreatment for density, 262
 resistance heating for, 96
 sorption of surfactants in, 306, 307
 surface charge of, 229
 vapor diffusion rate in, 328
 volumetric water content of, 325
 water content of, 328
Cleaning frequency, 335

Cleanup time, 319–320
Climatic conditions, effect on bioremediation rates, 39
Clogging, see also Plugging
 of activated carbon systems, 20
 avoidance with soil fracturing, 107
 of biofilters, 378
 minimization with microemulsions, 305
 prevention with bioreactors, 51
Clogging of aquifers by ozonation, 236
Cloning, 183
Closed-loop treatment system, 85
 for bioaugmentation, 280
 BioPurge[SM]/BioSparge[SM], 106
 Detoxifier™, 9
Clouding behavior, 305
Cloud point-phase separation of nonionic surfactants, 305
C:N (carbon-to-nitrogen) ratio, 250, 258
C:N:P (carbon-to-nitrogen-to-phosphorus) ratio, 143, 250, 258
Coal tar-contaminated soil, treatment of
 conversion to carbonaceous adsorbent, 20
 with CROW process, 87
 by encapsulation, 12
 ultrasound-enhanced soil washing, 10
Cobbles for immobilization, 59
Coculture for anaerobic degradation, 167
COD (chemical oxygen demand), 116
 for assessment of bioremediation feasibility, 199
 biodegradation indicator, 401
 range of treatment, 22
Coenzyme A, 156
Coenzyme M, 173, 266; see also Factor$_{420}$ Coenzymes, 298
Cofactors, 298
Coimmobilization, of nutrients and microorganisms, 290
Coke support, 58
Coke tray aerators/countercurrent packed columns/air stripping, 344
Cold climate/weather soil treatment, 10, 34
Colloidal gas aphrons (CGA), 241–242
 formation and foam size, 164
 for LNAPL treatment, 164
Colloids removal, 29
Colony hybridization, 397
Column reactors, 20, 58
Combination reactors, 51, 67–68
Combined aerobic/anaerobic treatment, 67–68, 76–78, 169, 248
Combined air/water flushing, 244
Combined oxidation, 79
Combined technologies, 3–4, 5, 76–78, 109–110, 417–428; see also Treatment trains
Combustion, 363–369; see also Incineration
Cometabolism, 2, 140, 155, 156, 209, 286–287, 292–297
 accumulation of metabolites from, 195
 anaerobic, 167, 170, 176, 294
 bacterial counts with, 293
 biotin requirement for, 195
 of branched and cyclic hydrocarbons, 292
 by-product degradation, 195, 294
 of condensed ring aromatics, 130
 of contaminants and surfactant, 305
 in denitrification, 171, 176, 294
 and diauxie effect, 297
 effect of concentration on, 214
 enzyme competition in, 196, 294
 enzyme induction with, 195
 examples of, 293–294
 of four- or more ring PAHs, 292
 inhibition of, 146, 196
 of organics and products, 296

oxygenases in, 292
    for PAH degradation, 145, 146, 151, 152, 154, 155, 292, 294
    for recalcitrant molecules, 258, 292, 294
    in sequential anaerobic/aerobic treatment, 76
    species exhibiting, 295
    substrates for, 292
    threshold contaminant concentration for, 293
Cometabolism products/microorganisms/analogs, 295–296
Cometabolites, 296–297
    with denitrification, 248
Commensalism, 146, 154, 159–160, 286–287
    for condensed ring aromatics, 130
    for four- or more ring PAHs, 292
    using metabolic intermediates, 195
Commensalistic symbiosis, with cycloalkanes, 130
Commercial products, 278, 280
Compaction, 248
Competition of microorganisms, reduction of, 148
Competition of surfactants, reduction of, 306
Competitive inhibition, 141
Competitive sorption, 224
Complexing agents, 14
Composition of fuel oils, 115–124
Compost
    as biofilter packing material, 376
    effect on PAH biodegradation, 143
    reuse for landscaping, 13
Compostable organics, 43, 45
Composting, 41–46, 227
    anaerobiosis, 44
    bacteria in, 44
    mobile composting unit, 45
    as pretreatment for landfarming, 34
    reactors, 46
Comprehensive Environmental Response, Compensation, and Liability Act (CERCLA), 89
Compressibility, soil, 47
Concentration of contaminants, 213–217
    biodegradable concentrations, 215–216
    effect
        on adaptation time, 213
        on biodegradation/rates, 142–143, 151, 409, 410–411
        on rate of emission, 327
        of temperature on, 371
    as measure of biodegradation, 401
    for monitoring bioremediation, 401
Condensation for VOCs, 354, 357, 369, 371–372, 373
Condensation solvent recovery system, 373
Condensed PAHs, 211
Condensed ring aromatics, degradability of, 130
Cone penetrometer system, 115
Congo red, 273
Conjugation for detoxification, 212
Conservative tracers, 405
Consortia, microbial, 70, 286–287; see also Anaerobic bacteria; Microbial consortia; Mixed cultures
Contact stabilization, 56
Contained recovery of oily wastes (CROW) process, 87
Contaminant alteration, 300–315
    by surfactants, 300–313
Contaminant analysis methods, 406
Contaminant hot spots, 225
Contaminated soil, microbial counts in, 393–396
Continuous-feed rotary kiln, 6
Continuous-flow mode of trickling filters, 59
Convective transport, 288

Conventional anaerobic digester, 71–72
Cooke's aureomycin-rose bengal medium, 388
Cooling towers/air stripping, 344
Cooxidation, 2, 140, 292
Cooxidizers, 293
Copper-contaminated aqueous stream, treatment of by chemical precipitation, 27
Copper quenching, 304
Copper resistance, 66, 192
Corexit 0600, 304
Corncobs
    for PAH degradation, 150, 155
    as carbon source for fungi, 149, 271, 273
Corona, *in situ*, 79
Corrugation irrigation, 226
Corynebacterium, 2
Coryneforms, as oligotrophs, 267
Cost of Remediation Model (CORA), EPA, 5
Costs for remediation treatment technologies, 37
Countercurrent packed columns/air stripping, 344, 345
Counts of microorganisms, see also Bioremediation monitoring methods
    in Antarctic soils, 396
    in biofiltration, 378
    in bioslurry reactors, 48
    with cometabolism, 293
    in contaminated soil, 393–396
    effect of biostimulation on, 255, 394–396
    effect of dormancy on, 384
    effect on biodegradation rate, 409
    in uncontaminated soil, 392, 393, 394
Covalent binding to soil humus, 51
Cover materials, 333; see also Synthethic membrane covers
Covers
    for biopiles, 51
    for lagoons, 335
Crankcase oils biodegradation, by landtreatment, 40
Creosote
    accumulation in landtreatment, 37
    effect on soil wettability, 164
Creosote-contaminated soil, treatment of
    by CROW process, 87
    by HRUBOUT®, 96
    by soil washing, 10
Cresols, anaerobic degradation of, 174
Critical micelle concentration, 149, 303, 304; see also Surfactant micelles
    ranges for surfactants, 311
Critical number of microorganisms for biodegradation, 140, 279, 393
Cross-adaptation for degradation, 177
Cross-flow towers/air stripping, 344
CROW (contained recovery of oily wastes) process, 87
Crude oil biodegradation, 158
    by landtreatment, 40
Crude oil composition, 115
Crude oil-contaminated soil, treatment of
    by encapsulation, 12
    by HRUBOUT® process, 96
    by landfarming, 31, 34
    by soil washing, 10
    by surfactants, 85
Cryogenic unit, 95
Cryo-scanning electron microscopy (cryo-SEM)/X-ray analysis for NAPLs, 165
Crystallization, 17, 28

INDEX

Crystal violet, 274
Cultivation, 31
Cunninghamella elegans, 268
    PAH degradation by, 147
    for PAH degradation end products, 194
Current treatment technologies, 5
Customblen, 256
Cyanide-contaminated soil, treatment of, by Chapman soil-washing process, 11
Cyanidium caldarium
    chromium removal by, 188
    heavy metals removal by, 284
Cyanobacteria, 2
    bioaccumulation and sorption of heavy metals by, 275
    bioconcentration of nickel by, 190
    lead toxicity to, 188
    metabolism of recalcitrant compounds by, 275
    metallothioneins and, 179
    PAH degradation by, 147, 150, 151
    phototrophs, 274
Cyanobacteria/bacteria biofilms, 275Cyanobacteria photozone, 275
Cyclic alkanes, biodegradation of, 130
Cyclic hydrocarbons, cometabolism of, 292
Cyclic or surge pumping, 104
Cyclic steam injection/vacuum extraction, for JP-5-contaminated soil, 98
Cycloalkane degradation by cometabolism and commensalism, 130
Cycloalkanes, toxicity of, 130, 211
Cyclodextrins, 388
Cyclohexanone, 160
Cyclohexane biodegradation
    by aerobic nonphotosynthetic strains, 178
    by anaerobic degradation, 178
    by anaerobic photosynthesis, 178
    by cometabolism and commensalism, 159–160
    enzyme induction for, 178
Cyclohexanol, 160
Cycloheximide for protozoa, 299
Cyst-forming amoeba, 276
Cysts, 224
Cytochrome oxidases, 126
Cytochrome P-450, 126
    fungal dependence on, 149, 268
    monooxygenases, 194
Cytophagic protozoans, 276, 299
Cytoplasmic inclusions, 300
Cytotoxicity of hydrogenperoxide, 238

D

Database for bioremediation, 117, 122–123
DDT, anaerobic composting of, 44
Dead-end metabolites
    accumulation of, 195
    overcoming intracellular accumulation of, 285
Deamination, 212
Decantation, 355
Decarboxylation of aromatics, 131
Decision framework for remediation technologies, 418
Decision making guide for SVE, EPA, 89
Deep soil fracture bioinjection, 109
    for bioaugmentation, 109
    for nutrient addition, 254
    for soil aeration, 246–247

DEFT (direct epifluorescence filtration technique), 386
Defusing, 193
Degradation, aerobic, 123–165
Degradation of alkanes, 128–130
Degradation of alkenes, 131
Degradation, anaerobic, 165–178
Degradation of aromatics, 131
Degradation of specific compounds, 131–141, 173–178
Degradation pathways, 126
Dehydration inactivation, 289
Dehydrogenase activity, 405
Dehydrogenase-coupled respiratory activity, 391
Demethylation of aromatics, 131
    detoxification by, 212
    of mercury, 189
Denitrification, 2, 169–171, 174, 175, 176, 177
    anaerobic expanded/fluidized for, 76
    acclimation with, 169
    alternate carbon sources for, 170–171
    oxygen concentration for, 166, 169, 247
    fluidized-bed reactor for, 171
    sequential, 175, 176
Denitrifiers, 166, 169, 171, 175, 176, 177
    in cometabolism, 294
    counts of, 387
    in pneumatic fracturing, 108
Dense nonaqueous-phase liquid (DNAPL)-contaminated soil, treatment of by CROW process, 87
Dense nonaqueous-phase liquid (DNAPL) ganglia, 164
Dense nonaqueous-phase liquid (DNAPL) mobilization, 164
Dense nonaqueous-phase liquid (DNAPL) treatment with foam, 164 165
Dense nonaqueous-phase liquid (DNAPL) (dense nonaqueous-phase liquid), 160–165
Dense nonaqueous-phase liquids (DNAPLs), 160, 161; see also Light nonaqueous-phase liquids; Nonaqueous-phase liquids
Densifiers, for hydrogen peroxide mobility control, 79
Density
    of contaminant, 208, 330
    of soil effect on contaminant migration, 203, 208
        effect on soil moisture, 221
Density separation, 17, 28
Deodorization of waste gases, 47
Depth, soil, 326, 384
Desorption, 145, 205
    Freundlich isotherm, 205
    in landtreatment, 30, 34
    leachate production, 207
    from organic matter, 258
    rates, 303
    thermal, 359
Desulfovibrio vulgaris, in anaerobic fixed-film systems, 74
Detergents, 33, 302; see also Surfactants
Detoxification
    by deamination, 212
    by demethylation, 189, 212; see also Demethylation
    by fungi, 51, 141, 142, 268
    of heavy metals, 178, 180, 186, 189, 191, 212
    by hydrolysis, 212
    by hydroxylation, 212
    in landtreatment, 30
    of PAHs, 141, 142
    processes, 212
    soil characteristics for, 219
Detoxifier™, 93–95

for soil aeration, 247
treatment train, 427
for VOC-contaminated soil, 363
Dewatering activated sludge, 56
Dewatering/bioslurping, 165
Dialysis, see also Electrodialysis
for leachate treatment by-products, 18
removal of inorganic salts by, 29
Diatomaceous earth, 65
Diatoms, 275
benzo(a)pyrene degradation by, 151
naphthalene degradation by, 150
PAH degradation by, 147
Diauxie and cometabolism, 297
Diauxie effect, 297; see also Sparing
repression, 158
Diauxie/sequential anaerobic-aerobic treatment, 76
Dielectric constant, 202, 330
Dielectric heating, 98
Diesel-contaminated soil and groundwater, treatment of, by electrobioremediation, 111
Diesel fuel-contaminated aqueous stream, treatment of
by BIOPUR®, 66
by immobilized cells, 156
Diesel fuel-contaminated soil, treatment of
by BIOFAST, 52
by bioremediation, 256
by blade-mixing reactors, 50
by encapsulation, 12
by fixed-film systems, 58
by HRUBOUT® process, 96
by liming, 148
by soil flushing/washing, 10 30, 85
by steam extraction, 12
by wet or dry steam/vacuum, 98
Diesel fuel #2
composition of, 115, 116
treatment of, by prepared-bed reactors, 50
Diesel fuel #2–#6-contaminated soil, treatment of, by steam injection, 96
Diesel fuel, treatment, by landfarming, 38
Diesel oil, 156
Diesel oil-contaminated soil, treatment of, by environmental modifications, 148
Difco nitrate agar, 387
Diffused aeration/activated sludge, 15
Diffused aeration/air stripping, 344
Diffused-air aeration, 347
Diffusion coefficient, 332
of oxygen, 331
in soil pores, 328
Diffusion of contaminants, 203
Diffusion travel times, 332
Diking, 248
Dimethylsulfoxide (DMSO), 386
Diol, 126, 127
Diol epoxide, 127
Dioxygenases, 125
in aromatics biodegradation, 131
in PAH biodegradation, 141
Diphasic systems, 202
Diphenyl carbazide, 86
Dip slides, 385
Direct cell counts, 385
Direct DNA extraction, 397
Direct electric heating, 99
Direct epifluorescence filtration technique (DEFT), 386
Direct gene probe detection, 397

Direct microscopic counts, 385
Direct photodegradation for VOC-contaminated soil, 363
Direct viable counts, 385
Disagglomeration of soil, 50
Dispersion of contaminants, 203
Dissimilatory denitrification, 169
Dissimilatory pathway, 169
Dissimilatory sulfate-reducing bacteria, 172
Dissimilatory sulfate reduction, 172
Dissociation constant (pKa) for determining sorption, 205
Dissolution flux with NAPLs, 163
Dissolved air flotation, 15
treatment of groundwater with, 346
Distillation, 29, 354
for leachate treatment by-products, 18
for VOCs, 342, 372
for VOC recycle, 380
Distillation/steam stripping/solvent extraction, 423
Disulfonates, 149
Diterminal oxidation of aliphatics, 129
Diterminal oxidation of alkanes, 129, 130
dmpN, 399
DMSO (dimethylsulfoxide), 386
DNA extraction, direct, 397
DNA gene probes, 397
DNAPLs (dense nonaqueous-phase liquids), see Dense nonaqueous-phase liquids; Light nonaqueous-phase liquids; Nonaqueous-phase liquids
DNA probe techniques, 398
Dobanols, 306
biodegradability of, 307
in landfarming, 33
DOC (dissolved organic compounds), biodegradation indicator, 401
Dodecane biodegradation, 159
"Do nothing" approach for leachate treatment, 15
Dormancy of microorganisms, 384
Dowfax C10L, 305
Dowfax 8390, 305
for chromium-contaminated soil, 86, 188
Drainage, 36, 225, 226
Draining frequency, 335
Drum reactor, composting, 46
Dredging frequency, 335
Drizit-immobilized cells, 156
Drop count method, 389
Droplette counting method, 389
DSM-Deutsche Sammlung von Mikroorganismen und Zellkulturen GmbH database, 117, 122–123
Dual-beam γ densiometer for LNAPLs, 165
Dual vacuum extraction for benzene-contaminated soil, 92
Dual injected turbulent suspension (DITS) reactor, 49
Dual-phase extraction of NAPLs, 165
Dual sorbant, 328
Dual vacuum extraction system, 92
Duckweed for heavy metal sorption, 183
Dust control measures, 326
Dworkin Foster mineral medium, 240, 387
for hydrogen peroxide stabilization, 240
Dyes, fungal screening, 273

E

Earthworms, 30, 275, 276, 299
Ebiox vacuum heap™ system, 53
Ecological approach to bioremediation, 219
Ecotoxicity of surfactants, 301
EDTA (ethylenediaminetetraacetic acid), 14

INDEX 517

Effective depth, 326
Effective diffusion coefficient, 328
Effectiveness of treatments, 320
Effluent levels, 389
Effluent polishing, 19
Efflux mechanisms, 186, 188
Efflux system, 191
Eh adjustment for aerobic/anaerobic treatment, 249
EIMCO™ reactors, 47, 49
Elasticized polyolefin membrane liner material, 338
Electrical corona process, 79
Electric current for contaminant immobilization, 80
Electric fields, 79, 99
Electrobioreclamation, 110–112
    effect on microbial transport, 289
    for soil aeration, 244
Electrobioreclamation/bioventing, 112
Electrobioreclamation/hydraulic injection, 111
Electrobioreclamation/SVE, 112
Electrochemical remediation, 110–112
Electrodialysis, 18, 29
Electrokinetic-enhanced passive *in situ* biotreatment PISB system, 111
Electrokinetic potential reversal, 181
Electrokinetic remediation process, 110
Electrokinetic soil flushing for lead contaminated soil, 86
Electrokinetics, 110–112
Electrokinetic soil flushing, 86
Electrolytes, 328
Electromagnetic techniques, 402
Electromigration, 99
Electron acceptors, 166, 402
Electron acceptors, alternative, 168, 169, 241
Electron donor, 168
Electron sinks, 173
Electrobioremediation, 111, 112
Electro-osmosis, 110–112, 244
Electroosmotic dewatering/pneumatic fracturing, 108
Electro-oxidation, 11
Electrostatic interactions, 288
ELISA (enzyme-linked immunosorbent assay), 322, 406
Elution times for hydrocarbons, 13
Elutriate recycle system, 81, 82
Emission rate, 320
Emission reduction method, 320
Emissions, see also Volatile organic compounds; Volatilization
    in composting, 46
    control of, 320, 333
    effect of depth on, 326
    effect of operating surface area on, 325
    effect of wind/barometric pressure on, 325, 333–334
    from landtreatment, 321
    from soil contamination, 317–323
    from TSDFs, 318
Emissions control, 332–381
    synthetic membranes, 335–340
Emissions isolation flex chamber, 317
Emulsification, 158, 302
Emulsifiers, 283, 300, 301, 302
    biosurfactants, 307
Emulsion formation with bioslurping, 106
Encapsulation, 12, 290–292; see also Immobilization
    activated carbon for, 290
    for bioaugmentation, 147–148
    of cell-free enzymes, 299
    cell storage and survival by, 290
    coimmobilization by, 290
    contaminant immobilization by, 12, 80, 306

FDA assay for, 292
    with foams, 296, 306
    of fungi, 290
    of hydrogen peroxide, 108
    liposomes for, 291–292
    of nutrients, 255
    oxygen mass transfer with, 291
    diesel contamination treatment by, 12
    by vitrification, 99
Enclosed mechanical aeration systems, 12
Enclosed reactors, 50
End products, see also By-products; Metabolites
    of anaerobic degradation, 167, 193, 403
    of biodegradation, 192–198
    of fungal degradation, 124, 194
    of PAH degradation, 194
Engineered bioremediation, 101
Enhanced natural degradation, 246
Enrichment culturing, 387–389
    for seed organisms, 279, 387
Enterics, 259
Enterobacteriaceae, 285
Enumeration of microorganisms, 383–396
Environmental factors, see also Individual soil factors; Soil characteristics; Soil factors
    affecting biodegradation, 217–218
    control of in bioreactors, 46
    modification for
        bioaugmentation, 278, 285
        biodegradation, 219–262
        heavy metals, 180, 181
Environmental risk assessment for surfactants, 301
Environmental stress, 285
Enzymatic oxidative coupling 298
Enzyme activity, effect of soil depth on, 384
Enzyme assays, 405
Enzyme competition in cometabolism, 294
Enzyme extracts, 298
Enzyme immobilization, 66, 298
Enzyme induction, 140, 160, 178, see also Inducible enzymes
    beads for microorganisms for, 297
    with cometabolism, 195
    cyclohexane degradation by, 160
    PAH degradation by, 145, 152, 146, 155, 282
    for recalcitrant compounds, 131
    repression by nonhydrocarbons, 297
Enzyme-linked immunosorbent assay (ELISA), 406
Enzymes, 2, 298; see also Cell-free enzymes; Extracellular enzymes
    in cometabolism, 2
    fungal, 272
    immobilized in fixed-film systems, 66
    preparation of, 298
Enzyme substrate specificity, 131
Enzymic biodegradative reactions, 131
Epifluorescence
    filtration technique, direct (DEFT), 386
    microcolony technique, 386
Epoxidation, 193
Equilibrium distribution data for air stripping, 345
Equilibrium partitioning, 344
Esters for resin adsorption, 20
Ethane cometabolism, 156
Ether cleavage for detoxification, 212
Ethoxylated alkyl phenol, 85
Ethoxylated (EO) surfactant, 84
Ethoxylated fatty acids, 85
Ethylbenzene, 172, 177

Ethylene propylene rubber membrane liner material, 338
Eukaryotic inhibitors, 299
Eukaryots, 142
Eutrophication, 241
Eutrophs, 285
Evaporation, 28
   abiotic processes, 317
   effect on vaporization, 329
   for leachate treatment by-products, 18
   as pretreatment for VOCs, 358
   VOC recycle by, 380
   for VOCs, 329, 357
Evaporation/biofilm filtration, 70
Evaporation/carbon adsorption, 358
Evaporation retardants, 226
Evaporation, thin-film, 380
Excavation, 1
   problems with 5, 6
   treatment of soil from, 5–78, 418–419
Exoenzymes, see Extracellular enzymes
Exopolysaccharide (EPS), 248, 186–187
Expanded-bed anaerobic systems, see Anaerobic fixed-film systems
Expanded/fluidized bed, anaerobic, 75–76
Expanded upflow systems, 64
Ex situ biological systems, limitations of, 417
Ex situ electrobioreclamation, 110
Ex situ remediation
   with Detoxifier™
      limitations of, 417
      treatment trains for, 417–419
   treatment processes, 5–78
Extracellular biosurfactants, 307
Extracellular enzymes, 298, 299; see also Cell-free enzymes; Enzymes; Extracellular peroxidases
   binding by soil organic matter, 297–298
   in biofilters, 377
   fungal, 272
   FyreZyme™, 299
   with NAPLs, 163
   for PAH degradation, 147, 154
Extracellular membrane vesicles, 312
Extracellular peroxidases, 298
   of *Bjerkandera* sp. strain BOS55, 153
   in white-rot fungi, 299
Extracellular surfactant with NAPLs, 163
Extrachromosomal plasmids, 179
Extraction/bioventing, 104
Extraction, soil, 81–87, 358; see also Soil flushing; Soil washing; Pump-and-treat

F

Factors affecting biodegradation, 199–218
   biological, 217
   chemical/physical, 199–217
   environmental, 218
Factors F_{420} and F_{430}, 173, 266
Facultative anaerobes, 168, 232
Fastidious anaerobes, 166
Fatty acid analysis, 391
Fatty acid ethoxylate surfactants, biodegradation of, 313
Fatty acid formation
   from alkane degradation, 129, 155
   from alkene degradation, 131, 158
   from biodegradation, 193
Fatty acid metabolizers, 193
Fatty acids
   anaerobic degradation of, 168, 178
   anaerobic fermentation of, 193
   as bioemulsifiers, 159
   incorporation into cell lipid, 155, 159
Fatty acids, biodegradation of, 126, 129, 131, 159, 193
   by β oxidation pathway, 159
   by oxidative decarboxylation, 193
   by sulfate reduction, 172, 178
FDA (fluorescein diacetate-hydrolyzing activity), 292
Fences, 326
Fenton chemistry, 239
Fenton's reagent, 239
   for oxidation of multiring aromatics, 48
Fermentation, 173
Fermentation, methanogenic metabolism, 166
Fermenters, 51, 166
Ferric ion, electron acceptor, 166
Fertilization
   in landtreatment, 35
   manure amendments, 258
   plan/program, 254
Fertilizers, 253–254, 256; see also Nutrients
F-400 GAC regeneration, 19
FID (flame ionization detector), 401
Field capacity, 221, 222, 223, 225
Filamentous fungi, 267, 268, 269–270; see also Fungi
Filter counts, membrane, 390
Filterpak media in trickling filters, 59
Filters/vacuum extraction/oil-water separator/steam stripping/carbon adsorption, 422
Filter system, anaerobic, 74–75
Filtration, leachate, 18, 29
Filtration of microbes by soil, 287
Filtration/precipitation, 422
Filtration/sedimentation, 422
Fingerprinting, 406
   BIOLOG® system for, 407
   GC traces, 406
First-order kinetics, 408–409
   in landtreatment, 39
First-order regression, 222
FITC (fluorescein isothiocyanate)/flow cytometry, 407
Fixed-bed carbon adsorption, 351–353
Fixed-bed reactors, 57
Fixed-film, fixed-bed bioreactor, Allied Signal immobilized cell bioreactor (ICB), 65
Fixed-film reactors, 58
Fixed-film systems, 57–66
   activated biofilters (ABSs), 65
   aerobic, fixed-film, upflow reactor, 58
   airlift bioreactors, 57
   Allied Signal immobilized cell bioreactor (ICB), 65
   biofilms in, 57, 58
   biological activated carbon (BAC) systems, 64–65
      BAC-fluidized bed, 65
      GAC-FBR, 64
      with GAC filters, 64
      for heavy metals, 65
      PAC/activated sludge system, 65
      PAC in aeration basins, 64
      retention times in, 65
      for solvents, VOCs, 65
      with wet air oxidation, 65
   biological towers, 57, 59
   BIOPUR® and RBC, 66
   cell concentrations in, 57
   cell detachment in, 58
   cell retention time in, 57

INDEX

continuous-flow, fixed-bed reactor, 58
for diesel fuel-contaminated soil, 58
fixed-bed reactors, 57
fluidized-bed digester, 63
fluidized-bed reactors, 57, 60–64
  with activated carbon systems, 64
  advantages of, 356
  aerobic Oxitron system, 61, 62
  anaerobic Anitron system, 61
  with anaerobic filter, 70
  anaerobic pretreatment of leachates in, 63
  with brown coal coke adsorption, 21
  with carbon adsorption, 353, 356–357
  denitrification in, 171
  with digester, 63
  effect of contaminant concentration on, 61
  with GAC medium, 63
  hydraulic retention times, 63
  with immobilized bacteria, 58
  with incinerators, 10, 359
  nitrification in, 61
  with resin adsorption, 21
  series operation, 63
  VOC destruction in, 363
GAC–BAC, 64
granular activated carbon fluidized-bed reactor (GAC–FBR), 63
Henry's law constant in, 58
immobilization
  of algae in carrageenan, 65
  of enzymes, 66
  of living cells, 58
  media, 65, 58, 165
  of nonliving biomass, 58
  of *Trichoderma harzianum*, 65
innovative fixed-film processes, 65
loading rate for, 58
membrane biological reactors (MBR), 60–64
membrane aerobic/anaerobic reactor system (MARS), 62–63
with PAC medium (MBR PAC), 63
microorganisms in, 58
organic limit, 57
organic removal efficiency, 58
for PAH-contaminated aqueous streams, 58
predator grazing in, 58
PUF (porous polyurethane foams), immobilized enzymes or living cells, 66
rotating biological contactors (RBCs), 57, 59–60
  Beggiatoa in, 60
substrate removal rate, 58
suspended detached cells, 58
trickling filters, 57, 59
with upflow anaerobic sludge blanket (UASB), 73, 422
Fixed-film systems, anaerobic, 74; see also Anaerobic fixed-film systems
Fixed/fluidized-bed regeneration, 353
Flagella, 207
Flame ionization detector (FID), 401
Flameless thermal oxidizer, 9
Flares, 333, 363, 369
Flavobacterium, 2
Floating oil, 211
Flocculants biosurfactants, 308
  in upflow floc digester, 73
Flocculation, 18, 28
Flocculation/coagulation/sedimentation, 83
Flocculation/flotation, 28
Flocculation/sedimentation, 28

Flocor media, 59
Flooding, 81, 93
Flood irrigation, 225
Flotation, 10, 28
Flow cytometry, 407–408
Fluidized-bed, see Fixed-film systems
Fluid wall reactors, 359
Fluoranthene biodegradation, 43, 155
Fluorene biodegradation, 155
Fluorescein diacetate-hydrolyzing activity (FDA), 292
Fluorescein isothiocyanate (FITC)/flow cytometry, 407
Fluorescence probe, 305
Fluorescence spectroscopy, surfactants, 145, 304
Fluorescent and nonfluorescent pseudomonads, 152
Flushing, 332; see also Hot air flushing; Steam flushing; Soil flushing combined air-water flushing, 244
Fly ash, 12, 80
Foam flusing, 164
Foam fractionation, 83
Foam front advancement, 165
Foams
  for channeling, 164, 242
  for clogging, 305
  encapsulation of low-permeability lenses by, 306
  formation and size of, 164
  microbubbles, 306
  for NAPL treatment, 164, 165, 306
  support in fixed-film systems, 58
Food-to-microorganism (F/M) ratio, 56
Forced aeration
  in biopiles, 52, 245
  in composting, 42
Fourier transform infrared spectroscopy (FTIS), 383
Four or more rings, degradation of, 48, 49, 145, 146, 147, 153, 211, 292
Fracturing/bioremediation, 109
Fracturing, soil, see Hydraulic fracturing; Pneumatic fracturing
Freeboard depth, 333–334
Free product recovery, 7, 8, 105–106, 165
Freeze-drying cultures, 289, 290
Freezing cultures, 284
Freundlich isotherm, 205
Froth flotation, 11, 48
FTIR (Fourier transform infrared spectroscopy), 383
Fuel components/hydrocarbons, biodegradation by specific microorgansisms, 132–139, 263–277
Fuel-contaminated soil, treatment of
  by bioremediation, 100
  by bioventing/soil warming, 103
Fuel oil-contaminated soil, treatment of
  by RF heating, 98
  by soil washing, 10
Fuels, carbon adsorption of, 20
Fulvates, 257
Fulvic acids, 145, 154, 211, 258
Fume incineration/air stripping, 87
Fume plate method, 389
Fungal bacteriophages, 276
Fungal biodegradation
  counts with Calcofluor W®, 391
  monitoring, 399
Fungal biomass, 186
Fungal compost bioreactors, 51
Fungal degradation, 267–274
  cytochrome P-450-dependency, 268
  enzyme immobilization for, 299
  hydroxylation in, 168
  lignocellulosic substrate for, 149

manganese peroxidase in, 270
microorganisms in, 268–269
of PAHs, 150
trans-dihydrodiols from, 268
Fungal metabolites, 194, 195
Fungal mycelia, 259, 268, 273
Fungal spores, 267, 268, 271
Fungi, 2, 267–274; see also Filamentous fungi; Yeasts
  at acidic pH, 230
  arene oxides from, 124, 127, 194, 230
  beneficial metabolites from, 194
  bioaccumulation in, 275
  for biphenyl degradation, 150
  broth cultures for isolation of, 389
  carbon sources for, 271, 273
  in composting, 44
  counting methods for, 387, 391
  cytochrome P-450 monooxygenases in, 194
  cytoplasmic inclusions in, 300
  detoxifications with, 141, 142, 268
  effect of soil moisture on, 224
  effect of temperature on, 228, 270
  encapsulation of, 290
  enrichment culturing of, 388
  extracellular enzymes in, 272
  lead toxicity to, 188
  mercury accumulation of, 189
  mutagens from, 142
  for metal binding, 179, 183
  oxygenases in, 194, 230
  PAH biodegradation by, 49, 50, 51, 54
  screening for, 273
  selenium biomethylation by, 191
  soil inoculation of, 273
  trans-dihydrodiols from, 194
Fungi carpophores, 190
Fungizone, 387
Furrow irrigation, 226FyreZyme™, 299

G

GAC–FBR (granular activated carbon fluidized-bed reactor), 63
GAC (granular activated carbon), see Granular activated carbon (GAC)
Gallionella, 416
Gas chromatography/mass spectrometry (GC/MS), 405–406
Gaseous carbon adsorption, 350
Gases in subsurface, 317
Gas–liquid chromatography (GLC)/column chromatography, 39
Gas masks, 351
Gasoline
  biodegradation of, 132–139, 263–277
    by mutualism, 132, 134
  composition of, 115–121
  migration of, 208
Gasoline and diesel fuel-contaminated soil, treatment of, by bioaugmentation with indigenous bacteria, 280
Gasoline-contaminated aqueous stream, treatment of, by BIOPUR®, 66
Gasoline-contaminated soil and groundwater, treatment of, by electrobioremediation, 111
Gasoline-contaminated soil, leaching in, 13
Gasoline-contaminated soil, treatment of
  by encapsulation, 12
  by HRUBOUT® process, 96
  by pneumatic fracturing, 108
  by surfactant/cosurfactant–water solution, 83
  by SVE, 89, 91
  by wet or dry steam and vacuum, 98
Gasoline-utilizing microorganisms, counting method for, 387
Gasoline vapors, 91, 234, 322, 323, 379
Gas-permeable-membrane-supported (BPMS) reactor, 76
Gas phase adsorbents, 15
Gas-suspended biomass systems, 59
GAS 3D model for SVE, 91
Gas turbines, 74
GC/MS (gas chromatography/mass spectrometry), 405–406
Gene fusion, 182–183, 398
Gene probe detection, direct, 397
Gene probes, 284, 398
Genes, 283
  for heavy metal resistance, 182
Gene sequences, 398
Genetically engineered microorganisms, 3, 285, 286, 397, 398; see also Mutant microorganisms
Genetic engineering, 279, 283, 284, 294
Genotypes, 398
Geranyl–coenzyme A carboxylase, 126
Glass fibers, hollow, 65
GLC (gas–liquid chromatography)/column chromatography, 39
Glutaraldehyde, 58
Glycolipids, production of, 154
Gram-negative bacteria
  attachment to soil, 287
  increase during bioremediation, 391
  PAH degradation by, 144, 287
  preservation of, 290
Gram-positive bacteria
  attachment to soil, 287
  PAH degradation by, 146, 154
Granular activated carbon/biofilters, 379, 424
Granular activated carbon/clarification/air stripping/reinjection/bioremediation, 421
Granular activated carbon fluidized-bed reactor (GAC–FBR), 63
Granular activated carbon (GAC), 19, 64
  in activated sludge process, 421
  with biofilters, 424
  in biological activated carbon systems, 64
  by-products from leachate treatment with, 17
  in fluidized-bed reactors (FBR), 63
  for soil washing, 11
  for VOCs, 89, 351
Granular activated carbon/ozonation, 24, 422
Gravel slurry reactors, 49
Gravity sedimentation, 15
Grazing, 58, 275, 276–277, 299, 415; see also Predation; Protozoa
  antibiotics against, 299
  effect on acclimation period, 276, 277
Grease-contaminated aqueous stream, treatment of
  in MARS, 63
  by flotation, 28
Grease, half-life, 39
Green algae, 180, 181, 190, 150, 275
Green sulfur bacteria, 286
Groundwater contamination, treatment of
  by BioSparge℠, 107, 246
  by composting, 46
  by dissolved air flotation, 346
  by landfilling, 37
  technology for, 101
  treatment applicability, 8
Groundwater recovery technologies, 8
Growth-decoupled enzymic metal removal, 183
Guar gum, 107

H

Half-life
  of alkanes, 39
  of anthracene, 152
  calculation by Monod kinetic analysis, 383
  effect on remediation time, 408, 409
  formula for estimating, 314
  of grease, 39
  of naphthalene, 39, 412
  of oil, total, 39
  ozone, 237
  of PAHs, 39, 40, 144
  of phenanthrene, 152
Half-lives, contaminant, 412
  from first-order kinetics, 39
  in landtreatment, 38, 39
  of PAHs, 144
Halophilic methanogenic bacteria, 266
Hazardous waste treatment storage and disposal facilities (TSDFs), 317, 318
Heap leaching, 11
Heating oil degradation, 158
  by landtreatment, 38, 40
Heating soil, 95–98
  with bioventing, 423
  with soil vapor extraction, 91
Heavy fuel-contaminated soil, treatment of, with bacteria, 97
Heavy fuel oil-contaminated soil, treatment of, by steam injection, 96
Heavy metal binding/bioaccumulation/biosorption, 66–67, 178–192
  acclimation of microbes for, 66
  by activated carbon/BAC systems, 65
  by algae, 180, 183, 188, 284
  by bacteria, 183, 190, 275
  by bark, 255
  by biofilms, 59
  biological/metal interactions in, 180
  biomass source for, 183
  biomethylation in, 178, 181, 189
  biosorbant regeneration for, 184
  B_{12}-dependence of, 181
  by cell walls, 181, 186, 192
  control measures with hydrogen peroxide for, 240
  by earthworms, 276
  effect of pH on, 180, 181, 190
  environmental modifications for, 180, 181
  by exopolysaccharide, 186–187
  factors affecting, 179
  by fungi, 189
  by growth-decoupled enzymic process, 183
  by heat killed bacteria, 184
  indicator organisms in bioassays for, 180
  by intracellular traps, 178, 179, 188, 190
  by isoelectric focusing, 99
  metal recovery by, 66
  by metallothioneins, 179, 185
  methylation products from, 181
  by microbial products, 179
  organometallic compounds from, 181
  by phytoremediation, 113
  plant bioaccumulation, 53, 113, 183, 277
  precipitation at cell surface with, 27, 180
  by redox chemistry, 182
  by siderophores, 179
  from wet-scrubber blowdown streams, 66
Heavy metal-contaminated aqueous stream, treatment of
  biosorbent regeneration for, 184
  by chemical precipitation, 27
  by flocculation/sedimentation, 28
  by immobilized living cells, 183
  by immobilized nonliving biomass, 183
  by organic materials, 259
  by precipitation, 27
  removal approaches for, 183
  by sewage treatment, 183
Heavy metal-contaminated biomass, disposal of, 184
Heavy metal-contaminated soil, treatment of
  by electrobioremediation, 111, 112
  by electrokinetic soil flushing, 86
  by electromigration, 99
  by microbial oxidation-reduction reactions, 183
  by quartz furnace, 9
  by sequestering, 183
  by soil washing, 11, 183
  by soil washing/landtreatment, 84
  by steam injection/bioremediation, 98
  by thermal treatment, 9
  by volatilization, 183
Heavy metal resistance, 178–184, 185, 190–191
  by efflux mechanisms, 186, 188, 191
  by electrokinetic potential reversal on cell walls, 181
  by gene fusions, 182–183
  by metal-binding proteins, 178
  by microbial adaptation, 178, 182
  microorganisms exhibiting, 182, 186, 188, 189, 192
  by phytoremediation, 179, 259
  by plasmids, 179, 182, 185, 187, 190, 191
  test for, 180
Heavy metals, 178–192
  detoxification of, 178, 180, 182, 189
  effect of pH on, 188
  entrapment in soil from isoelectric focusing, 99
  hydrogen sulfide production, 179
  in leachates, 54
  leaching of, 182
  metalloids, 181
  migration controls for, 36
  mobility in wastes of, 180, 258
  recovery of, 66, 184
  in sewage systems, 54, 258–259
  sulfate precipitation of, 180
  sulfate reduction of, 181
  toxicity of, 180, 181, 187, 188, 189, 190
  uptake of, 179
Heavy metals immobilization
  by bioscreens, 112
  in landfarming, 32
  by organic matter and pH adjustment, 259
Heavy metal sorption, competition with contaminants, 260
Heavy metals removal, see Heavy metal binding/bioaccumulation/biosorption
Helber counting chamber, 385
Henry's law constant, 81
  with air stripping VOCs, 345
  application of, 345
  effect on vapor phase transport, 327, 330
  in fixed-film reactors, 58
  for liquid alkanes, 331
  for several compounds, 330
Heptane biodegradation, 159
Heterocycle biodegradation
  in bioslurry reactors, 47
  in EIMCO™ reactors, 49

Heterogeneity of soil, 200
Heterotrophic microorganisms
   bacteria, 264
      denitrifier counts, 387
      fungi, 2
      methods for obtaining energy, 264
Heterotrophic potential, 264
Heterotrophs, 264
Heterotrophy with phototrophs, 274
Hexadecane biodegradation
   in anaerobic lagoon, 73
   by fungi, 158
   as indicator for biodegradation potential, 158
Hexadecane-contaminated soil, treatment of, by composting, 45
Hexane biodegradation, 156, 158
Higher life forms, 276–277
High-frequency field heating, 98
High-temperature thermal soil treatment, 10
Homoacetogenic bacteria, 266
Homogenation of soil, 50
Homologous series of molecules, 282
Homologous substrate, 293
Homologues, surfactants, 301
Hopanes
   as biomarkers, 400
   persistence of, 130
Hopane series, 156
Hopanoic acids, as biomarkers, 400
Horseradish peroxidase, 298
Hot air flushing, 96
Hot air injection, 95, 96
Hot spots of contaminants, 225
HRUBOUT® process, 96
Human health criteria, 322
Humic acids
   deactivation of peroxidase by, 298
   effect on microbial membrane permeability, 302
   solubilization of during leaching, 14
Humic material
   attenuation with, 257
   complex formation with, 194, 204
   effect of pH on, 257
   percent of soil organic matter, 257
   sorption to phenol oxidases in, 205
Humidity, see also Relative humidity
   effect on sorption, 327
   effect on VOCs, 333
   retardation coefficients in, 224
Humification in composting, 45
Hybridization, 399
Hybridization to colony DNA, 396
Hybridization to community DNA, 396
Hybridization to community RNA, 397
Hybridization to DNA from enrichments, 396
Hydratable polymeric materials, 79
Hydrated lime in bioreactors, 47
Hydraulic conductivity, 10
Hydraulic fracturing, 107–109, 244; see also Pneumatic fracturing
Hydraulic injection/electrobioreclamation, 111
Hydraulic loading in landtreatment, 39
Hydraulic/pneumatic fracturing/bioaugmentation, 108
Hydraulic retention times in reactors, 63, 66
Hydroautomatic biofilter in ABFs, 65
Hydrocarbon biodegradation, sequence of, 127
Hydrocarbon concentration for monitoring bioremediation, 401
Hydrocarbon degraders, see Hydrocarbonoclastic microorganisms

Hydrocarbonoclastic microorganisms, 132–139, 263–277; see also Anthropogenic hydrocarbons; Aromatic hydrocarbons, biodegradation of; Bioaugmentation microorganisms
   addition to activated sludge, 54
   culture collection, 278
   fungi, isolation of, 389
   genera, 263
   for monitoring hydrocarbon contamination and remediation, 394
   percentage of population, 263
   superbugs, 101, 103, 278, 283
   seed organisms, 387
   yeasts, isolation of, 388
Hydrocarbon uptake, models for, 163
Hydrocyclone, 11
   in bioreactors, 47, 48
Hydrogen peroxide, 24, 237–241
   for aromatics in wastewaters, 298
   encapsulation for hydraulic fracturing, 108
   hydrogen peroxide:organism ratio, 238
   for mobility control, 79
   mobilization of metals by, 240
Hydrogen peroxide/ozone, 238
Hydrogen peroxide/UV light, 24, 27, 422
Hydrogen peroxide/Vyrodex method, 240
Hydrogen-producing, acetogenic bacteria, 266
Hydrogen sulfide production, 179, 189
Hydrogen-utilizing methanogens, 70
Hydrolysis, 80, 212–213
   detoxification by, 212
   in landtreatment, 30
   for PAHs, 141
   separation of surfactant from contaminant by, 83
Hydrolytic bacteria, 166, 266
Hydrophobic fertilizer, 256
Hydrophobic interactions, microbial sorption to soil, 288
Hydroxylases, 126
Hydroxylation
   of aromatics, 131
   detoxification by, 212
   of fungi, 268
Hydroxypropyl-β-cyclodextrin (HPCD), 84
Hyonic NP-90, 305
Hypersaline environments, 143
Hypochlorite, 241
HYPOL FHP 2000/3000, prepolymer, PUF from, 66

I

ICB (Allied Signal immobilized cell bioreactor), 65
ICP/MS (inductively coupled plasma/mass spectrometry), 406
Igepal CA-720, 149, 304
Igepal CO-603, 305
Ignition of oil, 212
Immobilization of contaminants, 80, 99, 300
   of heavy metals, 32, 36, 259
   by RF heating, 80
Immobilization of microorganisms, 290; see also Encapsulation
   aerobic/anaerobic treatment, 169, 249
   with agar, 58
   with alginates, 58, 65, 291
   with BAC, 58
   beads for, 291, 297
   for bioaugmentation, 147–148
   with Biofix and Drizit, 156

INDEX

of biomass in alginate, 183
diatomaceous earth for, 65
Diauxie with, 297
for diesel-contamination, 12, 156
in fixed-film systems, 65
for heavy metal-contaminated aqueous streams, 183
in landtreatment, 30
of living and nonliving biomass, 58
for PAH degradation, 146, 287
with PUF, 66
Immobilized cell bioreactor (ICB), 65
Immobilized enzymes, 66, 298, 299
Immobilized nutrients, 255
Immunofluorescence microscopy, 386
Incineration, see also Combustion
costs, 37
limitations of, 415
as soil treatment, 9–10
for VOC treatment, 363, 364–369, 370, 371
Incinerators, 9, 10, 364–369, 370, 371
Incomplete oxidation products, 195
Indigenous hydrocarbonoclastic microorganisms, 263
Indigenous microorganisms, 85, 100, 101
advantages of, 263
bioaugmentation with, 280
with bioventing, 104
competition with bioaugmented microorganisms, 285, 277
counts in bioslurry reactors, 48
in sewage sludge, 258
stimulation of, 280
Indole degradation, 64
Inducer molecules, 293
Inducible enzymes, 274, 297; see also Enzyme induction
Inductively coupled plasma/mass spectrometry (ICP/MS), 406
Inert gas condensation solvent recovery system, 373
Inflow/outflow drainage pipe locations, 333, 334
Infrared furnaces, 359
Infrared gas analyzer, 403
Infrared spectroscopy, 401
Infrared transmitting optical fibers, 383
Inhibition of biodegradation, 145
Inipol EAP-22, 256, 306
Injection
of air, 242, 96
of liquified gases, 242
of oxygen-releasing compounds/nutrients, 242–243
at shallow injection sites, 243
of steam, 96–98; see also Steam injection
of water, 241
Injection aerators, 347
Injection–extraction process, 87
Injection pipes, 81, 93
Innovative fixed-film processes, 65
Inoculant formulations, 279, 280, 187; see also Bioaugmentation; Soil inoculation
Inorganic indicators of bioremediation, 403–404
Insects, 275, 276, 299
In situ bioremediation
applicability of, 7
kinetics parameters for, 413
in landfarming, 31
limitations of, 415–417
processes, 100
site and soil characteristics for, 220
In situ controls, 335–342
In situ electrical corona process, 79

In situ electrobioreclamation/electro-osmosis/electrokinetics/electrochemical remediation, 110–112, 244
In situ percolation, 7
In situ remediation
with Detoxifier™, 93, 94
processes, 78–113, 358–359, 362–363
treatment trains for, 419–420
INT activity test, 385–386
Interactions of microbes in soil, 286
Interface
effect on degradation rate, 201
modifiers, 79
NAPL/water, 163
solvent–water, 202
Interfacial tension, 287, 311
Intermediary metabolites
from biodegradation, 192–198
in methanogenesis, 173
for monitoring bioremediation, 405
Internal cell structure analysis, WALS, 407
Internal combustion engines with SVE, 92
Interpartical porosity, 377
Intracellular accumulation of metabolites, 195, 285
Intracellular electron transport inclusions, 158
Intracellular enzymes, 297
Intracellular polyester (PHB), 112
Intracellular precipitation of heavy metals, 179
Intracellular traps for heavy metals, 178, 179, 185, 190
Intraparticle porosity, 377
Intrinsic biodegration/bioremediation, 101, 417
In vessel (reactor) systems, composting, 44
Ion exchange
for leachate treatment, 16
leachate treatment by-products from, 18
as polishing step, 29
Ion exchange/air stripping/carbon sorption, 426
Ion exchange resins, 21
Iron
as alternate electron acceptor, 241
bioaccumulation of, 66
resistance to, 188
Iron bacteria, 139, 240, 241
Iron–sulfide precipitates, 168
Irrigation, 225–226, 229
Isoalkanes, degradability of, 130
Isoelectric focusing, 99
Isomers of surfactants, 301
Isoprenoids, 156, 400
Isothermal conditions, 328
Isotherms, 20
Isotope fractionation, 404
Isotopic labeling of contaminants, 405

J

Jet fuel contaminant, *Cladosporium resinae*, 158
Jet fuel-contaminated soil, treatment of, by HRUBOUT® process, 96
JP-4, composition of, 115, 122–124
JP-4 jet fuel-contaminated soil, treatment of
by bioreactor, 34
by bioventing, 104
by denitrification, 169
by *in situ* remediation, 34
by landfarming, 34
JP-5, composition of, 115, 121–122
JP-5-contaminated soil, treatment of

by bioslurping, 106
by cyclic steam injection, 98
by hydrogen peroxide, 238
JP-5-utilizing bacteria, counting method for, 389

K

KAX-50, KAX-100, 12
Kerosene-contaminated aqueous stream, treatment of, by BIOPUR®, 66
Kerosene-contaminated soil and groundwater, treatment of by electrobioremediation, 111
Ketones for carbon and resin adsorption, 20
Kiln dust for stabilization, 12
Kinetics of biodegradation, 411, 412, 413
Klebsiella aerogenes, capsular material for cadmium binding, 186
KLF 2000 for aromatic solvents, 68
K_{oc}, soil-water partition coefficent, 84, 206
  sorption onto organic matter, 204, 206
K_{ow} (octanol/water partion coefficient), 84, 206
K_p (linear partition coefficient), 206
K (soil adsorption constant), 205, 206

L

Labeled contaminants, 404
Lactose utilization gene (*lacYZ*), 398
Lagoons, 57, 160, 333, 334
  aeration of, 245
  anaerobic, 73
  bioaugmentation of, 57
  biomass recycle in, 57
  bioslurry reactor, 57
  cover system on, 335
  leachate by-product treatment in, 17
  mixflo process with, 246
  oily waste and petroleum treatment in, 57
Lag period/phase, 207, 227, 278, 281
Land disposal restrictions, EPA, 13
Landfarming, see also Bioremediation; Landtreatment
Landtreatment, 29–41
  abiotic losses in, 40
  anaerobic degradation in, 36
  anaerobiosis in, 34
  application rates in, 36, 38, 39, 40
  of aviation gasoline-contaminated soil, 34
  bioaccumulation in 37
  bioaugmentation in, 31
  biodegradation rates in, 38, 39, 40
  carbon-to-nitrogen ratio in, 35
  of crude oil-contaminated soil, 34
  desorption in, 34
  of diesel fuel, 38
  environmental factors, 34–36
  half-life of waste in, 38, 39, 144; see also Half-lives
  of heating oil, 38
  immobilization of metals in, 32
  of JP-4 jet fuel-contaminated soil, 34
  limitations of, 415
  methods of waste application in, 320
  monitoring, 32, 36, 39, 41
  of oil-contaminated soil, 33
  of oily wastes, 36
  of PAH-contaminated refinery effluent sludge, 34
  of PAHs, 40

  of petroleum-contaminated soil, 34
  surfactants in, 33
  VOCs in, 38, 317, 319, 320, 332
Landtreatment/precipitation, 422
"Lasagna" process, 108
Laser-induced fluorometry, 115
Leachates
  from biopiles, 52
  characteristics of, 13
  collection in Detoxifier™, 93
  control in bioreactors, 47
  control of production of, 419
  from desorption, 207
  factors affecting quality of, 14
  heavy metal concentrations in, 54
  profile, 14
  recycling of, 14
  separation of surfactant from contaminant in, 83
  treatability of, 15
Leachate treatment, 13–29, 54–78
  by air stripping, 28
  by-products of, 14, 15, 17
  control measures for emissions from, 15
  by flotation, 28
  by landfarming, 34
  by ozonation/UV, 26
  polishing by activated carbon, 83
  pretreatment of, 14, 63
  process applicability for, 16
  process train for, 425
  by sedimentation/chemical precipitation, 28
  by steam stripping, 343
  systems
    biological, 54–78
    chemical, 13–29
  VOC release from, 54
  by wet air oxidation, 22
Leaching 14, 15
  effect of adsorption on, 203
  effect of contaminant immobilization on, 80
  in landtreatment, 30
  of PAHs, 141
  with rainfall, 326
  of VOCs, 13
Lead, 66, 80, 188–189
Lead additives, 406
Lead biosorption, 189
Lead-contaminated aqueous stream, treatment of, by chemical precipitation, 27
Lead-contaminated soil, major compounds in, 188
Lead-contaminated soil, treatment of
  by electrokinetic soil flushing, 86
  by soil washing, 86
  by stabilization processes, 12
Lead methylation, 189
LEL (lower explosive limit), 364
Lenses foam encapsulation of, 306
  NAPL, 160, 161, 165
Lewis acids, 180
Lichenysin A, 311
Light nonaqueous-phase liquid (LNAPL)-contaminated soil and groundwater, treatment of, by bioslurping, 105–106
Light nonaqueous-phase liquid (LNAPL) contamination, treatment of
  by bioslurping/bioventing, 165
  by colloidal gas aphrons, 164

Light nonaqueous-phase liquids (LNAPLs), 160–165; see also Dense nonaqueous-phase liquids (DNAPLs); Nonaqueous-phase liquids (NAPLs)
  dual-beam gamma densiometer for, 165
  free phase recovery of, 165
  transport of, 161
Lignin, 270
Ligninases, 272
Ligninolysis, 270, 272
Lignin peroxidase (LiP), 147, 270, 272
Lignite adsorption, 21
Lignocellulosic substrates, 150, 155
Liming, 80, 148, 231, 232
Limitations of soil treatment systems, 415–417
Linear partition coefficient, (K_p), 206
Linear primary alcohol ethoxylate surfactants, 313
Lipid biomarkers, 391
LiP (lignin peroxidase), 147, 270, 272
Lipopeptide surfactant, 311
Liposomes
  application of, 311
  for bioremediation, 311
  for encapsulation, 291–292
  formation *in situ*, 312
  nutrient encapsulation by, 255
Lipotin, 305
Liquid hydrocarbons, 202
Liquid injection incinerators, 10, 363
Liquid-phase carbon, 8
Liquid-phase hydrocarbon (LPH), 7
Liquid/solid contactors, 11
Liquified gases
Lithotrophs, 264
LNAPLs (light nonaqueous-phase liquids), see Light nonaqueous-phase liquids)
Loading capacity
  of activated carbon systems, 20
  with denitrification in fluidized-beds, 171
Loading rates, see also Application rates
  with activated sludge process, 56
  in anaerobic digester, 71
  with anaerobic fixed-film systems, 74
  with biofiltration, 378
  in landtreatment, 36, 38, 39, 40; see also Application rate
  with RBCs, 60Loam, 36
Logarithm of octanol/water partition coefficient (log P or log K_{ow}), 84, 162, 202, 206, 212
Log $K_{OH}°$ for various organic compounds, 314
Log K_{ow} (logarithm of octanol/water partition coefficient), 162, 202, 206, 212
Log P (logarithm of octanol/water partition coefficient), 84, 206
London–van der Waals forces, 288
Long-chain alkanes, 127
Lower explosive limit (LEL), 364
Low-molecular-weight PAH/high molecular-weight PAH (LPAH/HPAH) ratios, 406
Low-permeability lenses, 306
Low-permeability soils, treatment of
  by electrobioreclamation, 110, 111
  by *in situ* electrical corona process, 79
Low-temperature soil roasting, 6
Low-temperature thermal stripping, 6, 12, 359, 360, 361, 362
LPAH/HPAH (low-molecular-weight PAH/high-molecular weight PAH), ratios, 406
Lubricant-contaminated soil, treatment of, by HRUBOUT® process, 96

Lubricating oil, 44
Lubricating oil-contaminated soil, treatment of, by soil washing, 10
luxAB (bioluminescence gene), 398
Lysimeter samples, 403

M

Macroemulsions, 164
Macrofauna, 275, 276
Macroorganisms, 30
Macropore, 288
Macroreticular resins, 21
Magnetic resonance imaging (MRI), for DNAPL-soil interactions, 164
Maintenance energy, 214
Malodorous compounds, 233
Malt agar, 388
Manganese bacteria, 240
Manganese-contaminated aqueous streams, treatment of, by chemical precipitation, 27
Manganese-dependent peroxidase (MnP), 272
Manganese-inhibited peroxidase, of *Bjerkandera* sp. BOS55, 271
Manganese peroxidase (MnP) 147, 151, 270, 271
Manure amendments, 52, 258
Marker genes, 398
MARS (membrane aerobic/anaerobic reactor system), 62–63
Mass spectrometry (MS), 405–406
Mass transfer, 207
  in air stripping, 344
  with biofilters, 379
  in bioslurry reactors, 48
  coefficient, 13, 59, 326
  driving force for, 326
  effect of depth of soil on, 326
  in enclosed reactors, 5–51
  limiting factor in field, 200
  of NAPLs, 161
  of PAHs, 143, 145
  of PAHs in NAPLs, 161, 162
  in steam stripping, 98
  in trickling filters, 59
  in water treatment, 344
Maturing process of crude oil, 400
Mechanical surface aeration, 347
Medium-distillate fuel-contaminated soil, treatment of, by bioremediation, 103
Melanic materials, 14
Membrane aerobic/anaerobic reactor system (MARS) 62–63; see also Fixed-film systems
Membrane biological reactors, 60–64; see also Fixed-film systems
  MBR-PAC, 63
Membrane covers, 335–340
Membrane filter counts, 390
Membrane liner materials, 335–340
Membrane lipid, 159
Membrane permeability, 130, 302
Membrane vesicles, extracellular, 312
Mercury, 189–190
Mercury-contaminated aqueous stream, treatment of, by chemical precipitation, 27
Mesocosm, 33
Mesophiles, 227, 228
Metabolic pathways, 126
Metabolic potential of microbial community, 406

Metabolites, 192–198, 285; see also By-products; End products
 heavy metal removal by, 179
Meta cleavage, 131
Metal-binding proteins, 178
Metalloids, 181
Metallothioneines, 179, 185
Metals, heavy, see Heavy metals
Meta substitutions, 209
Methane, 2
 anaerobic degradation indicator, 403
 biodegradation of, 156, 173
 from blue-green algae, 266
Methane bacteria, 172
Methane monooxygenase (sMMO), 299, 399
Methanogenesis, 172–173
 inhibition of, 78
 monitoring, 406
Methanogenic bacteria, 70, 166, 173, 266
 coenzyme M in, 266
 competition with sulidogens, 167
 counting method for, 390
 factors F_{420} and F_{430} in, 173, 266
 fluorescent molecules in, 407
 halophiles, 266
 nickel requirement of, 70, 173, 266
 in pneumatic fracturing, 108
 rate-limiting factor in reactors, 173, 266
 sequestration of electron flow by, 167
 stable isotope analysis for, 404
 thermophiles, 266
Methanol, 159, 170
Methanosarcina barkeri, in anaerobic fixed-film systems, 74
Methanothrix-like microorganisms, in UASB digesters, 73
Methanotrophs, 173, 186, 188, 192
Methylation, 189, 190, 191, 212
Methyl ethyl ketone (MEK) biodegradation, 74
Methylmercury, 190
Methylotrophs, 76, 78, 156
Methylquinone biodegradation, 64
Micellar partitioning, 304
Micelle-phase/aqueous-phase partition coefficient, Km, 303
Micelles, see Critical micelle concentration; Surfactant micelles
Michaelis–Menten kinetics, 408
Microarthropods, 276
Microautoradiography, 391
Microbeads, 290
Microbial accumulation of metals, 66–67
Microbial activity, 407
Microbial adsorption/aerobic reactors, 67, 68
Microbial community analysis, 405, 407
Microbial consortia, see also Mixed cultures anaerobic, 166, 265–266
 for bioaugmentation, 286–287
 metabolic potential of, 406
 PAH degradation by, 146, 287
 surfactant biodegradation by, 313
Microbial counting methods, 383–396, 404, 408; see also Bioremediation monitoring methods
Microbial counts, see Bioremediation monitoring methods; Counts of microorganisms
Microbial ecology approach, 101–102
Microbial enzymes, 298; see also Enzymes; Cell-free enzymes
Microbial interactions, 264, 286
Microbial leaching, 66
Microbial nutrient, 237
Microbial predators, 276, 299

Microbial preservation, 284, 289–290; see also Encapsulation; Immobilization
Microbial respiration, 232
Microbial respiration quotient, 403
Microbial seeding, 263
Microbial surfactants, 84, 149, 154, 163, 302, 307–312
Microbial transport, 287–289
Microbiological approach, 101–102
Microbiological diffusion, 288
Microbodies, 300
Microbubbles, 164, 306
Microcalorimetry, 407
Microcapsule technique, 291
Micrococcus, 2
Micrococcus varians, in fixed-film systems, 58
Microcolony formation, 384
Microcolony epifluorescence technique, 386
Microcosms, 101, 412
Microemulsion, 304, 312
Microencapsulation, 290
Microfiltration (MF), 29
Microorganisms/analog enrichment, 423
Microorganisms, degradative, for specific fuel components/hydrocarbons, 132–139, 263–277
Microorganisms in bioremediation, see Aromatic hydrocarbons, biodegradation of; Anthropogenic hydrocarbons; Bioaugmentation microorganisms; Hydrocarbonoclastic microorganisms
Micropores, soil, 34, 161
Microscopic counts, 385–386, 391
Microspheres for encapsulation, 291
Microsphere tracers, 290
Migration
 contaminant, 202–203, 207, 208, 220, 228
 heavy metal, 36, 180, 182, 258; see also Immobilization
 microbial, 221, 287, 288, 289
Mineralization, 2, 115, 214, 217, 292
 of heavy metals, 178–198
 of organic compounds, 115–178
 rates, 411, 412
Mineral matter, 329
Mineral soil, 219
Mixed cultures, see also Microbial consortia
 in anaerobic degradation, 166–167, 196, 265–266; see also Anaerobic bacteria;
 for anthracene biodegradation, 152
 bacteria and fungi, 195
 for bioaugmentation, 278
 in biofilters, 377
 breakdown of toxic products by, 212
 cometabolism with, 294
 for complete degradation, 196, 248
 denitrification in, 174, 175
 diauxie effect on, 297
 metabolic potential of, 406
 for PAH biodegradation, 144, 152, 155, 158, 287
 for petroleum biodegradation, 263
 for pristane biodegradation, 156
 for sequential anaerobic/aerobic treatment, 76
 starter culture, 287, 289
 surfactant biodegradation by, 313
Mixed liquor–volatile suspended solids (MLVSS)
 in activated sludge process, 56
 in fluidized-bed reactors, 61
Mixed oils, 400
Mixed oxygen/nitrate conditions, 171, 174, 175, 176, 177
Mixflow process, 246

INDEX   527

Mixing trommels, 48
Mixture toxicity index, 212
MLVSS (mixed liquor-volatile suspended solids), see Mixed liquor-volatile suspended solids (MLVSS)
sMMO (methane monooxygenase), 299
MnP (manganese-dependent peroxidase), 272
MnP-based lipid peroxidation systems, 151
Mobile systems
    composting unit, 45
    Detoxifier™, 93, 94
    fluidized-bed incinerator, 359
    fluid wall reactors, 359
    incinerator, EPA, 9
    infrared furnaces, 359
    revolving tubular reactor, 50
    SCF oxidation unit, 12
    soils washer, EPA, 10
    steam stripping systems, 98
    thermal treatment systems, 359
Mobility of contaminants, 37, 79, 80, 81, 164, 202
Mobilization, 80
Modeling, 408
Models
    for biofiltration, 378
    for bioremediation monitoring, 408
    cost of remediation (CORA), 5
    for effect on vapor-phase sorption, 326
    gas 3D for SVE, 91
    for evaporation rate, 317
    for hydrocarbon uptake, 163
    for immobilized cells, 297
    for micellar partitioning, 304
    for NAPL retention, 162
    for NAPLs, 165, 304
    for SVE systems, 92
    for toxicity, 212
    for VOCs, 91–92
Modified activated sludge, 58
Moisture, see Soil moisture
Molds, 388
Molecular diffusion, 288, 326, 331
Molecular weight, 331
Mole fraction of diffusing compound, 327
Molybdenum, 66
Moniliales, 268
Monitoring anaerobic biodegradation, 403; see also Bioremediation monitoring methods
Monitoring bioremediation, see Bioremediation monitoring methods
Monoclonal antibody immunoassay, for VOCs and SVOCs, 322
Monod kinetic analysis, 383
Mononuclear aromatic hydrocarbons, biodegradation of, 134, 139–141, 173–177
Monooxygenase, 129
Monoterminal oxidation, 128
Montmorillonite, 204, 248, 262
Most-probable-number (MPN) method, 389–390, 391, 403, 404
Motile bacteria, 288
Motor oil-contaminated soil, treatment of, 11
    by encapsulation, 12
Moving-bed biofilm reactor, 65
Moving-bed reactors, 64
MPN (most-probable-number) methods, 389–390, 391, 403, 404
mRNA extraction, 399
MS (mass spectrometry), 405–406
Mucorales, 268
Mulchs, 228, 229

Multiphase, multicompositional simulator (UTCHEM), 304
Multiphase system with NAPLs, 162
Multiple hearth incinerator, 10, 363
Multiple plasmid transfer, 149
Multiple technologies, 420
Multiring aromatics, oxidation of, 48
Mushroom spawn, 273
Mutagens, 141, 142, 194
Mutant bacterial formulations, 285
Mutant microorganisms, 283–286
Mutation, induced, 279
Mutualistic relationships, 286; see also Microbial consortia
Mycelia, 268, 273
Mycobacteria, 267, 276
Myo-inositol, 290
Myxobacteria, 276

N

nahA, 399
NAH plasmids, 48, 284, 398
NALS (narrow angle light scatter), 407
Naphthalene
    biosurfactant production on, 154
    glycolipid production, 154
    half-life of, 39, 412
    solubility of, 148
    toxicity of, 148
Naphthalene biodegradation, 148–150
    anaerobic, 177
    with biosurfactants, 149, 154
    composting of, 43
    emissions in bioslurry reactors, 49
    in fixed-film reactors, 58
    by landfarming, 37
    under mixed oxygen/nitrate conditions, 177
Naphthalene-contaminated aqueous stream, treatment of, by BIOPUR®, 66
Naphthalene solubility, 149
NAPLs (nonaqueous-phase liquids), see Nonaqueous-phase liquids (NAPLs)
Natural attenuation, 1, 101, 417
Natural gas-fired, batch, rotary kiln, 6
Natural soil bacteria, 2
ndoB (naphthalene degradation) gene probes, 284
Nematodes, 276, 299
Neoprene membrane liner material, 338
Neutralization, 80, 93
Nickel, 190–191
    removal by green algae, 180
Nickel-contaminated aqueous waste, treatment of, by chemical precipitation, 27
Nickel-contaminated soil, treatment of by stabilization processes, 12
Nitrate
    for biodegradation, 171, 241
    for denitrification, 169
    as electron acceptor, 241
    as polutant, 171, 241
    as terminal electron acceptor, 169
Nitrate limitation, 170
Nitrate/oxygen treatment, 171
Nitrate reducers, 168
Nitrification
    in fluidized-bed reactors, 61, 76
    in RBCs, 60
Nitrifying bacteria

in anaerobic lagoons, 73
  biosorption with, 300
Nitrogen fixation, 252
Nitrogen limitation, 252
Nitrogen/oxygen treatment, 171
Nitrous oxide for monitoring bioremediation, 403
Nocardia, 2
NOCs (nonionic organic compounds), 303
Nonaqueous-phase liquid (NAPL)-contaminated soil, treatment of
  by soil heating, 95
  by SVE, 92–93
  thermal gradients for, 164–165
Nonaqueous-phase liquid (NAPL) lenses, 160, 161, 165
Nonaqueous-phase liquid (NAPL) soil residuals, treatment of, by steam injection, 97
Nonaqueous-phase liquid (NAPL)-soil-water system, 161
Nonaqueous-phase liquid (NAPL) sorption, 162
Nonaqueous-phase liquid (NAPL)-water interface, 163
Nonaqueous-phase liquids (NAPLs), see also Dense nonaqueous-phase liquids (LNAPLs); Light nonaqueous-phase liquids
  bioaugmentation for, 163
  biodegradation of, 160–165
  bioemulsification of, 163
  model for migration and remediation of, 304
  photomicrographs of, 165
  pseudosolubilization with, 163
  reaction with surfactants, 303
  sequestration, 161, 217
  toxicity of, 212
Nongrowth substrate, 292
Nonindigenous degradative organisms, 290
Nonionic organic compounds (NOCs), 303
Nonionic surfactants, 153, 254, 301, 302, 303
  for NAPLs, 163, 164
  reduction of oxygen transfer by, 305
  sorption of, 307
Nonligninolytic conditions, 147, 270
Nonmethanogenic bacteria, 70
Nonpolar adsorbents, 21
Nonradioactive reporter genes, 399
Nonvolatile organic compound (NVOC)-contaminated aqueous stream, treatment of, by BIOPUR®, 66
Nonvolatile organic compound (NVOC)-contaminated soil, treatment of
  by resistance heating, 96
  by soil washing, 10
Nonvolatile organic compounds (NVOCs), 10, 66, 96
Novel II 1412-56, 303
NRV (nutrient requirement value), 251
Nucleic acid
  content of bacteria, 407
  hybridization, 284, 398
Nucleic acid-based monitoring methods, 396–400
Numbers of microorganisms, see Microbial counts
Nutrient application, gaseous, with bioventing, 424
Nutrient requirement value (NRV), 251
Nutrients, 249–257
  addition by
    BioPurge^SM, 254
    deep soil fracture bioinjection™, 254
    hydraulic fracturing, 108
  carbon-to-nitrogen (C:N) ratio, 250
  carbon-to-nitrogen-to phosphorus (C:N:P) ratio, 143, 250, 258
  for composting, 45
  control in bioreactors, 46

dispersal, 108, 109
encapsulation of, 255
fertilizers, 253–254, 256; see also Fertilizers food-to-microorganism (F/M ratio), 56
in landtreatment, 35
nutrient requirement value (NRV), 251
phosphate precipitation, 251, 255
plugging with, 252–253
slow-release, 259NVOCs (nonvolatile organic compounds), see Nonvolatile organic compounds (NVOCs)
Nystatin for protozoa, 299

O

Obligate anaerobes, 168, 172, 173, 265
OCA (oil-core aphrons), 306
Octacosane biodegradation, 227
Octadecane, 304
Octane biodegradation, 159
Octanol/water partition coefficient (K_{ow}), 84
  applications of, 206
  effect of sorption by organic matter on, 206
  for NAPLs, 161
  for predicting contaminant mobility, 202
Off-gas treatment, 346, 363
  biofiltration for, 376
  with Detoxifier™, 93
  technology applicability for, 8
Off-gas treatment/soil venting/air injection 423
Office of Research and Development network retrieval system, EPA, 5
Office of Safety and Health Administration (OSHA), 322
Office of Technology Assessment (OTA), 3
Oil agar no. 2, 387
Oil biodegradation by pleomorphs, 396
Oil-contaminated aqueous stream, treatment of by flocculation/flotation, 28
  by flotation, 28
  in MARS, 63
  by ultrafiltration, 29
Oil-contaminated soil, treatment of
  by hot water systems, 10
  by landfarming, 33
  by supercritical fluid (SCF) oxidation, 11
  by surfactant washing, 83
Oil-core aphron (OCA), 164, 306
Oil degradation products, 127
Oil dispersants, 301
Oil, heavy, landtreatment of, 40
Oil recovery, 311
Oil-sand aggregates, treatment of, 11
Oil tar-contaminated soil, treatment of, with *Phanerochaete chrysosporium*, 270
Oil, total, half-life of, 39
Oily sludges, biodegradation of, 38, 59
Oily wastes, treatment of
  by bioaugmentation of activated sludge, 54
  landtreatment of, 36, 37, 40
  in sewage lagoons, 57
Oil-water interface, 202, 300
Oil-water separation, 422
Oleophilic fertilizer, 149, 256, 302, 310
Oleophilic/liphobic activated carbon, 19
Oligonucleotide probes, 397
Oligotrophic conditions, for production of ultramicrobacteria, 266, 289
Oligotrophs, 259, 266, 267, 384

INDEX

On-site remediation
  biological systems, 417
  combined technologies for, 417–419
  limitations of, 417
On-site treatment processes, 5–78
Opacity tube method, 391
Operating practices, 334
Operating surface area, 325
Operon on NAH plasmids, 48
Optimization of bioremediation, 219–315
  biological enhancement, 263–300
  contaminant alteration, 300–315
  variation in soil factors, 219–262
Orange II, 273, 274
Organic aqueous streams, pretreatment of, 342–357
Organic matrix, 257
Organic matter, 257–260
  adsorption in landfarming, 34
  binding of extracellular enzymes by, 297–298
  effect on volatilization, 329
  K_{oc} value, 204, 206
  for metal immobilization, 259–260
Organocadmium compounds, 186
Organometal complexes, 181, 259
OR (oxidation-reduction potential), 260
"Ortho" cleavage pathway, 131
"Ortho" pathway, 131, 264
Oscillatoria, 275
OSHA (Office of Safety and Health Administration), 322
Overloading, 38, 39, 248, 252
Oxidation
  in Detoxifier™, 93
  for groundwater treatment, 8
  in landtreatment, 30
  types of, 126
α-Oxidation, 129
β-Oxidation, 126, 129, 131, 193
ω-Oxidation, 126
Oxidation of contaminants, 212
Oxidation/reduction, 79–80
Oxidation/reduction (OR) potential, see Redox potential
Oxidative cleavage of aromatic rings, 126
Oxidizing agents, 235
Oxitron aerobic fluidized-bed system, 62
Oxygen, see Soil oxygen
Oxygenases, 125, 126, 127, 232
  in cometabolism, 292
  in cycloalkane degradation, 130
  of fungi, 194, 230
Oxygenating cakes, 235
Oxygenation, 232–247
Oxygen mass transfer, with encapsulation, 291
Oxygen/nitrate
  electron acceptor conditions, 249
  treatment, 171, 174, 175, 176, 177
Oxygen-releasing compounds/nutrients, 242–243
Ozonation, 16, 24, 79, 236
Ozonation/granular activated carbon, 24, 422
Ozonation/UV, 24, 26–27, 422
Ozonation/UV/bioremediation, 422
Ozone, 24, 236–237
  half-life of, 237
  with PAHs, 236
  reduction of TOC toxicity by, 27
Ozone oxidation
  intermediate products of, 27
  off-gases treatment by, 27

P

Packed absorption column, 374
Packed and plate columns, comparison, 375
Packed-bed reactors, 49
Packed beds, 58
Packed columns, 372, 374, 375
Packed towers, 347, 350, 375
Packing material, 47, 50
  for air stripping, 345
  for biofilter, 376–377
PAHs (polycyclic aromatic hydrocarbons), see Polycyclic aromatic hydrocarbons (PAHs)
Para–Bac, 287
Partition coefficient, 34, 326
Partitioning of bacteria, 202
Partitioning of hydrocarbons, 302
Passive biosorption of heavy metals, 179, 180, 183
Passive soil vapor extraction, 89
Pathogenic microorganisms, 416
Pathways, 142
PCB-contaminated soil, treatment of, by encapsulation, 12
PCR (polymerase chain reaction), 284, 397, 398, 400
Peanut hulls, 259
Peat
  for cell encapsulation, 290
  as packing material, 47, 50, 376
Pellet formation, avoiding in reactors, 47
Pentacyclic terpanes, biomarkers, 400
Perched water table, 225
Permanganate oxidation, 27
Permeability, cell, 302, 303
Permeability, soil
  effect on remediation time, 320
  fracturing to enhance, 107
Permitting, 47, 380
Peroxidases
  deactivation of, 298
  for enhanced bound residue formation, 51
  in fungal compost bioreactors, 51
  in fungal degradation, 270
  for oxygen production, 235
  for pyrene biodegradation, 154
Peroxide spraying, 11
Peroxyacids, 235
Persistence of contaminants, 409
PETROBAC^R, 228
Petrochemical-contaminated soil, treatment of
  with bacteria/thermal treatment, 97
  by steam injection, 96
Petrochemical sludge, treatment of
  carbon-to-nitrogen ratio for, 35
  by landtreatment, 40
Petroleum, biodegradation, 115
  anaerobic, 168
  bioaccumulation, 115
  rate in soil, 412
  in sewage lagoons, 57
Petroleum-contaminated soil, treatment of
  in bioslurry reactors, 48
  by bioventing, 104
  by CROW process, 87
  by landfarming and prescreening soil, 34
  with surfactants, 82
Petroleum contamination, treatment of with filamentous fungi, 267
  rate of biodegradation of, 412

Petroleum decision framework (API), 5
Petroleum-degrading microorganisms, counting methods for, 383–392; see also Bioremediation monitoring methods
  fungi, 387
  petroleum-degraders:to total heterotrophic bacteria ratio, 405
Petroleum fraction, heavy, treatment of
  by landfarming, 34
  by vacuum heap biostimulation system, 52
Petroleum hydrocarbon oxidation products, 196
Petroleum hydrocarbons, treatment selection, 5
Petroleum refinery wastewater, treatment of, by modified RBC, 60
Petroleum-sulfonate oil (PSO), 149, 304
Petroleum wastes, treatment of, by landfarming, 31
Petronates, 149, 153
PETRO RISc soil, 322
pH, see Soil pH
Phanerochaete chrysosporium, 140, 270
  in biofilters, 377
  extracellular enzymes of, 151, 154
  in fungal compost bioreactors, 51
  normal habitat, 273
  for PAH biodegradation, 51, 147, 151, 153, 154
  peroxidases, 151
  as soil inoculum, 292
Phanerochaete chrysosporium BKM-1767, 271
  for PAH biodegradation, 150, 153
  metabolites, 194
Phanerochaete laevis HHB-1625, 153
PHB (intracellular polyester), 112
Phenanthrene, 151–154, 400
  as biomarker, 400
  composting of, 43
  degradation inhibition by surfactants, 153
  degradation pathway of, 131
  emissions in bioslurry reactors, 49
  half-life of, 152
  soil flushing of, 86
PHENOBAC®, 302
Phenol, 65
  aerobic degradation of, 134
  anaerobic degradation of, 173–174
  biodegradation in bioslurry reactors, 47
Phenol-contaminated aqueous stream, treatment of, with BIOPUR®, 66
Phenol-contaminated soil, treatment of, by conversion into carbonaceous adsorbent, 20
Phenolics 34
Phenol oxidases, 205
Phenol red, 387
Phenols, biodegradation of
  by Allied Signal immobilized cell bioreactor (ICB)
  in anaerobic reactors, 74
  in bioslurry reactors, 47
  candidates for carbon adsorption, 20
  in EIMCO™ reactors, 49
  by fluidized-bed digester, 64
  rates of biodegradation, 40
Phenotypes, 398
Phosphate precipitation, 251, 255
Phospholipids analysis, 391
Phosphorus limitation, 252
Photoautotrophs, 147, 150
Photocatalyists, 19
Photocatalyzed hydrogen peroxide, 27
Photocatalyzed ozone, 27
Photochemical oxidation, 23–27
Photochemical reactions, 424

Photochemical reactivity, 315; see also Photolysis
Photodecomposition, 30, 274
Photodegradation, 314, 315, 363
Photo-efficiency, 19
Photoionization detector (PID), 401
Photolysis, 53, 313–315
  of PAHs, 141
  of VOCs, 380
Photolysis/reverse osmosis/bioremediation/surfactant scrubbing, 421
Photometabolism, anaerobic, 173, 177
Photomicrographs for NAPL assessment, 165
Photo-oxidation of VOCs, 380
Photoreactivity, 315
Photosensitive stilbazolium (SbQ) groups, 291
Photosynthesis, 178, 264
Photosynthetic bacteria, 274
Photosynthetic microorganisms, 68
Phototrophic purple nonsulfur bacteria, 173
Phototrophs, 68, 147, 274–275
Phthalate esters, 20
Phylogeny of microorganisms, 399
Physical/chemical treatment processes
  leachates/wastewaters, 13–29
  soil, 6–13, 79–99
Physical factors affecting biodegradation, 200–217
Physical soil treatment processes, 6–13, 79–100
  costs, 37
  ex situ, 6
  limitations of, 415
  on site, 6
Phytane, 126, 156
  as biomarker, 401
Phytochelatins, 179
Phytoplankton, 73
Phytoremediation, 113
Phytotoxicity, 34, 193, 233
Phytotoxic structures, 211
PID (photoionization detector), 401
Pile turning, 42
PISB (electrokinetic-enhanced passive *in situ* biotreatment (PISB) system, 111
pKa (dissociation constant), 84, 205
Plant roots, 259, 277
Plants, 53; see also Vegetation
  effect of death on C:N:P ratio 258
Plasma arc technology, 99, 359
Plasma remediation of *in situ* materials (PRISM), 99
Plasmid fusion, 283
Plasmids, 141, 185, 187, 189, 190, 191, 283, 284, 290, 306
  for heavy metal resistance, 179, 182, 190
  multiplasmid *Pseudomonas putida* (MPP), 284
Plasmodium, 267
Plate columns, 372, 374, 375, 376
Plate counts, 386–387, 389, 390
Plate towers, 375
Platinum wire electrodes (PWEs), 36
Pleomorphs, 396
Pleurotus ostreatus, in biofilters, 377
Plowing, 243
Plugging, 14, 248; see also Clogging
  with biopolymer shields, 112
  with electrobioreclamation, 111
  by *Gallionella*, 416
  with hydrogen peroxide, 240
  by nutrients, 252–253, 255
Plugging microorganisms, 416
Plume formation, 208

INDEX

PM500 membrane, 84
pMOL 28 plasmids, 182, 190, 191
pMOL 30 plasmids, 182, 190
pMOL 85 plasmids, 182
PNAs (polynuclear aromatics); see Polycyclic aromatic hydrocarbons (PAHs)
Pneumatic fracturing, 107–109, 244; see also Hydraulic fracturing
    denitrifiers in, 108
    "Lasagna Process", 108
Pneumatic fracturing/bioremediation, 108, 421
Pneumatic soil permeability, 91
Polar adsorbents, 21
Polishing techniques, 57, 76, 184
Polyacrylamide, 58, 65, 290
Polyad FB, 376
Polyalkenes, 158
POLYBAC® E Biodegradable Emulsifier, 302
POLYBAC® N biodegradable nutrients, 250
Poly B-411, Poly R-481, Poly Y-606, 273
Polybutylene, 338
Polychlorinated biphenyls (PCBs), 9
Polyclonal antisera for ELISA, 406
Polycyclic aromatic hydrocarbon (PAH), biodegradation
    by acclimated microorganisms, 142, 144
    by algae, 147, 150
    by algal/bacterial mats, 275
    anaerobic, 167, 177
    by bioaugmentation, 148
    in bioreactors, 46, 47, 49, 419
    by cometabolism, 145, 146, 152, 154, 155, 211, 294
    by commensalism, 146, 292
    competition with, 146
    compost enhancement for, 43, 143
    counts with, 390
    by cyanobacteria, 147
    degradability database, 117, 122–123
    by denitrification, 169
    by diatoms, 147
    dioxygenases in, 141
    effect
        of concentration on, 142–143, 202
        of interface on, 201
        of NAPLs on, 162
        of plants on, 53, 146, 277
        of salinity on, 143
        of soil type on, 143
        of solubility on, 201
        of sorption on, 145
        of structure on, 144, 209
        of surface area on, 213
        of temperature on, 143
        of vegetation on, 146
    end products of, 194
    by enzyme induction, 145, 146, 152, 255, 282
    by extracellular enzymes, 147, 154
    of four rings or higher, 48, 49, 145, 211, 292
    by fungi, 147, 149, 153, 194, 271
    by gram-negative bacteria, 144, 146, 287
    by gram-positive pacteria, 146–147, 151
    by immobilized cells, 287
    metabolite inhibition of, 146
    metabolites from, 194, 195
    microorganisms for, 141, 146
    by mixed cultures, 144, 152, 155, 287
    optimum C:N:P ratios for, 143
    optimum temperature for, 143
    overcoming mixture inhibition of, 212
    pathways of, 126, 131, 141, 142
    by photoautotrophs, 147, 274–275
    by plants, 277
    plasmids in, 141, 284
    rapid screening test for, 142
    redox potential for, 143
    soil moisture for, 143, 223
    substrate antagonism in, 146
    synergism in, 146
    vegetation in, 277
Polycyclic aromatic hydrocarbon (PAH)-contaminated aqueous stream, treatment of, by fixed-film systems, 58
Polycyclic aromatic hydrocarbon (PAH)-contaminated sludge, treatment of
    by bioslurry reactor, 148
    by landfarming, 34
Polycyclic aromatic hydrocarbon (PAH)-contaminated soil, treatment of
    by bioaugmentation, 148
    in bioreactors, 46, 47, 48, 49
    by bioremediation, 100
    in bioslurry reactors, 48, 49
    by bioventing, 104
    by chemical extraction, 11
    by coimmobilized nutrients and microorganisms, 290
    by CROW process, 87
    by electrical corona process, 79
    with encapsulated cells, 290, 296
    by fungal compost bioreactors, 51
    by landtreatment, 40
    with nonionic surfactant, 83
    by photolysis, 53
    in prepared-bed reactors, 50
    by ReTeC screw auger process, 9
    by SCF oxidation, 11
    by soil washing, 10
    by steam extraction, 12
    by thermal desorption, 9
    by vacuum heap biostimulation system, 52
    with vegetation, 53
    by WES screw auger-based process, 9
Polycyclic aromatic hydrocarbon (PAH) contamination, treatment of
    by bioslurry reactors, 49
    by EIMCO™ reactors, 49
    by fixed-film reactors, 58
    by hydrolysis, 141
    by leaching, 141
    by photolytic breakdown, 27
    with surfactants, 310–311
    by volatilization, 141
Polycyclic aromatic hydrocarbon (PAH) desorption, 145
Polycyclic aromatic hydrocarbon (PAH) persistance, 34, 144
Polycyclic aromatic hydrocarbons (PAHs), 141–155
    acclimation to, 144
    adsorption to soil by, 141
    bioaccumulation of, 141
    bioavailability of, 303
    candidates for carbon adsorption, 20
    composting of, 43
    desorption rates of, 303
    detoxification by fungi, 141, 142, 194, 268
    effect on biomass production, 251
    effect of fulvic acids on, 145
    half-lives in soil of, 39, 40, 144
    mass transfer from NAPLs, 161
    partitioning from NAPLs, 162
    solubility of, 145, 202
    sorption to soil organic matter by, 258

threshold concentration for acclimation to, 282
toxicity of, 145, 148
Polycyclic sterane, 156
Polyester elastomer membrane liner material, 339
Polyethylene, 59
Polyethylene glycol surfactants, 313
Polyethylene membrane liner material, 339
Poly-3-hydroxybutyrate (PHB), 112
Polymerase chain reaction (PCR), 284, 397, 398, 400
Polymer-based adsorbent, 376
Polymeric adsorbents, 21
Polymeric membrane liners, 336
Polymers for membrane liners, 336
Polynuclear aromatics (PNAs), see Polycyclic aromatic hydrocarbons (PAHs)
Polypropylene membrane liner material, 339
Poly R-478, 274
Polysaccharide matrix, 288
Polystyrene, 59
Polystyrene foam, 65, 148, 290
Polyurethane-encapsulated cells, 290
Polyurethane foam, 48, 65, 148
Polyvinylalcohol (PVA), 291
Polyvinyl chloride, 59
    membrane liner material, 339
Ponding, 81, 93
Ponds, 245
Population turnover, 253
Pore neck diameters, 233
Pores, see also Soil pores
Pore size, 221, 262, 328
    distribution, 288, 332
    reduction, 247
Pore spaces, 232, 326; see also Soil pores
    DNAPL ganglia distribution in, 164
    effect of oxygen on, 232
    effect of spraying on, 326
    foams in, 164
    protection for bacteria, 276–277
    vapor pressure of contaminants in, 327
    VOC migration in, 328
Pore tree model, 262
Porosity of soil, 328
Porous glass, 376
Porous polyurethane foams (PUF), 66
Porous tube reactor, 50, 412
Portland cement, 12
Posttreatment techniques, VOCs, 363–380
Potassium permanganate, 80
Powdered activated carbon, 19
    in activated sludge system, 56, 65
    in biocatalyst beads, 70
    leachate treatment by, 17
    in membrane biological reactors (MBR), 63
    regeneration of, 22, 421
    for soil washing, 11
Pozzolanic materials, 12
Prairie grasses, 53
Preacidification, 74
Predation, 275, 276–277; see also Grazing; Protozoa
    of bioaugmented organisms, 279, 276–277
    effect on substrate disappearance curves, 409
    in fixed-film systems, 58
    protection by encapsulation, 290
    in RBCs, 60
Predators, microbial, 276, 299
Precipitation, 28, 80

Precipitation/filtration, 422
Precipitation/landtreatment, 422
Precipitation/sedimentation, 422
Precipitators, 15
Prehumidification of off-gas, 377
Prepared-bed reactors, 50
Preservation, see Microbial preservation
Pressure drop, 332
Pressure filtration, 56
Pressure reactors, 51
Pretreatment processes
    for aqueous streams, 342–357
    for biological treatment, 50
    for bioslurry reactors, 47, 48
    for carbon or resin adsorption, 28
    composting, 34
    for disagglomeration of soil, 50
    for fixed-film systems, 65
    fluidized-bed/membrane bioreactors, 63
    fungal compost bioreactors, 51
    for *in situ* biodegradation, 80
    for metals removal, 344–345
    microfiltration, 29
    for sludges, 358
    for soil, 19, 47, 358–363
    soil washing, 48
    for VOCs in organic liquids, 342–357
    for wastewaters, 24
Primary degradation, 131
Prior exposure, 282
PRISM (plasma remediation of *in situ* materials), 99
Pristane, 126, 156
    as biomarker, 401
Process heater, 363, 369
Prokaryotes, 142, 180
Propane cometabolism, 156
Propidium iodide, 407
Proppants, 107, 108
Protocooperation, 286
Proton reducers, 166
Proton-reducing organisms, 70
Protozoa, 275, 415; see also Grazing; Predation
    antibiotics for, 299
    bioaccumulation by, 300
    counting methods for, 390, 391
    effect on acclimation period by, 276, 299
    effect on substrate disappearance curves by, 409
    grazing by, 279, 276–277
    hydrocarbon accumulation in, 380
    lead toxicity to, 188
Pseudomonadaceae, 285
Pseudomonads, 2, 43
Pseudomonas strains, 131
    adsorption with BAC, 64
    in biofilters, 379
    in bioreactors, 68
    biosurfactants from, 149, 154, 302
    for degradation of recalcitrant compounds, 117, 126
    in electrobioreclamation, 111
    in fixed-film systems, 65
    mercury-resistant, 189
    multiple plasmid transfer in, 149
Pseudosolubilization, NAPLs, 163
PSO (petroleum sulfonate-oil) surfactant, 149, 153, 304
Psychrotrophic microorganisms, 139, 149, 158, 159, 227
PUF (porous polyurethane foams), 66
Pug mills, 48

INDEX

Pulsed air injection, 104
Pulverization, 243–244
Pump-and-treat methods, 81–87, 164, 340, 420; see also Soil flushing; Soil washing; Extraction
Pump-and-treat/bioventing, 422
Purple nonsulfur bacteria (Rhodospirilliaceae), 274, 286
PUR (reticulated polyurethane), 66
PVA (polyvinylalcohol), 291
Pyrene, biodegradation of, 43, 154
Pyrene-contaminated soil, treatment of
   by chemical extraction, 11
   by SCF extraction, 13
Pyrene, soil transport of, 84
Pyrolysis of contaminants, 98, 99
QSAR (quantitative structure activity relationship), 212
Q_{10} effect, 227
Quantitative structure activity relationship (QSAR), 212
Quartz furnace, 9
Quaternary ammonium salts, biodegradation of, 313
Quaternary carbon atoms, 130
Quinone degradation, 64

R

R2A medium, 387
Radiochemical techniques, 404
Radiofrequency heating/soil vapor extraction, 421, 423
Radiofrequency (RF) heating, 80, 95, 98
Radioisotope-labeled probes, 399
Radiolabeled carbon dioxide, 403, 404
Radiolabeled hydrocarbons, 404
Radiorespirometry, 407
Radium, 66
Randomly amplified polymorphic DNA (RAPD), 396
Rainfall, 225, 326, 331
Rankine cycle, 23
RAPD (randomlly amplified polymorphic DNA), 396
Rapid automated counting methods, 390–391
RAS (return activated sludge), 56
Rate of emissions, 320, 327
Rate of mineralization, 214
Rates of biodegradation, see Biodegradation rate; Remediation time
RBC (rotating biological contactor), 57, 59–60
RBCAs (risk-based corrective actions), 417
RCRA (Resource Conservation and Recovery Act), 89
RCRA wastes, 416
Reactors/aerobic, anaerobic, chemotrophic and/or phototrophic microorganisms, 422
Recalcitrant compounds, see also Refractory compounds
   anaerobic degradation of, 47
   cometabolism for, 258, 292, 294
   effect of chemical structure on, 208–210
   effect of solubility on, 144
   enzyme induction for, 131
   from humic fraction complexes, 194
   hydrocarbons, 127, 128
   landfarming for, 33, 40
   microorganisms for, 117, 275, 286–287
   NAPLs, 160–165
   PAHs, 40, 144
   superbugs for, 101, 103, 278, 283
   tarry materials, 127
Recalcitrant compounds, treatment of
   by anaerobic degradation, 167
   by ATA MBR, 70
Reciprocating engines, 74

Recombinant *Escherichia coli*, 285
Recovery of chemicals, 376
Recycle of contaminants, 97, 98, 380
Recycling, composting, 43, 44
Red algae, 150, 275
Redox chemistry, 182
Redox conditions, 167
Redox potential, 220, 260–261
   for aerobes, 143
   for anaerobes, 143, 166, 265
   effect on electron acceptor selection, 166
   in landtreatment, 36
   for microbial processes, 125
   for monitoring landtreatment, 36
   for PAH degradation, 143
Reducible sulfur compounds, 172
Reductive dehalogenation, 68
Redwood, 59
Redwood slat aerators, 349
Refinery oily waste, landtreatment of, 33, 40
Refractility of internal cellular structure, 407
Refractory aromatics, treatment of
   with GAC–BAC
   hydrogen peroxide and ozone for, 238
Refractory compounds, see also Recalcitrant compounds
   bioaccumulation of, 275
Refrigeration, 357
Regeneration
   of carbon systems, 19, 353
   of photocatalysts, 19
Regenerative thermal incinerator, 368
Reinjection/clarification/granular activated carbon/air stripping/bioremediation, 421
Relative humidity, see also Humidity
   in vapor-phase carbon adsorption system, 325
Remazol brilliant blue R, 274
Remedial action assessment system (RAAS), 5
Remediation costs, 100
Remediation guidelines, 417
Remediation monitoring methods, see Bioremediation monitoring methods
Remediation strategies, 417
Remediation systems, 5
Remediation technologies
   decision framework for, 418
   applicability, 7–8
   system integration, 419
Remediation time, 319–320, 408–413; see also Biodegradation rate
   effect of half-life on, 408, 409
Removal efficiencies (RE), 205
Reporter genes, 398–399
RE (removal efficiencies), 205
Residues/residuals, 10, 34, 46, 87, 381
Resin adsorption, 18, 20–21
Resin adsorption/sedimentation, 28, 422
Resin, regenerative/*in situ* bioventing, 423
Resins, 21, 60
Resistance heating, 95–96, 98, 99
Respiration, 105, 264
RESOL 30, 85
Respirometry, 407
Resource Conservation and Recovery Act (RCRA) of 1976, 30, 89
Resource Conservation Company solvent extraction process, 11
Retardation coefficients, 224
ReTeC screw auger process, 9

Retention, 332
Retention capacity, 162
Retention times, 46, 65, 66, 70, 71, 72, 74
Reticulated polyurethane (PUR), 66
Return activated sludge (RAS), 56
Reuse, treated soil, 13
Reverse osmosis, 16, 18, 22, 29
Reverse osmosis/bioremediation/surfactant scrubbing/photo-
    lysis, 421
Reverse osmosis/wet air oxidation, 421
Reverse transcription-coupled-PCR (RT–PCR) analysis, 399
Rexophos 25/97, 305
RF (radiofrequency) heating, 98
Rhamnolipid, 154, 302
Rhizosphere, 53, 146, 277, 313
Rhodococcus sp. Q15, 227
Rhodospirilliaceae, 173, 274
Ribonuclease protection assay, 397
Ring cleavage mechanism, 131
Risk-based corrective actions (RBCAs), 417
Risk Reduction Engineering Laboratory (RREL), EPA, 5
Risk Reduction Engineering Laboratory stabilization process, 12
Rock, crushed, 58
Romicon Model HF-Lab-5 ultrafiltration unit, 84
Rose bengal, 385
Rotary kiln, 10, 359, 363
Rotary reactor (BSRR), 6
Rotating biological contactors (RBCs), 57, 59–60
Rotating drum bioreactor, 48
Rototilling, 12, 31, 276
rRNA, bacterial, 407
rRNA monitoring methods, 399
RT–PCR (reverse transcription-coupled-PCR) analysis, 399
Rubber particulates, 12
Rubber tires, 11

S

Saccharomyces cerevisiae, for heavy metal sorption, 183
Safe levels of carcinogens, 322
Salicylate, 48
Saline soil pH, 230
Salinity, 143, 411
Sampling, *in situ*, 335
Sand, 36, 58, 60
Sapindus mukorossi surfactant, 84
Sapromat, 201, 403
Saprophytic macrocytes, 190
Saturated compounds in soil, treatment of by landfarming, 34
Saturated zone treatment applicability, 7
Saturates, landtreatment of, 34
Sawdust
    biofilter packing material, 376
    in biopiles, 52
    for heavy metal removal, 259
SbQ (photosensitive stilbazolium) groups, 291
SBR (sequencing batch reactor), 69
Scenedesmus, in fixed-film processes, 65
SCF (self-cycling fermenter), 69
Screening techniques
    for crude-oil degraders, 281
    for ligninolytic fungi, 273–274
    for PAH degradation, 142
Screens, biological, 112
Scrubbers, 15
Seasonal fluctuations, 411
Sedimentation, 28
Sedimentation/carbon or resin adsorption, 422

Sedimentation/chemical precipitation, 422
Sedimentation/filtration, 422
Sedimentation/flocculation, 28
Sediments, emissions, 335
Seeding microorganisms, 263, 277, 278, 279, 387; see also Bio-
    augmentation; Inoculant formulations; Soil inoculation
    Cladosporium resinae, 269
    effect on lag period, 278
    induced mutations for, 279
Seeding, raw wastes, 43
Selective enrichments techniques, 387–389
Selective plate counts, 386
Selective pressure, 282
Selenium, 191
Self-cycling fermenter (SCF), 69
Semivolatile organic compound-contaminated soil, treatment of
    by bioventing, 104, 105
    by Detoxifier™, 97
    by hot air/steam extraction, 98
    by resistance heating, 96
    by RF heating, 98
    by soil washing, 10
    by steam stripping, 97
Semivolatile organic compounds (SVOCs)
    emissions in bioslurry reactors from, 49
    EPA method 8270 (GC/MS) for, 406
    removal by thermal desorption, 6
Sensitized photo-oxidation, 313, 363
Serqua 710, 306
Sequence of hydrocarbon biodegradation, 127
Sequencing aerobic/anaerobic conditions, 169, 248
    with cometabolism/diauxie growth, 76
    substrate inhibition, 788
Sequencing batch reactor (SBR), 69
Sequencing soil manipulation for heavy metals, 181
Sequential anaerobic/aerobic treatment, 76, 78
Sequential enrichment techniques, 388
Sequestration
    of electron flow, 167
    of heavy metals, 183
    by NAPLs, 161, 217
    by surfactants, 149
Sewage systems, heavy metals in, 54, 258–259
Sewage treatment, 183
Shallow soil gas sampling, 402
Shallow soil mixing (SSM), 79
Shallow soil mixing/soil vacuum extraction, 421
Shallow soil mixing/soil vapor extraction, 79, 362
Shearing, 61
Shields, biopolymer, 112
Shock loading
    with activated sludge process, 56
    of nutrients, 253
Siallon process, 12
Siderophores, 179
Signature microbial lipid biomarker (SLB), 391
Silica gel medium, 58, 387
Silica sand
    for immobilization in trickling filters, 59
    as proppants, 107
Silo reactor for composting, 46
Silts for bioslurry reactors, 48
Silver, 191–192
Single particle reactor (SPR), 6
Site cleanup, 94
SITE ETP (Superfund Innovative Technologies Evaluation
    Emerging Technologies Program), 48
Site/soil properties, 219–262

INDEX **535**

Site-specific remediation plan, 219
Skim-milk, 290
SLB (signature microbial lipid biomarker), 391
Slime-forming bacteria, 112
Slime layer of RBCs, 59
Slime molds, 276
Slop oil biodegradation, 396
Sludge biotreatment/wet air oxidation, 421
Sludge digestion/anaerobic reactors, 71
Sludge emissions, 335
Sludge pretreatment for VOCs, 358
Sludge systems, 245
Slugs, 275
Slurper spear, 105, 106
Slurry bioremediation, 47
Slurry reactor, 148
Soda ash, 27
Sodic soil pH, 230
Sodium alginates, 112
Sodium carbonate, 11
Sodium dodecyl sulfate (SDS), 83
    competition as growth substrate, 306
Soil adsorption constant (K), 84, 205, 206
Soil aeration, see Aeration, soil
Soil amendments, 47
Soil bacteria, 288
Soil biofilter packing material, 376
Soil characteristics, 314; see also Environmental factors; Soil factors
    aggregates, 287
    air-filled porosity, 328, 331
    bulk density, 328
    capacity to degrade contaminants, 221
    cation exchange capacity, 231
    for detoxification, 219
    effect on vapor pressure, 329
    pore size, 221, 262
Soil conditioners, 52
Soil dewatering, 48
Soil environment, 217–218, 219–262
Soil extraction, 81
Soil factors, see also Soil characteristics
    affecting biodegradation, 217–218, 220, 225
    modification for bioremediation, 219–262; see also Environmental factors
Soil filtering of bacteria, 221
Soil flushing, 81–87, 415; see also Soil washing; Extraction; Pump-and-treat
    for NAPLs, 163
    for PAHs, 83
    prevention of channeling with, 164
Soil gas humidity, 224
Soil gas monitoring, 402–403
Soil heating, 95–98
Soil heating/soil venting, 95
Soil inoculation, see also Bioaugmentation; Inoculant formulations; Seeding microorganisms
    encapsulation for, 147–148, 290–292
    of fungi, 273
    inoculant formulations, 279, 280, 287
Soil inoculation/bulking/tilling, 273
Soil maps, 220
Soil matter, composition, 329
Soil matrix, 112, 288
Soil mixing/photochemical microorganisms, 424
Soil moisture, 221–226, 228, 325–326
    with air stripping, 327
    for anaerobiosis, 224, 248

    bioventing for, 224
    in biopiles, 52
    in bioreactors, 46
    in composting, 44
    effect
        on biofiltration, 378, 379
        of density on, 221
        on fungi, 224
        on photolysis, 314
        on SVE, 91
        on thermal desorption, 6
        on VOC sorption, 91, 325
    for emissions control, 333
    in landtreatment, 35, 39
    model for effect on vapor-phase sorption, 326
    for PAH biodegradation, 143, 223
    and temperature correlations, 35
    VOC removal indicator, 325
Soil moisture/electrobioreclamation, 110–111, 112
Soil organic content, 257–259, 314
Soil oxygen, 232–246
    for biodegradation, 249
    concentration for denitrification, 166
    content of air and water, 234, 242
    control in bioreactors, 46
    delivery approaches, 241–247
    effect on biodegradation rates, 411
    effect of contaminant on, 233
    effect on biofiltration, 378
    in landfarming, 36
    for monitoring bioremediation, 402–403
    sand and, 36
    supply alternatives, 234
Soil permeability, 52, 91; see also Permeability
Soil pH, 229–232
    for actinomycetes, 230
    in bioreactors, 46
    for biodegradation, 35–36, 229–232
    control with liming, 193
    for directing pathway for PAH degradation, 195
    effect
        on fungi, 270
        on fulvates and humates, 257
        on heavy metals, 180, 181, 188
        of humic substances on, 257
        on hydrogen peroxide, 240
        on hydrolysis, 80
        of lime on, 148, 231; see also Liming
        on nutrients, 253
        on soil flushing, 82
    in landtreatment, 35–36
    optimum range for biodegradation, 230
    range for biodegradation, 220
Soil pigment content, 314
Soil pores, see also Pore spaces
    air-filled porosity of, 331
    diffusion coefficient in, 328, 331
    effect on molecular diffusion, 288, 331
    effect on oxygen, 232
    size, 221, 247, 262, 288, 332
    volatiles in, 327, 328, 402
Soil pretreatment processes for VOCs, 358–363
Soil properties, 219, 220, 328
Soil retention times, 104
Soil slurry reactor, 412
Soil structure, 261–262
    effect on microbial transport, 288
    effect on soil pores and oxygen, 233

Soil surface charge, 229, 230
Soil temperature, 227–229; see also Temperature
    albedo effect on, 228
    in bioreactors, 46
    for bioremediation, 227–229
    in bioventing, 104
    in cold climates, 227, 228
    in composting, 45, 227
    control of, 46, 228, 229
    effect
        on acclimation period, 227
        on adsorption, 227
        on air stripping, 345
        on biodegradation rate, 412
        on biofiltration, 378, 379
        on contaminant concentration, 371
        on contaminant solubility, 227
        on contaminant viscosity, 227
        on fungi, 270
        on lag period, 227
        on landtreatment biodegradation rates, 34, 39
        of mulches on, 228
        on soil washing, 86
        on SVE, 91
        on vapor pressure, 324–325
        on VOC removal, 91
        on volatilization, 324–325, 330
    gradient, 228, 324, 325
    high temperatures, 227, 228
    in landtreatment, 34
    low temperatures, 227
    mesophiles and, 228
    modification by moisture control, 35, 228
    mutant microorganisms and, 228
    optimum, 227
    for PAH degradation, 143
    PETROBAC® mutant bacterial hydrocarbon degrader, 228
    psychrophilic microorganisms and, 227
    Q_{10} effect, 227
    range for hydrocarbon utilizers, 227
    seasonal variations, 228
    in thermal desorption, 6
    thermophiles and, 227
Soil texture, 261–262, 328
    effect on biodegradation rates, 39
    effect on soil moisture, 221
    effect on soil pores and oxygen, 233
    improvement in biopiles, 52
Soil treatability database, EPA, 5
Soil treatment systems
    biological, 29–53, 100–113
    chemical, 6–13
    in landtreatment, 29–41
    limitations of, 415–417
Soil type, effect on
    biodegradation in bioslurry reactors, 48
    contaminants, 261–262
    distribution of biomass, 384
    enzyme activity, 384
    microbial activity with depth, 262
    microbial populations, 262
    NAPL retention, 162
    PAH biodegradation, 143
Soil types in U.S., 220
Soil vacuum extraction, 89; see also Soil vapor extraction
    prevention of channeling with, 164
Soil vacuum extraction/shallow soil mixing, 421

Soil vapor extraction (SVE), 89–93, 376
    applicability, 7
    effect of heat on, 91
    guide, EPA, 89
    model for VOCs–NAPLs, 340
    modified, 104
    with NAPLs, 92–93, 161
    permeability measurements for remediation time, 320
    variable gas flow rates in, 342
    for VOC-contaminated soil, 89, 340, 362
Soil vapor extraction/air sparging, 91, 422
Soil vapor extraction/electrobioremediation, 112
    off-gas treatment by biofiltration, 363, 379
Soil vapor extraction/radio-frequency heating, 421, 423
Soil vapor extraction/shallow soil mixing (SSM), 79, 362
Soil vapor extraction/thermal-catalytic oxidizer vapor treatment system, 422
Soil vapor extraction treatment technology resource matrix, EPA, 89
Soil vapor extraction/vapor-phase biofilters/bioremediation, 421
Soil vapor stripping, 89; see also Soil vapor extraction
Soil venting, 89, 359; see also Soil vapor extraction
Soil venting/air injection/off-gas treatment, 423
Soil venting/injection of liquid nitrogen, 242
Soil wafer reactor, 412
Soils washer, mobile, 10
Soil washing, 10–11, 81–87; see also Soil flushing; Pump-and-treat; Extraction
    by biosurfactant, 84
    for heavy metals, 11, 84, 86, 183
    pretreatment for bioslurry reactors, 48
    surfactant plugging, 85
    for VOC-contaminated soil, 358
Soil water-holding capacity, 221
Soil-water isotherms, 224
Soil water potential, 220, 225
Solar drying, 335
Solar radiation, 327
Sole carbon source utilization, 391
Solidification, 12, 80
Solidification/stabilization, 80, 93
Solids/liquids separation, 28
Solubility of contaminants, 200–203
    degradation of low solubility compounds, 162
    effect on biodegradation rate, 412
    effect on fungal metabolism, 268
    effect on recalcitrance, 144
    effect on toxicity, 211
    inhibition of PAH degradation by, 145
    K_{oc}, K_{ow}, K_p, 206
    number of rings and, 24
    octanol–water partition coefficient and, 202, 206–207
    partitioning and, 206
    soil adsorption constants (K), 205, 206
Solubilization, 301
Solvation, 211
Solvent-contaminated soil, treatment of
    by HRUBOUT® process, 96
    by steam extraction, 12
Solvent extraction, 18, 29, 83, 342, 344
Solvent extraction/steam stripping, 344
Solvent extraction/steam stripping/distillation, 423
Solvent hydrophobicity, 162
Solvent recovery, 354
Solvent refluxing, 333
Solvents, treatment of

INDEX

in bioreactor, 68
in EIMCO™ reactors, 49
Solvent-water interface, 202
Sorbant adsorption, 83
Sorption, 203–207; see also Absorption; Adsorption
   Baker's yeast for, 259
   competition between heavy metals and contaminants for, 260
   of contaminants to organic matter, 257–258
   displacement of organics, 327
   dissociation constant for determining, 205
   effect of cosolvent on, 145
   effect of flagella on, 207
   by fulvic acids, 154
   effect of humic material on, 204, 262
   effect of humidity on, 327
   effect on leaching, 203
   effect of surfactants on, 307
   of enzymes to organic matter, 298
   fungal mycelia for, 259
   for immobilization of contaminants, 80
   of microorganisms to soil, 288
   of PAHs, 143, 144, 145, 204, 258
   by phototrophs, 88
   of surfactants, 306, 307
   van der Waals forces in, 204, 288, 329
   of VOCs, 328, 329
Sparging, see also Air sparging
   applicability of, 7
Sparging/bioventing, 104
Sparging/injection–extraction, 87
Sparing, 154, 297; see also Diauxie effect
Spatial partitioning of heavy metals, 180
Specific gravities of hydrocarbons, 208
Spectrophotometer, 391
Spent carbon, 353
Sphagnum peat moss
   as biofilter packing, 376, 379
   for heavy metal sorption, 183
Sponges, 58
Spores, 224, 267, 268
Spray aerators, 349
Spray-drying cultures, 289
Spreading velocity of surfactants, 302
Spread-plate technique, 387
Sprinkler irrigation system, 226
SSM (shallow soil mixing), see Shallow soil mixing (SSM)
Stabilization, 12, 13, 80
Stabilization/solidification, 80, 93
Stable carbon isotope analysis, 404
Staphylococcus saprophyticus, in fixed-film systems, 58
Starch amendments in bioreactors, 47
Starter culture, 287, 289; see also Bioaugmentation; Soil inoculation
Starvation, 285
Static windrow systems, 44
Stationary transformation techniques, 297
Steam cleaning process, 96
Steam-enhanced recovery process, 96–98
Steam extraction, 12
Steam flushing, 96–98
Steam injection, 95, 96–98
   with bioremediation, 98
Steam injection, cyclic/vacuum extraction, 422
Steam regeneration of activated carbon, 353
Steam stripping, 18, 96–98, 343–344, 380
   pretreatment for VOCs, 28, 342, 343, 358
Steam stripping/air stripping, 422

Steam stripping/chemical precipitation/bioremediation, 419
Steam stripping/solvent extraction/distillation, 423
Steam stripping/vacuum extraction/oil–water separation/filtration/carbon adsorption, 422
Steam stripping/wet air oxidation, 22, 28, 421
Steam, wet or dry/vacuum, for gasoline-contaminated site, 98
Steranes, biomarkers, 400
Still bottoms, 22, 49
Still bottoms/wet air oxidation, 421
Storage ponds, 335
Straight-chain aliphatics
   anaerobic degradation of, 177
   biodegradation of, 156, 158
Straight-chain alkanes, biodegradation of, 156, 158
Straw
   in biopiles, 52
   in bioreactors, 47
   for composting, 42, 44
   for heavy metal removal, 259
Stressed cells, 285
Stress, environmental, 291
Strict anaerobes, 172, 265
Strict anaerobic conditions, 166
Stripping, 16, 28, 87–88, 342, 380; see also Air stripping; Steam stripping
Stripping tower/biofilter, 423
Structure
   chemical, 208–210
   soil, see Soil structure
Subirrigation sytem, 225
Substituted aromatic compounds, degradation pathway, 126
Substrate antagonism, 146
Substrate disappearance curves, 409
Substrate inhibition, 196, 286
Subsurface drip irrigation system, 224
Subsurface science program, 100
Subterminal oxidation, 126, 127
   of alkanes, 130
Succinyl-Co-A, 156
Sulfate, as electron acceptor, 172
Sulfate reducers, 168
Sulfate-reducing bacteria, 166, 167, 172
Sulfate reduction, 172, 175, 176, 177
   by heavy metals, 181
Sulfide precipitation
   of heavy metals, 180
Sulfidogenic conditions, 172
Sulfidogens, 172
Superbugs, 101, 103, 278, 283
Supercritical carbon dioxide, 13
Supercritical fluid extraction, 13
Supercritical fluid reactors, 25, 26
Supercritical fluids as solvents, 22
Supercritical fluid (SCF) oxidation, 11–12, 22–23
Supercritical water, 11, 23
Superfund Innovative Treatment Evaluation (SITE) Program, EPA, 5
Superfund sites, 317, 318, 416
Superoxides, 235
Surface area minimization, 333
Surface area, operating, 325
Surface diffusion, 288
Surface impoundment, 37
Surface irrigation methods, 225
Surface seepage, 81, 93
Surface-wetting properties, 207
Surfactant hydrolysis/neutralization, 83

Surfactant hydrolysis/phase separation, 83
Surfactant internalization, 311
Surfactant micelles, 149, 303; see also Critical micelle concentration
   fluorescence spectroscopy for, 304
   formation of, 149
   ingestion by microorganisms, 300
   partitioning model for, 304
   with phenanthrene, 153
   plugging by, 85
Surfactant partitioning, 302
   micelle-phase/aqueous-phase partition coefficient, Km, 303
Surfactants, 300–313
   aeration coefficient with, 305
   antibiotics as, 311
   bioavailability of, 145
   biodegradation inhibition by, 153, 306
   biodegradation of, 277, 312–313
   in bioreactors, 46
   for carbon adsorption, 20
   for CGAs, 242
   channeling of, 306
   chemical, 302–307
   cometabolism of, 305
   competition as growth substrate, 306
   competition with soil organic matter, 302
   concentration of, 153, 303, 304
   for contaminant alteration, 300–313
   for contaminant mobilization, 80
   for copper quenching protection, 304
   critical micelle concentration of, 149, 303, 304
   with Detoxifier™, 93
   effect on cell permeability, 303
   evaluation of potential of, 84
   foams, 83, 164, 305, 306; see also Foams
   for heavy metals removal, 86, 304
   homologues of, 301
   lipotin, 305
   microbial, 84, 307–312; see also Microbial surfactants
   for NAPLs, 303, 306
   nutrient–emulsifier mixtures, 305
   for oil–core aphrons, 306
   oil–sand aggregates, 11
   oleophilic fertilizer, 302
   for PAHs, 145, 303
   plasmids and, 306
   separation from contaminant, 83, 84
   for soil flushing, 82, 85
   for soil moisture control, 226
   for soil property modification, 226
   for solubilization during leaching, 14
   sorption of, 307
   toxicity of, 301
Surfactant/water process, for oil spills, 302
Surfactin, 311, 249
Suspended growth systems, 56–57, 65, 71–76
SVE (soil vapor extraction), see Soil vapor extraction (SVE)
SVOCs (semivolatile organic compounds), 322, see also Semivolatile organic compounds (SVOCs)
Symbiotic relationships, 70, 286
Synergism, 286
   in anaerobic degradation, 167; see also Anaerobic bacteria
   using metabolic intermediates, 195
   in PAH degradation, 146
   with phenanthrene degradation, 152
   physical/chemical/biochemical, 1
   of treatment techniques, 420

Synergistic effects of contamination, 212
Synthetic membrane covers, 335–340
Syntrophic associations, 70
   in biofilms, 288
   methanogens and fermenters, 173
   in sulfate reduction, 172, 177
System integration, 419

T

TA (total acidity), 400
TAD (thermophilic aerobic disgestion), 56–57
TAH (total aliphatic hydrocarbons, 406
TAH/total PAH (total aliphatic hydrocarbons/total PAHs) ratios, 406
Tank reactor for composting, 46
Tar-contaminated soil, treatment of, by soil washing, 11
Tarry materials, 127
TBA (tertiary butyl alcohol) biodegradation, 159
TCLP for VOCs, 13
TEL (tetra-ethyl lead), 304–305
Temperature (see also Soil temperature), in anaerobic fixed-film systems, 74
   of influent, 334
Tensiometric barriers, 99
Tensiometric barrier/SVE, 91
Tensiometric barrier/vacuum extraction, 99
Teratogen, 185
Tergitol 15-S-9, 305
Tergitol NP-10, 153, 304
Tergitol NPX, 145, 149
Terminal branching, 130
Terminal electron acceptors
   effect of redox potential on, 166
   nitrate, 169
   sulfate, 172
Terminal oxidation, 126
Termites, 276
Terraferm biosystem soil enclosed reactor, 50
Tertiary butyl alcohol (TBA) biodegradation, 159
Tertiary treatment, 65
Test plan and technical protocol for bioslurping, 106
Tetracyclic aromatic hydrocarbons degradation, 48, 146
Tetracyclic compounds, biodegradation of, 144, 145, 158, 209
Tetradecane biodegradation, 158
Tetradecene-contaminated soil, treatment of, in rotating drum bioreactor/bioaugmentation, 48
Tetra-ethyl lead-contaminated soil, treatment of, with surfactant/cosurfactant/water solution, 83
Tetra-ethyl lead (TEL), 304–305
Texture, see Soil texture
Thauera selenatis, 191
Thermal-catalytic oxidizer vapor treatment system/vapor extraction system, 422
Thermal desorption
   mobile system for VOCs, 6, 359
   of PAH-contaminated soil, 9
   for VOC-contaminated soil, 359, 360–362
Thermal desorption/Thermatrix flameless oxidation process, 9, 421
Thermal incineration, 363–369, 370, 371
Thermally enhanced soil vapor extraction (TESVE), 92
Thermal oxidation, 8, 15, 363
Thermal reactors, 6
Thermal regeneration of activated carbon, 353
Thermal stripping, 359, 360–362
   low-temperature, 12

INDEX 539

Thermal treatment, 6, 9–10, 363–372
Thermatrix flameless oxidation process, 9, 79, 421
Thermatrix flameless oxidation process/thermal desorption, 9, 79
Thermogram, 407
Thermophiles, 311
Thermophilic aerobic digestion (TAD), 56–57
Thermophilic fungi, 270
Thermophilic methanogens, 266
Thermophilic microorganisms, 42, 70, 117, 227
Thin-film evaporation, 380
Thin-layer chromatography-flame ionization detection (TLC–FID), 406
Three-ring PAHs, biodegradation of, 147, 153
Threshold contaminant concentrations, 214, 282, 293, 384
Threshold level of microorganisms, 279
Threshold limit value (TLV) for gasolines, 322
TIDE process, 12
Tilling, soil
　for aeration, 243, 246
　for bioaugmentation enhancement, 290
　for diesel oil-contaminated soil, 148
　effect on soil temperature, 229
　for fungal enhancement, 273
　in landfarming, 32, 35, 36
　for run-on and runoff control, 259
　for volatilization, 317, 319, 363
TLC–FID (thin-layer chromatography-flame ionization detection), 406
TLV (threshold limit value) for gasolines, 322
TOC (total organic compounds), biodegradation indicator, 401
TOL plasmids, 104, 284, 389, 398
Toluene biodegradation, 139–140
　anaerobic, 174–175
　in bioreactors, 68
　emissions in bioslurry reactors, 49
　in trickling filters, 59
Toluene monooxygenase (TMO) gene, 285
Total acidity (TA), 400
Total aliphatic hydrocarbons (TAH), 406
Total constituent analysis, 13
Total heterotroph, aliphatic hydrocarbon-degrading microorganism counts, 390
Total petroleum hydrocarbon (TPH), 5, 49
　analyzer, 93
　biodegradation indicator, 401–402
　with bioventing, 104
　levels with landfarming, 34
Total petroleum hydrocarbon (TPH) removal
　by landfarming, 34
　by open-system slurry reactor, 48
Total petroleum hydrocarbon (TPH) treatment, by dual vacuum extraction, 92
Toxic hydrocarbon concentrations, 213
Toxicity, 210–212; see also Phytotoxicity
　detoxification processes, 212
　effect
　　of attenuation on, 261
　　of contaminant concentration on, 211, 213
　　of contaminant solubility on, 145, 211
　　of contaminant sorption on, 258
　of fulvic acids, 211, 154, 258
　of heavy metals, 180, 181, 188
　log K_{ow} values and, 212
　of landtreatment overloading, 38
　of PAHs, 148
　of surfactants, 301

Toxicity characteristic leaching procedure (TCLP), 12
Toxic substances
　biodegradation by-products, 146, 193, 212
　　monitoring, 258
　cadmium, 185
　chromium, 148
　defusing, 193
　effect on substrate disappearance curves, 409
　lead, 188
　naphthalenes, 148
　processes affecting behavior of, 199
　volatiles, 207
Toxic units, 212
Trace concentrations, 214
Trace elements, 121, 124, 213, 253, 258–259
Trametes versicolor
　in biofilters, 377
　as soil inoclum, 292
　T. versicolor Paprican 52, 150, 153, 194, 271
Transcriptional fusion reporter gene, 399
Trans-dihydrodiols, 124–125, 126, 147, 194, 268
Transformation, 2, 416
　of PAHs, 141
　by phototrophs, 275
　product accumulation from, 294
Transport
　of contaminant, 161, 202
　　into cells, 130, 307
　of microbes 287–289
　of vapor phase, 327
Treatment efficiency, landtreatment, 39
Treatment processes
　residuals, 381
　suitability, 6
Treatment technologies, 5
Treatment trains, 415–428; see also Combined technologies
　in bioreactors, 417–419
　with Detoxifier™, 93, 427
Treatment zone, 31, 32
Trichoderma harzianum, 65
Trickle-bed reactor for biofilters, 378
Trickle irrigation, 225
Trickling filters, 57, 59
Tricyclic aromatics, biodegradation of, 146
Tricyclic PAHs, biodegradation of, 144, 145
Tricyclic-to-pentacyclic ratio, biodegradation parameter, 400
Tridecane biodegradation, 159
Triterpane C_{30} 17α(H)21β(H)hopane, 400
Triton N101, 153, 304
Triton X-100, 145, 149, 153, 163, 304, 305, 306
　protection from copper quenching, 149
　for soil washing, 11
Triton X-114, 11, 304, 305
Tropaeolin O, 273
TPH (Total petroleum hydrocarbon), see Total petroleum hydrocarbon (TPH)
TSDFs (hazardous waste treatment storage and disposal facilities), 317, 318
Tubular bioreactors, 49–50
Tumorigens, 142, 151, 194
Turbidimetric measurements, 391
Turned windrow systems for composting, 44
Turnover time
　for hydrocarbons, 412, 413
　for lead, 188
Tween-20-80, 302
Tween-80, 149, 303

competition for nutrients, 140
effect on organic matter, 306
enhancement of fungal degradation, 150
Tween-type surfactants, inhibition of biodegradation, 153

U

UASB (upflow anaerobic sludge blanket), 70, 72–73
Ultrafiltration
   for leachate treatment, 29
   leachate treatment by-products, 18
   separation of surfactant and contaminant, 83
Ultrafiltration/incineration, 29
Ultrafiltration/wet oxidation, 22, 29, 421
Ultramicrobacteria (UMB), 289
Ultrasound-enhanced soil washing, 11
Ultraviolet (UV)/air stripping, 346
Ultraviolet (UV)/hydrogen peroxide, 24, 27, 422
Ultraviolet (UV)/ozonation, 24, 26, 422
UMB (ultramicrobacteria), 289
Underground storage tanks, (UST), 89
Unicelles, 14
Unicellular fungi, 267
Unilamellar vesicles, 312
Upflow anaerobic sludge blanket (UASB), 70, 72–73
Upflow anaerobic sludge blanket (UASB)/fixed-film system, 73, 422
Upflow floc digester, 73
Uranium bioaccumulation, 66
Urea–peroxide, 235
U.S. Air Force Bioventing Initiative, 104
U.S. Air Force Center for Environmental Excellence (AFCEE) Technology Transfer Division Bioslurper Initiative, 106
U.S. Environmental Protection Agency (EPA) priority pollutant list, 141
U.S. Soil Conservation Service, 220
UST (underground storage tanks), 89
UTCHEM (multiphase, multicompositional simulator), 304

V

Vacuum-enhanced NAPL recovery, 165
Vacuum extraction, 165
Vacuum extraction/cyclic steam injection, 422
Vacuum extraction/steam stripping/oil-water separation/filters/carbon adsorption, 422
Vacuum heap biostimulation systems, 52–53, 247
Vacuum pump oil-contaminated soil, treatment of, by RF heating, 98
Vacuum regeneration of activated carbon, 353
van der Waals forces, 204, 210, 288, 329
Vapor diffusion, 328, 331
Vapor extraction, 415; see also Soil vapor extraction
Vapor extraction process layout, 90
Vapor extraction rates, 104
Vapor extraction systems, 340–342
Vapor incinerators, 365
Vaporization, 317, 328, 329
Vapor/liquid equilibria, 344
Vapor-phase adsorption, 350, 363
Vapor-phase biofilters/SVE/bioremediation, 421
Vapor-phase carbon adsorption, 8, 87, 351
Vapor-phase diffusion, 331
Vapor pressure, effect
   on boiling point 330
   on contaminant volatilization, 325

   of electrolytes on, 328
   on emissions, 327
   of gasoline, 234
   in soil pores, 328
   for selected compounds, 328
   of soil composition on, 329
   on SVE VOC removal, 91
   of temperature on, 324–325, 327–328
   on transport of vapor phase, 327
Vapors, gasoline, 322
Variation of soil factors for bioremediation 219–262
Vegetation, 300; see also Plants
   alfalfa plants, 300
   effect on PAH degradation, 53, 146, 277
   effect on soil temperature, 228
   heavy metal accumulation by, 53
   in landtreatment, 30, 34
   for PAH degradation, 53
   prairie grasses, 53
Venting systems, 329; see also Bioventing
   for aeration, 245
Veratryl alcohol, 154, 194, 272, 273
Vermiculite, 148, 290
Viable counting methods, 385, 386, 387, 389, 390, 391, 404, 408
Vibrating screens, 48
Victim substrate, 293
Viruses, 276
Viscosity, 203, 207, 287, 330
Vitrification, 80, 95, 99
Volatile organic compound-contaminated aqueous stream, treatment of
   by biological treatment, 357
   with BIOPUR®, 66
   by carbon adsorption, 350
   by condensation, 357, 369
   by evaporation, 357
   by steam stripping, 343
   with UV/hydrogen peroxide, 24
   with UV/ozone, 24
Volatile organic compound-contaminated sludges, treatment of
   by air stripping/carbon adsorption, 358
   by evaporation/carbon adsorption, 358
   by steam stripping, 358
Volatile organic compound-contaminated soil, treatment of
   by air stripping, 88
   by biofiltration, 376–380
   by bioventing, 104, 105
   with Detoxifier™, 97, 363
   by hot air/steam stripping, 98
   by low-temperature thermal stripping systems, 360, 361, 362
   by photodegradation, 363
   posttreatment techniques for, 363–380
   pretreatment processes for, 342–363
   by resistance heating, 96
   by RF heating, 98
   by soil washing/extraction, 358
   by steam stripping, 97
   by SVE, 362
   by SVE/SSM, 362
   by thermal desorption, 359
Volatile organic compounds (VOCs), 317–381; see also Emissions
   absorption, 372
   adsorption isotherms, 20
   adsorption to soil, 329
   analysis of, 406

INDEX 541

bacteriostatic effect of, 207
chemisorption of, 329
condensation of, 372
containment in bioreactors, 47
control of, 50, 319, 332–381, 419
detection of, 322
diffusion travel times of, 232
effect of clay on, 328
effect of evaporation on, 329
effect of humidity on, 333
effect on lag phase, 207
effect on remediation time, 319–320
emission rate, 317
evaporation rate models for, 91–92, 317
on Hazardous Substance List (HSL), 319
landtreatment of, 38, 317, 319, 320, 332
leaching potential of, 13
migration in soil of, 328
in petroleum products, 317–381
from petroleum refining industry, 317–318
recovery by decantation, 355
recycle of, 380
reduction of, 332
release rates in landtreatment, 319
removal efficiency for, 325, 326
removal techniques for, 6, 65, 334, 350
sorption of, 325
sources of, 317
technologies for removing, 333
toxicity of, 207
transformation in soil of, 317
treatment, storage, and disposal of, 321
VES removal of, 340
Volatile organic compounds (VOCs), removal/treatment of, 332–381
by activated sludge/air stripping, 56
with afterburners, 365
by air stripping, 56, 344–350
by aquaculture, 333
with BAC systems, 65, 333
with biofilters, 87, 89, 377, 378
biological, 357
with biopiles, 51–52
with BIOPUR®, 380
in bioslurry reactors, 47, 49
with boilers, 262, 269
by carbon adsorption, 350–357
in cement kilns, 363
destruction of, 363, 364, 365, 369, 371
by distillation, 342
by evaporation, 357
with flares, 333, 363, 369
in situ bioremediation, air sparging and/or vapor extraction, 420
by liquid injection incinerators, 363
by multiple hearths, 363
by packed/plate columns, 372
by photodegradation, 363
by photo-oxidation, 380
posttreatment techniques, 363–380
pretreatment techniques, 342–363
by refrigeration/condensation, 357
with rotary kiln, 359, 363
by solvent extraction, 344
by steam stripping, 343–344
by SVE/catalytic combustion, 89
by SVE/GAC, 89

by thermal incinerators, 15, 363, 364, 366, 368
by Thermatrix/thermal desorption, 9
in trickle-bed reactors, 378
in tubular reactors, 49
by vapor incinerators, 365
by VES, 340–342
Volatility, 207
Volatilization, 317; see also Emissions
 with bioventing, 104
 effect
  of adsorption on, 203
  of barometric pressure on, 325
  of boiling point on, 325, 330
  of soil depth on, 326
  of temperature on, 324
 of heavy metal-contaminated soil, 183
 enhancement of, 12
 factors affecting, 323–332
 mechanical, 12
 with NAPLs, 162
 of PAHs, 141
 rates of, 326, 333
Volumetric water content of soil, 325–326
Vyrodex method, 240

W

Wafer reactors, 51
WALS (wide angle light scatter), 407
WAO (wet air oxidation) process, see Wet air oxidation (WAO) process
Washing, see Soil Washing
Waste oil-contaminated gravel, treatment of, by gravel slurry reactor, 49
Waste stabilization ponds, 57, 245
Wastewater treatment systems
 biological, 54–78
 chemical, 13–29
Water-holding capacity, soil, 221, 262
Waterlogging, 39, 224, 226
Water potential for biodegradation, 220, 225
Water quality criteria, EPA, 324
Water solubility, 329
Water table, 165, 244–245
Water treatment process, biological, 54
Wax-impregnated graphite electrodes (WIGEs), 36
Weathering of hydrocarbons, 234, 297, 305
Well points, 243
WES process, 12
Wet air oxidation (WAO), 22, 23, 79
 for leachate treatment, 16, 18
 with regeneration of PAC, 421
Wet air oxidation/BAC, 65, 422
Wet air oxidation/reverse osmosis, 421
Wet air oxidation/sludge biotreatment, 421
Wet air oxidation/steam stripping, 421
Wet air oxidation/still bottoms, 421
Wet air oxidation/ultrafiltration, 421
Wet scrubber blowdown streams, 66
Wettability, 164
Wheat straw, 273; see also Straw
White-rot fungi, 270, 271, 273
 in biofilters, 377
 Bjerkandera adjusta CBS 595.78, 153
 B. sp. strain BOS55, 153
 extracellular peroxidases, 299
 for PAH degradation, 149, 150, 153, 155, 194

Phanerochaete chrysosporium BKM-F-1767, 153
Trametes versicolor Paprican 52, 153
Wickerham yeast-nitrogen base medium, 388
Wicking, 326
Wind, 325
Windbreakers, 334
Wind-powered bioventing systems, 105
Wind speed, 326
Wood bark, 44
Wood chips
   as biofilter packing, 376
   carbon source for fungi, 271
   in composting, 44, 42
   in fixed-film systems, 58
   for fungi soil inoculation, 273
Wood slat and tray aerators, 349
Worms, 30, 275, 276, 299

X

Xanthan gum, 112
Xenobiotic catabolic genes, 400
Xenobiotic degradation
   anaerobic, 166
   by extracellular enzymes, 298
   by mixed cultures, 286
Xenobiotics, 115, 208, 213, 292, 294
Xenon plasma flashlamp, 380
XM50 membrane, 84
X-ray dot maps, 165
X TRAX process, 9
Xylene-contaminated soil, treatment of, by steam stripping, 97
Xylenes
   aerobic degradation of, 140
   anaerobic degradation of, 175–176
   in bioslurry reactors, 49
xylE (toluene, xylene degradation)
   gene probe, 284
   gene, TOL plasmid, 399

Y

Yeast malt agar, 388
Yeast–nitrogen base medium, 388
Yeasts, 269; see also Fungi
   bioaccumulation in, 275
   for biphenyl degradation, 150
   counting methods , 387, 391
   cytoplasmic inclusions in, 300
   for heavy metal sorption, 183
   hydrocarbonoclastic, isolation of, 388
   malt agar for, 388
   nonspecific enzyme systems, 268
   for PAH degradation, 147, 150
   for silver accumulation, 192

Z

Zinc, 39, 192
Zinc-contaminated soil, treatment of, by stabilization processes, 12
Zone of incorporation, 31
Zoogleal sludge, 56
Zwittergent, 386
Zymogenous organisms, 283